Vassil N. Alexandrov   Jack J. Dongarra
Benjoe A. Juliano   René S. Renner
C. J. Kenneth Tan (Eds.)

KU-318-789

Computational
Science – ICCS 2001

International Conference
San Francisco, CA, USA, May 28-30, 2001
Proceedings, Part II

Springer

Volume Editors

Vassil N. Alexandrov
University of Reading
School of Computer Science, Cybernetics and Electronic Engineering
Whiteknights, PO Box 225, Reading RG 6 6AY, UK
E-mail: V.N.Alexandrov@rdg.ac.uk

Jack J. Dongarra
University of Tennessee
Innovative Computing Lab, Computer Sciences Department
1122 Volunteer Blvd, Knoxville TN 37996-3450, USA
E-mail: dongarra@cs.utk.edu

Benjoe A. Juliano
René S. Renner
Computer Science Department, California State University
Chico, CA 95929-0410, USA
E-mail: {Juliano/renner}@ecst.csuchico.edu

C.J. Kenneth Tan
The Queen's University of Belfast
School of Computer Science
Belfast BT7 1NN, Northern Ireland, UK
E-mail: cjtan@acm.org

QUEEN MARY
UNIVERSITY OF LONDON
LIBRARY

Cataloging-in-Publication Data applied for

Die Deutsche Bibliothek - CIP-Einheitsaufnahme

Computational science : international conference ; proceedings / ICCS
2001, San Francisco, CA, USA, May 28 - 30, 2001. Vassil N. Alexandrov
... (ed.). - Berlin ; Heidelberg ; New York ; Barcelona ; Hong Kong ;
London ; Milan ; Paris ; Singapore ; Tokyo : Springer
Pt. 2 . - (2001)
  (Lecture notes in computer science ; Vol. 2074)
  ISBN 3-540-42233-1

CR Subject Classification (1998):D, F, G. H. I, J

ISSN 0302-9743
ISBN 3-540-42233-1 Springer-Verlag Berlin Heidelberg New York

This work is subject to copyright. All rights are reserved, whether the whole or part of the material is
concerned, specifically the rights of translation, reprinting, re-use of illustrations, recitation, broadcasting,
reproduction on microfilms or in any other way, and storage in data banks. Duplication of this publication
or parts thereof is permitted only under the provisions of the German Copyright Law of September 9, 1965,
in its current version, and permission for use must always be obtained from Springer-Verlag. Violations are
liable for prosecution under the German Copyright Law.

Springer-Verlag Berlin Heidelberg New York
a member of BertelsmannSpringer Science+Business Media GmbH

http://www.springer.de

© Springer-Verlag Berlin Heidelberg 2001
Printed in Germany

Typesetting: Camera-ready by author
Printed on acid-free paper   SPIN 10781771   06/3142   5 4 3 2 1 0

Preface

Computational Science is becoming a vital part of many scientific investigations, affecting researchers and practitioners in areas ranging from aerospace and automotive, to chemistry, electronics, geosciences, to mathematics, and physics. Due to the sheer size of many challenges in computational science, the use of high performance computing, parallel processing, and sophisticated algorithms, is inevitable.

These two volumes (Lecture Notes in Computer Science volumes 2073 and 2074) contain the proceedings of the 2001 International Conference on Computational Science (ICCS 2001), held in San Francisco, California, USA, May 27–31, 2001. These two volumes consist of more than 230 contributed and invited papers presented at the meeting. The papers presented here reflect the aims of the program committee to bring together researchers and scientists from mathematics and computer science as basic computing disciplines, researchers from various application areas who are pioneering advanced applications of computational methods to sciences such as physics, chemistry, life sciences, and engineering, arts and humanitarian fields, along with software developers and vendors, to discuss problems and solutions in the area, to identify new issues, and to shape future directions for research, as well as to help industrial users apply various advanced computational techniques. The aim was also to outline a variety of large-scale problems requiring interdisciplinary approaches and vast computational efforts, and to promote interdisciplinary collaboration.

The conference was organized by the Department of Computer Science at California State University at Chico, the School of Computer Science at The Queen's University of Belfast, the High Performance Computing and Communication group from the Department of Computer Science, The University of Reading, and the Innovative Computing Laboratory at the University of Tennessee. This is the first such meeting and we expect a series of annual conferences in Computational Science.

The conference included 4 tutorials, 12 invited talks, and over 230 contributed oral presentations. The 4 tutorials were "Cluster Computing" given by Stephen L. Scott, "Linear Algebra with Recursive Algorithms (LAWRA)" given by Jerzy Waśniewski, "Monte Carlo Numerical Methods" given by Vassil Alexandrov and Kenneth Tan, and "Problem Solving Environments" given by David Walker. The constitution of the interesting program was due to the invaluable suggestions of the members of the ICCS 2001 Program Committee. Each contributed paper was refereed by at least two referees. We are deeply indebted to the members of the program committee and all those in the community who helped us form a successful program. Thanks also to Charmaine Birchmore, James Pascoe, Robin Wolff, and Oliver Otto whose help was invaluable.

We would like to thank our sponsors and partner organizations, for their support, which went well beyond our expectations. The conference was sponsored by Sun Microsystems (USA), IBM (UK), FECIT (Fujitsu European Center for Information Technology) Ltd. (UK), American Mathematical Society (USA), Pacific Institute for the Mathematical Sciences (Canada), Springer-Verlag GmbH,

California State University at Chico (USA), The Queen's University of Belfast (UK), and The University of Reading (UK).

ICCS 2001 would not have been possible without the enthusiastic support of our sponsors and our colleagues from Oak Ridge National Laboratory, University of Tennessee and California State University at Chico. Warm thanks to James Pascoe, Robin Wolff, Oliver Otto, and Nia Alexandrov for their invaluable work in editing the proceedings; to Charmaine Birchmore for dealing with the financial side of the conference; and to Harold Esche and Rod Blais for providing us with a Web site at the University of Calgary. Finally, we would like to express our gratitude to our colleagues from the School of Computer Science at The Queen's University of Belfast and the Department of Computer Science at The University of Reading, who assisted in the organization of ICCS 2001.

May 2001   Vassil N. Alexandrov
                                                    Jack J. Dongarra
                                                    Benjoe A. Juliano
                                                    Reneé S. Renner
                                                  C. J. Kenneth Tan

Organization

The 2001 International Conference on Computational Science was organized jointly by The University of Reading (Department of Computer Science), The University of Tennesse (Department of Computer Science), and The Queen's University of Belfast (School of Computer Science).

Organizing Committee

Conference Chairs:

Vassil N. Alexandrov, *Department of Computer Science, The University of Reading*
Jack J. Dongarra, *Department of Computer Science, University of Tennessee*
C. J. Kenneth Tan, *School of Computer Science, The Queen's University of Belfast*

Local Organizing Chairs:

Benjoe A. Juliano (*California State University at Chico, USA*)
Reneé S. Renner (*California State University at Chico, USA*)

Local Organizing Committee

Larry Davis (*Department of Defense HPC Modernization Program, USA*)
Benjoe A. Juliano (*California State University at Chico, USA*)
Cathy McDonald (*Department of Defense HPC Modernization Program, USA*)
Reneé S. Renner (*California State University at Chico, USA*)
C. J. Kenneth Tan (*The Queen's University of Belfast, UK*)
Valerie B. Thomas (*Department of Defense HPC Modernization Program, USA*)

Steering Committee

Vassil N. Alexandrov (*The University of Reading, UK*)
Marian Bubak (*AGH, Poland*)
Jack J. Dongarra (*Oak Ridge National Laboratory, USA*)
C. J. Kenneth Tan (*The Queen's University of Belfast, UK*)
Jerzy Waśniewski (*Danish Computing Center for Research and Education, DK*)

Special Events Committee

Vassil N. Alexandrov (*The University of Reading, UK*)
J. A. Rod Blais (*University of Calgary, Canada*)
Peter M. A. Sloot (*University of Amsterdam, The Netherlands*)
Marina L. Gavrilova (*University of Calgary, Canada*)

Program Committee

Vassil N. Alexandrov (*The University of Reading*, UK)
Hamid Arabnia (*University of Georgia*, USA)
J. A. Rod Blais (*University of Calgary*, Canada)
Alexander V. Bogdanov (*IHPCDB*)
Marian Bubak (*AGH*, Poland)
Toni Cortes (*Universidad de Catalunya, Barcelona*, Spain)
Brian J. d'Auriol (*University of Texas at El Paso*, USA)
Larry Davis (*Department of Defense HPC Modernization Program*, USA)
Ivan T. Dimov (*Bulgarian Academy of Science*, Bulgaria)
Jack J. Dongarra (*Oak Ridge National Laboratory*, USA)
Harold Esche (*University of Calgary*, Canada)
Marina L. Gavrilova (*University of Calgary*, Canada)
Ken Hawick (*University of Wales, Bangor*, UK)
Bob Hertzberger (*University of Amsterdam*, The Netherlands)
Michael J. Hobbs (*HP Labs, Palo Alto*, USA)
Caroline Isaac (*IBM UK*, UK)
Heath James (*University of Adelaide*, Australia)
Benjoe A. Juliano (*California State University at Chico*, USA)
Aneta Karaivanova (*Florida State University*, USA)
Antonio Laganà (*University of Perugia*, Italy)
Christiane Lemieux (*University of Calgary*, Canada)
Jiri Nedoma (*Academy of Sciences of the Czech Republic*, Czech Republic)
Cathy McDonald (*Department of Defense HPC Modernization Program*, USA)
Graham M. Megson (*The University of Reading*, UK)
Peter Parsons (*Sun Microsystems*, UK)
James S. Pascoe (*The University of Reading*, UK)
William R. Pulleyblank (*IBM T. J. Watson Research Center*, USA)
Andrew Rau-Chaplin (*Dalhousie University*, Canada)
Reneé S. Renner (*California State University at Chico*, USA)
Paul Roe (*Queensland University of Technology*, Australia)
Laura A. Salter (*University of New Mexico*, USA)
Peter M. A. Sloot (*University of Amsterdam*, The Netherlands)
David Snelling (*Fujitsu European Center for Information Technology*, UK)
Lois Steenman-Clarke (*The University of Reading*, UK)
C. J. Kenneth Tan (*The Queen's University of Belfast*, UK)
Philip Tannenbaum (*NEC/HNSX*, USA)
Valerie B. Thomas (*Department of Defense HPC Modernization Program*, USA)
Koichi Wada *University of Tsukuba*, Japan)
Jerzy Wasniewski (*Danish Computing Center for Research and Education*, DK)
Roy Williams (*California Institute of Technology*, USA)
Zahari Zlatev (*Danish Environmental Research Institute*, Denmark)
Elena Zudilova (*Corning Scientific Center*, Russia)

Sponsoring Organizations

American Mathematical Society, USA
Fujitsu European Center for Information Technology, UK
International Business Machines, USA
Pacific Institute for the Mathematical Sciences, Canada
Springer-Verlag, Germany
Sun Microsystems, USA
California State University at Chico, USA
The Queen's University of Belfast, UK
The University of Reading, UK

Table of Contents, Part II

Table of Contents, Part I

Computational Methods

Digital Imaging Applications

Session chairs:

J. A. Rod Blais (University of Calgary, Canada)
Gary F. Margrave (University of Calgary, Canada)
Hilary Alto (University of Calgary, Canada)

Densification of Digital Terrain Elevations Using Shape from Shading with Single Satellite Imagery

Mohammad A. Rajabi [1], J. A. Rod Blais [1]

Dept. of Geomatics Eng., The University of Calgary, 2500, University Dr., NW, Calgary, Alberta, Canada, T2N 1N4

{marajabi, blais}@ucalgary.ca

Abstract. Numerous geoscience and engineering applications need denser and more accurate Digital Terrain Model (DTM) height data. Collecting additional height data in the field, if not impossible, is either expensive or time consuming or both. Stereo aerial or satellite imagery is often unavailable and very expensive to acquire. Interpolation techniques are fast and cheap, but have their own inherent difficulties and problems, especially in rough terrain. Advanced space technology has provided much single (if not stereo) high-resolution satellite imageries almost worldwide. This paper discusses the idea of using Shape From Shading (SFS) methods with single high resolution imagery to densify regular grids of heights. Preliminary results are very encouraging and the methodology is going to be implemented with real satellite imagery and parallel computations.

1 Introduction

In the present context, Digital Terrain Models (DTMs) are simply regular grids of elevation measurements over the land surface. They are used for the analysis of topographical features in GISs and numerous engineering computations. The National Center for Geographic Information and Analysis (NCGIA) provides a list of global DTMs with different resolutions (from 30 arc minutes to 30 arc seconds) which are freely available on the Internet [1]. NCGIA also provides a list of freeware and shareware programs to handle different formats of DTM files. DTMs can be used for numerous applications such as, e.g.,

- Calculating cut-and-fill requirements for earth works engineering, such as road construction or the area that would be flooded by a hydroelectric dam.
- Analyzing intervisibility to plan route location of radar antennas or microwave towers, and to define viewsheds.
- Displaying 3D landforms for military purposes and for landscape design and planning.
- Analyzing and comparing different kinds of terrain.

- Computing slope maps, aspect maps and slope profiles that can be used to prepare shaded relief maps, assist geomorphological studies, or estimate erosion or run-off.
- Computing of geophysical terrain corrections for gravity and other field observations.
- Displaying thematic information or for combining relief data such as soils, land use or vegetation.
- Providing data for image simulation models of landscapes and geographical processes.
- Investigating of erosion processes, land slides, contamination by dangerous chemicals, etc.

Traditionally, photogrammetry has been used as the main method of producing DTMs, however, there are cases for which no aerial photography is available. Recently, with the rapid improvement in remote sensing technology, automated analysis of stereo satellite data has been used to derive DTM data. Automatic generation of elevation data from SPOT stereo satellite imagery has been discussed in [2], [3], and [4]. But even in this case, there are some places that due to some reasons such as cloud coverage, technical and/or political limitations are not covered by stereo satellite imagery.

On the other hand, the availability of single satellite imagery for nearly all of the Earth is taken for granted nowadays. But unfortunately, reconstruction of objects from monocular images is very difficult, and in some cases, not possible at all. Inverse rendering or the procedure of recovering three-dimensional surfaces of unknown objects from two-dimensional images is an important task in computer vision research. A robust procedure, which can correctly reconstruct surfaces of an object, is important in various applications such as visual inspection, autonomous land vehicle navigation, and surveillance and robot control. In the past two decades, there have been extensive studies of this topic. Shape from defocusing [5], [6], shape from stereopsis [7], shape from motion [8], shape from texture (SFT) [9], [10], and shape from shading (SFS) [11] [12], [13] are examples of techniques used for inverse rendering.

This paper is the result of investigating the use of shape from shading techniques to improve the quality of the interpolated DTM data with single satellite imagery of better resolution than the DTM data. The idea is highly motivated by the wide availability of satellite remotely sensed imagery such as Landsat TM and SPOT HRV imagery.

2 Shape from Shading

Shape recovery in computer vision is an inverse problem which transforms single or stereo 2D images to a 3D scene. Shape from stereo, shape from motion, shape from texture, and shape from shading are examples of methods used for shape recovery in computer vision. Shape from shading (SFS) recovers the surface shape from gradual variations of shading in the image. The recovered shape can be expressed either in

terrain   height   z(x,y)   or   surface   normal $\vec{N}$ or   surface   gradient
$(p,q)=(\partial z / \partial x, \partial z / \partial y)$.

To solve the SFS problem, the first step is to study the image formation process. A Lambertian model is the simplest one in which it is assumed that the gray level at each pixel depends on light source direction and surface normal. Assuming that the surface is illuminated by a distant point source, we have the following equation for the image intensity:

$$R(x, y) = \rho \ \vec{N} . \vec{L} = \rho \frac{pl_1 + ql_2 + l_3}{\sqrt{p^2 + q^2 + 1}} \tag{1}$$

where ρ is the surface albedo, $\vec{N}$ is the normal to the surface and $\vec{L} = (l_1, l_2, l_3)$ is the light source direction. Even if ρ and $\vec{L}$ are known, the SFS problem will still be a challenging subject, as this is one nonlinear equation with two unknowns for each pixel in the image. Therefore, SFS is an underdetermined problem in nature and in order to get a unique solution, if there is any at all, we need to have some constraints.

Based on the conceptual differences in the algorithms, there are three different strategies to solve the SFS problem: 1. Minimization approaches 2. Propagation approaches, and 3. Local approaches. The following subsections briefly review these approaches. A more detailed survey of SFS methods can be found in [14].

2.1 Minimization Approaches

Based on one of the earliest minimization methods, the SFS problem is formulated as a function of surface gradients, while brightness and smoothness constraints are added to overcome the underdeterminedness condition [15]. The brightness constraint ensures the reconstructed shape produce the same brightness as the input image. The smoothness constraint in terms of second order surface gradients helps in reconstruction of a smooth surface. The shape at the occluding boundary was used for initialization to imply a correct convergence [15].

Other minimization approaches use more or less the same logic to deal with the SFS problem. However, some variations in either formulation of the problem or the constraints (or both) are usually implemented.

2.2 Propagation Approaches

In this approach there is a characteristic strip along which one can compute the surface height (object depth) and gradient, provided these quantities are known at the starting point of the strip. Singular points where the intensity is either maximum or minimum are usually the starting points. At singular points the shape of the surface is either known or can uniquely be determined. The first attempt for solving SFS problems using this method seems to be mentioned in Horn [16]. In this technique, shape information is propagated outward along the characteristic strips. The direction of characteristic strips is toward the intensity gradient.

2.3 Local Approaches

The basic assumption in this approach is that the surface is locally spherical at each pixel point. Intensity and its first and second derivatives [17] or intensity and only its first derivative [18] have been used to estimates the shape information. All local methods suffer from the local spherical assumption which may not be correct in all cases.

Another local approach uses linearized approximations of reflectance maps to solve the SFS problem [19]. In this method the reflectance map is formulated in terms of surface gradient and then the Fourier Transform is applied to the linear function and a closed form solution for the height (depth) at each point is obtained.

Another method makes use of computing the discrete approximation of the gradient [20]. Then the linear approximation of reflectance map as a function of height (depth) is employed. Using a Jacobi iterative scheme, this technique can estimate the height (depth) at each point. The main problem of the two latter approaches is with the linear approximation of the reflectance map. If the nonlinear terms are large, it will diverge.

3 Improvement of DTM Interpolation Using SFS Techniques

The problem under investigation is to improve the accuracy of the interpolated DTM grid data by applying SFS techniques on the corresponding single satellite imagery. The assumption is that the satellite imagery has better resolution than the original DTM data. To keep the problem simple, the resolution difference between the original DTM grid data and satellite imagery is considered to be one dyadic order. The original DTM data are used as boundary constraints in the SFS problem.

The assumptions made for this research are: 1) the surface is Lambertian, 2) the surface albedo is known, 3) the surface is illuminated by a distant point source, and finally 4) the position of the light source is known.

It is worth mentioning that in our special application of SFS where we are dealing with satellite imageries, the first assumption is the only one which is questionable. However, dealing with multispectral satellite imagery, one can get a good estimate of surface albedo by applying classification techniques. Meanwhile, the light source for satellite imageries is the Sun which is not only far away from the scene, but also has a known direction knowing the local time when the satellite is passing over the scene (which is assumed to be the case).

Our approach deals with a patch located at pixel (i,j) (see Figure 1) with nine points at a time out of which there are four grid points with known heights (squares) and five unknown points (circles) for which we want to improve the accuracy of the interpolation. The method mainly consists of three stages: 1) preprocessing, 2) processing, and 3) postprocessing. The preprocessing stage itself has two steps. In the first step, using interpolation (bilinear) techniques, the heights of the unknown points in the patch are estimated.

In the second step, using the known grid points, the relative orientation of the patch with respect to the light source is estimated. If this relative orientation implies that the patch is in the shadow, then there would be no useful shading information to improve the accuracy of the interpolated heights. Therefore, in this case the interpolated heights are considered the final height values.

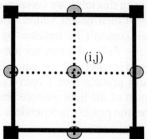

Fig. 1. A patch: squares are known grid points and circles are interpolated points.

Otherwise, the processing stage for each patch consists in solving an overdetermined system of equations of type (1) in terms of heights (z).

This is simply done by approximating p and q with finite differences. Writing the Lambertian reflectance model as a function of heights (z), we have nine nonlinear equations corresponding to the image intensity (I) of each point in the patch with four known heights (the grid points) and five unknown heights.

The nonlinear system of equations is solved using nonlinear least squares with the interpolated heights as the first approximation for the unknowns. Again it is worth mentioning that in most engineering applications, the DTM grid spacing is based on terrain roughness analysis, therefore, the interpolated heights are usually good estimates of heights for initializing the nonlinear solution. Moreover, the nonlinear least squares solution was constrained by the $\pm 3\sigma$ condition where σ is the expected standard deviation of the interpolated heights and comes from the DTM specifications.

The last stage, postprocessing, consists of taking arithmetic means of two solutions for the unknown heights located on the boundary of patches coming from the neighboring patches, except for the outsides of the peripheral patches.

4 Numerical Examples

The methodology described in Section 3 has been tested using a number of numerical examples. One synthetic object and one real DTM data set with their corresponding synthetic imageries were used in these experiments. The orientation of the light source was considered as a variable to investigate the effects of relative positions of the object and the light source on the solution.

The synthetic object under investigation is a convex hemisphere with a radius of 6 units, sampled at each 0.5 unit, i.e., a 33 by 33 pixel object. The corresponding DTM (one dyadic order less than the object, i.e., 17 by 17 pixels)) was extracted out from the object. Meanwhile, the corresponding image of the object was created using Lambertian reflectance model with a much (5 times) denser version of the object. The resulting image was passed through a smoothing filter to get an image with the same density as the original object under study.

The differences between the original object, the Interpolated Grid Solution (IGS) and the SFS solution were analyzed. Table 1 summarizes these results. The statistics shown in this table are computed with those pixels which our SFS method was able to updated their height values. The last row of the table shows the number of patches which couldn't be updated by this SFS method. This can be mainly due to two reasons. Either the patches are located with respect to the light source such that the average intensity of the patch is zero (patches in the shadow), or the nonlinear least squares process didn't converge. In Table 1 nothing has been mentioned about the azimuth of the light source. The symmetry of the synthetic object under investigation makes the process independent of the light source azimuth.

Table 1. The convex hemishpere

Elevation		$30°$	$35°$	$40°$	$45°$	$50°$	$55°$	$60°$
Object–IGS	Mean	-0.04	-0.04	-0.01	-0.01	0.00	-0.02	0.01
	Std	0.31	0.29	0.28	0.28	0.31	0.35	0.33
Object – SFS	Mean	-0.05	-0.07	-0.06	-0.05	-0.05	-0.06	-0.03
	Std	0.18	0.20	0.18	0.17	0.17	0.23	0.17
Patches not updated (out of 196)		41	39	28	26	19	15	20

Table 1 shows that in comparison to the IGS, the SFS solution has slightly worsened the mean differences. A detailed examination of the problem showed that this deterioration comes from the abrupt change of height values along the base of the hemisphere. Apart from these sections, the SFS solution has been able to improve the mean differences by an average of 30%. However, the improvement in the standard deviations of the differences is approximately 40% which is quite considerable. Meanwhile, for this synthetic object it is seen that the optimum elevation angle of the light source is about 50 degrees where we have about 45% improvement in standard deviation. As an example, Figure (2) shows the wiremesh of the object, the corresponding image and the wiremesh of differences of the original object and IGS, and the SFS solution where the light source is located at azimuth and elevation angle of $135°$ and $45°$ respectively.

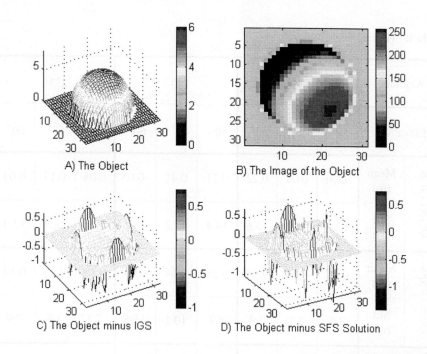

Fig. 2. A) The Original object; B) The corresponding image; The differences between the original object and the IGS in (C) and the SFS solution in (D).

The second test object is a real height data set from southern Alberta, Canada, measured with 25 metre spacing. It is a 1000 by 1000 grid with more than 1300 metre height difference which was down sampled to 125 metres (i.e., a 200 by 200 grid). This sampled height data was considered as the object under investigation and the corresponding DTM (one dyadic order wider, i.e., 250 metre grid) was extracted out from it. Meanwhile, the corresponding image of the object was created using a Lambertian reflectance model and the original 25 metre height data. The resulting image was passed through a smoothing filter to get an image with the same resolution as the object under study, i.e., 125 metres.

Table 2 summarizes the results of our experiment with this data set, which shows results similar to the previous case. The SFS solution has slightly worsened the absolute value of the mean differences and has improved the standard deviation of the mean differences with the average amount of 40%. It seems that during the data preparation process the original characteristics of the real data set has been lost. However, the results show a clear dependency of the SFS solution to the relative position of the object and the light source. As it can be seen, the worst results have been obtained with the light source at the azimuth of 180 degrees. Figure (3) shows the wiremesh of the object, the corresponding image and the wiremesh of differences between the original object, IGS, and the SFS solution, where the light source is located at azimuth and elevation angle of 135° and 45° respectively.

Table 2. The real height data set

Azimuth		135°			180°			225°		
Elevation		30°	45°	60°	30°	45°	60°	30°	45°	60°
Object - IGS	Mean (m)	-0.01	0.09	0.22	0.71	0.41	-0.03	0.08	0.17	0.09
	Std (m)	12.9	13.2	13.6	14.4	17.2	18.8	12.8	12.9	13.4
Object - SFS	Mean (m)	0.15	-0.04	0.03	0.26	0.34	0.05	0.13	0.13	0.12
	Std (m)	7.5	7.7	7.9	8.7	10.1	11.5	7.7	7.6	7.9
Patches not updated (out of 9409)		483	872	1484	4397	3922	3709	493	735	1239

5 Remarks and Conclusions

The use of a SFS technique to improve the accuracy of the interpolated DTMs has been investigated and preliminary results are very encouraging with simulated imageries. The next step is obviously to experiment with real satellite imageries. Georeferencing, clouds, shadowing, unknown surface albedo, and noise in addition to non dyadic scale factors between satellite imageries and DTMs seem to be problems that one has to deal with before solving the SFS problem.

This SFS method is a local method that can be easily implemented by parallel processing techniques. Considering the resolution of satellite imageries and the area that they cover, one can see that using local SFS techniques and parallel processing in our application is a must.

Looking at the results, one also realizes that the success of this SFS technique is highly dependent on the relative orientation of the patches and the light source. If the relative orientations of the object and the light source are such that there is no shading information, the SFS technique cannot improve the interpolation results. Obviously, one should not forget the morphology of the terrain. This may be another good criterion in addition to the resolution of the images, to be taken into consideration when one has different options in selecting satellite imagery for the DTM densification purposes.

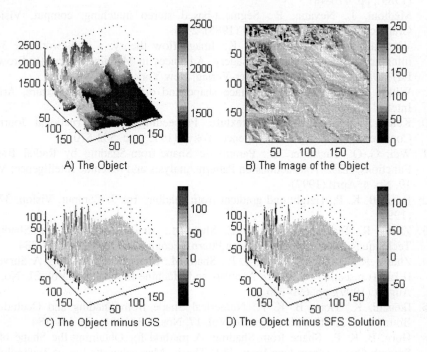

Fig. 3. A) The Original object; B) The corresponding image; The differences between the original object and the IGS in (C) and the SFS solution in (D).

References

1. The National Center for Geographic Information and Analysis (NCGIA): **www.ncgia.ucsb.edu/education/curricula/cctp/units**
2. Gugan, D. J., Dowman, I. J.: Topographic Mapping from Spot Imagery. Photgrammetric Engineering and Remote Sensing 54(10) (1988):1409-1414
3. Simard, R., Rochon, G., Leclerc, A.: Mapping with SPOT Imagery and Integrated Data Sets. Invited paper presented at the 16th congress of the International Society for Photogrammetry and Remote Sensing held July 1988 in Kyoto, Japan

4. Tam, A. P.: Terrain Information Extraction from Digital SPOT Satellite Imagery. Dept. of Geomatics Eng., The University of Calgary (1990)
5. Pentland, A. P.: A new sense for depth of field. IEEE. Trans. Pattern Anal. Match. Intell. PAMI-9(4) (1987), 523-531
6. Hwang T., Clark, J. J., Yuille, A. L.: A depth recovery algorithm using defocus information. IEEE Conference on computer vision and pattern recognition (1989), pp. 476-481
7. Medioni, J., Nevatia, R.: Segment-based stereo matching. comput. Vision Graphics Image Process, 31 July (1985), 2-18
8. Waxman, A. M., Gurumoothy, K.: Image flow theory: a framework for 3-D inference from time-varying imagery. Advances in Machine Vision (C Brown, Ed) (1988), pp. 165-224, Springer-Verlag, New York
9. Witkin, A. P.: Recovering surface shape and orientation from texture, Artif. Intell. 17 (1981), 17-47
10. Kender, J. R.: Shape from texture. Proceedings, 6th International Journal Conference on Aritficial Intelligence, Tokyo (1979), pp. 562-570
11. Wei, G.-Q., Hirzinger, G.: Parametric Shape-from-Shading by Radial Basis Functions. IEEE Transactions on Pattern Analysis and Machine Intelligence, Vol 19, No. 4, April (1997)
12. Horn, B. K. P.: Height and gradient from shading. Int. J. Comput. Vision, 37-5 (1990)
13. Zhang, R., Tsai, P. S., Cryer, J. E., Shah, M.: Analysis of Shape from Shading Techniques. Proc. Computer Vision Pttern Recognition (1994), pp. 377-384
14. Zhang, R., Tsai, P.-S., Cryer, J. E., Shah, M.: Shape from Shading: A Survey. IEEE Transcation on Pattern Analysis and Machine Intelligence, Vol. 21, No. 8, August (1999), pp. 690-706
15. Ikeuchi, K., Horn, B. K. P.: Numerical Shape from Shading and Occluding Boundaries. Artificial Intelligence, Vol. 17, Nos. 1-3 (1981), pp. 141-184
16. Horn, B. K. P.: Shape from Shading: A method for Obtaining the Shape of a Smooth Opaque from One View. PhD Thesis, Massachusetts Ins. of Technology (1970)
17. Pentland, A. P.: Local Shading Analysis, IEEE Trans. Pattern Analysis and Machine Intelligence Vol. 6 (1984), pp. 170-187
18. Lee, C. H., Rosenfeld, A.: Improved Methods of Estimating Shape from Shading Using the Light Source Coordinate System. Artificial Intelligence, Vol. 26 (1985), pp. 125-143
19. Pentland, A.: Shape information from Shading: A theory about Human Preception. Proc. Int'l Conf. Computer Vision (1988), pp. 404-413
20. Tsai, P. S., Shah, M.: Shape from Shading Using Linear Approximation, Image and Vision Computing J., Vol 12, No. 8 (1994), pp. 487-498

PC-Based System for Calibration, Reconstruction, Processing and Visualization of 3D Ultrasound Data Based on a Magnetic-Field Position and Orientation Sensing System

Emad Boctor[1, 2], A. Saad[2], Dar-Jen Chang[2], K. Kamel[2], A. M. Youssef[1]

[1] Biomedical Engineering Department, Cairo University Cairo, Egypt
eboctor@ieee.org
[2] CECS Department, University of Louisville KY, USA
{eboctor, ashraf, kamel, chang}@louisville.edu

Abstract. 3D-ultrasound can become a new, fast, non-radiative, non-invasive, and inexpensive tomographic medical imaging technique with unique advantages for the localization of vessels and tumors in soft tissue (spleen, kidneys, liver, breast etc.). In general, unlike the usual 2D-ultrasound, in the 3D-case a complete volume is covered with a whole series of cuts, which would enable a 3D reconstruction and visualization.

In the last two decades, many researchers have attempted to produce systems that would allow the construction and visualization of three-dimensional (3-D) images from ultrasound data. There is a general agreement that this development represents a positive step forward in medical imaging, and clinical applications have been suggested in many different areas. However, it is clear that 3-D ultrasound has not yet gained widespread clinical acceptance, and that there are still important problems to solve before it becomes a common tool.

1 Introduction

Conventional diagnostic ultrasound imaging is performed with a hand-held probe, which transmits ultrasound pulses into the body and receives the echoes. The magnitude and timing of the echoes are used to create a 2D gray scale image (B-scan) of a cross-section of the body in the scan plane.

Conventional freehand 3D ultrasound is a multi-stage process [1, 2, 3]. First, the clinician scans the anatomical organ of interest. Next, the ultrasound data is used to construct a 3D voxel array, which can then be visualized by, for example, arbitrary-plane slicing. A 3D freehand examination can be broken into three stages: scanning, reconstruction and visualization.

Before scanning, some sort of position sensor is attached to the probe. This is typically the receiver of an electromagnetic position sensor [4, 5, 6, 7, 8], as illustrated in *Figure 1*, although alternatives include acoustic spark gaps [9], mechanical arms [10] and optical sensors [11, 12]. Measurements from the electromagnetic position sensor are used to determine the positions and orientations of the B-scans with respect to a fixed datum, usually the transmitter of the electromagnetic position sensor.

In the next stage, the set of acquired B-scans and their relative positions are used to fill a regular voxel array. Finally, this voxel array is visualized using, for example, arbitrary-plane slicing, volume rendering or surface rendering (after 3D segmentation).

Freehand systems can be used to obtain arbitrary volumes of data, since the motion of the ultrasound probe is unconstrained. They are also cheap, requiring only existing, conventional ultrasound systems and relatively inexpensive additional components. For these reasons, research into freehand systems is very active.

1.1 Advantages of 3D Ultrasound System Over Conventional 2D Ultrasound:

There will be many advantages for developing 3D Ultrasound systems in comparison with usual 2D-imaging some of the suggested advantages are:

- The conventional 2D Ultrasound exam is subjective, i.e. it depends on the experience and knowledge of the diagnostician to manipulate the ultrasound transducer, mentally transform the 2D images into 3D tissue structures and make the diagnosis. While 3D Ultrasound systems will standardize the diagnosis procedure and minimize the dependence on diagnostician experience.

- 2D Ultrasound-guided therapeutic procedures are particularly affected because the process of quantifying and monitoring small changes during the procedure or over the course of time is severely limited by the 2D restrictions of the conventional exam. This practice is time consuming and inefficient and may lead to incorrect decisions regarding diagnosis and staging, and during surgery. The goal of 3D Ultrasound imaging is to provide an imaging technique that reduces the variability of the conventional technique, and it seems to play an important role in surgical planning.

- It is difficult to localize the thin 2D Ultrasound image plane in the organ, and difficult to reproduce a particular image location at a later time making the conventional 2D exam a poor modality for quantitative prospective or follow-up studies, while 3D Ultrasound imaging modality will overcome these restrictions.

- In 2D Ultrasound imaging calculating distances and volumes depends on formulas that approximate the body organs to regular geometrical shapes and also depends on the 2D view that the sonographer calculates the needed distance or volume for a more accurate measurement of distances and volumes.

- Sometimes patient's anatomy or orientation restricts the image angle, resulting in inaccessibility of the optimal image plane necessary for diagnosis. 3D Ultrasound imaging will facilitate an arbitrary selection of a cutting plane and the possibility of reconstructing a 3D-image from the data with several visualization tools.

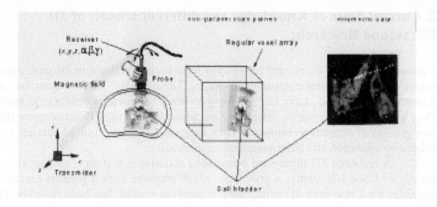

Figure 1: Conventional freehand 3D ultrasound imaging. This is a multi-stage process, involving scanning, reconstruction and visualization. The figure illustrates an examination of a gall bladder

1.2 3D Ultrasound vs. Other 3D Imaging Modalities: MRI, CT.

Recently, tomographic techniques such as CT, MRI, nuclear medicine, etc., experienced a breakthrough in several domains of diagnostic medical imaging.

3D ultrasound tomography is probably one of the most interesting applications within this area. Several attempts in the past have shown that 3D-ultrasound images usually cannot be volume rendered as a 3D-model by using methods known from CT and MRI datasets. The main difficulties originate in:
- The significant amount of noise and speckle in Ultrasound images.
- The regions representing boundaries are not sharp but show a width of several pixels.
- Intensity fluctuations resulting from surface curvature and orientation.
- Self-shadowing of deeper structures.

But in order to gain acceptance, 3-D ultrasound must have clear advantages over the other imaging modalities, e.g. computed x-ray tomography (CT), magnetic resonance imaging (MRI), positron emission tomography (PET) or conventional B-mode and Doppler 2D ultrasound. The main advantages that have been suggested for 3D ultrasound can be grouped into five categories: -

1. Ultrasound is a real time imaging modality, and 3D Ultrasound has the potential for displaying information in near real time too.

2. Extension of ultrasound to 3D provides new images, which would be impossible to visualize otherwise.

3. The reconstruction of 3D Ultrasound by computer potentially brings greater standardization and repeatability to conventional examinations.

2. Current State of Knowledge and different schools of 3D Ultrasound Research:

Various forms of 3D ultrasound echography have been developed in the past. Among recent examples, several companies, including Tomographic Technologies, Inc. and Acoustic Imaging, Inc., have been working to bring commercial products to market. They acquire 3D echography images by mechanically moving 2D echography slices over periods of seconds to minutes. They then visualize the acquired datasets using volume visualization and other visualization methods.

A real-time 3D ultrasound echography acquisition system is coming close to reality. At Duke University, a prototype, which employs such a parallel processing technique for a real-time 3D echography acquisition system, has been developed[35, 36].

Until now, too many universities and institutes have been working in 3D Ultrasound research area but the most powerful and promising systems exist in the following universities: -

- *University of Cambridge, Department of Engineering:* This team has developed "Stradx freehand 3D Ultrasound system" [20]. The most important features of this system are: -
Acquisition System: they use electromagnetic position sensors attached to the conventional ultrasound probe to label each B-scan image with its relative position and orientation, Stradx system is characterized in its well-calibrated acquisition system [15, 16]. Stradx uses a novel, state-of-the-art spatial calibration technique [17].
Visualization Tools: Stradx provides two visualization tools, both of which are available immediately after the data has been acquired, without having to wait for a voxel array to be constructed.
- *University of North Carolina at Chapel Hill, Department of computer science:* The team at North Carolina has been concentrating on two areas:
I- Real-time 3D ultrasound system: they have been working toward an ultimate 3D ultrasound system, which acquires and displays 3D volume data in real time [21]. Real-time display can be crucial for applications such as cardiac diagnosis, which need to detect certain kinetic features.
II- Augmented reality systems: with see-through head-mounted displays. They explored possibilities for in-place, real-time 3D ultrasound visualization.
- *Fraunhofer Institute for Computer Graphics, Germany:* The team there developed the first commercial 3D Ultrasound visualization package "In ViVo". It's an interactive system for the fast visualization of 3D Ultrasound datasets on general-purpose platforms [22]. "In ViVo" supports many visualization tools like:
- Iso-surfaces
- Contour wire frames
- Gradient shading
- Semi-transparent gels
- Max. Intensity projections
- Oblique cuts

3. Our Proposed Free Hand 3D Ultrasound System:

We started our research in 3D Ultrasound imaging about 3 years ago. At the first stage we made a survey over nearly all the existing schools working in 3D Ultrasound, we recognize that 3D Ultrasound system is a multi-phase system starts with choosing the hardware suitable for the intended application and ends with the best visualization technique that fits the required target see *Figure 3*. We found that every research group in this field concentrates on some areas but none of them intends to achieve a whole system for 3D Ultrasound imaging that is suitable to be clinically applied, every school has its own points of power and also has some lag in some phases, for example at UNC Chapel Hill they concentrate to build the whole system for augmented reality environment but they are not concerning with construction and visualization techniques for this system, also at Cambridge University they applied the simplest visualization methods although they have the best calibrated acquisition system for 3D Ultrasound.

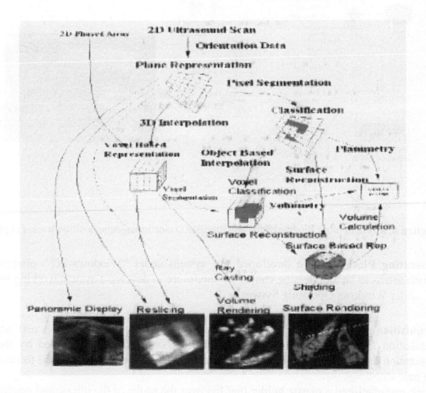

Figure 3: 3D Ultrasound different systems, Data processing and display

The cost and flexibility of free-hand imaging ensure that it will remain a popular choice for 3D Ultrasound systems. For these reasons, we decided to focus on building the infrastructure for a free-hand 3D Ultrasound system. In the next section we will discuss the infrastructure we built in details to indicate the current state of our project.

3.1 Infrastructure Prototype for 3D Ultrasound System:

Hardware Setup: The following is a list of hardware components that constitute our prototype free hand 3D Ultrasound system:

1- Pentium 200 MHz PC with 128 Mbytes RAM, 4 GBytes SCSI Hard Disk.

2- An electromagnetic position tracker "miniBIRD", see *Figure 4*, that is a six degree of freedom measuring device that is used to measure the position and orientation of a small receiver with respect to a transmitter.

3 - A digitizer card "Matrox Card" which is real time frame grabber capable of virtually capturing 25 frames/sec.

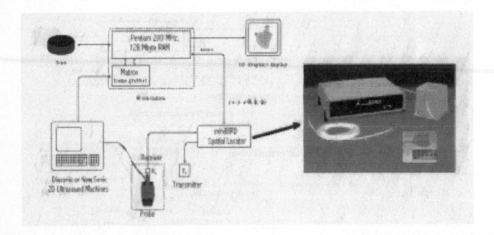

Figure 4: a system block diagram (left) and miniBIRD electromagnetic position tracker (right)

Operating Platform: We developed our system under "Windows NT" platform, which seems to be a promising choice for the future to develop a system working on a PC under Windows Operating System.

Acquisition System: The first step in the project was to implement a real time acquisition system to capture a set of randomly oriented images guided by their orientation data acquired from the spatial locator. To achieve these goals we followed the next steps:

1- we manufactured a plastic holder that fits over the probe of the ultrasound machine and has a holding position for the receiver of the spatial locator. Some precautions had to be taken as:

> -The material of the holder was chosen so that it would not affect the performance of the spatial locator i.e. it should contain no conducting materials.

- The design should not obstacle the free hand movement of the ultrasound probe as the physician was used before.
- The local axes of the receiver should not be in a special orientation with the local axis of the 2D ultrasound image like a perpendicular or parallel case, this condition is necessary for proper calibration of the spatial locator as will be indicated later.
- When removing the holder and reinstalling it again over the probe it should stay in the same location, this condition is also necessary for proper calibration of the spatial locator as will be indicated later.

2 - A key requirement of all free-hand imaging systems is the calibration of the spatial locator, there are two sources of errors in the readings of the electromagnetic tracker that should be taken into account to improve the registration of the free hand 2D slices.

- The first error in the readings comes from the effect of the proximity of metallic objects and induced noise from power cables exist in the ultrasound clinic's environment, we applied a look-up table technique to calibrate both position and orientation readings.
- The second error originates from the fact that the readings representing the position and orientation of the receiver's center relative to the center of the transmitter (reference point) are not the actual readings that we need, which are the readings representing the position and orientation of the corner of the B-scans relative to the center of the transmitter.

A typical calibration process, which often needs repeating every time a sensor is mounted on a probe, takes several hours for a skilled technician to perform. At Cambridge University, they presented a novel calibration technique which can be performed in a few minutes with a very good accuracy, but they invent a special phantom for this purpose which will add cost to the system and would not be available everywhere. We applied a novel technique to calibrate the spatial locator in a few minutes with a very good accuracy, but we used a very popular phantom for that purpose, which is the "AIUM Tissue Cyst Phantom".

Our technique is based on imaging the phantom from several orientations then entering this images into a program that automatically detect the coordinates of the phantom reflectors, this program was implemented by the authors in a previous project concerning calibration of ultrasound machines [18, 19].

3 - After performing the spatial locator calibration, we implemented the acquisition phase of the program that interfaces the digitizer card and the locator (via the serial interface) to the PC. A time synchronization process has been developed to achieve a minimal delay between the acquired 2D- ultrasound images and the associated orientation data.

Compounding and Reconstruction Techniques: The acquired B-images are randomly oriented so we applied a compounding technique that fits the images into an empty volume of voxels, the resolution of the voxel dataset could be changed upon user interface, then we applied different reconstruction techniques that translate the irregular voxel volume into a regular one, that is suitable for visualization algorithms. The reconstruction algorithms, we chose, have the following properties:

1- they are *local* so that their computational cost can be bounded,

2- they produce approximations (not interpolations) so that the *noise in samples can be suppressed*,

3- they produce *smooth approximation* results, so that the visualization results are also smooth without noisy artifacts,

4- they are *insensitive to the order in which the samples arrive*, so that the reconstruction results of identical objects are identical regardless of different paths of sweep upon acquisition.

Image Processing Library: Due to the speckle nature of ultrasound images, the visualization techniques that are valid for CT and MRI cannot be directly applied to 3D Ultrasound datasets. Image preprocessing algorithms should be applied first to the reconstructed datasets before visualization. While implementing this phase we put in our mind some goals to achieve:

- The main target of applying image processing routines is to suppress noise and speckle in the volume in conjunction with preserving the edges of the target organ.

- The applied algorithms should be very fast to achieve the global requirement of implementing an interactive system suitable for routine clinical times.

- These algorithms should not alter the dimensions and shape of the organs so that a quantitative analysis could be performed later on the volume.

- These algorithms should be applied to 3D volumes, not repeatedly applied to a set of 2D images, so that continuity is achieved in all the three dimensions.

Visualization Techniques: We have implemented some visualization techniques as follows:

- We have implemented *"Z shading"* technique to visualize iso-valued surfaces after performing filtering algorithms necessary for depicting meaningful iso-valued surfaces from ultrasound data. *Figure 6* shows a rendered scene using "Z shading" for the gall bladder of one of the authors. The main drawback of this method is its limitation for imaging particular structures.

- After that we took the direction of *"Volume Rendering"*, We implemented a Volume Rendering pipeline using "Voxelator Library" which is a prototype volume rendering software as an extension to the OpenGL library. *Figure 6* shows a screen of a rendered image by Voxelator library. The main drawback of using OpenGL library for volume rendering is the time consumed to update the rendered scene.

- Finally we have implemented a fast volume rendering technique, developed by Philippe Lacroute at Stanford University. This technique uses a Shear-Warp Factorization of the viewing transformation to accelerate the rendering process. *Figure 6* shows a rendered scene for the hepatic veins belonging to one of the authors using this technique.

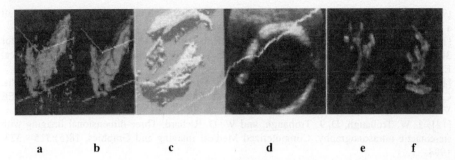

a   b   c   d   e   f

Figure 6: From the left: Z shading using iso-surfaces for the Gall Bladder of One of the authors a) Without Image Processing. b) With Image Processing. c) Volume rendered scene for the forehead part of a fetus using OpenGL Library. d) A cross section of the forehead of the fetus. e) and f) Two different views of volume rendered Hepatic veins of the author Using Fast Volume Rendering Algorithm.

4. Conclusion

In this paper, we have developed a prototype free-hand 3D ultrasound system that is capable of acquiring, constructing, and visualizing 3D ultrasound data sets. The main goal was to achieve high quality rendered images for anatomical structures within a reasonable time compared to the conventional ultrasound diagnostic time. The future research directions of the authors will be targeted on extracting useful quantitative information from the acquired 3D data, which could include 3D segmentation techniques and voxel classification methods.

References

[1] J. Deng, J. E. Gardener, C. H. Rodeck, and W. R. Lees. Fetal echocardiography in 3-dimensions and 4-dimensions. Ultrasound in Medicine and Biology, 22(8):979--986, 1996.

[2] R. N. Rankin, A. Fenster, D. Downey, P. L. Munk, M. F. Levin, and A. D. Vellet. Three-dimensional sonographic reconstruction: techniques and diagnostic applications. American J. of Roentgenology, 161(4):695-702, 1993.

[3] H. Steiner, A. Staudach, D. Spitzer, and H. Schaffer. Three-dimensional ultrasound in obstetrics and gynaecology: technique, possibilities and limitations. Human Reproduction, 9(9):1773--1778, 1994.

[4] C. Barry, C. Allot, N. John, P. M. Mellor, P. Arundel, D. Thomson, and J. C. Waterton. Three-dimensional freehand ultrasound: Image reconstruction and volume analysis. Ultrasound in Medicine and Biology, 23(8):1209--1224, 1997.

[5] P. R. Detmer, G. Bashein, T. Hodges, K. W. Beach, E. P. Filer, D. H. Burns, and D.E. Strandness Jr. 3D ultrasonic image feature localization based on magnetic scan- head tracking: in vitro calibration and validation. Ultrasound in Medicine and Biology, 20(9):923--936, 1994.

[6] S. W. Hughes, T. J. D'Arcy, D. J. Maxwell, W. Chiu, A. Milner, J. E. Saunders, and R. J. Sheppard. Volume estimation from multiplanar 2D ultrasound images using a remote electromagnetic position and orientation sensor. Ultrasound in Medicine and Biology, 22(5):561--572, 1996.

[7] D. F. Leotta, P. R. Detmer, and R. W. Martin. Performance of a miniature magnetic position sensor for three-dimensional ultrasound imaging. Ultrasound in Medicine and Biology, 24(4):597--609, 1997.

[8] T. R. Nelson and T. T. Elvins. Visualization of 3D ultrasound data. IEEE Computer Graphics and Applications, pages 50--57, November 1993.

[9] D. L. King, D. L. King Jr., and M. Y. Shao. Evaluation of in vitro measurement accuracy of a three-dimensional ultrasound scanner. Journal of Ultrasound in Medicine, 10:77--82, 1991.

[10] R. Ohbuchi, D. Chen, and H. Fuchs. Incremental volume reconstruction and rendering for 3D ultrasound imaging. Proceedings of SPIE --- The International Society for Optical Engineering, pages 312--323, 1992.

[11] A. State, D. T. Chen, C. Tector, A. Brandt, H. Chen, R. Ohbuchi, M. Bajura, and H. Fuchs. Case study: Observing a volume rendered fetus within a pregnant patient. In Proc. IEEE Visualization, 1994, pages 364--368, 1994.

[12] J. W. Trobaugh, D. J. Trobaugh, and W. D. Richard. Three-dimensional imaging with stereotactic ultrasonography. Computerized Medical Imaging and Graphics, 18(5):315-- 323, 1994.

[13] Shattuck, D. P., Weishenker, M.D., Smith, S.W., and von Ramm, O.T. ``Explososcan: A Parallel Processing Technique for High Speed Ultrasound Imaging with Linear Phased Arrays." JASA. 75(4): 1273-1282.

[14] Smith, S. W., Pavy, Jr., S.G., and von Ramm, O.T.``High-Speed Ultrasound Volumetric Imaging System - Part I: Transducer Design and Beam Steering." IEEE Transaction on Ultrasonics, Ferro., and Freq. Control. 38(2): 100-108.

[15] P. R. Detmer, G. Bashein, T. Hodges, K. W. Beach, E. P. Filer, D. H. Burns, and D.E. Strandness Jr. 3D ultrasonic image feature localization based on magnetic scanhead tracking: in vitro calibration and validation. Ultrasound in Medicine and Biology, 20(9):923-936, 1994.

[16] D. F. Leotta, P. R. Detmer, and R. W. Martin. Performance of a miniature magnetic position sensor for three-dimensional ultrasound imaging. Ultrasound in Medicine and Biology, 24(4):597–609, 1997.

[17] R. W. Prager, R. Rohling, A. Gee, and L. Berman. Rapid calibration for 3-D freehand ultrasound. Ultrasound in Medicine and Biology, in press.

[18] Emad M. Boctor, Ashraf A. Saad, Prof A. M. Youseef, and Prof James Graham "*Heuristic Based Approach For Extracting Calibration Parameters of Ultrasound Equipment* ", presented in the ISCA International Conference, June 11-13, 1997, Boston, Massachusetts, USA.

[19] Ashraf A. Saad, Emad M. Boctor and Prof Abo Bakr Youssef(PhD/MD), "*Statistical Based Automated Ultrasound Imaging Quality Control and Procedure* ", presented in the Fourth IEEE International Conference on Electronics, Circuits, and Systems ICECS'97, December 15-18, 1997, Cairo, Egypt.

[20] R. W. Prager, A. Gee, and L. Berman. STRADX: Real-Time Acquisition and Visualization of Free-Hand 3D Ultrasound CUED/F-INFENG/TR 319, Cambridge University Department of Engineering, April 1998.

[21] R. Ohbuchi, D. Chen, and H. Fuchs. Incremental volume reconstruction and rendering for 3D ultrasound imaging. Proceedings of SPIE - The International Society for Optical Engineering, 1808:312-323, 1992.

[22] G. Sakas, L-A. Schreyer, and M. Grimm. Preprocessing and volume rendering of 3D ultrasonic data. IEEE Computer Graphics and Applications, 15(4):47--54, July 1995.

Automatic Real-Time XRII Local Distortion Correction Method for Digital Linear Tomography

Christian Forlani, Giancarlo Ferrigno

Dipartimento di Bioingegneria, Politecnico di Milano, Piazza Leonardo da Vinci 32, 20133 Milano (Italia)
{forlani, ferrigno}@biomed.polimi.it
http://www.biomed.polimi.it

Abstract. An x-ray image intensifier (XRII) has many applications in diagnostic imaging, especially in real time. Unfortunately the inherent and external distortions (pincushion, S-distortion and local distortion) hinder any quantitative analysis of an image. In this paper an automatic real-time local distortion correction method for improving the quality of digital linear tomography images is presented. Using a digital signal processing (DSP), this method can work for an image up to 1Kx1Kx12 bit at 30fps. A local correction method has been used because it allows distortions such as those caused by poor axial alignment between the X-Ray cone beam and the XRII input surface and local distortions to be resolved that are generally neglected by global methods.

Introduction

Linear tomography is a helpful diagnostic exam that permits the acquisition of an x-ray relative to any plane, parallel to the x-ray table. However, inherent and external disturbances on the acquisition system chain cause several distortions when images are acquired by means of a digital system (XRII coupled with a CCD camera) [1][2].
The aim of this work is the development of an automatic real-time method for the correction of geometric distortions in digital linear tomography applications. A local correction algorithm has been used because it guarantees an efficient distortion correction only where needed. Geometric distortions considered and corrected here are: S-distortion, pincushion distortion, asymmetrical and misaligned axes, and local distortion.

Material and methods

The method needs a preliminary calibration setup based on the acquisition of a radio opaque square metal grid. For this work a 5x5 mm^2 cell size and a 2 mm metal wire thickness have been chosen, but no restrictions are imposed by the method. The acquisition-system consisted of 16" XRII tube coupled with a 1Kx1Kx 12bit CCD mounted on GMM Opera X-Ray table. The KV and mAs have been selected for guaranteeing an optimal contrast between grid and background with no image saturation. The DSP used for real-time application was a Matrox Genesis© PCI board working under Windows NT O.S..

Automatic grid detection and classification

Unlike other wire intersection point detection methods [3], this method uses as control points, the grid cells' Centers Of Gravity (COG) calculated from the acquired grid image by using a hardware blob analysis technique. This results in a reduction in the sensitivity to noise during control point set up. In order to exclude imperfect information about image cells across a radiographic circle, a preliminary circular region threshold has been applied. After thresholding, the method begins to classify the principal distorted axes, starting from the central COG. Cell classification has been obtained by an iterative procedure which is able to find the COG closest to a classified one along a desired direction, starting from a classified one. Nearest neighborhood COG can be found even if its position is shifted up to half a side of an expected square. The minimum distorted axis length is used to define the maximal squared region size where correction is applied. Once the principal axes are calculated, their information are used to classify all the COGs belonging to each quadrant region included in the square-shaped regions. Due to the presence of the radiographic circle and in order to guarantee the maximal dimension of the correction area, external circle COG cells are estimated by means of bilinear interpolation. Classified and estimated COG coordinates are ordered into a matrix. The central cell's area and COG are used to estimate the default cell and wire size parameters useful for the correction algorithm.

Correction algorithm

The correction algorithm uses two matrixes of COG points in order to create two bi-Dimensional Look Up Tables (2D LUT) used by a DSP to perform a real-time distortion correction. The first matrix (DM) containing distorted COG coordinates is obtained from the above described grid classification procedure. The second matrix (CM) is filled with corresponding corrected COG coordinates calculated using the default parameters (default cell and wire sizes), starting from the central distorted COG position. Each polygon formed by four adjacent elements (COG points) contained in the first matrix defines each single distorted local area. In the second matrix, the corresponding corrected square local area is defined by four elements located in the same matrix coordinates (see black points in Fig.1.). Each local area is corrected by using a 2D polynomial function characterized by 20 parameters. In order to calculate all parameters, sixteen COG points have been used: four concerning the local area and twelve belonging to the boundary of the square region. The COG information of the boundaries has been used to improve image continuity across neighboring local areas. The parameters obtained for each local area are used to create a 2D LUT that is used by the DSP image's real-time correction.

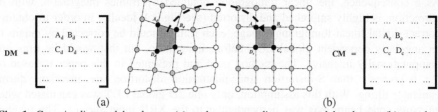

(a)   (b)

Fig. 1. Generic distorted local area (a) and corresponding corrected local area (b), used to correct distortion. Local area is delimited by four COG (black points), while boundary region (gray COG points) is used to guarantee continuity across neighboring local areas.

Linear tomography correction

The acquisition procedure, when traditional films are used, consists of continuous image integration during generator tilting (±20 degrees) and film shifting. The result is an x-ray image focused on a plane that is parallel to the table at the center of rotation. When digital equipment is used, the film is replaced by an XRII and a CCD and continuous integration is replaced by a discrete image integration. As noted before, due to XRII disturbance, each frame's results are affected by both typical XRII distortions as well as by angular distortions (see Fig.2a).

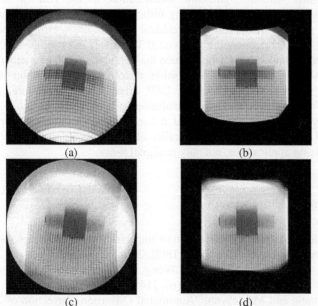

(a)   (b)

(c)   (d)

Fig. 2. Single frame distorted (a) and corrected (b). Linear Tomography distorted (c) and corrected (d).

As a consequence, the linear tomography obtained by frames integration, with no correction, is highly smoothed and distorted (see Fig.2c). Ideally, in order to obtain a correct digital linear tomography image, each frame should be corrected by means of an angularly dependent global LUT before integration, but that would be time and computationally intensive. The method used and presented in this paper is based on the hypothesis that S-distortion and pincushion distortion are invariant during generator tilting. With this assumption, only one 2D global LUT was calculated when an x-ray cone beam axis was perpendicular to the XRII input surface. The residual angular warping is corrected by fast global cosine correction method. In Fig.2b a corrected single frame, where the distortion and warping are substantially reduced, is shown. Comparing Fig.2c and Fig.2d, it is possible to note how the XRII disturbance, especially the S-distortion and pincushion distortions, deform and rotate the image reducing its quality.

Conclusion

An automatic local method to correct XRII image distortion in order to increase digital linear tomography image quality has been developed. Neither a grid cell size limit, or distortion information are needed in order to create 2D global LUT, with the exception of neighbor shift limit condition. Local correction allows the reduction of local distortions without significant modification to the rest of image. Image continuity has been guaranteed by using information about neighbor bounding cells. In simulated and real image tests, the residual distance mean errors were about 2.5% of the initial ones, corresponding to about 0.38 mm. It is expected that the use of a higher grid cell number can further reduce the residual error. The method is based on the hypothesis that the S and pincushion distortions are invariant during linear tomography acquisition. By using a DSP after set up procedure, this method guarantees a real time operation with resolution up to 1024x1024x12 bits. The method is completely self-setting, meaning that it can be directly hardware implemented as part of CCD camera. Its notable generality also permits its use in a wide range of image processing applications, not only in digital x-ray.

References

1. J.M. Boone, "Analysis and correction of imperfections in the image intensifier-TV-digitizer imaging chain" Med. Phys. 18(2), 236-242(1991).
2. S. Rudin, D.R. Bednarek and R. Wong, "Accurate characterization of image intensifier distortion", Med Phys. 18(6), 1145-1151(1991).
3. D.A. Reimann and M.J. Flynn, "Automated Distortion Correction of X-ray Image Intensifier Imager" Nuclear Science Symposium and Medical Imaging Conference, 2, 1339-1341(1992).

Meeting the Computational Demands of Nuclear Medical Imaging Using Commodity Clusters

Wolfgang Karl[1], Martin Schulz[1], Martin Völk[2], and Sibylle Ziegler[2]

{karlw, schulzm, voelk}@in.tum.de, s.ziegler@lrz.tum.de

[1] Lehrstuhl für Rechnertechnik und Rechnerorganisation, LRR–TUM
Institut für Informatik, Technische Universität München, Germany
[2] Nuklearmedizinische Klinik und Poliklinik, MRI-NM,
Klinikum Rechts der Isar, Technische Universität München, Germany

Abstract. Even though Positron Emission Tomography (PET) is a relatively young technique within Nuclear Medical Imaging, it has already reached a high level of acceptance. However, in order to fully exploit its capabilities, computational intensive transformations have to be applied to the raw data acquired from the scanners in order to reach a satisfying image quality. One way to provide the required computational power in a cost–effective and efficient way, is to use parallel processing based on commodity clusters.

These architectures are traditionally programmed using message passing. This, however, leads to a low–level style of programming not suited for the general user. In this work, a new programming environment based on a graphical representation of the application's behavior has been successfully deployed. The result is an image transformation application, which is both easy to program and fulfills the computational demands of this challenging application field.

1 Motivation

Over the last few years, Positron Emission Tomography (PET) has become a very important instrument in medical diagnosis procedures. However, in order to reach the level of image quality needed, computational intensive algorithms need to be deployed for the conversion of raw scanner data to humanly readable images, the so called PET image reconstruction. Quite a bit of work has been invested in increasing the image quality [5], but the computational demands remain high. One way to match these requirements in order to keep the time needed for image reconstruction process at an acceptable level and therefore to make the application of these improved algorithms in daily clinical routine feasible, is the deployment of parallel computing.

An attractive platform for such an approach are clusters built from commodity parts. They are cost effective and easy to build and maintain. Due to these very favorable properties, this class of architectures has recently earned a lot of attention and has started to replace traditional large–scale tightly coupled parallel systems. Clusters are generally programmed using message passing, mostly in the form of a standard library like PVM [1] or MPI [6], as this directly matches their distributed memory organization. These message passing APIs, however, are in most cases quite complex and cumbersome to apply. The user has to worry about many implementation details related to communication and data management, which do not belong to the actual problem to solve. This discourages many potential users, especially those directly from application areas without a formal computer science background, and therefore hinders the wide deployment of cluster architectures outside of computer science research.

One approach to overcome this problem of complex programmability is to deploy a programming environment which offers a high level of abstraction to the user and hides most of the low–level complexity of the underlying architecture. Such an approach has been taken within the NEPHEW project [9], which this work is part of, by applying a graphical programming environment, called Peak-Ware [8], for the implementation of several real–world applications including the reconstruction of PET images discussed in this paper.

In PeakWare, any communication within the whole system is abstracted into a modularized graphical representation which is easily comprehendible for the application programmer. The actual implementation of the communication, as well as the mapping of the application onto the target architecture, is automatically taken care of by the system. This creates a framework that enables an efficient implementation of a PET image reconstruction software on top of a Windows based PC cluster without having to deal with most of the complex issues normally involved in parallel and distributed processing. The resulting system is able to perform the reconstruction of a whole body scan in about 3 minutes. This is within the limits that allow an interactive on-line diagnosis by doctors while the patient remains in the clinic and within the scanner.

The remainder of this paper is organized as follows: Section 2 introduces PET imaging and its computational demands, followed by a discussion on how to match those using cluster computing in Section 3. Section 4 then presents the graphical programming environment used within this work and Section 5 provides details about the implementation of the PET reconstruction algorithm within this environment. The performance of the system is then evaluated in Section 6. The paper is rounded up by some concluding remarks and a brief outlook in Section 7.

2   Nuclear medical imaging using PET

Positron Emission Tomography is a nuclear medicine technique which allows to measure quantitative activity distributions in vivo. It is based on the tracer principle: A biological substance, for instance sugar or a receptor ligand, is labeled with a positron emitter and a small amount is injected intravenously. Thus, it is possible to measure functional parameters, such as glucose metabolism, blood flow or receptor density. During radioactive decay, a positron is emitted, which annihilates with an electron. This process results in two collinear high energy gamma rays. The simultaneous detection of these gamma rays defines lines-of-response along which the decay occurred. Typically, a positron tomograph consists of several detector rings covering an axial volume of 10 to 16 cm. The individual detectors are very small, since their size defines the spatial resolution. The raw data are the line integrals of the activity distribution along the lines-of-response. They are stored in matrices (sinograms) according to their angle and distance from the tomograph's center. Therefore, each detector plane corresponds to one sinogram. Image reconstruction algorithms are designed to retrieve the original activity distribution from the measured line integrals. From each sinogram, a transverse image is reconstructed, with the group of all images representing the data volume.

Ignoring the measurement of noise leads to the classical filtered backprojection (FBP) algorithm [3]. Reconstruction with FBP is done in two steps: Each projection is convolved with a shift invariant kernel to emphasize small structures but reduce frequencies above a certain limit. Typically, a Hamming filter is used for PET reconstruction. Then the filtered projection value is redistributed uniformly along the straight line. This approach has several disadvantages: Due to the filtering step it yields negative values, particular if the data is noisy, although intensity is known to be non-negative. Also the method causes streak artifacts and high frequency noise is accentuated during the filtering step.

Iterative methods were introduced to overcome the disadvantages of FBP. They are based on the discrete nature of data and try to improve image quality step by step after starting with an estimate. It is possible to incorporate physical phenomena such as scatter or attenuation directly into the models. On the down side, however, these iterative methods are very computational intensive. For a long time, this was the major drawback for clinical use of these methods, although they yield improved image quality.

One of the steps in iterative image reconstruction is the projection of measured PET data, just as in FBP. Expectation Maximization (EM) algorithms, however, forward project the images to compare the generated to the measured data. The result of this comparison is used to continually update the estimation of the image. The vector to update the image is calculated by using Poisson variables modeling the pixel emissions. With the Likelihood Function, an image is estimated for which the measured PET data would have been most likely to occur.

Following scheme shows one iteration step of an Expectation Maximization (EM) algorithm:

1. Estimation of distribution
2. Forward-projection
3. Compare estimated projection with measured projection
4. End loop when error estimate under predefined value
5. Back-projection of error estimate
6. New estimation of distribution

Since this method converges very slowly, this iteration step has to be performed many times (e.g. 100) for each pixel in each plane to converge. To reduce the number of iterations, a "divide and conquer" strategy is used. With

OSEM (Ordered Subset Expectation Maximization) [4] the events registered by the PET-scanner are divided into subsets. In one OSEM-iteration the typical steps of projection and back projection are done for each of theses subsets. The start-value for an iteration for each subset is gained from the result of the back projection of the previous subset. With ordering the subsets in a way that there is a maximum of information between each of them, the speed is further improved. Still, due to the iterative nature a substantial amount of time is required for the reconstruction of one image plane.

3 Meeting the computational demands with the help of clusters

In order to meet these computational demands for PET image reconstruction and still keep the total reconstruction time at an acceptable level, it is necessary to apply parallel processing. It is not only suited to speed-up the iterative reconstruction, but is also likely to guarantee image reconstruction times useful for a semi real–time, on–line diagnosis of a patients PET scan results while the patient is still within the PET scanner. This increases the medical accuracy as scans can easily be repeated without extra scanner time and therefore shortens the turnaround time for medical treatment. A typical upper bound for the reconstruction of a PET image that allows such guarantees is about four to five minutes.

An architecture well suited for this endeavor are clusters of commodity PCs. They provide an excellent price/performance ratio, are easy to build and maintain, and easily scalable to match the concrete computational demands. They are already used in many real–world application scenarios, including Nuclear Medical Imaging [7]. The main problem connected with this approach, however, is their difficult programmability. The most common programming approaches for clusters are based on message passing libraries, like MPI [6] or PVM [1]. With these libraries, however, the user is forced to a very low–level style of programming and is required to take care of additional tasks including the partitioning of code and data, the mapping of code onto individual nodes within the cluster, and the complete communication setup and management. This introduces a significant amount of additional complexity, which is not related to the concrete problem.

In order to make cluster architectures attractive to users not especially trained in parallel processing, a different programming environment has to be provided at a much higher level of abstraction. Such an environment has to be capable to hide the implementation complexity of the underlying architecture without sacrificing performance and/or functionality.

4 Easing the Programmability using a Visual Approach

In order to establish such an environment for the implementation of the PET image reconstruction, discussed in this work, a graphical tool, called PeakWare [8], is used. This tool was originally developed by Matra Systems & Information for real-time multiprocessor systems and has been adopted for Windows 2000 based clusters within the NEPHEW project. It completely hides the implementation of the communication framework from the user by automatically generating it from a graphical representation. This includes any startup procedures, notification mechanisms, as well as the actual data transport itself. The user only has to code the actual application functionality and PeakWare then automatically combines the individual parts into a full application.

Any development in PeakWare is generally done in five steps. First the application has to be decomposed into individual functional units, called modules, one of the central concepts of PeakWare. The separation into modules has to be done in a way that all communication between them can cleanly be specified in the form of a data flow graph.

The information gained through this analysis is then used to graphically describe the communication behavior in the so called software graph. An example with two modules and bidirectional communication is shown in Figure 1. Peak-Ware also offers the ability to scale individual modules, i.e. to replicate and distribute them among the cluster. This concept offers an easy way to introduce data parallelism into an application and allows the easy scaling to potentially arbitrary numbers of nodes.

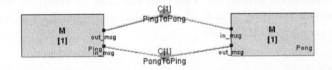

Fig. 1. PeakWare software graph (simple ping-pong communication)

Each module consists of several functions with the global communication channels as input and output arguments. The implementation of the functions themselves is done in external source files using conventional sequential programming in C. In the final application, these functions are then triggered automatically by PeakWare at corresponding communication events without requiring any further user intervention. This has to be seen in contrast to typical message passing models which require explicit receive calls and an explicit binding of incoming messages to their processing functions within the code.

The next step is the definition of the hardware that is supposed to be used for the application. This is again done with a graphical description, the hardware graph. An example of such a graph can be seen in Figure 2. It shows a small cluster of two compute and one host node connected by Fast Ethernet. This concept of an independent hardware graph enables the user to change the hardware in terms of node description and/or number of nodes without requiring changes in the software graph or the application itself.

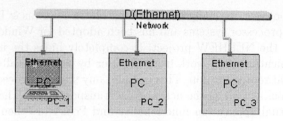

Fig. 2. PeakWare hardware graph (2 node cluster, with external development host)

Once the hard- and software graph have been completed, PeakWare gives the user the option to specify a mapping between modules (from the software graph) and nodes (as specified in the hardware graph). This mapping defines which module is executed on which node and hence represents the connection between the two graphs. It also allows the easy retargeting of applications to new hardware configurations as well as simple mechanisms for static load balancing.

The last step, after the mapping has been done and all routines have been implemented in external source files, is the code generation. In this process Peak-Ware uses the information from the graphical description of the software and hardware graphs and generates C source code that includes all communication and data distribution primitives. This code can then be compiled with conventional compilers resulting in a final executable and a shell script to start the application on all specified nodes.

5 PET image reconstruction using PeakWare

Due to their regular data layout, the decomposition for the PET image reconstruction can be achieved in a quite straightforward manner. Each input volume consists of a number of image planes (typically 47 or 63 per scan, depending on the scanner type). The reconstruction for each of these planes can be done independently. Therefore, a parallelization at the granularity of individual image planes is most promising.

This basic scheme was implemented in PeakWare using three different modules: a sender-module reads the raw input data and distributes the read image planes to a scaled consumer module in a round-robin fashion, one plane at a time. The consumer performs the actual reconstruction and after its completion forwards the resulting data to a receiver module, which is responsible for storing the final image. In addition the sender is informed, that a subsequent plane can be sent. The sender distributes the planes of the image until the reconstruction of all image planes has been acknowledged by the receiver. If no planes are left, planes might be resubmitted to idle consumers resulting in an easy, yet efficient fault tolerance scheme with respect to the consumer modules.

The software graph based on this design is depicted in Figure 3. It shows the three modules, with the consumer module scaled to $\$(SCALE)$ instances.

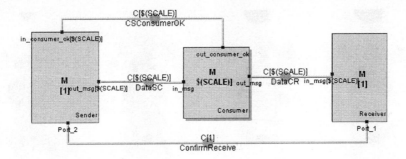

Fig. 3. Software graph for the PET reconstruction application.

In addition, the main data paths form the sender through the consumer to the receiver is visible in the middle augmented by two acknowledgment paths leading back to the sender.

The software-graph with its three modules can be mapped onto arbitrary hardware graphs. This enables the easy distribution of consumer module instances across arbitrarily scaled clusters. It is also possible to change the number of nodes without changing the software-graph by simply remapping it to a different hardware-graph. This independence of hardware and software description drastically eases the port of application between different hardware configuration and allows for an easy-to-handle scalability for the PET reconstruction application without the need for any code modifications.

6   Experimental Setup and Results

For the evaluation of the approach presented here, a research cluster consisting of four equally configured Dual processor PC nodes has been used. Each node is based on Intel's Xeon processors running at 450 MHz and is equipped with 512 MB main memory each. The interconnection fabric between the nodes is based on switched Fast Ethernet, which is used for both the external connection to the campus backbone, as well as for inter–node communication during the PET image reconstruction. All nodes run Windows 2000 enhanced only by a separate remote shell daemon.

For the evaluation, two different data sets have been used: one small data set of a human lung and one large data set of a whole human body. While the first one is acquired with a single scan, the latter resembles several consecutive scans of the different body sections, which are then merged into one full image. The complete size and resolution data for both data sets is shown in Table 1 and some examples of resulting image slices for the whole body case are shown in Figure 4. These two sets represent the two extreme cases of data sets currently available in the clinical routine. With improving scanner quality and availability of increased processing power, however, the trend is certainly towards larger data set sizes.

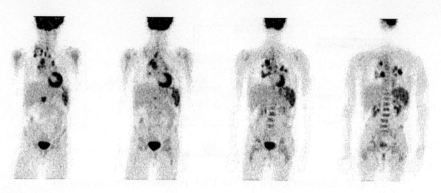

Fig. 4. Reconstructed and post–processed images of a human body (coronal slices, Nuklearmedizin TUM).

Description	Parameters					Execution times		
	Size	Planes	Scan Res.	Image Res.	Scans	Seq.	4 CPUs	8 CPUs
Human lung	14 MB	31	256x192	128x128	1	2m 54s s	1m 13s	1m 2s
Whole body	130 MB	282	256x192	128x128	6	15m 41s	4m 41s	3m 0s

Table 1. Evaluation data sets and the absolut exection times in various configurations

Table 1 also includes the absolute execution times of the code using the two data sets on various numbers of CPUs and Figure 5 (left) shows the same data transformed into speed-ups over the runtime time of a sequential execution without PeakWare involved. While the smaller data set performs rather purely, not exceeding a speed-up of 2.8 on 8 CPUs, the larger data set exhibits a much better performance with a speed-up of over 5.2. More important for the usability of the system in daily clinical routine is that the absolute execution times for any data set does not exceed the 3 minutes on 8 CPUs. This is short enough to allow a medical diagnosis based on a scan while the patient is still inside the PET scanner. Scans can therefore be repeated or adjusted to new target areas immediately without having to reschedule the patient for additional scans.

For a closer evaluation of the performance details, Figure 5 (middle and right) shows the aggregated absolute execution times over all CPUs in a breakdown into the different program phases: the startup time needed by the PeakWare environment to launch the program on all nodes, the file I/O time needed to read the initial data set and to store the final image volume, the time spent in the communication subroutines, and the time needed for the actual image reconstruction. The latter one is further broken down in a phase called *Weights*, a sequential preprocessing step, and *Rekon* the execution of the iterative reconstruction process. In this kind of graph bars of equal heights indicate perfect scalability, as the total execution time or work spent on any number of nodes is

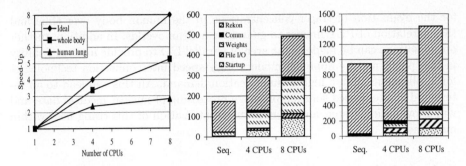

Fig. 5. Speedup (left) and Aggregated execution times (in sec) — middle: data set human lung, right: data set whole body.

equal showing that no additional overhead is introduced during the parallelization.

This behavior is clearly visible for the actual reconstruction phase for both data sets. In the small data set, however, where the total execution time on larger number of nodes is dominated by the sequential and therefore not scalable phases of the code, especially the preprocessing phase *Weights*, the overall speed-up is severely limited. In the case of the larger data set, the reconstruction time clearly outweighs any other phase, which translated into the high speed-up values shown above. It can be noted, however, that with increasing data set sizes, the I/O phase becomes more relevant demanding a new parallel I/O solution for the future with even larger data sets envisioned.

7 Conclusions and Future work

Nuclear imaging using Positron Emission Tomography is establishing itself as an important and very useful method for medical diagnosis. It is, however, connected with large computational demands for the image preparation. In order to match those, parallel processing is required and clusters built from commodity components provide a very cost–effective platform for this application domain. The problem connected with this approach is its complex and cumbersome programmability, making it difficult for scientists to use them at their full capacity.

In order to overcome this problem, the approach presented here deploys a graphical tool, which allows the specification of independent modules along with the communication between them. This raises the level of abstraction significantly and therefore eases the implementation process. The result is a very efficient and easy–to–implement and –use PET image reconstruction system, which satisfies the requirements for a use in daily clinical routine. It is already used in a production environment and has found acceptance with both doctors and medical personnel.

Based on this success, the approach is likely to be widened to more applications within the domain of nuclear medical imaging, like the spectral analysis for the evaluation of tracer concentrations in the human body over time and the correlation of PET images to guarantee a clean overlay. In addition, further optimizations can be made in the area of parallel I/O to provide more scalability in this phase and by applying modern high–speed interconnection technologies for applications with a high communication demand. All this will lead to an integrated cluster based solution of image processing of nuclear medical data and will allow easier, faster, and more accurate utilization of this rising field within medicine.

Acknowledgments

This work was supported by the European Commission in the Fourth Framework Programme in the context of the ESPRIT Project 29907, NEPHEW (*NE*twork of *PC*s *HE*terogeneous *W*indows-NT Engineering Toolset). The reconstruction algorithms have been implemented at the University of Michigan by Prof. Fessler and are distributed in the form of a library, called ASPIRE [2].

References

1. A. Beguelin, J. Dongarra, A. Geist, R. Manchek, and V. Sunderam. *A User's Guide to PVM Parallel Virtual Machine.* Oak Ridge National Laboratory, Oak Ridge, TN 37831-8083, July 1991.
2. J. Fessler. Aspire 3.0 user's guide: A sparse iterative reconstruction library. Technical Report TR–95–293, Communications & Signal Processing Laboratory, Department of Electrical Engineering and Computer Science, The University of Michigan Ann Arbor, Michigan 48109-2122, November 2000. Revised version.
3. G.T. Herman. *Image Reconstruction from Projections.* Springer-Verlag, Berlin Heidelberg New York, 1979.
4. H. Hudson and R. Larkin. Accelerated image reconstruction using ordered subsets of projection data. *IEEE Transactions on Medical Imaging*, 13:601–609, 1994.
5. R. Leahy and C. Byrne. Recent developments in iterative image reconstruction for PET and SPECT. *IEEE Transactions on Nuclear Sciences*, 19:257–260, 2000.
6. Message Passing Interface Forum (MPIF). MPI: A Message-Passing Interface Standard. Technical Report, University of Tennessee, Knoxville, June 1995. http://www.mpi-forum.org/.
7. S. Vollmar, M. Lercher, C. Knöss, C. Michael, K. Wienhard, and W. Heiss. BeeHive: Cluster Reconstruction of 3-D PET Data in a Windows NT network using FORE. In *In proceedings of the Nuclear Science Symposium and Medical Imaging Conference,* October 2000.
8. WWW:. Peakware — Matra Systems & Information . http://www.matra-msi.com/ang/savoir_infor_peakware_d.htm, January 2000.
9. WWW:. SMiLE: Nephew (Esprit project 29907) . http://wwwbode.in.tum.de/Par/arch/smile/nephew, June 2000.

An Image Registration Algorithm Based on Cylindrical Prototype Model

Joong-Jae Lee, Gye-Young Kim, and Hyung-Il Choi

Soongsil University, 1-1, Sangdo-5 Dong, Dong-Jak Ku, Seoul, Korea
ljjhop@vision.soongsil.ac.kr, {gykim,hic}@computing.soongsil.ac.kr

Abstract. We propose an image registration algorithm based on cylindrical prototype model to generate a face texture for a realistic 3D face model. This is a block matching algorithm which aligns 2D images of a 3D cylindrical model. While matching blocks it doesn't use same sized blocks but variable sized blocks with considering a curvature of 3D model. And we make a texture of aligned images using an image mosaic technique. For this purpose, we stitch them with assigning linear weights according to the overlapped region and using the cross-dissolve technique.

1 Introduction

The realistic 3D modeling provides user with more friendly interface and is also very useful in virtual reality, broadcasting, video teleconferencing, interface agent and so on. We can obtain the 3D model of human through the hardware equipment like 3D scanner [1]. It is very easy and simple solution, but it is quite an expense. A number of techniques have been developed for resolving this problem. One way is to use front and side views of face, which creates a model as texture mapping after modifying a generic model [2][3]. However, it only uses two face images so it is less realistic at other views except front and side views. Another way is to use face images of four or eight views, which produces a face texture using panoramic techniques and mapping [4]. Through it, we may have the natural result. But it requires images photographed in exact angle, because it uses only image stitching in panoramic techniques. Mapping to model with the images that weren't photographed in accurate angles makes an unnatural texture, and also we wouldn't create a realistic 3D model. To resolve such situations, we propose the image registration algorithm based on cylindrical prototype model. It enables us to align images precisely using the correlation between them, even though exact view angles would not be met when a face is photographed. And the natural face texture is generated by stitching the aligned images with the cross-dissolving. For this purpose, we adjust weights according to the overlapped region. So, more realistic 3D face model can be created.

2 Image registration with variable sized blocks

The process of block matching is to partition each of two images to same sized blocks. And then it finds a candidate block, within a search area in another image, which is most similar to the source block in one image, according to a predetermined criterion [5]. But the curvature of 3D cylindrical model causes surface patches of same size to be projected on 2D image as blocks of different size. While matching blocks, the proposed algorithm doesn't use same sized blocks but variable sized blocks with considering a curvature of 3D model. We denote block widths of 2D image of 3D cylindrical prototype model as in (1)

$$
\begin{cases}
BW = k_n \\
k_0 = 2R\sin\left(\frac{\theta_{Div}}{2}\right) \\
k_n = k_0 \cos\left[\left(|n| - \frac{1}{2}\right) \times \theta_{Div}\right]
\end{cases}
\tag{1}
$$

$$
where\ 0 < \theta_{Div} < \frac{\pi}{4},\ 1 \le |n| \le \frac{\pi}{2\theta}
$$

where θ_{Div} denotes the division angle partitioning a 3D model into same portions, R denotes a radius of model. And k_0 denotes the base length of an inscribed triangle when the 3D model is divided by θ_{Div}, k_n except k_0 is the size of each block projected on 2D image. In (1), n represents column index of blocks projected on the same column of 2D image. It is designed to be positive for right hand side block and negative for left hand side block. It gets larger magnitude as its block lies closer to the left and right border of 2D image. Fig. 1 shows us the result of producing variable sized blocks. BW becomes small when the absolute value of index n grows.

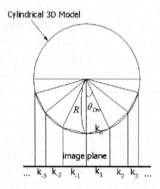

Fig. 1. Variable sized blocks depending on the curvature of cylindrical model

Fig. 2 shows us the different block widths of the proposed approach against the same block widths of the typical approach. In particular, we can see the variable sized blocks depending on the curvature of model. We partition an image to variable sized blocks to perform block matching. That is, we changeably adjust size of candidate blocks to be matched against the source block.

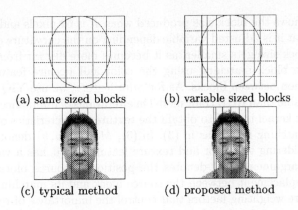

(a) same sized blocks   (b) variable sized blocks

(c) typical method   (d) proposed method

Fig. 2. Comparison of block size as block paritioning with front face view

Fig. 3(a) illustrates how to determine a block size of $BW(I_p)$ and the closed form of its equation is shown in (2). In (2), $BW(I_p)$ denotes the length between I_p and $I_{p\prime}$. It is obtained by looking for $I_{p\prime}$ which is projected on 2D image from $p\prime$ on 3D model. Here, $p\prime$ is separated from p on 3D model by division angle θ_{Div}.

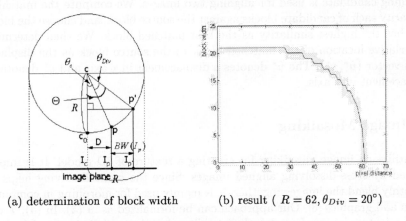

(a) determination of block width   (b) result ($R = 62, \theta_{Div} = 20°$)

Fig. 3. Variable adjustment of block width

$$BW(I_p) = |I_{p\prime} - I_p| = |I_{p\prime} - I_c| - |I_p - I_c| \qquad (2)$$
$$= Rsin\Theta - D$$
$$= Rsin\left(\theta_{Div} + \theta_c\right) - D$$
$$where \ \theta_c = sin^{-1}\left(\frac{D}{R}\right), \ D = |I_p - I_c|$$

Fig. 3(b) shows the block size produced when R is 62 pixels and θ_{Div} is 20°, we can note that it is adjusted variable depending on the curvature of 3D model. In fact, the block size gets smaller, as it becomes more distant from center. We accomplish the block matching using the color and texture feature that can reflect correlation between blocks. As for color feature, we use YIQ color model that is more stable than RGB color model in alterations of luminance [6]. Besides, Gabor Wavelet kernel is used to obtain the texture characteristics of a block [7]. We define a matching metric as in (3). In (3), $MF(i, j; u, v)$ denotes matching similarity considering the color and texture feature and it has a value between 0 and 1. The argument (i, j) denotes the position of source block, and (u, v) denotes the displacement between the source block and its matching candidate. The α and β are weighting factors that control the importance of related terms. We assign 0.4 to α and 0.6 to β experimentally.

$$\mathbf{d}(i, j) = (u^*, v^*) \ where \ MF(i, j; u, v) \ is \ maximized \qquad (3)$$
$$MF(i, j; u, v) = 1 - [\alpha \cdot M_{color}(i, j; u, v) + \beta \cdot M_{texture}(i, j; u, v)]$$
$$where \ 0 < \alpha, \beta < 1, \alpha + \beta = 1, 0 \leq MF(i, j; u, v) \leq 1$$

In (3), the displacement of (u^*, v^*) between the source block and its best matching candidate is used for aligning two images. We compute the matching similarity each of candidate blocks against the source block, and choose the block that has the highest similarity as the best matched block. We then determine the relative location of the matched block to the source block as the displacement vector (u^*, v^*). The u^* denotes a displacement in x axis and v^* denotes a displacement in y axis.

3 Image Mosaiking

We utilize an image mosaiking for creating a texture of 3D model. It is implemented by cross-dissolving aligned images. Since the cross-dissolving helps to smoothly blend the images together, it is mainly used for morphing in computer vision and graphics [8]. Our approach can be formalized as in (4). In (4), T denotes a texture image, I_1 and I_2 depicts aligned 2D images. P_i denotes ith pixel, w_1 and w_2 denotes weights for I_1 and I_2 to stitch them.

$$T = \begin{cases} I_1, & if \ Visible(P_i, I_1) \ and \ \neg Visible(P_i, I_2) \\ I_2, & if \ \neg Visible(P_i, I_1) \ and \ Visible(P_i, I_2) \\ w_1 \cdot I_1 + w_2 \cdot I_2, & if \ Visible(P_i, I_1) \ and \ Visible(P_i, I_2) \end{cases} \qquad (4)$$

Especially for overlapped region, images are stitched by assigning weights w_1 and w_2 to I_1 and I_2, respectively, according to rules defined in (5). Fig. 4 shows three types of overlapping occurred when aligning two images with displacement of D.

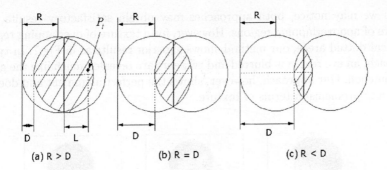

(a) R > D   (b) R = D   (c) R < D

Fig. 4. Three types of overlaps occurring when image stitching

Weights are determined as in (5) by considering the amount of overlap and displacement. In (5), R is radius of a model, D is a displacement between two aligned images and L denotes the distance from the center of an overlapped region to any pixel P_i as denoted in Fig. 4(a).

$$
w_1 = \begin{cases} \frac{2L}{2R-D}, & \text{if } D > R \ or \ D < R \\[2mm] \frac{2L}{R}, & \text{if} D = R \end{cases}
$$

$$
w_2 = 1 - w_1
$$

$$
where \ 0 < w_1, w_2 \leq 1, w_1 + w_2 = 1
$$
(5)

4 Experimental results and disccusions

To evaluate the proposed approach, we have used face images photographed in each of four and eight directions. Fig. 5 shows the result of generated texture from those images. To compare the performance of our approach with typical method, we experiment with face images having photographic errors of about 10° of viewing angle as shown in Fig. 6.

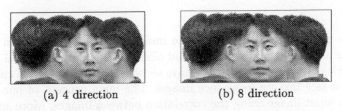

(a) 4 direction   (b) 8 direction

Fig. 5. Results of generated texture with face imges

As we may notice, both approaches may obtain satisfactory results for a texture of nonoverlapping regions. However, for a texture of overlapping regions, elliptical dotted areas, our method shows superior results as in Fig.7. In typical approach, an eye region is blurred and two ears are revealed much like the ghosts phenomenon. Our approach, however, shows the perfect alignment and does not make any problems in terms of texture.

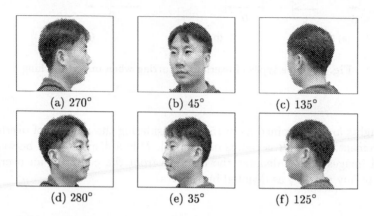

(a) 270°   (b) 45°   (c) 135°

(d) 280°   (e) 35°   (f) 125°

Fig. 6. Face images photographed in exact and inexact directions

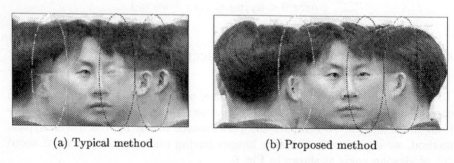

(a) Typical method   (b) Proposed method

Fig. 7. Comparison of results produced with the photographic errors

In this paper, we have proposed an image registration algorithm based on cylindrical prototype model which could align the 2D photographical images of 3D cylindrical prototype model. We have showed that our approach can be used to obtain a face texture using face images. The main concept of our approach is to align input image using the correlation between images. More specifically, when block matching, it uses variable sized blocks with considering a curvature of 3D model instead of same sized blocks. The proposed approach can solve the problems like the blurring effect and ghosts phenomenon, occurred when a texture is generated with images photographed at inaccurate angles.

Acknowledgement

This work was partially supported by the KOSEF through the AITrc and BK21 program (E-0075)

References

1. Yuencheng Lee, Demetri Terzopoulos, and Keith Waters.: Realistic Modeling for Facial Animation. Proceedings of SIGGRAPH 95. In Computer Graphics (1995) 55–62
2. Takaaki Akimoto, Yasuhito Suenaga.: 3D Facial Model Creation Using Generic Model and Side View of Face. In IEICE TRANS. On INF and SYST., Vol E75-D, No.2 (1992): 191–197
3. Takaaki Akimoto, Yasuhito Suenaga.: Automatic Creation of 3D Facial Models. In IEEE Computer Graphics and Applications (1993) 16–22
4. Han Tae-Woo.: 3D face modeling system for realistic facial expression animation. MD Thesis, Department of Computer Science, Korea Advanced Institute of Science Technology (1998)
5. Kyoung Won Lim, Byung Cheol Song, and Jong Beom Ra.: Fast Hierarchical Block Matching algorithm Utilizing Spatial Motion Vector Correlation. Proceedings of Visual Communications and Image processing, Vol. 3024 (1997) 284–291
6. Yu-Ich Ohta, Takep Kanade, Toshiyuki Sakai.: Color Information for Region Segmentation. In Comuter Graphics and Image Processing, Vol. 13 (1980) 222–241
7. W. Y. Ma and B. S. Manjunath.: Texture Features and Learning Similarity. IEEE International Conference on Computer Vision and Pattern Recognition, San Francisco, CA (1996)
8. G. Wolberg.: Image morphing: a survey. The Visual Computer, 14(8/9) (1998) 360–372

An Area-Based Stereo Matching Using Adaptive Search Range and Window Size[1]

Han-Suh Koo and Chang-Sung Jeong

Department of Electronics Engineering, Korea University
1-5ka, Anam-dong, Sungbuk-ku, Seoul 136-701, Korea
E-mail: {esprit@snoopy, csjeong@chalie}.korea.ac.kr

Abstract. In area-based stereo matching algorithm, the proper determination of search range and window size are two important factors to improve the overall performance of the algorithm. In this paper we present a novel technique for area-based stereo matching algorithm which provides more accurate and error-prone matching capabilities by using adaptive search range and window size. We propose two new strategies (1) for determining search range adaptively from the disparity map and multiresolutional images of region segments obtained by applying feature-based algorithm, and (2) for changing the window size adaptively according to the edge information derived from the wavelet transform such that the combination of two adaptive methods in search range and window size greatly enhances accuracy while reducing errors. We test our matching algorithms for various types of images, and shall show the outperformance of our stereo matching algorithm.

Keywords. Visualization, Stereo Vision, Image Processing

1 Introduction

The stereo matching algorithm is a technique that analyses two or more images captured at diverse view points in order to find positions in real 3D space for the pixels of 2D image. The stereo matching methods have been used in various fields such as drawing the topographical map from aerial photograph and finding the depth information of objects in machine vision system. Nowdays, optical motion capture techniques using stereo matching algorithms are being developed for visual applications such as virtual reality or 3D graphics.

Stereo matching algorithms can be generally classified into two methods: feature-based and area-based ones. The feature-based method matches the feature elements between two images, and uses the interpolation to obtain the disparity information for the pixels other than the feature elements, while the area-based method performs matching between pixels in two images by calculating the correlations of the pixels residing in the search window. Area-based method cannot match feature element with more accuracy than feature-based method even though it can make more dense disparity map [1, 2]. Moreover, it has more possibility of error in the area of insufficient texture information or depth discontinuities. In this paper we present a novel technique for area-based stereo matching algorithm which provides more accurate and error-prone

---

[1] This work was supported by KISTEP and BK21 Project.

matching capabilities by using adaptive search range and window size. We propose two new strategies (1) for determining search range adaptively from the disparity map and multiresolutional images of region segments obtained by applying feature-based algorithm, and (2) for changing the window size adaptively according to the edge information derived from the wavelet transform such that the combination of two adaptive methods in search range and window size greatly enhance accuracy while reducing errors. We test our matching algorithms for various types of images, and shall show the outperformance of our stereo matching algorithm.

The paper is organized as follows. In section 2, we briefly describe about the existing stereo matching algorithms and review the merits and defects of each approach. In section 3, we present our area-based stereo matching algorithm. In section 4, we explain the experimental results of our matching algorithm. Finally, in section 5, we give a conclusion.

2   Stereo Matching Methods

In this section we classify stereo matching methods into area-based and feature-based matching methods and briefly describe about them. Area-based method uses the correlation of intensity between patches of two images to find, for each pixel in one image, its corresponding pixel in the other image based on the assumption that the disparities for the pixels in the neighborhood region of one pixel are nearly constant. It can match each pixel in the right image to its corresponding pixel in the left image with maximum correlation or minimum SSD(Sum of squared difference) between two windows in the left and right images[3, 4]. However, most of the area-based methods are sensitive to noise and variation between images, and have a high probability of error in the area where there are insufficient texture information or depth discontinuities. Generally, if the window is too small, it give a poor disparity estimate, because the signal(intensity variation) to noise ratio is low. If, on the other hand, the window is too large, then the position of maximum correlation or minimum SSD may not represent correct matching due to the different projective distortions in the left and right images. To overcome this defect, Kanade and Okutomi[5] proposed an adaptive window method in an iterative stereo matching algorithm. They selected the size and shape of a window by iteratively updating the disparity estimate in order to obtain the least uncertainty for each pixel of an image. Also, Sun[6] proposed a multiresolutional stereo algorithm which makes use of dynamic programming technique. Previous works with adaptive window require so much processing time because of the iterative routines, and restrict the matching window to a perfect square. Our algorithm provides more simple and flexible method for selecting the window adaptively.

Feature-based method extracts primitives such as corners, junctions, edges or curve segments as candidate features for matching by finding zero-crossings or gradient peaks. Many researchers use line segments frequently in the feature-based stereo matching algorithms[7, 8] because they are abundant and easy to

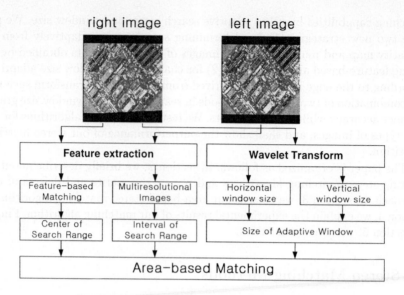

Fig. 1. Process flow diagram of our algorithm

find in an image, and their characteristics such as length, gradient and orientation are easy to match compared with the other primitives. Hoff and Ahuja[9] integrate feature matching and interpolation by using local parametric fitting in a hypothesis-test manner. Their algorithm is computationally expensive, because it considers too many matching possibilities when fitting for edges by Hough transform. Maripane and Trivedi[10] uses multiple primitives such as regions, lines, and edges in order to reduce the ambiguity among the features. Feature-based method is more accurate and less sensitive to photometric variations than area-based method since feature elements are used for matching. However, it requires additional interpolation for sparse feature elements, and has some difficulties in deriving a general algorithm applicable to all type of images.

3 Algorithms

In this section, we present a new area-based stereo matching algorithm using adaptive search range and adaptive window. Our algorithm consists of three stages. In the first stage, we compute, for each pixel in the right image, its search range in the left image. The search range in the left image is a one dimensional range along x axis which is defined by its center and interval. The center of the search range is obtained by executing feature-based matching with the region segments as feature elements, and the interval of the search range by investigating the region segments in multiresolutional images. In the second stage, we determine, for each pixel in the right image, its search window based on the edge information derived from the wavelet transform. In the third stage, we try to match each pixel in the right image with the pixels in the search range by moving the window along the pixels in the search range, and determine the disparity map by finding the pixel in the search range with the maximum correlation between two windows in the left and right images respectively. The diagram of the overall process flow of our algorithm is illustrated in figure 1. We shall describe about each stage more in detail in the following subsequent subsections.

3.1 Adaptive Search Range

In this subsection we shall show how to find the center and interval of the search range.

The center of the search range: The center of the search range is determined from the disparity map by executing feature-based matching with region segments as features. Most of the existing feature-based algorithms use edges as feature elements since they can obtain a relatively precise matching result due to the abundant appearance of edges in an image. Instead, our algorithm adopts region segments as feature elements since the smaller number of regions segments in an image, along with the region-based primitive's high discrimination power, reduces the number of false targets and increases the accuracy of matching. Moreover, the region segments can provide information about the search range for larger portion of the image when executing area-based matching algorithm.

1). Feature extraction: Region segments are extracted by finding threshold values from gray-level histogram[11]. Three threshold values are calculated by executing threshold-based segmentation method iteratively, and the region segments are divided into four groups according to the threshold values. Then, a series of high, medium, and low resolution images are generated by reductions in size by a factor of two using a Gaussian convolution filter for each group. Each group is represented by a binary map where each pixel indicates whether it belongs to the group or not. The number of groups is determined from the experiment so that the size of region segments may not be too large or too small. Although thresholding is unstable against photometric variations, the results in this step are acceptable, because it helps to find candidate matches but not correct matches.

2). Morphology filter: There may exist excessively small region segments generated during the feature extraction. Therefore, we apply the opening operation in morphology filter to the low resolution images in four groups in order to prevent the errors which may arise from those small region segments.

3). Feature-based matching: The disparity map is obtained by executing feature-based matching with region segments as feature elements. However, it is not the final disparity map, but the one which shall be used to compute the center of the search range and hence to find the more accurate disparity map by area-based matching. Therefore, we do not need the exact disparity map at this point. Thus, in our feature-based matching, low resolution images in four groups are used in order to ignore the small region segments and speed up the

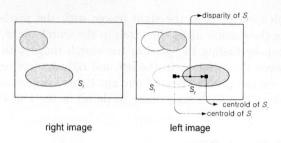

Fig. 2. Disparity of region segment

matching process. We up-sample the low resolution images to the image with original size in order to calculate the disparities in actual distance. After up-sampling, we match region segments using geometric properties such as position, size, eccentricity, and orientation, and find, for each pixel in the right image, its disparity and center of the search range as follows: Let S_l be a region segment in the right image and S_r be its matched region segment in the left image. If a pixel belongs to any region segment S_l, its disparity is defined as the difference in x value between the centroids of S_l and S_r; otherwise its disparity is set to 0. (See figure 2.) Then, for each pixel in the right image, the center of its search range is defined as the pixel in the left image which is apart from itself by its disparity. The center of the search range computed from the disparity map allows the more exact area-based matching through the enhanced selection of the search range.

The Interval of Search Range The interval of search range for a pixel in the right image is determined according to its closeness to the boundary of the region segment. The more a pixel is close to the boundary of the region segment to which it belongs, the larger the interval of its search range becomes to reflect the abrupt change in the boundary of the region segments. The closeness of a pixel to the boundary of the region segment is determined by investigating the inclusion property in the region segments of multiresolutional images as follows: First, we remove region segments which do not appear in low resolution image in order to eliminate small region segments which have insignificant influence but might incur errors. (See Figure 3.) Then, we up-sample each of multiresolutional images to the original high resolution size. For a region segment R_i, let U_l, U_m, and U_h be its up-sampled region segments from low, medium, and high resolution images respectively. Then, the relation $U_l \subset U_m \subset U_h$ usually holds, since during up-sampling, one pixel is expanded to its corresponding four neighbor pixels in the next higher resolution image. For the region segment R_i, we can classify three subregions: $SR_1 = U_h$, $SR_2 = U_m - U_h$, $SR_3 = U_l - U_m$. SR_1 is the innermost subregion which corresponds to the original region segment, and SR_3 is the outermost subregion. SR_2 is the subregion which lies between SR_1 and SR_3. Let I_i be the interval of the search range for a pixel p which lies in SR_i. Then, we determine, for each pixel p, its interval of the search range according to in which subregion it lies while satisfying $I_1 < I_2 < I_3$. In other words, the interval of the search range for a pixel becomes larger if it gets more close to the outer boundary of the region segments to which it belongs, since the edges in

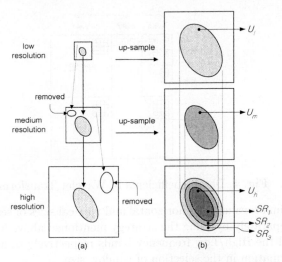

Fig. 3. The Interval of Search Range (a) Elimination of small region segment (b) Up-sampled regions U_l, U_m, U_h and subregions SR_1, SR_2, SR_3

the boundary the region segments are liable to change abruptly. If pixels are not detected in any up-sampled region segments, the interval of its search range is set to maximum since no disparity information is acquired from region segment matching process.

3.2 Adaptive Search Window

The proper size of matching window is one of the important factors to improve the performance of the area-based matching algorithm. A large window is proper to flat surface, but it blurs disparity edge. On the other hand, a small window preserves a disparity edge well, but it produces much error in flat surface. For this reason, a window size need to be selected adaptively depending on edge information. We propose a new strategy to select an appropriate window by evaluating edge information derived from wavelet transform[12]. Edge existence can be determined by applying a threshold to the detail coefficients of Low/High and High/Low frequency band since high frequency component appears in the detail coefficients. (See figure 4.) Therefore the vertical and horizontal sizes of the matching window are determined as follows: First, we apply 3-level 2D wavelet transform to the right image, and then determine whether a pixel is an edge component or not at each level of 2D wavelet transform by comparing the detail coefficient with the threshold value which is obtained from the experiment. The edge strength of a pixel can be regarded as more strong if the pixel is detected as edge component at lower levels. If a pixel is detected as edge component at all levels, the edge strength becomes the largest at the pixel, and the size of the search window is set to minimum value in order to capture the disparity edge more precisely. If a pixel is not detected as edge component at higher levels, the edge strength becomes smaller at the pixel, and the size of the search window increases. If an edge is not detected at any level, the size of search window is

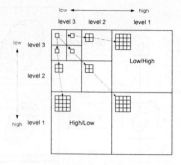

Fig. 4. Detail coefficients of Wavelet Transform

set to to maximum value. The horizontal and vertical sizes of search window are determined separately by using the strategy mentioned above for the Low/High frequency and the High/Low frequency bands respectively to adaptively reflect the edge information in the selection of window size.

3.3 Area-based Stereo Matching

After computing search range and window size for each pixel in the right image in the first and second stages respectively, we now attempt to execute area-based matching in order to find the final disparity map in the third stage. We try to match each pixel in the right image with the pixels residing in its search range of the left image using the appropriate window selected during the second stage, and compute the disparity map by finding the pixel in the left image with the maximum correlation between two image pairs in the corresponding windows. Finally, we perform the refinement of the disparity map by applying median filter in order to satisfy the continuity constraint[6]. The continuity constraint need to be satisfied since a disparity of single pixel which is excessively different from its neighborhood can be regarded as an error rather than correct value[3].

4 Experimental Results

We tested our stereo-matching algorithms presented in this paper for three real stereo images: apple, fruit and pentagon. Figure 5 illustrates stereo matching analysis for apple stereo image. Figure 5(a) shows a 512×512 right image for apple, and figure 5(b) shows the search range map where each pixel is represented by intensity according to its search interval. The brighter pixel represents the shorter search interval. Figure 5(c) shows the edge component map obtained by applying thresholding to the detail coefficients of wavelet transform for the right image, where the bright pixels at each level represent the existence of edge components. We used the daubechies basis vectors for wavelet transform. Based on the edge information, we determined the horizontal and vertical sizes of the search window. Figure 5(d) through (f) compare the disparity maps under various conditions. In the disparity map, the brighter pixel is the more close one to the camera. Figure 5(d) shows the disparity map obtained by applying general area-based matching algorithm with search range and the search window

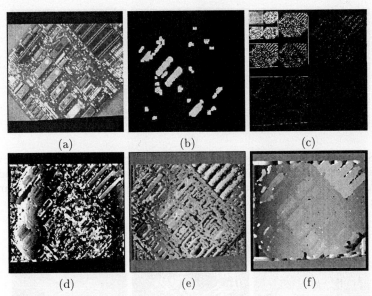

(a) (b) (c)

(d) (e) (f)

Fig. 5. Stereo matching analysis for apple image (a) Right image (b) Search range map (c) Edge component map obtained from wavelet transform (d) Disparity map of area-based matching algorithm using search range of 6 pixels and 9×9 search window (e) Disparity map of area-based matching algorithm using adaptive search range algorithm but without adaptive window algorithm (f) Disparity map of our area-based matching algorithm using adaptive search range and window

fixed to 6 pixels and 9×9 respectively, and figure 5(e) shows the disparity map obtained by area-based matching algorithm with adaptive search range but with 3×3 fixed window size. Figure 5(f) is the disparity map obtained by using our area-based matching algorithm using adaptive search range and window. In our algorithm, the search range varies from 6 pixels to 42 pixels, and the window size from 3 pixels to 21 pixels adaptively. Compared with figure 5(e) and (f), figure 5(d) shows more errors in disparities around the center of the image. Figure 5(e) clearly shows the improvement over figure 5(d) by using adaptive search range. However, figure 5(e) still has some errors around the center since 3×3 search window is too small to get the accurate result for the flat surface around the center. Figure 5(f) shows the further improvement over figure 5(e) by making use of adaptive search range and window together.

Figure 6 illustrates stereo matching analysis for 512×512 fruit image shown in figure 6(a). Figure 6(b) shows the disparity map obtained by area-based matching algorithm with adaptive window and fixed search range of 20 pixels. Compared with figure 6(c), figure 6(b) shows more errors in a tablecloth and an apple behind. This fact tells that too large and uniform search range is not adequate for distant objects in the image.

Figure 7 illustrates stereo matching analysis for 512×512 pentagon image shown in figure 7(a). We compare the disparity map for two cases: one when applying area-based matching algorithm using fixed search range and window size,

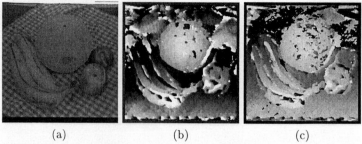

(a)   (b)   (c)

Fig. 6. Stereo matching analysis for fruit image (a) Right image (b) Disparity map of area-based matching algorithm using adaptive window algorithm but without adaptive search range algorithm (c) Disparity map of our area-based matching algorithm using adaptive search range and window

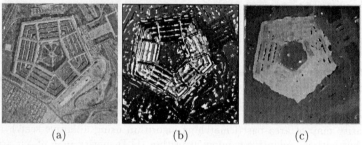

(a)   (b)   (c)

Fig. 7. Stereo matching analysis for pentagon image (a) Right image (b) Disparity map of area-based matching algorithm using search range of 20 pixels and 9×9 window (c) Disparity map of our area-based matching algorithm using adaptive search range and window

and the other using adaptive search range and window size as in our matching algorithm. Figure 7(b) shows the disparity map when using fixed search range of 20 pixels and fixed 9×9 window. In figure 7(b) we can easily see that some errors occurs in upper portion of the pentagon building due to the large window size. Figure 7(c) shows the clear improvement over figure 7(b) by using adaptive search range and window size. The comparison of disparity maps shows that the combination of adaptive search range and window as in our matching algorithm greatly enhance the matching capability by appropriately changing the search range and window size according to the various surface conditions.

5 Conclusion

In this paper we have presented the new technique for area-based stereo matching algorithm which provides more accurate and error-prone matching capabilities by using adaptive search range and window size. We have proposed two new strategies for finding adaptive search range and window size respectively.

The center of the search range is obtained by executing feature-based matching with the region segments as feature elements. Adopting region segments as feature elements allows reduction in the number of false targets and increase in the accuracy of matching due to the smaller number of region segments in an image, along with the region-based primitive's high discrimination power while providing information about the search range for larger portion of the image for area-based matching. The center of the search range computed from the disparity map using feature-based matching allows the significant enhancement in the selection of the search range. The interval of the search range is determined by investigating the up-sampled multiresolution images. Classifying the region segments in multiresolutional images into several subregions provides the elaborated criteria to the closeness to the boundary of the region segment and allows the more refined care for the abrupt change in the edge boundary. The appropriate window is selected adaptively according to the edge information derived from the wavelet transform by reflecting the edge information appearing in the detail coefficients of the wavelet transform. We have tested our matching algorithms for various types of images, and have shown that the combination of two adaptive methods in search range and window size greatly improves the performance of area-based matching algorithm.

References

1. S. D. Cochran, G. Medioni, "3-D Surface Description From Binocular Stereo," IEEE Trans. PAMI, vol. 14, no. 10, pp. 981-994, Oct. 1992.
2. G. Wei, W. Brauer, and G. Hirzinger, "Intensity- and Gradient-Based Stereo Matching Using Hierarchical Gaussian Basis Functions," IEEE Trans. PAMI, vol. 20, no. 11, pp. 1143-1160, Nov. 1998.
3. O. Faugeras, "Three-Dimensional Computer Vision; A Geometric Viewpoint," pp. 189-196, Massachusetts Institute of Technology, 1993.
4. M. Okutomi and T. Kanade, "A Multiple-Baseline Stereo," IEEE Trans. PAMI, vol. 15, no. 4, pp. 353-363, Apr. 1993.
5. T. Kanade and M. Okutomi, "A Stereo Matching Algorithm with an Adaptive Window: Theory and Experiment," IEEE Trans. PAMI, vol. 16, no. 9, pp. 920-932, Sep. 1994.
6. C. Sun, "A Fast Stereo Matching Method,", pp. 95-100, Digital Image Computing: Techniques and Applications, Massey University, Auckland, New Zealand, Dec. 1997.
7. K. L. Boyer and A. C. Kak, "Structural Stereopsis for 3-D Vision," IEEE Trans. PAMI, vol. 10, no. 2, pp. 144-166, Mar. 1988.
8. Shing-Huan Lee and Jin-Jang Leou, "A dynamic programming approach to line segment matching in stereo vision," Pattern Recognition, vol. 27, no. 8, pp. 961-986, 1994.
9. W. Hoff and N. Ahuja, "Surface From Stereo: Integrating Feature Matching, Disparity Estimation and Contour Detection," IEEE Trans. PAMI, vol. 11, no. 2, pp. 121-136, Feb. 1989.
10. S. B. Maripane and M. M. Trivedi, "Multi-Primitive Hierarchical(MPH) Stereo Analysis," IEEE Trans. PAMI, vol. 16, no. 3, pp. 227-240, Mar. 1994.
11. J. R. Parker, "Algorithms for Image Processing and Computer Vision," pp. 119-120, John Wiley & Sons, 1997.
12. S. E. Umbaugh, "Computer Vision and Image Processing," pp. 125-130, Prentice Hall, 1998.

Environmental Modeling

Session chair:

Zahari Zlatev (Danish Environmental Research Institute, Denmark)

Methods of Sensitivity Theory
and Inverse Modeling
for Estimation of Source Parameters and
Risk/Vulnerability Areas

Vladimir Penenko[1] and Alexander Baklanov[2]

[1]Institute of Computational Mathematics and Mathematical Geophysics, Siberian Division of Russian Academy of Sciences, Lavrentiev av., 6, RUS-630090, Novosibirsk, Russia
penenko@ommfao.sscc.ru
[2]Danish Meteorological Institute, Lyngbyvej, 100, DK-2100 Copenhagen, Denmark
alb@dmi.dk

Abstract. The source parameters estimation, based on environment pollution monitoring, and assessment of regions with high potential risk and vulnerability from nuclear sites are the two important problems for nuclear emergency preparedness systems and for long-term planning of socio-economical development of territories. For the discussed problems, most of modelers use the common back-trajectory techniques, suitable only for Lagrangian models. This paper discusses another approach for inverse modeling, based on variational principles and adjoint equations, and applicable for Eulerian and Lagrangian models. The presented methodology is based on both direct and inverse modeling techniques. Variational principles combined with decomposition, splitting and optimization techniques are used for construction of numerical algorithms. The novel aspects are the sensitivity theory and inverse modeling for environmental problems which use the solution of the corresponding adjoint problems for the given set of functionals. The methodology proposed provides optimal estimations for objective functionals, which are criterion of the atmospheric quality and informative content of measurements. Some applications of the suggested methods for source parameters and vulnerability zone estimations are discussed for important regions with environmental risk sites.

1 Introduction

Estimation of source parameters based on environmental pollution monitoring is a very important issue for national emergency response systems. For example, after the Algeciras accident in Spain (30 May 1998) many European monitors measured peaks of air radioactive contamination, but during several days the reason was unknown. Similar situations had happend after the Chernobyl and many others "man-made" catastrophes. Revealing regions with a high potential risk and vulnerability from nuclear or industrial sites is also important in the long-term planning of socio-economical development of territories and emergency systems.

A combination of the direct and inverse modelling approaches allows to solve some environmental and nuclear risk problems much more effectively compared with traditional ways based on a direct modelling. In this article we describe some aspects of application and development of the simulation technique proposed in [1–13]. It should be noted that most of researchers, which use for the discussed problems inverse methods, exploit the common back-trajectory techniques, suitable only for Lagrangian models [14-16]. The inverse modeling procedure described in the paper is based on variational principles with splitting and adjoint equations and it is applicable for Eulerian and Lagrangian models [7,9].

Advanced mathematical models of hydrodynamics and models of transport and transformation of pollutants in the gaseous and aerosol states are rather complicated, because they should take into account both natural and anthropogenic factors which affect the processes considered. In one paper it is impossible to give an extensive description of the simulation technique for a whole set of models. Therefore, we present as an example the ideas and main constructional elements of the models of transport and transformation of pollutants. It is assumed that the meteorological characteristics of the atmospheric system are known and prescribed as input information in the transport models.

2 Formulation of the Problems for Transport and Transformation of Pollutants

A dual description of the models as shown in [3, 5], is used to construct numerical schemes and their algorithms for direct, adjoint, and inverse problems: (1) as differential equations of transport and transformation of multi-species pollutants, and (2) in the variational form with a help of an integral identity.

The basic equation system of the model is written in the following form [4-6]:

$$(\Lambda\vec{\varphi})_i \equiv \frac{\partial \pi\varphi_i}{\partial t} + L(\pi\varphi_i) + (H(\vec{\varphi}))_i - f_i - r_i = 0, \quad i = \overline{1,n}, \quad n \geq 1, \quad (1)$$

where $\vec{\varphi} = \{\varphi_i(\vec{x},t), i = \overline{1,n}\} \in Q(D_t)$ is the vector-function of the state, φ_i - concentration of the ith pollutant, n - number of different substances, $\vec{f} = \{f_i(\vec{x},t), i = \overline{1,n}\}$ - source functions, r_i - functions which describe uncertainties and errors of the model, $L(\pi\varphi_i) = \text{div}[\pi(\vec{u}\varphi_i - \mu\,\text{grad}\,\varphi_i)]$ - advective-diffusive operator, $\vec{u} = (u_1, u_2, u_3)$ - velocity vector, $\mu = \{\mu_1, \mu_2, \mu_3\}$ - coefficients of turbulent transfer along the coordinates $\vec{x} = \{x_i, i = \overline{1,3}\}$, $H(\vec{\varphi})$ - nonlinear matrix operator of an algebraic form, which describes the processes of transformation of pollutants, π - pressure function, whose form depends on the chosen coordinate system, $D_t = D \times [0,\bar{t}]$, D - domain of variation of the spatial coordinates $\vec{x}$, $[0,\bar{t}]$ - time interval, and $Q(D_t)$ - space of the state functions, which satisfy the conditions at the boundary of the domain D_t. The transport operator is made antisymmetric using the continuity equation of [3].

The initial conditions at $t = 0$ and model parameters can be written in the following form:

$$\vec{\varphi}^0 = \vec{\varphi}_a^0 + \vec{\xi}_0(\vec{x}), \qquad \vec{Y} = \vec{Y}_a + \vec{\zeta}(\vec{x}, t), \tag{2}$$

where $\vec{\varphi}_a^0$ and $\vec{Y}_a$ are a-priori estimates of the initial fields $\vec{\varphi}^0$ and vector of parameters $\vec{Y}$; $\vec{\xi}_0(\vec{x})$, $\vec{\zeta}(\vec{x}, t)$ are the errors of the initial state and parameters. If we suppose that the model and input data are exact, the error terms in (1)-(2) should be omitted. The boundary conditions of the model are consequences of the physical content of the problem under investigation.

If the model implies the presence of errors, the expressions which describe the latter are formally included into the source functions as additional components.

The variational formulation of the model has the form [3,4]

$$I(\vec{\varphi}, \vec{\varphi}^*, \vec{Y}) \equiv \int_{D_t} (\Lambda \vec{\varphi} - \vec{f}) \vec{\varphi}^* \, dD \, dt = 0, \tag{3}$$

where $\vec{\varphi} \in (D_t)$ is the state function, $\vec{\varphi}^* \in Q^*(D_t)$ - adjoint or co-state function, $Q^*(D_t)$ - space of functions adjoint to $Q(D_t)$, $\vec{Y} = \{Y_i, i = \overline{1, n}\} \in R(D_t)$ - vector of parameters of the model, and $R(D_t)$ - region of their admissible values. The integral identity (3) is constructed taking into account the boundary conditions. The integrand in (3) is made symmetric, which automatically (i.e., without additional operations of differentiation and integration) ensures energy balance of functional (3) with the substitution $\vec{\varphi}^* = \vec{\varphi}$.

Now let us describe one more essential element of the construction. We mean the data of measurements. In order to include them into the model processing, it is necessary to formulate the functional relationship between the measurements themselves and state functions. Let this relation take the form

$$\vec{\Psi}_m = \vec{M}(\vec{\varphi}) + \vec{\gamma}(\vec{x}, t), \tag{4}$$

where $\vec{\Psi}_m$ is the set of observed values; $\vec{M}(\vec{\varphi})$ - set of measurement models; $\vec{\gamma}(\vec{x}, t)$ - errors of these models. The values of $\vec{\Psi}_m$ are defined on the set of points $D_t^m \in D_t$.

From point of view of the computational technology, the methods of inverse modeling are more suited to work with global (integral) characteristics of the models and processes than to work with the local ones. This is why we determine the set of such objects in a form of functionals.

For the purposes of monitoring, predicting, controlling, designing, and constructing algorithms of inverse simulation, we introduce a set of functionals

$$\Phi_k(\vec{\varphi}) = \int F_k(\vec{\varphi}) \chi_k(\vec{x}, t) \, dD \, dt, \qquad k = \overline{1, K}, \quad K \geq 1, \tag{5}$$

where $F_k(\vec{\varphi})$ are the prescribed functions on a set of the state functions, which are differentiable with respect to $\vec{\varphi}$, $\chi_k \geq 0$ - weight functions, and $\chi_k \, dD \, dt$

- corresponding Radon or Dirac measures in D_t. Using a functional (5) and an appropriate choice of the functions $F_k(\vec{\varphi})$ and χ_k, we can find generalized estimates of the behavior of the system, ecological restrictions on the quality of the environment, results of observations of various types, purpose control criteria, criteria of the quality of the models, etc. [5, 6]. For example, measured data (4) can be presented as

$$\Phi_0(\vec{\varphi}) = \left(\left(\vec{\Psi}_m - \vec{M}(\vec{\varphi}) \right)^T \chi_o S \left(\vec{\Psi}_m - \vec{M}(\vec{\varphi}) \right) \right)_{D_t^m}, \qquad (6)$$

where the index T denotes the operation of transposition. For source estimation, in addition to the functionals (6), it is necessary to consider a set of functionals in the form (5). Each of them describes an individual measurement.

Variational formulation (3) is used for the construction of the discrete approximations of the model. For these purposes, a grid D_t^h is introduced into the domain D_t, and discrete analogs $Q^h(D_t^h)$, $Q^{*h}(D_t^h)$, $R^h(D_t^h)$ of the corresponding functional spaces are defined on it. Then, the integral identity (3) is approximated by its sum analog

$$I^h(\vec{\varphi}, \vec{Y}, \vec{\varphi}^*) = 0, \quad \vec{\varphi} \in Q^h(D_t^h), \quad \vec{\varphi}^* \in Q^{*h}(D_t^h), \quad \vec{Y} \in R^h(D_t^h). \qquad (7)$$

The superscript h denotes a discrete analog of the corresponding object. Numerical schemes for model (1) are obtained from the stationarity conditions of the functional $I^h(\vec{\varphi}, \vec{Y}, \vec{\varphi}^*)$ with respect to arbitrary and independent variations at the grid nodes D_t^h of the grid functions $\vec{\varphi}^* \in Q^{*h}(D_t^h)$ for direct problems and $\vec{\varphi} \in Q^h(D_t^h)$ for adjoint problems [3].

3 The basic algorithm of inverse modeling and sensitivity studies

Let us use the ideas of the optimization theory and variational technique for the statement of the inverse problems and construction of methods for their solution. In this case, all approximations are defined by the structure of the quality functional and way of its minimization on the set of values of the state functions, parameters, and errors of the model discrete formulation [5].

The basic functional is formulated so that all the available real data, errors of the numerical model, and input parameters are taken into account:

$$
\begin{aligned}
J(\vec{\varphi}) = {} & \Phi_k(\vec{\varphi}) + \left(\vec{r}^T W_1 \vec{r} \right)_{D_t^h} + \left(\left(\vec{\varphi}^0 - \vec{\varphi}_a^0 \right)^T W_2 \left(\vec{\varphi}^0 - \vec{\varphi}_a^0 \right) \right)_{D^h} \\
& + \left(\left(\vec{Y} - \vec{Y}_a \right)^T W_3 \left(\vec{Y} - \vec{Y}_a \right) \right)_{R^h(D_t^h)} + I^h(\vec{\varphi}, \vec{Y}, \vec{\varphi}^*).
\end{aligned}
\qquad (8)
$$

Here, the first term is given by (5),(6), the second term takes into account the model errors, the third term describes errors in the initial data, the fourth term is responsible for the errors of the parameters, and the fifth term is a numerical model of the processes in a variational form, W_i, ($i = 1, 2, 3$ are weight matrices). The stationarity conditions for the functional (8) gives us the following system of equations:

$$\frac{\partial I^h(\vec{\varphi}, \vec{\varphi}^*, \vec{Y})}{\partial \vec{\varphi}^*} = B\Lambda_t \vec{\varphi} + G^h(\vec{\varphi}, \vec{Y}) - \vec{f} - \vec{r} = 0, \tag{9}$$

$$\frac{\partial I^h(\vec{\varphi}, \vec{\varphi}_k^*, \vec{Y})}{\partial \vec{\varphi}} + \frac{\partial \Phi_k^h}{\partial \vec{\varphi}} = (B\Lambda_t)^T \vec{\varphi}_k^* + A^T(\vec{\varphi}, \vec{Y})\vec{\varphi}_k^* - \vec{\eta}_k = 0, \tag{10}$$

$$\vec{\varphi}_k^*(\vec{x})\mid_{t=\bar{t}} = 0, \tag{11}$$

$$\vec{\eta}_k(\vec{x}, t) = grad_{\vec{\varphi}} \Phi_k^h(\vec{\varphi}) \equiv \frac{\partial \Phi_k^h(\vec{\varphi})}{\partial \vec{\varphi}}, \tag{12}$$

$$\vec{\varphi}^0 = \vec{\varphi}_a^0 + M_0^{-1} \vec{\varphi}_k^*(0), \qquad t = 0, \tag{13}$$

$$\vec{r}(\vec{x}, t) = R^{-1}(\vec{x}, t)\vec{\varphi}_k^*(\vec{x}, t), \tag{14}$$

$$\vec{Y} = \vec{Y}_a + M_1^{-1} \zeta_k, \tag{15}$$

$$\vec{\Gamma}_k = \frac{\partial}{\partial \vec{Y}} I^h(\vec{\varphi}, \vec{Y}, \vec{\varphi}_k^*), \tag{16}$$

$$A(\vec{\varphi}, \vec{Y})\vec{\varphi}' = \frac{\partial}{\partial \alpha} \left[G^h(\vec{\varphi} + \alpha \vec{\varphi}', \vec{Y}) \right]] \mid_{\alpha=0}, \tag{17}$$

where Λ_t is a discrete approximation of the time differential operator, B - diagonal matrix (some diagonal elements of which can be zero), $G(\vec{\varphi}, \vec{Y})$ - nonlinear matrix space operator depending on the state function and parameters, $A^T(\vec{\varphi}, \vec{Y})$ - space operator of the adjoint problem, $\vec{\Gamma}_k$ - functions of model sensitivity to the variations of parameters, and α - real parameter. The operations of differentiation (9),(10),(12),(16) are performed for all grid components of the state function, adjoint functions, and parameters. For temporal approximation of functional (8), we use the method of weak approximation with fractional steps in time [3]. This is why the equations (9) and (10) are the numerical splitting schemes. The system of equations (9)-(17) is solved with respect to $\vec{r}, \vec{\varphi}^0, \vec{Y}$ by the iterative procedures beginning with the initial approximations for the sought functions

$$\vec{r}^{(0)} = 0; \qquad \vec{\varphi}^{0(0)} = \vec{\varphi}_a^0; \qquad \vec{Y}^{(0)} = \vec{Y}_a. \tag{18}$$

Three basic elements are necessary for the realization of the method: 1) algorithm for the solution of the direct problem (9),(13)-(15); 2) algorithm for the solution of the adjoint problem (10),(12); 3) algorithm for the calculation of the sensitivity functions $\vec{\Gamma}_k$ with respect to variations of parameters (16). Then, the main sensitivity relations are constructed [3,5]:

$$\delta \Phi_k(\vec{\varphi}) = (\vec{\Gamma}_k, \delta \vec{Y}) = \frac{\partial}{\partial \alpha} I^h(\vec{\varphi}, \vec{Y}_a + \alpha \delta \vec{Y}, \vec{\varphi}_k^*) \mid_{\alpha=0}, \quad k = \overline{0, K}. \tag{19}$$

where the symbol δ indicates variations of the corresponding objects, $\delta \vec{Y} = \{\delta Y_i, i = \overline{1, N}\}$ is the vector of variations of the parameters $\vec{Y}$.

Inverse problems and methods of inverse simulation are formulated from the conditions of minimization of functionals (5) in the space of parameters or from the estimates of sensitivity to variations of these parameters. It includes algorithms for solving problems (9)–(17) and implementation of the feedback from the functionals to the parameters, which are derived from the relations (16) and (19):

$$\frac{dY_\alpha}{dt} = -\eta_\alpha \Gamma_{k\alpha}, \qquad \alpha = \overline{1, N_\alpha}, \qquad N_\alpha \leq N, \qquad (20)$$

where η_α are the coefficients of proportionality, which are found in the course of solving the problem, and N_α is the number of refined parameters.

It should be noted that the proposed variational principle, being combined with the splitting technique, allows to derive the hybrid set of Eulerian and Lagrangian transport models [7,9,8]. Such types of models complement each other so that their merits increase and demerits decrease in some respect.

In constructing numerical schemes for the transport model, the hydrothermodynamic components of the function of atmosphere state are assumed to be known and can be prescribed by different methods. In particular, they can be calculated from the models of atmospheric dynamics, which are integrated simultaneously with the transport models. For diagnostic studies, evaluation of information quality of observation systems, and implementation of scenarios of ecological prospects, transport models may be used together with models of the informational type, which generate characteristics of atmospheric circulation on the basis of retrospective hydrometeorological information [17]. In this case, for the construction of the directing phase space, we use Reanalysis data [18] for the global and large scales, and DMI-HIRLAM archives [19] for the meso- and regional scales.

The suggested approach enable to solve a broad class of environmental and nuclear risk problems more effectively. This approach includes: (i) development of a methodology to reveal the prerequisites of ecological disasters; (ii) environmental quality modeling; (iii) detection of the potential pollution sources; (iv) evaluation of their intensity; (v) estimation of the informative quality of monitoring systems; (vi) calculation of sensitivity and "danger" functions for protected areas; (vii) territorial zoning with respect to the danger of being polluted by known and potential sources of contamination; (viii) interconnections between source and recipient of pollution (direct and feedback relations); (ix) estimation of risk and vulnerability of zones/regions from nuclear and other dangerous sites, and (x) estimation of sensitivity functions for selected areas with respect to risk sites.

The number of applied problems were solved with the help of different modifications of the described methodology. Application of the direct and inverse modeling procedures, including sensitivity investigation, are given in [7,9,10,13]. Some specific studies of nuclear risk and vulnerability were presented in [11,12].

4 Algorithm for source estimation

Source term estimation is based on the use of monitoring data and inverse modeling. It is a particular case of the general inverse modeling algorithm [5,6]. The source function is considered as one of the parameters in $\vec{Y} \in R(D_t)$. The proposed computational scheme of the algorithm is as follows.

1) Suppose that the source term in (1) can be presented as a function

$$f = \sum_{\alpha=1}^{\alpha_N} Q_\alpha(t)\omega_\alpha(\vec{x}), \tag{21}$$

where Q_α and ω_α are the power and space pattern of the source with the number $\alpha = \overline{1, \alpha_N}$.

2) Choose a set of measurements Ψ_m and design the form of the functionals $\Phi_k(\vec{\varphi}, \Psi_m), k = \overline{1, K}$ for description of the measuremens. Note that the state function have to be included into the functionals for adjoint problems.

3) Calculate the sensitivity functions which can show us if the sought-for source can be observed with a given set of measurements. It means that one has to solve as many adjoint problems as number of measurements. Note, that such problems can be effectively run in parallel computing. Then the obtaned solutions of the adjoint problems should be investigated to find the observability regions where supports of the sensisitivity functions overlap in space and time. One should search for a source namely within these areas. The greater is the number of overlapped supports, the more exact source placement can be isolated.

4) Formulate the sensitivity relations for the chosen set of functionals

$$\delta\Phi_k(\vec{\varphi}, \Psi_m) = (\varphi_k^*(\vec{x}, t), \sum_{\alpha=1}^{\alpha_N} \delta Q_\alpha \omega_\alpha(\vec{x})), \quad k = \overline{1, K}, \tag{22}$$

which give us the possibility to define the source power variations. In a linear case, (22) gives the relation for power itself. For more precise identification of the source parameters the general algorithm of inverse modeling should be used.

5 An example to detect the source location

A test problem is in the detection of the source location using measurement data. As observational data, the solution of the 3D hemispheric direct problem on pollutant transport from the point source in the Central Europe was taken [10]. The scenario holds a 10-day period (24 March - 02 April 1999). Atmospheric circulation was reconstracted by means of numerical model [17] using Reanalysis data [18] in assimilation mode. The 2D section of concentration field at the surface level is presented in Fig.1 (left).

Following a criteria of relative significance level of about 10^{-4} with respect to the maximum concentration value at 2 April 1999, ten monitoring points, located at the Earth surface, were choosen. The "observations" were provided during 1 day. For this monitoring system, the 11 inverse scenarios were created for estimation of the functionals (5) with the linear F_k and Dirac measures [7,9,13]. The first scenario takes into account the entire monitoring set. In Fig.1 (right), the surface level of the sources' sensitivity function is displayed. The local maximums are obtained near the monitoring sites. As it is seen, there is no direct evidence for detection of the source. The generalized characteristic of the next 10 scenarios is presented in Fig.2 (left). Each of the scenarios gives the sensitivity functions for the functional that takes into account just one observation. The generalized characteristic is the function which describes the overlap of the sensitivity function supports. The value of this function gives the number of overlapped areas. The domains with the great values is preferable to search a source. The value 10 shows the domain where 10 supports overlap. In Fig. 2 (right), which is extraction from Fig.2 (left), this domain is displayed.

Several other examples of source term estimation, based on the direct and inverse local-scale modelling in a case of accidental contamination from an unknown release, and estimation of regional risk and vulnerability from various nuclear and industrial dangerous sites are discussed in [8,12,13].

Thus, the suggested methodology, based on a combination of direct and inverse modeling, extends the area of application of mathematical models. It allows us to find key parameters which are responsible for the processes' behavior. Variational principle provides relations between measured data and models, and gives the construction for a whole set of algorithms to solve environmental problems.

Acknowledgments. V.P.'s work is partly supported by the Grants of the RFBR (01-05-65313 and 00-15-98543) for the Leading Scientific Schools of Russia, Russian Ministry of Industry and Science (0201.06.269/349), and European Commission (ICA2-CT-2000-10024).

References

[1] Marchuk, G.I.: Mathematical modeling in the environmental problems. Moscow, Nauka (1982) (in Russian)

[2] Marchuk, G. I.: Adjoint equations and analysis of complex systems. Kluber Academic Publication (1995)

[3] Penenko, V.V.: Methods of numerical modeling of the atmospheric processes. Leningrad, Gidrometeoizdat (1981) (in Russian)

[4] Penenko, V.V. and Aloyan, A.E.: Models and methods for environment protection problems. Nauka, Novosibirsk (1985)

[5] Penenko, V.V.: Some Aspects of Mathematical Modeling Using the Models Together with Observational Data. Bull. Nov. Comp. Cent. 4 (1996) 31-52

[6] Penenko, V.V.: Numerical models and methods for the solution of the problems of ecological forecast and design. Survey of the Applied and Industry Mathematics. Vol. 1, 6 (1994) 917–941

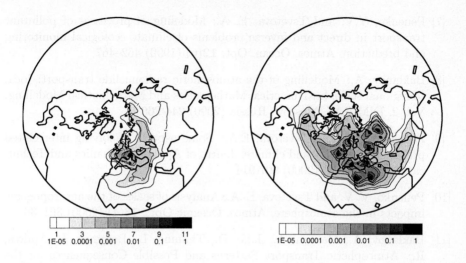

Figure 1: Solution of the direct problem: concentration fields (left), and the sensitivity function based on the inverse simulation for the entire set of observations (right)

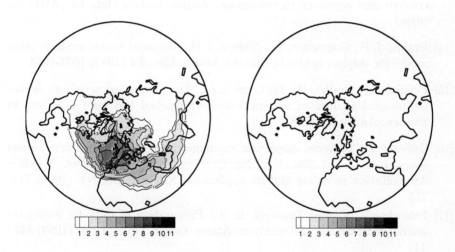

Figure 2: The overlapped supports of the sensitivity functions for 10 individual observations (left) and detected sourse location area (right)

[7] Penenko, V.V. and Tsvetova, E. A.: Modeling of processes of pollutant transport in direct and inverse problems of climato-ecological monitoring and prediction. Atmos. Ocean. Opt. **12**(6) (1999) 462–467

[8] Baklanov, A.: Modelling of the atmospheric radionuclide transport: local to regional scale. In: Numerical Mathematics and Mathematical Modelling, Vol. 2. INM RAS, Moscow, Russia (2000) 244-266

[9] Penenko, V.V. and Tsvetova, E.A.: Some aspects of solving interrelated problems of ecology and climate. Journ. of Applied Mechanics and Technical Physics, **41**(5) (2000) 907-914

[10] Penenko, V. V. and Tsvetova, E. A.: Analysis of scales of the anthropogenic impacts in the atmosphere. Atmos. Oceanic Opt., **13**(4) (2000) 361–365

[11] Baklanov, A., Mahura, A., Jaffe, D., Thaning, L., Bergman, R. Andres, R.: Atmospheric Transport Patterns and Possible Consequences for the European North after a Nuclear Accident. Journal of Environmental Radioactivity (2001) (accepted)

[12] Rigina, O. and Baklanov, A.: Regional radiation risk and vulnerability assessment by integration of mathematical modelling and GIS-analysis. J. Environment International (2001) (submitted)

[13] Penenko, V.V.: Revealing the areas of high ecological vulnerability: the concept and approach to realization. Atmos. Ocean. Opt. **14** (2001) (in press)

[14] Prahm, L.P., Conradsen, K., Nielsen, L.B.: Regional sourse quantification model for sulphur oxides in Europe. Atmos. Env. **14** (1980) 1027-1054

[15] Persson, C., Rodhe, H., De Geer, L.E.: The Chernobyl accident: A meteorological analysis of how radionuclides reached and were deposited in Sweden. Ambio, 16 (1987) 20-31

[16] Seibert, P.: Inverse dispersion modelling based on trajectory-derived sourse-receptor relationsships. In: Gryning, S.E., Chaumerliac, N.(eds.): Air pollution modeling and its application XII, Plenum, NY (1997) 711-713

[17] Penenko, V. V. and Tsvetova, E. A.: Preparation of data for ecological studies with the use of Reanalysis. Atmos. Ocean. Opt., **12**(5) (1999) 447–449

[18] Kalnay, E., Kanamits, M., Kistler, R. et al.: The NCEP/NCAR 40-year reanalysis project. Bull. Amer. Meteorol. Soc. **77**, (1996) 437–471

[19] Sørensen, J.H., Laursen, L., Rasmussen; A.: Use of DMI-HIRLAM for operational dispersion calculations. In: Gryning, S.E., Millan, M.M.(eds.): Air pollution modeling and its application X, Plenum, NY (1994) 373-381

The Simulation of Photochemical Smog Episodes in Hungary and Central Europe Using Adaptive Gridding Models

István Lagzi[1], Alison S. Tomlin[2], Tamás Turányi[1,3], László Haszpra[4], Róbert Mészáros[5], Martin Berzins[6],

[1]Department of Physical Chemistry, Eötvös University (ELTE), H-1518 Budapest, P.O.Box 32, Hungary
[2]Department of Fuel and Energy, The University of Leeds, Leeds, LS2 9JT, UK
fueast@leeds.ac.uk
[3]Chemical Research Center, H-1525 Budapest, P.O.Box 17, Hungary
[4]Institute for Atmospheric Physics, Hungarian Meteorological Service, H-1675 Budapest, P.O.Box 39
[5]Department of Meteorology, Eötvös University (ELTE), H-1518 Budapest, P.O.Box 32, Hungary
[6]School of Computer Studies, The University of Leeds, Leeds, LS2 9JT, UK

Abstract. An important tool in the management of photochemical smog episodes is a computational model which can be used to test the effect of possible emission control strategies. High spatial resolution of such a model is important to reduce the impact of numerical errors on predictions and to allow better comparison of the model with experimental data during validation. This paper therefore presents the development of an adaptive grid model for the Central European Region describing the formation of photochemical oxidants based on unstructured grids. Using adaptive methods, grid resolutions of less than 20 km can be achieved in a computationally effective way. Initial simulation of the photochemical episode of August 1998 indicates that the model captures the spatial and temporal tendencies of ozone production and demonstrates the effectiveness of adaptive methods for achieving high resolution model predictions.

1 Introduction

Previous EUROTRAC investigations have shown that some of the highest regional ozone concentrations in Europe can be observed in Central Europe, including Hungary. During summer ozone episodes, the ozone burden of natural and agricultural vegetation is often well beyond tolerable levels. Budapest is one of the biggest cities in this region, emitting significant amounts of ozone precursor substances. An important tool in the management of photochemical smog episodes is a computational model which can be used to test the effect of possible emission control strategies. High spatial resolution of such a model is important to reduce the impact of numerical errors on predictions and to allow better comparison of the model with experimental data during validation. The review paper of Peters et al.[1] highlights the importance of developing more efficient grid systems for the next generation of air pollution models in order to "capture important smaller scale atmospheric phenomena". This paper therefore presents the development of an adaptive grid model for the Central European

Region describing the formation of photochemical oxidants based on unstructured grids. The initial base grid of the model uses a nested approach with a coarse grid covering the wider central European region and a finer resolution grid over Hungary. Further refinement or de-refinement is then invoked using indicators based on the comparison of high and low order numerical solution of the atmospheric diffusion equation. Using this method, grid resolutions of less than 20 km can be achieved in a computationally effective way.

2 Model Description

The model describes the spread of reactive air pollutants within a 2D unstructured triangular based grid representing layers within the troposphere over the Central European region including Hungary. The model describes the horizontal domain using a Cartesian coordinate system through the stereographic polar projection of a curved surface onto a flat plane. The total horizontal domain size is 1540 km × 1500 km. Vertical resolution of pollutants is approximated by the application of four layers representing the surface, mixing, reservoir layers and the free troposphere. Reactive dispersion in the horizontal domain is described by the atmospheric diffusion equation in two space dimensions:

$$\frac{\partial c_s}{\partial t} = -\frac{\partial(u c_s)}{\partial x} - \frac{\partial(w c_s)}{\partial y} + \frac{\partial}{\partial x}\left(K_x \frac{\partial c_s}{\partial x}\right) + \frac{\partial}{\partial y}\left(K_y \frac{\partial c_s}{\partial y}\right)$$

$$+ R_s(c_1, c_2, ..., c_q) + E_s - (\kappa_{1s} + \kappa_{2s})c_s, \tag{1}$$

where c_s is the concentration of the s'th compound, u,w, are horizontal wind velocities, K_x and K_y are eddy diffusivity coefficients and κ_{1s} and κ_{2s} are dry and wet deposition velocities respectively. E_s describes the distribution of emission sources for the s'th compound and R_s is the chemical reaction term which may contain nonlinear terms in c_s. For n chemical species an n-dimensional set of partial differential equations (p.d.e.s) is formed describing the rates of change of species concentration over time and space, where each is coupled through the nonlinear chemical reaction terms.

The four vertical layers of the model are defined as; the surface layer extending to 50m, the mixing layer, a reservoir layer and the free troposphere (upper) layer. The mixing layer extends to a height determined by radiosonde data at 12am, but is modelled to rise smoothly to a height determined by radiosonde data at 12pm during the day[2]. The reservoir layer, if it exists above the height of the mixing layer, extends from the top of the mixing layer to an altitude of 1000 m. Vertical mixing and deposition are parameterised according to the vertical stratification presented by VanLoon[3]. Deposition velocities are assumed to be constant across the whole domain. The eddy diffusivity coefficients for the x and y directions were set to 50 m²s⁻¹ for all species.

Relative humidity and temperature data were determined by the meteorological model ALADIN with a time resolution of 6 hours and spatial resolution of 0.1 × 0.15 degrees[4]. The local wind speed and direction were obtained from the ECMWF database ensuring conservation properties of the original data[5]. The ECMWF data has a time resolution of 6 hours and a spatial resolution of 2.5×2.5 degrees. These data were interpolated to obtain data relevant to a given space and time point on the unstructured grid using conservative methods as described in Ghorai et al.[6].

The emission of precursor species into the domain was described by the EMEP emissions inventory for 1997 based on a 50km $\times$ 50km grid[7]. The emissions data have to be interpolated onto the unstructured mesh following each change to the mesh during refinement. This is achieved using the mass conservative method of overlapping triangles. The EMEP data is therefore split into triangular grid cells and the proportion of each EMEP triangle that overlaps each mesh triangle calculated. Point sources, although not included in the current simulations, can be included in the model by averaging them into the appropriate triangle. It follows that as the mesh resolution is improved, the description of point source emissions will also improve. During the initial simulations as described here the GRS chemical scheme was used[8] to enable fast turn around times. Photolysis rate constants have been chosen to be in agreement with those used by Derwent and Jenkin [9] and are expressed as m'th order rate constants with units (molecule cm^{-3})$^{1-m}$ s^{-1}. The photolysis rates were parameterised by the function $J_q = a_q \exp(-b_q \sec \theta)$, where θ is the solar zenith angle and q is the reaction number. The solar zenith angle is calculated from:

$$\cos \theta = \cos(\text{lha}) \cos(\text{dec}) \cos(\text{lat}) + \sin(\text{dec}) \sin(\text{lat}),$$

where lha is the local hour angle (i.e. a function of the time of day), dec is the solar declination (i.e. a function of the time of year) and lat is the latitude. Temperature dependent rate constants are represented by standard Arrhenius expressions.

3 Solution Method

The basis of the numerical method is the space discretisation of the p.d.e.s derived from the atmospheric diffusion equation on unstructured triangular meshes using the software SPRINT2D[10–12]. This approach, (known as the "Method of Lines"), reduces the set of p.d.e.s in three independent variables to a system of ordinary differential equations (o.d.e.s) in one independent variable, time. The system of o.d.e.s can then be solved as an initial value problem, and a variety of powerful software tools exist for this purpose[13]. For advection dominated problems it is important to choose a discretisation scheme which preserves the physical range of the solution. A more in-depth discussion of the methods can be found in previous references[14, 10–12].

3.1 Spatial Discretisation and Time Integration

Unstructured triangular meshes are commonly used in finite volume/element applications because of their ability to deal with general geometries. In terms of application to multi-scale atmospheric problems, we are not dealing with complex physical geometries, but unstructured meshes provide a good method of resolving the complex structures formed by the interaction of chemistry and flow in the atmosphere and by the varying types of emission sources. The term unstructured represents the fact that each node in the mesh may be surrounded by any number of triangles whereas in a structured mesh this number would be fixed. In the present work, a flux limited, cell-centered, finite volume discretization scheme of Berzins and Ware[10, 11] was chosen on an unstructured triangular mesh. This method enables accurate solutions to be determined for both smooth and discontinuous flows by making use of the local Riemann solver flux techniques (originally developed for the Euler equations) for the advective parts of the fluxes, and centered schemes for the diffusive part. The scheme used for the treatment of the advective terms is an extension to irregular triangular meshes of the nonlinear scheme described by Spekreijse[15] for regular Cartesian meshes. The scheme of Berzins and Ware has the desirable properties, see Chock[16], of preserving positivity eliminating spurious oscillations and restricting the amount of diffusion by the use of a nonlinear limiter function. The advection scheme has been shown to be of second order accuracy[11, 17]. For details of the application of the method to the atmospheric diffusion equation in two-space dimensions see Tomlin et al.[12]. The diffusion terms are discretised by using a finite volume approach to reduce the integrals of second derivatives to the evaluation of first derivatives at the midpoints of edges. These first derivatives are then evaluated by differentiating a bilinear interpolant based on four mid-point values[10]. The boundary conditions are implemented by including them in the definitions of the advective and diffusive fluxes at the boundary.

A method of lines approach with the above spatial discretization scheme results in a system of o.d.e.s in time which are integrated using the code SPRINT[13] with the Theta option which is specially designed for the solution of stiff systems with moderate accuracy and automatic control of the local error in time. Operator splitting is carried out at the level of the nonlinear equations formed from the method of lines by approximating the Jacobian matrix, and the method is described fully in Tomlin et al.[12]. The approach introduces a second-order splitting error but fortunately this error only alters the rate of convergence of the iteration as the residual being reduced is still that of the full o.d.e. system. This provides significant advantages over other splitting routines such as Strang splitting.

3.2 Mesh Generation and Adaptivity

The initial unstructured meshes used in SPRINT2D are created from a geometry description using the Geompack mesh generator[18]. These meshes are then refined and coarsened by the Triad adaptivity module which uses tree like data structures to enable efficient mesh adaptation by providing the necessary connectivity. A method of refinement based on the regular subdivision of triangles has been chosen. Here an original triangle is split into four similar triangles by connecting the midpoints of the edges as shown in Fig. 1. These may later be co-alesced into the parent triangle when coarsening the mesh. This process is called local h-refinement, since the nodes of the original mesh do not move and we are simply subdividing the original elements. In order to implement the adaptivity

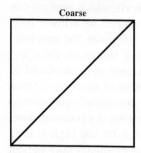 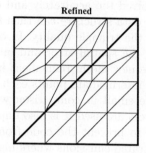

Fig. 1. Method of local refinement based on the subdivision of triangles

module a suitable criterion must be chosen. The ideal situation would be that the decision to refine or derefine would be made on a fully automatic basis with no user input necessary. In practice a combination of an automatic technique and some knowledge of the physical properties of the system is used. The technique used in this work is based on the calculation of spatial error estimates. Low and high order solutions are obtained for each species and the difference between them gives a measure of the spatial error. The algorithm can then choose to re-fine in regions of high spatial error by comparison with a user defined tolerance for one or the sum of several species. For the ith p.d.e. component on the jth triangle, a local error estimate $e_{i,j}(t)$ is calculated from the difference between the solution using a first order method and that using a second order method. For time dependent p.d.e.s this estimate shows how the spatial error grows lo-cally over a time step. A refinement indicator for the jth triangle is defined by an average scaled error ($serr_j$) measurement over all $npde$ p.d.e.s using supplied absolute and relative tolerances:

$$serr_j = \sum_{i=1}^{npde} \frac{e_{i,j}(t)}{atol_i/A_j + rtol_i \times C_{i,j}}, \tag{2}$$

where $atol$ and $rtol$ are the absolute and relative error tolerances and $C_{i,j}$ the computed concentration of species i in triangle j. This formulation for the scaled error provides a flexible way to weight the refinement towards any p.d.e. error. In these calculations a combination of errors in species NO and NO_2 were used as refinement indicators.

An integer refinement level indicator is calculated from this scaled error to give the number of times the triangle should be refined or derefined. Since the error estimate is applied at the end of a time-step it is too late to make the refinement decision. Methods are therefore used for the prediction of the growth of the spatial error using linear or quadratic interpolants. The decision about whether to refine a triangle is based on these predictions, and the estimate made at the end of the time- step can be used to predict errors at future time-steps. Generally it is found that large spatial errors coincide with regions of steep spatial gradients. The spatial error estimate can also be used to indicate when the solution is being solved too accurately and can indicate which regions can be coarsened. The tree data structure can then be used to restore a lower level mesh which maintains the triangle quality. In order to minimise the overhead involved in calculating the spatial error and interpolating the input data and concentrations onto the new mesh, refinement indicators are not calculated at each time point but at intermediate points each 20 minutes of simulation time. The errors induced are small in comparison with the computational savings made[19]. If point sources are used it is also important to set a maximum level of refinement in order to prevent the code from adapting to too high a level in regions with concentrated emissions where high spatial gradients may persist down to very high levels of refinement.

4   Simulation of Photochemical Smog Episode

The model was tested via the simulation of a photochemical oxidant episode that took place in Hungary in August, 1998. During almost the whole month wind speeds were low limiting dispersion, and strong sunshine resulted in high photo-oxidant levels over most of Europe. The simulation period was from midnight 1 August 1998 to midnight 5th August 1998. Both fixed and adaptive mesh simulations were carried out in order to investigate the influence of mesh resolution on predicted concentrations. The coarse level 0 grid covered Central Europe as seen in Fig. 2 and is defined by a maximum edge length of 228 km. For

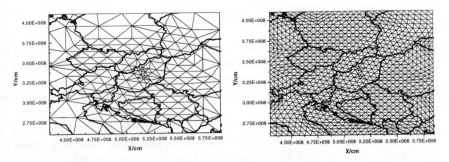

Fig. 2. The structure of level 0 and level 2 base meshes

comparison, simulations where also carried out using a level 2 refined base grid as shown in Fig. 2 with a maximum edge length of 56km. For both simulations extra mesh refinement to 3 levels (edge length 28km) was carried out around the K-puszta monitoring station of the Hungarian Meteorological Service, located 70 km south-southeast from Budapest, in order to carry out comparisons of measured and simulated ozone concentrations. The initial conditions for the most important species were as a follows: 1.9 ppb for NO_2, 3 ppb for NO, ppb 20.3 O_3, 4 ppb for VOC and were constant across the domain. A certain amount of spin up time was therefore required for the model in order to eliminate the impact of the initial conditions. Fig. 3 shows the simulated ozone concentrations after 4 days of simulation on 5th August, 1998 at 15.00 for the two base grids where no transient refinement has taken place. The impact of mesh resolution can be seen and the higher resolution simulation predicts higher peak concentrations over the region South of Hungary.

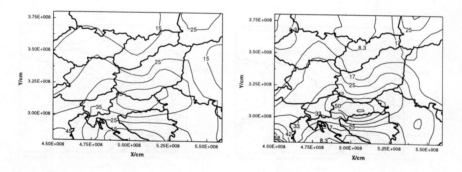

Fig. 3. Calculated ozone concentration after 4 days of simulation on 5th August, 1998 at 15.00, level 0 vs level 2 results

An adaptive simulation was also carried out where transient refinement was limited to 4 levels below the base mesh, in this case leading to a minimum edge length of approximately 20 km. Fig. 4 shows the grid at 18.00 on the third simulation day where refinement has effectively taken place around regions of high gradients in NO and NO_2 concentrations. The refinement region is seen to cover a large part of Hungary and the surrounding region where steep concentration gradients occur. The simulated ozone concentrations for a level 2 base mesh and the level 4 adaptive mesh shown in Fig. 4 at the simulated time of 18.00 on 5th August, 1998 are shown in Fig. 5. The high ozone concentration area is found to be in the Western region of Hungary where the grid is also dense for the adaptive solution, and to Southwest of Hungary. There are significant differences in the predicted peak ozone concentrations for the base mesh and the adaptive simu-

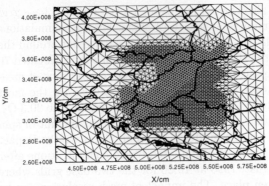

Fig. 4. The adaptive mesh at the simulated time of 18.00 on 5th August, 1998

lation with the peak not rising above 50ppb for the coarse mesh and reaching 90ppb in the South West region of the mesh for the adaptive solution.

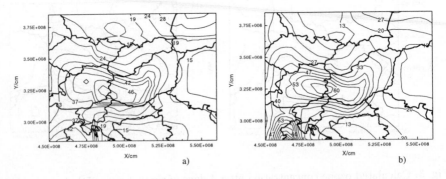

Fig. 5. Comparison of ozone predictions , at the simulated time of 18.00 on 5th August, 1998 using a) a level 2 base mesh and b) a level 4 adaptive mesh

Fig. 6 shows the simulated and measured ozone concentrations at the K-puszta station as a function of time during the simulation. As expected the initial comparisons are poor during the model spin up period reflecting the need for more accurate initial conditions. During the latter part of the simulation the simulated ozone levels increase but are still considerably lower than the ozone levels measured at K-puszta. There are several possible reasons for this. The meteorological data used for the simulations was quite coarse and therefore the region of predicted high ozone levels may be spatially inaccurate as a result. High ozone concentrations are predicted in the Southwest region of the mesh but lower concentrations in the area of K-puszta. It is also clear that the assumption of constant initial conditions across the mesh is not accurate and affects the early part of the simulation. The amount of ozone generated over the 5 days is quite well represented by the model, again leading to the requirement for better initial conditions. The emissions data is also based on a fairly coarse grid. Future improvements to the model will therefore be to include higher spatial resolution meteorological data and emissions data for the Budapest region. The incorporation of measured concentrations for initial conditions where possible would also improve the simulations.

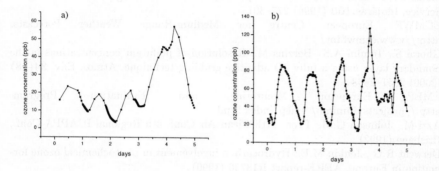

Fig. 6. Comparison of a) Model predictions with b) measured ozone concentrations at the K-puszta station

5 Conclusions

An adaptive grid model describing the formation and transformation of photochemical oxidants based on triangular unstructured grids has been developed for the Central European Region, which includes Hungary. The model automatically places a finer resolution grid in regions characterised by high concentration gradients and therefore by higher numerical error. Using an adaptive method it is therefore possible to achieve grid resolutions of the order of 20 km without excessive computational effort. Initial simulations of the photochemical episode of August 1998 indicate that the model under predicts ozone concentrations, but captures the spatial and temporal tendencies. The comparison of level 0 and level 2 base mesh simulations with a level 4 adaptive simulation demonstrates the impact of mesh refinement on predicted ozone concentrations and indicates that using a coarse mesh may lead to underpredictions of peak values.

Acknowledgements The authors acknowledge the support of OTKA grant T025875 and UK-Hungarian co-operation grant GB50/98.

References

1. Peters L.K., Berkovitz C.M., Carmichael G.R., Easter R.C., Fairweather G., Ghan S.J., Hales J.M., Leung L.R., Pennell W.R., Potra F.A., Saylor R.D. and Tsang T.T.: The current and future direction of Eulerian Models in simulating the tropospheric chemistry and transport of trace species: a review. Atmos. Env. **29** (1995) 189–222.
2. Matyasovszky I., Weidinger T.: Characterizing air pollution potential over Budapest using macrocirculation types. Idojárás **102** (1998) 219–237
3. VanLoon M.: Numerical methods in smog prediction. PhD. Thesis, GWI Amsterdam (1996)
4. Hornyi A., Ihsz I., Radnti G.: ARPEGE/ALADIN: A numerical Weather prediction model for Central-Europe with the participation of the Hungarian Meteorological Service. Idojárás **100** (1996) 277–301
5. ECMWF: European Centre for Medium-Range Weather Forecasts. http://www.ecmwf.int/
6. Ghorai S., Tomlin A.S., Berzins M.: Resolution of pollutant concentrations in the boundary layer using a fully 3D adaptive gridding technique. Atmos. Env. **34(18)** (2000) 2851–2863
7. EMEP: European Monitoring and Evaluation Program. http://projects.dnmi.no/ emep/index.html
8. Azzi M., Johnson G.M.: Proc. 11th Clean Air Conf. 4th Regional IUAPPA Conf., Brisbane (1992)
9. Derwent R.G., Jenkin M.E.: Hydrocarbon involvement in photochemical ozone formation in Europe. AERE-report-R13736 (1990)
10. Berzins M., Ware J.M.: Reliable Finite Volume Methods for the Navier Stokes Equations. In: Hebeker F-K, Rannacher R, Wittum G (eds) Notes on numerical Fluid Mechanics. Viewg, Wiesbaden (1994) 1–8
11. Berzins M., Ware J.: Positive cell-centered finite volume discretization methods for hyperbolic equations on irregular meshes. Appl. Num. Math. **16** (1995) 17–438
12. Tomlin A., Berzins M., Ware J., Smith J., Pilling M.J.: On the use adaptive gridding methods for modelling chemical transport from multi-scale sources. Atmos. Env. **31** (1997) 2945–2959
13. Berzins M., Dew P.M., Furzeland R.M.: Developing software for time-dependent problems using the method of lines and differential algebraic integrators. Appl. Numer. Math. (1989) 5375–390
14. Berzins M., Lawson J., Ware J.: Spatial and Temporal Error Control in the Adaptive Solution of Systems of Conversation Laws. Advances in Computer Methods for Partial Differential Equations, IMACS VII (1992) 60–66
15. Spekreijse S.: Multigrid solution of monotone second order discretizations of hyperbolic conservation laws. Math. Comp. **47** (1987) 135–155
16. Chock D.P.: A comparison of numerical methods for solving the advection equation III. Atmos. Env. **25A** (1991) 553–571
17. Ware J. and Berzins M.: Adaptive Finite Volume Methods for Time-dependent P.D.E.s. In: Modeling, Mesh Generation and Adaptive Numerical Methods for PDEs. (eds.) I. Babuska et.al. Volume 75 in IMA Volumes in Mathematics and Its Applications Series, Springer Verlag, (1995) 417–430
18. Joe B. and Simpson R.B.: Triangular meshes for regions of complicated shape. Int. J. Numer. Meth. Eng. **23** (1991) 987–997
19. Hart G.J.: Multi-scale atmospheric dispersion modelling by use of adaptive gridding techniques. PhD. thesis, Leeds, UK (1999)

Numerical Solution of the Aerosol Condensation/Evaporation Equation

Khoi Nguyen and Donald Dabdub

Department of Mechanical & Aerospace Engineering
University of California, Irvine
Irvine, CA 92612, USA
ddabdub@uci.edu

Abstract. A new method is developed to solve the condensation equation as it relates to Air Quality Models using both semi-Lagrangian and Lagrangian fluxes to increase resolution and perform accurately under stringent conditions that occur in the atmosphere. The new method, Partitioned Flux Integrated Semi-Lagrangian Method (PFISLM), can be used with lower-order interpolators and they produce highly accurate results. PFISLM is positive definite, peak retentive, mass conservative, and suppresses oscillations. Research indicates the differences between PFISLM and other traditional flux integrated semi-Lagrangian methods are significant when solving the aerosol condensation/evaporation equation. PFISLM is created to handle specific difficulties associated with the time and space discretization of the aerosol operator in air quality models.

1 Introduction

Air quality models that include aerosol dynamics often employ the continuous distribution approach (Pilinis, 1990) to represent aerosols undergoing evaporation and condensation. The fundamental equation governing the condensation process for an internally mixed aerosol is derived in Pilinis (1990). The equation is given as

$$\frac{\partial p_i}{\partial t} = H_i p - \frac{1}{3}\frac{\partial H p_i}{\partial \mu}, \tag{1}$$

where p_i, H_i, μ, $p = \sum_i^n p_i$, and $H = \sum_i^n H_i$ are the mass distribution of species i, mass transfer rate of species i, log of the diameter of the particle, total concentrations, and total mass transfer rates respectively with n being the number of aerosol compounds.

Typically, Equation (1) is solved by means of operator splitting into two parts: the growth,

$$\frac{\partial p_i}{\partial t} = H_i p, \tag{2}$$

and the redistribution,

$$\frac{\partial p_i}{\partial t} = -\frac{1}{3}\frac{\partial H p_i}{\partial \mu}. \tag{3}$$

Section 2 describes a new algorithm to solve the aerosol condensation/evaporation equation. Numerical tests are presented in Section 3.

2 The Algorithm

2.1 Aerosol Redistribution

PFISLM uses both semi-Lagrangian fluxes and Lagrangian positions to resolve the space discretization. A graphical representation of the new scheme is shown in Figure 1. A typical semi-Lagrangian flux is the mass that is contained from the semi-Lagrangian position of the cell interface, (μ_1) to the cell interface, (μ_2).

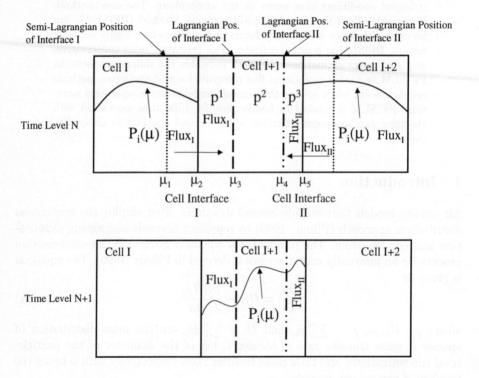

Fig. 1. Partitioned Flux Integrated Semi-Lagrangian Method uses both the Semi-Lagrangian and Lagrangian position of cell interfaces in its computation. The flux leaving cell interface I is not averaged into cell I+1 but is recorded into the partition defined from μ_2 and μ_3.

The evacuated semi-Lagrangian flux enters the partition defined by the cell interface, (μ_2) and the Lagrangian position, (μ_3). This area from (μ_2) to (μ_3) is known as a partition. Since, the total mass in cell $I + 1$ is known, the distribution inside each cell is further resolved. Namely, the mass occupied by the first partition, μ_2 to μ_3, is $p^1 = F_I$, the mass occupied by the second partition, μ_3 to μ_4, is $p^2 = p_i(t)$, and the mass occupied by the third partition from μ_4 to μ_5 is $p^3 = F_{II}$. In the next time step, the mass distribution inside cell $I + 1$ is interpolated (using both linear and quadratic interpolators) more accurately because more information is available to describe the mass partitioning inside the cell due to the Lagrangian partitioning. Depending on the mass transfer rate, each cell can have up to three partitions.

2.2 Aerosol Growth

The growth term given in Equation (2), is computed from a mass balance and the growth law. The growth is computed for the entire cell and is distributed to the enclosed partitions accordingly. The growth for the entire cell is found by a direct integration of the growth equations,

$$\frac{\partial p_i}{\partial t} = H_i p \tag{4}$$

$$\frac{\partial p}{\partial t} = Hp. \tag{5}$$

The solution to those two simultaneous equations is given as

$$p_i(t + \Delta t, \mu) = p_i(t, \mu) + \frac{H_i}{H} p(t, \mu) \left[e^{H \Delta t} - 1 \right]. \tag{6}$$

The mass increase or decrease in each partition is computed by solving a simple set of equations that represents mass conservation, the solution to the growth equation, and the contribution of the cell growth from each partition in accordance to its relative mass to the entire cell. That set of equations is given as

$$p_i \Delta \mu = p^1 \Delta \mu_1 + p^2 \Delta \mu_2 + p^3 \Delta \mu_3 \tag{7}$$

$$growth = \frac{H_i}{H} p(t, \mu) \left[e^{H \Delta t} - 1 \right] \tag{8}$$

$$massgrowth_j = massgrowth \frac{mass_j}{mass} \tag{9}$$

Here the $massgrowth$ is the growth of the mass in the entire cell, $growth$ is the growth of concentration in the entire cell, $mass_j$ is the mass occupied by partition j, $massgrowth_j$ is the growth of the mass in partition j, p_i is the concentration in the entire cell, and p^j is the concentration in partition j. Substituting $massgrowth_j = growth_j \Delta \mu_j$, $massgrowth = growth \Delta \mu$, and $mass_j = p^j \Delta \mu_j$ into Equation (12), the solution to the growth term in each partition is obtained as,

$$growth_j = growth \frac{p^j}{p_i}. \tag{10}$$

3   Numerical Test

There were hundreds of test perform in order to gauge the accuracy of the solver. For the sake of brevity, here we will report on one of them. The test consists on 200 iterations of back and forth condensation and evaporations of a given aerosol distribution using 12 fixed bins. The rate of evaporation and condensations is constant. Therefore, the final solution is expected to equal the initial conditions. Figure 2 shows that results are significantly less diffusive than those produced by Bott's solver (Bott, 1989) which is currently used in several air quality models.

The numerical test uses the following initial conditions:

$$p_i(\mu) = \begin{cases} \frac{i}{20}(1 + \cos[\frac{\mu-5}{4}]) & \text{if } |\mu - 5| \le 4 \\ 0 & \text{otherwise} \end{cases}$$

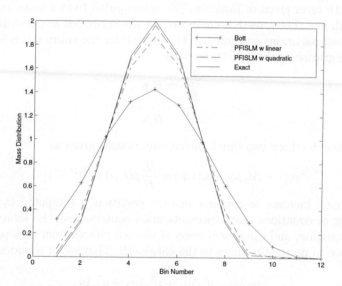

Fig. 2. Numerical tests show that the Bott solver is diffusive while the PFISLM with linear and quadratic interpolators preserve the peaks within 10%.

4   Conclusions

A new approach to solving the condensation equation is developed to handle these difficult characteristics of the aerosol operator. The new method is mass conservative, positive definite, peak retentive, and suppresses oscillations. The approach is a super set of flux integrated semi-Lagrangian methods using both semi-Lagrangian and Lagrangian fluxes. It effectively increases the resolution of the bins by keeping information about the partitions inside each bin. This increased resolution permits accurate representation of the actual dynamics of aerosols, especially with limited number of bins.

References

Pilinis, C.: Derivation and numerical solution of the species mass distribution equations for multicomponent particulate systems. Atmos. Environ. **24** (1990) 1923–1928

Bott, A.: A positive definite advection scheme obtained by nonlinear renormalization of the advective fluxes. Mon. Wea. Rev. **117** (1989) 1006–1015

Efficient Treatment of Large-Scale
Air Pollution Models on Supercomputers

Zahari Zlatev

National Environmental Research Institute
Frederiksborgvej 399, P. O. Box 358, DK-4000 Roskilde, Denmark

Abstract. The air pollution, and especially the reduction of the air pollution to some acceptable levels, is an important environmental problem, which will become even more important in the next two-three decades. This problem can successfully be studied only when high-resolution comprehensive models are developed and used on a routinely basis. However, such models are very time-consuming, also when modern high-speed computers are available. Indeed, if an air pollution model is to be applied on a large space domain by using fine grids, then its discretization will always lead to huge computational problems. Assume, for example, that the space domain is discretized by using a (288x288) grid and that the number of chemical species studied by the model is 35. Then ODE systems containing 2903040 equations have to be treated at every time-step (the number of time-steps being typically several thousand). If a three-dimensional version of the air pollution model is to be used, then the number of equations must be multiplied by the number of layers. The treatment of large air pollution models on modern parallel computers by using efficient numerical algorithms will be discussed in this paper.

1   Large-scale air pollution models

Mathematical models are indispensable tools when different air pollution phenomena are to be studied on a large space domain (say, the whole of Europe). Such models can be used to find out when and where certain critical levels are exceeded or will soon be exceeded. All important physical and chemical processes must be taken into account in order to obtain reliable results. This leads to huge computational tasks, which have to be handled on big high-speed computers.

A short description of the particular air pollution model used, the Danish Eulerian Model, [11], will be given in this section in order to illustrate the numerical difficulties that are to be overcome when large air pollution models are treated on modern computers. We shall concentrate our attention on the mathematical backgrounds of this model. However, many details about the physical and chemical processes involved in the models as well as many of the validation tests and different air pollution studies that were carried out with the Danish Eulerian Model are documented in [2], [4], [5], [9], [11]-[16]. Many comparisons of model results with measurements can be seen in web-site of this model, [8].

The Danish Eulerian Model is described by a system of partial differential equations (PDE's):

$$\frac{\partial c_s}{\partial t} = -\frac{\partial(uc_s)}{\partial x} - \frac{\partial(vc_s)}{\partial y} - \frac{\partial(wc_s)}{\partial z}$$

$$+\frac{\partial}{\partial x}\left(K_x\frac{\partial c_s}{\partial x}\right) + \frac{\partial}{\partial y}\left(K_y\frac{\partial c_s}{\partial y}\right) + \frac{\partial}{\partial z}\left(K_z\frac{\partial c_s}{\partial z}\right)$$

$$+E_s - (\kappa_{1s} + \kappa_{2s})c_s + Q_s(c_1, c_2, \dots, c_q), \quad s = 1, 2, \dots, q. \tag{1}$$

The notation used in the system of PDEs (1) can be explained as follows: (i) the concentrations of the chemical species are denoted by c_s, (ii) the variables u, v and w are wind velocities, (iii) K_x, K_y and Kz are diffusion coefficients, (iv) the emission sources are described by E_s, (v) k_{1s} and k_{2s} are deposition coefficients (dry deposition and wet deposition, respectively) and (vi) the chemical reactions are represented by the non-linear functions $Q_s(c_1, c_2, \dots, c_q)$.

2 Splitting procedures

It is difficult to treat the system of PDE's (1) directly. This is the reason for using different kinds of splitting. A splitting procedure, based on ideas proposed in [6] and [7], leads to five sub-models, representing respectively: (i) the horizontal transport (the advection) of the pollutants in the atmosphere, (ii) the horizontal diffusion, (iii) the chemistry (together with the emission terms), (iv) the deposition and (v) the vertical exchange. These sub-models are transformed to systems of ordinary differential equations, ODEs, by using different kinds of space discretization. The splitting procedure, actually used when the Danish Eulerian Model is handled on computers, is discussed in detail in [11] and [14]. The most time- consuming sub-models are the horizontal transport and the chemistry. The numerical methods used in these sub-models are shortly discussed in the next two sections.

3 Horizontal Transport

The horizontal transport (the advection) is described mathematically by the following system of q independent PDEs:

$$\frac{\partial c_s}{\partial t} = -\frac{\partial(uc_s)}{\partial x} - \frac{\partial(vc_s)}{\partial y}, \quad s = 1, 2, \dots, q. \tag{2}$$

Several numerical algorithms have been tried to handle the horizontal transport sub-model ([5], [11], [14], [15]). All numerical methods, which have been tried, have both advantages and drawbacks. Two of the methods gave good results: the pseudo-spectral algorithm and a finite element method.

3.1 The pseudo-spectral algorithm

If only the accuracy is considered, then the pseudo-spectral algorithm ([9], [11]) seems to be the best one among the tested methods. However, it requires periodical boundary conditions, which is a serious drawback especially when grid-refinement on some parts of the space domain is to be used.

The pseudo-spectral algorithm applied in a version of the Danish Eulerian Model is fully described in [11]. Approximations of the first order spatial derivatives are obtained by expanding the unknown functions c_s in Fourier series containing *sine* and *cosine* functions. The truncation of this series leads to a trigonometric polynomial. The derivative of this polynomial is the required approximation of the first order spatial derivative under consideration. The use of Fourier series implies a requirement for periodicity of the concentrations on the boundaries. If high accuracy is to be achieved by keeping not too many terms in the Fourier series, then an additional requirement for periodicity of the derivatives of the concentrations up to some order n has also to be satisfied. The last requirement is a serious drawback of this algorithm, but the results are rather good if the space domain is very large and if some care is taken in order to obtain periodicity on the boundaries. The method is not suitable for modern procedures based on local refinement in some parts of the space domain.

3.2 Application of a finite element approach

The finite element approach, [5], [11], is in general less accurate (unless very complicated finite elements are used), but it is much more flexible when grid-refinement is needed in some parts of the space domain. Moreover, it is easier to develop parallel codes for the advection sub-model when some kind of finite elements are used. The finite element approach can be considered, as many other numerical methods for solving systems of PDEs, as a method based on expanding the unknown functions, the concentrations, in Taylor series. Truncating these series one arrives at algebraic polynomials. The derivatives of these polynomials are then used to obtain approximations to the first order spatial derivatives of the concentrations. In the following part of this paper it will be assumed that the finite elements algorithm from [3] is used.

4 Treatment of the Chemical Part

The chemical part of an air pollution model (including the emissions) is described mathematically by a non-linear and stiff system of ODEs, which can be written in the following form:

$$dg/dt = f(g,t), \quad g \in \mathbf{R}^{N_x \times N_y \times N_z \times q}, \quad f \in \mathbf{R}^{N_x \times N_y \times N_z \times q}, \tag{3}$$

where N_x, N_x and N_x are the numbers of grid-points along the coordinate axes, g contains approcximations of the concentrations at the grid-points and f depends on the chemical reactions.

This is normally the most time-consuming part. Also here different numerical algorithms have been tried. A detailed discussion of the algorithms used until now is given in [1]. Three of the used until now numerical methods will be shortly discussed here: (i) the QSSA (Quasi-Steady-State-Approximation) algorithm, (ii) some classical methods for solving systems of ODEs and (iii) numerical methods based on partitioning the system of ODEs and using the Backward Euler Method to treat the partitioned system.

4.1   The QSSA algorithm

The QSSA algorithm is based on a division of the chemical reactions into three groups and trying to use different devices for every group. It is simple and fast. Moreover, it is easy to implement this algorithm on parallel computers. The low accuracy of the algorithm is its major drawback. There are several improvements of the algorithm. More details about the QSSA algorithm and its performance can be found in [1].

An improved version of the QSSA algorithm, which is also discussed in [1], will be used in this paper. An attempt to improve the performance of the algorithm is made in the improved version.

4.2   Using classical numerical methods

Classical numerical methods for solving systems of ODEs are also applicable. The methods used until now were: (i) the Backward Euler Method, (ii) the Trapezoidal Rule and (iii) a second order Runge-Kutta algorithm.

If these methods are carefully implemented they can become competitive with the QSSA algorithm when performance is considered and, furthermore, the classical numerical methods are more accurate than the QSSA algorithm. There are, however, still some open problems when the classical methods are to be used on parallel architecture. The major difficulties are connected with the quasi-Newton iterative methods, which have to be used, because all considered methods are implicit. The iterative procedure converges with different rates in different parts of the space domain. This leads to lack of loading balance and, thus, deteriorates the performance of the code on the parallel computers. It should be mentioned, however, that some progress has been achieved recently (see, for example, [3]).

Some more details about the classical methods for solving systems of ODEs, which have been used until now, can be found in [1].

4.3   Using partitioning

Some of the chemical reactions used in a large-scale air pollution model are slow, while the others are fast. This fact can be exploited in the solution of the system of ODEs arising in the chemical sub-model. A method based on this idea has been developed and tested in [1]. The system of ODEs is first partitioned and then the Backward Euler Method is used to handle the partitioned system. The results were very encouraging. However, if the method has to be applied on a parallel computer, then difficulties, which are quite similar to those arising when classical numerical methods are used, have to be overcome. The conclusion is that the partitioning methods can successfully be used on a sequential computer, but some more work is needed in order to make them more efficient when a parallel computer is to be applied.

5 Major Modules of the Code

When a code for running a large-scale air pollution model on computers is developed, then it is important to have different easily exchangeable modules for the different parts of the computational work. Two two-dimensional versions of the Danish Eulerian model, which are discussed in this paper, consists of the following five major modules:

- **Advection module.** This module treats the horizontal transport. The diffusion subroutines have been added to this module.
- **Chemistry module.** This module handles the chemical reactions involved in the model. The deposition subroutines have been added to this module.
- **Initialization module.** All operations needed to initialize arrays and to perform the first readings of input data are united in this module.
- **Input operations module.** Input data are read in this module when this is necessary; otherwise simple interpolation rules are used, both in space and in time, to produce the needed data.
- **Output operations module.** This module prepares output data. The total number of output files prepared in the 2-D versions of the model is at present eight. The total amount of the output data is about 675 Mbytes when the 2-D model is discretized on a (288x288) grid and run over a time-period of one month (but it must be emphasized here that the model is normally run over a time-period of several, up to ten, years). For some of the concentrations (ozone and nitrogen dioxide) hourly mean values are calculated. For other concentrations and some depositions, diurnal mean values are prepared. For all concentrations and depositions studied by the model, monthly mean values are calculated in one of the files.

The initialization module is called only once (in the beginning of the computations), while all other modules are called at every time-step.

6 Use of Parallel Computers

The performance of two 2-D versions of the Danish Eulerian Model will be discussed in this section. The first version is discretized on a (96x96) grid, while the second one is discretized on a (288x288) grid. It is assumed that the model is run over a time-period of one month. In practice, many such runs, several hundreds or even several thousands, are to be performed. The computational tasks that are to be handled in these two versions consist of four systems of ODEs that have to be treated during more than 10 thousand time-steps. The two versions have been run on several large supercomputers (such as Fujitsu, SGI Origin 2000, IBM SP, etc.). The performance of the code on IBM SMP nodes will be demonstrated in this section.

6.1 The IBM SMP Nodes

Every IBM SMP nodes of the available architecture at the Danish Computing Centre contains 8 processors. There are two nodes at present. Each node can be considered as a shared memory computer. A combination of several nodes can be considered as a distributed memory machine. Thus, there are two levels of parallelism: (i) a shared memory mode within each node of the IBM SMP computer and (ii) a distributed memory mode across the nodes of the IBM SMP computer.

6.2 Shared Memory Mode

Assume first that the computer is to be used as a shared memory machine. It is important to identify large parallel tasks in the two most time-consuming parts of the code: (i) the advection module and (ii) the chemistry module.

6.2.1. Parallel Tasks in the Advection Module. In the advection module, the horizontal transport of each pollutant can be considered as a parallel task. The number of pollutants in the Danish Eulerian Model is 35. This means that the loading balance is not perfect. This is partly compensated by the fact that the parallel tasks so selected are very large.

6.2.2. Parallel Tasks in the Chemical Module. In the chemical module, the chemical reactions at a given grid-point can be considered as parallel tasks. This means that the number of parallel tasks is very large (i.e. the loading balance is perfect). This number is (i) 82944 in the version of the model that is discretized on a (288x288) grid and (ii) 9216 in the version of the model that is discretized on a (96x96) grid. However, the tasks are not large. Therefore it is convenient to group several grid-points as a bigger task. It will be shown that the performance can be improved considerably by a suitable choice of the groups of grid-points.

6.3 Distributed Memory Mode

Consider now the case where parallelism across the nodes is to be utilized. Then we need two extra modules: (i) a pre-processing module and (ii) a post-processing module.

6.3.1. Pre-processing Module. The first of these modules performs a pre-processing procedure. In this procedure the data is divided into several parts (according to the nodes that are to be applied) and sent to the assigned for the job nodes. In this way each node will work on its own data during the whole run.

6.3.2. Post-processing Module. The second module performs a post-processing procedure. This means that every node prepares during the run its own output files. At the end of the job, the post-processing module is activated (i) to collect the output data on one of the nodes and (ii) to prepare them for future applications (visualizations, animations, etc.).

6.3.3. Benefits of Using the Additional Modules. By the preparation of the two additional modules, one avoids some excessive communications during the run. However, it should also be stressed that not all communications are avoided. The use of the pre-processor and the post-processor is in fact equivalent to the application of a kind of domain decomposition. Therefore, it is clear that some communications are needed along the inner boundaries of the domains. Such communications are to be carried out only once per step and only a few data are to be communicated. Thus, the actual communications that are to be carried out during the computational process are very cheap.

6.4 Numerical results

Many hundreds of runs have been carried out in order to check the performance under different conditions. The most important results are summarized below.

6.4.1. Choosing the size of the parallel tasks. One of the crucial issues is the proper division of the grid-points into a suitable number of parallel tasks in the chemical module. Tests with more than 20 numbers of grid-points per parallel task (i.e. with different sizes of the parallel tasks) have been carried out. Some results are given in Table 1.

Size	Advection	Chemistry	Total time
1	62	586	730
24	63	276	432
48	63	272	424
96	64	288	447
576	61	389	612

Table 1

Computing times (given in seconds) obtained in runs of the version discretized on a (96x96) grid on the IBM SMP computer with different numbers of grid-points per parallel task in the chemical part. The numbers of grid-points per parallel task in the chemical part are given in the first column (under "Size"). The number of processors used in these runs is 16.

Three conclusions can be drawn from the tests with different sizes of the parallel tasks: (i) if the parallel tasks are too small, then the computing times are large, (ii) if the parallel tasks are too large, then the computing times are also large and (iii) for tasks with medium sizes the performance is best. It is difficult to explain this behaviour. The following explanation should be further tested and justified. If the tasks are very small (1,2 and 4 grid- points per parallel task), then the overhead needed to start the large number of parallel tasks becomes considerable. On the other hand, if the parallel tasks are too large (576 grid-points per parallel task), then the data needed to carry out them are too big and cannot stay in cache. Parallel tasks of medium sizes seem to be a good compromise.

The maximum number of grid-points per parallel tasks when the version of the model discretized on a (96x96) grid is 9216/p, where p is the number of processors used. The maximum number is 576 when 16 processors are used

6.4.2. Runs on different numbers of processors. The version of the model, which is discretized on a (96x96) grid, has been run on 1, 2, 4, 8 and 16 processors. When the number of processors is less than 16, only one node of the IBM SMP computer has been used. This means that in this case the code works in a shared memory mode only. Two nodes of the IBM SMP computer are used when 16 processors are selected. In this case a shared memory mode is used within each node, while the code performs in a distributed memory mode across the nodes. Some results are given in Table 2 (computing times) and Table 3 (speed-ups and efficiency of the parallel computations).

The results show clearly that good performance can be achieved in both cases: (i) when the code is running in a shared memory mode during the whole run (this happens when the number of processors is less than 16) and (ii) when the code is running by utilizing both modes (this is the case when the number of processors is 16). OPEN MP instructions are used when the code distributed memory mode. runs in a shared memory mode, while MPI subroutines are called when the code is run in a distributed memory mode.

6.4.3. Scalability of the code. It is important to be able to run efficiently the code when the grid is refined. Some tests with a version of the code, which is discretized on a (288x288) grid, have been carried out. There are plans to carry out some runs with a version of the code discretized on a (480x480) grid in the near future.

The computational work in the advection part is increased by a factor of 27 when the (288x288) grid is used, because the number of grid-points is increased by a factor of 9, while the time-step is decreased 3 times.

The computational work in the chemical part is increased only by a factor of 9 when the (288x288) grid is used, because the number of grid points is increased by a factor of 9, while the time-step used in the chemical part remains unchanged.

A few runs with the version discretized on a (288x288) grid have been carried out and compared with the corresponding runs for he version discretized on a (96x96) grid. The purpose was to check if the times spent are changed as should be expected (about 27 times for the advection part and about 9 time for the chemical part) when the same number of processors are used. Some results are given in Table 4. It is seen that the results are rather closed to the

Processors	Advection	Chemistry	Total time
1	933	4185	5225
2	478	1878	2427
4	244	1099	1405
8	144	521	799
16	62	272	424

Table 2

Computing times (given in seconds) obtained in runs of the version discretized on a (96x96) grid on the IBM SMP computer with different number of processors. The number of grid-points per parallel task in the chemical part is 48.

Processors	Advection	Chemistry	Total time
2	1.95 (98%)	2.23 (112%)	2.15 (108%)
4	3.82 (96%)	3.81 (95%)	3.72 (93%)
8	6.48 (81%)	8.03 (100%)	6.54 (82%)
16	15.01 (94%)	15.39 (96%)	12.32 (77%)

Table 3

Speed-ups and efficiency (given in brackets) obtained in runs of the version discretized on a (96x96) grid on the IBM SMP computer with different number of processors. The number of grid-points per parallel task in the chemical part is 48. expected results. Some more experiments are needed.

Process	(288x288)	(96x96)	Ratio
Advection	1523	63	24.6
Chemistry	2883	288	10.0
Total	6209	432	14.4

Table 4

Computing times (given in seconds) obtained in runs of the two versions of the code, discretized on a (96x96) grid and on a (288x288) grid, on the IBM SMP computer by using 16 processors. The number of grid-points per parallel task in the chemical part is 96.

7 Conclusions

Many important environmental studies can successfully be performed only if (i) efficient numerical methods are implemented in the mathematical models used and (ii) the great potential power of the modern supercomputers is utilized in an optimal way.

Some results obtained in the effort to solve these two tasks have been presented. There are still a lot of unresolved problems. The efforts to find more efficient numerical methods for the large air pollution models and to optimize further the code for runs on parallel machines are continuing.

References

1. Alexandrov, V., Sameh, A., Siddique, Y. and Zlatev, Z., Numerical integration of chemical ODE problems arising in air pollution models, Environmental Modelling and Assessment, Vol. 2, 1997, 365–377.
2. Bastrup-Birk, A., Brandt, J., Uria, I. and Zlatev, Z., Studying cumulative ozone exposures in Europe during a seven-year period, Journal of Geophysical Research, Vol. 102, 1997, 23917-23935.
3. Georgiev, K. and Zlatev, Z., Parallel sparse matrix algorithms for air pollution models, Parallel and Distributed Computing Practices, to appear.
4. Harrison, R. M., Zlatev, Z. and Ottley, C. J., A comparison of the predictions of an Eulerian atmospheric transport chemistry model with measurements over the North Sea, Atmospheric Environment, Vol. 28, 1994, 497-516.
5. Hov, Ø., Zlatev, Z., Berkowicz, R., Eliassen, A. and Prahm, L. P., Comparison of numerical techniques for use in air pollution models with non-linear chemical reactions , Atmospheric Environment, Vol. 23 (1988), 967-983.
6. Marchuk, G. I., Mathematical modeling for the problem of the environment, Studies in Mathematics and Applications, No. 16, North-Holland, Amsterdam, 1985.
7. McRae G. J., Goodin, W. R. and Seinfeld, J. H., Numerical solution of the atmospheric diffusion equations for chemically reacting flows, Journal of Computational Physics, Vol. 45 (1984), 1-42.
8. WEB-site for the Danish Eurlerian Model, http://www.dmu.dk/AtmosphericEnvironment/DEM.
9. Zlatev, Z., Application of predictor-corrector schemes with several correctors in solving air pollution problems, BIT, Vol. 24, 1984, 700-715.
10. Zlatev, Z., Computational methods for general sparse matrices, Kluwer Academic Publishers, Dordrecht-Boston-London, 1991.
11. Zlatev, Z., Computer treatment of large air pollution models, Kluwer Academic Publishers, Dordrecht-Boston-London, 1995.
12. Zlatev, Z., Christensen, J. and Eliassen, A., Studying high ozone concentrations by using the Danish Eulerian Model, Atmospheric Environment, Vol. 27A, 1993, 845-865.
13. Zlatev, Z., Christensen, C. and Hov,Ø., A Eulerian air pollution model for Europe with non-linear chemistry, Journal of Atmospheric Chemistry, Vol. 15 (1992), 1-37.
14. Zlatev, Z., Dimov, I. and Georgiev K., Studying long-range transport of air pollutants, Computational Science and Engineering, Vol. 1, No. 3, (1994), 45-52.
15. Zlatev, Z., Dimov, I. and Georgiev K., Three-dimensional version of the Danish Eulerian Model, Zeitschrift für Angewandte Mathematik und Mechanik, Vol. 76, S4, (1996), 473-476.
16. Zlatev, Z., Fenger, J. and Mortensen, L., Relationships between emission sources and excess ozone concentrations, Computers and Mathematics with Applications, Vol. 32, No. 11 (1996), 101-123.

High Performance Computational Tools and Environments

Session chair:

Dale Shires (U.S. Army Research Laboratory, USA)

Pattern Search Methods
for Use-Provided Points[*]

Pedro Alberto[1], Fernando Nogueira[1], Humberto Rocha[2], and Luís N. Vicente[3]

[1] Departamento de Física, Universidade de Coimbra, 3004-516 Coimbra, Portugal
[2] Universidade Católica, Pólo de Viseu, 3504-505 Viseu, Portugal
[3] Departamento de Matemática, Universidade de Coimbra, 3001-454 Coimbra,
Portugal

Abstract. We show how pattern search methods can be adapted to the
optimization problem contexts where there are ways to provide points
that can lead to an objective function decrease. The paradigm here is
that it is the user and the optimization algorithm together, and not the
optimization algorithm alone, that lead the calculation of new points.
We are especially concerned with problems where objective function evaluations are expensive and for which parallel computing is available.

In this short paper we describe how pattern search methods for unconstrained
optimization problems of the form

$$\min f(x), \quad x \in \mathbb{R}^n$$

can be applied when the user can and wishes to provide a routine to compute
new points. An example of this situation arises in molecular geometry optimization, where new points can be provided by the user by applying physically
relevant geometrical transformations to the current configuration. These geometrical transformations can potentially lead to a decrease in the objective function,
i.e., in the total energy of the cluster of atoms that is being considered [1].

Pattern search methods exhibit enough flexibility to accommodate the user-
provided point calculation. The main idea is to use patterns that fill the space
surrounding the current iterate with a reasonable distribution of pattern points
and pattern directions. In this way, a new point calculated by the user can be
projected onto the pattern in such a way that the projected pattern point is
reasonably close to the point provided by the user. The objective function is
only evaluated at the projected pattern point and not at the user-provided point
– an important requirement to regularize the overall algorithm and guarantee
convergence properties [1]. The pattern must also be defined so that the linear
algebra involved in the projection can be cheaply computed.

We will provide a very brief introduction to pattern search methods, introducing only the notation necessary to describe the accommodation of user-provided

[*] This work was supported by FCT (under the grant Praxis/P/FIS/14195/1998), by
Centro de Física Computacional, and by Centro de Matemática da Universidade de
Coimbra.

points. The reader is referred to the papers [2, 3, 5, 7] for motivation, theory, and other related material on pattern search methods.

If the iteration k of a pattern search method is *successful*, the next iterate must provide a decrease in the objective function: $f(x_{k+1}) < f(x_k)$. Pattern search methods iteratively generate points in an integer lattice (pattern), visiting at each iteration k a subset of the pattern called the *mesh* M_k. The mesh M_k can be defined through a set $\mathcal{V}$ of m positive bases[1] and a mesh size parameter $\Delta_k > 0$ in the following way:

$$M_k = \{x_k + \Delta_k \mathcal{V} z : \ z \in W \subseteq \mathbb{Z}^{|\mathcal{V}|}\},$$

where $|\mathcal{V}|$ is the sum of the number of vectors in all positive bases. The choice we actually made in our implementation of the pattern search methods for user-provided points is

$$W = \{n e_i : n \in \mathbb{N}, \ i = 1, \dots, |\mathcal{V}|\},$$

where e_i is the i-th column of the identity matrix of order $|\mathcal{V}|$. Choices for $\mathcal{V}$ are described and discussed in [1].

The mechanism of pattern search methods consists of two steps at every iteration. In the first step, called the *search step*, a finite search is performed on the mesh, with the goal of finding a new iterate that decreases the value of the objective function at the current iterate. This step, called the *search step*, searches only a finite number of points in the mesh. The search step provides the flexibility for a global search, and influences the quality of the local minimizer or stationary point found by the method [1, 4, 6]. If the search step is unsuccessful, a second step, called the *poll step*, is performed around the current iterate with the goal of decreasing the objective function.

The poll step follows stricter rules and appeals to the concept of positive bases. In this step the candidate for a new iterate x_{k+1} is chosen in the *mesh neighborhood* around x_k

$$\mathcal{N}(x_k) = \{x_k + \Delta_k v : \ \text{for all } v \in V_k(x_k)\},$$

where $V_k(x_k)$ is a positive basis chosen from the finite set $\mathcal{V}$ of positive bases. The poll step attempts to perform a local search in a mesh neighborhood that, for a sufficient small mesh parameter Δ_k, is guaranteed to provide an objective function reduction, unless the current iterate is at a stationary point. If the poll step also fails, then the mesh parameter Δ_k must be decreased.

Pattern search methods for user-provided points can now be described for use in a parallel environment where, say, N_p processors are available.

Algorithm 1 (Pattern search methods for user-provided points)

---

[1] A positive basis for $\mathbb{R}^n$ can be defined as a set of nonzero vectors of $\mathbb{R}^n$ whose nonnegative combinations span $\mathbb{R}^n$, but no proper set does. It can be shown that every positive basis has between $n + 1$ and $2n$ elements.

0. Initialization Choose a rational number $\tau > 1$ and an integer number $m_{max} \geq 1$. Choose $x_0 \in \mathbb{R}^n$ and $\Delta_0 \in \mathbb{R}_+$. Set $k = 0$.

1. Search step (in current mesh)

   1. For each processor p in $\{1, \ldots, N_p\}$:

      (a) Obtain a point u_{k+1}^p from the user.

      (b) Compute $x_{k+1}^p = x_k + \Delta_k \mathcal{V} z^p$, where z^p is the optimal solution of the integer programming problem

$$\min_{z \in W} \| u_{k+1}^p - (x_k + \Delta_k \mathcal{V} z) \|. \tag{1}$$

      (c) Evaluate f on the mesh point x_{k+1}^p.

   2. If

$$\min_{p \in \{1, \ldots, N_p\}} f(x_{k+1}^p) < f(x_k),$$

   then set

$$x_{k+1} = \mathrm{argmin}_{x_{k+1}^p} f(x_{k+1}^p),$$

   and go to step **3**, expanding M_k (search step and iteration are declared successful).

2. Poll step (in mesh neighborhood given by the positive basis)

   This step is reached only if the search step is unsuccessful.

   1. Obtain a point u_{k+1} from the user.

   2. Determine $v_k \in \mathcal{V}$ such that

$$\frac{\langle u_{k+1} - x_k, v_k \rangle}{\| u_{k+1} - x_k \|} = \max_{v \in \mathcal{V}} \frac{\langle u_{k+1} - x_k, v \rangle}{\| u_{k+1} - x_k \|}. \tag{2}$$

   3. Set $V_k(x_k)$ to the positive basis in $\mathcal{V}$ that contains v_k, and then set $\mathcal{N}(x_k) = \{x_k + \Delta_k v : \text{for all } v \in V_k(x_k)\}$.

   4. List the points in $\mathcal{N}(x_k)$ by increasing order of the values of the angles between $u_{k+1} - x_k$ and the corresponding vectors in $V_k(x_k)$.

   5. Following the list given above, divided in groups of N_p points, start evaluating in parallel the function f in $\mathcal{N}(x_k)$.
Stop if a point $x_{k+1} \in \mathcal{N}(x_k)$ is found such that $f(x_{k+1}) < f(x_k)$. In this case go to step **3**, expanding M_k (poll step and iteration are declared successful).
If $f(x_k) \leq f(x)$ for every x in the mesh neighborhood $\mathcal{N}(x_k)$, go to step **4**, shrinking M_k (poll step and iteration are declared unsuccessful).

3. Mesh expansion (at successful iterations) Let $\Delta_{k+1} = \tau^{m_k^+} \Delta_k$ (with $0 \leq m_k^+ \leq m_{max}$). Increase k by one, and move to step **1** for a new iteration. (The value of $\tau^{m_k^+}$ can be chosen according to user-provided information.)

4. Mesh reduction (at unsuccessful iterations) Let $\Delta_{k+1} = \tau^{m_k^-} \Delta_k$ (with $-m_{max} \leq m_k^- \leq -1$). Increase k by one, and move to step **1** for a new iteration. (The value of $\tau^{m_k^-}$ can be chosen according to user-provided information.)

We point out that due to our choice of W, problems (1) and (2) are easily solved in the order of $|\mathcal{V}|n$ floating point operations.

It is also important to note that by listing the points in $\mathcal{N}(x_k)$ using the order suggested in step 2.4, the poll step starts by evaluating f in the points of $\mathcal{N}(x_k)$ closer to u_{k+1}.

Although we have described a parallel algorithm, a serial version of the algorithm is a straightforward adaptation of the parallel version. Both versions have been implemented in Fortran 95. The parallel version uses the parallelization protocol MPI. The codes and their documentation can be downloaded from the web site:

http://www.mat.uc.pt/~lvicente/psm/

The user must provide a routine to compute new points, and a routine to evaluate f at points specified by the algorithm. The calling sequences of these two routines are currently coded in the following form:

```
SUBROUTINE func( n, xk, f )

SUBROUTINE trial_point( n, xk, n_i_userpar, i_userpar, &
                        n_r_userpar, r_userpar, xtrial )
```

The output parameters are f and xtrial. In the routine trial_point, the user is given some information, stored in the integer vector i_userpar and in the real vector r_userpar, to indicate the amount of effort that can be put in the calculation. For instance, it is natural in the search step to ask the user to make a trial point calculation greedier or less conservative than in the the poll step.

References

1. P. Alberto, F. Nogueira, H. Rocha, and L. N. Vicente. Pattern search methods for molecular geometry problems. Technical Report 00-20, Departamento de Matemática, Universidade de Coimbra, 2000.
2. C. Audet and J. E. Dennis. Analysis of generalized pattern searches. Technical Report TR00-07, Departament of Computational and Applied Mathematics, Rice University, 2000.
3. C. Audet and J. E. Dennis. Pattern search algorithms for mixed variable programming. *SIAM J. Optim.*, 11:573–594, 2001.
4. W. E. Hart. Comparing evolutionary programs and evolutionary pattern search algorithms: A drug docking application. In *Proc. Genetic and Evolutionary Computation Conf.*, 1999.
5. R. M. Lewis and V. Torczon. Rank ordering and positive bases in pattern search algorithms. Technical Report TR96-71, ICASE, 1999.
6. J. C. Meza and M. L. Martinez. On the use of direct search methods for the molecular conformation problem. *Journal of Computational Chemistry*, 15:627–632, 1994.
7. V. Torczon. On the convergence of pattern search algorithms. *SIAM J. Optim.*, 7:1–25, 1997.

In-situ Bioremediation: Advantages of Parallel Computing and Graphical Investigating Techniques

M.C.Baracca, G.Clai, P.Ornelli

ENEA HPCN Via Martiri di Montesole 4 40129-Bologna, ITALY
email:[baracca,ornelli]@bologna.enea.it

Abstract. The mathematical modelling and the simulation of the bioremediation process are fairly complex and very computational demanding. The use of parallel computers allows the prediction of the effects of bioremediation interventions safely and cheaply in a realistic time-scale. Nevertheless, the data amount resulting from a real field simulation is so large that the only way to analyze it, in an effective way, is to work with its graphical interpretation, converting the huge numerical data volumes into 3D animation.

1 Introduction

The quality, duration and cost of a bioremediation intervention depend on the success of laboratory or pilot plant experiments, which are time consuming and not always reliable. Since the mathematical modelling and the simulation of the bioremediation process are fairly complex and very computational demanding, they can be used to predict in a realistic time-scale the effects of their actual application safely and cheaply, only through the exploitation of parallel computers.

Nevertheless, the data amount resulting from a real field simulation is so large that the only way to analyze it, in an effective way, is to work with its graphical interpretation, converting the huge numerical data volumes into 3D animation.

The two-years collaborative COLOMBO Project funded by the European Union has succesfully dealt with the application of both parallel computing and graphical investigating techniques to the simulation of the *in situ* bioremediation of contaminated soil.

2 The Bioremediation Model

The decontamination process works by stimulating the growth of indigenous bacteria and the model predicts the effects of bioremediation starting from geological, chemical and microbiological data as well as experimental results from the pilot plants designed and realized in the framework of the project. In order to solve succesfully soil pollution problems, researchers must simulate contaminant transport, diffusion and transformation processes and, at the same time, estimate the many parameters defining the bioremediation strategy.

The cellular automata (CA) modelling technique[1] was adopted in order to describe the phenomena taking place at the intermediate scale inside a portion of soil rather then the microscopic phenomena involved. The three dimensional bioremediation model[2], composed of mesoscopic size cells, describes real field interventions referred to heterogeneous regions simulating the fluid flow with diffusion and transport of pollutant inside the soil. The rule governing the system evolution has a "layered" structure, where the first layer is related to the water and pollutant flow through a porous saturated or unsaturated soil, the second describes the chemical behaviour and the third deals with the interaction between chemicals and biomass.

The model was tested on two real fields cases in Germany, the first at the US-Depot in Germersheim and the second near the Deutsche Bahn in Frankfurt.

3 The Bioremediation Simulation Code

The theory of CA lends itself well to the bioremediation simulation. In our case the automaton is related to a finite three-dimensional space and consists of a number of cells. Each of them is associates with a set of 6 neighbouring cells and its state is defined by a set of attributes, which are known as substates. The automaton evolves through a specified rule that decides the state of each cell in the next time-step based on the current state of the cell in question and those of its neighbours. The locality of the considered phenomena (bacteria and pollutants interact gradually over time and only within their neighbourhood area) benefits from parallel computers since one can decompose the automaton domain and map the components to cooperating computing processors with relative acceptable communication overhead, due to the boundary data exchange. Moreover, the CA approach is appropriate since it allows both to represent system heterogeneities, simply setting proper parameter values, and to tailor the cell sizes according to the problem needs and the available computing resources. The code has two modes and its execution is managed by a computational steering mechanism switching from the fluid dynamical to the biochemical layers simulation (and viceversa) when proper condition are satisfied. A real intervention simulation requires a very large number of iterations, it can run from several hours to some days, depending on whether soil conditions are saturated or unsaturated.

The benchmarks on CrayT3E-900 [3] and Beowulf cluster, have established that the code execution scales well with the number of available processors, as long as the size of the per processor data is more than half of the exchanged boundary data. In the pictures below, the benchmark results for the large model (256x128x13 cells) on the Cray platform, performed by the Edinburgh Parallel Computing Center researchers, are provided toghether with analogous benchmark results for the medium model (128x96x13 cells) on a Beowulf cluster.

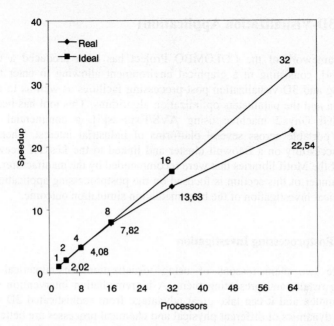

Fig 1: Speedup (blue) and optimum speedup (red) scaling curves for the large model on the Cray T3E (2-64 processors)

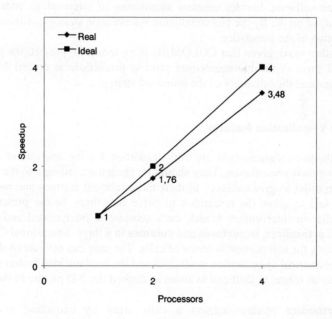

Fig 2: Speedup (blue) and optimum speedup (red) scaling curves for the medium model on Beowulf cluster (1-4 processors)

4 The 3D-Visualization Application

In the framework of the COLOMBO Project has been produced a user oriented interface[4], consisting in a graphical environment allowing to enter the data pre-processing and 3D visualization post-processing facilities as well as to run the batch simulation and the parameters optimization algorithms. This tool has been developed on an SGI Onyx2 machine using AVS/Express[5], a commercial visualization package, portable across several platforms of industrial interest. Then it has been ported succesfully on a Beowulf cluster and linked to the LessTif freeware libraries instead of the Motif libraries that were recommended by the manifacturer.

The remainder of this section is focused on the postprocessing application devoted to the graphical investigation of the bioremediation simulation outcome.

4.1 The Postprocessing Investigation

There are two main reasons of using visualization for numerical simulations: analyzing results and presenting them. A bioremediation intervention simulation is fairly complex and it can take great advantage from sophisticated 3D visualization, since the dynamics of different physical and chemical processes are better observed in graphical form. Starting from raw data gathered during the simulation, the researcher may analyze, study and discover useful features and numerical results that cannot be appreciated during the simulation. The interplay between the visualization and the simulation software, besides constant monitoring of intermediate states, enables the user to adjust on the fly the bioremediation intervention strategy, without the complete re-execution of the simulation.

On the other hand, given that COLOMBO is an industrial project, the presentation in graphical form of the bioremediation process predictions is crucial for proving the advantages and the reliability of the proposed strategy.

4.2 The Visualization features

The methods of visualization are very important for the analysis of the numerical results and their presentation. They should be chosen accordingly to the phenomena in object, in order to give evidence to their more relevant features and more interesting aspects, and to allow the researcher to delve into them. In the present case of the bioremediation intervention model, each substate is interpolated and visualized by means of **orthoslices**, **isosurfaces** and **volumes** in a three dimensional Cartesian space representing the soil portion in terms of cells. The user can activate or deactivate each of the geometrical visualization modalities and the implemented features. Moreover, it is possible to rotate, to shift and to zoom as desired the 3-D picture in the visualization space.

Each **orthoslice** module subsets a cells array by extracting one slice plane perpendicular to one of the Cartesian axis. The user can move interactively the orthoslices in the range from 0 to the axis maximum dimension. When using this modality, with a mouse click, it is possible to retrieve the substate value and the coordinates of a cell by means of the **probe** feature, cutting out the interpolated data produced by the visualization process.

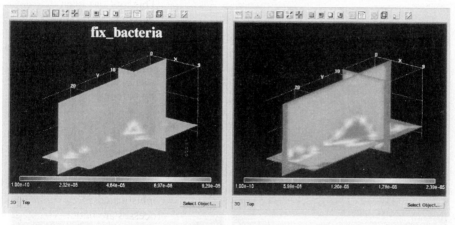

Fig 3   Fig 4

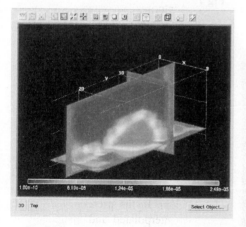

Fig 5

Figures 3, 4, 5 show the 3D automaton, representing the Frankfurt real field test case. In this example, the visualization is performed by means of orthoslices. Specifically, the bacteria able to degrade the contaminants are investigated: the pictures sequence shows three time steps of the biomass growth, starting from the beginning of the remediation intervention.

In order to outline correlated substates or threshold effects, the **isosurface** module, that creates a surface with a given constant value level, has been improved including the option to map on a substate isosurface the cell values of an other substate.

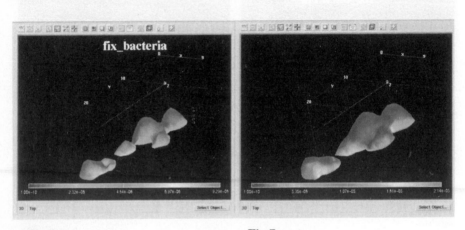

Fig 6   **Fig 7**

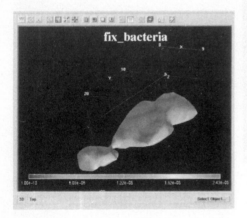

Fig 8

In figures 6, 7, 8, the 3D automaton, representing the Frankfurt real field test case, is visualized by means of the isosurface feature. This example shows two correlated substates, the oxigen concetration in water and the biomass. Given a specific value of the oxigen concentration, the related isosurface shape is created while the corresponding bacteria distribution is mapped onto it. The color coding scheme is presented at the bottom of each window. The pictures show three subsequent time steps of the biomass growth corresponding to the same oxigen concentration level.

The **isovolume** module that displays the interpolated volume of the cells whose values are greater or less than a specified level, as well as the **bounded volume,** displaying the substate volume included beetween two chosen cell values, are the bases for quantitative and statistical evaluations.

Vectorial fields visualization has been provided, in order to show in an effective way the air, water and pollutant fluxes. The vector fields are visualised by arrows, each one oriented as the local flux direction; the arrow color, mapped within a suitable color scale, represents the local flux value. Vector field planes can be visualized one at a time: the user can change the orthogonal plane and move it along the selected direction while the visualization is being performed.

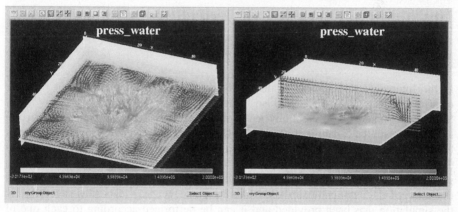

Fig 9   **Fig 10**

Figures 9 and 10 refer to the real field test case located in Germersheim. In the pictures the water flux is analyzed, shifting the attention from an horizontal to a vertical flux plane superposed to the water pressure distribution, at the injection and extraction wells level.

From the analysis of the bioremediation model applied to real field emerged that, usually, the surface dimensions are much larger than the field depth, so that the automaton cells look like a parallelepiped and not a cube. In order to properly visualize the automaton in this case, the **Scale** facility allows to choose and set a different scale for each axis. The Scale factor affects at the same time the scalar substate as well as the vector fields, so that in case of a superposition of a substate and a vector field, a change in the Scale factor results in a rescaling of the whole image.

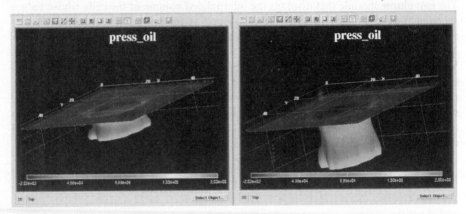

| Fig 11 | Fig 12 |

In the figures 11 and 12, the Scale facility is applied to the contaminant pressure distribution mapped on a potential isosurface.

Since the dynamics of different physical and chemical processes are better observed in graphical form, the capability to produce **3D Animation** visualizing the evolution of a bioremediation process, carried on according to a chosen intervention strategy, is crucial for demonstrating the reliability of the postprocessing visualization as a powerful tool for planning and designing actual interventions. Therefore, the temporal evolution of the automaton substates, based on a sequence of data files saved during the simulation, has been provided and it can be performed according to each one of the implemented geometrical modalities, as well as for vectorial fields.

5 Conclusions

The bioremediation model reliability in evaluating the degradation rates, the code scalability allowing the prediction of the interventions duration in a realistic time-scale and the advantages of the visualization tool have determined the success of the COLOMBO Project.

The 3D animation capability to present the interventions details and the remediation processes evolution, in an effective way, was essential for the end-user to win support of the Italian Ministry for Scientific Research for the study and the application of new bioremediation techniques to polluted sites located in Italy.

REFERENCES

1. J.von Neumann, "Theory of Self Reproducing Automata", Univ.Illinois Press, Champaign, Ill.,1966.
2. M.Villani, M.Mazzanti, R.Serra, M.Andretta, S.Di Gregorio, "Simulation model implementation description", Deliverable D11 of COLOMBO Project, July 1999.
3. K.Kavoussanakis et al., "CAMELot Implementation and User Guide", Deliverable D9 of COLOMBO Project, September 2000.
4. M.C.Baracca, G.Clai, P. Ornelli, "Pre/Post Processor Description", Deliverable D12 of COLOMBO Project, July 1998.
5. Advanced Visual System Inc. , "Using AVS/Express", July 1998.

Adaptive Load Balancing for MPI Programs*

Milind Bhandarkar, L. V. Kalé, Eric de Sturler, and Jay Hoeflinger

Center for Simulation of Advanced Rockets
University of Illinois at Urbana-Champaign
{bhandark,l-kale1,sturler,hoefling}@uiuc.edu

Abstract. Parallel Computational Science and Engineering (CSE) applications often exhibit irregular structure and dynamic load patterns. Many such applications have been developed using MPI. Incorporating dynamic load balancing techniques at the application-level involves significant changes to the design and structure of applications. On the other hand, traditional run-time systems for MPI do not support dynamic load balancing. Object-based parallel programming languages, such as Charm++ support efficient dynamic load balancing using object migration. However, converting legacy MPI applications to such object-based paradigms is cumbersome. This paper describes an implementation of MPI, called Adaptive MPI (AMPI) that supports dynamic load balancing for MPI applications. Conversion from MPI to this platform is straightforward even for large legacy codes. We describe our positive experience in converting the component codes ROCFLO and ROCSOLID of a Rocket Simulation application to AMPI.

1 Introduction

Many Computational Science and Engineering (CSE) applications under development today exhibit dynamic behavior. Computational domains are irregular to begin with, making it difficult to subdivide the problem such that every partition has equal computational load, while optimizing communication. In addition, computational load requirements of each partition may vary as computation progresses. For example, applications that use Adaptive Mesh Refinement (AMR) techniques increase the resolution of spatial discretization in a few partitions, where interesting physical phenomena occur. This increases the computational load of those partitions drastically. In applications such as the simulation of pressure-driven crack propagation using Finite Element Method (FEM), extra elements are inserted near the crack dynamically as it propagates through structures, thus leading to severe load imbalance. Another type of dynamic load variance can be seen where non-dedicated platforms such as clusters of workstations are used to carry out even regular applications [BK99]. In such cases, the availability of individual workstations changes dynamically.

---

* Research funded by the U.S. Department of Energy through the University of California under Subcontract number B341494.

Such load imbalance can be reduced by decomposing the problem into several smaller partitions (many more than the available physical processors) and then mapping and re-mapping these partitions to physical processors in response to variation in load conditions. One cannot expect the application programmer to pay attention to dynamic variations in computational load and communication patterns, in addition to programming an already complex CSE application. Therefore, the parallel programming environment needs to provide for dynamic load balancing *under the hood*. To do this effectively, it needs to know the precise load conditions at runtime. Thus, it needs to be supported by the runtime system of the parallel language. Also, it needs to predict the future load patterns based on current and past runtime conditions to provide an appropriate re-mapping of partitions.

Fortunately, an empirical observation of several such CSE applications suggests that such changes occur slowly over the life of a running application, thus leading to the *principle of persistent computation and communication structure* [KBB00]. Even when load changes are dramatic, such as in the case of adaptive refinement, they are infrequent. Therefore, by measuring variations of load and communication patterns, the runtime system can accurately forecast future load conditions, and can effectively load balance the application.

Charm++[KK96] is an object-oriented parallel programming language that provides dynamic load balancing capabilities using runtime measurements of computational loads and communication patterns, and employs object migration to achieve load balance. However, many CSE applications are written in languages such as FORTRAN, using MPI (Message Passing Interface) [GLS94] for communication. It can be very cumbersome to convert such legacy applications to newer paradigms such as Charm++ since the machine models of these paradigms are very different. Essentially, such attempts result in complete rewrite of applications.

Frameworks for computational steering and automatic resource management, such as AutoPilot [RVSR98], provide ways to instrument parallel programs for collecting load information at runtime, and a fuzzy-logic based decision engine that advises the parallel program regarding resource management. But it is left to the parallel program to implement this advice. Thus, load balancing is not transparent to the parallel program, since the runtime system of the parallel language does not actively participate in carrying out the resource management decisions. Similarly, systems such as CARMI [PL95] simply inform the parallel program of load imbalance, and leave it to the application processes to explicitly move to a new processor. Other frameworks with automatic load balancing such as the FEM framework [BK00], and the framework for Adaptive Mesh Refinement codes [BN99] are specific to certain application domains, and do not apply to a general programming paradigm such as message-passing or to a general purpose messaging library such as MPI. TOMPI [Dem97] and TMPI [TSY99] are thread-level implementations of MPI. The techniques they use to convert legacy MPI codes to run on their implementations are similar to our approach. However, they do not support FORTRAN. Also, they do not provide automated dynamic load balancing. TOMPI is a single processor simulation tool for MPI programs, while TMPI attempts to implement MPI efficiently on shared memory multiprocessors.

In this paper, we describe a path we have taken to solve the problem of load imbalance in existing FORTRAN90-MPI applications by using the dynamic load balancing capabilities of Charm++ with minimal effort. The next section describes the load-balancing framework of Charm++ and its multi-partitioning approach. In section 3, we describe the implementation of Adaptive MPI, which uses user-level migrating threads, along with message-driven objects. We show that it is indeed simple to convert existing MPI code to use AMPI. We discuss the methods used and efforts needed to convert actual application codes to use AMPI, and performance implications in section 4.

2 Charm++ Multi-partitioned Approach

Charm++ is a parallel object-oriented language. A Charm++ program consists of a set of communicating objects. These objects are mapped to available processors by the message-driven runtime system of Charm++. Communication between Charm++ objects is through asynchronous object-method invocations. Charm++ object-methods (called *entry* methods) are atomic. Once invoked, they complete without waiting for more data (such as by issuing a blocking receive). Charm++ tracks execution times (computational loads) and communication patterns of individual objects. Also, method execution is directed at objects, not processors. Therefore, the runtime system can migrate objects transparently.

Charm++ incorporates a dynamic load balancing (LB) framework, that acts as a gateway between the Charm++ runtime system and several "plug-in" load balancing strategies. The object-load and communication data be viewed as a weighted communication graph, where the connected vertices represent communicating objects. Load balancing strategies produce a new mapping of these objects in order to balance the load. This is an NP-hard multidimensional optimization problem, and producing optimal solution is not feasible. We have experimented with several heuristic strategies, and they have been shown to achieve good load balance [KBB00]. The new mapping produced by the LB strategy is communicated to the runtime system, which carries out object migrations.

NAMD [KSB+99], a molecular dynamics application, is developed using Charm++. As simulation of a complex molecule progresses, atoms may move into neighboring partitions. This leads to load imbalance. Charm++ LB framework is shown to be very effective in NAMD and has allowed NAMD to scale to thousands of processors achieving unprecedented speedups among molecular dynamics applications (1252 on 2000 processors). Another application that simulates pressure-driven crack propagation in structures has been implemented using the Charm++ LB framework [BK00], and has been shown to effectively deal with dynamically varying load conditions (Figure 1.) As a crack develops in a structure discretized as a finite element mesh, extra elements are added near the crack, resulting in severe load imbalance. Charm++ LB framework responds to this load imbalance by migrating objects, thus improving load balance, as can be seen from increased throughput measured in terms of number of iterations per second.

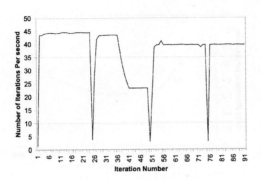

Fig. 1. Performance of the Crack-Propagation application using Charm++ load-balancing framework.

The key to effectively using the Charm++ LB framework is to split the computational domain into several small partitions. These smaller partitions, called virtual processors, (or *chunks*) are then mapped and re-mapped by the LB framework in order to balance the load across physical processors.

Having more chunks to map and re-map results in better load balance. Large number of chunks result in smaller partitions that utilize the cache better. Also, having more independent pieces of computation per processor results in better latency tolerance with computation/communication overlap. However, mapping several chunks on a single physical processor reduces the granularity of parallel tasks and the computation to communication ratio. Thus, multi-partitioning present a tradeoff in the overhead of virtualization and effective load balance.

In order to study this tradeoff, we carried out an experiment using a Finite Element Method application that does structural simulation on an FEM mesh with 300K elements. We ran this application on 8 processor Origin2000 (250 MHz MIPS R10K) with different number of partitions of the same mesh mapped to each processor. Results are presented in figure 2. It shows that increasing number of chunks is beneficial up to 16 chunks per physical processor, as number of elements per chunk decreases from 300K to about 20 K. This increase in performance is caused by better cache behavior of smaller partitions, and overlap of computation and communication (latency tolerance). Further, the overhead introduced for 32 and 64 chunks (with 10K and 5K elements per chunk, respectively) per physical processor is very small. Though these numbers may vary depending on the application, we expect similar behavior for many applications that deal with large data sets and have near-neighbor communication.

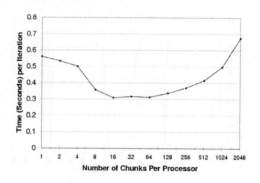

Fig. 2. Effects of multi-partitioning in an FEM application.

One way to convert existing MPI applications to the multi-partitioned approach is to represent an MPI process by a Charm++ object. However, this is not trivial, since MPI processes contain blocking receives and collective operations that do not satisfy the atomicity requirements of entry methods.

3   Adaptive MPI

AMPI implements MPI processes with user-level threads so as to enable them to issue blocking receives. Alternatives are to use processes or kernel-level threads. However, process context switching is costly, because it means switching the page table, and flushing out cache-lines etc. Process migration is also costlier than thread migration. Kernel-threads are typically preemptive. Accessing any shared variable would mean use of locks or mutexes, thus increasing the overhead. Also, one needs OS support for migrating kernel threads from one process to another. With user-level threads, one has complete control over scheduling, and also one can track the communication pattern among chunks, as well as their computational and memory requirements.

It is difficult to migrate threads because any references to the stack have to be valid after migration to a different address space. Note that a chunk may migrate anytime when it is blocking for messages. At that time, if the thread's local variables refer to other local variables on the stack, these references may not be valid upon migration, because the stack may be located in a different location in memory. Thus, we need a mechanism for making sure that these references remain valid across processors. In the absence of any compiler support, this means that the thread-stacks should span the same range of virtual addresses on any processor where it may migrate.

Our first solution to this problem was based on a stack-copy mechanism, where all threads execute with the same system stack, and contents of the thread-stack were copied at every context-switch between two threads. If the system stack is located at the same virtual address on all processors, then the stack references will remain valid even after migration. This scheme's main drawback is the copy overhead on every context-switch (figure 3). In order to increase efficiency with this mechanism, one has to keep the stack size as low as possible at the time of a context-switch.

Our current implementation of migratable threads uses a variant of the *iso-malloc* functionality of PM^2 [ABN99]. In this implementation, each thread's stack is allocated such that it spans the same reserved virtual addresses across all processors. This is achieved by splitting the unused virtual address space among physical processors. When a thread is created, its stack is allocated from a portion of the virtual address space assigned to the creating processor. This ensures that no thread encroaches upon addresses spanned by others' stacks on any processor. Allocation and deallocation within the assigned portion of virtual address space is done using the mmap and munmap functionality of Unix. Since we use isomalloc for fixed size thread stacks only, we can eliminate several overheads associated with PM^2 implementation of isomalloc. This results in context-switching overheads as low as non-migrating threads, irrespective of the stack-size, while allowing migration of threads. However, it is still more efficient to keep the stack size down at the time of migration to reduce the thread migration overheads.

3.1 Conversion to AMPI

Since multiple threads may be resident within a process, variables that were originally process-private will now be shared among the threads. Thus, to convert an MPI program to use AMPI, we need to privatize these global variables, which typically fall in three classes.

1. Global variables that are "read-only". These are either *parameters* that are set at compile-time, or other variables that are read as input or set at the beginning of the program and do not change during execution. It is not necessary to privatize such variables.
2. Global variables that are used as temporary buffers. These are variables that are used temporarily to store values to be accessible across subroutines. These variables have a characteristic that there is no blocking call such as MPI_recv between the time the variable is set and the time it is ever used. It is not necessary to privatize such variables either.
3. True global variables. These are used across subroutines that contain blocking receives and therefore there is a possibility of a context-switch between the definition and use of the variable. These variables need to be privatized.

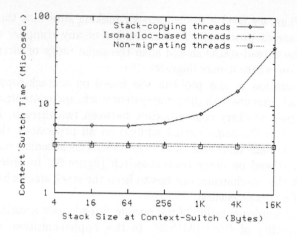

Fig. 3. Comparison of context-switching times of stack-copying and isomalloc-based migrating threads with non-migrating threads.

To systematically privatize all global variables, we create a FORTRAN 90 type, and make all the global variables members of that type. In the main program, we allocate an instance of that type, and then pass a pointer to that instance to every subroutine that makes use of global variables. Access to the members of this type have to be made through this pointer.

A source-to-source translator can recognize all global variables and automatically make such modifications to the program. We are currently working on modifying the front-end of a parallelizing compiler [BEF⁺94] to incorporate this translation. However, currently, this has to be done by hand. The privatization requirement is not unique to AMPI. Other thread-based implementations of MPI such as TMPI [TSY99] and TOMPI [Dem97] also need such privatization.

4   Case Studies

We have compared AMPI with the original message-driven multi-partitioning approach to evaluate overheads associated with each of them using a typical Computational Fluid Dynamics (CFD) kernel that performs Jacobi relaxation on large grids (where each partition contains 1000 grid points.) We ran this application on a single 250 MHz MIPS R10K processor, with different number of chunks, keeping the chunk-size constant. Two different decompositions, 1-D and 3-D, were used. These decompositions vary in number of context-switches (blocking receives) per chunk. While the 1-D chunks have 2 blocking receive calls per chunk per iteration, the 3-D chunks have 6 blocking receive calls per chunk per iteration. However, in both cases, only half of these calls actually block waiting for data, resulting in 1 and 3 context switches per chunk per iteration respectively. As can be seen from figure 4, the optimization due to availability of local variables across blocking calls, as well as larger subroutines in the AMPI version neutralizes thread context-switching overheads for a reasonable number of chunks per processor. Thus, the load balancing framework can be effectively used with user-level threads without incurring any significant overheads.

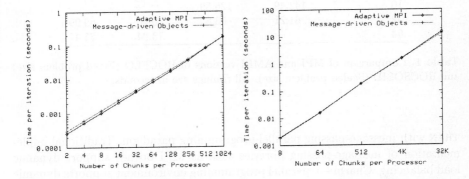

Fig. 4. Performance a Jacobi relaxation application. (Left) with 1-D decomposition. (Right) with 3-D decomposition.

Encouraged by these results, we converted some large MPI applications developed as part of the Center for Simulation of Advanced Rockets (CSAR) at University of Illinois. The goal of CSAR is to produce a detailed multi-physics rocket simulation [HD98]. GEN1, the first generation integrated simulation code, is composed of three coupled modules: ROCFLO (an explicit fluid dynamics code), ROCSOLID (an implicit structural simulation code), and ROCFACE (a parallel interface between ROCFLO and ROCSOLID) [PAN+99]. ROCFACE and ROCSOLID have been written using FORTRAN 90-MPI (about 10K and 12K lines respectively.) We converted each of these codes to AMPI. This conversion, using the techniques described in the last section, resulted in very few changes to original code. The converted codes can still link and run with original MPI. In addition, the overhead of using AMPI instead of MPI is shown (table 1) to be minimal, even with the original decomposition of one partition per processor. We expect the performance of AMPI to be better when multiple partitions are mapped per processor, similar to the situation depicted in figure 2. Also, the ability of AMPI to respond to dynamic load variations outweighs these overheads.

5 Conclusion

Efficient implementations of an increasing number of dynamic and irregular computational science and engineering applications require dynamic load balancing. Many such applications have been written in procedural languages such as FOR-

No. Of	ROCFLO		ROCSOLID	
Processors	MPI	AMPI	MPI	AMPI
1	1637.55	1679.91	67.19	63.42
2	957.94	916.73	–	–
4	450.13	437.64	–	–
8	234.90	278.93	69.81	71.09
16	142.49	126.59	–	–
32	61.21	63.82	70.70	69.99
64	–	–	73.94	75.47

Table 1. Comparison of MPI and AMPI versions of ROCFLO (Fixed problem size) and ROCSOLID (Scaled problem size). All timings are in seconds.

TRAN with message-passing parallel programming paradigm. Traditional implementation of message-passing libraries such as MPI do not support dynamic load balancing. Charm++ parallel programming environment supports dynamic load balancing using object-migration. Applications developed using Charm++ have been shown to adaptively balance load in presence of dynamically changing load conditions caused even by factors external to the application, such as in timeshared clusters of workstations. However, converting existing procedural message-passing codes to use object-based Charm++ can be cumbersome. We have developed Adaptive MPI, an implementation of MPI on a message-driven object-based runtime system, and user-level threads, that run existing MPI applications with minimal change, and insignificant overhead. Conversion of legacy MPI programs to Adaptive MPI does not need significant changes to the original code structure; the changes that are needed are mechanical and can be fully automated. We have converted two large scientific applications to use Adaptive MPI and the dynamic load-balancing framework, and have shown that for these applications, the overhead of AMPI, if any, is very small. We are currently working on reducing the messaging overhead of AMPI, and also automating the code conversion methods.

References

[ABN99] Gabriel Antoniu, Luc Bouge, and Raymond Namyst. An Efficient and Transparent Thread Migration Scheme in the PM^2 Runtime System. In *Proc. 3rd Workshop on Runtime Systems for Parallel Programming (RTSPP) San Juan, Puerto Rico, Held in conjunction with the 13th Intl Parallel Processing Symp. (IPPS/SPDP 1999), IEEE/ACM. Lecture Notes in Computer Science 1586*, pages 496–510. Springer-Verlag, April 1999.

[BEF+94] William Blume, Rudolf Eigenmann, Keith Faigin, John Grout, Jay Hoeflinger, David Padua, Paul Petersen, Bill Pottenger, Lawrence Rauchwerger, Peng Tu, and Stephen Weatherford. Polaris: Improving the Effectiveness of Parallelizing Compilers. In *Proceedings of 7th International Workshop on Languages and Compilers for Parallel Computing*, number 892 in Lecture

Notes in Computer Science, pages 141–154, Ithaca, NY, USA, August 1994. Springer-Verlag.

[BK99]   Robert K. Brunner and Laxmikant V. Kalé. Adapting to Load on Workstation Clusters. In *The Seventh Symposium on the Frontiers of Massively Parallel Computation*, pages 106–112. IEEE Computer Society Press, February 1999.

[BK00]   Milind Bhandarkar and L. V. Kalé. A Parallel Framework for Explicit FEM. In *Proceedings of the International Conference on High Performance Computing*, Bangalore, India, December 2000.

[BN99]   D. S. Balsara and C. D. Norton. Innovative Language-Based and Object-Oriented Structured AMR using Fortran 90 and OpenMP. In Y. Deng, O. Yasar, and M. Leuze, editors, *New Trends in High Performance Computing Conference (HPCU'99)*, Stony Brook, NY, August 1999.

[Dem97]   Erik D. Demaine. A Threads-Only MPI Implementation for the Development of Parallel Programs. In *Proceedings of the 11th International Symposium on High Performance Computing Systems (HPCS'97), Winnipeg, Manitoba, Canada*, pages 153–163, July 1997.

[GLS94]   W. Gropp, E. Lusk, and A. Skjellum. *Using MPI: Portable Parallel Programming with the Message-Passing Interface*. MIT Press, 1994.

[HD98]   M. T. Heath and W. A. Dick. Virtual Rocketry: Rocket Science meets Computer Science. *IEEE Comptational Science and Engineering*, 5(1):16–26, 1998.

[KBB00]   L. V. Kale, Milind Bhandarkar, and Robert Brunner. Run-time Support for Adaptive Load Balancing. In *Proceedings of 4th Workshop on Runtime Systems for Parallel Programming (RTSPP) Cancun - Mexico*, March 2000.

[KK96]   Laxmikant V. Kale and Sanjeev Krishnan. Charm++: Parallel Programming with Message-Driven Objects. In Gregory V. Wilson and Paul Lu, editor, *Parallel Programming using C++*, pages 175–213. MIT Press, 1996.

[KSB+99] Laxmikant Kalé, Robert Skeel, Milind Bhandarkar, Robert Brunner, Attila Gursoy, Neal Krawetz, James Phillips, Aritomo Shinozaki, Krishnan Varadarajan, and Klaus Schulten. NAMD2: Greater Scalability for Parallel Molecular Dynamics. *Journal of Computational Physics*, 151:283–312, 1999.

[PAN+99] I. D. Parsons, P. V. S. Alavilli, A. Namazifard, J. Hales, A. Acharya, F. Najjar, D. Tafti, and X. Jiao. Loosely Coupled Simulation of Solid Rocket Moters. In *Fifth National Congress on Computational Mechanics*, Boulder, Colorado, August 1999.

[PL95]   J. Pruyne and M. Livny. Parallel Processing on Dynamic Resources with CARMI. *Lecture Notes in Computer Science*, 949, 1995.

[RVSR98] Randy L. Ribler, Jeffrey S. Vetter, Huseyin Simitci, and Daniel A. Reed. Autopilot: Adaptive Control of Distributed Applications. In *Proc. 7th IEEE Symp. on High Performance Distributed Computing*, Chicago, IL, July 1998.

[TSY99]   H. Tang, K. Shen, and T. Yang. Compile/Run-time Support for Threaded MPI Execution on Multiprogrammed Shared Memory Machines. In *Proceedings of 7th ACM SIGPLAN Symposium on Principles and Practice of Parallel Programming (PPoPP'99)*, 1999.

Performance and Irregular Behavior of Adaptive Task Partitioning [*]

Elise de Doncker, Rodger Zanny, Karlis Kaugars, and Laurentiu Cucos

Department of Computer Science
Western Michigan University
Kalamazoo, MI 49008, USA
{elise,rrzanny,kkaugars,lcucos}@cs.wmich.edu

Abstract. We study the effect of irregular function behavior and dynamic task partitioning on the parallel performance of the adaptive multivariate integration algorithm currently incorporated in PARINT. In view of the implicit hot spots in the computations, load balancing is essential to maintain parallel efficiency. A convergence model is given for a class of singular functions. Results are included for the computation of the cross section of a particle interaction. The adaptive meshes produced by PARINT for these problems are represented using the PARVIS visualization tool.

1 Introduction

We focus on the irregularity of adaptive multivariate integration, resulting from irregular function behavior as well as the dynamic nature of the adaptive subdivision process. The underlying problem is to obtain an approximation Q to the multivariate integral $I = \int_D f(\mathbf{x})d\mathbf{x}$ and an absolute error bound E_a such that $E = |I - Q| \le E_a \le \varepsilon = \max\{\varepsilon_a, \varepsilon_r|I|\}$, for given absolute and relative error tolerances ε_a and ε_r, respectively. The integration domain D is a $d-$dimensional hyper-rectangular region.

For cases where the number of dimensions is relatively low, our software package PARINT (available at [2]) provides a parallel adaptive subdivision algorithm. Particularly for functions with singularities, necessarily leading to *hot-spots* in the computations, load balancing is crucial in keeping high error subregions distributed over the processors, and henceforth avoiding work anomalies caused by unnecessary subdivisions or idle times. We examine the effects of irregular behavior on algorithm convergence and scalability (using an isoefficiency model) for several types of centralized and decentralized load balancing. Furthermore we use a tool, PARVIS [3], for visualizing these effects by displaying properties of the domain decomposition performed during the computations.

In Section 2 we describe the adaptive algorithm; load balancing strategies are outlined in Section 3. In Section 4 we examine scalability and algorithm convergence. Results (in Section 5) include execution times and the visualization of region subdivision patterns for an application in particle physics.

---

[*] Supported in part by the National Science Foundation under grant ACR-0000442

2 Adaptive Task Partitioning

Adaptive task partitioning is a general technique for dynamically subdividing a problem domain into smaller pieces, automatically focusing on the more difficult parts of the domain. Adaptive task partitioning is used here in the context of region partitioning for multivariate integration. The ideas also apply to problems in areas such as adaptive optimization, branch and bound strategies, progressive (hierarchical) radiosity for image rendering, adaptive mesh refinement and finite element strategies.

The adaptive algorithm for numerical integration tends to concentrate the integration points in areas of the domain $\mathcal{D}$ where the integrand is the least well-behaved. The bulk of the computational work is in the calculation of the function f at the integration points. The granularity of the problem primarily depends on the time needed to evaluate f and on the dimension; an expensive f generally results in a problem of large granularity, and in higher dimensions we evaluate more points per integration rule (i.e., per subregion or task).

Initially the integration rules are applied over the entire region, and the region is placed on a priority queue ordered by the estimated error of the regions. The regular loop iteration consists of: removing the region with the highest estimated error from the priority queue; splitting this region in half; evaluating the two new subregions; updating the overall result and error estimate; and inserting the new subregions into the queue. This process continues until the estimated error drops below the user's desired threshold, or, a user-specified limit on the number of function evaluations is reached.

In the PARINT distributed, asynchronous implementation of the adaptive partitioning algorithm, all processes act as *integration worker* processes; one process additionally assumes the role of *integration controller*. The initial region is divided up among the workers. Each executes the adaptive integration algorithm on their own portion of the initial region, largely independent of the other workers, while maintaining a local priority queue of regions (stored as a heap). All workers periodically send updates of their results to the controller; in turn, the controller provides the workers with updated values of the estimated tolerated error τ, calculated as $\tau = \max\{\varepsilon_a, \varepsilon_r|Q|\}$. A worker becomes *idle* if the ratio R_E of its total local error to the total tolerated error τ falls below its fraction R_V of the total volume (of the original domain $\mathcal{D}$). We define the *error ratio* as the ratio R_E/R_V; thus a worker becomes idle when its error ratio reaches 1. To maintain efficiency, a dynamic load balancing technique is employed to move work to the idle workers, and to generally keep the load distributed over all the workers.

A useful model for analyzing the behavior of an adaptive partitioning algorithm is the *region subdivision tree*. Each node in the tree corresponds to a region that was evaluated during the execution of the algorithm. The root represents the initial region $\mathcal{D}$; each other node has a parent node corresponding to the region from which it was formed and either zero or two children. The leaf nodes correspond to the regions on the priority queue(s) at the end of execution, and the number of nodes in the tree is equivalent to the number of region evaluations during execution [1]. Note that, as any adaptive partitioning algorithm will have methods for prioritizing, selecting, and partitioning work, the notion of a domain subdivision tree exists in any problem domain for which some sort of dynamic, progressive partitioning can be utilized.

3 Load Balancing

3.1 Scheduler Based (SB)

Consider a *receiver initiated, scheduler based* technique where the controller acts as the scheduler and keeps an IDLE-STATUS list of the workers.

The controller is kept aware of the idle status of the workers via the workers' update messages. When a worker i informs the controller via a regular update message of its non-idle status, the controller selects (in a round-robin fashion) an idle worker j and sends i a message containing the id of j. Worker i will then send a work message to j containing either new work or an indication that it has no work available. Worker j receives this message and either resumes working or informs the controller that it is still idle. This strategy was implemented in PARINT1.0.

Apart from the disadvantage of a possible bottleneck at the controller, the amount of load balancing performed by the implementation appears limited for certain test cases. In order to tailor the strategy to increase the amount of load balancing, we consider sending larger amounts of work in individual load balancing steps. The controller could furthermore select multiple idle workers to be matched up with busy workers within a single load balancing step. As described subsequently, we will also investigate decentralizations of the strategy.

3.2 Controller Info, Decentralized (CID)

In addition to the updated tolerated error τ, in this technique the controller provides the workers with its current IDLE-STATUS information. Once workers have this information, they can perform load balancing independently of the controller.

An idle worker j issues a work request to a busy worker i (selected round-robin from the list). Upon receiving the request, i either sends j work, while setting the status of j locally to non-idle, or informs j that it has no work available. In the latter case, j will change its status of i to idle. If as a result of the load balancing step, j transfers from idle to non-idle status, it informs the controller (in a regular update message), and the latter updates its IDLE-STATUS list. Note this is a receiver-initiated strategy.

3.3 Random Polling (RP) / Allocation (RA)

RP and RA are potentially simpler decentralized strategies. In RA, a busy processor periodically sends off work to a randomly selected processor. This results in a random distributing of the load and has the advantage that each load balancing step requires only one communication. As a disadvantage, no idle-status information of the target processor is used. The latter could be remedied by providing the workers with an IDLE-STATUS list (resulting in a sender-initiated version of CID), however, this may result in a large number of collisions of work allocations to potentially few idle targets.

In RP, an idle processor requests work from a randomly selected processor. Compared to RA, two communications are needed per load balancing step. However, the first is a small request message. Compared to RA, work gets transferred only if warranted according to the target processor's (non-)idle status.

4 Simple Scalability Analysis

We will use the isoefficiency model of [4] to address scalability issues, particularly with respect to load balancing. In this model, all of the load W is initially in one processor; the analysis focuses on the work needed to distribute the load over all the processors. Note that, for our application, this may be thought of as the type of situation in the case of a point singularity, where one processor contains the singularity.

For our application, the work W can be characterized as an initial amount of error of the problem (or total estimated error in the adaptive algorithm). In a load balancing step, a fraction of a worker's error w will be transferred to another worker. The model assumes that when work w is partitioned in two parts, ψw and $(1 - \psi)w$, there is a constant $\alpha > 0$ (which may be arbitrarily small) such that $1 - \alpha \geq \psi \geq \alpha$; $\alpha \leq 0.5$, so that the load balancing step leaves both workers with a portion bounded by $(1 - \alpha)w$.

With respect to receiver-initiated load balancing, let $V(p)$ denote the number of requests needed for each worker to receive at least one request, as in [4]. If we suppose that the total error would remain constant, then as a result of load balancing, the amount of error at any processor does not exceed $(1 - \alpha)W$ after $V(p)$ load balancing steps; this is used in [4] to estimate the number of steps needed to attain a certain threshold, since under the given assumptions the work at any processor does not exceed δ after $(\log_{\frac{1}{1-\alpha}} \frac{W}{\delta})V(p)$ steps. Note that this only models spreading out the work, while the total work remains constant.

In order to allow taking numerical convergence into account, let us assume that during a load balancing cycle (of $V(p)$ steps), the total error has decreased by an amount of βW as a result of improved approximations. For the sake of simplicity consider that we can treat the effect of load balancing separately from that of the corresponding integration computations, by first replacing W by $W - \beta W = (1 - \beta)W$, then as a result of load balancing the amount of error at any processor does not exceed $(1 - \alpha)(1 - \beta)W$ at the end of the cycle. This characterizes a load balancing phase (until the load has settled) in the course of which the total load decreases. Realistically speaking, it is fair to assume that the global error will decrease monotonically (although not always accurate, for example when a narrow peak is "discovered").

Our current implementations do not send a fraction of the error in a load balancing step, but rather one or more regions (which account for a certain amount of error). In the case of a local singularity, the affected processor's net decrease in error greatly depends on whether it sends off its singular region. Furthermore our scheme is driven more by the attempt to supply idle or nearly idle processors with new work, rather than achieving a balanced load.

In the next subsection we study the numerical rate of convergence for the case of a singular function class.

4.1 Algorithm Convergence Ratio for Radial Singularities

Consider an integrand function f_ρ which is *homogeneous* of degree ρ around the origin [5], characterized by the property $f_\rho(\lambda \mathbf{x}) = \lambda^\rho f_\rho(\mathbf{x})$ (for $\lambda > 0$ and all $\mathbf{x}$). For example, $f(\mathbf{x}) = r^\rho$ where r represents the radius $\sqrt{\sum_{j=1}^{d} x_j^2}$ is of this form, as well as $f(\mathbf{x}) = (\sum_{j=1}^{d} x_j)^\rho$. Note that an integrand function of the form $f(\mathbf{x}) = r^\rho g(\mathbf{x})$, where $g(\mathbf{x})$ is a smooth function, can be handled via a Taylor expansion of $g(\mathbf{x})$ around the origin.

Let us focus on the representative function $f(\mathbf{x}) = r^\rho$, integrated over the d−dimensional unit hypercube. Our objective is to estimate the factor $(1 - \beta)$ by which the error decreases due to algorithm convergence. We assume that the error is dominated by that of the region containing the singularity at the origin. Since our adaptive algorithm uses region bisection, it takes at least d bisections to decrease the size of the subregion containing the origin by a factor of two in each coordinate direction. We will refer to this sequence of subdivisions as a *stage*.

The error associated with this type of region is given by

$$E(k, \rho) = \int_0^{\frac{1}{2^k}} \cdots \int_0^{\frac{1}{2^k}} r^\rho d\mathbf{x} - \sum_{i=1}^{q} w_i r_i^\rho,$$

where q is the number of points in the cubature formula and r_i is the value of r at the i-th cubature point. Then $E(k + 1, \rho) = 2^{-(d+\rho)} E(k, \rho)$.

Recall that a load balancing step corresponds to an update, which is done every n_s subdivisions (n_s is a parameter in our implementation). On the average $\mathcal{O}(p)$ load balancing steps are needed for each processor to do an update, i.e., get a chance to either be put on the IDLE-STATUS list as idle, or act as a donor. Note that $\sim p$ updates corresponds to $\sim n_s p$ bisections. At the processor with the singularity this corresponds to $\sim \frac{n_s p}{d}$ stages (assuming that it continually subdivides toward the singularity). Since each stage changes the error by a factor of $\frac{1}{2^{d+\rho}}$ and there are $\sim \frac{n_s p}{d}$ stages, we estimate $1 - \beta \sim (\frac{1}{2^{d+\rho}})^{\frac{n_s p}{d}}$.

Note that $d + \rho > 0$ for the integral to exist. For a *very* singular problem, $d + \rho$ may only be slightly larger than zero, thus $\frac{1}{2^{d+\rho}}$ only slightly smaller than 1 (i.e., there will be slow convergence).

4.2 Isoefficiency of SB

In our SB strategy, the work requests are driven by the worker updates (at the controller). An update from a busy processor will result in an idle processor requesting work from the busy processor (if there are idle processors). So, ignoring communication delays, we consider an update by a processor, done when there are idle processors, as a request for work to that processor (via the controller). This assumes that the computation time in between a worker's successive updates are large compared to the communication times. If we assume that each worker sends an update after a fixed number (n_s) of subdivisions (and the system is considered homogeneous), the update from a specific worker will be received within $\mathcal{O}(p)$ updates on the average. Therefore, $V(p) = \mathcal{O}(p)$.

Under the above conditions, this scheme behaves like the DONOR based SB and GRR in [4]. Consequently, the isoefficiencies for communication and contention on a network of workstations (NOW) are $\mathcal{O}(p \log p)$ and $\mathcal{O}(p^2 \log p)$, respectively (dominated by $\mathcal{O}(p^2 \log p)$).

4.3 Isoefficiency of CID

Without the IDLE-STATUS list at each worker, this scheme would be as Asynchronous Round Robin (ARR) in [4]). However, with (current) versions of the IDLE-STATUS list available at the workers, it inherits properties of SB, but avoiding algorithm contention, and attempting to avoid network contention. Further analysis is required to determine the isoefficiency of this load balancing technique.

4.4 Isoefficiency of RP

The isoefficiencies for communication and network contention on a NOW are known to be $\mathcal{O}(p \log^2 p)$ and $\mathcal{O}(p^2 \log^2 p)$, respectively (dominated by $\mathcal{O}(p^2 \log^2 p)$).

5 PARVIS and Results

PARVIS [3] is an interactive graphical tool for analyzing region subdivision trees for adaptive partitioning algorithms. Point singularities give rise to typical subdivision structures which can be recognized using PARVIS, by displays of the subdivision tree or projections of the subdivisions on coordinate planes. The data within the nodes of the tree further indicate the location of the singularity. Non-localized singularities generally span large portions of the tree, but can often be detected in the region view projections on the coordinate planes. Error ratios between a node and its descendants provide information on asymptotic error behavior and possible conformity to established error models [1].

Figure 1 shows the subdivision tree and the projection of the subdivisions on the x_0, x_1-plane for the function

$$f(\mathbf{x}) = \frac{1}{(x_0^2 + x_1^2 + x_2^2 + x_3^2)^\alpha (x_0^2 + x_1^2 + (1 - x_2)^2 + (1 - x_3)^2)^\beta},$$

with $\alpha = .9$, $\beta = .7$, over the 4–dimensional unit cube, to a relative error tolerance of 10^{-6}. The two pruned subtrees and the clustered subdivisions correspond to the integrand singularities at the origin and at $(0,0,1,1)$.

PARVIS can color the tree nodes by the range of the error estimate. With regard to performance characteristics, nodes can also be colored according to the time of generation or by the processor that owns the region. The latter gives information on the efficacy of load balancing.

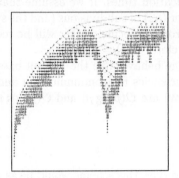 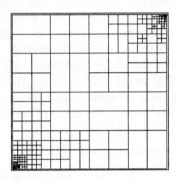

Fig. 1. Left: Subdivision tree for $f(\mathbf{x})$;   Right: Region projection on x_0, x_1–plane

When the adaptive partitioning technique encounters a narrow peak in a function, the estimated error can momentarily increase as the peak is discovered, before being reduced again as the peak is adequately subdivided. We tested this behavior by creating a parameterized integrand function, where the parameters can be varied randomly to define a "family" of similar integrand functions. Each function is over a certain dimension, and contains a number of random peaks, each of some varying height and width. The function definition is

$$f(\mathbf{x}) = \sum_{i=0}^{n-1}(((\gamma_i(\sum_{j=0}^{d-1}(x_j - p_{i,j})^2))^{\rho/2}) + \frac{1}{\mu_i})$$

where n is the number of peaks and d is the dimension. The values γ_i, ρ_i, and, μ_i determine the height and "width" of peak i, and the jth coordinate of the ith peak is given by $p_{i,j}$. Figure 2 shows the log of the error ratio as a function of the current number of iterations. The fluctuations illustrate the temporary error increases incurred as the algorithm subdivides around the peak.

Figure 3 displays the subregions produced when one of these functions is integrated; the function parameters used specify 100 peaks for $d = 2$. The clusters of smaller subdivisions indicate the locations of the peaks. Figure 4 gives the obtained speedup vs. the number of processors, from the adaptive PARINT algorithm using our SB load balancing scheme. These runs were done on *Athena*, our Beowulf cluster, which consists of 32 (800MHz) Athlon processors connected via a fast ethernet switch.

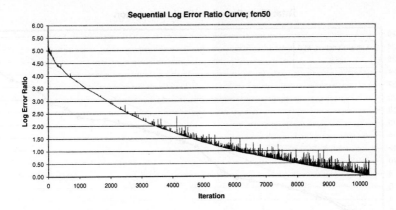

Fig. 2. Estimated error ratio log curve

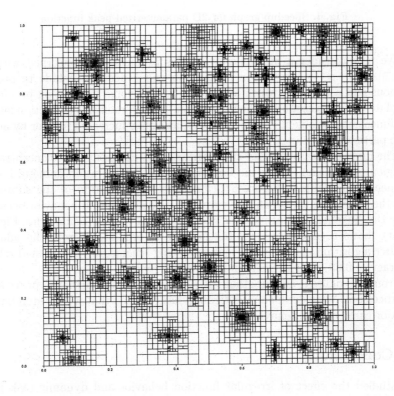

Fig. 3. Plot of subregions produced during integration of parameterized peak function

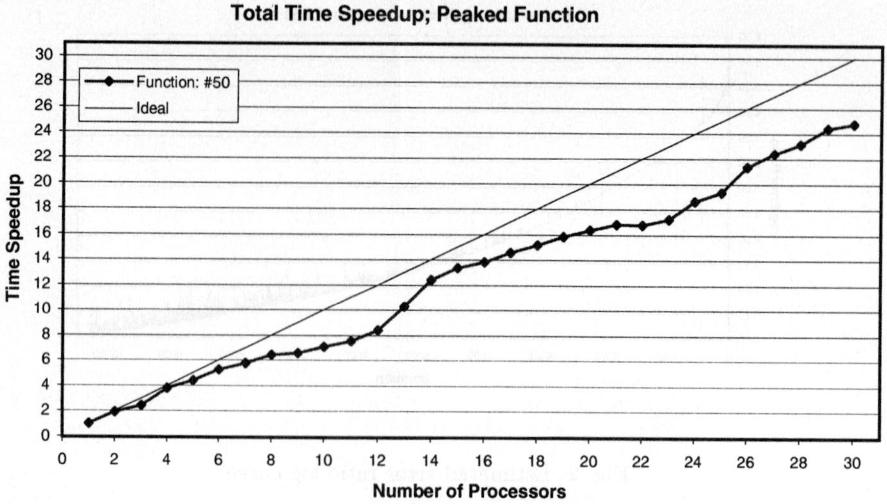

Total Time Speedup; Peaked Function

Fig. 4. Speedup graph for 2D parameterized peak function

We now consider a high energy physics problem [6]) concerned with the $e^+ e^- \rightarrow \mu^+ \mu^- \gamma$ interaction (1 electron and 1 positron collide to produce 2 muons and a photon), which leads to a 4-dimensional integration problem to calculate the *cross section* of the interaction. Adding radiative correction leads to a 6-dimensional problem. The problem dimension is increased further by adding more particles.

The function has boundary singularities, as is apparent from concentrations of region subdivisions along the boundaries of the integration domain. For the 4-dimensional problem, Figure 5 (left) depicts a projection of the subregions onto the x_1, x_2–plane, which also reveals a ridged integrand behavior occurring along the diagonals of that plane as well as along the region boundaries. Figure 5 (right) shows a time graph vs. the number of processors, from the adaptive PARINT algorithm using our SB load balancing scheme, for a requested relative accuracy of 0.05 (which results in fairly small runs).

From this figure it appears that the run-time does not further improve above 10 processors. More detailed investigation has revealed that work redundancy is a major cause of this behavior.

6   Conclusions

We studied the effect of irregular function behavior and dynamic task partitioning on the parallel performance of an adaptive multivariate integration algorithm. In view of the singular/peaked behavior of the integrands, cross sections for collisions such as $e^+ e^- \rightarrow \mu^+ \mu^- \gamma$ in high energy physics are very computational intensive and require supercomputing to obtain reasonable accuracy.

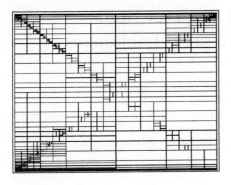 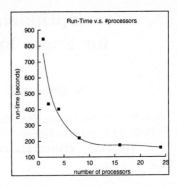

Fig. 5. Left: Region projection on x_1, x_2−plane; Right: Time graph on *Athena*

Adaptive subdivision methods hold promise on distributed computing systems, providing that the load generated by the singularities can be balanced effectively. We use the adaptive subdivision technique in PARINT to manage the singularities, in combination with the PARVIS visualization tool for a graphical integrand analysis.

Further work is needed for a detailed analysis of the interactions of particular aspects of the parallel adaptive strategy (e.g., of the balancing method) with the singular problem behavior.

Acknowledgement. The authors thank Denis Perret-Gallix (CERN) (CNRS Bureau Director, Japan) for his cooperation.

References

1. E. de Doncker and A. Gupta. Multivariate integration on hypercubic and mesh networks. *Parallel Computing*, 24:1223–1244, 1998.
2. Elise de Doncker, Ajay Gupta, Alan Genz, and Rodger Zanny. http://www.cs.wmich.edu/~parint, PARINT Web Site.
3. K. Kaugars, E. de Doncker, and R. Zanny. PARVIS: Visualizing distributed dynamic partitioning algorithms. In *Proceedings of the International Conference on Parallel and Distributed Processing Techniques and Applications (PDPTA'00)*, pages 1215–1221, 2000.
4. V. Kumar, A. Y. Grama, and N. R. Vempaty. Scalable load balancing techniques for parallel computers. *Journal of Parallel and Distributed Computing*, 22(1):60–79, 1994.
5. J. N. Lyness. Applications of extrapolation techniques to multidimensional quadrature of some integrand functions with a singularity. *Journal of Computational Physics*, 20:346–364, 1976.
6. K. Tobimatsu and S. Kawabata. Multi-dimensional integration routine DICE. Technical Report 85, Kogakuin University, 1998.

Optimizing Register Spills
for Eager Functional Languages

S. Mishra[1], K. Sikdar[1], and M. Satpathy[2]

[1] Stat-Math Unit, Indian Statistical Institute, 203 B. T. Road, Calcutta-700 035
[2] Dept. of Computer Science, University of Reading, Reading RG6 6AY, UK
(res9513, sikdar)@isical.ac.in, M.Satpathy@reading.ac.uk

Abstract. Functional programs are *referentially transparent* in the sense that the order of evaluation of subexpressions in an expression does not matter. Any order of evaluation leads to the same result. In the context of a compilation strategy for eager functional languages, we discuss an optimization problem, which we call the *Optimal Call Ordering Problem*. We shaow that the problem is NP-complete and discuss heuristics to solve this problem.

Keywords: *NP-Completeness; Functional Languages; Register Utilization*

1 Introduction

Functional programs are *referentially transparent* [3]. This means, the value of an expression never changes throughout its computational context. As a consequence, we can evaluate an expression in any order. Consider the expression $f(x, y) + g(x, y)$. If it is an expression in a functional program, the values of x and y remain the same throughout the computation of the expression. So, we can do any of the following to get the final result: (i) first make the call to f, then make the call to g and add their results, or alternatively, (ii) first evaluate the call to g, then the call to f and add their results.

So far as evaluation order is concerned, functional languages are divided into *eager functional languages* and the *lazy functional languages*. In an eager functional language, the arguments of a function are evaluated before the call to the function is made, whereas, in a lazy functional language, an argument of a function is evaluated only if it is needed. If two operands are present in machine registers, operations like additions and subtraction usually take one machine cycle. On the other hand, if the operands are present in memory, they are first brought to registers, and then the addition or the subtraction operation is performed. Thus memory operations are usually far more costlier than the register operations. So it is always preferable that, as much operations as possible should be done in registers. But it is not always possible since the number of registers is small, and the number of values which are *live* at a given point of time could be high [2]. Further, function calls and programming constructs like *recursion* complicate the matter. When we cannot accommodate a *live* value in a register, we need to store it in memory. Storing a register value in memory and retrieving it later to a register constitute a *register spill* or simply a *spill*. It is desired that a program should be computed with minimum number of spills.

In this paper, we will discuss an optimization problem which aims at minimizing the number of register spills in the context of a compilation strategy for eager functional programs. Section 2 introduces the problem. Section 3 discusses the context of the problem. Section 4 shows that the problem is NP-complete. In Section 5, we discus about heuristics for this problem. Section 6 concludes the paper.

2   Our Problem

We make the following assumptions: (i) The language is an eager functional language, and (ii) a function call may destroy the contents of the live registers. The later assumption is necessary because we shall have to deal with recursion. We now illustrate our problem through an example.

Example 1: Figure 1 shows the *directed acyclic graph* (DAG) for the following expression. The arguments of calls F_1 and F_2 share subexpressions.

$$F_1(1 + (x * (2 - (y * z)))) + F_2((x * (2 - y * z)) + ((2 - y * z) + (y * z/3)))$$

We can evaluate calls F_1 and F_2 in any order. For evaluating F_1, we need to evaluate the argument of F_1 and hence the expression DAG enclosed within **LP2**. Similarly, before making the call to F_2, we have to evaluate the argument of F_2 which is the expression DAG enclosed within **LP3**. LP2 and LP3 have many shared subexpressions between them. Now let us see the relative merits and demerits of the two possible orders of evaluation. Note that we are only considering the number of spills which result due to the shared nodes between the argument DAGs of F_1 and F_2.

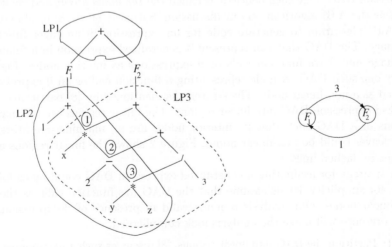

Fig. 1. Sharing between the arguments of two calls F_1 and F_2

– *Evaluation of F_1 is followed by F_2:* We have to first evaluate LP_2. Then the call to F_1 will be made. Next, LP_3 will be evaluated and it will be followed by the call to F_2. Observe that, during the evaluation of the DAG within LP_2, it will evaluate three shared computations, marked by circled 1, 2 and 3 in the figure. Immediately after the evaluation of LP_2, they are in registers. But following our assumptions, the call to F_2 may destroy them. So when LP_3 will be evaluated they may not in the registers in which they got evaluated. So we need to store them in memory. In conclusion, we need to make 3 spills.

– *Evaluation of F_2 is followed by F_1:* We have to first evaluate LP_3. Then the call to F_2 will be made. After that LP_2 will be evaluated and it will be followed by the call to F_1. Observe that, during the course of LP_3's evaluation, it will evaluate only one shared computations, marked by circled 1 in the figure, which will be needed by LP_2. So in memory we need to store its value and retrieve it when LP_2 needs it. In conclusion, we will make 1 spill.

We observed that the second approach is more efficient than the first approach since we need to make lesser number of spills. Note that we are not just saving two or three spills. Such calls may occur in a recursive environment. In such a case the number of spills that we save by choosing a better evaluation order could be very very large. So, choosing the better evaluation order is important.

3   The Problem Context

For computing an arithmetic expression, it is usually represented as a DAG [2]. Such a DAG is simple in the sense that the interior nodes are machine operators and the leaf-nodes are either literals or memory locations. The problem of computing a simple expression DAG with minimum number of instructions (or minimum number of registers such that no spill occurs) is NP-complete [1]. Aho, Johnsson and Ullman have discussed various heuristics for computing an expression DAG. One such heuristic is called the *top down greedy* and we will refer it as the AJU algorithm in our discussion. Satpathy et al. [8] have extended the AJU algorithm to generate code for an expression in an eager functional language. The DAG that can represent a generalized expression in a functional language may have function calls or if-expressions as interior nodes. Figure 2 shows one such DAG. A node representing a function call or an if-expression is termed as a non-linear node. The strategy as mentioned in [8] partitions a generalized expression DAG into linear regions. The linear regions are the maximal regions of a DAG such that its interior nodes are all machine operators, and their leaves could be non-linear nodes. Figure 2 shows the linear regions as the regions in dashed lines.

A strategy for evaluating a generalized expression DAG could be as follows. Here, for simplicity, let us assume that the DAG has function calls as the only non-linear nodes. Our analysis remains valid in presence of if-expressions but their presence will make the analysis look complicated.

Step 1: Partition the DAG into linear regions. [8] discusses such a partitioning algorithm.

Step 2: Select any linear region that lies at the leaf-level of the DAG (such a linear region do not have any non-linear node at any of its leaves). Evaluate this non-linear region using the AJU algorithm and store its result in memory. Now replace this non-linear region in the DAG by a leaf node labeled with the above memory location.

Step 3: Evaluate all function calls whose arguments are available in memory locations. Replace the calls in the DAG by their results in memory locations.

Step 4: Continue steps 2 and 3 till the result of the original DAG is available in a single memory location.

Alternatively, we can express the above strategy as follows:

− First evaluate all the nonlinear nodes lying at the leaves of a linear region one by one and spill their results. Then evaluate rest of the linear region using the AJU algorithm assuming that the results of all the nonlinear nodes at the leaves are in memory.

Note that when all the non-linear nodes of a linear region are already evaluated, their results are in memory. So, at this point of time, the linear region under evaluation is a simple expression DAG and so it could be evaluated using the AJU algorithm. Further notice that, this is a bottom-up strategy. We elaborate it through the evaluation of the DAG in figure 2. In the figure, the linear regions LR1 till LR4 are shown as the regions in dotted lines. For evaluating such a DAG, evaluation always starts with a linear region such that none of its leaf-level nodes is a non-linear node. In the present figure, LR2 and LR4 are the only candidates for evaluation. We can evaluate any of them first. Let us choose to evaluate LR4 (the argument of function J). After evaluation of LR4 (its result is in memory), a call to J is made since it satisfies the step 3 of the evaluation strategy. Let the result of J be stored in memory location m_2. Now LR3 is one linear region whose only non-linear node is available in memory. So LR3 could be evaluated by using AJU algorithm. After the evaluation of LR3, a call to H is made and let us store its result in m_3. Next the linear region which is a candidate for evaluation is LR2. It is evaluated and a call to G is made. Let the result of this call be stored in m_1. At this stage, LR1 could be evaluated using AJU algorithm since the results of all of its non-linear nodes are in memory. Let us store the result of LR1 in location m_0. And this the desired result.

3.1   Our problem in general form

In Figure 3, function calls $F_1, F_2, \ldots, F_n$ lie at the leaf-levels of the linear region LP. $\text{DAG}_1, \ldots, \text{DAG}_n$ respectively represent the arguments of such calls. For simplicity, we have assumed that all such functions are single argument functions. Further, we can assume that there are no non-linear nodes present in such argument DAGs (if they are present, they will have already been evaluated and their results will be residing in memory). So, all such argument DAGs are simple expression DAGs and they could be evaluated using the AJU algorithm in any order.

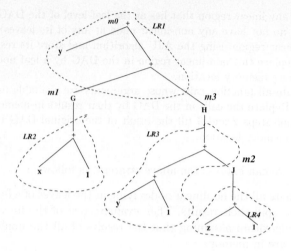

Fig. 2. Evaluation of a generalized DAG through a bottom-up strategy

$F_1, F_2, \ldots, F_n$ will induce, due to the sharing between the argument DAGs, certain number of spills which are as discussed in the previous section. So now the problem is: how to choose an evaluation order of the calls to $F_1, \ldots, F_n$ such that the number of spills due to the sharing among argument DAGs is minimum. From now on we will refer to this problem as the *optimal evaluation ordering problem (OEOP)*. Formally, OEOP can be defined as follows:

Instance : A linear region whose leaves have function calls $F_1, F_2, \ldots, F_n$ at its

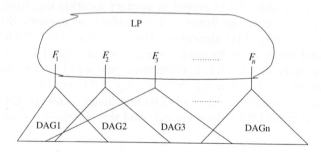

Fig. 3. Sharing between arguments of n calls $F_1, F_2, \ldots, F_n$

leaf-level, and the calls respectively have DAG_1, ..., DAG_n as argument DAGs. All such DAGs are simple expression DAGs (refer to Figure 3).

Solution : A permutation $\pi = (\pi_1, \pi_2, \ldots, \pi_n)$ of $(1, 2, \ldots, n)$.

Cost : $w(\pi)$ = number of shared nodes to be spilled to evaluate n function calls (due to sharing between the argument DAGs) in the order $F_{\pi_1}, F_{\pi_2}, \ldots, F_{\pi_n}$.

Goal: Minimize the cost function.

3.2 Assumptions and Problem Formulation

To analyse the OEOP, we will make the following simplifying assumptions.

- *All the functions are of one arguments:* This is to make the analysis simpler.
- *Assumption of simple-sharing:* An expression is shared between at most two function DAGs. Further, if a subexpression is shared by two calls F_i and F_j then no subexpression of it is shared by any function F_k $(k \neq i, j)$.

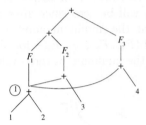

Fig. 4. DAGs of A, B and C not satisfying the assumption of simple sharing

In Figure 4, the node marked 1 is shared by the three functions A, B and C. Let B be evaluated before C. Now, whether B will spill node 1 for C depends on whether A has been evaluated already or not (i.e. whether A has spilled the node 1 or not). In other words, the spilling decisions between B and C depends on the context in which they are evaluated. To make such spilling decision between any two function calls independent of the context in which they are evaluated, we have introduced the assumption of simple sharing. We shall denote the OEOP satisfying above assumptions as OEOP(S). Obviously, OEOP(S) is in class NP.

4 NP-completeness of OEOP

Let Figure 3 represent an instance of OEOP(S). Given n calls as in the figure, we can obtain a weighted symmetric digraph $G = (V_n, A, w)$ which will preserve all the sharing information. In the digraph, the node set $V_n = \{1, 2, \ldots, n\}$, where node i corresponds to the function F_i. There will be a directed edge from node i to node j with weight w_{ij} if the argument DAG of F_i *cuts* the argument DAG of F_j at w_{ij} points. What it means is that if the call to F_i is evaluated ahead of the call to F_j, then w_{ij} shared nodes will have to be spilled to memory. Note that the weight of the edge from i to j may be different from the weight of the edge from j to i. For instance, if we construct a digraph for the functions A and B in the figure 1, then there will be an edge from node A to B with weight 3 and the reverse edge will have weight 1.

Let us evaluate the calls $F_1, \ldots F_n$ in the figure in the order $F_{\pi_1}, F_{\pi_2} \ldots F_{\pi_n}$, where $\pi = (\pi_1, \ldots \pi_n)$ is a permutation of $(1, 2, \ldots n)$. Then the number of spills we will have to make immediately after the evaluation of F_{π_1} is the sum of the weights of all the out-going edges from π_1. The number of spills we will make immediately after the evaluation of F_{π_2} is the sum of weights of the out-going edges from π_2 to all nodes excepting π_1. This is because F_{π_1} has already been evaluated, and π_2 should not bother about F_{π_1}. Continuing in this manner, the number of spills we will make after the evaluation of F_{π_j} is the sum of weights of all outgoing edges from π_j other than the edges to $\pi_1, \pi_2, \ldots \pi_{j-1}$. Let us call the number of spills that are made after the evaluation of F_{π_i} be SPILL_i. So the evaluation order $F_{\pi_1}, F_{\pi_2}, \ldots F_{\pi_n}$ will make it necessary to make $\text{SPILL}(F_{\pi_1}, F_{\pi_2}, \ldots F_{\pi_n}) = \text{SPILL}\pi_1 + \text{SPILL}\pi_2 + \ldots + \text{SPILL}\pi_n$ spills. Note that the value of $\text{SPILL}\pi_n$ will be zero since after its evaluation we need not make any spill. To find out the minimum number of spills we shall have to find out the value of $\text{SPILL}(F_{\pi_1}, F_{\pi_2}, \ldots, F_{\pi_n})$ for all possible evaluation order $F_{\pi_1}, F_{\pi_2}, \ldots, F_{\pi_n}$ and take the minimum of these values.

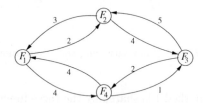

Fig. 5. Graph showing sharing information of an instance of OEOP(S).

Example: Let A, B, C and D be functions satisfying the simple sharing assumption (Figure 5). Let the evaluation order be $B, D, C,$ and A. Then:

$$\text{SPILL}(B, C, D, A) = \text{SPILL}_B + \text{SPILL}_C + \text{SPILL}_D + \text{SPILL}_A = 13.$$

We shall convert this arc weighted symmetric digraph G to a nonnegative arc weighted complete digraph $G_n = (V_n, A_n, w)$ by adding new arcs to G with zero arc weights.

Lemma 1. *The number of spills in the complete graph G_n due to the evaluation order $F_{\pi_1}, F_{\pi_2}, \ldots, F_{\pi_n}$ is same as the number of spills in G (the n function calls with sharing as represented by G) under the same evaluation order.* □

For computing the total number of spills for the order $F_{\pi_1}, F_{\pi_2}, \ldots, F_{\pi_n}$; the weights of the arcs of the form (π_i, π_j) with $j > i$ are considered in deciding on the number of spills. It is easy to see that these arcs corresponding to the above evaluation order form an acyclic tournament on V_n in G_n [7]. This we shall refer to as the acyclic tournament $T_\pi = \{(\pi_i, \pi_j) | i < j\}$ on V_n in G_n represented by the permutation π. A permutation of nodes of V_n defines a unique acyclic tournament on V_n and *vice versa*. In conclusion, for obtaining the optimal evaluation order in relation to the graph G, it would be enough if we obtain an acyclic tournament on V_n in $G_n = (V_n, A_n, w)$ with minimum total arc weight. This problem is known as MINLOP and it is known to be NP-complete [6]. MINLOP$_{12}$ is same as MINLOP with only difference that weight of an arc is either 1 or 2. We will show that MINLOP$_{12}$ is NP-complete by reducing it from MINFAS. In MINFAS, for a given digraph $G = (V_n, A)$ we are asked to find a subset B of A with minimum cardinality such that $(V_n, A - B)$ is acyclic. MINFAS is known to be NP-complete [5].

Lemma 2. *Let $G = (V_n, A)$ be a digraph and $B \subseteq A$ be a FAS (feedback arc set) of G. Then there exists a $C \subseteq B$ such that (V_n, C) is acyclic and C is a FAS of G.*

Proof If (V_n, B) is acyclic then $C = B$. Otherwise, (V_n, B) has at least one directed cycle. Since B is a FAS, $(V_n, A - B)$ is acyclic. Now construct an acyclic tournament (V_n, T) of G_n with $(A - B) \subseteq T$. We define $C = T^c \cap B$ where $T^c = \{(j, i) | (i, j) \in T\}$. C is FAS of G because $A - C = A \cap T$ and $A \cap T$ is acyclic (see Figure 6). □

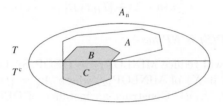

Fig. 6. $C \subseteq B \subseteq A$ and $T \cup T^c = A_n$

Given an FAS B of G the set C in the Lemma 2 can be constructed in polynomial time as follows. Let $(v_1, v_2, \ldots, v_n)$ be a permutation of V_n for which v_1 is a node in $(V_n, A - B)$ with zero indegree, v_2 is a node with zero indegree in the acyclic subdigraph of $(V_n, A - B)$ induced by the node set $V_n - \{v_1\}$ and so on. Note that $v_1, v_2, \ldots, v_n$ can be chosen in polynomial time. Let $T = \{(v_i, v_j) | i < j\}$. Clearly $A - B \subseteq T$ and T is an acyclic tournament on V_n. $C = T^c \cap B$ can be constructed in polynomial time. In conclusion, every FAS induces another FAS which is acyclic. *From now on, by FAS we will mean this induced acyclic FAS.*

Lemma 3. *The no. of arcs in an acyclic tournament on V_n in G_n is $\frac{1}{2}n(n-1)$.*

Theorem 1. *MINLOP$_{12}$ is NP-complete.*

Proof - MINLOP$_{12}$ is in NP since MINLOP is. To show that MINLOP$_{12}$ is NP-hard we shall reduce MINFAS to MINLOP$_{12}$.

Let $G = (V_n, A)$ be an instance of MINFAS. From this we shall construct an instance $G_n = (V_n, A_n, w)$ of MINLOP$_{12}$ with $w(e) = 1$ if $e \notin A$ and $w(e) = 2$ if $e \in A$. Let (V_n, T) be an acyclic tournament of G_n. From the proof of Lemma

2, it can be easily seen that $T \cap A$ is a FAS of G. By Lemma 3 and from the definition of the weight function w it follows that $w(T) = \frac{1}{2}n(n-1) + |T \cap A|$.

Next we shall show that, B is a FAS of G with $|B| \leq k$ if and only if there exists an acyclic tournament (V_n, T_B) of G_n with $w(T_B) \leq \frac{1}{2}n(n-1) + k$.

$\Leftarrow$: Let (V_n, T_B) be an acyclic tournament of G_n with $w(T_B) \leq \frac{1}{2}n(n-1) + k$.

$$w(T_B) = \frac{1}{2}n(n-1) + |T_B \cap A| \leq \frac{1}{2}n(n-1) + k$$

$$i.e. \ |T_B \cap A| \leq k$$

Choose $B = T_B \cap A$. Since B is an FAS of G we are done.

$\Rightarrow$: Let B be a FAS with $|B| \leq k$. By Lemma 2 (V_n, B) is acyclic. From definition of feedback arc set, $(V_n, A - B)$ is acyclic. Let (V_n, T) be an acyclic tournament with $(A - B) \subseteq T$. This arc set T can be constructed in polynomial time. Define $T_B = T^c$. Clearly $T_B \cap B \subseteq B$ and $(T_B \cap B) = (T_B \cap A)$.

$$i.e. \ \ w(T_B) = \frac{1}{2}n(n-1) + |T_B \cap A|$$

$$= \frac{1}{2}n(n-1) + |T_B \cap B| \leq \frac{1}{2}n(n-1) + k.$$

Theorem 2. *OEOP(S) is NP-hard.*

Proof sketch: We will reduce MINLOP$_{12}$ to OEOP(S). Let $G_n = (V_n, A_n, w)$ be an instance (call it X) of MINLOP$_{12}$, where $A_n = \{(i,j)|i \neq j$ and $i,j \in V_n\}$. Let $V_n = \{1, 2, \ldots, n\}$. Construct an instance of OEOP(S) from X. The subdigraph induced by any pair of nodes i and j will into one of:
Case 1: $w(i,j) = w(j,i) = 1$; **Case 2:** $w(i,j) = w(j,i) = 2$ and
Case 3: either $w(i,j) = 1$ and $w(j,i) = 2$ or $w(i,j) = 2$ and $w(j,i) = 1$.
For Case 1, we include the subexpressions $t_k^1 = a_k * (b_k + c_k)$ and $t_k^2 = (b_k + c_k) + d_k$ in the functions F_i and F_j respectively. It can be seen easily that if we evaluate t_k^1 and t_k^2 in any order only one memory spill will be required and it fits with the arc weights. Similarly the expressions t_k^1 and t_k^2 are chosen for Case 2 and Case 3. The idea is that the weights between nodes i and j in X fall into one of the three cases. The corresponding expressions will be linked to the arguments of the function calls F_i and F_j such that the simple sharing assumption is not violated. Doing this for all pair of nodes in the instance X, we can obtain Y, the instance of OEOP(S). Y will have function calls F_1 till F_n and their arguments will be shared. And then it can be established that for any permutation $\pi = (\pi_1, \pi_2, \ldots \pi_n)$ of $(1, 2, \ldots n)$, $\sum_{i=1}^{n-1} \sum_{j=i+1}^{n} w(F_{\pi_i}, F_{\pi_j})$ is same as the number of spills that will be made if the functions of Y are evaluated according to the same permutation. $\square$

5 Heuristics for OEOP

The amount of sharing between the arguments of functions in user programs is usually not high. However, such programs, for optimization purposes, are transformed into equivalent programs. Function unfolding is one such important transformation technique, and this technique introduces sharing [4]. Our observation is that the number of function call nodes at the leaf-level of an expression DAG (after unfolding) does not go beyond 8 or 9, and further, the number of shared nodes between the argument DAGs of two functions also hardly goes beyond 8 or 9. So, the graph that we will construct will not have nodes more than 9 and further the weights of the edges in it will be within 0 to 9. With such restrictions, the following greedy heuristic will give satisfactory results.

Construct the weighted graph from the DAG of the linear region. Then find the node such that the sum of the outgoing arc weights for this node is minimum. Evaluate the corresponding call. Then the call is removed from the linear region and it is replaced with a memory location. Get the new weighted graph, and repeat the process till a single call remains. This is the last call to be evaluated.

6 Conclusion

We have discussed an optimization problem that occurs while following a compilation technique for eager functional languages. We have shown that the problem is NP–Complete. The context in which problem arises, keeps the dimension of the problem small. Our experiments show that greedy heuristics provide satisfactory results.

Acknowledgements: The authors would like to thank A. Sanyal, A. Diwan and C.R. Subramanian for useful discussions.

References

1. Aho A.V., Johnson S.C. & Ullman J.D., Code generation for expressions with common subexpressions, JACM, Vol. 24(1), January 1977, pp. 146-160.
2. Aho A.V., Sethi R. & Ullman J.D.,*Compilers : Principles, Techniques and Tools*, Addison Wesley, 1986.
3. Bird R. & Wadler P., *Introduction to Functional Programming*, Printice Hall, 1988.
4. Davidson J.W. & Holler A.M., Subprogram Inlining: A study of its effect on program execution time, IEEE Tr. on *Software Engineering*, Vol. 18(2), Feb. 1992, pp. 89-102.
5. Garey M.R. & Johnson D.S., *Computers and Intractability: A Guide to the Theory of NP-Completeness*, Freeman and Company, New York, 1979.
6. Grotschel M., Junger M. & Reinelt G., On the acyclic subgraph polytope, *Math. Programming* 33 (1985), pp. 28-42.
7. Harary F., *Graph Theory*, Addition-Wesley, Reading, MA, 1969.
8. Satpathy M., Sanyal A & Venkatesh G., Improved Register Usage of Functional Programs through multiple function versions, *Journal of Functional and Logic Programming*, December 1998, MIT Press.

A Protocol for Multi-Threaded Processes with Choice in π-Calculus

Kazunori Iwata, Shingo Itabashi, Naohiro Ishii

Dept. of Intelligence and Computer Science, Nagoya Institute of Technology,
Gokiso-cho, Showa-ku, Nagoya, 466-8555, Japan
kazunori@egg.ics.nitech.ac.jp
shingo@egg.ics.nitech.ac.jp
ishii@egg.ics.nitech.ac.jp

Abstract. We have proposed a new protocol for the multi-threaded processes with choice written in π-calculus[1, 2]. This protocol frees the multi-threaded processes from deadlock. It has been defined as transition rules. We have shown why the protocol avoids the deadlock in the multi-threaded processes.

1 Introduction

We propose a new protocol for multi-threaded processes with choice written in π-calculus. π-calculus is a process calculus which can describe a channel-based communication among distributed agents. Agents communicate with each other in π-calculus according to the following rules:

1. A message is successfully delivered when two processes attempt an output and an input at the same time.
2. Agents are allowed to attempt outputs and inputs at multiple channels simultaneously, with only one actually succeeding.

This process of communication has been identified as a promising concurrency primitive[3–7].

In π-calculus agents have a property to choose one process from concurrent processes. In order to choose one process, agents get a mutex-lock and execute the process. The other processes are blocked by the lock and will be stopped, if the process will be successfully executed. This chosen process may be easily executed, if agents execute concurrent processes only by itself. However, these processes are executed by the communication among agents. The mutex-lock is very complex to avoid deadlock. Hence, we adjust the situations in the communication and define the protocol to avoid deadlock[7–12]. In the protocol, we appropriate each process in π-calculus to the thread.

2 π-calculus

π-calculus is a process calculus which is able to describe dynamically changing networks of concurrent processes. π-calculus contains just two kinds of entities: process and channels. Processes, sometimes called agents, are the active components of a system. The syntax of defining a process is as follows:

$$
\begin{aligned}
P ::= \; & \overline{x}y.P && /* \text{ Output } */ \\
& x(z).P && /* \text{ Input } */ \\
& P \mid Q && /* \text{ Parallel composition } */ \\
& (\nu x)P && /* \text{ Restriction } */ \\
& P + Q && /* \text{ Summation } */ \\
& 0 && /* \text{ Nil } */ \\
& !P && /* \text{ Replication } */ \\
& [x = y]P && /* \text{ Matching } */
\end{aligned}
$$

Processes interact by synchronous rendezvous on channels, (also called names or ports). When two processes synchronize, they exchange a single value, which is itself a channel.

The output process $\overline{x}y.P_1$ sends a value y along a channel named x and then, after the output has completed, continues to be as a new process P_1. Conversely, the input process $x(z).P_2$ waits until a value is received along a channel named x, substitutes it for the bound variable z, and continues to be as a new process $P_2\{y/z\}$ where, y/z means to substitute the variable z in P_2 with the received value y. The parallel composition of the above two processes, denoted as $\overline{x}y.P_1 \mid x(z).P_2$, may thus synchronize on x, and reduce to $P_1 \mid P_2\{y/z\}$.

Fresh channels are introduced by restriction operator ν. The expression $(\nu x)P$ creates a fresh channel x with scope P. For example, expression $(\nu x)(\overline{x}y.P_1 \mid x(z).P_2)$ localizes the channel x, it means that no other process can interfere with the communication between $\overline{x}y.P_1$ and $x(z).P_2$ through the channel x.

The expression $P_1 + P_2$ denotes an external choice between P_1 and P_2: either P_1 is allowed to proceed and P_2 is discarded, or converse case. Here, external choice means that which process is chosen is determined by some external input. For example, the process $\overline{x}y.P_1 \mid (x(z).P_2 + x(w).P_3)$ can reduce to either $P_1 \mid P_2\{y/z\}$ or $P_1 \mid P_3\{y/w\}$. The null process is denoted by 0. If output process (or input process) is $\overline{x}y.0$ (or $x(z).0$), we abbreviate it to $\overline{x}y$ (or $x(z)$).

Infinite behavior is allowed in π-calculus. It is denoted by the replication operator $!P$, which informally means an arbitrary number of copies of P running in parallel. This operator is similar to the equivalent mechanism, but more complex, of mutually-recursive process definitions.

π-calculus includes also a matching operator $[x = y]P$, which allows P to proceed if x and y are the same channel.

$\rightarrow$.

3 The Protocol for Multi-Threaded Processes with Choice

In this section, we propose a protocol for multi-threaded processes with choice. The processes concurrently communicate each other. First, we introduce basic concepts concerning the communication.

3.1 Basic Concepts

Agent: Agents are units of concurrent execution of our concurrent and distributed system. Agents are implemented as threads. If agents meet the choice process, they make new threads for each process in choice process.

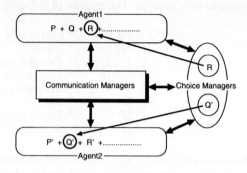

Fig. 1. Relationship

Communication Manager: Communication Managers(CMs) manage communication requests on channels from agents. They make possible for agents to communicate with one another. They have queues which consist of the communication requests from agents.

Choice Manager: Choice Managers(CHMs) manage choice processes on agents. They observe the threads made from the choice, and decide which process should be remained.

The relationship among these elements(Agents, CMs,CHMs) is in Fig. 1.

3.2 The Behavior of Agents

States of an Agent: An agent have the variable to store its state. The set of possible states of agent is {*init, wait-CMout, wait-CMin, wait-CHMid, wait-CHM, wait-Res, done, stopped* }.

Tab. 1 describes, in terms of state transition rules, the behavior of the agent.

3.3 The Behavior of CHMs

States of a CHM: CHMs have variables named flag and queue. The variable flag stores the status of one choice process, *suspend* means the choice process is suspended now, *try* means the choice process is tried to execute, *done* means the choice process is executed. The variable queue stores the processes in choice process that are tried to execute by CMs.

Tab. 2 describes the behavior of the CHM.

3.4 The Behavior of CMs

States of a CM: Each CM has two queues named in-*xxx* and out-*xxx* (*xxx* means arbitrary strings) The queue store the request from agents according to the kind of process. $\stackrel{div}{=}$ denotes the process which divides the queue into the first element and the others.

Tab. 3, 4 describes the behavior of the CM.

4 Free Processes from Deadlock

The processes with choice are nondeterministic, thus the executions have various results. Hence, if the executions are in deadlock, it is difficult to find the cause. In this section, we show why the protocol frees the processes from the deadlock.

To show the freedom from the deadlock, we consider four cases.

1. There is no choice process and the only one paired agents(input and output) use a channel.
2. There is no choice process and the agents use channels. It means some input and output agents use the same channel.
3. The first process, which a CHM determine to execute, in the choice process can be executed.
4. The first process, which a CHM determine to execute, in the choice process cannot be executed.

Table 1. Transition Rules of an Agent

Rule	State	Input	Process	Next State	Output	Other Actions
R1	*init*	-	$\overline{x}[\overrightarrow{y}]$	*wait-CMout*	Send $(in, x, \overrightarrow{y}, aid, 0, 0)$ to CM.	-
R2	*init*	-	$x(\overrightarrow{z})$	*wait-CMin*	Send $(out, x, \overrightarrow{z}, aid, 0, 0)$ to CM.	-
R3	*init*	-	$P \equiv Q + \ldots$	*wait-CHMid*	Send (P, aid) to CHM.	The current process is not changed.
R4	*wait-CHMid*	Get (Cid) from CHM.	$P \equiv Q + \ldots$	*wait-CHM*	-	Each process in choice is divided into one agent and each agent has the state *wait-CHM*.
R5	*wait-CHM*	-	$\overline{x}[\overrightarrow{y}]$	*wait-Res*	Send $(in, x, \overrightarrow{y}, aid, Cid, pid)$ to CM.	-
R6	*wait-CHM*	-	$x(\overrightarrow{z})$	*wait-Res*	Send $(out, x, \overrightarrow{z}, aid, Cid, pid)$ to CM.	-
R7	*wait-Res*	Get $(resume)$ from CHM.	$\overline{x}[\overrightarrow{y}]$	*wait-CMout*	-	This thread is selected to execute.
R8	*wait-Res*	Get $(resume)$ from CHM.	$x(\overrightarrow{z})$	*wait-CMin*	-	This thread is selected to execute.
R9	*wait-Res*	Get $(stop)$ from CHM.	-	*stopped*	-	This thread is stopped.
R10	*wait-CMout*	Get $(output)$ from CM.	$\overline{x}[\overrightarrow{y}]$	*init* if there are more processes otherwise *done*.	-	The state *done* means the process is finished.
R11	*wait-CMin*	Get $(\overrightarrow{v})$ from CM.	$x(\overrightarrow{z})$	*init* if there are more processes otherwise *done*.	-	The state *done* means the process is finished.

Table 2. Transition Rules of an CHM

Rule	State	Input	Next State	Output	Other Actions
R1	-	Get (P, aid) from Agent.	flag = *suspend*	Send (Cid) to *aid*.	Numbering each process.
R2	flag = *suspend*	Get (aid, pid). from CM	flag = *try*	Send *yes* to CM.	-
R3	flag = *try*	Get (aid, pid) from CM.	queue = queue $+ (aid, pid)$	-	-
R4	flag = *try* queue = $\emptyset$	Get $(suspend, aid, pid)$ from CM.	flag = *suspend*	-	-
R5	flag = *try* queue $\stackrel{div}{\rightarrow} (aid', pid')$ +queue'	Get $(suspend, aid, pid)$ from CM.	queue = queue' $+ (aid, pid)$	Send *yes* to CM which sent (aid', pid').	-
R6	flag = *try*	Get $(executed, aid, pid)$ from CM.	flag = *done*	Send $(resume)$ to *aid* with *pid* and $(stop)$ to *aid* without *pid*.	-
R7	flag = *done*	Get (aid, pid) from CM.	-	Send *no* to CM.	-

Table 3. Transition Rules of a CM(1)

Rule	State	Input	Next State	Output
R1	in-x = $\emptyset$	Get $(out, x, \overline{z}', aid, 0, 0)$ from agent.	out-x = out-x $+ (aid, \overline{z}', 0, 0)$	-
R2	out-x = $\emptyset$	Get $(in, x, \overline{y}', aid, 0, 0)$ from agent.	in-x = in-x $+ (aid, 0, 0)$	-
R3	in-x $\neq \emptyset$ in-x $\stackrel{div}{\rightarrow} (aid', 0, 0)$ $+$ in-x'	Get $(out, x, \overline{z}', aid, 0, 0)$ from agent.	in-x = in-x'	Send $(output)$ to *aid* and $(\overline{z}')$ to *aid'*.
R4	out-x $\neq \emptyset$ out-x $\stackrel{div}{\rightarrow} (aid', \overline{z}', 0, 0)$ $+$ out-x'	Get $(in, x, \overline{y}', aid, 0, 0)$ from agent.	out-x = out-x'	Send $(output)$ to *aid'* and $(\overline{z}')$ to *aid*.
R5	in-x = $\emptyset$	Get $(out, x, \overline{z}', aid, Cid, pid)$ from agent.	out-x = out-x $+ (aid, \overline{z}', Cid, pid)$	-
R6	out-x = $\emptyset$	Get $(in, x, \overline{y}', aid, Cid, pid)$ from agent.	in-x = in-x $+ (aid, Cid, pid)$	-
R7	in-x $\neq \emptyset$ in-x $\stackrel{div}{\rightarrow} (aid', 0, 0)$ $+$ in-x'	Get $(out, x, \overline{z}', aid, Cid, pid)$ from agent.	Send (aid, pid) to Cid and if get *yes* from Cid then: in-x = in-x'	Send $(output)$ to *aid* $(\overline{z}')$ to *aid'* $(execute, aid, pid)$ to Cid.
			if get *no* from Cid then: Ignore this input.	-
R8	out-x $\neq \emptyset$ out-x $\stackrel{div}{\rightarrow} (aid', \overline{z}', 0, 0)$ $+$ out-x'	Get $(in, x, \overline{y}', aid, Cid, pid)$ from agent.	Send (aid, pid) to Cid and if get *yes* from Cid then: out-x = out-x'	Send $(output)$ to *aid'* $(\overline{z}')$ to *aid'* $(execute, aid, pid)$ to Cid.
			if get *no* from Cid then: Ignore this input.	-

Table 4. Transition Rules of an CM(2)

Rule	State	Input	Next State	Output
R9	in-x $\neq \emptyset$ in-x $\xrightarrow{div} (aid', Cid', pid')$ + in-x'	Get $(out, x, \overline{z}', aid, Cid, pid)$ from agent.	Send (aid, pid) to Cid and (aid', pid') to Cid' and if get *yes* from Cid and Cid' then:	
			in-x = in-x'	Send $(output)$ to aid $(\overline{z})$ to aid' $(execute, aid, pid)$ to Cid $(execute, aid', pid')$ to Cid'.
			if get *yes* from Cid and *no* from Cid' then:	
			in-x = in-x' Apply these rules again.	-
			if get *no* from Cid and *yes* from Cid' then:	
			Ignore this input.	Send $(suspend, aid', pid')$ to Cid'.
			if get *no* from Cid and Cid' then:	
			in-x = in-x' Ignore this input.	-
R10	out-x $\neq \emptyset$ out-x $\xrightarrow{div} (aid', Cid', pid')$ + out-x'	Get $(in, x, \overline{y}', aid, Cid, pid)$ from agent.	Send (aid, pid) to Cid and (aid', pid') to Cid' and if get *yes* from Cid and Cid' then:	
			out-x = out-x'	Send $(output)$ to aid' $(\overline{z})$ to aid $(execute, aid, pid)$ to Cid $(execute, aid', pid')$ to Cid'.
			if get *yes* from Cid and *no* from Cid' then:	
			out-x = out-x' Apply these rules again.	-
			if get *no* from Cid and *yes* from Cid' then:	
			Ignore this input.	Send $(suspend, aid', pid')$ to Cid'.
			if get *no* from Cid and Cid' then:	
			out-x = out-x' Ignore this input.	-
R11	in-x $\neq \emptyset$ in-x $\xrightarrow{div} (aid', Cid, pid)$ + in-x'	Get $(out, x, \overline{z}', aid, 0, 0)$ from agent.	Send (aid, pid) to Cid and if get *yes* from Cid then:	
			in-x = in-x'	Send $(output)$ to aid $(\overline{z})$ to aid'.
			if get *no* from Cid then:	
			in-x = in-x' Apply these rules again.	-
R12	out-x $\neq \emptyset$ out-x $\xrightarrow{div} (aid', \overline{z}', Cid, pid)$ + out-x'	Get $(in, x, \overline{y}', aid, 0, 0)$ from agent.	Send (aid, pid) to Cid and if get *yes* from Cid then:	
			out-x = out-x'	Send $(output)$ to aid' $(\overline{z})$ to aid.
			if get *no* from Cid then:	
			out-x = out-x' Apply these rules again.	-

Case 1 Let the input process be A_{in} and the output process A_{out}. Each process uses the same link. We consider the process A_{in} is registered to CMs before A_{out}.

1. By R2 in Tab. 3, the id of A_{in} is registered to in-x. Then, the state of A_{in} is changed to *wait-CMin* by R2 in Tab. 1

2. If A_{out} is registered to a CM, by R3 in Tab. 3, the value in A_{out} is output to the process indicated by the id in the top of in-x. Then, A_{in} gets the value from the CM and executes next process by R11 in Tab. 1. The state of A_{out} gets the results from the CM and executes next process by R10 in Tab. 1.

Hence, the process A_{in} and A_{out} can communicate each other.

Case 2 Let the nth input and the nth output processes which use the same channel exist.

1. The mth input processes have already registered to CMs.

   (a) If m $==$ 1 (the length of in-x is 1) then
       This condition is same as the case 1. Hence the communication succeeds.

   (b) Assuming that the communications succeed on m $==$ k (the length of in-x is k) then considering the condition as m $==$ k + 1 (the length of in-x is k + 1) :

       When the condition on m $==$ k + 1,
       i. Let the next registered process be the output process.

          By R3 in Tab. 3, the value in the output process is sent to the process indicated by id in in-x. The output process proceeds to the next process through the state *wait-CMout* by R10 in Tab. 1. The process, which gets the value by R11 in Tab. 1, proceeds to the next process.
          The process in the top on in-x and the output process communicates with each other. The length of in-x is changed to m - 1, that means m $==$ k.

       ii. Let the next registered process be the input process.

          The length of in-x is changed to m + 1, then the communication succeeds by the previous case.
       Then by the assumption of the induction(b), the communication succeeds in any cases.

Case 3 We consider about the choice process $A_1 + A_2 + \cdots + A_n$ and $B_1 + B_2 + \cdots + B_n$.

Let the process A_1 be able to communicate with B_1 and the process A_2 be able to communicate with B_2 and so on. It means the different process uses a different channel.

The choice process $A_1 + A_2 + \cdots + A_n$ is divided into the process $A_1, A_2, \ldots$ and A_n and are registered to a CHM, by R3 and R4 in Tab. 1. Each process proceeds independently but has the state *wait-CHM*. The choice process $B_1 + B_2 + \cdots + B_n$ is done like as the choice process $A_1 + A_2 + \cdots + A_n$, but different CHM.

There are many combination to execute these processes. Before explaining it, we explain the actions of CHMs.

A CHM gets the processes and commits them to memory (see R1 in Tab. 2). It does not know the channel in the processes and checks the process which is requested to execute by CMs (see R2 and R3 in Tab. 2). The requests means a possibility to use the channel which the process want to use and the process can be executed if the CHM answers *yes*. When the CHM gets the first request, it returns the answer *yes*(see R2 in Tab. 2). When the CHM gets the second request or more requests, it store the requests in the queue and check the head request in the queue if the first request cannot be executed (see R3, R4 and R5 in Tab. 2).

We consider only the case that the process A_1 and B_1 can communicate with each other. However, there are many cases on the communication. These cases are distinguished by the order of registration to CMs. The kind of order is as follows:

1. $A_1 \rightarrow B_1 \rightarrow$ the other processes

   or

   $B_1 \rightarrow A_1 \rightarrow$ the other processes

   These cases means the process A_1 and B_1 registered before the others, and communicate with each other.

   We explain the first case in them.

   The process A_1 registers to a CM by R5 or R6 in Tab. 3. The process B_1 registers to the CM and the CM requests CHMs to execute A_1 and B_1 by R9 or R10 in Tab.4. CHMs answer *yes* to the CM, because the requests is the first request for each CHM(see R2 in Tab.2).

   The CM allow to communicate A_1 and B_1 by R9 or R10 in Tab.4 and CHMs send *stop* to the others by R6 in Tab.2. The other processes which get stop the execution.

   If the other processes register to a CM (before getting the signal *stop*), CHMs answer *no* to the CM by R7 in Tab.2.

2. Some processes do not include the pair A_i and $B_i \rightarrow A_1 \rightarrow$ Some processes do not include the pair to the processes which have already registered $\rightarrow B_1 \rightarrow$ the others

   or

   Some processes do not include the pair A_i and $B_i \rightarrow B_1 \rightarrow$ Some processes do not include the pair to the processes which have already registered $\rightarrow A_1 \rightarrow$ the others

   These cases means the some processes registered before the process A_1(or B_1) registering. When the process B_1 (or A_1) registers to a CM before the pair to the processes which have already registered registers, The CM requests to CHMs to execute it and CHMs answer *yes* to the CM. When the CM gets the answer *yes*, it sends CHMs the signal of execution(see R9 and R10 in Tab. 4). In this condition, if the pair to the processes registers to a CM before CHMs get the signal from the CM, the CM requests to CHMs and the CHM blocks this process(see R4 in Tab. 2). However, whether this process is blocked by the CHM or not, the process A_1 and B_1 communicate with each other. Because the CM has already sent the signal of execution to CHMs and CHMs send which process should be executed to Agent (in this case A_1 and B_1).

In this case CHMs block a process, but it does not generate a dead lock. Because blocking the process is generated by the determination which process should be executed. Hence, the blocked process has no chance to be execute and the block has no influence.

In this case, we consider only two choice processes in which each process uses a different channel. However, if many choice processes in which each process uses same channel, they do not generate a dead lock. Because, the CM considers the processes according to their channel and one CHM manage one choice process and the CHM blocks a process only if the other process which is managed by the CHM can communicate.

Case 4 We consider about these choice processes $A_1 + A_m$, $B_1 + B_n$.

In this case, we consider the condition that CMs sends CHMs the signal of suspension(see R9 and R10 in Tab. 4).

Let the process A_1 be able to communicate with B_1 and the process A_m and B_n be able to communicate other processes(M and N).

If all processes have finished to register CHMs, the condition that CMs sends CHMs the signal of suspension is generated by the order of requests to CHMs from CMs and answers from CHMs.

The order is as follows:

The CM sends the requests to execute the pair A_1 and B_1 B_n and N(see R9 and R10 in Tab. 4). In this condition, if the CHM which manages $A_1 + A_2$ arrows to execute A_1 and the CHM which manages B_1+B_n arrows to execute B_n, the communication between B_n and N succeed. Then B_1 cannot be executed and the CHM answers *no* to the CM(see R9 and R10 in Tab. 4). The CM sends the signal of suspension to the CHM which manages $A_1 + A_2$ The CHM removes the process A_1 from the queue and waits for new request from CMs.

In this case, the process A_1 blocks other processes in the same choice process. However, the process A_1 releases the block if the process B_1 cannot communicate.

If the number of the process in one choice process, CHMs consider only the first process in the queue which stores the requests from CMs. Hence, the condition is the same as this condition.

We consider the all possible condition and show the freedom from deadlock. Hence, by using the protocol, we avoid to deadlock in π-calculus.

5 Conclusion and Future Work

In this paper, we have proposed a new protocol for multi-threaded processes with choice written in π-calculus. In the protocol, we have defined the three elements: agents, Communication Managers(CMs) and Choice Managers(CHMs). Agents are units of concurrent execution of our concurrent and distributed system. CMs manage communication requests on channels from agents. CHMs manage choice processes on agents.

We have shown why the protocol frees the processes from the deadlock. If the agents have no choice process, any process do not be blocked. Hence, the deadlock is avoided. If the agents have choice processes, CHMs order the requests from CMs and manage the block to avoid deadlocks. Hence, the protocol frees the processes from the deadlock.

One of the future works is to implement the compiler for π-calculus and build it into any system like as agent systems, distributed object systems and so on.

6 Acknowledgement

A part of this research result is by the science research cost of the Ministry of Education.

References

1. R. Milner. Polyadic π-calculus:a Tutorial. LFCS Report Series ECS-LFCS-91-180, Laboratory for Foundation of Computer Science, 1991.
2. Milner, R., Parrow, J.G., and Walker, D.J. A calculus of mobile processes. In *Information and Computation,100(1)*, pages 1–77, 1992.
3. R. Bagrodia. Synchronization of asynchronous processes in CSP. In *ACM Transaction on Programming Languages and Systems*, volume 11, No. 4, pages 585–597, 1989.
4. M. Ben-Ari. Principles of Concurrent and Distributed Programing. In *Prentice-Hall International(UK) Limited*, 1989.
5. R.Milner D. Berry and D.N. Turner. A semantics for ML concurrency primitive. In *POPL'92*, pages 119–129, 1992.
6. Benjamin C. Pierce and David N. Turner. Concurrnet Objects in a Process Calculus. LNCS 907, pp.187–215, proc. TPPP'95, 1995.
7. MUNINDAR P. SINGH. Applying the Mu-Calculus in Planning and Reasoning about Action. *Journal of Logic and Computation*, 8:425–445, 1998.
8. G.N.Buckley and A. Silberschatz. An effective implementation for the generalized input-output construct of CSP. In *ACM Transactions on Programming Languages and Systems*, volume 5, No.2, pages 223–235, 1983.
9. C.A.R. Hoare. Communicating sequential processes. In *Communications of the ACM*, volume 21, No.8, pages 666–677, 1985.
10. E. Horita and K. Mano. Nepi: a network programming language based on the π-calculus. In *Proceedings of the 1st International Conference on Coordination Models, Languages adn Applicationos 1996*, volume 1061 of *LNAI*, pages 424–427. Springer, 1996.
11. E. Horita and K. Mano. Nepi^2: a two-level calculus for network programming based on the π-calculus. In *IPSJ SIG Notes, 96-PRO-8*, pages 43–48, 1996.
12. E. Horita and K. Mano. Nepi^2: a Two-Level Calculus for Network Programming Based on the π-calculus. ECL Technical Report, NTT Software Laboratories, 1997.

Mapping Parallel Programs onto Distributed Computer Systems with Faulty Elements

Mikhail S. Tarkov[1], Youngsong Mun[2], Jaeyoung Choi[2], and Hyung-Il Choi[2]

[1] Fault-tolerant computer systems department, Institute of Semiconductor Physics,
Siberian branch, Russian Academy of Sciences,
13, Lavrentieva avenue, Novosibirsk, 630090, Russia
tarkov@isp.nsc.ru
[2] School of Computing, Soongsil University,
1-1, Sang-do 5 dong, DongJak-gu, Seoul, Korea
{mun, choi, hic}@ computing.soongsil.ac.kr

Abstract. Mapping one-measured ("line", "ring") and two-measured ("mesh", "torus") parallel program structures onto a distributed computer system (DCS) regular structures ("torus", "two-measured circulant", "hypercube") with faulty elements (processor nodes and links) is investigated. It is shown that: 1) one-measured program structures mapped better than two-measured structures; 2) when failures are injected to the DCS structure the one-measured structures mapping aggravated very lesser than the two-measured structures mapping. The smaller a node degree of a program graph and the greater a node degree of a DCS graph the better the mapping quality. Thus, the one-measured program structure's mappings are more fault-tolerant than two-measured structure's one and more preferable for organization of computations in the DCS with faulty elements.

1 Introduction

A Distributed Computer System (DCS) [1-4] is a set of Elementary Machines [EM] that are connected by a network to be program-controlled from these EM. Every EM includes Computing Unit (CU) (a processor with a memory) and a System Device (SD). A functioning of the SD is controlled by the CU and the SD has input and output poles connected by links to output and input poles of v neighbor EM. A structure of the DCS is described by a graph $G_s(V_s, E_s)$, V_s is a set of EM and $E_s \subseteq V_s * V_s$ is a set of links between EM. For the DCS a graph $G_p(V_p, P_p)$ of parallel program is determined usually as a set V_p of the program branches (virtual elementary machines) that communicate with each other by "point-to-point" principle by transmission of messages across logical (virtual) channels (one- or two-directed) from a set $E_p \subseteq V_p * V_p$. In general case numbers (weights) that characterize computing complexities of branches and intensities of communications between neighbor branches are corresponded to nodes $x, y \in V_p$ and edges (or arcs) $(x, y) \in E_p$ accordingly.

A problem of mapping structure of the parallel program onto structures of DCS is equivalent to the graph isomorphism problem that is NP-complete [5,6]. Now efforts of researchers are directed to search effective heuristics suitable for most cases. In many cases observed in practice of parallel programming the weights of all nodes (and all edges) of program graph can be considered equal to each other and can be neglected .In this case a problem of mapping structure of parallel program onto structure of DCS has the following form [5]. The graph $G_p(V_p, P_p)$ of parallel program is considered as a set V_p and a function

$$G_p: V_p * V_p \to \{0,1\},$$

satisfying

$$G_p(x, y) = G_p(y, x), \quad G_p(x, x) = 0$$

for any $x, y \in V_p$. The equation $G_p(x, y) = 1$ means that there is an edge between nodes x and y, that is $(x, y) \in E_p$. Analogously the graph $G_s = (V_s, E_s)$ is determined as a set of nodes (elementary machines (EM)) V_s and a function

$$G_s: V_s * V_s \to \{0,1\}.$$

Here E_s is a set of edges (of links between EM). Suppose that $|V_p| = |V_s| = n$. Let's designate the mapping of parallel program branches to EM by one-to-one function $f_m: V_p \to V_s$.

The mapping quality can be evaluated by the number of program graph edges coinciding with edges of DCS graph. Let's call this number as a cardinality $|f_m|$ of mapping f_m and define it by the expression [5] (the maximum criterion of the mapping quality):

$$|f_m| = (1/2) \sum_{x \in V_p, y \in V_p} G_p(x, y) * G_s(f_m(x), f_m(y)). \tag{1}$$

The minimum criterion of the mapping quality has the form

$$|f_m| = (1/2) \sum_{x \in V_p, y \in V_p} G_p(x,y) * L_s(f_m(x), f_m(y)). \qquad (2)$$

Here $L_s(f_m(x), f_m(y))$ is equal to the distance between nodes $f_m(x)$ and $f_m(y)$ on the graph G_s.

2 Mapping Algorithm

Let's consider the following approach to the mapping problem [7]. Let initially $f_m(x) = x$. Let $e_p(x)$ be an environment (a set of neighbors) of a node x on the graph G_p and $e_s(x)$ be its environment on the graph G_s. For each node $x \in V_p$ we shall test twin exchanges of all nodes i and j if these nodes are satisfied to the condition

$$i \in e_p(x) \& i \notin e_s(x) \& j \in e_s(x) \& state(j) = 0,$$

where $state(j) = 0$, if a node j has not been exchanged in $e_s(x)$ and $state(j) = 1$ otherwise. Exchanges that don't make worse the performance of mapping f_m will be fixed. This approach is based on assumption about high probability of situation when such exchange increasing the number of nodes $i \in e_p(x)$ in the environment $e_s(x)$ improves (or at least doesn't make worse) the value of performance criterion $|f_m|$. It is obvious that the number of the analyzed exchanges in one evaluation of all nodes $x \in V_p$ doesn't exceed the value $v_p v_s n, n = |V_p|$, where v_p and v_s are maximum degrees of nodes of graphs G_p and G_s accordingly. With $v_p v_s < n$ this approach provides the reduction of computations with respect to the well-known Bokhari algorithm (B-algorithm) [5] and with increase of n the effect of reduction is also increased. On the base of suggested approach the procedure Search has been developed for search of local extremum of function $|f_m|$ in the mapping algorithm (MB-algorithm) that is a modification of the B-algorithm. Besides the MB-algorithm includes following modifications . Firstly initial value of $|f_m|$ is compared with E_p and if the equality is carried out then the algorithm is completed. Secondly check-up of the equality $|f_m| = |E_p|$ executed after every completion of the Search. These examinations also lead very often to a cutting down of the algorithm implementation time.

The MB-algorithm is presented below for the criterion (1). Here: TASK is the description of the graph G_p, VS is the current mapping f_m, BEST is the best found mapping f_m, card(f) is the performance criterion of the mapping f, card $(TASK)$ is the performance of the graph G_p selfmapping, BEST ←VS is the creation of the copy BEST of the mapping VS, Rand-interchange (VS) is the random exchange of n node pairs in the mapping VS, Output(BEST) is the output of the best mapping f_m.

```
Procedure MB;
begin
    done=false; BEST ← VS;
    if (card(VS)=card(TASK)) done=true;
        while (done ≠ true)    do
            begin Search(VS,TASK);
                if (card(VS) > card(BEST))
                    begin BEST ← VS; Rand_exchange(VS) end
            else       done=true;
        end
    Output(BEST)
end.
```

The performance of the MB-algorithm has been investigated [7] on the mapping standard parallel program structures ("line", "ring", "mesh") onto regular graphs of DCS (E_2-graph i.e. torus, optimal D_2-graph i.e. two-measured circulant, hypercube) with a number of nodes n varying from 8 to 256. Results of MB-algorithm tests show that the cardinality achieved by the algorithm of mapping graph G_p onto graph G_s with number of nodes $n=|V_p|=|V_s| \le 256$ is no less than 90% of the number of nodes in G_p for cases when G_p is the "line" or the "ring" and no less than 80% for the case when the G_p is the "mesh". The MB-algorithm reduces the time of mapping with respect to B-algorithm in two exponents for mapping one-measured structures ("line", "ring") onto E_2- and D_2-graphs and in exponent for mapping two-measured structures ("mesh") onto hypercube. For all that the MB-algorithm doesn't make worse the performance of the mapping with respect to the B-algorithm.

In this paper we investigate the MB-algorithm quality for mapping parallel program structures onto structures of distributed computer systems (DCS) with faulty elements. The mapping quality is investigated for different DCS network topologies (structure types) and for different numbers of faulty processors and links (interconnections).

Usually the DCS graphs initially are regular (torus, hypercube etc) but during the DCS functioning the failures in processors and links can be arised. As a result the parallel program must be remapped onto the nonregular connected functioning part of the DCS. The decentralized operating system of the DCS provides the equality of program branches count to the DCS functioning processor count, i.e. $|Vs|=|Vp|$ and the computational load is evenly distributed among functioning processors [3,4]. In other words, the parallel program can be adjusted to a number of processors in the DCS.

3 Conditions for Mapping Parallel Programs onto DCS with Faulty Processors and Links

How many failures can be tolerant in a DCS for mapping parallel program? There are two types of bounds for the tolerant multiplicity t of failures.

First, a number of faulty nodes must be no more than $(n-1)/2$, $n=|V|$. This bound is determined by selfdiagnosability condition [2]. In accordance with the bound, selfdiagnostics algorithm works correctly only when the nonfaulty part of the system has no less than half of the nodes number.

Second, the graph with faulty nodes and edges must be connected. This property is necessary for communications between branches of the parallel program. The probability p_t of the DCS graph disconnection and the probability $p_c = 1 - p_t$ of the disconnection absence were investigated with respect to the number of faulty nodes and edges. Investigation was realized by Monte Carlo simulation of regular graphs with faulty nodes and edges and the simulation results are presented below.

Figs.1-6 show the simulation plots of p_t versus the percentage of failed nodes (Figs.1,3,5) and edges (Figs.2,4,6) for a torus (Figs.1,2), a circulant (Figs.3,4) and hypercube (binary cube) (Figs.5,6) with n=64 nodes.

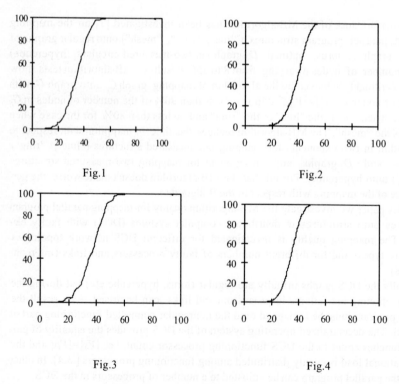

Fig.1

Fig.2

Fig.3

Fig.4

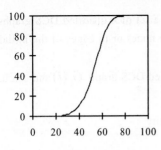

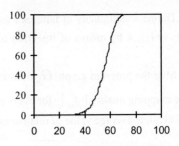

Fig.5   Fig.6

Simulation results show that:

1) for tori and two-measured circulants $p_t \approx 1$ for $t>n/2$, and for hypercubes $p_t \approx 1$ for $t>3n/4$;

2) for tori and two-measured circulants $p_c \approx 1$ for $t \leq 0.2n$ (see Figs.1-4), and for hypercubes $p_c \approx 1$ for $t \leq 0.3n$ (see Figs.5,6);

These results are true both for graphs with faulty nodes and for graphs with faulty edges. In the second case n is a number of the DCS edges. They show that graph topologies with higher node degree allow the network to sustain a larger number of failures before disconnection. Thus for $t \leq 0.2n < (n-1)/2$ all considered topologies are practically connected and we can apply the above mapping algorithm.

4 Investigation of the Mapping Algorithm

Four types of parallel programming graphs G_p were considered: "line", "ring", "mesh" and "torus". These graphs with the number of nodes from $n=8$ to 256 were mapped onto the DCS structure (torus, two-measured circulant (D2-graph), hypercube) with faulty nodes or faulty edges.

The plan of the mapping algorithm investigation is:

1. Create initial graph G_s with a given regular topology and a given number of nodes.

2. Define a multiplicity of failures t and generate a set of k distorted DCS graphs $G_s(i)$, $i=1,...,k$ by means of injection of failures to nodes or to edges of the initial graph.

3. Map the program graph G_p onto every distorted DCS graph $G_s(i)$ and calculate the mapping quality $|f_m|_i$ for every mapping .

4. Calculate average value of relative mapping quality

$$r_m = \frac{1}{kE_p} \sum_{i=1}^{k} |f_m|$$

Results of the mapping algorithm simulation can be distinguished on two groupes:
1. mapping "line" and "ring";
2. mapping "torus" and "mesh".

Table 1. Minimal values $r_m(V) / r_m(E)$ of average mapping qualities for faulty nodes and for faulty edges

G_p G_s	Torus	Circulant	Hypercube
Line	0.802 / 0.803	0.801 / 0.802	0.868 / 0.867
Ring	0.793 / 0.796	0.797 / 0.790	0.859 / 0.850
Mesh	0.571 / 0.792	0.538 / 0.790	0.633 / 0.613
Torus	0.500 / 0.595	0.495 / 0.468	0.586 / 0.586

For the first group (Table.1), if the number of faulty nodes $t < 0.2n$ than the average mapping quality is no less than 0.8 when the DCS graph has topology of torus or circulant and is no less than 0.86 when the DCS graph has topology of hypercube.

For the second group, if the number of faulty nodes $t < 0.2n$ than the average mapping quality is no less than 0.46 when the DCS graph has topology of torus or circulant and is no less than 0.58 when the DCS graph has topology of hypercube.

Figs.7-10 show the simulation plots of the r_m versus the number n of the DCS nodes for several values of the percentage of failed nodes (Tables. 2-5).

Table 2. Mapping quality, "line → torus" mapping

	16	36	64	100	144	169	225
0%	1	0.971	0.905	0.909	0.923	0.929	0.938
10%	0.937	0.864	0.828	0.842	0.836	0.843	0.856
20%	0.926	0.839	0.804	0.802	0.813	0.810	0.813

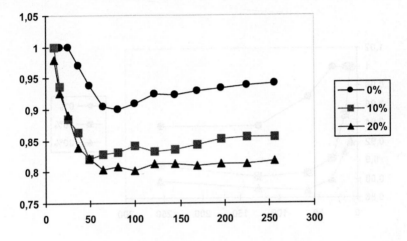

Fig.7. Mapping quality, "line → torus" mapping

Table 3. Mapping quality, "mesh → torus" mapping

	16	36	64	100	144	196	256
10%	0.865	0.698	0.674	0.639	0.633	0.641	0.628
20%	0.783	0.652	0.626	0.577	0.575	0.566	0.558

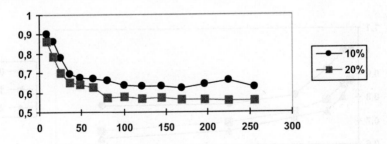

Fig.8. Mapping quality, "mesh → torus" mapping (r_m=1 for 0% percentage of failed nodes)

Table 4. Mapping quality, "line → hypercube" mapping

	8	16	32	64	128	256
0%	1	1	1	0.968	0.937	0.937
10%	1	0.948	0.902	0.887	0.883	0.895
20%	0.919	0.909	0.880	0.868	0.870	0.878

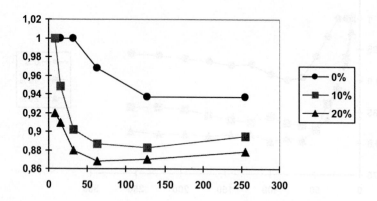

Fig.9. Mapping quality, "line → hypercube" mapping

Table 5. Mapping quality, "mesh → hypercube" mapping

	8	16	32	64	128	256
0%	1	1	0.923	0.812	0.776	0.771
10%	0.875	0.857	0.794	0.743	0.699	0.665
20%	0.875	0.805	0.742	0.719	0.661	0.633

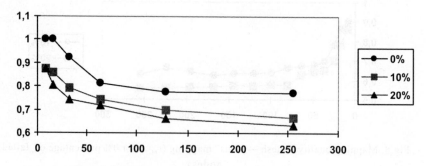

Fig.10. Mapping quality, "mesh → hypercube" mapping

5 Conclusion

Mapping typical graphs ("line","ring","mesh","torus") of parallel programs onto regular graphs ("torus", "two-measured circulant", "hypercube") of distributed computer system (DCS) is considered when the DCS has failures such as faulty processors (nodes) or faulty links (edges). It is shown by the mapping algorithm simulation that:

1) one-measured program structures (line and ring) are mapped better than two-measured structures (mesh and torus);

2) the one-measured structures mapping is changed to worse very less (6 – 15% for 20% faulty elements) than the two-measured structures mapping(15-45%).

For hypercube the mapping quality values are greater than for torus and two-measured circulant . The smaller node degree of a program graph and the greater node degree of a DCS graph the greater the mapping quality. Thus, the one-measured program structure's mappings are more fault-tolerant than two-measured structure's one and more preferable for organization of computations in the DCS with faulty elements.

Acknowledgement

This work was supported by the BK21 program (E-0075).

Referencez

1. Korneev, V.V.: Architecture of Computer Systems with Programmable Structure.Novosibirsk, Nauka (1985)(in Russian).
2. Dimitriev, Yu.K.: Selfdiagnostics of modular computer systems. Novosibirsk, Nauka (1994) (in Russian)
3. Korneev, V.V., Tarkov, M.S.: Operating System of Microcomputer System with Programmable Structure MICROS, Microprocessornie sredstva i sistemy (Microprocessor means and systems). 4 (1988) 41-44 (in Russian)
4. Tarkov, M.S.:Parallel Fault Tolerant Image Processing in Transputer System MICROS-T. Nauchnoe priborostroenie. Vol.5. 3-4 (1995) 74-80 (in Russian)
5. Bokhari, S.H.: On the Mapping Problem, IEEE Trans. Comput., C-30(3), (1981) 207-214
6. Lee, S.-Y., Aggarval, J.K.:A Mapping Strategy for Parallel Processing, IEEE Trans. Comput., C-36(4), (1987) 433-442
7. Tarkov, M.S.: Mapping Parallel Programs Onto Distributed Robust Computer Systems. Proceed. of the 15th IMACS World Congress on Scientific Computation, Modelling and Applied Mathematics, Berlin, August 1997. V.6. Application in Modelling and Simulation. Proc. Ed. By Achim Sydow. (1997) 365-370

Enabling Interoperation of High Performance, Scientific Computing Applications:

Modeling Scientific Data With The Sets & Fields (SAF) Modeling System

Mark C. Miller[1], James F. Reus[1], Robb P. Matzke[1], William J. Arrighi[1],
Larry A. Schoof[2], Ray T. Hitt[2] and Peter K. Espen[2]

[1]Lawrence Livermore National Laboratory, 7000 East Ave, Livermore, CA. 94550-9234
`{miller86, reus1, matzke1, arrighi2}@llnl.gov`
[2]Sandia National Laboratories, PO Box 5800 Albuquerque, NM 87185
`{laschoo, rthitt, pkespen}@sandia.gov`

Abstract. This paper describes the *Sets and Fields* (SAF) scientific data modeling system; a revolutionary approach to interoperation of high performance, scientific computing applications based upon rigorous, math-oriented data modeling principles. Previous technologies have required all applications to use the same data structures and/or meshes to represent scientific data or lead to an ever expanding set of incrementally different data structures and/or meshes. SAF addresses this problem by providing a small set of mathematical building blocks—sets, relations and fields—out of which a wide variety of scientific data can be characterized. Applications literally *model* their data by assembling these building blocks. A short historical perspective, a conceptual model and an overview of SAF along with preliminary results from its use in a few ASCI codes are discussed.

1 Introduction

This paper describes the *Sets and Fields* (SAF, pronounced "safe") scientific data modeling system: a revolutionary approach to storage and exchange of data between high performance, scientific computing applications. It is being developed as part of the *Data Models and Formats* (DMF) component of the *Accelerated Strategic Computing Initiative* (ASCI) [15].

ASCI-DMF is chartered with developing a suite of software products for reading, writing, exchanging and browsing scientific data and for facilitating interoperation of a diverse and continually expanding collection of scientific computing software applications. This amounts to solving a very large scale scientific software integration problem. Small scale integration has been addressed by a number of existing products; SEACAS[20], PACT[4], Silo[18], Exodus II[19], CDMLib[1], netCDF[16], CDF[12], HDF[11],[3] and PDBLib[4] to name a few. There are many others. Numerous isolated successes with these products on the small scale lead many to believe any one is sufficient for the large scale. It is merely a matter of comming to agreement. However, because these products force applications to express their data in terms of handfuls of data structures and/or mesh types, this has never been practical. Consequently, these approaches have succeeded in integration on the small scale but hold little promise for the large scale.

The key to integration and sharing of data on the large scale is to develop a small set of building blocks out of which descriptions for a wide variety of scientific data can be constructed. Each new and slightly different kind of data involves the use of the same building blocks to form a slightly different *assembly*, to literally *model* scientific data.

To achieve this, the building blocks must be at once, primitive and abstract. They must be primitive enough to model a wide variety of scientific data. They must be abstract enough to model the data in terms of *what* it represents in a mathematical or physical sense independent of *how* it is represented in an implementation sense. For example, while there are many ways to represent the airflow over the wing of a supersonic aircraft in a computer program, there is only one mathematical/physical interpretation: a field of 3D velocity vectors over a 2D surface. This latter description is immutable. It is independent of any particular representation or implementation choices. Understanding this *what* versus *how* relationship, that is *what* is represented versus *how* it is represented, is key to developing a solution for large scale integration of scientific software.

These are the principles upon which SAF is being developed. In the remaining sections, we present a brief history of scientific data exchange, an abstract conceptual view of scientific data, the data modeling methodology developed for and implemented in SAF and preliminary results from integration of SAF with some ASCI applications.

2 Historical Perspective

In the early days of scientific computing, roughly 1950 - 1980, simulation software development at many labs invariably took the form of a number of software "stovepipes". Each big code effort included sub-efforts to develop supporting tools for visualization, data differencing, browsing and management. Developers working in a particular stovepipe designed every piece of software they wrote, simulation code and tools alike, to conform to a common representation for the data. In a sense, all software in a particular stovepipe was really just one big, monolithic application, typically held together by a common, binary or ASCII file format. Data exchanges across stovepipes were achieved by employing one or more computer scientists whose sole task in life was to write a conversion tool called a *linker* and keep it up to date as changes were made to one or the other codes that it linked. In short, there was no integration. Furthermore, there was duplication of effort in the development of similar tools for each code.

Between 1980 and 2000, an important innovation emerged, the general purpose I/O library. In fact, two variants emerged each working at a different level of abstraction. One focused on computer science objects such as arrays, structs and linked lists. The other focused on computational modeling objects such as structured and unstructured meshes and zone- and node-centered variables. Examples of the former are CDF, HDF and PDBLib (early 80's). Examples of the latter are EXODUS (1982), Silo (1988) and CDMLib (1998).

Unfortunately, both kinds of I/O libraries focused more upon the *hows* of scientific data representation than the *whats*. That is, they characterize *how* the data is represented in terms of arrays, structs and linked-lists or in terms of structured and unstructured meshes with zone- and node-centered variables. To use these products across the gamut of scientific data, a set of conventions in *how* these objects are applied to represent various kinds of data must invariably be adopted. And, without a lot of cooperative effort, different organizations wind up using a different set of conventions for the same kinds of data. Worse still, for many kinds of scientific data, both qualitative and quantitative information can be completely lost in the resultant representations.

A related innovation that emerged during this same time is also worth mentioning, the visualization tool-kit. Examples are AVS[21], DX[14] and Khoros[13]. While these products were developed primarily for visualization, they included functionality to exchange scientific data between software modules via common data representations. From this point of view, most notable among these is DX for its implementation of a data model based upon abstract, mathematical building blocks [14], [5].

Large scale integration of scientific computing software requires a *common abstraction* capable of supporting a wide variety of scientific computing applications. An appropriate abstraction is the continuous mathematical setting from which the majority of scientific computing software is ultimately derived.

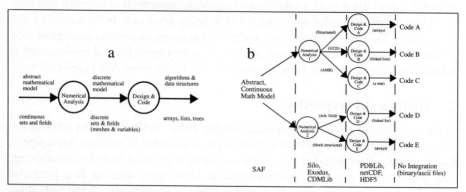

Fig. 1 a) typical scientific computing application design process, b) example of different codes that result from different decisions in each phase of the design

In Fig. 1a[8], we illustrate the basic stages in the development of a scientific computing application. The process begins with an abstract mathematical characterization of the problem or class of problems to be solved. From there, a thorough numerical analysis is performed resulting in a discrete approximation to the original continuous problem or problem class. Invariably, this results in particular kinds of meshes and interpolation schemes for dependent variables defined over them. Next, the software design and coding stage is entered resulting in a set of data structures and software components implementing the numerical model. At each stage, there are numerous design and implementation decisions to be made, each resulting in a different implementation of a solution for the same problem or problem class. This is illustrated in Fig. 1b[8].

Of course, the most important point of Fig. 1b is that the only salient feature of the data that is truly a *common abstraction* is the abstract mathematical setting from which all the different implementations are derived. This is the only context which is immutable across all design and implementation decisions. Unfortunately, historically all of the rich mathematical information necessary to support this abstraction is routinely lost, left in design documents, programmer's notebooks or the heads of the computational engineers. This is so because there has never existed an I/O library that allowed software developers to model their data in these fundamental, abstract, mathematical terms.

3 The Abstract Mathematical Setting: Fields

SAF is designed to provide the abstract mathematical setting alluded to in the previous section. It is based upon representing *fields*[8]. Fields serve as the standards for real (or imagined) observables in the real, continuous world. In terms of scientific computing, fields are the dependent variables of a simulation. In fact, a field is really just a generalization of a *function*. Like functions, fields are specified in three logically distinct parts; something akin to the domain of a function called the *base-space*, something akin to the range of a function called the *fiber-space* and a *map* that relates points in the base-space to points in the fiber-space. Furthermore, both the base-space and fiber-space are infinite point-sets with topological dimensions called the *base-dimension* and *fiber-dimension*. This terminology comes from the mathematics of *fiber-bundles*[5].

Next, if a field is intended to represent a real (or imagined) observable in the real, continuous world, then it has infinitely precise value at an infinite number of points. How do we represent it on a computer? More importantly, how do we formulate a representation that is common across many scientific computing applications? One common representation is given in Eq.1[9]

$$F(x) = \sum_i f_i b_i(x) \qquad (1)$$

where we represent a continuous field, $F(x)$, by a *basis-set*, $\{b_i(x)\}$ and a *dof-set*, $\{f_i\}$. We call the f_is *degrees of freedom* or *dofs*[8]. To call them *values* would imply that they are in fact *equal* to the field for certain points, x, and this is only true when the basis functions are *interpolating*. They are most appropriately viewed as the degrees of freedom in the representation of the field.

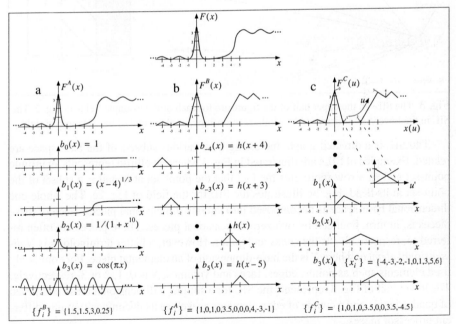

Fig. 2 An arbitrary one dimensional field and possible representations.

Three examples are illustrated in Fig. 2. In this figure, the horizontal axes represent the base-space. The vertical axes represent the fiber-space and the wiggly lines we've plotted represent the map between the two. While there is much to be learned by comparing these examples in detail, we will only touch on a few key points.

First, Fig. 2c introduces a level of indirection by re-parameterizing on a new base-space, u. In some sense, u represents the infinite point-set *along* the wiggly line itself, as opposed to either axis. On u, even the independent variable, x, is a field. It is a special field because it is a *coordinate field*. Nonetheless, this illustrates the point that even the independent variable(s) of a simulation are fields like any other field. This also means that apart from the fields F and x, u is nothing more than a topological space. And, statements about u and its various subsets are concerned solely with set-algebraic relationships of infinite point-sets. Next, observing that u is broken into subsets each of which is independently mapped onto an elemental base-space, u', leads to another key point in modeling a field; the *subset inclusion lattice* or *SIL*[8].

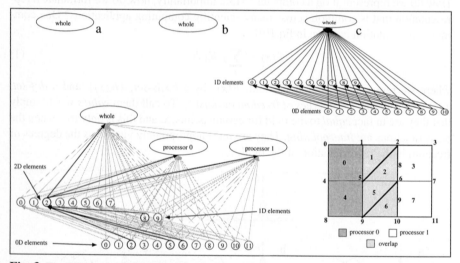

Fig. 3 The SILs in the upper half of the figure are for each of the example fields in Fig. 2. The SIL in the lower half is drawn to highlight relationships for element 2 of the 2D mesh at right.

The SIL is a directed graph, indicating how various subsets of the base-space are related. Examples of SILs are illustrated in Fig. 3 where an ellipse represents an infinite point-set and an arrow represents the fact that the point-set at its tail is a subset of the point-set at its head. Fig. 3c illustrates the SIL for the field of Fig. 2c. The whole one dimensional base-space is decomposed into ten one dimensional pieces. Each of these pieces is, in turn, bounded by two zero dimensional pieces. These pieces are often referred to, respectively, as the *zones* and *nodes*. However, a better terminology is *N dimensional element* where *N* is the base-dimension of an elemental piece. For commonly used elements such as points, edges, faces and volumes, *N* is 0, 1, 2 or 3, respectively. The lower half of Fig. 3 represents the SIL for a simple, 2 dimensional mesh composed of quad, triangle and a couple of edge elements as well as its decomposition into different processor pieces.

Across all possible representations for a field, there are subtle trade-offs between how a representation is split between the dof-set, basis-set and SIL. There is probably some *conservation of information* principle at work. To see this, observe that the basis-set in Fig. 2a is rather complicated, involving rational polynomials and trigonometric functions while the dof-set is small. The basis-set in Fig. 2b involves only a single function, $h(x)$, shifted about integral positions on the x base-space. The dof-set is larger. And, in both of these cases, the SIL contains only a single set. Further comparison of Fig. 2b with Fig. 2c. reveals that while both are piecewise linear representations, in Fig. 2c, we have an additional set of numbers, $\{x_i^C\}$. These can be viewed either as an entirely new dof-set for the independent variable, x, as a field defined on the u base-space or as parameters that control the shape of the member functions of a basis-set for the field F defined on the x base-space. Ultimately, the difference is determined by how the basis-set is defined. If it is defined on the x base-space with member functions such as $b_1(x)$, $b_2(x)$ and $b_3(x)$ in Fig. 2c, then the SIL contains just one set for the whole base-space. If, on the other hand, the u base-space is decomposed into elemental subsets, u', then the basis-set contains just two functions, u' and $1-u'$, and the SIL is more involved because it must specify how these pieces fit together. This latter view is illustrated in Fig. 3c. Finally, the field in Fig. 2b could be characterized as a special case of that of Fig. 2c when $u=x$ and $\{x_i^B\} = \{i\}$. Again, the difference is in the basis-set.

3.1 The Data Model Enables Development of Set and Field Operators

There is much more to modeling fields than there is room to discuss here. Furthermore, the ability to model fields is only the first, small step. Of much greater importance is the plethora of operations on sets and fields such a data model enables. Therefore, we conclude this section by postulating a number of these operations[8].

With such a model, we have all the information necessary to evaluate a field at any point in its base-space. We can efficiently describe fields that are defined on only a subset of a mesh instead of having to define them with a bunch of zeros over the whole mesh. We can easily postulate a *restriction* operator which restricts a field to a particular subset of its base-space. Such an operation is invaluable when we are interested in looking at only a portion of some large data set. Likewise, we can postulate a number of set operations such as *union, intersection, difference, boundary-of, copy-of,* and *neighbor-of.*

Looking, again, at the SILs in Fig. 3, we see that different types of mesh structure can be captured in terms of different patterns of subset relations (arrows) between the N dimensional elements. Only the pattern need be captured. With this approach, various kinds of structured grids such as rectangular, triangle-strips, hexagonal, could be easily and efficiently represented. In a completely arbitrarily connected grid, there is no such pattern and the subset relations (arrows) must be explicitly characterized.

We can represent *inhomogeneous* fields. For example, we can represent a symmetric tensor field which is a 2D tensor over an infinitely thin, 2D wing but is a 3D tensor over a 3D fuselage. Furthermore, we can characterize multiple, independent decompositions of a base-space into element-blocks, materials, processor domains, parts in an assembly, etc. In the case of processor decompositions, the SIL provides the information to enable operators that seamlessly change between different decompositions to run a problem on 32 processors, for example, and restart it on 48. Finally, we can describe multiple, different representations for the same field and seamlessly exchange between them. Operations such as these are fundamental to large scale, scientific software integration.

4 The Sets and Fields (SAF) Scientific Data Modeling System

SAF represents a first cut at a portable, parallel, high performance application programming interface for reading and writing shareable scientific data files based upon abstract data modeling principles. Our primary goal has been to demonstrate that we can apply this technology to do the same job previously achieved with mesh-object I/O libraries like Silo and EXODUS, that is to simply describe and store our data to files. Our belief is that if we cannot demonstrate this basic capability, there is little point in trying to address more advanced features of scientific data management made possible with this technology.

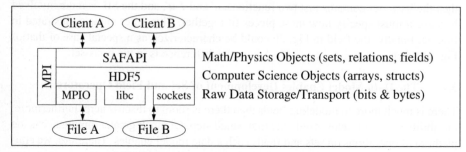

Fig. 4 Overall SAF System Software Architecture

A coarse view of the architecture is illustrated in Fig. 4. SAF is built upon the Hierarchical Data Format, version 5 (HDF5) and the Message Passing Interface (MPI). It is a procedural implementation of a design based upon object oriented techniques. Based on the conceptual model, *set*, *relation* and *field* objects were defined along with the following operators:

- **declare**: create a handle to a new object and define its parameters.
- **describe**: get the declaration parameters of an object.
- **write**: put out "raw" data (typically problem-sized) that populates an object
- **read**: get "raw" data of an object.
- **find**: retrieve objects based on matching criteria and/or traversing the SIL.

4.1 Sets and Collections

Sets are used to define the base-space for fields. In theory, every subset of the base-space, even every node, edge, face and volume element is a full-fledged set. However, in the implementation of SAF, we have had to differentiate between two kinds of sets; *aggregate* and *primitive*. Aggregate sets are the union of other sets in the base-space. As such, aggregate sets are used to establish *collections* of primitive sets as well as other aggregate sets. Primitive sets are not the union of any other sets in the base-space. Typically, primitive sets represent computational elements; nodes, edges, faces and volumes. Primitive sets are never instantiated as first class sets in SAF. Instead, they only ever appear as members of collections.

4.2 Relations

A relation describes how members of a *domain* collection are set-algebraically related to members of a *range* collection. There are numerous possibilities for different kinds of relations, but currently only two have been implemented; *subset relations* and *topology relations*. A subset relation defines a subset relationship between two aggregate sets by identifying the members in a range collection, on the superset, that are in the domain collection on the subset. For example, a processor subset is specified by a subset relation which identifies all the elements in the whole that are on the processor. A topology relation defines how members of a domain collection are knitted together to form a network or mesh by virtue of the fact that they share members of a range collection. For example a collection of zones is knitted together by identifying the nodes each zone shares.

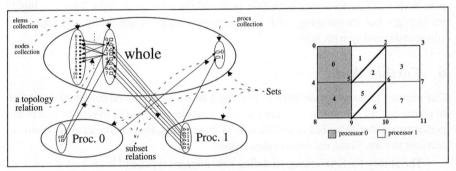

Fig. 5 An example of sets, collections, subset and topology relations.

4.3 Field-Templates and Fields

Fields are defined in SAF in two steps. First, a field-template is defined by attaching a fiber-space to a set. A fiber-space is specified by its fiber-dimension (e.g. the number of components in the field), an algebraic-type such as scalar, vector or tensor, a basis-type such as Euclidean and a physical quantity such as length, time or charge. In fiber-bundle theory, the cross-product of the fiber-space and the base-space it is attached to defines a *section-spaced*[5], the space of all possible fields of the given fiber type. An instance of a field is a slice or cross-section through the section-space. A field-template defines a section-space.

In the second step, a field is defined as an instance of a field-template. The basis-set is defined along with the specific units of measure such as meters, seconds or coulombs. Finally, the dof-set is defined by specifying a mapping of dofs to pieces of the base-space. Currently, SAF supports mappings of dof-sets which are *n:1* associated with members of collections. That is, for every member there are *n* dofs that influence the field over it. These mappings are a generalization of the older notions of node- and zone-centered variables commonly used in I/O libraries and visualization tools.

5 Results

A key result and milestone for SAF is that sub-organizations in ASCI who currently use EXODUS or Silo, have demonstrated that SAF is general enough to replace *both* of these older scientific database technologies. This is a critical result. It validates our approach to large scale integration. In addition, SAF is currently being integrated with several ASCI applications and support tools including SNL's SIERRA[10] and LLNL's Ale3d[22] and MeshTV[17]. The results in Fig. 6 com-

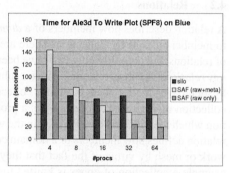

Fig. 6 Preliminary performance results

pare SAF's I/O performance with Silo in the Ale3d simulation code. They are highly preliminary but encouraging. SAF's performance suggests scalability and appears to compare favorably with Silo.

6 Further Work

Our primary focus so far has been to develop a sufficiently general data model and then to prove that we can produce an easy to use, scalable, parallel I/O library of acceptable performance that describes scientific data in these terms. From here, there are many directions to take. Some are enumerated below.

- The ability for SAF clients to define new element types and basis-sets at run time.
- The ability to register specific assemblies of SAF objects in an object registry.
- The ability to characterize derived fields along with the specific derivation rules.
- A publish/subscribe paradigm for exchanging data in-situ instead of by I/O to a file.
- MPI-like messaging pitched in terms of sets and fields instead of arrays and structs.
- The addition of set and field operators.
- A query language analagous to SQL used in relational database systems.

SAF is currently being prepared for a full public release around the end of March, 2001. At that time, there will be publicly viewable web pages at or near http://www.ca.sandia.gov/asci-sdm/ (follow the "Data Models and Formats" link)

7 Acknowledgements

Special thanks to David Butler for his early and important contributions in defining the data model [7]. In addition, we thank our alpha users and early adopters: Nancy Collins, Doug Speck, Rob Neely and Greg Sjaardema. Our sincere thanks, also, to numerous Livermore and Sandia management advocates who have supported this work: Linnea Cook, Celeste Matarazzo, Tom Adams, Terri Quinn, Charles McMillan, John Zepper, James Peery and Constantine Pavlakos. Thanks also to Mike Folk and the HDF team for the HDF5 product, upon which SAF is built, and continued enthusiastic collaboration in ASCI-DMF. This work was performed under the auspices of the U.S. Department of Energy by the University of California, Lawrence Livermore National Laboratory under Contract No. W-7405-Eng-48.

References

[1]   Ambrosiano, J., Holten, J., Medvick, P., Parker J., Reed, T., "CDMlib: Simplifying Data Exchange Among Simulation Codes", to be published report of the Los Alamos National Laboratory (1998).

[2]   Baum, J. D., "Elements of Point-Set Topology", Dover Publications Inc., 1991.

[3]   Brown, S., Folk, M., Goucher, G., Rew, R.,"Software for Portable Scientific Data Management," *Computers in Physics*, Vol. 7, No. 3, pp. 304-308, (1993).

[4]   Brown, S. A., "PACT User's Manual", University of California Research Lab Report, Lawrence Livermore National Laboratory, UCRL-MA-112087 (1999).

[5]   Butler, D. M., Pendley, M. H., "A Visualization Model Based on the Mathematics of Fiber Bundles", *Computers in Physics*, V3 N5, pp. 45-51 (1989).

[6]   Butler, D. M., Bryson, S., "Vector Bundle Classes Form Powerful Tool for Scientific Visualization", *Computers in Physics*, V6 N6, pp. 576-584 (1992).

[7]   Butler, D. M., "ASCI Data Models and Formats Committee Array API Specification" aka "VB-1.0", working draft, Limit Point Systems and Sandia National Laboratories, (1997).

[8]   Butler, D. M., various consultations on the ASCI-DMF project (1997-1999)

[9]   Cheney, E. W., "Introduction to Approximation Theory", McGraw-Hill, 1966.

[10]   Edwards, H. C. "A Look at SIERRA: A Software Environment for Developing Complex Multi-physics Applications", http://www.cfd.sandia.gov/sierra.html (2000).

[11]   Fortner, Brand, "HDF: the Hierarchical Data Format. " (Technology Tutorial) Dr. Dobb's Journal V23, N5, pp. 42-48 (1998).

[12]   Goucher, G., Mathews, J., "The National Space Science Data Center Common Data Format," *R&T Report*, NASA/Goddard Space Flight Center Publication, (December 1994).

[13]   Grygo, G., "Khoros public domain environment integrates stages of data visualization. (product announcement)", *Digital Review* v8, n34 (Nov 4, 1991)

[14]   Haber, R., Lucas, B. and Collins., N. "A Data Model for Scientific Visualization with Provisions for Regular and Irregular Grids", *Proceedings IEEE Visualization '91 Conference*, pp. 298-305, October, 1991.

[15]   Larzelere, A.R., II, "Creating simulation capabilities," *IEEE Computational Science and Engineering,* Jan.-March, 5, 1 (1998).

[16]   Rew, R. K., Davis, G., "NetCDF: An Interface for Scientific Data Access" *IEEE Computer Graphics and Applications*, V4, pp. 72-82, July (1990).

[17]   Roberts, L., Brugger, E. S.,Wookey, S. G., "MeshTV: scientific visualization and graphical analysis software", University of California Research Lab Report, Lawrence Livermore National Laboratory, UCRL-JC-118600 (1999).

[18]   Roberts, L. J., "Silo User's Guide", University of California Research Lab Report, Lawrence Livermore National Laboratory, UCRL-MA-118751-REV-1 (2000).

[19]   Schoof, L. A., Yarberry, V. R., "EXODUS II: A Finite Element Data Model," Tech. Rep. SAND92-2137, Sandia National Laboratories (1994).

[20]   Sjaardema, G. D., "Overview of the Sandia National Laboratories Engineering Analysis Code Access System", SAND92-2292, Sandia National Laboratories (1993).

[21]   Upson, C., et. al., "The Application Visualization System: a computational environment for scientific visualization", *IEEE Computer Graphics and Applications*, July, 1989.

[22]   Wallin, B. K., Nichols III, A. L., Chow, E., Tong, C., "Large multi-physics simulations in ALE3D", *Tenth Society for Industrial and Applied Mathematics Conference on Parallel Processing and Scientific Computing*, Portsmouth, VA, March 12-14, 2001.

Intelligent Systems Design and Applications

Session chairs:

Ajith Abraham (Monash University, Australia)
Baikunth Nath (Monash University, Australia)

ALEC: An Adaptive Learning Framework for Optimizing Artificial Neural Networks

Ajith Abraham and Baikunth Nath

School of Computing & Information Technology
Monash University (Gippsland Campus), Churchill 3842, Australia
{Email: Ajith.Abraham, Baikunth.Nath@infotech.monash.edu.au}

Abstract: In this paper we present ALEC (Adaptive Learning by Evolutionary Computation), an automatic computational framework for optimizing neural networks wherein the neural network architecture, activation function, weights and learning algorithms are adapted according to the problem. We explored the performance of ALEC and artificial neural networks for function approximation problems. To evaluate the comparative performance, we used three different well-known chaotic time series. We also report some experimentation results related to convergence speed and generalization performance of four different neural network-learning algorithms. Performances of the different learning algorithms were evaluated when the activation functions and architecture were changed. We further demonstrate how effective and inevitable is ALEC to design a neural network, which is smaller, faster and with a better generalization performance.

1. Introduction

In Artificial Neural Network (ANN) terminology, function approximation is simply to find a mapping $f: R^m \Rightarrow R^n$, given a set of training data. Even then, finding a global approximation (applying to the entire state space) is often a challenging task. The important drawback with the conventional design of ANN is that the designer has to specify the number of neurons, their distribution over several layers and interconnection between them. In this paper, we investigated the speed of convergence and generalization performance of backpropagation algorithm, conjugate gradient algorithm, quasi Newton algorithm and Levenberg-Marquardt algorithm. Our experiments show that architecture and node activation functions can significantly affect the speed of convergence of the different learning algorithms. We finally present the evolutionary search procedures wherein ANN design can evolve towards the optimal architecture without outside interference, thus eliminating the tedious trial and error work of manually finding an optimal network [1]. Experimentation results, discussions and conclusions are provided towards the end.

2. Artificial Neural Network Learning Algorithms

If we consider a network with differentiable activation functions, then the activation functions of the output units become differentiable functions of both the input variables and of the weights and biases. If we define an error function (E), such as sum of squares function, which is a differentiable function of the network outputs, then this error function is itself a differentiable function of the weights. We can therefore evaluate the derivatives of the error with respect to the weights, and these derivatives can then be used to find weight values, which minimize the error function, by using one of the following learning algorithms:

Backpropagation Algorithm (BP)

BP is a gradient descent technique to minimize the error E for a particular training pattern. For adjusting the weight (w_k), in the batched mode variant the descent is based on the gradient ∇E ($\frac{\delta E}{\delta w_k}$) for the total training set:

$$\Delta w_k(n) = -\epsilon * \frac{\delta E}{\delta w_k} + \alpha * \Delta w_k(n-1) \tag{1}$$

The gradient gives the direction of error E. The parameters ε and α are the learning rate and momentum respectively. A good choice of both the parameters is required for training success and speed of the ANN.

Scaled Conjugate Gradient Algorithm (SCGA)

Moller [5] introduced the scaled conjugate gradient algorithm as a way of avoiding the complicated line search procedure of conventional conjugate gradient algorithm (CGA). According to the SCGA, the Hessian matrix is approximated by

$$E''(w_k)p_k = \frac{E'(w_k + \sigma_k p_k) - E'(w_k)}{\sigma_k} + \lambda_k p_k \tag{2}$$

where E' and E'' are the first and second derivative information of global error function $E(w_k)$. The other terms p_k, σ_k and λ_k represent the weights, search direction, parameter controlling the change in weight for second derivative approximation and parameter for regulating the indefiniteness of the Hessian. In order to get a good quadratic approximation of E, a mechanism to raise and lower λ_k is needed when the Hessian is positive definite. Detailed step-by-step description can be found in [5].

Quasi - Newton Algorithm (QNA)

Quasi-Newton method involves generating a sequence of matrices $G^{(k)}$ which represents increasingly accurate approximations to the inverse Hessian (H^{-1}). Using only the first derivative information of E [4], the updated expression is as follows:

$$G^{(k+1)} = G^{(k)} + \frac{pp^T}{p^T v} - \frac{(G^{(k)}v)\,v^T\,G^{(k)}}{v^T G^{(k)} v} + (v^T G^{(k)} v)uu^T \tag{3}$$

where $p = w^{(k+1)} - w^{(k)}$, $v = g^{(k+1)} - g^{(k)}$, $u = \frac{p}{p^T v} - \frac{G^{(k)} v}{v^T G^{(k)} v}$ and T represents transpose of a matrix. A significant advantage of the QNA over the CGA is that the line search does not need to be performed with such great accuracy.

Levenberg-Marquardt (LM) algorithm

When the performance function has the form of a sum of squares, then the Hessian matrix can be approximated to $H = J^T J$; and the gradient can be computed as $g = J^T e$, where J is the Jacobian matrix, which contains first derivatives of the network errors with respect to the weights, and e is a vector of network errors. The Jacobian matrix can be computed through a standard backpropagation technique that is less complex than computing the Hessian matrix. The LM algorithm uses this approximation to the Hessian matrix in the following Newton-like update:

$$x_{k+1} = x_k - [J^T J + \mu I]^{-1} J^T e \qquad (4)$$

When the scalar μ is zero, this is just Newton's method, using the approximate Hessian matrix. When μ is large, this becomes gradient descent with a small step size. As Newton's method is more accurate, μ is decreased after each successful step (reduction in performance function) and is increased only when a tentative step would increase the performance function. By doing this, the performance function will always be reduced at each iteration of the algorithm [4].

3. Experimental Setup Using Artificial Neural Networks

We used a feedforward network with 1 hidden layer and the training was performed for 2500 epochs. The numbers of hidden neurons were varied (14,16,18,20,24) and the speed of convergence and generalization error for each of the four learning algorithms was observed. The effect of node activation functions, Log-Sigmoidal Activation Function (LSAF) and Tanh-Sigmoidal Activation Function (TSAF), keeping 24 hidden neurons for the four learning algorithms was also studied. Computational complexities of the different learning algorithms were also noted during each event. In our experiments, we used the following 3 different time series for training the ALEC/ANN and evaluating the performance. The experiments were replicated 3 times and the worst errors are reported.

a) Waste Water Flow Prediction

The problem is to predict the wastewater flow into a sewage plant [6]. The water flow was measured every hour. It is important to be able to predict the volume of flow $f(t+1)$ as the collecting tank has a limited capacity and a sudden increase in flow will cause to overflow excess water. The water flow prediction is to assist an adaptive online controller. The data set is represented as $[f(t), f(t-1), a(t), b(t), f(t+1)]$ where $f(t), f(t-1)$ and $f(t+1)$ are the water flows at time $t, t-1, and t+1$ (hours) respectively. $a(t)$ and $b(t)$ are the moving averages for 12 hours and 24 hours. The time series consists of 475 data points. The first 240 data sets were used for training and remaining data for testing.

b) Mackey-Glass Chaotic Time Series

The Mackey-Glass differential equation is a chaotic time series for some values of the parameters $x(0)$ and τ [9].

$$\frac{dx(t)}{dt} = \frac{0.2x(t - \tau)}{1 + x^{10}(t - \tau)} - 0.1 x(t). \tag{5}$$

We used the value $x(t-18)$, $x(t-12)$, $x(t-6)$, $x(t)$ to predict $x(t+6)$. Fourth order Runge-Kutta method was used to generate 1000 data series. The time step used in the method is 0.1 and initial condition were $x(0)=1.2$, $\tau=17$, $x(t)=0$ for $t<0$. First 500 data sets were used for training and remaining data for testing.

c) Gas Furnace Time Series Data

This time series was used to predict the CO_2 (carbon dioxide) concentration $y(t+1)$ [8]. In a gas furnace system, air and methane are combined to form a mixture of gases containing CO_2. Air fed into the gas furnace is kept constant, while the methane feed rate $u(t)$ can be varied in any desired manner. After that, the resulting CO_2 concentration y(t) is measured in the exhaust gases at the outlet of the furnace. Data is represented as $[u(t), y(t), y(t+1)]$ The time series consists of 292 pairs of observation and 50% of data was used for training and remaining for testing.

3.1 Experimentation results using ANNs

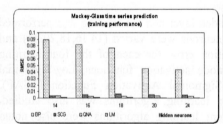
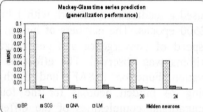

Figure 1. Mackey-Glass time series (training and generalization performance)

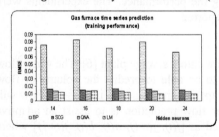
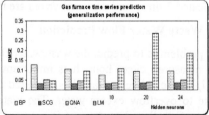

Figure 2. Gas furnace time series (training and generalization performance)

Figures 1, 2 and 3 illustrate the training error and generalization performance of the 4 learning algorithms (BP, SCG, QNA and LM) for the three different time series. Figures 4 (a, b, c) reveal the convergence characteristics of the different learning algorithms for the 3 time series predictions when the hidden neurons activation functions are varied. Figure 5 shows the plot of approximate computational complexity (in Billion Floating Operations - BFlops) for the four different learning algorithms using TSAF and varying the number of hidden neurons.

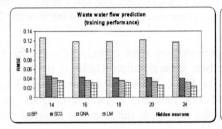

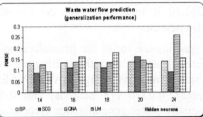

Figure 3. Waste water flow prediction (training and generalization performance)

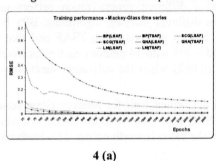

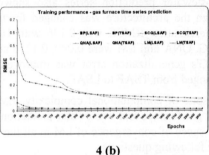

4 (a)   **4 (b)**

Figure 4 (a), (b), (c). Convergence characteristics due to change of hidden neuron activation functions (Mackey Glass, gas furnace and waste water prediction)

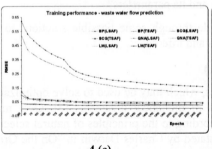

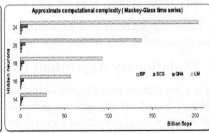

4 (c)   **Figure 5.** Computational complexity

3.2 Discussion of results obtained using artificial neural networks

Our experiments and the following discussion highlight the difficulty in finding an optimal ANN which is smaller in size, faster in convergence and with the best generalization error.

For Mackey Glass series all the 4 learning algorithms tend to generalize well as the hidden neurons were increased. However the generalization was better when the hidden neurons were using TSAF. LM showed the fastest convergence regardless of architecture and node activation function. However, the figures depicting computational complexity of LM are very amazing. For Mackey glass series (with 14 hidden neurons), when BP was using 0.625 BFlops, LM used 29.4 BFlops. When the hidden neurons were increased to 24, BP used 1.064 BFlops and LM's share jumped to 203.10 BFlops. LM gave the lowest generalization RMSE of 0.0009 with 24 hidden neurons.

For gas furnace series the generalization performance were entirely different for the different learning algorithms. BP gave the best generalization RMSE of 0.0766 with 18 hidden neurons using TSAF. RMSE for SCG, QNA and LM were 0.033 (16 neurons), 0.0376 (18 neurons) and 0.045 (14 neurons) respectively. QNA gave marginally better generalization error when the activation function was changed from TSAF to LSAF.

Wastewater prediction series also showed a different generalization performance when the architecture was changed for the different learning algorithms. BP's best generalization RMSE was 0.136 with 16 hidden neurons using TSAF and that of SCG, QNA and LM were 0.090, 0.1276 and 0.095 with 14 neurons each respectively. SCG's generalization error was improved (0.082) when the activation function was changed from TSAF to LSAF.

In spite of computational complexity, LM performed well for Mackey Glass series. For gas furnace and wastewater prediction SCG algorithm performed better. However the speed of convergence of LM in all the three cases is worth noting. This leads us to the following questions:

> What is the optimal architecture for a given problem?
> What activation function should one choose?
> What is the optimal learning algorithm and its parameters?

The following sections will hopefully guide through some possible solutions to the above questions via our further experiments and inferences.

4. Evolutionary Algorithms (EA)

EAs are population based adaptive methods, which may be used to solve optimization problems, based on the genetic processes of biological organisms [3]. Over many generations, natural populations evolve according to the principles of natural selection and "Survival of the Fittest", first clearly stated by Charles Darwin in "On the Origin of Species". By mimicking this process, EAs are able to "evolve" solutions to real world problems, if they have been suitably encoded. The procedure may be written as the difference equation:

$$x[t + 1] = s(v(x[t])) \tag{6}$$

$x[t]$ is the population at time t, v is a random variation operator, and s is the selection operator.

5. Adaptive Learning by Evolutionary Computation (ALEC)

Evolutionary computation has been widely used for training and automatically designing ANNs. However, not much work has been reported regarding evolution of learning mechanisms, which is surprisingly the most challenging part [2]. ALEC mainly focuses on the adaptive search of learning algorithms according to the problem. An optimal design of an ANN can only be achieved by the adaptive evolution of connection weights, architecture and learning rules which progress on different time scales [1]. Figure 6 illustrates the general interaction mechanism with the learning mechanism of the ANN evolving at the highest level on the slowest time scale. All the randomly generated architecture of the initial population are trained by four different learning algorithms and evolved in a parallel environment. Parameters of the learning algorithm will be adapted (example, learning rate and momentum for BP) according to the problem. Figure 7 depicts the basic algorithm of ALEC.

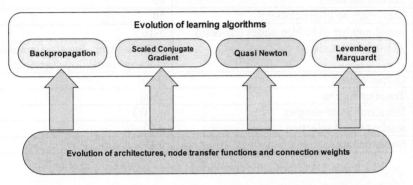

Figure 6. Interaction of various evolutionary search mechanisms in ALEC

1. *Set t=0 and randomly generate an initial population of neural networks with architectures, node transfer functions and connection weights assigned at random.*

2. *In a parallel mode, evaluate fitness of each ANN using BP/SCG/QNA and LM*

3. *Based on fitness value, select parents for reproduction*

4. *Apply mutation to the parents and produce offspring (s) for next generation. Refill the population back to the defined size.*

5. *Repeat step 2*

6. *STOP when the required solution is found or number of iterations has reached the required limit.*

Figure 7. ALEC algorithm for evolutionary design of artificial neural networks

6. ALEC: Experimentation Setup and Results

We have applied ALEC to the three-time series prediction problems mentioned in Section 3. For performance comparison, we used the same set of training and test data that were used for ANNs. The parameters used in our experiments were set to be the same for all the 3 problems. Fitness value is calculated based on the RMSE achieved on the test set. The best-evolved ANN is taken to be the best individual in the last generation. As the learning process is evolved separately, user has the option to pick the best ANN (e.g. less RMSE, fast convergence, less computational expensive etc.) among the four learning algorithms. Genotypes were represented using binary coding and the initial populations of network architectures were randomly created based on the following ALEC parameters.

Table 1. Parameters used for evolutionary design of artificial neural networks

Population size	40
Maximum no of generations	50
Initial number of hidden nodes (random)	5-16
Activation functions	tanh, logistic, linear, sigmoidal, tanh-sigmoidal, log-sigmoidal
Output neuron	linear
Training epochs	2500
Initialization of weights	+/- 0.3
Ranked based selection	0.50
Mutation rate	0.1

Table 2. Performance comparison between ALEC and ANN

Time series	Learning algorithm	ALEC		ANN	
		‡RMSE	*Architecture	‡RMSE	*Architecture
Mackey Glass	BP	0.0077	7(T), 3(LS)	0.0437	24(TS)
	SCG	0.0031	11(T)	0.0045	24(TS)
	QNA	0.0027	6(T),4(TS)	0.0034	24(TS)
	LM	0.0004	8(T),2(TS),1(LS)	0.0009	24(TS)
Gas Furnace	BP	0.0358	8(T)	0.0766	18(TS)
	SCG	0.0210	8(T),2(TS)	0.0330	16(TS)
	QNA	0.0256	7(T),2(LS)	0.0376	18(TS)
	LM	0.0223	6(T),1(LS),1(TS)	0.0451	14(TS)
Waste Water	BP	0.0547	6(T),5(TS),1(LS)	0.1360	16(TS)
	SCG	0.0579	6(T),4(LS)	0.0820	14(LS)
	QNA	0.0823	5(T),5(TS)	0.1276	14(TS)
	LM	0.0521	8(T),1(LS)	0.0951	14(TS)

* Architecture = hidden neurons distribution and activation functions.
† Activation function (T=tanh, S=sigmoidal, LS-log-sigmoidal, TS=tanh-sigmoidal)
‡ RMSE on test set

We used a learning rate of (0.2-0.05) and a momentum of (0.2-0.05) for BP algorithm. For SCG the parameter controlling change in weight for second derivative approximation was chosen as 5e-05(+/-100%) and the parameter for regulating the indefiniteness of the Hessian as 5e-07(+/- 100%). In QNA we varied the scaling factors and step sizes after each generation. For LM we used 1 as the factor for memory/speed trade off to converge faster, adaptive learning rate of 0.001 (+/- 100%) and learning rate increasing and decreasing factor of 10 and 0.1 respectively. The experiments were repeated three times and the worst RMSE values are reported.

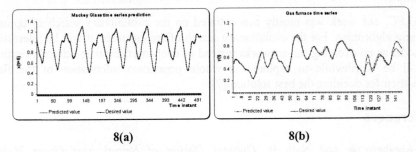

8(a)   8(b)

Figure 8(a). ALEC test results using BP algorithm for Mackey Glass series **(b)** Gas furnace series **(c)** Waste water flow series

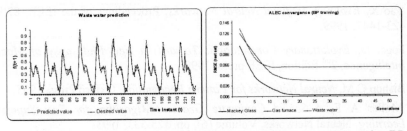

8 (c)   **Figure 9.** ALEC convergence using BP
                                learning after 50 generations

7. Discussions

Table 2 shows comparative performance between ALEC and ANN. For BP algorithm RMSE error was reduced by 82.4% for Mackey Glass series while it was 52.3% for gas furnace and 60.3% for wastewater prediction. At the same time number of hidden neurons got reduced by 58.4% (Mackey Glass), 55.5% (Gas furnace) and 31.25 (wastewater) respectively. The percentage savings in RMSE and hidden neurons are very much similar for all the four algorithms. LM algorithm gave the best results for Mackey Glass and wastewater prediction and SCG performed well for gas furnace series. Overall LM algorithm gave the best RMSE error even though it is highly computational expensive. We deliberately terminated the training after 2500 epochs (regardless of early stopping in some cases) for ANN and ALEC problems, just to have a generalization performance comparison.

8. Conclusion

Selection of the architecture (number of layers, hidden neurons, activation functions and connection weights) of a network and correct learning algorithm is a tedious task for designing an optimal artificial neural network. Moreover, for critical applications and hardware implementations optimal design often becomes a necessity. In this paper, we have formulated and explored; ALEC: an adaptive computational framework based on evolutionary computation for automatic design of optimal artificial neural networks. Empirical results are promising and show the importance and efficacy of the technique.

In ALEC, our work was mostly concentrated on the evolutionary search of optimal learning algorithms. For the evolutionary search of architectures, it will be interesting to model as co-evolving sub-networks instead of evolving the whole network. Further, it will be worthwhile to explore the whole population information of the final generation for deciding the best solution [7].

References

[1] Abraham A and Nath B, *Optimal Design of Neural Nets Using Hybrid Algorithms*, In proceedings of 6th Pacific Rim International Conference on Artificial Intelligence (PRICAI 2000), pp. 510-520, 2000.

[2] Yao X, *Evolving Artificial Neural Networks*, Proceedings of the IEEE, 87(9):1, 423-1447, 1999.

[3] Fogel D, *Evolutionary Computation: Towards a New Philosophy of Machine Intelligence*, 2nd Edition, IEEE press, 1999.

[4] Bishop C M, *Neural Networks for Pattern Recognition*, Oxford Press, 1995.

[5] Moller A F, *A Scaled Conjugate Gradient Algorithm for Fast Supervised Learning*, Neural Networks, Volume (6), pp. 525-533, 1993.

[6] Kasabov N, Foundations of Neural Networks, Fuzzy Systems and Knowledge Engineering, The MIT Press, 1996.

[7] Yao X and Liu Y, *Making Use of Population Information in Evolutionary Artificial Neural Networks*, IEEE Transactions on Systems, Man and Cybernetics, Part B: Cybernetics, 28(3): 417-425, 1998.

[8] Box G E P and Jenkins G M, *Time Series Analysis, Forecasting and Control*, San Francisco: Holden Day, 1970.

[9] Mackey MC, Glass L, *Oscillation and Chaos in Physiological Control Systems*, Science Vol 197, pp.287-289, 1977.

Solving Nonlinear Differential Equations by a Neural Network Method

Lucie P. Aarts[1] and Peter Van der Veer[1]

[1] Delft University of Technology, Faculty of Civilengineering and Geosciences,
Section of Civilengineering Informatics,
Stevinweg 1, 2628 CN Delft, The Netherlands
l.aarts@citg.tudelft.nl, p.vdveer@ct.tudelft.nl

Abstract. In this paper we demonstrate a neural network method to solve non-linear differential equations and its boundary conditions. The idea of our method is to incorporate knowledge about the differential equation and its boundary conditions into neural networks and the training sets. Hereby we obtain specifically structured neural networks. To solve the nonlinear differential equation and its boundary conditions we have to train all obtained neural networks simultaneously. This is realized by applying an evolutionary algorithm.

1 Introduction

In this paper we present a neural network method to solve a nonlinear differential equation and its boundary conditions. In [1] we have already demonstrated how we could solve linear differential and linear partial differential equations by our neural network method. In [2] we showed how to use our neural network method to solve systems of coupled first order linear differential equations. In this paper we demonstrate how we incorporate knowledge about the nonlinear differential equation and its boundary conditions into the structure of the neural networks and the training sets. Training the obtained neural networks simultaneously now solves the nonlinear differential equation and its boundary conditions. Since several of the obtained neural networks are specifically structured, the training of the networks is accomplished by applying an evolutionary algorithm. An evolutionary algorithm tries to find the minimum of a given function. Normally one deals with an evolutionary algorithm working on a single population, i.e. a set of elements of the solution space. We however use an evolutionary algorithm working on multiple subpopulations to obtain results more efficiently. At last we graphically illustrate the obtained results of solving the nonlinear differential equation and its boundary conditions by our neural network method.

2 Problem Statement

Many of the general laws of nature, like in physics, chemistry, biology and astronomy, find their most natural expression in the language of differential equations. Applications also abound in mathematics itself, especially in geometry, and in engineering, economics, and many other fields of applied science.

In [10] one derives the following nonlinear differential equation and its boundary conditions for the description of the problem of finding the shape assumed by a flexible chain suspended between two points and hanging under its own weight. Further the y-axis pass through the lowest point of the chain.

$$\frac{d^2 y}{dx^2} = \sqrt{1 + \frac{dy}{dx} \cdot \frac{dy}{dx}} , \tag{1}$$

$$y(0) = 1, \tag{2}$$

$$\frac{dy}{dx}(0) = 0 . \tag{3}$$

Here the linear density of the chain is assumed to be a constant value. In [10] the analytical solution is derived for system (1), (2) and (3) and is given by

$$y(x) = \frac{1}{2}\left(e^x + e^{-x}\right). \tag{4}$$

In this paper we consider the system (1), (2) and (3) on the interval $x \in [-1, 2]$.

3 Outline of the method

By knowing the analytical solution of (1), (2) and (3), we may assume that $y(x)$ and its first two derivatives are continuous mappings. Further we define the logsigmoid function f as

$$f(x) = \frac{1}{1 + \exp(-x)}. \tag{5}$$

By results in [8] we can find such real values of α_i, w_i and b_i that for a certain natural number m the following mappings

$$\varphi(x) = \sum_{i=1}^{m} \alpha_i f(w_i x + b_i), \tag{6}$$

$$\frac{d\varphi}{dx}(x) = \sum_{i=1}^{m} \alpha_i w_i \frac{df}{dx}(w_i x + b_i), \tag{7}$$

$$\frac{d^2\varphi}{dx^2}(x) = \sum_{i=1}^{m} \alpha_i w_i^2 \frac{d^2 f}{dx^2}(w_i x + b_i), \tag{8}$$

respectively approximate $y(x), \dfrac{dy}{dx}$ and $\dfrac{d^2 y}{dx^2}$ arbitrarily well. The networks represented by (6), (7) and (8) have one hidden layer containing m neurons and a linear output layer. Further we define the DE-neural network of system (1), (2) and (3) as the not fully connected neural network which is constructed as follows. The output of the network represented by (7) is the input of a layer having the function $g(x) = x^2$ as transfer function. The layer contains one neuron and has no bias. The connection weight between the network represented by (7) and the layer is 1. The output of the layer is the input of a layer with the function $h(x) = \sqrt{x}$ as transfer function. The layer contains one neuron and has a bias with value 1. The connection weight between the two layers is 1.

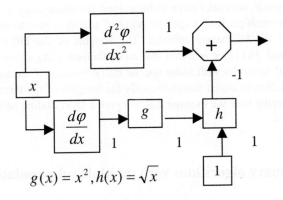

Fig. 1. The DE-neural network for system (1), (2) and (3)

The output of the last layer is subtracted from the output of the network represented by (8). A sketch of the DE-neural network of system (1), (2) and (3) is given in Fig. 1.

Since the learnability of neural networks to simultaneously approximate a given function and its unknown derivatives is made plausible in [5], we observe the following. Assume that we have found such values of the weights that the networks represented by (6), (7) and (8) respectively approximate $y(x)$ and its first two derivatives arbitrarily well on a certain interval. By considering the nonlinear differential equation given by (1) it then follows that the DE-neural network must have a number arbitrarily close to zero as output for any input of the interval. In [6] it is already stated that any network suitably trained to approximate a mapping satisfying some nonlinear partial differential equations will have an output function that itself approximately satisfies the partial differential equations by virtue of its approximation of the mapping's derivatives. Further the network represented by (6) must have for input $x = 0$ an output arbitrarily close to one and the network represented by (7) must give for the same input an output arbitrarily close to zero.

The idea of our neural network method is based on the observation that if we want to fulfil a system like (1), (2) and (3) the DE-neural network should have zero as output for any input of the considered interval $[-1,2]$. Therefore we train the DE-neural network to have zero as output for any input of a training set with inputs $x \in [-1,2]$. Further we have the following restrictions on the values of the weights. The neural network represented by (6) must be trained to have one as output for input $x = 0$ and for the same input the neural network represented by (7) must be trained to have zero as output. If the training of the three networks has well succeeded the mapping φ and its first two derivatives should respectively approximate y and its first two derivatives.

Note that we still have to choose the number of neurons in the hidden layers of the networks represented by (6), (7) and (8), i.e. the natural number m by trial and error.

The three neural networks have to be trained simultaneously as a consequence of their inter-relationships. It is a specific point of attention how to adjust the values of the weights of the DE-neural network. The weights of the DE-neural network are highly correlated. In [11] it is stated that an evolutionary algorithm makes it easier to generate neural networks with some special characteristics. Therefore we use an evolutionary algorithm to adjust simultaneously the weights of the three neural networks. Before we describe how we manage this, we give a short outline of what an evolutionary algorithm is.

4 Evolutionary algorithms with multiple subpopulations

Evolutionary algorithms work on a set of elements of the solution space of the function we would like to minimize. The set of elements is called a population and the elements of the set are called individuals. The main idea of evolutionary algorithms is that they explore all regions of the solution space and exploit promising areas through applying recombination, mutation, selection and reinsertion operations to the individuals of a population. In this way one hopefully finds the minimum of the given function. Every time all procedures are applied to a population, a new generation is created. Normally one works with a single population. In [9] Pohlheim however states that results are more efficiently obtained when we are working with multiple subpopulations instead of just a single population. Every subpopulation evolves over a few generations isolated (like with a single population evolutionary algorithm) before one or more individuals are exchanged between the subpopulations. To apply an evolutionary algorithm in our case, we define e_1, e_2 and e_3 as the means of the sum-of-squares error on the training sets of respectively the DE-neural network, the network represented by (6) and the network given by (7). Here we mean by the mean of the sum-of-squares error on the training set of a certain network, that the square of the difference between the target and the output of the network is summed for all inputs and that this sum is divided by the number of inputs. To simultaneously train the DE-neural network and the networks represented by φ and $\dfrac{d\varphi}{dx}$ we minimize the expression

$$e_1 + e_2 + e_3, \tag{9}$$

by using an evolutionary algorithm. Here equation (9) is a function of the variables α_i, w_i and b_i.

5 Results

In this section we show the results of applying our neural network method to the system (1), (2) and (3). Some practical aspects of training neural networks that are well known in literature also hold for our method. In e.g. [3] and [7], it is stated that if we want to approximate an arbitrary mapping with a neural network represented by (6), it is advantageous for the training of the neural networks to scale the inputs and targets so that they fall within a specified range. In this way we can impose fixed limits on the values of the weights. This prevents that we get stuck too far away from a good optimum during the training process. By training the networks with scaled data all weights can remain in small predictable ranges. In [2] more can be found about scaling the variable where the unknown of the differential equation depend on and scaling the function values of the unknown of the differential equation, to improve the training process of the neural networks.

To make sure that in our case the weights of the networks can remain in small predictable ranges, we scale the function values of the unknown of the nonlinear differential equation. Since we normally do not know much about the function values of the unknown we have to guess a good scaling of the function values of the unknown. For solving the system (1), (2) and (3) on the considered interval $[-1,2]$ we decide to scale y in the following way:

$$y_M = \frac{y}{2}. \tag{10}$$

Hereby the system (1), (2) and (3) becomes

$$2\frac{d^2 y_M}{dx^2} = \sqrt{1 + 4\frac{dy_M}{dx} \cdot \frac{dy_M}{dx}}, \tag{11}$$

$$y_M(0) = \frac{1}{2}, \tag{12}$$

$$\frac{dy_M}{dx}(0) = 0. \tag{13}$$

We now solve the system (11), (12) and (13) by applying the neural network method described in Sect. 3. A sketch of the DE-neural network for system (11), (12) and (13) is given in Fig. 2.

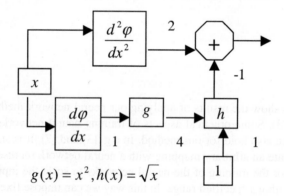

Fig. 2. The DE-neural network for system (11), (12) and (13)

We implemented the neural networks by using the Neural Network Toolbox of Matlab 5.3 ([4]). Further we used the evolutionary algorithm implemented in the GEATbx toolbox ([9]). When we work with the evolutionary algorithms implemented in the GEATbx toolbox, the values of the unknown variables α_i, w_i and b_i have to fall within a specified range. By experiments we noticed that we obtain good results if we restrict the values of the variables α_i, w_i and b_i to the interval $[-5,5]$. The DE-neural network is trained by a training set with inputs $x \in \{-1, -0.9, .., 1.9, 2\}$ and the corresponding targets of all inputs are zero. Further we have to train the neural network represented by φ to have one as output for input $x = 0$ and for the same input the neural network represented by $\dfrac{d\varphi}{dx}$ must have zero as output. The number of neurons in the hidden layer of the neural networks represented by (6), (7) and (8) is taken equal to 6. Therefore the number of variables which have to be adapted is equal to 18. After running the chosen evolutionary algorithm for 1500 generations with 160 individuals divided over 8 subpopulations we take the set $x \in \{-1, -0.95, 0.9, ..., 1.95, 2\}$ as input to compute the output of the neural networks represented by $2\varphi, 2\dfrac{d\varphi}{dx}$ and $2\dfrac{d^2\varphi}{dx^2}$. We also compute the analytical solution of (1), (2) and (3) and its first two derivatives for $x \in \{-1, -0.95, 0.9, ..., 1.95, 2\}$. By comparing the results we conclude that the approximation of y and its first two derivatives by respectively 2φ and its first two derivatives are very good. Both the neural network method solution of (1), (2) and (3) and its first two derivatives as the analytical solution of (1), (2) and (3) and its first two derivatives are graphically illustrated in Fig. 3 and Fig. 4. The errors between the neural network method solution of (1), (2) and (3) and its first two derivatives on the one hand and the analytical solution of (1), (2) and (3) and its first two derivatives on the other hand are graphically illustrated in Fig. 5. The difference between the target of the DE-neural network of the system (1), (2) and (3), i.e. zero for any input x of the set $\{-1, -0.95, 0.9, ..., 1.95, 2\}$ and its actual output is also illustrated in Fig. 5. Considering Fig. 5, we can conclude that the approximations of the solution of (1), (2) and (3) and its first derivative are somewhat better than the approximation of the second derivative of the solution of (1), (2) and (3). Since we are however in most numerical solving methods for differential equations interested in the approximation of just the solution itself, our results are really satisfying.

6 Concluding Remarks

In this paper we used our neural network method to solve a system consisting of a nonlinear differential equation and its two boundary conditions. The obtained results are very promising and the concept of the method appears to be feasible. In further research more attention will be paid to practical aspects like the choice of the evolutionary algorithm that is used to train the networks simultaneously. We will also do more extensive experiments on scaling issues in practical situations, especially the scaling of the variable where the unknown of the differential equation depends on.

References

1. Aarts, L.P., Van der Veer, P.: Neural Network Method for Solving Partial Differential Equations. Accepted for publication in Neural Processing Letters (200?)
2. Aarts, L.P., Van der Veer, P.: Solving Systems of First Order Linear Differential Equations by a Neural Network Method. Submitted for publication December 2000
3. Bishop, C.M.: Neural Networks for Pattern Recognition. Clarendon Press, Oxford (1995)
4. Demuth, H., Beale, M.: Neural Networks Toolbox For Use with Matlab, User's Guide Version 3. The Math Works, Inc., Natick Ma (1998)
5. Gallant, R.A., White H.: On Learning the Derivatives of an Unknown Mapping With Multilayer Feedforward Networks. Neural Networks 5 (1992) 129-138
6. Hornik, K., Stinchcombe, M., White, H.: Universal Approximation of an Unknown Mapping and Its Derivatives Using Multilayer Feedforward Networks. Neural Networks 3 (1990) 551-560
7. Masters, T.: Practical Neural Networks Recipes in C++. Academic Press, Inc. San Diego (1993)
8. Li, X.: Simultaneous approximations of multivariate functions and their derivatives by neural networks with one hidden layer. Neurocomputing 12 (1996), 327-343
9. Pohlheim, H., Documentation for Genetic and Evolutionary Algorithm Toolbox for use with Matlab (GEATbx): version 1.92, more information on http://www.geatbx.com (1999)
10. Simmons, G.F.: Differential equations with applications and historical notes. 2nd ed. McGraw-Hill, Inc., New York (1991)
11. Yao, X.: Evolving Artificial Neural Networks. Proceedings of the IEEE 87(9) (1999) 1423-1447

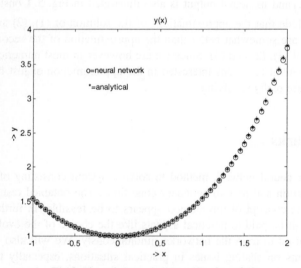

Fig.3. The solution of system (1), (2) and (3)

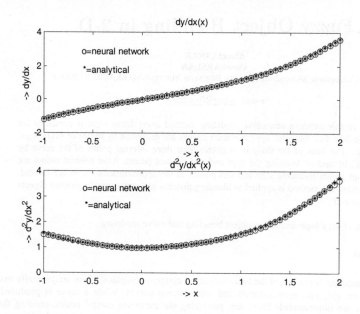

Fig.4. The first two derivatives of the solution of system (1), (2) and (3)

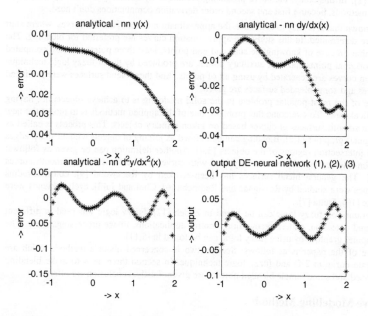

Fig. 5. The errors between the analytical solutions of (1), (2) and (3) and its first two derivatives on the one hand and the neural network method solution of (1), (2) and (3) and its first two derivatives on the other hand. Also the output of the DE-neural network of (1), (2) and (3) is illustrated

Fuzzy Object Blending in 2-D

Ahmet ÇINAR
Ahmet ARSLAN
Fırat Universitesi, Muhendislik Fakultesi Bilgisayar Muhendisligi ELAZIG, TURKEY

e-mail: acinar@firat.edu.tr

Abstract: In this paper, a new curve modeling method using fuzzy logic is presented for blending in 2-D. For this goal, approximation curves are produced by means of fuzzy logic techniques. The base of the study is to obtain least three internal points of the curve by fuzzy logic, in case of knowing the start and end control points. After internal points are obtained, all points including start and end points of the approximation curve are created. As application, the method is applied to blending process and some sample blended objects are given in 2-D.

Key words: Fuzzy logic technique, object blending and curve modeling

1 Introduction

The curve modeling is the one of the basic subject in computer graphics. There are generally two curves in computer graphics: approximation and interpolation curves. While a curve is produced, different methods are implemented. These are; producing the curve via control points, creating the curve to be satisfied C^1 continuous at starting and ending points and generating curve by knowing curvature radius [1]. In this study, the curve producing technique by control points is improved. It is easier than other methods, because first and second order derivation computations don't need.

It is well - known that it is easy to construct the approximation or interpolation curves where start and end points are determined. In this study, the approximation curves are preferred for blending. The basis of work is that: in case of knowing the start and end points, least three points, that are computed by the means of original points and their auxiliary points, are produced by the of fuzzy logic technique and approximation curves are generated by using all of points. And then, blend surfaces were produced by means of curves and some blended surfaces are given.

The one of the most popular problem in the solid modeling is to achieve objects not having certain mathematical form. To overcome this problem, one of the applied methods is to model the new objects by joining smooth surfaces or curves based on given primary objects. This process is called as blending on computer graphics [1,2]. Blending can be defined as joining of objects without sharp edges or corners in the intersection regions of the objects [3,4]. Another definition can be given as follows; intersecting or non-intersecting objects can be joined with curves and surfaces having smooth curves and surfaces [8]. The general blend surfaces has been worked by Rockwood [9], constant-radius blending techniques were studied by Rossignac and Requicha [6], Choi and Ju[3], cyclide blends were studied by Schene [10], Dutta [7].

The first literature on fuzzy logic can be found in Zadeh [12]. Fuzzy logic was used in different areas such as sound detection, pattern recognition, control applications, image processing,...., etc. The works about computer graphics by using fuzzy logic can be found in [5,11].

The structure of the paper is as follows; Section two is concerned about a method, which are developed to create curve in 2-D and fuzzy logic technique. In section three, how to make blending process is mentioned and some sample blended objects are given. Lastly, conclusion are offered.

2 Fuzzy Curve Modelling Method

The control points have to be known, while the approximation and interpolation curves are created by means of control points. In this study, when start and end points are known, the curve is produced by means of other points, which are calculated by these points. The basis of study is to compute the interval points by fuzzy logic technique, where start and end points are known. While that computation was done, at least three points except start and end points are used to create curve. The reason is that, local changes on the created curve must be able to satisfy at least three points. Figure 1 depicts this effect. Let $A(u, v)$ and $B(u, v)$ be start and end points given in 2-D. Infinite curves can be created by means of these two points. If a controlled curve is wanted to be created, then the location of auxiliary points (second points) is important.

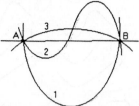

Figure 1. $A(u,v)$, $B(u,v)$ are two points and infinite curves can be produced by means of these points.

So, some parameters will be needed to create a controlled curve as well. Therefore, the shape of curve can be satisfied. It is required to use different control points to create curve 1 on Figure 1. Similarly, curve 2 and 3 require different control points. To compute location of these points by fuzzy logic, we use points called as auxiliary points. The reason of using auxiliary points is that, the control mechanism, that will satisfy local changing on the curve, is to obtain. These points will determine shape of curve and set the basis of blending.

2.1 Developed Method

Let $A(u,v)$ and $B(u,v)$ be two points in 2-D. Infinite curves can pass from these two points. Let us define two other points to create curve that will satisfy property to be requested and passing from these points. Let $A'(u,v)$ and $B'(u,v)$ be two auxiliary points. Auxiliary points are points at distance δ of original $A(u,v)$ and $B(u,v)$ points. Let δ be very small such as $\delta << 1$. Let α and β be two angles. (Figure 2.)

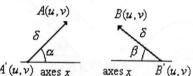

Figure 2. Case of $A(u,v)$, $B(u,v)$, $A'(u,v)$, and $B'(u,v)$ points

There may be different cases on location of points. Because angles can change interval $[0^0, 360^0]$. Figure 3 depicts created curves for different α and β values.

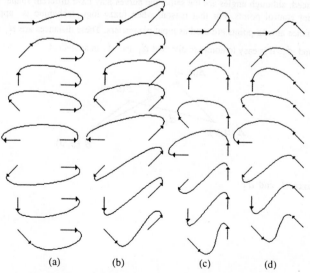

(a)	(b)	(c)	(d)

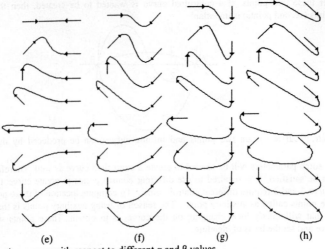

(e)   (f)   (g)   (h)

Figure 3. Creating curves with respect to different α and β values.

Having α and β determined, it is not enough to take only these angles to be able to use fuzzy logic technique. The reason is that, if the distance among points is not determined and only angle is evaluated then, undesired cases may occur. So, the curves can not be produced. This case was graphically depicted in figure 3.

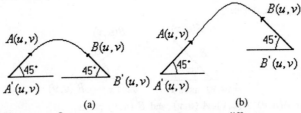

(a)   (b)

Figure 4. The angles α and β are same, but the produced curves are different.

If noticed, although angles are the same, the curves may have different shape. Because curves require different control points. For that reason, when fuzzy logic technique is applied to, angles α, β and distance among points are used as input parameters. These distances are d_1 and d_2. Figure 5 depicts d_1 and d_2. It is easy to compute distance d_1 and d_2 in analytical.

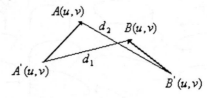

Figure 5. Distance d_1 and d_2.

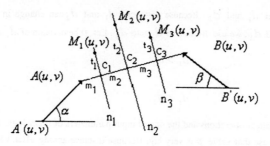

Figure 6. Cases of m_1, m_2, m_3, t_1, t_2 and t_3.

If it is noticed to figure 6, the points $M_1(u,v)$, $M_2(u,v)$ and $M_3(u,v)$ are points to be computed. While locations of these points are computed, lines n_1, n_2 and n_3 are used. These lines are located as perpendicular to the line connecting the points $A(u,v)$ and $B(u,v)$. The distances m_1, m_2, m_3 and n_1, n_2 and t_1, t_2 and t_3 are important. Simply, distances m_1, m_2, m_3 and n_1, n_2, n_3 are computed by means of fuzzy logic technique. If these distances and points C_1, C_2, C_3 are determined then, to compute exact points $M_1(u,v)$, $M_2(u,v)$ and $M_3(u,v)$ can be computed in fuzzy inference machine. Center of sums is used as inference machine.(Equation 1).

$\mu(u_i)$: membership functions.

u_i: The value of membership functions.

$$u^* = \frac{\sum_{i=1}^{l} u_i \cdot \sum_{k=1}^{n} \mu(u_i)}{\sum_{i=1}^{l} \sum_{k=1}^{n} \mu(u_i)} \qquad (1)$$

The fourth degree approximation curve can be created by equation (2). The $M_1(u,v)$, $M_2(u,v)$ and $M_3(u,v)$ points are intermediate points computing by the means of developed method.

$$P(u,v) = (1-u)^4 A(u,v) + 4(1-u)^3 M_1(u,v) +$$
$$6(1-u)^2 u^2 M_2(u,v) + 4u^3(1-u)M_3(u,v) + u^4 B(u,v) \qquad (2)$$

Figure 7 shows curves produced by equation (2).

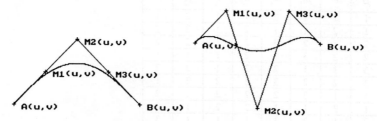

Figure 7. Curves produced by equation (2)

2.2 Rules for Fuzzy Logic and Construction of Membership Functions.

In this section, membership functions selected by the means of parameters determined in section 2.1 and produced rule sets by these functions are presented. It is known that, the parameters that will be evaluated for fuzzy logic technique are angle α, β and distances d_1, d_2. However, there is a problem to select d_1 and d_2. Because distances d_1 and d_2 can change in a very large interval. Instead of d_1 and d_2, we use division of d_1 into d_2. Let p be division of d_1 into d_2.

$$p = \frac{d_1}{d_2} \qquad\qquad d_1 > d_2$$

$$p = \frac{d_2}{d_1} \qquad\qquad d_1 < d_2 \qquad\qquad (3)$$

So, both mathematical operations and the one of input parameters of fuzzy logic are decreased. Noticed that, there is no case that value p is very big. Because distance among $A(u,v)$ and $A'(u,v)$ is very small and distance among $B(u,v)$ and $B'(u,v)$ is very small, too. For that reason, distance d_1 is almost equal to distance d_2, and so, it will be selected in interval $[0-1.25]$ of p. Figure 7 shows membership functions used as input parameters and figure 8 shows membership functions selected as output parameters.

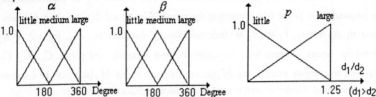

Figure 8. Membership functions used as input parameters.

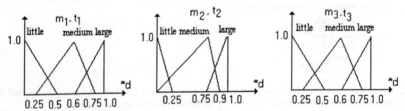

Figure 9. Membership functions used as output parameters

Table 1. Fuzzy control rules

Inputs			Outputs					
α	β	p	m_1	m_2	m_3	t_1	t_2	t_3
1	1	1	1	2	1	1	2	1
1	1	2	1	1	2	1	2	2
1	2	1	2	3	2	1	2	2
1	2	2	3	1	2	2	1	1
1	3	1	1	2	2	1	3	2
1	3	2	2	3	3	2	3	1
2	1	1	3	2	2	2	1	2
2	1	2	1	1	2	2	1	2
2	2	1	2	3	1	1	1	3
2	2	2	1	2	2	2	2	2
2	3	1	2	2	1	1	2	3
2	3	2	2	1	2	3	1	2
3	1	1	3	2	3	3	1	2
3	1	2	2	1	3	2	3	1
3	2	1	1	2	2	1	2	2
3	2	2	1	2	2	1	3	2
3	3	1	3	1	1	2	1	1
3	3	2	1	3	1	1	2	1

Note 1. For α and β; 1, 2 and 3 values show little, medium and large, respectively.

    2. For p; the 1, 2 values show little and large, respectively.

    3. For m_1, m_2, m_3 and t_1, t_2 t_3; the 1 2, 3 values show little, medium and large, respectively.

3 Object Blending in 2-D

Applying the method presented in section 2 to blending process require to know exactly some parameters and points $A(u,v)$, $A^{'}(u,v)$ and $B(u,v)$, $B^{'}(u,v)$. The method can apply to blending process, after these points and parameters are known.

Let $f(u,v)$ and $g(u,v)$ be two surfaces in 2-D and let $f^{'}(u,v)$ and $g^{'}(u,v)$ be their offset surfaces. It is known that an offset surface is a translated surface at distance d to the surface normal direction (Figure 10).

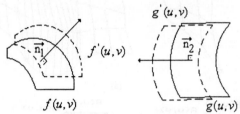

Figure 10. Original surfaces; $f(u,v)$ and $g(u,v)$, their offset surfaces; $f^{'}(u,v)$ and $g^{'}(u,v)$ and surface normals $\vec{n_1}$ and $\vec{n_2}$.

According to developed method, it is not important whether surfaces intersect or not.

$$f(u,v) - g^{'}(u,v) = 0 \qquad (4)$$

$$f^{'}(u,v) - g(u,v) = 0 \qquad (5)$$

If equation (4) and (5) is solved by means of Newton-Raphson method, then intersection curves, namely points $A(u,v)$ and $B(u,v)$, can be found. By the help of these points, $A^{'}(u,v)$ and $B^{'}(u,v)$ can be computed. For this purpose, offset surfaces are translated to distance δ. Hence, the points $A^{'}(u,v)$ and $B^{'}(u,v)$ are obtained. However, δ must be very little. Because δ gives only the information on shape of original surface $f(u,v)$ and $g(u,v)$.

δ should be very little value. Because δ must be able to give an information about shape of original surfaces $f(u,v)$ and $g(u,v)$. If it is selected big value, then it won't give an information shape of original surfaces. Therefore, while blending process is done, blending process having smooth curves and surfaces can not realized. For that reason, δ must be little enough.

In this study, δ is selected in range $0.0001 < \delta < 0.1$, especially for $\delta = 0.02$. Figure 11 shows the computation steps.

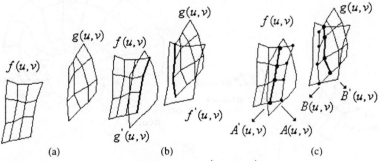

(a)   (b)   (c)

Figure 11. Computing of points $A(u,v)$, $B(u,v)$ and $A^{'}(u,v)$, $B^{'}(u,v)$.

After the points $A(u,v)$, $B(u,v)$ and $A^{'}(u,v)$, $B^{'}(u,v)$ are computed, three points should be computed. At the beginning, $f(u,v)$ and $g(u,v)$ are translated toward to each for other by their surface normal. This process is maintained by arising intersection. Intersection curves namely $A^{'}(u,v)$ and $B^{'}(u,v)$ are computed by means of equations 4 and 5. It can be solved by the help of Newton—Raphson method.

Figure 12 depicts some blended objects and their blending points. The case before blending process is on the left side, and the case after blending process is on the right side.

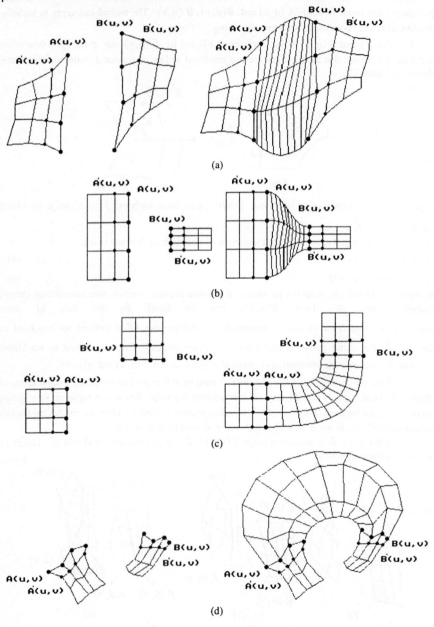

(a)

(b)

(c)

(d)

Figure 12.Some blended objects produced by means of developed method and their blending points.

4 Conclusion

In this paper, a new technique to create an approximation curve has been presented. The basic of the developed method is to create an approximation curve by the help of fuzzy logic technique. Produced curves have been used for blending in 2-D. As a future work, the method will be tried to be applied on blending in 3-D.

5 References

[1] **Ahmet ÇINAR , Ahmet ARSLAN**, "Two Approaches On Blending", ISCIS XI The Eleventh International Symposium On Computer and Information Sciences, November 6-8 Antalya, Türkiye, 1996.

[2] **Farin G.**, Curves and Surfaces for Computer Aided Geometric Design, Academic Press, Boston, 1990.

[3] **Choi K.B. , Ju Y. S.**, 'Constant - Radius blending in surface modelling', Computer Aided Design', Vol . 21 , No . 4 , May, 1989.

[4] **Filkins P.C., Tuohy S.T., Patrikalakis M. N.,** 'Computational methods for blending approximations' ,Engineering With Computer' , 9 , pp 49 - 62, 1993.

[5] **Gao Lun Shan, Kawarada Hideo,** ' Applications of fuzzy average to curve and surfaces fitting', 2nd Fuzzy Eng. Sym.IEEE Inter. Con. on Fuzzy systems, Vol. 2, p 971-978 , 1995.

[6] **Rossignac J. R., Requicha A. G.** 'Constant radius blending in solid modeling', 'Comput. . Mech . Eng . , July , pp 65 – 73, 1984.

[7] **Srinivas L. V. , Dutta Debasisih,** 'Blending and joining using cyclides', ASME Advances In Design Automation, Vol. 2, pp 231-236, 1992.

[8] **Jung-Hong Chuang, Wei-Chung Wang,** Variable-radius blending by constrained spline generation, Visual Computer, Vol: 13, N: 7, 316-329, 1997.

[9] **Rockwood . A. P., Owen J. C.,** Blending surfaces in solid modelling, in: Farin G., ed., Geometric Modeling, Algorithms and New Trends, SIAM, Philadelphia, PA, 367-383, 1994.

[10] **Seth Allen, Debasish Dutta,** Cyclides in pure blending I, Computer Aided, Geometric Design, 14, 51-75, 1997.

[11] **Saga Sato, Makino Hiromi**, A Method for Modelling Freehand Curves-the Fuzzy Spline Interpolation, Systems and Computers in Japan , Vol: 26, No:19, pp:1610-1619, 1995.

[12] **Zadeh L.** "Fuzzy sets ", Inf. Control, Vol:8, pp:338-353, 1965.

An Adaptive Neuro-Fuzzy Approach for Modeling and Control of Nonlinear Systems

Otman M. Ahtiwash and Mohd Zaki Abdulmuin

Center of Factory and Process Automation, Faculty of Engineering,
University of Malaya
50603 Kuala Lumpur, MALAYSIA.
ahtiwash@tm.net.my

Abstract. Fuzzy Inference Systems (FISs) and Artificial Neural Networks (ANNs), as two branches of Soft Computing Systems (SCSs) that pose a human-like inference and adaptation ability, have already proved their usefulness and have been found valuable for many applications [1],[2]. They share a common framework of trying to mimic the human way of thinking and provide an effective promising means of capturing the approximate, inexact nature of the real world process. In this paper we propose an Adaptive Neuro-Fuzzy Logic Control approach (ANFLC) based on the neural network learning capability and the fuzzy logic modeling ability. The approach combines the merits of the both systems, which can handle quantitative (numerical) and qualitative (linguistic) knowledge. The development of this system will be carried out in two phases: The first phase involves training a multi-layered Neuro-Emulator network (NE) for the forward dynamics of the plant to be controlled; the second phase involves on-line learning of the Neuro-Fuzzy Logic Controller (NFLC). Extensive simulation studies of nonlinear dynamic systems are carreid out to illustrate the effectiveness and applicability of the proposed scheme.

1 Introduction

In recent years, Fuzzy Inference Systems (FISs) and Artificial Neural Networks (ANNs) have attracted considerable attention as candidates for novel computational systems because of the variety of the advantages that they offer over conventional computational systems [1]-[5]. Unlike other classical control methods, Fuzzy Logic Control (FLC) and ANNs are more model free controllers, i.e. they do not require exact mathematical model of the controlled system. Moreover, they are becoming well-recognized tools of designing identifiers and controllers capable of perceiving the operating environment and imitating a human operator with high performance.

FLC has the strengths of linguistic control, parallelism, relaxation, flexibility, and robustness. But there has been no systematic approach in implementing the adaptive fuzzy control system. For example, the shape and location of membership function for each fuzzy variable must be obtained by trail-error or heuristic approach. Also, when an expert cannot easily express his knowledge or experience with the linguistic form of (*If-Then*) control rule, it is not easy to construct an efficient control rule base [6].

ANNs have the characteristics of high parallelism, fault-tolerance, and adaptive and learning abilities. But there exist some problems in the neural control; firstly, it is not easy to decide the optimal number of layers and neurons; secondly, the learning algorithm of the neural network has slow convergence speed and thirdly, the neural networks take numerical (quantitative) computations rather than symbolic or linguistic (qualitative) computations [7].

In order to overcome such problems, there have been considerable research efforts to integrate FLC and ANNs for developing what is known as neuro-fuzzy control systems [5]-[10]. The fusion of the two approaches, which can enlarge their individual strengths and overcome their drawbacks, will produce a powerful representation flexibility and numerical processing capability [11]-[13], [14]-[17].

In this paper we present another approach of an adaptive neuro-fuzzy logic control scheme (ANFLC) using the hybrid combination of fuzzy logic and neural networks. The proposed control scheme consists of a neuro-fuzzy logic controller (NFLC) and a neuro-emulator (NE). In the NFLC, the antecedent and consequent parts of the fuzzy rules are constructed using a multi-layered neural network with clustering method. The NFLC is trained to refine its parameters adaptively using error backpropagation learning algorithm (EBP). After constructing the adaptive neuro-fuzzy control system by NFLC and NE, the effectiveness of the proposed scheme will be demonstrated and evaluated by different nonlinear dynamic cases.

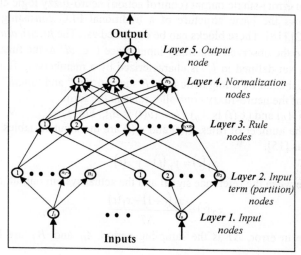

Fig. 1. Topology of the Neuro-Fuzzy model

2 Neuro-Fuzzy Logic Controller

Both the FLC and ANNs have been employed together to design a neuro-fuzzy logic controller (NFLC). A fuzzy system with learning ability has been constructed and is trained directly from the input-output data of the plant. Since the NFLC has the property of learning; membership functions and fuzzy rules of the controller can be tuned automatically by the learning algorithm [6], [13], [15]. Learning is based on the performance error, which is evaluated by comparing the process output with the desired or required output.

The NFLC presented here is based on a self-learning FLC. The learning method is basically a special form of the error backpropagation (EPB), which is used for training ANNs. To train the controller, the EBP method is employed to propagate the plant output error signal through different stages in time.

The NFLC architecture is composed of five layers as shown in Fig. 1, where the layers are functionally described as: the input layer (L_1), the fuzzy partition layer (L_2), the firing strength layer (L_3), the normalized firing strength layer (L_4) and the output layer (L_5). The first four layers perform the fuzzy partition of the input space and construct the antecedent part while the last layer together with the weights and the results obtained by the partition carry out the implementation task of control and learning. This structure can update membership function and rule base parameters according to the gradient descent update procedure.

2.1 Fuzzy Elements of the Neuro-Fuzzy Logic Controller

Since a Multi-Input-Multi-Output (MIMO) system can always be separated into group of a Multi-Input-Single-Output (MISO) systems, we only consider a multi-input (error and change in error)-single output (control action) neuro-fuzzy logic controller here.

Fig.2 shows the basic structure of a traditional FLC consisting of four major blocks [1],[12],[18]. These blocks can be described as: The *fuzzification interface* is a mapping from the observed non-fuzzy input space $U \subset \mathcal{R}^n$ to the fuzzy set defined in U. The fuzzy set defined in U is characterized by a membership function $\mu_F : U \to [0,1]$, and is labelled by a linguistic term F such as "*big*" and "*small*". The *fuzzy rule base* is a set of the neuro-fuzzy controller rules in the form:

R_j: If $(x_1(t)$ is $A_{1j})$ and $(x_2(t)$ is $A_{2j})$ Then $(u(t)$ is $B_j)$, For $j = 1, \ldots, N$, and, $t = 1,2,\ldots$
Where N is the number of rules, $x_1(t)$ and $x_2(t)$ are the input variables to the NFLC at time t given as [15]:

$$x_1(t) = y_r(t) - y_p(t) \tag{1}$$

is the error between the reference signal and the actual system output, and

$$x_2(t) = \frac{x_1(t+1) - x_1(t)}{\Delta T} \tag{2}$$

is the change in error. ΔT is the sampling period, A_{ij} and B_j are linguistic terms characterized by the membership functions $\mu_{A_{ij}}(x_i(t))$ and $\mu_{B_j}(u(t))$ respectively.

Throughout this study, the A_{ij} uses the Gaussian shaped membership function, defined by [10],[17]:

$$\mu_{A_{ij}}(x_i(t)) = \exp\left\{-\left(\frac{(x_i(t) - c_{ij})}{\sigma_{ij}}\right)^2\right\} \tag{3}$$

The *fuzzy inference machine* is a decision making logic which employs fuzzy rules from the fuzzy rule base to determine the fuzzy outputs corresponding to its fuzzified inputs. Using the centroid defuzzification method, the *defuzzification interface* determines the non-fuzzy output of the neuro-fuzzy controller in the form [3]:

$$u(t) = \frac{\sum\limits_{j=1}^{N} \bar{y}_j \left(\mu_{A_{1j}}(x_1(t)) \, \mu_{A_{2j}}(x_2(t))\right)}{\sum\limits_{j=1}^{N} \left(\mu_{A_{1j}}(x_1(t)) \, \mu_{A_{2j}}(x_2(t))\right)} \tag{4}$$

In the previous equation, x_i is the input of node i, A_{ij} is linguistic label associated with this node, c_{ij} and σ_{ij} are adjustable real valued parameters representing the centre and width of the the Gaussian-shaped function; $\bar{y}_j$ is the point in the output space at which μ_{B_j} achieve its maximum.

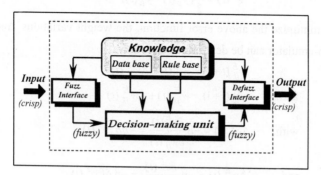

Fig. 2. General Structure of Fuzzy Systems

2.2 Neuro-Emulator Model

In this paper, we consider the dynamic system goverened by the following relationship [2],[4]:

$$y(t) = f\{y(t-1),...,y(t-n_y), u(t),...,u(t-n_u)\} \tag{5}$$

where, $y(t)$ and $u(t)$ are the system output and input repectively; n_y, n_u are the corresponding lags in the output and input, $f(\cdot)$ is a nonlinear function. The task is to approximate and generalize $f(\cdot)$ using the multi-layer neuro-emulator with,

$$X(t) = [y(t-1) \ldots y(t-n_y) \ u(t) u(t-n_u)]^T \tag{6}$$

We may formulate the problem using the multi-layer neuro-emulator with the input-output response

$$\hat{y}(t) = \hat{f}(X(t)) \tag{7}$$

Training is done by using EBP algorithm whose MISO relationship forward stage is

$$net_i = \sum_{j=1}^{n_1} w_{ij}^E x_j + b_i \tag{8}$$

and, $\quad o_i(t) = g(net_i) \tag{9}$

where, the superscript E stands for the neuro-emulator; w_{ij}^E and b_i are the hidden-input node connection weights and the threshold respectively, for $j = 1,...,n_1$ and $i =$

$1,...,n_2$; n_1 is the number of nodes in the input layer and n_2 is the number nodes in the hidden layer. Furthermore, $g(.)$ is the sigmoid activation function described as [4]:

$$g(\bullet) = \frac{1}{1+\exp(-(\bullet))} \tag{10}$$

$$y_E(t) = \sum_{j=1}^{n_2} w_{jk}^E \, o_j(t) \tag{11}$$

where, w_{jk}^E are the output-hidden node weights. The error function $e^E(t)$ utilized during the learning period can be defined as

$$e^E(t) = \frac{1}{2}(y_p(t) - y_E(t))^2 \tag{12}$$

In order to minimize the above error function, the weight variations Δw_{ij}^E and Δw_{jk}^E of the neuro-emulator can be determined as follows

$$w_{ij}^E(t+1) = w_{ij}^E(t) + \Delta w_{ij}^E(t) \tag{13}$$

and,

$$w_{jk}^E(t+1) = w_{jk}^E(t) + \Delta w_{jk}^E(t) \tag{14}$$

with,

$$\Delta w_{ij}^E(t) = -\eta \frac{\partial e^E(t)}{\partial w_{ij}^E(t)} = -\eta \delta_j^E x_j(t) \tag{15}$$

and,

$$\Delta w_{jk}^E(t) = -\eta \frac{\partial e^E(t)}{\partial w_{jk}^E(t)} = -\eta \delta_k^E o_j^E(t) \tag{16}$$

where,

$$\delta_k^E = y_p(t) - y_E(t) \tag{17}$$

and,

$$\delta_j^E = g'(X(t)) \sum_j \delta_k^E w_{jk}^E \tag{18}$$

where, $g'(\bullet)$ denotes the activation function derivation. After the NE has been trained to emulate the plant exactly, the plant output $y_p(t)$ is replaced with the NE output $y_E(t)$. Then the error signal δ^c of the NFLC can be obtained as follows:

$$\delta_k'^E = y_r(t) - y_E(t) \tag{19}$$

$$\delta_j'^E = g'(X(t)) \sum_j \delta_k'^E w_{jk}^E \tag{20}$$

Thus the performance error at the output of the NFLC can be obtained as [2],[14]:

$$\delta^c = \sum_j \delta_j'^E w_{jk}^E \tag{21}$$

where, the superscript c stands for the neuro-fuzzy controller.

3 Adaptive Neuro-Fuzzy Logic Control System

When there exist some variations in the internal or external environments of the controlled plant, it will be required for the controller to possess the adaptive ability to deal with such changes. Thus, in this section, an adaptive neuro-fuzzy logic control system (ANFLC) will be developed by using the NFLC described earlier in section 2.

But, when applying the previous NFLC, there are some difficulties in obtaining the performance error signal. To overcome this problem we use the neuro-emulator (NE) presented in the previous section, which can emulate the plant dynamics and backpropagate, the errors between the actual and desired outputs through the NE. Fig. 3 shows the proposed scheme constructed by the two input-single output NFLC and the NE, where k_1, k_2 and k_3 are the scaling gains for x_1, x_2 and u respectively.

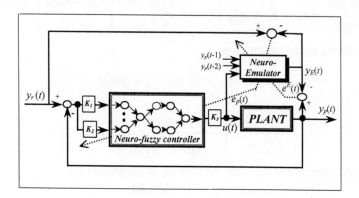

Fig. 3. The Proposed Adaptive Neuro-Fuzzy Logic Controller Scheme

3.1 Learning Mechanism

At each time step we adjust the parameters of the NE before updating the controller. For this purpose, the EBP training algorithm is used to minimize the performance error e_p as follows:

$$e_p(t) = \frac{1}{2}\{(y_r(t) - y_E(t))^2\} \tag{21}$$

From equation (4), we train $\bar{y}_j$ as follows [6],[9]:

$$\bar{y}_j(t) = \bar{y}_j(t-1) + \Delta\bar{y}_j(t) \tag{22}$$

$$\Delta\bar{y}_j(t) = -\eta \frac{\partial e_p(t)}{\partial c_{ij}(t)} + \alpha \Delta\bar{y}_j(t-1) \tag{23}$$

where, η is the learning rate and α is the constant momentum term. Using the chain rule, we have:

$$\frac{\partial e_p(t)}{\partial \bar{y}_j(t)} = e_p(t) \frac{\mu_j}{\sum_j^N \mu_j} \sum_{i=1}^{n_2} w_{ij}^c w_{jk}^c g'(net_i) \tag{24}$$

Thus the training algorithm for $\bar{y}_j$

$$\bar{y}_j(t) = \bar{y}_j(t-1) - \eta \, [e_p(t) \frac{\mu_j}{\sum_j^N \mu_j} \sum_{i=1}^{n_2} w_{ij}^c w_{jk}^c g'(net_i)] + \alpha \, \Delta\bar{y}_j(t-1) \tag{25}$$

In order to train c_{ij} in (3) and (4), we use

$$c_{ij}(t) = c_{ij}(t-1) + \Delta c_{ij}(t) \tag{26}$$

$$\Delta c_{ij}(t) = -\eta \frac{\partial e_p(t)}{\partial c_{ij}(t)} + \alpha \, \Delta c_{ij}(t-1) \tag{27}$$

Again, using the chain rule, we have

$$\frac{\partial e_p(t)}{\partial c_{ij}(t)} = e_p(t) \sum_{i=1}^{n_2} w_{ij}^c \, w_{jk}^c \, g'(net_i) \left(\frac{\bar{y}_j(t) - u(t)}{\sum_{j=1}^{N} \mu_j} \right) \left(\frac{x_i(t) - c_{ij}(t)}{(\sigma_{ij})^2} \right) \tag{28}$$

and the training algorithm for c_{ij} is

$$c_{ij}(t) = c_{ij}(t-1) - \eta \left[e_p(t) \sum_{i=1}^{n_2} w_{ij}^c \, w_{jk}^c \, g'(net_i) \right.$$

$$\left. \left(\frac{\bar{y}_j(t) - u(t)}{\sum_{j=1}^{N} \mu_j} \right) \left(\frac{x_i(t) - c_{ij}(t)}{(\sigma_{ij})^2} \right) \right] + \alpha \, \Delta c_{ij}(t-1) \tag{29}$$

By using the same above methods, we train σ_{ij} in (3) and (4) as:

$$\sigma_{ij}(t) = \sigma_{ij}(t-1) + \Delta \sigma_{ij}(t) \tag{30}$$

Thus the training algorithm for σ_{ij} is

$$\sigma_{ij}(t) = \sigma_{ij}(t-1) - \eta \left[e_p(t) \sum_{i=1}^{n_2} w_{ij}^c \, w_{jk}^c \, g'(net_i) \right.$$

$$\left. \left(\frac{\bar{y}_j(t) - u(t)}{\sum_{j=1}^{N} \mu_j} \right) \left(\frac{(x_i(t) - c_{ij}(t))^2}{(\sigma_{ij})^3} \right) \right] + \alpha \, \Delta \sigma_{ij}(t-1) \tag{31}$$

4   Simulations

In this work, the proposed model is examined using two different applications. Firstly, the well known example of a nonlinear dynamic system given in [6],[19] was used to test for the predicted plant output. This is governed by the following difference equation:

$$y(t+1) = a_1 \, y(t) + a_2 \, y(t-1) + f[u(t)] \, ;$$

The nonlinear function has the form:

$f(u) = 0.6 \sin(\pi u) + 0.3 \sin(3 \pi u) + 0.1 \sin(5 \pi u)$;   In order to predict the plant outputs, the following difference equation was used:

$$\hat{y}(k+1) = a_1 \, y(t) + a_2 \, y(t-1) + \hat{f}[u(t)] \, ;$$

where $a_1 = 0.3$ and $a_2 = 0.6$. For the prediction stage we select $\eta = 0.01$, $n_1 = 3$, $n_2 = 12$, and $\alpha = 0.7$. The parameters w_{ij} and w_{kj} are initialized randomly and uniformly distributed over [-0.5, 0.5]. Training data of 500 samples are generated from the plant model and used to train w_{ij} and w_{kj} in order to minimize the performance error at each time interval. Fig.4(a) shows that the model is approximated and converged to the plant output after only a few iterations. For the control stage, after the learning process is finished, the model is tested utilizing the same intilized parameters and the trained weights were being used to train NFLC in (25), (29) and (31) using (21). Fig.4(b) shows the improved results obtained using the combined scheme. Moreover, the adaptive capabilities and generalization abilities of the proposed scheme are further investigated by changing the input functions to: $f(u) = 0.3\cos(u(t)) + 0.25\sin(u(t))$ for $250 \le t \le 325$ and $f(u) = 0.45\sin(u(t))$ for $325 < t \le 500$, where $u(t) = \sin(2\pi t/100)$, the results are also shown in Fig. 4(b). the ANFLC model has a good match with the actual model with $E_p = 0.02638$, obviously, the proposed scheme can commendably identify the plant and assured its tracking performance.

Secondly, for further invetigations, the scheme was tested on a highly nonlinear system used in [20]:

$$y(t) = 0.8\sin(y(t-1)) + \frac{0.4y(t-1)\sin(y(t-2))}{1+y^2(t-2)} + 0.5u(t),$$

with the desired input sequence signal chosen as $[\sin(t/20) + \sin(t/50)]$.

After extensive testing and simulations, the ANFLC model proved a good performance in forecasting the output of the complex-dynamic plant, it has a good match with the actual model where the performance error minimized from $E_p = 0.355$ to $E_p = 0.00138$, The results of the prediction and control stages of this nonlinear system are presented in Fig. 5(a) and (b), respectively. Comparable performance to the first plant were obtained.

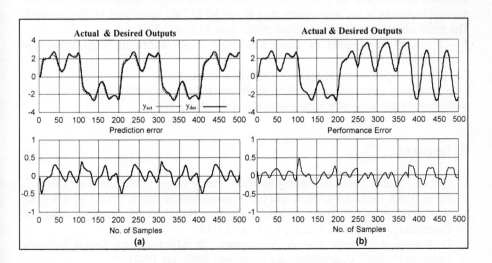

Fig. 4. Actual and Desired Outputs for 1st model: (a) For Prediction Stage: ➔ $E_p = 0.2310$; (b) For Control Stage: ➔ $E_p = 2.638e\text{-}02$

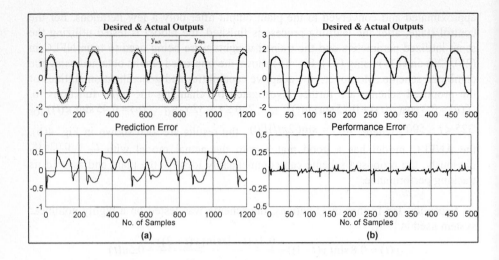

Fig. 5. Actual and Desired Outputs for 2nd model: (a) For Prediction Stage: ➔ $E_p = 0.355$; (b) For Control Stage: ➔ $E_p = 1.38e\text{-}03$

5   Conclusions

A flexible, adaptive neuro-fuzzy controller scheme (ANFLC) using the integration of FLC and ANNs has been proposed in this paper. The main features of the proposed control is that it does not require a reference model and it can be used to identify the unknown dynamics of the plant. The membership functions in the antecedent part and the real numbers in the consequent part of the NFLC are optimized by this method. The main advantages of the proposed model over FLC & ANNs are:

- A neural net mapping, of "Black Box" type, which is difficult to interpreted , is avoided.
- The tuning problem of fuzzy controllers is eliminated.

Two nonlinear examples are treated to demonstrate the potential applicability and usefulness of this approach in nonlinear dynamic processes.

References

1. C. Lee, "Fuzzy Logic in Control Systems: Fuzzy Logic Controller, Part I & II ", *IEEE Trans. on Systems, Man, and Cybernetics*, Vol.20, No.2, 1990, pp. 404-418.
2. V. Kasparian and C. Batur, "Model Reference Based Neural Network Adaptive Controller", *ISA Trans.*, Vol.37, 1998, pp. 21-39.
3. D. Psaltis; A. Sideris; and A. Yamamura, "A Multi-layered Neural Network Controller", *IEEE Control Systems Magazine*, 1988, pp. 17-21.
4. J. Tanomaru and S. Omatu, "Process Control by On-Line Trained Neural Controllers", *IEEE Transactions on Industrial Electronics*, Vol.39, No.6, 1992, pp. 511-521.

5. L-X. Wang, *"Adaptive Fuzzy Systems and Control: Design and Stability Analysis"*, Prentice Hall, NJ, 1994.
6. L-X. Wang and J. Mendel, "Back-propagation fuzzy system as nonlinear dynamic system identifiers", *IEEE International Conference on Fuzzy System.* 1992, pp. 1409 –1418.
7. C. Lin and C. Lee, "Neural-network-based fuzzy logic control and decision system", *IEEE Trans. Computers,* vol. 40, no. 12, 1991, pp. 1320-1336,
8. C-T. Lin; C-F Juang and C-P Li, "Water bath temperature control with a neural fuzzy inference network", *Fuzzy Sets And Systems,* Vol.111, Issue 2. 2000, pp. 285-306.
9. D. Kaur and B. Lin, "On the Design of Neural-Fuzzy Control System", *Inter. Journal of Intelligent Systems,* Vol.13, 1998, pp.11-26.
10. J. Kim, and N. Kasabov, "HyFIS: adaptive neuro-fuzzy inference systems and their application to nonlinear dynamical systems", *Neural Networks,* Vol. 12, Issue 9, 1999, pp. 1301-1319.
11. P. Dash; S. Panda; T. Lee; J. Xu and A. Routray, "Fuzzy and Neural Controllers for Dynamic Systems: an Overview", *IEEE Proceedings.* 1997, pp. 810-816.
12. R-J Jang and T. Sun, "Neuro-Fuzzy Modeling and Control", *Proceeding of the IEEE,* Vol.83, No.3, 1995, pp. 378-404.
13. S. Horikawa; T. Furahashi; and Y. Uchikawa, "On fuzzy modeling using fuzzy neural networks with the backpropagation algorithm", *IEEE Trans. on Neural Networks,* vol.3, 1992, pp. 801-806.
14. T. Hasegawa; S-I Horikawa; T. Furahashi and Y. Uchikawa, "An Application of Fuzzy Neural Networks to Design Adaptive Fuzzy Controllers", *Proceedings of International joint Conference on Neural Networks,* 1993, pp.1761-1764.
15. Y. Hayashi; E. Czogala and J. Buckley, "Fuzzy neural controller", *IEEE Intelligent. Conf. on Fuzzy Systems*, 1992, pp. 197-202.
16. Y. Shi & M. Mizumoto, "Some considerations on conventional neuro-fuzzy learning algorithms by gradient descent method", *Fuzzy Sets and Systems.* Vol.112, Issue 1, 2000, pp. 51-63.
17. Y. Wang and G. Rong, "A Self-Organizing Neural Network Based Fuzzy System", *Fuzzy Sets and Systems*, No.103, 1999, pp. 1-11.
18. P. Lindskog, *"Fuzzy Identification from a Grey Box Modeling Point of View"*, Technical Report, Department of Electrical Engineering, Linkoping University, Linkoping, Sweden, 1996.
19. K. S. Narendra and K. Parthsarathy, "Identification and control of dynamical systems using neural networks", *IEEE Trans. on Neural Networks,* Vol.1. No.1, 1990, pp. 4-27.
20. L. Pavel and M. Chelaru "Neural Fuzzy Architecture for Adaptive Control", *Proceedings of IEEE*, pp. 1115-1122. 1992.

The Match Fit Algorithm: A Testbed for the Computational Motivation of Attention

Joseph G. Billock[1], Demetri Psaltis[1], and Christof Koch[1]

California Institute of Technology
Pasadena, CA 91125, USA
billgr@sunoptics.caltech.edu

Abstract. We present an assessment of the performance of a new on-line bin packing algorithm, which can interpolate smoothly from the Next Fit to Best Fit algorithms, as well as encompassing a new class of heuristic which packs multiple blocks at once. The performance of this novel $O(n)$ on-line algorithm can be better than that of the Best Fit algorithm. The new algorithm runs about an order of magnitude slower than Next Fit, and about two orders of magnitude faster than Best Fit, on large sample problems. It can be tuned for optimality in performance by adjusting parameters which set its working memory usage, and exhibits a sharp threshold in this optimal parameter space as time constraint is varied. These optimality concerns provide a testbed for the investigation of the value of memory and attention-like properties to algorithms.

1 Introduction

The computational abilities of humans and computers are in many ways complementary. Computers are good at routine, serialized tasks where high degrees of precision are required. Humans are good at dealing with novel situations in cases where highly precise operation is less important. One contributing factor to this is the computational architecture used by computers and humans. Computer algorithms usually approach problems with the goal of seeking exact solutions, or at least solutions which are optimal in some sense. To do this, they use as much information about the problem domain as possible. Human computation, we argue, is bound by a different set of constraints. The computational architecture of the brain is set up in a way which sharply limits the amount of information reaching its planning areas. This bottleneck (which mandates the contents of "awareness") makes humans very good at generalizing, dealing with novel situations, and responding quickly, but less good at finding exact solutions.

We would like to explore the area where the demands placed on computer algorithms are more similar to those which the brain handles–situations where highly complex problems defy exact solutions, and where external time pressure forces rapid response– and investigate how computer algorithms deal with these constraints.

The bin packing problem is a promising testbed. It is a known NP hard problem ([2], [3]), and is very general, with applications to cutting stock, machine and job scheduling, parallel processing scheduling, FPGA layout, loading problems, and more ([4], [1]). In its most basic form, the problem is phrased thus:

Definition 1. *Given a set S of real numbers in $(0, 1]$, we wish to find the smallest possible number k such that there are k subsets of S, s_i, $i = 1..k$, with $s_i \cap s_j = \emptyset$ for $i \neq j$, $\cup s_i = S$, and $\forall s_i \sum \{s_i\} \leq 1$.*

One can think of this problem as an assortment of blocks of varying sizes being fit into bins of unit size. The goal is to fit the blocks into as few bins as possible without overfilling them. The elegance of the bin-packing problem has attracted much attention. Since finding exact solutions for NP problems is computationally intractable, researchers have generally attempted to find heuristics which perform well, and to analyze this performance. A large number of heuristic approaches have been suggested. These can be classified into on-line and meta-heuristic approaches. The on-line approaches (such as the Best Fit algorithm) are in general much, much faster than the meta-heuristic approaches (such as genetic algorithms or simulated annealing). In this paper, we will present and characterize a new online heuristic for the bin packing problem.

The algorithm we describe is motivated by some considerations on the Best Fit algorithm. Given that the worst case performance ratio of on-line algorithms is quite low, from 1.7 for Best Fit to 2 for Next Fit ([5], [6]), there is not much room for algorithms which perform dramatically better than these. When we consider a time-constrained problem, however, where algorithms are under time pressure to produce the best possible packing, then we can consider algorithms which have comparable or slightly improved performance than Best Fit (BF), but which perform at speeds nearer to that of Next Fit (NF). The inspiration for our algorithm is a characteristic of the way in which Best Fit treats bins in its interim solutions. Bins below a certain level can be thought of as "in progress," that is, actively being used by the algorithm to pack new blocks. Above a certain bin level, however, the flow of incoming bins is enough to make the algorithm unlikely to work any more with a particular bin at that level. Furthermore, the number of "in progress" bins remains quite constant even for very large problems. We have designed an algorithm which operates in linear time (as Next Fit), but which uses approximately those bins and blocks which Best Fit would use, and whose performance is thus very close to that of Best Fit.

2 Match Fit Algorithm

The operation of the algorithm (see Fig. 1) maintains in memory those bins which are actively involved in the solution of the problem. It does this by limiting the size of its "short-term memory" according to parameter specification, and taking full bins out of this short-term memory to keep within that bound. We have introduced a memory for blocks, as well. The algorithm matches blocks and bins from its memory, which is usually quite small compared to the problem as a whole. The criteria used for matching produces bins which are nearly full (and will not be re-examined). Thus, the algorithm takes advantage of the relatively

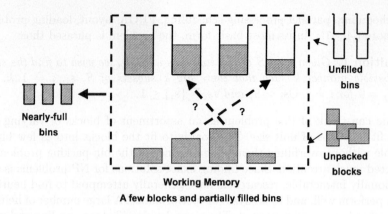

Working Memory
A few blocks and partially filled bins

Fig. 1. The Match Fit algorithm illustrated. The algorithm maintains a memory of bins and blocks. On each cycle it attempts to find matches between bin/block pairs which are within a small threshold of making a full bin. When it finds such a match, it removes the nearly-full bin from its memory and refreshes its memory from the queue of waiting blocks and/or new, empty bins. If no such match is found, blocks are put into bins in which they don't match exactly, but the bin is not removed from memory.

small number of bins in the "in progress" category for excellent time-performance even for large problems (the algorithm is $O(n)$) with a relatively small ($O(1)$) memory. After each iteration, the memory is replenished from any blocks left unpacked. The bin memory is replenished with empty bins. If no suitable matches in memory can be found, the algorithm forces placement of blocks into bins anyway, and then replenishes memory.

We have run Match Fit on the Falkenauer [7] test sets and compared its performance to Best Fit (Table 1).

	Best Fit	Match Fit 3 Bins, 3 Blocks	Match Fit 10 Blocks, 6 Bins	Match Fit unlimited memory
Performance Ratio (mean)	0.96	0.90	0.96	> 0.99

Table 1. Falkenauer data test comparison results

Statistically, Match Fit (MF) can outperform the Best Fit algorithm. For working memory sizes of only 10 blocks and 6 bins, or of 6 blocks and 8 bins, the average performance of Match Fit was 0.96, which is what Best Fit performs on the test set. For this test, the 1000 block probems were used, which is composed of problems with a uniform normalized block distribution on integers in [20,100] with bin size 150. The working memory size, then, is about 2% of the problem size for performance at parity with Best Fit. For very large working memory sizes (comparable to the size of the problem), Match Fit very often yeilds optimal solutions (which are known for these test problems), with an average performance of 0.995.

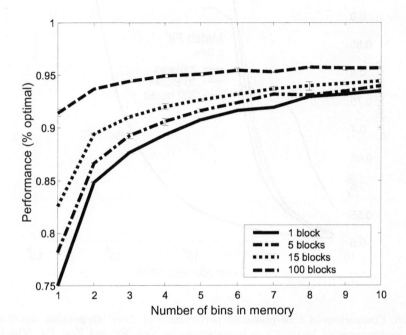

Fig. 2. Performance of MF algorithm on large test problem. 10,000 blocks; uniform distribution (0, 1] on size. The bins are of size unity. The performance increases as the working memory of the algorithm increases slowly towards the performance of Best Fit (which is 98%).

In Fig. 2, the algorithms performance on a very large problem (10,000 blocks; uniformly distributed in size) is shown. The early saturation of performance suggests that Match Fit will do well under time pressure. The reason is that since the MF algorithm takes a shorter time to operate, and still can produce competitive performance with BF, then when there is not much time to operate, the MF algorithm will be able to keep up and perform better than BF. The saturation is comparable when either bin or block memory is increased. Of the two, increasing block memory offers slightly better marginal performance. This suggests that bin packing algorithms which operate in very resource-limited environments would do well to expand the number of blocks they consider simultaneously alongside, or even before, they expand the number of partially-filled bins they consider.

3 Time-pressured performance characteristics

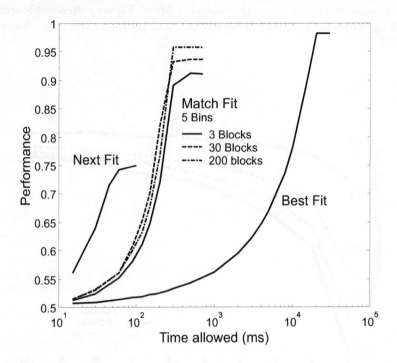

Fig. 3. Comparison of time-pressured performance of three bin-packing algorithms. The performance of Match-Fit is intermediate to Next Fit and Best Fit. The time allowed for the problem solution is shown in milliseconds, but will scale if a different processor is used for the problem, while maintaining the general shape of the curves. The problem is packing 10,000 blocks uniformly distributed in size over $(0, 1]$ into unity-sized bins.

Our primary focus in this investigation is the performance under time pressure of these various algorithms. By time pressure we mean the performance of the algorithm in a situation where the best solution is demanded from the algorithm after a particular length of time. The time constraint is enforced by an external controller, which allows the algorithms to run on a problem for a fixed amount of time, and then allocates any remaining blocks unpacked by the algorithm at one block per bin. This is equivalent to blocks passing a real-life packing machine operated by one of these algorithms. If the algorithm could not consider a particular block as it passed (that is, if the blocks passed too quickly) then that block would pass outside the working area of the machine and be placed into its own bin. Fig. 3 shows a performance comparison between the BF, NF, and a few configurations of the MF algorithm (with 5 bins and a varying numbers of blocks available to its working memory). The problems being solved by the algorithms are the packing of the 10,000-block problem (uniform (0,1] distribution on block size) into bins of size unity. As can be seen, the performance of the MF algorithm is intermediate to BF and NF performance.

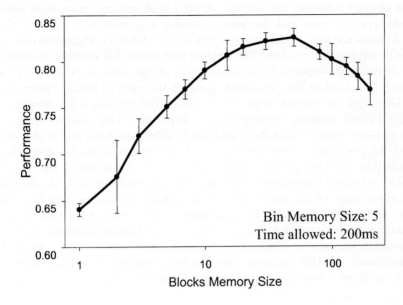

Fig. 4. Optimality in memory size for time-pressured MF algorithm. Under time pressure, there is a performance optimum for the MF algorithm. Too many items in memory slows the algorithm down too much, whereas with too few items, it does not perform as well. The problem is packing 10,000 blocks uniformly distributed in size over (0, 1] into unity-sized bins.

An examination of the performance characteristics for the MF algorithm indicates that when the algorithm is in its most interesting performance region in terms of its time-pressured performance–that is, performing well, but not yet at its optimum where it would be best to choose the most possible memory– there is an optimum in the amount of working memory the algorithm uses. This is shown more explicitly in Fig. 4. The optimal for this case (with 5 bins) is about 25-30 blocks in memory. When fewer blocks are used, the performance is less because the algorithm doesn't have as good a chance of finding good packings for blocks. When the memory uses more blocks, the performance also decreases, because although good packings are being found, it takes the algorithm longer to find them and runs out of time. In the case of the BF algorithm, which is equivalent to a limiting case of the MF algorithm where an almost unlimited bin memory is allowed (but using a single block in memory at a time), the packing is superior, but the time taken is roughly two orders of magnitude more.

There is another interesting effect in how the optimal strategy changes as the time pressure eases. Fig. 5 illustrates that there is a quite abrupt change in the optimal approach to solving the problem as the time pressure is slowly varied. In this figure, the memory parameters of the algorithm were varied widely (from 1 bin and 1 block to 200 blocks in memory and 20 bins in memory). For each value of time pressure, the various Match Fit algorithms using their memory parameters were run, and the best performer was examined. The top plot of Fig. 5 shows the best performance of any of the Match Fit algorithm configurations in solving the problem. The bottom plot shows the working memory size (the addition of blocks and bins in memory) at this optimal point. We observe a sharp threshold in the parameter space of the optimal configuration. Below this threshold, the optimal approach to solving the problem is for the algorithm to use a small working memory to best advantage. This remains true as the time pressure eases off and the algorithm is able to perform better and better. When the performance becomes close to its asymptotic limit, however, there is a transition. For time pressures less than this transitional value, the algorithm is better off to use basically as much memory as is available to it (the saturation in working memory size shows reflects the maximum sizes used in the simulations). Before the threshold, the performance curves exhibit the clear optimum we anticipate for a system solving a demanding problem in real-time: there is an optimum in the amount of resources it should dedicate to the task. As the time pressure eases off, this optimum becomes less pronounced, and the approaches which use more resources start to become attractive.

4 Conclusions

The Match Fit algorithm provides an interpolation between two of the most interesting on-line approaches to solving the bin packing problem. The bin packing problem has the charactistic, shared with many difficult problems, that only part of the problem is important at any one time, and neglecting much of the problem state in the functioning of the algorithm need not lead to great performance deficits. We then devised an algorithm which would take advantage of this idea by utilizing a user-defined amount of computational resources in its solution of the problem.

Since we are able to vary the size of the memory available to the MF algorithm smoothly, we can observe the impacts of strategies which are resource-intensive and those which do not require so many resources. When not under time pressure, we observe asymptotic performance behavior with respect to the amount of resources used. When the algorithm is under time pressure, however, there is an optimum in the computational resources used by the algorithm. This corresponds to the difference between the usual computer and human computational strengths. When there is less time pressure, and when exact solutions are wanted, the approach which uses the most information about the problem is favored. Under conditions where time pressures are important, it is best to

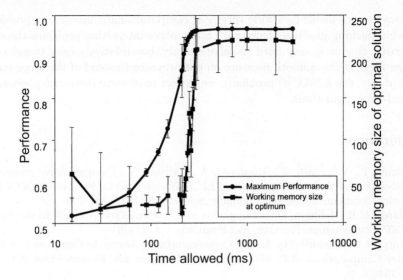

Fig. 5. The performance of MF as a funtion of time pressure and the corresponding sudden change in strategy needed to achieve optimal performance. For time pressures such that the algorithm cannot perform at its asymptotic level, the optimal strategy is to use a relatively small working memory. The transition between this regime-that where it is optimal to use a very small working memory and that where it is optimal to use a very large working memory-is extremely sharp. The error bars in the top plot show the standard deviation in performance of the memory configuration with the highest mean performance over 10 runs of the simulation. The error bars in the bottom plot indicate the standard deviations for those values of working memory size for which the performance at a given time pressure was ever the best in any simulation run, and so are quite pessimistic.

severely restrict the amount of information considered by the planning parts of the algorithm.

Under increasing time pressure, we observe that there is a very sharp threshold between the optimal performance of the two approaches. When time pressure is severe-that is, there is not enough time to quite get to near-asymptotic performance, it is advantageous to select a strategy which uses very few computational resources. When the time pressure is not so severe, it very quickly becomes advantageous to use very large (relative to the previous case) amounts of computational resources. The reason for this is that under intense time pressure, the system switches from performance being asymptotic in the amount of resources used to having an optimum. Instead of providing a performance boost, having more resources available simply "distracts" the algorithm and slows it down as it has to take time to take the extra information into account.

These two lessons-an optimum in the performance vs. resource utilization curve when the algorithm operates in time constrained environment, and the sudden transition from a limited-memory optimum to a large-memory optimum-we believe are extensible to a wide variety of computationally interesting problems. The bin packing problem shares with many other interesting problems the characteristic that it is very hard to solve exactly, but relatively easy to get close. When there is this smooth measure on performance (instead of the more binary case of, say, the k-SAT [8] problem), we expect to observe these two phenomena in real-time algorithms.

References

1. Izumi, T., Yokomaru, T., Takahashi, A., Kajitani, Y.: Computational complexity analysis of Set-Bin-Packing problem. IEICE Transactions on Fundamentals Of Electronics Communications and Computer Sciences: **5** (1998) 842–849
2. Garey, M.R., Johnson, D.S.: Computers and Intractability: A Guide to the Theory of NP-Completeness. Freeman, San Francisco, CA. (1979)
3. Karp, R.M.: Reducibility Among Combinatorial Problems. In Complexity of Computer Computations, R.E. Miller and J.W. Thatcher eds. Plenum Press, NY. 1972 85–104
4. Johnson, D.S.: Fast Algorithms for Bin-Packing. Journal of Computer Systems Science **8** (1974) 272–314
5. Johnson, D.S., Demers, A., Ullman, J.D., Garey, M.R., Graham,R.L.: Worst-Case Performance Bounds for Simple One-Dimensional Packing Algorithms. SIAM Journal of Computing **3** (1974) 299–326
6. Mao, W.: Tight Worst-case Performance Bounds for Next-k-Fit Bin Packing. SIAM Journal on Computing **22(1)** (1993) 46–56
7. Falkenauer, E.: A Hybrid Grouping Genetic Algorithm for Bin Packing. Working paper CRIF Industrial Management and Automation, CP 106 - P4, 50 av. F.D.Roosevelt, B-1050 Brussels, Belgium. (1996)
8. Monasson, R., Zecchina, R., Kirkpatrick, S., Selman, B., Troyansky, L.: Determining computational complexity from characteristic 'phase transitions'. Nature **400(6740)** (1999) 133–137

Automatic Implementation and Simulation of Qualitative Cognitive Maps[1]

João Paulo Carvalho   José Alberto Tomé

IST – Instituto Superior Técnico
INESC – Instituto de Engenharia de Sistemas e Computadores
R. Alves Redol, 9, 1000 LISBOA, PORTUGAL
joao.carvalho@inesc.pt   jose.tome@inesc.pt

Abstract: This short paper presents the overview of an ongoing project which goal is to obtain and simulate the dynamics of qualitative systems through the combination of the properties of Fuzzy Boolean Networks and Fuzzy Rule Based Cognitive Maps.

1. Introduction

Decision makers, whether they are social scientists, politicians or economists, usually face serious difficulties when approaching significant, real-world dynamic systems. Such systems are composed of a number of dynamic concepts interrelated in complex ways, usually including feedback links that propagate influences in complicated chains. Axelrod work on Cognitive Maps (CMs) [1] introduced a way to represent real-world qualitative dynamic systems, and several methods and tools [2,3] have been developed to analyze the structure of CMs. However, complete, efficient and practical mechanisms to analyze and predict the evolution of data in CMs are necessary but not yet available for several reasons [4]. System Dynamics tools like those developed by J.W.Forrester [5] could be a solution, but since in CMs numerical data may be uncertain or hard to come by, and the formulation of a mathematical model may be difficult, costly or even impossible, then efforts to introduce knowledge on these systems should rely on natural language arguments in the absence of formal models. Fuzzy Cognitive Maps (FCM), as introduced by Kosko [6], are an alternative approach to system dynamics. However, in most applications, a FCM is indeed a man-trained Neural Network that doesn't share the qualitative properties of other fuzzy systems [7], and the modeled system ends up being represented by a quantitative matrix without any qualitative knowledge. Therefore, we can say that there are currently no tools available to adequately represent the dynamics of qualitative systems.

Even if we had the tools to represent and analyze those systems, a different issue would be the building of a qualitative CM. The standard methodology for acquisition of the necessary data relies usually on lengthy processes of individual and simplistic

---

[1] This work is partially supported by the program POSI (under the EU 3rd framework program) and by the Portuguese Foundation for Science and Technology (FCT)

information collection: the result ends up being a list of quantitative relevant world concepts displayed as a matrix where weights represent the relations between those concepts [2,3,4,6]. Therefore we end up with quantitative CMs where any vestiges of a system with real-world rich qualitative entities and rich qualitative relations were simply ignored. Besides, there is the additional problem of combining the views of different analysts into one single map, which is usually done by simple math operations between the resulting matrixes.

This paper presents a general overview of an ongoing work that pretends to potentiate two original research lines to solve the problems presented above.

The first research line is in the potential use of really Qualitative Cognitive Maps as a tool not only to describe but also to simulate scenarios in qualitative real world systems like Social Conflicts, Political Science cases or Economic real-world problems. Fuzzy Rulebased Cognitive Maps (RB-FCM) [7,8,9,10] were introduced in previous works and are still being developed. They provide new Fuzzy Operators and a complete methodology for analysis and design of the dynamics of qualitative systems.

The second research line develops a new class of Boolean Neural Nets - Fuzzy Boolean Networks (FBN) [11,12,13] - which behave like fuzzy systems and are capable of non supervised learning of rules and of using the learned rules in qualitative reasoning.

2. Automatic Implementation and Simulation of Cognitive Maps

This paper purposes and introduces the overall architecture of AISCMap (Automatic Implementation and Simulation of Cognitive Maps). AISCMap should automatically generate a RB-FCM using FBN for the acquisition of the rules, concepts and relations that compose the map, allowing the simulation of what-if scenarios in the qualitative system that it models. The following sections provide a simplified description of AISCMap.

2.1 Inputs for AISCMAP

The inputs for AISCMap consist of the output provided by expert's analysis of the system to be modeled. Even after debate, each expert might have its own view of the problem, and a preferred way of representing the system. They can use natural language to express concepts and the rules relating them; they can provide examples of possible relations and their effect; they can define what is the type of relation involving two or more concepts; they can describe the system using matrixes, etc.

2.2 The Fuzzy Boolean Network Module

The FBN module extracts RB-FCM concepts, rules and relations from the expert raw data combining them into a single qualitative cognitive map. The map is described using RB-FCMSyntax, which is a language developed to provide a complete description of a RB-FCM.

Neural Networks are known for their capability in apprehending relationships among variables, but is also well known their difficulty in expliciting these relations into human comprehensive rules [15]. Fuzzy logic based systems, are very useful in explaining behaviors by a set of qualitative rules [14], but are not so good in which concerns learning of those rules. A synergetic cooperation between these two paradigms leads usually to hybrid systems that can learn and also explain the relationships by rules. However, these hybrid systems are usually an artificial engineered human work, where algorithms as well as fuzzy and neural components are "inserted" into the system. Although being based on common grounds, FBN are a new class of neural fuzzy systems that don't have many things in common with other neuro-fuzzy systems. Fuzzy qualitative reasoning is a natural and emergent property of Boolean networks[11,12]. These nets present many similarities with natural systems. They are capable of learning rules from non-supervised experiments in an Hebbian-like manner and they can automatically adapt to the granularity[13] of the antecedent variables and select the relevant ones. FBN are Boolean Nets with Macroscopic Emergent Fuzzy Reasoning, which allows them to extract qualitative rules from experimental data [11,12]. Here are some of the main FBN features:

- Variables/Concepts are associated with neural areas
- Values are given by neural activation
- Local Random Neural Connections
- Macro Structured (Links connecting Areas)
- Hebbian like Nonsupervised Learning by experiments
- Embedded Variable Selection and Granular adaptation

2.3 From FBN to RB-FCM

In AISCMap it is necessary to extend the working principles of the FBN to the problem of RB-FCM. This involves the adequate definition of the relationship between concepts and structure of the Maps and the neural areas and sub-areas as defined on the FBN.

2.4 Representing and Modeling Real World Qualitative Dynamic Systems

The RB-FCM allow a representation of complex real-world dynamic qualitative systems and the simulation of events and their influence in the system. RB-FCM are essentially composed of fuzzy Concepts and fuzzy Relations. Concepts are fuzzy variables that represent the value or the change in Actors, or Social, Political, Economical or Abstract entities that compose the system we are trying to model [7,8,9,10].

Relations between concepts are defined using fuzzy rule bases and can be of several types: Fuzzy Causal; Fuzzy Influence; Possibilistic; Invariant and Time variant Probabilistic; Possibilistic and Probabilistic; Fuzzy Similarity; Level/Change [10].

RB-FCM are iterative and provide mechanisms to simulate the dynamics of the system it represents (including fuzzy mechanisms to simulate time, which is not a given issue when relations are not time dependent mathematical equations as in system dynamics tools). The evolution of the system through time might converge in a single state or a cycle of states under certain conditions [7].

One of the main important uses for a dynamic model of a system, is the analysis of "WHAT-IF" scenarios ("What happens to the system when some event occurs?"), since the change of state in one or more concepts or relations affect the system in ways that are usually difficult or impossible to predict due to complex feedback links. RB-FCM are well adapted to these analysis, since introduction or removal of concepts and/or relations are possible and easily done without the hassle of the exponential increase of rules usually associated with fuzzy rule based systems [7,9].

3. Conclusions

AISCMAP is an ongoing work that is the result of a synergic approach of two different areas. Most of its modules are on an advanced stage of evolution, and the obtained results are promising, but there is still a lot of work to be done especially on the interaction between the expert results and the inputs to the system; and on the interaction FBN/RB-FCM.

4. References

[1]Axelrod,R.,The Structure of Decision: Cognitive Maps of Political Elites, Princeton University Press, 1976

[2]Laukkanen,M.,Comparative Cause Mapping of Management Cognition:A computer Database For Natural Data,Helsinki School of Economics and Business Publications,1992

[3]Decision Explorer, Banxia Software

[4]Laukkanen,M.Conducting Causal Mapping Research:Opportunities and Challenges, in Managerial and Organisational Cognition,edited by Eden, C.and Spender, Sage,1998

[5]Forrester, J.W., several papers available in http://sysdyn.mit.edu/sd-intro/home.html

[6]Kosko,B., "Fuzzy Cognitive Maps", Int. Journal of Man-Machine Studies, 1986

[7]Carvalho,J.P.,Tomé, J.A.,"Rule Based Fuzzy Cognitive Maps and Fuzzy Cognitive Maps - A Comparative Study", Proceedings of the 18th International Conference of the North American Fuzzy Information Processing Society, NAFIPS99, New York

[8]Carvalho, J.P. , Tomé, J.A.,"Rule Based Fuzzy Cognitive Maps- Fuzzy Causal Relations", Computational Intelligence for Modelling, Control and Automation, Edited by M. Mohammadian, 1999

[9]Carvalho,J.P.,Tomé,J.A., "Fuzzy Mechanisms For Causal Relations", Proceedings of the Eighth International Fuzzy Systems Association World Congress, IFSA'99, Taiwan

[10]Carvalho,J.P.,Tomé, J.A.,"Rule Based Fuzzy Cognitive Maps – Qualitative Systems Dynamics", Proceedings of the 19th International Conference of the North American Fuzzy Information Processing Society, NAFIPS2000, Atlanta

[11]Tomé, J.A., "Counting Boolean Networks are Universal Approximators", Proceedings of the 1998 Conference of the North American Fuzzy Information Processing Society, Florida

[12]Tomé,J.A., "Neural Activation ratio based Fuzzy Reasoning", Proceedings of the 1998 IEEE World Congress on Computational Inteligence, Anchorage, Alaska

[13]Tomé, J.A., "Automatic Variable Selection and Granular Adaptation in Fuzzy Boolean Nets", Proceedings of the 18th International Conference of the North American Fuzzy Information Processing Society, NAFIPS99, New York

[14]Mamdani,E.and Gaines,B.,Fuzzy "Reasoning and its Applications".London: Academic Press, 1981.

[15]Rumelhart,D.,G.Hinton,R.Williams "Learning Internal representations by error propagation" Parallel Distributed Processing,volI,chap.8.Cambridge,MA,MIT Press, 1986

Inclusion–Based Approximate Reasoning

Chris Cornelis[1] and Etienne E. Kerre[1]

Department of Mathematics and Computer Science, Ghent University
Fuzziness and Uncertainty Modelling Research Unit
Krijgslaan 281 (S9), B-9000 Gent, Belgium
{chris.cornelis,etienne.kerre}@rug.ac.be
WWW homepage: http://allserv.rug.ac.be/~ekerre

Abstract. We present a kind of fuzzy reasoning, dependent on a measure of *fulfilment* of the antecedent clause, that captures all the expressiveness of the traditional approximate reasoning methodology based on the Compositional Rule of Inference (CRI) and at the same time rules out a good deal of its inherent complexity. We also argue why this approach offers a more genuine solution to the implementation of analogical reasoning than the classically proposed similarity measures.

1 Introduction and Preliminaries

Reasoning with imprecise information expressed as fuzzy sets has received an enormous amount of attention over the last 30 years. More specifically, researchers have undertaken various attempts to model the following reasoning scheme (an extension of the modus ponens logical deduction rule), known as Generalized Modus Ponens (GMP):

$$\textbf{IF } X \text{ is } A \textbf{ THEN } Y \text{ is } B \quad (1)$$
$$\frac{X \text{ is } A' \quad\quad\quad\quad\quad\quad (2)}{Y \text{ is } B' \quad (3)}$$

where X and Y are assumed to be variables taking values in the respective universes U and V; furthermore $A, A' \in \mathcal{F}(U)$ and $B, B' \in \mathcal{F}(V)$[1].

Zadeh suggested to model the if–then rule (1) as a fuzzy relation R (a fuzzy set on $U \times V$) and to apply the Compositional Rule of Inference (CRI) to yield an inference about Y. The CRI is the following inference pattern:

$$\frac{X \text{ and } Y \text{ are } R}{\frac{X \text{ is } A'}{Y \text{ is } R \circ_T A'}}$$

where $\circ_T$ represents the fuzzy composition of R and A' by the t–norm[2] T, i. e. for every $v \in V$ we have:

---

[1] By $\mathcal{F}(U)$ we denote all fuzzy sets in a universe U.
[2] A t–norm is any symmetric, associative, increasing $[0,1] \times [0,1] \to [0,1]$ mapping T satisfying $T(1,x) = x$ for every $x \in [0,1]$

$$B'(v) = \sup_{u \in U} T(A'(u), R(u,v)) \ . \tag{1}$$

We will refer to the above approach as CRI–GMP, i.e. an implementation of GMP by CRI.

It is obvious that in order for CRI–GMP inference to be reasonable, R and T can not be just any combination of a fuzzy relation and a t–norm. One important way of determining the soundness of an inference strategy is to check whether it complies with certain desirable properties of inference in classical logic. For example, we might impose the following (non-exhaustive) list of criteria:

A.1 $B \subseteq B'$ (nothing better than B can be inferred)
A.2 $A'_1 \subseteq A'_2 \Rightarrow B'_1 \subseteq B'_2$ (monotonicity)
A.3 $A' = A \Rightarrow B' = B$ (compatibility with modus ponens)
A.4 $A' \subseteq A \Rightarrow B' = B$ (fulfilment of A implies fulfilment of B)

The first three are all standard in the approximate reasoning literature (see e.g. [1] [6]); we have added the fourth[3] as a straightforward extension of a commonsense principle that stems from the following intuition: the antecedent of a rule represents a restriction enforced on a variable, and the fulfilment of this restriction (this happens whenever $A' \subseteq A$) is a *sufficient* condition for the fulfilment of the consequent. This is exemplified by the crisp rule "Any Belgian older than 18 is entitled to get his driving licence". To conclude that a person may apply for a licence, it suffices to know that his age is restricted to belong to a *subset* of $[18, +\infty[$.

Noticing the link with classical logic, authors have proposed various fuzzifications of the implication operator to model R. A very general definition of a so-called implicator is the following:

Definition 1.1. (Implicator) *Any $[0,1]^2 \to [0,1]$–operator $\mathcal{I}$ of which the first and second partial mappings are decreasing, respectively increasing, and so that $\mathcal{I}(0,0) = \mathcal{I}(0,1) = \mathcal{I}(1,1) = 1$ and $\mathcal{I}(1,0) = 0$ is called an implicator.*

Table 1 lists some common implicators. After choosing an implicator $\mathcal{I}$, we put $R(u,v) = \mathcal{I}(A(u), B(v))$ for all $(u,v) \in U \times V$.

The suitability of a given $(T, \mathcal{I})$ pair to implement the CRI–GMP can then be evaluated with respect to the listed criteria; extensive studies exist on this issue (see e.g. [6]).

One particularly unfortunate aspect of the CRI is its high complexity. In general, for finite universes U and V so that $|U| = m$ and $|V| = n$, an inference requires $O(mn)$ operations. Mainly for this reason, researchers have explored

---

[3] Strictly speaking, A.4 is superfluous, as it is a direct consequence of A.1, A.2 and A.3 combined. We explicitly mention it here because of its intuitive logical appeal.

Table 1. Implicators on the unit interval ($(x, y) \in [0,1]^2$)

Symbol	Name	Definition
$\mathcal{I}_m$	Zadeh	$\mathcal{I}_m(x,y) = max(1-x, min(x,y))$
$\mathcal{I}_a$	Łukasiewicz	$\mathcal{I}_a(x,y) = min(1, 1-x+y)$
$\mathcal{I}_g$	Gödel	$\mathcal{I}_g(x,y) = \begin{cases} 1 & \text{if } x \leq y \\ y & \text{otherwise} \end{cases}$
$\mathcal{I}_b$	Kleene–Dienes	$\mathcal{I}_b(x,y) = max(1-x, y)$
$\mathcal{I}_\Delta$	Goguen	$\mathcal{I}_\Delta(x,y) = \begin{cases} 1 & \text{if } x \leq y \\ \frac{y}{x} & \text{otherwise} \end{cases}$

other ways of performing fuzzy inference while preserving its useful characteristics. In the next sections, we will propose an alternative based on a fuzzification of crisp inclusion, and argue why it is better than the existing methods.

2 Inclusion–Based Approach

From the criteria listed in the introduction, it is clear that when the observation A' of X is a subset of A, the restriction on Y should be exactly B by A.4. On the other hand, as A' moves away from ("is less included in", "less fulfills") A, by A.1 the imposed restriction on Y can only get weaker, meaning that we will be able to infer less and less information. Bearing in mind the close relationship between fulfilment and inclusion, we might capture this behaviour provided we can somehow measure the degree of inclusion of A' into A.

Indeed, if we have such a measure (say $Inc(A', A)$) at our disposal, we can use it to transform the consequent fuzzy set B into an appropriate B'. Schematically, this amounts to the following:

$$\frac{\textbf{IF } X \text{ is } A \textbf{ THEN } Y \text{ is } B \\ X \text{ is } A'}{Y \text{ is } f(Inc(A, A'), B)}$$

Good candidates for the (f, Inc) pair will preferably be such that A.1 through A.4 hold with as little extra conditions added as possible. In addition, we would like to have $Inc(B', B) = Inc(A', A)$, in order that a kind of symmetry between the fulfilment of B' by B and that of A' by A is respected. In this section and the next, we will consider each of these problems in turn.

2.1 Fuzzification of Set Inclusion

As has been done for just about every aspect of classical set theory and binary logic, several authors have introduced frameworks for extending the subsethood notion to fuzzy sets. For a long time Zadeh's definition of inclusion, which reads

$$A \subseteq B \iff (\forall u \in U)(A(u) \le B(u)) , \qquad (2)$$

remained unquestioned by the fuzzy community. Bandler and Kohout were among the first to contest this naive view, terming it "an unconscious step backward in the realm of dichotomy". They commented on the relationship between implication and inclusion, arguing that in classical set theory A is a subset of B (A, B defined in a universe U) if and only if

$$(\forall u \in U)(u \in A \Rightarrow u \in B) \qquad (3)$$

For each element u, we check whether $u \in A$. If so, then u must also be in B. Otherwise, we enforce no constraint on the membership of u to B. Bandler and Kohout suggested the following fuzzy extension of this mathematical principle ($\mathcal{I}$ being an implicator):

$$Inc_\mathcal{I}(A,B) = \inf_{u \in U} \mathcal{I}(A(u), B(u)) . \qquad (4)$$

As there are various implicators with varying suitability, we also have the choice between a multitude of inclusion operators which have or have not certain desired properties. We have adapted a list of selected properties taken from [10]:

I.1 $A \subseteq B \iff Inc_\mathcal{I}(A,B) = 1$

I.2 $(\exists u \in U)(A(u) = 1 \text{ and } B(v) = 0) \iff Inc_\mathcal{I}(A,B) = 0$

I.3 $B \subseteq C \Rightarrow Inc_\mathcal{I}(C,A) \le Inc_\mathcal{I}(B,A)$

I.4 $B \subseteq C \Rightarrow Inc_\mathcal{I}(A,B) \le Inc_\mathcal{I}(A,C)$

I.5 $Inc_\mathcal{I}(A,B) = Inc_\mathcal{I}(coB, coA)$

I.6 $Inc_\mathcal{I}(B \cup C, A) = \min(Inc_\mathcal{I}(B,A), Inc_\mathcal{I}(C,A))$

I.7 $Inc_\mathcal{I}(A, B \cap C) = \min(Inc_\mathcal{I}(A,B), Inc_\mathcal{I}(A,C))$

I.8 $Inc_\mathcal{I}(A,C) \ge \min(Inc_\mathcal{I}(A,B), Inc_\mathcal{I}(B,C))$

Criterion I.2 might at first glance seem harsh (e.g. Wilmott [12] and Young [13] preferred to leave it out in favour of more compensating operators), but as Sinha and Dougherty [8] proved, it is indispensible if we want $Inc_\mathcal{I}$ to be a faithful extension of the classical inclusion, that is, $Inc_\mathcal{I}(A,B) \in \{0,1\}$ if A and B are crisp sets.

Regarding these conditions we may prove the following theorem: [10]

Theorem 2.1. (Properties of $Inc_\mathcal{I}$)

1. $Inc_\mathcal{I}$ always satisfies I.3, I.4, I.6, I.7 and the sufficient part of I.2
2. If $\mathcal{I}$ satisfies $(\forall(x,y) \in [0,1]^2)(x \le y \iff \mathcal{I}(x,y) = 1)$, then $Inc_\mathcal{I}$ satisfies I.1
3. If $\mathcal{I}$ satisfies $(\forall(x,y) \in [0,1]^2)(\mathcal{I}(x,y) = \mathcal{I}(1-y,1-x))$, then $Inc_\mathcal{I}$ satisfies I.5
4. If $\mathcal{I}$ is transitive, then $Inc_\mathcal{I}$ satisfies I.8

It turns out that we can test $Inc_{\mathcal{I}}$ for suitability by checking just the discriminating conditions I.1, (the necessary part of) I.2, I.5 and I.8. For instance, $Inc_{\mathcal{I}_a}$ satisfies all conditions (including the necessary part of I.2 provided the universe is finite) [10] [11], while $Inc_{\mathcal{I}_b}$ does not since the Kleene–Dienes implicator is neither contrapositive nor transitive.

2.2 Modification of the Consequent

After evaluating $\alpha = Inc(A', A)$ by a chosen inclusion measure, we proceed by modifying B into a suitable output B', i.e. $B'(v) = f(\alpha, B)(v)$ for all v in V. To comply with condition A.1, it is clear that $f(\alpha, B)(v) \geq B(v)$. On the other hand, assuming Inc satisfies the monotonicity condition I.4, f ought to be increasing w.r.t. its first argument to fulfil A.2. Lastly, to have A.3 and A.4 it is mandatory that $f(1, B) = B$, whatever $B \in \mathcal{F}(V)$.

The need for modification mappings also arises in similarity–based reasoning, where instead α will result from a similarity measurement, so we can "borrow", so to speak, some of the work that has been done in that field. For instance, Türkşen and Zhong [9] use the following forms, for all $v \in V$:

$$f_1(\alpha, B)(v) = \begin{cases} \min\left(1, \frac{B(v)}{\alpha}\right) & \text{if } \alpha > 0 \\ 1 & \text{otherwise} \end{cases}$$

$$f_2(\alpha, B)(v) = \alpha B(v)$$

f_2 drops out immediately because it violates A.1. f_1 on the other hand does obey our postulates, and so does the following mapping:

$$f_3(\alpha, B)(v) = B(v)^\alpha$$

for all $v \in V$. Another alternative, adopted from Bouchon–Meunier and Valverde [4], introduces a level of *uncertainty* proportional to $1 - \alpha$, thus making inference results easy to interpret: the higher this value gets, the more the original B is "flooded" by letting the minimal membership grade in B' for every v in V become at least $1 - \alpha$:

$$f_4(\alpha, B)(v) = \max(1 - \alpha, B(v))$$

We thus observe that several modification mappings serve our cause; deciding which one to choose depends largely on the application at hand. Nevertheless, in a situation where we would like the inference result to be in accordance somehow with the output of a given CRI–GMP system, one mapping might be considered more eligible than the next one. This will be clarified in the next section.

3 Link with the Compositional Rule of Inference

In the previous sections we showed how inclusion could be related to implication. It is not surprising, then, to discover that the behaviour of the CRI–GMP based on particular t–norm/implicator pairs can be linked to our inclusion approach. In fact, it can serve as a benchmark for testing the quality as well as the soundness of our proposed inference system.

In particular, we may show that for a residuated[4] implicator generated by a continuous t–norm T, the following theorem holds:

Theorem 3.1. *Let T be a continuous t–norm. If B' represents the result obtained with CRI–GMP based on the $(T, \mathcal{I}_T)$ pair, i.e. for all $v \in V$*

$$B'(v) = \sup_{u \in U} T(A'(u), \mathcal{I}_T(A(u), B(v))),$$

then

$$Inc_{\mathcal{I}_T}(B', B) \geq Inc_{\mathcal{I}_T}(A', A)$$

Additionally, if $(\forall \alpha \in [0,1])(\exists v \in V)(B(v) = \alpha)$, then

$$Inc_{\mathcal{I}_T}(B', B) = Inc_{\mathcal{I}_T}(A', A)$$

Proof. This proof is based on properties of residuated implicators generated by continuous t–norms as listed by Klir and Yuan: [7]

I $\quad \mathcal{I}_T \left(\sup_{j \in J} a_j, b \right) = \inf_{j \in J} \mathcal{I}_T(a_j, b)$

II $\quad \mathcal{I}_T(T(a,b), d) = \mathcal{I}_T(a, \mathcal{I}_T(b, d))$

III $\quad \mathcal{I}_T(\mathcal{I}_T(a, b), b) \geq a$

IV $\quad T(a, b) \leq d \iff \mathcal{I}_T(a, d) \geq b$

where J is an arbitrary index set, $(a_j)_{j \in J}$ a family in $[0, 1]$, $a, b, d \in [0, 1]$.

$$
\begin{aligned}
Inc_{\mathcal{I}_T}(B', B) &= \inf_{v \in V} \mathcal{I}_T(B'(v), B(v)) \\
&= \inf_{v \in V} \mathcal{I}_T \left(\sup_{u \in U} T(A'(u), \mathcal{I}_T(A(u), B(v))), B(v) \right) \\
&= \inf_{v \in V} \inf_{u \in U} \mathcal{I}_T(T(A'(u), \mathcal{I}_T(A(u), B(v))), B(v)) \quad \text{(Property I)} \\
&= \inf_{v \in V} \inf_{u \in U} \mathcal{I}_T(A'(u), \mathcal{I}_T(\mathcal{I}_T(A(u), B(v)), B(v))) \quad \text{(Property II)} \\
&\geq \inf_{v \in V} \inf_{u \in U} \mathcal{I}_T(A'(u), A(u)) \quad \text{(Property III)} \\
&= Inc_{\mathcal{I}_T}(A', A)
\end{aligned}
$$

In the one but last step of this proof, if we can guarantee that there exists a $v \in V$ so that $B(v) = A(u)$, then for this value the infimum over all $v \in V$ is reached (considering that $\mathcal{I}_T(\mathcal{I}_T(A(u), A(u)), A(u)) = \mathcal{I}_T(1, A(u)) = A(u)$ and that the second partial mapping of $\mathcal{I}_T$ is increasing). $\qquad \square$

---

[4] Any t–norm T generates a so–called residuated implicator $\mathcal{I}_T$ by the formula $\mathcal{I}_T(x, y) = \sup\{\gamma | \gamma \in [0, 1]$ and $T(x, \gamma) \leq y\}$ [10]

Corollary 3.1. *For every $v \in V$, the inference result $B'(v)$ obtained with CRI–GMP based on the $(T, \mathcal{I}_T)$ pair, where T is a continuous t–norm, is bounded above by the expression $\mathcal{I}_T(Inc_{\mathcal{I}_T}(A', A), B(v))$.*

Proof. Suppose $Inc_{\mathcal{I}_T}(A', A) = \alpha$. From theorem 3.1 we conclude that $B' \in \left\{ B^* \in \mathcal{F}(V) | \inf_{v \in V} \mathcal{I}_T(B^*(v), B(v)) \geq \alpha \right\}$. Hence, for all $v \in V$, we obtain successively:

$$
\begin{aligned}
\mathcal{I}_T(B'(v), B(v)) \geq \alpha &\iff T(B'(v), \alpha) \leq B(v) \quad \text{(Property IV)} \\
&\iff T(\alpha, B'(v)) \leq B(v) \quad \text{(Symmetry of } T) \\
&\iff \mathcal{I}_T(\alpha, B(v)) \geq B'(v) \quad \text{(Property IV)}
\end{aligned}
$$

This completes the proof. □

In effect, this shows that if we put $f(\alpha, B)(v) = \mathcal{I}_T(\alpha, B(v))$ for every v in V, a conclusion entailed by our algorithm is a superset (not necessarily a proper one) of the according CRI–GMP result, which can be regarded as a justification of its soundness: when we replace the output of the CRI–GMP by a less specific fuzzy set, the result will never contain more information than we were allowed to deduce with the original (reliable) inference mechanism.

We will make these results concrete by applying them to the Lukasiewicz implicator[5], of which we already know (from section 2) that the associated inclusion measure excels w.r.t. the proposed axioms. In this case, $\alpha = Inc_{\mathcal{I}_a}(A', A) = \inf_{u \in U} \min(1, 1 - A'(u) + A(u))$ and according to corollary 3.1 $f(\alpha, B)(v) = \min(1, 1 - \alpha + B(v))$ imposes itself as the proper choice of modification mapping.

Example 3.1. Consider the rule

IF dirt level is *low* **THEN** washing time is *medium*

which could be part of the implementation of a simple fuzzy washing machine. We agree to express X (dirt level) in percentages, while the universe of Y (washing time) is taken to be the real interval $[0, 90]$. For simplicity, all membership functions have triangular shape.

We illustrate the results of our approach to several fuzzy inputs. Figure 1 shows the membership functions for "low" and "medium". The first input in figure 2 could be interpreted to mean "somewhat low", while the second may represent the concept "around 20". It can be checked that $Inc_{\mathcal{I}_a}$(somewhat low,low) $= 0.66$ and $Inc_{\mathcal{I}_a}$(around 20,low) $= 0.60$.

In figure 3 we show the inference result obtained with the inclusion–based approach. Each picture also shows, as a dashed line, the output generated by the corresponding CRI–GMP method (i.e. the one based on the $(W, \mathcal{I}_a)$ pair).

---

[5] $\mathcal{I}_a$ is in fact a residual implicator generated by the t–norm W defined as $W(x, y) = max(0, x + y - 1)$, $(x, y) \in [0, 1]^2$

Notice that, while the maximal difference between the CRI–GMP result and ours is bounded above by $1 - Inc_{\mathcal{I}_a}(A', A)$ (since $B(v) \leq B'(v) \leq f(\alpha, B)(v)$), we generally get a fairly tight upper approximation and in a lot of cases, as 3.b exemplifies, the results are equal.

As far as efficiency is concerned, the inclusion method clearly outperforms the CRI–GMP: our technique requires only one inf–calculation (one min–calculation in a finite universe) per inference while in the latter case we need to perform this tedious operation for *every* $v \in V$.

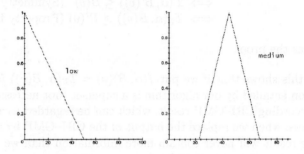

Fig. 1. a) concept *"low"* b) concept *"medium"*

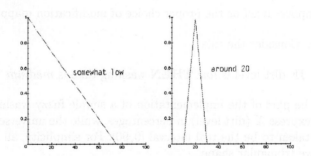

Fig. 2. a) concept *"somewhat low"* b) concept *"around 20"*

4 Relationship to Analogical Reasoning

Several authors have pursued the strategy of analogical reasoning in their quest to model human behaviour in various cognitive tasks, such as classification and decision making. As Bouchon et al. [3] state, "it amounts to inferring, from a

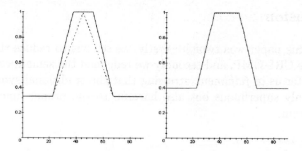

Fig. 3. a) inference for *"somewhat low"* b) inference for *"around 20"*

known case which is similar enough to the encountered situation, that what is true in the known case might still be true or approximately true in this situation".

A panoply of *similarity measures* have been reviewed in the literature as a means to draw analogies between situations (fuzzy sets). Just like our inclusion measures, these operators yield a [0,1]–valued degree for any two fuzzy sets in a given universe, which can likewise be used to set up appropriate modification mappings. Similarity is however not a uniquely defined notion, and care should be taken when adopting such or such interpretation for use in an application. The predominant characteristics that most people look for in a similarity measure S seem to be those proposed by Zadeh [14], i.e.[6]

S.1 $S(A, A) = 1$ (reflexivity)
S.2 $S(A, B) = S(B, A)$ (symmetry)
S.3 $\min(S(A, B), S(B, C)) \leq S(A, C)$ (min–transitivity)

where A, B and C are fuzzy sets in a given universe U.

For our purposes, i.e. the modelling of approximate analogical reasoning, symmetry is actually both counterintuitive and harmful! Counterintuitive, because we compare an observation A' to a reference A and not the other way round; harmful, because imposing symmetry inevitably clashes with the soundness condition A.4 (from $A' \subseteq A$ infer $B' = B$), and renders inference absolutely useless (just imagine a *symmetrical* measure S that satisfies $S(A, B) = 1$ if $A \subseteq B$). Still, most of the research done so far did not take heed of this evident pitfall, with the notable exception of Bouchon–Meunier and Valverde [4]: they are also inclined to drop the symmetry requirement, but they failed to link their proposed operators to the intuitive notion of fulfilment that we have developed in the course of this paper.

So while we see that the idea of analogical reasoning is in itself very useful, its execution so far has often been subject to essential misconceptions that have put it, from a logical perspective, on the wrong track.

---

[6] The transitivity condition S.3 can be relaxed by allowing a general t–norm to replace min

5 Conclusion

Our aim in this paper was twofold: firstly, we set out to reduce the complexity present in the CRI–GMP, and secondly we redefined the semantics of analogical reasoning in terms of *fulfilment*, stressing that the traditional symmetry condition is not only superfluous but also harmful to the proper functioning of an inference system.

6 Acknowledgements

Chris Cornelis would like to acknowledge the Fund for Scientific Research Flanders (FWO) for supporting the research elaborated on in this paper.

References

1. Baldwin J.F., Pilsworth, B.: Axiomatic approach to implication for approximate reasoning using fuzzy logic. Fuzzy Sets and Systems **3** (1980) 193–219
2. Bandler W., Kohout, L.: Fuzzy power sets and fuzzy implication operators. Fuzzy Sets and Systems **4** (1980), 13–30
3. Bouchon-Meunier B., Dubois D., Godo L., Prade, H.: Fuzzy sets and possibility theory in approximate and plausible reasoning. Fuzzy sets in approximate reasoning and information systems, Kluwer Academic Publishers. (1999) 15–190
4. Bouchon–Meunier B., Valverde L.: A fuzzy approach to analogical reasoning. Soft Computing **3** (1999) 141–147
5. Cornelis C., De Cock M., Kerre E.E.: The generalized modus ponens in a fuzzy set theoretical framework. Fuzzy IF-THEN Rules in Computational Intelligence, Theory and Applications (eds. D. Ruan en E. E. Kerre), Kluwer Academic Publishers (2000) 37–59
6. Fukami,S., Mizumoto, M., Tanaka, T.: Some considerations on fuzzy conditional inference. Fuzzy Sets and Systems **4** (1981) 243–273
7. Klir, G.J, Yuan, B.: Fuzzy sets and fuzzy logic, theory and applications. Prentice Hall PTR (1995)
8. Sinha D., Dougherty, E.R.: Fuzzification of set inclusion: theory and applications. Fuzzy Sets and Systems **55** (1993) 15–42
9. Türkşen, I.B., Zhong, Z.: An approximate analogical reasoning scheme based on similarity measures and interval-valued fuzzy sets. Fuzzy Sets and Systems **34** (1990) 323–346
10. Van Der Donck, C.: A study of various fuzzy inclusions. Masters' thesis (in Dutch). University of Ghent (1998)
11. Van Der Donck, C., Kerre, E.E.: Sinha–Dougherty approach to the fuzzification of set inclusion revised. Submitted to Fuzzy Sets and Systems (2000)
12. Willmott, R.: Mean measures of containment and equality between fuzzy sets. Proc. of the 11th Int. Symp. on Multiple–Valued Logic, Oklahoma City (1981) 183–190
13. Young, V.R.: Fuzzy subsethood. Fuzzy Sets and Systems **77** (1996) 371–384
14. Zadeh, L.A.: Similarity relations and fuzzy orderings. Information Sciences **3** (1971) 177–200

Attractor Density Models with Application to Analyzing the Stability of Biological Neural Networks

Christian Storm and Walter J. Freeman

University of California at Berkeley, Graduate Group in Biophysics,
9 Wellman Court, Berkeley, CA USA 94720
raystorm@uclink4.berkeley.edu, wfreeman@socrates.berkeley.edu
http://sulcus.berkeley.edu

Abstract. An attractor modeling algorithm is introduced which draws upon techniques found in nonlinear dynamics and pattern recognition. The technique is motivated by the need for quantitative measures that are able to assess the stability of biological neural networks which utilize nonlinear dynamics to process information.

1 Introduction

In formulating neural networks which rely on nonlinear dynamics as an information processing medium [1] [2], two pertinent questions arise when the stability of the network is considered: 1) How stable is the network dynamics to input? 2) If one or more of the parameters in the network are altered, to what degree are the dynamics modified? These questions are central to understanding the utility of chaotic neural networks since their answers form the foundation for computational stability and robustness or, ultimately, the degree in which such systems can be relied on to perform extended parallel computations. However, of the present tools offered to dynamicists none are capable of answering these questions on a quantitative basis.

1.1 Nonlinear System Identification

To date a suite of nonlinear system identification methods have been introduced for dynamical systems, each geared toward solving a particular identification problem [3] [4]. However, of the current methods offered in the literature, none are comparative in nature in the sense that approximative models are created for the sole purpose of comparison. In this paper a novel nonlinear system identification method is presented which relies on tools found in nonlinear dynamics and pattern recognition to form approximative models that lend themselves to comparison on a statistical basis. The density of states approach is founded upon the notion that a dynamical system has a corresponding set of attractors where the dynamical evolution of the system in a particular volume of phase space is governed by the attractor or its associated basin which occupies that space [5].

2 Attractor Density Models

Using a set of trajectories originating from an independent and identically distributed initial condition set **S** intersecting a collection of attractors **A** or their associated basins **B**, a particular region of state space can be independently explored. From a probabilistic reinterpretation of these independent state space trajectories a probability density model may be defined. Specifically, given a realization set $\{y\}$ generated from a known model $F(p)$ with unknown parameters or inputs p acting on a known domain $\{x\}$, $F(p) : \{x\} \to \{y\}$, a probability function may be estimated from a transformation of the set $\{y\}$, $P(T(\{y\}))$, such that $P(T(\{y\}))$ corresponds with $F(p) : \{x\}$.

2.1 Feature Selection and Segmentation

The methodology used to form attractor density models is motivated by the two tenets of pattern recognition- Feature selection and segmentation of the feature set. Feature selection is carried out by decomposing independent explorations of state space, generated from a fixed domain, into reconstruction vectors. Segmentation of the feature vectors relies on exploiting the variance, covariance structure of a reconstruction vector set.

Given a time series $h(\mathbf{a}_t) = y_t$ where A is a compact finite-dimensional set of states $\mathbf{a} \in A$ observed with a generic function h, $h : A \to \mathbf{R}$, the states $\mathbf{a}$ may be reproduced by n-dimensional vectors $\mathbf{b}$ produced by a linear operation on the time series values

$$\mathbf{b} = M[h(\mathbf{a}), h(\mathbf{a}_{-\tau}), ..., h(\mathbf{a}_{-(n-1)\tau})]^T \tag{1}$$

where τ is a positive real number and $n \geq 2d + 1$ with box counting dimension d [6]. In the present context of feature selection in a nonstationary process, the reconstruction operator $M = M_4 M_3 M_2 M_1$ is comprised of three terms: M_1 finds and removes the mean from the realization, M_2 finds and normalizes the demeaned realization to unit energy, M_3 finds the spectral coefficients through the convolution of the remaining realization with a Morlet filter bank [7]. This has the effect of breaking each realization of the chaotic process into its dynamical modes. To segment each set of vectors, the dominant modes are extracted using singular value decomposition, M_4.

2.2 Density Estimation

Each dynamical state vector represents a state $\mathbf{a}$ of one of the attractors $A \in \mathbf{A}$. The density of trajectories in the neighborhood of state $\mathbf{a}$ can be viewed statistically in terms of having a certain probability of being visited. Using this notion an attractor representation may be built by calculating the density of states of an attractor in reconstruction space.

Under the assumption that each dynamical state vector is a sample from an unknown probability density function f, a kernel function K provides a means of generating a nonparametric estimate $\hat{f}$ of f. The kernel density estimate $\hat{f}$ in dimension d at point $\mathbf{a}'$ with samples $\mathbf{a}_i$ is defined as [4]

$$\hat{f}(\mathbf{a}') = \frac{1}{nh^d} \sum_{i=1}^{n} K\left(\frac{(\mathbf{a}' - \mathbf{a}_i)^T (\mathbf{a}' - \mathbf{a}_i)}{h^2}\right) \qquad (2)$$

where K is defined as the multivariate Gaussian joint probability density function.

3 A Test of Segmentation

An interesting demonstration of this methodology is to see how well attractor density models are segmented for a range of parameter values in the Mackey-Glass equation [8]. First, attractor density models are built for each parameterization ($\tau = 6, 8, ..., 200, a = 0.16, 0.18, ..., 0.24$) using forty realizations.

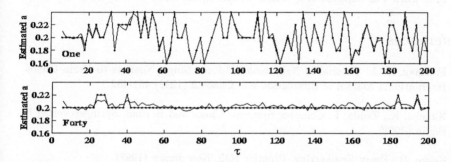

Fig. 1. Rounded estimates (*dotted lines*) of a given one and forty realizations. The raw estimates are denoted by *solid lines*. The true value is $a = 0.2$ for all values of τ

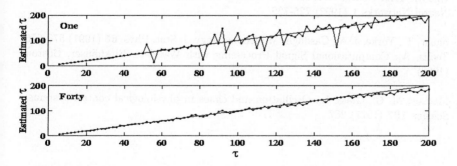

Fig. 2. Rounded estimates (*dotted lines*) of τ given one and forty realizations. The true values are denoted by the 45° *solid lines*

Given a set of one and forty realizations generated with unknown parameters $\tau = 6, 8, ..., 200$ and $a = 0.20$, which model best describes the data? This question is answered by finding the center of mass of the density values found for each model using the points in each realization set. It is evident from the estimations of a and τ in Figure 1 and 2, respectively, that increased sample sizes reduce the estimate variances indicating that with modest sample sizes a good degree of segmentation can be expected. The variability in the estimates is attributed to statistical fluctuations.

4 Conclusion

In this paper, the basis for creating attractor density models is described. The aim in formulating this technique lies in the need for nonlinear system identification methods which promote the comparison between models. With such methods, the ability to analyze the stability of chaotic neural networks to changes in internal parameters or input will be greatly improved.

This work was supported by ARO MURI grant ARO DAAH 04-96-0341.

References

1. Freeman, W.J.: Tutorial in Neurobiology: From Single Neurons to Brain Chaos. International Journal of Bifurcation and Chaos **2** (1992) 451-482

2. Kaneko, K., Tsuda, I.: Complex Systems: Chaos and Beyond. Springer-Verlag b, Berlin (2001)

3. Kosko, B.: Fuzzy Engineering. Prentice Hall, New Jersey (1997)

4. Silverman, B. W.: Density Estimation for Statistics and Data Analysis. Chapman and Hall, New York (1986)

5. Storm, C.: A Novel Dynamical Invariant Measure Addresses the Stability of the Chaotic KIII Neural Network. Proceedings of International Joint Conference on Neural Networks **1** (1999) 725-729

6. Sauer, T., Yorke, J. A., Casdagli, M.: Embedology. J. Stat. Phys. **65** (1991) 579-616
7. Teolis, A.: Computational Signal Processing with Wavelets. Birkhäuser, Boston (1998)

8. Mackey, M. C., Glass, L.: Oscillation and chaos in physiological control systems. Science **197** (1977) 287

MARS: Still an Alien Planet in Soft Computing?

Ajith Abraham and Dan Steinberg[t]

School of Computing and Information Technology
Monash University (Gippsland Campus), Churchill 3842, Australia
Email: ajith.abraham@infotech.monash.edu.au

[t] Salford Systems Inc
8880 Rio San Diego, CA 92108, USA
Email: dstein@salford-systems.com

Abstract: The past few years have witnessed a growing recognition of soft computing technologies that underlie the conception, design and utilization of intelligent systems. According to Zadeh [1], soft computing consists of artificial neural networks, fuzzy inference system, approximate reasoning and derivative free optimization techniques. In this paper, we report a performance analysis among Multivariate Adaptive Regression Splines (MARS), neural networks and neuro-fuzzy systems. The MARS procedure builds flexible regression models by fitting separate splines to distinct intervals of the predictor variables. For performance evaluation purposes, we consider the famous Box and Jenkins gas furnace time series benchmark. Simulation results show that MARS is a promising regression technique compared to other soft computing techniques.

1. Introduction

Soft Computing is an innovative approach to construct computationally intelligent systems that are supposed to possess humanlike expertise within a specific domain, adapt themselves and learn to do better in changing environments, and explain how they make decisions. Neurocomputing and neuro-fuzzy computing are well-established soft computing techniques for function approximation problems. MARS is a fully automated method, based on a divide and conquer strategy, partitions the training data into separate regions, each with its own regression line or hyperplane [2]. MARS strengths are its flexible framework capable of tracking the most complex relationships, combined with speed and the summarizing capabilities of local regression. This paper investigates the performance of neural networks, neuro-fuzzy systems and MARS for predicting the well-known Box and Jenkins time series, a benchmark problem used by several connectionist researchers. We begin with some theoretical background about MARS, artificial neural networks and neuro-fuzzy systems. In Section 6 we present experimentation setup using MARS and soft computing models followed by results and conclusions.

2. What are Splines?

Splines can be considered as an innovative mathematical process for complicated curve drawings and function approximation. Splines find ever-increasing application in the numerical methods, computer-aided design, and computer graphics areas. Mathematical formulae for circles, parabolas, or sine waves are easy to construct, but how does one develop a formula to trace the shape of share value fluctuations or any time series prediction problems? The answer is to break the complex shape into simpler pieces, and then use a stock formula for each piece [4]. To develop a spline the X-axis is broken into a convenient number of regions. The boundary between regions is also known as a knot. With a sufficiently large number of knots virtually any shape can be well approximated. While it is easy to draw a spline in 2-dimensions by keying on knot locations (approximating using linear, quadratic or cubic polynomial etc.), manipulating the mathematics in higher dimensions is best accomplished using basis functions.

3. Multivariate Adaptive Regression Splines (MARS)

The MARS model is a spline regression model that uses a specific class of basis functions as predictors in place of the original data. The MARS basis function transform makes it possible to selectively blank out certain regions of a variable by making them zero, allowing MARS to focus on specific sub-regions of the data. MARS excels at finding optimal variable transformations and interactions, as well as the complex data structure that often hides in high-dimensional data [3].

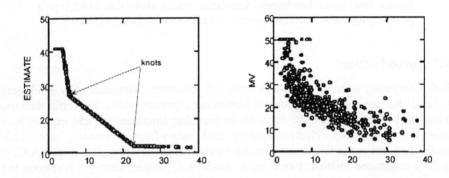

Figure 1. MARS data estimation using spines and knots (actual data on the right)

Given the number of predictors in most data mining applications, it is infeasible to approximate the function $y=f(x)$ in a generalization of splines by summarizing y in each distinct region of x. Even if we could assume that each predictor x had only two distinct regions, a database with just 35 predictors would contain more than 34 billion regions. Given that neither the number of regions nor the knot locations can be specified a priori, a procedure is needed that accomplishes the following:

- judicious selection of which regions to look at and their boundaries, and
- judicious determination of how many intervals are needed for each variable

A successful method of region selection will need to be adaptive to the characteristics of the data. Such a solution will probably reject quite a few variables (accomplishing variable selection) and will take into account only a few variables at a time (also reducing the number of regions). Even if the method selects 30 variables for the model, it will not look at all 30 simultaneously. Such simplification is accomplished by a decision tree (e.g., at a single node, only ancestor splits are being considered; thus, at a depth of six levels in the tree, only six variables are being used to define the node).

MARS Smoothing, Splines, Knots Selection and Basis Functions

A key concept underlying the spline is the knot. A knot marks the end of one region of data and the beginning of another. Thus, the knot is where the behavior of the function changes. Between knots, the model could be global (e.g., linear regression). In a classical spline, the knots are predetermined and evenly spaced, whereas in MARS, the knots are determined by a search procedure. Only as many knots as needed are included in a MARS model. If a straight line is a good fit, there will be no interior knots. In MARS, however, there is always at least one "pseudo" knot that corresponds to the smallest observed value of the predictor. Figure 1 depicts a MARS spline with three knots.

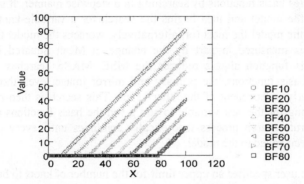

Figure 2. Variations of basis functions for c = 10 to 80

Finding the one best knot in a simple regression is a straightforward search problem: simply examine a large number of potential knots and choose the one with the best R^2. However, finding the best pair of knots requires far more computation, and finding the best set of knots when the actual number needed is unknown is an even more challenging task. MARS finds the location and number of needed knots in a forward/backward stepwise fashion. A model which is clearly overfit with too many knots is generated first, then, those knots that contribute least to the overall fit are removed. Thus, the forward knot selection will include many incorrect knot locations, but these erroneous knots will eventually, be deleted from the model in the backwards pruning step (although this is not guaranteed).

Thinking in terms of knot selection works very well to illustrate splines in one dimension; however, this context is unwieldy for working with a large number of variables simultaneously. Both concise notation and easy to manipulate programming expressions are required. It is also not clear how to construct or represent interactions using knot locations. In MARS, Basis Functions (BFs) are the machinery used for generalizing the search for knots. BFs are a set of functions used to represent the information contained in one or more variables. Much like principal components, BFs essentially re-express the relationship of the predictor variables with the target variable. The hockey stick BF, the core building block of the MARS model is often applied to a single variable multiple times. The hockey stick function maps variable X to new variable X^*:

max $(0, X - c)$, or
max $(0, c - X)$

where X^* is set to 0 for all values of X up to some threshold value c and X^* is equal to X for all values of X greater than c. (Actually X^* is equal to the amount by which X exceeds threshold c). The second form generates a mirror image of the first. Figure 2 illustrates the variation in BFs for changes of c values (in steps of 10) for predictor variable X, ranging from 0 to 100.

MARS generates basis functions by searching in a stepwise manner. It starts with just a constant in the model and then begins the search for a variable-knot combination that improves the model the most (or, alternatively, worsens the model the least). The improvement is measured in part by the change in Mean Squared Error (MSE). Adding a basis function always reduces the MSE. MARS searches for a pair of hockey stick basis functions, the primary and mirror image, even though only one might be linearly independent of the other terms. This search is then repeated, with MARS searching for the best variable to add given the basis functions already in the model. The brute search process theoretically continues until every possible basis function has been added to the model.

In practice, the user specifies an upper limit for the number of knots to be generated in the forward stage. The limit should be large enough to ensure that the true model can be captured. A good rule of thumb for determining the minimum number is three to four times the number of basis functions in the optimal model. This limit may have to be set by trial and error.

4. Artificial Neural Network (ANN)

ANN is an information-processing paradigm inspired by the way the densely interconnected, parallel structure of the mammalian brain processes information. Learning in biological systems involves adjustments to the synaptic connections that exist between the neurons [7]. Learning typically occurs by example through training, where the training algorithm iteratively adjusts the connection weights (synapses). These connection weights store the knowledge necessary to solve specific problems. A typical three-layer feedforward neural network is illustrated in Figure 3. Backpropagation (BP) is one of the most famous training algorithms for multilayer perceptrons. BP is a gradient descent technique to minimize the error E for a particular training pattern. For adjusting the weight (w_{ij}) from the i-th input unit to the j-th output, in the batched mode variant the descent is based on the gradient ∇E ($\frac{\delta E}{\delta w_{ij}}$) for the total training set:

$$\Delta w_{ij}(n) = -\epsilon * \frac{\delta E}{\delta w_{ij}} + \alpha * \Delta w_{ij}(n-1)$$

The gradient gives the direction of error E. The parameters ε and α are the learning rate and momentum respectively.

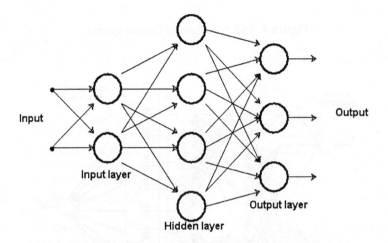

Figure 3. Typical three-layer feedforward network architecture

5. Neuro-Fuzzy (NF) System

We define a NF [6] system as a combination of ANN and Fuzzy Inference System (FIS) [9] in such a way that neural network learning algorithms are used to determine the parameters of FIS. As shown in Table 1, to a large extent, the drawbacks pertaining to these two approaches seem largely complementary.

Table 1. Complementary features of ANN and FIS

Black box	Interpretable
Learning from scratch	Making use of linguistic knowledge

In our simulation, we used ANFIS: Adaptive Network Based Fuzzy Inference System [5] as shown in Figure 5, which implements a Takagi Sugeno Kang (TSK) fuzzy inference system (Figure 4) in which the conclusion of a fuzzy rule is constituted by a weighted linear combination of the crisp inputs rather than a fuzzy set.

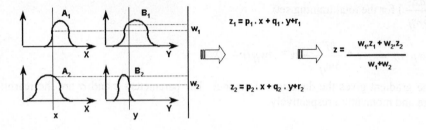

Figure 4. TSK type fuzzy inference system

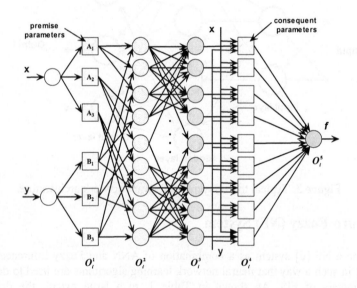

Figure 5. Architecture of the ANFIS

Architecture of ANFIS and the functionality of each layer is as follows:

Layer-1 Every node in this layer has a node function

$$O_i^1 = \mu_{A_i}(x), \text{ for } i = 1, 2$$

or

$$O_i^1 = \mu_{B_{i-2}}(y), \text{ for } i = 3, 4, \ldots$$

O_i^1 is the membership grade of a fuzzy set A ($= A_1$, A_2, B_1 or B_2), specifies the degree to which the given input x (or y) satisfies the quantifier A. Usually the node function can be any parameterized function.. A gaussian membership function is specified by two parameters c (membership function center) and σ (membership function width)

$$\text{guassian } (x, c, \sigma) = e^{-\frac{1}{2}\left(\frac{x-c}{\sigma}\right)^2}$$

Parameters in this layer are referred to premise parameters.

Layer-2 Every node in this layer multiplies the incoming signals and sends the product out. Each node output represents the firing strength of a rule.

$$O_i^2 = w_i = \mu_{A_i}(x) \times \mu_{B_i}(y), i = 1,2.......$$

In general any T-norm operators that perform fuzzy AND can be used as the node function in this layer.

Layer-3 Every i-th node in this layer calculates the ratio of the i-th rule's firing strength to the sum of all rules firing strength.

$$O_i^3 = \overline{w_i} = \frac{w_i}{w_1 + w_2}, i = 1,2.....$$

Layer-4 Every node i in this layer has a node function

$$O_1^4 = \overline{w_i} f_i = \overline{w_i}(p_i x + q_i y + r_i),$$

where $\overline{w_i}$ is the output of layer3, and $\{p_i, q_i, r_i\}$ is the parameter set. Parameters in this layer will be referred to as consequent parameters.

Layer-5 The single node in this layer labeled Σ computes the overall output as the summation of all incoming signals: $O_1^5 = Overall\ output = \sum_i \overline{w_i} f_i = \frac{\sum_i w_i f_i}{\sum_i w_i}$.

ANFIS makes use of a mixture of backpropagation to learn the premise parameters and least mean square estimation to determine the consequent parameters. A step in the learning procedure has two parts: In the first part the input patterns are propagated, and the optimal conclusion parameters are estimated by an iterative least mean square procedure, while the antecedent parameters (membership functions) are assumed to be fixed for the current cycle through the training set. In the second part the patterns are propagated again, and in this epoch, backpropagation is used to modify the antecedent parameters, while the conclusion parameters remain fixed. This procedure is then iterated.

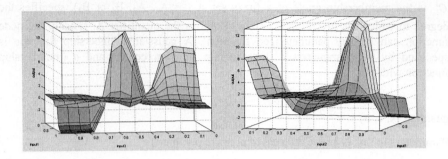

Figure 6. 3D view of Gas furnace time series training data (I/O relationship)

6. Experimental Setup Using Soft Computing Models and MARS

Gas Furnace Time Series Data: This time series was used to predict the CO_2 (carbon dioxide) concentration $y(t+1)$ [10]. In a gas furnace system, air and methane are combined to form a mixture of gases containing CO_2. Air fed into the gas furnace is kept constant, while the methane feed rate $u(t)$ can be varied in any desired manner. After that, the resulting CO_2 concentration y(t) is measured in the exhaust gases at the outlet of the furnace. Data is represented as $[u(t), y(t), y(t+1)]$. The time series consists of 292 pairs of observation and 50% was used for training and remaining for testing purposes. Figure 6 shows the complexity of the input / output relationship in two different angles. Our experiments were carried out on a PII, 450MHz Machine and the codes were executed using MATLAB and C++.

- **ANN training**

  We used a feedforward neural network with 1 hidden layer consisting of 24 neurons (tanh-sigmoidal node transfer function). The training was terminated after 6000 epochs. Initial learning rate was set at 0.05.

- **ANFIS training**

  In the ANFIS network, we used 4 Gaussian membership functions for each input parameter variable. Sixteen rules were learned based on the training data. The training was terminated after 60 epochs.

- **MARS**

  We used 5 basis functions and selected 1 as the setting of minimum observation between knots. To obtain the best possible prediction results (lowest RMSE), we sacrificed the speed (minimum completion time).

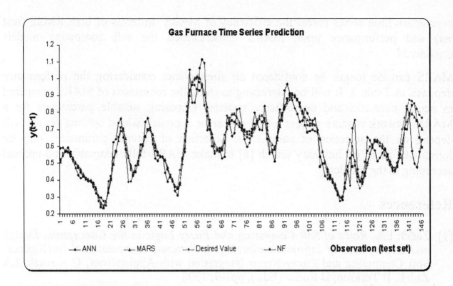

Figure 7. Gas furnace series prediction using soft computing models and MARS

- **Performance and results achieved**

    Figure 7 illustrates the test results achieved for the gas furnace time series. Table 2 summarizes the comparative performances of the different soft computing models and MARS in terms of performance time, training error and testing error obtained.

Table 2. Results showing performance comparison between MARS and soft computing models for gas furnace series prediction

Model	Root Mean Squared Error		BFlops	Epochs	Training time (seconds)
	Training Set	**Test Set**			
MARS	0.0185	0.0413	-	-	1
ANN	0.0565	0.0897	0.5802	6000	250
NF	0.0137	0.0570	0.0005	60	40

*Computational load in billion flops (BFlops)

7. Conclusion

In this paper we have investigated the performance of MARS and compared the performance with artificial neural networks and neuro-fuzzy systems (ANFIS), which are well-established function approximators. Our experiments to predict the benchmark time series reveal the efficiency of MARS. In terms of both RMSE (test set) and performance time, MARS outperformed the soft computing models considered.

MARS can no longer be considered an alien planet considering the performance depicted in Table 2. It will be interesting to study the robustness of MARS compared to neural networks and neuro-fuzzy systems. Choosing suitable parameters for a MARS network is more or less a trial and error approach where optimal results will depend on the selection of parameters. Selection of optimal parameters may be formulated as an evolutionary search [8] to make MARS fully adaptable and optimal according to the problem.

References

[1] Zadeh LA, *Roles of Soft Computing and Fuzzy Logic in the Conception, Design and Deployment of Information/Intelligent Systems*, Computational Intelligence: Soft Computing and Fuzzy-Neuro Integration with Applications, O Kaynak, LA Zadeh, B Turksen, IJ Rudas (Eds.), pp1-9, 1998.

[2] Friedman, J. H, *Multivariate Adaptive Regression Splines*, Annals of Statistics, Vol 19, 1-141, 1991.

[3] Steinberg, D, Colla, P. L., and Kerry Martin (1999), *MARS User Guide*, San Diego, CA: Salford Systems, 1999.

[4] Shikin E V and Plis A I, *Handbook on Splines for the User*, CRC Press, 1995.

[5] Jang J S R, *Neuro-Fuzzy Modeling: Architectures, Analyses and Applications*, PhD Thesis, University of California, Berkeley, July 1992.

[6] Abraham A and Nath B, *Designing Optimal Neuro-Fuzzy Systems for Intelligent Control*, The Sixth International Conference on Control, Automation, Robotics and Vision, (ICARCV 2000), December 2000.

[7] Abraham A and Nath B, *Optimal Design of Neural Nets Using Hybrid Algorithms*, In proceedings of 6th Pacific Rim International Conference on Artificial Intelligence (PRICAI 2000), pp. 510-520, 2000.

[8] Fogel D, *Evolutionary Computation: Towards a New Philosophy of Machine Intelligence*, 2nd Edition, IEEE press, 1999.

[9] Cherkassky V, *Fuzzy Inference Systems: A Critical Review*, Computational Intelligence: Soft Computing and Fuzzy-Neuro Integration with Applications, Kayak O, Zadeh LA et al (Eds.), Springer, pp.177-197, 1998.

[10] Box G E P and Jenkins G M, *Time Series Analysis, Forecasting and Control*, San Francisco: Holden Day, 1970.

Data Reduction Based on Spatial Partitioning

Gongde Guo, Hui Wang, David Bell and Qingxiang Wu

School of Information and Software Engineering, University of Ulster
Newtownabbey, BT37 0QB, N.Ireland, UK
{G.Guo, H. Wang, DA.Bell, Q.Wu}@ulst.ac.uk

Abstract. The function of data reduction is to make data sets smaller, while preserving classification structures of interest. A novel approach to data reduction based on spatial partitioning is proposed in this paper. This algorithm projects conventional database relations into multidimensional data space. The advantage of this approach is to change the data reduction process into a spatial merging process of data in the same class, as well as a spatial partitioning process of data in different classes, in multidimensional data space. A series of partitioned regions are eventually obtained and can easily be used in data classification. The proposed method was evaluated using 7 real world data sets. The results were quite remarkable compared with those obtained by C4.5 and DR. The efficiency of the proposed algorithm was better than DR without loss of test accuracy and reduction ratio.

1 Introduction

Data reduction is a process used to transform raw data into a more condensed form without losing significant semantic information. In data mining, data reduction in a stricter sense refers to feature selection and data sampling [1], but in a broader sense, data reduction is regarded as a main task of data mining [2]. Data mining techniques can thus, in this broad sense, be regarded as a method for data reduction. Data reduction is interpreted as a process to reduce the size of data sets while preserving their classification structures. Wang, *et al* [3] propose a novel method of data reduction based on lattices and hyper relations. The advantage of this is that raw data and reduced data can be both represented by hyper relations. The collection of hyper relations can be made into a complete Boolean algebra in a natural way, and so for any collection of hyper tuples its unique least upper bound (lub) can be found, as a reduction.

According to the method proposed in [3], the process of data reduction is to find the least upper bound of the raw data and to reduce it. The process of data reduction can be regarded as a merging process of simple tuples (raw data) and hyper tuples that have been generated in the same class to generate new hyper tuples. The success of each merging operation depends on whether the new hyper tuple generated from this merging operation covers in same sense a simple tuple of another class. If the new hyper tuple does cover a simple tuple of another class, the operation is cancelled. The merging operation repeats recursively until all the data including hyper tuples and same class simple tuples cannot be merged again. The main drawback of the method proposed in [3] is its efficiency, since much time is spent in trying probable merge. In this paper, we introduce the *complementary* operation of hyper tuples and attempt to use the irregular regions to represent reduced data. The main goal of the proposed algorithm is to improve its efficiency and reduction ratio whilst preserving its classification accuracy.

The remainder of the paper is organized as follows. Section 2 introduces the definitions and notation. Section 3 describes the data reduction and classification algorithm based on spatial partitioning, in which the execution process of the algorithm is demonstrated by graphical illustration. The experimental results are described and the evaluation is given in Section 4. Section 5 ends the paper with a discussion and an indication of proposed future work.

2 Definitions and Notation

In the context of the paper, *Hyper relations* are a generalization of conventional database relations in the sense that it allows sets of values as tuple entries. A *hyper tuple* is a tuple where entries are sets instead of single values. A hyper tuple is called a *simple tuple*, if all its entries have a cardinality of 1. Obviously a simple tuple is a special case of hyper tuple.

Consider two points p_i, p_j denoted as $p_i=(p_{i1}, p_{i2}, ..., p_{in})$, $p_j=(p_{j1}, p_{j2}, ..., p_{jn})$ and two spatial regions a_i, a_j denoted as $a_i=([t_{i11}, t_{i12}], [t_{i21}, t_{i22}], ..., [t_{in1}, t_{in2}])$, $a_j=([t_{j11}, t_{j12}], [t_{j21}, t_{j22}], ..., [t_{jn1}, t_{jn2}])$ in multidimensional data space, in which $[t_{il1}, t_{il2}]$ is the projection of a_i to its *l-th* component, and $t_{il2} \geq t_{il1}$, $l=1, 2, ..., n$. In this paper, for simplicity and uniformity, any point p_i is represented as a spatial region in multidimensional data space, viz. $p_i=([p_{i1}, p_{i1}], [p_{i2}, p_{i2}], ..., [p_{in}, p_{in}])$. This is often a convenient and uniform representation for analysis.

Definition 1 Given two regions a_i, a_j in multidimensional data space, the *merging operation* of two regions denoted by ' $\cup$ ' can be defined as: $a_i \cup a_j=([\min(t_{i11}, t_{j11}), \max(t_{i12}, t_{j12})], [\min(t_{i21}, t_{j21}), \max(t_{i22}, t_{j22})], ..., [\min(t_{in1}, t_{jn1}), \max(t_{in2}, t_{jn2})])$.

The *intersection operation* ' $\cap$ ' of two regions in multidimensional data space can be defined as:
$a_i \cap a_j =([\max(t_{i11}, t_{j11}), \min(t_{i12}, t_{j12})], [\max(t_{i21}, t_{j21}), \min(t_{i22}, t_{j22})], ..., [\max(t_{in1}, t_{jn1}), \min(t_{in2}, t_{jn2})])$. $a_i \cap a_j$ is empty, if and only if there exists a value of l such that $\max(t_{il1}, t_{jl1}) > \min(t_{il2}, t_{jl2})$, where $l=1,2,..., n$.

A point merging (or intersecting) with a region can be regarded as a special case according to above definition.

Definition 2 Given a region a_j in multidimensional data space denoted as $a_j = ([t_{j11}, t_{j12}], [t_{j21}, t_{j22}], ..., [t_{jn1}, t_{jn2}])$, the *complementary operation* of a_j is defined as: $\overline{a_j} =$
$([\overline{[t_{j11}, t_{j12}]}, [t_{j21}, t_{j22}], ..., [t_{jn1}, t_{jn2}]) \cup ([t_{j11}, t_{j12}], \overline{[t_{j21}, t_{j22}]}, ..., [t_{jn1}, t_{jn2}]) \cup ... \cup ([t_{j11},$
$t_{j12}], [t_{j21}, t_{j22}], ..., \overline{[t_{jn1}, t_{jn2}]}) \cup (\overline{[t_{j11}, t_{j12}]}, \overline{[t_{j21}, t_{j22}]}, [t_{j31}, t_{j32}], ..., [t_{jn1}, t_{jn2}]) \cup ([\overline{t_{j11}, t_{j12}}],$
$\overline{[t_{j21}, t_{j22}]}, \overline{[t_{j31}, t_{j32}]}, ..., [t_{jn1}, t_{jn2}]) \cup ... \cup (\overline{[t_{j11}, t_{j12}]}, ..., \overline{[t_{j(n-1)1}, t_{j(n-1)2}]}, \overline{[t_{jn1}, t_{jn2}]}) \cup ... \cup$
$(\overline{[t_{j11}, t_{j12}]}, \overline{[t_{j21}, t_{j22}]}, ..., \overline{[t_{jn1}, t_{jn2}]})$), is the region in the multidimensional data space complementary region a_j.

Definition 3 Given a point p_i denoted as $p_i=(p_{i1}, p_{i2}, ..., p_{in})$ and a region a_j denoted as $a_j=([t_{j11}, t_{j12}], [t_{j21}, t_{j22}], ..., [t_{jn1}, t_{jn2}])$ in multidimensional data space, the *hyper similarity* of p_i, a_j denoted as $S(p_i, a_j)$ is defined as follows:

If a_j is a regular region, $a_j=([t_{j11}, t_{j12}], [t_{j21}, t_{j22}], ..., [t_{jn1}, t_{jn2}])$, the hyper similarity $S(p_i, a_j)$ is equal to the number of l which satisfies $t_{jl1}{\leq}p_{il}{\leq}t_{jl2}$, in which $l=1, 2, ..., n$.

If a_j is an irregular region, consisting of h regular regions, denoted as $a_j =\{ a_{j1}, a_{j2}, ..., a_{jh} \}$, the *hyper similarity* $S(p_i, a_j)$ equals the value of $\max(S(p_i, a_{j1}),$ $S(p_i, a_{j2}), ..., S(p_i, a_{jh}))$.

Definition 4 Given a point $p_i =(p_{i1}, p_{i2}, ..., p_{in})$ and a region a_j in multidimensional data space, the universal *hyper relation* '$\leq$' is defined as: $p_i \leq a_j$, if and only if the point p_i falls into the spatial region of a_j.

If a_j is a regular region denoted as $a_j =([t_{j11}, t_{j12}], [t_{j21}, t_{j22}], ..., [t_{jn1}, t_{jn2}])$. $p_i{\leq}a_j$ if for all values of l, $t_{jl1}{\leq} p_{il}{\leq} t_{jl2}$, where $l=1, 2, ..., n$.

If a_j is an irregular region, consisting of h regular regions, $a_j =\{ a_{j1}, a_{j2}, ..., a_{jn} \}$. $p_i{\leq} a_j$, if and only if there exists a regular region a_{jl} where $p_i{\leq}a_{jl}$, in which, $l=1,2,..., h$.

For simplicity, all the data attributes used in this paper for data reduction and classification are numerical. Set union operation (respectively intersection and complementation operations) can be used to replace the '$\cup$' operation ('$\cap$' and '-' operations respectively) defined above for categorical data or binary data. In addition, the standard set inclusion operation and subset operation can be used to replace the hyper similarity operation and universal hyper relation operation respectively.

3 Data Reduction & Classification Algorithm

Let a training data set $D_i=\{d_1, d_2, ..., d_m\}$, where $d_i =(d_{i1}, d_{i2}, ..., d_{in})$. d_i is represented as $d_i=([d_{i1}, d_{i1}], [d_{i2}, d_{i2}], ..., [d_{in}, d_{in}])$ using spatial region representation as a point in multidimensional data space. Supposing that there are k classes in the training data set and d_{in} is a decision attribute, value $d_{in}{\in}\{t_1, t_2, ..., t_k\}$, the spatial partitioning algorithm is as follows:

1. $t=0$

2. $M_i^t = \bigcup_{d_{jn}=t_i}\{d_j\}$, $d_j \in D_t$, $i=1, 2, ..., k$, $j=1, 2, ..., m$

3. $M_{i,j}^t = M_i^t \cap M_j^t$, $i{\neq}j$, $i=1, 2, ..., k$, $j=1, 2, ..., k$

4. $S_i^t = M_i^t \cap \overline{M_{i,1}^t} \cap ... \cap \overline{M_{i,j-1}^t} \cap \overline{M_{i,j+1}^t} \cap ... \cap \overline{M_{i,k}^t}$, $i=1, 2, ..., k$

5. $D_{t+1} =\{d_i \mid d_i \in M_{i,j}^t, i{\neq}j, i=1, 2, ..., k, j=1, 2, ..., k\}$

6. If $(D_{t+1}{=}\phi)$ go to 8

7. $t=t+1$, go to 2

8. $R_i =\{ S_i^0, S_i^1, \cdots, S_i^t \}$, $i=1, 2, ..., k$.

Some symbols used in the above algorithm are: S_i^t -the biggest irregular region of t_i class obtained in the t-th cycle; D_t -the training data set used in the t-th cycle; M_i^t - the merging region of t_i class data obtained in the t-th cycle; $M_{i,j}^t$ - the intersection of M_i^t and M_j^t obtained in the t-th cycle. R_i - the results obtained via running the algorithm are a series of distributive regions of t_i class data, in which, $i=1, 2,..., k$.

Given the training data set shown in Figure 2-1, the running process of the algorithm in 2-dimensional space is illustrated graphically below.

The training data set includes 30 data points and is divided into three classes of black, grey and white. The distribution of data points in 2-dimensional data space is shown in Figure 2-1.

At the beginning of the running process, we merge all the data in the same class for each individual class and gain three data spatial regions M_1^0, M_2^0, M_3^0 represented by bold line, fine line and broken line respectively. The intersections of $M_{1,2}^0, M_{1,3}^0, M_{2,3}^0$ are also represented by bold line, fine line and broken line in Figure 2-2 and Figure 2-3, in which, $M_{1,2}^0 = M_1^0 \cap M_2^0$, $M_{1,3}^0 = M_1^0 \cap M_3^0$, $M_{2,3}^0 = M_2^0 \cap M_3^0$. In the first cycle, three partitioning regions S_1^0, S_2^0, S_3^0 shown in Figure 2-4 are obtained.

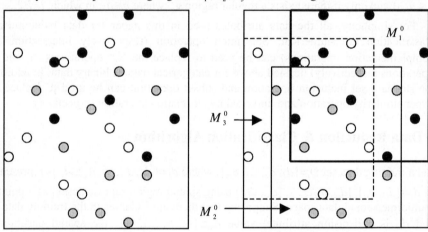

Figure 2-1. The distribution of data points *Figure 2-2.* Three merging regions

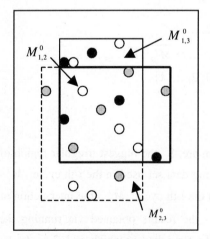

Figure 2-3. Three intersection regions

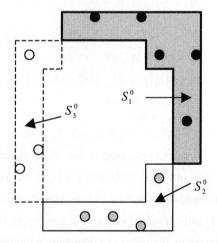

Figure 2-4. Three partitioning regions

In which, $S_1^0 = M_1^0 \cap \overline{M_{1,2}^0} \cap \overline{M_{1,3}^0}$, $S_2^0 = M_2^0 \cap \overline{M_{1,2}^0} \cap \overline{M_{2,3}^0}$, $S_3^0 = M_3^0 \cap \overline{M_{1,3}^0} \cap \overline{M_{2,3}^0}$. Obviously, if test data falls into S_1^0 (or S_2^0, S_3^0), it belongs to the black class (or the grey, white class respectively). If it falls into none of S_1^0, S_2^0 and S_3^0, it should fall into $M_{1,2}^0$, or $M_{1,3}^0$ or $M_{2,3}^0$. If so, it can not be determined which class it belongs to. All the data in the original data set which belong to $M_{1,2}^0$ or $M_{1,3}^0$ or $M_{2,3}^0$ are taken out and form a new training data set. This new training data set is partitioned again and another three merging regions: M_1^1, M_2^1, M_3^1 as well as another three intersection regions: $M_{1,2}^1, M_{1,3}^1, M_{2,3}^1$ are obtained in the second cycle. The process of merging and partitioning is executed recursively until there is no data in the new training set. This process is illustrated graphically below from Figure 2-4 to Figure 2-9.

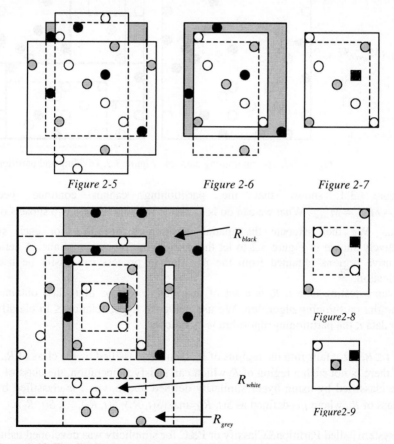

Figure 2-5   *Figure 2-6*   *Figure 2-7*

Figure2-8

Figure2-9

Figure 2-10. The distributive regions of black data

A series of distributive regions of data of each class are obtained via learning from the training data set. The distributive regions of data of the black class are represented against a grey background in Figure 2-10.

Using irregular regions representing spatial distributive regions of different classes can give higher data reduction efficiency. The term 'irregular regions' in this paper means the projection of the spatial region to each dimension might be a series of discrete intervals.

It is probable that the partitioning process cannot continue for some data distributions because equal merging regions could be obtained in the partitioning process. See Figure 3-1 for instance. The three merging regions of M_{black}, M_{grey} and M_{white} are totally equal to each other.

In this situation, one resolution is to select an attribute as a partitioning attribute to divide the data set into two subsets and then for each subset according to the above algorithm, continue partitioning until all the data has been partitioned.

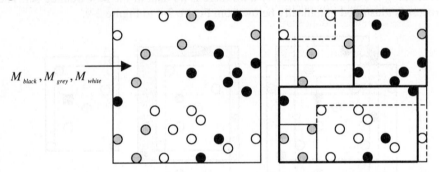

Figure 3-1. Special training data set   *Figure 3-2.* Dividing and partitioning

Figure 3-1 shows that the partitioning cannot continue because $M_{black} = M_{grey} = M_{white}$. What we can do is to divide the data set into two subsets on the Y-axis. We then execute the spatial partitioning operation for each subset respectively shown in Figure 3.2 to let the partitioning process continue. A series of distributive regions obtained from the partitioning process can easily be used in classification.

Given a testing data t, R_i is a set of distributive regions of t_i class obtained by running the partitioning algorithm. We use universal hyper relation $\leq$ to classify the testing data t, the partitioning algorithm is as follows:

- If $t \leq R_i$ viz. t falls into the regions of R_i, then t is classified by the class of R_i.
- If there is not such a region of R_i which can satisfy $\leq$ operation, the class of t can be classified by using hyper similarity defined above, viz. t is classified by the class of R_j, where j is defined as $S(t, R_j)=\max(S(t, R_1), S(t, R_2), \ldots, S(t, R_k))$.

A system called Partition&Classify or P&C for simplicity was developed using the proposed method, it can classify unlabeled testing data effectively. The algorithm was evaluated for classification using some real world data sets and the results are quite remarkable. The experimental results are reported in next section.

This sort of data reduction and classification is very helpful for large databases and data mining based on some of the following reasons [3]:

- It reduces the storage requirements of data used mainly for classification;
- It offers better understandability for the knowledge discovered;
- It allows feature selection and continuous attribute discretization to be achieved as by-products of data reduction.

4 Experiment and Evaluation

The ultimate goal of data reduction is to improve the performance of learning, hence the main goal of our experiment is set to evaluate how well our proposed method performs for data reduction and to calculate its accuracy of prediction and performance for some real world data sets. We use the 5-fold cross validation method to evaluate its prediction accuracy and compare the results obtained from experiment with some of standard data mining methods.

Seven public databases are chosen from the UCI machine learning repository. Some information about these databases is listed in Table 1.

Data set	NA	NN	NO	NB	NE	CD
Aust	14	4	6	4	690	383:307
Diab	8	0	8	0	768	268:500
Hear	13	3	7	3	270	120:150
Iris	4	0	4	0	150	50:50:50
Germ	20	11	7	2	1000	700:300
TTT	9	9	0	0	958	332:626
Vote	18	0	0	18	232	108:124

Table 1. General information about the data sets

In Table 1, the meaning of the title in each column is follows: NA-Number of attributes, NN-Number of Nominal attributes, NO-Number of Ordinal attributes, NB-Number of Binary attributes, NE-Number of Examples, and CD-Class Distribution.

We also selected the C4.5 algorithm installed in the Clementine' software package as our benchmark for comparison and the DR algorithm [3] as a reference to data reduction. A 5-fold cross validation method was used to evaluate the performance of C4.5, DR and the P&C algorithm, the classification accuracy and the data reduction ratio were obtained and shown in Table 2. The reduction ratio we used is defined as follows:

(The number of tuples in the original data set - The number of the biggest irregular regions in the model) / (The number of tuples in the original data set).

The experimental results in Table 2 show that P&C outperforms C4.5 with respect to the cross validation test accuracy for the data sets but Vote. For the data sets with more numerical attributes (e.g. Iris and Diab data sets) P&C excels in ratio of data reduction compared to DR while preserving the accuracy of classification. Both DR and P&C were tested on the same PC with Pentium(r) III Processor, experiments show that DR has the highest testing accuracy among the three tested algorithms and

P&C has more higher reduction ratio than DR. In particular, on average P&C is about 2 times faster than DR.

Data set	TA:C4.5	TA:DR	TA:P&C	RR:DR	RR:P&C
Aust	85.2	87.0	86.9	70.6	69.1
Diab	72.9	78.6	77.4	68.6	71.1
Hear	77.1	83.3	82.5	74.1	74.2
Iris	94.0	96.7	96.7	94.0	97.1
Germ	72.5	78.0	77.2	73.1	73.2
TTT	86.2	86.9	86.1	81.5	80.7
Vote	96.1	87.0	86.3	99.1	98.8
Average	83.4	85.4	84.7	80.1	80.6

Table 2. A comparison of C4.5, DR and P&C in testing accuracy and reduction ratio
(TA-Testing Accuracy, RR-Reduction Ratio).

5 Conclusion

In this paper, we have presented a novel approach to data reduction and classification (and so data mining) based on spatial partitioning. The reduced data can be regarded as a model of the raw data. We have shown that data reduction can be viewed as a process to find the biggest irregular region to represent the data in the same class. It executes union, intersection and complement operations in each dimension using the projection of spatial regions in multidimensional data space and represents the raw data of the same class by the local biggest irregular regions to realize the goal of data reduction. A series of spatial regions obtained from the learning process can be used in classification. Further research is required into how to eliminate noise and resolve the marginal problem to improve testing accuracy as current P&C is sensitive to noise data and data in marginal areas has lower testing accuracy.

References

1. Weiss, S. M., and Indurkhya, N. (1997). *Predictive Data Mining: A Practical Guide.* Morgan Kaufmann Publishers, Inc.
2. Fayyad, U. M. (1997). Editorial. *Data Mining and Knowledge Discovery – An International Journal* 1(3).
3. Hui Wang, Ivo Duntsch, David Bell. (1998). *Data reduction based on hyper relations.* In proceedings of KDD98, New York, pages 349-353.
4. Duntsch,I., and Gediga, G. (1997). *Algebraic aspects of attribute dependencies in information systems.* Fundamenta Informaticae 29:119-133.
5. Gratzer, G. (1978). *General Lattice Theory.* Basel: Birkhauser.
6. Ullman, J. D. (1983). *Principles of Database Systems.* Computer Science Press, 2 edition.
7. Wolpert, D. H. (1990). *The relationship between Occam's Razor and convergent guessing. Complex Systerms* 4:319-368.
8. Gongde Guo, Hui Wang and David Bell. (2000). *Data ranking based on spatial partitioning.* In proceedings of IDEAL2000, HongKong, pages78-84. Springer-Verlag Berlin Heidelberg 2000.

Alternate Methods in Reservoir Simulation

Guadalupe I. Janoski, Andrew H. Sung

silfalco@cs.nmt.edu, sung@cs.nmt.edu
Department of Computer Science
New Mexico Institute of Mining and Technology
Socorro, New Mexico 87801

Abstract. As time progresses, more and more oil fields and reservoirs are reaching maturity; consequently, secondary and tertiary methods of oil recovery have become increasingly important in the petroleum industry. This significance has added to the industry's interest in using simulation as a tool for reservoir evaluation and management to minimize costs and increase efficiency. This paper presents results of several experiments using soft computing algorithms and techniques to perform history matching, a well-known important simulation task usually performed to calibrate reservoir simulators.

1 Introduction

An important step in calibrating a petroleum reservoir simulator for modeling a particular reservoir or oil field is to perform *history matching*, where the simulation engineer attempts to match the simulator-generated production curves (consisting of the output of oil, gas, and water) against the reservoir's actual production over a period of time. Once properly calibrated, the simulator can be used to predict future production and perform other tasks for reservoir evaluation.

Since the reservoir physics is well understood in the small scale (See Janoski et al, 2000 for more information) history matching appears to be fairly simple. In actuality, however, satisfactory results for history matching are usually difficult to achieve, and have mostly been obtained by the use of *ad hoc* methods. A typical method involves a petroleum simulation engineer familiar with the reservoir running the simulator with tentative input values; manually inspecting results after each run, and adjusting the input parameters according to their knowledge or experience; and repeating the simulation–adjustment process until satisfactory results are obtained; at which point the final set of input values is adopted to "calibrate" the simulator. This results in high costs in machine usage, labor, and time.

For example, while in this study of our version of the problem we are only attempting a history match with a small 8 production well section, we must deal with a search space of over 2^{12954} different possible solutions even under a greatly simplified version of the reservoir model for simulation. The reason for this large solution space is that we must include 17 wells in the surrounding area for environmental data, and use a multi-layer gridded cube consisting of 7 layers, each layer having 256 grid blocks, with each grid block having over 32 parameters with real number ranges. This is illustrated with the single level map in Fig. 1 of the well layout (Chang & Grigg, 1998).

In short, three compounding factors contribute to the difficulty of history matching: problem size, inadequate resources, and the necessity for human intervention in the simulation process.

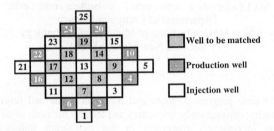

Fig. 1. A well map of the simulation area. The interior 8 production wells {7, 8, 9, 12, 14, 17, 18, 19} were the wells to be matched, while the remaining wells were provided for environmental data.

1.1 Problem Size

We have an enormous data space in which to locate the best solution; the shear size makes finding the optimal solution a true "grand challenge" problem. Needless to say it is impossible to try each solution, and some intelligent search algorithm must be used. Since the simulation is based on a highly simplified model of the reservoir to make the computation tractable, we aim to search for optimal solutions as exact solutions may never be located.

1.2 Inadequate Resources

The second problem originates directly from the first, in that even supercomputers would be hard pressed to find solutions in a history matching problem, except in the most trivial cases. Further, the high cost of supercomputers would preclude the smaller oil companies from performing history matching, while in fact the smaller companies–the majority of them operating on mature fields–may benefit the most from doing it. This makes the idea of developing smart algorithms for searching the data space even more appealing.

1.3 Human intervention

This problem is the hardest to deal with since it is not apparent how one can automate the handling of the inter-simulation parameter adjustment without any human intervention. We show, however, that several soft computing based methods can be used to inspect the simulation results and perform parameter adjustment with satisfactory results.

2 The Reservoir

The historical data was acquired from the East Vacuum Grayburg/San Andres Unit (EVGSAU) reservoir, which is owned and operated by the Phillips Petroleum Company. The reservoir covers 7025 acres in Lea County, New Mexico.

The data gathered from this reservoir includes three phases: primary depletion from 1959-1979, secondary depletion that consisted of water-flood injection from 1980-1985, and finally a tertiary phase which applied a CO_2-flood WAG (water alternating gas) injection, from 1985-1992.

In 1990 a pilot area of the EVGSAU reservoir was selected as a site for a foam field trial to comprehensively evaluate the use of foam for improving the effectiveness of CO_2 injection projects, so as to advance the CO_2-foam technology for improved oil recovery. Specifically, the prime directive of the foam field trial was to prove that a foam could be generated and that it could aid in suppressing the rapid CO_2 breakthrough by reducing the mobility of CO_2 in the reservoir. Operation of the foam field trial began in 1991 and ended in 1993. The response from the foam field trial was very positive; it successfully demonstrated (Martin, et al., 1995) that a strong foam could be formed *in situ* at reservoir conditions and that the diversion of CO_2 to previously bypassed zones/areas due to foam resulted in increased oil production and dramatically decreased CO_2 production. This foam trial period formed the tertiary period of the reservoir.

3 The Simulator

As part of the CO_2 project, the multi-component pseudo-miscible simulator MASTER (Miscible Applied Simulation Techniques for Energy Recovery), which was developed by the U.S. Department of Energy, was modified by incorporating a foam model and used to conduct a history match study on the pilot area at EVGSAU to understand the process mechanisms and sensitive parameters. The ultimate purpose was to establish a foam predictive model for CO_2-foam processes. Details of a manually performed history match and results are reported in (Chang & Grigg, 1998).

In doing our experiments a modified version of the simulator was used. The most important modification was a simplification of the simulator input data. Instead of the dozens of input parameters required, we used only permeability and relative permeability values. A secondary modification was also made as we decided that input values for only a single reservoir layer will be specified, and proportional adjustment will be used to give the inputs for the remaining layers. These modifications brought our simulation input parameters down to 32 parameters, consisting of 25 permeability parameters and 7 relative permeability parameters. This leaves us with a simpler, but still nearly untenable problem.

4 Expert System and Fuzzy Control

Building a simple expert system and fuzzy controller proved to be of invaluable use in understanding the complex well interactions as permeability values were adjusted on the well map during the simulation process.

4.1 Expert System

Our initial studies began with the construction of a simple expert system (ES) for parameter adjustment, which would later form the basis of the fuzzy controller. The rules of the controller were formulated empirically from a study of more than 200 simulation runs that formed an initial parameter study. The ES was composed of 25 IF-THEN rule groups, one for each well. These rules used a combination of actual well error values and predicted well error values. See Fig. 2 for an example. For ease of use the error values were divided into one of nine ranges: {EL Extremely Low, VL Very Low, L Low, SL Slightly Low, K within tolerance, SH Slightly High, H High, VH Very High, EH Extremely High}. Each rule set was run in sequence and the resulting predicted set passed to the next rule to be used as an actual set.

The ES proved very useful as it allowed for rapid prototyping, and quick reduction of error in the match. The primary problem was the granularity induced by having only 9 error ranges. Standardized parameter alteration values tended to cause oscillation as error ranges would tend to bounce between two opposing sets (such as H to L, and L to H), in later runs. Refer to Fig. 3. Due to the fact that wells 8 and 12 tended to work inversely of each other, oscillation tended to occur, i.e., as the match of one well improves, the other's would worsen. Despite this, reductions in error by over 800% by the fourth or fifth iteration of the ES were not uncommon.

```
If well error value for well 8 is SH Slightly High and
well error value for well 12 is SL Slightly Low then
decrease parameter 3 by 30. Change predicted set for
well 8 and 12 to K.
```

Fig. 2. A partial IF-THEN rule for well 3. The underlined denotes well error ranges.

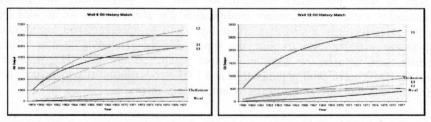

Fig. 3. Results of a match on the primary depletion period, for wells 8 and 12. As can be seen, initial error of the match decreases rapidly, towards the case being matched (actual historical data in this simulation).

4.2 Fuzzy Controller

This section describes a fuzzy controller for automatic parameter adjustment in using MASTER for history matching. The purpose of a controller is to carry out the parameter adjustment automatically and thus eliminate human intervention. The benefits of fuzzy control in this application are the ability to get around the problems of complexity in formulating exact rules and to deal with situations where there are multiple meta rules that may be applicable under similar circumstances. For example, consideration of fluid flow physics leads to the development of three "meta-rules" for permeability adjustment:

1.   If both wells' outputs are too high, then choose those blocks whose reduction in permeability leads to lower outputs.

2.   If wells' outputs are too low, then choose those blocks whose increase in permeability leads to higher outputs.

3.   If one well's output is too high and the other's is too low, then choose those blocks whose alteration in permeability leads to proportional, corrective shifts of outputs.

Rules of the third type are most difficult to obtain, even for experts or simulation engineers familiar with the reservoir, since many factors need to be considered before a decision is made regarding which blocks' permeabilities to increase and which blocks' to decrease; thus the need for developing the rules empirically.

The fuzzy controller consists of sections:

1.   Fuzzification Module: Accepts condition/Input and calculated membership grades to express measurement uncertainties.

2.   Fuzzy Inference Engine: Uses the fuzzified measurements and the rules in the rule base to evaluate the measurements.

2.   Fuzzy Rule Base: contains the list of fuzzy rules.

3.   Defuzzification Module: converts the conclusion reached by the inference engine, into a single real number answer

The primary benefits of using fuzzy control is that it is easy to design and tune, and it avoids the difficulty of formulating exact rules for control actions. The fuzzy controller's rules are empirically obtained, based on a parameter study in which a single well's permeability value was altered while the rest of the 24 permeability values were held constant. The fuzzy controller (Klir and Yuan, 1995; Jang et al., 1997) implemented for permeability adjustment is of the simplest kind in that percentage errors and control actions are fuzzified, but only rarely will more than one rule fire. The control action applied is thus usually only scaled by the membership grade of the percentage error in the error fuzzy set. The adaptive controller works as follows.

Fuzzification is accomplished by usage of membership functions.   After a simulation is run, an error calculation is made from the simulated and the synthetic case or historical data based on a percent error formula. This value is then used to determine error values membership in each fuzzy set: {EL Extremely Low, VL Very Low, L Low, SL Slightly Low, K within tolerance, SH Slightly High, H High, VH Very High, EH Extremely High}. The corresponding fuzzy set values are -4, -3, -2, -1, 0, 1, 2, 3, 4, respectively.

Inference begins once the membership grades are calculated. It assigns the fuzzy set with the highest membership value for each well. If an equilibrium condition is reached between two sets, the set value closest to K is chosen.

Rule Firing is our next step. Within the fuzzy rule base there are 3 types of rules: I (increase production rules), D (decrease production rules), and P (shift production from one well to another). Based on the fuzzy set assigned to each well, we can decide the rule type that needs to be applied.. Based on the fuzzy set value assigned to each well, we can calculate the average set distance from K and decide the change degree (firing strength) of a rule that needs to be applied.

The final step is application of the control action. The action taken depends on the chosen rule type and the degree change needed. The parameters for the next simulation run are now altered.

Many experiments have been conducted (Janoski, 1999). The fuzzy controller's performance depends, naturally, on the definition of fuzzy sets for error and the definition of the fuzzy sets for control actions; therefore, the rule base needs to be fine tuned for optimal performance. Since the rules must be based on empirical observations, other factors, such as scaling factors of the controller (Palm, 1995), may not be quite as critical. The basic idea of using a fuzzy controller for automatic parameter adjustment in history matching, however, has been validated by using a specific controller with crisp control actions. In this case we were able to obtain very good matches within 5 iterations for the two wells over their primary production period of 18 years. Previously, with manual adjustment, such close matches would easily take several weeks to achieve.

5 Neural Networks

The neural network method we chose uses a group of predicted oil production output as training data. The data was acquired by running the MASTER simulator with a set of ranged values so as to cover the hypothetical case's production output curve. This resulted in a set of curves, which bracketed the synthetic case. We found that it was necessary for the training data to cover the synthetic case history so to restrict the range of the problems. The figure below shows a very small set of cases that cover the history. The solid line in is the synthetic case history.

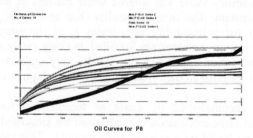

Oil Curves for P8

Fig. 4. The bracketing curves for the synthetic case (solid line).

Once the network is well trained, we feed the network with the historical data to get 25 permeability parameters and 7 relative permeability parameters. We then feed these parameters into the MASTER simulator to check if these parameters are acceptable and to create an estimate of the errors for these parameters.

The Network we built for this study is a three-layer feedforward network, with 27 input units (historical data), 30 hidden units and 32 outputs (permeabilities). The scaled conjugate descent learning algorithm (Möller, 1993) is used in training. Fig. 5 shows the comparisons between the desired output and the output from a trained network. We can see good matches between predicted value and desired value except for one pair; however, this mismatch can likely be attributed to the fact that certain permeability values have little effect on the output history. (For example, during the first 20 years, the 25th permeability value causes less than a 1% change across its complete value range.) Furthermore, Fig. 5 shows an experimental result in using the neural network to match the chosen hypothetical case, displaying a very close match. Currently the training data and testing data are all simulation results from MASTER simulator. In future experiments real historical data will be used as the number of parameters are increased.

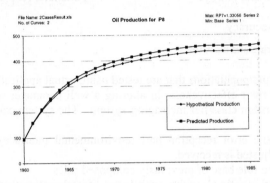

Fig. 5. Synthetic case match using the neural network.

6 Genetic Algorithms

Initial genetic algorithm (GA) trials were run using differing crossover methods. These studies proved interesting in that little information was needed in creating the GA system, but at the same time proved to have a huge drawback. As simulation times could range up to 45 minutes on even a 600MHz PC, creating initial populations and simulating future generations became extremely costly. As a result smaller populations for initial testing were used, thus limiting the GA, as large degrees of similarity occurred between population members in succeeding generations.

In doing these studies we worked with a data vector composed of the parameter set. Primary studies were done using multipoint crossover. Population improvement tended toward only a 1-3% change in the initial few generations. In using standard crossover the best results were found using a complete generational replacement scheme with small random number of crossover points for each new child.

7 Hybrid Systems: A Classifier System

ARIA (Automatic Recombinant Input Approach) uses a non-standard genetic based learning machine (GBML) to create a set of rules that may be used for atom rule creation for control in the history matching.

Each rule consists of a set of atom-like subsections. For example in rule 1 below each line is an atom section that we work with:

Rule 1:	**Classifier 1:**
Error Environment Match:	Error Calculations:
Error Well X=range	Error for well N= Error
Actions:	Parameter List:
Change Parameter N by X	Parameter N=pN
Statistics:	
Age, Uses, Accuracy	

Fig. 6. ARIA example rule and classifier.

ARIA consists of rule populations that are tested using actual application data, and are tracked based on their effectiveness, in altering a well parameter set. It uses a standard genetic algorithm classifier messaging system. The system consists of four parts:

1. Part one is the error reception part (environmental interface) in which a parameter set to be adjusted is received.
2. Part two is the rule base of previously composed rules.
3. Part three is a genetic algorithm with fuzzy control section that creates new rules when a rule is not available from the database.
4. The final part is the messaging system that tracks a rule's effectiveness, matches a rule to a parameter set and takes relevant action.

The first part of the ARIA, the environmental interface, creates error estimates from the actual historical production data. This error calculation is done by using a Sum Squared Error (SSE) calculation between the predicted output of each well and its historical values. Once the SSE has been calculated for each well, these 8 values become the environmental pattern (classifier) that will be used in the messaging system for rule matching. Fig. 6. is an example of an environmental classifier. It has two parts the error calculations, and the list of parameters that belonged to the simulation data.

These error values in the classifier are then matched based on a set of fuzzy rules in the messaging system to appropriate action rules. The fuzzy control in this section is very basic and consists of simplistic rules that determine if a rule exists within range, a tolerance factor, and rates each rule by its statistical information, and then determines which is the most appropriate rule. This is done by attempting to find a rule whose Error Environmental Match ranges for each well bracket the classifiers error calculations. Since there do exist 8 error calculations the idea of a tolerance factor was introduced, in which not all 8 error calculations must be in range. This calculation is done by using an averaging function to calculate how out of range the error set is. If an adequate rule is found, it is then used and statistical success or failure data of its use on the simulation parameters is average together. On the other hand, if an appropriate rule cannot be located the ARIA system invokes the genetic fuzzy logic creation algorithm to create a new rule, which is then used.

This method has shown some promise in application, and is merely an extension off of the previous work for the MASTER WEB project in which fuzzy control was applied resulting in convergence and error control within ten generation, and 200% error ranges, proving to be a very quick and accurate system. Currently the system has been running small numbers of iterations, as tests are being run to determine the best initial rule population. Currently small numbers of changes, that rely on being able to affect parameters within the lower third of their value ranges, without causing parameters to go out side of their allowed values, have shown the most successful ability to converge to a solution. They have been able to come within a 30 to 70% error within approximately 15 iterations.

The size of the rule base has also been shown to have a significant effect on the number of iterations, as the larger the size the more likely an appropriate rule will be found. Created rules have are extremely dependent on the genetic algorithm used, as wells have complex interactions.

8 Conclusion

In this paper, we have proposed several soft computing based methods to assist in the challenging task of reservoir simulation and presented experimental results.

While simulation experts have traditionally preformed history matching semi-automatically, soft computing algorithms offer a great deal of promise in the field of reservoir simulation, particularly history matching. These algorithms have performed very well in our experiments in minimizing human intervention in the simulation process, thereby resulting in great savings.

Preliminary results indicate that the soft-computing-enhanced simulation on the EVGSAU reservoir is capable of producing satisfactory matches within hours. These results would have taken weeks to achieve previously. We believe, therefore, that the potential of our approach is clearly demonstrated and further research and development efforts are warranted to achieve a cost effective and fully automatic history matching.

8 Acknowledgement

We would like to gratefully acknowledge that support for this research was received from Sandia National Laboratories and the State of New Mexico. Dr. Eric Chang (previously with the Computer Science Department of New Mexico Tech) and Dr. Reid Grigg (of the Petroleum Recovery Research Center of New Mexico Tech) initiated our interest in this project and made invaluable contributions during its development. Mr. F.S Li helped perform many neural network modeling experiments.

References

1. Ammer, J.R., Brummert, A.C., and Sams, W.N. "Miscible Applied Simulation Techniques for Energy Recovery – Version 2.0." Report to U.S. Department of Energy, Contract No. DOE/BC–91/2/SP, February 1991.
2. Aziz, K., and A. Settari. Petroleum Reservoir Simulation. London: Applied Science, 1979.
3. Chang, S. -H. and Grigg, R. B. "History Matching and Modeling the CO2-Foam Pilot Test at EVGSAU." Paper SPE 39793 presented at the 1998 SPE Permian Basin Oil and Gas Recovery Conference, Midland, Texas.
4. Jang, J.-S. R., Sun, C.-T., and Mizutani, E. Neural-Fuzzy and Soft Computing. New Jersey: Prentice-Hall, 1997.
5. Janoski, G., Pietrzyk, M., Sung, A. H., Chang, S.-H., Grigg, R. B. "MASTER Web: A Petroleum Reservoir Simulation Tool." Proceedings of the International Conference on Web-Based Modeling and Simulation & Virtual Worlds and Simulation Conference. San Diego: Simulation Councils Inc., 2000.
6. G.J. Klir, and B. Yuan. Fuzzy Sets and Fuzzy Logic: Theory and Applications. New Jersey: Prentice Hall, 1995.
7. Janoski, G., Pietrzyk, M., Sung, A. H., Chang, S.-H., Grigg, R. B. "Advanced Reservoir Simulation Using Soft Computing." Intelligent Problem Solving: Methodologies and Approaches, Lecture Notes in Artificial Intelligence 1821, (2000): 623-628.
8. Kosko B. Fuzzy Engineering. New Jersey: Prentice Hall, 1997.
9. Li, H.J., Li, F.S., Sung, A.H., and Weiss, W.W. "A Fuzzy Inference Algorithm for Lithology Analysis in Formation Evaluation." Intelligent Problem Solving: Methodologies and Approaches, Lecture Notes in Artificial Intelligence 1821, (2000): 623-628.
10. Martin, F.D., Stevens, J.E., and Harpole, K.J. "CO2-Foam Field Test at the East Vacuum Grayburg/San Andres Unit." SPERE (Nov. 1995) 266.
11. Möller M.F. "A Scaled Conjugate Gradient Algorithm for Fast Learning." Neural Networks, 6, (1993): 525-533.
12. Nghiern, L. "An Integral Approach for Discretizing the Reservoir Flow Equations." SPE Reservoir Engineering 3 no. 2(May 1988):685-690.
13. Palm R. "Scaling of Fuzzy Controllers Using the Cross-Correlation." IEEE Tran. Fuzzy Systems 3, no. 1, (1995) : 116-123.
14. Peaceman, D. W. Fundamentals of Numerical Reservoir Simulation. New York: Elsevier, 1977.
15. Pruess, K., and G. S. Bodvarsson. "A Seven-Point Finite Difference Method for Improved Grid Orientation Performance in Pattern Steamfloods." SPE Paper 12252. Richardson, Texas: Society of Petroleum Engineers. 1983.
16. Xiong, et al. "An Investigation into the Application of Fuzzy Logic to Well Stimulation Treatment Design." SPE 27672. SPE Computer Applications. Texas: Society of Petroleum Engineers. Feb. 1995.

Intuitionistic Fuzzy Sets in Intelligent Data Analysis for Medical Diagnosis

Eulalia Szmidt and Janusz Kacprzyk

Systems Research Institute, Polish Academy of Sciences
ul. Newelska 6, 01–447 Warsaw, Poland
E-mail: {szmidt, kacprzyk}@ibspan.waw.pl

Abstract

We propose a new approach for medical diagnosis by employing intuitionistic fuzzy sets [cf. Atanassov [1, 2]]. Solution is obtained by looking for the smallest distance [cf. Szmidt and Kacprzyk [7, 8]] between symptoms that are characteristic for a patient and symptoms describing illnesses considered. We point out advantages of this new concept over the method proposed by De, Biswas and Roy [4] where intuitionistic fuzzy sets were also applied, but the max-min-max rule was used instead of taking into account all, unchanged symptom values as proposed in this article.

1 Introduction

For many real world problems, imperfect, imprecise information is a vital part of the problem itself, and a continuing reasoning without proper modelling tools may lead to generating inaccurate inferences.

Traditional methods of analysis are oriented toward the use of numerical techniques. By contrast, much of human reasoning involves the use of variables whose values are not numerical. This observation is a basis for the concept of a linguistic variable, that is, a variable whose values are words rather than numbers, in turn represented by fuzzy sets.

The use of linguistic variables represents a significant paradigm shift in system analysis. More specifically, in the linguistic approach the focus of attention in the representation of dependencies shifts from difference and differential equations to fuzzy $IF - THEN$ rules in the form $IF\ X\ is\ A\ THEN\ Y\ is\ B$, where X and Y are linguistic variables and A and B are their linguistic values, e.g. $IF\ Pressure\ is\ high\ THEN\ Volume\ is\ low$.

Description of system behaviour in the language of fuzzy rules lowers the need for precision in data gathering and data manipulation, and in effect may be viewed as a form of data compression.

But there are situations when description by a (fuzzy) linguistic variable, given in terms of a membership function only, seems too rough. For example, in decision making problems, particularly in the case of medial diagnosis, sales analysis, new product marketing, financial services, etc. there is a fair chance of the existence of a non-null hesitation part at each moment of evaluation of an unknown object.

Intuitionistic fuzzy sets (Atanassov [1, 2]) can be viewed in this context as a proper tool for representing hesitancy concerning both the membership and non-membership of an element to a set. To be more precise, a basic assumption of fuzzy set theory that if we specify the degree of membership of an element in a fuzzy set as a real number from [0, 1], say a, then the degree of its non-membership is automatically determined as $1 - a$, need not hold for intuitionistic fuzzy sets. In intuitionistic fuzzy set theory it is assumed that the non-membership should not be more than $1 - a$. The difference let us express a lack of knowledge (hesitancy concerning both the membership and non-membership of an element to a set). In this way we can better model imperfect information – for example, we can express the fact that the temperature of a patient changes, and other symptoms are not quite clear.

In this article we will present intuitionistic fuzzy sets as a tool for reasoning in the presence of imperfect facts and imprecise knowledge. An example of medical diagnosis will be presented assuming there is a database, i.e. description of a set of symptoms S, and a set of diagnoses D. We will describe a state of a patient knowing results of his/her medical tests. Description of the problem uses the notion of an intuitionistic fuzzy set. The proposed method of diagnosis involves intuitionistic fuzzy distances as introduced in (Szmidt and Kacprzyk [7, 8]). Advantages of such an approach are pointed out in comparison with the method presented in (De, Biswas and Roy [4]) in which the max-min-max composition rule was applied.

The material in the article is organized as follows. In Section 2 we briefly overview intuitionistic fuzzy sets. Section 3 presents De, Biswas and Roy's [4] approach for medical diagnosis via intuitionistic fuzzy relations, or – to be more precise – via the max-min-max composition. In Section 4 we propose a new approach for solving the same problem – we also use intuitionistic fuzzy sets but the final diagnosis is pointed out by the smallest distance between symptoms characteristic for a patient and symptoms decsribing considered illnesses. Finally, we finish with some conclusions in Section 5.

2 Brief introduction to intuitionistic fuzzy sets

As opposed to a fuzzy set (Zadeh [9]) in $X = x$, given by

$$A^{'} = \{< x, \mu_{A'}(x) > | x \in X\} \tag{1}$$

where $\mu_{A'} : X \to [0, 1]$ is the membership function of the fuzzy set $A^{'} : \mu_{A'}(x) \in [0, 1]$; is the membership of $x \in X$ in $A^{'}$, an intuitionistic fuzzy set (Atanassov [2]) $A \in X$ is given by

$$A = \{< x, \mu_A(x), \nu_A(x) > | x \in X\} \tag{2}$$

where: $\mu_A : X \to [0, 1]$ and $\nu_A : X \to [0, 1]$ such that

$$0 \le \mu_A(x) + \nu_A(x) \le 1 \tag{3}$$

and $\mu_A(x), \nu_A(x) \in [0, 1]$ denote the degree of membership and non-membership of $x \in A$, respectively.

Obviously, each fuzzy set corresponds to the following intuitionistic fuzzy set

$$A = \{< x, \mu_{A'}(x), 1 - \mu_{A'}(x) > | x \in X\} \qquad (4)$$

For each intuitionistic fuzzy set in X, we will call

$$\pi_A(x) = 1 - \mu_A(x) - \nu_A(x) \qquad (5)$$

a *hesitation margin* (or *intuitionistic fuzzy index*) of $x \in A$, and it is a hesitation degree of whether x belongs to A or not [cf. Atanassov [2]]. It is obvious that $0 \le \pi_A(x) \le 1$, for each $x \in X$.

On the other hand, for each fuzzy set A' in X, we evidently have

$$\pi_{A'}(x) = 1 - \mu_{A'}(x) - [1 - \mu_{A'}(x)] = 0, \text{ for each } x \in X \qquad (6)$$

Therefore, we can state that if we want to fully describe an intuitionistic fuzzy set, we must use any two functions from the triplet:

- membership function,

- non-membership function, and

- hesitation margin.

In other words, the application of intuitionistic fuzzy sets instead of fuzzy sets means the introduction of another degree of freedom into a set description (i.e. in addition to μ_A we also have ν_A or π_A).

3 An intuitionistic fuzzy sets approach to medical diagnosis due to De, Biswas and Roy [4]

By following the reasoning of De, Biswas and Roy [4] (which is an extension of Sanchez's approach [5, 6]), we will now consecutively recall their approach to medical diagnosis via intuitionistic fuzzy sets, or to be more precise – via intuitionistic fuzzy relations that in effect boils down to applying the max-min-max composition [3].

The approach presented by De, Biswas and Roy [4] involves the following three steps:

- determination of symptoms,

- formulation of medical knowledge based on intuitionistic fuzzy relations, and

- determination of diagnosis on the basis of composition of intuitionistic fuzzy relations.

Table 1:

Q	$Temperature$	$Headache$	$Stomach$ $pain$	$Cough$	$Chest$ $pain$
Al	$(0.8, 0.1)$	$(0.6, 0.1)$	$(0.2, 0.8)$	$(0.6, 0.1)$	$(0.1, 0.6)$
Bob	$(0.0, 0.8)$	$(0.4, 0.4)$	$(0.6, 0.1)$	$(0.1, 0.7)$	$(0.1, 0.8)$
Joe	$(0.8, 0.1)$	$(0.8, 0.1)$	$(0.0, 0.6)$	$(0.2, 0.7)$	$(0.0, 0.5)$
Ted	$(0.6, 0.1)$	$(0.5, 0.4)$	$(0.3, 0.4)$	$(0.7, 0.2)$	$(0.3, 0.4)$

A set of n patients is considered. For each patient p_i, $i = 1, \ldots, n$, a set of symptoms S is given. As a result, an intuitionistic fuzzy relation Q is given from the set of patients to the set of symptoms S.

Next, it is assumed that another intuitionistic fuzzy relation R is given – from a set of symptoms S to the set of diagnoses D. The composition T of intuitionistic fuzzy relations R and Q describes the state of a patient given in terms of a membership function, $\mu_T(p_i, d_k)$, and a non-membership function, $\nu_T(p_i, d_k)$, for each patient p_i and each diagnosis d_k.

The functions are calculated in the following way [4]:

$$\mu_T(p_i, d_k) = \bigvee_{s \in S} [\mu_Q(p_i, s) \wedge \mu_R(s, d_k)] \tag{7}$$

and

$$\nu_T(p_i, d_k) = \bigwedge_{s \in S} [\nu_Q(p_i, s) \vee \nu_R(s, d_k)] \tag{8}$$

where $\bigvee = \max$ and $\bigwedge = \min$.

Example 1 [4] Let there be four patients: Al, Bob, Joe and Ted, i.e. $P = \{Al, Bob, Joe, Ted\}$. The set of symptoms considered is $S = \{temperature, headache, stomach pain, cough, chest-pain\}$. The intuitionistic fuzzy relation $Q(P \rightarrow S)$ is given in Table 1.

Let the set of diagnoses be $D = \{Viral fever, Malaria, Typhoid, Stomach problem, Chest problem\}$. The intuitionistic fuzzy relation $R(S \rightarrow D)$ is given in Table 2.

Therefore, the composition T (7)–(8) is given in Table 3.

But as the max-min-max composition was used when looking for T, "dominating" symptoms were in fact only taken into account. So, in the next step an improved version of R is calculated for which the following holds [4]:

- $S_R = \mu_R - \nu_R \pi_R$ is the greatest, and

- equations (7)–(8) are retained.

Effects of the presented improvements [4] are given in Table 4.

□

Table 2:

R	Viral fever	Malaria	Typhoid	Stomach problem	Chest problem
Temperature	(0.4, 0.0)	(0.7, 0.0)	(0.3, 0.3)	(0.1, 0.7)	(0.1, 0.8)
Headache	(0.3, 0.5)	(0.2, 0.6)	(0.6, 0.1)	(0.2, 0.4)	(0.0, 0.8)
Stomach pain	(0.1, 0.7)	(0.0, 0.9)	(0.2, 0.7)	(0.8, 0.0)	(0.2, 0.8)
Cough	(0.4, 0.3)	(0.7, 0.0)	(0.2, 0.6)	(0.2, 0.7)	(0.2, 0.8)
Chest pain	(0.1, 0.7)	(0.1, 0.8)	(0.1, 0.9)	(0.2, 0.7)	(0.8, 0.1)

Table 3:

T	Viral fever	Malaria	Typhoid	Stomach problem	Chest problem
Al	(0.4, 0.1)	(0.7, 0.1)	(0.6, 0.1)	(0.2, 0.4)	(0.2, 0.6)
Bob	(0.3, 0.5)	(0.2, 0.6)	(0.4, 0.4)	(0.6, 0.1)	(0.1, 0.7)
Joe	(0.4, 0.1)	(0.7, 0.1)	(0.6, 0.1)	(0.2, 0.4)	(0.2, 0.5)
Ted	(0.4, 0.1)	(0.7, 0.1)	(0.5, 0.3)	(0.3, 0.4)	(0.3, 0.4)

Table 4:

	Viral fever	Malaria	Typhoid	Stomach problem	Chest problem
Al	0.35	0.68	0.57	0.04	0.08
Bob	0.20	0.08	0.32	0.57	0.04
Joe	0.35	0.68	0.57	0.04	0.05
Ted	0.32	0.68	0.44	0.18	0.18

It seems that the approach proposed in [4] has some drawbacks. First, the max-min-max rule alone (Table 3) does not give a solution. To obtain a solution, the authors [4] propose some changes in medical knowledge $R(S \to D)$. But it is difficult to justify the proposed changes as medical knowledge $R(S \to D)$ is (at least, should be without a doubt) based on many cases, and knowledge of physicians, so it is difficult to understand sudden, arbitral changes in it.

Next, the type of changes: $S_R = \mu_R - \nu_R \pi_R$ means that the membership function describing relation R (medical knowledge) is weakened. But, in the idea of intuitionistic fuzzy sets there is nowhere assumed that the membership function can decrease because of the hesitation margin or the non-membership function. The hesitation margin (or part of it) can be split between the membership and non-membership functions (so, in fact, it can be added to, not substracted from the membership function). Summing up, the proposed improvements, although leading to some solutions, are difficult to understand because of arbitral (both from practical (doctors' knowledge) and theoretical (theory of intuitionistic fuzzy sets)) points of view.

4   Medical diagnosis via distances for intuitionistic fuzzy sets

To solve the same problem as in [4], but without manipulations in medical knowledge base, and with taking into account all the symptoms characteristic for each patient, we propose a new method based on calculating distances between diagnoses and patient tests.

As in [4], to make a proper diagnosis D for a patient with given values of tested symptoms S, a medical knowledge base is necessary. In our case a knowledge base is formulated in terms of intuitionistic fuzzy sets.

To compare the approach proposed in this article with the method of De, Biswas and Roy [4], and described shortly in Section 3, we consider just the same data. Let the set of diagnoses be $D = \{$ *Viral fever, Malaria, Typhoid, Stomach problem, Chest problem*$\}$. The considered set of symptoms is $S = \{$ *temperature, headache, stomach pain, cough, chest-pain*$\}$.

The data are given in Table 5 – each symptom is described by three numbers: membership μ, non-membership ν, hesition margin π. For example, for malaria: the temperature is high ($\mu = 0.7$, $\nu = 0$, $\pi = 0.3$), whereas for the chest problem: temperature is low ($\mu = 0.1$, $\nu = 0.8$, $\pi = 0.1$). In fact data in Table 2 and Table 5 are exactly the same (due to (5)) but by involving in an explicit way the hesitation margin too, we want to stress that the values of all three parameters are necessary in our approach.

The considered set of patients is $P = \{$ *Al, Bob, Joe, Ted*$\}$. The symptoms characteristic for the patients are given in Table 6 – as before, we need all three parameters (μ,ν,π) describing each symptom but the data are the same (due to (5)) as in Table 1.

Our task is to make a proper diagnosis for each patient p_i, $i = 1,\ldots,4$.

Table 5:

	Viral fever	Malaria	Typhoid	Stomach problem	Chest problem
Temperature	(0.4, 0.0, 0.6)	(0.7, 0.0, 0.3)	(0.3, 0.3, 0.4)	(0.1, 0.7, 0.2)	(0.1, 0.8, 0.1)
Headache	(0.3, 0.5, 0.2)	(0.2, 0.6, 0.2)	(0.6, 0.1, 0.3)	(0.2, 0.4, 0.4)	(0.0, 0.8, 0.2)
Stomach pain	(0.1, 0.7, 0.2)	(0.0, 0.9, 0.1)	(0.2, 0.7, 0.1)	(0.8, 0.0, 0.2)	(0.2, 0.8, 0.0)
Cough	(0.4, 0.3, 0.3)	(0.7, 0.0, 0.3)	(0.2, 0.6, 0.2)	(0.2, 0.7, 0.1)	(0.2, 0.8, 0.0)
Chest pain	(0.1, 0.7, 0.2)	(0.1, 0.8, 0.1)	(0.1, 0.9, 0.0)	(0.2, 0.7, 0.1)	(0.8, 0.1, 0.1)

Table 6:

	Temperature	Headache	Stomach pain	Cough	Chest pain
Al	(0.8, 0.1, 0.1)	(0.6, 0.1, 0.3)	(0.2, 0.8, 0.0)	(0.6, 0.1, 0.3)	(0.1, 0.6, 0.3)
Bob	(0.0, 0.8, 0.2)	(0.4, 0.4, 0.2)	(0.6, 0.1, 0.3)	(0.1, 0.7, 0.2)	(0.1, 0.8, 0.1)
Joe	(0.8, 0.1, 0.1)	(0.8, 0.1, 0.1)	(0.0, 0.6, 0.4)	(0.2, 0.7, 0.1)	(0.0, 0.5, 0.5)
Ted	(0.6, 0.1, 0.3)	(0.5, 0.4, 0.1)	(0.3, 0.4, 0.3)	(0.7, 0.2, 0.1)	(0.3, 0.4, 0.3)

To fulfill the task we propose to calculate for each patient p_i a distance of his symptoms (Table 6) from a set of symptoms s_j, $j = 1, \ldots, 5$ characteristic for each diagnosis d_k, $k = 1, \ldots, 5$ (Table 5). The lowest obtained distance points out a proper diagnosis.

In (Szmidt and Kacprzyk [7, 8]) we proved that the only proper way of calculating the most widely used distances for intuitionistic fuzzy sets is to take into account all three parameters: the membership function, the non-membership function, and the hesitation margin. To be more precise, the normalised Hamming distance for all the symptoms of the i-th patient from the k-th diagnosis is equal to

$$l(s(p_i), d_k) = \frac{1}{10} \sum_{j=1}^{5} (|\mu_j(p_i) - \mu_j(d_k)| + |\nu_j(p_i) - \nu_j(d_k)| +$$
$$+ |\pi_j(p_i) - \pi_j(d_k)|) \tag{9}$$

The distances (9) for each patient from the considered set of possible diagnoses are given in Table 7. The lowest distance points out a proper diagnosis: Al suffers from malaria, Bob from stomach problem, Joe from typhoid, whereas Ted from fever.

We obtained the same results, i.e. the same quality diagnosis for each patient when looking for the solution while applying the normalized Euclidean distance [cf. Szmidt and Kacprzyk [7, 8]]:

$$q(s(p_i), d_k) = (\frac{1}{10} \sum_{j=1}^{10} (\mu_j(p_i) - \mu_j(d_k))^2 + (\nu_j(p_i) - \nu_j(d_k))^2 +$$

Table 7:

	Viral fever	Malaria	Typhoid	Stomach problem	Chest problem
Al	0.28	0.24	0.28	0.54	0.56
Bob	0.40	0.50	0.31	0.14	0.42
Joe	0.38	0.44	0.32	0.50	0.55
Ted	0.28	0.30	0.38	0.44	0.54

Table 8:

	Viral fever	Malaria	Typhoid	Stomach problem	Chest problem
Al	0.29	0.25	0.32	0.53	0.58
Bob	0.43	0.56	0.33	0.14	0.46
Joe	0.36	0.41	0.32	0.52	0.57
Ted	0.25	0.29	0.35	0.43	0.50

$$+ \quad (\pi_j(p_i) - \pi_j(d_k))^2)^{\frac{1}{2}} \tag{10}$$

The results are given in Table 8 – the lowest distance for each patient p_i from possible diagnosis D points out a solution. As before, Al suffers from malaria, Bob from stomach problem, Joe from typhoid, whereas Ted from fever.

5 Conclusions

By employing intuitionistic fuzzy sets in databases we can express a hesitation concerning examined objects. The method proposed in this article, performing diagnosis on the basis of the calculation of distances from a considered case to all considered illnesses, takes into account values of all symptoms. As a result, our approach makes it possible to introduce weights for all symptoms (for some illnesses some symptoms can be more important). Such an approach is impossible in the method described in (De, Biswas and Roy [4]) because the max-min-max rule "neglects" in fact most values except for extreme ones.

References

[1] Atanassov K. (1986) Intuitionistic fuzzy sets, Fuzzy Sets and Systems, 20 (1986) 87-96.

[2] Atanassov K. (1999) Intuitionistic Fuzzy Sets: Theory and Applications. Physica-Verlag.

[3] Biswas R. (1997) Intuitionistic fuzzy relations. Bull. Sous. Ens. Flous.Appl. (BUSEFAL), Vol. 70, pp. 22–29.

[4] De S.K., Biswas R. and Roy A.R. (2001) An application of intuitionistic fuzzy sets in medical diagnosis. Fuzzy Sets and Systems, Vol. 117, No.2, pp. 209 – 213.

[5] Sanchez E. (1977) Solutions in composite fuzzy relation equation. Application to medical diagnosis Brouwerian Logic. In: M.M. Gupta, G.N. Saridis, B.R. Gaines (Eds.), Fuzzy Automata and Decision Process, Elsevier, North-Holland.

[6] Sanchez E. (1976) Resolution of composition fuzzy relation equations. Inform. Control, Vol. 30, pp. 38–48.

[7] Szmidt E. and Kacprzyk J. (1997) On measuring distances between intuitionistic fuzzy sets, Notes on IFS, Vol.3, No.4, pp.1 - 13.

[8] Szmidt E. and J. Kacprzyk J. (2000) Distances between intuitionistic fuzzy sets, Fuzzy Sets and Systems, Vol.114, No.3, pp.505 - 518.

[9] Zadeh L.A. (1965) Fuzzy sets. Information and Control 8, pp. 338 - 353.

Design of a Fuzzy Controller Using a Genetic Algorithm for Stator Flux Estimation

Mehmet KARAKOSE   Mehmet KAYA   Erhan AKIN

Firat University Department of Computer Engineering, 23119,Elazig, TURKEY
e-mail: {mkarakose,mekaya,eakin}@firat.edu.tr

Abstract: Performance of the field orientation in induction motors depends on the accurate estimation of the flux vector. The voltage model used for field orientation has in the flux calculation process an open integration problem, which is generally solved with a feedback loop. In this paper, a new method is developed for the feedback loop of the integrator. The method, as apart from studies in the literature, uses a fuzzy controller determined membership functions using a genetic algorithm (GA). For this purpose, a fuzzy controller is designed and tested on various motors of different power ratings. The proposed method is simulated by using MATLAB-SIMULINK and implemented on an experimental system using a TMS320C31 digital signal processor.

1 Introduction

The performance of the vector control is related to the accuracy of the resultant flux phase and magnitude information of the induction motor. There are two common methods to estimate the stator or rotor flux vector of the induction motor. These are called the current model and the voltage model. The current model does not contain an open integration, but requires the rotor parameters information and motor speed measurement during the flux calculation process. In this method, the flux vector is also estimated at standstill due to the lack of an open integration process. On the other hand, the voltage model is sensitive to stator parameters and requires voltage and current measurements, but it also does not require speed measurement. In sensorless vector control, the voltage model is most preferred. There are various methods of stator flux estimation using the voltage model. Rotor flux can also be estimated by using stator flux and motor parameters. The stator flux equations in the stator reference frame known as the voltage model are as follows:

$$\psi_{s\alpha} = \int (u_{s\alpha} - i_{s\alpha} R_s)\,dt \qquad (1a)$$

$$\psi_{s\beta} = \int (u_{s\beta} - i_{s\beta} R_s)\,dt \qquad (1b)$$

By using Equation 1a-1b, $\psi_{s\alpha}$ and $\psi_{s\beta}$ can be estimated. The angle of the stator flux orientation can be calculated by using these equations. This angle is used for axes transformation between $\alpha\beta$ to dq and called as transformation angle θ_s.

$$\theta_s = arctg\frac{\psi_{s\beta}}{\psi_{s\alpha}} \qquad (2)$$

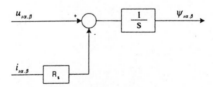

Fig. 1. Block diagram of the voltage model

The block diagram in Fig. 1 shows the flux calculation process described in Equations 1a-1b. In this process, the integrator has no feedback. This open integration process causes drift in the integration. This integration process can be analog or digital. Digital flux integration has integration method and integration step size, errors in the current and voltage measurements, variations in the stator resistance due to temperature and frequency of the current, initial value of the integrator, error introduced by the finite bit length of the processor, execution time of the commands etc.

It is clear from the above open integrator that the performance of the voltage model depends on the stator resistance. This resistance gains further importance in low-speed operation where the voltage drop on this resistance becomes significant. At zero speed, it is not possible to estimate the stator flux by using this algorithm.

Numerous papers exist in the literature on the integration problem, especially sensorless type direct vector-controlled drives. For example, Ohtani [1] proposed to add a feedback to the integration algorithm to solve the problem. In Ohtani's study, a steady-state analysis of the issue is presented. In [2] another approach is studied. This is based on two-phase voltage and current measurements, and stability analysis is made under dynamic conditions. In this study, flux magnitude and flux derivative are used in the feedback loop. However, the gain of the flux derivative feedback is a function of the load and speed. Therefore, the performance of the system depends on the correct value of these feedback gains. A study by Patel [3] employs a cascade connected and automatically adjusted low-pass filter instead of the integrator. An algorithm proposed by Akın [4] eliminates the DC component of the integrator output, but it only works in the steady state.

Some of the algorithms mentioned here are a δ feedback integration algorithm to solve the stability problem of the integrator and a PI feedback integration algorithm. The block diagram and transfer function of the δ feedback integrator are given in Fig. 2 and Equation 3 respectively.

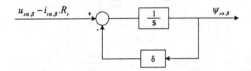

Fig. 2. Block diagram of δ feedback integrator

$$F(s) = \frac{1}{s+\delta} \tag{3}$$

In this integrator, the δ feedback path is used for stability of the integrator. When the frequency is high $(jw >> \delta)$, the integrator function behaves well. However, when jw becomes comparable to δ, both magnitude and phase errors appear and affect the performance both at steady state and in transient conditions. The δ feedback integrator, shown in Fig. 2, eliminates offset substantially. However, integration output still involves phase shift and error in magnitude [5].

In this study, a fuzzy controller is developed for the feedback loop of the integrator. This fuzzy controller has the advantages of robustness, ease of design and good transient response.

One of the main problems of a fuzzy controller design is the process of determining membership functions. Usually such a determination is obtained from human experts. However, the approach adopted for acquiring the shape of any particular membership function often depends on the application. For most fuzzy logic control problems, the membership functions are assumed to be linear and usually triangular in shape. So, typically the sets that describe various factors of importance in the application and the issues to be determined are the parameters that define the triangles. These parameters are usually based on the control engineer's experience and/or are generated automatically. However, for many other applications, triangular membership functions are not appropriate as they do not represent accurately the linguistic terms being modelled, and the shape of the membership functions have to be elicited directly from the expert, by a statistical approach or by automatic generation of the shapes.

GA was employed first by Karr [6] in determination of membership functions. Karr has applied GA to the design of a fuzzy logic controller (FLC) for the cart pole problem. Meredith [7] has applied GA to the fine tuning of membership functions in a FLC for a helicopter. Initial guesses for the membership functions are made by a control engineer, and the GA adjusts the defining parameters by through use in order to minimize the movement of a hovering helicopter. Again, triangular membership functions are used. Lee and Takagi [8] have also tackled the cart problem. They have taken a holistic approach by using GA to design the whole system determining the optimal number of rules as well as the membership function, which are again triangular. Recently, Arslan and Kaya [9] have proposed a new method for determination of fuzzy logic membership functions using GAs.

2 A Survey of Fuzzy Controller and Genetic Algorithms

Fuzzy logic is a technology based on engineering experience and observations. In fuzzy logic, an exact mathematical model is not necessary, because linguistic variables are used in fuzzy logic to define system behavior rapidly. Fuzzy logic is a very recent technology relative to conventional controllers; its areas of application are increasing very quickly. Fuzzy PID, fuzzy PI, fuzzy PD and fuzzy mixed controllers are fuzzy controller design approaches, but unlike conventional controllers the focus is not in the modeling [10].

Some of the problems, such as stability and performance, are encountered both in fuzzy controllers and conventional controllers. Unlike conventional control design, where mathematical models are used to solve these problems, fuzzy controller design involves IF-THEN rules defined by an expert to tackle these problems.

There are two methods that are commonly used to design fuzzy controllers: trial and error method and the theoretical method. In trial and error, IF-THEN rules are defined by using expert knowledge and experience. Then, these rules are applied to the actual system. Unlike the theoretical approach where the parameters are adjusted to guarantee the desired performance, in the fuzzy method the IF-THEN rules are modified until the desired performance is achieved. In practice, both methods can be used to obtain better performance [11].

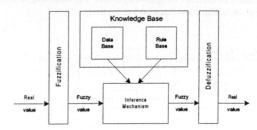

Fig. 3. Block diagram of fuzzy control architecture

The fuzzy controller has four components as shown in Fig. 3. These are:

a- Fuzzifier: The input values are scaled and grouped into fuzzy sets. In other words, the input values labeled and transformed into linguistic variables.

b- Inference mechanism: The inference mechanism uses a database and a rule base. The database involves membership functions that are used by the inference mechanism to make fuzzy decisions.

b- Rule Base: Rule base is a set of IF-THEN rules defined by an expert. The inference mechanism uses these rules.

c- Defuzzifier: The linguistic variables manipulated by the inference mechanism are converted back to real values.

In a fuzzy controller design, the knowledge and observations of an expert are more important than the underlying mathematical model. This expert knowledge and observation is used while the system is being designed. This kind of approach provides an opportunity to easily embed experience into a controller, which has been gained over a long time. However, it is not possible to obtain automation during controller design.

A GA is an iterative procedure that consists of a constant-size population of individuals, each one represented by a finite string of symbols, known as the genome, encoding a possible solution in a given problem space. This space, referred to as the search space, comprises all possible solutions to the problem at hand. Generally speaking, the GA is applied to spaces that are too large to be exhaustively searched. The standard GA proceeds as follows: an initial population of individuals is generated at random or heuristically. Every evolutionary step, known as a generation, the individuals in the current population are decoded and evaluated according to some predefined quality criterion, referred to as the fitness function. To form a new population individuals are selected according to their fitness. Many selection procedures are currently in use one of the simplest being Holland's original fitness-proportionate selection [12]. Selection alone cannot introduce any new individuals into the population. These are generated by genetically inspired operators, of which the best known are crossover and mutation.

3 Design of a Fuzzy Controller Using Genetic Algorithms

In this paper, a fuzzy controller is used in the feedback loop of the integrator in the voltage model as shown in Fig. 4. The proposed fuzzy controller is based on rules and can be adopted to different machines easily. The membership functions of the fuzzy controller used are determined using GAs. Unlike conventional controllers, fuzzy controllers are less sensitive to sensor errors and small variations of the parameters.

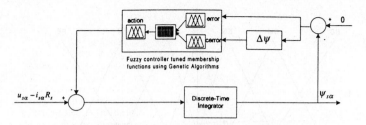

Fig. 4. Block diagram of the fuzzy controller for stator flux estimation

As shown in Fig. 4, the core of the estimation of the stator flux is the discrete integration of the difference between $(u_{s\alpha} - i_{s\alpha}.R_s)$ and the feedback signal. R_s is assumed constant during the simulation. In this figure, first the flux is compared to zero reference in the feedback loop. Next, this difference and the derivative of the difference are given as inputs to the fuzzy logic controller tuned membership functions using GAs. Each variable of the fuzzy controller is represented by using 7 membership functions at the input, as shown in Fig. 5a-5b, and 9 membership functions at the output in Fig. 5c. Initially, the base values and intersection points are chosen randomly. The ranges of the input and output variables are assumed to be [-1,1], [-1.5, 1.5] and [-1.5, 1.5], respectively. The fuzzy rule base which resulted in the most efficient process for this fuzzy controller is as shown in Fig. 5d. For better precision, the number of membership functions can be increased at the expense of computational cost.

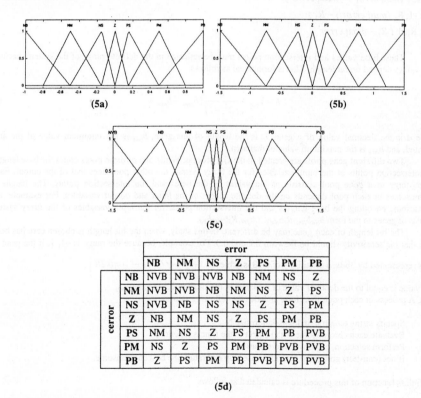

Fig. 5. Membership functions, rule table and surface viewer of fuzzy controller **(a).** Initial membership functions of input variable "*error*" **(b).** Initial membership functions of input variable "*cerror*" **(c).** Initial membership functions of output variable "*action*" **(d).** Rule table of fuzzy controller.

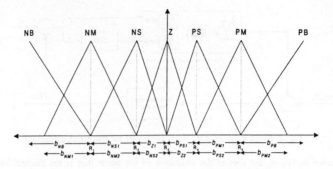

Fig. 6. The base lengths of the membership functions for input variable *"error"*

The goal expected from the GA is to find the base lengths and intersection points of triangles corresponding to the input data. Each base length has minimum and maximum values available. For example, the GA searches base value b_{NB} between the minimum value of *error* (i.e. -1) and the maximum value of *ac* (i.e. 1). The search intervals of some the base values and intersection points for variable *error* are as follows:

b_{NB}: -1, 1 *(min(error) – max(error))*

R_1: -1, 1 *(min(error) – max(error))*

b_{NM1}: -1, R_1 *(min(error) – R_1)*

b_{NM2}: R_1, 1 *(R_1 – max(error))*

These base values and intersection points must be reflected in the definite range of the system, because their values depend on bit length. This is formulated as follows.

$$b = b_{min} + \frac{d}{(2^L - 1)} (b_{max} - b_{min}) \qquad (4)$$

where d is the decimal value of a gene, L is the length of this gene, b_{min} is the minimum value of the area reflected, and b_{max} is the maximum value of that area.

Two different gene pools are created in the GA process. While one of these pools codes the base lengths and intersection points of the input variables for the fuzzy system, the other pool does that of the output. Each chromosome in a gene pool consists of the values of the bases and the intersection points. The length of chromosomes in each pool depends on the definite ranges of the input and output variables. For example, the chromosome encoding the base lengths and the intersection points for the input variables of the fuzzy system consists of genes in the form $b_{NB}b_{NM1}R_1b_{NM2}....b_{PM1}R_5b_{PM2}b_{PB}$.

The bit length of each gene may be different. In this study, when the bit length is chosen care has been taken that the sensitivity should be between 0.2 and 0.3. For example, because the range is –1, 1, if the gene of b_{NB} is represented by 3bits, a sensitivity of 0.28 is achieved. $\left(\frac{1}{2^3 - 1} * (1 - (-1)) = 0.28 \right)$

This value is equal to the decimal value of a change in the smallest bit of gene b_{NB}.
The GA process in each pool, therefore, includes the following steps:

(i)   Specify string length l_s and population size N,
(ii)   Evaluate each chromosome with respect to the fitness function,
(iii)   Perform selection, crossover and mutation,
(iv)   If not (end-test) go to step (ii), otherwise stop and return the best chromosome.

The fitness function of this procedure is calculated as follows:

$$Fitness\ function = Max.\ error - Total\ error \qquad (5)$$

The maximum error is made large enough to prevent the value of the fitness function from being negative. The maximum error is found as follows:

$$\sum_{i=1}^{9}\left(action_{GA_i}-1.5\right)^2 \qquad (6)$$

From the equation above, the maximum error is equal to 20.25. Total error is calculated as follows:

$$\sum_{i=1}^{9}\left(action_i-action_{GA_i}\right)^2 \qquad (7)$$

where $action_{GA_i}$ is the output found by GA in current cycle and $action_i$ is the output obtained in previous cycle.

To find the desired results, first of all, corresponding membership functions are found for the ith values of the input variables. Then, it is determined whether or not these two membership functions may be put in a rule. If there is a rule between these two membership functions, the grades of the membership functions are calculated and analysed to determine if the rule contains AND or OR. If these two membership functions are ANDed, outputs are determined from lower grades of membership functions, but if two membership functions are ORed, outputs are determined from higher grades of membership functions.

If there is more than one membership function intersecting with the ith input of any input variable, then the outputs of these membership functions are both evaluated, and the one which has less error is used.

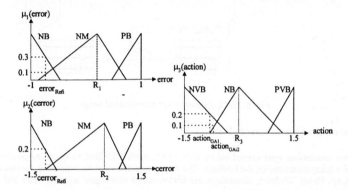

Fig. 7. The method of finding appropriate outputs of GA for the ith inputs

The ith inputs intersect with *NB* and *NM* for *error* and with *NB* for *dd*. After obtaining the intersection situation from Fig. 4, grades of membership are determined for each membership function. Then, from the rule bases that were obtained from both *small* and *slightly dirty*, the outputs are calculated from *short* and *warm*. While doing calculations, μ_2 is taken as 0.2, because the rule involves AND ($\mu_2 < \mu_1$). Since there is also a rule between *medium* and *slightly dirty*, the outputs are calculated for $\mu_1 = 0.1$. The outputs are calculated from both rules, and the one which has less error compared to the desired output is used. This situation is depicted for 3 membership functions in Fig. 7. In other words,

$$if\ [(\ action_i - action_{GA_{i1}}\ )^2 < (\ action_i - action_{GA_{i2}}\ )^2\ ] then$$

$$action_{GA_i} = action_{GA_{i1}}\quad else\ action_{GA_i} = wt_{GA_{i2}}$$

Two important points should be noted here. First, if the intersection between membership function and reference input occurs on the left-hand side of the intersection point (R_1) in a non-right triangle, and the output occurs in the range of non-right triangle for the rule, then the output is also taken from the left-hand side of the intersection point. The same rule is valid for the right hand side as well. The second important point is that if there is no rule between any two membership functions for input variables. However, The rule base used here has complete rules.

4 Experimental Setup

To verify the proposed compensation algorithm, an experimental setup with an induction motor drive has been constructed. Fig. 8 shows the block diagram of the drive system where a conventional field oriented control is implemented. An IGBT inverter, which is controlled by a digital hysteresis control algorithm, is used to drive the motor. The controller board is a DS1102 from dSPACE GmbH. The processor on the controller board is a Texas Instruments TMS320C31 32-bit floating-point processor with a 60 ns instruction cycle. The DS1102 is also equipped with a four channel analog-to-digital converter (two 16-bit and two 12-bit channels), a four channel 12-bit DAC and two incremental encoders.

In this experimental setup, LEM sensors are used to measure two-phase currents. The voltage information is obtained from the inverter switching position including dead time effects. The vector control algorithm has an execution cycle time of 35 μs. The new algorithm with the fuzzy controller consumes an additional 80 μs of execution time. In this experimental setup the speed control is implemented by using speed estimation. However, the speed estimation algorithm has been checked with encoder output for confirmation.

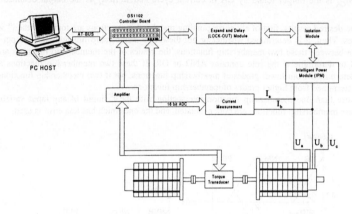

Fig. 8. Block diagram of the experimental setup

5 Simulation and Experimental Results

Various simulations were carried out by using MATLAB/SIMULINK to assess the performance of the integrator with a fuzzy controller on the feedback. The fuzzy controller used for estimating the flux is developed by the MATLAB Fuzzy Toolbox. Simulations are performed to investigate transient state and steady state performance of the proposed flux estimator.

In the simulations, we tested the fuzzy controller by using two induction motors of different power rating. The stator flux waveform of the fuzzy-controlled for a 3-HP motor is given in Fig. 9. In Fig. 10 experimental result of the fuzzy controller is given. The difference between these results is negligible. The same simulations mentioned above are also performed for a 500-HP motor as shown Fig. 11.

Experimental results mentioned above are obtained by TRACE31 software. Experiments and simulations are also performed in different torque reference and load conditions for the three motors used in the previous simulations. In these simulations the fuzzy controller performed better than other algorithm in most of the torque load conditions.

The membership functions of the fuzzy controller are determined by using GA in offline run. Although the GA decreases a little the performance of the system, the user's effect disappears on the system.

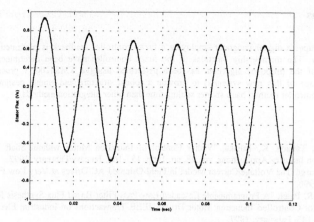

Fig. 9. Simulation results of the fuzzy controller for 3 hp motor.

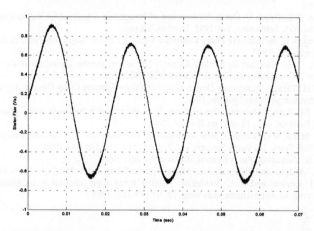

Fig. 10. Experimental results of the fuzzy controller for 3 hp motor.

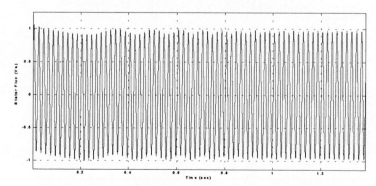

Fig. 11. Simulation results of the fuzzy controller for 500 hp motor.

6 Conclusions

In this paper a digital integrator employing a fuzzy controller feedback is presented to calculate the stator flux vector. The membership functions of the fuzzy controller have been determined by using GA. Implementation of the proposed algorithm has been performed using a TMS320C31 processor. The most important advantage of this new system is to provide a robust structure and simple design. Moreover, the proposed fuzzy controller method has a shorter settling time then other integration methods.

References

1. T. Ohtani, N. Takada, K. Tanaka, "Vector Control of Induction Motor without Shaft Encoder", IEEE Transactions on Industry Applications, vol. 28, no. 1, pp. 157-164, January/February, 1992.
2. Jos van der Burgt, The Voltage/Current Model in Field-Oriented AC Drives at Very Low Flux Frequencies, PhD Thesis, 1996.
3. B.K. Bose, N.R. Patel, "A Programmable Cascaded Low-Pass Filter-Based Flux Synthesis for a Stator Flux-Oriented Vector-Controlled Induction Motor Drive", IEEE Transactions on Industrial Electronics, vol. 44, no. 1, pp. 140-143, February, 1997.
4. Akın E., A New Method for Rotor Flux Orientation of Induction Motor via Stator Fluxes, Firat University, PhD Thesis, 1994.
5. E. Akın, H. Can, H.B. Ertan, Y. Üçtuğ, "Comparison of Integration Algoritms for Vector Control", ICEM98, Vol. 3, pp. 1626-1631, September, 2-4, 1998, İstanbul, Turkey.
6. C. L. Karr, "Design of an Adaptive Fuzzy Controller Using a Genetic Algorithm Proc. of the 4th Intl. Conf. on Genetic Algorithms", 1991.
7. D. L. Meredith, C. L. Karr and K. Krishna Kamur, "The use of genetic algorithms in the design of fuzzy logic controllers", 3rd workshop on Neural Network WNN'92, 1992.
8. M. A. Lee and H. Takagi, "Integrating Design Stages of Fuzzy Systems Using Genetic Algorithms", Second IEEE Intl. Conference on Fuzzy Systems, 1993.
9. A. Arslan and M. Kaya, "Determination of fuzzy logic membership functions using genetic algorithms", Fuzzy Sets and Systems, vol:118 no:2, pp:297-306, 2001.
10. B. Hu, G.K.I. Mann, R.G. Gosine, "New Methodology for Analytical and Optimal Design of Fuzzy PID Controllers", IEEE Trans. On Fuzzy Systems, vol. 7, no. 5, 521-539, October 1999.
11. K.M. Passino, S. Yurkovich, Fuzzy Control, Addison-Wesley, 1998.
12. J. H. Holland, "Adaptation in Natural and Artificial Systems" Ann Arbor, MI: Univ. Mich. Press, 1975.
13. F. Herrara, M. Lozano and J. L. Verdegay, "Tuning fuzzy logic controllers by genetic algorithms", International Journal of Approximate Reasoning, 12(3/):299-315, 1995.

Object Based Image Ranking Using Neural Networks

Gour C Karmakar[†], Syed M Rahman[‡], and Laurence S Dooley[†]

[†]Gippsland School of Computing & Information Technology, Monash University
Gippsland Campus, Churchill VIC 3842, Australia
E-mail: {Gour.Karmakar, Laurence.Dooley}@infotech.monash.edu.au

[‡]Department of Computer and Information Sciences, Minnesota State University
Mankato, MN56001, USA
E-mail: mahbubur.syed@mnsu.edu

Abstract. In this paper an object-based image ranking is performed using both supervised and unsupervised neural networks. The features are extracted based on the moment invariants, the run length, and a composite method. This paper also introduces a likeness parameter, namely a similarity measure using the weights of the neural networks. The experimental results show that the performance of image retrieval depends on the method of feature extraction, types of learning, the values of the parameters of the neural networks, and the databases including query set. The best performance is achieved using supervised neural networks for internal query set.

1   Introduction

With the emergence of the Internet and multimedia technology the development of effective content-based image retrieval systems has become a major research area. The Internet, through the World Wide Web, has given users the capability to access vast amounts of information stored on image databases across the network [1]. Object ranking is defined as the listing of objects in decreasing order of similarity to a query object, and is performed by retrieving objects similar to the query reference. The similarity is measured by matching the two objects. There are two types of matching – exact matching and partial matching [2]. Exact matching retrieves the items that are perfectly matched with the user request while partial matching retrieves those items where there are some similarities, but it is not a perfect match. Exact matching is usually utilized in relational databases to perform a query [3], while partial matching, because of the huge variance in pictorial information is used in image retrieval systems. Ranking therefore represents the degree of the resemblance of objects in the database with the query object. A higher order of ranking denotes a higher degree of likeness.

In this paper a model is introduced using neural networks to measure the similarity between the objects in a database and a query object. The ranking is performed using the actual value of the output node of the neural network [1]. Features were extracted from the objects by utilizing three types of feature extraction methods: the moment invariants [4], the run length [5], and the composite [6] methods. Experiments were conducted to compare the discriminating capability of the features obtained from the composite, the run length and the moment invariants methods. A comparative study was also performed to examine the performance of supervised and unsupervised modes of training for neural networks for image based ranking applications.

The organization of the paper is as follows. A discussion of the proposed models for similarity measurement based on neural networks is given section 2. Section 3 explains the techniques of object ranking using neural networks. The methods for retrieval performance evaluation for object ranking and experimental results are illustrated in sections 4 and 5 respectively. Finally some concluding remarks are presented in section 6.

2 Similarity Measure Using Neural Networks

Various object features are used to train the neural networks. During training, the weights of the networks are adjusted, representing the knowledge about particular object' s features. The actual value of an output unit is determined by applying an activation function to the values of the weights connected to that output unit and this determines the class into which the object is classified. The value of the network's output unit can be used to measure the similarity between the two objects. The similarity measurement is described by the following formula:-

$$Sim(O_0,Q_0) = \left(1 - \frac{|O_0 - Q_0|}{Q_0}\right) \times 100\% \tag{1}$$

where O_0 is the actual value of network's output unit for the database object;
Q_0 is the actual value of network's output unit for the query object;
$Sim(O_0,Q_0)$ is the similarity measure between O_0 and Q_0.

This approach is simple, computationally cost effective and can be easily implemented since the neural network has been chosen as the classifier to perform the ranking. It also gives good performance in measuring the similarity but the main disadvantage is that its performance depends on the neural network correctly classifying the object. If the object is misclassified the output value is totally irrelevant.

3 Techniques of Object Ranking Using Neural Networks

Artificial neural networks may be employed for classification of objects. In the first step the database objects are classified using a neural network. Then the class to which the query object belongs is identified and the ranking of the objects within that class with respect to the query object is computed. The detailed steps for ranking by an artificial neural network are shown in Fig. 1. Query object(s) are presented to the system in order to identify similar objects and their ranking. Features are obtained from both the query object(s) and the database objects by the feature extracting methods. Both types of neural networks (supervised and unsupervised) are used for ranking of objects. Neural networks are trained using the features set from the database objects and the classification of the objects contained in the database is performed using the same features set used in training. The feature set computed from the query

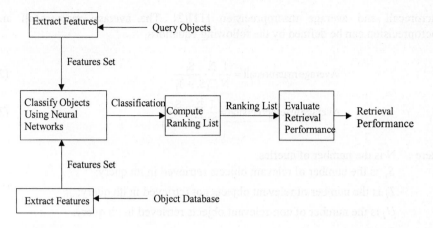

Fig. 1. Block diagram of steps necessary for object ranking using neural networks approach

object(s) is applied to the neural networks in order to get the class(es) to which the query object(s) belong. The similarity between the query object(s) and the database objects is computed using the neural network based and city block distance function [7] approaches. The rank order among the objects is derived using the similarity values. The performance of retrieval for ranking purposes is evaluated using the retrieval performance evaluation approaches described in Sect. 4.

4 Methods of Retrieval Performance

Performance evaluation mainly focuses on two aspects, namely effectiveness and efficiency [2]. Effectiveness reflects the capability of the system to satisfy the user whereas efficiency measures the cost and the effort required to solve a specific problem [9]. Cleverdon gave six measurable quantities in order to measure the performance of information retrieval systems [8]. They are the coverage of collection, the average retrieval time (time lag) i.e. the time between the search request and the retrieved objects, the presentation of the output, the endeavor needed for the user to get output to his search request, the recall and the precision of the system. The first four criteria can be easily measured. Recall and precision are most commonly used to measure the effectiveness of a retrieval system. The relevance of the image is determined by the user's satisfaction level based on the significance of the image to a query image. Recall is defined as the ratio between the number of relevant images retrieved and the total number of relevant images in the image database. Precision is defined as the ratio between the number of relevant images retrieved and the total number of images retrieved [10]. The output of the retrieval system can be presented in various ways such as rank position or co-ordination level (the number of common terms between the query and images) [8]. The precision-recall value can be represented by an ordered pair (R, P), if the output of the retrieval system is various ranks as cut-off points. In order to measure the overall performance of the information retrieval system, the average values of recall and precision are computed from all the results for a set of queries. We can compute the average values in two main ways, average macrorecall and average macroprecision [11][2]. The average macrorecall and macroprecision can be defined by the following equations,

$$\text{Average macrorecall} = \frac{1}{N}\sum_{i=1}^{N}\frac{S_i}{S_i + T_i} \qquad (2)$$

$$\text{Average macroprecision} = \frac{1}{N}\sum_{i=1}^{N}\frac{S_i}{S_i + U_i} \qquad (3)$$

where   N is the number of queries;

S_i is the number of relevant objects retrieved in ith query;

T_i is the number of relevant objects not retrieved in ith query;

U_i is the number of non-relevant objects retrieved in ith query.

The average values of macro recall-precision pairs obtained from the above pair of equations are used to draw the average curve for recall-precision pairs.

5   Experimental Results

The object database contains around 200 different real objects having variation in shape, scale and orientation. The objects are classified into nine manual classes based on the model of the object. Four query sets were used in the experiments. Each query set consists of nine real objects and each object is chosen from each manual class. A representative sample of query sets (query set 1) after rotation normalization used in object ranking based on neural networks is shown in Fig. 2. The experiments were conducted using both supervised and unsupervised neural networks.

Fig. 2. Query set 1 after rotation normalization

5.1   Ranking Using Supervised Neural Networks

Internal and external queries were used for supervised neural networks. The internal query set (query set 1) is a subset of learning set whereas the external query set is taken from the outside of the learning set. Back propagation network having three slabs in the hidden layer with different activation functions have been applied for supervised leaning. The architecture of the network and its parameters used in training are the following.

Parameters used in the run length and the composite method for internal query set:

Number of neurons in each hidden slab=30
Minimum average error=0.00005

Learning epochs taken (run length) =3662
Learning epochs taken (Composite) =152
Learning rate = 0.1
Momentum=0.1
Initial weights=0.3
Activation functions linear [-1,1], Gaussian, Tanh, Gaussian-complement and logistic
are used for input, three hidden and output units respectively.

Parameters used in the moment invariants method for internal query set:

Number of neurons in each hidden slab=8
Minimum average error=0.0016541
Learning epochs taken =307384
All other parameters are the same as the run length and composite methods.

The macro recall-precision curves obtained from the experiments are shown in Fig. 3. The experimental results show that the composite method clearly outperforms both the run length and moment invariants methods. The composite method gave 100% precision for all values of recall, which demonstrates that features obtained from this method, have a good discriminatory power. The main drawback of this method is that it is not fully automatic, as the user needs to add the query set into the object database and assign the class to which the query belongs and then perform the query. The similarity values for this method were computed utilizing the proposed neural networks approach (equation 1) i.e. the actual output values of the neural network.

Another series of experiments were performed using three different external query sets, each of which contained nine different real objects. These query sets were not subsets of the learning set and different numbers of neurons in each hidden slab were used for the run length and the composite methods. A representative samples of the experimental results and parameters used in training are shown in Tables 1 and 2. Minimum average error was taken as 0.0005 and the similarity values were computed using the neural networks approach. The remaining parameters were exactly the same as the parameters used in the run length and composite methods for an internal query set. The experimental results show that the values of recall precision vary with the query set even for the same number of neurons in each hidden slab and other parameters. The selection of query set and number of neurons in each hidden slab have an impact on the precision. If the correlation coefficient of the learning set tends to 1, the precision will be determined only by the correct classification of the query set and it remains constant for all recall values. For this region, the precision values of the composite method are constant for some neurons in Table 2. In the composite method, the maximum precision obtained was 0.67 for data set 2 with 40 neurons in each hidden slab. In the run length method, it was 0.65 for data set 3 with 15 neurons in each hidden slab. The computational time increases with an increase in the number of neurons in each hidden slab. The results also show that the required number of epochs and time during learning for the composite method is much less than the run length method. Thus it may be concluded that the composite method is better than the run length method with respect to precision, learning time and total number of epochs required. If users require the same precision irrespective of recall then this method may be applied in practical applications but it is very hard to obtain 100 percent precision even for a low level (i.e. 10 percent) of recall.

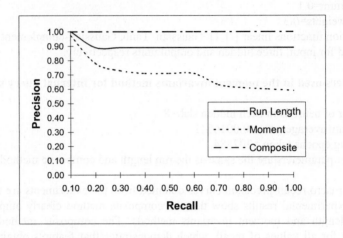

Fig. 3: Macro recall-precision curve averaged for an internal query set (query set 1) using supervised neural networks

Table 1. The precision values at various levels of recall for query set 1 for the run length method using supervised neural networks

#neurons / hidden slab	Number of epochs	Learning time hh:mm:ss	Precision at various levels of recall (in percentage)									
			10	20	30	40	50	60	70	80	90	100
15	2051	00:28:39	.35	.33	.33	.33	.34	.34	.33	.34	.34	.34
20	1820	00:32:32	.52	.46	.46	.46	.46	.45	.46	.45	.45	.45
25	1674	00:37:08	.52	.54	.54	.54	.55	.55	.55	.54	.54	.54
30	1827	00:47:22	.44	.44	.43	.43	.44	.44	.44	.43	.43	.43
40	1732	00.59:09	.22	.20	.21	.21	.21	.21	.22	.22	.22	.22

Table 2: The precision values at various levels of recall for query set 1 for the composite method using supervised neural networks

#neurons / hidden slab	Number of epochs	Learning time hh:mm:ss	Precision at various levels of recall (in percentage)									
			10	20	30	40	50	60	70	80	90	100
15	66	00:01:00	.35	.35	.35	.35	.35	.35	.35	.35	.34	.34
20	74	00:01:28	.34	.34	.34	.34	.34	.34	.34	.34	.34	.34
25	68	00:01:42	.34	.34	.34	.34	.34	.34	.34	.34	.34	.34
30	75	00:02:15	.28	.28	.24	.24	.24	.24	.24	.24	.24	.24
40	68	00:02:41	.44	.44	.44	.44	.44	.44	.44	.44	.44	.44

5.2 Ranking Using Unsupervised Neural Networks

We have applied unsupervised neural networks, called self-organising maps (SOM), in order to retrieve similar objects and rank them with respect to the user's query. The experiments have been conducted having considered three different numbers of output units i.e. 9, 10 and 15 as the classification results vary with the number of output units of the self-organising map. The values of the parameters used during training of the SOM are the following:

- Initial weight = 0.50, learning rate = 0.50, and neighbourhood radius is taken as the number of output neurons – 1
- The distance function used was Vanilla (Euclidean) and feature selection method is Rotation
- A missing value is regarded as error condition and the total number of epochs is 10000.
- Winning neuron = 1 and all others = 0

Objects are classified by applying both learning and a test set to the SOM. Three different query sets were used for this purpose. The similarity values were calculated using the city block distance function. The average values of macro-precision were computed using equation 3 for 10% to 100% recall levels. Fig. 4 shows the effectiveness of object retrieval in terms of macro recall-precision of the composite, run length and moment invariants methods for 9 output neurons. The same experiments were performed using 10 and 15 output neurons, but this has not been included in this paper.

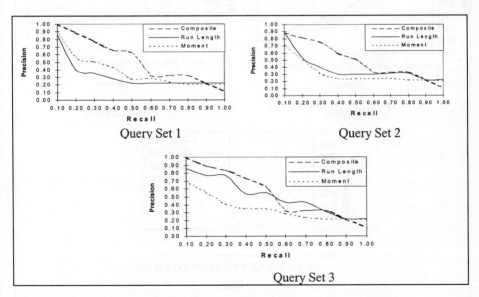

Fig. 4. Average macro recall precision curves for 3 differ

The macro recall-precision curves show that the composite method gave significantly improved precision over the run length and moment invariants methods up to approximately 60% of recall for almost all query sets and for different numbers of output units. The composite method gave poor performance for a higher level of recall, between 80% and 100% of recall compared to the run length and moment invariants methods. This is due to the change in orientation angle due to the variations of the pixels of the objects of the same class, resulting in changes in the distribution of horizontal chord lengths. Besides, the shapes of the objects in the same class are also different to some extent and manual segmentation adds to this. However, the overall classification performance of the composite method was better than that of both the run length and the moment invariants methods. As a result the composite method only gave poor performance for higher levels of recall, but overall the composite method based on SOM achieved better performance up to 50% of recall when the number of output units is equal to the manual classes (i.e. 9). The run length also outperformed the moment invariants method. We also analyzed the experimental results for each method considering different sets of query objects and the number of output units (9, 10 and 15). The results for 15 output neurons are shown in Fig. 5. The graphs showed the sensitivity of the different feature extraction methods to variations in the query sets and the number of output units used.

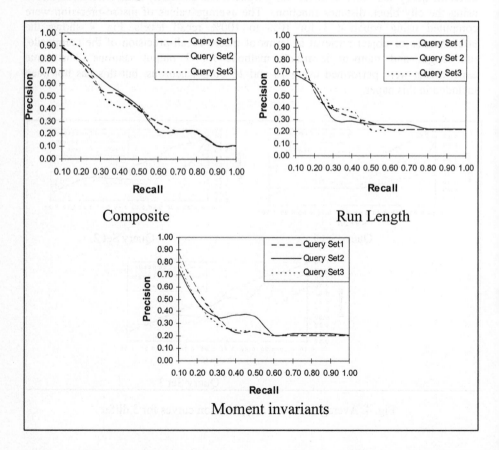

Fig. 5. Average macro recall precision curves for 3 feature extraction methods for 15 output neurons

The recall precision curves show that the composite method can cope better with variation in the query objects than the run length and moment invariants methods. The run length based features are more sensitive to variations in the query set compared to the moment invariants method. So it can be concluded that the composite features have significant discriminating potential in object recognition. Unsupervised neural based object ranking will give promising results for practical applications when users need to retrieve up to 50% of the total relevant objects contained in a database. It is not practically feasible to rank the objects using the traditional computational method without grouping the objects or retrieving some of the objects for a given criterion, for a database containing a large number of objects.

6 Conclusion and Discussions

In this paper we have investigated three feature extraction methods: namely the composite, the run length and the moment invariants methods. Four query sets have been utilized for ranking purposes. Each query set consisted of nine different real objects and was selected from the object database. Ranking was performed using the achieved classification output of the neural networks. We have experimented by applying two types of neural networks, i.e. supervised and unsupervised, to rank the objects matched to a user's query. The similarity values were computed by applying the distance function method (city block distance) and proposed neural networks approach. The neural networks approach gave good similarity values with respect to the query object. Supervised neural networks achieved better performance when the query set is a subset of the learning set but the major drawback is that it needs manual intervention to assign classes to all query objects and add them to the database before performing the query. The precision values (Tables 1 and 2) vary with the number of neurons in each hidden slab and the query set used for supervised neural networks based ranking when the query set is not a subset of the learning set. For some experiments, the classification correlation coefficient of the learning set approached unity and in these cases the precision remains constant irrespective of recall levels. The time and total number of epochs required for training in the composite method was very small compared to the run length method. The experiments on ranking have also been conducted for each feature extraction method using unsupervised neural networks. In this case, the ranking performance was promising and suitable in practice for large databases. The composite method gave a poorer performance for high level of recalls compared with the run length and the moment invariants methods due to errors in rotation normalization, manual segmentation and differences in object shape in a particular manual class. In conclusion, the composite method gives the best performance of the three feature extraction methods and the overall performance of the run length method is better than the moment invariants method.

References

1. Karmakar, G. C.: Shape And Texture Based Feature Extraction For Object Ranking Using Neural Networks. Masters Thesis, Gippsland School of Computing and Information Technology, Monash University, Australia, September 1998.

2. Chen, C.: Image Retrieval Using Multiresolution Analysis and Wavelet Transform. Masters Thesis, Department of Computer Science, RMIT University, Australia, March 1997.

3. Gudivada, V. N., Raghavan, V. V.: Content Based Image Retrieval Systems. IEEE Computer Magazine, pp. 18-22, September 1995.

4. Ming-Kuei Hu. Visual Pattern Recognition by Moment Invariants. IRE transactions on Information Theory, Vol: IT-8, Feb 1962.

5. Karmakar, G. C., Rahman, S. M., Bignall, R.J.: Object Ranking Using Run Length Invariant Features. In the Proceedings of the International Symposium on Audio, Video, Image Processing and Intelligent Application (ISAVIIA 98), Baden-Baden, Germany, pp: 52-56, August 1998.

6. Karmakar, G. C., Rahman, S. M., Bignall, R.J.: Composite Features Extraction and Object Ranking. In the Proceedings of the International Conference on Computational Intelligence for Modeling, Control and Automation (CIMCA'99), Vienna, Austria, pp: 134 - 139, 17-19 February 1999.

7. Rahman, S. M., Haque, N.: Image Ranking Using Shifted Difference. In Proceedings of the ISCA 12th International Conference on Computers and Their Applications. Tempe, Arizona, USA, March, pp. 110-113, 1997.

8. Rijsbergen, V. C. J.: Information Retrieval. Butterworths, London, Second Edition, 1997. http://www.dcs.glasgow.ac.uk/Keith/Chapter.7/Ch.7.html.

9. Salton, G., McGill, M. J.: Introduction to Modern Information Retrieval. McGraw-Hill, New York, 1983.

10. Sajjanhar, A., Lu, G.: A Grid-Based Shape Indexing and Retrieval Method. The Australian Computer Journal, Vol: 29, No: 4, pp: 131-140, November, 1997.

11. Salton , G.: Automatic Information Organisation and Retrieval. McGraw-Hill, New York, 1968.

A Genetic Approach
for Two Dimensional Packing with Constraints

Wee Sng Khoo, P. Saratchandran and N. Sundararajan

School of Electrical and Electronic Engineering,
Nanyang Technological University, Singapore.
Email: epsarat@ntu.edu.sg

Abstract. In this paper, a new genetic algorithm based method is proposed for packing rectangular cargos of different sizes into a given loading area in a two dimensional framework. A novel penalty function method is proposed for checking the solution strings that have violated the loading area's aspect constraints. This method penalizes the solution strings based on the extent by which they violate the length and breadth constraints. The results from the proposed genetic algorithm are compared with the popular heuristic method of Gehring et. al. to show the advantages of this method.

Keywords: Genetic Algorithm, Two dimensional packing, Penalty function, Sentry point

1. Introduction

The field of the genetic algorithms [1-4] has been growing and has proved successful in many real-world applications like floorplan design in VLSI [5-7], cutting [8-11] and packing [12-20]. The reason for the popular usage of genetic algorithm is its robustness. If an artificial system can be made more robust, costly redesigns can be reduced or eliminated. Other attractive features of genetic algorithm are their ability to interface with other approaches for solving problems and it is very easy to modify for further enhancement as well as adding constraints. In this paper, a genetic algorithm is proposed to maximize the number of cargos that can be loaded into the given area without violating loading area's aspect constraints. The paper is organized as follows. Section 2 describes the representation of the problem and all the genetic operations. Section 3 gives results for cargo loading examples from the proposed method as well the heuristic method of Gehring et. al.[22]. Section 4 gives conclusions for this study.

2. Proposed Genetic Algorithm for 2-Dimensional Packing

2.1 Model Representation

A slicing tree structure is used to model the relative arrangement of the cargos (which is a solution to the problem). A slicing tree is an oriented rooted binary tree. Each terminal node of the tree is the identification number of the cargo. Each internal node is labeled v or h. v corresponds to a vertical joint or cut and h corresponds to a horizontal joint or cut. If the tree is viewed from top to bottom, it specifies how a big rectangle is cut into smaller rectangles. If it is viewed from bottom to top, it indicates how the smaller rectangles are joined together. In this problem, the packing method used will be similar to guillotine cuts [18-19]. That is the tree will be viewed from bottom to top, from left to right. At each internal node where two sub-rectangles are joined, it is considered as a new rectangle to the upper node. In addition, each node will be assigned an orientation symbol of '→' or '↑' except for the topmost node. The '→' symbol indicates a 90 degree turn of the rectangle and the '↑' symbol will indicates that the rectangle is kept upright (that is no change of orientation). As the tree is traversed from bottom to the top and left to right, a post-order polish notation is most suitable to represent the tree. Using a post-order polish notation would make it easier for the definition of objective function and coding of the strings. The string is coded in terms of the nodes and their orientation symbol ('→' or '↑'). In the coded string, the orientation symbol will follow the corresponding node. Hence, for n cargos, the strings will have length of $(4*n-3)$. All these are shown more explicitly in Fig. 1.

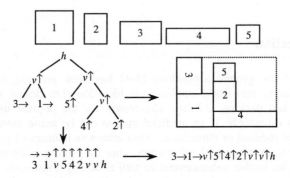

Fig. 1. Tree representation of cargo layout

2.2 Objective Function

After defining the model, an objective function for evaluating the string fitness is required. The density of the packing will be used as the fitness value for this objective function. The packing density is defined as the ratio of the sum of the area of all the cargos to the total area used by the solution string for packing all these cargos. The fitness value itself is in normalized form, a value of 1 will indicate 100% efficiency of packing (i.e., no wastage of area). For a given number of cargos, the sum of the area of all cargos is always fixed. The total area used by the solution string for packing is calculated as follows. As the string is in post-order polish form, these calculations need only sum, max and swap operation together with a stack. If a rectangle's identification number is encountered, the length and breadth of this rectangle is read from the rectangle's length and breadth and pushed into the stack. If a '$\rightarrow$' symbol is encountered, the length and the breadth of the rectangle on the top of the stack is swapped. If a '$\uparrow$' symbol is encountered, no swapping is required. If a 'v' or a 'h' is encountered, the stack will pop two rectangles' length and breadth. A new meta-rectangle is created from these two rectangles. The length and breadth of the two rectangles are pushed back into the stack. The length of the new meta-rectangle is calculated as follows. If the node has a label 'v', the length of the meta-rectangle is the maximum of the two rectangles' length, and breadth of the meta-rectangle is the sum of the two rectangles' breadth. If it has a label 'h', the length of the meta-rectangle is the sum of the two rectangles' length, and breadth of the meta-rectangle is the maximum of the two rectangles' breadth. After the last symbol of the string, only a rectangle's length and breadth will be in the stack. The product of this length and breadth will give the total area required by the solution string for packing the cargos.

2.3 Selection

In this algorithm, Stochastic universal sampling selection [1,3] is used for the selection process. As the population size is small, Stochastic universal sampling is preferred as it would give a better selection [2].

2.4 Crossover

The tree structure with the identification number of the cargos renders normal crossovers like single-point, multi-point and even uniform not applicable. Hence, a new crossover operation has to be defined specifically to suit the tree structure. While forming new crossover, the following constraints have to be maintained so that after crossover the solution string is still a valid one. That is, after crossover, the child strings should have the exact numbers of cargo identification number, there should not be any repeat cargo identification number in the same string and the child strings should still be able to form valid tree (i.e., no missing terminal nodes or broken branches). In order to satisfy these constraints, the proposed crossover will choose only the common terminal nodes of the trees and interchange the nodes with their respective orientation symbol. This can be done by randomly choosing a cargo identification number in one parent string and interchange this cargo identification number and its orientation symbol with the same cargo identification number and its corresponding orientation symbol in the other parent string. This crossover operation will be controlled by the probability of crossover, p_c.

2.5　Mutation

In this genetic algorithm, three types of mutation operations are defined. They are mutation of joint, orientation and interchange.

2.5.1　Joint

Mutation of joint is to invert the joint symbol in the string from 'v' to 'h' and vice versa. This operation is controlled by the probability of joint, p_j. It will determine the chances of the mutation of a joint.

2.5.2　Orientation

Mutation of orientation is to invert the orientation symbol in the string from '$\rightarrow$' to '$\uparrow$' and vice versa. This operation is controlled by the probability of orientation, p_o that determines the chances of the mutation of orientations.

2.5.3　Interchange

Interchange mutation is to swap the position of any two random cargo identification numbers in the string. This operation is controlled by the probability of interchange, p_i that will determine the chances of the mutation due to an interchange.

2.6　Elitism

Elitism has been found to improve the genetic algorithm's performance significantly. Hence, in this genetic algorithm, elitism is also implemented to prevent the loss of highly fit strings missed out by the selection process or destroyed by crossover or mutation. The proposed elitism method selects the weakest string in the offspring's generation and replaces it with the fittest string from the parents' generation. The reason for choosing a low replacement is to let the population explore wider in the search domain. At the same time, it does not let the population to converge too early to a local optimum and miss other better solutions.

2.7　Area Constraint

The area constraint is handled as follows. First, the total area of all the given cargos is calculated to see if it is less than or equal to the loading area. If the total area is greater than the loading area, a cargo is randomly selected and withdrawn from the cargo list, so that the number of cargos to be packed is reduced by one. Then the total area of these reduced cargos is checked again. This procedure is repeated until the total cargo area is less than or equal to the loading area. After this condition is satisfied, the genetic algorithm described earlier is applied. After each generation, every string in the population is checked to see whether it satisfies two constraints. First one is that the solution's required packing area should be less than or equal to the loading area. Second one is that the required packing area's length and breadth do not exceed the loading area's length and breadth. If a string does not fulfill these two conditions that means that string has violated the constraints. In every generation, all the strings in the population are checked against these two constraints. If after n generations, there is still no single string that satisfies these two constraints, then it means that it is not possible to find a valid solution using the current set of cargos. Hence, a cargo is again randomly selected and withdrawn from the cargo list. The genetic algorithm will reset, and start from the first generation with a new cargo list. This process will repeat until a valid string is obtained. The genetic algorithm will then proceed beyond the nth generation and will continue until the stopping criterion is met. The choice of generation 'n' as a 'sentry point' is arbitrary. If the sentry point is set too high, there could be wastage in computation. If it is set to too low, there would not be enough evolutions for a valid solution to come about. There would be a lot of cargos excluded from packing resulting in solution strings that are not the optimum.

2.8 Penalty Function

In the process of handling constraints, whenever a violated string is encountered, the objective function will be modified with a penalty function to suppress the fitness value of that string such that the violated string will have a weak fitness value. Different types of penalty functions could be used to handle the violated strings. One is to give all those violated strings a zero value for their fitness. Another is to reduce the fitness by an arbitrary fraction. Assigning a zero value to the fitness is too drastic. Although the strings may be violating the constraints, it may also have some good schema inside it. Hence assigning a zero value is not a good penalty function. Reducing the fitness by an arbitrary fraction will not differentiate between solution strings of extreme violation and slight violation of the constraints. In the proposed penalty function, if the length of the solution has violated the loading area length, then fitness of the string is reduced by a factor of L_r. If the breadth is violated, the fitness is then reduced by a factor of B_r. If both are violated, then its fitness is reduced by both factors. In this way, the penalized fitness depends on the extent by which a solution string violates the constraints. The penalized fitness is calculated as:

$$\text{Penalized Fitness} = \text{Fitness} \times L_r \times B_r \tag{1}$$

$$\text{where} \quad L_r = \begin{cases} CL/SL & \text{if } CL/SL < 1 \\ 1 & \text{if otherwise} \end{cases},$$

$$\text{and} \quad B_r = \begin{cases} CB/SB & \text{if } CB/SB < 1 \\ 1 & \text{if otherwise} \end{cases}$$

CL & CB are the length and breadth of loading area and SL & SB are the length and breadth of the solution string.

3. Simulation Results

3.1 Optimal values of crossover and mutations probabilities

Extensive simulations studies [21] were carried out to obtain the optimal values for the crossover and mutations probabilities. These are $p_c = 0.99$, $p_i = 0.51$, $p_j = 0.02$ and $p_o = 0.01$.

3.2 Results of comparison with the heuristic method

A comparison between our proposed genetic algorithm and the heuristic method [22] proposed by Gehring et.al. is carried out for two examples. In both examples, there are 21 cargos that need to be loaded. The dimensions of the cargos used in the comparison study are listed in Appendix A. In example 1, the loading area is 700 cm x 300 cm. Fig. 2 shows the results of the two methods. It can be seen from the figure that our genetic algorithm is able to pack all the 21 cargos while the heuristic method is able to pack only 18 cargos.

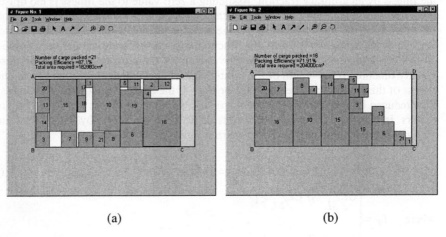

(a)   (b)

Fig. 2. Packing results from the two methods for Example 1. (a: proposed GA, b: Heuristic)

In example 2, the given loading area is 500 cm x 300 cm. This is less than the sum of the area of all the 21 cargos. Hence it is impossible to pack all the cargos within the loading area. Fig. 3 shows the results obtained form our genetic algorithm and the heuristic method. It can be seen from the figure that our genetic algorithm is able to pack 17 cargos while the heuristic method is able to pack only 13 cargos.

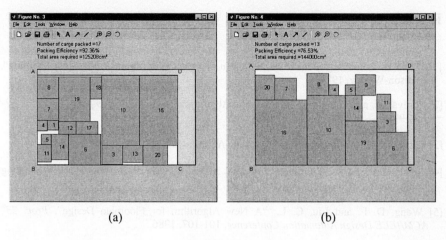

(a)   (b)

Fig. 3. Packing results from the two methods for Example 2. (a: proposed GA, b: Heuristic)

Table 1 summarizes the results obtained for the two examples. The packing efficiency referred in the table is the ratio of sum of the area of the cargos packed to the area required by the algorithms (area ABCD in Fig. 2 & 3) to pack these cargos.

	Number of Cargos Packed		Packing Efficiency	
	Genetic Algorithm	Heuristic Method	Genetic Algorithm	Heuristic Method
Example 1	21	18	87.10 %	71.91 %
Example 2	17	13	92.36 %	76.53 %

Table 1. Results of the two sets of comparisons

4. Conclusion

In this paper, a new genetic algorithm for solving the cargo loading problem in two-dimension is proposed. The algorithm makes use of a novel penalty function method to handle the solution strings that violate the length and breadth constraints imposed by the loading area. Comparisons with a well-known heuristic method have shown that the proposed genetic algorithm is superior in terms of number of cargos packed and packing efficiency.

Reference

[1] David E. Goldberg, "Genetic Algorithm in Search, Optimization & Machines Learning", Addison-Wesley, 1989.

[2] Srinivas, E. and Patnaik, L. M., "Genetic Algorithms: A Survey", *Computer* Vol. 24/6, 17-26, 1994.

[3] Mitchell, M., "An Introduction To Genetic Algorithms", MIT Press, 1997.

[4] Ribeiro Filho, J. L., Treleaven, P. C. and Alippi, C., "Genetic-Algorithms Programming Environments", *Computer* Vol. 24/6, 28-43, 1994.

[5] Wong, D. F. and Liu, C. L., "A New Algorithm for Floorplan Design", *Proc. 23th ACM/IEEE Design Automation Conference*, 101-107, 1986.

[6] Wong, D. F. and Sakhamuri, P. S., "Efficient Floorplan Area Optimization", *Proc. 26th ACM/IEEE Design Automation Conference*, 586-589, 1989.

[7] Wang, T. C. and Wong, D. F., "An Optimal Algorithm for Floorplan Area Optimization", *Proc. 27th ACM/IEEE Design Automation Conference*, 180-186, 1990.

[8] Grinde, R. B. and Cavalier, T. M., "A new algorithm for the minimal-area convex enclosure problem.", *European Journal of Operation Research* 84, 522-538, 1995.

[9] Li, Z. and Milenkovic, V., "Compaction and separation algorithms for non-convex polygons and their applications.", *European Journal of Operation Research* 84, 539-561, 1995.

[10] Valério de Carvalho, J. M. and Guimarães Rodrigues, A. J., "An LP-based approach to a two-stage cutting stock problem.", *European Journal of Operation Research* 84, 580-589, 1995.

[11] Arenales, M. and Morabito, R., "An AND/OR-graph approach to the solution of two-dimensional non-guillotine cutting problems.", *European Journal of Operation Research* 84, 599-617, 1995.

[12] Bischoff, E. E. and Wäscher, G., "Cutting and Packing", *European Journal of Operational Research* 84, 503-505, 1995.

[13] Darrell Whitley, V Scott Gordan and A. P. Willem Böhm, "Knapsack problems", *Handbook of Evolutionary Computation* 97/1, G9.7:1-G9.7:7, 1997.

[14] Dowsland, K. A. and Dowsland, W. B., "Packing problem.", *European Journal of Operation Research* 56, 2-14, 1992.

[15] Neolißen, J., "How to use structural constraints to compute an upper bound for the pallet loading problem.", *European Journal of Operation Research* 84, 662-680, 1995.

[16] Bischoff, E. E., Janetz, F. and Ratcliff, M. S. W., "Loading pallet with non-identical items.", *European Journal of Operation Research* 84, 681-692, 1995.

[17] George, J. A., George, J. M. and Lamar, B. W., "Packing different-sized circles into a rectangular container.", *European Journal of Operation Research* 84, 693-712, 1995.

[18] Kröger, B., "Guillotineable bin packing: A genetic approach.", *European Journal of Operation Research* 84, 645-661, 1995.

[19] Hwang, S.-M., Kao, C.-Y. and Horng, J.-T., "On Solving Rectangle Bin Packing Problems Using Genetic Algorithms.", *IEE Transactions on Systems, Man, and Cybernetics,* Vol. 2, 1583-1590, 1994.

[20] Lin, J.-L., Footie, B., Pulat, S., Chang, C.-H. and Cheung J. Y., "Hybrid Genetic Algorithm for Container Packing in Three Dimensions.", *Proc. 9th Conference on Artificial Intelligence for Applications*, 353-359, 1993.

[21] Khoo, W. S., "Genetic Algorithms Based Resource Allocation Methods", Technical Report EEE4/038/00, School of Electrical and Electronic Engineering, Nanyang Technological University, Singapore, Dec 2000.

[22] Gehring, H., Menschner, K. and Meyer, M., "A computer-based heuristic for packing pooled shipment containers.", *European Journal of Operation Research* 44, 277-288, 1990.

Appendix A Cargo Details

Dimensions of the 21 cargos are listed in Table 2 (Length by Breadth in cm).

Cargo Number.	Length x Breadth(cm)	Cargo Number.	Length x Breadth(cm)	Cargo Number.	Length x Breadth(cm)
1	30 x 35	8	68 x 68	15	120 x 220
2	46 x 66	9	64 x 64	16	163 x 205
3	60 x 66	10	120 x 220	17	40 x 69
4	32 x 33	11	44 x 55	18	36 x 69
5	32 x 33	12	40 x 55	19	100 x 140
6	97 x 103	13	54 x 64	20	62 x 78
7	68 x 68	14	54 x 80	21	50 x 62

Table 2. Dimensions of the cargos used in Examples 1 and 2

Task Environments for the Dynamic Development of Behavior

Derek Harter and Robert Kozma

Department of Mathematical Sciences; University of Memphis
Memphis, TN 38152 USA
{dharter, rkozma}@memphis.edu
http://www.psyc.memphis.edu/{∼harterd, ∼kozmar}

Abstract. The development of complex, adaptive behavior in biological organisms represents vast improvement over current methods of learning for artificial autonomous systems. Dynamical and embodied models of cognition [1–13] are beginning to provide new insights into how the chaotic, non-linear dynamics of heterogeneous neural structures may self-organize in order to develop effective patterns of behavior. We are interested in creating models of ontogenetic development that capture some of the flexibility and power of biological systems. In this paper we present a testbed for the creation and testing of models of development. We present some results on standard neural networks in learning to perform this task and discuss future plans for developmental models in this environment.

1 Introduction

1.1 Development and Non-Linear Dynamics

The development of behavior in biological organisms is primarily a self-organizing phenomenon. Organisms are born with a basic repertoire of motor skills and instinctive needs. These are often tied to simple action-loops [1], which provide a basic repertoire of simple pattern completion and instinctive behaviors that can begin to satisfy the intrinsic drives of the organism. As the organism develops both physically and behaviorally, however, these instinctive behavior patterns begin to be associated with more general sensory stimuli. The organism learns to recognize patterns in the environment that are important and useful affordances for beneficial behaviors [14]. Increasingly complex patterns of behavior are organized around the solutions that are discovered at earlier stages of development.

Thelen and Smith [13, 15] view development as a shifting ontogenetic landscape of attractor basins. As physical and behavioral patterns develop the landscape is continually reformed and reshaped. Each developed behavior opens up many possibilities for new more complex patterns of behavior, while closing off possibilities for others. Even relatively simple tasks can provide opportunities for the development of increasingly complex strategies in order to improve performance. For example in the simple task we present in the next section, humans develop higher level strategies for improving their performance.

Many theories of the development of behavior in biological organisms are beginning to view it in terms of a self-organizing dynamical system [13, 9, 8]. The organization of patterns of behavior is viewed, in some sense, as the formation and evolution of attractor landscapes. Some research [12, 5, 4, 6, 7, 10] also indicates that chaotic dynamics may play an essential part in the formation of perception and behavior in biological organisms.

1.2 Category Formation through Aperiodic Chaotic Dynamics

Following experimental evidence and theoretical modeling [5, 10], categories are associated with localized wings of a complex, high-dimensional chaotic attractor. The attractor landscape is formed in a flexible way reflecting past experiences and continuously modified based on the actual information the system receives from the environment.

The system typically resides in a high-dimensional basal state. It can be kicked-off from this state in response to external factors or internal developments. As the result, the state of the system is switched to a low-dimensional wing, which might represent a previously learnt memory pattern or an elementary action. The system might reside in this wing or it may visit sequentially various wings, thus representing a possible behavioral pattern. In either case, the system moves back to the basal state upon completing the sensory input-induced identification task or performing the desired action. The advantage of such a dynamical approach is the flexibility and robustness of the selection of the behavior and action as it is seen in biological systems. We attempt to implement this dynamical strategy to solve action selection tasks as outlined in this work.

1.3 Task Environments for Testing Models of Development

Biological organisms are capable of marvelously complex patterns of behavior in pursuit of the satisfaction of their endogenous drives. However, it is not always apparent how much of the complexity of their behavior is internally generated, and how much emerges from the interaction of simple response patterns within a complex task environment. It seems doubtful that true progress in understanding the properties of intelligent behavior can be made by studying disembodied, syntactic systems [16, 17, 1, 8, 18]. Intelligent behavior, at least in biological organisms, seems built upon a foundation of fast and robust pattern recognition and completion, both of static and temporally extended, often vague and noisy patterns. This observation is suggestive of several features that may be necessary in the developmental processes of biological organisms.

Of course it is desirable to develop models of behavior in realistic and complex environments, but it is not always possible. The question then becomes: what are the features of real world environments that are necessary for the development of complex behavior in biological organisms. Can we begin to study the phenomenon of the dynamic development of behavior in such a way that our results are valid in more realistic, complex environments. The following are a few features of environments that seem critical in creating models of the development of behavior.

Real Time Constraints Invariably biological organisms face critical time pressures that constrain their behavior. It is not simply a matter of doing the right thing, biological organisms must do the right thing and do it in a timely manner. It is better to do something, even something not optimal, on time, than to be too late and become someone else's dinner.

Developmental processes occur within the real time interactions of the organism with the environment. A testbed for developmental processes should be able to support tasks with real time constraints on the behavior of the developing system. However, in order to begin creating models it is desirable to loosen the restrictions of real time constraints. Therefore the testbed should have configurations with and without real time constraints on behavior. The task should be similar enough in both configurations so that mechanisms developed without constraints can eventually be tried in the more demanding situation of tasks with real time constraints.

Multiple Sensory Modalities The task environment should support the simulation of multiple sensory modalities, so that associations can be formed between purely reactive (intrinsic) behaviors, and more complex senses and behavior. This type of learned association between events in disparate sensory modalities seems to be crucial to many types of category formation and learned behavior. For example, in classical conditioned learning, the co-occurrence of an auditory stimuli (bell) with onset of reward results in a conditioned response being developed [19–22].

Exploitable Environmental Regularities Even though an environment is complex, it still must posses statistically significant, exploitable regularities in order to be a viable, survivable niche. Such regularities may co-occur in spatial location and time, or be more spatially or temporally extended. A major part of developmental processes is discovering such regularities and learning to exploit them.

2 Packing Task

Towards the end of studying and creating models of development, we have begun work on creating appropriate tasks with the previous properties. We describe a packing task here which is one such environment, and some work on standard machine learning tools in this environment.

2.1 Description

In the packing task, the behaving system is presented with a series of shapes, one shape at a time. In this packing task, which is a simplified form of the Tetris game [23], the system can be presented with one of 3 shapes as shown in figure 1. The goal of the task is to move and rotate a shape before allowing it to drop onto a playing field in such a way as to end up with as compact of a packing as possible. An example of a packing trial in progress can be found in figure 2.

Fig. 1. The shapes used in the packing task.

Fig. 2. An example packing task trial. Shapes enter from the top and must be positioned and rotated before they are dropped. Performance is evaluated by the height and the density of the packing of the shapes.

The behaving system does not know in advance what sequence of shapes it will be given. In our version of the packing task, the system is given random sequences of 10 shapes. The performance of the system on the packing task is evaluated by examining the density of their packing and by examining the total height of the resulting packing.

The system can produce two types of behavior. It must specify where to position (or move) the shape in the playing field, and how to rotate the shape. Once the system has specified the position and rotation of the shape, it is allowed to fall down onto the playing field. The shape settles into place and the next shape is presented to the behaving system to be positioned and rotated.

2.2 Encoding

We now present an example of a standard neural network that learns to perform the simple packing task. The neural network needs to be given some sense of the current state of the environment. For the experiments performed here, two pieces of input were given to the network: the type of shape that has appeared, and a perception of the contours of the current playing field.

The encoding of the type of shape is relatively simple. In the packing task environment there are 3 different shape types. We used 2 bits to encode the type of shape. The shapes in figure 1 were given numbers (from left to right) of 0, 1 and 2 and were encoded as 00, 01 and 10 respectively.

The perception of the state of the playing field (the environment) is necessary in order to produce good behavior on where and how to position the shape before dropping it. In our reduced packing task, the playing field consisted of 5 columns. We sensed the height of each column currently in the environment, and encoded this for training and testing the networks. The lowest point in the playing field is used as a baseline and is encoded as having a height of 0. All other heights are calculated from the baseline depth. We used 2 bits to encode the height of each column, and simply ignored perception of columns that were greater than 3 units above the baseline.

For example in figure 2, the leftmost column has the lowest depth in the playing field, and would be encoded with height 0. The next column to the right has a height 2 units above the baseline. So from left to right, the height of the columns in figure 2 would be encoded as 0, 2, 1, 2, 2. The type of shape shown in figure 2 about to be dropped is shape number 1. As stated before we used 2 bits to encode the shape type, and 2 bits for each of the column heights, for a total of 12 bits of input. The situation shown in figure 2 would be encoded as:

```
Type Col1 Col2 Col3 Col4 Col5
0 1  0 0  1 0  0 1  1 0  1 0
```

For the output of the system we developed the following encoding. We encoded the position to place the shape from the left edge in 3 bits. We need to be able to specify up to 5 units of displacement, thus we needed 3 bits to encode the 5 possibilities. The shapes can be rotated in increments of 90 degrees. Shape 2 (the L shape) can be rotated into 4 different distinct orientations. Therefore we also needed 2 bits to encode all possible specifications of rotation.

2.3 Training

We trained standard backpropogation networks using the encoding described above. For training data we had a human perform 50 packing trials, and we captured and encoded the input and the output of the behavior that the human produced when performing the packing task. We also captured a similar set of data for testing. We trained and tested the networks with many different configurations of number of hidden nodes and epochs trained. We then chose the best configurations in order to evaluate the performance of the networks on the packing task as discussed in the next section. The neural network performed best with 50 hidden nodes.

2.4 Experiments

We then used our packing task testbed in order to evaluate the performance of the networks on simulated packing trials. We gave the networks 100 random trials and measured their performance by calculating the packing density and height that they achieved. Packing height is simply a measure of the highest column of blocks in the playing field. Packing density is measured by looking at the ratio of the number of filled spaces in the packing to the total area of the packing. In figure 2 the packing has a height of 4 and a density of 17 / 20 or 0.85. The lower the height of the packing is the better the performance and similarly the denser the packing is the better the performance.

2.5 Results

A human learned the packing task and was asked to perform the task for 100 trials. Similarly the resulting neural networks were run on 100 trials of the packing task. Table 1 shows a comparison of the average performance on the 100 trials of the neural network and the human. Figure 3 shows a histogram of the performance of the human and the neural network rated by height and density.

Table 1. Comparison of average height and density performance measures on 100 simulated packing tasks

	Height	Density
Human	7.62	0.8748
Neural Network	8.18	0.8261

Basic neural networks perform adequately on the packing task, but obviously are not quite as good at packing as humans, even for this simplified task domain. Humans, when performing this task, alter their strategies as the task progresses. Early on in a packing trial, a human is willing to leave open opportunities for particular shapes. People know intuitively that, even though they see shape types at random, they are likely to see the particular shape type needed if it is still early in the trial. However, as the trial progresses, strategies shift to those that will simply minimize the height of their packing.

This shift in strategies causes confusion for simple backpropogation networks. They see this as conflicting output patterns for the same input. Strategies that would possibly correct this deficiency for basic neural networks and other solutions will be discussed in the next section.

2.6 Discussion of Development in the Packing Task

The development of differing strategies given the context of the problem is a prime example of the development of skills in biological organisms. People are not given explicit examples of appropriate shifts in strategies. They develop such strategies by interacting with the task environment, and guided by their previous experience with the constraints of the problem. They seem to quickly and intuitively embody the opportunities that situations afford for good behaviors, and how such opportunities change with the changing situation. In other words, they develop a set of skills and strategies for improving their performance on the problem simply through interaction and experience in the task domain.

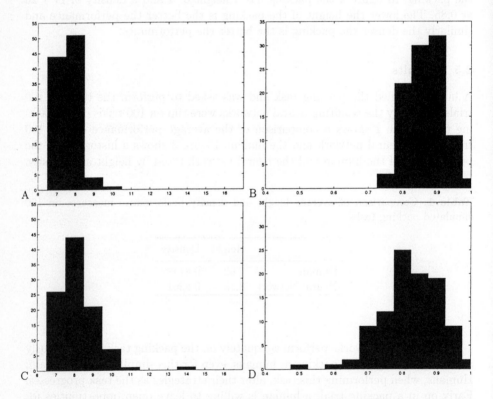

Fig. 3. Histograms of performance on 100 trials of the packing task. The top two figures (A and B) show the performance of a human subject in the packing task, while the bottom two (C and D) display the performance of a neural network. In the left column we are measuring performance by the height of the packing. On the right we show performance by the density of the packing.

Even in our simple environment we see that people develop differing strategies for behavior based on the context of the progress of the trial. For example, humans first learn a basic set of good contours and correspondance with different shape types that provide for efficient packings. From this basic set of behavior, they begin to develop preferences for patterns that keep open future opportunities. For example, some contours will naturally accommodate more than one shape type, and are preferred over other patterns that limit good packing to a single shape type. Even further, people begin to develop higher level strategies at this point. For example, if it is early in the trial they wait for more optimal packings, but later on they simply try and minimize the height. The challenge in creating models of development is in capturing this ability to, not only softly assemble solutions through a repertoire of learned and innate skills, but to also develop new skill and effective higher level strategies for the problem domain.

3 Future Directions

The basic neural networks presented here are not quite capable of human level performance in the packing task. The primary reason for this deficiency is an inability to perceive the changes in circumstances that cause a shift in the behavior of the human trainers. We have no doubt that adding on more contextual input (such as the current height of the packing, or a count of the number of shapes packed so far) would improve the performance of the basic network, though it remains to be seen if it could equal human performance. Also, other methods such as recurrent, dynamical neural networks, or genetic algorithm optimizations, should be capable of bringing standard methods of machine learning up to human level performance on this simple task.

The point is not to equal human performance in this simplified domain, but to begin to create models that can develop behavior on their own in a cognitively plausible manner, and that display some of the flexibility of biological development. Most standard methods of machine learning should be able to competently handle the packing task environment in its simplified form but inevitably will break down as we add complexity and real time constraints to the task.

KIII is a dynamical memory device, which has been used successfully to solve difficult classification problems in vague, and noisy environments [10]. The KIII model incorporates several KII sets, which can be interpreted as units generating limit cycle oscillations in an autonomous regime. High-dimensional aperiodic and chaotic behavior does not emerge until the complete KIII system is formed. KIII has a multi-layer architecture with excitatory and inhibitory lateral, feed-forward, and feedback connections. KIII models can grasp the essence of the observed dynamic behavior in certain biological neural networks. It seems feasible to build a simplified version of KIII for the action selection task addressed in this work. We call it 3*KII model, as it consists of 3 mutually interconnected KII sets. Each KII set has a well-defined oscillation frequency. The complete 3*KII model, however, may exhibit high-dimensional, aperiodic oscillations as the result of competing, incommensurate frequencies of the KII components.

The advantage of 3*KII is that it allows a self-organized encoding of behavioral patterns into localized wings of a high-dimensional attractor. Therefore, we can obtain flexible and noise-resistant transitions among the states of the system, self-organized into a sequence of elementary actions of phase transitions. It is expected that defining a more challenging packing task with larger number and more complicated block patterns and also larger play field the application of dynamical encoding and action selection mechanism as 3*KII would prove to be beneficial. Also the emergence of self-organized action patterns would be imminent and complex behavioral patterns could be studied.

4 Conclusion

The development of behavior, even in a simplified environment such as the packing task, can shed light on the mechanisms of biological development and learning. Biological organisms are able to effectively develop increasingly complex skills and strategies simply by interacting with and solving problems in their environment. The dynamic, self-organization of behavior in biological organisms is a powerful model of learning that, if better understood, would provide great opportunities for improved artificial behaving and learning systems. Development of behavior in biological organisms can be viewed as a self-organizing dynamical system. Some research also indicates the importance of chaotic modes of organization in the development of behavior. We can begin to study models of development even in simplified ways as long as we are aware of the essential properties of the environments that are exploited by biological organisms during developmental process. Some of these properties include critical real time constraints, a rich sensory modality and environmental regularities and exploitable features.

References

1. Andy Clark. *Being There: Putting Brain, Body, and World Together Again.* The MIT Press, Cambridge, MA, 1997.
2. Gerald M. Edelman and Giulio Tononi. *A Universe of Consciousness: How Matter Becomes Imagination.* Basic Books, New York, NY, 2000.
3. Stanley P. Franklin. *Artificial Minds.* The MIT Press, Cambridge, MA, 1995.
4. Walter J. Freeman. Consciousness, intentionality and causality. In Núñez and Freeman [24], pages 143–172.
5. Walter J. Freeman. *How Brains Make Up Their Minds.* Weidenfeld & Nicolson, London, 1999.
6. Walter J. Freeman and Robert Kozma. Local-global interactions and the role of mesoscopic (intermediate-range) elements in brain dynamics. *Behavioral and Brain Sciences*, 23(3):401, 2000.
7. Walter J. Freeman, Robert Kozma, and Paul J. Werbos. Biocomplexity: Adaptive behavior in complex stochastic dynamical systems. *BioSystems*, 2000.
8. Horst Hendriks-Jansen. *Catching Ourselves in the Act: Situated Activity, Interactive Emergence, Evolution and Human Thought.* The MIT Press, Cambridge, MA, 1996.

9. J. A. Scott Kelso. *Dynamic Patterns: The Self-organization of Brain and Behavior*. The MIT Press, Cambridge, MA, 1995.
10. Robert Kozma and Walter J. Freeman. Chaotic resonance - methods and applications for robust classification of noisy and variable patterns. *International Journal of Bifurcation and Chaos*, 11(6), 2001.
11. Robert F. Port and Timothy van Gelder, editors. *Mind as Motion: Explorations in the Dynamics of Cognition*. The MIT Press, Cambridge, MA, 1995.
12. Christine A. Skarda and Walter J. Freeman. How brains make chaos in order to make sense of the world. *Behavioral and Brain Sciences*, 10:161–195, 1987.
13. Esther Thelen and Linda B. Smith. *A Dynamic Systems Approach to the Development of Cognition and Action*. The MIT Press, Cambridge, MA, 1994.
14. J. J. Gibson. *The Ecological Approach to Visual Perception*. Houghton Mifflin, 1979.
15. Esther Thelen. Time-scale dynamics and the development of an embodied cognition. In Port and van Gelder [11], chapter 3, pages 69–100.
16. Rodney A. Brooks. Elephants don't play chess. *Robotics and Autonomous Systems*, 6:3–15, 1990.
17. Rodney A. Brooks. A robot that walks: Emergent behaviors from a carefully evolved network. In Randall Beer, R. Ritzmann, and T. McKenna, editors, *Biological Neural Networks in Invertebrate Neuroethology and Robotics*. Academic Press, 1993.
18. Rolf Pfeifer and C. Scheier. *Understanding Intelligence*. The MIT Press, Cambridge, MA, 1998.
19. Paul F. M. J. Verschure, B. Kröse, and Rolf Pfeifer. Distributed adaptive control: The self-organization of behavior. *Robotics and Autonomous Systems*, 9:181–196, 1992.
20. Paul F. M. J. Verschure, J. Wray, Olaf Sporns, Giulio Tononi, and Gerald M. Edelman. Multilevel analysis of classical conditioning in a behaving real world artifact. *Robotics and Autonomous Systems*, 16:247–265, 1995.
21. Paul F. M. J. Verschure. Distributed adaptive control: Explorations in robotics and the biology of learning. *Informatik/Informatique*, 1:25–29, 1998.
22. I. P. Pavlov. *Conditioned Reflexes*. Oxford University Press Press, Oxford, 1927.
23. David Kirsh and Paul Maglio. Reaction and reflection in tetris. In J. Hendler, editor, *Artificial Intelligence Planning Systems: Proceedings of the First Annual International Conference (AIPS92)*, Morgan Kaufman, San Mateo, CA, 1992.
24. Rafael Núñez and Walter J. Freeman, editors. *Reclaiming Cognition: The Primacy of Action, Intention and Emotion*. Imprint Academic, Bowling Green, OH, 1999.

Wavelet Packet Multi-Layer Perceptron for Chaotic Time Series Prediction: Effects of Weight Initialization

Kok Keong Teo, Lipo Wang* and Zhiping Lin

School of Electrical and Electronic Engineering
Nanyang Technological University
Block S2, Nanyang Avenue
Singapore 639798
http://www.ntu.edu.sg/home/elpwang
{p7309881e, elpwang, ezplin}@ntu.edu.sg
*To whom all correspondence should be addressed

Abstract. We train the wavelet packet multi-layer perceptron neural network (WP-MLP) by backpropagation for time series prediction. Weights in the backpropagation algorithm are usually initialized with small random values. If the random initial weights happen to be far from a good solution or they are near a poor local optimum, training may take a long time or get trap in the local optimum. Proper weights initialization will place the weights close to a good solution with reduced training time and increased the possibility of reaching a good solution. In this paper, we investigate the effect of weight initialization on WP-MLP using two clustering algorithms. We test the initialization methods on WP-MLP with the sunspots and Mackey-Glass benchmark time series. We show that with proper weight initialization, better prediction performance can be attained.

1 Introduction

Neural networks have demonstrated great potential for time-series prediction where system dynamics is nonlinear. Lapedes and Farber [1] first proposed to use a multi-layer perceptron neural network (MLP) for nonlinear time series prediction. Neural networks are developed to emulate the human brain that is powerful, flexible and efficient. However, conventional neural networks process signals only on their finest resolutions. The introduction of wavelet decomposition [2]-[6] provides a new tool for approximation. It produces a good local representation of the signal in both the time and the frequency domains. Inspired by both the MLP and wavelet decomposition, Zhang and Benveniste [7] proposed wavelet network. This has led to rapid development of neural network models integrated with wavelets. Most researchers used wavelets as basis functions that allow for hierarchical, multiresolution learning of input-output maps from data. The wavelet packet multi-layer perceptron (WP-MLP) neural network is an MLP with the wavelet packet as a feature extraction method to obtain time-frequency information. The WP-MLP had a lot of success in classification applications in acoustic, biomedical, image and speech.

Kolen and Pollack [8] shown that feedforward network with the backpropagation technique is very sensitive to the initial weight selection. Prototype pattern [18] and the orthogonal least square algorithm [19] can be used to initialize the weights. Geva [12] proposed to initialized the weights by clustering algorithm that is based on mean local density (MLD).

In this paper, we apply the WP-MLP neural network for temporal sequence prediction and study the effects of weight initialization on the performance of the WP-MLP. The paper is organized as follows. In Section 2, we review some background on WP-MLP. In Section 3, we describe the initialization methods. In Section 4, we present the simulation results of time series prediction for two benchmark cases, i.e. the sunspots and Mackey-Glass time series. Finally, Section 5 presents the conclusion.

2 Wavelet Packet MLP

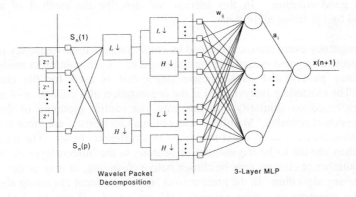

Fig. 1. Model of WP-MLP

Fig. 1 shows the WP-MLP used in this paper. It consists of three stages, i.e., input with a tapped delay line, wavelet packet decomposition, and an MLP. The output $\hat{x}(n + 1)$ is the value of the time series at time $n+1$ and is a function of the values of the time series at previous time steps:

$$s_n = [x(n), x(n-1), \ldots, x(n-p)]^T \tag{1}$$

Coifman, Meyer and Wickerhauser [9] introduced wavelet packets by relating multiresolution approximations with wavelets. Recursive splitting of vector spaces is represented by a binary tree. We employed the wavelet packet transformation to produce time-frequency atoms. These atoms provide us with both time and frequency information with varying resolution through out the time-frequency plane. The time-frequency atoms can be expanded in a tree-like structure to create arbitrary titling, which is useful for signals with complex structure. A tree algorithm [10] can be employed for computing wavelet packet transform by using the wavelet coefficients as filter coefficients. The next sample is predicted using a conventional MLP with k input units, one hidden layer with m sigmoid neurons and one linear output neuron, as shown in Fig. 1. The architecture of the WP-MLP is defined as, $[p:l(wlet):h]$ where p is the number of tapped delays, l is the number of decomposition level, *wlet* is the type of wavelet packet used, and h is the number of hidden neurons. In the present work, the MLP is trained by the backpropagation algorithm using the Levenberg-Marquadt method [11].

3 Weight Initialization by Clustering Algorithms

The initialization of the weights and biases has a great impact on the network training time and generalization performance. Usually, the weights and biases are initialized with small random values. If the random initial weights happen to be far from a good solution or they are near a poor local optimum, training may take a long time or get trap in the local optimum [8]. Proper weights initialization will place the weights close to a good solution, which reduces training time and increases the possibility of reaching a good solution. In this section, we describe the method of weights initialization by clustering algorithms.

The time-frequency event matrix is defined by combining the time-frequency patterns and its respective targeted outputs of the training data. Each column is made up of time-frequency pattern and its respective target that is used for the clustering algorithm. The clustering analysis [13] is the organization of a collection of patterns into clusters based on similarity. In this way, the natural grouping of the time-frequency events is revealed. Member event within a valid cluster are more similar to each other than they are to an event belonging to a different cluster. The number of clusters is then chosen to be the number of neurons in the hidden layer of the WP-MLP. The number of clusters may be chosen before clustering, or may be determined by the clustering algorithm. In the present work, the hierarchical clustering algorithm [14] and the counterpropagation network [15] are tested. Hierarchical clustering algorithm consist of three simple steps:

- Compute the Euclidean distance between every pair of events in the data set.
- Group the events into a binary, hierarchical cluster tree by linking together pairs of events that are in close proximity. As the events are paired into binary clusters, the newly formed clusters are grouped into larger cluster until a hierarchical tree is formed.
- Cut the hierarchical tree to form the clusters according to the number of hidden neurons chosen based on a prior knowledge acquired.

Counterpropagation network is a combination of a competitive layer and another layer employing Grossberg learning. For the purpose for this paper, the forward-only variant of the counterpropagation network shall be used. Let us consider a competitive learning network consisting of a layer of N_n competitive neurons and a layer of $N+1$ input nodes, N being the dimension of the input pattern.

$$h_k = \sum_{j}^{N+1} g_{kj} x_j \qquad (2)$$

where g_{kj} is the weight connecting k neuron to all the inputs, h_k is the total input into neuron k and x_j is j element of the input pattern. If neuron k has the largest total input, it will win the competition and becoming the sole neuron to response to the input pattern. The input patterns that win the competition in the same neuron are considered to be in the same cluster. The neuron that win the competition will have it weight adjust according with the equation shows below.

$$w_{kj}^{new} = (1 - \alpha(t))w_{kj}^{old} + \alpha(t)x_j \tag{3}$$

where learning constant $\alpha(t)$ is a function of time. The other neurons do not adjust their weights. Learning constant will change according to (13) to ensure that each training pattern has equal statistical importance and independence of presentation order [16] [17].

$$\alpha(\tau_k) = \frac{\alpha(1)}{1 + (\tau_k - 1)\alpha(1)} \tag{4}$$

where $\tau_k \geq 1$ is the number of times neuron k has modified its weights including the current update. At the end of the training, those neurons that never had it weights modified are discarded. In this way, we had found the number of clusters and the number of hidden neurons of the WP-MLP. Both clustering algorithms had formed clusters and we can proceed to initialize the WP-MLP. The weights of the hidden neuron are assigned to be the centroid of each cluster.

4 Simulation Results

We have chosen two time series often found in the literature as a benchmark, i.e., the Mackey-Glass delay-differential equation and the yearly sunspot reading. For both data sets are divided into three sections for training, validation and testing. When there is an increase in the validation error, training stops. In order to have a fair comparison, simulation is carried out for each network over 100 trials (weight initialization can be different from trial to trial, since the clustering result can depend on the random initial starting point for a cluster).

4.1 Mackey-Glass chaotic time series

The Mackey-Glass time-delay differential equation is defined by

$$\frac{dx(t)}{dt} = \frac{ax(t - \tau)}{1 + x(t - \tau)^{10}} - bx(t) \tag{5}$$

where $x(t)$ represents the concentration of blood at time t when the blood is produced. In patients with different pathologies, such as leukemia, the delay time τ may become very large, and the concentration of blood will oscillate, becoming chaotic when $\tau \geq 17$. In order to test the learning ability of our algorithm, we choose the value $\tau = 17$ so that the time series is chaotic. The values of a and b are chosen 0.2 and 0.1

respectively. The training data sequences are of length 400, followed by validation and testing data sequences of length 100 and 500 respectively. Tests were performed on random initialization, hierarchical clustering algorithm and counterpropagation network on varying network architecture. Eight neurons in the hidden layer are chosen and this is a prior knowledge required by hierarchical clustering algorithm to cut the tree i.e. there are eight clusters. The counterpropagation network is given ten neurons to process the data to search for the optimal clustering.

Table 1. Validation and test MSE of WP-MLP with random initialization on Mackey-Glass chaotic time series

Network Architecture	Validation MSE			Test MSE		
	Mean	Std	Min	Mean	Std	Min
[14:1(Db2): 8]	4.17×10^{-4}	1.89×10^{-4}	2.36×10^{-4}	1.80×10^{-5}	8.13×10^{-6}	1.02×10^{-5}
[14:2(Db2): 8]	7.86×10^{-5}	3.71×10^{-4}	2.94×10^{-5}	3.73×10^{-6}	1.86×10^{-5}	1.25×10^{-6}
[14:3(Db2): 8]	3.62×10^{-4}	3.62×10^{-4}	2.24×10^{-4}	1.56×10^{-5}	4.51×10^{-6}	9.65×10^{-6}
*[14: 3(Db2):8]	1.15×10^{-6}	2.94×10^{-6}	1.15×10^{-6}	1.12×10^{-5}	3.83×10^{-6}	1.49×10^{-6}

*Wavelet Decomposition

Table 2. Validation and test MSE of WP-MLP initialized with hierarchical clustering algorithm on Mackey-Glass chaotic time series

Network Architecture	Validation MSE			Test MSE		
	Mean	Std	Min.	Mean	Std	Min.
[14: 1(Db2): 8]	5.66×10^{-6}	1.42×10^{-5}	2.16×10^{-6}	5.65×10^{-6}	1.42×10^{-5}	2.16×10^{-6}
[14:2(Db2): 8]	2.74×10^{-6}	2.16×10^{-7}	2.31×10^{-6}	2.75×10^{-6}	3.24×10^{-7}	2.31×10^{-6}
[14:3(Db2): 8]	3.46×10^{-6}	1.50×10^{-6}	1.53×10^{-6}	3.45×10^{-6}	1.65×10^{-6}	1.53×10^{-6}
*[14: 3(Db2):8]	2.45×10^{-5}	5.54×10^{-5}	9.06×10^{-6}	2.81×10^{-5}	5.54×10^{-5}	1.38×10^{-5}

*Wavelet Decomposition

Table 3. Validation and test MSE of WP-MLP initialized with counterpropagation clustering algorithm on Mackey-Glass chaotic time series

Network Architecture	Validation MSE			Test MSE		
	Mean	Std	Min.	Mean	Std	Min.
[14: 1(Db2):10]	1.935×10^{-5}	3.69×10^{-5}	1.78×10^{-7}	2.33×10^{-5}	4.30×10^{-5}	3.54×10^{-7}
[14:2(Db2):10]	1.02×10^{-5}	1.06×10^{-5}	1.73×10^{-7}	1.22×10^{-5}	1.21×10^{-5}	2.42×10^{-7}
[14:3(Db2):10]	2.303×10^{-5}	4.21×10^{-5}	4.62×10^{-7}	2.96×10^{-5}	3.52×10^{-5}	9.82×10^{-7}
*[14: 3(Db2):10]	1.96×10^{-5}	1.95×10^{-5}	6.20×10^{-6}	2.34×10^{-5}	2.24×10^{-5}	7.86×10^{-6}

*Wavelet Decomposition

Table 1 shows the results on prediction errors for the Mackey-Glass time series using different architecture of the WP-MLP, i.e. the mean, the standard deviation, and the minimum Mean Squared Error (MSE) of the prediction errors over 100 simulations. Table 2 and 3 show the result on prediction error for network initialized by hierarchical and counterpropagation respectively.

Hierarchical clustering algorithm provides consistent performance for the various network architecture except the network with wavelet decomposition. Counterpropagation network managed to attain the lowest minimum MSE with various network architecture. Thus we can say that it had indeed found the best clustering to place the weights near the optimum solution. The higher mean, larger standard deviation and lower minimum suggested that the probability of getting the optimal clustering is low. Therefore, the counterpropagation network needs more tuning to find the optimal clustering frequently.

4.2 Sunspots time series

Sunspots are large blotches on the sun that is often larger in diameter than the earth. The yearly average of sunspot areas has been recorded since 1700. The sunspots of years 1700 to 1920 are chosen to be the training set, 1921 to 1955 as the validation set, while the test set is taken from 1956 to 1979. Tests were performed on random initialization, hierarchical clustering algorithm and counterpropagation network on varying network architecture. Eight neurons in the hidden layer are chosen and this is a prior knowledge required by hierarchical clustering algorithm to cut the tree i.e. there are eight clusters. The counterpropagation network is given ten neurons to process the data to search for the optimal clustering.

Table 4. Validation and test NMSE of WP-MLP with random initialization on Sunspots time series

Network	Validation NMSE			Test NMSE		
Architecture	Mean	Std	Min	Mean	Std	Min
[12: 2(Db1): 8]	0.0879	0.300	0.0486	0.3831	0.2643	0.1249
[12:1(Db2):8]	0.0876	0.0182	0.059	0.4599	0.7785	0.1256
[12:2(Db2):8]	0.0876	0.0190	0.0515	1.0586	6.2196	0.1708
*[12:3(Db2):8]	0.0876	0.0182	0.0590	0.5063	1.0504	0.1303

*Wavelet Decomposition

Table 5. Valdiation and test NMSE of WP-MLP with initialized with hierarchical clustering algorithm on Sunspots time series

Network	Validation NMSE			Test NMSE		
Architecture	Mean	Std	Min	Mean	Std	Min
[12: 2(Db1): 8]	0.0674	0.0123	0.0516	0.2682	0.1716	0.1305
[12:1(Db2):8]	0.0714	0.0103	0.0499	0.2906	0.1942	0.1426
[12:2(Db2):8]	0.0629	0.0100	0.0508	0.2126	0.0624	0.1565
*[12:3(Db2):8]	0.0691	0.0069	0.0523	0.2147	0.0585	0.1430

*Wavelet Decomposition

Table 6. Validation and test NMSE of WP-MLP with counterpropagation clustering algorithm on Sunspots time series

Network	Validation NMSE			Test NMSE		
Architecture	Mean	Std	Min	Mean	Std	Min
[12: 2(Db1): 10]	0.079	0.0121	0.050	0.277	0.0819	0.1383
[12:1(Db2):10]	0.078	0.012	0.047	0.271	0.0880	0.1451
[12:2(Db2):10]	0.078	0.0130	0.0485	0.2722	0.0722	0.1442
*[12:3(Db2):10]	0.071	0.011	0.048	0.252	0.0631	0.1324

*Wavelet Decomposition

Table 5 shows the results on prediction errors for the Sunspot time series using different architecture of the WP-MLP, i.e. the mean, the standard deviation, and the minimum Normalized Mean Squared Error (NMSE) of the prediction errors over 100 simulations. Table 6 and 7 show the result on prediction error for network initialized by hierarchical and counterpropagation respectively. Both initialization methods shown superior performance in all aspect compared to the WP-MLP that is initialized randomly. Thus we can say that for the problem of sunspots, the initialization methods place the weights close to a good solution frequently.

5 Conclusion

In this paper, we used hierarchical and counterpropagation clustering algorithms to initialize WP-MLP for time series prediction. Usually the weights are initialized with small random values that maybe far from or close to a good solution. Therefore the probability of being located near a good solution depends on the complexity of the solution space. In a relatively simpler problem such as Mackey-Glass time series, it is expected to have less poor local optimum. Thus the chance of being placed near a good solution is higher and initialization of weights may not be important. When dealing with difficult problem like Sunspot time series, it is expected to have a complex solution space that has lot of local optima. Thus random initialization cannot provide the consistent performance attained by the both initialization methods. Therefore, the weights of the WP-MLP must be initialized to achieve better performance.

REFERENCE

1. Lapedes, A., Farber, R.: Nonlinear Signal Processing Using Neural Network: Prediction and System Modeling. Los Alamos Nat. Lab Tech. Rep, LA-UR-872662 (1987)
2. Mallat, S.G.: Multifrequency channel decompositions of images and wavelet models. IEEE Transactions on Acoustics, Speech and Signal Processing, Vol. 37, No. 12 (1989) 2091 – 2110
3. Mallat, S.G.: A Theory for Multiresolution Signal Decomposition: The Wavelet Representation. IEEE Transactions on Pattern Recognition, Vol. 11, No. 7 (1989) 674 –693

4. Vetterli, M., Herley, C.: Wavelets and filter banks: theory and design. IEEE Transactions on Signal Processing, Vol. 40, No.9 (1992) 2207 –2232
5. Mallat, S.G.: A Wavelet Tour of Signal Processing. Academic Press (1998)
6. Strang, G., Nguyen, T.: Wavelet and Filter Banks. Wellesly-Cambridge Press (1996)
7. Zhang, Q., Benveniste, A.: Wavelet Networks. IEEE Transactions on Neural Networks, Vol. 3, No. 6. (1992) 889 -898
8. Kolen, J.F., Pollack, J.B.: Back Propagation is Sensitive to Initial Conditions. Tech. Rep. TR 90-JK-BPSIC, Laboratory for Artifical Intelligence Research, Computer and Information Science Department, (1990)
9. Coifman, R.R., Meyer, Y., Wickerhauser, M.V.: Entropy-Based Algorithm for Best Basis Selection. IEEE Transactions on Information Theory, Vol. 38, No. 2 Part 2 (1992) 713-718
10. Vetterli, M., Herley, C.: Wavelets and filter banks: theory and design. IEEE Transactions on Signal Processing, Vol. 40, No.9 (1992) 2207–2232
11. Hagan, M.T., Menhaj, M.B.: Training Feedforward Networks with the Marquardt Algorithm. IEEE Transactions on Neural Networks, Vol. 5, No. 6 (1994) 989 -993
12. Geva, A.B.: ScaleNet-Multiscale Neural-Network Architecture for Time Series Prediction. IEEE Transactions on Neural Networks, Vol. 9, No. 9 (1998) 1471 –1482
13. Jain, A.K., Murty, M.N., Flynn, P.J.: Data Clustering: A Review. ACM Computing Surveys, Vol. 31, No. 3 (1999) 264-323
14. Aldenderfer, M.S., Blashfield, R.K.: Cluster Analysis. Sage Publications (1984)
15. Nielsen, R.H.: Neurocomputing. Addisson-Wesley Publishing Company. (1990)
16. Wang, L.: On competitive learning. IEEE Transaction on Neural Network, Vol. 8,No. 5 (1997) 1214-1217
17. Wang, L.: Oscillatory and Chaotic dynamic in neural networks under varying operating conditions. IEEE Transactions on Neural Network, Vol. 7, No. 6 (1996) 1382-1388
18. Denoeux, T., Lengelle, R.: Initializing back propagation network with prototypes. Neural Computation, Vol. 6 (1993) 351-363
19. Lehtokangas, M., Saarinen, J., Kaski, K., Huuhtanen, P.: Initialization weights of a multiplayer perceptron by using the orthogonal least squares algorithm. Neural Comput., Vol. 7, (1995) 982-999

Genetic Line Search

S. Lozano[1], J.J. Domínguez[2], F. Guerrero[1] and K. Smith[3]

[1]Escuela Superior de Ingenieros, Camino de los Descubrimientos, s/n, 41092 Sevilla, Spain
slozano@cica.es, fergue@esi.us.es
[2]Dpto. Lenguajes y Sistemas Informáticos - University of Cadiz, Spain
juanjose.dominguez@uca.es
[3]School of Business Systems, Monash University, Clayton, Victoria 3800, Australia
kate.smith@infotech.monash.edu.au

Abstract. All unconstrained and many constrained optimization problems involve line searches, i.e. minimizing the value of a certain function along a properly chosen direction. There are several methods for performing such one-dimensional optimization but all of them require that the function be unimodal along the search interval. That may force small step sizes and in any case convergence to the closest local optimum. For multimodal functions a line search along any direction is likely to have multiple valleys. We propose using a Genetic Line Search with scalar-coded individuals, convex linear combination crossover and niche formation. Computational experiences show that this approach is more robust with respect to the starting point and that a fewer number of line searches is usually required.

1. Introduction

Unconstrained optimization deals with the problem of minimizing (or maximizing) a function in the absence of any restrictions, i.e.

$$\underset{\bar{x}\in\Re^n}{Min} \quad f(\bar{x}) \ . \tag{1}$$

Most solution methods consist in repeating a basic two-phase process: determining a search direction $\bar{d}^k$ and a step length α_k^* such that

$$\bar{x}^{k+1} \ \leftarrow \ \bar{x}^k + \alpha_k^* \, \bar{d}^k \ . \tag{2}$$

How to carry out such iterative process depends on the differentiability or not of function f. The Cyclic Coordinate Method, the Method of Hooke and Jeeves and the Rosenbrock Method determine the search direction without need of derivatives information [1]. On the contrary, the Gradient Method, Newton's Method, Quasi-Newton and Conjugate Gradient Methods all require that function f is at least once (more often twice) continuously differentiable. All these methods require performing a number of line searches. Thus, line searches can be considered the backbone of all these unconstrained optimization approaches. It can even be argued that they also play a basic role in constrained optimization since many algorithms solve a constrained problem through a sequence of unconstrained problems via Lagrange relaxation or penalty and barrier functions.

A line search can be exact or inexact. The first case is equivalent to a one-dimensional optimization problem

$$Min \qquad F(\alpha_k) \equiv f(\overline{x} + \alpha_k \overline{d}^k) \ .$$
$$\alpha_k \in (0, \alpha_k^{max}) \tag{3}$$

where the search interval $[0, V_k^{max}]$ depends on the maximum stepsize, which guarantees a unique minimum in the interval of uncertainty. Most exact line search methods use such strict unimodality requirement to iteratively reduce the interval of uncertainty by excluding portions of it that do not contain the minimum. They differ in whether or not they use derivative information. Line search without using derivatives include Dichotomous Search, Fibonacci Search and Golden Section Method [1]. Line search using derivatives include Bisection Search and Newton's Method. Polynomial (quadratic or cubic) Interpolation Methods may or may not use derivative information [1].

Very often, in practice, the expense of excessive function and gradient evaluations forces a finish of the line search before finding the true minimum of $F(V_k)$ in (3). Since that may impair the convergence of the overall algorithm that employs the line search, different conditions may be imposed for terminating the search provided that a significant reduction in the objective function f has been achieved [2].

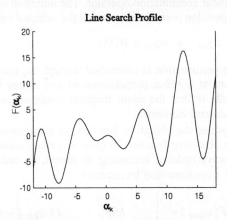

Fig. 1. Typical multi-modal function profile along a search direction

The requirement that the search interval has a unique minimum implies that only the first valley along the search direction is explored and that the overall optimization algorithm cannot converge to a local minimum other than the one at the bottom of the valley in which the starting point $\overline{x}$ lies. For multi-modal functions, many valleys are likely to happen along any search direction as shown in Figure 1. Some of those valleys may be deeper than the first one. It would be sensible to move to a deeper valley since that would give a bigger reduction in f. In other words, freeing the line search from restricting itself to the neighborhood of $\overline{x}$ allows for jumping across valleys, thus accelerating the optimization process.

To implement such a look-further-ahead strategy a robust and efficient procedure must be used to solve a one-dimensional, multi-modal optimization problem. Genetic Algorithms (GA) are well-suited for this task since they do not require the function to be differentiable (not even continuous) and can be parallel-processed. In [3] combining GA and local optimization of a continuous function is suggested. It is proposed the use of the conventional methods for unconstrained optimization that we have mentioned above, starting with the solution obtained by the GA or, better yet, augmenting the GA with a problem-specific hill-climbing operator yielding a hybrid GA. Our approach is completely different. We embed the GA in the overall optimization process replacing traditional line search methods.

The rest of this paper has the following structure. Section 2 describes the application of GA to the line search task while Section 3 show some computational experiences from which, in Section 4, conclusions are drawn.

2. Genetic Line Search

The Genetic Line Search (GLS) algorithm we propose has the following features:
- Individuals are scalars, coded as floating-point real numbers. Let ν_{knt} the n-th individual of the population after t iterations in the k-th line search.
- Two operators are used, one for crossover and another for mutation. The crossover operator is a convex linear combination operator. The mutation operator consists in generating a random gaussian perturbation around the selected individual.

$$\alpha_{knt} \leftarrow \alpha_{knt} + N(0,\sigma) . \tag{4}$$

The magnitude of the perturbation is controlled through the standard deviation σ, which is kept constant at a value proportional to one minus the probability of mutation. The rationale is that the more frequent mutations are performed, the smaller their magnitude and vice versa.
- Fitness is defined as minus the objective function for a minimization problem. It would be equal to the objective function in case of a maximization problem. The population is kept always ordered according to fitness. In order to keep fitness values positive adding a constant may be required

$$F(\alpha_{knt}) \leftarrow F(\alpha_{knt}) + \left| \underset{j \in Population(t)}{Min} F(\alpha_{kjt}) \right| + 1 . \tag{5}$$

- A steady state GA has been implemented. In each iteration either mutation (with probability p_m) or crossover (with probability $p_c = 1 - p_m$) is applied. Mutation requires just one individual and produces another one. Crossover requires two parents and generates two off-springs of which the least fit is discarded. Therefore, the population is renewed at a rate of one individual per iteration.
- The individual to be deleted from the population is selected independently of the individuals to reproduce. Individuals to reproduce through mutation or crossover are selected with uniform probability, i.e. selection for reproduction is not fitness-

biased. On the contrary, selection for deletion is rank-based with a (non-linear) geometric probability distribution [4].

- In order to promote diversification, a sharing function scheme which encourages niche formation is used [3]. It modifies the fitness landscape penalizing the accumulation of individuals near one another. When several individuals are very similar to each other they mutually reduce their fitness. There is a distance threshold, called the niche size, which sets the minimum distance of separation of two individuals for them not to interfere each other. Instead of directly fixing that parameter, the user is asked for a desired number of solutions per niche. The population size divided by the number of solutions per niche gives the expected number of niches and from that the estimated niche size is derived.

- The parameters that have been introduced so far are: population size, crossover and mutation probabilities and number of solutions per niche. Another parameter to be fixed is the number of iterations after which the GLS stops, returning the best individual in the final population. Since the selection for deletion is population-elitist there is no need to distinguish between on-line and off-line performance. In case a different GA were used, off-line performance should be monitored.

There is yet another parameter and an important one for that matter. It is the problem scale S which will be used for the different line searches. This parameter specifies whether the magnitudes of the decision variables are of order unity or of the order of thousands or millions. This helps in computing the maximum and minimum stepsizes in each line search. But before we express how stepsizes are bounded, some considerations are in order:

- Even in the case of unconstrained optimization it is always possible for real-world problems to set upper and lower limits on most if not all variables. Let l_i and u_i respectively the lower and upper bounds on variable i.

- Limits on the decision variables translate, through the components d_i^k of the search direction, into limits on the allowed stepsize for line search k.

- The stepsize scale in line search k is related to the problem scale S through the norm of the search direction $\left\| \overline{d}^k \right\|$. Thus, if the search direction is not normalized, problem scale must be adjusted to give the proper stepsize scale.

- Besides the previous limits on the stepsize, the user may impose an overall maximum stepsize v^{max} valid for all line searches.

- Conventional line search techniques do not explore negative values for the stepsize since the search direction is normally chosen so as to decrease the objective function locally in that direction. However, our approach is not restricted to the neighborhood of the line search origin and although it can be expected that small negative values increase the objective function value, further away along the negative axis there may be deep valleys worth exploring. Therefore, our approach sets a negative lower bound on the step size. In the end we have

$$\alpha_k^{min} \leq \alpha_k \leq \alpha_k^{max} . \tag{6}$$

where

$$0 \leq \alpha_k^{max} \leq Min \left\{ \underset{i:d_i^k>0}{Min} \ \frac{u_i-x_i^k}{d_i^k}, \ \underset{i:d_i^k<0}{Min} \ \frac{l_i-x_i^k}{d_i^k}, \ \frac{S}{\left\|\overline{d}^k\right\|}, \alpha^{max} \right\}. \qquad (7)$$

$$0 \geq \alpha_k^{min} \leq Max \left\{ \underset{i:d_i^k<0}{Max} \ \frac{u_i-x_i^k}{d_i^k}, \ \underset{i:d_i^k>0}{Max} \ \frac{l_i-x_i^k}{d_i^k}, \ -\frac{S}{\left\|\overline{d}^k\right\|}, -\alpha^{max} \right\}. \qquad (8)$$

Whenever a mutation causes an individual to step out of this interval, it is repaired by setting its value equal to the closest feasible value. Since the convex crossover operator guarantees that if the two parents fall inside a closed interval, so does every off-spring generated in such a way, no reparation is ever required after applying crossover.

The stepsize bounding mechanism described above may seem a bit complicated but a proper determination of the search interval prior to starting each line search can significantly improve the overall performance of the optimization algorithm. Too narrow a search interval is a waste from the GLS point of view. On the other side, too wide a search interval may cause the GLS to be stretched too thin. Note, finally, that GLS explores a $[V_k^{min}, V_k^{max}]$ search interval that resembles but differs from the interval $[0, V_k^{max}]$ in equation (1). Not only negative values are allowed but also the interval scale is larger, usually involving multiple valleys in which to move into. This feature is crucial if better performance than conventional methods is aimed.

A pseudocode description of the GLS algorithm follows:

```
procedure GLS
begin
       t 7 1
       initialize Population(1)
       evaluate adjusted fitness and order Population(1)
       while t < number-of-iterations-allowed do
       begin
              t 7 t+1
              select operator (crossover or mutation)
              perform selection for reproduction
              apply operator
              perform selection for deletion
              insert new individual
              readjust fitness and reorder Population(t)
       end
end
```

As it has been said before, GLS is embedded in an overall optimization algorithm whose pseudocode description is:

```
procedure optimization
begin
```

```
    input starting point x̄¹
    input problem scale S
    input (if any) bounds lᵢ and uᵢ
    input (if any) maximum stepsize αᵐᵃˣ
    k 7 1
    select search direction d̄ᵏ
    perform GLS from x̄ᵏ along d̄ᵏ obtaining x̄ᵏ⁺¹
    while significant-reduction-obtained do
    begin
            k 7 k+1
            select search direction d̄ᵏ
            perform GLS from x̄ᵏ along d̄ᵏ obtaining x̄ᵏ⁺¹
    end
end
```

Note that the proposed termination condition is based on the amount of reduction in the objective function obtained in the last line search, i.e.

$$\frac{\left| f(\bar{x}^k) - f(\bar{x}^{k+1}) \right|}{\left| f(\bar{x}^k) \right|}. \tag{9}$$

3. Computational Experiences

There exists in the literature a set of standard problems [5] to help compare the relative performance of different line search methods. A thorough study to benchmark GLS will be the subject of another paper. What we report here are the results obtained with the proposed approach for just two of the problems tested. Results of using GLS to train feed-forward Neural Networks have been reported in [6].

The first test is the trigonometric function (TRIG) which in the interval under consideration $[0,100]^n$ has many local minima and a global minimum in the vicinity of the point $(100,100,...,100)$. The second one is the Rastrigin function (RAST) which in the interval under consideration $[-50,50]^n$ has many local minima and a global minimum at $(0,0,...,0)$.

Figure 2 shows the successive line searches performed by GLS and a standard Golden Section algorithm [1] for RAST function starting at $(-49, 47)$ and using Cyclic Coordinate Method for selecting the direction of descent. While GOLDEN requires multiple small-step line searches and gets stuck at a near-by local minimum, GLS require just two line searches to reach the global minimum.

324

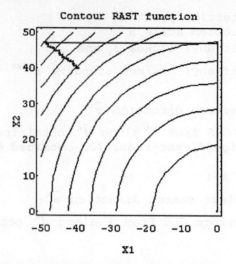

Fig. 2. Succesive line searches starting at (-49,47)

Figures 3 and 4 show the results for minimizing RAST function starting from 100 random points. GLS always requires more computing time but, independently of the starting point, almost always reaches the global minimum. Golden, on the other hand, is faster but it converges to the optimum only when the starting point is close.

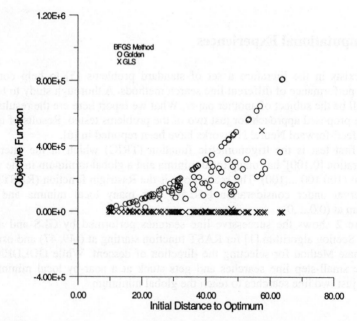

Fig. 3. Initial Distance to Optimum vs final RAST Obj. Funct. for 100 random starting points

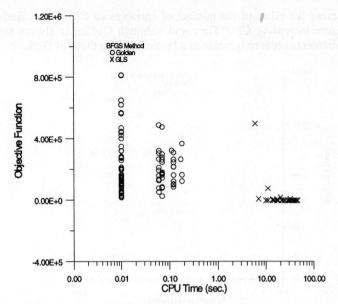

Fig. 4. CPU Time vs final RAST Objective Function for 100 random starting points

The results shown so far correspond to the two-variables case. To test the scalability of the methods, we considered an increasing number of variables: 5, 10, 20, 25, 50, 75 and 100. Figure 5 shows how both Golden and GLS find lower values of TRIG Objective Function as the number of variables increases although GLS always finds better values than Golden. The gap seems to widen with the number of variables.

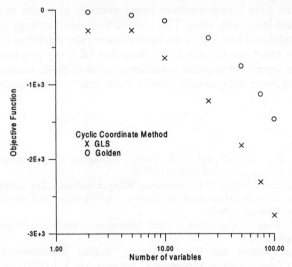

Fig. 5. TRIG Objective Function vs number of variables

Figure 6 shows the effect of the number of variables on CPU Time. Both Golden and GLS require increasing CPU Time and although Golden is always faster than GLS, its requirements seem to increase at a faster rate than those of GLS.

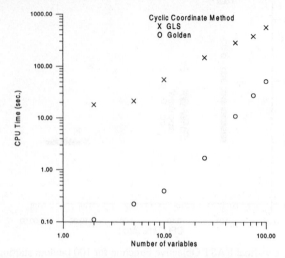

Fig. 6. CPU Time vs number of variables

4. Summary

This paper deals with a new approach to unconstrained optimization of multi-modal functions, consisting on performing successive line searches using a Genetic Algorithm. Genetic Line Search explores larger intervals along the search direction which translate into larger step sizes. This look-further-ahead strategy overcomes the local nature of traditional line search methods allowing the algorithm to escape from local optima. The experiments carried out show that GLS requires more computing time for each line search but requires a smaller number of them, is more robust with respect the starting point and generally obtains better values of the objective function.

References

1. Bazaraa, M.S., H.D. Sherali and C.M. Shetty, *Nonlinear Programming. Theory and Algorithms*, Wiley, Singapore, 2nd edition (1993)
2. Fletcher, R., *Practical Methods of Optimization*, Wiley, Chichester, 2nd edition (1987)
3. Goldberg, D.E., *Genetic Algorithms in Search, Optimization and Machine Learning*, Addison-Wesley, Reading (1989)
4. Michalewicz, Z., *Genetic Algorithms + Data Structures = Evolution Programs*, Springer-Verlag, New York (1992)
5. More, J.J., B.S. Garbow and K.E. Hillstrom, "Testing Unconstrained Optimization Software", *ACM Transactions on Mathematical Software*, vol. 7, 1 (1981) pp. 17-41
6. Lozano, S., J.J. Dominguez, F. Guerrero, L. Onieva and J. Larrañeta, "Training Feedforward Neural Networks Using a Genetic Line Search", in *Smart Engineering System: Neural Networks, Fuzzy Logic, Data Mining and Evolutionary Programming*, C.H. Dagli, M. Akay, O. Ersoy, B.R. Fernández and A. Smith (eds.), ASME Press, New York (1997) pp. 119-124

HARPIC, an Hybrid Architecture Based on Representations, Perception and Intelligent Control: A Way to Provide Autonomy to Robots

Dominique Luzeaux and André Dalgalarrondo

DGA/Centre Technique d'Arcueil,
16bis av. Prieur de la Côte d'Or, 94114 Arcueil Cedex, France,
luzeaux@etca.fr,dalga@etca.fr,
WWW home page: http://www.etca.fr/CTA/gip/Publis/Luzeaux/

Abstract. In this paper we discuss an hybrid architecture, including reactive and deliberative behaviors, which we have developed to confer autonomy to unmanned robotics systems. Two main features characterize our work: on the one hand the ability for the robot to control its own autonomy, and on the other hand the capacity to evolve and to learn.

1 Introduction

As was mentioned recently in a call for participation to a special issue on intelligent systems design, complex intelligent systems are getting to the point where it almost feels as if "someone" is there behind the interface. This impression comes across most strongly in the field of robotics because these agents are physically embodied, much as humans are. There are several primary components to this phenomenon. First, the system must be capable of action in some reasonably complicated domain: a non-trivial environment within which the system has to evolve, and a rather elaborate task which the system should fulfill. Second, the system must be capable of communicating with other systems and even humans using specified possibly language-like modalities, i.e. not a mere succession of binary data, but some degree of symbolic representations. Third, the system should be able to reason about its actions, with the aim of ultimately adapting them. Finally, the system should be able to learn and adapt to changing conditions to some extent, either on the basis of external feedback or relying on its own reasoning capacities.

These remarks have guided our research and in order to bring up some answers and to build an unmanned ground robot that could perform well in ill-known environments, we have focused on the robot control architecture, which is the core of the intelligence, as it binds together and manages all the components. In the next sections, we will first discuss shortly what we understand under autonomy for robots and control architectures. Then we will describe the architecture we propose and give some implementation details and results. The final sections are then dedicated to the possibility for the system first to control its own autonomy depending on external input and second to learn and adapt.

1.1 Autonomy in robotic systems

In order to tackle autonomous robots, one has first to delimit the scope of expected results; this calls for a temptative definition of autonomy. A first necessary condition for a system to be called autonomous is to be able to fire reactions when faced with external disturbances: this yields a concept obviously parametrized by the nature and diversity of disturbances one can act against. However mere reaction to disturbances cannot be truly accepted as autonomy as it does not encompass longer-term decision abilities. A more demanding definition includes the ability to change the interaction modes with the environment: this captures the idea that an autonomous organization is not static in its functioning ways and can "adapt". Looking closely at the implications, one sees that such an organization necessarily has to internalize external constraints, which means the ability to *integrate knowledge of its own dynamics and representation of the exterior*. To sum up, an interaction without explicit representation of both an internal world corresponding to the system and an external world relative to the environment cannot be called autonomous (consider a painting industrial robot with simple contact feedback as a counterexample). Notice that this does not mean the representations have to be entirely different (on the contrary, efficient sensorimotor closed loops require an integration of the various representations!). Concluding these paragraphs on autonomy, we see that although there are epistemological necessary conditions for autonomy, there is no absolute autonomy: a system can reasonably only be said "more autonomous" than another.

In our approach to autonomous systems, we have proceeded in a bottom-up fashion, handling first the control and perception issues, and looking for adequate representations which could integrate both these issues, leading to sensorimotor i/o behavior [4, 10, 11]. Since we are not interested in robot wandering aimlessly within corridors, possibly avoiding scientists strolling through the lab, but have to deal with changing situations, even rapidly changing, ranging from slight modification of the environment to its thorough transformation (or at least a transformation of the stored model: e.g. discrepancies between the cartographic memory of the world and the current location, due to stale information or possible destruction of infrastructures), we have incorporated deliberative and decision capacities in the system, that do not run necessarily at the same temporal rate than the i/o behavior execution.

In order to scale up to system level, we must then turn to control architectures, i.e. we have to detail how the various functions related to perception and action have to be organized in order for the whole system to fulfill a given objective. To achieve this, we work with *hybrid control architectures*, integrating a lower level focusing on intelligent control and active perception, and a higher level provided through the mission planning.

1.2 Robot control architectures

As a complex system collocating sensors, actuators, electronic and mechanical devices, computing resources, a robot has to be provided ways to organize these various heterogeneous components in order to fulfill its prescribed mission, which furthermore may evolve in time. This is all the more important when additional constraints, such as real-time and cost issues – nowadays a major issue for operational systems – are involved. The control architecture deals with these problems and brings answers to the following questions:

– how is the system built from its basic components?
– how do the parts build up a whole?
– how should components be (re)organized to fulfill missions changing in time?

By basic components, one has to understand mechanical, electronical and software aspects, sensors, actuators, but also the ways to relate these elements and the interfaces between the various subsystems.

For a general overview of existing control architectures, see [1]. The first architectures historically introduced in mobile robots derive from the sense-plan-act paradigm taken from hard artificial intelligence, and follow a top-down approach relying on a recursive functional decomposition of the problem into subproblems down to a grain level where an explicit solution to the problem is given. Such architectures have been shown to suffer from the symbol-grounding [5], the frame and the brittleness problems [6]. In other words, they manipulate symbols which cannot be related in a constructive way to features of the environment and they have to rely on a model of the environment which has to be complete and redefined hierarchically in order to cope with the top-down functional decomposition. While this works for static environments, any unplanned situation will have dramatic impact on the robot! [3]

In reaction to that approach, bottom-up approaches have been proposed, inspired by biology and ethology. They do not rely on explicit models of the environment but on input-output reactive behaviors, which may be aggregated together to solve a more complex task. One of the most famous bottom-up architectures is Brook's subsumption architecture.

However it is generally admitted that both approaches have failed, mainly because of their radical positions: top-down approaches lead to awkward robots unable to cope with any unforeseen change of the environment or the mission, while bottom-up approaches have led to promising animal-like robots which unfortunately could not solve complex problems or missions. Building on their respective advantages (and hopefully not cumulating their drawbacks!), hybrid architectures have been investigated in recent years. They try to have on the one hand a reactive component and on the other hand a decision or planning module; the difficulty is of course the interface between these layers and here lies the diversity of the current approaches.

2 HARPIC

2.1 General description

We propose an hybrid architecture (cf. figure 1) which consists in four blocks organized around a fifth: *perception processes*, an *attention manager*, a *behavior selector* and *action processes*. The core of the architecture relies on *representations*.

Sensors yield data to perception processes which create representations of the environment. These perception processes are activated or inhibited by the attention manager and receive also an information on the current executed behavior. This information is used to foresee and check the consistency of the representation. The attention manager updates representations (on a periodical or on an exceptional basis), supervises the environment (detection of new events) and the algorithms (prediction/feedback control) and guarantees an efficient use of the computing resources. The action selection module chooses the robot's behavior depending on the predefined goal(s), the current action, the representations and their estimated reliability. Finally, the behaviors control the robot's actuators in closed loop with the associated perception processes.

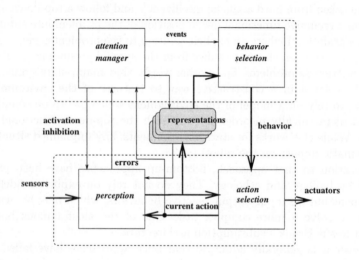

Fig. 1. Functional diagram of the HARPIC architecture.

The key ideas of that architecture are:

• *The use of sensorimotor behaviors linking internally and externally perceptions and low-level actions*: the internal coupling allows to compare a prediction of the next perception, estimated from the previous perception and the current control, with the perception obtained after application of the control, in order to decide whether the current behavior runs normally or should be changed.

• *Use of perception processes* with the aim of creating local situated representations of the environment. No global model of the environment is used; however less local and higher level representations can be built from the instantaneous local representations.

• *Quantitative assessment of every representation* [13]: every algorithm has associated evaluation metrics which assign to every constructed representation a numerical value that expresses the confidence which can be given to it. This is important, because any processing algorithm has a domain of validity and its internal parameters are best suited for some situations: there is no perfect algorithm that always yields "good" results.

• *Use of an attention manager*: it supervises the executions of the perception processing algorithms independently from the current actions. It takes into account the processing time needed for each perception process, as well as the cost in terms of needed computational resources. It looks also for new events due to the dynamics of the environment, which may signify a new danger or opportunities leading to a change of behavior. It may also fire processes in order to check whether sensors function nominally and can receive error signals coming from current perception processes. In practice, for instance with a vision sensor, the attention will focus on the illumination conditions, on the consistency between the movement of the robot and the temporal consistency of the representations, and on error signals sent by perception processes. With this information it is then possible to invalidate representations due to malfunctioning sensors or misused processes.

• *The behavior selection module* chooses the sensorimotor behaviors to be activated or inhibited depending on the predefined goal, the available representations and the events issued from the attention manager. This module is the highest level of the architecture. It should be noted that the quantitative assessment of the representations plays a key role in the decision process of the behavior selection: on the one hand a representation might be more or less adapted to the current situation, depending for instance on the sensor used or on the conditions of the perception acquisition (e.g. a day camera used during night operating conditions will yield representations to which a lower confidence should be assigned a priori; the same holds also for instance for a detector relying on translational invariance while the robot has such a motion that this assumption is incorrect) ; on the other hand some representations might be more interesting for some behaviors or might provide an improved help to choose between several behaviors (e.g. a wall-following behavior needs information on contours more than velocity vectors, while a tracking behavior has the opposite needs). Therefore every behavior weights also each representation depending on its direct usability, and this weight is combined with the intrinsic assessment of the representation.

• *The action selection module* regroups the lower-level controllers operating on the actuators. It uses valid representations in order to compute the control laws.

This modular architecture allows to develop independently the various processes belonging to each of the four entities, before integrating them together. Its originality relies in both the lower-level loop between perception and action, necessary for instance for active vision and any situated approach of perception, and the decoupling of the perception and the action processes, which become behaviors during their execution. The latter avoids the redundancy of common components, saves computational resources when representations are common to several behaviors and limits the conflicts when accessing the hardware resources of the robot. Compared to other hybrid architectures (the so-called three-layer approach which develops the three following levels: symbolic level, reactive behaviors, lower-level control), we have tried to focus more on the relations between these three levels in order to take into account the heterogeneous loop aspect characterizing a complex robot.

Furthermore, the proposed architecture with both loops between perception and action, one at a lower level, the other one relying on representations, seems a plausible model from a biological point of view [2]: whereas the lower-level sensorimotor closed loop is characteristic of simpler organisms, the cognitive capacities of higher organisms, like mammals, can be explained by another loop relying on a dynamical model of the environment and the organism, running in feedforward mode with a slower time scale. This higher-level loop allows also open loop planning in order to check the validity of some representations and can trigger their update (e.g. the racing pilot who concentrates on the track and the race, when rehearsing in his head before he actually starts the race).

Finally, the attention manager is a notion used in biological vision [7], and provides a very efficient supervision concept for artificial systems, firing batch processes or reacting on discrete events. The asynchronous property of the control architecture is due to that attention manager, and we think this property is a keystone in complex systems, which have to deal with unpredictable environments and limited resources.

2.2   Implementation and experimentation

Fundamental capacities of our architecture encompass modularity, encapsulation, scalability and parallel execution. For these needs we decided to use a multi-agent formalism that fits naturally our need for encapsulation in independent, asynchronous and heterogeneous modules. The communication between agents is realized by messages. Object oriented language are therefore absolutely suited for programming agents: we chose C++. We use POSIX Threads to obtain parallelism: each agent is represented by a thread in the overall process. We do not aim at promoting multi-agent techniques, for us it is merely an interesting formalism and our architecture could be implemented without them.

All the processing is done by a PC bi-PIII 500 MHz running Linux, with a frame grabber card. We did not use any specialized hardware or real-time operating system. The robots used both in indoor and outdoor experiments are Pioneer AT from RWI equipped with a monochrome camera, with motorized lens, mounted on a pan-tilt unit (cf. figure 2). The robot links to the computer are a radio modem at 9600 bauds and a video link which transmits 768×512 pixel images at a 25 Hz rate. The robot displacement speed is about 400 mm/s.

We have developed perception agents, for lane, obstacle and target detection, and action agents for lane following, obstacle avoiding, target tracking and lens aperture adjusting. The processing time of the perception agents varies between 5 and 30 ms according to their algorithms.

Fig. 2. Outdoor experiments with the robot (structured roads, unstructured tracks).

3 Computational intelligence and controlled autonomy

Through a parallel execution of a wide range of algorithms and after the choice of the "best action" depending on the evaluation of the most adequate representation given a current situation and behavior, it is easily understandable how a high autonomy is achieved. Namely there is a straightforward scheduling of contextual modes: for instance, for an outdoor robot, there is no need to tell explicitly the robot when it drives on structured roads and when it is off-road, since the right representation computed by the various vision algorithms should be selected by the navigation behavior. As an illustration, when dealing with our robot in indoor environments with competing behaviors such as wall-following, obstacle avoidance, navigation with given heading, we have analyzed the sequence of behaviors as selected by the architecture, and there was a natural transition from wall-following to navigation with given heading and back to wall-following when a door occurred.

This is all the more important in structured outdoor environments, where curbs or road marks can suddenly disappear, either because of a crossroad or more often because the reference structure is partially damaged. Truly, we observed [13] that, while driving on roads delimited by sidewalks, detectors based on contour detection had higher assessment marks than other representations based for instance on multifractal texture indices, whereas the contrary happened when arriving on unstructured lanes such as off-road tracks.

Whereas usual roboticists would then ask for more robust perception algorithms, we prefer a more intelligent design that does not require perfect perception or action to work.

Of course, a critic arises then concerning the intrinsic stability of the behavior selector: one could fear a new behavior would be chosen every time step. In order to avoid this, we have established a dichotomy between behavior selection and action selection. A behavior has thus a different time scale than the firing of an action or the data acquisition process; a consequence is a temporal smoothing. Besides, the behavior selector uses a knowledge base which defines which behaviors might follow a given behavior. This reduces the combinatorics and introduces some determinism in the behavior scheduling. One should note that in order not to have too much determinism, we allow the attention manager to interrupt the current behavior whenever it is necessary, depending on unforeseen detected events. Our different experiments up to now have shown that this balancing mechanism works well. Such intrinsic instability within the architecture should not be judged too severely, since we introduced intentionally asynchronism and non-determinism in order to be able to cope with unforeseen and changing situations: solving the frame problem and the brittleness problem cannot be done without introducing some structural instability, but one should take care to provide some means to control that unstability as best as possible.

As a consequence of what has been discussed up to now, a most interesting property is the fact that the robot can control its own autonomy by relying on external features, such as perceptual landmarks that can trigger special events or behaviors once they have been detected by the perception processes, transformed into representations and interpreted by the attention manager.

A typical straightforward application consists in an autonomous robot looking for traffic signs along the road, that limit its freedom of decision. A similar example is an autonomous robot, to which instructions are given for instance through sign language or even speech (speech processing algorithms are then needed among the perception processes), which yield either information on the environment or provide orders to be followed, like modification of the current behavior.

Such ability to switch from autonomous to semi-autonomous modes is from our point of view a true sign of computational intelligence: while autonomy is often conceived for an artificial system as not needing external rules to choose its own behavior, being able to follow suddenly rules while you were evolving freely just relying on yourself is a proof of intelligent design!... and a prerequisite for robots to be accepted among humans.

More elaborate control modes of the robot's own autonomy obviously can be conceived which do not necessarily rely on a priori semantics to be discovered by the perception processes. Such less explicit control of autonomy is exerted in a multirobot scenario [16], where we have implemented several robots with the previous architecture after having provided it with an additional communication process which allows the exchange of individual representations among different robots upon request. The autonomy of every robot is then implicitly limited by the autonomy of its partners, as the completion of the task is evaluated at the group's level and what is actually developed is the multirobot group's autonomy.

4 Computational intelligence and learning

In previous work [9, 10, 14], we have presented methodologies that allow to learn the sensorimotor lower-level loop. They proceed in two phases: identification of a qualitative model of the dynamical system and inversion of that model to deduce a fuzzy-like control law. Such research has shown that furthermore higher-level representations could emerge relative to the topology of the environment or the sequencing of behaviors [12, 17].

It would be interesting to see where and how similar learning mechanisms could be introduced within the control architectures. The lower reactive level which binds the perception and the action selection modules have already been dealt with as was recalled previously. Of course it is always possible to improve the already available methods: for instance, learning control laws up to now does not include any explicit prediction, as there is no quantitative analysis or comparison of the actual effect of the input with respect to the predicted effect. If this was achieved, it would be possible to include corrective terms within the learned control law (as is done in predictive feedforward control) and to have reasoning mechanisms on the consistency of the identified qualitative model relatively to the observed effects (techniques on qualitative model verification from [8] could be used for that purpose).

But it would also be interesting to add learning mechanisms at the representation level, especially concerning their update: as in biological systems, forgetting or short-term and long-term memory mechanisms could be implemented).

Concerning the introduction of learning mechanisms within action selection algorithms, the literature is rather rich [15], and most techniques described in the reference could be advantageously implemented in our work.

On the contrary, introducing learning at the attention manager level is an open problem: it could be done in order to interpret either the awaited events as a function of the current behavior and the potential behaviors that could follow it, or unexpected events that can trigger emergency answers. Another direction could be to have adaptive sequencing of the behaviors.

5   Conclusion

We have discussed robot control architectures and proposed an architecture which aims at filling the gap between reactive behavioral and deliberative decision systems, while keeping a close eye on the dynamic management of all the resources available to the robot.

We have also shown how it was possible to propose various autonomy modes, since the autonomy of the robot can be easily controlled by feeding the environment the landmarks that, after interpretation, provide selective behaviors.

Finally, we have discussed the introduction of learning techniques at the various levels of the architecture, and we have argued how such learning mechanisms, if successfully implemented, could yield not only adaptive, but also evolutive control architectures. All of this would then achieve a degree of autonomy, about which we may only dream for now...

References

1. R.C. Arkin. *Behavior-based robotics*. A Bradford Book, The MIT Press, 1998.

2. A. Berthoz. *Le sens du mouvement.* Éditions Odile Jacob, 1997.

3. M.A. Boden, editor. *The philosophy of artificial life.* Oxford Readings in Philosophy, 1996.

4. A. Dalgalarrondo and D. Luzeaux. Rule-based incremental control within an active vision framework. In 4th *Int. Conf. on Control, Automation, Robotics and Vision,* Westin Stamford, Singapore, 1996.

5. S. Harnad. The symbol grounding problem. *Physica D,* 42:335–346, 1990.

6. P.J. Hayes. The frame problem and related problems in artificial intelligence. In J. Allen, J. Hendler, and A. Tate, editors, *Readings in Planning,* pages 588–595. Morgan Kaufmann Publishers, Inc., 1990.

7. S.M. Kosslyn. *Image and Brain.* A Bradford Book, The MIT Press, 1992.

8. B. Kuipers. *Qualitative Reasoning: Modeling and Simulation with incomplete knowledge.* The MIT Press, 1994.

9. D. Luzeaux. Let's learn to control a system! In *IEEE International Conference on Systems Man Cybernetics,* Le Touquet, France, 1993.

10. D. Luzeaux. Learning knowledge-based systems and control of complex systems. In *15th IMACS World Congress,* Berlin, Germany, 1997.

11. D. Luzeaux. Mixed filtering and intelligent control for target tracking with mobile sensors. In 29th *Southeastern Symposium on System Theory,* Cookeville, TN, USA, 1997.

12. D. Luzeaux. Catastrophes as a way to build up knowledge for learning robots. In *16th IMACS World Congress,* Lausanne, Switzerland, 2000.

13. D. Luzeaux and A. Dalgalarrondo. Assessment of image processing algorithms as the keystone of autonomous robot control architectures. In J. Blanc-Talon and D. Popescu, editors, *Imaging and Vision Systems: Theory, Assessment and Applications.* NOVA Science Books, New York, 2000.

14. D. Luzeaux and B. Zavidovique. Process control and machine learning: rule-based incremental control. *IEEE Transactions on Automatic Control,* 39(6), 1994.

15. P. Pirjanian. Behavior coordination mechanisms: state-of-the-art. Technical report, University of Southern California, TR IRIS-99-375, 1999.

16. P. Sellem, E. Amram, and D. Luzeaux. Open multi-agent architecture extended to distributed autonomous robotic systems. In *SPIE Aerosense'00, Conference on Unmanned Ground Vehicle Technology II,* Orlando, FL, USA, 2000.

17. O. Sigaud and D. Luzeaux. Learning hierarchical controllers for situated agents. In *Proceedings of the 14th International Congress on Cybernetics,* Bruxelles, Belgium, 1995.

Hybrid Intelligent Systems for Stock Market Analysis

Ajith Abraham[1], Baikunth Nath[1] and Mahanti P K[2]

[1]School of Computing & Information Technology
Monash University (Gippsland Campus), Churchill 3842, Australia
Email: {Ajith.Abraham, Baikunth.Nath}@infotech.monash.edu.au

[2]Department of Computer Science and Engineering
Birla Institute of Technology, Mesra-835 215, India
Email: deptcom@birlatech.org

Abstract: The use of intelligent systems for stock market predictions has been widely established. This paper deals with the application of hybridized soft computing techniques for automated stock market forecasting and trend analysis. We make use of a neural network for one day ahead stock forecasting and a neuro-fuzzy system for analyzing the trend of the predicted stock values. To demonstrate the proposed technique, we considered the popular Nasdaq-100 index of Nasdaq Stock Market[SM]. We analyzed the 24 months stock data for Nasdaq-100 main index as well as six of the companies listed in the Nasdaq-100 index. Input data were preprocessed using principal component analysis and fed to an artificial neural network for stock forecasting. The predicted stock values are further fed to a neuro-fuzzy system to analyze the trend of the market. The forecasting and trend prediction results using the proposed hybrid system are promising and certainly warrant further research and analysis.

1. Introduction

During the last decade, stocks and futures traders have come to rely upon various types of intelligent systems to make trading decisions. Several hybrid intelligent systems have in recent years been developed for modeling expertise, decision support, complicated automation tasks etc. We present a hybrid system fusing neural networks and neuro-fuzzy systems aided with a few well-known analytical techniques for stock market analysis.

Nasdaq-100 index reflects Nasdaq's largest companies across major industry groups, including computer hardware and software, telecommunications, retail/wholesale trade and biotechnology [1]. The Nasdaq-100 index is a modified capitalization-weighted index, which is designed to limit domination of the Index by a few large stocks while generally retaining the capitalization ranking of companies. Through an investment in Nasdaq-100 index tracking stock, investors can participate in the collective performance of many of the Nasdaq stocks that are often in the news or have become household names. In this paper we attempt to forecast the values of six individual stocks and group index as well as the trend analysis of the different stocks. Individual stock forecasts and group trend analysis might give some insights of the actual performance of the whole index in detail. To demonstrate the efficiency of the proposed hybrid system we considered the two years stock chart information (ending 13 March 2001) of six major industry groups listed on the national market tier of the Nasdaq Stock Market (Nasdaq-100 index).

Neural networks are excellent forecasting tools and can learn from scratch by adjusting the interconnections between layers. Fuzzy inference systems are excellent for decision making under uncertainty. Neuro-fuzzy computing is a popular framework wherein neural network training algorithms are used to fine-tune the parameters of fuzzy inference systems. For the stock forecasting purpose we made use of a neural network trained using scaled conjugate gradient algorithm. However, the forecasted stock values might deviate from the actual values. We modeled the deviation of the predicted value from the required value as a fuzzy variable and used a fuzzy inference system to account for the uncertainty and decision-making. In section 2 we explain the details of the proposed hybrid system and its components followed by experimentation setup and results in section 3. Conclusions and further work are provided towards the end.

2. A Soft Computing Framework for Stock Market Analysis

Soft computing introduced by Lotfi Zadeh is an innovative approach to construct computationally intelligent hybrid systems consisting of neural network, fuzzy inference system, approximate reasoning and derivative free optimization techniques [4]. In contrast to conventional artificial intelligence, which only deals with precision, certainty and rigor the guiding principle of soft computing is to exploit the tolerance for imprecision, uncertainty, low solution cost, robustness, partial truth to achieve tractability, and better rapport with reality.

Figure 1 depicts the hybrid intelligent system model for stock market analysis. We start with data preprocessing, which consists of all the actions taken before the actual data analysis process starts. It is essentially a transformation T that transforms the raw real world data vectors X_{ik}, to a set of new data vectors Y_{ij}. In our experiments, we used Principal Component Analysis (PCA) [3], which involves a mathematical procedure that transforms a number of (possibly) correlated variables into a (smaller) number of uncorrelated variables called principal components. In other words, PCA performs feature extraction. The first principal component accounts for as much of the variability in the data as possible, and each succeeding component accounts for as much of the remaining variability as possible. The preprocessed data is fed into the Artificial Neural Network (ANN) for forecasting the stock outputs. ANN "learns" by adjusting the interconnections (called weights) between layers. When the network is adequately trained, it is able to generalize relevant output for a set of input data. Learning typically occurs by example through training, where the training algorithm iteratively adjusts the connection weights. We trained the ANN using a Scaled Conjugate Gradient Algorithm (SCGA) [6]. In the Conjugate Gradient Algorithm (CGA) a search is performed along conjugate directions, which produces generally faster convergence than steepest descent. A search is made along the conjugate gradient direction to determine the step size, which will minimize the performance function along that line. A line search is performed to determine the optimal distance to move along the current search direction. Then the next search direction is determined so that it is conjugate to previous search direction. The general procedure for determining the new search direction is to combine the new steepest descent direction with the previous search direction. An important feature of the CGA is that the minimization performed in one step is not partially undone by the next, as it is the case with gradient descent methods. However, a drawback of CGA is the requirement of a line search, which is computationally expensive. The SCGA is basically designed to avoid the time-consuming line search at each iteration. SCGA combine the model-trust region approach, which is used in the Levenberg-Marquardt algorithm with the CGA. Detailed step-by-step descriptions of the algorithm can be found in Moller[6].

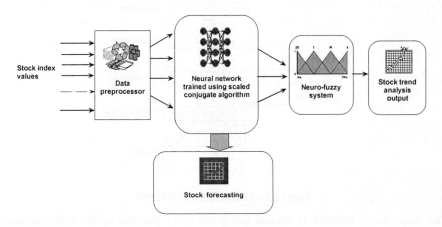

Figure 1. Block diagram showing hybrid intelligent system for stock market analysis

The forecasted outputs by the neural network are further analyzed using a neuro-fuzzy system. This time our aim is to analyze the upward and downward trends of the different forecasted stocks. Since the forecasted values will deviate from the desired value (depending upon the prediction efficiency of ANN), we propose to make use of the uncertainty modeling capability of Fuzzy Inference System (FIS)[7]. We define a neuro-fuzzy system as a combination of ANN and FIS in such a way that neural network learning algorithms are used to determine the parameters of FIS [5]. We used an Evolving Fuzzy Neural Network (EFuNN) implementing a Mamdani type FIS and all nodes are created during learning. EFuNN has a five-layer structure as shown in Figure 2. The input layer followed by a second layer of nodes representing fuzzy quantification of each input variable space. Each input variable is represented here by a group of spatially arranged neurons to represent a fuzzy quantization of this variable. Different Membership Functions (MFs) can be attached to these neurons (triangular, Gaussian, etc.). The nodes representing MFs can be modified during learning. New neurons can evolve in this layer if, for a given input vector, the corresponding variable value does not belong to any of the existing MF to a degree greater than a membership threshold. The third layer contains rule nodes that evolve through hybrid supervised/unsupervised learning. The rule nodes represent prototypes of input-output data associations, graphically represented as an association of hyper-spheres from the fuzzy input and fuzzy output spaces. Each rule node r is defined by two vectors of connection weights – W_1 (r) and W_2 (r), the latter being adjusted through supervised learning based on the output error, and the former being adjusted through unsupervised learning based on similarity measure within a local area of the input problem space. The fourth layer of neurons represents fuzzy quantification for the output variables. The fifth layer represents the real values for the output variables. EFuNN evolving algorithm used in our experimentation was adapted from [2].

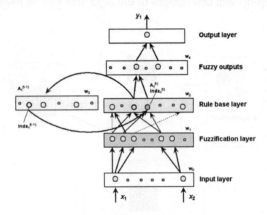

Figure 2. Architecture of EFuNN

The neuro-fuzzy network is trained using the trend patterns of the different stock values. The difference between the day's stock value and the previous day was calculated and used for training the NF system. If all the stock values were increasing we classified it as positive trend "1" and "0" otherwise. The proposed NF system is capable of providing detailed trend analysis of individual stocks and also interdependencies of various stocks and how they affect the overall index.

3. Experimentation setup and test results

We considered 24 months stock data [1] for training and analyzing the efficiency of the proposed hybrid intelligent system. In our experiments, we used Nasdaq-100 main index values and six other companies listed in the Nasdaq-100 index. Apart from the Nasdaq-100 index (IXNDX); the other companies considered were Microsoft Corporation (MSFT), Yahoo! Inc. (YHOO), Cisco Systems Inc. (CSCO), Sun Microsystems Inc. (SUNW), Oracle Corporation (ORCL) and Intel Corporation (INTC). Figures 3 and 4 depict the variation of stock values for a 24 months period from 22 March 1999 to 20 March 2001.

For each *t*, the stock values *x (t)* were first standardized and *x (t-1), x (t-2), x (t-3)* were computed. The data was then passed through the data pre-processor to ensure that the input vectors were uncorrelated. The PCA analysis also revealed that by providing only *t* and *x (t)* the neural networks could be trained within the required accuracy. 80% of the data was used for training and remaining was used for testing and validation. The same set of data was used for training and testing the neuro-fuzzy system. While the proposed neuro-fuzzy system is capable of evolving the architecture by itself, we had to perform some initial experiments to decide the architecture of the ANN. More details are reported in the following sections. Experiments were carried out on a Pentium II 450MHz machine and the codes were executed using MATLAB. Test data was presented to the network and the output from the network was compared with the actual stock values in the time series.

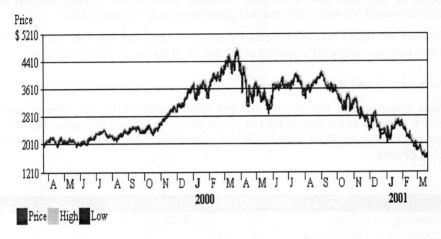

Figure 3. 24 months data of Nasdaq-100 index adapted from [1]

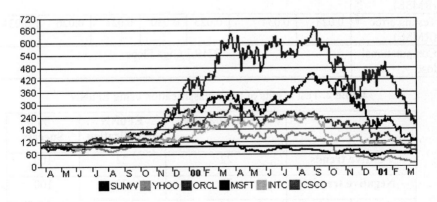

Figure 4. 24 months data of 6 companies adapted from [1]

ANN – SCG Algorithm

We used a feedforward neural network with 8 input nodes and two hidden layers consisting of 20 neurons each. We used tanh-sigmoidal activation function for the hidden neurons. The training was terminated after 2000 epochs. The test data was passed through the network after the training was completed.

EFuNN Training

Each of the input variables consists of the difference in the stock value (example. today's value – yesterday's value). For training the EFuNN, we had 8 input variables, the standardized stock value differences and the time factor. We used 4 membership functions for each of the 8 input variable and the following EFuNN parameters: sensitivity threshold $Sthr$=0.99, error threshold $Errthr$=0.001. Online learning of EFuNN created 503 rule nodes and the training error achieved was 6.5E-05.

However we report only the collective trend of all the seven stock values. If all the trends were increasing we classified as "1" and "0" if the trends were going down.

Performance and Results Achieved

Table 1 summarizes the training and test results achieved for the different stock values. Figure 5 and 6 depicts the test results for the prediction of Nasdaq-100 index and other company stock values. Table 2 summarizes the trend prediction results using EFuNN.

Table 1. Training and testing results using neural network

	Nasdaq	Microsoft	Sun	Cisco	Yahoo	Oracle	Intel
Learning epochs	2000						
Training error (RMSE)	0.0256						
Testing error (RMSE)	0.028	0.034	0.023	0.030	0.021	0.026	0.034
Computational load	7219.2 Giga Flops						

Table 2. Test results of trend classification using EFuNN

	Actual quantity	EFuNN classification	% Success
Positive trends	22	22	100
Negative trends	78	78	100
Computational load	18.25 Giga Flops		

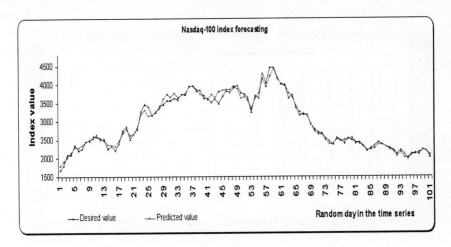

Figure 5. Forecast test results for Nasdaq-100 index

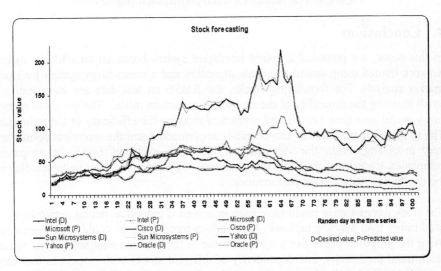

Figure 6. Forecast test results for the six companies listed under Nasdaq-100 index

Figure 7 shows the test results for the collective trend prediction of Nasdaq-100 index and the six 6 company stock values using EFuNN.

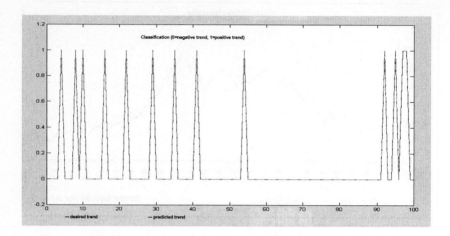

Figure 7. Test results for trend prediction using EFuNN

4. Conclusions

In this paper, we proposed a hybrid intelligent system based on an artificial neural network trained using scaled conjugate algorithm and a neuro-fuzzy system for stock market analysis. For forecasting stocks, the RMSE on test data are comparatively small showing the reliability of the developed prediction model. The proposed neuro-fuzzy model also gave 100% trend prediction showing the efficiency of the technique. The proposed hybrid system can be easily implemented and the empirical results are very promising. From the viewpoint of the stock exchange owner, participating companies, traders and investors the technique might help for better understanding of the day-to-day stock market performance.

The stock forecast error could have been improved if individual neural networks were used rather than a single network. Also various trend analyses could have been done using the proposed neuro-fuzzy system. Some of the possible analyses are individual stock trend predictions, interdependency of different stocks with respect to the main index as well as individual companies. Our future works will be organized in that direction.

References

[1]   Nasdaq Stock Market[SM]: http://www.nasdaq.com

[2]   Kasabov N and Qun Song, *Dynamic Evolving Fuzzy Neural Networks with 'm-out-of-n' Activation Nodes for On-line Adaptive Systems*, Technical Report TR99/04, Department of information science, University of Otago, 1999.

[3]   Duszak Z and Loczkodaj W W, *Using Principal Component Transformation in Machine Learning*, Proceedings of International Conference on Systems Research, Informatics and Cybernetics, Baden-Baden Germany, p.p 125-129, 1994.

[4]   Zadeh LA, *Roles of Soft Computing and Fuzzy Logic in the Conception, Design and Deployment of Information/Intelligent Systems*, Computational Intelligence: Soft Computing and Fuzzy-Neuro Integration with Applications, O Kaynak, LA Zadeh, B Turksen, IJ Rudas (Eds.), pp1-9, 1998.

[5]   Abraham A & Nath B, *Designing Optimal Neuro-Fuzzy Systems for Intelligent Control*, In proceedings of the Sixth International Conference on Control Automation Robotics Computer Vision, (ICARCV 2000), Singapore, December 2000.

[6]   Moller A F, *A Scaled Conjugate Gradient Algorithm for Fast Supervised Learning*, Neural Networks, Volume (6), pp. 525-533, 1993.

[7]   Cherkassky V, *Fuzzy Inference Systems: A Critical Review*, Computational Intelligence: Soft Computing and Fuzzy-Neuro Integration with Applications, Kayak O, Zadeh LA et al (Eds.), Springer, pp.177-197, 1998.

On the Emulation of Kohonen's Self-organization via Single-Map Metropolis-Hastings Algorithms

Jorge Muruzábal

Statistics and Decision Sciences Group

University Rey Juan Carlos, 28936 Móstoles, Spain

j.muruzabal@escet.urjc.es

Abstract. As data sets get larger and larger, the need for exploratory methods that allow some visualization of the overall structure in the data is becoming more important. The self-organizing map (SOM) introduced by Kohonen is a powerful tool for precisely this purpose. In recent years, SOM-based methodology has been refined and deployed with success in various high-dimensional problems. Still, our understanding of the properties of SOMs fitted by Kohonen's original algorithm is not complete, and several statistical models and alternative fitting algorithms have been devised in the literature. This paper presents a new Metropolis-Hastings Markov chain Monte Carlo algorithm designed for SOM fitting. The method stems from both the previous success of bayesian machinery in neural models and the uprise of computer-intensive, simulation-based algorithms in bayesian inference. Experimental results suggest the feasibility as well as the limitations of the approach in its current form. Since the method is based on a few extremely simple chain transition kernels, the framework may well accommodate the more sophisticated constructs needed for a full emulation of the self-organization treat.

1 Introduction

Kohonen's self-organizing map (SOM) [10] provides a fast, scalable and easy-to-interpret visualization tool. Complemented with Sammon's mapping [11] and several diagnostic statistics [1], it has proved useful for several data analysis tasks, see e. g. [12,13]. While this applied success is stimulating, the relative theoretical opacity of the original fitting *algorithm* has made it hard to define the particular state of affairs that the SOM *structure* should converge to [4]. One of the frequently cited problems of the standard fitting algorithm is that it is not supported by any statistical model when perhaps it should, for it attempts after all to carry out a density estimation task.

Several bayesian models have been proposed to bear on this issue. Utsugi's prior [16] places a direct smoothing constraint on the set **w** of SOM pointers. The generative topographic mapping (GTM) approach [2] introduces a latent covariate and tries to hard-wire the SOM's smoothness by means of a non-linear map from latent to data space. This map effectively links the set of pointers beyond the SOM topology. As a result of this modelling effort, several EM-like techniques are now available as alternative fitting algorithms for the SOM structure. However, the practical usefulness of these algorithms is to be fully demonstrated yet [9]. Meanwhile, there clearly remains room for analysis of further algorithms that, on the basis of these sensible models, pursue the emulation of the self-organizing ability.

In recent years, analysis of more complex bayesian models has been made possible thanks to fundamental advances in Markov chain Monte Carlo (MCMC) theory, see e.g. [5, 15]. MCMC methods are based on iterative *sampling* algorithms: given a target posterior distribution π on some space of structures of interest, they provide a collection of structures that can be approximately taken as *iid* draws from π. Estimates of a variety of posterior features, together with their corresponding standard errors, can be built in principle from these values. In the context of neural networks, the most general models involve network architecture parameters γ, network weights or pointers $\mathbf{w} = \mathbf{w}(\gamma)$ and other objects like prior hyperparameters α and likelihood parameters β (collectively denoted by $h = \{\alpha, \beta\}$). The most sophisticated MCMC methods rely on *reversible jump* MCMC theory [6] to explore full posteriors of the form $\pi = \pi(\gamma, \mathbf{w}, h/\mathbf{x})$, where $\mathbf{x}$ is the training data. For example, Rios Insua and Müller [14] derive an approximation to the posterior of the number of units in the hidden layer of a standard feed-forward network. Similar contributions in other neural models have been hindered by the difficulty in elliciting sensible prior distributions over the huge space of possible $(\gamma, \mathbf{w}, h)$. For example, in the SOM model the map size would become a variable parameter. Neither Utsugi [16, 17] nor Bishop and coworkers [2, 3] consider indeed MCMC methods for the SOM model.

The first MCMC method in the SOM context has been proposed in [18]. Utsugi essentially formulates a *Gibbs* sampler [5, 15] for posteriors of the form $\pi = \pi(\mathbf{w}, h/\gamma, \mathbf{x})$. In practice, the SOM is typically planar, and we know that the choices of topology, shape and size are not crucial to obtain useful results. Hence, conditioning on γ substantially reduces the complexity of the problem while imposing only a mild limitation on the overall scope. In this paper I consider an even simpler MCMC method (developed independently of [18]) based on posteriors of the form $\pi(\mathbf{w}/h, \gamma, \mathbf{x})$, where α and β are scalar quantities. Since the choice of conditioning h may modify the posterior landscape substantially, the approach involves some pretesting with various α and β and is thus quite exploratory in its present form. Still, the class of *Metropolis-Hastings* (MH) algorithms [5, 15] reviewed below is rather flexible and permits to explore how far can one go by replacing the conditional distributions in Gibbs samplers with simple transition kernels inspired by the SOM's original fitting algorithm and bayesian assumptions. Note also that the two MCMC algorithms just discussed always maintain a single SOM in memory. Such algorithms are thus markedly different from *multiple* MCMC samplers, see e.g. [8]; these maintain a population of networks from which transition proposals based on several networks can be made.

The organization is as follows. The basic notation and assumptions in bayesian SOM modelling are provided in section 2. Section 3 summarizes some relevant MCMC theory. Section 4 presents the new class of MH algorithms and section

5 reports on their performance in some data sets. Section 6 summarizes and provides directions for future research.

2 Bayesian Self-Organizing Maps

The self-organizing map is a biologically-inspired network of interconnected neurons or units s, each endowed with an associated pointer $w_s \in \mathbb{R}^m$ [10]. Let us focus for simplicity on the 2-D case and consider squared SOMs with $r = k^2$ units and standard connectivity. A data matrix $\mathbf{x}$ containing n exchangeable vectors $x^{(l)} \in \mathbb{R}^m$ is used for training. A trained SOM (fitted by the standard algorithm or otherwise) should satisfy two key desiderata: (i) the "density" of the pointer cloud should resemble the underlying distribution of the data; and (ii) pointers should exhibit topological order or self-organization, a notion unfortunately hard to pin down precisely [4]. The standard fitting procedure tends to achieve these two goals at a reasonable computational cost. For inspection of trained SOMs, we usually project the map $\mathbf{w} = \{w_s, s = 1, .., r\}$ onto 2-D space via Sammon's mapping [11]. Since the set of pointers "inherits" the connectivity pattern, pointers can be linked to its immediate neighbours on these images and we can evaluate informally the amount of organization in the fitted SOM, see Figures 1 and 2 below.

Statistical models recently introduced for the SOM agree to set a Gaussian mixture sampling (or *generative*) model $P(\mathbf{x}/\mathbf{w}, \beta) = \prod\limits_{l=1}^{n} \sum\limits_{s=1}^{r} \frac{1}{r} f(x^{(l)}/w_s, \beta)$, where $f(x/w_s, \beta) = \left(\frac{\beta}{2\pi}\right)^{\frac{m}{2}} \exp\{-\frac{\beta}{2} \|x - w_s\|^2\}$ and $\beta > 0$ controls the dispersion of the data "generated" by any given unit [2, 16]. As regards the prior $P(\mathbf{w}/\alpha)$, it is customary to assume independent coordinates. A general choice for $P(\mathbf{w}/\alpha)$ is the Gaussian process prior $P(\mathbf{w}/\alpha) = \left(\frac{1}{2\pi}\right)^{\frac{rm}{2}} |\alpha|^{-\frac{m}{2}} \prod\limits_{j=1}^{m} \exp\{-\frac{1}{2} w_{(j)}^T \alpha^{-1} w_{(j)}\}$, where α is, in principle, a dispersion matrix expressing the desired bias towards smoothness in some way and $w_{(j)} \in \mathbb{R}^r$ collects the j-th coordinates from all pointers. The full model is completed by the second-stage prior $P(\alpha, \beta)$, from which the key joint distribution $P(\mathbf{x}, \mathbf{w}, \alpha, \beta) = P(\mathbf{x}/\mathbf{w}, \beta)P(\mathbf{w}/\alpha)P(\alpha, \beta)$ follows.

In their GTM model [2], Bishop, Svensén and Williams extend the previous formulation. They assume a latent variable $z \in \mathbb{R}^L$, $L \ll m$, and a prior density on latent space $P(z)$ which is discrete uniform on some fixed $\mathbf{z} = \{z_s, s = 1, ..., r\}$ (typically an evenly-spaced, rather arbitrary collection of latent vectors). A nonlinear, one-to-many map τ, parametrized by a $m \times M$ matrix Λ, formally links the latent and data spaces, so that each pointer w_s becomes $w_s = \tau(z_s, \Lambda) = \Lambda\Phi(z_s)$ for a set Φ of M fixed basis functions defined on $\mathbb{R}^L$. As shown by the authors, the emerging log-likelihood $L(\Lambda, \beta; \mathbf{x}) = \sum\limits_{l=1}^{n} \log\left\{\frac{1}{r} \sum\limits_{s=1}^{r} f(x^{(l)}/\tau(z_s, \Lambda), \beta)\right\}$ can be maximized by a variant of the EM algorithm. Note that each z_s plays the role of a single neuron and each $\tau(z_s, \Lambda)$ plays the role of a single pointer, so that the number of degrees of freedom of the map $\mathbf{w}$ is reduced substantially. Further, since τ is continuous and smooth, the $\tau(z_s, \Lambda)$ centroids will be automatically organized, see [2] for details on Φ, M and L.

Returning to our previous bayesian model, in [3] GTM's latent structure is used to reduce the complexity in $P(\mathbf{w}/\alpha)$ by taking $\alpha_{st} \propto \exp\{-\frac{\lambda}{2} \|z_s - z_t\|^2\}$, where z_s and z_t are the latent vectors generating map units s and t respectively and $\lambda > 0$ is a single scalar expressing correlation decay that naturally plays the role of α in $P(\mathbf{x}, \mathbf{w}, \alpha, \beta)$. Utsugi [18] dismisses this idea and prefers to achieve a similar level of simplicity via the alternative choice $\alpha = (\lambda D^T D + \mu E^T E)^{-1}$, where D and E are fixed matrices and λ and μ are (scalar) weight factors. Utsugi assigns the main role to the D matrix (E is just included to guarantee a proper prior). If we set $\mu = 0$ for simplicity, we are led to $P(\mathbf{w}/\alpha) = P(\mathbf{w}/\lambda) = \prod_{j=1}^{m} \left(\frac{\lambda}{2\pi}\right)^{\frac{R}{2}} \sqrt{\Delta} \exp\left\{-\frac{\lambda}{2} \|Dw_{(j)}\|^2\right\}$, where R and Δ are the rank and product of positive eigenvalues of $D^T D$ respectively. This is the prior used below for the usual D smoothing matrix, the so-called five-point star approximation to the Laplacian operator in 2-D [16]. Specifically, D has $R = (k-2)^2$ rows (one for each interior unit), and the row addressing pointer (u, v) presents a -4 at location (u, v), ones at its four neighbours' locations $\{(u - 1, v), (u + 1, v), (u, v - 1), (u, v + 1)\}$ and zeros elsewhere. The resulting conditional log posterior function is $\log P(\mathbf{w}/\mathbf{x}, \lambda, \beta) = \sum_{l=1}^{n} \log \sum_{s=1}^{r} \exp\left\{-\frac{\beta}{2} \|x^{(l)} - w_s\|^2\right\} - \frac{\lambda}{2} \sum_{j=1}^{m} \|Dw_{(j)}\|^2$. The following MCMC computations are based on $P(\mathbf{w}/\mathbf{x}, \lambda, \beta)$ playing the role of $\pi(\mathbf{w}/h, \gamma, \mathbf{x})$ as discussed in the Introduction (it will simply be written $\pi(\mathbf{w})$ below). I sometimes refer to the two summands in this expression as the fit and smoothness components respectively.

3 Some MCMC background

We now briefly review the basic aspects of the Metropolis-Hastings (MH) class of algorithms [5, 15]. The key result states that an invariant distribution π of a time-homogeneous Markov chain G with transition kernel Γ is also its limiting distribution provided (G, Γ) is aperiodic and irreducible. Intuitively, for the chain to be irreducible any state should be reachable from any other state. Likewise, an aperiodic chain is not forced to visit certain (subsets of) states in any systematic way. The main idea in bayesian analysis is to simulate a suitable chain (G, Γ) having the posterior of interest π as invariant distribution. The limiting (long-run) behaviour of the chain is then taken as an approximation to iid sampling from π.

The class of MH algorithms easily yields kernels Γ guaranteing the stationarity of any given π as follows. Let $q(a, b)$ denote a proposal density configuring the MH transition kernel Γ. Then, given that the chain is at state a, a random proposal b is made according to $q(a, \cdot)$ and accepted with probability $\psi(a, b) = \min\{1, \frac{\pi(b)q(b,a)}{\pi(a)q(a,b)}\}$; if the proposal is not accepted, then the chain stays at state a and a new b is drawn from $q(a, \cdot)$, etc. This procedure guarantees the stationarity of π. In practice, q is to be devised to ensure that the chain is also aperiodic and irreducible.

An easy way to do this is to decompose q as a (finite) mixture of several densities q_θ (with respective activation probabilities p_θ), all of which have π as stationary distribution and one of which, say θ_0, does possess the desired properties of aperiodicity and irreducibility in a trivial way. The mixture chain defined by $q = \sum_\theta p_\theta q_\theta$ first selects some θ and then makes a proposal according to $q_\theta(a, \cdot)$. As long as the corresponding p_{θ_0} is strictly positive, this mixture q inherits the target properties and hence provides a means to simulate the posterior π according to the MH strategy. The basis density θ_0 is usually quite simple; it is the remaining q_θ in the mixture which provide adequate scope for the strategy.

In the case of SOMs $\mathbf{w} \in \mathbb{R}^{rm}$, it is straightforward to see that the role of θ_0 can be played by either a joint spherical Gaussian random walk $\tilde{\mathbf{w}} \sim N_{rm}(\mathbf{w}, \sigma_1^2 I)$ (updating all pointers w_s at once) or else a uniform mixture density made up by the lower-dimensional, single-pointer spherical Gaussian random walks $N_m(w_s, \sigma_s^2 I)$ (updating a single w_s at a time). Here typically the σ_s^2 are all equal to some σ_2^2. These "background" processes are referred to below as (B1) and (B2) respectively. Note that, in either case, $q(\mathbf{w}, \tilde{\mathbf{w}}) = q(\tilde{\mathbf{w}}, \mathbf{w})$, so $\psi(\mathbf{w}, \tilde{\mathbf{w}})$ boils down to $\min\{1, \frac{\pi(\tilde{\mathbf{w}})}{\pi(\mathbf{w})}\}$.

Consider now the case of more general block transitions q_θ in our SOM context. Now index θ refers to (possibly overlapping) subsets of coordinates of $\mathbf{w}$, for example, those associated to one or several pointers w_s. The activation probabilities p_θ correspond to random drawing among all possible θ. If we decompose, with an obvious notation, $\mathbf{w} = \{\mathbf{w}_\theta, \mathbf{w}_{(\theta)}\}$, we typically use $q_\theta(\mathbf{w}, \tilde{\mathbf{w}}) = q(\mathbf{w}, \tilde{\mathbf{w}}_\theta)$ for all θ, that is, at each transition step proposals are made to update the θ portion only (but these proposals are always made in the same way as in the case of (B2) above). It follows that $\mathbf{w}_{(\theta)} = \tilde{\mathbf{w}}_{(\theta)}$ and hence $\psi(\mathbf{w}, \tilde{\mathbf{w}}) = \min\{1, \frac{\pi(\tilde{\mathbf{w}}_\theta/\tilde{\mathbf{w}}_{(\theta)})q(\tilde{\mathbf{w}}, w_\theta)}{\pi(\mathbf{w}_\theta/\mathbf{w}_{(\theta)})q(\mathbf{w}, \tilde{\mathbf{w}}_\theta)}\}$. An important particular case occurs then when $q(\mathbf{w}, \tilde{\mathbf{w}}_\theta) = q(\mathbf{w}_{(\theta)}, \tilde{\mathbf{w}}_\theta)$ equals the conditional posterior density $\pi(\tilde{\mathbf{w}}_\theta/\mathbf{w}_{(\theta)})$, the so-called Gibbs sampler. In this case, $\psi(\mathbf{w}, \tilde{\mathbf{w}}) \equiv 1$, that is, all proposed transitions are automatically carried out. Of course, depending on the complexity of θ and π, it may not be always straightforward to find the conditionals $\pi(\cdot/\mathbf{w}_{(\theta)})$ required for sampling.

Utsugi's [18] Gibbs sampler involves chains that have *entire* SOMs $\mathbf{w}$ together with hyperparameters λ, β and membership dummies $\mathbf{y}$ as state space. Thus, his sampler alternates between $\pi(\mathbf{y}/\mathbf{x}, \mathbf{w}, \beta)$, $\pi(\mathbf{w}/\mathbf{x}, \mathbf{y}, \lambda)$, $\pi(\lambda/\mathbf{x}, \mathbf{y}, \mathbf{w})$ and $\pi(\beta/\mathbf{x}, \mathbf{y}, \mathbf{w})$. Here, in contrast, we are envisaging a sampler oriented to SOM portions θ (for fixed choice of hyperparameters and using no latent structure). In the next section, we consider MH algorithms based on simple proposal densities $q(\mathbf{w}, \tilde{\mathbf{w}}_\theta)$ for certain types of subsets θ. This simplicity comes of course at the price of having to evaluate the posterior ratios $\frac{\pi(\tilde{\mathbf{w}})}{\pi(\mathbf{w})}$ at each tentative $\tilde{\mathbf{w}}_\theta$.

4 SOM fitting via MH samplers

It is clear that the traditional SOM algorithm, although Markovian in nature, is far from the MH family. For example, in the standard sequential implementation of this algorithm all pointers in some neighbourhood are linearly shifted at each time step towards the current data item $x^{(l)}$, the neighbourhood width being a user-input decreasing function. This shifting inspires nonetheless one of the proposal densities $q(\mathbf{w}, \tilde{\mathbf{w}}_\theta)$ discussed below. The other stems from the smoothing character of the assumed prior $P(\mathbf{w}/\lambda)$.

Four MH algorithms for SOM fitting are explored. The first two, (B1) and (B2) (given above), are meant as simple reference algorithms of scant value on their own; they include a single tunable scalar each. The other two algorithms, (A1) and (A2), are mixture chains based on (B1) or (B2) as background processes that incorporate smoothing and block-shift transitions. Let us describe these kernels first.

Smoothing transitions apply to individual interior units s only. They exploit the SOM topology in order to provide a simple smoothing counterpart to the conditional density discussed in the previous section. Specifically, let ν_s denote the standard 8-unit neighbourhood of unit s (excluding s itself) and write ∂w_s for the usual average (mean) vector of w_s, $s \in \nu_s$. A transition is then proposed from w_s to some $\tilde{w}_s$ drawn from a $N_m(\partial w_s, \tau_1^2 I)$ distribution. The associated ratio $\frac{q(\tilde{\mathbf{w}}_{(s)}, w_s)}{q(\mathbf{w}_{(s)}, \tilde{w}_s)}$ becomes $\exp\{-\frac{1}{2\tau_1^2}(\|w_s - \partial w_s\|^2 - \|\tilde{w}_s - \partial w_s\|^2)\}$, so that $\psi(\mathbf{w}, \tilde{\mathbf{w}}) = \exp\{\min[0, \Psi(\mathbf{w}, \tilde{\mathbf{w}})]\}$ with $\Psi(\mathbf{w}, \tilde{\mathbf{w}}) = [\log \pi(\tilde{\mathbf{w}}) - \log \pi(\mathbf{w})] - \frac{1}{2\tau_1^2}(\|w_s - \partial w_s\|^2 - \|\tilde{w}_s - \partial w_s\|^2)$. Note that the intended smoothing effect occurs only if τ_1^2 is relatively small, in which case the last term in $\Psi(\mathbf{w}, \tilde{\mathbf{w}})$ may downplay any log posterior improvement introduced by the new $\tilde{w}_s$. Thus, care is needed when setting the value of the shift variance τ_1^2.

Block-shift transitions act on edges of the net. These transitions are intended to facilitate the higher mobility that such units are expected to need. Specifically, given a unit s on some edge of the network, let ν_s denote now unit s together with its 5 immediate neighbours (3 in the case of corners). A single δ is drawn from a $N_m(0, \tau_2^2 I)$ distribution, and $\tilde{w}_s = w_s + \delta$ for all $s \in \nu_s$. Clearly, in this case $\psi(\mathbf{w}, \tilde{\mathbf{w}})$ simplifies again as in (B1) or (B2). Note the difference with respect to the standard training algorithm whereby δ is different for each s.

We can now describe the remaining MH algorithms; they implement two possible combinations of the above ideas and are defined as follows. Algorithm (A1): at each time step a coin with probability of heads χ is tossed. If heads, a global transition (B1) is attempted. If tails, a unit is randomly selected from the network. If interior, then a smoothing transition is proposed, otherwise a local (B2) transition is attempted. Algorithm (A2): at each time step a similar coin is tossed. If heads, a local (B2) transition is attempted. If tails, a unit is selected as before. If interior, a smoothing transition is proposed, otherwise a block-shift is attempted. Note that both (A1) and (A2) present four tunable scalars each.

5 Experimental results

In this paper I concentrate on the issue of how effective the proposed kernels are with regard to the goal of emulating the standard fitting algorithm [11]. SOMs obtained by these algorithms are contrasted in several toy problems for various choices of hyperparameters λ and β. Relatively small SOMs of 6×6 and 7×7 neurons are used (so we have approximately the same number of interior and edge units). Before we actually dwell into the experiments, a couple of remarks are in order.

It is a fact that MH samplers tend to struggle with multimodality. In practice, this entails that convergence to a suboptimal posterior mode is expected in nearly all cases. This is not so critical though, for there are clearly many useful modes in our SOM context. Hence, the proposed algorithms are evaluated on the basis of the individual SOMs obtained after a fixed number of iterations. Of course, random allocation of pointers is likely to require exceedingly long runs, and alternative initialization procedures are almost compulsory. A simple idea (similar to initialization based on principal components [11] and used below) is to place all initial SOM pointers regularly on a random hyperplane (going through m randomly chosen rows of $\mathbf{x}$). The result typically will not fit the data well, yet it is a flat structure that should free samplers from the ackward phase of early organization.

As regards selection of tunable scalars, the heads probability χ was simply set to $\frac{1}{2}$ throughout. As usual, all sampling variances were tuned on the basis of the acceptance rate ξ of their proposed transitions. This selection process is somewhat tricky since acceptance rates typically decrease along the run and not much is known about optimal values in general. The following simple strategy was based on the anticipated run length. Background processes (B1) and (B2) were first run separately in order to select values for σ_1^2 and σ_2^2 leading to ξ's around 30% after a few thousand trials. These selected values were then maintained in (A1) and (A2), and the remaining τ_1^2 and τ_2^2 were tuned so that the overall ξ's remained between 15 and 25% after 10,000 trials.

Two artificial test cases are considered: a four-cluster Y-shaped data set ($n = 100$, $m = 3$) and a cigar-shaped data set ($n = 100$, $m = 10$). The four-cluster data were generated as a balanced mixture of four Gaussians with small spherical spread and centers located at the corners of a folded Y, see Figure 2. Dimensionality is kept low in this case to allow for direct inspection; the 3-D folding guarantees an interesting structure for the 2-D SOM to capture. The cigar-shaped data (see Figure 1) consists of an elongated bulk with point-mass contamination. The bulk (80%) was derived from a Gaussian distribution with zero mean and equicorrelation dispersion matrix with correlation .9. The remaining 20% were generated by another Gaussian with small spherical spread and mean far away from the bulk.

Let us begin with the cigar-shaped data. The fit is done under $\lambda = 100$ and $\beta = 1$ (thus placing a strong emphasis on the smoothness of the final map). Algorithm (A1) was executed 5 times for 10,000 trials under standard deviations $\sigma_1 = .005$, $\sigma_2 = .045$ and $\tau_1 = .025$; these led consistently to acceptance rates

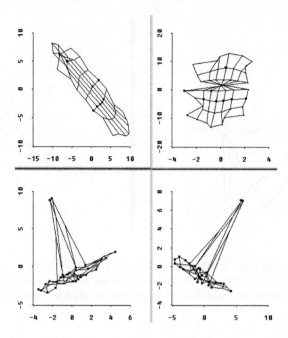

Fig. 1. Projections (via Sammon's map) of SOMs fitted to the cigar-shaped data by MH algorithm (A1) (top row) and by the standard algorithm (bottom row). In all cases solid squares cover the elongated bulk and circles cover the outliers (dots are empty).

by their associated proposals of about 15%, 3% and 58%, with a combined rate of 23%. Figure 1 portrays two SOMs obtained by this MH sampler (second and worst with regard to log posterior values) together with two SOMs fitted by the standard algorithm. As expected, the larger value of λ translates into rather flat structures by (A1), yet we can recover the structure in the data by focusing on the pattern of nonempty units. On the other hand, the standard algorithm arranges pointers more faithfully according to the underlying data density. While not fully organized, these SOMs undoubtedly provide a more accurate description of the data.

Consider next the four-cluster data. We now examine performance by (A2) under $\lambda = 10$, $\beta = 1,000$ (thus priming heavily the fit component) and $\sigma_2 = .2$, $\tau_1 = .15$ and $\tau_2 = .35$ (leading respectively to partial acceptance rates of 35%, 10% and 5%, with an overall rate of about 21%). Five runs were again conducted, and this time the median log posterior SOM was selected for comparison. Figure 2 shows this SOM together with another map fitted by the standard algorithm. The differences are again outstanding as regards visual appearance and log density values. Specifically, the standard algorithm scores about $-6,100$ and -530 in the smoothness and fit log density scale respectively, whereas (A2) yields -50 and $-1,710$ respectively. Hence, the standard algorithm is clearly willing to sacrifice a good deal of smoothness in order to arrange pointers closer to the data.

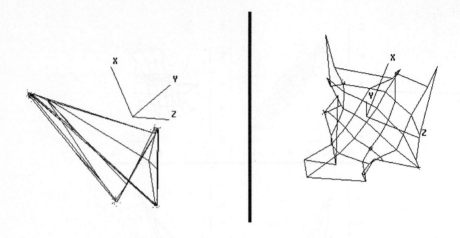

Fig. 2. 3-D rotating plots for the four-cluster data: SOMs fitted by the standard algorithm (left) and by the MH algorithm (A2). Data are superimposed for reference. Axes are included to highlight the different viewpoints adopted in each case.

We conclude that, even under the favorable β used in this run, the MH algorithm can not produce the clusters of pointers needed to improve the fit beyond the smoothness requirement.

6   Summary and concluding remarks

A new class of MH algorithms for SOM fitting has been presented and some preliminary experiments reported. Following Utsugi's [16] model choice for smoothing prior and gaussian mixture likelihood, it has been shown that it is relatively easy to emulate the smoothness property of the trained SOM via MH algorithms. A useful analysis may then proceed on the basis of the pattern of non-empty units on the network. However, it is the SOM's density estimating goal which remains ellusive and may require the design of additional transition kernels. Specifically, kernels that somehow home in on detected concentrations of the data should be most useful here. Furthermore, alternative prior distributions should be devised in order to set up a more flexible sampling scheme; the prior used here may be too strict in penalizing slight but useful departures from very smooth arrangements of pointers. Overall, it is hoped that the reviewed MH approach proves useful for the future development of new kinds of SOM samplers including the forementioned reversible jump samplers [6, 14], multiple-map samplers [8] and Gibbs samplers [18]. In addition, the class of *adaptive* samplers presented in [7] may be useful to cope with the issue of kernel variance tuning.

Acknowledgement. The author is supported by grants HID98-0379-C02-01 and TIC98-0272-C02-01 from the spanish CICYT agency.

References

1. Bauer, H.-U., Herrmann, M., and Villmann, T. (1999). Neural Maps and Topographic Vector Quantization. *Neural Networks*, Vol. 12, 659–676.
2. Bishop, C. M., Svensén, M., and Williams, K. I. W. (1998). GTM: The Generative Topographic Mapping. *Neural Computation*, Vol. 10, No. 1, 215–235.
3. Bishop, C. M., Svensén, M., and Williams, K. I. W. (1998). Developments of the Generative Topographic Mapping. *Neurocomputing*, Vol. 21, 203–224.
4. Cottrell, M., Fort, J. C., and Pagès, G. (1998). Theoretical aspects of the SOM algorithm. *Neurocomputing*, Vol. 21, 119–138.
5. Gilks, W. R., Richardson, S., and Spiegelhalter, D. J. (1996). *Markov Chain Monte Carlo in Practice*. Chapman and Hall.
6. Green, P. J. (1995). Reversible Jump Markov Chain Monte Carlo Computation and Bayesian Model Determination. *Biometrika*, Vol. 82, 711–732.
7. Haario, H., Saksman, E., and Tamminen, J. (1999). Adaptive Proposal Distribution for Random Walk Metropolis Algorithm. *Computational Statistics* Vol. 14, No. 3, 375–395.
8. Holmes, C. C. and Mallick, B. K. (1998). Parallel Markov chain Monte Carlo Sampling: an Evolutionary Based Approach. Manuscript available from the MCMC Preprint Service, see http://www.statslab.cam.ac.uk/~mcmc/
9. Kiviluoto, K. and Oja, E. (1998). S-map: A Network with a Simple Self-Organization Algorithm for Generative Topographic Mappings. In *Advances in Neural Information Processing Systems* (M. I. Jordan, M. J. Kearns, and S. A. Solla, Eds.), vol. 10, 549–555. MIT Press.
10. Kohonen, T. (1997). *Self-Organizing Maps* (2nd Ed.). Springer-Verlag.
11. Kohonen, T., Hynninen, J., Kangas, L., and Laaksonen, J. (1995). SOM_PAK. The Self-Organizing Map Program Package. Technical Report, Helsinki University of Technology, Finland.
12. Kohonen, T., Kaski, S., Lagus, K., Salojärvi, J., Paatero, V., and Saarela, A. (2000). Self Organization of a Massive Document Collection. *IEEE Transactions on Neural Networks*, Vol. 11, No. 3, 574–585.
13. Muruzábal, J. and Muñoz, A. (1997). On the Visualization of Outliers via Self-Organizing Maps. *Journal of Computational and Graphical Statistics*, Vol. 6, No. 4, 355–382.
14. Rios Insua, D. and Müller, P. (1998). Feed-Forward Neural Networks for Nonparametric Regression. In *Practical Nonparametric and Semiparametric Bayesian Statistics* (D. Dey, P. Müller and D. Sinha, Eds.), 181–193. Springer-Verlag.
15. Tierney, L. (1994). Markov Chains for Exploring Posterior Distributions. *Annals of Statistics*, Vol. 22, 1701–1762.
16. Utsugi, A. (1997). Hyperparameter Selection for Self-Organizing Maps. *Neural Computation*, Vol. 9, No. 3, 623–635.
17. Utsugi, A. (1998). Density estimation by mixture models with smoothing priors. *Neural Computation*, Vol. 10, No. 8, 2115–2135.
18. Utsugi, A. (2000). Bayesian Sampling and Ensemble Learning in Generative Topographic Mapping. *Neural Processing Letters*, Vol. 12, No. 3, 277–290.

Quasi Analog Formal Neuron
and Its Learning Algorithm Hardware

Karen Nazaryan

Division of Microelectronics and Biomedical Devices,
State Engineering University of Armenia,
375009, Terian Str. 105, Yerevan, Armenia
nakar@freenet.am

Abstract. A version of learning algorithm hardware implementation for a new neuron model – quasi analog formal neuron (QAFN) is considered in this paper. Due to the presynaptic interaction of "AND" type, wide functional class (including all Boolean functions) for the QAFN operating is provided based on only one neuron. There exist two main approaches of neurons, neural networks (NN) and their learning algorithm hardware implementations: analog and digital. The QAFN and its learning algorithm hardware are based on those two approaches simultaneously. Weight reprogrammability is realized based on EEPROM technique that is compatible with CMOS technology. The QAFN and its learning algorithm hardware are suitable to implement in VLSI technology.

1 Introduction

An interest in the artificial neural networks and neurons has been greatly increased especially in the last years. Research show, that the computing machines based on NNs can easily solve such difficult problems which require not only standard algorithmic approaches. These neural networks and neurons are based on models of biological neurons [1, 2].

Thus, a biological neuron model is presented in fig.1 (below, the word *neuron* is implied as the artificial model of biological neuron). A neuron, has one or more inputs x_1, x_2, x_3, ... x_n and one output y. Inputs are weighted, i.e. input x_j is multiplied by weight w_j. Weighted inputs of the neuron s_1, s_2, s_3, ... s_n are summed by resulting algebraic sum S. In general, a neuron has a threshold value, which also is summed algebraically with S. The result of this whole sum is the argument of f activation function of the neuron.

One of the most attractive features of NNs is their capability to be learned and adapted to solve various problems. That adaptability is provided by using learning algorithms, the aim of which is to find the optimum weight sets for the neurons. There exist two main approaches of learning algorithm performance: hardware and software.

In general, neural models differ by the output activation function, input and weight values. For instance, activation function of classical formal neuron (FN) is Boolean one, inputs accept binary and weights - integer values [1, 3]. In analog neuron model, the output function, inputs and weights accept analog values, etc. [4, 5, 6].

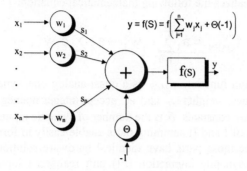

$$y = f(S) = f\left(\sum_{j=1}^{n} w_j x_j + \Theta(-1)\right)$$

Fig. 1. Mathematical model of biological neuron

2 The QAFN model

Currently, there exist many neural models and variants of their hardware implementations. This work considers a learning algorithm hardware for the new neuron model named QAFN (fig.2) [7, 8], which is based on classical FN and enables the increase of FN functional possibilities.

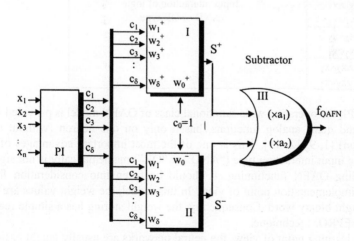

Fig. 2. Quasi analog formal neuron

On the other hand, research shows [8, 9, 10, 11] that the proposed QAFN model simplifies ways of hardware implementation, with simultaneous improvement of some technical parameters and features: operating speed, accuracy, consuming area in semiconductor crystal, etc.

Thus, QAFN realizes the following mathematical equation:

$$f_{QAFN} = a_1 \sum_{j=0}^{\delta} c_j w_j^+ - a_2 \sum_{j=0}^{\delta} c_j w_j^-, \tag{1}$$

where the activation function f_{QAFN} is a quasi analog one, synaptic c_j inputs accept logic 1 and 0 values, weights w_j^+ and w_j^- accept integer nonnegative $(0,+1,+2,+3\ldots)$ values, a_1 and a_2 are constants, δ is the number of synaptic inputs (fig.3).

The separation of I and II summing parts enable easily to form positive and negative weights, since those parts have identical hardware solutions [9, 10, 11]. Input interaction or presynaptic interaction (PI) unit realizes a logical input interaction between values of vector $X=(x_1, x_2, \ldots x_n)$, by forming logical values of vector $C=(c_1, c_2,\ldots c_\delta)$. Values of c-s are all possible variants of logical interactions between input values, including all inputs without interactions. Consequently, $\delta=2^n-1$ (where n is the number of information inputs x). For instance, in $n=3$ case, values of c-s are the followings:

N=3, δ=7,	$(c_1,c_2,\ldots c_7)$ (x_1,x_2,x_3)
$c_1=x_1$	
$c_2=x_2$	Without input interaction
$c_3=x_3$	
$c_4=x_1 \wedge x_2$	
$c_4=x_1 \vee x_2$	
$c_5=x_1 \wedge x_3$	
$c_5=x_1 \vee x_3$	Input interaction of logic
$c_6=x_2 \wedge x_3$	AND or OR type
$c_6=x_2 \vee x_3$	
$c_7=x_1 \wedge x_2 \wedge x_3$	
$c_7=x_1 \vee x_2 \vee x_3$	

Due to PI, the operating wide functional class of QAFN model is provided for both Boolean and quasi analog functions, based only on one neuron (without network construction) [1, 9, 12, 13]. This is one of the most important advantages of QAFN model. The input interaction type ("AND" or "OR") mathematically is not significant for providing QAFN functioning. It should be taken into consideration from the hardware implementation point of view. In this model, the weight values are defined by the weight binary word. Consequently, the weight storing has a simple realization by using EEPROM technique.

From a learning point of view, the neural networks are usually taught easier if the weights of the neurons are allowed to assume both positive and negative values [5]. One of the developed versions of QAFN [9], which provides that opportunity for all weights, is illustrated in fig.3. In this case $a_1=a_2=1$ and only one of the summing parts (I) or (II) are used, and positive or negative weighted influence is defined by switching the weighted inputs s_1, s_2, s_3 ... either to positive or negative input of the subtractor:

$$s_j = c_j w_j = c_j (w_j^+ - w_j^-) = \begin{cases} c_j w_j^+ & \text{if } w_j^- = 0; \\ -c_j w_j^- & \text{if } w_j^+ = 0. \end{cases} \quad (2)$$

The most significant bit (MSB) of *weight binary word* defines the sign of the given weight, as it is done in the known computing systems. MSB controls the output of a switch (IV) to the corresponding positive or negative input of the subtractor (III), depending on the weight sign value.

For example, if the number of bits in the weight binary word is eight, the values of w_j are the following:

$w_j^{(7)}$	$w_j^{(6)}$	.	.	.	.	.	$w_j^{(0)}$	w_j
0	1	1	1	1	1	1	1	+127
0	1	1	1	1	1	1	0	+126
:								:
0	0	0	0	0	0	1	0	+2
0	0	0	0	0	0	0	1	+1
0	0	0	0	0	0	0	0	0
1	1	1	1	1	1	1	1	−1
1	1	1	1	1	1	1	0	−2
:								:
1	0	0	0	0	0	0	1	−127
1	0	0	0	0	0	0	0	−128

Thus, QAFN enables to overcome functional drawbacks of classical neurons, increase functional possibilities and adaptability, and improve technical parameters (speed, features, used area etc.).

3 The QAFN learning

3.1 The Learning Algorithm

A brief description of a learning algorithm for QAFN will be considered here. The output function $f(k)$ for the k-th combination of input x-s, generating $C(k)=(c_1(k), c_2(k), ... c_\delta(k))$ synaptic inputs, is given by:

$$f(k) = S(k) = \sum_{j=0}^{\delta} w_j(k) \cdot c_j(k) \quad (3)$$

If $f(k)$ is equal to the desired value for the k-*th* combination of input variables, we will say that $C(k)$ belongs to the ω_1 functional class, otherwise – $C(k)$ belongs to the ω_2 functional class [12, 14]. Due to the presynaptic interaction unit of QAFN, ω_1 and ω_2 functional classes may be overlapped [9].

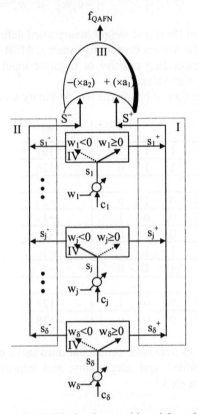

Fig. 3. QAFN block scheme with weight switches

Let $W(k)$ be an arbitrarily selected initial weight vector. Two learning multitudes are given that represent ω_1 and ω_2 functional classes (they may be linearly unseparable), respectively. In this case the k-th step of the neuron learning is the following [12, 14]:

1. If $X(k) \in \omega_1$ and $W(k) \cdot C(k) < 0$, then the $W(k)$ weight vector is replaced by $W(k+1) = W(k) + \varphi \cdot C(k)$ vector;

2. If $X(k) \in \omega_2$ and $W(k) \cdot C(k) \geq 0$, then the $W(k)$ weight vector is replaced by $W(k+1) = W(k) - \varphi \cdot C(k)$ vector;

3. Otherwise, the $W(k)$ weight vector is not changed, i.e. $W(k+1) = W(k)$.

where φ is the correcting factor. φ is a real positive constant value. It influences on the learning rate. It is advised to choose value φ depending on the given problem.

The learning process is accomplished at the $k_r + \gamma$ step, that is

$$W(k_r) = W(k_r + 1) = W(k_r + 2) = ... = W(k_r + \gamma), \tag{4}$$

where γ is the number of training multitudes [14].

The above described neuron learning algorithm converges after finite number of iterations that is proved in [14].

The brief description of the learning algorithm for QAFN is considered below.

Assigning the output error

$$E = T - A, \tag{5}$$

where T and A are target (desired) and actual outputs, respectively. Considering that T and A accept quasi analog values and x-s, consequently c-s – logic 0 or 1, the learning algorithm for the QAFN, based on the algorithms described in [1, 12, 13, 14], can be written in the following way:

$$W(k+1) = W(k) + \Delta(k) \cdot C(k), \tag{6}$$

where

$$\Delta(k) = \begin{cases} -\varphi & \text{if } E < 0 \ (T(k) < A(k)) & \text{(2nd item)}; \\ 0 & \text{if } E = 0 \ (T(k) = A(k)) & \text{(3th item)}; \\ +\varphi & \text{if } E > 0 \ (T(k) > A(k)) & \text{(1st item)}. \end{cases} \tag{7}$$

Here, φ corresponds to the unit weight value.

It is obvious that E/φ is always equal to an integer value.

Schematically, it can be presented as shown in fig.4. The *output error generator* (OEG) generates the $\Delta(k)$, comparing target (T) and actual (A) values (see equation (7)).

3.2 The Learning Hardware

Computation of synaptic weight values is included within the chip by the digital and analog circuit blocks, and the synaptic weight values are stored in the digital memory.

Before the illustration of the hardware implementation details, a technical particularity should be taken into consideration. The problem is that from the engineering point of view, it is very difficult to provide the exact analog value equal to the target one because of the technological mismatch of the semiconductor components. In order to overcome that problem, the following is recommended to take into consideration:

$$U_{target} = \frac{U^+ + U^-}{2} \tag{8}$$

$$\Delta U = U^+ - U^- \tag{9}$$

where U^+, U^- are the bounds of higher and lower voltage levels, between which the actual output value of QAFN is assumed equal to the target one. Of course, ΔU

should be as small as possible considering the hardware implementation properties, in order to reach the ideal case. It is obvious that ΔU and unit weight should have the same order of their values.

Thus, learning target value is given by the analog values of U^+ and U^- considering that the ΔU difference should be the same for all desired output values and input combinations. A voltage level-shifting circuit block (controlled by only U^+) to generate U^+ and U^- values can solve that problem.

The hardware of QAFN learning algorithm is illustrated in fig.5. A simple OEG is designed to generate *up/down* signals. A couple of the voltage comparators generate a digital voltage code comparing the target and actual analog values. That code is then used in the *up/down* signal generation during clock signal appearance. The corresponding digital weight (for which $c_j=1$) is *increased/decreased* by the up/down counter, accordingly. Up/Down signal generation is shown in Table 1. Each value of the weight word bits is stored in a memory cell and is allowed to renew due to EEPROM (electrically erasable-programmable read only memory) technique [15, 16]. To refresh the weights, an access (EN - enable) to the triggers of the corresponding counter, is required. The operational modes of the neuron are presented in Table 2.

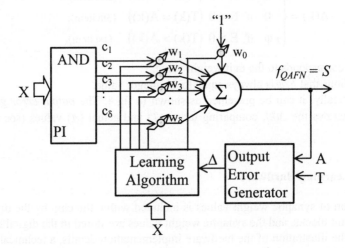

Fig. 4. Schematic representation of QAFN learning

Table 1. Up/Down signal generation

Clk	U_{actual}, U_{target}	c_j	up	down	Learning Algorithm
↑	-	0	0	0	item 3
↑	$U_{actual} < U^-$	1	↑	0	item 1
↑	$U_{actual} > U^+$	1	0	↑	item 2
↑	$U^- < U_{actual} < U^+$	1	0	0	item 3

Table 2. The neuron operational modes

Terminal			Weight storage	Counter (cnt.)		Comments
CS	OE	U_{PR}		access to trigger inputs	Trigger outputs	
Working Mode						
1	1	5v	read	-	R_{off}	
Training Mode						
1	1	5v	read	On	R_{off}	Previous weight setting in counter
0	0	5v	R_{off}	Off	cnt.	Learning algorithm realization
0	1	18v pulse	-	Off	cnt.	Previous weight erasing
0	0	18v pulse	write	Off	cnt.	New weight storing

4 Conclusions

The neuron suggested in this paper could be used in complex digital-analog computational systems, in digital filters, data analyzers, controllers, complex Boolean and quasi – analog functioning, digital – analog signal processing systems, etc., where high speed and adaptability are crucial. Weight discretization is the most reliable way of eliminating leakage problems associated with, for instance, capacitive analog storage. Using static digital storage is a convenient solution, it is, however, quite area consuming. Boolean full-functionality and wide functional class for quasi – analog functions are provided by only one QAFN, without network construction, which is one of the most important advantages of the neuron model. A convenient hardware is designed for the neuron learning algorithm performance. The fabrication mismatches of the semiconductor components are taken into consideration, designing a convenient output error generator, based on some engineering approaches. Since the weight correcting value is fixed compatibly with the QAFN hardware and the learning algorithm is performed for only one neuron, the learning rate is relatively high, which depends only on clock signal frequency and possibilities of hardware implementation limitations. The hardware approach of the QAFN learning algorithm implementation provides a substantial high speed for the learning process. Reprogrammability of the weights is realized due to EEPROM technique. The CMOS QAFN and its learning algorithm are suitable to implement in VLSI technology.

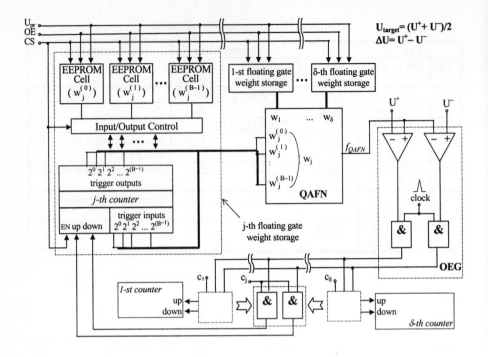

Fig. 5. Schematic of QAFN learning algorithm hardware with analog/digital circuit blocks and digital weight storage.

References

1. Mkrtichyan, S.: Computer Logical Devices Design on the Neural Elements. Moscow: Energia (1977), (in Russian).
2. Mkrtichyan, S.: Neurons and Neural Nets. Moscow: Energia, (1971) (in Russian).
3. Mkrtichyan, S., Mkrtichyan, A.S., Lazaryan, A.F., Nazaryan, K.M. and others. Binary Neurotriggers for Digital Neurocomputers. Proc of NEUREL 97, Belgrade, (1997) 34-36.
4. Mead, C.: Analogue VLSI Implementation of Neural Systems. Kluwer, Boston (1989).
5. Haykin, S.: Neural Networks. A Comprehensive Foundation. Macmillan, New York, (1994)
6. Chua, L., and Yang, L.: Cellular Neural Networks: Theory. IEEE Trans. Circuits and Systems, vol 35, N 5, (1988). 1257-1290.
7. Mkrtichyan, S., and Nazaryan, K.: Synthesis of Quasi Analog Formal Neuron. Report of Science Academy of Armenia, N3, Yerevan: (1997). 262-266 (in Armenian).
8. Nazaryan, K.: Research and Design of Formal Neurons (MSc thesis). State Engineering University of Armenia, Yerevan (1997) (in Armenian).
9. Nazaryan, K..: Circuitry Realizations of Neural Networks (PhD thesis). State Engineering University of Armenia, Yerevan (1999) (in Armenian).
10.Nazaryan, K.: Circuit Realization of a Formal Neuron: Quasi Analog Formal Neuron. In Proc of NEUREL 97. Belgrade, (1997) 94-98.

365

11.Nazaryan, K.,: The Quasi Analog Formal Neuron Implementation Using a Lateral Weight Transistor. Proc of NEURAP 98. Marseilles, (1998). 205-209.
12.Mkrtichyan, S., Lazaryan, A.: Learning Algorithm of Neuron and Its Hardware Implementation. In Proc of NEUREL 97. Belgrade, (1997) 37-42.
13.Mkrtichyan, S., Navasardyan, A., Lazaryan, A., Nazaryan, K. and others.: Adaptive Threshold Element. Patent AR#472, Yerevan, (1997). (in Armenian).
14.Tu, J. and Gonzalev, R.: Pattern Recognition Principles, Mir, Moscow (1978). (in Russian).
15.Holler, M., Tam, S., Castro, H., Benson, R.: An Electrically Trainable Artificial Neural Network (ETANN) with 10240 Floating Gate Synapses. In Proc of IEEE/INNS Int. Joint Conf. of Neural Networks 2, (1989) 191-196.
16.Lee, B., Yang, H., Sheu, B.: Analog Floating-Gate Synapses for General-Purpose VLSI Neural Computation. IEEE Trans. on Circuit and Systems 38, (1991) 654-658.

Producing Non-verbal Output for an Embodied Agent in an Intelligent Tutoring System

Roger Nkambou[1] and Yan Laporte[2]

[1]Département d'Informatique
Université du Québec à Montréal, Montréal, H3C 3P8, Canada
Nkambou.roger@uqam.ca
[2]Département de Mathématiques et Informatique
Université de Sherbrooke, Sherbrooke, J1K 2R1, Canada
ylaporte@acm.org

Abstract. This paper discusses how we can generate non-verbal output through an embodied agent from a user's actions in an ITS. The most part of the present work is related to maintaining an emotional state for a virtual character. We present the basic emotional model we used for internal emotions management for the character. Then we follow with an overview of the agent's environment followed by the role he is designed to play using our own system as a reference. We then give an overview of its internal architecture before we move on to what are the inputs taken by the system and how those are treated to modify the emotional model of the agent.

1 Introduction

Intensive researches have been done over the years on the development of believable agents, embodied agents or real time animated characters [1, 2]. These works have so far achieved very convincing results and impressive demonstrations. Some of them has been done in the field of ITS while the others are more in the interest of cognition theory and adaptive interfaces. Both of those fields are tightly linked to the development of ITS. The current work aimed at making agents being somewhat empathic by giving them the capability to guess the mood and emotional state of the user. For this purpose different models have been created to represent the emotions of the user.

This paper discusses how we can generate non-verbal output though an embodied agent from a user's actions in an ITS. The most part of the present work is related to maintaining an emotional state for a virtual character. We present the basic emotional model we used for internal emotions management for the character. Then an overview of the agent's environment is presented followed by the role the agent is designed to play using our own system as a reference. We then give detail of its internal architecture before move on to what are the inputs taken by the system and how those inputs are computed in order to modify the emotional model of the agent.

2 A Model to Represent Emotions

Among the various cognitive models proposed to represent emotions, one that has been used many times (including in the Oz project [3]) is the Ortony, Clore & Collins (OCC) model [4]. This model has been used by Elliot [5] to become the model presented in table 1. This model was usually used to depict the emotional state of a

Table 1. The OCC Model as described by Clark Elliot 1997

Group	Specification	Category Label and Emotion type
Well-Being	Appraisal of a situation as an event	**Joy**: pleased about an event **Distress**: displeased about an event
Fortunes-of-Others	Presumed value of a situation as an event affecting another	**Happy-for**: pleased about an event desirable for another **Gloating**: pleased about an event undesirable for another **Resentment**: displeased about an event desirable for another **Jealousy**: resentment over a desired mutually exclusive goal **Envy**: resentment over a desired non-exclusive goal **Sorry-for**: displeased about an event undesirable for another
Prospect-based	Appraisal of a situation as a prospective event	**Hope**: pleased about a prospective desirable event **Fear**: displeased about a prospective undesirable event
Confirmation	Appraisal of a situation as confirming or disconfirming an expectation	**Satisfaction**: pleased about a confirmed desirable event **Relief**: pleased about a disconfirmed undesirable event **Fears-confirmed**: displeased about a confirmed undesirable event **Disappointment**: displeased about a disconfirmed desirable event
Attribution	Appraisal of a situation as an accountable act of some agent	**Pride**: approving of one's own act **Admiration**: approving of another's act **Shame**: disapproving of one's own act **Reproach**: disapproving or another's act
Attraction	Appraisal of a situation as containing an attractive or unattractive object	**Liking**: finding an object appealing **Disliking**: finding an object unappealing
Well-being /Attribution	Compound emotions	**Gratitude**: admiration + joy **Anger**: reproach + distress **Gratification**: pride + joy **Remorse**: shame + distress
Attraction/Attribution	Compound emotion extensions	**Love**: admiration + liking **Hate**: reproach + disliking

user and sometime try to infer what his state will become after a given episode. From this model, we can easily define couples of contradictory emotions [6] as indicated in table 2.

What we intended to do in our system is to use this model as an internal model for our agent and not for the user [7]. This practice has been discussed and judged flawed by the original authors [3] mainly because of the limited perception and judgment of a computer system compared to a human being. While this note was stated about inverting a model established from human behavior, which is unlike in our system where the model is driven by arbitrary rules, it is partly addressed in our approach by using the model in a defined environment and with a specific role to play. While this is obviously imperfect, it provides us with a unified representation of the emotion status of our agent. Some work has been done in this regard by morphing different faces together.

Each couple of emotions is represented by a single numerical value that can vary around a central point 0 being the absence of any imbalance in that emotion couple. Increasing the value of a given emotion will, at the same time, decrease the value associated to its counterpart.

Table 2. The Emotion Couples derived from the OCC model

Joy/Distress	Happy-for/Resentment	Sorry-for/Gloating	Hope/Fear
Satisfaction/ Disappointment	Relief/Fears-Confirmed	Admiration/Reproach	Pride/Shame
Liking/Disliking	Gratitude/Anger	Gratification/Remorse	Love/Hate
Jealousy/-Jealousy			

3 Defining the Environment

The Intelligent Tutoring System used as a test environment is comprised of a collection of virtual laboratories (some of them in 3D) where the student must perform tasks and solve problems using interactive simulations [8]. The learners are undergraduate university students and all the activities are part of different scientific curriculum. The emphasis of this system is on simulations lab equipment to accomplish tasks that would otherwise need to be done in a laboratory. This is done in order to allow students to virtually manipulate equipment by themselves and this practice provides a way to overcome problems related to the availability and high cost of the equipment.

The system activities range from 3D simulations of the equipment to more standard interfaces to accomplish tasks such as solving Logic problems in a way where it can be followed by the system. The user also uses the system to view the documentation associated with the activities. The realm of observable events for the agent is contained within those boundaries of consulting documentation and manipulating the virtual laboratories to accomplish given activities and tasks. A structure annotated with domain knowledge or at least a detailed description of what is available to the agent is associated with each activity.

All the information contained in the user model is available as well. The user model [9] is comprised in three parts: the cognitive model, the behavioral or affective model and an inference engine. The entire curriculum is mapped with a structure of underlying concepts which make it possible to specify the required concepts of a given activity and the ones acquired through a given step of the activity. Even particular steps or actions in an activity can be mapped. In the student model, each concept is associated with a mastering level and the source of this belief (the system, the inference engine, the user himself etc.).

The agent doesn't have any perception of anything outside this closed realm. While this limits its capacity to provide appropriate feedback, the user expects it to be that way because the agent is the agent for the tutoring system and nothing else.

4 Defining the Role of the Agent

Eventually, the agent will play the role of a coach in the system providing comments, critics, help and insights. Right now since the system is still in its first version, the goal set for the agent is to provide continual non-verbal feedback on the user's actions (or inactions). It acts more as a companion providing a dynamic element not to replace but to temporarily palliate to the lack of human participation which is a strong element of laboratory activities in the real world. This companion is a companion in the sense of another student [10] but rather a silent coach or maybe a strong student [11] nodding or twitching to the learner's doing. This allows mostly unobtrusive feedback to occur. The agent is also reflects the internal state of the system trying to track what the user is trying to do.

There are a number of concerns that are usually part of designing an embodied agent that must be addressed before defining the system. Most of these are given fixed value in the context of an ITS. The motivation of the agent will not change in our system, it remains at all times that the users acquires new knowledge and completes activities successfully. The "another" mentioned in the definition of the different emotions will always correspond to the learner. Also the attractive and unattractive objects will always be associated with learner's success and learner's failure in doing an action. The different event will however include actions that cannot lead directly to failure or success such as consulting a piece of media or the intermediate steps between two points where we know if the user is successful or unsuccessful. Events can also be constituted of user's inaction.

Also, the agent acts in his own reserved space and doesn't have to do any direct manipulation in the world at this point[2].

5 Architecture Overview

The agent (called Emilie) is designed in three layers each abstracting a part of the problem of generating movement in the agent to provide a response to events in the system. These events are exchanged in the form of messages. The interface is written in Java using the Java3D library while the other layers are implemented in an expert system. An expert system seemed appropriate because it has been used to model emotions in the past [12] and is relatively easy to program.

The interface layer is concerned with managing the 3D geometry and generating the movements from the actions issued from the motor layer. These commands are simple but still carry a lot of geometric information related to the virtual world. Examples of such commands are *lookAt()*, *setSmileLeft()*, *setSmileRight*, *setMouthOpeningCenter()*, *setLeftEyelid()* and so on. While most of them operate with relative parameters which are independent from the geometrical world, some, such as *lookAt()* must provide much more dependent information, in this case, a point in 3D space to look at.

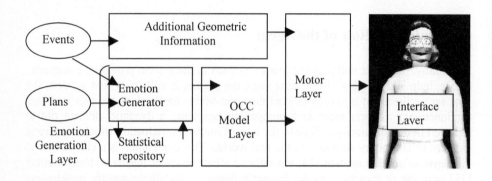

Fig. 1. Overview of the system

5.1 The OCC Model

The OCC model is the emotional model presented earlier in this article. It is implemented as a set of slots in the expert system with rules making sure it has both a ceiling and a floor so the values can't go off the scale. This is done to prevent a user to make the agent so angry or so happy that any changes afterward would be nearly imperceptible. The last set of rules is fired on periodical time intervals to bring the values closer to zero every time. These mechanisms are in place because no matter by which event the emotion was caused, the agent should eventually "get over it" and resume his earlier attitude with time.

5.2 The Statistical Repository

It contains useful information to generate an appropriate response. It stores information such as how many times was the user out of a known to be right path in a row. How many times was the user out of the known to be right path in a row the last time? Is the user on a streak correcting his past mistakes? How wrong the user usually is? What is the ratio of the user being right so far? Has the system considered a past stream of events from the student false while it led to the right solution earlier in this activity? These are important information to achieve coherent emotional response. If the user often uses path to the solution that are unknown of the system all the time and the agent looks angry every time, it is not a desirable behavior. An acceptable behavior would be to look confused or interested in the user's doing until we can evaluate what he is doing. Also if the user has a long streak of events where he is correct, we can't suddenly be angry or have high disapproval on a single mistake.

5.3 The Emotion Generator

The emotion generator is the part of the system that takes events and the planned actions in the system and generates modifications to emotion model. It takes into account information in the statistical repository to adjust his variations before sending them to the OCC model. The modifications made to the OCC model are in the form of decreases or increases of the values for the different emotion couples.

5.4 Additional Geometric Information

This part of the agent content information such as position on the screen of the latest actions. This information can be directly sent to the motor layer and is not related to the other layers anyway since their contribution is essentially in the real of the geometrical representation of the situation.

6 Choosing and Treating the Input

Since the information treated inside an ITS is very varied and comes in very large quantity, we had to define what was relevant to know to be able to generate changes to the emotional model. A large part of this effort is to reduce the amount of information to be taken into account to come up with a character showing satisfying behavior. First, all the information for each event is not relevant to the embodied agent. Domain tied knowledge is nearly unusable for his purpose. The information we retained for all the events coming from the user's are eventually brought down to two parameters: the degree of difficulty of the action accomplished (or to be accomplished in a case of failure) and the degree of "wrongness " of the student or rather how far his action is compared to the expected one (the right action having a value of zero). Another parameter carried in an event but unused by the emotion generator layer is the geometrical coordinates of where the event occurred, this is sent directly to the motor layer as an additional geometric information. Another type of event but which is treated differently is when the user undoes his last actions. This last category of action doesn't carry any parameters.

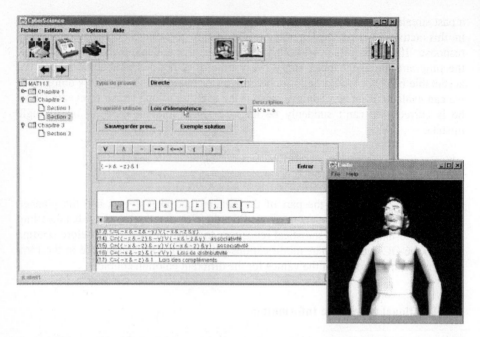

Fig. 2. The user just did something which is not part of the solution

The plans received by the emotion generation layer are merely a list of the difficulty factors that should be encountered by the student before he completes the activity. This information is to be used to produce hope, fear Satisfaction, Relief, Fears-confirmed, Disappointment.

The system also uses a lot of information that comes from the user model. This information is used to create a realistic degree of difficulty for a step or an event. The user model contains the information about the concepts mastered by the learner. In the curriculum we have prerequisites and concepts associated to the different steps in an activity and for the activity itself.

To determine the difficulty level associated with an activity or a step within this activity, we first retrieve an initial difficulty factor set by the designer of the activity. Since arbitrary values or values obtained by statistic about how difficult is an activity are unlikely to reflect the difficulty encountered by any given learner. To come to a better evaluation, we make adjustments based on information available in the curriculum and in the user model. If the student masters all the associated concepts well over the degree required to perform the step or the activity, then we set the associated difficulty level to a lower value. If the episode (the context in which the degree of mastery of a student on a concept was determined by the system) is judged similar to the current one, the difficulty level is set even lower because we consider that the user has already performed a similar task. If however the user has borderline or even incomplete requirements to be able to succeed in the activity, the difficulty factor is raised to a higher value. Also if the activity leads to a large number of new concepts, the degree of difficulty is set higher because we assume that the activity involves a lot of thinking and reasoning for the student to discover these new concepts.

The retrieval of a value to the parameter indicating how far is the student from the expected action is of course very domain related. There are however a few guidelines that can be established. First if the user is doing the expected action, the value should be of zero. If the user is doing something unexpected but that is not identified as a mistake, the value should be small. If the user makes a known mistake then the value should be higher.

7 Converting Events in Emotional Changes

After we have extracted the information about how difficult is the step associated with the incoming event and how wrong was the student was we use a set of predetermined rules to fire procedures that will adjust the OCC model. As an example, an event said to be "good" will generate joy and happy-for emotions, these emotions will be stronger if the degree of difficulty associated was higher. But the rules can be richer and lead to a much more sophisticated emotional feedback. If the user was in a streak of "bad" events, we should introduce satisfaction or relief. If the user just accomplished successfully a step that was judged as the most difficult in the activity we should raise satisfaction and relief only if the user encountered difficulties. If this was the last step in an activity, similar variations should be made to the model. Relief could be introduced if the user took a long time before taking action. Admiration value can be raised if the user succeeded on the first try to a difficult step of activity. It is said the user will prefer be flattered as to be reprimanded [13] so responses to "bad" events should be generally weaker.

When modifications are made to the OCC model, the motor layer converts them in movements or expression for the agent. This conversion includes very simple rules like setting the smile in relation with the joy value. But can also allow more sophisticated actions such as playing little scenes or series of actions when an emotion factor had a strong variation in a short period or modifying the amount of time the agent is directly looking at the user or to the action on the screen. In Figure 2 for example, the user just performed an action that we know is a mistake by selecting an item in the drop down list while it should have chosen another one, to show this the user looks at the source of the event while having an expression cause by the sudden rise of disappointment and distress.

Another example of a result is visible on figure 3 where the sudden rise of relief, joy and admiration led to the visible expression. The facial expression is in majority due to the joy and happy-for values while the hand movement is a result of the important variation in the admiration and/or relief levels.

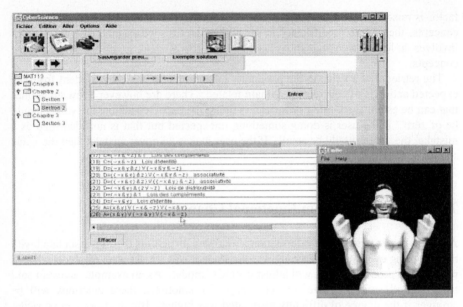

Fig. 3. The user just finished a difficult activity

8 Future Improvements

The current system uses the OCC model, which makes some expressions difficult to represent. For example it is still unclear how the system should represent confusion when the user accomplishes an unexpected action that is of an unknown value. Maybe there are more suitable models already existing or maybe we will have to tailor our own.

Numerous improvements could be made to the visual interface of the agent. Among them is the creation of a new 3D model which would be more cartoon like and require less polygons to draw, therefore offer better performances. Also another beneficial improvement would be that the agent would be in the same window where the activities take place. Adding transparency so the agent would look as it is drawn over the workspace would also make the system more usable. Better animation would also help to offer a better experience for the user.

Of course such a system would always beneficiate of a larger set of rules. Providing an authoring tool would most likely make the task of enlarging the current base easier. It would also offer better chances for our system to be reused in other contexts.

Some functionality will have to be integrated as well for the agent to really act as a coach. The agent should be able to converse with the user in some way. A text-to-speech feature would also improve the effect of the agent's feedback. For these features to be implemented, the agent will need to make a more important use of the other parts of the system such as the curriculum [14] or the representation of contexts.

It should also be able to choose appropriate materials from the media available when the user asks for help.

9 Conclusion

We proposed a way to generate simulated emotional responses to an embodied agent. Those visible responses are triggered by changes in an internal emotional model based on cognition theory. We have given examples of how information available in an ITS can be used to modify this model in response to events from the user. Such a system could be implemented in conjunction with other sources of actions to raise the quality of the interaction with embodied agents in Intelligent Tutoring Systems. We will continue to enhance this model and the implementation we are using. We are considering to eventually release the project and it's source code in hope that other people will use and enhance it.

10 References

1. Rickel, J. and Johnson, L. 2000. Task-Oriented Collaboration eith Embodied Agents in Virtual Worlds. In J. Cassel, J. Sullivan and S. Prevost (Eds.), *Embodied Conversational Agents*, MIT Press 2000.
2. Johnson, L., Rickel, J. and Lester, J.C. (2000). "Anumated Pedagogical Agents: Face-to-face interaction in interactive learning environments. *International Journal of Artificial Intelligence in Education*. 11: 47-78.
3. Ortony A., Clore G., and Collins A. *The Cognitive Structure of Emotions*, Cambridge: Cambridge University Press, 1988.
4. Bates J. The Role of Emotion in Believable Agents. Communications of the ACM, Speical Issue on Agents, July 1994
5. Elliott, C. 1992. *The Affective Reasoner: A Process Model of Emotions in a Multi-agent System*. Ph.D. Dissertation, Northwestern University. The Institute for the Learning Sciences, Technical Report No. 32.
6. Sassine Abou-Jaoude and Claude Frasson. *Emotion computing in competitve learning environments*. In Working Notes of the ITS '98 Workshop on Pedagogical Agents, pages 33--39, San Antonio, Texas, 1998.
7. Elliott, C.; Rickel, J.; and Lester, J. 1997. Integrating affective computing into animated tutoring agents. In *Proceedings of the IJCAI97 workshop, Animated Interface Agents: Making them Intelligent*, 113-121.
8. Nkambou, R., Laporte, Y. 2000. Integrating Learning Agents in Virtual Laboratory. In *Proceeding of World Conference on Educational Multimedia, Hypermedia & Telecommunication*, pp. 1669-1671. AACE.
9. Nkambou, R. 1999. Managing inference process in student modelling for intelligent tutoring systems. In Proceeding of the *IEEE International Conference on Tools with Artificial Intelligence, pp. 16-21*. IEEE press.
10. Aimeur, E., Dufort, H., Leibu, D., and Frasson, C. (1997). Some justifications for the learning by disturbing paradigm. In DuBoulay, B., Mizoguchi, R. (eds.), *Artificial Intelligence in Education: Knowledge and Media in Learning Systems. The Proceedings of AI-ED 97, Kobe, Japan, August 1997, 119-126.*

11. Hietala, P., Niemirepo, T. (1998). "Current Trends and Applications of Artificial Intelligence in Education. Presented at *The Fourth World Congress on Expert Systems*, Mexico City, Mexico.
12. Scherer, K. 1993. Studying the emotion-antecedent appraisal process: An expert system approach. *Cognition & Emotion* 7(3):325-356.
13. Fogg, B. J., Nass, C. 1997. Silicon sycophants: The effects of computers that flatter. *International journal of human-computer studies* 46(5):551-561.
14. Nkambou, R., Frasson, C. and Gauthier, G. 1998. A new approach to ITS-curriculum and course authoring: the authoring environment. *Computers & Education*. 31: 105-130.

Co-evolving a Neural-Net Evaluation Function for Othello by Combining Genetic Algorithms and Reinforcement Learning

Joshua A. Singer

Stanford University

Abstract. The neural network has been used extensively as a vehicle for both genetic algorithms and reinforcement learning. This paper shows a natural way to combine the two methods and suggests that reinforcement learning may be superior to random mutation as an engine for the discovery of useful substructures. The paper also describes a software experiment that applies this technique to produce an Othello-playing computer program. The experiment subjects a pool of Othello-playing programs to a regime of successive adaptation cycles, where each cycle consists of an evolutionary phase, based on the genetic algorithm, followed by a learning phase, based on reinforcement learning. A key idea of the genetic implementation is the concept of feature-level crossover. The regime was run for three months through 900,000 individual matches of Othello. It ultimately yielded a program that is competitive with a human-designed Othello-program that plays at roughly intermediate level.

1 Background

This paper describes a way to search the space of neural-net functions using an approach that combines the genetic algorithm with reinforcement learning. The hybrid approach is founded on the intuition that crossover and gradient descent exploit different properties of the fitness terrain and can thus be synergistically combined. A secondary idea is that of *feature-level crossover*: each hidden node of a three-layer neural-net may be thought of as extracting a feature from the net's input. It is these features that are mixed by the crossover operation.

The paper also describes a software experiment that employs this hybrid approach to evolve a pool of Othello-playing programs. The best evolved program played about as well as *kreversi*, the hand-crafted Othello program written by Mats Luthman and ported to KDE/LINUX by Mario Weilguni. (See www.kde.org for more information.) *Othello*, also known as *Reversi*, is a popular two-player board-game whose rules are described briefly in section 1.3.

1.1 Neural Nets As Game-Players

Many people have used neural-nets as the basis for playing games, notably Tesauro [1], whose TD Gammon program applied reinforcement learning to a neural-net in self-play and greatly surpassed all previous efforts at computer backgammon, and recently Chellapilla and Fogel [2], who applied a mutation-only genetic algorithm to a pool of neural-nets, producing a net that could play checkers at near-expert level. Neural-net approaches to Othello in particular include that of Moriarty and Miikkulainen [3], who applied the genetic algorithm to a flexible encoding representation that allowed them to evolve multi-layer, recurrent neural-net topologies.

1.2   The Impetus for a Combined Approach

The goal of all learning methods can be roughly characterized as the search for an optimal point within some space, where optimality usually means maximizing a fitness function or minimizing a cost function on the space. In the case of Othello, the search space is a parameterized space of static evaluation functions, and the point searched for, $\mathbf{u}^*$, is the parameter whose associated function $f_{\mathbf{u}^*}(\mathbf{x})$ most closely approximates the true value function of the game. In general, a global optimum cannot be found, and we content ourselves with methods that can find points that are highly fit (very close to optimal).

There are at least two reasonable ways to search for points with high fitness. One is to use gradient-descent to find a local optimum. This is the method used by reinforcement learning. The other is to use crossover, and in so doing to make what we might call the *decomposability assumption*: that the fitness of a point in the space can be approximated as the sum of fitness-values associated with substructures of the point. For example, suppose the search space is $\mathbf{R}^n$ with associated fitness function F, and that we can decompose $\mathbf{R}^n$ into the cross-product of subspaces $\mathbf{R}^m$ and $\mathbf{R}^k$, each with their own fitness-functions, F_1, and F_2, in such a way that when we decompose $\mathbf{u} \in \mathbf{R}^n$ as $\mathbf{u} = (\mathbf{u}_1, \mathbf{u}_2)$, $\mathbf{u}_1 \in \mathbf{R}^m$, $\mathbf{u}_2 \in \mathbf{R}^k$, we find that $F(\mathbf{u})$ is approximately equal to $F_1(\mathbf{u}_1) + F_2(\mathbf{u}_2)$. Then the original fitness terrain, $(\mathbf{R}^n, F)$, is a decomposable fitness terrain.

As an example of how crossover works to improve fitness in a decomposable fitness terrain, consider two parameters, $\mathbf{u} = (\mathbf{u}_1, \mathbf{u}_2)$ with fitness-values $(50, 1)$ for a total fitness of approximately 51, and $\mathbf{z} = (\mathbf{z}_1, \mathbf{z}_2)$ with fitness-values $(2, 60)$, for a total fitness of approximately 62. If these are crossed to yield the two child parameters $\mathbf{c}_1 = (\mathbf{u}_1, \mathbf{z}_2)$ and $\mathbf{c}_2 = (\mathbf{z}_1, \mathbf{u}_2)$, then $\mathbf{c}_1$ will have a fitness of approximately 110.

Crossover has a weakness though: it can't make something from nothing. Given a pool of points with low-fitness, possessed of no useful substructures, crossover is no more likely than macromutation to yield a point with improved fitness [4].

Gradient-descent does not possess this weakness of crossover. While, in the standard genetic algorithm, mutation is the motive force for discovering useful substructures, in this combined approach, gradient-descent can be used to find useful substructures much more quickly. No matter how unfit a particular point is, it can be used as the starting point for gradient-descent, ultimately leading to a locally optimal point. Furthermore, it is a fairly safe guess that this point, being a local optimum, possesses at least one good substructure!

Consider then the following scenario: we start with ten randomly chosen points in the fitness terrain. The fitness of each being poor, we don't expect crossover to yield any interesting results. So instead we perform ten gradient-descent searches, yielding ten different local optima. Presumably each of these local optima possesses at least one good substructure, and, since they come from different portions of the fitness terrain, it may well be that they possess ten *different* good substructures. Now crossover has a lot to work with. If we cross pairs of points from this pool of local optima, we are very likely to produce new points of higher fitness than any in the original pool.

We therefore have the following situation: given any point that is not already a local optimum, we can use gradient-descent to find a local optimum, yielding a new point with improved fitness. And, given any two points with relatively high fitness, we can use crossover to yield, with high probability, a new point with improved fitness. But if this new point truly has improved fitness over its parents, then it can't reside in the search space near either of its parents, since its parents are local optima. Therefore, we can use gradient-descent again on this new point to yield a new local optimum with still better fitness. By interleaving the two methods, we hope to produce a sequence of points with ever-increasing fitness.

The impetus then for the combined approach is to achieve a synergistic interplay between reinforcement learning, which finds local optima, and the genetic algorithm, which can combine local optima to produce new points with still higher fitness.

1.3 How Othello Is Played

Othello is played on an 8x8 board. One player plays as White, the other as Black. Players take turns placing a stone of their color on an unoccupied square until all the squares have a stone on them, or until neither player has a legal move. The player with the most stones on the board at the end of the game wins.

Every time a player places a stone, he causes some of the stones that his opponent has already played to "flip" colors. So, for example, every time White plays, some black stones on the board change to white stones, and every time Black plays, some white stones on the board change to black stones. The exact way in which this stone-flipping occurs is covered in the rules for Othello, which can be found at *http://www.othello.org.hk/tutorials/eng-tu1.html*.

The game's final score is $v = \frac{W}{W+B}$. Under this scoring method, .5 is a tie. Scores higher than .5 are wins for White, and scores lower than .5 are wins for Black. In typical game-theoretic fashion, White tries to maximize the score, while Black tries to minimize it.

1.4 Game-Playing and Evaluation Functions

Virtually all game-playing programs contain a subroutine called an *evaluation function*. The evaluation function is an approximation to its game-theoretic counterpart: the *value function*. To each possible game-state, the value function assigns a value that indicates the outcome of the game if play were to continue optimally, both players making their best possible moves from that state onward. In Othello, a game-state consists of two pieces of information: the board-position, and whose move it is.

At any point during a game, the player whose move it is has several possible legal moves, each leading to a different successor game-state. We denote the set of successor game states of a given state $\mathbf{x}$ by $Succ(\mathbf{x})$. If White knew the value function, he would select the move leading to the highest-valued successor state. Similarly, if Black knew the value function, he would select the move leading to the lowest-valued successor state.

The value function is not discoverable in practice, so instead we approximate it with what we call an *evaluation function*. The evaluation function is obtained by applying the minimax algorithm to what we call a *static evaluation function*: a parametrized function from some function space, $f_{\mathbf{u}}(\mathbf{x}) \in \mathcal{F}$. Here, $\mathbf{x}$ is a vector representing a game-state, and $\mathbf{u}$, called a *parameter*, is a vector of values that selects a particular function from the family $\mathcal{F}$. For the experiment, we use neural-nets as static evaluation functions, which means that the parameter $\mathbf{u}$ consists of the weight-values and bias-values that uniquely determine the net.

The evaluation function, which we'll denote $h(\mathbf{x}; d, f)$, evaluates a game-state $\mathbf{x}$ by expanding the game tree beneath $\mathbf{x}$ to some specified search-depth, d, applying f to the terminal nodes of the expansion, and then using the minimax algorithm to trickle these values back up to the top to yield a derived value for $\mathbf{x}$. Specifically, for a given static evaluation function f and a fixed depth d, the evaluation of a game state $\mathbf{x}$ is

$$h(\mathbf{x}; d, f) \triangleq \begin{cases} f(\mathbf{x}), & d = 0 \\ \\ \max_{\mathbf{x}' \in Succ(\mathbf{x})} h(\mathbf{x}', d-1; f), & d \neq 0, \\ & \text{White-to-play} \\ \\ \min_{\mathbf{x}' \in Succ(\mathbf{x})} h(\mathbf{x}', d-1; f), & d \neq 0, \\ & \text{Black-to-play} \end{cases}$$

1.5 Choosing a Move

When it is a player's turn to move, it must decide which of the successors of the current state $\mathbf{x}$ it should move to. In the experiment, a player makes a weighted random choice, based on the relative values of each move, using the following exponential weighting scheme: let $\mathbf{x}'_1 \ldots \mathbf{x}'_n$ be the successors $\mathbf{x}$. Then the probability of selecting the move that leads to successor $\mathbf{x}'_i$ is:

$$p_i = \begin{cases} \dfrac{e^{Th(\mathbf{x}'_i; d, f)}}{\sum_j e^{Th(\mathbf{x}'_j; d, f)}} & \mathbf{x} \text{ is a White-to-play state} \\ \\ \dfrac{e^{-Th(\mathbf{x}'_i; d, f)}}{\sum_j e^{-Th(\mathbf{x}'_j; d, f)}} & \mathbf{x} \text{ is a Black-to-play state} \end{cases}$$

where T is a tuning parameter that controls how tight or dispersed the probability distribution is. Higher values of T make it more likely that the best-rated move will actually be selected. The experiment uses a value of $T = 10$. At $T = 0$, all moves are equally likely to be chosen, regardless of their evaluation. At $T = 10$, if move A has an evaluation .1 greater than move B, then move A is $e^{10..1} = e$ times more likely to be chosen than move B.

1.6 Neural-Networks as Evaluation Functions

This project uses neural-networks as static evaluation functions. Each neural-network has a 192-node input layer, three hidden nodes, and a single output node. A network in learning mode will use backpropagation to modify its weights. Network nodes have standard sigmoid functions $g(z) \triangleq (1 + e^{-z})^{-1}$.

The input layer accepts a 192-bit encoding of the game state. When an Othello-player wants to evaluate a game-state, it passes the properly-encoded game-state to its neural-network and interprets the neural-network's output as the result of the evaluation.

A game state $\mathbf{x}$ is encoded as the concatenation of three 64 bit vectors, $\mathbf{q} = q_1 \ldots q_{64}$, $\mathbf{r} = r_1 \ldots r_{64}$, and $\mathbf{s} = s_1 \ldots s_{64}$. That is, $Encode(\mathbf{x}) = \mathbf{qrs}$. These 64 bit vectors are determined as follows:

$$q_j = \begin{cases} 1 & \text{square } j \text{ of } \mathbf{x} \text{ has a white stone} \\ 0 & \text{otherwise} \end{cases}$$

$$r_j = \begin{cases} 1 & \text{square } j \text{ of } \mathbf{x} \text{ has a black stone} \\ 0 & \text{otherwise} \end{cases}$$

$$s_j = \begin{cases} 1 & \text{square } j \text{ of } \mathbf{x} \text{ is empty} \\ 0 & \text{otherwise} \end{cases}$$

For the above scheme, it is inconsequential how one numbers the squares of the Othello board, so long as they each have a unique number from 1 to 64.

2 Experiment Design

We begin with a pool of players, each with its own randomly drawn neural-network. These players then undergo a regime of alternating evolutionary and learning phases.

In the evolutionary phase, the current generation of players gives rise to a next generation of players through the genetic algorithm; that is, through the mechanisms of mutation, crossover, and replication. Fitness values are assigned to the players based on their rankings in a single round-robin tournament.

In the learning phase, each player in the newly created generation augments the talent it has inherited from its parents by playing successive round-robin Othello tournaments, modifying its static evaluation function $f_\mathbf{u}(\mathbf{x})$ as it learns from experience, using the reinforcement learning algorithm.

A round-robin tournament is simply a tournament in which each player plays a game against every other player exactly once. In the evolutionary phase, the result of a single round-robin tournament determines the fitness of the players. In the learning phase, the outcomes of the successive round-robin tournaments are irrelevant. The players are in effect just playing practice games.

In both phases, when it is a player's turn to move, the player determines which move to make as described in section 1.5. That is, it evaluates the successors of the current state and chooses one of the successors randomly, based on relative value. The evaluation is performed as described in section 1.4, with the player's neural-net acting as the static evaluation function f, and using a search-depth $d = 1$.

At the end of each learning phase, the weights of the neural-nets may be substantially different than they were at the beginning of the learning phase. In the subsequent evolutionary phase it is these *new* weights that are carried over into the next generation via crossover and replication operations. Thus, this regime is Lamarckian: the experience gained by the nets during their "lifetimes" is passed on to their progeny.

2.1 Evolutionary Phase

The evolutionary method used in this project is the genetic algorithm. The evolutionary phase pits the players against each other in a round-robin tournament, assigning to each player as a fitness value the fraction of games that it won in the tournament. To be precise, the fitness measure is:

$$fitness = \frac{GamesWon}{TotalGames} + \frac{\frac{1}{2}GamesTied}{TotalGames}$$

which gives some credit for ties and always falls in $[0, 1]$. Players in this generation are then mated to produce the next generation: a new set of players, produced as the children of players in the current generation via the operations of crossover, mutation, and replication. The experiment used a generation size of 10.

In *crossover*, two players from the current generation are selected at random, but based on fitness, so that players with higher fitness values are more likely to be selected, and their neural-network evaluation functions are merged—in a way to be described in detail later—to produce two new child players in the new generation. In *replication*, a single player from the current generation is selected at random, but based on fitness, and copied verbatim into the new generation. Every child of the new generation thus created is subject to *mutation*: with a certain probability, some of the weights or biases on its neural-net are randomly perturbed.

Because of the selection pressure the genetic algorithm imposes, poorly-performing players are likely to get culled from the population. More importantly, if two different players discover different features of the game-state that are useful for approximating the value function, these features can be combined via crossover into a child whose neural-net employs both features in its evaluation function.

Feature-Level Crossover

Given the topology of the neural nets in the experiment, the function they represent is always of the form:

$$f_{\mathbf{u}}(\mathbf{x}) = g(b + \sum_{i=1}^{3} \omega_i g(b_i + \sum_{j=1}^{192} w_{ij} a_j))$$

where a_j is the activation of the jth input node, w_{ij} is the weight to the ith hidden node from the jth input node, b_i is the bias of the ith hidden node, g is the sigmoid function, ω_i is the weight from the ith hidden node to the output node, and b is the bias of the output node.

One way to view the above function is as follows:

$$f_{\mathbf{u}}(\mathbf{x}) = g(b + \sum_{i=1}^{3} \omega_i \phi_i(\mathbf{x}))$$

where

$$\phi_i(\mathbf{x}) \overset{\triangle}{=} g(b_i + \sum_{j=1}^{192} w_{ij} a_j)$$

The ϕ_i can be thought of as features of the raw game state. Their normalized, weighted sum (plus a bias term) then constitutes the evaluation function. Under this view of the neural-net's structure, the hidden nodes, which implement the functions ϕ_i above, each extract a feature of the raw game-state. Depending on the initial weight-settings of the nets before learning begins, the hidden nodes of various nets may learn to represent quite different features of the raw game state, and some of these features may be good predictors of the value function.

If we assume that the value function can be approximated by a roughly linear combination of more or less independent features of the game state, and if a hidden node is capable of representing such features, then any net possessing even one such feature will very likely be more fit than a net possessing none, and a net possessing two different features will be more fit still.

Given the above assumptions, the fitness terrain is decomposable, with a point $\mathbf{u} \in \mathcal{F}$ decomposing as

$$\mathbf{u} = (\phi_1, \phi_2, \phi_3, \omega_1, \omega_2, \omega_3, b)$$

where the ϕ_i are the above feature functions, expressed as vectors whose components are the bias and all incoming weights of the ith hidden node. That is, the components of ϕ_i are the elements $\{w_{ij}, b_i\}$, $j = 1 \ldots 192$.

The crossover operation is structured to take advantage of the presumed decomposability of the fitness terrain. It combines nets at the feature-level, rather than at the lowest-, or weight-level. When two nets are selected to mate in crossover, their ϕ_i are pooled to create a set of 6 features, from which three are drawn at random to be the features of the first child. The remaining three become the features of the second child. In addition, the ω_i of each child are set to that of the corresponding inherited ϕ_i. For example, if a child inherits ϕ_2 from parent 1, then it also inherits ω_2 from parent 1. Finally, the b value of each child is randomly inherited from one of its parents. The important point is that the set of values that determine a feature are never disturbed through crossover. All the crossover operation does is determine which combination of the parents' features the children will inherit.

Mutation

For every new child created, either through crossover or reproduction, there is a $1/10$ probability that the child will have a mutation in precisely one of its weights or node-biases. When a mutation occurs, the current weight or bias is thrown away and replaced with a new value drawn uniformly on $[-.7, .7]$. This is a very low mutation rate, and, unless the mutation is sufficient to knock the neural net's parameter out of the local bowl in which it resides into a qualitatively different part of the fitness terrain, the next learning-phase will probably just pull it right back to very nearly the same local optimum it occupied before the mutation. Mutation plays a minor role in the experiment because the gradient-descent behavior of reinforcement learning, rather than mutation, is supposed to provide the impetus for the discovery of useful substructures.

2.2 Learning Phase

The learning method used in this project is reinforcement learning. During the learning phase, the Othello players modify the weights and biases of their neural-nets with every move of every game they play, in accordance with the reinforcement learning algorithm. The method is simple. At each game state, $\mathbf{x}$, $Player_1$ passes the training pair $(\mathbf{x}, h(\mathbf{x}; f_{\mathbf{u}_1}, d))$ to its neural-net, and $\mathbf{u}_1$ is updated accordingly. Similarly, $Player_2$ passes the training pair $(\mathbf{x}, h(\mathbf{x}; f_{\mathbf{u}_2}, d))$ to its neural-net. After both players have updated their nets based on game state $\mathbf{x}$, the player whose turn to move it is chooses a move using the fitness-based random-selection method described in section 1.5, and play continues.

A round-robin tournament pits every player against every other exactly once. So, with a pool of 10 players, a tournament consists of 45 games. This experiment conducts 200 tournaments in each learning phase. The first 100 are conducted with the learning parameter of the nets set to .1 so that their weights will quickly hone-in on the bottom of the local error bowl. (Not to be confused with the weight vector parameter $\mathbf{u}$ of the net, the learning parameter, λ, is a scalar that determines how much the net alters $\mathbf{u}$ as the result of a single training pair. Higher values of λ result in greater changes to $\mathbf{u}$. Lower values of λ are used for fine-tuning, when $\mathbf{u}$ is already close to optimal.) Then the next 100 tournaments are conducted with the learning parameter of the nets set to .01, for fine tuning.

Endgame Calculation

Othello has a nice property: with each move made, the number of empty squares left on the board decreases by one. Therefore, once the game reaches the point where there are only a handful of empty squares left on the board, the remaining game-tree can be completely calculated.

During the learning phase, the players perform endgame calculation when there are 10 empty squares left. They then perform one final learning-step apiece, using as a target value $v(\mathbf{x})$, *the true value of the game-state*, obtained from the endgame calculation. That is, when $\mathbf{x}$ represents a game state that has only 10 empty squares left, then the final learning pair $(\mathbf{x}, v(\mathbf{x}))$ is passed to the neural-nets of both $Player_1$ and $Player_2$ and the game ends. (Note that which player wins the game is irrelevant to the learning phase.)

It is important to stress that whereas all previous learning-steps only use the approximation $h(\mathbf{x}; f_{\mathbf{u}}, d)$ for a target, the final learning-step uses the true game-state value, $v(\mathbf{x})$, obtained by performing a *complete* game-tree expansion from $\mathbf{x}$ and calculating $\frac{W}{W+B}$ at each of the terminal nodes of the expansion. Specifically, we have

$$
v(\mathbf{x}) \stackrel{\triangle}{=} \begin{cases} \frac{W}{W+B} & \mathbf{x} \text{ is a terminal game-state} \\[2ex] \max_{\mathbf{x}' \in Succ(\mathbf{x})} v(\mathbf{x}') & \text{not terminal and White-to-play} \\[2ex] \min_{\mathbf{x}' \in Succ(\mathbf{x})} v(\mathbf{x}') & \text{not terminal and Black-to-play} \end{cases}
$$

3   Results and Analysis

The experiment ran for 100 generations, taking three months to complete, during which time $900,000$ individual games were played at a rate of approximately one game every 9 seconds. At the end of the experiment, a final round-robin tournament was played among the 100 best players of each generation, assigning to each player as a final fitness rating the fraction of games won, modified to account for ties:

$$
fitness = \frac{GamesWon}{TotalGames} + \frac{\frac{1}{2}GamesTied}{TotalGames}
$$

Unlike while in learning mode, when the experiment used a search-depth of 1, the final tournament used a search-depth of 4. A search-depth of 4 is still very modest—the nets can calculate that in a fraction of a second—and probably gives a more robust appraisal of the evaluation functions.

Figure 4 at the end of this paper shows the fitness of these best-of-gen players as a function of generation number. One can clearly see that fitness steadily improves for the first twelve generations, after which it levels off at just a little above .5. The highest-fitness player is $BestOfGen_{29}$, with a rating of of .68.

For some comparison with a human-written program, $BestOfGen_{29}$ was played against *kreversi*, a well-known program available with KDE for LINUX. kreversi plays a modified positional game. A purely positional game assigns different values to each square of the board and evaluates a game state as the sum of the values of the squares occupied by white minus the the sum of the values of the squares occupied by black. kreversi uses a convex combination of position and piece-differential to evaluate its game states, with piece-differential receiving higher and higher weight as the game progresses. kreversi also uses a dynamic search-depth: states near the end of the game are searched more deeply.

$BestOfGen_{29}$ played a twenty game tournament against kreversi on its level three setting, which uses a minimum search-depth of 4, with the following results: as white, $BestOfGen_{29}$ won 7 games and lost 3. As black, $BestOfGen_{29}$ won 3 games, lost 6, and tied 1.

3.1   Diverse Features Failed to Develop

Casual analysis of the best-of-gen nets fails to reveal that any interesting crossover took place. In net after net, the three hidden nodes end up with the same general pattern of weight values. Figures 2 and 3 illustrate this similarity. It is possible that, given the fixed three-layer topology, there is only one useful feature to be discovered. On the other hand, it is possible that the three hidden nodes are different in subtle ways not so easily detectable to the eye. They may in fact provide a useful redundancy, allowing the nets to be more robust. In this sense, the three hidden nodes are like three slightly different Othello-evaluators that are "bagged" together by the overall network. (*Bagging* is a method in statistics of averaging the results of different estimators in order to obtain a lower-variance estimate.)

How to Read Figures 1—3

Figure 1 on the last page shows the values of the weights incoming to the hidden layer of neural-net $BestOfGen_1$. Each of the net's hidden nodes has 192 incoming weights. This set of incoming weights is represented graphically as a column of Othello boards: a white board, a black board, and an empty board ($3 \cdot 64 = 192$). Since the net has three hidden nodes, the figure has three columns, one for each node. The squares of the boards are drawn with five different gray levels; white codes the highest value range, $[0.6, 1.0]$, and black codes the lowest value range, $[-1.0, -0.6]$ The color of the square represents the value of the associated incoming weight. Above each column of boards is printed the weight of the outgoing link on that hidden node. A hidden node with a relatively higher outgoing weight than the other two hidden nodes will participate more significantly in determining the final output value of the net.

Figures 2 and 3 are drawn in exactly the same format. Note that the biases of the nets are not shown on any of the figures.

Evaluation Functions Learned to Like Corners and Dislike Corner-Neighbors

Probably the single most important piece of strategic advice to give any beginning Othello player is: *get the corners*. An immediate corollary is to avoid placing a stone on any square adjacent to a corner, since this usually enables one's opponent to grab the corner.

Reviewing the feature plots, one can see that, by generation 29, the nets have learned to value the corners. This is reflected by high-values (light shades of gray) in the corner squares of the white board and low-values (dark shades of gray) in the corner squares of the black board. Remember that the evaluation functions always assess the board-position from White's point of view, so a white-stone in the corner will be good (high value), whereas a black-stone in the corner will be bad (low value).

By generation 29, not only have the corners of the white-board become high values, but the adjacent squares have become low values. Thus, at the very least, we can conclude that the nets have learned the fundamentals.

The Fitness Plateau

It is perhaps somewhat surprising that the fitness levelled-off so quickly. One might have expected fitness to increase steadily throughout all 100 generations. But one must remember that each generation comprises not only a genetic-mixing step, but also 200 learning tournaments, consisting of 45 games each. Thus, by generation 30, 9000 games are played. It may be that this was enough for the reinforcement learning phases to reach convergence. Had a more complex net topology been chosen, it is possible that the nets would have learned more and longer.

However, it is not surprising in itself that a plateau occurred. Indeed, this is almost inevitable. Once a fairly high-level of fitness is achieved, subsequent genetic pairings are likely to either leave fitness unchanged or reduce it, since there are many ways to go down, and few ways left to climb, the fitness terrain. And, given that a plateau occurs, the fitness level of those players in the plateau must perforce be around .5. Since, by assumption, they all have about the same fitness, the probability that one player from the plateau group beats another from the plateau group must be roughly .5.

4 Conclusions And Future Directions

The central idea of this paper is that, in domains where it can be applied, reinforcement learning may be better than random mutation as an engine for the discovery of useful substructures. A secondary idea is that of feature-level crossover: that the substructures we wish to discover may be identified with the hidden nodes of a neural net. Feature-level crossover is merely a way to do crossover on neural-nets. It needn't be coupled with reinforcement learning, nor restricted to fixed-topology nets.

As a side note, the distinction between the terms *substructure* and *feature* is actually one of genotype and phenotype. In the space of neural-net functions, the substructures that confer high fitness are simply those that lead to useful feature-extractors in the phenotype.

Whether or not the Othello fitness terrain is a decomposable terrain is an open question. If it is not, then the crossover operation should perform no better than macromutation in finding points of improved fitness [4]. It is however reasonable to suspect that the Othello fitness terrain can be characterized by more or less independent features, because this is how humans analyze board positions. It is also an open question whether reinforcement learning applied to Othello can lead to the discovery of features along the lines suggested by this paper. Although the experiment yielded a fairly good player, there is little indication that diverse features were discovered, or that crossover played a significant role.

Further, any hybrid strategy, being inherently more complex than a single method, requires justification based on its practical worth, the more so because it seems that mutation alone can still be very effective in finding points of high fitness [2, 7]. While this paper does not investigate the efficacy of the hybrid method relative to other methods, that is an important direction for future research.

Acknowledgments

I would like to thank John Koza for his many helpful comments and suggestions.

References

1. G. Tesauro, "Temporal Difference Learning and TD-Gammon". *Communications of the ACM*, vol. 38, no. 3, pp. 58-68, 1995
2. K. Chellapilla and D. B. Fogel, "Evolution, Neural Networks, Games, And Intelligence". *Proceedings of the IEEE*, vol. 87, no. 9, pp. 1471-96, 1999
3. D. E. Moriarty and R. Miikkulainen, "Discovering Complex Othello Strategies Through Evolutionary Neural Networks". *Connection Science*, vol. 7, no. 3-4, pp. 195-209, 1995
4. Terry Jones, "Crossover, Macromutation, and Population-Based Search", *Proceedings of the Sixth International Conference on Genetic Algorithms*
5. Dimitri Bertsekas and John Tsitsiklis, *Neuro-Dynamic Programming*. Belmont, Massachusetts: Athena Scientific, 1996.
6. John Koza, *Genetic Programming*. MIT, 1996
7. J. B. Pollack and A. D. Blair, "Co-Evolution in the Successful Learning of Backgammon Strategy", *Machine Learning*, vol. 32, no. 3, pp. 225-40, 1998
8. Brian D. Ripley, *Pattern Recognition And Neural Networks*. Cambridge University Press, 1996.
9. Richard Sutton and Andrew Bartow, *Reinforcement Learning*. Windfall Software, 1999.

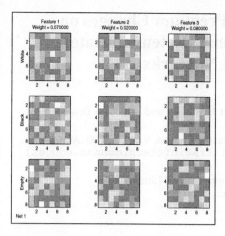

Fig. 1. The best player of generation 1–a random neural-net

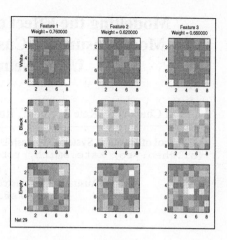

Fig. 2. the best-of-run player, from generation 29. Note the high value placed on the corners

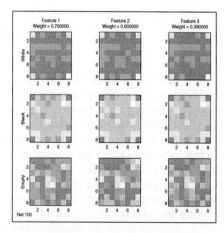

Fig. 3. The best player of generation 100. This neural-net is nearly as good as net 29, winning 60 percent of its games, and its structure is very similar to that of net 29

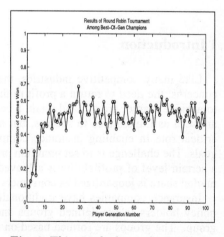

Fig. 4. This graph shows the fraction of games won by each generation's best player in a round-robin tournament against the other generations' best players. The graph reveals a general trend of increasing fitness through generation 29, at which point fitness levels off

Modelling the Effect of Premium Changes on Motor Insurance Customer Retention Rates Using Neural Networks

Ai Cheo Yeo[1], Kate A. Smith[1], Robert J. Willis[1], and Malcolm Brooks[2]

[1]School of Business Systems, Monash University, Clayton, Victoria 3800, Australia
{ai.cheo.yeo, kate.smith, rob.willis@infotech.monash.edu.au}

[2]Australian Associated Motor Insurers Limited
mbrooks@aami.com.au

Abstract. This paper describes a neural network modelling approach to premium price sensitivity of insurance policy holders. Clustering is used to classify policy holders into homogeneous risk groups. Within each cluster a neural network is then used to predict retention rates given demographic and policy information, including the premium change from one year to the next. It is shown that the prediction results are significantly improved by further dividing each cluster according to premium change. This work is part of a larger data mining framework proposed to determine optimal premium prices in a data-driven manner.

1 Introduction

Like many competitive industries, the insurance industry is driven by two main concerns: the need to return a profit to their shareholders and investors, and the need to achieve market growth and retain a certain level of market share. These two goals are seen as imperatives to success, but are often conflicting. Premium prices play a critical role in enabling insurance companies to find a balance between these two goals. The challenge is to set premium prices so that expected claims are covered and a certain level of profitability is achieved, yet not to set premium prices so high that market share is jeopardized as consumers exercise their rights to choose their insurers.

Insurance companies have traditionally determined premium prices by assigning policy holders to pre-defined groups and observing the average behaviour of each group. The groups are formed based on industry experience about the perceived risk of different demographic groups of policy holders. With the advent of data warehouses and data mining however comes an opportunity to consider a different approach to premium pricing: one based on data-driven methods. By using data mining techniques, the aim is to determine optimal premiums that more closely reflect the genuine risk of individual policy holders as indicated by behaviours recorded in the data warehouse.

In previous work we have proposed a data mining framework for tackling this problem (Smith *et al.*, 2000) (Yeo *et al.*, 2001). This framework comprises components for determining risk groups, determining the sensitivity of policy holders to premium change, and combining this information to determine optimal premium prices that appropriately balance profitability and market share. Recently we have presented the results of the first component where clustering techniques are used to define risk groups (Yeo *et al.*, 2001). The purpose of this paper is to investigate the second component of the data mining framework: modelling the effect of premium price changes on the customer retention rate.

In Section 2 we review the data mining framework employed in this study. A case study approach utilising a database of over 330,000 policy holders is used to evaluate the effectiveness of various techniques within this framework. Section 3 summarises the previously published results for risk group classification. The use of neural networks for modelling retention rates under various premium changes is then discussed in Section 4. A strategy for improving the retention rate prediction by dividing the data into more homogeneous groups and using separate neural network models for each group is presented, and the results are compared to a single neural network model. Computational results of prediction error rates are presented in Section 5 for all risk classification groups. Conclusions are drawn in Section 6, where future research areas are identified.

2 A Data Mining Framework

Figure 1 presents a data mining framework for determining the appropriate pricing of policies based upon the interaction of growth, claims and profitability. The framework consists of four main components: identifying risk classifications, predicting claim costs, determining retention rates, and combining this information to arrive at optimal premiums. Firstly, the estimated risk of policy holders must be calculated, and used to determine optimal premium values. The total premiums charged must be sufficient to cover all claims made against the policies, and return a desired level of profit. The levels of predicted claims can also be used to forecast profits, when coupled with premium information. However premiums cannot be set at too high a level as customers may terminate their policies, affecting market share. Sales forecasting is determined by marketing information as well as models that predict customer retention or "churn" rates. When integrated, this information provides a methodology for achieving the two goals of market growth and profitability.

For optimal premiums to be set, the insurance company thus needs to determine estimated claim costs and the effect of changes in premiums on retention rates. The estimation of claim cost requires an accurate assessment of risk, discussed in the next section.

3 Risk Classification

Insurance companies group policy holders into various risk groups based on factors which are considered predictors of claims. For example younger drivers are considered to be a higher risk and so they are charged a higher premium. In designing the risk classification structure, insurance companies attempt to ensure maximum homogeneity within each risk group and maximum heterogeneity between the risk groups. This can be achieved through clustering. In previous work (Yeo *et al.*, 2001) we have shown that the data-driven k-means clustering approach to risk classification can yield better quality predictions of expected claim costs compared to a previously published heuristic approach.

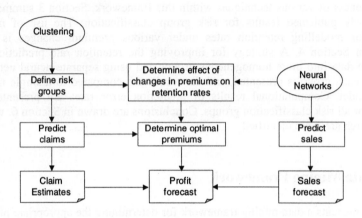

Fig. 1. A framework for data mining within the insurance industry to determine optimal premium prices. It includes components for ascertaining the perceived risk of policy holders, their sensitivity to premium changes, and a mechanism for combining this information to arrive at premiums that try to balance profitability and market share.

The data for this study were supplied by an Australian motor insurance company. Two data sets (training set and test set), each consisting of 12-months of comprehensive motor insurance policies and claim information were extracted. The training set consisted of 146,326 policies with due dates from 1 January to 31 December 1998 while the test set consisted of 186,658 policies with due dates from 1 July 1998 to 30 June 1999. Thirteen inputs were used to cluster policy holders.

The k means clustering model was used to generate a total of 30 risk categories. Figure 2 is a graphical representation of the claim frequency and average claim cost of the 30 clusters.

Fig. 2. Claim frequency and average claim cost of the 30 risk groups (clusters)

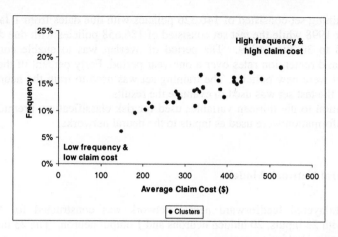

In the insurance industry, risk is measured by frequency or the probability of a claim and the amount of claims. The higher the frequency, the higher the risk. The higher the amount of claims, the higher the risk. The k-means clustering model was able to find clusters which have significantly different claim frequency and claim cost, without being provided with any claim information as input variables. In other words, clustering is able to distinguish between low and high risk groups.

Besides the k means clustering we also experimented with a previously published heuristic model (Samson *et al.*, 1987) and fuzzy c-means model, but the k-means clustering model produced the best results on the test set with a weighted absolute deviation of 8.3% compared to the 15.6% for the fuzzy c-means model and 13.3% for the heuristic model (Yeo *et al.*, 2001).

Now that the policy holders have been classified into 30 risk groups based on their demographic and policy information, we can examine the pricing sensitivity within each cluster. Neural networks are used in the following section to model the effect of premium price change on retention rates.

4   Modelling Retention Rates

Neural networks have been used in this paper to learn to distinguish policy holders who are likely to terminate their policies from those who are likely to renew. They are an ideal tool for solving this problem due to their proven ability to learn to distinguish between classes, and to generalise their learning to unseen data. Prediction of termination rates or "churn" prediction is a significant area of research, particularly in the telecommunications and insurance industries. Several researchers have successfully applied neural networks to churn prediction problems, and to better understand the factors affecting a customer's decision to churn (Smith *et al.*, 2000) (Mozer *et al.*, 2000) (Madden *et al.*, 1999) (Behara *et al.*, 1994).

4.1 Data

The training set consisted of 146,326 policies with due dates from 1 January to 31 December 1998 while the test set consisted of 186,658 policies with due dates from 1 July 1998 to 30 June 1999. The period of overlap was to enable comparison of exposure and rentention rates over a one-year period. Forty percent of the policies in the test set were new policies. The training set was used to train the neural networks and while the test set was used to evaluate the results.

In addition to the thirteen variables used for risk classification, premium and sum insured information were used as inputs to the neural networks.

4.2 Neural Network Models

A multilayered feedforward neural network was constructed for each of the clusters with 23 inputs, 20 hidden neurons and 1 output neuron. The 23 inputs were:
1. policy holder's age,
2. policy holder's gender,
3. area in which the vehicle was garaged,
4. rating of policy holder,
5. years on current rating,
6. years on rating one,
7. number of years policy held,
8. category of vehicle,
9. sum insured,
10. total excess,
11. vehicle use,
12. vehicle age,
13. whether or not the vehicle is under finance.
14. old premium,
15. new premium,
16. old sum insured,
17. new sum insured,
18. change in premium,
19. change in sum insured,
20. percentage change in premium,
21. percentage change in sum insured,
22. ratio of old premium to old sum insured,
23. ratio of new premium to new sum insured.

The hyberbolic tangent activation function was used. Input variables which were skewed were log transformed. The neural network produces output between zero and one, which is the probability that that a policy holder will terminate his policy. Figure 3 shows the probability of termination of Cluster 11. A threshold value is used to decide how to categorise the output data. For example a threshold of 0.5 means that if the probability of termination is more than 0.5, then the policy will be classified as terminated. In our case, we have set a threshold so that the predicted termination rate is equivalent to the actual termination rate for the whole cluster. For cluster 11, the termination rate is 14.7% which means a threshold of 0.204 is needed to produce the same predicted termination rate, as shown in Figure 3.

4.3 Prediction Accuracy

To determine how well the neural networks were able to predict termination rates for varying amounts of premium changes, the clusters were then divided into various bands of premium as follows: decrease in premiums of less than 22.5%, premium decrease between 17.5% and 22.5%, premium decrease between 12.5% and 17.5% etc. The predicted termination rates were then compared to the actual termination rates. For all the clusters the prediction accuracy of the neural networks starts to deteriorate when premium increases are between 10% to 20%. Figure 4 shows the actual and predicted termination rates for one of the clusters (Cluster 24).

Fig. 3. Determining the threshold value of the neural network output

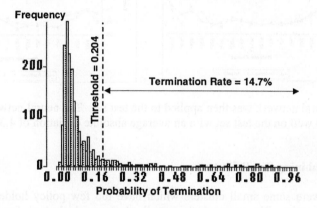

Fig. 4. Prediction accuracy for one neural network model of cluster 24

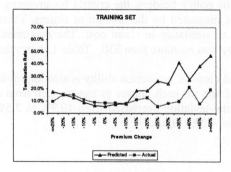

4.4 Generating more homogeneous models

In order to improve the prediction accuracy, the cluster was then split at the point when prediction accuracy starts to deteriorate. Two separate neural networks were trained for each cluster. The prediction accuracy improved significantly with two neural networks as can be see from Figure 5. The average absolute deviation decreased from 10.3% to 2.4%.

Fig. 5. Prediction accuracy for two networks model of cluster 24 (training set)

Fig. 6. Prediction accuracy for two networks model of cluster 24 (test set)

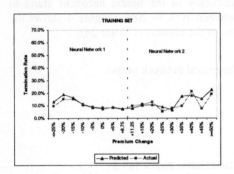

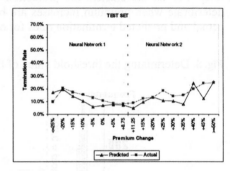

The neural network was then applied to the test set. The neural network performed reasonably well on the test set with an average absolute deviation of 4.3% (Figure 6).

4.5 Combining Small Clusters

There were some small clusters which have too few policy holders to train the neural networks. We grouped the small clusters which had fewer than 7,000 policies.Since the objective is to ultimately determine the optimal premium which reflect the risk of the policy holders, the criteria for grouping has to be similarity in risk. Risk in turn is measured by the amount of claims. Therefore the clusters were grouped according to similarity in claim cost. The maximum difference in average claim cost per policy was no more than $50. Table 1 shows the grouping of the small clusters.

For the combined clusters, prediction ability is also improved by having two neural networks instead of one for each cluster as can be seen from Figure 7 and Figure 8. The average absolute deviation decreased from 10.3% to 3.5%. The test set has an absolute deviation of 4.2% (Figure 9).

5. Results

Table 2 presents a summary of the results for all clusters. It shows clearly that the average absolute deviation between the actual and predicted termination rates is significantly reduced by employing two neural networks per cluster rather than a single neural network. It appears that a single neural network is unable to simultaneously learn the characteristics of policy holders and their behaviours under different premium changes. This is perhaps due to the fact that many of the large premium increases are due to an upgrade of vehicle. Since these policy holders may well expect an increase in premium when their vehicle is upgraded they may have different sensitivities to premium change compared to the rest of the cluster. Attempting to isolate these policy holders and modelling their behaviours results in a better prediction ability.

Table 1. Grouping of Small Clusters

Cluster	No of Policies	Average Claim Cost	Difference
13	2,726	344	
14	2,714	343	16
28	6,441	328	
12	1,422	285	
2	3,988	280	7
30	2,366	278	
5	5,606	270	
26	1,460	262	14
23	3,374	256	
3	1,595	249	
10	1,610	248	
8	2,132	247	15
7	1,601	235	
18	1,446	234	
1	4,101	231	
15	1,621	217	48
16	1,505	194	
6	2,346	183	
21	3,566	138	
25	1,445	125	40
22	1,401	116	
17	1,411	98	

Fig. 7. Prediction accuracy for one network model of combined clusters 5, 23 and 26

Fig. 8. Prediction accuracy for two networks model of combined clusters 5, 23 and 26 (training set)

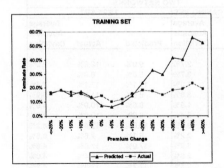

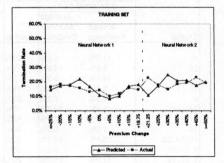

6. Conclusions and Future Research

This paper has examined the use of neural networks for modelling customer retention rates within homogeneous groups. The work is part of a data mining framework for determining optimal premium prices. Clustering is used to arrive at homogeneous groups of policy holders based on demographic information. This information is then supplemented with premium details, and a neural network is used to model termination rates given premium changes. We have shown that significant improvements in prediction accuracy can be obtained by further dividing each cluster to isolate those policy holders with a significant increase in premium. It is believed that these policy holders behave differently due to the greater number of these policy holders who have upgraded their vehicles.

Fig. 9. Prediction accuracy for two networks model of combined clusters 5, 23 and 26 (test set)

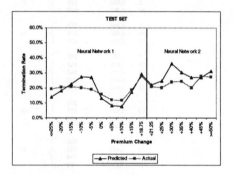

Table 2. Summary of Results

	ONE NETWORK			TWO NETWORKS			
	TRAINING SET			TRANING SET	TEST SET		
Cluster	Actual/ Predicted	Average Absolute Deviation	Split at	Average Absolute Deviation	Predicted	Actual	Average Absolute Deviation
4	11.2%	5.5%	20%	2.7%	9.9%	12.6%	2.9%
20	6.8%	10.8%	15%	3.7%	5.0%	6.9%	3.4%
11	14.7%	8.3%	10%	5.0%	17.1%	17.1%	7.4%
29	9.5%	8.8%	15%	3.8%	8.5%	11.3%	5.9%
9	8.8%	7.2%	20%	1.9%	8.6%	10.0%	4.5%
24	9.5%	10.3%	10%	2.4%	8.2%	10.9%	4.3%
27	11.6%	12.4%	15%	3.1%	10.8%	12.6%	4.2%
19	7.5%	6.3%	15%	2.2%	8.1%	7.6%	3.6%
13, 14, 28	15.1%	10.9%	10%	3.8%	17.9%	17.9%	4.8%
2, 12 ,30	13.5%	9.2%	10%	2.8%	13.1%	15.4%	4.0%
5, 23, 26	14.7%	10.3%	20%	3.5%	17.7%	17.3%	4.2%
3, 7, 8, 10, 18	11.7%	6.7%	20%	3.1%	12.3%	14.4%	4.4%
1, 6, 15, 16	10.9%	5.2%	20%	2.2%	11.1%	12.9%	4.0%
17, 21, 22, 25	8.7%	5.4%	15%	2.9%	8.4%	9.8%	3.4%

References

Behara, R.S., Lemmink, J. (1994): Modeling the impact of service quality on customer loyalty and retention: a neural network approach. 1994 Proceedings Decision Sciences Institute. 3, 1883-5

Madden, G., Savage, S., Coble-Neal, G. (1999): Subscriber churn in the Australian ISP market. Information Economics & Policy 11(2), 195-207

Mozer, M.C., Wolniewicz, R., Grimes, D.B., Johnson, E., Kaushansky, H. (2000): Predicting subscriber dissatisfaction and improving retention in the wireless telecommunication. IEEE Transactions on Neural Networks 11(3), 690-6

Samson, D., Thomas, H. (1987): Linear models as aids in insurance decision making: the estimation of automobile insurance claims. Journal of Business Research 15, 247-256

Smith, K.A., Willis, R.J., Brooks, M. (2000): An analysis of customer retention and insurance claim patterns using data mining: a case study. Journal of the Operational Research Society 51, 532-541

Yeo, A.C., Smith, K.A., Brooks, M. (2001): A comparison of soft computing and traditional approaches for risk classification and claim cost prediction in the automobile insurance industry. In: Reznik, L., Kreinovich, V. (Eds.): Soft computing in measurement and information acquisition. Springer-Verlag, Heidelberg , accepted for publication.

On the Predictability of Rainfall in Kerala
An Application of ABF Neural Network

Ninan Sajeeth Philip[1] and K. Babu Joseph[1]

Cochin University of Science and Technology, Kochi-682022, Kerala State, India.
nspp@iucaa.ernet.in

Abstract. Rainfall in Kerala State, the southern part of Indian Peninsula in particular is caused by the two monsoons and the two cyclones every year. In general, climate and rainfall are highly nonlinear phenomena in nature giving rise to what is known as the 'butterfly effect'. We however attempt to train an ABF neural network on the time series rainfall data and show for the first time that in spite of the fluctuations resulting from the nonlinearity in the system, the trends in the rainfall pattern in this corner of the globe have remained unaffected over the past 87 years from 1893 to 1980. We also successfully filter out the chaotic part of the system and illustrate that its effects are marginal over long term predictions.

1   Introduction

Although technology has taken us a long way towards better living standards, we still have a significant dependence on nature. Rain is one of nature's greatest gifts and in third world countries like India, the entire agriculture depends upon rain. It is thus a major concern to identify any trends for rainfall to deviate from its periodicity, which would disrupt the economy of the country. This fear has been aggravated due to the threat by the global warming and green house effect. The present study has a soothing effect since it concludes that in spite of short term fluctuations, the general pattern of rainfall in Kerala has not undergone major deviations from its pattern in the past.

The geographical configuration of India with the three oceans, namely Indian Ocean, Bay of Bengal and the Arabian sea bordering the peninsula gives her a climate system with two monsoon seasons and two cyclones inter-spersed with hot and cold weather seasons. The parameters that are required to predict the rainfall are enormously complex and subtle so that the uncertainty in a prediction using all these parameters is enormous even for a short period. The period over which a prediction may be made is generally termed the event horizon and in best results, this is not more than a week's time. Thus it is generally said that the fluttering wings of a butterfly at one corner of the globe may cause it to produce a tornado at another place geographically far away. This phenomenon is known as the *butterfly effect.*

The objective of this study is to find out how well the periodicity in these patterns may be understood using a neural network so that long term predictions can be made. This would help one to anticipate with some degree of confidence the general pattern of rainfall for the coming years. To evaluate the performance of the network, we train it on the rainfall data corresponding to a certain period in the past and cross validate the prediction made by the network over some other period. A difference diagram[1] is plotted to estimate the extent of deviation between the predicted and actual rainfall. However, in some cases, the cyclone may be either delayed or rushed along due to some hidden perturbations on the system, for example an increase in solar activity. (See figure 2.) These effects would appear as spikes in the difference diagram. The information from the difference diagram is insufficient to identify the exact source of such spike formation. It might have resulted from slight perturbations from unknown sources or could be due to an inaccurate modeling of the system using the neural network. We thus use a standard procedure in statistical mechanics that can quantitatively estimate the fluctuations[1]. This is to estimate the difference in the Fourier Power Spectra (FPS) of the predicted and actual sequences. In the FPS, the power corresponding to each frequency interval, referred to as the *spectral density* gives a quantitative estimate of the deviations of the model from reality. Rapid variations would contribute to high frequency terms and slowly varying quantities would correspond to low frequency terms in the power spectra.

The degree of information that may be extracted from the FPS is of great significance. If the model agrees with reality, the difference of the power spectra (*hereafter referred to as the residual FPS*) should enclose minimum power in the entire frequency range. An exact model would produce no difference and thus no residual FPS pattern. If there are some prominent frequency components in the residue, that could indicate two possibilities; either the network has failed to comprehend the periodicity, or that there is a new trend in the real world which did not exist in the past. One can test whether the same pattern exists in the residual FPS produced on the training set and confirm whether it is a new trend or is the drawback of the model.

A random fluctuation would be indicated in the residual FPS by amplitudes at all frequency values as in the case of 'white' noise spectra. Here again, how much power is enclosed within the FPS gives a quantitative estimate of the perturbations. A low amplitude fluctuation can happen due to so many reasons. But its effect on the overall predictability of the system would be minimal. If however, the residual FPS encloses a substantial power from the actual rainfall spectrum, the fluctuations could be catastrophic.

In this study, the results indicate that the perturbations produced by the environment on the rainfall pattern of Kerala state in India is minimal and that there is no evidence to envisage a significant deviation from the rainfall patterns prevailing here.

The general outline of the paper is as follows. In section 2 we present a brief outline of the Adaptive Basis Function Neural Network (ABFNN), a variant of

---

[1] This is the diagram showing the deviation of the predicted sequence from the actual rainfall pattern.

the popular Back-Propagation algorithm. In section 3, the experimental set up is explained followed by the results and concluding remarks in section 4.

2 Adaptive Basis Function Neural Networks

It was shown in [2] that a variant of the Back-Propagation algorithm (backprop) known as the Adaptive Basis Function Neural Network performs better than the standard backprop networks in complex problems. The ABFNN works on the principle that the neural network always attempts to map the target space in terms of its basis functions or node functions. In standard backprop networks, this function is a fixed sigmoid function that can map between zero and plus one or between minus one and plus one the input applied to it from minus infinity to plus infinity. It has many attractive properties that make the backprop an efficient tool in a wide variety of applications. However serious studies conducted on the backprop algorithm have shown that in spite of its widespread acceptance, it systematically outperforms other classification procedures *only* when the targeted space has a sigmoidal shape [3]. This implies that one should *choose* a basis function such that the network may represent the target space as a nested sum of products of the input parameters in terms of the basis function. The ABFNN thus starts with the standard sigmoid basis function and alters its nonlinearity by an algorithm similar to the weight update algorithm used in backprop.

Instead of the standard sigmoid function, ABFNN opts for a variable sigmoid function defined as

$$O_f = \frac{a + tanh(x)}{1 + a} \tag{1}$$

Here a is a control parameter that is initially set to unity and is modified along with the connection weights along the negative gradient of the error function. It is claimed in [2] that such a modification could improve the speed and accuracy with which the network could approximate the target space.

The error function is computed as:

$$E = \sum_k \frac{(O_k - O_k^*)^2}{2} \tag{2}$$

with O_k representing the network output and O_k^* representing the target output value.

With the introduction of the control parameter, the learning algorithm may be summarized by the following update rules. It is assumed that each node, k, has an independent node function represented by a_k. For the output layer nodes, the updating is done by means of the equation:

$$\Delta a_k = -\beta \left(O_k - O_k^*\right) \frac{1 - O_k}{1 + a_k} \tag{3}$$

where β is a constant which is identical to the learning parameter in the weight update procedure used by backprop.

For the first hidden layer nodes, the updating is done in accordance with the equation:

$$\Delta a_{k-1} = -\sum_i w_{ij}\beta\left(O_k - O_k^*\right)\left(1 - O_k\right)\left[(1 + a_k) + O_k\left(1 + a_k\right)\right]\frac{\partial O_k}{\partial O_{k-1}} \quad (4)$$

Here w_{ij} is the connection weight for the propagation of the output from node i to node j in the subsequent layer in the network.

The introduction of the control parameter results in a slight modification to the weight update rule (Δw_{ij}) in the computation of the second partial derivative term $\frac{\partial O_j}{\partial I_j}$ in:

$$\Delta w_{ij} = -\beta\frac{dE}{dw_{ij}} = -\beta\frac{\partial E}{\partial O_j}\frac{\partial O_j}{\partial I_j}\frac{\partial I_j}{\partial w_{ij}} \quad (5)$$

as:

$$\frac{\partial O_j}{\partial I_j} = \left[(1 - a_j) + (1 + a_j)O_j\right]\left(1 - O_j\right) \quad (6)$$

The algorithm does not impose any limiting values for the parameter a and it is assumed that care is taken to avoid division by zero.

3 Experimental Setup

In pace with the global interest in climatology, there has been a rapid updating of resources in India also to access and process climatological database. There are various data acquisition centers in the country that record daily rainfall along with other measures such as sea surface pressure, temperature etc. that are of interest to climatological processing. These centers are also associated to the World Meteorological Organization (WMO). The database used for this study was provided by the Department of Atmospheric Sciences of Cochin University of Science and Technology, a leading partner in the nation wide climatological study centers.

The database consists data corresponding to the total rainfall in each month of Trivandrum city in Kerala, situated at latitude-longitude pairs ($8^o29'$ N - $76^o57'$ E). Although the rainfall data from 1842 were recorded, there were many missing values in between that we had to restrict to periods for which a continuous time series was available. This was obtained for the period from 1893 to 1980.

For training our network, the monthly rainfall database from 1893 to 1933 was used. This corresponds to the first 480 data samples in the database. Since rainfall has an yearly periodicity, we started with a network having 12 input nodes each with the corresponding month's total rainfall as input. It was observed that the network accuracy increases systematically as we increased the number of input nodes from 12 to 48 covering the data of 4 years. The training RMS error in this case went as low as 0.0001 in about 3000 iterations. Any increase in input nodes resulted in poorer representations. However, the performance of the network on the remaining test data was found to be poor due to over fitting in the training data. This happened because of the large number of free parameters in the network compared to the size of the training set. Further experimentation showed that it is not necessary to input the data of all the twelve months of the previous 4 years, but, four sets of 3 months data centered over the predicted month of the fifth year in the 4 previous years will give good generalization properties. We thus finalized our network with 12 input nodes each for the 3 months input data over 4 years, 7 hidden nodes and one output node. In this way, based on the information from the four previous years around the targeted month, the network is to predict the amount of rain to be expected in the same month of the fifth year.

The training was carried out until the RMS error stabilized to around 0.085 over the training data(40 years from 1893-1933). The ABFNN converged to this value in around 350 iterations starting from random weight values.

After training, testing was carried out independently on the remaining test data set. For generating the difference diagram, the data corresponding to the entire 87 years of rainfall in Trivandrum city was presented to the network. The difference of the output produced by the network to the actual rainfall in each month was computed and plotted. This diagram enables one to make an easy estimate of the error levels in both the training and test periods. For a quantitative estimate of generalization error on the independent test set, the power spectrum of the output produced by the network was subtracted from the power spectra of the actual rainfall data in those months. This residual spectra was then compared with the spectral differences between the actual patterns in the training and test periods. In a dynamic system, there will always be some residual spectra. But if the network could generalize well and capture the dynamics in the training data, the output produced by it with the input from the test period will be more identical to the actual pattern. This means, the residual spectra will contain lesser residual power. If this power is less than the power contained in the difference between the spectra of the training and test periods of the actual rainfall patterns, it indicates that the deviations are not due to any new phenomenon, but are due to the trends that existed from the training period itself. Now the concern is the power content of the residual spectra. The lower the power enclosed in this residual spectra, the better is the learning and the generalization. The advantage of the scheme is that it provides a platform for comparing the performance of various networks with ease and allows one to visualize the spectral components that have actually deviated.

4   Results and Discussion

In figure 1 we show the difference pattern produced by the network over the entire 87 year period on the Trivandrum database. The training period is also shown. It may be noted that the deviations of the predicted signal from the actual in the entire dataset falls within the same magnitude range. The root mean square error on the independent test dataset was found to be around 0.09. The two positive spikes visible in the plot corresponding to the two instances

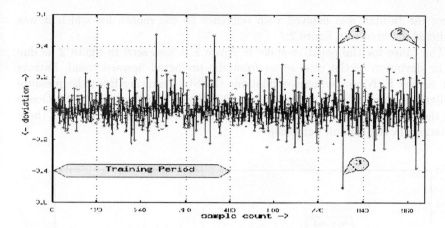

Fig. 1. The deviation of the predicted sequence from the actual time series data is shown. It may be noted that the deviations in the training period and the rest of the series are comparable in magnitude. The larger spikes are mainly due to the delay in the actual commencement of the monsoon or cyclone. Label 1 indicates an example where the monsoon extended into the following month giving more rainfall than anticipated. Label 2 in the figure represent an example where the monsoon was actually delayed by one month than was anticipated thus producing two spikes, one negative and the other positive. Label 3 corresponds to a negative spike caused by the change in trend from the predicted increase in rain in the following year.

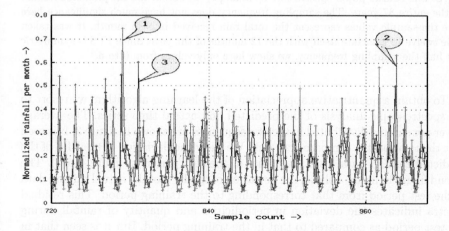

Fig. 2. An overlap of the predicted and the actual time series data in the region corresponding to the spikes in figure 1 is shown. The labels indicate the reason for the spikes seen in figure 1.

where the rainfall was delayed with reference to the month for which it was predicted are shown in figure 2.

The cause for the negative spike in figure 1 is also seen in figure 2 as due to the deviation of the time series from the predicted upward trend. Factors such as the El-Nino southern oscillations (ENSO) resulting from the pressure oscillations between the tropical Indian Ocean and the tropical Pacific Ocean and their quasi periodic oscillations are observed to be the major cause of these deviations in the rainfall patterns[4].

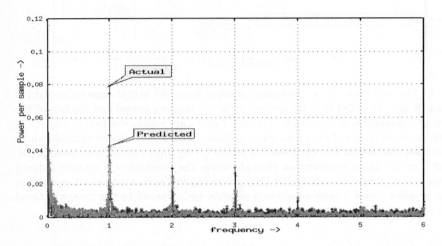

Fig. 3. The Fourier power spectra corresponding to the actual and the predicted rainfall for the entire 87 years. The sampling frequency does not have much significance here since the sample values represent the total rain received in each month. It was taken to be twelve to associate itself with the 12 months of the year. Since FPS is symmetric over half the sampling frequency, we show here only values from 0 to 6.

To obtain a quantitative appreciation of the learning algorithm, we resort to the spectroscopic analysis of the predicted and actual time series. The average power per sample distributed in the two sequences are shown in figure 3. It is seen that the two spectra compare very well. We show the deviation of the FPS of the predicted rainfall data from the actual in the test period in figure 4(top). The second figure in figure 4 shows the deviation of the FPS of the actual rainfall data in the test period from that corresponding to the training period. This residual spectra indicates the deviation in periodicities and quantity of rainfall during the test period as compared to that in the training period. But it is seen that in the top figure most of the new frequency components visible in the second figure are missing. From the scaling property of Fourier spectra, one knows that power deviations in the spectra indicates deviations in the quanta of rainfall. Such deviations are indicators of the error associated with the prediction and are caused by the nonlinearity in the global factors that affect rainfall. New trends, or deviations of the existing rainfall pattern are identified by new frequency

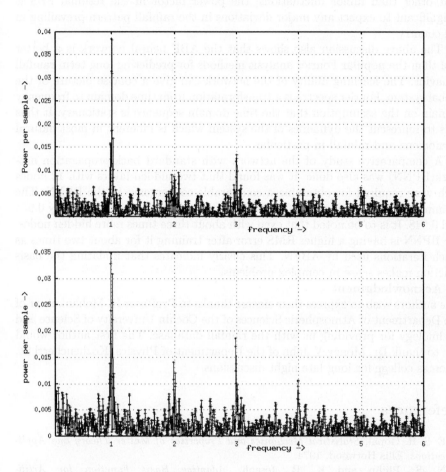

Fig. 4. The residual FPS corresponding to the predicted and actual rainfall in the years following the training period is shown on top. The spectral difference from a Fourier analysis perspective (the difference of the FPS in the training period and the period corresponding to the top figure) is presented in the bottom figure. Note that the spectral differences are minimum in the top figure. This is to say that the deviations observed in the actual spectra in the bottom figure followed from the trends that existed in the training period.

components and not by the amplitudes of the component frequencies. The fact that there are no major new frequency components in the top figure suggests that the network could capture those trends seen in the second figure from the training data itself and that the rainfall trend in the test period has not changed considerably from the training period. In other words, it is to be concluded that other than minor fluctuations, the power factor in the residual FPS is insignificant to expect any major deviations in the rainfall pattern prevailing in the country.

The above discussion also shows that the ABF neural network is a better tool than the popular Fourier analysis methods for predicting long term rainfall behavior. The learning ability of the network can give a concise picture of the actual system. Fourier spectra is a transformation from time domain to frequency domain on the assumption that the time domain sequence is stationary. It thus fails to represent the dynamics of the system which is inherent in most natural phenomena and rainfall in particular.

A comparative study of the network with standard back-propagation network(BPNN) was also done. It was found that two hidden layers with 12 nodes each were required for obtaining a comparable convergence in the BPNN. The training and test errors after 1000 iterations were found to be respectively 0.954 and 0.948. It is to be noted that even with about three times more hidden nodes, the BPNN is having a higher RMS error after training it for about two times as much iterations used by ABFN. This clearly indicates that updating the basis function makes sense in complex problems.

Acknowledgment

The authors wish to express their sincere thanks to Professor K. Mohankumar of the Department of Atmospheric Sciences of the Cochin University of Science and Technology for providing us with the rainfall database. The first author would like to thank Dr. Moncy V John of the Department of Physics, Kozhencherri St. Thomas college for long late night discussions.

References

1. E. S. R. Gopal, *Statistical Mechanics and Properties of Matter Theory and Applications*, Ellis Horwood, 1974.
2. N. S. Philip and K. B. Joseph, *Adaptive Basis Function for Artificial Neural Networks*, (*in press*) Neurocomputing Journal, also available at http://www.geocities.com/sajithphilip/research.htm
3. M. A. Kraaijveld *Small sample behaviour of multi-layer feed forward network classifiers: Theoretical and Practical Aspects* (Delft University Press, 1993).
4. A. Chowdhury and S. V Mhasawade, *Variations in meteorological floods during summer monsoon over India*, Mausam, **42**, 2, 167-170, 1991.

A Job-Shop Scheduling Problem with Fuzzy Processing Times

Feng-Tse Lin

Department of Applied Mathematics, Chinese Culture University
YangminShan, Taipei, Taiwan, ROC
ftlin@staff.pccu.edu.tw

Abstract. Job-shop scheduling is a difficult problem, both theoretically and practically. This problem is a combinatorial optimization of considerable industrial importance. Although the job-shop scheduling problem has often been investigated, very little of this research is concerned with the uncertainty characterized by the imprecision in problem variables. In this paper, we investigate the job-shop scheduling problem with imprecise processing times. We first use triangular fuzzy numbers to represent imprecise processing times, and then construct a fuzzy job-shop scheduling model. Our work intends to extend the deterministic job-shop scheduling problem into a more generalized problem that would be useful in practical situations.

1 Introduction

The job-shop scheduling problem is concerned with allocating limited resources to operations over time [3]. Between the operations precedence constraints for a job can be defined. Although job-shop scheduling has always had an important role in the field of production and operations management, it is a difficult problem in combinatorial optimization [12]. The difficulty is due to the high number of constraints, unfortunately unavoidable in the real-world applications [9]. The job-shop scheduling problem can be described as follows. We are given n jobs and m machines. Each job consists of a sequence of operations that must be processed on m machines in a given order. Each operation must be executed uninterrupted on a given machine for a given period of time and each machine can only handle at most one operation at a time. The problem is to find a schedule, an allocation of the operations of n jobs to certain time intervals on m machines, with a minimum overall time.

Solving the job-shop scheduling problem requires a high computational effort and considerable sophistication [2, 4]. Instead of investigating using optimal algorithms, it is often preferred to use approximation algorithms such as heuristics and meta-heuristics, e.g. simulated annealing, genetic algorithms, and tabu search [1, 5, 6-9, 15, 17, 19]. However, most of the methods proposed in the literature required the assumption that all time parameters are known exactly. This is a strong assumption, which may cause severe difficulties in practice. An example is the difficulty in estimating the exact processing times for all jobs on the machines. In fact, there are many vaguely formulated relations and imprecisely quantified physical data values in real world descriptions since precise details are simply not known in advance. Stochastic methods exist, but not many address imprecise uncertainty. Although the job-shop scheduling problem has often been investigated, very few of these studies take uncertainty, typified by the imprecision or vagueness in time estimates, into account [10, 14, 15, 16, 18].

In this study, we investigate a job-shop scheduling problem with imprecise processing time and use fuzzy numbers to represent imprecise processing times in this problem. The main interest of our approach is that the fuzzy schedules obtained from Property 1 are the same type as those in the crisp job-shop scheduling problem. The fuzzy job-shop scheduling model, in the case of imprecise processing times, is then an extension of the crisp problem.

2 Job-Shop Scheduling Problem

The deterministic job-shop scheduling problem is stated as follows. There are n jobs to be scheduled on m machines. Each job consists of a sequence of operations that must be processed on m machines in a given order [13]. Each operation is characterized by specifying both the required machine and the fixed processing time. Several constraints on jobs and machines, which are listed as follows [11]:

(1) Each job must pass through each machine once and only once.
(2) Each job should be processed through the machine in a particular order.
(3) Each operation must be executed uninterrupted on a given machine.
(4) Each machine can only handle at most one operation at a time.

The problem is to find a schedule to determine the operation sequences on the machines in order to minimize the total completion time. Let c_{ik} denote the completion time of job i on machine k, and t_{ik} denote the processing time of job i on machine k. For a job i, if the processing on machine h precedes that on machine k, we need the following constraint:

$$c_{ik} - t_{ik} \geq c_{ih}$$

On the other hand, if the processing on machine k comes first, the constraint becomes

$$c_{ih} - t_{ih} \geq c_{ik}$$

Thus, we need to define an indicator variable x_{ihk} as follows:

$$x_{ihk} = \begin{cases} 1, \text{processing on machine } h \text{ precedes that on machine } k \text{ for job } i \\ 0, \text{otherwise} \end{cases}$$

We can then rewrite the above constraints as follows:

$$c_{ik} - t_{ik} + L(1 - x_{ihk}) \geq c_{ih}, \quad i = 1, 2, ..., n, \quad h, k = 1, 2, ..., m$$

where L is a large positive number. Consider two jobs, i and j, that are to be processed on machine k. If job i comes before job j, we need the following constraint:

$$c_{jk} - c_{ik} \geq t_{jk}$$

Otherwise, if job j comes first, the constraint becomes

$$c_{ik} - c_{jk} \geq t_{ik}$$

Therefore, we also need to define another indicator variable y_{ijk} as follows.

$$y_{ijk} = \begin{cases} 1, \text{if job } i \text{ precedes job } j \text{ on machine } k \\ 0, \text{otherwise} \end{cases}$$

We can then rewrite the above constraints as follows:

$$c_{jk} - c_{ik} + L(1 - y_{ijk}) \geq t_{jk}, \quad i, j = 1, 2, ..., n, \quad k = 1, 2, ..., m$$

The job-shop scheduling problem with a makespan objective can be formulated as follows:

$$\min \max_{1 \leq k \leq m} [\max_{1 \leq i \leq n} [c_{ik}]] \tag{1}$$

$$\text{s.t.} \quad c_{ik} - t_{ik} + L(1 - x_{ihk}) \geq c_{ih}, \quad i = 1, 2, ..., n, \quad h, k = 1, 2, ..., m \tag{2}$$

$$c_{jk} - c_{ik} + L(1 - y_{ijk}) \geq t_{jk}, \quad i, j = 1, 2, ..., n, \quad k = 1, 2, ..., m \tag{3}$$

$$c_{ik} \geq 0, \quad i = 1, 2, ..., n, \quad k = 1, 2, ..., m \tag{4}$$

$$x_{ihk} = 0 \text{ or } 1, \quad i = 1, 2, ..., n, \quad h, k = 1, 2, ..., m \tag{5}$$

$$y_{ijk} = 0 \text{ or } 1, \quad i, j = 1, 2, ..., n, \quad k = 1, 2, ..., m$$

3   Constructing A Fuzzy Job-Shop Scheduling Model

3.1   Imprecise Processing Time

As in real life situations, some unexpected events may occur, resulting in small changes to the processing time of each job. Therefore, in many situations, the processing time can only be estimated as being within a certain interval. Let us illustrate this point using one example. The example taken from [10] is an industrial case in a chemical environment. The chemical reaction depends not only on the pressure and the temperature but also on the quality of the components. It is very difficult to control this dependence. Therefore, the uncertainties of those reaction times must also be modeled. Because of this interval estimation feature, the representation of processing time for a job can be more realistically and naturally achieved through the use of a fuzzy number. As a result, decision-makers (DM) do not need to give a single precise number to represent the processing time of a job. We use t_{jk}, the processing time for job j on machine k, to denote the process time for each job in this paper. However, t_{jk} is just an estimate and its exact value is actually unknown.

3.2 A Fuzzy Job-Shop Sequencing Model

Consider the schedule for the job-shop problem is performed several times in practical situations. Obviously, the processing time for this schedule at different execution times is not necessarily the same. Therefore, an estimated processing time interval, i.e. $[t_{jk} - \Delta_{jk1}, t_{jk} + \Delta_{jk2}]$, should be given to represent the possible range of values for the processing time. Thus the use of interval $[t_{jk} - \Delta_{jk1}, t_{jk} + \Delta_{jk2}]$ is more appropriate than the use of a single estimate, t_{jk}, in practical situations. The DM should carefully determine the parameters Δ_{jk1} and Δ_{jk2}, which satisfy $0 < \Delta_{jk1} < t_{jk}$ and $0 < \Delta_{jk2}$, for defining an acceptable processing time range for any particular problem. After that, the DM can choose an appropriate value from the interval $[t_{jk} - \Delta_{jk1}, t_{jk} + \Delta_{jk2}]$ as an estimate for the processing time for job j on machine k. Obviously, when the estimate happens to be t_{jk}, which is the crisp processing time, the error rate is zero. When the estimate deviates from t_{jk}, the error rate will become larger. In fact, we can use the term "confidence level" instead of "error rate" while we consider processing time interval based on the fuzzy viewpoint. We can therefore say that the confidence level is one if the processing time estimate equals t_{jk}. Otherwise, when the processing time estimate deviates from t_{jk}, the confidence level will become smaller. Finally, when the estimate approaches one of the two ends of the interval, i.e. $t_{jk} - \Delta_{jk1}$ or $t_{jk} + \Delta_{jk2}$, the confidence level will be close to zero.

A level 1 triangular fuzzy number corresponding to the above interval $[t_{jk} - \Delta_{jk1}, t_{jk} + \Delta_{jk2}]$ is given as follows:

$$\tilde{t}_{jk} = (t_{jk} - \Delta_{jk1}, t_{jk}, t_{jk} + \Delta_{jk2}; 1) \in F_N(1), \tag{6}$$

$$0 < \Delta_{jk1} < t_{jk}, \quad 0 < \Delta_{jk2}, \quad j = 1, 2, ..., n, \quad k = 1, 2, ..., m.$$

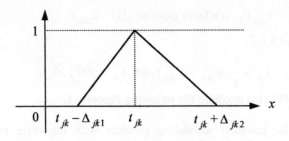

Fig. 1. A triangular fuzzy number $\tilde{t}_{jk}$.

Fig. 1 shows a level 1 triangular fuzzy number $\tilde{t}_{jk}$. From fig. 1 we can see that the membership grade at t_{jk} in $\tilde{t}_{jk}$ is 1. However, the more the fuzzy number deviates from the t_{jk} position, the lesser the fuzzy numbers membership grade in $\tilde{t}_{jk}$. The membership grade at $t_{jk} - \Delta_{jk1}$ or $t_{jk} + \Delta_{jk2}$ is zero. Of course, we can see that the confidence level for an estimate in the interval corresponds to the membership grade of a fuzzy number in the fuzzy sets. This concept naturally leads to the use of fuzzy numbers for the processing time in the job-shop scheduling problem.

Let $\tilde{c}_{ik} = (c_{ik}, c_{ik}, c_{ik}; 1) = (\tilde{c}_{ik})_1 \in F_N(1)$. (7)

The signed distance of level λ fuzzy number is defined as follows. For each $\lambda \in [0,1]$ and $\tilde{A} = (a,b,c;\lambda) \in F_N(\lambda)$. The signed distance of $\tilde{A}$ measured from $\tilde{0}_1$ is defined by $d(\tilde{A}, \tilde{0}_1) = \frac{1}{4}(2b + a + c)$. Then $d(\tilde{c}_{ik}, \tilde{0}_1) = c_{ik}$ is the signed distance from $\tilde{0}_1$ to $\tilde{c}_{ik}$. Since $d(\tilde{c}_{ik}, \tilde{0}_1) = c_{ik} > 0$, it is a positive distance from $\tilde{0}_1$ to $\tilde{c}_{ik}$. Therefore $c_{ik} = d(\tilde{c}_{ik}, \tilde{0}_1)$ is the completion time measured from the origin 0.

The relations $\succ, \approx$ are the ranking as defined on $F_N(1)$. We then fuzzify (1) - (4) using (6) and (7) to obtain the following fuzzy job-shop scheduling problem:

$$\min \max_{1 \leq k \leq m} \left[\max_{1 \leq i \leq n} [d(\tilde{c}_{ik}, \tilde{0}_1)] \right] \tag{8}$$

s.t. $\tilde{c}_{ik} \ominus \tilde{t}_{ik} \oplus (L(\tilde{1} - x_{ihk}))_1 \gtrapprox \tilde{c}_{ih}, \quad i = 1,2,...,n, \quad h,k = 1,2,...,m$ (9)

$\tilde{c}_{jk} \ominus \tilde{c}_{ik} \oplus (L(\tilde{1} - y_{ijk}))_1 \gtrapprox \tilde{t}_{jk}, \quad i,j = 1,2,...,n, \quad k = 1,2,...,m$ (10)

$\tilde{c}_{ik} \succ \tilde{0}_1, \gtrapprox j = 1,2,...,n, k = 1,2,...,m$ (11)

$x_{ihk} = 0 \text{ or } 1, \quad i = 1,2,...,n, k = 1,2,...,m$ (12)

$y_{ijk} = 0 \text{ or } 1, \quad i,j = 1,2,...,n, \quad k = 1,2,...,m$

Note that $(L(\tilde{1} - x_{ihk}))_1$ is a fuzzy point at $L(1 - x_{ihk})$.

Since $d(\tilde{c}_{ik}, \tilde{0}_1) = c_{ik}$, $\qquad\qquad\qquad\qquad\qquad\qquad\qquad\qquad$ (13)

and $d(\tilde{t}_{jk}, \tilde{0}_1) = t_{jk} + \dfrac{1}{4}(\Delta_{jk2} - \Delta_{jk1})$, let $t_{jk}^* = d(\tilde{t}_{jk}, \tilde{0}_1)$. $\qquad\qquad$ (14)

We summarize (9)-(14) to obtain the following Property 1.

Property 1. The job-shop scheduling problem with imprecise processing time modeled by fuzzy number is as follows:

$$\min \max_{1 \le k \le m} [\max_{1 \le i \le n} [c_{ik}]] \qquad\qquad\qquad\qquad\qquad\qquad (15)$$

s.t. $\quad c_{ik} - t_{ik}^* + L(1 - x_{ihk}) \ge c_{ih}$, $\quad i = 1, 2, ..., n$, $\quad h, k = 1, 2, ..., m$ $\qquad$ (16)

$\qquad c_{jk} - c_{ik} + L(1 - y_{ijk}) \ge t_{jk}^*$, $\quad i, j = 1, 2, ..., n$, $\quad k = 1, 2, ..., m$ $\qquad$ (17)

$\qquad c_{ik} \ge 0$, $\quad i = 1, 2, ..., n$, $\quad k = 1, 2, ..., m$ $\qquad\qquad\qquad\qquad\qquad$ (18)

$\qquad x_{ihk} = 0$ or 1, $\quad i = 1, 2, ..., n$, $\quad h, k = 1, 2, ..., m$ $\qquad\qquad\qquad$ (19)

$\qquad y_{ijk} = 0$ or 1, $\quad i, j = 1, 2, ..., n$, $\quad k = 1, 2, ..., m$

4 Computational Results

Consider an example from [11]. We compare the results obtained from Property 1 with that of the crisp job-shop scheduling problem using the Johnson's constructive algorithm. Johnson's algorithm for, $n/2/G/F_{max}$ (i.e. n jobs, two machines, general job-shop problem, to minimize makespan) problem is stated briefly as follows:

Suppose that the set of n jobs $\{J_1, J_2, ..., J_n\}$ may be partitioned into four types of jobs.

Type A: Those to be processed on machine M_1.

Type B: Those to be processed on machine M_2.

Type C: Those to be processed on both machines in the order M_1 then M_2.

Type D: Those to be processed on the machine in the order M_2 then M_1.

Then the construction of an optimal schedule is straightforward, as follows:

(1) Schedule the jobs of type A in any order to give the sequence S_A.

(2) Schedule the jobs of type B in any order to give the sequence S_B.

(3) Schedule the jobs of type C according to Johnson's algorithm for $n/2/F/F_{max}$ (n jobs, 2 machines, flow-shop problem, to minimize makespan) problems to give the sequence S_C.

(4) Schedule the jobs of type D according to Johnson's algorithm for $n/2/F/F_{max}$ problems to give the sequence S_D.

The example is the $9/2/G/F_{max}$, nine jobs and two machines, with times and processing order as given in Table 1.

Table 1. Processing order and times.

Job	First Machine	Second Machine
1	M_1 $8(=t_{11})$	M_2 $2(=t_{12})$
2	M_1 $7(=t_{21})$	M_2 $5(=t_{22})$
3	M_1 $9(=t_{31})$	M_2 $8(=t_{32})$
4	M_1 $4(=t_{41})$	M_2 $7(=t_{42})$
5	M_2 $6(=t_{52})$	M_1 $4(=t_{51})$
6	M_2 $5(=t_{62})$	M_1 $3(=t_{61})$
7	M_1 $9(=t_{71})$	$-$
8	M_2 $1(=t_{82})$	$-$
9	M_2 $5(=t_{92})$	$-$

An optimal crisp sequence for Table 1, obtained by Johnson's algorithm is

	Processing Sequence of jobs
Machine M_1	(4, 3, 2, 1, 7, 5, 6)
Machine M_2	(5, 6, 8, 9, 4, 3, 2, 1)

We can see that the total time F_{max} = 44 for the optimal sequence. From Property 1, we consider the following fuzzification for processing time t_{jk}.

$$\tilde{t}_{11} = (8-2, 8, 8+4), \tilde{t}_{21} = (7-1, 7, 7+4), \tilde{t}_{31} = (9-3, 9, 9+1), \tilde{t}_{41} = (4-1, 4, 4+2),$$

$$\tilde{t}_{51} = (4-0.5, 4, 4+1), \tilde{t}_{61} = (3-0.6, 3, 3+1.5), \tilde{t}_{71} = (9-3, 9, 9+2),$$

$$\tilde{t}_{12} = (2-0.8, 2, 2+1), \tilde{t}_{22} = (5-1, 5, 5+2), \tilde{t}_{32} = (8-1, 8, 8+4), \tilde{t}_{42} = (7-2, 7, 7+3),$$

$$\tilde{t}_{52} = (6-2, 6, 6+1), \tilde{t}_{62} = (5-0.9, 5, 5+2), \tilde{t}_{82} = (1-0.5, 1, 1+0.6), \tilde{t}_{92} = (5-2, 5, 5+1)$$

By (14) we obtain the following fuzzy processing times.

$$t_{11}^* = 8.5, t_{21}^* = 7.75, t_{31}^* = 8.5, t_{41}^* = 4.25, t_{51}^* = 4.125, t_{61}^* = 3.225, t_{71}^* = 8.75, t_{12}^* = 2.05,$$

$$t_{22}^* = 5.25, t_{32}^* = 8.75, t_{42}^* = 7.25, t_{52}^* = 5.75, t_{62}^* = 5.275, t_{82}^* = 1.025, t_{92}^* = 4.75.$$

Then we find an optimal schedule as follows:

Type A jobs: job 7 on M_1.

Type B jobs: jobs 8 and 9 on M_2 in an arbitrary order (8,9).

Type C jobs: jobs 1, 2, 3, and 4 require M_1 first and then M_2. Using Johnson's algorithm we obtain the relations $t_{31}^* < t_{32}^*$, $t_{41}^* < t_{42}^*$, and $t_{41}^* < t_{31}^*$; $t_{11}^* > t_{12}^*$, $t_{21}^* > t_{22}^*$, and $t_{12}^* < t_{22}^*$. Thus the sequence obviously is (4, 3, 2, 1).

Type D jobs: jobs 5 and 6 require M_2 first and then M_1. Again, using Johnson's algorithm we obtain the relations $t_{51}^* < t_{52}^*$, $t_{61}^* < t_{62}^*$, and $t_{61}^* < t_{51}^*$. Note that M_1 now becomes the second machine. The sequence is therefore (5, 6).

Finally, an optimal sequence for Property 1 is

	Processing Sequence of Jobs
Machine M_1	(4, 3, 2, 1, 7, 5, 6)
Machine M_2	(5, 6, 8, 9, 4, 3, 2, 1)

The processing sequence for jobs obtained from the fuzzy case is the same as that for the crisp case. The total time is $F_{max}^* = 45.1$ for the optimal sequence of Property 1. We compare the result of Property 1 with that of crisp case as follows:

$$\frac{F_{max}^* - F_{max}}{F_{max}} \times 100 = 2.5\%.$$

5 Conclusion

In this paper, we have investigated the fuzzy job-shop scheduling problem with imprecise processing times. In conclusion, we point out that our work has produced the following main results for the job-shop scheduling problem with fuzzy processing times modeled as fuzzy numbers. In Section 3, we presented a fuzzy job-shop scheduling problem based on triangular fuzzy numbers. In (14), if $\Delta_{jk2} = \Delta_{jk1}, \forall j, k$, obviously, Fig. 1 is an isosceles triangle and we obtain $t_{jk}^* = t_{jk}, \forall j, k$. Thus, the equations (15)-(19) will become the same as (1)-(5). If $\Delta_{jk2} = \Delta_{jk1} = 0$, $\forall j, k$, the job-shop scheduling in the fuzzy sense of Property 1 will become the crisp job-shop scheduling of equations (1)-(5). Therefore, the job-shop scheduling in the fuzzy sense of Property 1 is an extension of the crisp scheduling of (1)-(5).

The interpretation of Fig. 1 is as follows: When $\Delta_{jk2} > \Delta_{jk1}, \forall j, k$, the triangle is skewed to the right-hand side, thus obtaining $t^*_{jk} > t_{jk}, \forall j, k$. This means that the completion time in the fuzzy sense is longer than in the crisp case. Conversely, when $\Delta_{jk2} < \Delta_{jk1}, \forall j, k$, the triangle is skewed to the left-hand side, thus obtaining $t^*_{jk} < t_{jk}, \forall j, k$, indicating that the completion time in the fuzzy sense is shorter than the crisp case. From (14), we can see that $t^*_{jk} = t_{jk} + \frac{1}{4}(\Delta_{jk2} - \Delta_{jk1})$ is the processing time in the fuzzy sense, where t^*_{jk} equals the crisp processing time, t_{jk}, plus some variants due to the inclusion of fuzzy values.

The comparison of [10] with our work is as follows. Fortemps used six-point fuzzy numbers to represent fuzzy durations and fuzzy makespan. In his approach, the resulting fuzzy framework was not an extension of the original crisp problem. He used a simulated annealing technique to solve the fuzzy model obtaining the optimization sequence of the problem. In our approach, however, the crisp duration x become an interval $[x - \Delta_1, x + \Delta_2]$, $0 < \Delta_1 < x$, $0 < \Delta_2$, represents an allowable range of duration. We let the fuzzy number $\tilde{x} = (x - \Delta_1, x, x + \Delta_2; 1)$ correspond to the interval $[x - \Delta_1, x + \Delta_2]$. Then we used a signed distance method for ranking fuzzy numbers to obtain a fuzzy job-shop scheduling problem. After defuzzifying the fuzzy problem using the signed distance ranking method, we obtained a job-shop scheduling problem in the fuzzy sense. The resulting fuzzy framework in our approach is an extension of the original crisp problem. Therefore, the algorithms that were used for solving the crisp job-shop sequencing problem can also be used for solving the fuzzy problem.

References

1. Adams, J., Balas, E., and Zawack, D.: The Shifting Bottleneck Procedure for Job Shop Scheduling. International Journal of Flexible Manufacturing Systems, Vol.34, No. 3, (1987) 391-401
2. Applegate, D. and Cook, W.: A Computational Study of The Job-Shop Scheduling Problem. ORSA Journalon Computing, Vol. 3, No. 2, (1991) 149-156
3. Baker, K. R.: Introduction to Sequencing and Scheduling. John Wiley & Sons, Inc., New York (1974)
4. Bellman, R. Esogbue, A., and Nabeshima, I.: Mathematical Aspects of Scheduling and Applications. Pergamon Press, Oxford (1982)
5. Blackstone, J., Phillips D., and Hogg, G.: A State-of-The-Art Survey of Dispatching Rules for Manufacturing Job Shop Operations. International Journal of Production Research, Vol. 20, (1982) 26-45

6. Cheng, R. Gen, M. and Tsujimura, Y.: A Tutorial Survey of Job-Shop Scheduling Problems Using Genetic Algorithms: Part I. Representation. International Journal of Computers and Industrial Engineering, Vol. 30, No. 4, (1996) 983-997

7. Croce, F., Tadei, R., and Volta, G.: A genetic algorithm for the job shop problem. Computers and Operations Research, Vol. 22, (1995) 15-24

8. Coffman, E. G. Jr.: Computer and Job-Shop Scheduling Theory. John Wiley & Sons, Inc., New York (1976)

9. Falkenauer, E. and Bouffoix, S.: A Genetic Algorithm for Job Shop. Proceedings of the IEEE International Conference on Robotics and Automation, (1991) 824-829

10. Fortemps, P.: Jobshop Scheduling with Imprecise Durations: A Fuzzy Approach. IEEE Transactions on Fuzzy Systems, Vol. 5, No. 4, (1997) 557-569

11. French, S.: Sequencing and Scheduling: An introduction to the Mathematics of the Job-Shop. John Wiley & Sons, Inc., New York (1982).

12. Garey, M., Johnson, D., and Sethi, R.: The Complexity of Flowshop and Jobshop Scheduling. Mathematics of Operations Research, Vol. 1, (1976) 117-129

13. Gen M., and Cheng, R.: Genetic Algorithms and Engineering Design, John Wiley & Sons, Inc., New York (1997)

14. Ishii, H., Tada, M., and Masuda, T.: Two Scheduling Problems with Fuzzy Due Dates. Fuzzy Sets and Systems, Vol. 46, (1992) 339-347

15. Ishibuchi, H., Yamamoto, N., Murata, T., and Tanaka, H.: Genetic Algorithms and Neighborhood Search Algorithms for Fuzzy Flowshop Scheduling Problems. Fuzzy Sets and Systems, Vol. 67, (1994) 81-100

16. Kerr, R. M. and Slany, W.: Research issues and challenges in fuzzy scheduling. CD-Technical Report 94/68, Christian Doppler Laboratory for Expert Systems, Technical University of Vienna, Austria (1994)

17. Krishna, K., Ganeshan, K. and Janaki Ram, D.: Distributed Simulated Annealing Algorithms for Job Shop Scheduling. IEEE Transactions on Systems, Man, and Cybernetics, Vol. 25, No. 7, (1995) 1102-1109

18. McCahon, C. S. and Lee, E. S.: Fuzzy Job Sequencing for A Flow Shop. European Journal of Operational Research, Vol. 62, (1992) 294-301

19. van Laarhoven, P. J. M., Aarts, E. H. L, and Lenstra, J. K.: Job Shop Scheduling by Simulated Annealing. Operations Research, Vol. 40, No. 1, (1992) 113-125

Speech Synthesis Using Neural Networks Trained by an Evolutionary Algorithm

Prof. Dr. Trandafir MOISA[1], Dan ONTANU[2], Adrian Horia DEDIU[3]

[1] CSE department, Politehnica University, Splaiul Independentei 313, 77206, Bucharest, Romania
tmoisa@cs.pub.ro
[2] CST, Tolstoi 4B, 712893, Bucharest, Romania,
dano@starnets.ro
[3] CST, Tolstoi 4B, 712893, Bucharest, Romania
hd@ziplip.com

Abstract. This paper presents some results of our research regarding the speech processing systems based on Neural Networks (NN). The technology we are developing uses Evolutionary Algorithms for NN supervised training. Our current work is focused on Speech Synthesis and we are using a Genetic Algorithm to train and optimize the structure of a hyper-sphere type Neural Network classifier. These Neural Networks are employed at different stages of the Speech Synthesis process: recognition of syllables and stress in the high level synthesizer, text to phonemes translation, and control parameters generation for the low level synthesizer. Our research is included in a pilot project for the development of a bilingual (English and Romanian) engine for text to speech / automatic speech recognition and other applications like spoken e-mail, help assistant, etc.

Keywords: Evolutionary Algorithms, Neural Networks, Supervised Training, Speech Synthesis

1 Introduction

In order to generate faster and more accurate results in different applications, many researchers use Neural Networks (NN) in conjunction with Evolutionary Algorithms (EA). Thus, evolving weights and thresholds, the network topology and node transfer functions, as well as evolving learning rules or finding a near optimal training set are examples of using EA in the NN training process. A "State of the Art" presentation of Evolving Neural Networks can be found in [Yao 1999].

Our approach is based on developing techniques for simultaneous optimization of both weights and network topology.

It is well-known that the coding scheme used by an EA is crucial for successfully solving a given problem. There have been some debates regarding the cardinal of the alphabet used for the coding scheme. According to Goldberg [Goldberg 1989], the lower the number of symbols in the alphabet, the higher implicit parallelism can be achieved. This way, a genetic algorithm should work with binary genes. However, other studies revealed unexpected advantages of using high cardinality alphabets [Michalewicz 1992]. Genetic operators (average crossover) dedicated to solve specific problems can benefit from such a coding scheme.

Among the EA used for NN training, such as - Genetic Algorithms (GA), Evolutionary Strategies (ES), Evolutionary Programming (EP), etc. – we focused on GA with real number chromosomes and special genetic operators like average crossover [Beasley 1993] and rank-space selection [Winston 1992].

We are using NN trained by GA for solving two practical problems: prediction and text-to-speech (TTS) engine.

As known, a text-to-speech engine is a system aimed to convert a source text into its spoken (i.e. voiced) equivalent [Moisa 1999a], [Moisa 1999b], [Moisa 2000]. The source text is generally supposed to be ASCII encoded, preferably without any spelling or lexical restrictions. Its spoken counterpart is a voice stream, that is, a sequence of voice samples, with a specified sampling frequency and dynamic range. We have developed a complete text-to-speech engine for the English and Romanian languages. The general structure of the TTS engine is outlined in **Fig. 1**:

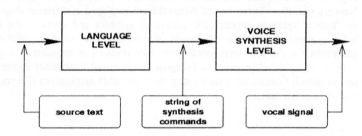

Fig. 1. General structure of the TTS engine

2 Parallel Training of Neural Networks Using Genetic Algorithms

We used the Soft Computing Genetic Tool [Dediu 1996], [Dediu 1999], for the parallel training of the neural networks. Having an Excel type fitness function evaluation interface, we used on a given row the input data set, the neural network architecture, the output computed value, the desired output value, and the error for the current data set. The neural network uses the genes values as weights and thresholds. Using the copy formulas capability, we extended the described structure for the whole data set available. Normally, when referring a data element, we used a relative addressing mode, and when we referred to a gene value, we used an absolute address mode. The fitness function computes the average error for the training data set available. The algorithm works and minimizes the fitness function, thus the global error for the whole data set.

From formal point of view, we modified the well-known "Back Propagation" training algorithm [Lippman 1987]:

Repeat
  select one input data set;
  compute NN output;
  adjust weights and thresholds;
until stop criteria achieved;

And we obtained the following algorithm:

Repeat
  parallel compute NN outputs for whole input data set;
  adjust weights and thresholds;
until stop criteria achieved;

The stop criteria might be err < ε or we can stop after a certain number of training iterations or after a given amount of time.

The following picture (**Fig. 2**) might create an idea of what is happening during the training phase of the neural network using Soft Computing Genetic Tool.

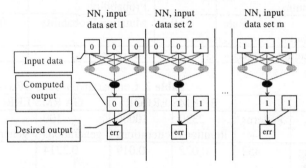

Fig. 2. Parallel training of neural networks using genetic algorithms

Behind the interface of the Soft Computing Genetic Tool, there is a genetic engine that proposes values for genes, than the interface evaluates the fitness function and sends back to the genetic engine the fitness function value.

We should remark the higher degree of parallelism achieved by our algorithm. Thus, the GA contains a parallel population of individuals, every individual having a parallel set of neural network structures, every such structure being, as known, a massively parallel computing system.

3 Neural Networks Trained by Genetic Algorithms - First Results

We tested our method on several data sets and we studied the behavior of neural networks depending on different transfer functions. We realized that the genetic algorithm needs some hints coming from the neural processing elements transfer functions, in order to be able to find out weights and thresholds. Thus, a step transfer function is practically useless since, after a slightly modification of a gene (weight or threshold) only the luckiest mutations can reduce the global error.

We used for tests the "Game of life" problem where a cell surrounded by eight neighbors survives if it has between 2 and 3 living neighbors. Our neural network structure has 9 inputs, one output and 19 hidden neurons fully interconnected. We run the tests and we compare the results of our GA method with a BP training algorithm. We used as genetic operators shuffle crossover, average mutation [Michalewicz 1992] and rank-space selection [Winston 1992]. The following tables summarize algorithms' parameters and comparative RMS errors for GA and BP methods (**Table 1** and **Table 2**). Analyzing the results, we can conclude that the GA used for NN training can find fast a satisfying solution, not so well tuned but well suited for generalization.

Table 1. Algorithms' training parameters

BP parameters		GA parameters	
Max Iterations:	10000	Max Generations:	1000
Learn Rate (Eta):	0.01	Population size:	10
Learn Rate Minimum:	0.001	Crossover Probability	80%
Learn Rate Maximum:	0.3	Mutation Probability	0.48%
Momentum (Alpha):	0.8	Nr. of genes:	210

Table 2. Comparative results BP versus GA

		BP RMS ERROR		GA RMS ERROR	
	Patterns	2330 iterations	10000 iterations	100 generations	1000 generations
Training data set:	452	0.022	0.019	0.2214	0.147
Test data set	60	0.028	0.026	0.016	0.0012

4 Neural Networks Trained by Genetic Algorithms - a Text-To-Speech Engine

The language level comprises several natural language processing stages ([Moisa 1999a], [Ontanu 1998]) such as: brief lexical analysis, exception treatment, conversion to phonological notation (at both rough and fine scale), detection of prosodic features responsible for rhythm and intonation, as well as the generation of the sequence of commands for driving a voice synthesis machine. This processing chain eventually translates the source text in a sequence of lexical tokens, each of these being, in turn, a sequence of phonemes, labeled with prosodic markers. On this basis is finally generated the input for the next stage, namely the sequence of commands. At the voice synthesis level we find a concrete voice generation system, based either on a model for the human vocal tract or, on concatenation of small voice units, such as diphones, half-syllables and the like, extracted by processing human voice recordings. Our TTS engine employs the latter approach, one based on a TD-PSOLA recent development ([Dutoit 1996]). Such a synthesis machine has the advantage of providing a simple command which, essentially specifies the fundamental frequency and duration for every phoneme that is to be played.

4.1 The neural model for the TTS engine

The neural structure presented in **Fig. 3** is a 2-layer hypersphere classifier, as described in [Ontanu 1993] or [Ontanu 1998]. Each node on the hidden layer (H_i) is related to a n-dimensionally hypersphere in the problem space, centered in C_i and having the radius r_i. If an input pattern belongs to such a domain (say, to the one of center C_k and radius r_k), the associated node (H_k) will fire.

The decision region for a binary (2-class) problem is the union (set theoretic) of all the domains represented by the nodes H_i. This is achieved by putting the final neuron (Y) to perform a "logical OR" among all H_i outputs. The training algorithm is basically a covering procedure in the problem space, trying to approximate the distribution of "positive answer" patterns. As opposed to [Reilly 1982], such a procedure, described in section 2 of the paper, effectively builds the structure by growing the hidden layer accordingly to problem complexity. When compared to multi-layer perceptrons, trained by backpropagation, the overall advantages of such networks are: the speed of the training process, and better generalization capabilities

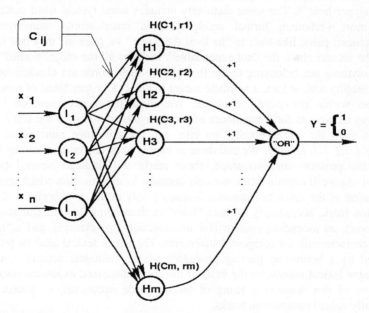

Fig. 3. The structure of a 2-layer binary hypersphere classifier: Hi - active nodes associated to hyperspheres of center Ci and radius ri in the problem space, Ii – passive input nodes

4.2 Lexical analysis and exception processing

This process first looks for the segmentation of the source text in two roughly defined categories: words and punctuation marks. By "word" we mean in this context any sequence of alphanumeric characters and/or usual punctuation marks, separated by white spaces (i.e. tabs, carriage return, line feed, or the space character itself). Next, every scanned word is classified as "expandable" and "not expandable". The first class comprises various numeric expressions, i.e. numbers (*23, 3.14,* etc), date and time marking conventions (*07-26-1999, 13:23,* etc), acronyms like *IBM* or *HTML,* short-cuts for usual words like *ASAP* ("as soon as possible"), *etc* ("etcaetera"), measurement units and so on. The class also includes geographical terms (cities, countries, rivers, mountains), names for well-known companies and institutions (*FMI, Microsoft, Xerox*), foreign currencies (*Euro, USD, DM*), etc. Every such word is replaced by an appropriate expansion, which is a conversion to its spoken form. For instance, *23* becomes, after expansion, *twenty-three.* We say that some of the words are directly "expandable", meaning that the spoken form is found by using some appropriate procedure; these are the various numerical expressions we've talked above. On the contrary, for any other terms, we consider using an exception dictionary, in fact, an associative and continuously extendable database, containing the proper expansion for every entry it has. For some word categories, the dictionary gathers not only the corresponding spoken string, but also items like the plural and singular forms or the word's gender. This helps solving situations like *30 Km/h,* which obviously require the plural form for *km/h* ("kilometers per hour" and not "kilometer per hour"). The same dictionary includes some typical word contexts for solving most verb/noun formal ambiguities; in other words, homonymic (or homographical) pairs, like *lives* in "he lives the house" vs. *lives* in "they fear for their lives". The second class, the "not expandable", includes all the other "normal" words (that is, anything not belonging to the first class). Such words are checked against a pronounciability test, in fact, a heuristic meant to detect a proper blend of vowels and consonants within the word's structure. We allow for a maximum of 3 vowel successions as well as for a maximum of 4 consonants in a row. If the test fails, we choose to spell the word. Finally, we take into account some punctuation marks, namely: *{.}, {,}, {:}, {;}, {?}.* We call these as having prosodic value, meaning the fact that, at the prosodic analysis stage, these marks will indicate several types of intonation. As we'll explain latter, we only consider local intonation which consists in the alteration of the voice fundamental frequency only in the neighborhood of such a punctuation mark, accordingly to three "laws": a descending contour (for statement-type clauses), an ascending contour (for interrogation-type clauses), and a "neutral" contour, characteristic for compound statements. The whole lexical analysis process is performed by a bottom-up parsing procedure whose "semantic actions" consist in creating new lexical tokens, on the basis of the above discussed expansion operations. The output of this stage is a string of tokens, made exclusively of letters and/or prosodically valued punctuation marks.

4.3 Translation to phonemic notation

As seen from figure **Fig. 4**, this process involves two distinct stages. The first one (the primary translation), converts each token produced by the lexical analyzer to a string of phonemes. This is accomplished by means of using a neural network (set of hypersphere classifiers as illustrated in **Fig. 5,** trained over a 120,000 items pronunciation dictionary. As a result, a letter is usually replaced by a sound (phoneme) or a pair of sounds. Stressed vowels are also marked accordingly. The notation employs one character per phoneme and is inspired by a convention proposed at the LDC (Linguistic Data Consortium). The second stage takes into account some allophonic variations, like the aspired counterparts for [t], [p], [k] (cf. *pit, cut, task*) or the [o] variant in words like *sort*. The output of either two levels is a string of converted tokens. The main difference is that the "fine" process outputs also the inherited prosodic markers (i.e., for rhythm and intonation).

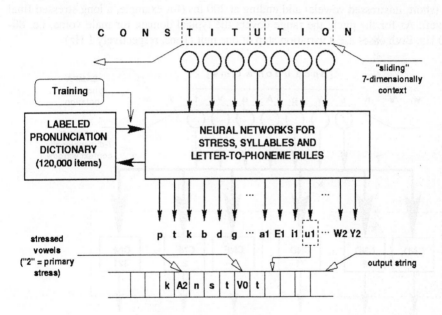

Fig. 4. Translation to phonetic notation and detection of prosodic features (such as stress) using neural networks (hypersphere classifiers)

4.4 Driving the voice synthesizer. Neural networks for generating duration and pitch

In this section we briefly present the main aspects of a neural principle for generating the command in TD-PSOLA like voice synthesizers. As known, these voice generators are driven at phoneme level, by means of specifying the duration and the pitch (fundamental frequency) curve. Specifying fundamental frequency amounts to giving a discrete approximation of it, that is several "location-value" pairs spread over the phoneme duration. The TD-PSOLA algorithm uses then interpolation in order to achieve a continuous pitch variation. This forms the basis for the intonation control. Given the above hypotheses, we consider labeling vowel-centered phonemic contexts with duration and pitch tags. These contexts are obtained by sliding a 7-characters window over the string of phoneme tokens (interspersed with stress markers), representing the output of the former text processing stages. We gather only those windows centered on a vowel. The net eventually recognizes the attributes for the central vowel of the current context, namely its duration (in *ms*) and the fundamental frequency value, say, in the middle of its length (50%), expressed in Hertz. **Fig. 5** reveals a set of binary neural classifiers, trained and built using the algorithm presented in this paper. The duration and pitch variation ranges are conveniently sampled into a set of discrete values. Thus, we allow for vowel duration starting at 30 ms (short, unstressed vowels) and ending at 200 ms (for example, a long stressed final vowel). As for the pitch, the range covers the typical domain for male voice, i.e. 80-400 Hz. Both cases use a variation step of one unit (1 ms, respectively 1 Hz).

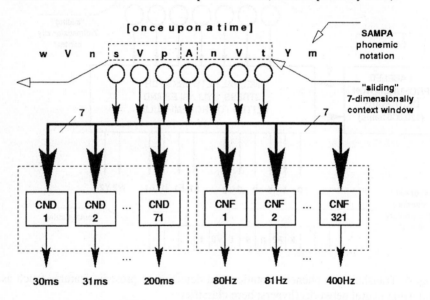

Pitch and duration for current vowel: /A/ (e.g. 95ms and 120Hz)

Fig. 5. Set of neural classifiers for recognizing duration and pitch.
(CND = Duration Neural Classifier, CNF = Frequency Neural Classifier)

For every discrete value, within the specified ranges, we associate a 2-layer hypersphere neural classifier aimed at recognizing contexts having that value as its attribute. We come, therefore, to a set of 71 binary networks for duration and 321 for frequencies, as pointed out in the figure below.

It becomes obvious that the training process for the above nets requires the evaluation (measurement) of both parameters for a large set of vowel contexts. We recorded and used a dictionary of about 4000 English basic words. From the corresponding voice recordings, we further extracted by hand more than 8000 phonemic contexts and measured for each one the duration and the pitch values at three different knots (15%, 50%, 85%). Training sets were created for each of these measurement points. Then, the learning process built the appropriate net structures, as presented above. The low-levels synthesis engine is based on diphone concatenation. A diphone can be viewed as the transition region between two successive phonemes. A large base of such elements is used for the synthesis process. For the English language, it roughly includes about 1600 diphones and amounts to 7 MB of storage. The output of the **CNDs** and **CNFs** (**Fig. 5**) is translated into commands with the following syntax:

<phoneme_name> <duration> {<position> <pitch value>}.

Curly brackets indicate that the synthesizer accepts several pairs "duration-pitch", as mentioned earlier, but we confine ourselves to a simpler command structure, comprising a single pitch specification. The synthesizer generates vocal signal at 16,000 samples per second, with 16 bits per sample. These values express, in fact, the recording properties of the diphones database.

5 Concluding Remarks

Our experience in using NN models (particularly hyperspehere classifiers) to develop TTS and ASR engines, revealed the strength of the method and the benefits of this approach.

To further improve the results, we investigated different NN training methods using EA. In this respect, we have developed a GA employing high cardinality alphabets for the coding scheme and a rank space selection operator. This way, a higher parallelism has been achieved, speeding up the training process.

The results of this approach have been emphasized by an increased efficiency of our bilingual (English and Romanian) engine for text to speech / automatic speech recognition, as well as related applications such as: spoken e-mail, help assistant, etc.

References

[Beasley 1993] David Beasley, David R. Bull, Ralph R. Martin, *An Overview of Genetic Algorithms, Part 2, Research Topics,* Univerity Computing, 1993, 15(4) 170-181
[Box 1970] G.E. Box and G.M. Jenkins, *Time Series Analysis: Forecasting and Control,* Holden Day, 1970.

[Dediu 1996] A.H. Dediu, D. Mihaila, *Soft Computing Genetic Tool, Proceedings of First European Congress on Intelligent Techniques and oft Computing EUFIT'96*, Aachen 2-5 September 1996, Vol. 2,pp. 415-418.

[Dediu 1999] Dediu Adrian Horia, Agapie Alexandru, Varachiu Nicolae, *Soft Computing Genetic Tool V3.0 – Applications*, 6-th Fuzzy Days in Dortmund-Conference on Computational Intelligence; 1999

[Dutoit 1996] T. Dutoit, V. Pagel, N. Pierret, F. Bataille, O. Van der Vrecken, The MBROLA project: *Towards a Set of High Quality Speech Synthesizers* Free of Use for Non Commercial Purposes Proc. ICSLP 96, Fourth International Conference on Spoken Language Processing, vol.3, p. 1393-1396, Philadelphia, 1996

[Goldberg 1989] D. E. Goldberg. *Genetic Algorithms in search, optimization and machine learning*. Addison-Wesley, 1989

[Gori 1998] M. Gori and F. Scarselli, *Are Multilayer Perceptrons Adequate for Pattern Recognition and Verification?*, IEEE Transactions on Patern Analysis and Machine Intelligence, November 1998, Volume 20, Number 11, pp. 1121-1132.

[Lippman 1987] Richard P. Lippman, *An Introduction to Computing with Neural Nets*, IEEE ASSP MAGAZINE APRIL 1987.

[Michalewicz 1992] Z. Michalewicz. *Genetic Algorithms + Data Structures = Evolution Programs*. Springer Verlag, 1992

[Moisa 1999a] Trandafir Moisa, Dan Ontanu, *Learning Romanian Language Using Speech Synthesis*, Advanced Research in Computers and Communications in Education, New Human Abilities for the Networked Society, Edited by Geoff Cumming, Tashio Okamoto, Louis Gomez, Proceedings of ICCE'99, 7th International Conference on Computers in Education, Chiba, Japan, pp. 808-815, IOS Press, Volume 55, 1999

[Moisa 1999b] Trandafir Moisa, Dan Ontanu, *A Spoken E-mail System*, 12th International Conference on Control Systems and Computer Science – CSCS12, May 26 – 29, 1999, Bucharest, Romania.

[Moisa 2000] Trandafir MOISA, Adrian Horia DEDIU, Dan ONTANU, *Conversational System Based on Evolutionary Agents*, CEC 2000, International Conference on Evolutionary Computation USA 2000

[Ontanu 1993] Onțanu, D.-M. *"Learning by Evolution. A New Class of General Classifier Neural Networks and Their Training Algorithm"*, Advances in Modelling & Analysis, AMSE Press, Vol. 26, nr. 2/1993, pp. 27-30

[Ontanu 1998] Dan Ontanu, Voice Synthesis, referate for the PhD. thesis *"Neural Models for Specch Recognition and Synthesis"*, under the direction of prof. dr. Trandafir Moisa, author, "Politehnica" University, Bucharest, July 1998

[Reilly 1982] Reilly, Cooper, Elbaum, *"A Neural Model for Category Learning"*, Biological Cybernetics, 45, p. 35-41, 1982

[Winston 1992] Patrick Henry Winston, *"Artificial Intelligence"* third ed., Addison-Wesley, June 1992

[Weigend 1994] A.S. Weigend, N.A. Gershenfeld (Eds.), *Time Series Prediction: Forecasting the Future and Understanding the Past*, Addison Wesley, 1994.

[Yao 1999] X. Yao: *Evolving artificial neural networks*, Proceedings of IEEE 87(9), pp. 1423-1447, September 1999.

A Two-Phase Fuzzy Mining and Learning Algorithm for Adaptive Learning Environment

Chang-Jiun Tsai   S. S. Tseng*   Chih-Yang Lin

Department of Computer and Information Science
National Chiao Tung University
Hsinchu 300, Taiwan, R.O.C.
E-mail: sstseng@cis.nctu.edu.tw

Abstract. As computer-assisted instruction environment becomes more popular over the world, the analysis of historical learning records of students becomes more important. In this work, we propose a Two-Phase Fuzzy Mining and Learning Algorithm, integrating data mining algorithm, fuzzy set theory, and look ahead mechanism, to find the embedded information, which can be provided to teachers for further analyzing, refining or reorganizing the teaching materials and tests, from historical learning records.

Keyword: Fuzzy set theory, Data Mining, Machine Learning, CAI

1. Introduction

As Internet becomes more popular over the world, the amounts of teaching materials on WWW are increasing rapidly and many students learn knowledge through WWW. Therefore, how to design and construct computer-assisted instruction environment together with its teaching materials is of much concern. In recent years, an adaptive learning environment [13], [14], [15] has been proposed to offer different teaching materials for different students in accordance with their aptitudes and evaluation results. After students learn the teaching materials through the adaptive learning environment, the teachers can further analyze the historical learning records and refine or reorganize the teaching materials and tests.

In this work, we propose a Two-Phase Fuzzy Mining and Learning Algorithm to find the embedded information within the historical learning records for teachers to further analyze, reorganize and refine the learning path of teaching materials or tests. The first phase of proposed algorithm uses a fuzzy data mining algorithm, *L*ook Ahead *F*uzzy *M*ining Association Rule *Alg*orithm (*LFMAlg*), integrating association rule mining algorithm, Apriori [1], fuzzy set theory, and look ahead mechanism, to find the embedded association rules from the historical learning records of students. The output of this phase can be fed back to teachers for reorganizing the tests, and will be treated as the input of the second phase. The second phase uses an inductive learning algorithm, AQR algorithm, to find the concept descriptions indicating the missing concepts during students learning. The results of this phase can be provided to teachers for further analyzing, refining or reorganizing the learning path.

*Corresponding author

Some related works and our motivation are first described in Section 2. Section 3 presents the framework of analyzing the historical learning records. Section 4 shows the algorithms. Concluding remarks are given in Section 5.

2. Related Works and Motivation

The web-based educational systems are becoming more and more popular over the world. Several approaches, which can be used to organize the teaching materials, have been developed in the past ten years [2], [3], [7], [10], [12]. As to the evaluation, [7] provides the evaluation mechanism to find out what instructional objectives in some sections the students do not learn well. However, the students always need to learn all teaching materials in each section for the first time no matter how the teaching materials are suitable for them or not. Therefore, an adaptive learning environment was proposed in [13] and [14], which can offer different teaching materials for different students in accordance with their aptitudes and evaluation results. The idea is to segment the teaching materials into teaching objects that is called Object-Oriented Course Framework as shown in Figure 1 The teaching materials can be constructed by organizing these teaching objects according to learning path, which can be defined and provided by teachers or teaching materials editors for students learning.

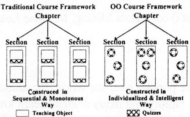

Figure 1. The comparison between traditional and OO course framework

The architecture of adaptive learning environment is shown in Fig. 2. All teaching objects are stored in Teaching Object Resource Database. When teachers want to construct the teaching materials about some subjects, they can retrieve the teaching objects from Teaching Object Resource Database, and define the learning path about these teaching objects.

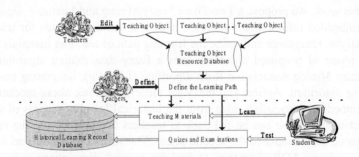

Figure 2. The architecture of adaptive learning environment

As we know, the learning path is usually defined as tree structure containing some teaching materials, and each teaching object of teaching materials may contain the teaching content and quizzes or tests to discriminate the students' learning performance. Therefore, students can learn these teaching materials by following the learning path. For the example of learning path shown in Fig. 3, there is a specific subject of teaching material including four teaching objects, A, B, C, and D. That means A is the pre-requisite knowledge of B and C. In other words, students should first learn teaching object A, learn teaching objects B and C, and finally learn teaching object D.

Figure 3. An example of learning path

All learning records, which are stored into Historical Learning Record Database, would be used in the adaptive learning environment. In other words, the system would reconstruct the course framework according to these learning records for students learning again, if students cannot learn well about the teaching material [14]. Besides, the learning records also can be used to refine the learning path by teachers. Because each quiz or test may consist of more than one concept, some embedded information about the relationships among the high grades of quizzes and low grades of quizzes can be used to determine whether the previously defined learning path is reasonable or not. Therefore, we propose a Two-Phase Fuzzy Mining and Learning Algorithm to find the embedded information about relationships among the learning records.

3. The Flow of Two-Phase Fuzzy Mining and Learning Algorithm

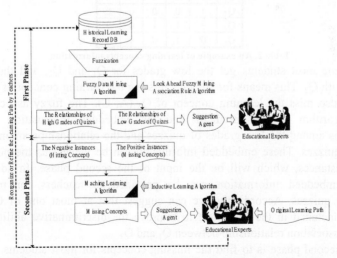

Figure 4. The flow of Two-Phase Fuzzy Mining and Learning Algorithm

Fig. 4 shows the flow of our Two-Phase Fuzzy Mining and Learning Algorithm, which can provide teachers the some embedded information for further analyzing, refining and reorganizing the learning path and the tests.

Two-Phase Fuzzy mining and Learning Algorithm

Input: The learning records of students from Historical Learning Record Database.

Output: The information of missing concepts.

Phase1: Use Look Ahead Fuzzy Mining Association Rule Algorithm to find the fuzzy association rules of quizzes from the historical learning records.

Phase2: Use Inductive Learning Algorithm to find the missing concept during students learning.

The first phase, data mining phase, applies fuzzy association rule mining algorithm to find the associated relationship information, which is embedded in learning records of students. In Table1, there are ten learning records of students, and each record has the grades of 5 quizzes, where the highest grade of each quiz is 20.

Student ID	Q_1	Q_2	Q_3	Q_4	Q_5	Total
1	12	14	18	3	9	56/100
2	10	6	12	6	7	41/100
3	3	6	6	1	5	21/100
4	8	10	8	2	8	36/100
5	16	18	20	20	20	94/100
6	0	3	3	1	4	11/100
7	1	8	6	4	10	29/100
8	2	3	3	0	3	11/100
9	12	16	14	4	14	60/100
10	6	8	12	2	10	38/100

Table 1. An example of students' learning record

Assume each quiz may contain one or more learning concepts. As shown in Table 2, related learning concepts A, B, C, D, and E may be contained in these five quizzes, where "1" indicates the quiz contains this concept, and "0" indicates not.

	A	B	C	D	E
Q_1	1	0	0	1	0
Q_2	1	0	1	0	0
Q_3	1	0	0	0	0
Q_4	0	1	1	0	0
Q_5	0	1	0	0	1

Table 2. An example of learning concepts information

Assume most students get the low grades for Q_1 and Q_2, and thus Q_1 may associate with Q_2. This means for students missing the learning concept of Q_1 (A, D), they may also miss the learning concept of Q_2 (A, C). The fuzzy association rule mining algorithm is then used to find the embedded information about the relationships among the low grades of quizzes and the relationships among the high grades of quizzes. These embedded information can be viewed as the positive and negative instances, which will be the input of the second phase. In addition, the obtained embedded information can be provided to teachers for making up appropriate quizzes. Accordingly, we can suggest that at most one of Q_1 and Q_2 should be included in the same test to improve the discriminative ability, if there exists the association relationship between Q_1 and Q_2.

The second phase is to find the missing concepts for most students. We apply inductive learning strategy, AQR learning algorithm [8] which is suitable for symbolic learning, to include all positive instances and exclude all negative instances. Then the teachers can refine learning path according to these information and original learning path. As mentioned above, in Table 2, the Q_1 (A, D) and Q_2 (A, C) belong to the set of low grades of quizzes. Thus the set of missing concepts, (A, C, D), can be regarded as a positive instance for AQR learning algorithm. On the contrary, the set of hitting concepts can be regarded as the negative instance. By using AQR algorithm for these training instances, some rules which can include all positive instances and exclude all negative instances can be learned.

4. Algorithm

In this section, *L*ook Ahead *F*uzzy *M*ining Association Rule *Alg*orithm (*LFMAlg*) used in the first phase and AQR algorithm used in the second phase will be described.

4.1 Fuzzy Data Mining Algorithm

IBM Almaden Research Center proposed the association rule mining algorithm, Apriori algorithm [1], to find the embedded information within a large number of transactions. The famous example of supermarket shows trend of buying behavior of customers. Unfortunately, the Apriori algorithm only works in ideal domains where all data are symbolic and no fuzzy data are present. However, real-world applications sometimes contain some numeric information, which need to be transformed into symbolic. For example, if the grade of quiz is 78, different experts may have different opinions of the concept "high grade". Thus, how to transform the linguistic data into the symbolic data and how to let Apriori algorithm be able to handle numeric information of these linguistic data are of most importance. Our idea is to improve Apriori algorithm by applying fuzzy set theory to overcome the problem of existing fuzzy regions among the data.

A fuzzy set is an extension of a crisp set, which allows only full membership or no membership at all, whereas fuzzy set allows partial membership. To apply fuzzy concepts, the membership function of each quiz's grade, which can transform numeric data into fuzzy set, is defined in Fig. 5.

As we know, in the association rule mining algorithm, the support of association rule is larger than a *minimum support threshold*, defines as α_ℓ, and the confidence of association rule is larger than a *minimum confidence threshold*, defined as λ, in ℓ-large itemset. In Apriori algorithm, α_ℓ is a constant, and then the number of association rules may decrease when ℓ increases in ℓ-large itemset. That may cause losing of some association rules in larger itemsets. However the minimum support threshold may be too small to exhibit the meaning of association rule. Hence, we apply Look Ahead mechanism to generate Next Pass Large Sequence [6] to overcome the above problem.

Figure 5. The given membership function of each quiz's grade

As mentioned above, we proposed the Look Ahead Fuzzy Mining Association Rule Algorithm, integrating Apriori algorithm, Look Ahead mechanism, and Fuzzy Set Theory, to find the association rules within fuzzy data set as below.

Definition

$f_{ij}(k)$: indicates the kth student's fuzzy values for the ith quiz and the jth degree of fuzzy function.

F_{ij} : indicates the sum of all students' fuzzy value for the ith quiz and the jth degree of fuzzy function, i.e., $F_{ij} = \sum_{k=1}^{n} f_{ij}(k)$, where n indicates the number of students.

C_{ℓ} : indicates ℓ-candidate itemset.

$NPLS_{\ell}$: indicates ℓ-Next Pass Large Sequence ($NPLS$) [6].

L_{ℓ} : indicates ℓ-large itemset.

α_{ℓ} : indicates the minimum support threshold of ℓ-large itemset.

$support(x)$: indicates the support values of the element x for $x \in C_{\ell}$.

Look Ahead Fuzzy Mining Association Rule Algorithm

Input: The learning records of students from Historical Learning Record Database.
The minimum support threshold α_1 in the 1-large itemset, L_1.
The minimum confidence threshold λ.

Output: The fuzzy association rules of learning records of students.

STEP1: Transform the grades of each quiz into fuzzy value, $f_{ij}(k)$, for all students according to the fuzzy membership function.

STEP2: $C_1 = \{ F_{ij} / F_{ij} = \sum_{k=1}^{n} f_{ij}(k) \}, \quad and \quad \ell = 1$ (1)

STEP3: $L_{\ell} = \{ x / support(x) \geq \alpha_{\ell}, \quad for \quad x \in C_{\ell} \}$ (2)

STEP4: $\alpha_{\ell+1} = max(\dfrac{\alpha_1}{2} \quad , \quad \alpha_{\ell} - \dfrac{\alpha_1}{\ell * c}), \quad where\ c\ is\ a\ constant.$ (3)

STEP5: $NPLS_{\ell} = \{ x / support(x) \geq \alpha_{\ell+1}, \quad for \quad x \in C_{\ell} \}$ (4)

STEP6: If $NPLS_{\ell}$ is null,
         then stop the mining process and go to **STEP8**,
         else generate the ($\ell+1$)-candidate set, $C_{\ell+1}$, from $NPLS_{\ell}$.

STEP7: $\ell = \ell + 1$ and go to **STEP3**.

STEP8: Determine the association rules according to the given λ and all large itemsets.

Example 4.1

According to the given membership function defined in Fig. 5, Table 3 shows the fuzzication result of the data of Table 1. The "Low" denotes "Low grade ", "Mid" denotes "Middle grade", and "High" denotes "High grade". $Q_i.L$ denotes the low degree of fuzzy function of the ith quiz, $Q_i.M$ denotes the middle degree of fuzzy function of the ith quiz, and $Q_i.H$ denotes the high degree of fuzzy function of the ith quiz.

Student ID	Q_1			Q_2			Q_3			Q_4			Q_5		
	Low	Mid	High	Low	Mid	High	Low	Mid	High	Low	Mid	High	Low	Mid	High
1	0.0	0.0	0.7	0.0	0.0	1.0	0.0	0.0	1.0	0.8	0.0	0.0	0.0	0.8	0.2
2	0.0	0.5	0.3	0.3	0.5	0.0	0.0	0.0	0.7	0.3	0.5	0.0	0.2	0.8	0.0
3	0.8	0.0	0.0	0.3	0.5	0.0	0.3	0.5	0.0	1.0	0.0	0.0	0.5	0.3	0.0
4	0.0	1.0	0.0	0.0	0.5	0.3	0.0	1.0	0.0	1.0	0.0	0.0	0.0	1.0	0.0
5	0.0	0.0	1.0	0.0	0.0	1.0	0.0	0.0	1.0	0.0	0.0	1.0	0.0	0.0	1.0
6	1.0	0.0	0.0	0.8	0.0	0.0	0.8	0.0	0.0	1.0	0.0	0.0	0.7	0.0	0.0
7	1.0	0.0	0.0	0.0	1.0	0.0	0.3	0.5	0.0	0.7	0.0	0.0	0.0	0.5	0.3
8	1.0	0.0	0.0	0.8	0.0	0.0	0.8	0.0	0.0	1.0	0.0	0.0	0.8	0.0	0.0
9	0.0	0.0	0.7	0.0	0.0	1.0	0.0	0.0	1.0	0.7	0.0	0.0	0.0	0.0	1.0
10	0.3	0.5	0.0	0.0	1.0	0.0	0.0	0.0	0.7	1.0	0.0	0.0	0.0	0.5	0.3

Table 3. The fuzzication results of the data of Table 1

According to the fuzzy data shown in Table 3, Fig. 6 shows the process of finding the association rules for the original data shown in Table 1. In this case, assume $\alpha_1 = 2.1$ and $\lambda = 0.75$. Table 4 shows the process of calculating the confidence of 2-large itemset. For example, the confidence value of the itemset (Q_2.L, Q_1.L) is 0.86, means that many students get low grades of Q_1 and Q_2 simultaneously. The embedded information shows that Q_1 may have similar property or may contain similar learning concept with Q_2. Therefore, we may suggest that Q_1 and Q_2 do not appear at the same test.

Itemset	Confidence	Itemset	Confidence
(Q_1.L, Q_2.L)	0.46	(Q_1.H, Q_2.H)	0.89
(Q_2.L, Q_1.L)	0.86	(Q_2.H, Q_1.H)	0.72
(Q_1.L, Q_3.L)	0.54	(Q_1.H, Q_3.H)	1
(Q_3.L, Q_1.L)	1	(Q_3.H, Q_1.H)	0.61
(Q_1.L, Q_5.L)	0.49	(Q_1.H, Q_5.H)	0.7
(Q_5.L, Q_1.L)	0.91	(Q_5.H, Q_1.H)	0.68
(Q_2.L, Q_3.L)	0.86	(Q_2.H, Q_3.H)	0.9
(Q_3.L, Q_2.L)	0.86	(Q_3.H, Q_2.H)	0.68
(Q_2.L, Q_5.L)	0.9	(Q_2.H, Q_5.H)	0.67
(Q_5.L, Q_2.L)	0.9	(Q_5.H, Q_2.H)	0.79

Table 4. The process of calculating the confidence of 2-large itemset

$$\alpha_1(2.1), C_1 \rightarrow L_1$$
$$\rightarrow \alpha_2(1.89) \rightarrow NPLS_1 \rightarrow C_2 \rightarrow L_2$$
$$\rightarrow \alpha_3(1.79) \rightarrow NPLS_2 \rightarrow C_3 \rightarrow L_3$$
$$\rightarrow \alpha_4(1.72) \rightarrow NPLS_4 \rightarrow C_4 \rightarrow L_4$$
$$\rightarrow \alpha_5(1.66) \rightarrow NPLS_5(Null)$$

Figure 6. The process of mining association rules

In the process of data mining, only any two items both with high grades or both with low grades would be considered as a 2-large itemset, because if one with high grades and the other with low grades, the found embedded information is meaningless for teachers. However, teachers cannot explain why students do not learn well for some specific quizzes according to the output of first phase. In other words, we cannot intuitively induce what learning concepts the students miss in accordance with the information about association rules of the quizzes, although information about what concepts are contained in quizzes is known. Therefore, we apply inductive machine learning strategy, AQR algorithm [8], to find the missing learning concepts.

4.2 Inductive Learning Algorithm

AQR is one kind of the batch and inductive learning algorithm [8] that uses the basic AQ algorithm [11] to generate the concept description, which can include all positive instances and exclude all negative instances. When learning process is running, AQR performs a heuristic search through the hypothesis space to determine the description. The training instances, including the set of positive instances and the set of negative instances, are learned in stages; each stage generates a single concept description, and removes the instances it covers from the training set. The step is repeated until enough concept descriptions have been found to cover all positive instances.

In the first phase of the Two-Phase Fuzzy Mining and Learning Algorithm, LFMAlg is used to find the association rules embedded in the historical learning records of students. The large itemsets except 1-large itemset can be considered as the training instances for AQR algorithm. Because the large itemsets of low grades, the missing concepts, can be considered as positive instances of the training instances, and because the large itemsets of high grades can be considered as negative instances, the concept descriptions which *include/exclude* all *positive/negative* instances can show the more precise missing concepts of students. The association rule of the first phase's output consists of the left-hand-side (LHS) and right-hand-side (RHS). The LHS may consist of one or more items. To transform the LHS into the training instance, the union operator for the concepts contained by the items of LHS is used. For example, the item $(Q_1.L, Q_3.L, Q_4.L)$ is one of 3-large itemset, and the confidence λ is larger than the defined minimum confidence threshold. Then the itemset can be transformed into a positive instance by run the union operator for Q_1's and Q_3's learning concepts shown in Table 2, i.e., (10010, 10000) = (10010). To combine the learning concept of RHS of this item, Q_4, the positive instance can be expressed as (10010, 01100).

Before using AQR algorithm to induce the concept descriptions, which cover missing concepts, the pre-process of transforming the itemsets found by the first phase into the training instances should be done as mentioned above.

The concepts derived from AQR are represented as the multiple-valued logic calculus with typed variables, which can be represented as follows.

> If *<cover>* then predict *<class>*
>
> , where   *<cover>* = *<complex 1>* or ⋯ or *<complex m>*,
>
>   *<complex>* = *<selector 1>* and ⋯ and *<selector n>*,
>
>   *<selector>* = *<attributes r values>*,
>
>   *<r>* = *relation operator.*

A *selector* relates a variable to a value or a disjunction of values. A conjunction of *selectors* forms a *complex*. A *cover* is a disjunction of *complexes* describing all positive instances and none of the negative ones of the concept. The AQR algorithm is described as below.

AQR Algorithm [8]

Input: The set of positive instances and the set of negative instances.
Output: The information of missing concepts.

SETP1: Let *POS* be a set of positive instances and let *NEG* be a set of negative instances.

SETP2: Let *COVER* be the empty cover.

SETP3: While *COVER* does not cover all instances in *POS*, process the following steps. Otherwise, stop the procedure and return *COVER*.

SETP4: Select a *SEED*, i.e., a positive instance not covered by *COVER*.

SETP5: Call procedure *GENSTAR* to generate a set *STAR*, which is a set of complex that covers *SEED* but that covers no instances in *NEG*.

SETP6: Let *BEST* be the best complex in *STAR* according to the user-defined criteria.

SETP7: Add *BEST* as an extra disjunction of *COVER*.

GENSTAR procedure

SETP1: Let *STAR* be the set containing the empty complex.

SETP2: While any complex in STAR covers some negative instances in *NEG*, process the following steps. Otherwise, stop the procedure and return *STAR*.

SETP3: Select a negative instance E_{neg} covered by a complex in *STAR*

SETP4: Specialize complexes in STAR to exclude E_{neg} by:

Let *EXTENSION* be all selectors that cover *SEED*, but not E_{neg}.

Let *STAR* be the set $\{x \cap y \mid x \in STAR, y \in EXTENSION\}$

SETP5: Repeat this step until sizes of *STAR* $\leq$ max-star (a user-defined maximum). Remove the worst complex from *STAR*.

Example 4.2

Table 5 shows the training instances of transforming the 2-large itemsets, in which support value are larger than the defined minimum confidence threshold.

Itemset	Positive (+)/ Negative (-)	Instance	Itemset	Positive (+)/ Negative (-)	Instance
$(Q_2.L, Q_1.L)$	+	(10100, 10010)	$(Q_2.L, Q_5.L)$	+	(10100, 01001)
$(Q_3.L, Q_1.L)$	+	(10000, 10010)	$(Q_5.L, Q_2.L)$	+	(01001, 10100)
$(Q_5.L, Q_1.L)$	+	(01001, 10010)	$(Q_1.H, Q_2.H)$	-	(10010, 10100)
$(Q_2.L, Q_3.L)$	+	(10100, 10000)	$(Q_1.H, Q_3.H)$	-	(10010, 10000)
$(Q_3.L, Q_2.L)$	+	(10000, 10100)	$(Q_5.H, Q_2.H)$	-	(01001, 10100)

Table 5. An example of training instances

The three pairs, (11101, 10010), (00100, 10000), and (01001, 10100), which are the learning concepts students do not learn well, are found by the learning process of AQR algorithm. For example, (00100, 10000) means that the learning concept, (C), have high degree of relationship with (A), even if (C) may introduce (A). The teachers may analyze learning results of AQR algorithm as shown in Fig. 7.

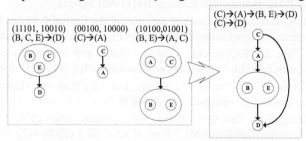

Figure 7. The analysis of learning results of AQR algorithm

5. Concluding Remarks

In this work, we proposed the Two-Phase Fuzzy Mining and Learning Algorithm. In the first phase, LMFAlg was proposed to find the embedded association rules from the historical learning records of students. In the second phase, the AQR algorithm was applied to find the concept descriptions indicating the missing concepts during students learning. The obtained results can be fed back to the teachers for analyzing, refining or reorganizing learning path of teaching materials and the tests. Now, we are trying to apply our algorithm to the virtual mathematics curriculum of senior high school in Taiwan.

Acknowledgement

This work was partially supported by the National Science Council of the Republic of China under Grant No. NSC 90-2511-S009-023.

Reference

1. Agrawal, R., Srikant, R.: Fast Algorithms for Mining Association Rules. Proc. Of the 20th Int'l Conference on Very Large Database. (1994)

2. Alessi, S.-M., Trollip, S.-R.: Computer-based Instruction: Methods and Development. 2nd. Englewood Cliffs, NJ: Prentice-Hall (1991)

3. Antal, P.: Animated Explanations Using Adaptive Student Models. Proc. of INES'97. (1997) 573-576

4. Au, W.-H., Chan, K.-C.-C.,: FARM: A Data Mining System for Discovering Fuzzy Association Rules. Proc. of FUZZ-IEEE'99. 3 (1999) 1217-1222

5. Chen, M.-S., Han, J., Yu, P.-S.: Data Mining: An Overview from A Database Perspective. IEEE Transaction on Knowledge and Data Engineering. 8(6) (1996) 866-883

6. Chen, S.-S., Hsu, P.-Y., Chen, Y.-L.: Mining Web Traversal Rules with Sequences. MIS Review 1(1999) 53-71

7. Chou, C.: A Computer Logging Method for Collecting Use-reported Inputs During Formative Evaluation of Computer Network-assisted Distance Learning. Proc. of ED-Media'96. (1996)

8. Clark, P., Niblett, T.: The CN2 Induction Algorithm. Machine Learning. 3 (1989) 261-283

9. Hong, T.-P., Kuo, C.-S., Chi, S.-C.: A Fuzzy Data Mining Algorithm for Quantitative Values. Proc. of Knowledge-based Intelligent Information Engineering Systems. (1999) 480-483

10. Hwang, G.-J.: A Tutoring Strategy Supporting System for Distance Learning on Computer Networks. IEEE Transactions on Education. 1(4) (1998) 343-343

11. Michalski, R.-S.: On The Quasi-minimal Solution of The General Covering Problem. Proc. of the 5th International Symposium on Information Processing. (1969) 125-128

12. Sun, C.-T., Chou, C.: Experiencing CORAL: Design and Implementation of Distance Cooperative Learning. IEEE Transactions on Education. 39(3) (1996)357-366

13. Su, G.-H., Tseng, S. S., Tsai, C.-J., Zheng, J.-R.: Building An Object-oriented and Individualized Learning Environment on the WWW. Proc. of ICCE'99. (1999) 728-735

14. Su, G.-H., Tseng, S. S., Tsai, C.-J., Zheng, J.-R.: Implementation of An Object-Oriented Learning Environment Based on XML," Proc. of ICCE'00. 2 (2000) 1120-1127

15. Tsai, C.-J., Tseng, S. S., Su, G.-H.: Design of An Object-oriented and Rule-based Virtual School. Proc. of GCCCE'00. (2000) 320-327

Applying Genetic Algorithms and Other Heuristic Methods to Handle PC Configuration Problems

Vincent Tam and K.T. Ma

{vtam,makengte}@comp.nus.edu.sg
Department of Computer Science
School of Computing, National University of Singapore
Lower Kent Ridge Road, Singapore 119260.

Abstract. Singapore is developing very fast as an Information Technology (IT) hub in which many people are keen to configure and build their own personal computers (PC). Like many real-life configuration problems, a well-designed PC configuration often represents a challenge in which given the wide diversity of hardware components, the ever-changing PC technology and the limited compatibility between some hardware components, we are interested to obtain an (sub-)optimal configuration for each specific usage restricted to a budget limit and other preferred criteria. In this paper, we formally defined these PC configuration problems as discrete optimization problems. Then we proposed a systematic and flexible framework in which we can integrate any heuristic search method for solving these difficult real-world discrete optimization problems. A possible advantage of our proposed framework is that users can flexibly add in or modify their specific requirements at any time. To demonstrate the feasibility of our proposal, we built a prototype of an intelligent Personal Computer Configuration Advisor available on the Web to assist the general users in configuring their own PCs. Interestingly, our work opens up many new directions for future investigation including the improvement of our optimizer to handle more complicated users' requirements, and the possible uses of efficient learning algorithms such as the ID3 algorithm [2] to classify different user-defined configurations into useful examples to guide the search during optimization.

Keywords : Personal Computer Configuration Problems, Genetic Algorithms, Intelligent Web Applications.

I. INTRODUCTION

Similar to many well-developed Asian countries like Hong Kong or Singapore, many people, due to the influence of the popular "Do-It-Yourself" (DIY) philosophy from western society, are keen in configuring and building their own personal computers (PC) to suit their own requirements, often as the first step to learn to use software or access the Internet. Clearly, the PC configuration problems are widely occurring decision problems faced by many people in which people are always interested to know the optimal, or possibly sub-optimal, PC configuration within their limited budget. However, since the PC technology nowadays are changing very quickly, the diversity of PC hardware components, such as the different types of processors, random-access memory (RAM) and the motherboards, and their limited compatibility between certain components due to the underlying proprietary manufacturers often complicates the whole decision problem, thus making it difficult to handle by the general public.

In fact, many configuration problems [2] as in the area of computer-aided design or manufacturing (CAD/CAM) [6] are well studied. In particular, there were some interesting research work [6] on formulating the machine configuration problems formally as discrete optimization problems, and then applying local search methods such as genetic algorithms [11,12] or simulated annealing [9,10] to handle these configuration problems successfully. The previous research experience reported in [6] already revealed that formulating the machine, or any general, configuration problems as constrained optimization problems (COPs) will definitely result in systematic handling of users' requirements as constraints or optimization criteria. At the same time, the users can flexibly add in or modify their requirements at any time to see the resulting configurations for planning or control [5]. Thus, in this paper, we firstly gave a formal definition of these PC configuration problems as discrete optimization problems. More importantly, based on this formal definition, we proposed a systematic and flexible framework for solving these difficult real-world discrete optimization problems.

Basically, a discrete optimization problem involves a set Z of variables, each variable V_i ($\in Z$) with a finite domain D_i of discrete values, a set C of constraints on some subsets of variables limiting the combination of values assigned to those involved variables, and a set of objective functions f_j for minimization/maximization. The challenge is to find a globally optimal solution which minimizes/maximizes all the objective functions f_j while satisfying all the constraints in C. Mathematically, the discrete minimization problem can be specified as follow.

$$\min \sum_{j=1}^{m} f_j \quad subject\ to \quad \forall c \in C \quad cf(c) = 0 \ \dots\dots\dots\dots\dots(1)$$

where m denotes the total number of objective functions in the problem, and $cf(c)$ *is an arbitrary function* which returns 0 when the specific constraint c is satisfied. Otherwise, it returns 1. Clearly, to solve maximization problems, we can simply negate all objective function f_j in the above formulation as $\min \sum_{j=1}^{m} - f_j$. In handling these optimization problems, which are always *NP-complete* [3,4], one of the frequently used heuristics is the branch-and-bound (B&B) heuristic [1,2] in which the exploration of any partial solution in a search tree will be abandoned immediately whenever the search cost of that partial solution, as represented by an arbitrary objective function, already exceeds the minimal cost for the optimal solution found so far. For instance, the B&B heuristic has been successfully applied to handle the famous traveling salesman problems [2,3] to guarantee the finding of the shortest path to transverse all the required cities in one round trip. However, in the worst cases, the B&B search method may still require exponential time to find out the optimal solution for any general COP. Besides, in some cases [10] even with the aid of useful heuristics, finding a feasible solution satisfying all the constraints in C is still very difficult. Thus, the users may relax some constraints in C, and ready to accept some "partial solutions" [5] representing the optimal or sub-optimal to the objective functions. On the other hands, there are some real-life COPs which are sparsely constrained, thus having too many feasible solutions. In those cases, it will then become difficult to guarantee global optimality of the resulting solution within a reasonable period of time. Thus, the users may also be willing to accept a sub-optimal solution as a trade-off for efficiency. In general, many real-life PC configuration problems belong to this latter case of discrete COPs in which given the diversity of possible PC configurations as feasible solutions, the users will often accept a sub-optimal solution when the resulting configuration can be returned quickly from the optimizer.

Therefore, based on the discrete COP formulation, we proposed a flexible and systematic framework to handle the PC configuration problems in which most of the useful information about the PC hardware components is often stored in some database files locally in different computer companies. Usually, these companies will also make use of the information already stored in those database files to advertise the products on their company Web pages. Accordingly, inside our proposed framework, we consider the optimization of options for PC configurations based on information provided from a centralized database system. It should be noted that the assumption of a centralized database system is generally valid since nowadays, most of the database systems we used in companies are often compatible with the de facto Open DataBase Connectivity (ODBC) standard [7]. In general, the heterogeneous database systems linked up via the Internet and the ODBC interface layer can be regarded a centralized database system when we ignore the latency time mainly due to the network communication and the possibly inconsistent data stored in these different database systems. Nevertheless, based on the presumably consistent data about PC components stored in a centralized database system, there are two major approaches we used to efficiently optimize the ultimate PC configurations to satisfy the users' requirements for handling these specific instances of real-life COPs. First, similar to the branch-and-bound method [1,2], we prune off any alternative choice which already exceeds the planned budget during each search step. In addition, we include a constant threshold value n to monitor the size of the possible PC configurations we will consider in each search step so as to avoid the combinatorial explosion problem frequently occurring in solving these PC configuration problems given the diversity of some particular PC components. Clearly, since we are not performing an exhaustive search, the approaches we used can be efficient but may not be able to guarantee the global optimality of the resulting configurations.

To demonstrate the feasibility of our proposed systematic framework, we firstly collected the actual data from some major computer shopping centers in Singapore to build our own centralized and local databases of PC components for investigation, and then used the widely available Practical Extraction and Report Language (PERL) [7, 8] Version 5.0 to implement the afore-mentioned preliminary search strategies to handle these practical PC configuration problems. The empirical experience of using our proposed approach to handle the PC configuration problems is fairly encouraging. Furthermore, the resulting optimizer is used to build a platform-independent prototype of the useful Web-based Personal Computer Configuration Advisor so as to facilitate the general users to quickly set up their required PC configurations. Obviously, our work opens up many new directions for future investigation such as improving our current optimizer to handle more complicated users' requirements, and the possible uses of efficient learning algorithms such as the ID3 algorithm [1,2] to classify different user-defined configurations into useful examples to guide the search during optimization.

This paper is organised as follows. Section 2 describes the PC configuration problems as specific instances of discrete COPs, thus forming a systematic framework for optimizing choices of PC components to satisfy the users' requirement. In Section 3, we detail and justify our proposed search strategies to handle these specific COPs according to their unique problem structures. Section 4 provides the empirical evaluation of our cross-platform prototype implementation of the proposed optimizer in terms of efficiency and costs of the resulting configurations, with an example application of our optimizer to build a Web-based PC Configuration Advisor. Lastly, we conclude our work in Section 5.

2. PC Configuration Problems

Basically, the PC configuration problem is to select a configuration of personal computer hardware parts to build a complete system, taking into account the *compatibility* issues between the different hardware components. For instance, Intel Pentium II CPU should be attached to a Slot-1 Motherboard. Definitely, one of the most important optimization criteria in many real-life situation is *price*. Thus, a general formulation of the PC configuration problems as discrete COPs can be as follows.

$$\min \sum_{j=1}^{m} \mathrm{cost}(Pj) \qquad subject\ to \qquad \forall comp(Pi, Pj) \in C \quad cf(comp(Pi, Pj)) = 0 \ldots\ldots\ldots(2)$$

In (2), cost(Pj) denotes the cost for the component Pj, C specifies all the compatible relations (constraints) between the components Pi and Pj, and the arbitrary function cf returns 0 when comp(Pi, Pj) is evaluated as true. For example, when the variables for the CPU and motherboards are : P_{CPU} = "Intel PII CPU" and P_{MB} = "Slot-1 Motherboard"respectively, comp(P_{CPU}, P_{MB}) = true, then cf(comp(P_{CPU}, P_{MB})) = 0. Clearly, given the above general COP formulation, it is flexible to add in or modify the users' requirements specified as constraints or optimization criteria. For instance, when new components P_X and P_Y are added into the PC market, it is easy to simply add a new constraint comp(P_X , P_Y) into C to store their compatibility information. [1] Besides, for more complicated real-life cases, we can simply extend the minimization function to consider other important factors such as a weighting value for each component to reflect the users' preference.

Variables

The variables $Pi's$ of interest for the PC configuration problem is the set of hardware components which will be assembled to build a working personal computer system. Below is the table of the variable names which we used to formulate a typical PC configuration problem and their short descriptions for clarity.

Variables	Short Descriptions
CPU	Central Processing Unit. The core of a computer. For simplicity, only uni-processor computers are considered in this problem. Needs to be compatible with Memory and Mainboard.
Memory	The hardware where data is temporary stored. Need to be compatible with CPU and MainBoard.
MainBoard	The PVC-type board which all the various hardware components are connected together. Most of the constraints are defined on this variable.
Casing	The covering of the computer. No constraint.
DisplayCard	The hardware component which handles the processing of the graphical display. No constraint.
HardDisk	Mass storage device. No constraint.
CD_ROM	Read-only input media. DVD and CD Writer/Rewriter do not belong to this variable.
Modem	The interface to telephone whereby the computer can be connected to other computers over the PBX-type modem. No constraint.
Monitor	The display unit. No constraint.
Printer	The printing device. No constraint.
Scanner	The scanning input device. No constraint.
SoundCard	The device which handles the processing of audio input/output. No constraint.

Table 1. **Typical Variables involved in PC Configuration Problems**

Clearly in the above table, there are about 12 common variables usually involved in configuring a PC nowadays. The first three variables, namely CPU, Memory and MainBoard, of the table represent the most important variables which are often highly constrained. The remaining nine variables are totally un-constrained. Thus, this special relationship between the PC components depicts the unique problem structure of the PC configuration problem : the constraints between the first 3 variables will leave very few combinations of options to be matched against a possibly large number of possible choices in a later stage of the search. It should be noted that this large possibility occurred in the later search stage may pose major difficulty to most optimizers such the branch-and-bound techniques since no other available information can be used to prune off any value during the search.

---

[1] Clearly, some readers may argue that these compatibility constraints are in fact no different from the logical relations like "father(john, mary)" to specify "john" is the father of "mary" in some logic programs. Of course, we would agree on that since it is generally true for most constraints. However, the "encapsulation" of these general relations between objects as a specific constraint will in general provide *a more systematic* way to handle that particular kind of constraints. As a result, we will discuss how a special-purpose search algorithm (as constraint solver) can be designed to handle those compatibility constraints *more efficiently* for our PC configuration problems in the next section.

Domain

Table 2 shows the actual domain size and the range of integer values stored for each variable.

Variables	Size	Range	Variables	Size	Range
CPU	40	[0..39]	CD_ROM	26	[0..25]
Memory	46	[0..45]	Modem	32	[0..31]
MainBoard	69	[0..68]	Monitor	105	[0..104]
Casing	14	[0..13]	Printer	34	[0..33]
DisplayCard	110	[0..109]	Scanner	42	[0..41]
HardDisk	82	[0..81]	SoundCard	25	[0..24]

Table 2. The Actual Domains for Variables in PC Configuration Problems

The average domain size for all the variables is 52. Thus, the size of the search space for all possible combinations of PC components is roughly of the order of 52^{12}. More importantly, this huge search space has lots of possible sub-trees which cannot be further pruned off by any heuristic. Accordingly, the conventional enumerative search [5] may simply give unsatisfactory performance to return the globally optimal PC configuration. Therefore, in the next section, we will discuss our proposed search strategy which sacrifices global optimality for efficiency to tackle these real-life COPs.

Furthermore, it should be noted that these PC, or possibly other, configuration problems are very much different from some famous routing problems such as the traveling salesman problem, on which the branch-and-bound (B&B) strategy performs reasonably well, in terms of the underlying domains for each variable. For a traveling salesman problem with *12* cities, the underlying domains for all the variables $V_1..V_{12}$, denoting the first to the last city to be visited on a trip, are the same. That is the range [1..12] to represent the **same** *12* cities. Thus, at each search step, we can use the B&B heuristic to prune off any alternative route starting and ending with the same cities, say "city 1 - ......- city 9", but with a higher total distance covered. However, in our PC configuration problems, the possible choices at each search step will be totally different from the previous choices already occurred in the previous search steps. Thus, the B&B heuristic may not necessarily be a useful heuristic in solving the PC configuration problems.

Constraints

The following list shows some example of the constraints frequently occurred in the PC configuration problems. The first one specifies that the total cost of the configuration must be less than or equal to the budget predefined by the user. The second and third constraints are obviously about the *CPU, Memory* and *MainBoard*. The second one states that the type of the *CPU* must match with the type of the *MainBoard* while the last one specifies that the speed of the *MainBoard* and the *Memory* must be the same.

- TotalPrice <= Budget
- CPU.Type = Mainboard.Type
- Mainboard.Bus_Speed = Memory.Bus_Speed

Clearly, it is straightforward to add in more constraints as users' new requirements. As in most PC configurations, the cost and performance are two common factors for consideration.

3. OUR PROPOSED FRAMEWORK & SEARCH STRATEGIES

In this section, we will firstly look into a systematic framework for integrating different optimizers to handle the PC configuration problems based on the formulation of discrete optimization problems which we discussed in the previous section. Figure 1 shows our proposed systematic framework for integration of different optimizers.

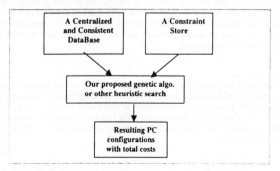

Fig. 1. Our Proposed Systematic Framework for Optimizing PC Configurations

It should be noted that in our systematic framework for constrained optimization to handle the PC configuration problems, we assume our search and optimization process have to performed on a set of consistent data from a centralized database system. In addition, there is a constraint store to provide useful information about the users' requirements as a collection of constraints. Based on these useful information, our optimizer with customizable search strategies will return a set of (sub-)optimal PC configurations with their total costs to the users according to the predefined optimization criteria. Clearly, other possible optimizers can also be added into our framework as some intelligent plug-in components for better efficiency or handling more complicated optimization criteria.

In the following, we will discuss our proposed optimizer with customizable search strategies which may not guarantee to always return the PC configuration with the globally minimal cost. However, as opposed to enumerative search, our proposal will definitely be more efficient. Table 3 shows the two main factors we used to control the search, together short descriptions to explain their roles. Both control parameters are predefined by the user of the search strategies.

Control	Short Descriptions
Budget	The budget for the whole personal computer system. Total prices of the components are not allowed to exceed the budget.
Threshold	Number of the best partial solutions to be considered in each search step

Table 3. Our Control Factors to Search for (Sub-)Optimal PC Configuration

Similar to the B&B heuristic, the predefined control parameter *Budget* will be used to filter out any partial configuration which already exceeds its allowed value. In addition, the *Threshold* value n will help to ensure the search will always return the best n partial configurations after sorting all possible partial configurations against their total prices at each search step. Based on these two strategies, we designed a beam-search based optimizer to solve these PC configuration problems as follows.

```
1. Retrieve and sort the values for the current variable by price.

2. Find the next matching value of the variable in the sorted value queue.
Check that all the constraints are satisfied for this value. This includes the
budget constraint (that is Total_price <= Budget).

3. If no constraint is violated, goto 6.

4. Else if queue is not empty, repeat 2. Else Fail.

5. Set the current solution set to include this value.

6. If number of partial solutions < threshold, repeat 2. Else 7

7. If variable queue is not empty, proceed to set current variable to next
variable in the queue. Repeat 1.

8. If number of solutions > 0 then report Solved. Else report Failure.
```

Our second proposal is a min-conflict heuristic (MCH) based micro-genetic algorithm (MGA) [6], that is an evolutionary algorithm with small population size of chromosomes, to handle the PC configuration problems. In general, when handling constraint satisfaction or optimization problems with evolutionary algorithms, a bit-string called chromosome will usually be used to represent a possible valuation for all the variables in the constrained problem. When the valuation represented by a particular chromosome violates no constraints in the problem, the chromosome denotes a solution. For solving constrained problems with a large number of variables (say > 500), a MGA with a small population size PZ in the range of 6-20 may often outperform the traditional evolutionary algorithms with larger population size [6]. This is probably because the computational cost can be minimized by focusing the search only on a reasonably small population of chromosomes without much impact on the search efficiency.

The min-conflict heuristic is to consider modifying only a single variable at a time, and to assign a value to that variable which is locally minimum in terms of constraint violations. When there are several local minima (ties) in terms of constraint violations, a value will randomly be selected among these ties. In our previous work [12], we proposed a mutation-based evolutionary search scheme (*MCH-MGA*) to efficiently solve constraint satisfaction problems (CSPs) [3] as follow.

```
MCH-MGA(CSP, fitness(), PZ, MAX_GENS) {

    initialize the 1st generation of PZ chromosomes randomly;

    set best_fit = best fitness value of the current population;
    set ngens = 1;

    while (best_fit != 0 && ngens < MAX_GENS) {
```

```
      uses selection scheme to choose the pivot_gene;

      applies MCH-mutate to pivot_gene;

      applies descent-mutate to other genes;

      if (population is in local minimum) then applies popu-learn;

      update best_fit;

      ngens++;

   }

   if(best_fit == 0) return solution;

   return failure;

}
```

Clearly, the *MCH-MGA* depends on the selection scheme, either *update-select* or *usage-select*, to pick up the most important variable/gene for the current search step to apply *MCH-mutate* in order to get the greatest improvement if possible. For the remaining variables, it generally applies the more efficient operator *descent-mutate*. The *MCH-mutate* is basically a MCH variable updating operator while the *descent-mutate* operator is a restrictive operator which allows only mutations resulting in decreases in the total number of constraint violations. The *popu-learn* operator is simply an adaptation of the heuristic learning mechanism discussed in [6] to a population of chromosomes rather than a single chromosome. For detail, refer to [6].

To handle the PC configuration problems, we have to firstly modify the original *fitness* function to combine the cost of constraint violations with actual *cost* of the different PC hardware components $P_j's$. Assume the average cost for each PC component P_j is about *1000*, we can adjust the *fitness* function to an arbitrary and relatively larger weight such as *10, 000* associated with the part for constraint violations as follows.

$$\min \sum_{j=1}^{m} (\text{cos} t(Pj) + 10,000 * \sum_{\forall comp(Pi,Pj) \in C} cf(comp(Pi,Pj)))................(3)$$

Thus, the ultimate effects produced by *(3)* on the evolutionary search will help to avoid any possible constraint violation. It should be noted that with this new formulation, it is straightforward to add in the users' preferences for certain PC components by adding another weighted term, for example *1000 * user_pref(Pj)*, into the above *fitness* function. Besides carefully adjusting the *fitness* function, we have to modify the testing condition for the *while-loop* in the *MCH-MGA* to : *while (ngens < MAX_GENS)* since we are now handling optimization problems instead of CSPs. Therefore, the unsatisfiability testing condition: *best_fit != 0* should be removed. Furthermore, the last two statements of the *MCH-MGA* procedure should also be modified to always: return (sub-)optimal solution for the optimization problems at hands. After putting into all these minor modifications to the above *MCH-MGA* procedure, we rename this new MCH-based evolutionary algorithm for solving optimization problems as *MCH-MGA-OPT*. In the following section, we are going to compare the performance of *MCH-MGA-OPT* against the beam-search based optimizer in solving the real-life PC configuration problems.

4. EMPIRICAL EVALUATION

Based on the above algorithms for our proposed optimizers, we built a prototype in PERL Version 5.0 since we planed to integrate our optimizer to build a Web-based PC Configuration Advisor to facilitate the general users to configure their own PC. To evaluate the performance of our prototype for the proposed optimizer to handle the PC configuration problems, we focused on two main aspects of the computational results returned by our optimizer in the following analysis. The first important factor to consider is the quality of the solution in term of the total costs of the PC configurations for minimization. The second factor of interest is the efficiency of our optimizer as reflected by the timings in CPU seconds. Accordingly, we run our prototype 10 times for different settings of threshold values on a DEC-Alpha workstation running Digital Unix Version 4.0.8.

The table below shows a sorted listing of the total costs of the different PC configurations, labeled from 01 to 30, returned by our beam-search optimizer for *Budget <= $3000* and *Threshold = 30*.

Cfg Cost	Cfg Cost	Cfg Cost	Cfg Cost	Cfg Cost
01 753	07 757	13 763	19 765	25 766
02 755	08 759	14 763	20 765	26 767
03 755	09 760	15 763	21 765	27 767
04 755	10 762	16 764	22 765	28 768
05 757	11 762	17 764	23 765	29 769
06 757	12 763	18 764	24 766	30 769

Clearly, there is a fair difference of SG $16 between the first minimal total cost and the last total cost of the configuration 30 as shown in the above table. In addition, it should be noted that from our initial experiments, the top 10 configuration-cost pairs usually remain unchanged for the same budget limit even though we changed the threshold value from 30 to 20 and then to 10. This demonstrated the stable performance of our optimizer in handling these PC configuration problems.

Table 4 shows the performance of our beam-search optimizer for handling the PC configuration problems with *Budget* <= $3000 and varying the *Threshold* value at *1, 5, 10, 20* and *30*. For each case, the reported data is the average CPU time in seconds over *10* runs.

Threshold	Average CPU time in seconds (10 runs)
1	3.18
5	3.19
10	3.32
20	3.40
30	3.51

Table 4. Performance of our Optimizer for Different Threshold Values

The above table clearly showed that our optimizer is fairly efficient in handling a typical PC configuration problem with different threshold values to control the search. Also, the performance of our optimizer is compact and stable since there is only slight increase in the average CPU time from 3.18 to 3.51 seconds when the corresponding threshold value is changed from *1* to *30*.

MAX_GEN	20	50	100
PZ = 1	0.35s (847.2)	0.71s(715)	1.28s(715)
5	1.26s(716.8)	3.10s(715)	6.04s(715)
10	2.38s(715)	5.99s(715)	12.03s(715)
20	4.76s(715)	11.97s(715)	24.07s(715)

The above table gives the performance of the *MCH-MGA-OPT* for *PZ = 1, 5, 10 and 20* and *MAX_GENS = 20, 50 and 100* for comparison. For each table entry, the first data represents the average CPU time in seconds while the second one

in parentheses denotes the cost in SG$. When we compare against the quality of solutions as returned by the beam-search based optimizer, it is clear that the *MCH-MGA-OPT* consistently outperformed the beam-search based optimizer with the optimal cost = SG $715 for most cases. In fact, as for *MCH-MGA-OPT*, the minimal cost over *10 runs* consistently stays at SG $715 for all the cases, which probably represents the global optimal solution[2]. In general, the beam-search based optimizer could only better the average cost returned by *MCH-MGA-OPT* in the weakest case where *PZ=1 and MAX_GEN = 20*. On the other hand, the *MCH-MGA-OPT* generally outperformed the beam-search based optimizer in terms of solution quality.

Since the performance of our proposed optimizers from the initial experiments is satisfactory, we integrated our optimizer as a plug-in component into our targeted Web-based PC Configuration Advisor for optimizing different PC configurations according to users' budget. Firstly, our Web-based PC Configuration Advisor supports rule-based information processing to actively anticipate or validate users' inputs, and also analyze hardware information stored in databases to provide useful advice on the best PC configurations for certain usage subject to a user-defined budget limit, and a pre-defined threshold limit used during the search. Specially, we allows the active uses of rules, as stated in the rule script to represent some domain experts' knowledge, to "dynamically" prune off options within the same page (intra-page) or between the pages (inter-page). For example, when the user selects the "Type of PC" to be configured is "Server" on the first page, the option of "17-inch monitor" will be automatically removed in the last page[3] with the active use of the domain experts' rules. Besides ensuring valid inputs from the users, it is mainly used to remove irrelevant choices so as to facilitate the search efficiency. After all possible pruning, our proposed optimizer, implemented as a easy-to-modify plug-in component inside our PC Configuration Advisor, then performs the controlled *(best-n)* search according to the predefined budget limit and threshold values during optimization of PC configurations. Thus, for a user's requirement like *budget ≤ 3000 and threshold = 20*, our Web-based PC Configuration Advisor will definitely return only the first *20* optimal PC configurations with the total cost less than *$3000*. In case where no PC configuration can meet the specified budget, an error message will be printed on the resulting Web page.

---

[2] Of course, this will require a complete solver to later verify the global optimality of our solution found.

[3] Technically, this special feature is achieved with JavaScript through the uses of cookies to remember some important values for each page.

5. CONCLUSION

It is undeniable that the PC configuration problems are both practical and interesting optimization problems widely occurring in many well-developed countries of the world. However, specific configuration problems are interesting since they have the unique problem structures that some components are highly constrained while the rest are totally unconstrained. This may pose challenge to some conventional enumerative search methods. As the PC technology is likely to be more complicated in the future, it will definitely become more difficult to handle these specific configuration problems. In this paper, we gave a formal definition of the PC configuration problems as discrete optimization problems. Based on this problem formulation, we proposed a systematic framework which allows the integration of different optimizers for optimizing the PC configurations according to the useful information stored in a centralized database and constraint store. More importantly, after analyzing the unique features of the PC configuration problems which may possibly lead to combinatorial explosion during the search, we proposed two useful search strategies, one evolutionary algorithm and one beam-search, to control the search and optimization process in our optimizer. The min-conflict heuristic based MGA optimizer compared favorably against the beam-search method in handling these real-life PC configuration problems. To demonstrate the feasibility of our proposals, we implemented prototypes of our proposed optimizers for some empirical evaluation, and later integrated it into a Web-based PC Configuration Advisor to facilitate the general public in configuring various PCs.

Clearly, there are several interesting directions for our future investigation. One obvious direction is to try our optimizer on many different cases, and also include other optimizers such as the B&B method for a complete comparison. Second, it would be interesting to improve our optimizer to handle more complicated constraints and optimization criteria to handle more complex real-life PC, or other general, configuration problems. Lastly, we are currently studying the possible uses of efficient learning algorithms such as the ID3 algorithm [2] to classify different user-defined configurations into useful examples so as to guide the search during the optimization process.

Acknowledgement

We are grateful to Mr. W.K. Foo for his valuable help in implementing the beam-search based optimizer. In addition, we would like to acknowledge the ARF grant (ref. no. RP3991612), the National University of Singapore.

6. REFERENCES

[1] "Artificial Intelligence : A Knowledge-Based Approach" by Morris W. Firebaugh, PWS-Kent Publishing Company, Boston, 1988.

[2] "Artificial Intelligence" by Elaine Rich and Kevin Knight, McGraw-Hill International Edition, 1991.

[3] "Discrete Mathematics – A Unified Approach" by Stephen A. Wiitala, McGraw-Hill International Edition, 1987.

[4] "Introduction to Algorithm" by Thomas H. Cormen, Charles E. Leiserson and Ronald L. Rivest, The MIT Press, 1994.

[5] "Foundations of Constraint Satisfaction" by Edward Tsang, Academic Press, 1993.

[6] "Genetic algorithms versus simulated annealing : Satisfaction of large sets of algebraic mechanical design constraints" by A.C. Thornton, in *Proceedings of Artificial Intelligence in Design*, pp. 381-398, 1994.

[7] "Discover PERL 5" by Naba Barkakati, IDG Books WorldWide Inc., 1997.

[8] "Programming in PERL" by Larry Wall, O'Reilly, 1995.

[9] "Boltzmann machines for traveling salesman problems" by E. Aarts and J. Korst, *European Journal of Operational Research*, 39:79-95, 1989.

[10] "Optimization by simulated annealing : an experimental evaluation; Part II, graph coloring and number partitioning" by D. Johnson, C. Aragon, L. McGeoch, and C. Schevon. Operations Research, 39(3)378-406, 1991.

[11] "Solving small and large scale constraint satisfaction problems using a heuristic-based microgenetic algorithm" by G.Dozier, J. Bowen and D. Bahler. *In Proceedings of the IEEE International Conference on Evolutionary Computation*, 1994.

[12] "Improving Evolutionary Algorithms for Efficient Constraint Satisfaction" by Vincent Tam and Peter Stuckey, *International Journal on Artificial Intelligence Tools*, Vol. 8, No. 2, World Scientific Publishers, December 1999.

Forecasting Stock Market Performance
Using Hybrid Intelligent System

Xiaodan Wu[1], Ming Fung[2], Andrew Flitman[3]

[1] jenny.wu@infotech.monash.edu.au
[2] mkfung@yahoo.com
[3] andrew.flitman@infotech.monash.edu.au
School of Business Systems, Faculty of Information Technology,
Monash University, Victoria 3800 Australia

Abstract: Predicting the future has always been one of mankind's desires. In recent years, artificial intelligent techniques such as Neural Networks, Fuzzy Logic, and Genetic Algorithms have gained popularity for this kind of applications. Much research effort has been made to improve the prediction accuracy and computational efficiency. In this paper, a hybridized neural networks and fuzzy logic system, namely the *FeedForward NeuroFuzzy* (FFNF) model, is proposed to tackle a financial forecasting problem. It is found that, by breaking down a large problem into manageable "chunks", the proposed FFNF model yields better performance in terms of computational efficiency, prediction accuracy and generalization ability. It also overcomes the black art approach in conventional NNs by incorporating "transparency" into the system.

1. Introduction

Over the years, economy forecasters have been striving to predict various financial activities, such as exchange rates, interest rates and stock prices. Successful prediction of the "economic cycle" helps them to maximize their gains in the market place and/or to prevent from business failure. Owing to the nonlinear, dynamic and chaotic nature of economic environment, financial problems are usually difficult to analyze and model using conventional techniques. Those techniques rely heavily on assumptions that often mask constraints in the initial problem domain. As the result, the models derived tend to have poor quality in applications. As the alternative or supplementary approaches, artificial intelligent (AI) techniques, such as neural networks (NNs), fuzzy logic (FL) and genetic algorithms (GAs), have drawn serious attention from both industrial and business sectors in recent years. Many encouraging results have been achieved when applying these techniques to financial market [1,3, 5, 6].

Depending on the scenario being modeled, a particular technique or a combination of techniques can be used to achieve certain goals. Because each technique has its own strengths and limitations in terms of performance, reliability, missing functionality and biases, some problems cannot be solved or efficiently solved by using single technique. In these cases, a hybrid approach, which employs a combination of techniques, may provide rich opportunities to successfully implement the problem modeling.

This paper explores the application of AI techniques to forecast the monthly S&P500 closing price. A hybrid *FeedForward NeuroFuzzy*) model is proposed. A basic 3-layer backpropropagation NN is used as a benchmark to gauge the performance of the hybrid NeuroFuzzy model. The trend prediction accuracy was used to evaluate the performance of the models.

2. Experimental Data Selection and Preprocessing

Data selection, which usually requires knowledge of domain experts, is an important part in building a successful prediction model. Appropriate data is the essential ingredient to start with the modeling process. Obtained from "http://www.economagic .com/popular.htm", the data used in this paper cover periods from January 1992 to April 2000. In light with the experts' suggestions, the following variables are used as the inputs and output of the forecasting model:

x_1 : Consumer Price Index $\qquad$ x_2 : Total Industrial Production Index
x_3 : Leading Economic Index $\qquad$ x_4 : Bank Prime Loan Rate
x_5 : Federal Funds Rate $\qquad$ x_6 : Unemployment Rate
x_7 : S&P500 lag(1) $\qquad$ x_8 : S&P500 lag(2)
x_9 : S&P500 lag(3) $\qquad$ y : S&P500 closing prices (output)

As stock market prices can move up and down quite frequently, it makes more sense to predict the trend of the financial data series rather than the actual value itself. Since most stocks tend to move in the same direction as the S&P500, it would be desirable to know which direction this indicator heads.

Data preprocessing facilitates efficiency and effectiveness, the computational cost reduction, and better generalization ability. In this paper, S&P500 lags(1-3) were preprocessed to be the previous three months' closing values. These historical data are considered having significant impact on the predicted one.

3. Neural Networks Approaches

Owing to the complex and unstructured characteristics of economic forecasting, neural networks (NNs) may be the appropriate approach to solve the problem under consideration. Compared with conventional statistical methods, neural networks are more robust in terms of coping with noisy and incomplete data. They can interrelate both linear and non-linear data as well as the important influential factors that are not based on historical numerical data. They may also eliminate the constraints of unrealistic assumptions masked by conventional techniques. These make NNs more flexible and effective in the domain of financial forecasting.

In experiment data was split into three sub-sets for model training, test, and validation. This is to ensure that the trained network will not simply memorize the data in the training set but learn the relationships amongst the variables, i.e. having good generalization ability.

There are many issues involved in NNs development. These include choice of input variables, number of hidden neurons and their activation functions, network architecture and parameters, etc. Based on a trial and error approach, different architectures and parameters have been experimented in order to obtain an optimal configuration. The performance of this optimal network is then used to comparing with the proposed FFNF system.

4. A NeuroFuzzy Approach

Fuzzy logic (FL) provides a means of representing uncertainty and is good at reasoning with imprecise data. In a fuzzy system variables are defined in terms of fuzzy sets, and fuzzy rules are specified to tie the fuzzy input(s) to the output fuzzy set. Such rules derived also make it easy to maintain and update the existing knowledge in the problem domain. Bearing these benefits, FL also has limitations. In most applications, knowledge is embedded in the data itself and thus requiring significant effort to derive the accurate rules and the representative membership functions. Further, the rules derived cannot adapt automatically to changes. They have to be manually altered to cater for change in conditions.

NNs technique is commonly referred to as the "black box" approach. The causes of certain NNs' behaviours can neither be interpreted nor can it be changed manually. That is, the final state of the NN model cannot be interpreted explicitly. Contrast to NNs' implicit knowledge acquisition, FL facilitates transparency in terms of explicit rule presentation. From this sense, FL may be used to improve the effectiveness of a NNs model used for trading, for example, by incorporating knowledge about financial markets with clear explanation of how the trading recommendations are derived. It may also make the model verification and optimization easy and efficient [1].

The advantages from FL point of view is that NNs' learning ability can be utilized to adjust and fine-tune the fuzzy membership functions. This solves the difficult and resource-consuming process of deriving the rule base manually. The resultant optimal rule base turns out to be an advantage for NNs as it solves the "black box" problem.

In short, NNs learn from data sets while FL solutions are easy to verify and optimize. The combination of both techniques results in a hybrid NeuroFuzzy model. With the complementary advantages offered by both sides, they work together to deliver value.

There are many ways to integrate NNs and FL. In this paper, NN is adopted as the learning tool to evolve a set of fuzzy rules, based solely on the data itself. The inferred fuzzy rules can be used for forecasting the monthly S&P500 trend. The developed system would have the following characteristics:

- Vague inputs can be used
- More general boundaries can be described
- Automated fuzzy rule extraction
- Learning the rules through training of NNs

4.1. System Connections

The proposed NeuroFuzzy system is a FL system trained by a NN. A simplified structure of the system connections, which involves two inputs with two and three fuzzy sets respectively, is shown in Figure 1.

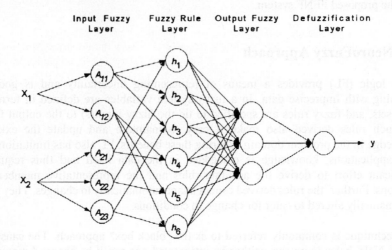

Figure 1. Connections between different layers in the NeuroFuzzy module

Where:

- The *input fuzzy layer* represents linguistic sets in antecedent fuzzy membership functions. The encoding of the shapes, centres, and the widths of membership functions is done in this layer. The total number of fuzzy sets for all input variables determines the number of neurons as follow, where Ni is the number of fuzzy sets for input xi, and m is the total number of input variables:

$$\sum_{i=1}^{m} N_i$$

- The *fuzzy rule layer* is built as fuzzy associative memories (FAM). A FAM is a fuzzy logic rule with an associated weight. Deboeck [2] has demonstrated how FAMs can be used to extract rules. With FAM, each rule is assigned a degree of support representing the individual importance of the rule. Rules themselves can be "fuzzy", meaning validity between 0 and 1. Each neuron in this layer represents a unique combination of input fuzzy sets. The links between this layer and the input fuzzy layer are designed in a way that all combinations among input fuzzy sets are represented as shown in table 1.

Table 1. Fuzzy rule table for x_1 and x_2

x_1 \ x_2	A_{21}	A_{22}	A_{23}
A_{11}	R_1	R_2	R_3
A_{12}	R_4	R_5	R_6

The "AND" association for each fuzzy set combination forms the antecedent relation established by these links, which build a relation between the input and output fuzzy membership functions and define the fuzzy rules. Each neuron in the module represents a fuzzy rule and the total number of neurons can be obtained by the product of all N_is in the input fuzzy layer as:

$$\prod_{i=1}^{m} N_i$$

- The *output fuzzy layer* represents the consequent fuzzy membership functions for outputs. The number of fuzzy sets for output fuzzy variables determines the number of neurons in this layer. Each neuron in the fuzzy rule layer is connected to all neurons in this layer. The weights between fuzzy rule layer and output fuzzy layer are rule matrices, which can be converted into a FAM matrix.
- The *defuzzification layer* is for rule evaluation. The weight of the output link from the neuron represents the center of maximum of each output membership function of the consequent. The final output value is then calculated using the center of maximum method.

4.2. The FeedForward NeuroFuzzy Model

The system structure identifies the fuzzy logic inference flow from the input to the output variables. Fuzzification in the input interface translates analog inputs into fuzzy values. The fuzzy inference takes place in the rule blocks which contain the linguistic control rules. The outputs of the rule blocks are linguistic variables. Defuzzification in the output interface converts the fuzzy variables back into crisp values. Figure 2 shows the structure of the FFNF system including input interface, rule blocks and output interface, where the connection lines symbolize the data flow.

Based on the nature of the corresponding economic indictors, the original inputs of the FFNF system $(x_1 - x_9)$ are categorized into three groups (Table 2) to form three sub-modules of the first tier of the rule blocks (see Figure 2). The second tier of the rule blocks has one processing module (the Final Module in Figure 2) designed to sum the rule evolving results from the first level modules. This layered structure helps to simplify the initial complex problem and to improve system efficiency. The size of the fuzzy rule base developed has been significantly reduced from a total of $3^{10} = 59,049$ rules (10 variables with 3 membership functions each) derived from a non-layered model to 324 rules (4 modules x 3^4 each) of the current structure.

Table 2. Grouping of input, intermediate and output variables

	Group 1 (indices)	Group 2 (rates)	Group 3 (historical data)	Final System
Inputs	x_1	x_4	x_7	y_1
	x_2	x_5	x_8	y_2
	x_3	x_6	x_9	y_3
Outputs	y_1	y_2	y_3	y

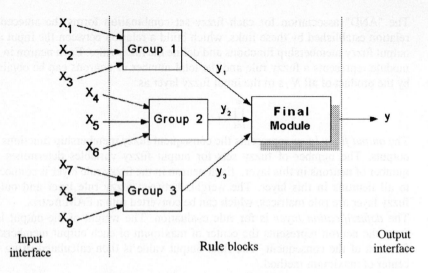

Figure 2. Structure of the FFNF System.

4.3. NeuroFuzzy Learning Algorithm

Considering a simple case of two input variables (x_1, x_2) and one output variable (y) for demonstration purpose, the following mathematics forms the basis of the learning algorithm [4].

$$x_1 : A_{1i} \ (i = 1, 2, \ldots, r)$$
$$x_2 : A_{2j} \ (j = 1, 2, \ldots, k)$$

where : r means the number of fuzzy partitions for x_1, k means for x_2

Gaussian type membership functions :

$$A_{1i}(x_1) = \exp(-(x_1 - a_{1i})^2 / 2\sigma^2_{1i})$$
$$A_{2j}(x_2) = \exp(-(x_2 - a_{2j})^2 / 2\sigma^2_{2j})$$

Where :
a_{1i} and σ_{1i} are respectively the centre and width of A_{1i}, assuming that the rule base is defined based on all the above combinations of A_{1i} and A_{2j} $(i = 1, ..r, j = 1, ..k)$ [4] :

Rule 1 : $\qquad A_{11}, A_{21} \Rightarrow R_1$

. . .

Rule k : $\qquad A_{11}, A_{2k} \Rightarrow R_k$

. . .

Rule $2k$: $\qquad A_{12}, A_{2k} \Rightarrow R_{2k}$

. . .

Rule $(i-1)k+j$: $\quad A_{1i}, A_{2j} \Rightarrow R_{(i-1)k+j}$

. . .

Rule rk : $\qquad A_{1r}, A_{2k} \Rightarrow R_{rk}$

The above fuzzy inference rules can be expressed in the form of rule table below:

Table 3. Fuzzy rule table for A_{1i} and A_{2j}

x_1 \ x_2	A_{21}	A_{22}	...	A_{2j}	...	A_{2k}
A_{11}	R_1	R_2	...	R_j	...	R_k
A_{12}	R_{k+1}	R_{k+2}	...	R_{k+j}	...	R_{2k}
$\vdots$	$\vdots$	$\vdots$	$\vdots$	$\vdots$	$\vdots$	$\vdots$
A_{1i}	...	...	...	$R_{(i-1)\,k+j}$	...	...
$\vdots$	$\vdots$	$\vdots$	$\vdots$		$\vdots$	$\vdots$
A_{1r}	$R_{(r-1)\,k+1}$	...	...	...	...	R_{rk}

Based on the above, a fuzzy inference y can be obtained using product-sum-gravity fuzzy reasoning method. The intersection (agreement) of A_{1i} and A_{2j} ($i = 1...r; j = 1 ...k$) at (x_1, x_2) is calculated as:

$$h_{(i-1)(k+j)} = A_{1i}(x_1)A_{2j}(x_2)$$

5. Experiment Results and Discussion

For the purpose of performance comparison of the two models (NNs vs. FFNF), the trend accuracy was calculated as shown below:

$$\text{Correct Trend} = \frac{100}{n}\sum_{i=1}^{n}a_i; \quad \text{where } a_i = \begin{cases} 1 & \text{if } (x_{t+1}-x_t)(\hat{x}_{t+1}-x_t) \geq 0 \\ 0 & \text{otherwise} \end{cases}$$

where n, x, $\hat{x}$ are the number of observations, the actual and predicted values respectively.

A correct trend of 100% means that the model can perfectly predict the directional changes, while a value of greater than 50% means that the model performs better than tossing a coin. The evaluation was performed based on the proposed FFNF model and the optimal NN configuration in Table 4. The results of the two models are given in Table 5.

For the NN model, the best result (in bold font) was obtained using a 3-layer architecture trained with the standard backpropagation (BP) algorithm. The transfer function was linear over the range [1, -1]. The tanh function was adopted as the activation function in the hidden layer. Both the learning rate and the momentum were at 0.1, and the initial weight was set at 0.3 (see Table 4).

Table 4. The optimal configuration obtained using a standard 3-layer BP NN

Input Variables	Hidden Neurons	Learning Rate	Momentum	Initial Weights	R^2 Training	R^2 Test	R^2 Pattern
6	22	0.6	0.9	0.6	0.9817	0.9836	0.9830
9	**13**	**0.1**	**0.1**	**0.3**	**0.9908**	**0.9884**	**0.9905**
12	30	0.7	0.8	0.3	0.9775	0.9454	0.9836

There are constraints that may limit the scope of the selection of input variables. Subjective opinions may also have significant influence on how successful the prediction model will be. The contribution factor of a variable can be considered as a rough measure for evaluating the significance of the variable, relative to the other variables in a same network, in predicting the output. In our experiment, the consumer price index (CPI) is the most significant input variable of the prediction model.

Based on initial values, the NeuroFuzzy model tunes rule weights and membership function definitions so that the system converges to the behaviour represented by the data sets. Changes in the weights during learning can be interpreted as changes in the membership functions and fuzzy rule.

Despite its flexibility in system tuning, the model requires substantial effort in optimization so as to arrive at the optimal combination of member functions, fuzzy rules, and fuzzy inference and defuzzification methods. If no rules were generated with a reasonable level of degree of support (DoS) for certain input variables, it is reasonable to assume that they either do not contribute or are correlating to another variable for which rules are generated. Examples of the rule bases evolved for the modules are shown in Appendix.

In the experiment, the original inputs are categorized into three groups to form three sub-modules of the first tier of the rule blocks (see Figure 2 above). The rules evolved from these sub-modules are expressed in a simple, clear-explained format. They would not represent the overall system behavior directly but can, however, provide valuable assistance in understanding the complex initial problem domain.

The second tier of the rule blocks has one processing "Final Module" designed to sum the rule evolving results from the first level modules. The inputs of it are the defuzzified crisp values of the three sub-modules in the previous stage. The rule base evolved from this Final Module cannot be used in a stand-alone manner. Instead, by feeding all the inputs through three rule bases derived in the first tier, the final output in the second tier is meaningful and therefore is applicable for prediction purpose.

At this stage, the system cannot operate in parallel as the results of the first tier modules have to be computed before feeding into the Final Module. However the intermediate outputs of the first tier are valuable and useful. They facilitate decision-making assistance depending on expert knowledge or subjective opinions. The final output when using all four modules as a single rule block provides high prediction accuracy. The layered structure helps simplify the initial complex problem and to improve system efficiency.

As shown in Table 5, the FFNF system is superior to the standard 3-layer BPNN in the financial forecasting problem under consideration. By breaking down a large, complex problem into manageable smaller modules, the FFNF design yields better performance in terms of computational efficiency (i.e. a significant reduction in the rule base), prediction accuracy and good generalization ability.

Table 5. Models performance

	3-Layer BPNN	FFNF
Correct Trend	96%	99%

6. Conclusions

Financial prediction in an unstructured, nonlinear and dynamic environment is a sophisticated task. It is worthy of considering the hybrid artificial intelligence approach. This paper proposes a *Feedforward NeruoFuzzy* (FFNF) system for financial application. The system integrates the fuzzy rule-based techniques and the NNs technique to accurately predict the monthly S&P500 trend. Experimental results showed that the proposed system has better performance than the BPNN model. By highlighting the advantages and overcoming the limitations of individual AI techniques, a hybrid approach can facilitate the development of more reliable and effective intelligent systems to model expert thinking and to support the decision-making processes. It has diverse potential and is a promising direction for both research and application development in the years to come.

References

1. Altrock, C.: Fuzzy Logic and NeuroFuzzy Applications in Business and Finance. Prentice PTR (1997)

2. Deboeck, G.J.: Why Use Fuzzy Modeling. In: Deboeck, G.J. (ed.): Trading on the Edge. John Wiley and Sons, New York (1994) 191-206

3. Goonatilake, S. and Khebbal S.: Intelligent Hybrid Systems: Issues, Classifications and Future Directions. In Goonatilake, S. and Khebbal S.(eds.): Intelligent Hybrid Systems. John Wiley and Sons, U.K. (1995) 1-19

4. Shi, Y., Mizumoto, M., Yubazaki, N. and Otani, M.: A Method of Generating Fuzzy Rules Based on the Neuro-fuzzy Learning Algorithm. Journal of Japan Society for Fuzzy Theory and Systems.Vol. 8, No.4 (1996) 695-705

5. Tano, S.: Fuzzy Logic for Financial Trading. In Goonatilake, S. and Treleaven, P. (eds.): Intelligent Systems for Finance and Business. John Wiley and Sons, Chichester (1995) 209-224

6. Yen, J.: Fuzzy Logic-A Modern Perspective. IEEE Transactions on Knowledge and Data Engineering. Vol. 11, No. 1 (1999) 153-165

Appendix: Examples of the Rule Base Evolved

Table A1. Rules of the rule block "Group 1"

IF			THEN	
x_1	x_2	x_3	DoS (weights)	y_1
low	low	low	1.00	low
medium	low	low	0.86	low
medium	medium	low	0.99	low
medium	high	medium	0.66	high
high	medium	medium	0.64	low
high	high	medium	0.90	high

Rule 1: IF (x_1 = low and x_2 = low and x_3 = low) OR (x_1 = medium and x_2 = low and x_3 = low) OR (x_1 = medium and x_2 = medium and x_3 = low) OR (x_1 = high and x_2 = medium and x_3 = medium), THEN y = low

Rule 2: IF (x_1 = medium and x_2 = high and x_3 = medium) OR (x_1 = high and x_2 = high and x_3 = medium), THEN y = high

Table A2. Rules of the rule block "Group 2"

IF			THEN	
x_4	x_5	x_6	DoS (weights)	y_2
low	low	medium	0.98	low
low	low	high	0.88	low
medium	medium	low	0.98	high
medium	medium	medium	0.76	low
high	medium	low	0.59	medium
high	high	low	0.55	high
high	high	medium	0.67	low

Rule 1: IF (x_1 = low and x_2 = low and x_3 = medium) OR (x_1 = low and x_2 = low and x_3 = high) OR (x_1 = medium and x_2 = medium and x_3 = medium) OR (x_1 = high and x_2 = high and x_3 = medium), THEN y = low

Rule 2: IF (x_1 = high and x_2 = medium and x_3 = low), THEN y = medium

Rule 3: IF (x_1 = medium and x_2 = medium and x_3 = low) OR (x_1 = high and x_2 = high and x_3 = low), THEN y = high

Table A3. Rules of the rule block "Final Module"

IF			THEN	
y_1	y_2	y_3	DoS (weights)	y
low	low	low	0.65	decrease
low	medium	low	0.72	decrease
medium	medium	low	0.59	decrease
high	medium	medium	0.82	increase
high	high	medium	0.98	increase

Rule 1: IF (x_1 = low and x_2 = low and x_3 = low) OR (x_1 = low and x_2 = medium and x_3 = low) OR (x_1 = medium and x_2 = medium and x_3 = low), THEN y = decrease

Rule 2: IF (x_1 = high and x_2 = medium and x_3 = medium) OR (x_1 = high and x_2 = high and x_3 = medium), THEN y = increase

Multimedia

Session chair:

Rachel McCrindle (University of Reading, UK)

The MultiMedia Maintenance Management (M⁴) System

Rachel J. McCrindle

Applied Software Engineering Research Group, Department of Computer Science, The University of Reading, Whiteknights, PO Box 225, Reading, Berkshire, RG6 6AY, UK
Tel: +44 118 931 6536, Fax: +44 118 975 1994
r.j.mccrindle@reading.ac.uk

Abstract. Although adoption of a software process model or method can realise significant benefits, there is generally a need to provide a level of computerised support if it is to be usefully applied to large real-world systems. This is particularly true in relation to the software maintenance discipline, where many of the problems to date have typically arisen from deficiencies in recording and being able to easily access any knowledge regained about a system during maintenance. The MultiMedia Maintenance Management (M⁴) system has been designed and prototyped as a meta-CASE framework in order to promote maximum process flexibility and product extensibility. As such, a variety of bespoke or host-resident tools for activities such as process management, information reclamation and documentation, configuration management, risk assessment etc. may be plugged into the M⁴ system and activated through a universal front-end.

1 Introduction

The last decade has witnessed an explosion in terms of the abundance, size and complexity of software systems being developed such that they now play a key role in almost every aspect of today's society [11]. Indeed, the software industry may be seen as one of continual growth: in the industrial sector as automatic control of many systems, previously unrealisable due to hardware limitations, becomes possible through the transference of complex algorithms to a software base [18]; and in the business arena where software plays an increasingly central role in business process optimisation and redesign [7]. Commensurate with the increasing reliance being placed on software is the need for cost effective, high quality software products that meet customer expectations, perform reliably and safely [3] and which can evolve to meet the changing requirements of a dynamic industry [1]. Additionally, recognition of software as a key corporate asset is now becoming evident and the importance of maintaining this asset is gaining momentum [16].

The importance of investing in the maintenance and control of software is further substantiated by the fact that in many of today's computer systems the hardware expenditure can no longer be considered the major factor in any costing scenario [17]. The high cost associated with software may be attributed to a number of factors linked not only to the human-intensive nature of the process but also to the characteristics of the software itself. The enormity of the task is evident when we consider that even a decade ago software systems were being described as the "*most intricate and complex of men's handiworks requiring the best use of proven engineering management methods*" [5]. Since then, the internal complexity of software has progressively risen as has the number and heterogeneity of components.

Other factors to contend with include the increased linkage of software with its environment, firmware and third party components, wider application bases, changes in architecture over the lifetime of a system, fragmentation of upgrade paths, the increasingly distributed nature of software products and the need to link them with supporting information [12].

This growing awareness of software has precipitated an increase in the research being conducted within the software process model arena and the co-requirement of developing automated tools to support the resultant models. This paper describes the M^4 system created to support the development, evolution and maintenance of large-scale software systems.

2 Automated lifecycle support

Various maintenance process models exist [1, 2, 4, 10, 12] which address the maintenance process from a number of different perspectives (e.g. economic, task-oriented, iterative, reuse, request driven, reverse engineering). Although these models describe the maintenance process in varying levels of detail, they all centre on the evolutionary nature of software. In addition other characteristics shown to be important for the production of high quality long-lasting software include the ability to enable effective communication, to support cost-effective maintenance, to facilitate a re-useable process, to support evolution by serving as a repository for modifications and to facilitate effective planning and increased understanding of the systems being maintained. As such the M^4 has been developed with a number of key features in mind:

Control: an underlying core set of facilities should be integrated into the toolset to satisfy the requirement to regain and keep control of a system in a defined and consistent manner.

Adaptability: the framework should enable the integration of different tools into the toolset to support the individual working practices and methods of different maintenance organisations.

Extensibility: information gathered from other sources or tools should be able to be brought into the framework and maintained as a coherent set of data.

Evolution: new technologies should be exploited as they come into the mainstream computing community.

3. Development of the M^4 system

The M^4 system has been developed primarily to provide semi-automated support for the ISCM (Inverse Software Configuration Management) process model [12] and its associated PISCES (Proforma Identification Scheme for Configurations of Existing Systems) method [13]. However, due to its development as a flexible and open meta-CASE framework rather than a rigidly defined more traditional CASE tool [9] it can be adapted to incorporate and support other process models if required. The components currently incorporated in the M^4 system are shown in Figure 1.

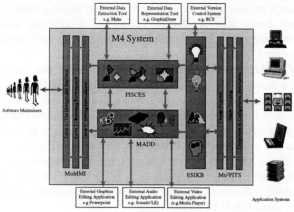

Figure 1. Overview of M^4 System

3.1 EISKB

The Extensible System Information Knowledge Base (EISKB) forms the core of the tool. The purpose of the EISKB is to stored the rules, component details and domain knowledge pertaining to a system configuration. The ESIKB as it is currently implemented is actually an amalgamation of technologies spanning simple flat files, relational database systems and the inherent storage mechanisms of the host system. This rather piecemeal approach has arisen due to the development of individual tools which handle data storage locally but whose data can be shared through in-built data exchange mechanisms or via simple parsing tools which extract relevant data from the different toolsets as required. Whilst this approach has worked successfully and can exploit fully the advantages of flexibility there are also arguments for having a central M^4 repository, with strictly defined data structures, for use by all tools within the M^4 framework. However whilst this would offer more seamless tool integration and would lessen the likelihood of maintaining any redundant data within the M^4 system it may also make the incorporation of proprietary tools difficult or even impossible.

3.2 PISCES

The activities of the PISCES (Proforma Identification Scheme for Configurations of Existing Systems) tool are primarily concerned with the extraction and recording of information regarding the components and configuration of an application and the overall environment in which they play a part. As such the tool primarily implements a series of defined templates for incrementally building-up software system configuration information, although close linkages are also formed with the information recorded by the Mu²PITS system and with any host system tools that aid the extraction of information about a system's components. The following facilities are offered by the PISCES tool [13].

Reclamation of configuration information: a series of templates are provided as an on-line guide to the consistent information collection and documentation of any software system being maintained. As maintenance occurs, information may be successively added and saved using one of two approaches: the templates may be updated and saved in the normal manner or a 'baseline' may be struck and the template saved as a new version. The point at which baselines are struck is definable by the maintenance organisation, for example, this may be on a temporal basis or on a per maintenance basis after a change or series of changes have been made. Progressive baselining of the templates enables an evolutionary history of an entire system to be recorded.

Creation of HTML links and web pages: by creating the templates within the Microsoft Office suite of programs, the facility to add HTML links and covert the templates to web pages is enabled. Maintenance across the web then becomes a possibility especially if maintainers can link up via a web communications system [8].

Dependency information extraction and viewing: linkage to tools such as awk, grep and Make for extraction of dependency information; access to the resultant components such as Makefiles; and study of the code when combined with the maintainers expertise enables dependency information to be recovered and documented. Additionally small bespoke utilities enable features such as the resultant dependency tree structures to be displayed.

File location information and maps: the physical file locations for components can be identified and displayed through the use of simple data extraction utilities on either a per component or per system basis.

Incremental annotation facility: provision of a simple editor enables notes to be made regarding the understanding of the system gained during the comprehension process, or as a means of providing a temporary note to the maintainer, for example, to flag an activity that still needs to be carried out or to warn other maintainers of a troublesome section of code etc.

3.3 MADD

The MADD (Multimedia Application Documentation Domain) environment supports the management and control of, and access to the underlying multimedia types. A 'natural-language' control centre provides the primary linkage mechanism within the environment enabling links to be established between the different project documents. The key facilities provided by the MADD environment are [14]:

Viewing and loading of stored documentation: this facility enables the multimedia documentation previously stored on a project within the M^4 system to be viewed. This includes the ability of the M^4 system to display recorded video, to play audio, to display 'pure' text or texts with incorporated animation or graphics, and to display statistics in a graphical or animated format. The system also enables concurrent viewing of different multimedia attributes, for example, it allows code to be viewed alongside an associated video recording.

Creation and saving of the documentation: this facility enables new multimedia documentation to be input into the M^4 system, stored within the M^4 environment as a project directory and subsequently accessed through the MuMMI interface.

Editing of stored documentation: this facility enables changes to be made to stored documentation. This is required to keep the documentation up-to-date and concurrent with the state of the software project or simply to make corrections to erroneous documentation. In the case of audio and video files, these may be externally edited and replaced in the system, or the editing tools may be loaded and the file edited through the M^4 system itself. Text documentation may be displayed as 'pure' text or may be associated with other multimedia files. For example, whilst in edit mode the user can use the audio tool to record and associate a voice-over with a piece of text, thereby adding value to and aiding the understanding of an otherwise less descriptive pure text file.

Appending to stored documentation: this facility enables new information to be appended to the existing versions of the documentation held within the M^4 system. Although closely allied to the editing function, implementation is handled differently as changes are sequential in nature thereby building up a change history of a particular component.

Deletion of documentation: this facility enables documentation to be deleted either as complete multimedia files or as linkages to files. However, this should be supported with the disciplined practice of archiving the documentation before allowing it to be deleted from the project environment.

Provision of security and access rights: this facility provides a security mechanism in the form of password access. This is important because only certain personnel within an organisation may have the rights to change project documentation.

Data storage and retrieval: this enables the multimedia documentation accessible by the M^4 system to be stored in the form of files in the user directory. There is no restriction on the organisation of these files, thereby allowing the user to adopt a preferred method of operation. This can involve the storage of independent files associated with each project in individual directories. Alternatively, storage can be via a link to a database application, as occurs in connection with the information stored within the Mu^2PITS tool. The advantage of the latter method is the provision of a more organised framework and the ability to conduct more complex queries and searches.

It is also acknowledged that a fully configured M^4 system will store large quantities of multimedia documentation, in the form of external files. The need for a high capacity storage medium is determined by the size of video and audio files. However this technical problem of storage is becoming less of an issue due to the development of efficient compression methods such as MPEG, increases in computation power even in low-end machines, and the continuing reduction in the price of hard-disk and removable storage.

3.4 Mu2PITS

The Mu^2PITS (MultiMedia Multi-Platform Identification and Tracking System) tool supports documentation of the identifying features of system components and their integration into software configurations. The nature of the tool is such that it enables attributes such as the relationships existing between different file types to be recorded and the location of these components to be tracked across distributed networks. In this respect Mu^2PITS differs from many of the traditional configuration management tools which tend to concentrate on text based products. Mu^2PITS also supports the production of change requests and tracking of the status of changes throughout the maintenance process. The key facilities (Figure 2) provided by the Mu^2PITS tool are [6].

Documentation of component attributes: this facility enables information about a particular component to be documented. Once created the resultant component records may be amended, versioned or deleted. The documentation supports all types of multimedia components as well as the more traditional text-based components.

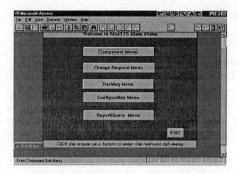 

Figure 2. Mu²PITS main menu   Figure 3. Process configuration form menu

Documentation of configuration composition: this facility enables the documented components to be associated with one or more configurations. In this way master configuration lists of the components of configurations can be built-up. Additionally, dependencies on a per component basis can be recorded. Once created the resultant configuration records may be amended, versioned or deleted (Figure 3).

Creation/amendment/deletion of a change request: this facility enables a change request to be generated for a particular component. In order to avoid possible conflicts and loss of valuable data the system only allows one active request per component at any one time. The resultant change request form may be updated if the change request has to be amended in some way after scrutiny by the change control board and is archived once the approved change has been completed or immediately if the change has been rejected.

Creation/amendment/deletion of a status tracking form: this facility enables the status of an accepted change request to be tracked during the maintenance process and it thereby allows the status of the change itself to be monitored. A status tracking form cannot be raised until the change request form has been completed and agreed. The status attribute of the tracking form is amended during the change process to reflect the status of the component and the form is archived once the change has been completed.

Creation of reports and queries: this facility enables information to be obtained regarding the data held within the Mu²PITS database. Information may be output to a printer as reports or to the screen as queries. The reports range from listing of component details to master configuration lists to management statistics, for example details of how many components are undergoing changes at a particular point in time.

3.5 MuMMI

The MuMMI (MultiMedia Maintenance Interface) is the front-end to the M^4 system, and is used to display the multimedia documentation [14]. This documentation is any information of relevance to understanding the high-level design or low-level code of a software product. It is routinely added to the system throughout maintenance thereby keeping the documentation up to date in relation to the underlying software product. The MuMMI co-operates closely with the MADD and is based on three levels of access:

Level 1 - MuMMI Manager: this level is used to select the particular project or system undergoing comprehension (Figure 4). Selection is via a graphical menu of multimedia attributes and determines whether it is the graphical, textual, audio or video documentation associated with a project that is initially displayed. Once a valid project file has been selected control is automatically switched to level-2 of the interface.

Level 2 - Browsing MuMMI Project Environment: this level is organised with respect to the selected project. Initially only the selected multimedia view of the project documentation and the control centre are displayed. The control centre uses a 'natural language' command interface in order to fulfil the requirements for fast operation by an experienced user group. The user communicates with the system through the entry of defined commands. A history of the commands entered within a session is maintained and can be invoked again by simple selection. As the program comprehension process proceeds the user can enter commands to display other multimedia views of the project related to the initially opened file.

Figure 4. Project selection window

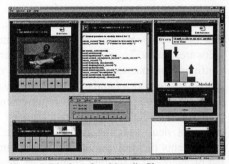

Figure 5. Display of multimedia files

The interaction between this level of the MuMMI and the MADD component of the M^4 system is one of read-only access. Thus the user is able to browse and display the multimedia documentation on accessible project files but is prevented from making any changes to stored material. The user can enter the editing levels of the M^4 system through a command in the control centre followed by access via a password protection mechanism.

Level 3 - Editing MuMMI Project Environment: the interaction between the third level of the interface and the MADD component and is that of read- and write-access. Thus at this level the user is able to edit the existing multimedia documentation and to create and store new documentation within the M^4 framework. This facility allows update or extension of the information stored on an existing product as it is recovered during the program comprehension process (Figure 5).

3.6 M^4

The Multimedia Maintenance Manager (M^4) provides the overall framework of the system, binding together the ESIKB information storage mechanism, the MuMMI front-end and the MADD back-end. It also provides hooks into the bespoke Mu^{2}PITS and PISCES tools as well as enabling links to be made to external point function tools such as those for version management and information extraction. The evolution of the M^4 system has also incorporated animated models of the maintenance process (Figures 6 & 7).

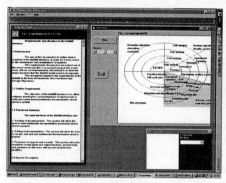
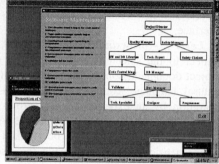

Figure 6. Animated view of the spiral process procedures & personnel

Figure 7. Animated company maintenance

Additionally with such a dynamic environment as the M^4 system and the emphasis that needs to be placed on keeping documentation concurrent with the state of the software, there is a need for rigorous change and version control. Software configuration management requirements must also extend to the ability of the M^4 system to support a family of concurrently released software product configurations to different clients. This latter requirement of recording component and product configuration details has been addressed in the Mu^2PITS tool. However Mu^2PITS does not deal with the issue of change to and subsequent storage of the actual components themselves. As the M^4 documentation is in different multimedia formats the problems of managing changes are more complex than for text management alone particularly if the changes are to be stored as a series of deltas rather than complete files. Currently, links have been provided to an external, but tailorable change and version control system for multi-platform code management and work is underway on a fully-distributed multimedia configuration management system [15].

4 M^4 System in Use

Section 3 has described and discussed the functionality offered by each of the main components of the M^4 system. It is also important to establish the use of the M^4 system within the task-oriented context of the way in which maintainers perform their role. It is intended that the M^4 system is used during the program comprehension process prior to each maintenance change according to the following procedures:

Browse: initially, the maintainer browses the program code and any associated documentation already present in the MuMMI system that appears to be of use to understanding the proposed change and the affected parts of the system.

Evoke: if appropriate, the maintainer evokes any data extraction tools etc. existing within or integrated into the M^4 framework in order to generate additional understanding or dependency information about the construction of the system.

Complete: the maintainer documents any additional findings about the affected parts of the system within the specific template generated for the system under study.

Update: as the comprehension process occurs the maintainer gains more information about the application itself, the application domain, the detailed functionality or design of the system etc. This understanding is incrementally recorded using an appropriate multimedia type and subsequently stored within the M^4 system.

In essence the M^4 system acts as a 'maintenance oracle' capturing all previous maintenance experience within a single environment. The effects of this system are three-fold:

Increased understanding: as domain, system and application knowledge is incrementally built-up and recorded so too is the understanding of the system and its importance in the context of the organisational and system domain better understood. This has the effect of ensuring that changes can be prioritised towards those of a key business nature and it also ensures that changes can be made more safely with reduced risk of the ripple effect occurring.

Faster changes: as more information is built-up about a system the time attributed to the comprehension process should become successively reduced. This is because the prior knowledge of the maintainer themselves and of any previous maintainer is available for use and hence provides a 'head start' so that the time taken to understand a change and its impact on the rest of the system can become progressively more rapid.

Foundation for re-engineering: the increased documentation and understanding of a system, as well as notes about a system or collection of metrics about errors etc. assists with identifying the areas of the system that will most benefit from being re-engineered. Additionally, identification of the components and records of how they are synthesised into complete configurations provides a solid foundation on which the actual re-engineering process can be based.

Additionally, although the use of the M^4 system has primarily being discussed within the context of its use during maintenance, maximum benefit of the M^4 system comes as a result of it being used from the inception of a green-field project. In these circumstances, the M^4 system can be populated with the complete set of components and documentation about an application. Additionally, all design decisions, key meetings or reasoning of an informal, semi-formal and formal nature can be captured from the outset of development through a combination of the different multimedia attributes. Used in this way, the M^4 system pre-empts the maintenance process and acts as a 'maintenance escort' during hand-over of the application [14].

5 Conclusions and Benefits

The M^4 system framework may be summarised as having a multimedia (MuMMI) front-end, an underlying repository (ESIKB) and a control and management back-end (MADD). In addition to this there is a data collection and collation mechanism (PISCES), support for the SCM activities of identification, control, status accounting and audit (Mu^2PITS), and the integration of other tools such as those for data extraction and version control. These tools plus their ability to link to other tools may be considered to provide a complete maintenance environment. In summary the benefits afforded by the M^4 system are:

Role: the M^4 system supports the ISCM process and PISCES method but can be readily adapted to incorporate other process models.

Flexible: adoption of a meta-CASE approach to the development of the M^4 system enables it to be very flexible in terms of its functionality, and readily extensible with regard to the range of tools that it can incorporate.

Generic: as well as being adaptable for different operational platforms, the system may be tailored to suit different application domains and programming languages through the on-line implementation of the PISCES templates.

Cost effective: many current maintenance tools are costly to buy and implement. The M^4 system can be integrated with tools already residing on the host system thereby reducing cost, and minimising the disruption associated with training maintainers to use new tools.

Easy to use: as the PISCES method provides guidance rather than prescription for the maintenance process, the different working practices of maintainers can be accommodated within the framework of the system whilst still providing the required consistency and control.

Multimedia: the M^4 system makes use of a hypermedia approach to enrich the maintenance and evolution of software systems. It also provides the MuMMI interface as a means of managing the recording, activation and dissemination of multimedia material pertaining to software system configurations.

Domain Knowledge: The M^4 system pays particular attention to being able to capture domain knowledge about a system undergoing maintenance. This is facilitated by exploitation of the multimedia capabilities described above.

Communicative: maintainers have expressed the need for better communication between the development and maintenance teams. This includes the need for the initial transfer of knowledge between the two teams to be as complete as possible, as well as the requirement for long term conservation, dissemination and refinement of expertise from one maintainer to another. The long term transfer facility is provided by using the M^4 system during maintenance whilst initial transfer of knowledge is facilitated if the M^4 system is used from the outset of the development process.

Transferability: a by-product of the M^4 system development is the transferability of the underlying framework into other realms of information management. Although the M^4 system has centred around providing an environment for the development and maintenance of software systems it has become evident that there is enormous potential for expansion of the ideas into many other application areas requiring production and control of a mixed media type.

Although a working prototype exhibiting the above characteristics has evolved during the course of the research, there are still a number of ways in which the M^4 system can be improved both in relation to the degree of functionality offered by the M^4 system and in relation to the quality of development of the prototype system. For example, work is continuing on various aspects of the tool associated with risk management, project documentation frameworks, enhanced web and multimedia support. The M^4 system was developed as a proof of concept meta-CASE system and features of the tool have attracted considerable interest. For example a specific version of the tool concentrating on enabling large-scale knowledge management and personnel communication both locally and across the Internet has been developed for a major international company.

Acknowledgements

Thanks must go to Stuart Doggett and Kirsty-Anne Dempsy, final year project students and Frode Sandnes, Research Assistant, on the VES-GI project for their help in implementing parts of the tool-set. Much appreciation is also due to Professor Malcolm Munro for his support in formulating the ISCM process model and PISCES method.

References

1. Basili, V.R., *Viewing Software Maintenance as Re-use Oriented Software Development*, IEEE Software, Vol. 7, pp19-25, Jan. 1990.
2. Bennett, K.H., Cornelius, B., Munro, M. and Robson, D., *Software Maintenance*, in J. McDermid, ed. Software Engineer's Reference Book, Chapter 20, Butterworth-Heinemann, 1991.
3. Bhansali, P.V., *Universal Safety Standard is Infeasible - for Now*, IEEE Software, pp. 8-10, March 1996.
4. Boehm, B.W., *Software Engineering*, IEEE Transactions Computers, pp. 1226-1241, Dec. 1976.
5. Brooks, F.P., *The Mythical Man Month: Essays on Software Engineering*, Reading, Mass., Addison-Wesley, 1982
6. Dempsey, K-A., McCrindle, R.J. and Williams, S., *Multimedia Multi-Platform, Identification and Tracking System (Mu^{2}PITS)*, Final Project Report, Supervised by R.J. McCrindle, Department of Computer Science, the University of Reading, 1996.
7. Georges, M., Message from the General Chair, Proceedings, *International Conference on Software Maintenance*, France, 1995, IEEE Computer Society Press, 1995.
8. Hill, S. and McCrindle, R.J., *The Virtual Body Project*, Draft Paper, March 2001.
9. Iivari, J, *Why are CASE Tools not Used?*, Communications of the ACM, Vol. 39, No.10, pp. 94-103, October 1996.
10. Lano, K. and Malic, N., *Reengineering Legacy Applications using Design Patterns*, In Proceedings, Eighth International Workshop on Software Technology and Engineering Practice, pp. 326-338, London, July 1997
11. Lehman, M.M., Software'e Future: Managing Evolution, *IEEE Software*, pp. 40-44, January-February, 1998.
12. McCrindle, R.J., *Inverse Software Configuration Management*, PhD Thesis, The University of Durham, 1998
13. McCrindle, R.J. (nee Kenning) and Munro, M., *PISCES - An Inverse Configuration Management System*, Chapter 17 in Reuse and Reverse Engineering In Practice, Ed. P.A.V. Hall, Chapman & Hall, 1992
14. McCrindle, R.J., and Doggett, S. *The Multimedia Maintenance Interface System*, in Proceedings COMPSAC 2000, Taipei, Taiwan.
15. O'Connell, P. and McCrindle R.J., Using SOAP to Clean Up Configuration Management, Draft Paper, March 2001.
16. Parnas, D.L., *Software Ageing*, In Proceedings 16th International Conference on Software Engineering, pp. 279-287, 1994.
17. Shapiro, S., *Splitting the Difference: the Historical Necessity of Synthesis in Software Engineering*, IEEE Annals of the History of Computing, Vol. 19, No. 1, pp. 20-54, 1997.
18. Warwick, K., *An Introduction to Control Systems*, 2nd Edition, Advanced Series in Electrical and Computer Engineering, Vol. 8, World Scientific, 1996.

Visualisations; Functionality and Interaction

Claire Knight and Malcolm Munro

Visualisation Research Group, Research Institute in Software Evolution,
University of Durham, Durham, DH1 3LE, UK.
Tel: +44 191 374 2554, Fax: +44 191 374 2560
{C.R.Knight, Malcolm.Munro}@durham.ac.uk
http://vrg.dur.ac.uk/

Abstract. The interface to any visualisation application can essentially be split into two; the visualisation itself, and the interface that is associated with the interface. Despite the fact that the data plays a major part in any visualisation application, this is generally accessed only through either the interface or the visualisation. An important issue then arises over the location of functionality in the application. To be able to provide a usable and effective visualisation application (including data integration) where are the various selections, filters, customisations, and interrogation mechanisms located; the visualisation or the interface? The interaction mechanisms involved play a part in guiding this decision, but the amount of coupling between the visualisation and interface is also important.

1 Introduction

Visualisations are very direct interfaces to the data that forms their basis. Because of the variety of display and representation, and the problems of multiply overloading functionality onto glyphs there is usually a requirement for a separate, more traditional, interface that works in conjunction with the visualisation graphics. This can be extremely beneficial from a usability perspective; it provides an easier, more familiar, route into using any visualisation, and there can even be replication of functionality to allow for different users. The problem comes when designing such visualisations and their interfaces; where is the best place for interaction and functionality?

The likelihood is that there will be a split of functionality between the interface and the visualisation, and that in order to facilitate this functionality in the visualisation suitable interaction will be required. There is also the issue of which functionality to place where. Whilst some of this may be dependent on the data, on the tasks being supported, and even the user base of the visualisation it may be that general guidelines can be developed.

This paper presents some initial guidelines that have stemmed from the process of working with three-dimensional system and software visualisations. These have required the use of standard interface techniques as well as the various procedures that are concerned with visualisations. Three dimensional interfaces may suffer more from the perceptual issues associated with interfaces because of the expectation that much human-computer interaction occurs within the desktop metaphor, and the more obvious separation of being in a virtual space compared to clicking on windows. Desktop visualisation and virtual reality may lose the two-dimensionality aspect through clever use of graphics but there is still the screen restriction and the normal user positioning and devices to use the systems. This means that it is necessary to have controls and input areas somewhere thus the need for the issues presented in this paper to be considered.

2 Concatenation or Combination?

An important subject when discussing visualisations, interfaces, and for wont of a better term, visualisation applications, is where the division actually lies. For the user of a visualisation application (visualisation and the interface) this distinction should be trivial and relatively transparent. Indeed it should not really concern them at all. However, it is vital that this is considered by those who design and create visualisation applications as this will have a direct impact on the visualisation and the visualisation application usability [2].

There is obviously some degree of coupling between the interface and the visualisation because they work together to create the visualisation application. There is also the need to tailor the interface options to work with the data (as any visualisation has to) and to allow change of those aspects considered to be parameters to the visualisation. From a more established aspect, there is the expectation of what various parts of the interface will do. This has a direct impact on the visualisation if that interface links to the visualisation. Slider bars are popular ways of controlling various aspects of visualisations because of their familiarity and ease of use. A novel use of these would likely cause interface and visualisation usability problems for this very reason. In this case the control would be better integrated in another aspect of the application.

Essentially the uncertainty lies with whether a complete visualisation application consists of:

visualisation + interface   *[Concatenation]*

Or whether is can be seen as:

visualisation• interface   *[Combination]*

This is a subtle difference, but the important distinction is whether there is a bolting together of the two components, or whether they are actually combined together with a higher degree of coupling and awareness than a standard API would provide.

The concatenation of two components can create an application, but as to whether it is a good one because of the lack of awareness is an area that would require more investigation. Combination provides a more coupled interface and visualisation and thus defeats some of the principles of reuse but may create a better overall application. Also, with component technologies much improved, there is no reason why major parts of the interface (for example) cannot be sub-components that can then be reused with some form of overall data and visualisation specific co-ordination and parameterisation.

3 Visualisation Interaction

Visualisations have more use if they are interactive [1] as they then allow users to be able to work with, investigate, browse, interrogate, and generally view the various aspects of the data without relying on predefined fixed views. Whilst it is necessary to provide one or more representations and then the views that are valid for this representation in combination with the data (essentially the visualisation) there is still much user freedom if interaction is possible.

This then solicits the question of what interaction is allowed for a given visualisation. This is in two main areas; (a) interrogation style interaction with the graphics and hence the data, and (b) navigational interaction within the visualisation. With current technologies there is the ability to be able to incorporate various interaction mechanisms, but there is need to evaluate which are necessary for usability and operation rather than because it is technically feasible.

The first difference that any visualisation is likely to have over a standard graphical user interface is the number of degrees of freedom for movement and thus interaction. Whilst a normal interface may involve several discrete windows, or some amount of scrolling to incorporate all of the necessary data, this is the most that a user has to deal with. The contents of the windows (such as in a word processor) are not covered by this definition. In the same way that a two dimensional visualisation is a possibly infinite canvas contained within a window, any data interface is the same. There is the ability to scroll (pan) around this space, to zoom, to (if provided) utilise overviews such as table of contents or high-level images of the complete visualisation, and to be able to interact with the data contained within this creative space. Three-dimensional visualisations provide not only an extra dimension to consider, but also the rotation around these three axes; roll, pitch, and yaw. There is not this issue with two-dimensional visualisations, because whilst there might be the ability to rotate the visualisation around one of the axis there is generally the requirement that the other axis is fixed in order the display is still visible to the user. Therefore three-dimensional visualisations actually add another four degrees of possible movement. The navigation around the space should therefore take this into account. For two-dimensional displays standard mouse and keyboard interaction should not pose any problems. On the other hand, three dimensions provides an added challenge to the process of orientation and movement. This implies that the use of navigational aids is necessary in order to prevent a feeling of being lost [5, 7]. Pettifer and West [10] also relate the problem to the systems and metaphors in use today.

> *"Losing a cursor on the desktop is one thing, losing yourself in cyberspace is quite another."*

Three-dimensional worlds are potentially infinite whereas desktops are of generally finite space even if current implementations are able to cover several screens.

The interaction with graphical objects is also based around a familiar interaction method. Generally most three-dimensional visualisations are integrated into desktop applications therefore the use of gloves, and virtual spaces such as a CAVE [8, 9], are not the largest concern. Such interfaces would require more consideration than this paper is able to provide, and also would likely be able to learn much from the Virtual Reality community over interaction and usability. The usual interaction with graphical objects (regardless of dimension) is through selection, usually with a mouse click, and then the selection of the required detail or data refinement. These can be chosen through a variety of mechanisms; interface interaction with the selected graphic, the use of drop-down menus, right or multiple clicking.

The various ways of integrating with the visualisation for navigation and data interrogation have been summarised, but the issue for anyone who designs visualisations is to make justified decisions as to which are integrated into a visualisation application. An instinctive answer would be that incorporating all variations of interaction would provide the most usable interface because it would cater with the different demands of different users. This may not actually be the case because there would reach a point where the interaction styles would have to be overloaded on order to provide all of the necessary interaction. In other words, the same actions would have different functions depending on the context. Whilst data context may provide this (such as what detail is available from a specific type of graphical object in the visualisation) to do this with pervasive operations like mouse clicks is likely to cause problems for users. There is the associated issue that this then makes the interface more confusing and it then has a steeper learning curve. If a visualisation is too complex, compared to even rudimentary other analysis or understanding techniques, then it is not going to be used. This means that the related benefits that visualisation can provide of augmentation and amplification will be lost because of unrelated issues relating to the interface.

4 Functionality Location

Regardless of the permitted interactions with the interface and the visualisation of the visualisation application there are decisions as to be made as to where it is best to locate the functionality. Without any functionality, it is irrelevant what interaction and interface niceties are present as they will never have any impact on the data being visualised. Because interaction may be duplicated between the visualisation and the interface then it means that the functionality that is located behind that interaction needs to be consistent. Without some degree of detailed coupling between interface and visualisation this could not be guaranteed. They would both need to utilise the same routines (at an implementation level) therefore encapsulating this and making it directly accessible to both visualisation and interface, with adequate linking to enable the impact of that interaction (such as filtering the data) to actually affect the visualisation and interface.

Functionality can also be only available from one of the two main parts of a visualisation application. In these cases the question becomes not one of implementation and ensuring consistency and accuracy, but of being able to locate the functionality in the best place for its use (usability) and most effectiveness based on the actions it actually encompasses. There are four possible choices:

1. *From the visualisation only*
2. *From the interface only*
3. *From a combination of the visualisation and interface (requiring both)*
4. *From both the visualisation and the interface (duplication)*

Since the effectiveness of a visualisation (application) has dependencies on both the underlying data and how well the representations reflect this, and on the tasks that it is intended to support [4], then this may dictate which of these options applies to each piece of functionality. It may also be there is user demand for the inclusion of functionality in a particular place. Since the interface is generally included to augment the visualisation, then there will probably be little functionality that is restricted to the interface, but the same cannot be said of the visualisation. As a rule of thumb to guide the decision making process the following applies:

1. *If the functionality has a local impact (such as interrogating a graphical object) then it should be attached to the visualisation. Duplication through interface provision is also feasible.*
2. *If the functionality has a global impact (such as over all of the visualisation) then it is best located on the interface but combined with the visualisation.*
3. *If the functionality is related to the application (such as preferences) then it should only be attached to the interface.*

These guidelines make the distinction between global and local. If would make little sense if one wanted to change all glyphs representation a particular type of data to have to do each one individually. Likewise, selecting one object in order to obtain more detail about it does not make sense if applied to the global scale of a visualisation. There is some crossover, such as when the detail of an object then dictates that this should then become the focus of a new query (for example). In cases such as this a local impact would move to having a global one and therefore the integrated functionality guideline above would come into play. Despite all of this, these are only guidelines and doubtless certain data sets will dictate that they cannot be followed directly.

5 Conclusions

This paper has examined the interaction between the visualisation and interface components of a visualisation application. This is important to be able to consider where is the best place to locate functionality in the application. This has impacts on usability and therefore use of the application, and also for implementation and design of visualisations.

Related work by the authors [3] involves the development of a component-oriented framework for visualisations and data sources. This allows for multiple data sources to be integrated with multiple visualisations. The development of the framework so far dictates that the various visualisation views that are applicable and their control panel (the interface in the context of this paper) are a complete component unit. This allows for the required level of coupling considered necessary to fulfill the various guidelines expressed in this paper.

This framework development provides a perfect opportunity for developing further these ideas, and for then investigation aspects such as linking between visualisation styles and views (that are not already linked), the impact that cross data and cross visualisation annotation has on the location of functionality, and the issues associated with dealing with multiple data sources and multiple visualisations. This paper provides much further work, and also seeks to try and add science to the creative process that currently surrounds visualisations [6]. There is always going to be a create element to any visualisation because of the graphic design involved, but there is not reason why solutions cannot be engineered through the other aspects of designing and creating visualisations.

Acknowledgements

This work has been supported by the EPSRC project VVSRE; Visualising Software in Virtual Reality Environments.

References

1. Chalmers, M., Design Perspectives in Visualising Complex Information, Proceedings IFIP 3rd Visual Databases Conference, March 1995.
2. Neilson, J. (1993). Usability Engineering, Academic Press Professional Publishing, San Francisco, 1993.
3. Knight, C., and Munro, M., m-by-n Visualisations, Submitted to IEEE Visualization, Information Visualization Symposium 2001, University of Durham, Department of Computer Science Technical Report...
4. Knight, C., Visualisation Effectiveness, submitted to Workshop on Fundamental Issues of Visualisation, Proceedings of CISST, 2001, University of Durham, Department of Computer Science Technical Report...
5. Ingram, R. and Benford, S., Legibility Enhancement for Information Visualisation, Proceedings of IEEE Visualization '95, October 30 - November 3 1995.
6. Card, S., Mackinlay, J., and Shneiderman, B., (Editors), Readings in Information Visualization: Using Vision to Think, Morgan Kaufmann, February 1999.
7. Dieberger, A., Providing spatial navigation for the World Wide Web, Spatial Information Theory, Proceedings of Cosit '95, pp93-106, September 1995.
8. Leigh, J., and Johnson, A. E., Supporting Transcontinental Collaborative Work in Persistent Virtual Environments, IEEE Computer Graphics & Applications, Vol. 16, No. 4, pp47-51, July 1996.
9. Leigh, J., Rajlich, P. J., Stein, R. J., Johnson, A. E., and DeFanti, T. A., LIMBO/VTK: A Tool for Rapid Tele-Immersive Visualization, Proceedings of IEEE Visualization '98, October 18-23, 1998.
10. Pettifer, S., and West, A., Deva: A coherent operating environment for large scale VR applications, Presented at the first Virtual Reality Universe conference, April 1997.

DMEFS Web Portal : A METOC Application

Avichal Mehra and Jim Corbin

Integrated Data Systems Laboratory
Engineering Research Center
Mississippi State University
Stennis Space Center
MS 39529
Mehra@erc.msstate.edu

Abstract. Distributed Marine Environment Forecast System (DMEFS) is a research testbed for demonstrating the integration of various technologies and components prior to operational use as well as a framework in which to operate validated meteorological and oceanic (METOC) numerical models. The focus of the DMEFS is to create an open computational web portal for a distributed system for describing and predicting the marine environment that will accelerate the evolution of timely and accurate forecasting. The primary goals are to first, focus the adaptation of distributed (scalable) computational technology into oceanic and meteorological predictions, and secondly, to shorten the model development time by expanding the collaboration among the model developers, the software engineering community and the operational end-users. The web portal provides a secure, seamless access to high performance resources, hiding their complexity. It is extensible and is designed for rapid prototyping, validation and deployment of legacy computational models as well as new models and tools by providing a set of common toolkits.

Background

Oceanic prediction in the littoral is vastly more complex than in the open ocean. The temporal and special scales of variability are much shorter; the parameters of interest, such as waves, storm surge, optical clarity, tide, sediment transport, beach trafficability, currents, temperature, salinity, etc., are very different. In addition, the real time or even historic observations of these parameters, in many areas of potential interest, are either very restricted or limited at best. The ability to simulate the littoral environment rapidly, accurately, and across many temporal and spatial scales poses very significant challenges to gathering, processing, and disseminating information. At the same time, underlying computer architectures have undergone an equally drastic change. No longer in the business of making vector supercomputers, U.S. computer manufacturers produce multiprocessor computers made from commodity processors with a high-speed interconnect.   While such architectures offer vast improvements in cost versus performance, software written for vector supercomputers must be extensively rewritten to run on these new architectures. Thus, there is an emerging demand for scalable parallel models that not only can run on current architectures, but also can be easily adapted to new ones [1]. Finally, the exponential growth in data communications, exemplified by the Internet infrastructure and web applications, are enabling dramatic advances in distributed computational environments.

The Distributed Marine-Environment Forecast System (DMEFS) [2] attempts to address three essential goals. First, it incorporates the latest advances in distributed computing. Second, it provides the means of substantially reducing the time to develop, prototype, test, and validate simulation models. Third, it supports genuine, synergistic collaboration among computer specialists, model developers, and operational users. Several critical components that such a system must have to advance these three goals are discussed in this paper.

Importance of DMEFS to the Navy

Currently, most models are developed within the R&D community as stove-pipe systems with unique requirements for control of execution, inputs and outputs. All of which are accomplished by unique execution of scripts that usually function properly only in the original research environment within which they were developed. When the model is transitioned to the operational community, the scripts have to be re-written to accommodate the new operating environment, new data inputs, and new product outputs. As a result, many man-hours are needed just to get the new model to function in the new operating environment.

DMEFS provides a virtual common computer operating environment within which to both develop the new models and to operate them once validated. Secondly, as new models have been introduced into the operational community, each one usually uses unique applications to process input and output data, such as quality control, data interpolation schemes, and graphics. Therefore an operator must know several different ways to do the same task depending on the model of interest. Within DMEFS, from the user's perspective, the same functions will be done the same way through a common graphical user interface. Thus enormous time is saved in training and indoctrination of new personnel or when introducing a new model. Thirdly, in the current way of doing business, when a program sponsor needs updated information regarding a project or status of model, he must ask for input from the developer or an operator. With the web portal access of DMEFS, any authorized user can come directly into the system to obtain information and graphics. And finally, while most computer software developments within the Navy research community are beginning to use industry standards, most METOC model development does not. Developing new models within the DMEFS infrastructure allows the model developers to exploit the latest in industry standard software engineering technology without significant changes to their underlying code.

DMEFS Web Portal

The DMEFS web portal targets the creation of an open, extensible framework, designed for rapid prototyping, validation, and deployment of new models and tools. It is operational over evolving heterogeneous platforms distributed over wide areas with web-based access for METOC tools and for forecast-derived information. The open framework is critical to facilitate the collaboration among the researchers within the community.

DMEFS web portal is a "meta-system" that provides the model developer with desktop access to geographically distributed high performance computational resources, both hardware and software. The portal is developed using enabling high performance and commodity technologies and services. It provides the foundation to facilitate the rapid development, integration, testing, and deployment of various meteorology-oceanography meta-applications. It simplifies the applications programming environment and nurtures a collaborative environment for users in a virtual workplace. The resources owned and managed by the stakeholders are made available in a controlled and secured manner.

DMEFS Portal Services

The DMEFS is a service oriented architecture to support high performance distributed computing.. Any METOC application embedded in DMEFS or developed using DMEFS may also take advantage of the various services provided by DMEFS.

A set of services provided by the DMEFS Web Portal currently under development are (see Figure 1):

Metacomputing Services: These services include a spectrum of high performance scalable computers (supercomputers) with sufficient computational power to perform the required calculations and an extensive set of software tools to aid in model development. These computers may be distributed across a local area network and/or a wide area network. The computational facilities include experimental and operational high speed networks, digital libraries, hierarchical storage devices, visualization hardware, etc. It is essential to provide services to facilitate their discovery and use. Security and access control services enable the resource providers to offer the use of the resources in a controlled manner using a flexible access control policy. Resource management services allow the registration and scheduling the availability of the resources in a metacomputing environment. A few preliminary glimpses of such an environment with screen-shots are presented in [3].

METOC Model Services: METOC Model Services is a suite of realistic numerical oceanic and atmospheric models optimized for the new scalable computer architectures, showing useful forecast skill on time and space scales of scientific interest. A suite also consists of model test cases and skill assessment procedures for model validation, comparison and testing.. Since the most skillful forecasts are made by coupled model-observation systems with data assimilation and real-time adaptive sampling, a forum for testing methods for assimilation of the mesoscale *in situ* and remote observations including radar observations will also be provided in future. Some other issues which can be studied include treatment of open boundary conditions including multi-scale nesting. Interfaces to the METOC Model Services enables the scheduling, execution, and validation of the model results.

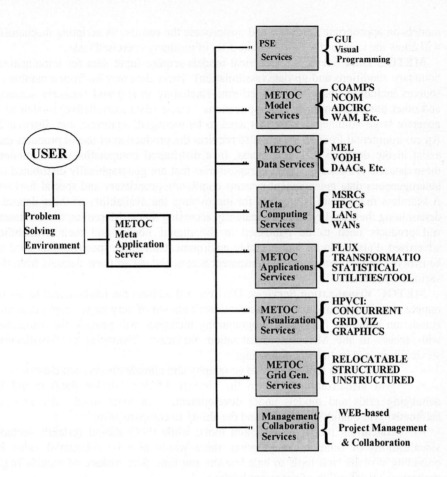

Fig. 1. DMEFS Portal Services

DMEFS Portal Services (Cont.)

A few more services to be provided by DMEFS web portal in the future are:

Problem Solving Environment (PSE): The DMEFS PSE will provide a visual programming paradigm similar to that employed by a number of scientific visualization software such as AVS, SGI/Explore, KHOROS, etc. It will allow the scientists and engineers to work comfortably and intuitively in an environment that allows the users to set up their task and data flow modules and connect them via data pipes to compose and execute applications. This model and/or the associated visual programming paradigm has been adopted by other metacomputing initiatives. The PSE in DMEFS will allow the users to authenticate their identity, discover the availability of numerical models and couple them appropriately; discover, acquire, and preprocess appropriate initialization and assimilation data; set-up and execute the models on appropriate hardware and postprocess the results. A scripting mechanism will allow the user to store and execute a set of routinely exercised tasks.

METOC Data Services: Numerical models require input data for initialization, boundary conditions and up-date (assimilation). These data may be from a number of sources including observational platforms (including in-situ and remotely sensed), and other numerical models and applications. These (data assimilative) models also generate large volumes of data that need to be managed, analyzed, and distributed. An environmental forecast system also requires the production of useful products that assist in the decision making process. In a distributed computational environment these datasets are usually stored in repositories that are geographically distributed on heterogeneous data management systems employing proprietary and special formats. A seamless mechanism is required for discovering the availability of these datasets, determining their quality, and transparently accessing them. Moreover, these datasets and products need to be published in the digital library and their availability advertised. Further, in an interoperable environment interface specifications need to be evolved and implemented for transparent access and use of these datasets from the various repositories.

METOC Visualization Services: DMEFS will address the fundamental issues of remote visualization and the management/manipulation of very large geophysical and simulation datasets. The visual programming metaphor will provide the researcher with access to the METOC Visualization Services. Examples of visualization services required include the following:

- visualization tools to browse geography and climate/observation databases
- grid visualization tools (in the absence of any solution data) to aid in debugging grids and models under development; such tools should also provide automated grid quality diagnostics and the ability to compare grids
- concurrent model visualization tools; while these should certainly include visualizations of solutions in progress, there would also be substantial value in providing visualization tools to examine the run-time performance of models (e.g., concurrent visualization of processor loads).

With distributed concurrent visualization, the user will be able to set up and run a simulation in a remote computing environment and then utilize the equipment in the remote compute environment to do part of the visualization task, potentially concurrently.

METOC Grid Generation Services: DEMFS will develop a robust, economical, and efficient structured/unstructured grid generation system tailored to support ocean and atmospheric models within the DTA framework. The METOC Grid Generation Services will support the generation of quality 2-D, 2.5-D and 3-D structured/unstructured grids. It will seamlessly obtain geometric and ocean characteristics by interrogating the DMEFS METOC Data Services.

Application Analysis Services: A spectrum of utilities and tools is required to extract information from the very large geophysical and simulation datasets for various reasons (e.g., data compression for transmission and storage, feature extraction for data steering and adaptation, spatial and time scaling or transformation for coupling different applications, and extract statistical characteristics of results). As an example, an analysis process may be run "adjacent" to a compute intensive storm surge simulation, perhaps concurrently, to extract information of the response (compressed and scalable in an appropriate "metacomputing" format) for transmission, storage, and subsequent transformation to the specific requirements for a receiving process for structural simulation at a distant location. Tools are also required to extract the macromodels for parametric simulations.

Management and Collaboration Services: Tools and data services are desirable for posting and reviewing program and project status, targeting two levels of access: internal for use by the DMEFS management and researchers and for appropriate external use by sponsors. A convenient WEB-based project management tool, which targets ease of use, is anticipated to provide means to quickly enter and modify goals at the individual and different organizational levels while providing appropriate access for cross-disciplinary research in a virtual organization. In addition, there is the need to establish real-time collaborative tools between remote researchers to facilitate hands-on demonstrations and communications (e.g., shared applications associated with desktop teleconferencing).

DMEFS Portal Users

Overall, three different groups of users of a fully functional DMEFS web portal can be identified. These are (as indicated in Figure 2) researchers, operators and managers.

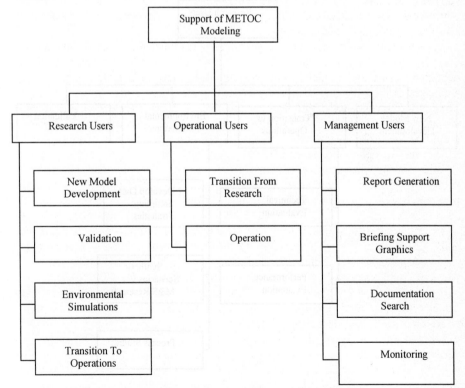

Fig. 2. High Level View of DMEFS Web Portal Users

The high level view of the major tasks performed by the three classes of users is shown in Figure 2. Figures 3, 4 and 5 present a further breakdown of those high level tasks in sub-tasks or objectives from each user perspective. The research support functions require the most flexibility and demand for real time system configuration. The basic tasks for a research user can be broadly defined under four categories: new model development, transition to operations, environmental simulations and validation. These tasks can be broken down into further sub-tasks as shown in Figure 3. The operational user is more focused on product generation and less on processes, but needs more automation and tailored environment for routine run streams for both operations (to run a model or to derive products from archived model runs) and transitions. These activities lead to other sub-tasks as shown in Figure 4.

The management user focus's more on monitoring status. He or she can be from research or operations, but can also represent project, program or sponsor management, or users from the public domain. These users are distinctly high-level users of the portal and would use web access primarily for information only. This information could be used for generating reports, making briefs, monitoring the status of the project and/or searching for relevant documents and references.

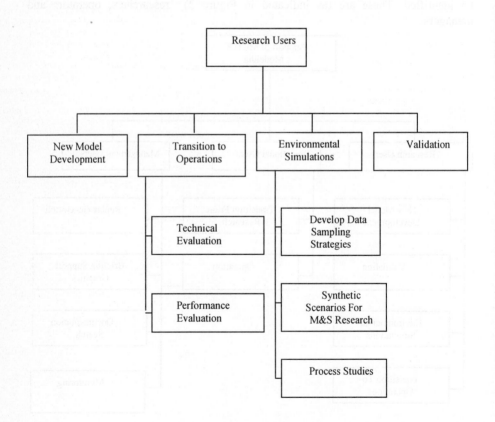

Fig. 3. Research Users for DMEFS Web Portal

483

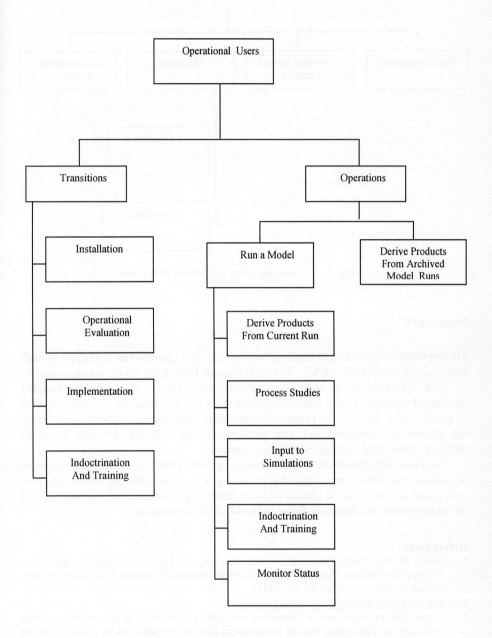

Fig. 4. Operational Users of DMEFS Web Portal

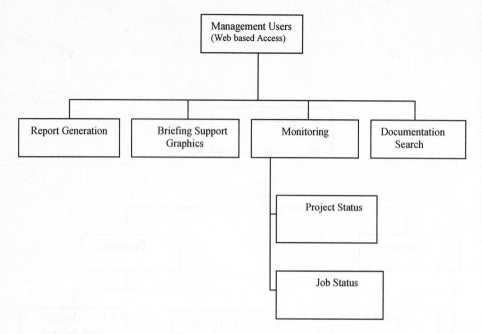

Fig. 5. Management Users of DMEFS Web Portal

Summary

The exponential growth in computing technologies has changed the landscape in high performance computing (HPC). The migration of HPC from vector supercomputers to both tightly-coupled shared-memory multiprocessors and loosely coupled distributed multiprocessors has complicated software development and maintenance of legacy codes. Second, the exponential growth in data communications, specifically the Internet infrastructure and web applications, are enabling dramatic advances related to distributed computational environments.

Distributed Marine-Environment Forecast System (DMEFS) web portal attempts to address the above two issues by providing a suite of services which are pre-configured for three varied classes of users namely the research, the operational and the management user typically found in the METOC community.

References:
1. Joseph W. McCaffrey, Donald L. Durham, and James K. Lewis, "A Vision for the Future of Naval Operational Oceanic Nowcast/Forecast Systems," Marine Technology Society Journal, Vol. 31, No. 3, pp. 83-84 (1997).
2. http://www.erc.msstate.edu/~haupt/DMEFS.
3. Tomasz Haupt, Purushotham Bangalore, and Gregory Henley, " A Computational Web Portal for the Distributed Marine Environment Forecast System," to be presented at the IEEE International Symposium on Cluster Computing and the Grid, 15-18 May, Brisbane Australia (2001).

Validation Web Site:
A Combustion Collaboratory over the Internet

Angela Violi[1], Xiaodong Chen[2], Gary Lindstrom[2], Eric Eddings[1],

and Adel F. Sarofim[1]

[1] Department of Chemical and Fuels Engineering, University of Utah
84112 Salt Lake City, Utah
[2] School of Computing, University of Utah
84112 Salt Lake City, Utah

Abstract. The Soot Model Development Web Site (SMDWS) is a project to develop collaborative technologies serving combustion researchers in DOE national laboratories, academia and research institutions throughout the world. The result is a shared forum for problem exploration in combustion research. Researchers collaborate over the Internet using SMDWS tools, which include a server for executing combustion models, web-accessible data storage for sharing experimental and modeling data, and graphical visualization of combustion effects. In this paper the authors describe the current status of the SMDWS project, as well as continuing goals for enhanced functionality, modes of collaboration, and community building.

1 Introduction

The emergence of the Internet has enabled interactions among people working in the same field around the globe. This has inspired many efforts to address the associated problems of effective group communication, and provide access to information sources that are geographically and administratively disparate. Collaboratories are one way to provide virtual shared workspace to users. A Collaboratory is a wide area distributed system supporting a rich set of services and tools that facilitate effective synchronous and asynchronous communication between two or more people who are not co-located [1]. To date, collaboratories have been developed in many areas including material science [2], natural science [3], telemedicine [4], space physics [5], and environmental molecular sciences [6]. In the combustion field, web-based collaboratories have been developed in the areas of turbulent diffusion flames [7] and diesel engines [8].

The University of Utah Center for the Simulation of Accidental Fires and Explosions (C-SAFE) [9] is an interdisciplinary project funded under the DOE Accelerated Strategic Computing Initiative Alliances program. This center focuses on providing large-scale high performance science-based tools for the numerical simulation of accidental fires and explosions, targeted at issues associated with safe handling and storage of highly flammable materials.

The Validation Web Site (SMDWS) [10] is in turn a research project under C-SAFE providing a shared problem exploration environment for combustion researchers. The SMDWS enables researchers to work together across geographic and organizational boundaries to solve complex interdisciplinary problems, and to securely share access to resources at participating web including specialized databases and computational servers. In short, the goal of the SMDWS project is to make science and information exchange more efficient, opportunistic and synergistic for combustion researchers at large.

This paper discusses the application area, the SMDWS requirements, and the SMDWS architecture and design.

2 Application: Combustion Modeling

Soot formation is a significant environmental problem associated with the operation of many practical combustion devices.

An accurate simulation of soot formation is critical to the C-SAFE goal of predicting the radiative heat transfer in fires, since soot is a dominant component of hydrocarbon fires. High heat flux and emissions from accidental fires pose significant human and property hazards.

The Soot Model Development Web Site has been developed to allow the C-SAFE group working on the development of soot models to collaborate with researchers at LLNL, LBL, SNLA, SNLL, UC Berkeley, Polytechnic of Milan, and the University of Naples. Model development is a complex task requiring a close interplay among experimentation, development of chemical submodels, model validation and formulation of new modeling concepts. Progress in this already difficult task is further complicated by the interdisciplinary nature of the required collaboration, as well as the geographical distribution of experts in the field throughout the national laboratories, industry, and universities. Effective model development requires validation of individual model components as well as strategies for their integration. For example, C-SAFE's soot modeling task has the specific goal of developing reliable, experimentally validated models of soot formation that can be used in a numerical simulation and optimization of combustion devices. Success in this task clearly requires the critical comparison and co-design of soot formation mechanisms being developed both within and outside of C-SAFE. Hence a coordinate effort between the many researchers is required to obtain the best soot formation mechanism possible. It was judged early in the C-SAFE project that development of an interactive, web-based tool for this purpose was essential to achieving this goal. The C-SAFE Validation group has adopted the role of leading this effort and is hosting an electronic interaction facility with the assistance of Computer Scientists within the C-SAFE team. We have collected several combustion modeling codes and made them available to the SMDWS over the Internet: *ChemkinII*, a chemical kinetics package for the analysis of gas-phase chemical kinetics; *ChemkinII/Soot* [11], which is *Chemkin* with kinetic modeling of soot formation; and *DSmoke* [12], which supports the analysis of different reactors in series or in parallel coupled with a mixer or splitter. Several reaction mechanisms are also available at the SMDWS: Lawrence Livermore National Laboratory's mechanisms for the formation of aromatic compounds and other species [13,14], kinetic mechanisms by Prof. Frenklach's group at the University of California at Berkeley [15,16], reaction schemes by Polytechnic of Milan, Italy [17], etc. The SMDWS also provides links to important public databases in the area (e.g., www.ecs.umass.edu/MBMS/) for browsing and comparison with data contributed to the SMDWS web site. Also provided is a database of premixed flames data collected from the literature. The modeling results obtained by Utah's own soot combustion group are reported on the web to invite critical feedback and solicit additional contributions from other research groups. The goal of the SMDWS is to make doing science, engineering and information exchange among research partners more efficient and synergistic. Researchers are able to tackle new problems with their existing methods and to access new methods through collaborations that were previously infeasible due data and software interchange obstacles.

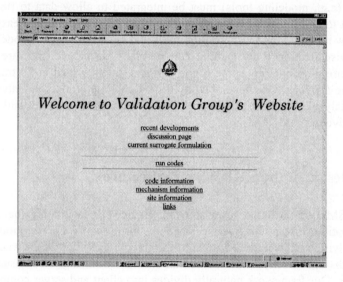

Fig. 1. Soot Model Development Web Site welcoming page

3 Soot Model Development Web Site Requirements

The SMDWS has defined a specific set of requirements in order to accomplish these goals. The researchers at different sites must be able to:

- Share graphical data easily using web browsers
- Discuss modeling strategies and easily share model descriptions between groups
- Archive collaborative information in a web-accessible electronic notebook
- Utilize a remote execution system to run combustion models at remote locations and
- Readily explore and analyze experimental data and modeling results through web browsers

At present a first approximation to these goals has been implemented, including use of our web server as the only computational host, and a rudimentary file based electronic notebook facility. Many of zero-dimensional chemistry simulations (such as *Premix* of the Chemkin software) can run for hours to days on high-end workstations and produce kilobytes of data. The resulting data sets are visualized using simultaneous 2-D plots of multiple variables. In addition to dealing with the practicalities of such scale issues, many of the collaborative tools (e.g., modeling tools) must be integrated to allow data to flow easily between tools. The SMDWS architecture must support rapid and easy integration of new modeling and visualization tools and newly developed models. Modeling and visualization tools include commercially developed software, as well as software developed by some of the partners, often written in Fortran (for example the *DSmoke* code developed by Prof. Ranzi's group at the University of Milan, Italy). Users must be able to locate and combine resources and data to form effective teams.

4 Design of the SMDWS Architecture

The SMDWS is based on a software framework facilitating the addition and integration of computing services and supporting tools. Careful thought as to the underlying architecture is critical to the smooth evolution of the web site to meet growth and diversification of collaborative, modeling, and visualization demands. Our framework naturally divides into client and server components. The server component has the following properties:

1. Support for running codes from local and remote contributors, including pre- and post-processors
2. Validation of required and optional input
3. Delivery of output files as specified by client

4. Browsing and query-driven inspection of arbitrarily large output files
5. Return of partial results upon failure, with diagnostic reports.

The SMDWS server architecture consists of two parts: a web server and database server. Both parts run under Window NT Server 4.0.

1. Our web server uses an Apache 1.3.14 web server, supplemented by an Apache JServ 1.0 servlet engine. One multipurpose Java servlet (under JSDK2.0) receives two kinds of requests from clients:

    a) Get the input parameters and input data files for simulation run from client side, then call a DOS shell script and an Expect script (Expect 5.21) to run the executable codes, and finally store the results in the database.
    b) Receive a database query from client side, then get the database response and return the results to the client. Client web pages provide input parameters through a Java Applet (JDK1.1.8) interface. In addition, we have provided an optional username and password protected version of our website where Raven SSL 1.4.1 provides Digest Authentication for web security.

2. Our database server runs on Microsoft SQL Server 7.0. The Java servlet communicates with SQL Server through an RmiJdbc server and a JDBC/ODBC bridge.

The client framework is Java code supporting the integration of contributed combustion modeling codes (rarely in Java) to be integrated into the SMDWS. Properties include:

*   Status monitoring of long running execution requests with periodic updates
*   Termination and subsequent recovery of intermediate results
*   Flexible invocation of all computational resources, including pre-processors and post-processors
*   Definition and upload of required and optional input files
*   Inspection and download of expected output files
*   Parameterizable visualization, currently limited to 2D plotting

Employing client and server frameworks allows new capabilities to be added easily, facilitates maintenance of SMDWS client and server code, maximizes code reuse and enables integration with third party components such as web and database server tools.

Future development of the SMDWS will support electronic notebooks and shared workspaces where researchers can annotate combustion codes, input data sets, and result files --- all accessible through their web browsers. The electronic notebooks will contain a rich variety of information – text and images, and links to information can be simultaneously viewed, edited and annotated on the pages of a notebook. Similarly, it is planned that shared workspaces will be made interactively accessible using a web browser, enabling users to dialog about modeling codes, input files, output files, data plots, and notes on the computational models.

A 2-D plotting package using the Java 2D Graph Package v2.4 has been integrated into the SMDWS modeling client template, such that users can interactively plot modeling results from several modeling runs. Users can then save plots as files or save plot files in a file repository.

5 Collaborative Tools in the SMDWS

The challenges involved in successful deployment of the SMDWS include development and implementation of an appropriate collaboratory infrastructure, integration of new interdisciplinary tools, and the deployment and adoption of the SMDWS among the research community. To accomplish this, we have planned a systematic approach to the development, integration, and testing of tools. The initial phase of deployment has been to directly involve the most interested combustion researchers in tool development and testing. The SMDWS has been found to have highest impact when geographically separated researchers are motivated to interact directly with each other to achieve their goals. The focus effort provides direct support to researchers as they confront some of the cultural changes inherent in electronic collaboration. These changes include the use of new technologies and the development of social behaviors in electronic communication and collaboration [1]. One challenge in particular is the balance of the increased access to each other data, resources, and presence with the desire to invoke electronic security to protect the same.

The SMDWS project will be broadening the target research to multiple projects including a broader range of collaborators and the dissemination of new technology to the other interested groups. The broadening of the target research areas will provide a greater test of the generality of our approach and an opportunity to incorporate some advanced tools.

5.1 Examples of SMDWS application

Descriptions of illustrative scenarios in the SMDWS serve to further describe its application in combustion research. Here we provide a few brief descriptions:

- A user can retrieve previous results generated by a collaborator from a shared workspace, interpret this data then rerun modeling code with new inputs, and put new results in the shared workspace that can be viewed by collaborators.
- A scientist can start a modeling run in collaboration with a distant colleague. The results will be stored in a repository for both to see.
- The combustion scientists upload input files from local machines for remote execution of the code using the respective SMDWS clients. Output files are returned to the scientist's machines and the modeling results could be immediately analyzed using the SMDWS client's interactive 2D plotting package (see Figure 2).

All the calculations are documented in web-based files that can be accessed by collaborators. As mentioned above, our ultimate goal is for the combustion scientist to use real-time collaboration tools to discuss results with colleagues.

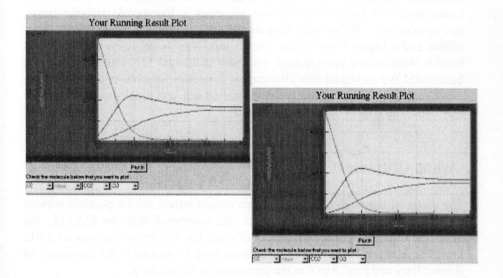

Fig. 2. Calculation results for some species concentrations

6 Status and Future Work

The SMDWS project is continuing to develop collaborative technologies for combustion researchers distributed throughout the national laboratories and academia. We have assembled an overall soot formation mechanism starting from a surrogate hydrocarbon fuel that has been experimentally developed to reproduce the features of jet fuels important for pool fires. This mechanism draws on experimental and *ab initio* approaches to obtaining rate data for the elementary reaction assembled from the literature. Validation of the mechanism has been performed by comparison of the soot formation mechanism with measurements from laboratory flames. Thus for the site permits the remote running of the soot models on a local server, thus providing a means for all collaborators work with the same version of mechanisms and models. The site also provides a Discussion Page for dialogue on the results of mechanism comparisons. User friendly and low maintenance security is a difficult challenge and will require continuous development with user feedback for some time. In the future we expect to offer a real-time conferencing capability. Ultimately, desktop video and data conferencing among researchers will be investigated as appropriate Internet connection bandwidth becomes available.

Collaboratory technology is having an impact, however. One of the most significant impacts of the SMDWS is to extend the researcher's capabilities to use new tools by making possible the use of such tools in collaboration with an expert at another location. The ability to handle such knowledge resources is a key to interdisciplinary collaboration and an important characteristic of the SMDWS.

Future work will build on the current successes in framework implementation, tool development. We are searching simultaneously both a greater range of capabilities and a higher level of user friendliness. New target research groups will involve international participation with new challenges for interoperability and bandwidth. We anticipate new client/server templates for multi-code simulations, new capabilities for collaborative visualization and group collaboration built on new tools, and improved commercial tools.

Acknowledgement

The scope of the work addressed in this project would not be possible without the efforts, support of the group of people associated with the C-SAFE, the School of Computing at the University of Utah, Dr. Westbrook's group at LLNL, Prof. Frenklach at University of California, Prof. D'Alessio at the University of Naples, Italy and Prof. Ranzi at the University of Milan, Italy.

This research is funded by the University of Utah Center for the Simulation of Accidental Fires and Explosions (C-SAFE), funded by the Department of Energy, Lawrence Livermore National Laboratory, under subcontract B341493. Additional support was provided by the University of Naples "Federico II", Department of Chemical Engineering.

References

1. Kouzes R.T., Myers J.D. and Wulf W.A., Collaboratories: Doing Science On the Internet, IEEE Computer, volume 29, number 8, August 1996

2. Parvin B., Taylor J. and Cong G., Deep view: a collaborative framework for distributed microscopy, Proceedings of SC 98, Orlando Florida, Nov.1998

3. Sunderam V.S., Cheung S.Y., Hirsh M. et al., CCF: collaboratory computing frameworks, Proceedings of SC 98, Orlando Florida, Nov.1998

4. Kilman D.G. and Forslund D.W., "An international collaboratory based on virtual patient records, Communications of the ACM, volume 40, n.8, (1997)

5. Subramanian S., Malan G.R., Sop Shim H. et al., "Software Architecture for the UARC web-based collaboratory", IEEE Internet Computing, volume 3, n.2, (1999), pg. 46-54

6. Payne D.A. and Myers J.D., The EMSL collaborative research environment (CORE) – collaboration via the World Wide Web, Proceedings of the Fifth Workshop on Enabling Technologies: infrastructure for collaborative enterprises (WET ICE'96), Stanford, California, June 1996

8. http://www.ca.sandia.gov/tdf/Workshop.html

8. http://www-collab.ca.sandia.gov/dcc/

9. http://www.C-SAFE.utah.edu/

10. http://panda.cs.utah.edu/~validate

11. Kenneth L. Revzan, Nancy J. Brown and Michael Frenklach, at http://www.me.berkeley.edu/soot/

12. Ranzi E., Faravelli T., Gaffuri P. and Sogaro A., *Combustion and Flame* 102: (1995), 179-192

13. Marinov, N.M., Pitz, W.J., Westbrook, C.K., Castaldi, M.J., and Senkan, S.M., *Combustion Science and Technologies,* 116-117, (1996), 211-287

14 Curran, H. J., Gaffuri, P., Pitz, W. J., and Westbrook, C. K., *Combustion and Flame* 114, (1998), 149-177

15. Hai Wang and Michael Frenklach, *Combustion and Flame* 110, (1997), 173-221

16. Joerg Appel, Henning Bockhorn, and Michael Frenklach, *Combusion and Flame* 121, (2000), 122-134

17. Ranzi E., Sogaro A., Gaffuri P., Pennati G., Westbrook C.K., Pitz W.J., *Combustion and Flame* 99, (1994), 201-211

The Policy Machine for Security Policy Management

Vincent C. Hu, Deborah A. Frincke, and David F. Ferraiolo
National Institute of Standards and Technology, 100 Bureau Dr. Stop 8930 Gaithersburg
Maryland 20899-8930, USA
{vhu, dferraiolo}@nist.gov
Department of Computer Science of the University of Idaho, Moscow Idaho 83844, USA
frincke@cs.uidaho.edu

Abstract. Many different access control policies and models have been developed to suit a variety of goals; these include **Role-Based Access Control**, **One-directional Information Flow**, **Chinese Wall**, **Clark-Wilson**, **N-person Control**, and **DAC**, in addition to more informal ad hoc policies. While each of these policies has a particular area of strength, the notational differences between these policies are substantial. As a result it is difficult to combine them, both in making formal statements about systems which are based on differing models and in using more than one access control policy model within a given system. Thus, there is a need for a unifying formalism which is general enough to encompass a range of these policies and models. In this paper, we propose an open security architecture called the *Policy Machine* (*PM*) that would meet this need. We also provide examples showing how the *PM* specifies and enforces access control polices.

1 Introduction

Access control is a critical component of most approaches to providing system security. Access control is used to achieve three primary objectives: (1), determining which subjects are entitled to have access to which objects (Authorization); (2) determining the access rights permitted (a combination of access modes such as read, write, execute, delete, and append); and (3) enforcing the access rights. An access control policy describes how to achieve these three goals; to be effective, this policy needs to be managed and enforced. There is a vast array of techniques that define and enforce access control policies within host operating systems and across heterogeneous bodies of data [NCSC98]. Although these techniques are successful in the specific situations for which they were developed, the current state of security technology has, to some extent, failed to address the needs of all systems [Spencer et al99, HGPS99] in a single notation. Access control policies can be as diverse as the applications that rely upon them, and are heavily dependent on the needs of a particular environment. Further, notations that easily express one collection of access control policies may be awkward (or incapable) in another venue. An example of this situation would be when a company's documents are under **One-direction Information Flow** [BL73, Biba77, Sand93] policy control at the development stage. When the development is finished, the documents that are available for use by employees, could then be required to be controlled by a role-based or **RBAC** [FCK95, SCFY96] policy. Most existing commercial technologies used to provide security to systems are restricted to a single policy model, rather than permitting a variety of models to be used [Spencer et al99]. For instance, Linux applies a **DAC** [NCSC87] policy, and it is difficult to implement **RBAC policy** (among others) in such a system. Further, if an organization decides to change from one policy model to another, it is quite likely that the new policy model will have to be implemented above the operating systems level, perhaps even as part of the application code or through an intermediary. This is inconvenient, subject to error, slow, and makes it difficult to identify or model the overall "policy" that is enforced by the system.

To meet this challenge, we have developed the *Policy Machine (PM)*. The underlying concept of the *PM* relies on the separation of the access control mechanism from the access control policy [JSS97, JSSB97]. This enables enforcement of multiple access control policies within a single, unified system. Although complete policy coverage is an elusive goal, the *PM* is capable of expressing a broad spectrum of well-known access control policies. Those we have tested so far include: **One-directional Information Flow, Chinese Wall** [BNCW89], **N-person Control** [NCSC91] **and DAC**. These were selected partly because they are so well known, and partly because they differ greatly from one another. A further advantage of *PM* is that it is highly extensible, since it can be augmented with any new policy that a specific application or user may require. This paper will demonstrate the functionalities of *PM*, and illustrate *PM*'s universal, compositional and combinational properties.

2 Policy Machine (PM)

In this section we describe our design of *PM*, and explain how we achieve our goal of a unified description and enforcement of policies. Our design of *PM* is related to the traditional **reference monitor** approach. A reference monitor is not necessarily a single piece of code that controls all accesses; rather, it is an abstraction or model of the collection of access controls [Ande72]. As we have applied the concept, *PM* does not dictate requirements for particular types of subject or object attributes nor the relationships of these attributes within its abstract database. Because of this generality *PM* can be used to model the implementation of a wide variety of access control policies, but *PM* is not itself a policy.

The structure of *PM* is based on the concept that all enforcement can be fundamentally characterized as either static (e.g., **BBAC**), dynamic (e.g., **Work Flow**), or historical (e.g., **Chinese Wall**) [GGF98, SZ97]. In the next five sections, we introduce the primary components of the *PM*: the *Policy Engine (PE)* with *PE* databases and the security policy operations, which is the *General Policy Operations (GPO)*. We then discuss the Universal and Composition properties of the *PM* following each introduction.

2.1 Policy Engine (PE)

The task of the *PE* is to receive access requests and determine whether they should be permitted. Access requests are of the form *<User Identity, Requested Operations, Requested Objects>*. To determine the acceptability of a request, the *PE* executes three separate phases (Figure 1): the request management phase, the subject-object mediation phase, and the access history recording and database management phase to determine whether access is granted or denied. The request management phase is primarily used to assign the requesting user to a *subject*. The subject-object mediation phase is used to decide if the *subject* is permitted to access the requested *object*. The access history recording and database management phase determines if the granted access will affect the state of the *PM* if a historical policy is implemented.

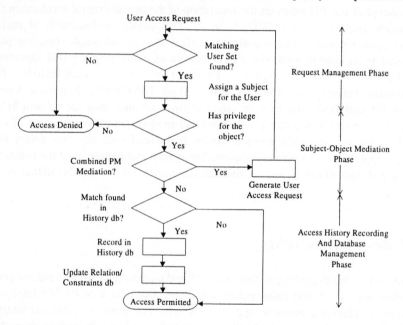

Fig. 1. *PM* phases

Request Management Phase (RMp)
This phase is activated by the detection of a user request. The *PE* responds by creating a *subject* along with *operation*, and *object* specified within the user's request. For example, consider the following request:

   (Mary; write; a_file) i.e., *user Mary* wants to write to *a_file*.
Currently, the *Relations database (Rdb)* has entries as follow:

   Managers = (Mary), Programmers = (Managers, Bob, Sue) i.e., the *Managers user set* contains *user Mary*, and *user set Programmers* contains *user set Managers* and *users Bob* and *Sue*.
Thus, *user Mary* is assigned as:

   subject = (Mary, Managers, Programmers); write; a_file
and is able to continue to the next phase.

Subject-Object Mediation Phase (SOMp)

The *SOMp* takes the input (*subject, operations* and *object*) created by the *RMp*, and the process authorizes access under two conditions. First, there is a match of the *subject*'s request and there is an entry in the *Privilege database (Pdb)*. In this case, the requested access is authorized. Second, there exists no further subject-object mediation check that is required under a different *PM*. For example, assume that, in the *SOMp*, PE generated the following message when in the *RMp:*

subject = (Mary, Managers, Programmers); write; a_file

The *SOMp* will then check *Pdb*, and find entries:

(Programmers; all; a_file)

This means that the *Programmers user set* has *all* privileges to the file *a_file*. Therefore the request

subject = (Mary, Managers, Programmers); write; a_file

is authorized.

Access History Recording and Database Management Phase (AHRDMp)

This phase evaluates the relevance of the authorized *event* with respect to the history-based policies that are stored in the *History Based Relations database (HBRdb)*. History-based policies are driven by an *event* and an *action* stored in the *HBRdb*. If the *event* received matches *events* stored in the *HBRdb* then PE in *AHRDMp* invokes the *action* associated with the *event*. The *action* either creates new constraints in the *Constraints database (Cdb)*, and/or updates relations in the *Pdb*. Consistency is checked whenever a relation or constraint is created. For example, *HBRdb* contains an entry:

event = (subject_a; all; x_file),   response = (Generate Constraints = (((subject_a; all; y_file) ⊕ (*:*:*)), ((subject_a; all;z_file) ⊕ (*:*:*)); Cdb)

This means that *subject_a* is prohibited from any kind of access (by excluding (⊕) all (*)) to *y_file* and *z_file* if *x_file* has been accessed by *subject_a*. Entries for the constraints will be added into the *Cdb*. This example can be used for policies that require **SOD** constrains, such as **Chinese Wall** [BNCW89] policies.

2.1.1 PE Databases

PE databases provide relation, privilege, constraint, and history information for *PE* processes. The databases are maintained either by *GPO* or by the *AHRDMp*. In this section, we will describe each of the four types of databases.

Relations database (Rdb) maintains the *inheritance* relations between sets. For example, *Rdb* is used in the *RMp* to establish the active *user set(s)* of a *subject*. An example is in Figure 2.

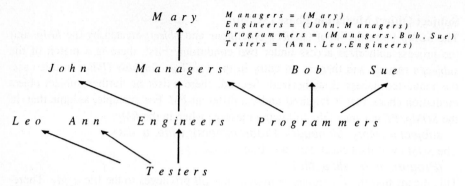

Fig. 2. Set Relations database example

Constraints database (*Cdb*) contains constraint relations of *subject (sets)*, *object (sets)*, and *operation(s)*. It is used by *RMp* as a reference to establish a *subject* from a *user's* access request. An example entry:

$(SS_1; P_1; OS_1)$, $(SS_2; P_2; OS_2)$ i.e., The combination of *subject (sets)* SS_1, *operation(s)* P_1 and *object (sets)* OS_1, and the combination of *subject (sets)* SS_2, *operation(s)* P_2 and *object (sets)* OS_2 are mutually excluded.

Privilege database (*Pdb*) is accessed by *SOMp* for searching the *subject (sets)*, *operation(s)* and *object (sets)* relations. A relation in *Pdb* defines the privilege for the *subject (sets)*. An example of *Pdb* entry is:

Programmers; read, write, execute; file_a i.e., *subject set Programmers* can *read*, *write*, and *execute file_a*.

History Based Relations database (*HBRdb*) contains information, which is used in the *AHRDMp* to perform state transaction of a historical-related policy embedded in the database with the structure:

st_i, *Event, Actions*, st_j, where st_i is the current state of the embedded historical access control policies, *Event* is an authorized access :(*subject (sets)*, *operation(s)*, *object (sets)*). *Actions* are sets of actions $(a_1.....a_n)$, each a_i = (*dbname, Updates*), where the database *dbname* is the name of one of the *Cdb* or *Pdb* will be updated with the information in *Updates*. st_j is the next historical state *PM* will be when the *Event* occurred.

2.1.2 Universal Property

PM can support most major access models, making it a possible to express the corresponding access control policies that these models represent. In order to reduce the support needed to express a multitude of models, we take advantage of previous work which has shown that certain model categories may be simulated or represented by other model categories. For instance, Sandhu has shown in [Sand93] that **Lattice-based** access control models such as the **Bell-Lapadula** model [BL73] and **Biba** model [Biba77] may simulate **Information Flow** polices (including **MAC** policies) and **Chinese Wall** policies. In *PM*, **Lattices** are represented by a **set** implemented by data records in a database and embedded in the *PE*'s *Rdb* as described in section *2.1.1 Rdb*. The relations of the elements of the **Lattices** are checked in *RMp* as described in section 2.1 *RMp*. This allows us to describe a wide variety of models with a minimum of *PM* features.

As a second example, *PM* maintains records for access events and updates databases to reflect changes of the states caused by the events according to the historical policy states that are embedded in *HRBdb*. As indicated in section 2.1.1 *HRBd*, the historical related policy is implemented by a table structure of a database, the process in the *AHRDMp* tracks the historical state of the policy by referring the information in the table as described in section 2.1 *AHRDMp*. Therefore, policy models which can be mapped to state transition machine models can also modeled by *PM*; this allows us to support **Clark Wilson** [CWAC87] and **N-person Control** policy models, for example.

Gligor, Gavrila and Ferraiolo [GGF98] showed that **SOD** policies can be enforced either statically or dynamically. We can use either method by employing *PE*'s *Cdb* as described in section 2.1.1 statically and dynamically. The static *Cdb* constrains relations between subjects, objects and/or operations, and provides the tool for statically separating users of their privileges. The dynamic *Cdb* can be generated by the process of *AHRDMp* according to *the actions* stored in the historical policy database and can separate the users' privileges dynamically. Therefore, through the *Cdb* the *PM* is able to enforce policy models which require either static or dynamic **SOD**. These policies include **Work Flow** [AH96] and **RBAC**.

DAC access control policy can be achieved by appropriately managing the *Pdb* tables associated with each user. The *Pdb* table for each *user* should be set as a controlled *object* of the administrator's *PE*. These users will then have sole and total control of their own *Pdb* table, thereby allowing them to delegate the access privilege to other *users*.

Although a formal proof of correspondence is beyond the scope of this paper, the above paragraphs indicate how the properties and components of *PM* allow *PM* to model some major access control models.

2.2 General Policy Operations (GPO)

GPO is a set of operations for expressing access control policies. It allows users to specify, together with the authorizations, the policy according to which access control decisions are to be made. *GPO* has the following components: *Basic sets* define the basic elements and sets for the *GPO*. *Database query functions* represent restrictions or relationships between elements. *Administrative operations* are used for the administration of the *PM* databases. *Rules* are used to keep the transitions in the states of *PM* during the system operation.

Basic Sets
A set of basic sets, defines the *users, objects, operations* and their relations.

Examples:

US = the set of *user sets*, $(US_1, ..., US_n)$.

US_i = the set of *users*, $(u_1, ..., u_n)$, where each *user* is assigned to *user set* US_i.

Database Query Functions

There are three types of database query functions:

1. *Set query* functions, for retrieving relations from *Rdb*. Examples:

   member_of (s) denotes the members of set s (s is inherited by the members).

   set_of (m) denotes the sets the member m is assigned to (inherit from).

   transitive_member_of (s) denotes the transitive members of set s.

   transitive_set_of (m) denotes the transitive sets the member m belongs to.

2. *Constraint query* functions, for retrieving constrains from *Cdb*. Examples:

   userSet_constrained (s) denotes the constraint relations of the *user (set)* s.

   objectSet_constrained (o) denotes the constraint relations of the *object (set)* o.

   userSet_permitted (s) denotes the permitted relations related to the *user (set)* s.

   objectSet_permitted (o) denotes the permitted relations of the *object (set)* o.

3. *Process mapping* functions, for retrieving information of current process. Examples:

   operation_request (u) denotes the operation requested by *user* u.

   object_request (u) denotes the *object* requested by *user u*.

   subject_of (u) denotes the subject associated with *user u*.

   active_userSets (s) denotes the set of active user(sets) associated with *subject s*.

   access(s,p,o) return 1 if *subject s* is authorized to access *object o* by *operation p*.

Administrative Operations

A set of administrative commands. Examples:

   addMember (x, y), *rmMember (x, y)*, new member x is added/removed to set y in *Rdb*.

   addConstraints(c), *rmConstraintse(c)*, constraints c is added/removed to *Cdb* database.

   addRelations(c), *rmRelations(c)* , relation c is added/removed to *Pdb*.

Rules

The rules represent assumptions about the *GPO*. Examples:

   The member assigned to a given set is exactly the member directly inheriting the given set in Rdb, and the sets assigned to a given member are exactly the sets directly inherited by the member in Rdb.

   Any two user sets assigned to a third user set do not inherit (directly or indirectly) one another in Rdb.

2.2.1 PM Composition

The *Administration Operations* of *GPO* allow users of *PM* to interact with *PM*; they are used to add, remove, and update a *PE* database entry. Before an *Administration*

Operation is executed, *GPO* will invoke the *Database Query Functions* built in the *GPO* to retrieve information required for the operation. *GPO* will then check the validation of the logics and formats governed by the *GPO's Rules*. The *Rules* of *GPO* guarantee the consistencies of the relations, constraints and privileges of the information stored in the *PE databases*.

Some of the functions for well-known access control policies can be built in the *GPO*. Through the *Administration Operations, PM* users/administrators can implement and manage an access control policy. For example, to add a relation for users in a Separation of Duty (**SOD**) policy, *PM* administrators can execute the command:

AddSODusers (x, y), i.e., add separation of duty (mutual exclusive) relation for *users x* and *users y*.

The *Administrative Operations* are a key feature of *GPO*, since they provide the *PM* administrator a tool for composing new access control policies. Through the *GPO's Rule*, the information stored in the *PE* database can be generated and modified without *PM* administrators having any known access control policy in mind.

The composition capability of *GPO* supports the important feature of *PM* that separates the security policies from the security mechanisms.

2.3 PM Combination

A key feature of *PM* is its ability to combine policies. Policies are established in different *PMs* by implementing their databases and *GPOs*. When policies are combined, *PMs* are essentially "chained together". In practice, this means that a "User Request" begins at the "first" *PM*, and it is passed along to subsequent *PMs* for examination. Each *PM* will in turn examine its own state of completion (i.e., ability to decide if there is enough information to process the request). If it has not reached a state of completion, it will pass the token as an input to the next *PM* (see Figure 3).

An *object* can be constrained under more than one policy, for example, a user may get past the **MAC** policy check, but may still need to be authorized again by other policies say, **RBAC** for allowing the requested operation to the object. For example, the sequence of checks could be: Can the user read secret information, is the user a doctor, is the patient assigned to the doctor for the operation/procedure.

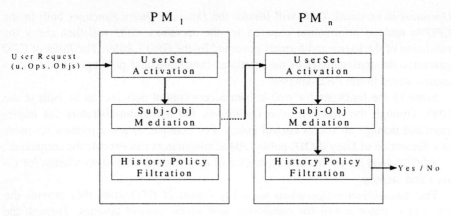

Fig. 3. *PM* combination

3 Conclusions

In this paper, we have demonstrated the concept of constructing universal policy architecture without detailed and formal specification. The issue of formal specification will be the subject of our planned future research. *PM* functions are executed through sequential phases (as described in Section 2) without recursive operations in the *PM* functions; therefore it is computable in polynomial time. Since *PM*'s database is always consistent (no conflict mapping of relations), users' access requests are handled conclusively; that is, they are either authorized or denied.

PM has significant theoretical and practical implications; it is an architecture that is capable of implementing virtually any access control policy. It also helps to promote interoperability and the use of innovative tools for policy generation and visualization that can be built on top of existing access control primitives and scaled to the largest virtual enterprise.

References

[AH96] Atluri V., Huang W., "An Authorization Model for Workflows". *Proceedings of the Fifth European Symposium on Research in Computer Security in Lecture Notes in Computer Science, No 1146,* September 1996.

[Ande72] Anderson J. P., "Computer Security Technology Planning Study," *ESD_TR_73-51, Vol. 1, Hanscom AFB, Mass.,* 1972.

[Bark97] Barkley J., "Comparing Simple Role Based Access Control Models and Access Control Lists", *Proceedings of the Second ACM Workshop on Role-Based Access Control,* November 1997, page 127-132.

[Biba77] Biba K. J., "Integrity Considerations for Secure Computer Systems," *ESD-TR-76-372, USAF Electronic Systems Division,* Bedford, Mass., April 1977.

503

[BL73] Bell D.E., and Lapadula L. J., "Secure Computer Systems: Mathematical Foundations and Model," *M74-244, MITRE Corp.*, Bedford, Mass., 1973.

[BNCW89] Brewer, D., Nash, M. "The Chinese Wall Security Policy." *Proc IEEE Symp Security & Privacy, IEEE Comp Soc Press,* 1989, pp 206-214.

[CW87] Clark D. D., and Wilson D. R., "A Comparison of Commercial and Military Security Policies," *Proc. of the 1987 IEEE Symposium on Security and Privacy*, Oakland, California, 1987, pp.184-194

[CWEO89] Clark D. D., and Wilson D. R., "Evolution of a Model for Computer Integrity", *NIST Special Publication 500-168, Appendix A*, September 1989

[FCK95] Ferraiolo D. F., Cugini J. A., Kuhn D. R., "Role-Based Access Control (RBAC): Features and Motivations", *Proc. of the 11th Annual Conference on Computer Security Applications. IEEE Computer Society Press*, Los Alamitos, CA. 1995.

[GGF98] Gligor V. D., Gavrila S. I., Ferraiolo D., "On the Formal Definition of Separation-of-Duty Policies and their Composition", In *IEEE Symposium on Computer Security and Privacy*, April 1998.

[HGPS99] Hale J., Galiasso P., Papa M., Shenoi S., "Security Policy Coordination for Heterogeneous Information Systems", *Proc. 15th Annual Computer Security Applications Conference, Applied Computer Security Associates*, December 1999.

[ISW97] Irvine C. E., Stemp R., Warren D. F., "Teaching Introductory Computer Security at a Department of Defense University", *Naval Postgraduate School* Monterey, California, NPS-CS-97-002, April 1997

[JSS97] Jajodia S., Samarati P., and Subrahmanian V. S., ``A Logical Language for Expressing Authorizations," *Proc. IEEE Symp,* Oakland, Calif., May 1997.

[JSSB97] Jajodia S., Sammarati P., Subrahmanian V. S., and Bertino E., "A Unified Frame Work for Enforcing Multiple Access Control Policies", *Proc. ACM SIGMOD Conf. On Management of Data*, Tucson, AZ, May 1997.

[NCSC87] National Computer Security Center (NCSC). "A GUIDE TO UNDERSTANDING DISCRETIONARY ACCESS CONTROL IN TRUSTED SYSTEM", *Report NSCD-TG-003 Version1*, 30 September 1987.

[NCSC91] National Computer Security Center, "Integrity in Automated information System", *C Technical Report 79-91, Library No. S237,254*, September 1991.

[NCSC98] National Computer Security Center, "1998 Evaluated Products List", *Washington, D.C., U.S. Government Printing Office.*

[Sand93] Sandhu R. S., "Lattice-Based Access Control Models", *IEEE Computer, Volume 26*, Number 11, November 1993, page 9-19.

[SCFY96] Sandhu R. S., Coyne E. J., Feinstein H. L., and Youman C. E., "Role-Based Access Control Models", *IEEE Computer, Volume 29, Number 2*, February 1996, page 38-47.

[Spencer et al99] Spencer R., Smalley S., Loscocco P., Hibler M., Andersen D., and Lepreau J., "The Flask Security Architecture: System Support for Diverse Security Policies", *http://www.cs.utah.edu/fluz/flask*, July 1999.

[SZ97] Simon R.T., and Zurko M. E., "Separation of Duty in Role-Based Environments," *Proc. of the Computer Security Foundations Workshop X*, Rockport, Massachusetts, June 1997.

Multi-spectral Scene Generation and Projection

Session chair:

James B. Johnson, Jr. (U.S Army Redstone Technical Test Center, USA)

The Javelin Integrated Flight Simulation

Charles Bates[1], Jeff Lucas[2], Joe Robinson[3]

[1]US Army AMCOM, AMSAM-RD-SS-SD, Redstone Arsenal, AL 35898-5000
Charlie.bates@rdec.redstone.army.mil
[2]Computer Sciences Corporation, 4090 South Memorial Pkwy, Huntsville, AL 35815
Jeff.lucas@rdec.redstone.army.mil
[3]CAS, Inc., 555 Sparkman Drive, Huntsville, AL 35814
Joe.robinson@rdec.redstone.army.mil

Abstract. The cornerstone of the all-digital simulation for the Javelin missile system is its accurately rendered high fidelity 8-12µm infrared imagery of targets, clutter, and countermeasure effects. The Javelin missile system is a medium range, manportable, shoulder-launched, fire-and-forget, anti-armor weapon system. Javelin has two major components: a reusable Command Launch Unit (CLU) and a missile sealed in a disposable launch tube assembly. The CLU incorporates an integrated day/night sight and provides target engagement capability in adverse weather and countermeasure environments. The Javelin missile incorporates an imaging infrared seeker with a fire-and-forget tracker that allows the gunner to fire the missile and immediately take cover. The on-board tracker guides the missile to the target until impact. The tracker and its ability to maintain lock on the target throughout flight emerged as the most critical aspect of the Javelin system. A cost-effective way to develop and test new tracker algorithms became a necessity. Additionally, an innovative way to determine system performance of a fire and forget imaging infrared system was needed. From these requirements came the development of the Javelin Integrated Flight Simulation (IFS). The Javelin IFS is a high fidelity all-digital simulation whose primary functions are tracker algorithm development, flight test predictions and reconstructions, and system performance assessment. The IFS contains six models: environment model, seeker model, tracker model, autopilot, six degrees-of-freedom (6-DOF) model, and gunner model. The most critical aspect of the IFS is feeding the tracker with as realistic imagery as possible. The seeker model working in conjunction with the environment model gives the IFS highly realistic and fully verified, validated, and accredited imagery of targets, terrain, and countermeasures.

1 Introduction

The Javelin Project Office undertook a comprehensive program for managing key system, subsystem, and component-level models and simulations, related databases, and test data. These models and simulations evolved from initial all-digital simulations supporting requirements analysis, and they have continued to support the program throughout its development life cycle. The primary simulation developed to represent the Javelin system was the Javelin Integrated Flight Simulation (Javelin IFS).

The Javelin IFS is a high-fidelity, all-digital simulation whose primary functions are tracker algorithm development, flight test predictions and reconstructions, and system-performance assessment. The Javelin IFS contains an environment model, a seeker model, a tracker model, a guidance model, a six degrees-of-freedom (6-DOF) model, and a gunner model. All of which are controlled by an executive and data from each model is passed via shared memory segments. The term Integrated Flight Simulation was coined to signify the use of Mission Critical Code Resources (i.e. Flight Software) as part of the simulation. This integration has proven to be the bedrock on which Javelin program successes have been built.

The Javelin's prime contractor and the U.S. Army's Aviation and Missile Research, Development and Engineering Center (AMRDEC) developed the Javelin IFS jointly during engineering and manufacturing development (EMD). Verification, validation, and accreditation (VV&A) of the Javelin IFS was performed by AMRDEC, and accreditation by the Army Materiel Systems Analysis Activity (AMSAA). The IFS was then used to assess the performance of the system through many thousands of monte-carlo simulation runs with various environmental and attack scenarios. The data derived from this performance assessment was used to support the Low-Rate Initial Production (LRIP) phase of the program.

2 The Javelin Integrated Flight Simulation

The IFS represents the first simulation from AMCOM AMRDEC of an imaging fire-and-forget missile system that included large sections of the Mission Critical Code Resource (MCCR) as a part of the simulation. The architecture of the IFS was unique at its creation (early 90's) in that it treated models or groups of models as independent applications that communicated through message queues and shared memory segments. This approach was driven primarily by the state of the art (at the time) of compilers, operating systems, and hardware. Initially, the Javelin system was modeled with software that was written in Fortran, ADA, C, and Pascal. Grouping each of these models into a monolithic simulation and interfacing the disparate programming languages became difficult in both implementation and maintenance.

A decision was made to create smaller components and interface them using inter-process communication features of the Unix operating system. Socket based communication could have been chosen but the overhead compared to shared memory access was significant enough to decide otherwise. This architecture, illustrated in Figure 1, is working well for the Javelin project but it isn't essential to simulating this system. There are many solutions to this problem and all would be valid. The important aspect of the IFS is the segmenting of the problem into smaller sets. Maintenance of the simulation became more manageable.

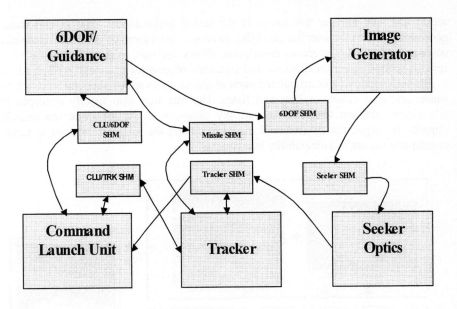

Fig. 1. IFS Block Diagram.

The IFS consists of five applications to provide modeling and interfaces for the 6DOF hardware simulation, guidance, tracking, seeker optics, environment, and gunner/launch unit modeling. An executive application coordinates the creation and removal of the message queues and shared memory segments, as well as the actual execution of the simulation applications. Each simulation application signals the other applications through message queues when another execution frame is needed.

As an imaging simulation, the IFS has struggled with the tradeoffs associated with simulation of an infrared scene. These tradeoffs include providing enough fidelity to adequately test the tracker software, the cost associated with collecting signatures that represents various objects, backgrounds, and phenomenon under a variety of conditions, and execution of enough simulation runs to provide valid statistical datasets for evaluation. These tradeoffs have resulted in developing over 25 square miles of terrain databases at 1 ft resolution with multiple signatures as well as over 50 target models with as many as 20000 polygons each and executing in excess of 200000 simulation runs.

2.1 The Environment Model

The Environment Model is in short the IFS scene generator. Its purpose is to create a high-resolution image for a specified scenario from the viewpoint of the missile seeker and then transfer this image to the seeker optics model. The Javelin scene generator provides the user the capability to create "environments" through a scenario description file. This scenario description allows the user to specify the use of a specific section of terrain elevation and signature, objects such as tanks, trees, bushes, events such as moving or articulated parts of objects and various phenomenon such as smoke and dust clouds or fires and flares. The user may also specify atmospheric attenuation, ambient temperature, and sky temperature. At the point the missile impacts the target, the environment model calculates the intersection point in target coordinates for use in vulnerability assessments.

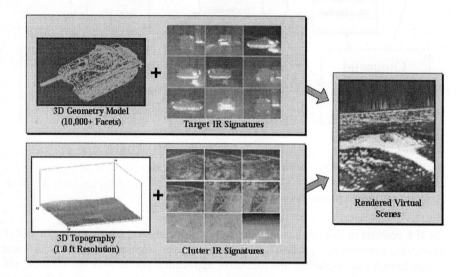

Fig. 2. Building a Scene.

From its inception, The environment model was developed for the Silicon Graphics platform. Originally, SGI provided the best platform for the creation of this type of imaging simulation as well as providing the cost/benefit ratio to afford to execute thousands of runs that might take as long as an hour each. As time and technology has progressed, the scene generator has been improved to take advantage of new features but the pace of adoption has been slow due to configuration management issues. Army acquisition rules require simulations to be accredited by an agency independent of the developers. This insures a good product but also necessarily, limits changes to the product to keep reaccredidation costs low. For this reason the Javelin target models are still rendered using shaded polygons and the terrain while texture still relies on a regularly gridded terrain skin. When the design for this software was completed, the Reality Engine series of graphics board was the best available but unfortunately 4 MBs of texture RAM was not enough for both target and terrain.

As the cost/benefit ratio becomes unbalanced due to technology shifts, the effort to modify the scene generation becomes justified and that is the case now. New thrusts will include converting targets and terrains to textured models in a commercially available format (i.e. OpenFlight®).

The targets and terrain databases used by the IFS are predominately characterized as empirical datasets. Early in the program, a decision was made to not depend on predictive signatures for either target or terrain but rather gather as much measured data as possible. Initial studies of the Javelin missile system indicated that the hit point of the missile on the target would be controlled overwhelmingly by the tracker. Now that seems an obvious statement but to clarify, when all other error sources of the missile are combined for Javelin, the system can still consistently hit a point on a target smaller than its diameter. Therefore, the controlling factor is the tracker error. Of course, the tracker error is controlled by the scene itself hence the use of measured signatures.

The process by which these models were created was laborious at first but as time passed and the process was refined, automated tools lowered the workload.

The process of building target models included:

1. Obtain accurate drawings or measurements of the target.
2. Obtain calibrated infrared signatures from the target.
3. Construct a polygonal model of the target using drawings and add additional vertices to capture thermal detail as necessary.
4. Render an image of the resulting model and compare back to the measured signature.

This process seems straightforward and it is but it does result in target models that can exceed 20,000 facets.

Terrain modeling follows a similar process but on a larger scale.

1. Gather topographic information in the form of elevation data and class descriptions (i.e. vegetation types and location). This was obtained very often through survey overflights.
2. Collect calibrated signature data from different class types.
3. Create a topographical description file for the elevation data.
4. Create a signature description from the class maps.
5. Render and compare the result with the original signature measurements.

This process required a great deal of infrastructure support but the Army has capitalized on its investment by making the datasets available to other programs. This community sharing has resulted in a variety of models that provide excellent evaluation and development capabilities to the imaging tracker community.

Everyone would like to have the ability to predict signatures and it will happen at some point in the near future. However, due to VV&A concerns at the time, the method of collecting the exact imagery for that point in time emerged as the preferred approach. No one could argue the validity of real target signatures and real terrain signatures. It is very likely that the datasets collected on this program will provide the means to validate predictive techniques in the future when they are available.

2.2 The Seeker Model

The seeker model provides an optics and signal-processing model of the Javelin system. The features include modeling for focus, focal plane array (FPA) non-linearity, bad pixels, gain and level processing, and electronic noise. The seeker application obtains a high-resolution scene image from the environment model and creates a low-resolution image by first applying an optical blurring function. This optical blurring function varies as a function of range to a surface and provides the blurring that occurs as the missile approaches the target. The parameters of the blurring function can be varied statistically to cover the manufacturing error of the optics. After the optically processed low-resolution image is created then the effects of FPA non-linearity, bad pixels and gain and level processing are added. This resulting image represents the image that would be presented to the Javelin tracker in the real hardware. The seeker application makes this image available through shared memory to the tracker model.

2.3 Tracker Model

The tracker model is how the Integrated Flight Simulation got its name. The original concept was to "integrate" the real tracker MCCR code (Ada) into the IFS. The same source code base for the Javelin tracker is compiled for the missile and compiled to run on SGI processors for the simulation. This means that the same program flow is occurring on both processors. The only sections of the MCCR code not included in the IFS are the built-in-test (BIT) and the array processor functions that run on specialized chip sets in real-time language (RTL). Because the core of the tracker is preserved in the IFS, it can be said that the tracker is not simulated.

Math operations, particularly floating point calculations in the array processor sometimes differ slightly. A matching process is rigorously undertaken for each tracker delivery to identify any of these types of differences and provide compensation for them in a simulation specific portion of the tracker code. This addresses such issues as rounding differences and truncation differences.

The tracker model receives seeker imagery from the seeker model through a shared memory interface. The tracker then performs its tracking algorithms in order to compute the desired aimpoint on the intended target. From the computed aimpoint, the track errors are calculated and then sent to the autopilot where the guidance electronics command the control actuator subsystem (CAS) to move the missile.

2.4 The Missile Hardware and Guidance Model

The 6DOF missile application models all the hardware of the missile. This hardware includes the seeker gimbal dynamics, flight computer process timing, launch and flight rocket motors, the fin actuator dynamics, and the missile aerodynamics. This application also models the earth, atmosphere, and wind. The user also has the capability of simulating hardware failures to support testing of range safety algorithms in the flight software. The guidance software is included in this application rather than separating as the tracker in order to support legacy studies. The 6DOF application can run in a standalone mode and provides monte-carlo simulation capability particularly for cases where the tracker is not of concern. When used in the IFS, this application provides the missile viewpoint to the environment model to render a scene for the seeker model. It accepts the tracker error from the tracker model to provide a closed loop simulation. At initialization, the 6DOF application interfaces with the Gunner Model to position the missile and to respond to flight software mode changes as directed by the gunner model.

2.5 The Gunner Model

The gunner model portion of the IFS simulates the interface between the gunner and the missile hardware.

The Command Launch Unit (CLU) is the Javelin sight. It allows the gunner to find the target on the battlefield, lock-on to the target, and fire the missile. The gunner identifies and centers the target in the CLU sight then switches to the missile seeker to size the track gates around the target. When the tracker has locked on, the gunner can fire the missile.

In the IFS, the gunner/CLU model performs this same function. The scene generator knows which pixels within the seeker image are target and which are background. This target truth is available through a shared memory interface to the gunner model via an image color-coded to identify target location. The gunner model uses this image to center and size the track gates roughly around the target. Moving the target within the image is accomplished by giving the scene generator a new pitch/yaw orientation. This mimics the gunner moving the CLU/missile to position the target near the center of the image. Sizing the track gates is accomplished by setting and clearing bits in the CLU/tracker interface. Since the actual Javelin Tracker is running within the IFS, these single bit commands are the same as those sent from an actual CLU. Monte Carlo errors are applied to track gate size and position. These error distributions are based on measurements of gunners in the field.

3 Next Generation IFS

As stated earlier, when the cost/benefit ratio is unbalanced enough then upgrading is justified. The users of this simulation (Javelin PMO, AMCOM AMRDEC, and Raytheon) have determined that case exist now. With the availability of stable object-oriented language compilers (i.e. C++) and new players in the high-end graphics market as well as new computing platforms, there are a myriad of opportunities to gain new efficiencies or attempt more complex problem sets.

3.1 New Simulation Framework

The first item of improvement has been to adapt to a new simulation framework developed and promoted by AMCOM AMRDEC and referred to as the Common Simulation Framework (CSF). CSF was developed to aid AMRDEC in promoting a common set of tools in its development and evaluation business. CSF is an object-oriented and shared object approach to building simulations. Users are encouraged to develop models as classes with interfaces that are similar to the interfaces of the real hardware. A shared object is created with this class definition and libraries of these classes are gathered to construct a simulation. The simulation is constructed by instantiating the classes and connecting through their open interfaces. The obvious benefit is the promotion of software reuse. As different types of models are created, eventually a library of missile parts will be developed. Software reuse, however, is not the prime motivator. Missile systems are never just created and then statically maintained for the lifetime of the missile. As new threats are realized, improvements are necessary and to aid in this effort a simulation that allows itself to freely adapt new models is desired. One example would be as countermeasures are created that defeat the current tracker, tests could be perform on other vendors trackers to keep abreast of new technologies.

The Javelin IFS has been transferred to this new framework and is being tested to insure that all capabilities are maintained as well as maintaining a valid system simulation of Javelin.

3.2 New Scene Generator

With the explosion of new 3D graphics capabilities, vendors are releasing new products at a surprising rate. This certainly opens up the opportunity to increase capability and lower costs at the same time. For this reason, a new scene generator is being developed to take advantage of all the scene generation techniques available. The new scene generator will use models and terrain databases retuned and converted to use the OpenFlight® format and will be rendered using scene graphs. The volumetric effects such as smoke and fire will be implemented using 3D texturing and the increasing availability of texture RAM. When it is possible, some phenomenon will be implemented using particle effects. All these techniques along with efficient use of levels-of-detail will provide efficiencies necessary to implement increasingly complex testing scenarios such as shadows and reflective processes.

4 Conclusion

The Javelin IFS is a fully verified, validated, and accredited model of the Javelin missile system and its sub-systems. Because of the fidelity and validity of the IFS, Javelin can with confidence use the IFS to determine system performance, execute flight predictions and reconstructions, and perform tracker design and analysis. The IFS is an example of a simulation that is an integral part of a current US Army missile program and is being used on future missile programs as well.

A Multi-spectral Test and Simulation Facility to Support Missile Development, Production, and Surveillance Programs

James B. Johnson, Jr.[1] and Jerry A. Ray[2]

[1] U.S. Army Developmental Test Command, Redstone Technical Test Center,
CSTE-DTC-RT-E-SA, Redstone Arsenal, AL 35898, USA
James.Johnson@rttc.army.mil

[2] U.S. Army Aviation and Missile Command, Research, Development and
Engineering Center, AMSAM-RD-SS-HW, Redstone Arsenal, AL 35898, USA
jerry.a.ray@redstone.army.mil

Abstract. The Multi-Spectral Missile Life-Cycle Test Facility (MSMLCTF) is a single hardware-in-the-loop (HWIL) simulation complex that will uniquely have the capability to perform life-cycle simulation/testing on next-generation multi-sensor missile systems. The facility will allow for simulated end-to-end flight tests in a non-destructive manner as well as complete evaluation of all missile electronics and software. This testing will be performed at both the All-Up-Round (AUR) level and the Guidance Section (GS) level. The MSMLCTF will be used to support the Army Transformation by evaluating operational flexibility before fielding, reducing life-cycle costs, improving system performance by evaluating technology insertions and evaluating interoperability impacts. This paper gives a general description of the MSMLCTF, cost tradeoffs between live flight testing and simulation, and an update on facility development status.

1 Introduction

In the past, missile systems were tested solely at an open-air flight range. This technique was relatively simple. In recent years, missile systems have become much more complex utilizing multiple sensors such as millimeter wave and infrared. Also in the past, weapon developers developed and used simulation facilities to develop a weapon and testers developed and used simulation facilities to test a weapon. This paper describes a new multi-spectral simulation facility which will be utilized to both develop and test next-generation multi-spectral missile systems.

2 Historical Perspective

The authors played an integral role in developing two previous simulation facilities – the Millimeter System Simulator One (MSS-1) and the Simulation/Test Acceptance

Facility (STAF). Lessons were learned in the development of these facilities which led us to the development of the Multi-Spectral facility. A brief description of each of these facilities is provided.

2.1 Millimeter System Simulator One (MSS-1)

The MSS-1 facility was developed by the U.S. Army Aviation and Missile Command's (AMCOM) Research, Development and Engineering Center (RDEC) in the mid-1980's primarily for the Longbow HELLFIRE missile system guidance electronics (no explosives). The Longbow HELLFIRE is a millimeter wave radar guided system. This facility featured a high fidelity target array, flight motion simulation, and signal injection. The principal utility of the facility was development of the weapon system in a repeatable edge-of-the-envelope environment. Capabilities included end-to-end hardware-in-the-loop simulations using multiple targets and countermeasures. Early prototype guidance systems were placed in the facility for development of weapon system software algorithms. Also, many guidance packages scheduled to be flown on a missile system were placed in the facility to judge flight readiness. The missile system is in production today; however, guidance packages are still inserted into the facility to periodically check edge-of-the-envelope performance. An artist's sketch of the facility is shown in Figure 1.

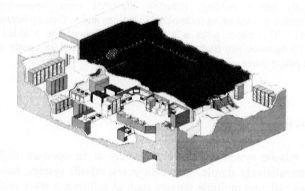

Fig. 1. Millimeter-wave System Simulator One (MSS-1) Facility

2.2 Simulation/Test Acceptance Facility (STAF)

The STAF facility was developed jointly by the U.S. Army Developmental Test Command's (DTC) Redstone Technical Test Center (RTTC) and AMCOM RDEC in the mid-1990's. The mission of STAF was to perform functional testing on Longbow

517

HELLFIRE missiles (complete all-up-rounds with explosives). The STAF operated in a similar fashion to the MSS-1 facility with the exceptions that the missile was complete with explosives installed and the target array was much more simplistic. The target array was deemed adequate for a production environment. The facility, still in use today, is used for production sampling, special engineering tests, and stockpile surveillance testing. An artist's sketch of this facility is shown in Figure 2.

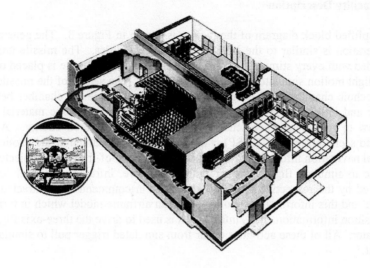

Fig. 2. Simulation/Test Acceptance Facility (STAF)

3 Next Generation Multi-Spectral Simulation Facility

Many lessons were learned in the development of the MSS-1 and STAF facilities. Emerging weapon system technologies indicate the requirement for a *Multi-Spectral* hardware-in-the-loop simulation facility. Multi-Spectral implies the incorporation of both Millimeter Wave and Imaging Infrared in the same guidance package. As such, the simulation facility must be capable of providing multi-spectral targets.

3.1 Concept

Taking advantage of the lessons learned on the MSS-1 and STAF facilities, the authors devised a concept for the Multi-Spectral facility which houses both a non-hazardous high-fidelity facility and a hazardous lower-fidelity facility. The facility will consist of two test chambers, one high fidelity and the other lower fidelity, with a common control building. The high fidelity chamber will be used primarily for missile development and will feature multi-spectral multiple targets, countermeasures, and edge-of-the-envelope evaluations. This facility will be used to design hardware and develop software algorithms. The lower-fidelity chamber will be used for pre-flight confidence testing, production lot sampling, stockpile reliability tests, and special engineering tests. The multi-spectral targets in this facility will be more simplistic relative to the high fidelity chamber.

3.2 Facility Description

A simplified block diagram of the facility is shown in Figure 3. The general concept of operation is similar to the MSS-1 and STAF facilities. The missile under test is provided with every stimuli it is capable of sensing. The missile is placed on a three-axis flight motion simulator (pitch, yaw, and roll). The sensor of the missile is inside an anechoic chamber. An exotic material is placed inside the chamber between the sensor and millimeter wave target generation horns. The exotic material acts as a window to millimeter wave energy and a mirror to infrared energy. A dynamic infrared scene projector (DIRSP) is placed off-axis and aimed at the exotic material. Inertial navigation information, driven by a six-degree-of-freedom, is injected into the missile to simulate flight. Proper millimeter wave, infrared, and laser targeting is supplied by the respective target generators. Fin commands are extracted from the missile and this information is computed in an airframe model which in turn supplies fin position information. This information is used to drive the three-axis flight motion simulator. All of these activities occur from simulated trigger pull to simulated target impact.

3.3 Facility Utilization

Multiple weapon systems are being targeted for test in the Multi-Spectral facility. During system development and demonstration phases, the facility will be used to aid in system design (hardware and software) and integration for system breadboards. During the final stages of this acquisition phase, the facility will be used for pre-flight risk reduction testing for all guidance sensors and all-up-rounds prior to live flight testing. The facility will be made available to operational testers to augment their operational test scenarios and possibly reduce operational flight test requirements. The facility will be used for lot acceptance testing, engineering evaluations/tests, and stockpile reliability tests during the Production and Deployment phases. During all phases, the facility will be used to evaluate operational flexibility before fielding, reducing life-cycle costs, improve system performance by evaluating technology insertions, and evaluating interoperability impacts.

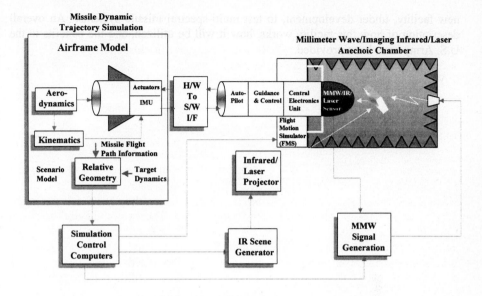

Fig. 3. Multi-Spectral Simulation Facility Block Diagram

3.4 Cost Avoidance

The authors performed a cost tradeoff of performing only a flight test program at a test range versus performing a combination of flight testing and hardware-in-the-loop simulation testing.

Some assumptions that went into this analysis for each phase of the weapon life cycle were: (1) the number of missiles that would be fired per year with no hardware-in-the-loop simulation, (2) the number of missiles undergoing hardware-in-the-loop simulation per year in the multi-spectral simulation facility as well as the number of missiles actually fired as augmentation. The results of the cost tradeoff using conservative values result in a cost avoidance of at least $90M over the life of a weapon system with potential of up to $100M. The tremendous cost avoidance is due to the non-destructive nature of the simulated flights which allows the rounds to be placed in inventory upon test completion. This same analysis indicates a facility cost payback period of less than five years.

4 Conclusion

This paper gave a historical perspective on the development of hardware-in-the-loop simulation facilities to test missile systems. Lessons learned have been applied to a new facility, under development, to test multi-spectral missile systems. An overall description of how this facility works, how it will be utilized, and the benefits to the U.S. Army have been provided.

Correlated, Real Time Multi-Spectral Sensor Test and Evaluation (T&E) in an Installed Systems Test Facility (ISTF) Using High Performance Computing

John Kriz[1], Tom Joyner[1], Ted Wilson[1], and Greg McGraner[1]

[1] Naval Air Warfare Center – Aircraft Division (NAWC-AD), Air Combat Environment Test & Evaluation Facility (ACETEF), Building 2109, Suite N223, Shaw Road, Patuxent River, MD 20670
{KrizJE, JoynerTW, WilsonJE2, McgranerGL}@navair.navy.mil

Abstract. Military aircraft are increasingly dependent on the use of installed RF and Electro-optical sensor systems and their data correlation/fusion to contend with the increasing demands and complexities of multi-mission air warfare. Therefore, simultaneous, and correlated, ground-based T&E of multiple installed RF and Electro-optical (EO) sensors and avionics systems/subsystems is used to optimize the use of scarce flight testing resources while also satisfying increasingly sophisticated test requirements. Accordingly, it is now an accepted, and highly cost-effective practice, to first conduct rigorous ground testing and evaluation (T&E) of the installed/integrated sensors' performance and their interaction using high fidelity modeling and simulation with real time scene rendering. Accordingly, valid ground-based T&E of installed multi-spectral (RF, IR, Visible) sensors requires that the simulations' spatial, spectral and temporal components be of sufficient fidelity and correlation to produce sensor responses that are indistinguishable from responses to "real-world" conditions. This paper discusses accomplishment of the foregoing goals, and fidelity challenges through innovative utilization of High Performance Computing (HPC) parallelization, visualization, shared memory optimization and scalability for multi-spectral modeling, simulation and real time rendering.

1 Introduction

Today's military aircraft are increasingly dependent on a diversity of electromagnetic sensors to provide aircrews and tactical commanders with unparalleled battle space awareness and engagement opportunity. The missions are equally diverse in such areas as reconnaissance, search and rescue, time vision navigation/evasion, target acquisition, target search and track, missile warning, and terminal missile homing. Achieving these missions requires employment of the most advanced sensors available to provide the war fighter with the greatest possible tactical advantage. Accordingly, multi-spectral sensors are an integral component of the flight and mission control avionics. Consequently, overall mission performance is directly linked to the combined performance of the onboard mission critical sensors

1.1 Installed Sensor Testing

The foregoing critical inter-relationship necessitates extensive developmental and operational sensor test and evaluation (T&E) of these integrated sensor/avionics environments prior to their release for operational use. The highest fidelity of systems performance testing occurs during flight-testing. However, it is becoming increasingly cost prohibitive to provide all mission representative environments during flight-testing. Therefore, the aviation community has developed ground-based dynamic, high fidelity instrumentation capabilities to test installed aircraft sensor subsystems. These ground-based capabilities are increasingly using correlated high fidelity real-time simulation and stimulation to conduct T&E of the installed sensor's performance and interaction with other sensors integrated within the weapons system platform. The computational challenge is to replicate the multi-spectral, temporal and geometric fidelity of the real-world problem-domain in a virtual warfare environment. Figure 1 illustrates the many attributes to be replicated, and correlated in real-time.

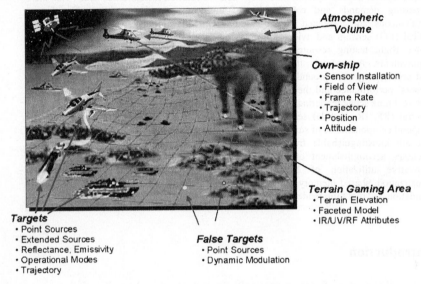

Fig. 1. The Computational Challenge: Accurately Replicate the Battle-space

1.2 Joint Installed System Test Facility (JISTF)

The combination of increasingly sophisticated test requirements and today's declining investments in test assets requires the coordinated and innovative development of modeling, simulation and emerging technologies to maximize use of scarce testing investment resources. In response to these emerging T&E needs, the Office of the Secretary of Defense (OSD), Central Test and Evaluation Investment Program (CTEIP), has sponsored development of three Joint Installed System Test Facility (JISTF) enhancements based on dynamic virtual reality simulation technology.

These enhancement projects are the Communications Navigation & Identification (CNI)/Joint Communications Simulator (JCS), Generic Radar Target Generator (GRTG), and Infrared Sensor Stimulator (IRSS). This paper addresses the capabilities of the JISTF GRTG and IRSS enhancements at the Air Combat Environment Test and Evaluation Facility (ACETEF), Naval Air Warfare Center, Aircraft Division (NAWC-AD), Patuxent River, MD, and their HPC requirements, and utilization. GRTG and IRSS provide the capability to simultaneously test multiple installed avionics and sensor subsystems in a ground test environment. Moreover, they are used to evaluate multi-sensor data fusion/correlation and subsystems' interoperability for Communications and Data Link subsystems, GPS, RADAR, and Infrared Sensors. Figure 2 illustrates the functional relationship of the three upgrades within the ACETEF architecture.

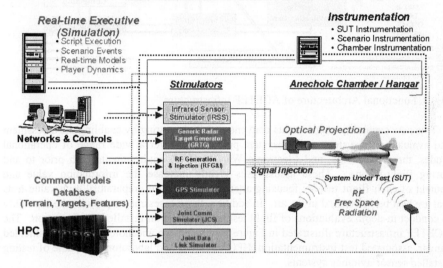

Fig. 2. Functional Block Diagram of Upgrades in ACETEF Architecture

1.3 Overview of the ACETEF Enhancements

The objective of the ACETEF enhancements is to develop, install, and operationally certify common core capability enhancements that will enable the effective testing of advanced CNI, RADAR, Infrared (IR) and Electronic Countermeasures (ECM) avionics subsystems. This enhanced test capability reduces the overall cost and schedule associated with the developmental and operational testing of existing and new avionics, sensor, and communication suites.

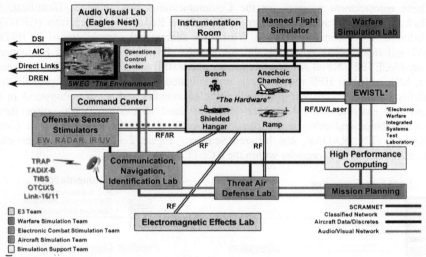

Fig. 3. Functional Architecture of ACETEF Infrastructure

This new capability supplements conventional flight testing by enabling the assessment and evaluation of system and subsystem performance over a wide range of operational modes, threat scenarios, and Electronic Warfare (EW) environments both prior to and during flight-testing. These assessments greatly increase the information value and content of each flight test by focusing attention on specific preparatory high value tests that are best be performed in the air. In addition, these assessments provide the capability to conduct in-depth evaluations of flight test anomalies in an installed configuration. The ACETEF infrastructure illustrated in Figure 3 includes two anechoic chambers, a shielded hangar, associated test instrumentation, HPC resources and laboratories capable of testing installed sensor/avionics systems.

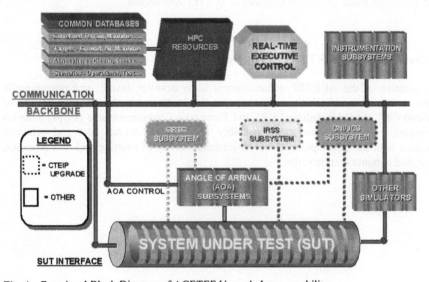

Fig. 4. Functional Block Diagram of ACETEF Upgrade Interoperability

Figure 4 illustrates the integration and interoperability of the GRTG and IRSS within the ACETEF infrastructure. Interoperability is achieved through a common communications backbone, which enables the real-time executive (RTEx) control to synchronize individual simulator/stimulator operations, the System Under Test (SUT) and the Instrumentation Subsystem to a common test scenario. Common data bases, standard communications interfaces, standard programming languages, and standard simulation models all contribute to the integrated operation of the ACETEF. Modular design principles and the use of an open architecture permit the addition of other simulators and a free space Angle Of Arrival (AOA) system in the future. The communications infrastructure also implements a number of standard interfaces to avionics buses through a SUT interface. The use of common test scenarios, common data bases and standard simulations used by hardware-in-the-loop (HITL) facilities and Open Air Ranges (OAR) provides the linkage and traceability needed to implement the Electronic Combat (EC) Test Process.

Additionally, their capabilities contribute to the evaluation of systems, and sensors, during development, and engineering model development (EMD) performance assessment. They also enable sensor test in a variety of environmental and warfare simulations that can not be replicated on the test range.

1.4 Generic Radar Target Generator (GRTG)

The GRTG is a modular and expandable radar environment stimulator for use with existing and emerging radar systems (e.g., Existing - F-22 APG-77, F/A-18 APG-73, and the JSF - Future capability). It is used to verify the functionality of air-to-air (A/A) and air-to-surface (A/S) modes (- Future capability) of radar systems. The GRTG generates radar skin returns and surface returns, and electronic countermeasures (ECM) against airborne radar's operating in the I/J bands (8-18GHz). It is capable of simulating a maximum of four targets in the main beam of the radar antenna. The modular design enables reconfiguration to incorporate enhancements in other spectra bands and radar modes as necessary to accommodate future systems. Additionally, 32 targets can be simulated in the radar field of regard (FOR).

The GRTG can be operated in a stand-alone mode or integrated with the ACETEF infrastructure. Signals are injected directly to the SUT radar receiver via RF cables or radiated free space from horn antennas to the radar antenna assembly. In the case of free space, the GRTG functions in conjunction with the angle of arrival (AOA) subsystem to provide multiple dynamic/static angles of arrival for radar/ECM targets. In the case of direct injection (RF), the GRTG replicates/models by-passed radar sub-assemblies by directly injecting the signal into the radar-processing avionics. The GRTG is capable of both autonomous (stand-alone) testing and correlated, real time, integrated multi-spectral testing with other simulators/stimulators (e.g., EW, IR, GPS). The latter form of testing is accomplished with the GRTG and other simulators/stimulators responding to an external ACETEF scenario simulation synchronized by the real-time executive controller. Figure 5 is a block diagram of the GRTG functional areas and relationships.

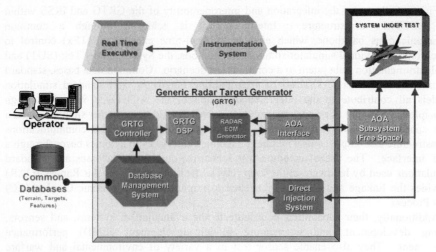

Fig. 5. GRTG Functional Block Diagram

1.5 Infrared Sensor Stimulator (IRSS)

The IRSS is a modular system used to generate high fidelity Infrared (IR) scenes for stimulation of installed IR Electro-Optic (EO) sensors installed on military aircraft undergoing integrated developmental and operational testing. It is capable of stimulating multiple types of sensors such as Forward Looking Infrared (FLIR), Missile Warning Systems (MWS), Infrared Search and Track (IRST) and Missile Seekers. It features include an open architecture, high fidelity physics-based rendering, repeatable, dynamic real-time IR and Ultraviolet (UV) simulation, free-space scene projection and direct image signal injection stimulation capability. The dedicated system configuration utilizes two SGI Onyx2 graphics computers and a proprietary scene rendering system (SRS) to provide three channels of real time sensor-specific scene simulation. With the parallel use of HPC resources, IRSS is scalable to ten channels of simulation output. The signal injection component also provides a sensor modeling capability for replicating sensor optics/detector modulation transfer function and evaluating performance characteristics and effectiveness of prototype designs (e.g., EMD). The IRSS system can operate in stand-alone mode, or in an integrated mode with the ACETEF infrastructure.

Figure 6 is a top-level functional block diagram of the IRSS and its six subsystems. The six subsystems are (1) Simulation Control Subsystem; (2) Scenario Development Subsystem; (3) Modeling and Database Development Subsystem; (4) Scene Generation Subsystem; (5) Point Source Projection Subsystem and (6) Signal Injection Subsystem/interface. The IRSS operator and/or the real-time executive control software provide inputs to the IRSS simulation control computer.

In the integrated mode, the ACETEF real time executive software controls IRSS SCRAMNET interfaces, timing, and execution priorities for system models and simulations to create a real-time virtual test environment. Scenario development inputs are provided by an operator and/or the executive software. Inputs includes (but are not limited to): aircraft data (position, airspeed, attitude), sensor data (FOV, FOR, modulation transfer function (MTF), and slew rates), and gaming area environment (targets, backgrounds and atmospherics effects). The IRSS functions in concert with the other stimulators to perform correlated, real time, integrated multi-sensor testing.

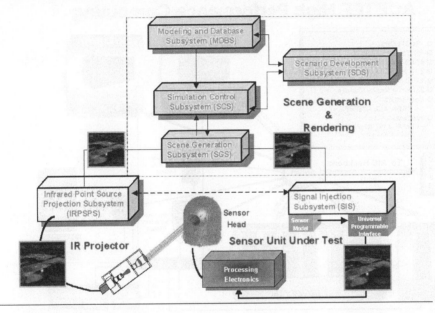

Fig. 4. IRSS Six Primary Functional Subsystems

2. The Need for High Performance Computing (HPC) Resources

The technical challenge for the ACETEF T&E enhancements is that they be valid test tools. Consequently, the spatial, spectral and temporal components of the computer-generated synthetic scenes/RF responses should be of sufficient fidelity to produce sensor responses that are indistinguishable from the tested sensor's "real-world" responses. Two fundamental capabilities are required to accomplish the foregoing validity objectives. First, the simulation generation equipment must have the capability to create a valid simulation of the real-world conditions. Second, the stimulators (RF emitters, IR scene projectors and signal injection modules) must be capable of accepting the simulated scene, retain sufficient scene fidelity, and interface appropriately to the sensor unit under test (UUT).

This section discusses the subtleties of accomplishing the foregoing goals, and provides an overview of how these T&E enhancements meet the fidelity challenge. The GRTG and IRSS have been designed to take advantage of present and future ACETEF HPC resources. When the complexity of GRTG and IRSS test requirements increases, HPC resources are used to augment dedicated computational systems. Areas such as increased number and/or types of sensors to be stimulated, increased frame rates, and scene complexity are illustrative of increased test requirements. Figure 7 is an architectural diagram of the ACETEF HPC configuration.

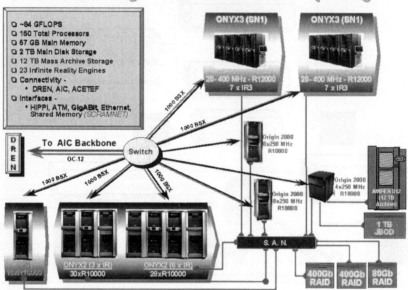

Fig. 7. ACETEF High Performance Computing Architectural Configuration

The GRTG configuration is capable of utilizing HPC resources for real-time stand-alone scenario control. Pre-planned product improvements to the GRTG includes a large area common terrain database server upgrade to support ultra-high-fidelity synthetic aperture radar (SAR) processing. The terrain database will be shared between the IRSS, and the GRTG to support multi-spectral testing. Figure 8 is an aerial view of the ACETEF Large Anechoic Chamber with a cutaway view of an F/A 18 and EA6B inside the chamber. Figure 9 illustrates an ACETEF multi-spectral testing configuration in one of the two anechoic chambers. The aircraft can be positioned on the ground or suspended from one of two 40 tons cranes located on the test chamber ceiling.

Fig. 8. ACETEF Large Anechoic Chamber

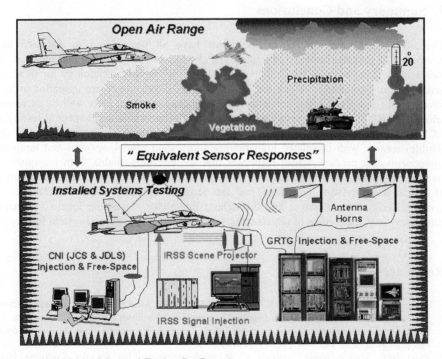

Fig. 9. ACETEF Multi-Spectral Testing Configuration

The IRSS performance and throughput requirements for dynamic real time rendering and physics-based accuracy necessitate use of an HPC solution. Accordingly, the IRSS architecture employs commercial-off-the-shelf (COTS) Silicon Graphics (SGI) fast symmetric multiprocessing hardware to minimize cost and development time. The following functional processes illustrate the high performance computational workload that the IRSS must satisfy.

1. IRSS provides three channels of digital video configured to support synchronized testing of multiple EO imaging sensors on a single aircraft or individual sensors on up to three aircraft. The architecture is scalable to a maximum of ten channels of output using a commensurate temporary allocation of HPC resources.

2. During real-time scene simulation, the multiprocessors are used to update polygon vertex locations and compute radiometrically correct floating-point radiance values for each waveband. Scene radiance is calculated on a frame by frame basis with each frame being rendered with the relevant contributions from the sky, sun, targets, terrain, and atmosphere. The individual frame calculations are accomplished as a function of the engagement geometry using existing validated high-fidelity IR models for phenomenology, terrain, and targets.

3. Summary and Conclusions

The JISTF IRSS and GRTG enhancements have all been carefully designed and developed to leverage current and future HPC resources. The IRSS enhancement is increasingly using more of HPC assets as it pushes the limits of technology for IR scene generation and rendering. When the GRTG and the CNI systems are upgraded to take advantage of a multi-spectral common terrain database systems, there will be increased reliance on HPC systems to provide the backbone for integrated multi-spectral testing at ACETEF. They provide the valid stimulator T&E of installed multi-spectral (RF, IR, Visible) sensors with correlated high fidelity stimulation spatial, spectral and temporal components that produce sensor responses that are indistinguishable from responses to "real-world" conditions. The availability and use of HPC resources has enabled resolution of critical issues relating to scaling, and the accommodation of additional channels of synchronized/correlated output and throughput limitations versus the level of scenario complexity (e.g. the content of targets and background descriptions, and system latency.)

References

1. T. Joyner, K. Thiem, R. Robinson, R. Makar, and R. Kinzly, "Joint Navy and Air Force Infrared Sensor Stimulator (IRSS) Program Installed Systems Test Facilities (ISTF)", Technologies for Synthetic Environments: Hardware-in-the-Loop Testing IV, Robert Lee Murrer, Editor, Proceedings of SPIE Vol. 3697, pp. 11-22, 1999.
2. DoD HPC Modernization Program FY96 Implementation Plan for the Common High Performance Computing (HPC) Software Support Initiative (CHSSI), dated October 1995

Infrared Scene Projector Digital Model Development

Mark A. Manzardo[1], Brett Gossage[1], J. Brent Spears[1], and Kenneth G. LeSueur[2]

[1] 555 Sparkman Drive, Executive Plaza, Suite 1622
Huntsville, AL 35816
{mmanzardo, bgossage, bspears}@rttc.army.mil
[2] US Army, Developmental Test Command
Redstone Technical Test Center
CSTE-DTC-RT-E-SA, Bldg. 4500
Redstone Arsenal, AL 35898-8052
klesueur@rttc.army.mil

Abstract. This paper describes the development of an Infrared Scene Projector Digital Model (IDM). The IDM is a product being developed for the Common High Performance Computing Software Support Initiative (CHSSI) program under the Integrated Modeling and Test Environments (IMT) Computational Technology Area (CTA). The primary purpose of the IDM is to provide a software model of an Infrared Scene Projector (IRSP). Initial utilization of the IDM will focus on developing non-uniformity correction algorithms, which will be implemented to correct the inherent non-uniformity of resistor array based IRSPs. After describing the basic components of an IRSP, emphasis is placed on implementing the IDM. Example processed imagery will be included, as well.

1 General Description of an Infrared Scene Projector (IRSP)

In general, an IRSP can be broken down into the interaction between the following subsystems/components: a Software Control Subsystem (SCS), a Computer Image Generator Subsystem (CIGS), a Control Electronics Subsystem (CES), an Environmental Conditioning Subsystem (ECS), an Infrared Emission Subsystem (IRES), a Projection Optics Subsystem (POS), a Non-Uniformity Correction Subsystem (NUCS), a Mounting Platform Subsystem (MPS) and a System Under Test (SUT). Figure 1 provides an overview block diagram illustration of an IRSP. This figure also illustrates the typical interaction between the subsystems. Not all of these interactions are required; some are for ease of use, configuration management, system safety, or other specific needs. It is unlikely that any given IRSP utilizes all of the illustrated interfaces.

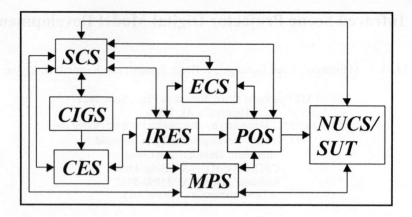

Fig. 1. Typical IRSP Block Diagram.

The following sections provide a brief description of each of the IRSP subsystems.

1.1 Software Control Subsystem (SCS)

The SCS can be one integrated program controlling all aspects of the IRSP from a single Graphical User Interface (GUI). The SCS can also have communication interfaces with other programs that are driving the overall simulation. The SCS can be hosted entirely on the CIGS, CES or it's own separate platform. In general, SCS components can be distributed throughout other IRSP subsystems. The form the SCS takes is usually driven by IRSP usage requirements tempered by cost restraints.

1.2 Computer Image Generator Subsystem (CIGS)

The CIGS is the hardware platform responsible for generating the dynamic input imagery used by the IRSP CES. The generation of this imagery is typically performed in real-time using 3-D target/terrain database(s), material database(s), etc. The result of the scene generation software (an IRSP SCS component) is a sequence of 2-D rendered scene images. The scene generator can either render these images in real-time for input to the IRSP CES or scene generation can be an initial step preparing the imagery for pre-projection processing. One such pre-projection processing step would be to perform software Real-time Non-Uniformity Correction (RNUC) on each image pixel. In some cases the scene generator is used to render extremely high fidelity scenes in non-real-time for storage and subsequent real-time playback.

1.3 Control Electronics Subsystem (CES)

The CES is used to provide all the necessary analog and digital signals to drive the resistor array. In addition to the image data these signals will also typically include bias, power, and address lines. In some cases, such as test patterns, image data can be generated on the CES hardware or uploaded across the bus interface of the CES hardware (e.g., VME, VXI, ...). However, more practical applications that require real-time dynamic image sequences use a high speed digital image interface (e.g., the SGI DD02). In some cases, pipeline image processing hardware is incorporated into the CES to simulate effects not incorporated in the scene generator and/or to correct for overall IRSP simulation latency. Examples of effects not incorporated in the scene generator that can be accommodated by dedicated hardware are atmospheric blur, missile seeker focus upon closure to a target, and RNUC of the resistor array. With or without dedicated hardware to perform the above special effects, the CES can impose lossy noise or digitization effects on the image data. The output of the CES is typically either high speed digital image data over a fiber optic interface or analog voltages from high speed digital to analog converters.

1.4 Environmental Conditioning Subsystem (ECS)

The ECS has the primary responsibility of providing an environment for proper operation of the resistor array. The ECS is responsible for creating (and sometimes maintaining) a vacuum environment around the array. The ECS is also responsible for creating and maintaining an operating temperature for the array. In addition, in some cases, the ECS is also extended to provide thermal conditioning of other IRSP components (e.g., the POS, NUCS and/or the SUT).

The ECS typically utilizes a contaminant free vacuum pump capable of pulling a vacuum less than 3E-04 torr. A turbo-molecular vacuum pumping station is typically used to create the vacuum. In some cases, the vacuum pump remains on continuously to maintain the vacuum environment. In other cases, the IRES has the ability to maintain a vacuum level for a sufficient amount of time for operation; the IRES is sometimes assisted by a small ION pump. The array's radiometric output is known to be a function of the vacuum level. This is primarily due to allowing a convection heat dissipation path. In an absolute vacuum, the array can only dissipate heat radiatively and conductively through it's legs to the substrate. When the vacuum pressure increases, providing for a convection heat dissipation path, the resistor does not heat up as much and the radiated energy is subsequently reduced. Studies indicate that a vacuum level below 3E-04 torr is sufficient to dramatically mitigate this effect.

Due to the large substrate heat dissipation requirements of micro-resistor arrays (typically in excess of 100 Watts), the heat removal is accomplished with a fluid recirculating chiller. The operating temperature of the resistor array affects the array's radiometric output. This directly affects the obtainable background radiometric floor and stability of the IRSP. In addition, the CMOS drive substrate changes it's characteristics as a function of temperature.

1.5 Infrared Emission Subsystem (IRES)

The IRES has the primary function of mounting the resistor array to allow it to interface to the rest of the IRSP. Usually, the IRES is a self-contained package providing electrical, mechanical, fluid, and vacuum interfaces. In some cases, the heat sink is external to the vacuum environment of the array, limiting the array operating temperature in this stand-alone package to something above the ambient dew point. Other arrays, are packaged in a full dewar configuration allowing the array operating temperature to go well below the dew point (some are designed to go to −35°C and others can go to 20K). In some configurations, the vacuum environment not only encloses the array, but also the POS and sometimes a detector array.

In any event, the primary component of the IRES is the resistor array itself. Some examples of characteristics, which are known to affect the array's input to output response are: material type, leg length, fill factor, etc. However, non-uniformities in resistor responses occur because of manufacturing inconsistencies and defects.

There are at least a couple of known scene based effects with some resistor arrays. The first is local heating near contiguously driven regions on the array. Another is an effect termed "bus-bar-robbing" which results from voltage drops by driving a large percentage of the array.

In addition to the scene based effects there are at least a couple of transient temporal effects. One is based on the real response time of the resistor. A similar effect is intra-frame droop.

1.6 Projection Optics Subsystem (POS)

The POS has the responsibility to properly project the IRES radiation into a Sensor or System Under Test (SUT). This is typically accomplished by collimating the IRES radiation. The focal length and pupil of the POS is chosen to allow for valid effective SUT evaluation. Being a real optical system, the POS will have spectral transmission losses, impose geometric and diffractive blur effects and impose geometric distortion effects on the projected image.

1.7 Non-Uniformity Correction Subsystem (NUCS)

The NUCS has the responsibility of being able to collect, reduce, and process resistor array imagery with the purpose of determining the proper correction coefficients necessary to remove the array non-uniformity's. The process has been implemented by isolating and measuring the output of each resistor as a function of input. This has been accomplished by commanding a sparse grid of resistors to multiple inputs and then stepping through the array until all resistors have been measured. The two main drawbacks to this process is the data collection time and the fact that the resistor arrays do not respond the same when commanded with a sparse grid pattern as they do when commanded with realistic scene imagery (local heating and what has been termed bus-bar robbing are considered to be the primary sources for the differences).

Some attempts have been made to drive contiguous blocks of resistors to collect non-uniformity data. Unfortunately, most are very dependent on optical coupling and distortion, but preliminary results are promising.

The primary component of the NUCS is the IR data collection camera. The camera acts as a transducer from IR radiation to a digital image. The camera detector(s) will have a limited resolution when digitizing the input radiation signal (e.g., 12-bit digital values representing a known max and min radiance/temperature). The primary effects the camera will impose on the input scene are spatial sampling artifacts, spatial and temporal noise, geometric distortion, and geometric and diffractive blur effects.

1.8 Mounting Platform Subsystem (MPS)

The MPS has the responsibility of aligning the IRES with the POS and the SUT (or NUCS camera). This can be a manual process using stages, rails, tripod(s), or equivalent. This can also be a fully computer controlled process using motorized stages and gimbals. In any event, there will always be some limited alignment error (at best on the order of the detector/resistor array spatial resolution).

1.9 System Under Test (SUT)

The SUT is the infrared imaging device being evaluated with the Infrared Scene Projector.

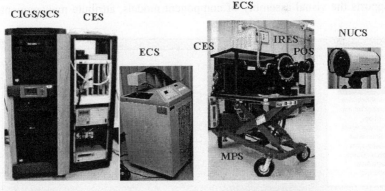

CIGS- SGI OnyxII w/ 8 processors
SCS - Resides on SGI
CES - VME Chassis
IRES - 544x672 array in Dewar

POS - Manual LWIR Zoom Lens
ECS - Recirculating Chiller & Vacuum
MPS - Pitch/Yaw Gimbal, Roll axis, and Scissor Jack
NUCS - AGEMA 1000LR & Tripod

Fig. 2. DEGA System (Example IRSP).

1.10 Example IRSP

For illustrative purposes, Figure 2 shows components of an example IRSP. Specifically, this is the Dynamic Infrared Scene Projector (DIRSP) Engineering Grade Array (DEGA) system used by the US Army Redstone Technical Test Center.

2 IDM Implementation and Example Imagery

The following sub-sections describe the simulation architecture used to implement the IDM and some example imagery. A detailed mathematical description of how the IDM is designed to accommodate all the IRSP effects is provided in a simultaneously published companion paper: "Infrared Scene Projector Digital Model Mathematical Description". In it's current state, the IDM incorporates the IRSP effects that tend to redistribute the image energy as detected by the SUT. These types of effects are: resistor fill factor, optical distortion, optical geometric aberrations, SUT diffraction, and SUT geometric aberrations. Over the next couple of years the IDM will grow to encompass the rest of the IRSP effects. However, the above effects are estimated to be the most computationally challenging, which justifies why they have been incorporated first.

2.1 Common Simulation Framework (CSF) Simulation Architecture

The IDM utilizes the Common Simulation Framework (CSF) to provide an IDM user with the capability to manage, store and execute the simulation. The CSF is a DoD-developed simulation architecture managed by US Army Aviation and Missile Research, Development and Engineering Center, Huntsville AL. The CSF-hosted IDM supports the visual assembly of component models, attribute management, and

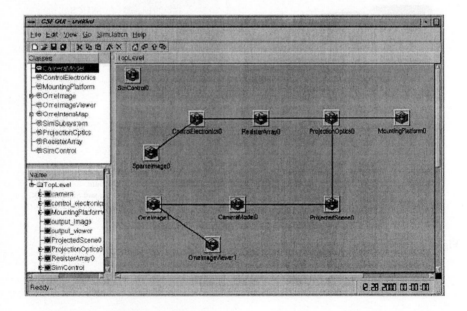

Fig. 3. CSF Implementation of the IDM.

model execution all within a unified application. The IDM software is built on the CSF framework libraries thus providing a seamless CSF/IDM integration. Since the CSF is written entirely in C++, run-time efficiency is maintained and multi-threading is easily applied. The object-oriented "plug-n-play" features of the CSF allow different models (e.g. cameras, resister arrays) to be substituted without changing the software. Figure 3 shows a screen-shot of the CSF-hosted IDM user interface.

In addition to using the CSF to host the IDM simulation, we have extended the CSF executive to include a real-time, frame-based scheduler. This allows the models to be executed in the same environment as the real-time NUC algorithms. The real-time executive is launched from within the CSF GUI shown above and allows a user to run, pause, reset, and stop the simulation. Pluggable image viewers are provided for viewing the results at various points in the processing chain.

2.1.1 Parallel Speedup of IDM

A parallel implementation of the IDM was performed for a 400(H)x200(V) resolution camera and a 672(H)x544(V) resolution resister array on a Silicon Graphics Incorporated (SGI) Onyx-2 with 8x300 MHz processors. Distortion effects calculations were not included in this parallel implementation. Figure 4 illustrates the comparison between the actual achieved run times versus the theoretical limit.

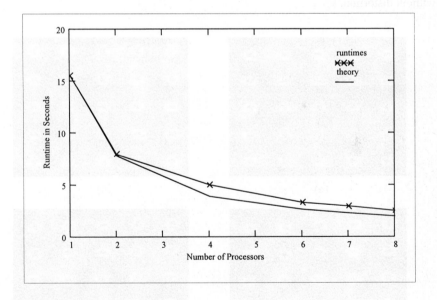

Fig. 4. Parallel Speedup of the IDM.

2.2 Example Imagery

The following example imagery illustrates usage of the IDM. The input represents the image driving a 672(H)x544(V) resistor array. The output represents the image captured by a 800(H)x405(V) camera. The input was a sparse resistor pattern, where every 16th row and every 14th column resistor was driven. One input image began the sparse pattern at resistor (0,0) the other began the pattern at resistor (1,1). Although the IDM processes the entire camera image, for illustrative purposes only a portion of the image is shown here. Figure 4 shows the upper 5x5 driven resistors.

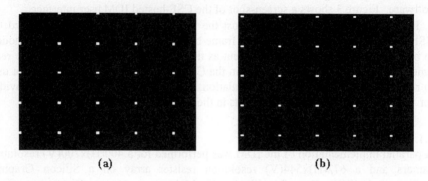

Fig. 5. 16x14 Sparse Resistor Input Images to the IDM: (a) start at (0,0), (b) start at (1,1).

Figure 5 illustrates the output of the IDM for the two input images with and without distortion.

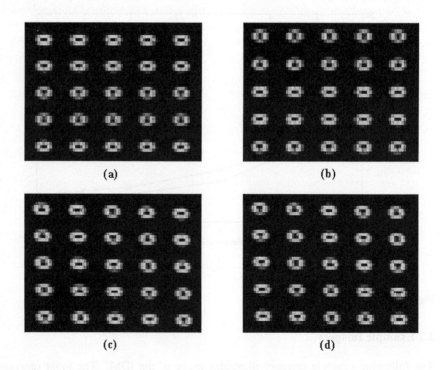

Fig. 6. Output IDM Images without (a, b) and with (c, d) Distortion.

Of interest in the IDM output imagery is the noticeable variation of detected energy distribution from resistor to resistor. This basically illustrates why a simple single convolution kernal for the whole image would be inaccurate. In general, one way to think of the process is that each detector uses a different convolution kernal of variable size and distribution. This is necessary to account for mis-registration between the resistors and detectors, optical distortion, and walk-off registration between the resistors and detectors. Walk-off registration occurs when the pitch of the detector samples does not match the pitch of the resistors. As the output imagery illustrates, the need for a variable convolution kernal holds true even without distortion.

2 Conclusions and Continuing Efforts

The IDM output has been validated from a "that looks about right" perspective. The output looks very reminiscent of real collected imagery. The goal for validation of the IDM is to eventually be able to compare real versus IDM imagery using objective image statistics and comparison metrics. The validation phase can begin after incorporating the rest of the IRSP effects.

Recall that in it's current state, the IDM incorporates the IRSP effects that tend to redistribute the image energy as detected by the SUT. These types of effects are: resistor fill factor, optical distortion, optical geometric aberrations, SUT diffraction, and SUT geometric aberrations. Over the next couple of years the IDM will grow to encompass the rest of the IRSP effects.

Efforts will continue to parallelize code for use on HPC multi-processor platforms.

Infrared Scene Projector Digital Model Mathematical Description

Mark A. Manzardo[1], Brett Gossage[1], J. Brent Spears[1], and Kenneth G. LeSueur[2]

[1] 555 Sparkman Drive, Executive Plaza, Suite 1622
Huntsville, AL 35816
{mmanzardo, bgossage, bspears}@rttc.army.mil

[2] US Army, Developmental Test Command
Redstone Technical Test Center
CSTE-DTC-RT-E-SA, Bldg. 4500
Redstone Arsenal, AL 35898-8052
klesueur@rttc.army.mil

Abstract. This paper describes the development of an Infrared Scene Projector Digital Model (IDM). The IDM is a product being developed for the Common High Performance Computing Software Support Initiative (CHSSI) program under the Integrated Modeling and Test Environments (IMT) Computational Technology Area (CTA). The primary purpose of the IDM is to provide a software model of an Infrared Scene Projector (IRSP). Initial utilization of the IDM will focus on developing non-uniformity correction algorithms, which will be implemented to correct the inherent non-uniformity of resistor array based IRSPs. Emphasis here is placed on how IRSP effects are modeled in the IDM.

1 General Description of an Infrared Scene Projector (IRSP)

In general, an IRSP can be broken down into the interaction between the following subsystems/components: a Software Control Subsystem (SCS), a Computer Image Generator Subsystem (CIGS), a Control Electronics Subsystem (CES), an Environmental Conditioning Subsystem (ECS), an Infrared Emission Subsystem (IRES), a Projection Optics Subsystem (POS), a Non-Uniformity Correction Subsystem (NUCS), a Mounting Platform Subsystem (MPS) and a System Under Test (SUT). Figure 1 provides an overview block diagram illustration of an IRSP. This figure also illustrates the typical interaction between the subsystems. Not all of these interactions are required; some are for ease of use, configuration management, system safety, or other specific needs. It is unlikely that any given IRSP utilizes all of the illustrated interfaces.

A more detailed description of each of these subsystems is provided in the simultaneously published companion paper: "Infrared Scene Projector Digital Model Development".

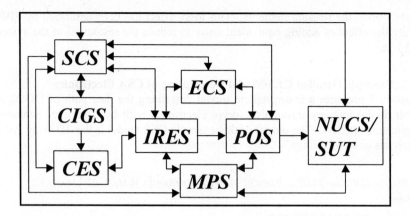

Fig. 1. Typical IRSP Block Diagram

2 IRSP Digital Model (IDM) Mathematical Description

The primary purpose of the IDM is to provide a digital model of the IRSP process allowing for NUC algorithm development while freeing up the actual hardware for utilization. The IDM will assume as input a temporal sequence of 2D digital image pixels with a 16-bit depth, the typical output of the CIGS. In other words, the IDM will not model the intricate process of actually generating the 2D rendered images. The IDM will provide as output a temporal sequence of 2D digital image pixels equivalent to the output of the NUCS camera or SUT. The bit depth of the output image will be dependent on the sensor. For example, the Agema 1000LR NUCS camera provides a 12-bit output digital image.

2.1 IDM CIGS Model

For IDM purposes, the CIGS is modeled simply as a sequence of two dimensional imagery. This is represented here as the image array DD02[,].

2.2 IDM CES Model

For drive electronics systems that are used to exercise the Honeywell type resistor array, the 16-bit digital image signal is converted to analog drive voltages. This conversion is not perfect and therefore comes with a loss in resolution. For drive electronics systems that are used to exercise the Santa Barbara Infrared (SBIR) type resistor array, the 16-bit digital image signal is driven directly into the digital input of the resistor array. This effectively provides a lossless data path to the array. It should be noted that on the chip these digital signals are converted to analog signals to actually drive the resistor elements. This lossy effect has been described by SBIR as having the effect of adding equivalent noise to reduce the resolution of the system to 14.5 bits.

2.2.2 Example Detailed CES Model Description of CSA Electronics

Equation 1 provides a mathematical model describing the data path of DDO2 input 16-bit digital imagery through the drive electronics built by Computer Science and Applications (CSA). The output of this example CES model is the 2D array AIE[,], which has units of voltage.

$$AIE[i,j] = DACNoise\{VLUT_{APB}[GainOff\{VLUT_{SIB}[DDO2[i,j]],i,j\},Mod(j,32)],Mod(j,32)\}$$

where :

(a)
$$\begin{bmatrix} \text{if}: & \text{direct to FIFO SIB mode} \\ & i = i_0 \ \& \ j = j_0 \\ \text{elseif}: & \text{A/B bank SIB mode} \\ & i = InvRowALUT[i_0,j_0] \ \& \ j = InvColALUT[i_0,j_0] \\ \text{where}: [i_0,j_0] \text{is the original row and column index in the DDO2 imagery} \end{bmatrix}$$

(b)
$$\begin{bmatrix} \text{if}: & SIB[i,j]*Gains[BPLUT[SIB[i,j]],i,j] <= 2^{\wedge}16-1 \quad -\text{and}- \\ & 0 <= \begin{cases} SIB[i,j]*Gains[BPLUT[SIB[i,j]],i,j] \\ +Offsets[BPLUT[SIB[i,j]],i,j] \end{cases} < 2^{\wedge}16 \\ & GainOff\{SIB[i,j],i,j\} = VLUT_{RNUC}\begin{bmatrix} SIB[i,j]*Gains[BPLUT[SIB[i,j]],i,j] \\ +Offsets[BPLUT[SIB[i,j]],i,j] \end{bmatrix} \\ \text{elseif}: & SIB[i,j]*Gains[BPLUT[SIB[i,j]],i,j] > 2^{\wedge}16-1 \\ & GainOff\{SIB[i,j],i,j\} = VLUT_{RNUC}[0xFFFF] \\ \text{elseif}: & SIB[i,j]*Gains[BPLUT[SIB[i,j]],i,j] <= 2^{\wedge}16-1 \quad -\text{and}- \\ & \begin{cases} SIB[i,j]*Gains[BPLUT[SIB[i,j]],i,j] \\ +Offsets[BPLUT[SIB[i,j]],i,j] \end{cases} > 2^{\wedge}16-1 \\ & GainOff\{SIB[i,j],i,j\} = VLUT_{RNUC}[0xFFFF] \\ \text{elseif}: & SIB[i,j]*Gains[BPLUT[SIB[i,j]],i,j] <= 2^{\wedge}16-1 \quad -\text{and}- \\ & 0 > \begin{cases} SIB[i,j]*Gains[BPLUT[SIB[i,j]],i,j] \\ +Offsets[BPLUT[SIB[i,j]],i,j] \end{cases} \\ & GainOff\{SIB[i,j]i,j\} = VLUT_{RNUC}[0] \\ \text{where}: & SIB[i,j] = VLUT_{SIB}[DDO2[i2,j2]] \end{bmatrix}$$

(1)

2.3 IDM ECS Model

As in a real IRSP, the IDM ECS acts primarily as a service unit. The ECS parameters (vacuum, IRES temperature, POS temperature, prevailing environment temperature ...) will be used to establish the necessary radiometric properties of the IRES, POS, and NUCS/SUT. The ECS set-points are considered to be temporally stationary (constant). However, due to the response and control loop(s) of real systems, there will be some drift and/or "noise" associated with the ECS parameters. To account for this the ECS parameters will be time varying values.

2.4 IDM MPS Model

As in a real IRSP, the IDM MPS acts primarily as a service unit for alignment. The MPS parameters will be used to establish the relative spatial location of the IRES, POS, and NUCS/SUT. The MPS set-points are considered to be temporally stationary (constant). However, due to vibrations, there will be some drift and/or "noise" associated with the MPS parameters. To account for this the MPS parameters will also be time varying values.

2.5 IDM IRES Model

2.5.1 IRES Bus-Bar Robbing Effect

Bus-Bar Robbing is an effect common in the micro-resistor array technology. Basically, as the high current drive voltage bus is tapped into, by each FET along the bus, a voltage drop occurs dependent on how much current is drawn. This effect is a scene dependent effect and may require knowledge of the entire image before adequate compensation can be implemented. Although the Bus-Bar Robbing operation is not fully modeled yet, the IDM is configured to accommodate this effect by processing on the output of the CES (see Equation 1 above) as follows:

$$PostBusBar[i, j] = BusBarRob\{AIE[\ ]\} \qquad (2)$$

The empty brackets [ ] in this relationship indicate that the operator $BusBarRob\{\ \}$ may need the complete AIE array (or some yet to be determined sub-array) to process the output for a given [i, j] resistor. For a given [i, j] resistor, the output of this operation is the discretely represented by array $PostBusBar[i, j]$.

2.5.2 IRES Substrate Heating Effect

IRES Substrate Heating is an effect which causes the local substrate to heat up when a large area of the array is driven at once. This is primarily due to the limitation of the heat sink to remove local heat. The local heat of a driven region will dissipate laterally across the heat sink to substrate boundary. This effect is a scene dependent effect. Although the IRES Substrate Heating operation is not fully modeled yet, the IDM is configured to accommodate this effect by processing on the output of the CES after the Bus-Bar Robbing effect as follows:

$$PreNU[i, j] = SubHeat\{PostBusBar[\ ]\} \qquad (3)$$

The empty brackets [ ] in this relationship indicate that the operator $SubHeat\{\ \}$ may need the complete $PostBusBar$ array (or some yet to be determined sub-array) to process the output for a given [i, j] resistor. For a given [i, j] resistor, the output of this operation is the discretely represented by array $PreNU[i, j]$.

NOTE: It may be alternatively sufficient to add an additional system blur to account for this effect. Adding addition blur effects are discussed later.

2.5.3 IRES Non-Uniformity Effect
The primary issue surrounding the utilization of an IRSP is the inherent output Non-Uniformity (NU) that exists between the resistors when driven to the same input. This NU is a complex unknown input/output relationship, which differs for each resistor. For IDM implementation purposes, each resistor is assigned a set of NU coefficients of a yet to be determined functional form. The modeled output after NU is represented by:

$$PostNU[i,j] = ArrayNU\{PreNU[i,j], NUCoefs[i,j,1...N]\} \tag{4}$$

The functional form used will be determined by empirical measurements. For a given [i, j] resistor, the output of this operation, $ArrayNU\{\ \}$, is the discretely represented array $PostNU[i,j]$.

2.5.4 IRES Fill Factor Effect
Each resistor does not emit energy as a point source. The energy comes from the distributed area of the resistive material. To represent this effect, $PostNU[i,j]$ is represented as a two dimensional sum of dirac delta functions that is convolved with the analytically continuous *FillFactor* function. NOTE: The exact form of the *FillFactor* function will differ between different array designs. Up until now, the IRSP image representation has been in a unitless discretely sampled coordinate system. It is advantageous at this point to represent the energy coming off the array in the angular coordinate system in collimated space after the POS. The description of signal after incorporating the Fill Factor effect becomes:

$$PostArray(\alpha,\beta) = \left(\sum_{i=0}^{N_i-1}\sum_{j=0}^{N_j-1} PostNU[i,j]\delta\left(\alpha - \frac{i-\delta i_{Array}}{iScale_{Array}}, \beta - \frac{j-\delta j_{Array}}{jScale_{Array}}\right)\right) ** FillFactor\left(\frac{\alpha}{\alpha Scale_{Array}}, \frac{\beta}{\beta Scale_{Array}}\right) \tag{5}$$

It is important to note that at this point the IRSP signal is represented as an analytically continuous function. There is no theoretical limit to the spatial resolution used to represent this as a discretely sampled image for viewing purposes. However, any attempt to represent this discretely is subject to sampling and aliasing artifacts. To mitigate or limit such artifacts appropriate reconstruction filtering techniques should be used. As it will be shown later, these artifacts are not issues for the IDM's intended use.

2.6 IDM POS Model

2.6.1 POS Transmission and Geometric Aberrations Effects
The POS will impose transmission, geometric aberrations, distortion, and diffraction effects. The diffraction effect is not implemented because diffraction will occur as a function of the POS and SUT optics combined. Since, an ideal POS will overfill or match the entrance pupil of an SUT, diffraction will occur at the SUT and is accounted for there. Similarly, the distortion effect is cumulative between the POS and SUT and is accounted for later at the system level. The POS may have a transmission that varies as a function of field of view. It is accounted for here as a multiplicative two dimensional function $\tau_{POS}(\ )$. The geometric aberrations effect is accounted for by convolution with another two dimensional analytical function $GPSF_{POS}(\ )$. The description of the IRSP signal coming out of the POS becomes:

$$PostPOS(\alpha,\beta) = \tau_{POS}(\alpha,\beta) PostArray(\alpha,\beta) ** GPSF_{POS}\left(\frac{\alpha}{\alpha Scale_{POS}}, \frac{\beta}{\beta Scale_{POS}}\right)$$

$$PostPOS(\alpha,\beta) = \tau_{POS}(\alpha,\beta) \sum_{i=0}^{N_i-1}\sum_{j=0}^{N_j-1} PostNU[i,j]\delta\left(\alpha - \frac{i - \delta i_{Array}}{iScale_{Array}}, \beta - \frac{j - \delta j_{Array}}{jScale_{Array}}\right) **$$

$$\left(FillFactor\left(\frac{\alpha}{\alpha Scale_{Array}}, \frac{\beta}{\beta Scale_{Array}}\right) ** \atop GPSF_{POS}\left(\frac{\alpha}{\alpha Scale_{POS}}, \frac{\beta}{\beta Scale_{POS}}\right) \right) \tag{6}$$

2.7 IDM SUT Model

2.7.1 SUT Optics Transmission, Geometric Aberrations, and Diffraction Effects
Similar to the POS, the SUT optics will have transmission, geometric aberration, distortion, and diffraction effects. As discussed earlier distortion effects will be handled later at the system level. Similar to the POS, transmission is accounted for here as a multiplicative two dimensional function $\tau_{SUT}(\ )$. The combined SUT geometric aberrations and diffraction effects are accounted for here by convolution with another two dimensional analytical function $PSF_{SUT}(\ )$. The description of the IRSP signal after SUT optics becomes:

$$PostSUTOptics(\alpha,\beta) = \tau_{System}(\alpha,\beta)PostPOS(\alpha,\beta) ** PSF_{SUT}\left(\frac{\alpha}{\alpha Scale_{SUT}},\frac{\beta}{\beta Scale_{SUT}}\right)$$

$$PostSUTOptics(\alpha,\beta) = \tau_{System}(\alpha,\beta)\left(\sum_{i=0}^{N_i-1}\sum_{j=0}^{N_j-1}PostNU[i,j]\delta\left(\alpha-\frac{i-\delta i_{Array}}{iScale_{Array}},\beta-\frac{j-\delta j_{Array}}{jScale_{Array}}\right)\right)**$$

$$\left(\begin{array}{l} FillFactor\left(\dfrac{\alpha}{\alpha Scale_{Array}},\dfrac{\beta}{\beta Scale_{Array}}\right)** \\[2ex] GPSF_{POS}\left(\dfrac{\alpha}{\alpha Scale_{POS}},\dfrac{\beta}{\beta Scale_{POS}}\right)** \\[2ex] PSF_{UUT}\left(\dfrac{\alpha}{\alpha Scale_{SUT}},\dfrac{\beta}{\beta Scale_{SUT}}\right) \end{array}\right) \tag{7}$$

where:

$$\tau_{System}(\alpha,\beta) = \tau_{SUT}(\alpha-\Delta\alpha_{POS2SUT},\beta-\Delta\beta_{POS2SUT})\tau_{POS}(\alpha,\beta) \tag{8}$$

NOTE: $\Delta\alpha_{POS2SUT}$ and $\Delta\beta_{POS2SUT}$ represent some potential registration shift between the Projection Optics and the SUT Optics. Assuming the SUT output image is centered on the SUT Optics, this will be the same equivalent shift of the detectors described later.

2.7.2 SUT Detector Active Area Effect

Similar to the resistor array fill factor effect, the detector does not detect at a point, but rather over some active area. The efficiency of this active area may not be continuous. It is therefore, represented here as a general two dimensional convolution function. The description of the IRSP signal just after detection becomes:

$$PostSUTDetector(\alpha,\beta) = PostSUTOptics(\alpha,\beta) ** DetectArea_{SUT}\left(\frac{\alpha}{\alpha Scale_{SUT}},\frac{\beta}{\beta Scale_{SUT}}\right)$$

$$PostSUTDetector(\alpha,\beta) = \tau_{System}(\alpha,\beta)\left(\sum_{i=0}^{N_i-1}\sum_{j=0}^{N_j-1}PostNU[i,j]\delta\left(\alpha-\frac{i-\delta i_{Array}}{iScale_{Array}},\beta-\frac{j-\delta j_{Array}}{jScale_{Array}}\right)\right)**$$

$$\left(\begin{array}{l} FillFactor\left(\dfrac{\alpha}{\alpha Scale_{Array}},\dfrac{\beta}{\beta Scale_{Array}}\right)** \\[2ex] GPSF_{POS}\left(\dfrac{\alpha}{\alpha Scale_{POS}},\dfrac{\beta}{\beta Scale_{POS}}\right)** \\[2ex] PSF_{UUT}\left(\dfrac{\alpha}{\alpha Scale_{SUT}},\dfrac{\beta}{\beta Scale_{SUT}}\right)** \\[2ex] DetectArea_{SUT}\left(\dfrac{\alpha}{\alpha Scale_{SUT}},\dfrac{\beta}{\beta Scale_{SUT}}\right) \end{array}\right) \tag{9}$$

$$PostSUTDetector(\alpha,\beta) = \tau_{System}(\alpha,\beta)\left(\sum_{i=0}^{N_i-1}\sum_{j=0}^{N_j-1}PostNU[i,j]\delta\left(\alpha-\frac{i-\delta i_{Array}}{iScale_{Array}},\beta-\frac{j-\delta j_{Array}}{jScale_{Array}}\right)\right)**$$

$$System(\alpha,\beta)$$

$$PostSUTDetector(\alpha,\beta) = \tau_{System}(\alpha,\beta)\left(\sum_{i=0}^{N_i-1}\sum_{j=0}^{N_j-1}PostNU[i,j]System\left(\alpha-\frac{i-\delta i_{Array}}{iScale_{Array}},\beta-\frac{j-\delta j_{Array}}{jScale_{Array}}\right)\right)$$

where:

$$System(\alpha,\beta)=\left(\begin{array}{c}FillFactor\left(\dfrac{\alpha}{\alpha Scale_{Array}},\dfrac{\beta}{\beta Scale_{Array}}\right)**GPSF_{POS}\left(\dfrac{\alpha}{\alpha Scale_{POS}},\dfrac{\beta}{\beta Scale_{POS}}\right)** \\ PSF_{SUT}\left(\dfrac{\alpha}{\alpha Scale_{SUT}},\dfrac{\beta}{\beta Scale_{SUT}}\right)**DetectArea_{SUT}\left(\dfrac{\alpha}{\alpha Scale_{SUT}},\dfrac{\beta}{\beta Scale_{SUT}}\right)\end{array}\right) \tag{10}$$

NOTE: Additional System Blur effects can easily be added to the *System* function as necessary via additional two dimensional convolution functions.

2.7.3 SUT Discrete Sampling Effect

Representing the image as discretely sampled by the SUT at the $[k,\ l]$ detector is accommodated by multiplication with an appropriately shifted dirac delta function. The description of the signal as discretely sampled by the SUT becomes:

$$PostSUTSamp[k,l]=\delta\left(\alpha-\frac{k-\delta k_{SUT}}{kScale_{SUT}},\beta-\frac{l-\delta l_{SUT}}{lScale_{SUT}}\right)PostSUTDetector(\alpha,\beta)$$

$$PostSUTSamp[k,l]=PostSUTDetector(\alpha,\beta)\Big|_{\alpha=\alpha_n[k]=\frac{k-\delta k_{SUT}}{kScale_{SUT}},\beta=\beta_n[l]=\frac{l-\delta l_{SUT}}{lScale_{SUT}}}$$

$$PostSUTSamp[k,l]=\tau_{System}(\alpha_n[k],\beta_n[l])\times \tag{11}$$

$$\left(\sum_{i=0}^{N_i-1}\sum_{j=0}^{N_j-1}PostNU[i,j]System\left(\alpha_n[k]-\frac{i-\delta i_{Array}}{iScale_{Array}},\beta_n[l]-\frac{j-\delta j_{Array}}{jScale_{Array}}\right)\right)$$

At this point the signal is once again represented as a discretely sampled image that can be readily calculated and viewed. However, it is advantageous to first account for full system distortion effects. In addition the above form requires a double summation extending over the entire resistor array for each $[k,l]$ detector. A simplification to this form is presented below as well.

2.7.4 SUT Noise Effect

The SUT will add noise to the output signal. In the IDM implementation this operation, *SUTNoise{ }*, remains to be developed. However, as a place holder the IDM will process the image as follows:

$$PostSUTNoise[k,l]=SUTNoise\{PostSUTSamp[\ ]\} \tag{12}$$

2.7.5 SUT Digitization Effect

Most SUTs of interest will also digitize the image in some form or fashion, even if this means external digitization of an RS170 image. In the IDM implementation this operation, *SUTDig{ }*, remains to be developed. However, as a place holder the IDM will process the image as follows:

$$Output[k,l]=SUTDig\{PostSUTNoise[\ ]\} \tag{13}$$

2.7 Full System Distortion Effect

The above representation is for an IRSP without any distortion, which is impractical. To incorporate overall system distortion effects, the nominal sampling location for the $[k,l]$ detector of $(\alpha=\alpha_n[k],\ \beta=\beta_n[l])$ can be replaced by the actual distortion effected sampling location of $(\alpha=\alpha_d[k,l],\ \beta=\beta_d[k,l])$. The following gives an example of how to incorporate overall IRSP distortion effect:

$$\alpha_d[k,l] = C_0 + C_1\alpha_n[k] + C_2\beta_n[l] + C_3r_n[k,l] + C_4\alpha_n^2[k] + C_5\beta_n^2[l] +$$
$$C_6\alpha_n[k]\beta_n[l] + C_7\alpha_n[k]r_n[k,l] + C_8\beta_n[l]r_n[k,l] + C_9r_n^2[k,l]$$
$$\beta_d[k,l] = D_0 + D_1\alpha_n[k] + D_2\beta_n[l] + D_3r_n[k,l] + D_4\alpha_n^2[k] + D_5\beta_n^2[l] +$$
$$D_6\alpha_n[k]\beta_n[l] + D_7\alpha_n[k]r_n[k,l] + D_8\beta_n[l]r_n[k,l] + D_9r_n^2[k,l]$$

$$\text{(14)}$$

$$where:$$
$$r_n[k,l] = \sqrt{\alpha_n^2[k] + \beta_n^2[l]},\ \alpha_n[k] = \frac{k - \delta k_{SUT}}{kScale_{SUT}},\ \beta_n[l] = \frac{l - \delta l_{SUT}}{lScale_{SUT}}$$

The C and D coefficients can be measured or modeled. With the incorporation of the distortion effect, the signal as discretely sampled by the SUT is:

$$PostSUTSamp[k,l] = PostSUTDetector(\alpha,\beta)\big|_{\alpha=\alpha_d[k,l],\beta=\beta_d[k,l]}$$
$$PostSUTSamp[k,l] = \tau_{System}(\alpha_d[k,l],\beta_d[k,l]) \times$$
$$\left(\sum_{i=0}^{N_i-1}\sum_{j=0}^{N_j-1} PostNU[i,j]System\left(\alpha_d[k,l] - \frac{i - \delta i_{Array}}{iScale_{Array}},\beta_d[k,l] - \frac{j - \delta j_{Array}}{jScale_{Array}}\right)\right)$$

$$\text{(15)}$$

2.8 Restriction of Summation Limits

The primary bottleneck in calculating the *PostSUTSamp* image above is the double summation extending over the entire resistor array for each $[k,\ l]$ detector. However, for a given $[k,\ l]$ detector there is only a need to sum up a limited number of i's (resistor array rows) and j's (resistor array columns) based on the fact that the rolled-up *System* function will be effectively zero outside the range $(\alpha_{min}<\alpha<\alpha_{max}, \beta_{min}<\beta<\beta_{max})$.

$$PostSUTSamp[k,l] = \tau_{System}(\alpha_d[k,l],\beta_d[k,l]) \times$$
$$\left(\sum_{i=i_{min}[k,l]}^{i_{max}[k,l]}\sum_{j=j_{min}[k,l]}^{j_{max}[k,l]} PostNU[i,j]System\left(\alpha_d[k,l] - \frac{i - \delta i_{Array}}{iScale_{Array}},\beta_d[k,l] - \frac{j - \delta j_{Array}}{jScale_{Array}}\right)\right)$$

$$\text{(16)}$$

In the above form, the output as discretely sampled by the SUT can be calculated and viewed as desired.

2.9 Complete IRSP System Model

The final modeled representation of the IRSP takes the form:

$$Output[k,l] = SUTDig\{SUTNoise\{PostSUTSamp[k,l]\}\} \tag{17}$$

where:

$$PostSUTSamp[k,l] = \tau_{System}(\alpha_d[k,l], \beta_d[k,l]) \times$$

$$\left(\sum_{i=i_{\min}[k,l]}^{i_{\max}[k,l]} \sum_{j=j_{\min}[k,l]}^{j_{\max}[k,l]} PostNU[i,j] System\left(\alpha_d[k,l] - \frac{i - \delta i_{Array}}{iScale_{Array}}, \beta_d[k,l] - \frac{j - \delta j_{Array}}{jScale_{Array}} \right) \right) \tag{18}$$

$$PostNU[i,j] = ArrayNU\{SubHeat\{BusBarRob\{AIE[\ ]\}\}, NUCoefs[i,j,1...N]\} \tag{19}$$

$$AIE[i,j] = DACNoise\{VLUT_{APB}[GainOff\{VLUT_{SIB}[DDO2[i,j]], i,j\}, Mod(j,32)] \tag{20}$$

3.0 Conclusion and Closing Remarks

In it's current state, the IDM incorporates the IRSP effects that tend to redistribute the image energy as detected by the SUT. These types of effects are: resistor fill factor, optical distortion, optical geometric aberrations, SUT diffraction, and SUT geometric aberrations. Over the next couple of years the IDM will grow to encompass the rest of the IRSP effects.

Using the equations described above requires detailed knowledge of the IRSP configuration and components. Example results of using the IDM based on a real IRSP can be found in the simultaneously published paper: "Infrared Scene Projector Digital Model Development". Unfortunately, publication limitations have precluded the derivations of the scale factors, limits, and other parameters used in the IDM. Interested parties are encouraged to contact the primary author for additional information.

Distributed Test Capability
Using Infrared Scene Projector Technology

David R. Anderson[1], Ken Allred[1], Kevin Dennen[1], Patrick Roberts[2], William R. Brown[3], Ellis E. Burroughs[3], Kenneth G. LeSueur[3], and Tim Clardy[3]

[1] ERC, Inc.
555 Sparkman Drive, Executive Plaza, Suite 1622
Huntsville, AL 35816
{danderson, kallred, kdennen}@rttc.army.mil
[2] Sparta, Inc.
4901 Corporate Dr NW
Huntsville, AL 35805
{proberts@rttc.army.mil}
[3] US Army, Developmental Test Command
Redstone Technical Test Center
CSTE-DTC-RT-E-SA, Bldg. 4500
Redstone Arsenal, AL 35898-8052
{rbrown, eburroughs, klesueur, tclardy}@rttc.army.mil

Abstract. This paper describes a distributed test capability developed at the Redstone Technical Test Center (RTTC) in support of the Javelin Fire-and-Forget missile system. Javelin has a target acquisition sensor for surveying the battlefield, and a missile seeker staring focal plane array (FPA) for target tracking. Individually, these systems can be tested using scene projector capabilities at RTTC. But when combined, using this distributed capability, complete system acquisition to missile performance can be tested with the gunner in the loop. This capability was demonstrated at the Association of the United States Army (AUSA) conference in October of 2000. The Defense Research and Engineering Network (DREN) was used to connect the tactical target acquisition sensor in Washington, D.C. to the missile seeker and guidance section in Redstone Arsenal, in Huntsville, AL. The RTTC High Performance Computer provided scene projector control, distributed architecture support and real-time visible and infrared scene generation.

1 Introduction

This paper describes a distributed test capability developed by RTTC in Redstone Arsenal, AL, in support of the Javelin missile system. Javelin is a man portable, fire-and-forget antitank missile employed to defeat current and future threat armored combat vehicles. The Javelin consists of a missile in a disposable launch tube and a reusable Command Launch Unit (CLU) with a trigger mechanism and day/night sighting device for surveillance, and target acquisition and built-in test capabilities. The night sight is a long wave infrared (LWIR) scanning sensor. The missile locks on to the target before launch using an LWIR staring focal plane array and on-board processing, which also maintains target track and guides the missile to the target after launch. The capability to test this system in a distributed environment was demonstrated at the AUSA Annual Conference in Washington D.C. in October 2000. Two RTTC scene projector systems utilizing micro-resister array technology were used to support this effort. The missile was located at the RTTC Electro-Optical Subsystem Flight Evaluation Laboratory (EOSFEL) on Redstone Arsenal, AL. In Washington D.C. the Dynamic Infrared Scene Projector (DIRSP) Engineering Grade Array (DEGA) was used to project dynamic scenes into the CLU. Two real-time infrared scene generators were synchronized and running simultaneously at both locations. These scene generators along with the IR scene projectors allowed both sensors to be submerged into the same virtual environment. Targets moving on the database did so on both databases at the same time. The Defense Research and Engineering Network (DREN) was used to pass data between the two facilities.

2 DIRSP Engineering Grade Array (DEGA)

The DEGA infrared scene projector system uses resister array technology that was developed by Honeywell under the RTTC DIRSP program. The arrays are made up of 544 x 672 micro-emitter pixels (365,568 total pixels). Pixel pitch in the 544 direction is designed to be 45 um and pixel pitch in the 672 direction is designed to be 61 um. This means that the physical size of the array format is 0.96 inches by 1.62 inches.

The resister array spectral characteristics accommodate the LWIR CLU target acquisition sensor of the Javelin system. The DEGA projects ground based scenes with an maximum apparent (effective blackbody) temperature of 373K, in the LWIR spectrum. The resister array is contained in a vacuum environment and mounted to a heat sink maintained at 238K. The micro-emitters are designed to ensure a temperature accuracy of 0.05K and a design goal for DEGA is to have 0.03K delta temperature accuracy and resolution.

The DEGA has a projection optics subsystem that contains optical elements necessary to project an image of the IR emitter array onto the entrance aperture of the sensor under test. Control electronics consist of all the processing and interface electronics as well as the software required to control the system. Finally, the computer image generator consists of the hardware and software necessary to render and display high resolution complex scene images in real-time at the required frame rates for testing.

3 EOSFEL

The EOSFEL is a state-of-the-art facility for testing EO missile guidance sections and control systems in a 6 degree-of-freedom (6-DOF), hardware-in-the-loop (HWIL) simulation environment. This closed-loop simulation capability expands current subsystem/component testing to include evaluation of system level performance under tactical flight scenarios. The EOSFEL provides a non-destructive environment for performing missile subsystem test throughout the weapon system life-cycle (i.e. design, development, qualification, production, and stockpile reliability).

The EOSFEL maintains state-of-the-art equipment for performing these various test/simulation exercises. A description of the major EOSFEL test/simulation equipment is presented in Table 1.

Table 1. Major EOSFEL Equipment

Simulation/Test Equipment	Description
Flight Motion Simulator (FMS) Mountable Infrared Scene Projector (FIRSP)	FMS Mounted WISP 512x512 Wideband IR Resistor Array, LWIR Optics, High Speed Drive Electronics, Real-Time Non-Uniformity Correction (NUC), 2D Image Playback Electronics, Vacuum & Temperature Control
3D Computer Image Generator (CIG)/ Real-time Computer	SGI Onyx2 InfiniteReality, 16 R12K Processors, 2 IR3 Graphics Pipes, 6 Raster Managers w/ 256 Mb Texture, Real-time Digital Video Port Interface to IRSP
Real-Time I/O Linkage System	DACs, ADCs, Discrete I/O, Bit3 & Scramnet Interfaces
Flight Motion Simulator (FMS)	Flight Motion Simulator (FMS) 8 inch dia. UUTs
Dynamic Fin Loader (DFL)	Electro-Servo Driven 4 Channel Dynamic Loader, Up to 14 inch dia. CASs
Environmental Conditioning	Temperature Conditioning of Seeker, CAS, & ESAF : -50 to +150F

4   Distributed Test Configuration

For this distributed exercise, two real-time computers were connected via a combination of an ATM Virtual Private Network supplied by the High Performance Computing Management Office (HPCMO) named DREN (Defense Research and Engineering Network) and commercial T1 leased lines for extending the connectivity to the metro Washington D.C. area. Figure 1 is a simple diagram that illustrates some of the components of the distributed test configuration. The real-time computer at the Redstone Technical Test Center operates as a distribution center for the HPCMO which provides extensive computer processing capabilities and high-bandwidth connections to remote sites. It was used to perform real-time scene generation, facility control, and network communications. The second real-time computer was transported to Washington D.C. along with the mobile DEGA scene projector and support electronics. A virtual prototype CLU was used to provide a gunner interface to control the Javelin system. A tactical CLU was mounted to stare into the DEGA scene projector. Using the gunner interface, the gunner had complete control of the missile system. The control switches are designed to slave the tactical CLU. Functions like FOV select, focus, and level and gain control on the gunner interface slave the tactical CLU. The gunner interface has laser ring gyros that monitor the motion of the gunner. Displacements in pitch and yaw are sensed and sent to the real-time computer. The real-time scene projected to the sensor under test, will move accordingly, based on the sensed motion of the gunner. When the gunner looks into the gunner interface eyepiece, he will see this motion.

There are several components that were modeled in the virtual world. A tank driving station was developed. It allowed a driver to use a joystick and drive around in the virtual scene. The tank position was continuously updated on both scene generators simultaneously. The station allowed the tank to be moved in any direction at any rate within the capabilities of the tank. As the tank moved around, the Javelin gunner, using the gunner interface, would engage the tank and fire the missile located in the EOSFEL. Tank position would update for the duration of the flight.

4.1 Daysight Mode

While in day mode, the gunner views the visible database. This scene is generated in real-time on the DEGA control computer. The gyros monitor gunner motion and send that information back to the real-time computer. The scene generator receives the displacement information and updates the look angles in the scene generator. At 30 Hz, the gunner gets an updated view of the virtual world from his new perspective.

4.2 CLU Nightsight

During target acquisition, the gunner interface allows the operator to choose day sight or night. While in night mode, the gunner is looking at actual FLIR video from the tactical CLU looking into the DEGA scene projector. This is made possible by a Javelin test set that is interfaced to the tactical CLU. The test set grabs digital CLU video in real-time and sends the imagery to the gunner interface with very little latency. In this mode, the gunner will continue to have the ability to pan the virtual database in search of a potential target.

4.3 Seeker Video

The gunner will center the target in the FOV, and if ready to engage the target, will switch to seeker mode. When the gunner pulls the seeker mode trigger on the gunner interface, it sends a command back to the missile system in Redstone, and powers on the missile and starts the initialization. Initialization includes software download and cryogenic cooling of the focal plane array. Once the missile starts to image the scene, that video is sent back to the gunner interface over the

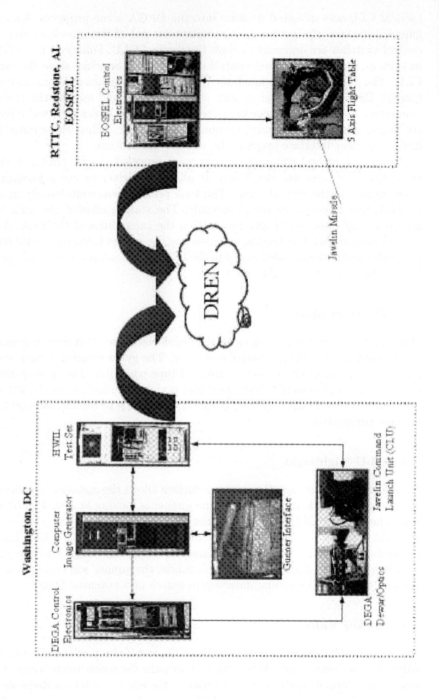

Fig. 1. Distributed Test Configuration

DREN. At this time, the gunner in Washington D.C. is completely slaving that missile in Redstone. The operator uses the gunner interface to size track gates, lock-on the target, and pull the fire trigger. When the missile is launched, it will go through a complete hardware-in-the-loop (HWIL) firing in the EOSFEL.

5 Network Infrastructure

For the AUSA demo, two separate UDP sockets were established for communication. One UDP port was dedicated for reception of missile video and EOSFEL lab status data and the other UDP socket was dedicated to transmitting mission and scenario feedback. Approximately 5000 bytes were transmitted by the EOSFEL and 164 bytes were received. The timing of the data transfers were scheduled at 30HZ but an effective rate of approximately 28 HZ was achieved during the demo. Several preliminary tests were conducted prior to the AUSA demo and the maximum latency was required to be less than 100 milliseconds in order to inject a smooth gunner motion into the HWIL missile track loop. Round trip latency from the EOSFEL at Redstone, to the AUSA demo was measured by a simple loop back test and was found to be approximately 20 milliseconds. The physical network connection was fairly extensive. The required data was transferred using a combination of the DREN and T1 networks. The EOSFEL ATM switch connected to an ATM switch in the RTTC HPC Distributed Center. From there, the ATM line runs to the CIC switch on Redstone. The signals then entered the DREN cloud. The Naval Research Lab in Washington, DC was the nearest DREN connection to the AUSA conference center. From that facility, the network goes through 2 leased T-1 lines and several routers to a switch that connects directly to the Onyx2 system controlling the DEGA scene projector.

6 Scene Generation and Databases

Scene Generation is the construction of radiometrically and visually calibrated digital images representing real-world situations that one would encounter on the test range or battlefield. These scenes may be based upon measured field-test imagery (empirical), physics-based 3-D computer models (predictive), or a combination of empirical and predictive sources. The scenes generated are then converted to a thermal or visual scene that is projected into the entrance aperture of a thermal imaging system.

Imagery that was projected into the CLU and the Javelin FPA was required to be generated in the Long Wave Infrared (LWIR) spectral bands. Injected day sight video is required to be in the visible spectrum. The LWIR imagery was projected into the tactical seeker and CLU from the FIRSP and the DEGA located at RTTC and Washington, D.C., respectively. The imagery from the tactical hardware was then acquired digitally and input into the viewing aperture of the gunner interface to give the gunner the appearance of tactical battlefield scenes. The visible spectrum was also generated and input digitally into the viewing aperture of the gunner interface. During the engagement exercise, the gunner could readily switch scenes between the visible, night sight, and seeker view. In order for the gunner not to perceive shifts in the presented imagery, the scenes presented to the gunner had to be correlated. In order to correlate the projected scenes, it was required that the visible and LWIR terrain and target databases be geometrically and geographically correlated. This was accomplished by placing the visible and LWIR textures on the same elevation map. Several databases have been correlated in the visible, LWIR and MWIR spectrums by RTTC under the scope of the DTC Virtual Proving Ground (VPG).

Fig. 2. TA-3 Visible Database and Target

7 Conclusion

Using multi-spectral scene generation and projection techniques provides earlier realistic testing which allows accelerated system design maturity and involves the user in early development. This testing technique provides a re-configurable cost effective supplement to destructive flight testing. The multi-spectral scene generation has also proven useful in combining Operational and Developmental test activities as well as rehearsals for live/field testing.

Another RTTC effort in the area of distributed multi-spectral scene generation is linking with weapon system developers. RTTC is the DoD sponsor in establishing a DREN connection at the Raytheon Javelin Lab in Tucson AZ.

Fig. 3. TA-3 LWIR Database and Target

The connection provides the developer with real-time High Performance Computing (HPC) resources and the EOSFEL and EOTASEL HWIL testing labs where multiple scene projectors are used for missile and target acquisition sensor testing.

The TOW Fire & Forget program will be tested using multiple scene generators, projectors, and IR databases. The system will employ a LWIR Target Acquisition Sensor (TAS) and a mid wave IR missile seeker. Upon missile power up the system will perform an automatic boresight on the two sensors and hand-off target information from the TAS to the missile. In order to perform qualification and pre-flight testing RTTC will use EOSFEL and EOTASEL scene generators and projectors to provide real-time, simultaneous, spatially correlated, in-band IR scenes to both sensors in their respective wave bands.

Development of Infrared and Millimeter Wave Scene Generators for the P3I BAT High Fidelity Flight Simulation

Jeremy R. Farris, Marsha Drake

System Simulation and Development Directorate,
U.S. Army Aviation and Missile Research and Development Center,
U.S. Army Aviation and Missile Command,
Redstone, Arsenal, Alabama 35898
Jeremy.Farris@rdec.redstone.army.mil
Simulation Technologies, Inc.,
3307 Bob Wallace Avenue,
Huntsville, Alabama 35805
Marsha.Drake@rdec.redstone.army.mil

The Pre-planned Product Improvement (P3I) Brilliant Anti-Armor BAT submunition High Fidelity Flight Simulation (HFS) is an all-digital, non-real-time simulation that combines the P3I BAT Six-Degrees-of-Freedom (6-DOF), a high fidelity dual mode seeker model, tactical flight software, and infrared (IR) / millimeter-wave (MMW) scene generation to produce a tool capable of system design, algorithm development and performance assessment. The HFS development methodology emphasizes the use of high fidelity IR and MMW imagery so that the seeker algorithms / tactical software can be incorporated to form a Software-In-The-Loop (SWIL) simulation. The IR and MMW scene generators use validated IR and MMW target signatures with correlated IR and MMW backgrounds to produce a realistic scene for a variety of battlefield scenarios. Both the IR and MMW scene generators have evolved with the availability high performance computational capability and system development to meet the design and performance assessment requirements of the P3I BAT program.

Introduction

The Pre-planned Product Improvement (P3I) Brilliant Anti-Armor BAT submunition High Fidelity Flight Simulation (HFS) is being developed by the U.S. Army Missile Research, Development, and Engineering Center (MRDEC), Systems Simulation Development Directorate (SS&DD) and the U.S. Army Tactical Missile System (ATACMS) / BAT Project Office support contractor as a tool to support system design evaluation, tactical algorithm development and performance assessment of the P3I BAT system. The HFS combines the P3I BAT 6-DOF model, a high fidelity dual mode seeker model, tactical flight software, and infrared (IR) / millimeter-wave (MMW) scene generation into an integrated, all-digital, Software-In-The-Loop (SWIL) simulation. As part of the P3I BAT suite of simulations, the HFS will be used to evaluate, analyze and assess the performance of the P3I BAT system.

Along with being a standalone analytical tool, the HFS is the all-digital companion tool for the Hardware-in-the-Loop (HWIL) simulation and flight testing, allowing for extended scenario capabilities including additional targets and countermeasures. The HFS development methodology emphasizes the use of high fidelity IR and MMW imagery and tactical flight software-in-the-loop to maximize the commonalities between the HFS, HWIL, and tactical submunition. This greatly simplifies verification and validation (V&V) activities, and places a strong emphasis on IR and MMW scene generation.

This paper addresses the IR and MMW scene generators currently being used in the P3I BAT HFS, and how they have evolved with the availability of high performance computational capability and system development to meet the design and performance assessment requirements of the P3I BAT program.

Background

Fig. 1. P3I BAT Submunition

P3I BAT Submunition

The P3I BAT is the Pre-Planned Product Improvement for the Base BAT submunition (Fig. 1). The BAT and P3I BAT systems, after being dispensed from the host vehicle such as the Army Tactical Missile System (ATACMS), perform as autonomously guided smart submunitions that are designed to search, detect, and defeat armored vehicles in many weather conditions. The Base BAT submunition whose primary mission is to attack moving armored target arrays, initially utilizes acoustics to determine a general location of the target, then employs an IR seeker for acquisition and terminal homing. The P3I BAT submunition has the same physical characteristics as the Base BAT configuration, but incorporates an improved dual-mode (IR/MMW) seeker design and an improved warhead that allows it to engage soft targets. These features greatly expand the possible engagement environments and target set. [1]

HFS

The P3I BAT HFS is a high-fidelity closed-loop simulation of the P3I BAT submunition that begins at dispense out of the carrier vehicle, and ends at submunition impact on the target. The methodology for the HFS development is primarily based on the methodology previously developed for the Javelin Antitank weapon system Integrated Flight Simulation (IFS). [2, 3] One of the major differences between these systems lies in the dual mode IR/MMW seeker used by the P3I BAT submunition. The Javelin seeker is an earlier generation IR-only seeker, and as such was not as complex to simulate and analyze as the dual mode P3I BAT seeker. To date, there are no other known dual mode simulations that approach the HFS in complexity and fidelity. [1]

The HFS was originally proposed as a tool to support the downselect between two competing seeker developers for the P3I BAT submunition. By integrating the seeker developer's detailed seeker model and algorithms into the HFS environment, the performance of each seeker design could be analyzed, and the effects of the seeker design on overall system performance could be quantified. The role of the HFS has since been expanded to include many areas, including providing the one (submunition)-on-many (targets) performance capability as an input to the many (submunitions)-on many (targets) system effectiveness simulation, STRIKE. [1]

Methodology

The HFS development methodology emphasizes four primary areas:

1. Reuse of Base BAT subsystem models common to P3I BAT
2. Use of tactical flight software-in-the-loop
3. Maximize use of common subsystem models between the HFS and HWIL.
4. Use of validated IR imagery and MMW measurement data for targets and backgrounds.

The primary goals behind the methodology are to:

1. Reduce development time.
2. Simplify verification and validation activities.
3. Provide an accurate assessment of the system's performance.

This approach to simulation development requires that the IR and MMW scene generators incorporate high fidelity target and clutter models, and the dual mode processing within the seeker requires that the MMW target and clutter models be generated in registration with the corresponding IR model. The use of high fidelity facet target models and backgrounds for the generation of IR imagery was a proven simulation methodology under the Javelin program. However, the critical test for the development of the P3I BAT HFS was to explore whether millimeter wave radar "imagery" could be incorporated into the simulation and utilized at the same time as the infrared imagery, with the same good results as those obtained using IR-only. No information was available as to whether this idea was feasible. Discussions were held with members of the MMW Longbow Hellfire Modular Missile System (LBHMMS) HWIL simulation team to obtain their expert opinions on the feasibility of the idea. They provided considerable information on the procedures and processes used to simulate MMW targets and clutter using the HWIL MMW radar arrays in real time. After these discussions, it was determined that these real-time HWIL modeling processes could be adapted and merged with IR clutter techniques to be run in a non-real-time all digital simulation. [1]

Model Development

The P3I BAT HFS IR and MMW scene generation development has evolved with the availability of high performance computational capability and the requirements of the P3I BAT program. To fully appreciate this evolution, the fidelity of the IR and MMW target and clutter models will be discussed along with rendering techniques applied for image generation.

IR Target Models

Under the Javelin program, the U.S. Army MRDEC Missile Guidance Directorate (MGD) developed a methodology for applying IR target measurement data to 3 dimensional target facet models. A target geometry model is developed with "enough" polygons to represent the resolution of the calibrated IR imagery. The vertices are then "painted" with the collected measurements and rendered applying gouraud shading. This target development and rendering methodology has proven to be a valid approach on the Javelin missile programs, and was chosen for the P3I BAT IR scene generator.

As the program matured and the P3I BAT HWIL facility was developed, it became evident that the computational load to render the high polygon target models would not meet the HWIL real-time requirements. So, the HWIL personnel developed a methodology to decimate the number of facets, and apply texture(s) that would adequately represent the resolution of the calibrated IR imagery. (see Fig. 2) [4]

Although the high polygon targets were not as much of an issue for the non-real-time HFS, the availability of low-cost high performance Silicon Graphics capable of rendering 12 bit textured polygons made the incorporation of algorithms to render the HWIL real-time targets possible, and the advantages offered by the approach include improved scene generator efficiency and assures commonality with HWIL, reducing cost and streamlining the verification and validation process.

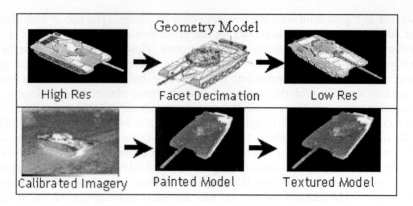

Fig. 2. P3I BAT HWIL Target Development Methodology [4]

MMW Target Models

Based on the LBHMMS HWIL experience, point scatterer models were developed for the P3I BAT program. The MMW target models developed have 150 scatterers per polarization per 0.5 degrees in azimuth and 5 degrees in depression. Each model has been verified and validated by comparing model frequency-averaged RCS and HRR aim-point statistics to those from the original measurement model [5]. The models are registered to the IR model by correlating the scatterer locations geometrically to the IR target coordinate frame. (see Fig. 3). [4]

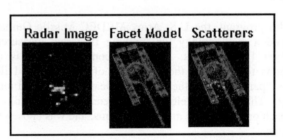

Fig. 3. P3I BAT MMW Target Development Methodology. [4]

In the utilization of these models, the number of scattering bins generated in target model development can approach 2,400 per polarization and more than 100 scattering centers thresholded from these bins are typically chosen. The computational requirements to process these scatterers are quite significant and a study has been made to quantify the effects of reducing the number of scattering centers to reduce run times and meet HWIL real-time requirements. Preliminary results indicate that the reduction of scatterers by half will generate RCS and aim-point statistics that are within the acceptable range of measurement statistics [5]. This reduction along with conventional parallel processing techniques will allow the HFS MMW scene generator to simulate complex battlefield scenarios within a reasonable amount of runtime.

IR Clutter Models

Under the Javelin program, the U.S. Army MRDEC Missile Guidance Directorate (MGD) developed a methodology for applying IR clutter measurement data to a class map describing the site to be simulated. Each class map consists of description of the features of the site and typically consists of 1-foot increment descriptions, with the classes characterized as shown in Table 1.

Table 1. Class Map Descriptions

Class	Description
0	Bushes
1	Tall Grass
2	Brush
3	Short Grass
4	Road Sand
5	Normal Sand
6	Powder Sand
7	Sparse Short Grass
8	Dirt
9	Gravel
10	Asphalt
11	Medium Veg
12	Mixed Forest
13	Deciduous Trees
14	Coniferous Trees
15	Snow
16	Road Snow
17	Traveled Road Snow
18	Snow over veg
19	Rock
22	Sky
59	Shadow

Initially, the P3I BAT IR scene generator rendered the clutter as a polygon mesh with each vertex "painted" to represent the measured class temperature data. If topography data was available, each vertex elevation was modified appropriately. At one foot resolution, the time to render the background was quite significant, and efforts to improve the scene generation efficiency were investigated. Again, the availability of low-cost high performance Silicon Graphics capable of rendering 12 bit textured polygons made it possible to represent the clutter with fewer polygons and still maintain the resolution of the class map. This approach was also chosen by the HWIL facility and provides one more area of commonality between the HFS and HWIL simulation domains.

MMW Clutter Models

The dual mode processing with the P3I BAT seeker requires that the MMW backgrounds be generated in registration with the corresponding IR model. This is controlled by the utilization of a class map (see IR Clutter Model description). Normalized sigma0 values obtained from collected measurement data are used in conjunction with the corresponding class map to generate the normalized even and odd polarization clutter map. This is obtained by applying the statistics reported to:

$$\sigma_o(\text{random, normalized}) = \sigma_o(\text{normalized}) + 20\log_{10}\mathbf{r} \tag{1}$$

where **r** is a Raleigh distribution with a mean of 1. The user then computes the sine of the depression angle to the given clutter patch and applies:

$$\sigma_o = \gamma\sin\theta \tag{2}$$

where θ is the depression angle and γ is the normalized sigma0. This model is considered valid in the range of depression angles of interest to this program. [6]

Currently, the BAT P3I HFS MMW scene generator has the option to read in pre-generated clutter maps or to dynamically generate the clutter maps on a run-to-run basis. Depth buffer data obtained from the IR scene generator provides information on which cells are "shadowed" by the target, and the model generates a list of sigma0 values to be processed. As can be imagined, the computational requirements are significant and efforts to optimize clutter cell processing are underway. This includes investigating the application of HWIL real-time optimized vector processing techniques.

Verification and Validation

Verification and validation of the P3I BAT IR and MMW scene generators is an ongoing process. The approach of utilizing IR and MMW target models that are developed with collected, not predicted, data simplifies the process but still requires that each model is validated with the calibrated measurement data. On the P3I BAT program, the target and clutter model developers have established and implemented verification and validation procedures to the approval and satisfaction of P3I BAT simulation community. In addition, the techniques by which the scene generators generate the simulated battlefield have been proven valid on other programs. The utilization of HWIL and flight test data has been critical in the verification and validation process and the commonality between the HFS and HWIL scene generation has streamlined these activities. Fig. 4 shows an example comparison between HFS and test IR imagery.

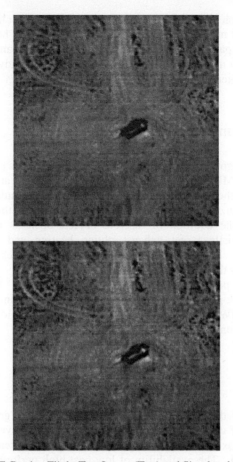

Fig. 4. P3I BAT Captive Flight Test Image (Top) and Simulated Image (Bottom)

Current Utilization

The P3I BAT HFS is currently being used in several studies. One study in particular is being performed to evaluate the tracker algorithms performance across a suite of target and clutter models. To support this study, upgrades to the IR and MMW scene generators were implemented to support the large Monte Carlo run matrices to be executed. One such upgrade has been proven to be instrumental in the timely execution of the analysis. Historically, IR and MMW clutter models are pre-generated and clutter variation is only obtained by reading in a new pre-generated model that is quite timely on a run-to-run basis. The P3I BAT signatures IPT developed a methodology in which collected class map statistics for particular sites, seasonal, and weather conditions were processed to form a clutter vector. For a Monte Carlo run set, both the IR and MMW scene generators utilize these clutter vectors to dynamically generate textures (IR) and normalized clutter maps (MMW) at the beginning of each run. In addition, a methodology was defined to apply a level term to the IR target model and allow for "yesterday's" target be placed in "today's" background. This capability provided the HFS an efficient means to Monte Carlo the IR and MMW clutter models, and allowed the evaluators to study the effects of clutter variation on the tracker algorithms.

Future Enhancements

There are several ongoing and planned activities to enhance the capabilities of the P3I BAT HFS IR and MMW scene generators. The activities are currently centered around increasing the fidelity and complexity of the simulated battlefield and implementing optimization techniques currently utilized in the HWIL simulation. Upgrades include the addition of discrete objects such as individual trees and buildings, and realistic target exhaust plume heating and obscuration effects. Battlefield effects such as burning targets and countermeasures (smoke, flares, etc.) will also be added as those models are developed.

In addition, the personnel responsible for the P3I BAT HWIL are currently responsible for developing a MMW scene generator for the Common Missile program that utilizes the HWIL real-time vector processing techniques. MMW scene generator optimization will be leveraged on this activity.

Conclusion

The development of the P3I BAT HFS IR and MMW scene generators is ongoing. The methodology of utilizing IR and MMW target models developed with collected measurement data has given the fidelity required to generate realistic imagery of the simulated battlefield. The availability of low-cost high performance computers has allowed for the implementation of real-time scene generation techniques and has increased the commonality between the HFS and HWIL simulation. As development progresses, the HFS will become an even more powerful tool for performance assessment, system design and algorithm development.

References

1. Kissell, Ann H., Williams, Joe E., Drake, V. Marsha: *"Development of an IR/MMW High Fidelity Flight Simulation (HFS) for the P3I BAT Submunition"*, presented at the American Institute of Aeronautics and Astronacics (AIAA) First Biennial National Forum on Weapon Systems Effectiveness. (1999)
2. TI/Martin Javelin Joint Venture: *"JAVELIN Integrated Flight Simulation User's Manual"*, Version 2.0, Document Number ND-0078. (1993)
3. Kissell, Ann H.: *"P3I BAT Simulation Transition Proposal"*, briefing. (1996)
4. Simulation Technologies: *"Simulation Technologies, Inc"*, briefing. (2001)
5. Saylor, Annie V.: *"BAT P3I Target Model Scatterer Number Study"*, Letter Report, RD-SS-99-04. (1999)
6. Saylor, Annie V.: *"MMW Background Models for P3I BAT"*, Letter Report. (2000)

Novel Models
for Parallel Computation

Session chair:

Frank Dehne (Carleton University, Canada)

A Cache Simulator for Shared Memory Systems

Florian Schintke, Jens Simon, Alexander Reinefeld

Konrad-Zuse-Zentrum für Informationstechnik Berlin (ZIB)
{schintke,simon,ar}@zib.de

Abstract. Due to the increasing gap between processor speed and memory access time, a large fraction of a program's execution time is spent in accesses to the various levels in the memory hierarchy. Hence, cache-aware programming is of prime importance. For efficiently utilizing the memory subsystem, many architecture-specific characteristics must be taken into account: cache size, replacement strategy, access latency, number of memory levels, etc.

In this paper, we present a simulator for the accurate performance prediction of sequential and parallel programs on shared memory systems. It assists the programmer in locating the critical parts of the code that have the greatest impact on the overall performance. Our simulator is based on the *Latency-of-Data-Access Model*, that focuses on the modeling of the access times to different memory levels.

We describe the design of our simulator, its configuration and its usage in an example application.

1 Introduction

It is a well-known fact that, over the past few decades, the processor performance has been increased much faster than the performance of the main memory. The resulting gap between processor speed and memory access time still grows at a rate of approximately 40% per year. The common solution to this problem is to introduce a hierarchy of memories with larger cache sizes and more sophisticated replacement strategies. At the top level, a very fast L1 cache is located near the CPU to speed up the memory accesses. Its short physical distance to the processor and its fast, more expensive design reduces the access latency if the data is found in the cache.

With the increasing number of memory levels, it becomes more important to optimize programs for temporal and spatial locality. Applications with a good locality in their memory accesses benefit most from the caches, because the higher hit rate reduces the average waiting time and stalls in arithmetic units.

In practice, it is often difficult to achieve optimal memory access locality, because of complex hardware characteristics and dynamic cache access patterns which are not known *a priori*. Hence, for a realistic cache simulation, two things are needed: a hardware description with an emphasis on the memory subsystem and the memory access pattern of the program code.

We have designed and implemented a simulator which is based on the *Latency-of-Data-Access Model* described in the next section. The simulator is execution-driven, i.e. it is triggered by function calls to the simulator in the target application. It determines a runtime profile by accumulating the access latencies to specified memory areas.

2   The Latency-of-Data-Access Model

The *Latency-of-Data-Access (LDA)* model [9] can be used to predict the execution characteristics of an application on a sequential or parallel architecture. It is based on the observation that the computation performance is dominated (and also strictly limited) by the speed of the memory system.

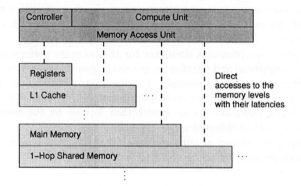

Fig. 1. Concept of the LDA model.

In LDA, data movements are modeled by direct memory accesses to the various levels in the memory hierarchy. The number of data movements combined with the costs of computations provide an accurate estimate on the expected runtime [9].

LDA models the execution of an application by a set of abstract instructions, which are classified into *arithmetic operations, flow control*, and *data movements* between the processing unit(s) and the memory system. The data movements are considered to be issued from an explicit load/store unit with direct access to each level of the memory hierarchy (see Fig. 1).

With this abstract model, complex protocols of today's memory hierarchies can be treated uniformly, while still giving very precise results. Cache misses caused by a load operation in a real machine are modeled by a single access to the next lower memory level that holds the data. This is done with the rationale that the time taken by the initial cache miss is much shorter than the time taken by the access to the next lower memory level. Hence, for each load/store instruction, LDA uses the execution cost of that access to the specific memory level.

Note that the precise timing depends on the access pattern: Consecutive memory accesses are much faster than random accesses. The LDA model reflects these effects by adding different costs to the individual accesses.

3 The LDA Simulator

Our simulator [8] gets two sources of input: the configuration of the memory architecture of the target machine, and the memory access pattern of the simulated program. The latter is obtained by manually inserting function calls to the simulator in the program source code.

We have verified the accuracy of our simulator by comparing simulated data to the timing of practical execution runs.

In the following sections we describe the features and architecture of the simulator in more detail and explain its usage with a simple example.

3.1 Features

The LDA simulator can be used to optimize programs, but also to help in the design of the memory hierarchy of a processor. It can be compiled on any system with a standard C++ compiler. The simulated architecture is independent of the computer that the simulator runs on.

The simulator shows the currently accumulated execution time in every time step of the simulation. It is also possible to get the number of memory accesses for a specific memory level.

The simulator allows to split the costs for user-specified memory areas to different cost accounts. With this feature, the execution time on subsets of the address space can be monitored (Sec. 4), which is useful for, e.g., optimizing the data structure for better cache usage.

3.2 Architecture

Figure 2 shows the simulator's architecture described in the *Unified Modeling Language UML* [5]. The actual implementation has been done with the C++ programming language. The simulator consists of two main parts:

- The classes `MachineDef`, `Protocol`, `CacheDef`, `RStrategy`, and `WStrategy` describe the static architecture of a memory hierarchy.
- The classes `Caches`, `CacheLevel`, and `Cacheline` describe the dynamic part. Their behavior depends on the static description.

The classes `Caches`, `Cachelevel`, and `Cacheline` determine which memory levels are accessed for each memory operation and then update their internal counters for calculating the execution time. For this purpose, it is necessary to simulate the behavior of the cache levels and to determine the currently cached addresses.

The user interface is defined in the two classes `MachineDef` and `Caches`.

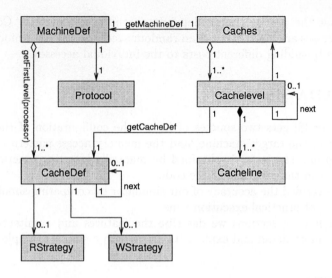

Fig. 2. UML class diagram of the simulator showing the software architecture of the simulator.

The simulator is execution-driven and the target source code must be instrumented by manually inserting function calls into the program. By the approach not to use an executable editing library like EEL [1], the user has the burden to modify the source code on the one hand, but on the other hand, it allows him to focus on the most relevant parts in the application, e.g. program kernels. Simulation overhead has only to be added where really necessary. An example of an instrumented program code is shown in Fig. 4.

3.3 Configuration

At startup time, a configuration file is read by the simulator that contains the static specification of the memory architecture. Many different architectures can be specified in the same configuration file and can be referenced by a symbolic name. With this feature, several system architectures can be simulated in the same run.

Memory hierarchies of arbitrary depths and sizes may be defined, thereby also allowing to simulate hierarchical shared memory systems. For each memory level, there may be different settings for

- the number of blocks,
- the blocksize,
- the associativity,
- the replacement strategy,
- the write strategy, and
- the read and write latency.

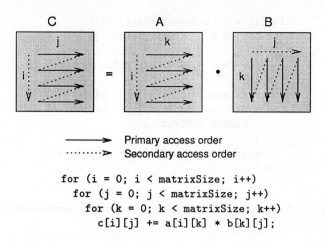

```
for (i = 0; i < matrixSize; i++)
  for (j = 0; j < matrixSize; j++)
    for (k = 0; k < matrixSize; k++)
      c[i][j] += a[i][k] * b[k][j];
```

Fig. 3. Memory access pattern of an unblocked matrix-matrix multiplication and its implementation (with C memory layout).

For the input configuration file, the hardware characteristics of the target architectures can be obtained either from the technical manuals of the system or by running a suite of microbenchmarks. For configurable systems, only microbenchmarks can be used to determine the actual configuration. In our experiments we used the method of McVoy and Staelin [3]. It measures the memory hierarchy characteristics by enforcing strided accesses to memory blocks of different sizes. Their scheme determines the cache size, the latency and the blocksize for each memory level.

However, some of the characteristics can not be determined by microbenchmarks. This is true for the cache associativity, the write strategy (write through or write back), and the replacement strategy (LRU, FIFO, random, ...) which must be specified by hand.

4   Example

In this section, we illustrate the use of the simulator with an unblocked matrix-matrix multiplication on a single and an SMP multiprocessor system.

4.1   Matrix Multiplication on a Single Processor

As an example we have simulated the matrix-matrix multiplication $A * B = C$. Figure 3 depicts the default memory layout of the matrices. The two-dimensional memory blocks of matrices A and C are mapped to the linear address space of the computer by storing them line by line. Hence, the system accesses the elements of A and C with consecutive memory accesses, resulting in a good spatial locality. The accesses to matrix B, however, occur strided over the memory. The lower spatial locality of matrix B slows down the whole computation.

```
for (i = 0; i < matrixSize; i++)
  for (j = 0; j < matrixSize; j++)
  {
    for (k = 0; k < matrixSize; k++)
    {
      caches.load(&a[i][k] - offset, 4);
      caches.load(&b[k][j] - offset, 4);
      caches.doFloatOps(2);
    }
    caches.store(&c[i][j] - offset, 4);
  }
```

Fig. 4. Instrumented code for the matrix-matrix multiplication.

Temporal locality is given only in the accesses to the result matrix C. The temporal locality increases with growing matrix size (Fig. 6).

In preparation of the simulation, all load and store operations must be searched in the source code and function calls to the simulator must be inserted. Figure 4 shows the instrumented code. Note that the function calls to the simulator caches.load, caches.store expect the (one-dimensional) array address and the data size as parameters.

In practice, the elements c[i][j] of the result matrix are stored only once to the main memory. Hence in the simulation we can save CPU time by moving the data assignment out of the inner loop as shown in Fig. 4.

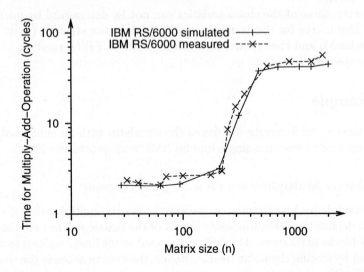

Fig. 5. Measured and simulated execution times of the matrix-matrix multiplication for different matrix sizes.

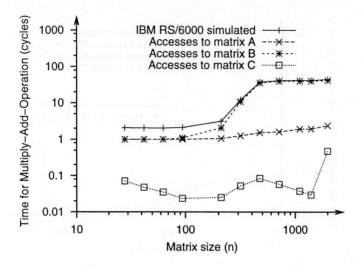

Fig. 6. Split-down of the execution times for each matrix involved in the multiplication.

4.2 Single Processor Results

Figure 5 shows the measured and simulated execution times of the matrix multiplication for different matrix sizes. In theory, we would expect an $O(n^3)$ execution effort, corresponding to a horizontal line in Fig. 5. In practice, the actual growth rate is not cubical, because of the memory hierarchy.

As can be seen in Fig. 5, our simulation closely matches the real execution time. The prediction is slightly more optimistic, because the simulator does not include side effects of the operating system like task switching and interrupts. In real life these effects invalidate a part of the cache and thereby produce additional overhead.

Figure 6 gives a more detailed view. It shows the access time for each single matrix. This data has been determined by counting the accesses to matrix A, B, and C separately. As can be seen now, for the small matrix sizes, matrix A and B are both stored in the first level cache. When the matrices get bigger, matrix B falls out of the cache, while A is still in. This is caused by the layout of matrix B in the memory, which does not support spatial access locality.

The locality of matrix B can be improved by storing it in the transposed form, giving a similar access pattern for both source matrices, A and B. Each of them takes about half of the cache capacity. Figure 7 shows the simulation results for this improved memory layout.

Another interesting observation in Fig. 6 is the fact that the accesses to matrix C seem to become faster with increasing matrix size. This is due to the increasing number of computations that are done in the inner loop before storing a result element in matrix C. A sudden jump in the graph occurs only

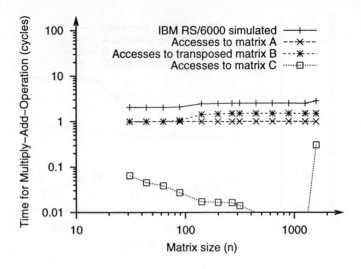

Fig. 7. Simulation of the optimized matrix multiplication with transposed matrix B.

when matrix C grows to such an extend that it cannot be completely contained in the cache.

4.3 Matrix Multiplication on an SMP

We have also simulated the matrix multiplication for a dual Pentium III processor system with 16 kB L1 and 256 kB L2 cache. Figure 8 gives the results of two versions of the algorithm, an alternating distribution where one processor computes all even and the other computes all odd matrix elements, and a blocked distribution where each processor computes a half consecutive block of matrix C. Clearly the blocked version is faster due to lower memory contention.

The memory contention has been simulated by including a simulation of the Pentium's MESI consistency protocol into the simulator. Note that this does not only simulate cache conflicts but also contention on the system bus.

5 Related Work

Several other simulators for caches exist, but they don't use the LDA model for their analysis and have another focus. Some of them are limited in the number of supported memory levels, others aim at simulating different hardware features.

The *Linux Memory Simulator (limes)* [2] has been designed for simulating SMP systems. It has been developed by modifying a compiler backend to generate instrumented code. This approach limits the flexibility of instrumentation and it also depends very much on the specific compiler. This execution-driven simulator is able to analyze parallel programs on a single host processor by creating threads for each target system.

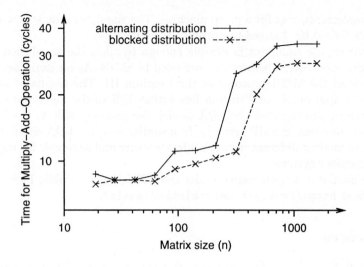

Fig. 8. Simulated execution times of the matrix multiplication on a dual processor SMP system.

The *Rice Simulator for ILP Multiprocessors (RSIM)* [6] has been designed with an emphasis on simulating instruction level parallelism. With this level of details, the internal processor behavior can be simulated more precisely, but at the cost of a higher simulation runtime. The *Wisconsin Wind Tunnel II* [4] in contrast tries to reduce the simulation time by executing instructions of the target machine in parallel on some host processors.

For analyzing new cache architectures the *Multi Lateral Cache Simulator (mlcache)* was developed [10]. It is used, for example to simulate cache that are split into temporal and spatial parts [7]. The depth of the memory hierarchy is limited to only a few levels.

6 Conclusion

We have presented a generic execution-driven simulator for the accurate prediction of memory accesses based on the *Latency-of-Data-Access Model* [9]. This model is motivated by the observation that the computation speed of modern computers is dominated by the access time to the various memory levels.

The simulator can be used to simulate a wide variety of target platforms. This is done with a configuration file that describes the memory characteristics in terms of the number of memory levels, cache sizes, access latency, and replacement strategy.

The simulator is triggered by the execution of the program, which must be instrumented manually before runtime. This allows the user to focus on the important code modules, e.g., the most time-consuming kernels. With the example of a matrix multiplication, we have shown how to use the simulator for deriving better implementations for a given machine. The simulator indicates which data structure should be changed to optimize the code.

Moreover, the simulator is flexible enough to allow the simulation of complex cache coherency schemes that are used in SMPs. As an example, we have implemented the MESI protocol of the Pentium III. The results of our simulation of a dual processor Pentium lies within 10% of the measured execution time, thereby proving that the LDA model also properly reflects the impact of memory contention in SMP systems. In a similar way, the LDA simulator could be used to analyze different network infrastructure and protocols for distributed shared memory systems.

The simulator is open source under the GNU General Public License and is available at http://www.zib.de/schintke/ldasim/.

References

1. James R. Larus and and Eric Schnarr. EEL: Machine-independent executable editing. In *Proceedings of the SIGPLAN'95 Conference on Programming Language Design and Implementation*, pages 291–300, 1995.
2. Davor Magdic. Limes: A multiprocessor environment for PC Platforms. In *IEEE Computer Society Technical Committee on Computer Architecture Newsletter*, March 1997.
3. Larry McVoy and Carl Staelin. lmbench: portable tools for performance analysis. In *USENIX 1996 Annual Technical Conference, January 22–26, 1996. San Diego, CA, USA*, pages 279–294, Berkeley, CA, USA, January 1996. USENIX.
4. Shubhendu S. Mukherjee, Steven K. Reinhardt, Babak Falsafi, Mike Litzkow, Mark D. Hill, David A. Wood, Steven Huss-Lederman, and James R. Larus. Wisconsin Wind Tunnel II: A fast, portable parallel architecture simulator. *IEEE Concurrency*, 8(4):12–20, October/December 2000.
5. OMG. *Unified Modeling Language Specification*. Open Management Group, Version 1.3 edition, June 1999.
6. Vijay S. Pai, Parthasarathy Ranganathan, and Serita V. Adve. RSIM: An Execution-Driven Simulator for IPL-Based Shared-Memory Multiprocessors ans Uniprocessors. In *IEEE Computer Society Technical Committee on Computer Architecture Newsletter*, March 1997.
7. M. Prvulović, D. Marinov, Z. Dimitrijević, and V. Milutinović. Split Temporal/Spatial Cache: A Survey and Reevaluation of Performance. In *IEEE Computer Society Technical Committee on Computer Architecture Newsletter*, July 1999.
8. Florian Schintke. Ermittlung von Programmlaufzeiten anhand von Speicherzugriffen, Microbenchmarks und Simulation von Speicherhierarchien. Technical Report ZR-00-33, Konrad-Zuse-Zentrum für Informationstechnik Berlin (ZIB), 2000.
9. Jens Simon and Jens-Michael Wierum. The Latency-of-Data-Access Model for Analyzing Parallel Computation. *Information Processing Letters*, 66(5):255–261, June 1998.
10. Edward Tam, Jude Rivers, Gary Tyson, and Edward S. Davidson. mlcache: A flexible multi-lateral cache simulator. Technical Report CSE-TR-363-98, Computer Science and Engineering, University of Michigan, May 1998.

On the Effectiveness of D-BSP as a Bridging Model of Parallel Computation[*]

Gianfranco Bilardi, Carlo Fantozzi, Andrea Pietracaprina, and Geppino Pucci

Dipartimento di Elettronica e Informatica, Università di Padova, Padova, Italy.
{bilardi,fantozzi,andrea,geppo}@artemide.dei.unipd.it

Abstract. This paper surveys and places into perspective a number of results concerning the D-BSP (Decomposable Bulk Synchronous Parallel) model of computation, a variant of the popular BSP model proposed by Valiant in the early nineties. D-BSP captures part of the proximity structure of the computing platform, modeling it by suitable decompositions into clusters, each characterized by its own bandwidth and latency parameters. Quantitative evidence is provided that, when modeling realistic parallel architectures, D-BSP achieves higher effectiveness and portability than BSP, without significantly affecting the ease of use. It is also shown that D-BSP avoids some of the shortcomings of BSP which motivated the definition of other variants of the model. Finally, the paper discusses how the aspects of network proximity incorporated in the model allow for a better management of network congestion and bank contention, when supporting a shared-memory abstraction in a distributed-memory environment.

1 Introduction

The use of parallel computers would be greatly enhanced by the availability of a model of computation that combines the following properties: *usability*, regarded as ease of algorithm design and analysis, *effectiveness*, so that efficiency of algorithms in the model translates into efficiency of execution on some given platform, and *portability*, which denotes the ability of achieving effectiveness with respect to a wide class of target platforms. These properties appear, to some extent, incompatible. For instance, effectiveness requires modeling a number of platform-specific aspects that affect performance (e.g., interconnection topology) at the expense of portability and usability. The formulation of a *bridging model* that balances among these conflicting requirements has proved a difficult task, as demonstrated by the proliferation of models in the literature over the years.

In the last decade, a number of bridging models have been proposed, which abstract a parallel platform as a set of processors and a set of either local or shared memory banks (or both) communicating through some interconnection. In order to ensure usability and portability over a large class of platforms, these models do not provide detailed characteristics of the interconnection but, rather, summarize its communication capabilities by a few parameters that broadly capture bandwidth and latency properties.

---

[*] This research was supported, in part, by the Italian CNR, and by the Italian MURST under Project *Algorithms for Large Data Sets: Science and Engineering*.

Perhaps the most popular example in this arena is Valiant's *BSP* (*Bulk Synchronous Parallel*) model [Val90]. A BSP machine is a set of n processors with local memory, communicating through a router, whose computations are sequences of *supersteps*. In a superstep, each processor (i) reads the messages received in the previous superstep; (ii) performs computation on locally available data; (iii) sends messages to other processors; and (iv) takes part in a global barrier synchronization. A superstep is charged a cost of $w + gh + \ell$, where w (resp., h) is the maximum number of operations performed (resp., messages sent/received) by any processor in the superstep, and g and ℓ are parameters with g inversely related to the router's bandwidth and ℓ capturing latency and synchronization delays.

Similar to BSP is the LogP model proposed by Culler et al. [CKP$^+$96] which, however, lacks explicit synchronization and imposes a more constrained message-passing discipline aimed at keeping the load of the underlying communication network below a specified capacity limit. A quantitative comparison developed in [BHP$^+$96,BHPP00] establishes a substantial equivalence between LogP and BSP as computational models for algorithm design guided by asymptotic analysis.

In recent years, a number of BSP variants have been formulated in the literature, whose definitions incorporate additional provisions aimed at improving the model's effectiveness relative to actual platforms without affecting its usability and portability significantly (see e.g., [BGMZ95,BDM95,JW96b,DK96]). Among these variants, the *E-BSP* (*Extended BSP*) by [JW96b] and the *D-BSP* (*Decomposable BSP*) by [DK96] are particularly relevant for this paper. E-BSP aims at predicting more accurately the cost of supersteps with *unbalanced* communication patterns, where the average number h_{ave} of messages sent/received by a processor is lower than the corresponding maximum number, h. Indeed, on many interconnections, routing time increases with h_{ave}, for fixed h, a phenomenon modeled in E-BSP is by adding a term depending upon h_{ave} to the cost of a superstep. However, the functional shape of this term varies with the topology of the intended target platform, making the model somewhat awkward.

D-BSP extends BSP by incorporating some aspects of network proximity into the model. Specifically, the set of n processor/memory pairs is viewed as partitionable as a collection of clusters, where each cluster is able to perform its own sequence of supersteps independently of the other ones and is characterized by its own g and ℓ parameters, typically increasing with the size of the cluster. The partition into clusters can change dynamically within a pre-specified set of legal partitions. The key advantage is that communication patterns where messages are confined within small clusters have small cost, like in realistic platforms and unlike in standard BSP. In fact, it can be shown quantitatively that this advantage translates into higher effectiveness and portability of D-BSP over BSP. Clustering also enables efficient routing of unbalanced communication patterns in D-BSP, making it unnecessary to further extend the cost model in the direction followed by E-BSP. Thus, D-BSP is an attractive candidate among BSP variants and, in general, among bandwidth-latency models, to strike a fair balance among the conflicting features sought in a bridging model of parallel computation.

In Section 2, we define a restricted version of D-BSP where clusters are defined according to a regular recursive structure, which greatly simplifies the use of the model without diminishing its power significantly. In Section 3, we employ the methodology based on cross-simulations proposed in [BPP99] to quantitatively assess the higher effectiveness of D-BSP with respect to BSP, relatively to the wide class of processor networks. Then, in Subsection 3.1 we show that, for certain relevant computations and prominent topologies, D-BSP exhibits a considerably higher effectiveness than the one guaranteed by the general result. In such cases, the effectiveness of D-BSP becomes close to optimal. Furthermore, we present a general strategy to exploit communication locality: one of the corollaries is a proof that D-BSP can be as effective as E-BSP in dealing with unbalanced communication patterns. Finally, in Section 4 we show how D-BSP can efficiently support a shared memory abstraction, a valuable provision for algorithm development in a distributed-memory environment. The results presented in the section clearly indicate that the network proximity modeled by D-BSP can be exploited to reduce network congestion and bank contention when implementing a shared address space both by randomized and by deterministic strategies.

2 The D-BSP Model

The D-BSP (Decomposable BSP) model was introduced in [DK96] as an extension of Valiant's BSP [Val90] aimed at capturing, in part, the proximity structure of the network. In its most general definition, the D-BSP is regarded as a set of n processor/memory pairs communicating through a router, which can be aggregated according to a predefined collection of submachines, each able to operate independently. For concreteness, we focus our attention on a restricted version of the model (referred to as *recursive D-BSP* in [DK96]) where the collection of submachines has the following regular structure. Let n be a power of two. For $0 \leq i \leq \log n$, the n processors are partitioned into 2^i fixed, disjoint *i-clusters* $C_0^{(i)}, C_1^{(i)}, \cdots, C_{2^i-1}^{(i)}$ of $n/2^i$ processors each, where the processors of a cluster are able to communicate among themselves independently of the other clusters. The clusters form a hierarchical, binary decomposition tree of the D-BSP machine: specifically, $C_j^{\log n}$ contains only processor P_j, for $0 \leq j < n$, and $C_j^{(i)} = C_{2j}^{(i+1)} \cup C_{2j+1}^{(i+1)}$, for $0 \leq i < \log n$ and $0 \leq j < 2^i$.

A D-BSP computation consists of a sequence of labelled supersteps. In an *i-superstep*, $0 \leq i \leq \log n$, each processor executes internal computation on locally held data and sends messages exclusively to processors within its *i*-cluster. The superstep is terminated by a barrier, which synchronizes processors within each *i*-cluster independently. It is assumed that messages sent in one superstep are available at the destinations only at the beginning of the subsequent superstep. Let $g = (g_0, g_1, \ldots, g_{\log n})$ and $\ell = (\ell_0, \ell_1, \ldots, \ell_{\log n})$. If each processor performs at most w local operations, and the messages sent in the superstep form an *h-relation* (i.e., each processor is source or destination of at most h messages), then, the cost of the *i*-superstep is upper bounded by $w + h g_i + \ell_i$, for $0 \leq i \leq \log n$. Parameters g_i and ℓ_i are related to the bandwidth and latency guaranteed by the router when communication occurs within *i*-clusters. We refer to such a model as a D-BSP (n, g, ℓ).

Note that the standard BSP(n, g, ℓ) defined by Valiant can be regarded as a D-BSP (n, g, ℓ) with $g_i = g$ and $\ell_i = \ell$ for every i, $0 \leq i \leq \log n$. In other words, D-BSP introduces the notion of proximity in BSP through clustering, and groups *h*-relations into specialized classes associated with different costs. This ensures full compatibility between the two models, which allows programs written according to one model to run on any machine supporting the other, the only difference being their estimated performance.

In this paper, we will often exemplify our considerations by focusing on a class of parameter values for D-BSP of particular significance. Namely, let α and β be two arbitrary constants, with $0 < \alpha, \beta < 1$. We will consider D-BSP $(n, g^{(\alpha)}, \ell^{(\beta)})$ machines with

$$\begin{cases} g_i^{(\alpha)} = G \cdot (n/2^i)^\alpha, \\ \ell_i^{(\beta)} = L \cdot (n/2^i)^\beta, \end{cases} \quad 0 \le i \le \log n,$$

where G and L are two arbitrary positive constants. Note that these parameters capture a wide family of machines whose clusters feature moderate bandwidth/latency properties, such as, for example, multidimensional arrays.

3 D-BSP and Processor Networks

Intuitively, a computational model M, where designers develop and analyze algorithms, is *effective* with respect to a platform M', on which algorithms are eventually implemented and executed, if the algorithmic choices based on M turn out to be the right choices in relation to algorithm performance on M'. In other words, one hopes that the relative performance of any two algorithms developed on M reflects the relative performance of their implementations on M'. In order to attain generality, a computational model abstracts specific architectural details (e.g., network topology), while it incorporates powerful features that simplify its use but may not be exhibited by certain platforms. Therefore, in order to evaluate the effectiveness of a model M with respect to a platform M', we must establish the performance loss incurred by running algorithms developed for M (which do not exploit platform-specific characteristics) on M', and, on the other hand, we must assess how efficiently features offered by M to the algorithm designer, can be supported on M'.

More precisely, let us regard M and M' as two different machines, and define $S(M, M')$ (resp., $S(M', M)$) as the minimum slowdown needed for simulating M on M' (resp., M' on M). In [BPP99] a quantitative measure of effectiveness of M with respect to M' is given, and it is shown that the product $\delta(M, M') = S(M, M')S(M', M)$ provides an upper bound to this measure. Namely, effectiveness decreases with increasing $\delta(M, M')$ and is highest for $\delta(M, M') = 1$. When maximized over all platforms M' in a given class, this quantity provides an upper measure of the portability of M with respect to the class.

We use this approach to evaluate the effectiveness of D-BSP with respect to the class of processor networks. Let G be a connected n-processor network, where in one *step* each processor executes a constant number of local operations and may send/receive one point-to-point message to/from each neighboring processor (multi-port regimen). As is the case for all relevant network topologies, we assume that G has a decomposition tree $\{G_0^{(i)}, G_1^{(i)}, \cdots, G_{2^i-1}^{(i)} : \forall i, 0 \le i \le \log n\}$, where each $G_j^{(i)}$ is a connected subnet (*i-subnet*) with $n/2^i$ processors and $G_j^{(i)} = G_{2j}^{(i+1)} \cup G_{2j+1}^{(i+1)}$. By combining the routing results of [LR99,LMR99] one can easily show that for every $0 \le i \le \log n$ there exist suitable values g_i and ℓ_i related, respectively, to the bandwidth and diameter characteristics of the i-subnets, such that an h-relation followed by a barrier synchronization within an i-subnet can be implemented in $O\left(hg_i + \ell_i\right)$ time. Let M be a D-BSP $(n, \boldsymbol{g}, \boldsymbol{\ell})$ with these particular g_i and ℓ_i values, for $0 \le i \le \log n$. Clearly, we have that

$$S(M, G) = O\left(1\right) . \tag{1}$$

Vice-versa, an upper bound to the slowdown incurred in simulating G on M is provided by the following theorem proved in [BPP99, Theorem 3].

Theorem 1. *Suppose that at most b_i links connect every i-subnet to the rest of the network, for $1 \le i \le \log n$. Then, one step of G can be simulated on the D-BSP $(n, \boldsymbol{g}, \boldsymbol{\ell})$ in time*

$$S(G, M) = O\left(\min_{n' \le n} \left\{ \frac{n}{n'} + \sum_{i=\log(n/n')}^{\log n - 1} (g_i \max\{h_i, h_{i+1}\} + \ell_i) \right\} \right) , \tag{2}$$

where $h_i = \lceil b_{i-\log(n/n')}/(n/2^i) \rceil$, for $0 \le i \le \log n$.

We can apply Equations 1 and 2 to quantitatively estimate the effectiveness of D-BSP with respect to specific network topologies. Consider, for instance, the case of an n-node d-dimensional array. Fix $g_i = \ell_i = (n/2^i)^{1/d}$, for $0 \le i \le \log n$. Such D-BSP $(n, \boldsymbol{g}, \boldsymbol{\ell})$ can be simulated on G with constant slowdown. Since G has a decomposition tree with subnets $G_j^{(i)}$ that have $b_i = O\left((n/2^i)^{(d-1)/d}\right)$, the D-BSP simulation quoted in Theorem 1 yields a slowdown of $S(G, M) = O\left(n^{1/(d+1)}\right)$ per step. In conclusion, letting M be the D-BSP with the above choice of parameters, we have that $\delta(M, G) = O\left(n^{1/(d+1)}\right)$. The upper bound on $S(G, M)$ can be made exponentially smaller when each array processor has constant-size memory. In this case, by employing a more sophisticated simulation strategy, we can get $S(G, M) = O\left(2^{\Theta(\sqrt{\log n})}\right)$ [BPP99], thus significantly improving D-BSP's effectiveness.

It is important to remark that the D-BSP clustered structure provides a crucial contribution to the model's effectiveness. Indeed, it can be shown that, if M' is a BSP(n, g, ℓ) and G is a d-dimensional array, then $\delta(M', G) = \Omega\left(n^{1/d}\right)$ independently of g, l and the size of the memory at each processor [BPP99]. This implies that, under the δ metric, D-BSP is asymptotically more effective than BSP with respect to multidimensional arrays.

3.1 Effectiveness of D-BSP with respect to specific computations

We note that non-constant slowdown for simulating an *arbitrary* computation of a processor network on a D-BSP is to be expected since the D-BSP disregards the fine structure of the network topology, and, consequently, it is unable to fully exploit topological locality. However, for several prominent topologies and several relevant computational problems arising in practical applications, the impact of such a loss of locality is much less than what the above simulation results may suggest, and, in many cases, it is negligible.

Consider, for example, the class of processor networks G whose topology has a recursive structure with bisection bandwidth $O\left(n^{1-\alpha}\right)$ and diameter $O\left(n^{\beta}\right)$, for arbitrary constants $0 < \alpha, \beta < 1$ (note that multidimensional arrays belong to this class).

In this case, the results of [LR99,LMR99] imply that a D-BSP $(n, \boldsymbol{g}^{(\alpha)}, \boldsymbol{\ell}^{(\beta)})$ M can be efficiently supported on G, so that $S(M, G) = O(1)$. Hence, algorithms devised on M can be implemented on G with at most constant loss of efficiency. Although the reverse slowdown $S(G, M)$ is in general non-constant, we now show that M allows us to develop algorithms for a number of relevant computational problems, which exhibit optimal performance when run on G. Hence, for these problems, the loss of locality of M with respect to G is negligible.

Let $N \geq n$ items be evenly distributed among the n processors. A parallel prefix on these items (N-*prefix*) requires time $\Omega\left(N/n + n^\beta\right)$ to be performed on G [Lei92]. On the D-BSP $(n, \boldsymbol{g}^{(\alpha)}, \boldsymbol{\ell}^{(\beta)})$ there is a parallel prefix algorithm that runs in time $O\left(N/n + n^\alpha + n^\beta\right)$ [DK96]. Clearly, when $\alpha \leq \beta$ the implementation of the D-BSP algorithm on G exhibits optimal performance. We remark that optimality cannot be obtained using the standard BSP model. Indeed, results in [Goo96] imply that the direct implementation on G of any BSP algorithm for N-prefix, runs in time $\Omega\left(n^\beta \log n/\log(1 + \lceil n^{\beta-\alpha}\rceil)\right)$, which is not optimal, for instance, when $\alpha = \beta$.

We call k-*sorting* a sorting problem in which k keys are initially assigned to each one of n processors and are to be redistributed so that the k smallest keys will be held by processor P_0, the next k smallest ones by processor P_1, and so on. It is easy to see that k-sorting requires time $\Omega\left(kn^\alpha + n^\beta\right)$ to be performed on G because of the bandwidth and diameter of the network. On the D-BSP $(n, \boldsymbol{g}^{(\alpha)}, \boldsymbol{\ell}^{(\beta)})$ there is an algorithm that performs k-sorting in time $O\left(kn^\alpha + n^\beta\right)$ [FPP01], which is clearly optimal when ported to G. Again, a similar result is not possible with standard BSP: the direct implementation on G of any BSP algorithm for k-sorting runs in time $\Omega\left((\log n/\log k)(kn^\alpha + n^\beta)\right)$, which is not optimal for small k.

As a final important example, consider a (k_1, k_2)-*routing* problem where each processor is the source of at most k_1 packets and the destination of at most k_2 packets. Observe that a greedy routing strategy where all packets are delivered in one superstep requires $\Theta\left(\max\{k_1, k_2\} \cdot n^\alpha + n^\beta\right)$ time on a D-BSP $(n, \boldsymbol{g}^{(\alpha)}, \boldsymbol{\ell}^{(\beta)})$, which is the best one could do on a BSP(n, n^α, n^β). However, a careful exploitation of the submachine locality exhibited by the D-BSP yields a better algorithm for (k_1, k_2)-routing, which runs in time $O\left(k_{\min}^\alpha k_{\max}^{1-\alpha} n^\alpha + n^\beta\right)$, where $k_{\min} = \min\{k_1, k_2\}$ and $k_{\max} = \max\{k_1, k_2\}$ [FPP01]. Indeed, standard lower bound arguments show that such a routing time is optimal for G [SK94].

As a corollary of the above routing result, we can show that, unlike the standard BSP model, D-BSP is also able to handle unbalanced communication patterns efficiently, which was the main objective that motivated the introduction of a BSP variant, called E-BSP, by [JW96a]. Let an (h, m)-relation be a routing instance where each processor sends/receives at most h messages, and a total of m messages are exchanged. Although a greedy routing strategy for an (h, m)-relation requires time $\Theta\left(hn^\alpha + n^\beta\right)$ on both D-BSP and BSP, the exploitation of submachine locality in D-BSP allows us to route any (h, m)-relation in time $O\left(\lceil m/n\rceil^\alpha h^{1-\alpha} n^\alpha + n^\beta\right)$, which is equal or smaller than the greedy routing time and it is optimal for G. Consequently, D-BSP can be as effective as E-BSP in dealing with unbalanced communication as E-BSP, where the treatment of unbalanced communication is a primitive of the model.

4 Providing Shared Memory on D-BSP

A very desirable feature of a distributed-memory model is the ability to support a shared memory abstraction efficiently. Among the other benefits, this feature allows porting the vast body of PRAM algorithms [JáJ92] to the model at the cost of a small time penalty. In this section we present a number of results that demonstrate that D-BSP can be endowed with an efficient shared memory abstraction.

Implementing shared memory calls for the development of a *scheme* to represent m shared cells (*variables*) among the n processor/memory pairs of a distributed-memory machine in such a way that any n-tuple of variables can be read/written efficiently by the processors. The time required by a parallel access to an arbitrary n-tuple of variables is often referred to as the *slowdown* of the scheme.

Numerous randomized and deterministic schemes have been developed in the literature for a number of specific processor networks. Randomized schemes (see e.g., [CMS95,Ran91]) usually distribute the variables randomly among the memory modules local to the processors. As a consequence of such a scattering, a simple routing strategy is sufficient to access any n-tuple of variables efficiently, with high probability. Following this line, we can give a simple, randomized scheme for shared memory access on D-BSP. Assume, for simplicity, that the variables be spread among the local memory modules by means of a totally random function. In fact, a polynomial hash function drawn from a $\log n$-universal class [CW79], suffices to achieve the same results [MV84], but it takes only poly$(\log n)$ rather than $O(n \log n)$ random bits to be generated and stored at the nodes. We have:

Theorem 2. *Any n-tuple of memory accesses on a D-BSP (n, g, ℓ) can be performed in time*

$$O\left(\sum_{i=0}^{\lfloor \log(n/\log n) \rfloor - 1} T_{\mathrm{pr}}(i) + g_{\lfloor \log(n/\log n) \rfloor} \frac{\log n}{\log \log n} + \ell_{\lfloor \log(n/\log n) \rfloor}\right) \tag{3}$$

with high probability, where $T_{\mathrm{pr}}(i)$ denotes the time of a prefix-sum operation within an i-cluster.

Proof (Sketch). Consider the case of write accesses. The algorithm consists of $\lfloor \log(n/\log n) \rfloor + 1$ steps. More specifically, in Step i, for $1 \leq i \leq \lfloor \log(n/\log n) \rfloor$, we send the messages containing the access requests to their destination i-clusters, so that each node in the cluster receives roughly the same number of messages. A standard occupancy argument [MR95] suffices to show that, with high probability, there will be no more than $\lambda n/2^i$ messages destined to the same i-cluster, for a given small constant $\lambda > 1$, hence each step requires a simple prefix and the routing of an $O(1)$-relation in i-clusters. In the last step, we simply send the messages to their final destinations, where the memory access is performed. Again, the same probabilistic argument implies that the degree of the relation in this case is $O(\log n/\log \log n)$, with high probability.

For read accesses, the return journey of the messages containing the accessed values can be performed by reversing the algorithm for writes, thus remaining within the same time bound.

By plugging in the time for prefix in Eq. (3) we obtain:

Corollary 1. *Any n-tuple of memory accesses can be performed in optimal time* $O\left(n^{\alpha} + n^{\beta}\right)$, *with high probability, on a D-BSP* $\left(n, \boldsymbol{g}^{(\alpha)}, \boldsymbol{\ell}^{(\beta)}\right)$.

Observe that under a uniform random distribution of the variables among the memory modules, $\Theta\left(\log n / \log \log n\right)$ out of *any* set of n variables will be stored in the same memory module, with high probability, hence any randomized access strategy without replication would require at least $\Omega\left(n^{\alpha} \log n / \log \log n + n^{\beta}\right)$ time on a $\mathrm{BSP}(n, n^{\alpha}, n^{\beta})$.

Let us now switch to deterministic schemes. In this case, achieving efficiency is much harder, since, in order to avoid the trivial worst-case where a few memory modules contain all of the requested data, we are forced to replicate each variable and manage replicated copies so to enforce consistency. A typical deterministic scheme replicates every variable into ρ copies, which are then distributed among the memory modules through a map exhibiting suitable expansion properties. Expansion is needed to guarantee that the copies relative to any n-tuple of variables be never confined within few nodes. The parameter ρ is referred to as the *redundancy* of the scheme. In order to achieve efficiency, the main idea, originally introduced in [UW87] and adopted in all subsequent works, is that any access (read or write) to a variable is satisfied by reaching only a subset of its copies, suitably chosen to maximize communication bandwidth while ensuring consistency (i.e., a read access must always return the most updated value of the variable).

A general deterministic scheme to implement a shared memory abstraction on a D-BSP is presented in [FPP01]. The scheme builds upon the one in [PPS00] for a number of processor networks, whose design exploits the recursive decomposition of the underlying topology to provide a hierarchical, redundant representation of the shared memory based on $k + 1$ levels of logical modules. Such an organization fits well with the structure of a D-BSP, which is hierarchical in nature. More specifically, each variable is replicated into $r = O\left(1\right)$ copies, and the copies are assigned to r logical modules of level 0. In general, the logical modules at the i-th level, $0 \leq i < k$ are replicated into three copies, which are assigned to three modules of level $i + 1$. This process eventually creates $r3^{k} = \Theta\left(3^{k}\right)$ copies of each variable, and 3^{k-i} replicas of each module at level i. The number (resp., size) of the logical modules decreases (resp., increases) with the level number, and their replicas are mapped to the D-BSP by assigning each distinct block to a distinct cluster of appropriate size, so that each of the sub-blocks contained within the block is recursively assigned to a sub-cluster.

The key ingredients of the above memory organization are represented by the bipartite graph that governs the distribution of the copies of the variables among the modules of the first level, and those that govern the distribution of the replicas of the modules at the subsequent levels. The former graph is required to exhibit some weak expansion property, and its existence can always be proved through combinatorial arguments although, for certain memory sizes, explicit constructions can be given. In contrast, all the other graphs employed in the scheme require expansion properties that can be obtained by suitable modifications of the BIBD graph [Hal86], and can always be explicitly constructed.

For an n-tuple of variables to be read/written, the selection of the copies to be accessed and the subsequent execution of the accesses of the selected copies are performed on the D-BSP through a protocol similar to the one in [PPS00], which can be implemented through a combination of prefix, sorting and (k_1, k_2)- routing primitives. By employing the efficient D-BSP implementations for these primitives discussed in Section 3, the following result is achieved on a D-BSP $(n, g^{(\alpha)}, \ell^{(\beta)})$ [FPP01, Corollary 1].

Theorem 3. *For any value m upper bounded by a polynomial in n there exists a scheme to implement a shared memory of size m on a D-BSP $(n, g^{(\alpha)}, \ell^{(\beta)})$ with optimal slowdown $\Theta\left(n^\beta\right)$ and constant redundancy, when $\alpha < \beta$, and slowdown $\Theta\left(n^\alpha \log n\right)$ and redundancy $O\left(\log^{1.59} n\right)$, when $\alpha \geq \beta$. The scheme requires only weakly expanding graphs of constant degree and can be made fully constructive for $m = O\left(n^{3/2}\right)$ and $\alpha \geq 1/2$.*

An interesting consequence of the above theorem is that it shows that optimal worst-case slowdowns for shared memory access are achievable with constant redundancy for machines where latency overheads dominate over those due to bandwidth limitations, as is often the case in network-based parallel machines. When this is not the case, it is shown in [FPP01] that the proposed scheme is not too far-off from being optimal.

Perhaps, the most important feature of the above scheme is that, unlike the other classical deterministic schemes in the literature, it solely relies on expander graphs of mild expansion, hence it can be made fully constructive for a significant range of the parameters involved. Such mild expanders, however, are only able to guarantee that the copies of an arbitrary n-tuple of variables be spread among $O\left(n^{1-\epsilon}\right)$ memory modules, for some constant $\epsilon < 1$. Hence the congestion at a single memory module can be as high as $O\left(n^\epsilon\right)$ and the clusterized structure of D-BSP is essential in order to achieve good slowdown. In fact, any deterministic strategy employing these graphs on a BSP(n, g, ℓ) could not achieve better than $\Theta\left(gn^\epsilon\right)$ slowdown.

References

[BDM95] A. Bäumker, W. Dittrich, and F. Meyer auf der Heide. Truly efficient parallel algorithms: c-optimal multisearch for and extension of the BSP model. In *Proc. of the 3rd European Symposium on Algorithms*, pages 17–30, 1995.

[BGMZ95] G.E. Blelloch, P.B. Gibbons, Y. Matias, and M. Zagha. Accounting for memory bank contention and delay in high-bandwidth multiprocessors. In *Proc. of the 7th ACM Symp. on Parallel Algorithms and Architectures*, pages 84–94, Santa Barbara, CA, July 1995.

[BHP+96] G. Bilardi, K.T. Herley, A. Pietracaprina, G. Pucci, and P. Spirakis. BSP vs LogP. In *Proc. of the 8th ACM Symp. on Parallel Algorithms and Architectures*, pages 25–32, 1996. To appear in *Algorithmica*, Special Issue on Coarse Grained Parallel Algorithms.

[BHPP00] G. Bilardi, K. Herley, A. Pietracaprina, and G. Pucci. On stalling in LogP. In *Proc. of the Workshop on Advances in Parallel and Distributed Computational Models*, LNCS 1800, pages 109–115, May 2000.

[BPP99] G. Bilardi, A. Pietracaprina, and G. Pucci. A quantitative measure of portability with application to bandwidth-latency models for parallel computing. In *Proc. of EUROPAR 99*, LNCS 1685, pages 543–551, September 1999.

[CKP+96] D.E. Culler, R. Karp, D. Patterson, A. Sahay, E. Santos, K.E. Schauser, R. Subramonian, and T.V. Eicken. LogP: A practical model of parallel computation. *Communications of the ACM*, 39(11):78–85, November 1996.

[CMS95] A. Czumaj, F. Meyer auf der Heide, and V. Stemann. Shared memory simulations with triple-logarithmic delay. In *Proc. of the 3rd European Symposium on Algorithms*, pages 46–59, 1995.

[CW79] J.L. Carter and M.N. Wegman. Universal classes of hash functions. *Journal of Computer and System Sciences*, 18:143–154, 1979.

[DK96] P. De la Torre and C.P. Kruskal. Submachine locality in the bulk synchronous setting. In *Proc. of EUROPAR 96*, LNCS 1124, pages 352–358, August 1996.

[FPP01] C. Fantozzi, A. Pietracaprina, and G. Pucci. Implementing shared memory on clustered machines. In *Proc. of 2nd International Parallel and Distributed Processing Symposium*, 2001. To appear.

[Goo96] M.T. Goodrich. Communication-efficient parallel sorting. In *Proc. of the 28th ACM Symp. on Theory of Computing*, pages 247–256, Philadelphia, Pennsylvania USA, May 1996.

[Hal86] M. Hall Jr. *Combinatorial Theory*. John Wiley & Sons, New York NY, second edition, 1986.

[JáJ92] J. JáJá. *An Introduction to Parallel Algorithms*. Addison Wesley, Reading MA, 1992.

[JW96a] B.H.H. Juurlink and H.A.G. Wijshoff. The E-BSP model: Incorporating general locality and unbalanced communication into the BSP model. In *Proc. of EUROPAR 96*, LNCS 1124, pages 339–347, August 1996.

[JW96b] B.H.H. Juurlink and H.A.G. Wijshoff. A quantitative comparison of paralle computation models. In *Proc. of the 8th ACM Symp. on Parallel Algorithms and Architectures*, pages 13–24, June 1996.

[Lei92] F.T. Leighton. *Introduction to Parallel Algorithms and Architectures: Arrays • Trees • Hypercubes*. Morgan Kaufmann, San Mateo, CA, 1992.

[LMR99] F.T. Leighton, B.M. Maggs, and A.W. Richa. Fast algorithms for finding $O(\text{congestion} + \text{dilation})$ packet routing schedules. *Combinatorica*, 19(3):375–401, 1999.

[LR99] F.T. Leighton and S. Rao. Multicommodity max-flow min-cut theorems and their use in designing approximation algorithms. *Journal of the ACM*, 46(6):787–832, 1999.

[MR95] R. Motwani and P. Raghavan. *Randomized Algorithms*. Cambridge University Press, Cambridge MA, 1995.

[MV84] K. Mehlhorn and U. Vishkin. Randomized and deterministic simulations of PRAMs by parallel machines with restricted granularity of parallel memories. *Acta Informatica*, 21:339–374, 1984.

[PPS00] A. Pietracaprina, G. Pucci, and J. Sibeyn. Constructive, deterministic implementation of shared memory on meshes. *SIAM Journal on Computing*, 30(2):625–648, 2000.

[Ran91] A.G. Ranade. How to emulate shared memory. *Journal of Computer and System Sciences*, 42:307–326, 1991.

[SK94] J.F. Sibeyn and M. Kaufmann. Deterministic $1\text{-}k$ routing on meshes, with application to hot-potato worm-hole routing. In *Proc. of the 11th Symp. on Theoretical Aspects of Computer Science*, LNCS 775, pages 237–248, 1994.

[UW87] E. Upfal and A. Widgerson. How to share memory in a distributed system. *Journal of the ACM*, 34(1):116–127, 1987.

[Val90] L.G. Valiant. A bridging model for parallel computation. *Communications of the ACM*, 33(8):103–111, August 1990.

Coarse Grained Parallel On-Line Analytical Processing (OLAP) for Data Mining

Frank Dehne[1], Todd Eavis[2], and Andrew Rau-Chaplin[2]

[1] Carleton University, Ottawa, Canada,
frank@dehne.net,
WWW home page: http://www.dehne.net
[2] Dalhousie University, Halifax, Canada,
eavis@cs.dal.ca, arc@cs.dal.ca,
WWW home page: http://www.cs.dal.ca/~arc

Abstract. We study the applicability of coarse grained parallel computing model (CGM) to on-line analytical processing (OLAP) for data mining. We present a general framework for the CGM which allows for the efficient parallelization of existing data cube construction algorithms for OLAP. Experimental data indicate that our approach yield optimal speedup, even when run on a simple processor cluster connected via a standard switch.

1 Introduction

During recent years, there has been tremendous growth in the data warehousing market. Despite the sophistication and maturity of conventional database technologies, the ever-increasing size of corporate databases, coupled with the emergence of the new global Internet "database", suggests that new computing models may soon be required to fully support many crucial data management tasks. In particular, the exploitation of parallel algorithms and architectures holds considerable promise, given their inherent capacity for both concurrent computation and data access.

Data warehouses can be described as *decision support systems* in that they allow users to assess the evolution of an organization in terms of a number of key data attributes or dimensions. Typically, these attributes are extracted from various operational sources (relational or otherwise), then cleaned and normalized before being loaded into a relational store. By exploiting multi-dimensional views of the underlying data warehouse, users can "drill down" or "roll up" on hierarchies, "slice and dice" particular attributes, or perform various statistical operations such as ranking and forecasting. This approach is referred to as Online Analytical Processing or OLAP.

2 Coarse Grained Parallel OLAP

Data cube queries represent an important class of On-Line Analytical Processing (OLAP) queries in decision support systems. The precomputation of the different group-bys of a data cube (i.e., the forming of aggregates for every combination of GROUP BY attributes) is critical to improving the response time of the queries [10]. The resulting data structures can then be used to dramatically accelerate visualization and query tasks associated with large information sets. Numerous solutions for generating the data cube have been proposed. One of the main differences between the many solutions is whether they are aimed at sparse or dense relations [3, 11, 13, 15, 17]. Solutions within a category can also differ considerably. For example, top-down data cube computations for dense relations based on sorting have different characteristics from those based on hashing. To meet the need for improved performance and to effectively handle the increase in data sizes, parallel solutions for generating the data cube are needed. In this paper we present a general framework for the CGM model ([5]) of parallel computation which allows for the efficient parallelization of existing data cube construction algorithms. We present load balanced and communication efficient partitioning strategies which generate a subcube computation for every processor. Subcube computations are then carried out using existing sequential, external memory data cube algorithms.

Balancing the load assigned to different processors and minimizing the communication overhead are the core problems in achieving high performance on parallel systems. At the heart of this paper are two partitioning strategies, one for top-down and one for bottom-up data cube construction algorithms. Good load balancing approaches generally make use of application specific characteristics. Our partitioning strategies assign loads to processors by using metrics known to be crucial to the performance of data cube algorithms [1, 3, 15]. The bottom-up partitioning strategy balances the number of single attribute external sorts made by each processor [3]. The top-down strategy partitions a weighted tree in which weights reflect algorithm specific cost measures such as estimated group-by sizes [1, 15].

The advantages of our load balancing methods compared to the previously published parallel data cube construction methods [8, 9] are as follows. Our methods reduce inter-processor communication overhead by partitioning the load in advance instead of computing each individual group-by in parallel (as proposed in [8, 9]). In fact, after our load distribution phase, each processor can compute its assigned subcube without any inter-processor communication. Our methods maximize code reuse from existing sequential data cube implementations by using existing sequential data cube algorithms for the subcube computations on each processor. This supports the transfer of optimized sequential data cube code to the parallel setting.

The following is a high-level outline of our coarse grained parallel top-down data cube construction method:

1. Construct a lattice for all 2^d views.
2. Estimate the size of each of the views in the lattice.
3. To determine the cost of using a given view to directly compute its children, use its estimated size to calculate (a) the cost of scanning the view and (b) the cost of sorting it.

4. Using the bipartite matching technique presented in [14], reduce the lattice to a spanning tree that identifies the appropriate set of prefix-ordered sort paths.
5. Partition the tree into $s \times p$ sub-trees (s = oversampling ratio).
6. Distribute the sub-trees over the p compute nodes.
7. On each node, use the sequential Pipesort algorithm to build the set of local views.

In the following, we provide a more detailed description of the implementation. We first describe the code base and supporting libraries.

The Code Base

In addition to MPI, we chose to employ a C++ platform in order to more efficiently support the growth of the project. With the expansion of the code base and the involvement of a number of independent developers, several of whom were in geographically distinct locations, it was important to employ an object-oriented language that allowed for data protection and class inheritance. One notable exception to the OOP model, however, was that the more familiar C interface to MPI functions was used.

The LEDA Libraries

A full implementation of our parallel datacube algorithms was very labour intensive. We chose to employ third-party software libraries so that we could focus our own efforts on the new research. After a review of existing packages, we selected the LEDA libraries because of the rich collection of fundamental data structures (including linked lists, hash tables, arrays, and graphs), the extensive implementation of supporting algorithms, and the C++ code base [12]. Although there is a slight learning curve associated with LEDA, the package has proven to be both efficient and reliable.

The OOP Framework

Having incorporated the LEDA libraries into our C++ code base, we were able to implement the lattice structure as a LEDA graph, thus allowing us to draw upon a large number of built-in graph support methods. In this case, we have sub-classed the graph template to permit the construction of algorithm-specific structures for node and edge objects. As such, a robust implementation base has been established; additional algorithms can be "plugged in" to the framework simply by sub-classing the lattice template and (a) over-riding or adding methods and (b) defining the new node and edge objects that should be used as template parameters.

In the current implementation, the base lattice has been sub-classed so as to augment the graph for the sort-based optimization. For each view/node, we estimate its construction cost in two formats: as a linear scan of its parent and as a complete resorting of its parent. Since these cost assessments depend upon accurate estimates of the sizes of the views themselves, the inclusion of a view estimator is required.

Probabilistic View Size Estimator

A number of inexpensive algorithms have been proposed for view size estimation [16]. The simplest merely entails using the product of the cardinalities of each dimension to place an upper bound on the size of each cuboid. A slightly more sophisticated technique computes a partial datacube on a randomly selected sample of the input set. The result is then "scaled up" to the appropriate size. although both approaches can give reasonable results on small, uniformly distributed datasets, they are not as reliable on real world data warehouses. Consequently, the use of probabilistic estimators that rely upon a single pass of the dataset have been suggested. As described in [16], our implementation builds upon the counting algorithm of Flajolet and Martin [7]. Essentially, we concatenate the d dimension fields into bit-vectors of length L and then hash the vectors into the range $0 \ldots 2^L - 1$. The algorithm then uses a probabilistic technique to count the number of distinct records (or hash values) that are likely to exist in the input set. To improve estimation accuracy, we employ a universal hashing function [4] to compute k hash functions, that in turn allows us to average the estimates across k counting vectors.

The probabilistic estimator was fairly accurate, producing estimation error in the range of 5–6 % with 256 hash functions. However, its running time on large problems was disappointing. The main problem is that, despite an asymptotic bound of $O(n * 2^d)$, the constants hidden inside the inner computing loops are enormous (i.e, greater than one million!). For the small problems described in previous papers, this is not an issue. In high dimension space, it is intractable. The running time of the estimator extends into weeks or even months. Considerable effort was expended in trying to optimize the algorithm. All of the more expensive LEDA structures (strings, arrays, lists, etc.) were replaced with efficient C-style data types. Despite a factor of 30 improvement in running time, the algorithm remained far too slow. We also experimented with the GNU-MP (multi-precision) libraries in an attempt to capitalize on more efficient operations for arbitrary length bit strings. Unfortunately, the resulting estimation phase was still many times slower than the construction of the views themselves. At this point, it seems unlikely that the Flajolet and Martin estimator is viable in high dimension space.

A simple view estimator

We needed a fast estimator that could be employed even in high dimension environments. We chose to use the technique that bounds view size as the product of dimension cardinalities. We also improved upon the basic estimate by exploiting the fact that a child view can be no bigger than the smallest of its potential parents. The estimated size for a node is the minimum of 1) the product of the cardinalities of the node's dimensions, and 2) the estimated size of the node's smallest parent. However, additional optimizations that incorporate intermediate results are not possible since a parallel implementation prevents us from sequentially passing estimates up and down the spanning tree. Section 3 discusses the results obtained using this version of the view estimator.

Computing the Edge Costs

As previously noted, the values produced by the estimator only represent the sizes of each view, not the final edge costs that are actually placed into the lattice. Instead, the algorithm uses the view estimate to calculate the potential cost of scanning and sorting any given cuboid. An appropriate metric must be experimentally developed for every architecture upon which the datacube algorithm is run. For example, on our own cluster, an in-memory multi-dimensional sort is represented as $(d + 2)/3) * (n \log n)$, where d is the current level in the lattice. At present, we are working on a module that will be used to automate this process so that appropriate parameters can be provided without manually testing every architecture.

Constructing the spanning tree

Once the lattice has been augmented with the appropriate costs, we apply a weighted bipartite matching algorithm that finds an appropriate set of sort paths within the lattice (as per [14]). Working bottom-up, matching is performed on each pair of contiguous levels in order to identify the most efficient distribution of sort and scan orders that can be used to join level i to level $i - 1$. The matching algorithm itself was provided by LEDA and required only minor modification for inclusion in our design.

Min-Max Partitioning

As soon as the bipartite matching algorithm has been executed, we partition the lattice into a set of k sub-trees using the min-max algorithm proposed by Becker, Schach and Perl [2]. The original algorithm is modified slightly since it was designed to work on a graph whose costs were assigned to the nodes, rather than the edges. Furthermore, a *false root* with zero cost must be added since the algorithm iterates until the root partition is no longer the smallest sub-tree. The min-max algorithm performs nicely in practice and, in conjunction with the over-sampling factor mentioned earlier, produces a well balanced collection of sub-trees (see [6] for a more complete analysis).

Once min-max terminates, the k sub-trees are collected into p sets by iteratively combining the largest and smallest trees (with respect to construction cost). Next, each sub-tree is partitioned into a set of distinct prefix-ordered sort paths, and then packaged and distributed to the individual network nodes.

The local processor decompresses the package into its composite sort paths and performs a pipesort on each pipeline in its assigned workload. No further communication with the root node is required from this point onward.

Local Pipesorts

Pipesort consists of two phases. In the first round, the root node in the list is sorted in a given multi-dimensional order. In phase two, we perform a linear pass through the sorted set, aggregating the most detailed records into new records that correspond to the granularity level of each cuboid in the sort path. As the new records are produced, they are written directly to disk. For example, if we sort the data in the order $ABCD$, we will subsequently create the $ABCD$, ABC, AB, and A views as we traverse the sorted set.

Although we originally exploited LEDA's array sorting mechanism to sort the root node in memory, we have since re-written the sort using the C library routines in order to to maximize performance. At present, all input sorting is performed in main memory. In the future, we expect to incorporate robust external memory sorting algorithms into the project.

3 Performance Evaluation

We now discuss the performance of our goarse grained parallel data cube implementation. As parallel hardware platform, we used a cluster consisting of a front-end machine and eight processors. The front-end machine is used to partition the lattice and distribute the work among the other 8 processors. The front-end machine is an IBM Netfinity server with two 9 GB SCSI disks, 512 MB of RAM and a 550-MHZ Pentium processor. The processors are 166 MHZ Pentiums with 2G IDE hard drives and 32 MB of RAM, except for one processor which is a 133 MHZ Pentium. The processors run LINUX and are connected via a 100 Mbit Fast Ethernet switch with full wire speed on all ports. Clearly, this is a very low end, older, hardware platform. However, for our main goal of studying the speedup obtained by our parallel method rather than absolute times, this platform is sufficient. In fact, the speedups measured on this low end cluster are lower bounds for the speedup that our software would achieve on newer and more powerful parallel machines.

Figure 1 shows the running time observed as a function of the number of processors used. For the same data set, we measured the sequential time (sequential pipesort [1]) and the parallel time obtained through our parallel data cube construction method, using an oversampling ratio of $s = 2$. The data set consisted of 1,000,000 records with dimension 7. Our test data values were uniformly distributed over 10 values in each dimension. Figure 1 shows the running times of the algorithm as we increase the number of processors. There are three curves shown. The *runtime* curve shows the time taken by the slowest processor (i.e. the processor that received the largest workload). The second curve shows the *average time* taken by the processors. The time taken by the front-end machine, to partition the lattice and distribute the work among the compute nodes, was insignificant. The *theoretical optimum* curve shown in Figure 1 is the sequential pipesort time divided by the number of processors used.

We observe that the *runtime* obtained by our code and the *theoretical optimum* are essentially identical. That is, for an oversampling ratio of $s = 2$, an optimal speedup of p is observed. (The anomaly in the *runtime* curve at $p = 4$ is due to the slower 133 MHZ Pentium processor.)

Interestingly, the *average time* curve is always below the *theoretical optimum* curve, and even the *runtime* curve is sometimes below the *theoretical optimum* curve. One would have expected that the *runtime* curve would always be above the *theoretical optimum* curve. We believe that this *superlinear speedup* is caused by another effect which benefits our parallel method: improved I/O. When sequential pipesort is applied to a 10 dimensional data set, the lattice is partitioned into pipes of length up to 10. In order to process a pipe of length 10, pipesort needs to write to 10 open files at the same time. It appears that under LINUX, the number of open files can have a considerable impact on performance. For 100,000 records, writing them to 4 files each took 8 seconds on our system. Writing them to 6 files each took 23 seconds, not 12, and writing them to 8 files each took 48 seconds, not 16. This benefits our parallel method, since we partition the lattice first and then apply pipesort to each part. Therefore, the pipes generated in the parallel method are considerably shorter.

Figure 2 shows the running time as a function of the oversampling ratio s. We observe that, for our test case, the parallel *runtime* (i.e. the time taken by the slowest processor) is best for $s = 3$. This is due to the following tradeoff. Clearly, the workload balance improves as s increases. However, as the total number of subtrees, $s \times p$, generated in the tree partitioning algorithm increases, we need to perform more sorts for the root nodes of these subtrees. The optimal tradeoff point for our test case is $s = 3$. It is important to note that the oversampling ratio s is a tunable parameter. The best value for s depends on a number of factors. What our experiments show is that $s = 3$ is sufficient for the load balancing. However, as the data set grows in size, the time for the sorts of the root nodes of the subtrees increases more than linear whereas the effect on the imbalance is linear. For substantially larger data sets, e.g. 1G rows, we expect the optimal value for s to be $s = 2$.

4 Conclusion

As data warehouses continue to grow in both size and complexity, so too will the opportunities for researchers and algorithm designers who can provide powerful, cost-effective OLAP solutions. In this paper we have discussed the implementation of a coarse grained parallel algorithm for the construction of a multidimensional data model known as the datacube. By exploiting the strengths of existing sequential algorithms, we can pre-compute all cuboids in a load balanced and communication efficient manner. Our experimental results have demonstrated that the technique is viable, even when implemented in a *shared nothing* cluster environment. In addition, we have suggested a number of opportunities for future work, including a parallel query model that utilizes packed r-trees. More significantly perhaps, given the relatively paucity of research currently being performed in the area of parallel OLAP, we believe that the ideas we have proposed represent just a fraction of the work that might lead to improved data warehousing solutions.

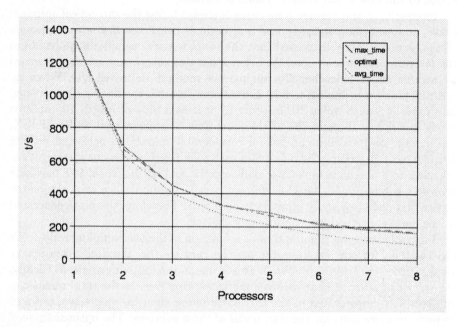

Fig. 1. Running Time In Seconds As A Function Of The Number Of Processors. (Fixed Parameters: Data Size = 1,000,000 Rows. Dimensions = 7. Experiments Per Data Point = 5.)

References

1. S. Agarwal, R. Agarwal, P.M. Deshpande, A. Gupta, J.F. Naughton, R. Ramakrishnan, and S. Srawagi. On the computation of multi-dimensional aggregates. In *Proc. 22nd VLDB Conf.*, pages 506–521, 1996.
2. R. Becker, S. Schach, and Y. Perl. A shifting algorithm for min-max tree partitioning. *Journal of the ACM*, 29:58–67, 1982.
3. K. Beyer and R. Ramakrishnan. Bottom-up computation of sparse and iceberg cubes. In *Proc. of 1999 ACM SIGMOD Conference on Management of data*, pages 359–370, 1999.
4. T. Cormen, C. Leiserson, and R. Rivest. *Introduction to Algorithms*. The MIT Press, 1996.
5. F. Dehne. Guest editor's introduction, special issue on "coarse grained parallel algorithms". *Algorithmica*, 24(3/4):173–176, 1999.

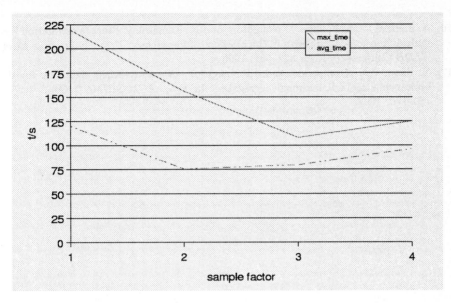

Fig. 2. Running Time In Seconds As A Function Of The Oversampling Ratio s. (Fixed Parameters: Data Size = 1,000,000 Rows. Number Of Processors = 8. Dimensions = 7. Experiments Per Data Point = 5.)

6. F. Dehne, T. Eavis, S. Hambrusch, and A. Rau-Chaplin. Parallelizing the datacube. *International Conference on Database Theory*, 2001.

7. P Flajolet and G. Martin. Probabilistic counting algorithms for database applications. *Journal of Computer and System Sciences*, 31(2):182–209, 1985.

8. S. Goil and A. Choudhary. High performance OLAP and data mining on parallel computers. *Journal of Data Mining and Knowledge Discovery*, 1(4), 1997.

9. S. Goil and A. Choudhary. A parallel scalable infrastructure for OLAP and data mining. In *Proc. International Data Engineering and Applications Symposium (IDEAS'99)*, Montreal, August 1999.

10. J. Gray, S. Chaudhuri, A. Bosworth, A. Layman, D. Reichart, M. Venkatrao, F. Pellow, and H. Pirahesh. Data cube: A relational aggregation operator generalizing group-by, cross-tab, and sub-totals. *J. Data Mining and Knowledge Discovery*, 1(1):29–53, April 1997.

11. V. Harinarayan, A. Rajaraman, and J.D. Ullman. Implementing data cubes efficiently. *SIGMOD Record (ACM Special Interest Group on Management of Data)*, 25(2):205–216, 1996.

12. Max Planck Institute. *LEDA*. http://www.mpi-sb.mpg.de/LEDA/.

13. K.A. Ross and D. Srivastava. Fast computation of sparse datacubes. In *Proc. 23rd VLDB Conference*, pages 116–125, 1997.

14. S. Sarawagi, R. Agrawal, and A.Gupta. On computing the data cube. Technical Report RJ10026, IBM Almaden Research Center, San Jose, California, 1996.

15. S. Sarawagi, R. Agrawal, and A. Gupta. On computing the data cube. Technical Report RJ10026, IBM Almaden Research Center, San Jose, CA, 1996.

16. A. Shukla, P. Deshpande, J. Naughton, and K. Ramasamy. Storage estimation for multidimensional aggregates in the presence of hierarchies. *Proceedings of the 22nd VLDB Conference*, pages 522–531, 1996.
17. Y. Zhao, P.M. Deshpande, and J.F.Naughton. An array-based algorithm for simultaneous multidimensional aggregates. In *Proc. ACM SIGMOD Conf.*, pages 159–170, 1997.

Architecture Independent Analysis of Parallel Programs

Ananth Grama[1], Vipin Kumar[2], Sanjay Ranka[3], and Vineet Singh[4]

[1] Dept. of Computer Sciences, Purdue University, W. Lafayette, IN 47907
ayg@cs.purdue.edu
[2] Dept. of Computer Sciences, University of Minnesota, Minneapolis, MN 55455
kumar@cs.umn.edu
[3] Dept. of Computer Sciences, University of Florida, Gainesville, FL 32611
ranka@cis.ufl.edu
[4] 10535 Cordova Road, Cupertino, CA 95014

Abstract. The presence of a universal machine model for serial algorithm design, namely the von Neumann model, has been one of the key ingredients of the success of uniprocessors. The presence of such a model makes it possible for an algorithm to be ported across a wide range of uniprocessors efficiently. Universal algorithm design models proposed for parallel computers however tend to be limited in the range of parallel platforms they can efficiently cover. Consequently, portability of parallel programs is attained at the expense of loss of efficiency. In this paper, we explore desirable and attainable properties of universal models of architecture independent parallel program design. We study various models that have been proposed, classify them based on important machine parameters and study their limitations.

1 Introduction

Conventional sequential computing is based on the von Neumann model. This role of this model is illustrated in Figure 1. Given problem P, an optimal algorithm O is designed for solving it for model I. A description of this algorithm, along with a description of the architecture of the target machine A is fed into the translator T. The translator generates machine code for the target machine. The set A spans the set of most conventional sequential computers and for this set, the von Neumann model fills in the role of model I.

In the above framework, the architecture independent algorithm design model I plays a very important role. It ensures that the same algorithm can be executed efficiently on a large variety of serial computers. Specifically, programs designed for this model have the property that they run in asymptotically the same time on every serial computer. The runtime of the algorithm on the said model is an accurate reflection of its asymptotic runtime on all computers subsumed by that model. The constants involved may not be reflected in the model but the model does provide a reasonable estimate. We refer to this as the **predictive** property of the model. Another interpretation of the predictive property is that

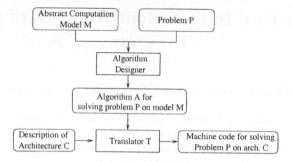

Fig. 1. The role of algorithm design model I in solving problem P on architecture A.

if two algorithms O1 and O2 are fed to the model, the model is able to predict which algorithm will have an asymptotically better performance on each computer subsumed by the model. We refer to this as the **ordering** property of the model. Finally, a translator exists that is capable of taking an algorithm specified for the model and converting it into a form that is executable by the target machine in asymptotically the same time as on model I. The translation process itself should not take an unreasonable amount of time. In the absence of such an architecture independent algorithm design model, we would have to design separate algorithms for each architecture.

The desirable properties of an architecture independent model do not change significantly in the context of parallel computers. However, due to added complexity, any such model can become difficult to program. It is therefore necessary to limit the complexity of the model. An architecture independent parallel programming model must have good coverage in architecture and algorithm space, be simple to work with, and have desirable performance prediction properties in the absolute, asymptotic, or relative (ordering) sense. It is unreasonable to expect accurate absolute (or even asymptotic) performance prediction from a model with good architecture and algorithm coverage. However, ordering properties are highly desirable for any proposed model.

2 Taxonomy of Parallel Algorithm Design Models

A number of models have been proposed that tradeoff accuracy, coverage, and simplicity to varying degrees. Whereas some models are relatively easy to design algorithms for, they may be difficult to realize in practice. In contrast, models that are closer to real machines tend to be less architecture independent and difficult to program. It is critical to identify various factors that differentiate the models and how they render models more or less realizable.

Global Memory Versus Seclusive Memories The strongest models of algorithm design allow each word in a part of the memory to be addressed independent of all others (*i.e.*, assume a global p-ported memory). Access to a certain word

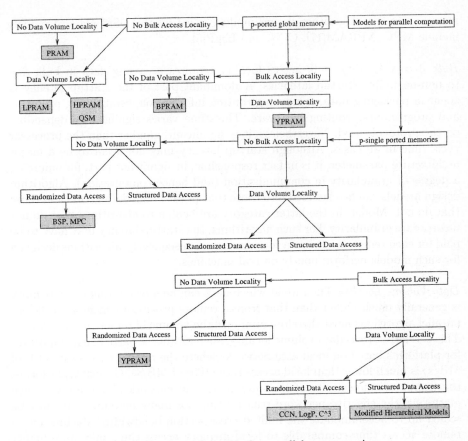

Fig. 2. Taxonomy of models for parallel programming.

in this global memory by a processor does not block access to any other word by another processor. Unfortunately, such a strong model comes with significant costs. If the total number of words in global memory is m and the number of processors is p, a completely non-blocking switch has a complexity $\Theta(mp)$. Furthermore, the presence of processor caches imposes stronger requirements of the global memory. Most PRAM models (PRAM, LPRAM [3], BPRAM [4]) assume a global memory.

The switch complexity can be reduced by organizing memory into banks (p banks of single-ported memories). When a processor is accessing a bank, access to all other words in the bank by other processors is blocked. If the switched memory is organized into b banks, the switch complexity of a non-blocking (at the bank level) switch is $\Theta(pb)$. This complexity is realizable at the current state of the art. Such a model is referred to as a Seclusive Memory model. When combined with appropriate communication cost models, seclusive memory models approximate machine architectures much better than global memory models. Algorithm design models based on the DSM class of models include MPC, XPRAM[17], CCN, and LogP[5].

Bulk Access Locality Bulk access locality requires accessing data in bulk due to non-negligible startup latencies. A dominant part of the startup latency is spent in packaging data, computing routing information, establishing protocols and programming routing hardware. This time varies significantly depending on the protocols used. Nevertheless, it is significantly higher than the processor cycle time in all cases. Although startup latency may not seem to be a major architecture parameter, it is in fact responsible, in significant part, for enforcing a degree of granularity in communication (and hence computation). Algorithm design models can be classified into those that reward bulk data access and those that do not. Models in the latter category are benign to algorithms that do not incorporate granularity. For such algorithms, the startup latency may have to be paid for each remote access on a real machine. Consequently, algorithms designed for such models perform poorly on real machines.

Data Volume Locality The rate at which data can be fetched from local memory is generally much higher than that from a remote processor's memory. Efficient parallel algorithms must therefore minimize the volume of remote data accessed. This is referred to as data volume locality. Such locality is particularly important for platforms based on local area networks where the remote access rate (10s of MB/s) is much lower than local access rate (100s of MB/s). On many platforms, the degree of connectivity of the interconnection network is limited. For such platforms, the effective per-word transfer time for many permutations must be scaled up to accommodate the limited cross-section bandwidth. Machines with remote access time comparable to local memory access time, may have higher effective remote access times when compensated for cross-section bandwidth.

Even if it were possible to construct machines with inter-processor communication bandwidth matching local memory bandwidth, the presence of caches complicates the situation. Assuming a high cache hit ratio on serial computations, the effective local access time is the time to access the cache (and not the main memory). The inter-processor communication bandwidth must therefore match the local cache bandwidth. This is impossible to achieve without making bulk accesses. Therefore, even models that assume communication bandwidth equal to local memory bandwidth must incorporate bulk-access to mask cache effects and startup latency.

Structured Communication Operations Typical applications can be viewed as a set of interacting subtasks executing on different processors. Interaction is by means of exchange of messages. It is often possible to formulate these tasks in such a way that the computation proceeds in interleaved steps of computation and communication. All processors (or subsets thereof) participate in the communication step. Since the communication involves more than a single pair of processors (aggregates or groups), such communication operations are also referred to as aggregate or collective communication operations.

Most current parallel computers are capable of exploiting the structure of aggregate operations. However, many models ignore this structure and assume that all redistributions take equal time. A very common assumption made is that if the maximum amount of data entering or leaving a processor in a particular redistribution is m then the communication time is proportional to m and is independent of the number of processors p. This can be a misleading assumption. Consider a many-to-many personalized communication with fixed data fan-in and fan-out. This communication operation is also referred to as the transportation primitive [15]. It is possible to perform this operation in $\Theta(m)$ time (m is the data fan-in or fan-out) only in the presence of slack and $O(p)$ cross-section bandwidth [15]. The accuracy of models making the above assumption can be poor and the class of real machines covered is relatively small.

There is a class of communication operations for which the optimal algorithms fit into a hierarchical framework. A hierarchical framework consists of hierarchies of redistributions. Redistributions involve only processors within a group. It is assumed that redistributions at various levels have no structure. Hierarchical models of parallel computers (YPRAM, PRAM) are capable of exploiting communication operations that fit into this hierarchical framework.

3 Existing Models for Parallel Programming

A variety of parallel algorithm design models have been proposed. They differ in their coverage, complexity, and accuracy.

LPRAM and BPRAM Models The LPRAM [3] model of computing is a global memory model. It assumes that the startup latency is negligible. Both of these parameters are unrealizable in real machines of reasonable sizes. Furthermore, it does not reward structured data accesses. Algorithms designed for this model may potentially pay significant startup overheads and make poor use of available communication bandwidth. The bandwidth for remote access is assumed to be less than that for local memory access. Therefore, algorithms designed for LPRAMs reward only data volume locality.

The BPRAM [4] model assumes non-negligible startup latencies and a communication bandwidth identical to local memory bandwidth. It therefore rewards bulk access locality but fails to capture data volume locality. Machines conforming to this model are difficult to construct because of the presence of a global memory. Furthermore, if the processing nodes have a hierarchical memory structure, the effective bandwidth for local access is the bandwidth to cache (assuming a high hit ratio). The communication bandwidth must therefore match the bandwidth to cache in this model. This is difficult to realize unless communication is highly clustered.

Hierarchical RAM models (HPRAM, YPRAM, HMM) The HPRAM [11, 10] and YPRAM [6] models of computation allow recursive decomposition of the machine until the algorithm uses only one processor and executes sequentially. For more than one processor, the communication cost charged is a function of the number of processors available. Assigning different cost to data redistributions in which communication is limited to smaller subsets of processors allows the design of algorithms that are more effective in utilizing locality especially for divide and conquer type applications. Despite the availability of variable cost redistributions, the formulations described in [11, 6] suffer from limitations similar to the BSP model (described subsequently) in terms of slack and worst case bandwidth requirements for every step. Analyzing time requirements using these models is difficult for non-recursively decomposable algorithms.

Bulk Synchronous Parallel and Related Models The XPRAM [17] model allows PRAM algorithms to run within a constant factor of optimal as long as there is sufficient slack. The XPRAM model forms the basis for Bulk Synchronous Parallelism (BSP). It assumes a seclusive memory and for efficient operation, assumes startup latency to be negligible and per-word access time to be similar to cycle time. Although such a model is not as difficult to construct as global memory models, the restrictions on startup latency and per-word transfer time are difficult to realize in real machines. The main feature of BSP is that it allows simulation of $p \log p$ PRAM processors on p seclusive memory processors in asymptotically the same time as on p PRAM processors.

Although BSP [16] is an attractive theoretical model, it has several drawbacks. The cost of making a bulk access of h words by a processor in a superstep is assumed to be $s + g'h$. Here, s is the startup cost and g' is the effective per-word transfer time. For efficient simulation, this cost must be comparable to the cost of h local memory accesses. This is difficult to achieve in realistic BSP simulations on actual platforms. The presence of hierarchical memory at processing nodes also requires bulk access locality. Even if the startup latency t_s of the network was somehow reduced to a negligible amount, the time for realizing a h relation must be similar to the cache access time for h words on a uniprocessor (and not local memory access time). This can be achieved only through bulk access of data from processors.

The CGM model, proposed by Dehne [7], is an extension of the BSP model. It addresses the undesirable possibility in BSP of a large number of small messages. A CGM algorithm consists of alternating phases of computation and global communication. A single computation-communication pair corresponds to a BSP superstep and the communication phase corresponds to a single h-relation in BSP. However, in the case of CGM, the size of the h-relation is given by $h = n/p$, where n corresponds to the problem size and p to the number of processors. With this restriction on permutation size, the communication cost of a parallel algorithm is simply captured in the number of communication phases. This communication cost model, rewards a small number of larger messages, thus capturing spatial locality better than the BSP model.

Another extension of the BSP model proposed by Baumker et al. [2] also attempts to increase the granularity of messages by adding a parameter B, which is the minimum message size for the message to be bandwidth bound as opposed to latency bound. Messages of size less than B are assigned the same cost as messages of size B, thus rewarding higher message granularity. The time taken by a BSP* algorithm is determined by the sum of computation and communication supersteps. A computation superstep with n local computations at each processor is assigned a cost $\max\{L, n\}$. A communication step with a h-relation of size s is assigned a cost $\max\{g \times h \times \lceil \frac{s}{B} \rceil, L\}$. Thus algorithms aggregate messages into larger h-relations to utilize all available bandwidth.

CCN, LogP and Related Models The Completely Connected Network (CCN) model of computing is a seclusive memory model. Although CCN has never been formally proposed as a model for architecture independent programming, its variants have been used extensively [12]. It assumes the presence of p processing elements connected over a completely connected network. The cost of sending a message of size m from one processor to any other is given by $t_s + t_w m$. Here, t_s is the startup latency and t_w is the bandwidth term. It therefore rewards bulk access and data volume locality.

In many architectures, small, fixed size messages can be sent with a small overhead using active messages. This forms the basis for the LogP model of computation. In the LogP model processors communicate using small fixed size messages of size m. The model rewards bulk access at the level of small messages (using parameter L) and data volume locality (using parameter g). Larger messages are sent in LogP by breaking them into smaller messages and pipelining them. For each of the smaller messages, the processor incurs an overhead of o at the sending end. The message can then be injected into the network. Subsequent messages can be injected at gaps of g. Here, g is determined by the network bandwidth. If a sequence of short messages needs to be sent, the gap of the first message can be overlapped with the preparation overhead o of the second message. Depending on values of g and o, two situations arise: $g > o$: in this case, the sending processor can prepare the messages faster than the network bandwidth. In this case, the network bandwidth will not be wasted. In the second case $o > g$: in this case, the processor cannot inject messages into the network fast enough to utilize available bandwidth. Therefore, it is desirable that the size of the small fixed message m be such that $g > o$. This implies that the model rewards bulk access to the level of m words. Indeed, if $g > o$ and there is no overlap of computation and communication, this bulk is sufficient to mask startup latencies.

The postal model [1], which predates the LogP model, is a simplification of the LogP model. The postal model assumes a message passing system of p processors with full connectivity (processors can send point-to-point messages to other processors), and simultaneous I/O (a processor can send one message and receive a message from another processor simultaneously). As is the case with LogP, messages have a predefined size. Larger messages are composed as aggregates of these smaller messages. In a postal model with latency λ, a (small) message sent at time t renders the sender busy for the time interval $[t, t+1]$ and the receiver busy for the interval $[t + \lambda - 1, t + \lambda]$. Notice that this corresponds to the overhead o of the LogP model being 1 and the latency λ being identical to L.

The LogP model relies on short messages. However, several machines handle long messages much better than a sequence of short messages. In such cases, the LogP communication model is not very accurate. The LogGP model aims to remedy this by adding a parameter G, which captures the bandwidth for large messages. G corresponds to the gap per byte (or data item) for large messages. In the LogP model, sending a k item message takes time $o + (k - 1) * \max\{g, o\} + L + o$ cycles. In contrast, sending the message in the LogGP model (assuming the message is large) is given by $o + (k - 1)G + L + o$ cycles. As before, the sending and receiving processors are busy only during the o cycles of overhead and the rest of the time can be overlapped with useful computation. By reducing the overhead associated with large messages, LogGP rewards increased communication granularity.

These models suffer from drawbacks that stem from their inability to exploit structure in underlying communication patterns. They assume that all redistributions can be accomplished in time proportional to message size ($\Theta(m)$ for a m word permutation). However, there are no known algorithms that can perform the permutation in this time even on $\Theta(p)$ cross-section bandwidth networks without slack. (The transportation primitive of Shankar and Ranka [15] performs this permutation in $\Theta(m)$ but it requires a total slack of $\Theta(p)$.) Therefore, the cost model is a lower bound for data redistributions on all architectures. Now consider an adaptation of these models for a lower connectivity network such as a 2-D mesh. The effective bandwidth is scaled down by the cross-section bandwidth to account for the limited bandwidth of the mesh while performing generalized redistributions. However, aggregate communication operations such as grid-communication and broadcasts consist entirely of redistributions that do not congest the network and are therefore capable of extracting the same network performance from a mesh as from a dense network such as a hypercube. In these cases, the communication cost of both models is an upper bound. The communication time predicted by the models is therefore neither an upper bound nor a lower bound. To remedy these problems the LogP model itself suggests use of different values of gap g for different communication patterns, however, it does not provide details on how to do so.

Contention Modeling Abstractions The Queuing Shared Memory Model (QSM) of Ramachandran et al. [13] also follows a bulk synchronous format with phases of shared memory reads, shared memory writes, and local computation. The performance model of QSM explicitly accounts for remote memory accesses, contention, and congestion. The primary parameters in the QSM model include the number of processors p and the remote data access rate in terms of local instruction cycles g. If in a phase, a parallel algorithm executes m_{op} local memory operations and m_{rw} remote operations, with a contention of κ, QSM associates a cost $\max\{m_{op}, g.m_{rw}, \kappa\}$ with it. Minimizing this maximizes local data access and minimizes contention to shared data. In addition to these, QSM also has a set of secondary parameters - latency l, barrier time L, message sending overhead o, memory bank contention h_r, and network congestion c. The process of algorithm design can be largely simplified by considering only the primary parameters and optimizing the secondary factors using known techniques such as pipelining for latency hiding, randomization for bank conflicts, etc.

The C^3 model, proposed by Hambrusch and Khokhar [9] abstracts a p-processor machine in terms of the communication packet size l, the setup cost for a message s, communication latency h, and bisection width b. A parallel algorithm is assumed to proceed in supersteps of computation and communication. The cost of the computation phase of a superstep is determined by the maximum number of data items accessed by any processor (expressed in terms of number of packets). The cost of the communication phase is determined by whether the communication is blocking or non-blocking. A blocking communication of m packets is charged a time $2(s + h) + m + h$ at the sender and $s + 2h + m$ at the receiver. A non-blocking communication of m packets is charged a time $s + m + h$ at the sender and m at the receiver. Communication congestion is modeled in terms of network congestion and processor congestion. This is determined by the number of processor pairs communicating c, and the average number of packets r being communicated by any pair. The network congestion C_l is given by $C_l = (r \times c)/b$ and the processor congestion C_p by $C_p = (r \times c \times h)/p$. The overall communication cost is the maximum of sender time, receiver time, processor congestion, and network congestion.

While both of these models are refinements of BSP and LogP models, they suffer from many of the same drawbacks as LogP and BSP models.

The A^3 model [8] expresses communication cost as a combination of the architecture and underlying permutation. It classifies commonly used communication operations into those that are sensitive to the bisection bandwidth, and those that are not. It then uses the bisection width of the network to derive appropriate costs for aggregate communication operations. Algorithm analysis is reduced to the process of a table lookup for accurate or approximate cost associated with each operation. It is shown that this model has good coverage in algorithm in architecture space and has excellent predictive properties.

4 Concluding Remarks

In this paper, we presented a brief survey of prominent parallel algorithm design models. As is evident, in spite of extensive research, the search for a universal parallel algorithm design model has had only limited success. Advances in this area have tremendous significance, not just for parallel architectures, but also for serial processors with deep memory hierarchies.

References

1. A. Bar-Noy and S. Kipnis. Designing Broadcasting Algorithms in the Postal Model for Message-Passing Systems. *Proceedings of 4-th ACM Symp. on Parallel Algorithms and Architectures*, pp. 13-22, 1992.

2. Armin Bumker, Wolfgang Dittrich, Friedhelm Meyer auf der Heide. Truly Efficient Parallel Algorithms: 1-optimal Multisearch for an Extension of the BSP Model. *Theoretical Computer Science*, 203(2): 175-203, 1998.

3. A. Agarwal, A. K. Chandra, and M. Snir. Communication complexity of PRAMs. Technical Report RC 14998 (No.64644), IBM T.J. Watson Research Center, Yorktown Heights, NY, 1989.

4. A. Agarwal, A. K. Chandra, and M. Snir. On communication latency in PRAM computations. Technical Report RC 14973 (No.66882), IBM T.J. Watson Research Center, Yorktown Heights, NY, 1989.

5. D. Culler, R. Karp, D. Patterson, A. Sahay, K.E. Schauser, E. Santos, R. Subramonian, and T. von Eicken. LogP: Towards a Realistic Model of Parallel Computation. *Proceedings of 4-th ACM SIGPLAN Symp. on Principles and Practices of Parallel Programming*, pp. 1-12, 1993.

6. Pilar de la Torre and Clyde P. Kruskal. Towards a single model of efficient computation in real parallel machines. In *Future Generation Computer Systems*, pages 395 – 408, 8(1992).

7. F. Dehne, A. Fabri, and A. Rau-Chaplin. Scalable parallel computational geometry for coarse grained multicomputers. *International Journal on Computational Geometry*, Vol. 6, No. 3, 1996, pp. 379 - 400.

8. Ananth Grama, Vipin Kumar, Sanjay Ranka, and Vineet Singh. A^3: A simple and asymptotically accurate model for parallel computation. In *Proceedings of Conference on Frontiers of Massively Parallel Computing*, page 8 pp, Annapolis, MD, 1996.

9. S. Hambrusch and A. Khokhar. C^3 : A parallel model for coarse-grained machines. *Journal of Parallel and Distributed Computing*, 32(2):139–154, Feb 1996.

10. T. Heywood and S. Ranka. A practical hierarchical model of parallel computation. i. the model. *Journal of Parallel and Distributed Computing*, 16(3):212–32, Nov. 1992.

11. T. Heywood and S. Ranka. A practical hierarchical model of parallel computation. ii. binary tree and fft algorithms. *Journal of Parallel and Distributed Computing*, 16(3):233–49, Nov. 1992.

12. Vipin Kumar, Ananth Grama, Anshul Gupta, and George Karypis. *Introduction to Parallel Computing: Algorithm Design and Analysis*. Benjamin Cummings/ Addison Wesley, Redwod City, 1994.

13. Vijaya Ramachandran. QSM: A General Purpose Shared-Memory Model for Parallel Computation. *FSTTCS* 1-5, 1997.

14. A. G. Ranade. How to emulate shared memory. In *Proceedings of the 28th IEEE Annual Symposium on Foundations of Computer Science*, pages 185 – 194, 1987.

15. R. Shankar and S. Ranka. Random data access on a coarse grained parallel machine ii. one-to-many and many-to-one mappings. Technical report, School of Computer and Information Science, Syracuse University, Syracuse, NY, 13244, 1995.

16. L. G. Valiant. A bridging model for parallel computation. *Communications of the ACM*, 33(8), 1990.

17. L. G. Valiant. General purpose parallel architectures. *Handbook of Theoretical Computer Science*, 1990.

18. J. S. Vitter and E. A. M. Shriver. Algorithms for Parallel Memory II: Hierarchical Multilevel Memories. newblock *Algorithmica*, 12(2-3), 1994, 148-169.

19. T. von Eicken, D. E. Culler, S. C. Goldstein, and K. E. Schauser. Active messages: a mechanism for integrated communication and computation. In *Proceedings of the 19th International Symposium on Computer Architecture*, 1992.

Strong Fault-Tolerance:
Parallel Routing in Networks with Faults[*]

Jianer Chen and Eunseuk Oh

Department of Computer Science, Texas A&M University,
College Station, TX 77843-3112, USA
Email: {chen,eunseuko}@cs.tamu.edu

Abstract. A d-regular network G is *strongly fault-tolerant* if for any copy G_f of G with at most $d - 2$ faulty nodes, every pair of non-faulty nodes u and v in G_f admits $\min\{d_f(u), d_f(v)\}$ node-disjoint paths in G_f from u to v, where $d_f(u)$ and $d_f(v)$ are the degrees of the nodes u and v in G_f. We show that popular network structures, such as hypercube and star networks, are strongly fault-tolerant. We develop efficient algorithms that construct the maximum number of node-disjoint paths of optimal or nearly optimal length in these networks with faulty nodes. Our algorithms are optimal in terms of their running time.

1 Introduction

Routing on large size networks with faults is an important issue in the study of computer interconnection networks. In this paper, we introduce a new measure for network fault tolerance: the *strong fault-tolerance*. Consider a network G and let G_f be a copy of G with a set S_f of faulty nodes. Let u and v be two non-faulty nodes in G_f. Based on local information, we know the degrees $d_f(u)$ and $d_f(v)$ of the nodes u and v in G_f and are interested in constructing the maximum number of node-disjoint paths between u and v in G_f. Obviously, the number of node-disjoint paths between u and v in G_f cannot be larger than $\min\{d_f(u), d_f(v)\}$. We are interested in knowing the precise bound on the size of the faulty node set S_f such that for any two non-faulty nodes u and v in G_f, there are $\min\{d_f(u), d_f(v)\}$ node-disjoint paths between u and v.

Clearly, if the network G is d-regular (i.e., all of its nodes have degree d), then in general the number of faulty nodes in the set S_f should not exceed $d - 2$ to ensure $\min\{d_f(u), d_f(v)\}$ node-disjoint paths between any two nodes u and v in G_f. This can be seen as follows. Let u and v be two nodes in G whose distance is larger than 3. Pick any neighbor u' of u and remove the $d - 1$ neighbors of u' that are not u. Note that no neighbor of u' can be a neighbor of v since the distance from u to v is at least 4. Let the resulting network be G_f. The degrees of the nodes u and v in G_f are d. However, there are obviously no d node-disjoint paths in G_f from u to v since one of the d neighbors of u in G_f, the node u', leads to a "dead-end". This motivates the following definition.

---

[*] This work is supported in part by NSF under Grant CCR-0000206.

Definition 1. A d-regular network G is *strongly fault-tolerant* if for any copy G_f of G with at most $d - 2$ faulty nodes, every pair of non-faulty nodes u and v admits $\min\{d_f(u), d_f(v)\}$ node-disjoint paths from u to v in G_f.

Strong fault-tolerance characterizes the property of parallel routing in a network with faulty nodes. Since one of the motivations of network parallel routing is to provide alternative routing paths when failures occur, strong fault-tolerance can also be regarded as the study of fault tolerance in networks with faults.

To authors' knowledge, there has not been a systematic study on parallel routing on networks with faults. In this paper, we will concentrate on the discussion on strong fault-tolerance of two extensively studied interconnection network structures, the hypercube networks and the star networks. We first give a brief review on the previous related research on these networks.

The hypercube networks are among the earliest and still remain as one of the most important and attractive network models. A large number of fault-tolerant algorithms dealing with single-path routing in the hypercube networks have been proposed (see, for example, [8, 7, 12]). Parallel routing on hypercube networks without faulty nodes was studied in [18]. The general fault tolerance properties of the star networks were first studied and analyzed in [1–3]. Algorithms for single-path routing in star networks with faults were developed in [4, 11, 17]. Parallel routing algorithms on the star networks were studied in [10, 9]. In particular, an efficient algorithm has been developed [6] that constructs the maximum number of node-disjoint paths of optimal length for any two nodes in the star networks. A randomized algorithm, based on the Information Dispersal Algorithm [15], for parallel routing in the star networks with faults was proposed in [16].

We will first study the strong fault-tolerance for the star networks. Taking the advantage of the *orthogonal decomposition* of the star networks, we develop an efficient algorithm that constructs node-disjoint paths between any two non-faulty nodes in the n-star network S_n with at most $n - 3$ faulty nodes: for any two non-faulty nodes u and v, our algorithm constructs $\min\{d_f(u), d_f(v)\}$ node-disjoint paths of minimum length plus a small constant between u and v.

Since the hypercube networks do not have a similar orthogonal decomposition structure, the techniques in parallel routing for the star networks with faults are not applicable to the hypercube networks. In order to effectively route parallel paths in the hypercube networks with faults, we develop a new technique called "neighbor pre-matching". Based on this technique, an algorithm is developed that constructs $\min\{d_f(u), d_f(v)\}$ node-disjoint paths of optimal length for any pair of nodes u and v in the n-cube network with at most $n - 2$ faulty nodes.

Our algorithms have optimal running time.

2 Strong fault-tolerance of the star networks

The *n-star network* S_n is an undirected graph consisting of $n!$ nodes labeled with the $n!$ permutations on symbols $\{1, 2, \ldots, n\}$. There is an edge between two nodes u and v in S_n if and only if the permutation v can be obtained from the permutation u by exchanging the positions of the first symbol and another symbol. The n-star network is $(n-1)$-regular. The star networks have received considerable attention as an attractive alternative to the widely used hypercube network model [2].

Each node u in the n-star network S_n, which is a permutation on the symbols $\{1, 2, \ldots, n\}$, can be given as a product of disjoint cycles (called the *cycle structure* for u) [5]. For each $2 \leq i \leq n$, an operation ρ_i on the permutations of $\{1, 2, \ldots, n\}$ is defined such that $\rho_i(u)$ is the permutation obtained from the permutation u by exchanging the first symbol and the ith symbol in u.

Denote by ε the identity permutation $\langle 12 \cdots n \rangle$. Since the n-star network S_n is vertex-symmetric [2], a set of node-disjoint paths from a node w to a node v can be easily mapped to a set of node-disjoint paths from a node u to ε. Therefore, we only need to concentrate on the construction of node-disjoint paths from u to ε in S_n. Let $dist(u)$ denote the distance, i.e., the length of the shortest paths, from u to ε in S_n. The value $dist(u)$ can be easily computed using the formula given in [2]. We say that an edge $[u, v]$ (in this direction) in S_n *does not follow the Shortest Path Rules* if $dist(u) \leq dist(v)$ [6]. It is also easy to check whether an edge follows the Shortest Path Rules [6].

Lemma 1. *If an edge $[u, v]$ in S_n does not follow the Shortest Path Rules, then $dist(v) = dist(u) + 1$. Consequently, let P be a path from u to ε in which exactly k edges do not follow the Shortest Path Rules, then the length of the path P is equal to $dist(u) + 2k$.*

For the n-star network S_n, let $S_n[i]$ be the set of nodes in which the symbol 1 is at the ith position. It is well-known [1] that the set $S_n[1]$ is an independent set, and the subgraph induced by the set $S_n[i]$ for $i \neq 1$ is an $(n-1)$-star network.

Our parallel routing algorithm is heavily based on the concept of *bridging paths* that connect a given node to a specific substar network in the n-star network. The following lemma will serve as a basic tool in our construction.

Lemma 2. *Let u be any non-faulty node in the substar $S_n[i]$ with $k_i \leq n - 3$ faulty nodes, $i \neq 1$. A fault-free path P from u to $\rho_i(\varepsilon)$ can be constructed in $S_n[i]$ in time $O(k_i n + n)$ such that at most two edges in P do not follow the Shortest Path Rules. In case u has a cycle of form $(i1)$, the constructed path P has at most one edge not following the Shortest Path Rules.*

Definition 2. Let u be a node in the n-star network S_n and u' be a neighbor of u in the substar $S_n[i]$, $i \neq 1$. For each neighbor v of u', $v \neq u$, a (u', j)-*bridging path* (of length at most 4) from u to the substar $S_n[j]$, $j \neq 1, i$, is defined as follows: if v is in $S_n[1]$ then the path is $[u, u', v, \rho_j(v)]$, while if v is in $S_n[i]$ then the path is $[u, u', v, \rho_i(v), \rho_j(\rho_i(v))]$.

Thus, from each neighbor u' in $S_n[i]$ of the node u, $i \neq 1$, there are $n - 2$ (u', j)-bridging paths of length bounded by 4 that connect the node u to the substar $S_n[j]$.

Since no two nodes in $S_n[i]$ share the same neighbor in $S_n[1]$ and no two nodes in $S_n[1]$ share the same neighbor in $S_n[j]$, for any neighbor u' of u, two

(u', j)-bridging paths from u to $S_n[j]$ have only the nodes u and u' in common. Moreover, for any two neighbors u' and u'' of u in $S_n[i]$ (in this case, the node u must itself be also in $S_n[i]$), since u' and u'' have no other common neighbor except u (see, for example, [9]), a (u', j)-bridging path from u to $S_n[j]$ and a (u'', j)-bridging path from u to $S_n[j]$ share no nodes except u.

Definition 3. Let u be a node in S_n and let u' be a neighbor of u in $S_n[i]$, $i \neq 1$. A (u', j)-bridging path P from the node u to the substar $S_n[j]$ is *divergent* if in the subpath of P from u to $S_n[1]$, there are three edges not following the Shortest Path Rules.

Note that the subpath from u to $S_n[1]$ of a (u', j)-bridging path P contains at most three edges. In particular, if the subpath contains only two edges, then the path P is automatically non-divergent.

In case there are no faulty nodes in the n-star network, each divergent (u', j)-bridging path can be efficiently extended into a path from u to $\rho_j(\varepsilon)$, as shown in the following lemma. A detailed proof for the lemma can be found in [13].

Star-PRouting
Input: a non-faulty node u in S_n with at most $n - 3$ faulty nodes.
Output: $\min\{d_f(u), d_f(\varepsilon)\}$ parallel paths of length $\leq dist(u) + 8$ from u to ε.
1. **if** the node u is in $S_n[1]$
1.1. **then for** each $j \neq 1$ with both $\rho_j(u)$ and $\rho_j(\varepsilon)$ non-faulty **do**
        construct a path P_j of length $\leq dist(u) + 6$ in $S_n[j]$ from u to ε;
1.2. **else** (* the node u is in a substar $S_n[i]$, $i \neq 1$ *)
1.2.1. **if** the node $\rho_i(\varepsilon)$ is non-faulty
    then pick a non-faulty neighbor v of u and construct a path P_v of
        length $\leq dist(u) + 4$ from u to ε such that all internal nodes
        of P_v are in $S_n[i]$ and P_v does not intersect a (u', j)-bridging
        path for any non-faulty neighbor $u' \neq v$ of u;
1.2.2. **if** the neighbor $u_1 = \rho_i(u)$ of u in $S_n[1]$ is non-faulty
    then find $j \neq 1, i$, such that both $\rho_j(u_1)$ and $\rho_j(\varepsilon)$ are non-faulty;
        extend the path $[u, u_1, \rho_j(u_1)]$ from $\rho_j(u_1)$ to $\rho_j(\varepsilon)$ in $S_n[j]$
        to make a path of length $\leq dist(u) + 8$ from u to ε;
2. maximally pair the non-faulty neighbors of u and ε that are not used
   in step 1: $(u'_1, \rho_{j_1}(\varepsilon)), \ldots, (u'_g, \rho_{j_g}(\varepsilon))$;
3. **for** each pair $(u', \rho_j(\varepsilon))$ constructed in step 2 **do**
3.1. **if** there is a non-divergent (u', j)-bridging path P with neither
        faulty nodes nor nodes used by other paths
    then pick this (u', j)-bridging path P
    else pick a divergent (u', j)-bridging path P with neither faulty
        nodes nor nodes used by other paths;
3.2. extend the path P into a fault-free path $P_{u'}$ of length $\leq dist(u) + 8$
    from u to ε such that the extended part is entirely in $S_n[j]$;

Fig. 1. Parallel routing on the star network with faulty nodes

Lemma 3. *There is an $O(n)$ time algorithm that, for a divergent (u', j)-bridging path P from a node u to a substar $S_n[j]$, extends P in $S_n[j]$ into a path Q from u to $\rho_j(\varepsilon)$, in which at most 4 edges do not follow the Shortest Path Rules. Moreover, for two divergent (u', j)-bridging paths P_1 and P_2, the two corresponding extended paths Q_1 and Q_2 have only the nodes u, u', and $\rho_j(\varepsilon)$ in common.*

Based on the above discussion, we are now ready to present our parallel routing algorithm on star networks with faults, which is given in Figure 1.

Step 1 of the algorithm constructs certain number of paths between non-faulty neighbors of the node u and non-faulty neighbors of the node ε, Step 2 of the algorithm maximally pairs the rest non-faulty neighbors of u with the rest non-faulty neighbors of ε. It is easy to see that the number of pairs constructed in Step 2 plus the number of paths constructed in Step 1 is exactly $\min\{d_f(u), d_f(\varepsilon)\}$. Since Step 3 of the algorithm constructs a path from u to ε for each pair constructed in Step 2, the algorithm **Star-PRouting** constructs exactly $\min\{d_f(u), d_f(\varepsilon)\}$ paths from u to ε. In fact, we can prove the following theorem (a detailed proof of the theorem can be found in [13]).

Theorem 1. *If the n-star network S_n has at most $n - 3$ faulty nodes and the node ε is non-faulty, then for any non-faulty node u in S_n, in time $O(n^2)$ the algorithm **Star-PRouting** constructs $\min\{d_f(u), d_f(\varepsilon)\}$ node-disjoint fault-free paths of length bounded by $dist(u) + 8$ from the node u to the node ε.*

The following example shows that the bound on the path length in Theorem 1 is actually almost optimal. Consider the n-star network S_n. Let the source node be $u = (21)$, here we have omitted the trivial cycles in the cycle structure. Then $dist(u) = 1$. Suppose that all neighbors of u and all neighbors of ε are non-faulty. By Theorem 1, there are $n - 1$ node-disjoint fault-free paths from u to ε. Thus, for each i, $3 \leq i \leq n$, the edge $[u, u_i]$ leads to one P_i of these node-disjoint paths from u to ε, where $u_i = (i21)$. Note that the edge $[u, u_i]$ does not follow the Shortest Path Rules. Now suppose that the node $(i2)(1)$ is faulty, for $i = 3, 4, \ldots, n - 1$ (so there are $n - 3$ faulty nodes). Then the third node on the path P_i must be $v_i = (ji21)$ for some $j \neq 1, 2, i$, and the edge $[u_i, v_i]$ does not follow the Shortest Path Rules. Since the only edge from v_i that follows the Shortest Path Rules is the edge $[v_i, u_i]$, the next edge $[v_i, w_i]$ on P_i again does not follow the Shortest Path Rules. Now since all the first three edges on P_i do not follow the Shortest Path Rules, by Lemma 1, $dist(w_i) = dist(u) + 3 = 4$, and the path P_i needs at least four more edges to reach ε. That is, the length of the path P_i is at least $7 = dist(u) + 6$. Thus, with $n - 3$ faulty nodes, among the $n - 1$ node-disjoint paths from u to ε, at least $n - 3$ of them must have length larger than or equal to $dist(u) + 6$.

The situation given above seems a little special since the distance $dist(u)$ from u to ε is very small. In fact, even for nodes u with large $dist(u)$, we can still construct many examples in which some of the node-disjoint fault-free paths connecting u and ε must have length at least $dist(u) + 6$ (see [13] for details).

3   Strong fault-tolerance of the hypercube networks

The construction of parallel routing paths in the star networks in the previous section heavily takes the advantage of the orthogonal decomposition structure $\{S_n[1], S_n[2], \ldots, S_n[n]\}$ of the star networks. In particular, the (u', j)-bridging paths route a node u through the substar $S_n[1]$ to the substar $S_n[j]$ so that extension of the paths can be recursively constructed in the substar $S_n[j]$. Unfortunately, the hypercube networks do not have a similar orthogonal decomposition structure so the above techniques are not applicable. We need to develop new techniques for parallel routing in the hypercube networks with faults. Because of space limit, some details are omitted. Interested readers are referred to [14].

Recall that an n-cube network Q_n is an undirected graph consisting of 2^n nodes labeled by the 2^n binary strings of length n. Two nodes are adjacent if they differ by exactly one bit. An edge is an i-edge if its two ends differ by the ith bit. The (Hamming) distance $dist(u, v)$, i.e., the length of the shortest paths, from u to v is equal to the number of the bits in which u and v differ. Since the n-cube network Q_n is vertex-symmetric, we can concentrate, without loss of generality, on the construction of node-disjoint paths from a node of form $u = 1^r 0^{n-r}$ to the node $\varepsilon = 0^n$. Define $dist(u) = dist(u, \varepsilon)$.

The node connected from the node u by an i-edge is denoted by u_i, and the node connected from the node u_i by a j-edge is denoted by $u_{i,j}$. A path P from u to v can be uniquely specified by a sequence of labels of the edges on P in the order of traversal, denoted by $u\langle i_1, \ldots, i_r \rangle v$. This notation can be further extended to a set of paths. Thus, for a set S of permutations, $u\langle S \rangle v$ is the set of paths of the form $u\langle i_1, \ldots, i_r \rangle v$, where $\langle i_1, \ldots, i_r \rangle$ is a permutation in S.

Our parallel routing algorithm is based on an effective pairing of the neighbors of the nodes u and ε. We first assume that the nodes u and ε have no faulty neighbors. We pair the neighbors of u and ε by the following strategy.

prematch-1
pair u_i with ε_{i-1} for $1 \leq i \leq r$,[1] and pair u_j with ε_j for $r + 1 \leq j \leq n$.

Under the pairing given by **prematch-1**, we construct parallel paths between the paired neighbors of u and ε using the following procedure.

procedure-1
1. for each $1 \leq i \leq r$, construct $n - 2$ paths from u_i to ε_{i-1}, $r - 2$ of them are of the form $u_i \langle S_1 \rangle \varepsilon_{i-1}$, where S_1 is the set of all cyclic permutations of the sequence $(i + 1, \ldots, r, 1, \ldots, i - 2)$; and $n - r$ of them are of the form $u_i \langle h, i + 1, \ldots, r, 1, \ldots, i - 2, h \rangle \varepsilon_{i-1}$, where $r + 1 \leq h \leq n$.
2. for each $r + 1 \leq j \leq n$, construct $n - 1$ paths from u_j to ε_j, r of them are of the form $u_j \langle S_2 \rangle \varepsilon_j$, where S_2 is the set of all cyclic permutations of the sequence $(1, 2, \ldots, r)$, and $n - r - 1$ of them are of the form $u_j \langle h, 1, 2, \ldots, r, h \rangle \varepsilon_j$, where $h \neq j$, and $r + 1 \leq h \leq n$.

---

[1] The operations on indices are by mod r. Thus, ε_0 is interpreted as ε_r.

The paths constructed by cyclic permutations of a sequence are pairwisely disjoint(see, for example [18]). It is easy to verify that for each pair of neighbors of u and ε, the paths constructed between them are pairwisely disjoint.

Lemma 4. *Let (u_h, ε_g) and (u_s, ε_t) be two pairs given by* **prematch-1**. *Then, there is at most one path in the path set constructed by* **precedure-1** *for the pair (u_h, ε_g) that shares common nodes with a path in the path set constructed by* **precedure-1** *for the pair (u_s, ε_t).*

Since the n-cube network Q_n may have up to $n - 2$ faulty nodes, there is a possibility that for a pair (u_i, ε_{i-1}) constructed by **prematch-1**, where $1 \leq i \leq r$, all $n - 2$ paths constructed by **precedure-1** from u_i to ε_{i-1} are blocked by faulty nodes. In this case, we can directly construct n node-disjoint paths from u to ε, as follows.

> **Assumption.** there is a pair (u_i, ε_{i-1}) given by **prematch-1**, $1 \leq i \leq r$, such that all $n - 2$ paths constructed by **prematch-1** for the pair are blocked by faulty nodes.
>
> **prematch-2**
> 1. pair u_i with ε_{i-2}, u_{i-1} with ε_i, and u_{i+1} with ε_{i-1};
> 2. for other neighbors of u and ε, pair them as in **prematch-1**;
>
> **procedure-2**
> 1. from u_i to ε_{i-2}: $u_i \langle i - 1, \ldots, r, 1, \ldots, i - 3 \rangle \varepsilon_{i-2}$;
> 2. from u_{i-1} to ε_i: $u_{i-1} \langle i + 1, \ldots r, 1, \ldots, i - 2 \rangle \varepsilon_i$;
> 3. from u_{i+1} to ε_{i-1}: $u_{i+1} \langle i + 2, \ldots, r, 1, \ldots, i - 2, i \rangle \varepsilon_{i-1}$;
> 4. for $g \neq i - 1, i, i + 1$, $1 \leq g \leq r$: $u_g \langle g + 1, \ldots, r, 1, \ldots, g - 2 \rangle \varepsilon_{g-1}$;
> 5. for $r + 1 \leq j \leq n$: if $i = 1$ then $u_j \langle 2, \ldots, r, 1 \rangle \varepsilon_j$ else $u_j \langle 1, \ldots, r \rangle \varepsilon_j$.

Lemma 5. *Under the conditions of* **prematch-2**, *the algorithm* **procedure-2** *constructs n fault-free parallel paths of length $\leq dist(u) + 4$ from u to ε.*

Now we consider the case in which the nodes u and ε have faulty neighbors. We apply the following pairing strategy, which pairs the edges incident on the neighbors of the nodes u and ε, instead of the neighbors of u and ε.

> **prematch-3**
> for each edge $[u_i, u_{i,i'}]$ where both u_i and $u_{i,i'}$ are non-faulty **do**
> 1. if $1 \leq i, i' \leq r$ and $i' = i + 1$, then pair $[u_i, u_{i,i'}]$ with the edge $[\varepsilon_{i-1,i-2}, \varepsilon_{i-1}]$, if $\varepsilon_{i-1,i-2}$ and ε_{i-1} are non-faulty;
> 2. if $1 \leq i, i' \leq r$ and $i' = i - 1$, then pair $[u_i, u_{i,i'}]$ with the edge $[\varepsilon_{i'-1,i'-2}, \varepsilon_{i'-1}]$, if $\varepsilon_{i'-1,i'-2}$ and $\varepsilon_{i'-1}$ are non-faulty;
> 3. otherwise, pair $[u_i, u_{i,i'}]$ with the edge $[\varepsilon_{j,j'}, \varepsilon_j]$ if $\varepsilon_{j,j'}$ and ε_j are non-faulty, where the indices j and j' are such that **prematch-1** pairs the node $u_{i'}$ with ε_j, and the node u_i with $\varepsilon_{j'}$.

Note that it is possible that an edge $[u_i, u_{i,i'}]$ with both u_i and $u_{i,i'}$ non-faulty is not paired with any edge because the corresponding edge in **prematch-3** contains faulty nodes. For each pair of edges given by **prematch-3**, we construct a path as follows.

procedure-3

1. for $1 \leq i, i' \leq r$, $i' = i+1$, and paired edges $[u_i, u_{i,i'}]$, $[\varepsilon_{i-1,i-2}, \varepsilon_{i-1}]$, construct the path $u_i \langle i', \ldots i - 2 \rangle \varepsilon_{i-1}$;

2. for $i \leq i, i' \leq r$, $i' = i-1$, and paired edges $[u_i, u_{i,i'}]$, $[\varepsilon_{i'-1,i'-2}, \varepsilon_{i'-1}]$, construct a path by flipping i and i' in the path $u_{i'} \langle i, \ldots, i' - 2 \rangle \varepsilon_{i'-1}$;

3. Otherwise, for paired edges $[u_i, u_{i,i'}]$, $[\varepsilon_{j,j'}, \varepsilon_j]$, if $i < i'$, construct a path by flipping j and j' in the path $u_i \langle i', \ldots j \rangle \varepsilon_{j'}$; if $i > i'$, construct a path by flipping i and i' in the path $u_{i'} \langle i, \ldots j' \rangle \varepsilon_j$.

Now we are ready to present our main algorithm for parallel routing in the hypercube networks with faults. The algorithm is given in Figure 2.

Cube-PRouting

input: non-faulty nodes $u = 1^r 0^{n-r}$ and $\varepsilon = 0^n$ in Q_n with $\leq n - 2$ faults.

output: $\min\{d_f(u), d_f(\varepsilon)\}$ parallel paths of length $\leq dist(u) + 4$ from u to ε.

1. **case 1.** u and ε have no faulty neighbors

    for each pair (u_i, ε_j) given by **prematch-1 do**

1.1.   **if** all paths for (u_i, ε_j) by **procedure-1** include faulty nodes
        then use **prematch-2** and **procedure-2** to construct n parallel
        paths from u to ε; STOP.

1.2.   **if** there is a fault-free unoccupied path from u_i to ε_j by **procedure-1**
        then mark the path as occupied by (u_i, ε_j);

1.3.   **if** all fault-free paths constructed for (u_i, ε_j) include occupied nodes
        then pick any fault-free path P for (u_i, ε_j), and for the pair $(u_{i'}, \varepsilon_{j'})$
        that occupies a node on P, find a new path;

2. **case 2.** there is at least one faulty neighbor of u or ε

    for each edge pair $([u_i, u_{i,i'}], [\varepsilon_{j,j'}, \varepsilon_j])$ by **prematch-3 do**

2.1.   **if** there is a fault-free unoccupied path from u_i to ε_j by **procedure-3**
        then mark the path as occupied by the pair $([u_i, u_{i,i'}], [\varepsilon_{j,j'}, \varepsilon_j])$;

2.2.   **if** all fault-free paths for the pair include occupied nodes
        then pick any fault-free path P for the edge pair, and for the edge
        pair that occupies a node on P, find a new path.

Fig. 2. Parallel routing on the hypercube network with faulty nodes

Lemma 5 guarantees that step 1.1 of the algorithm **Cube-PRouting** constructs n fault-free parallel paths of length $\leq dist(u) + 4$ from u to ε. Step 1.3 of the algorithm requires further explanation. In particular, we need to show that for the pair $(u_{i'}, \varepsilon_{j'})$, we can always construct a new fault-free path from $u_{i'}$ to $\varepsilon_{j'}$ in which no nodes are occupied by other paths. This is ensured by the following lemma.

Lemma 6. *Let (u_i, v_j) and $(u_{i'}, v_{j'})$ be two pairs given by **prematch-1** such that two paths constructed for (u_i, v_j) and $(u_{i'}, v_{j'})$ share a node. Then the algorithm **Cube-PRouting** can always find fault-free paths for (u_i, v_j) and $(u_{i'}, v_{j'})$, in which no nodes are occupied by other paths.*

A similar analysis shows that step 2.2 of the algorithm **Cube-PRouting** can always construct a new fault-free path without nodes occupied by other paths. Let us summarize all these discussions in the following theorem.

Theorem 2. *If the n-cube network Q_n has at most $n - 2$ faulty nodes, then for each pair of non-faulty nodes u and v in Q_n, in time $O(n^2)$ the algorithm* **Cube-PRouting** *constructs $\min\{d_f(u), d_f(v)\}$ node-disjoint fault-free paths of length bounded by $dist(u,v) + 4$ from u to v.*

4   Conclusion

Network strong fault-tolerance is a natural extension of the study of network fault tolerance and network parallel routing. In particular, it studies the fault tolerance of large size networks with faulty nodes. In this paper, we have demonstrated that the popular interconnection networks, such as the hypercube networks and the star networks, are strongly fault-tolerant. We developed algorithms of running time $O(n^2)$ that for two given non-faulty nodes u and v in the networks, constructs the maximum number (i.e., $\min\{d_f(u), d_f(v)\}$) of node-disjoint fault-free paths from u to v such that the length of the paths is bounded by $dist(u,v) + 8$ for the star networks, and bounded by $dist(u,v) + 4$ for the hypercube networks. The time complexity of our algorithms is optimal since each path from u to v in the network S_n or Q_n may have length as large as $\Theta(n)$, and there can be as many as $\Theta(n)$ node-disjoint paths from u to v. Thus, even printing these paths should take time $O(n^2)$. We have shown that the length of the paths constructed by our algorithm for the star networks is almost optimal. For the n-cube network Q_n, the length of the paths constructed by our algorithm is bounded by $dist(u,v) + 4$. It is not difficult to see that this is the best possible, since there are node pairs u and v in Q_n with $n - 2$ faulty nodes, for which any group of $\min\{d_f(u), d_f(v)\}$ parallel paths from u to v contains at least one path of length at least $dist(u,v) + 4$.

We should mention that Rescigno [16] recently developed a randomized parallel routing algorithm on star networks with faults, based on the Information Disersal Algorithm (IDA) [15]. The algorithm in [16] is randomized thus it does not always guarantee the maximum number of node-disjoint paths. Moreover, in terms of the length of the constructed paths and running time of the algorithms, our algorithms seem also to have provided significant improvements.

References

1. S. B. AKERS, D. HAREL, AND B. KRISHNAMURTHY, The star graph: an attractive alternative to the n-cube, *Proc. Intl. Conf. of Parallel Proc.*, (1987), pp. 393-400.
2. S. B. AKERS AND B. KRISHNAMURTHY, A group-theoretic model for symmetric interconnection networks, *IEEE Trans. on Computers 38*, (1989), pp. 555-565.
3. S. B. AKERS AND B. KRISHNAMURTHY, The fault tolerance of star graphs, *Proc. 2nd International Conference on Supercomputing*, (1987), pp. 270-276.

4. N. BAGHERZADEH, N. NASSIF, AND S. LATIFI, A routing and broadcasting scheme on faulty star graphs, *IEEE Trans. on Computers 42*, (1993), pp. 1398-1403.

5. G. BIRKHOFF AND S. MACLANE, *A Survey of Modern Algebra*, The Macmillan Company, New York, 1965.

6. C. C. CHEN AND J. CHEN, Optimal parallel routing in star networks, *IEEE Trans. on Computers 46*, (1997), pp. 1293-1303.

7. M.-S. CHEN AND K. G. SHIN, Adaptive fault-tolerant routing in hypercube multi-computers, *IEEE Trans. Computers 39* (1990), pp. 1406-1416.

8. G. -M. CHIU AND S. -P. WU, A fault-tolerant routing strategy in hypercube multicomputers, *IEEE Trans. Computers 45* (1996), pp. 143-154.

9. K. DAY AND A. TRIPATHI, A comparative study of topological properties of hypercubes and star graphs, *IEEE Trans. Parallel, Distrib. Syst. 5*, (1994), pp. 31-38.

10. M. DIETZFELBINGER, S. MADHAVAPEDDY, AND I. H. SUDBOROUGH, Three disjoint path paradigms in star networks, *Proc. 3rd IEEE Symposium on Parallel and Distributed Processing*, (1991), pp. 400-406.

11. Q.-P. GU AND S. PENG, Fault tolerant routing in hypercubes and star graphs, *Parallel Processing Letters 6*, (1996), pp. 127-136.

12. T. C. LEE AND J. P. HAYES, Routing and broadcasting in faulty hypercube computers, *Proc. 3rd Conf. Hypercube Concurrent Computers and Applications* (1988), pp. 625-630.

13. E. OH AND J. CHEN, Strong fault-tolerance: parallel routing in star networks with faults *Tech. Report*, Dept. Computer Science, Texas A&M University (2001).

14. E. OH AND J. CHEN, Parallel routing in hypercube networks with faulty nodes, *Tech. Report*, Dept. Computer Science, Texas A&M University (2001).

15. M. O. RABIN, Efficient dispersal of information for security, load balancing, and fault tolerance, *Journal of ACM 36*, (1989), pp. 335-348.

16. A. A. RESCIGNO, Fault-tolerant parallel communication in the star network, *Parallel Processing Letters 7*, (1997), pp. 57-68.

17. A. A. RESCIGNO AND U. VACCARO, Highly fault-tolerant routing in the star and hypercube interconnection networks, *Parallel Processing Letters 8*, (1998), pp. 221-230.

18. Y. SAAD, M. H. SCHULTZ, Topological properties of hypercubes, *IEEE Transactions on Computers, 37*, (1988), pp. 867-872.

Parallel Algorithm Design
with Coarse-Grained Synchronization

Vijaya Ramachandran*

Department of Computer Sciences
The University of Texas at Austin
Austin, TX 78712
vlr@cs.utexas.edu

Abstract. We describe the Queuing Shared-Memory (QSM) and Bulk-Synchronous Parallel (BSP) models of parallel computation. The former is shared-memory and the latter is distributed-memory. Both models use the 'bulk-synchronous' paradigm introduced by the BSP model. We describe the relationship of these two models to each other and to the 'LogP' model, and give an overview of algorithmic results on these models.

Keywords: Parallel algorithms; general-purpose parallel computation models; bulk-synchronous models.

1 Introduction

An important goal in parallel processing is the development of general-purpose parallel models and algorithms. However, this task has not been an easy one. The challenge here has been to find the right balance between simplicity, accuracy and broad applicability.

Most of the early work on parallel algorithm design has been on the simple and influential *Parallel Random Access Machine (PRAM)* model (see e.g., [29]). Most of the basic results on parallelism in algorithms for various fundamental problems were developed on this simple shared-memory model. However, this model ignores completely the latency and bandwidth limitations of real parallel machines, and it makes the unrealistic assumption of unit-time global synchronization after every fine-grained parallel step in the computation. Hence algorithms developed using the PRAM model typically do not map well to real machines. In view of this, the design of general-purpose models of parallel computation has been an important topic of study in recent years [3, 5–7, 10, 13, 15, 18, 25, 30, 32, 35, 36, 39, 44]. However, due to the diversity of architectures among parallel machines, this has also proved to be a very challenging task. The challenge here has been to find a model that is general enough to encompass the

---

* This work was supported in part by NSF Grant CCR-9988160 and Texas Advanced Research Program Grant 3658-0029-1999.

wide variety of parallel machines available, while retaining enough of the essential features of these diverse machines in order to serve as a reasonably faithful model of them.

In this paper we describe the approach taken by the *Queuing Shared-Memory (QSM) model* [22] and the *Bulk-Synchronous Parallel (BSP) model* [46]. The former is shared-memory while the latter is distributed-memory. The two models are distinguished by their use of *bulk synchronization* (which was proposed in [46]).

Bulk-synchronization moves away from the costly overhead of the highly synchronous PRAM models on one hand, and also away from the completely asynchronous nature of actual machines, which makes the design of correct algorithms highly nontrivial. The QSM and BSP are algorithmic/programming models that provide coarse-grained synchronization in a manner that facilitates the design of correct algorithms that achieve good performance.

In the following sections, we define the QSM and BSP models, and provide an overview of known relationships between the models (and between both models and the *LogP* model [14]) as well as some algorithmic results on these models.

2 Definitions of Models

In the following we review the definitions of the BSP [46], LogP [14], and QSM [22] models. These models attempt to capture the key features of real machines while retaining a reasonably high-level programming abstraction. Of these models, the QSM is the simplest because it has only 2 parameters and is shared-memory, which is generally more convenient than message passing for developing parallel algorithms. On the other hand the LogP is more of a performance evaluation model than a model for parallel algorithm design, but we include it here since it is quite similar to the BSP model.

BSP Model. The Bulk-Synchronous Parallel (BSP) model [46, 47] consists of p processor/memory components that communicate by sending point-to-point messages. The interconnection network supporting this communication is characterized by a bandwidth parameter g and a latency parameter L. A BSP computation consists of a sequence of "supersteps" separated by global synchronizations. Within each superstep, each processor can send and receive messages and perform local computation, subject to the constraint that messages are sent based only on the state of the processor at the start of the superstep.

Let w be the maximum amount of local work performed by andy processor in a given superstep and let h be the maximum number of messages sent or received by any processor in the superstep; the BSP is said to route an h-*relation* in this superstep. The *cost*, T, of the superstep is defined to be $T = \max(w, g \cdot h, L)$. The time taken by a BSP algorithm is the sum of the costs of the individual supersteps in the algorithm.

The (d, x)-BSP [11] is similar to the BSP, but it also attempts to model memory bank contention and delay in shared-memory systems. The (d, x)-BSP is parameterized by five parameters, p, g, L, d and x, where p, g and L are as in the original BSP model, the *delay* d is the 'gap' parameter at the memory banks, and the *expansion* x reflects the number of memory banks per processor (i.e., there are $x \cdot p$ memory banks).

The computation of a (d, x)-BSP proceeds in supersteps similar to the BSP. In a given superstep, let h_s be the maximum number of read/write requests made by any processor, and let h_r be the maximum number of read/write requests to any memory bank. Then the cost of the superstep is $\max(w, g \cdot h_s, d \cdot h_r, L)$, where w is as in the BSP.

The original BSP can be viewed as a (d, x)-BSP with $d = g$ and $x = 1$.

LogP Model. The LogP model [14] consists of p processor/memory components communicating through point-to-point messages, and has the following parameters: the *latency* l, which is the time taken by the network to transmit a message from one to processor to another; an *overhead* o, which is the time spend by a processor to transfer a message to or from the network interface, during which time it cannot perform any other operation; the *gap* g, where a processor can send or receive a message no faster than once every g units of time; and a *capacity constraint:* whereby a receiving processor can have no more than $\lceil l/g \rceil$ messages in transit to it. If the number of messages in transit to a destination processor π is $\lceil l/g \rceil$ then a processor that needs to send a message to π *stalls*, and does not perform any operation until the message can be sent. The *nonstalling LogP* is the LogP model in which it is not allowed to have more than $\lceil l/g \rceil$ messages in transit to any processor.

QSM and s-QSM models. The Queuing Shared Memory (QSM) model [22] consists of a number of identical processors, each with its own private memory, that communicate by reading and writing shared memory. Processors execute a sequence of synchronized phases, each consisting of an arbitrary interleaving of shared memory reads, shared memory writes, and local computation. The value returned by a shared-memory read can be used only in a subsequent phase. Concurrent reads or writes (but not both) to the same shared-memory location are permitted in a phase. In the case of multiple writers to a location x, an arbitrary write to x succeeds in writing the value present in x at the end of the phase. The *maximum contention* of a QSM phase is the maximum number of processors reading or writing any given memory location. A phase with no reads or writes is defined to have maximum contention one.

Consider a QSM phase with maximum contention κ. Let m_{op} be the maximum number of local operations performed by any processor in this phase, and let m_{rw} be the maximum number of read/write requests issued by any processor. Then the *time cost* for the phase is $\max(m_{op}, g \cdot m_{rw}, \kappa)$. The *time* of a QSM algorithm is the sum of the time costs for its phases. The *work* of a QSM algorithm is its processor-time product. Since the QSM model does not have a latency parameter, the effect of latency can be incorporated into the performance analysis by counting the number of phases in a QSM algorithm with the goal to minimizing that number.

The s-QSM (*Symmetric QSM*) is a QSM in which the time cost for a phase is $\max(m_{op}, g \cdot m_{rw}, g \cdot \kappa)$, i.e., the gap parameter is applied to the accesses at memory as well as to memory requests issued at processors.

The (g, d)-QSM is the most general version of the QSM. In the (g, d)-QSM, the time cost of a phase is $\max(m_{op}, g \cdot m_{rw}, d \cdot \kappa)$, i.e., the gap parameter g is applied to the memory requests issued at processors, and a different gap parameter d is applied to accesses at memory. Note that the QSM is a $(g, 1)$-QSM and an s-QSM is a (g, g)-QSM.

The special case of QSM and s-QSM where the gap parameter g equals 1, is the QRQW PRAM [19], a precursor to the QSM.

3 Relationships between Models

Table 1 presents recent research results on *work-preserving* emulations between QSM, BSP and LogP models [22, 42, 43].

An emulation of one model on another is work-preserving if the processor-time bound on the emulating machine is the same as that on the machine being emulated, to within a constant factor. The ratio of the running time on the emulating machine to the running time on the emulated machine is the *slowdown* of the emulation. Typically, the emulating machine has a smaller number of processors and takes proportionately longer to execute. For instance, consider the entry in Table 1 for the emulation of s-QSM on BSP. It states that there is a randomized work-preserving emulation of s-QSM on BSP with a slowdown of $O(L/g + \log p)$. This means that, given a p-processor s-QSM algorithm that runs in time t (and hence with work $w = p \cdot t$), the emulation algorithm will map the p-processor s-QSM algorithm on to a p'-processor BSP, for any $p' \leq p/((L/g) + \log p)$, to run on the BSP in time $t' = O(t \cdot (p/p'))$ w.h.p. in p.

Slowdown of Work-Preserving Emulations between Parallel Models				
Emulated Models (p procs)	Emulating Models			
	BSP	LogP (stalling)	s-QSM	QSM
BSP		$\log^4 p + (L/g) \log^2 p$	$\lceil \frac{g \log p}{L} \rceil$	$\lceil \frac{g \log p}{L} \rceil$
LogP (non-stalling)	L/l (det.)[1]	1 (det.)	$\lceil \frac{g \log p}{l} \rceil$	$\lceil \frac{g \log p}{l} \rceil$
s-QSM	$(L/g) + \log p$	$\log^4 p + (l/g) \log^2 p$		1 (det.)
QSM	$(L/g) + g \log p$	$\log^4 p + (l/g) \log^2 p + g \cdot \log p$	g (det.)	

Table 1. All results are randomized and hold w.h.p. except those marked as 'det.', which are deterministic emulations. These results are reported in [22, 42, 43]. Results are also available for the (d, x)-BSP and (g, d)-QSM.

[1] This result is presented in [9] but it is stated there erroneously that it holds for stalling LogP programs.

All emulations results listed in Table 1 are work-preserving, and the one mis-match is between stalling and non-stalling LogP, and here we do not know how to provide a work-preserving emulation with small slow-down of a stalling LogP on any of the other models. (Note that earlier it was stated erroneously in Bilardi et al. [9] that LogP is essentially equivalent to BSP, but this was refuted in Ramachandran et al. [43]).

This collection of work-preserving emulations with small slowdown between the three models – BSP, LogP and QSM — suggests that these three models are essentially interchangeable (except for the stalling versus nonstalling issue for LogP) in terms of the relevance of algorithm design and analysis on these models to real parallel machines.

4   Algorithms

QSM Algorithmic results. Efficient QSM algorithms for several basic problems follow from the following observations [22]. (An 'EREW' PRAM is the PRAM model in which each shared-memory read or write has at most one access to each memory location. A 'QRQW' PRAM [19] is the QSM model in which the gap parameter has value 1.)

1. *(Self-simulation)* A QSM algorithm that runs in time t using p processors can be made to run on a p'-processor QSM, where $p' < p$, in time $O(t \cdot p/p')$, i.e., while performing the same amount of work.
2. *(EREW and QRQW algorithms on QSM)*
   (a) An EREW or QRQW PRAM algorithm that runs in time t with p processors is a QSM algorithm that runs in time at most $t \cdot g$ with p processors.
   (b) An EREW or QRQW PRAM algorithm in the work-time framework that runs in time t while performing work w implies a QSM algorithm that runs in time at most $t \cdot g$ with w/t processors.
3. *(Simple lower bounds for QSM)* Consider a QSM with gap parameter g.
   (a) Any algorithm in which n distinct items need to be read from or written into global memory must perform work $\Omega(n \cdot g)$.
   (b) Any algorithm that needs to perform a read or write on n distinct global memory locations must perform work $\Omega(n \cdot g)$.

There is a large collection of logarithmic time, linear work EREW and QRQW PRAM algorithms available in the literature. By the second observation mentioned above these algorithms map on to the QSM with the time and work both increased by a factor of g. By the third observation above the resulting QSM algorithms are work-optimal (to within a constant factor). More generally, by working with the QSM model we can leverage on the extensive algorithmic results compiled for the PRAM model.

Some QSM algorithmic results for sorting and list ranking that focus on reducing the number of phases are given in [43]. Related work on minimizing the number of supersteps on a BSP using the notion of *rounds* is reported in [23] for sorting, and in [12] for graph problems. Several lower bounds for the number of phases needed for basic problems are presented in [34]. Some of these lower bounds are given in Table 2. The 'linear approximate compaction' problem mentioned in Table 2 is a useful subroutine for load-balancing; a simple algorithm for this problem on the QRQW PRAM (and hence the QSM) that improves on the obvious logarithmic time algorithm is given in [20].

problem (n=size of input)	Deterministic time l.b.	Randomized time l.b.	# of phases w/ p procs. and $O(n)$ work/phase
Lin. approx. compaction	$\Omega(g\sqrt{\frac{\log n}{\log\log n}})$	$\Omega(g\log\log n)$	$\Omega(\sqrt{\frac{\log n}{\log(n/p)}})$
OR	$\Omega(\frac{g\log n}{\log\log n})$	$\Omega(g\log^* n)$	$\Omega(\frac{\log n}{\log(n/p)})^\dagger$
Prefix sums, sorting	$\Omega(g\log n)^\dagger$	$\Omega(\frac{g\log n}{\log\log n})$	$\Omega(\frac{\log n}{\log(n/p)})^\dagger$

Table 2. Lower bounds for s-QSM [34]. ([34] also presents lower bounds for these problems for QSM and BSP.)
†This bound is tight since there is an algorithm that achieves this bound.

Some experimental results are presented in [24]. In this paper, the QSM algorithms for prefix sums, list ranking and sorting given in [43] were examined experimentally to evaluate the trade-offs made by the simplicity of the QSM model. The results in [24] indicate that analysis under the QSM model yields quite accurate results for reasonable input sizes.

BSP Algorithmic Results. By the emulation results of the previous section, any QSM algorithm can be mapped on to the BSP to run with equal efficiency and only a small slow-down (with high probability). Thus, all of the results obtained for QSM are effective algorithms for the BSP as well. Additionally, there has been a considerable amount of work on algorithms design for the BSP model. For instance, sorting and related problems are considered in [17, 23], list and graph problems are considered in [12], matrix multiplication and linear algebra problems are considered in [17, 46, 37], algorithms for dynamic data structures are considered in [8], to cite just a few.

LogP Algorithmic Results. Some basic algorithms for LogP are given in [14]. LogP algorithms for summing and broadcasting are analyzed in great detail in [28]. Several empirical results on performance evaluation of sorting and other algorithms are reported in the literature. However, there are not many other results on design and analysis of LogP algorithms. This appears to be due to the asynchronous nature of the model, and the capacity constraint requirement, which is quite stringent, especially in the nonstalling LogP.

References

1. M. Adler, J. Byer, R. M. Karp, Scheduling Parallel Communication: The h-relation Problem. In *Proc. MFCS*, 1995.
2. M. ADLER, P.B. GIBBONS, Y. MATIAS, AND V. RAMACHANDRAN, Modeling parallel bandwidth: Local vs. global restrictions, In *Proc. 9th ACM Symp. on Parallel Algorithms and Architectures*, 94–105, June 1997.
3. A. AGGARWAL, A.K. CHANDRA, AND M. SNIR, Communication complexity of PRAMs, *Theoretical Computer Science*, 71(1):3–28, 1990.

4. A. ALEXANDROV, M.F. IONESCU, K.E. SCHAUSER, AND C. SHEIMAN, LogGP: Incorporating long messages into the LogP model — one step closer towards a realistic model for parallel computation, In *Proc. 7th ACM Symp. on Parallel Algorithms and Architectures*, 95–105, July 1995.

5. B. ALPERN, L. CARTER, AND E. FEIG, Uniform memory hierarchies, In *Proc. 31st IEEE Symp. on Foundations of Computer Science*, 600–608, October 1990.

6. Y. AUMANN AND M.O. RABIN, Clock construction in fully asynchronous parallel systems and PRAM simulation, In *Proc. 33rd IEEE Symp. on Foundations of Computer Science*, 147–156, October 1992.

7. A. BAR-NOY AND S. KIPNIS, Designing broadcasting algorithms in the postal model for message-passing systems, In *Proc. 4th ACM Symp. on Parallel Algorithms and Architectures*, 13–22, 1992.

8. A. BAUMKER AND W. DITTRICH, Fully dynamic search trees for an extension of the BSP model, In *Proc. 8th ACM Symp. on Parallel Algorithms and Architectures*, 233–242, June 1996.

9. G. BILARDI, K. T. HERLEY, A. PIETRACAPRINA, G. PUCCI, P. SPIRAKIS. BSP vs LogP. In *Proc. ACM SPAA*, pp. 25–32, 1996.

10. G.E. BLELLOCH, *Vector Models for Data-Parallel Computing*, The MIT Press, Cambridge, MA, 1990.

11. G.E. BLELLOCH, P.B. GIBBONS, Y. MATIAS, AND M. ZAGHA, Accounting for memory bank contention and delay in high-bandwidth multiprocessors, In *Proc. 7th ACM Symp. on Parallel Algorithms and Architectures*, 84–94, July 1995.

12. E. CACERES, F. DEHNE, A. FERREIRA, P. FLOCCHINI, I. RIEPING, A. RONCATO, N. SANTORO, AND S. W. SONG. Efficient parallel graph algorithms for coarse grained multicomputers and BSP. In *Proc. ICALP*, LNCS 1256, pp. 390-400, 1997.

13. R. COLE AND O. ZAJICEK, The APRAM: Incorporating asynchrony into the PRAM model, In *Proc. 1st ACM Symp. on Parallel Algorithms and Architectures*, 169–178, June 1989.

14. D. CULLER, R. KARP, D. PATTERSON, A. SAHAY, K.E. SCHAUSER, E. SANTOS, R. SUBRAMONIAN, AND T. VON EICKEN, LogP: Towards a realistic model of parallel computation, In *Proc. 4th ACM SIGPLAN Symp. on Principles and Practices of Parallel Programming*, 1–12, May 1993.

15. C. DWORK, M. HERLIHY, AND O. WAARTS, Contention in shared memory algorithms, In *Proc. 25th ACM Symp. on Theory of Computing*, 174–183, May 1993.

16. S. FORTUNE AND J. WYLLIE, Parallelism in random access machines, In *Proc. 10th ACM Symp. on Theory of Computing*, 114–118, May 1978.

17. A.V. GERBESSIOTIS AND L. VALIANT, Direct bulk-synchronous parallel algorithms, *Journal of Parallel and Distributed Computing*, 22:251–267, 1994.

18. P.B. GIBBONS, A more practical PRAM model, In *Proc. 1st ACM Symp. on Parallel Algorithms and Architectures*, 158–168, June 1989.

19. P.B. GIBBONS, Y. MATIAS, AND V. RAMACHANDRAN, The Queue-Read Queue-Write PRAM model: Accounting for contention in parallel algorithms, *SIAM Journal on Computing*, vol. 28:733-769, 1999.

20. P.B. GIBBONS, Y. MATIAS, AND V. RAMACHANDRAN, Efficient low-contention parallel algorithms, *Journal of Computer and System Sciences*, 53(3):417–442, 1996.

21. P.B. GIBBONS, Y. MATIAS, AND V. RAMACHANDRAN, The Queue-Read Queue-Write Asynchronous PRAM model, *Theoretical Computer Science: Special Issue on Parallel Processing*, vol. 196, 1998, pp. 3-29.

22. P.B. GIBBONS, Y. MATIAS, AND V. RAMACHANDRAN, Can a shared-memory model serve as a bridging model for parallel computation? *Theory of Computing Systems* Special Issue on *SPAA '97*, 32:327-359, 1999.

23. M. GOODRICH, Communication-Efficient Parallel Sorting. In *Proc. STOC*, pp. 247–256, 1996.

24. B. GRAYSON, M. DAHLIN, V. RAMACHANDRAN, Experimental evaluation of QSM, a simple shared-memory model. In *Proc. IPPS/SPDP*, 1999.

25. T. HEYWOOD AND S. RANKA, A practical hierarchical model of parallel computation: I. The model, *Journal of Parallel and Distributed Computing*, 16:212–232, 1992.

26. B. H. H. JUURLINK AND H.A.G. WIJSHOFF, A quantitative comparison of parallel computation models, In *Proc. 8th ACM Symp. on Parallel Algorithms and Architectures*, pp. 13–24, 1996.

27. B.H.H. JUURLINK AND H.A.G. WIJSHOFF, The E-BSP Model: Incorporating general locality and unbalanced communication into the BSP Model, In *Proc. Euro-Par'96*, 339–347, August 1996.

28. R. KARP, A. SAHAY, E. SANTOS, AND K.E. SCHAUSER, Optimal broadcast and summation in the LogP model, In *Proc. 5th ACM Symp. on Parallel Algorithms and Architectures*, 142–153, June-July 1993.

29. R.M. KARP AND V. RAMACHANDRAN, Parallel algorithms for shared-memory machines, In J. van Leeuwen, editor, *Handbook of Theoretical Computer Science, Volume A*, 869–941. Elsevier Science Publishers B.V., Amsterdam, The Netherlands, 1990.

30. Z.M. KEDEM, K.V. PALEM, M.O. RABIN, AND A. RAGHUNATHAN, Efficient program transformations for resilient parallel computation via randomization, In *Proc. 24th ACM Symp. on Theory of Computing*, 306–317, May 1992.

31. K. KENNEDY, A research agenda for high performance computing software, In *Developing a Computer Science Agenda for High-Performance Computing*, 106–109. ACM Press, 1994.

32. P. LIU, W. AIELLO, AND S. BHATT, An atomic model for message-passing, In *Proc. 5th ACM Symp. on Parallel Algorithms and Architectures*, 154–163, June-July 1993.

33. P.D. MACKENZIE AND V. RAMACHANDRAN, ERCW PRAMs and optical communication, *Theoretical Computer Science: Special Issue on Parallel Processing*, vol. 196, 153–180, 1998.

34. P.D. MACKENZIE AND V. RAMACHANDRAN, Computational bounds for fundamental problems on general-purpose parallel models, In *ACM SPAA*, 1998, pp. 152-163.

35. B.M. MAGGS, L.R. MATHESON, AND R.E. TARJAN, Models of parallel computation: A survey and synthesis, In *Proc. 28th Hawaii International Conf. on System Sciences*, II: 61–70, January 1995.

36. Y. MANSOUR, N. NISAN, AND U. VISHKIN, Trade-offs between communication throughput and parallel time, In *Proc. 26th ACM Symp. on Theory of Computing*, 372–381, 1994.

37. W.F. MCCOLL, A BSP realization of Strassen's algorithm, Technical report, Oxford University Computing Laboratory, May 1995.

38. K. MEHLHORN AND U. VISHKIN, Randomized and deterministic simulations of PRAMs by parallel machines with restricted granularity of parallel memories, *Acta Informatica*, 21:339–374, 1984.

39. N. NISHIMURA, Asynchronous shared memory parallel computation, In *Proc. 2nd ACM Symp. on Parallel Algorithms and Architectures*, 76–84, July 1990.

40. S. PETTIE, V. RAMACHANDRAN, A time-work optimal parallel algorithm for minimum spanning forest. In Proc. Approx-Random'99, August 1999.
41. C. K. POON, V. RAMACHANDRAN, A randomized linear work EREW PRAM algorithm to find a minimum spanning forest. In Proc. 8th Intl. Symp. on Algorithms and Computation (ISAAC '97), Springer-Verlag LNCS vol. 1530, 1997, pp. 212-222.
42. V. RAMACHANDRAN, A general purpose shared-memory model for parallel computation, invited paper in *Algorithms for Parallel Processing*, Volume 105, IMA Volumes in Mathematics and its Applications, Springer-Verlag, pp. 1-17, 1999.
43. V. RAMACHANDRAN, B. GRAYSON, M. DAHLIN, Emulations between QSM, BSP and LogP: A framework for general-purpose parallel algorithm design. In *ACM-SIAM SODA'99*, 1999.
44. A.G. RANADE, *Fluent parallel computation*, PhD thesis, Department of Computer Science, Yale University, New Haven, CT, May 1989.
45. L. SNYDER, Type architecture, shared memory and the corollary of modest potential, *Annual Review of CS*, I:289–317, 1986.
46. L.G. VALIANT, A bridging model for parallel computation, *Communications of the ACM*, 33(8):103–111, 1990.
47. L.G. VALIANT, General purpose parallel architectures, In J. van Leeuwen, editor, *Handbook of Theoretical Computer Science, Volume A*, 943–972. Elsevier Science Publishers B.V., Amsterdam, The Netherlands, 1990.
48. H.A.G. WIJSHOFF AND B.H.H. JUURLINK, A quantitative comparison of parallel computation models, In *Proc. 8th ACM Symp. on Parallel Algorithms and Architectures*, 13–24, June 1996.

Parallel Bridging Models and Their Impact on Algorithm Design[*]

Friedhelm Meyer auf der Heide and Rolf Wanka

Dept. of Mathematics and Computer Science and Heinz Nixdorf Institute, Paderborn University, 33095 Paderborn, Germany. Email: {fmadh | wanka}@upb.de

Abstract. The aim of this paper is to demonstrate the impact of features of parallel computation models on the design of efficient parallel algorithms. For this purpose, we start with considering Valiant's BSP model and design an optimal multisearch algorithm. For a realistic extension of this model which takes the critical blocksize into account, namely the BSP* model due to Bäumker, Dittrich, and Meyer auf der Heide, this algorithm is far from optimal. We show how the critical blocksize can be taken into account by presenting a modified multisearch algorithm which is optimal in the BSP* model. Similarly, we consider the D-BSP model due to de la Torre and Kruskal which extends BSP by introducing a way to measure locality of communication. Its influence on algorithm design is demonstrated by considering the broadcast problem. Finally, we explain how our Paderborn University BSP (PUB) Library incorporates such BSP extensions.

1 Introduction

The theory of efficient parallel algorithms is very successful in developing new algorithmic ideas and analytical techniques to design and analyze efficient parallel algorithms. The *Parallel Random Access Machine* model (PRAM model) has proven to be very convenient for this purpose. On the other hand, the PRAM cost model (mis-)guides the algorithm designer to exploit a huge communication volume, and to use it in a fine-grained fashion. This happens because the PRAM cost model charges the same cost for computation and communication. In real parallel machines, however, communication is much more expensive than computation, and the cost for computation differs from machine to machine. Thus, it might happen that two algorithms for the same problem are incomparable in the sense that one is faster on machine A, the other is faster on machine B.

To overcome these problems, several proposals for so-called parallel bridging models have been developed: for example, the BSP model [16], the LogP model [6], the CGM model [7], and the QSM model [1].

A bridging model aims to meet the following goals: Its cost measure should guide the algorithm designer to develop efficient algorithms. It should be detailed enough to allow an accurate prediction of the algorithms' performance. It ought to provide an environment independent from a specific architecture and technology, yet reflecting the

---

[*] Partially supported by DFG SFB 376 "Massively Parallel Computation" and by the IST programme of the EU under contract number IST-1999-14186 (ALCOM-FT).

most important constraints of existing machines. This environment should also make it possible to write real portable programs that can be executed efficiently on various machines.

Valiant's BSP model [16] (*Bulk Synchronous model of Parallel computing*) is intended to bridge the gap between software and hardware needs in parallel computing. In this model, a parallel computer has three parameters that govern the runtime of algorithms: The number of processors, the latency, and the gap.

The aim of this paper is to demonstrate the impact of features of the bridging model used on the design of efficient algorithms. For this purpose, in Section 2, after a detailed description of the (plain) BSP model, we present an efficient algorithm for the multisearch problem. Then, in Section 3, we explain that it is sometimes worthwhile to consider additional parameters of the parallel machine. Examples for such parameters are the *critical block size* and the *locality function* of the machine. The effect of the critical block size is demonstrated by presenting a multisearch algorithm that is efficient also if the critical block size is considered. The benefit of taking the locality into account is shown by presenting an efficient algorithm for the broadcast problem. In Section 4, we report on the implementation of the PUB Lib, the Paderborn BSP Library.

2 The Plain BSP Model

The PRAM model is one of the most widely used parallel computation models in theoretical computer science. It consists of a number of sequential computers that have access to a shared memory of unlimited size. In every time step of a PRAM's computation, a processor may read from or write into a shared memory location, or it can perform a single local step. So it charges one time unit both for an internal step, and for accessing the shared memory though these tasks seem to be quite different. The PRAM model is very comfortable for algorithm design because it abstracts from communication bottlenecks like bandwidth and latency. However, this often leads to the design of communication intensive algorithms that usually show bad performance if implemented on an actual parallel computer. In order to overcome this mismatch between model and actual machine, Valiant identified some abstract parameters of parallel machines that enable algorithm designer to charge different cost for the different tasks, without to be committed to a special parallel computer, but with a hopefully reliable prediction of the algorithms' performance. It is called the BSP model and discussed next.

2.1 Definition

A BSP machine is a parallel computer that consists of p processors, each processor having its local memory. The processors are interconnected via an interconnection mechanism (see Fig. 1(a)).

Algorithms on a BSP machine proceed in *supersteps* (see Fig. 1(b)). We describe what happens in a superstep from the point of view of processor P_i: When a new superstep t starts, all $\mu_{i,t}$ messages sent to P_i during the previous superstep are available at P_i. P_i performs some local computation, or work, that takes time $w_{i,t}$ and creates $\lambda_{i,t}$ new messages that are put into the interconnection mechanism, but cannot be received

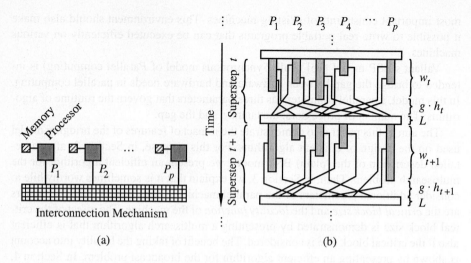

Fig. 1. (a) A BSP computer and (b) the execution of a BSP algorithm.

during the current superstep. Let $h_{i,t} = \max\{\mu_{i,t}, \lambda_{i,t}\}$. This superstep is finished by executing a barrier synchronization. Summarizing over all processors, we introduce the parameters $w_t = \max_i w_{i,t}$ and $h_t = \max_i h_{i,t}$. If an algorithm performs T supersteps, we use $W = \sum_{1 \leq t \leq T} w_t$ and $H = \sum_{1 \leq t \leq T} h_t$. W is called the local work of the algorithm, H its communication volume. Note that the size of the packets sent is not taken into account. In the course of analyzing a BSP algorithm, the task is to determine concrete values for T, H, and W.

The actual time that a BSP machine needs to execute the above BSP algorithm depends on the following machine parameters: L, the *latency*, and g, the *gap*, or *bandwidth inefficiency*. L is the maximum time that a message sent from a processor P_i needs to reach processor P_j, taken over all i and j. Likewise, L is also the time necessary for a single synchronization. Every processor can put a new packet into the interconnection mechanism after g units of time have been elapsed. Concrete values of g and L can be measured by experiments. For a different platforms catalogue, see [14]. The third machine parameter is p, the *number of processors*, or *the size of the machine*.

Hence, the runtime of the tth superstep is (at most) $w_t + g \cdot h_t + L$, and the overall runtime is (at most) $W + g \cdot H + L \cdot T$.

Note that we consider sums of times although it is often sufficient to only consider the maximum of the involved times when, e. g., pipelining can be used. This only results in a constant factor-deviation, whereas it simplifies many analyses considerably.

Let T_{seq} be the runtime of a best sequential algorithm known for a problem. Ideally, we are seeking BSP algorithms for this problem where $W = c \cdot T_{\text{seq}}/p$, $L \cdot T = o(T_{\text{seq}}/p)$, and $g \cdot H = o(T_{\text{seq}}/p)$, for a small constant $c \geq 1$. Such algorithms are called *c-optimal*. There are two popular ways to represent results of a BSP analysis. First, values of T, H, and W are given. For example, see Theorem 1 below. The other way is to state (with n denoting the input size) for which ranges of n/p, g, and L the algorithm is 1-optimal. Theorem 2 concluded from Theorem 1 is presented in this way.

2.2 An Example: The Multisearch Problem

As an instructive example, we outline an efficient BSP algorithm for the Multisearch Problem. Let U be a *universe* of objects, and $\Sigma = \{\sigma_1, \ldots, \sigma_m\}$ a partition of U into segments $\sigma_i \subseteq U$. The segments are *ordered*, in the sense that, for every $q \in U$ and segment s_i, it can be determined (in constant time) whether $q \in \sigma_1 \cup \cdots \cup \sigma_{i-1}$, or $q \in \sigma_i$, or $q \in \sigma_{i+1} \cup \cdots \cup \sigma_m$. In the (m,n)-multisearch problem, the goal is to determine, for n objects (called *queries*) $q_1, \ldots, q_n$ from U, their respective segments. Also the queries are ordered, but we are not allowed to conclude from $q \in \sigma_i$ that $q' \in \sigma_j$, for any pair i, j. Such problems arise in the context of algorithms in Computational Geometry (e. g., see Fig. 2).

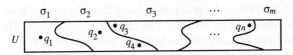

Fig. 2. A multisearch instance from Computational Geometry. Note that q_3 and q_4 are in different relative orderings with respect to each other and their respective segments.

For simplicity, we assume that $m = d^c$ for some constant c and that the segments are given in the form of a complete d-ary search tree $\mathcal{T}$ of height $\log_d m + 1$. d will be determined later. Every leaf represents a segment, the leftmost leaf σ_1, the rightmost σ_m. Every inner node contains $d - 1$ copies (called splitters) of the segments that split all its leaves into equally-sized intervals. A query q takes a path from the root (on level 0) of $\mathcal{T}$ to its respective leaf (on level $\log_d m$). In every inner node, it can be decided in time $O(\log d)$ by a binary search on the splitters to which child one has to go to resume the search.

In a seminal paper [13], Reif/Sen introduced a randomized PRAM algorithm that solves any $(O(n), n)$-multisearch instance on n processors in time $O(\log n)$, with high probability (w. h. p.), no shared memory cell being accessed at the same step more than once. A direct adaptation of this algorithm to the BSP model yields $T = W = H = O(\log n)$ which is far away from being optimal in the sense discussed above.

In the BSP setting, the parallel solution for the multisearch problem has to accomplish the following two tasks: (i) *Mapping*. In a preprocessing step, the nodes of $\mathcal{T}$ are mapped to the processors of the BSP machine. (ii) *Parallel multisearch*. After (i), the search for the segments of the n queries is performed in parallel.

The BSP algorithm works quite simple from a high-level point of view. The queries are distributed among the processors. Initially, they are all assigned to the root of $\mathcal{T}$, and they all must travel from the root to the correct leaf. In round t, it is determined for every query that is assigned to a node on level $t - 1$ to which node on level t it has to go, so the number of rounds is $\log_d m$. As it turns out, the number of supersteps per round can be constant for a proper choice of d and a broad range for the BSP parameters.

The crucial task of this algorithm is the way of how the information to which node v of $\mathcal{T}$ which is stored in some processor P_i query q has to go meets q which is stored in some processor P_j. Should v be sent to P_j, or should q be sent to P_i? We shall see that answering this question individually for pairs (q, v) is the main step in the design of an efficient BSP algorithm.

Two kinds of hot spots can emerge that, unfortunately, are not addressed in a PRAM setting. The first kind can happen at a *processor*: There can be many nodes stored in a single processor that will be accessed in one round, either by receiving queries, or by sending their information away. There is a surprisingly simple way to tackle this problem. As for the mapping used in our plain BSP algorithm, we assume that the $\Theta(m)$ nodes of $\mathcal{T}$ are mapped randomly to the processors. The important observation for this mapping is: If $m \geq p \log p$, every processor manages $(1 + o(1)) \cdot m/p$ nodes, with high probability. We shall see in the next section that this mapping cannot avoid some problems if the so-called critical block size is considered as an additional parameter. However, for the time being, this simple way of mapping avoids hot spots at the processors. Of course we shall avoid sending requested information more than once. This is the aim of the solution of the problems arising from the second kind of hot spots.

Namely, the other kind of hot spots can emerge at *nodes* of $\mathcal{T}$. There can be nodes which lie on the paths of many queries from the nodes to the leaves. E. g., the root is on the path of every query, and, hence, a hot spot. The idea is to handle the case that many paths go through a nodes differently from the other case.

This distinction and separate handling is the impact of the BSP model on the design of an efficient BSP algorithm for the multisearch problem.

For a node v, let $J(v)$ be the set of queries that go through v. $J(v)$ is called the job of v. $J(v)$ is called a *large* job, if $|J(v)| > r$, otherwise, it is called a *small* job. For our purposes, it is sufficient to choose $r = (n/p)^{1-\varepsilon}$ for some ε, $0 < \varepsilon < 1$.

Let $v_1, \ldots, v_{d^t}$ be the nodes on level t of $\mathcal{T}$. Let the jobs be distributed among the processors as shown in Figure 3. For small jobs $J(v_i)$, $J(v_i)$ is sent to the processor that

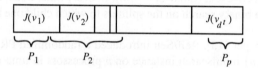

Fig. 3. Distribution of the jobs.

manages node v_i where the nodes of the next level are computed by binary search. This can be done easily.

Large jobs $J(v_i)$ are handled differently because we want to avoid to send too many queries to one processor. Here the strategy is to first distribute $J(v_i)$ evenly on a group of consecutive processors such that at most n/p queries are on each of these processors. Then the group's first processor receives the management information, i. e., the splitters, from the processor that manages v_i and broadcasts it to the rest of the group. This routine can be implemented by using integer sorting and a segmented parallel prefix for which efficient BSP algorithms have to be used. The development of such algorithms is a very interesting work in its own right, but for our purposes it suffices to state the following theorem. Its proof can be found in [2, 3]. Note that a simple consideration shows that in the case of small jobs, it can be bad to send many 'large' nodes to the presumably many small jobs on a single processor.

Theorem 1. *Let $T_{\text{bin}}(x, y)$ denote the sequential worst case time for x binary searches on a sequence of y segments. Let $c \geq 1$, $1 < k < p$ and $d = o((n/p)^{1/c})$.*

The multisearch algorithm presented above performs n searches on a d-ary tree of depth δ *in time* $\delta \cdot (W + g \cdot H + T \cdot L)$, *w.h.p., with*

$$W = (1 + o(1)) \cdot T_{bin}(n/p, d) + O(k \cdot ((n/p)^{1/c} + c + \log_k p)) \ ,$$
$$H = O(c \cdot (n/p)^{1/c} + k \cdot \log_k p) \ ,$$
$$T = O(c + \log_k p) \ .$$

With $T_{bin}(x, y) = O(x \log y)$ and, therefore, $T_{seq} = O(\delta \cdot n/p \cdot \log d)$ one can compute for which ranges of the machine parameters p, g and L the algorithm is 1-optimal.

Theorem 2. *Let* $c \geq 1$ *be an arbitrary constant and* $d = (n/p)^{1/c}$. *Then the algorithm solves the multisearch problem with n queries on a d-ary tree with depth* $\delta = \log_d p$ *in runtime* $\delta \cdot (1 + o(1)) \cdot T_{bin}(n/p, d)$ *for the following parameter constellations:*

- $n/p = \Omega((\log p)^c)$, $g = o(\log(n/p))$, *and* $L = o(n \log(n/p)/(p \log p))$.
- $n/p = \Omega(p^{\varepsilon})$, *for any* $\varepsilon > 0$, $g = o(\log(n/p))$, *and* $L = (n/p) \cdot \log(n/p)$.

3 Extensions of BSP

As mentioned in the introduction, the runtime computed with the BSP parameters should allow an accurate prediction of the algorithms' performance. However, quite often it can be observed that the prediction and the actual performance of an implementation deviate considerably despite a tight analysis. That means that at least in such cases, there are properties of existing machines that influence the performance heavily, but that are not visible in the BSP approach and therefore missing when algorithms are designed. In this section, two such properties are identified and incorporated into the BSP model. For both variations of the BSP model, we present algorithms that show the impact of the changes in the model on the algorithm design.

3.1 BSP*: Critical Block Size and the Multisearch Problem Revisited

Observations of existing interconnection mechanisms of real machines show that it is sometimes better to send large packets that contain many single information units than sending all these units separately. But – as mentioned above – the size of messages is not taken into account in the BSP model. In order to model this aspect, the *critical block size* B has been introduced as additional parameter [2, 8] resulting in the BSP* model. B is the minimum number of information units a message must have in order to fully exploit the bandwidth of the communication mechanism. This also leads to a modified definition of the gap: Every processor can put a new packet *of size* B into the communication mechanism after g^* units of time have been elapsed.

Now for superstep t, we count how many messages h_t are sent or received by a single processor, and how many information units s_t are parceled in these messages. For the runtime of the superstep, we charge $w_t + g^* \cdot (s_t + h_t \cdot B) + L$. Note that if h_t messages containing altogether $s_t = h_t \cdot B$ information units are sent, the runtime is $w_t + 2g^* \cdot s_t + L$. If h_t messages containing only $s_t = h_t$ information units are sent, the runtime is $w_t + g^* \cdot h_t \cdot (B + 1) + L$. That means that short messages are treated as if they were of size B. The number T of supersteps and the local work W of an BSP* algorithm are identical to those in the BSP model. Now, $H = \sum_t h_t$ is the *number of message start-ups*, and the *communication volume* is $S = \sum_t s_t$. The total runtime of a BSP* algorithm is $W + g^*(S + B \cdot H) + T \cdot L$.

Let H_{BSP} denote the number of messages of a BSP algorithm. In a bad case, it can be that every message contains only one information unit. That means the BSP* runtime is $W + g^*(1 + B) \cdot H_{BSP} + T \cdot L$. In this case, we are seeking for a new (or modified) algorithm for the BSP* machine that communicates (about) the same amount of information, but parceled into at most $H = H_{BSP}/B$ messages.

Indeed, the bad case mentioned above can occur in the BSP multisearch algorithm presented in Subsection 2.2. For example, suppose a very large search tree $\mathcal{T}$ and few queries that all belong to different leaves of $\mathcal{T}$. If we run our BSP algorithm with a random mapping of the nodes of $\mathcal{T}$, there will be a level (i. e., round) t_0, after that all jobs have size 1 and, due to the good distribution property of a random mapping, all jobs have to be sent from now on to different processors, w. h. p. That means that we cannot create large messages to exploit the bandwidth of the communication mechanism, and that almost all messages that have size 1 are charged with B. In fact, this computed runtime comes in total much closer to the actual performance of this algorithm on many real machines than the plain BSP runtime.

From the discussion above, it follows that the random mapping of the nodes of $\mathcal{T}$ causes the problems. In the following, we shall describe a different mapping that enables the algorithm also to parcel large messages after some critical level t_0 of $\mathcal{T}$ as described above has been reached. This mapping is the so-called z-mapping.

It is the impact of considering the critical block size in the BSP model on the design of an efficient algorithm for the multisearch problem.*

Note that also adapted BSP* algorithms for integer sorting, segmented parallel prefix and broadcast have to be designed and applied as subroutines.

Let $z \leq p$. The z-mapping of the nodes of a d-ary tree $\mathcal{T}$ with $\delta + 1$ levels works as follows, with t going from 0 through δ iteratively: On level t, $t \leq \log_d p$, there are at most p nodes which are randomly mapped to different processors. On level t, $t > \log_d p$, there are more than p nodes. For every processor P, let $R(P)$ be the subset of nodes in level $t - 1$ that has been mapped to P. All children of the nodes of $R(P)$ are distributed randomly among z randomly chosen processors.

In the random mapping, the children of $R(P)$ can be spread evenly among *all* processors, whereas in the z-mapping they are, so to speak, clustered to only z processors where they are spread evenly.

Now we observe how the multisearch algorithm has to behave in the case of small jobs if the z-mapping is applied. Large jobs are treated as before.

After the partition of jobs of level t-nodes has been completed, every small job $J(v)$ is stored in processor P that also stores v. P splits (by binary search) $J(v)$ into the jobs for the children nodes on level $t + 1$. The z-mapping ensures that these children are scattered only on at most z processors. Even if P has many small jobs, all children nodes are distributed among at most z processors. So P can create z large messages.

Of course, this description can only put some intuition across the success of this mapping. A detailed analysis of this algorithm [2, 3] leads to the following theorem that states when this algorithm is an optimal parallelization.

Theorem 3. *Let $c \geq 1$ be an arbitrary constant and choose $d = (n/p)^{1/c}$ and let $z = \omega(d \log p)$. Let $\mathcal{T}$ be a d-ary tree with depth δ that has been mapped onto the processors by using the z-mapping.*

W. h. p., the runtime of the above algorithm is $\delta \cdot (1 + o(1)) \cdot T_{\text{bin}}(n/p, d)$, i. e. 1-optimal, for the following parameter constellations:

- *$n/p = \omega((\log)^c)$, $g^* = o(\log(n/p))$, $B = o((n/p)^{1-1/c})$, and $L = o((n \log(n/p)/(p \log p))$.*
- *$n/p = \Omega(p^\varepsilon)$, for an arbitrary constant $\varepsilon > 0$, $g^* = o(\log(n/p))$, $B = o((n/p)^{1-1/c})$, and $L = (n/p) \log(n/p)$.*

3.2 D-BSP: Locality and the Broadcast Problem

Many parallel algorithms work in a recursive way, i. e., the parallel machine is split into two or more independent parts, or sets of processors, where similar subproblems of smaller size, have to be solved. It is often possible to partition a real parallel machine into independent parts such that we can consider the latency and the gap as functions $L(k)$ and $g(k)$, resp., of the size k of a set of processors because it can make sense that, e. g., the latency on a small machine is much smaller than on a huge machine. In the *decomposable* BSP machine model (D-BSP) introduced by de la Torre and Kruskal [15], in a step the machine can be partitioned into two equally-sized submachines. This partitioning process may be repeated on some submachines (i. e., submachines of different sizes are possible simultaneously). In the D-BSP model, the runtime of a superstep *on a single submachine* is $w_t + h_t \cdot g(p_t) + L(p_t)$, with p_t denoting the size of this submachine. The runtime of a superstep is the maximum runtime taken over the submachines.

In the following, we present an algorithm for the broadcast problem that can easily be implemented on the D-BSP machine. In the broadcast problem, processor P_1 holds an item α that has to be sent to the remaining $p - 1$ processors.

An obvious plain BSP algorithm for this problem works in $\log p$ supersteps. In superstep t, every processor P_i, $1 \leq i \leq 2^{t-1}$ sends α to processor P_{i+2^t} (if this processor exists). The BSP parameters are $T = W = H = O(\log p)$ so the BSP runtime is $O(\log p + g \cdot \log p + L \cdot \log p)$.

For example, suppose the interconnection mechanism to be a hypercube with Gray code numbering, guaranteeing $L(2^k) = O(k)$, and suppose $g(2^k) = O(1)$. Then the runtime of the plain BSP algorithm implemented directly on this D-BSP machine is $\Theta((\log p)^2)$.

The following recursive broadcast algorithm dubbed $\text{ROOT}(p)$ has been introduced by Juurlink *et al.* [12] (also see [14]): If there are more than two processors, $\text{ROOT}(\sqrt{p})$ is executed on $P_1, \ldots, P_{\sqrt{p}}$ (appropriate rounding provided). Then every processor P_i, $i \in \{1, \ldots, \sqrt{p}\}$, sends α to $P_{i \cdot \sqrt{p}}$. Now, the machine is partitioned in $\sqrt{p}$ groups of size $\sqrt{p}$ each, where finally $\text{ROOT}(\sqrt{p})$ is executed.

It is a nice exercise to write a D-BSP program that implements algorithm $\text{ROOT}(p)$ and to prove the following theorem.

The (recursive) partition of the problem into subproblems that have to be solved on 'compact' submachines is the impact of the locality function on the algorithm design.

Theorem 4. *On the D-BSP, algorithm* ROOT(p) *has a runtime of*

$$O\left(\sum_{i=0}^{\log\log p} 2^i \cdot \left(L(p^{1/2^i}) + g(p^{1/2^i}) \right) \right) .$$

For our hypercube example, the runtime of ROOT(p) is $O(\log p \log\log p)$.

Correspondingly to Theorem 4, Juurlink *et al.* prove a lower bound [12] that matches this upper bound for large ranges for $L(k)$ and $g(k)$.

Theorem 5. *Any algorithm for the broadcast problem on a p-processor D-BSP machine with latency $L(k)$ and gap $g(k)$ with $\log k / \log\log k \leq L(k) \leq (\log k)^2$ and $\log k / \log\log k \leq g(k) \leq (\log k)^2$ for all k, $1 \leq k \leq p$, has runtime*

$$\Omega\left(\sum_{i=0}^{\log\log p} 2^i \cdot \left(L(p^{1/2^i}) + g(p^{1/2^i}) \right) \right) .$$

4 The PUB Lib

In the previous sections, we showed how the BSP model and its extensions influence the design of efficient parallel algorithms. The other important aim of a bridging model like, in our case, BSP is to achieve portability of programs. That means there should be an environment on parallel machines that supports the coding of BSP algorithms independent from the underlying machine and that executes the code efficiently.

Such BSP environments are the Oxford BSPlib [10], the Green library [9], and the PUB Lib (Paderborn University BSP Library) [5].

The PUB Lib is a C Library to support the development and implementation of parallel algorithm designed for the BSP model. It provides the use of block-wise communication as suggested by the BSP* model, and it provides the use of locality as suggested in the D-BSP model by allowing to dynamically partition the machine into independent submachines.

It has a number of additional features:

- Collective communication operations like, e. g., broadcast and parallel prefix are provided and implemented in an architecture independent way. Furthermore, as they are non-synchronizing, they have very good runtimes.
- By providing partitioning and subset synchronization operations it is possible for users of the PUB Lib to implement and analyze algorithms for locality models as discussed in the previous section.
- Oblivious synchronization is provided.
- Virtual processors are incorporated into the PUB Lib. These processors can be used for solving problems of sizes that do not fit into main memory efficiently on a sequential machine. Furthermore, these virtual processors serve as a parallel machine simulator for easy debugging on standard sequential programming environments.

The PUB Lib has been successfully installed on the Cray T3E, Parsytec CC, GCel, GCPP, IBM SP/2, and workstation clusters, and on various single CPU systems.

All information for downloading and using the PUB Lib can be found on the project's webpage http://www.upb.de/~pub/.

Currently, the implementation of the migration of virtual processors from heavily loaded CPUs to less loaded CPUs is in progress.

Acknowledgments

The effort on BSP, BSP*, and on designing and analyzing algorithms for them and on implementing the PUB Lib in Paderborn is the work of many individuals: Armin Bäumker, Olaf Bonorden, Wolfgang Dittrich, Silvia Götz, Nicolas Hüppelshäuser, Ben Juurlink, Ingo von Otte, and Ingo Rieping who are certainly also authors of this paper.

References

1. M. Adler, P.B. Gibbons, Y. Matias, V. Ramachandran. Modeling parallel bandwidth: local versus global restrictions. *Algorithmica* 24 (1999) 381–404.
2. A. Bäumker. *Communication Efficient Parallel Searching*. Ph.D. Thesis, Paderborn University, 1997.
3. A. Bäumker, W. Dittrich, F. Meyer auf der Heide. Truly efficient parallel algorithms: 1-optimal multisearch for an extension of the BSP model. *TCS* 203 (1998) 175–203.
4. A. Bäumker, F. Meyer auf der Heide. Communication efficient parallel searching. In: *Proc. 4th Symp. on Solving Irregularly Structured Problems in Parallel (IRREGULAR)*, 1997, pp. 233–254.
5. O. Bonorden, B. Juurlink, I. von Otte, I. Rieping. The Paderborn University BSP (PUB) Library — Design, Implementation and Performance. In: *Proc. 13th International Parallel Processing Symposium & 10th Symposium on Parallel and Distributed Processing (IPPS/SPDP)*, 1999, pp. 99–104.
6. D. Culler, R. Karp, D. Patterson, A. Sahay, K. Schauser, E. Santos, R. Subramonian R., T. von Eicken. LogP: A practical model of parallel computation. *C.ACM* 39(11) (1996) 78–85.
7. F. Dehne, A. Fabri, A. Rau-Chaplin. Scalable parallel computational geometry for coarse grained multicomputers. *Int. J. Computational Geometry & Applications* 6 (1996) 379–400.
8. W. Dittrich. *Communication and I/O Efficient Parallel Data Structures*. Ph.D. Thesis, Paderborn University, 1997.
9. M.W. Goudreau, K. Lang, S.B. Rao, T. Suel, T. Tsantilas. Portable and efficient parallel computing using the BSP model. *IEEE Transactions on Computers* 48 (1999) 670–689.
10. J. Hill, B. McColl, D. Stefanescu, M. Goudreau, K. Lang, S. Rao, T. Suel, T. Tsantilas, R. Bisseling. BSPlib: The BSP programming library. *Parallel Computing* 24 (1998) 1947–1980.
11. B.H.H. Juurlink. *Computational Models for Parallel Computers*. Ph.D. Thesis, Leiden University, 1997.
12. B.H.H. Juurlink, P. Kolman, F. Meyer auf der Heide, I. Rieping. Optimal broadcast on parallel locality models. In: *Proc. 7th Coll. on Structural Information and Communication Complexity (SIROCCO)*, 2000, pp. 211–226.
13. J.H. Reif, S. Sen. Randomized algorithms for binary search and load balancing on fixed connection networks with geometric applications. *SIAM J. Computing* 23 (1994) 633–651.
14. I. Rieping. *Communication in Parallel Systems – Models, Algorithms and Implementations*. Ph.D. Thesis, Paderborn University, 2000.
15. P. de la Torre, C. P. Kruskal. Submachine locality in the bulk synchronous setting. In: *Proc. 2nd European Conference on Parallel Processing (Euro-Par)*, 1996, 352–358.
16. L. Valiant. A bridging model for parallel computation. *C.ACM* 33(8) (1990) 103–111.

A Coarse-Grained Parallel Algorithm for Maximal Cliques in Circle Graphs

E. N. Cáceres[1], S. W. Song[2], and J. L. Szwarcfiter[3]

[1] Universidade Federal do Mato Grosso do Sul
Departamento de Computação e Estatística
Campo Grande, MS, 79069, Brazil
edson@dct.ufms.br
[2] Universidade de São Paulo
Departamento de Ciência da Computação - IME
São Paulo, SP - 05508-900 - Brazil
song@ime.usp.br
[3] Universidade Federal do Rio de Janeiro
Instituto de Matemática and Núcleo de Computação Eletrônica
Rio de Janeiro, RJ, 21945-970, Brazil
jayme@nce.ufrj.br

Abstract. We present a parallel algorithm for generating the maximal cliques of a circle graph with n vertices and m edges. We consider the Coarse-Grained Multicomputer Model (CGM) and show that the proposed algorithm requires $O(\log p)$ communication rounds, where p is the number of processors, independent of n. The main contribution is the use of a new technique based on the unrestricted depth search for the design of CGM algorithms.

1 Introduction

In this paper we present a new CGM/BSP parallel algorithm for finding maximal cliques in circle graphs.

Circle graphs are a special kind of intersection graphs. Recognition of circle graphs has been an open problem for many years until the mid-eighties with solutions discovered independently by several researchers [1, 8, 10]. These algorithms basically makes use of a graph decomposition technique [5, 6]. The presented algorithms are sequential polynomial algorithms. a We consider the problem of generating all maximal cliques in a circle graph of n vertices and m edges. Szwarcfiter and Barroso have studied this problem and presented a sequential $O(n(m + \alpha))$ algorithm [11] where α is the number of maximal cliques of the graph. Cáceres and Szwarcfiter [3] present a PRAM algorithm of $O(\alpha \log^2 n)$ time using n^3 processors in a CREW PRAM.

To our knowledge there are no known parallel algorithms for finding maximal cliques in circle graphs under the coarse-grained parallel computing model. Based on [11] and [3] we present a CGM (Coarse-Grained Multicomputer) algorithm for this problem that requires $O(\log p)$ communication rounds. The main contribution of this paper is the introduction of the *unrestricted search* as a new technique for the design of CGM algorithms.

2 Coarse-Grained Multicomputer (CGM) Model

The PRAM model has been extensively utilized to produce many important theoretical results. Such results from PRAM algorithms, unfortunately, do not necessarily match the speedups observed on *real* parallel machines.

In this paper, we present a parallel algorithm that is based on a more practical parallel model. More precisely, we will use a version of the BSP model [13] referred to as the *Coarse Grained Multicomputer* (CGM) model [7]. In comparison to the BSP model, the CGM allows only bulk messages in order to minimize message overhead. A CGM consists of a set of p processors $P_1, \ldots, P_p$ with $O(N/p)$ local memory per processor and an arbitrary communication network (or shared memory). A CGM algorithm consists of alternating local computation and global communication rounds. Each communication round consists of routing a single h-relation with $h = O(N/p)$, i.e. each processor sends $O(N/p)$ data and receives $O(N/p)$ data. We require that all information sent from a given processor to another processor in one communication round is packed into one long message, thereby minimizing the message overhead. A CGM computation/communication round corresponds to a BSP superstep with communication cost $g\frac{N}{p}$. Finding an optimal algorithm in the coarse grained multicomputer model is equivalent to minimizing the number of communication rounds as well as the total local computation time. The CGM model has the advantage of producing results which correspond much better to the actual performance on commercially available parallel machines.

3 Notation and Terminology

Circle graphs are intersection graphs of a family of chords of a circle. Consider a family of n chords, numbered as $1 - 1', 2 - 2', \ldots, n - n'$, in a circle C (or equivalently in a square) (see Fig. 1 on the left). Assume that any two chords do not share a same endpoint. This corresponds to a circle graph $G = (V, E)$ where each chord corresponds to a vertex: $V = \{1, 2, \ldots, n\}$ and the edge set E is formed by edges (u, v) if chord u intersects chord v. The corresponding circle graph is shown on the right of Fig. 1.

A *circular sequence* S of G is the sequence of the $2n$ distinct endpoints of the chords in circle C, by traversing C in a chosen direction, starting at a given point in C. Denote by $S_1(v)$ and $S_2(v)$ respectively the first and second instances in S of the chord corresponding to $v \in V$ in C. Denote by $S_i(v) < S_j(w)$ (for $i, j = 1, 2$) when $S_i(v)$ precedes $S_j(w)$ in S. We have $S_1(v) < S_2(v)$.

Let $G = (V, E)$ be a circle graph, S the circular sequence of G. An S_1-*orientation G* of G is an orientation in which any directed edge $(v, w) \in E$ satisfies $S_1(v) < S_1(w)$. The S_1-orientation is an acyclic digraph (directed graph).

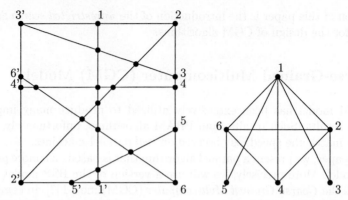

Fig. 1. Family of chords (left) and the corresponding circul graph on the right $G = (V, E)$

Let G denote an acyclic orientation of G. $A_v(G)$ and $A_v^{-1}(G)$ are the subsets of vertices that leave or enter v, respectively. For $v, w \in V$, v is an *ancestral* of w in G if the directed graph contains a path $v - w$. In this case, w is a *descendant* of v. Denote by $D_v(G)$ the set of descendants of v. If $w \in D_v(G)$ and $v \neq w$, then v is a *proper ancestral* of w and w a *proper descendant* of v. G is denominated a *transitive digraph* with respect to edges when $(v, w), (w, z) \in E$ implies $(v, z) \in E$. The *transitive reduction* G_R is the subgraph of G formed by the edges that are not motivated by transitivity. In other words, the transitive reduction of a directed graph G is the subgraph G_R with the smallest number of edges such for every path between vertices in G, G_R has a path between those vertices.

Let $G = (V, E)$ be a non-directed graph, $|V| > 1$ and G an acyclic orientation of G.

Let $v, w \in V$. We denote by $Z(v, w) \subset V$ the subset of vertices that are simultaneously descendants of v and ancestrals of w in G. An edge $(v, w) \in E$ *induces a local transitivity* when $G(Z(v, w))$ is a transitive digraph. Clearly, in this case the vertices of any path from v to w induce a clique in G. Furthermore, (v, w) *induces a maximal local transitivity* when there does not exist $(v', w') \in E$ different from (v, w) such that v' is simultaneously an ancestral of v and w' a descendant of w in G. (v, w) is denominated a *maximal edge*. The orientation G is *locally transitive* when each of its edges induces local transitivity. Fig. 2 shows an example of a locally transitive orientation.

Based on the following theorem, one can use locally transitive orientations for finding maximal cliques.

Theorem 1. *Let $G = (V, E)$ be a graph, G a locally transitive orientation of G and G_R the transitive reduction of G. Then there exists a one-to-one correspondence between the maximal cliques of G and paths $v - w$ in G_R, for all maximal edges $(v, w) \in E$.*

The proof can be found in [11].

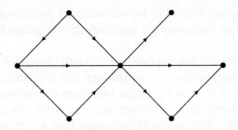

Fig. 2. A locally transitive digraph $G = (V, E)$

4 Algorithm for Unrestricted Search in Acyclic Digraphs

In a depth search of a graph, each edge is visited a constant number of times. Such a search process is known as a restricted search. An unrestricted search in a graph G is a systematic process of traversing G such that each edge is visited a finite number of times. The only restriction is that the process terminate.

The sequential unrestricted depth search of a connected graph $G = (V, E)$ is given as follows. The algorithm starts with a given vertex v chosen as root. It uses a stack Q.

Unrestricted Search (v)
Mark v
Push v onto stack Q
For each vertex w adjacent to v do
    if w is not in Q then
        visit edge (v, w)
        Unrestricted Search (w)
Pop v from Q
Unmark v

Now we will implement this algorithm in CGM. First we give some definitions. Given a digraph $G = (V, E)$, for each vertex $v \in V$ we define the *successor* of v ($\mathrm{suc}[v]$) as a fixed element of the adjacency list of v. Let $e = (u, v) \in E$. We define the *successor* of e as the edge $(v, \mathrm{suc}[v])$. A kl-path is a path C in G with initial edge $e = (k, l)$. An edge $e \in E$ belongs to kl-path C if e is the successor of some edge on the path C, and $e \notin C$. An edge (u, v) can belong to more than one kl-path.

For a undirected graph G, we consider each edge (u, v) as two distinct directed edges (u, v) and (v, u).

Since the successor of each vertex is fixed, all the edges on a kl-path incident with a vertex v have a same successor $(v, \mathrm{suc}[v])$. A kl-path can be a simple path, a cycle, a path together with a cycle or the union of two cycles. In the case when $G = (V, E)$ is an acyclic digraph, the kl-paths are formed by simple paths, with an initial vertex k and a final vertex t.

We present below an algorithm for unrestricted depth search in acyclic digraphs. The algorithm computes all maximal simple paths from a given root r.

The algorithm initially decomposes the digraph $G = (V, E)$ into a set of kl-paths. This decomposition is obtained through the definition of the successors of vertices and edges of G. The algorithm initializes the successor of each vertex $v \in V$ as the first element of its adjacency list of v. After the decomposition, the algorithm explores the edges of the kl-path such that $k = r$, where r is the root of the search and $l = \mathrm{suc}[r]$. The visited edges form a maximal simple path C. Once we have determined a maximal simple path $C = \{v_0, \cdots, v_p\}$, for $r = v_0$ on the kl-path, we can obtain a new maximal simple path, if it exists, as follows. We determine the last vertex $v_i \in C$ that has some vertex in its adjacency list that has not been visited yet. The successors of the vertices $v_j \notin \{v_0, \cdots, v_i\}$ are modified to be the first element of the adjacency list of each v_j, and the successor of v_i is altered to be the element of the adjacency list of v_i immediately following $\mathrm{suc}[v_i]$. The successors of the vertices $\{v_0, \cdots, v_{i-1}\}$ remain unaltered. The new definition of successors determines a new decomposition of the digraph into a set of kl-paths. We determine a simple path $C = \{v_i, \cdots, v_t\}$ on a kl-path in the new decomposition, with $k = v_i$ and $l = \mathrm{suc}[v_i]$. The path $C = \{v_0, \cdots, v_i, \cdots, v_t\}$ formed by the union of the paths $\{v_0, \cdots, v_i\}$ and $\{v_i, \cdots, v_t\}$ is a new maximal simple path. The remaining maximal simple path, if they exist, are computed analogously.

All the vertices w_i that are not reachable from the root vertex r are not included in the search.

Algorithm: Unrestricted Search

1. Define a root r, construct the adjacency lists of G and initialize the maximal simple path CM.
2. Decompose the digraph G into a set of kl-paths.
3. Determine a simple path C from r on the kl-path, with $k = r$ and $l = \mathrm{suc}[r]$.
4. Compute the maximal simple path $CM = CM \cup C$. Verify the existence of any vertex $v_i \in CM$ that has in its adjacency list a vertex that has not been visited yet.
5. In case v_i exists, $CM = \{v_0, \cdots, v_{i-1}\}$. Alter the successors of the vertices $v_i \notin CM$, and unmark the edges $e \in G$ that do not belong to the path CM.
6. Apply steps 2, 3, 4, and 5 to the set of unmarked edges of G for $r = v_i$ until all possible maximal simple paths from v_0 have been explored.

First, we rank the adjacency lists of all $v \in V$ and define the successor of v to be the first element of the adjacency list of v.

We start with the root r and with the defined successors, this represents a maximal path in G. Since our graph is acyclic, using list ranking, we can rank all vertices of this path. After this, we mark the last vertex of the path and change the successor (if there is one) of the tail of this path and compute another maximal path. Otherwise, using a proper data structure, we can backtrack in this path and visit a vertex that was not visited before.

The unrestricted search in this acyclic graph can be done in $O(\log p)$ communication rounds.

5 The Maximal Cliques Algorithm

An S_1-orientation G of a given circle graph G can be easily obtained through its circular sequence. We thus assume that a S_1-orientation is given as input. It can also be observed [11] that if G is locally transitive, then $(v, w) \in E$ is a maximal edge if and only if $A_v(G) \cap A_w(G) = A_v^{-1}(G) \cap A_w^{-1}(G) = \emptyset$.

Step 3 below is the main step of this algorithm and is based on the unrestricted search.

5.1 Algorithm Description

CGM Algorithm: Maximal Cliques

1. Construct the transitive reduction G_R.
2. Determine all maximal edges of G.
3. For each maximal edge $(v, w) \in E$, determine all paths $v - w$ in G_R.

5.2 An Illustrative Example

Consider the circle graph G of Fig. 1. Consider the S_1-orientation shown in Fig. 3 (a).

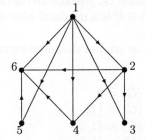 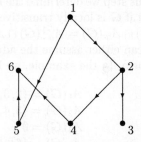

Fig. 3. (a) Graph G (S_1-orientation) and (b) Transitive Reduction Graph G_R

Step 1

Determine the transitive reduction G_R of G.

We can easily determine the adjacency lists $A_v^{-1}(G)$ for all the vertices of G. We can also obtain the lists $A_v^{-1}(G)$ by partitioning the edge lists into sublists, one for each vertex v. We use the array of lists ADJ to store the adjacency lists. $ADJ[v_i]$ contains the list of vertices v_j that arrives in v_i. Using recursive doubling, we remove from each adjacency list $ADJ[v_i]$ all the vertices that are in adjacency lists $ADJ[v_j]$, for all $v_j \in ADJ[v_i]$. This can be done as follows.

for all $v_i \in V$ in parallel do
    $ADJ[v_i] \leftarrow A_{v_i}^{-1}(G)$
    $BADJ[v_i] \leftarrow \cup_{v_j \in ADJ[v_i]} ADJ[v_j]$
repeat $\lceil \log n \rceil$ times
    for all $v_i \in V$ in parallel do
        $BADJ[v_i] \leftarrow BADJ[v_i] \cup \{\cup_{v_j \in BADJ[v_i]} BADJ[v_j]\}$
for all $v_i \in V$ in parallel do
    $ADJ[v_i] \leftarrow ADJ[v_i] - BADJ[v_i]$

The adjacency lists $A_{v_i}^{-1}(G_R)$ are given by $ADJ[v_i]$.

The main step that is repeated $O(\log n)$ times amounts to using the *pointer jumping* or *recursive doubling* technique. By this technique, it can be easily shown that the CGM algorithm requires $O(\log n)$ communication rounds. We can however do better by using the p^2-ruling set technique presented in [4] for the list ranking problem. An *r-ruling set* is defined as a subset of selected list elements that has the following properties: (1) No two neighboring elements are selected. (2) The distance of any unselected element to the next selected element is at most r. Because of the characteristics of circle graphs and the S_1-orientation, we have at most $O(n^2)$ partial lists in graph G. So this step can be done with $O(\log p)$ communication rounds.

For our example G_R is

$$\boxed{(1,2)(1,5)(2,3)(2,4)(4,6)(5,6)}$$

Step 2

In this step we determine the maximal edges of G. We base on the observation [11] that if G is locally transitive, then $(v, w) \in E$ is a maximal edge if and only if $A_v(G) \cap A_w(G) = A_v^{-1}(G) \cap A_w^{-1}(G) = \emptyset$.

We can either assume the adjacency lists as part of the input or easily computer them. In the example, we have

$$
\begin{array}{ll}
A_1(G) = \{2,3,4,5,6\} & A_1^{-1}(G) = \emptyset \\
A_2(G) = \{3,4,6\} & A_2^{-1}(G) = \{1\} \\
A_3(G) = \emptyset & A_3^{-1}(G) = \{1,2\} \\
A_4(G) = \{6\} & A_4^{-1}(G) = \{1,2\} \\
A_5(G) = \{6\} & A_5^{-1}(G) = \{1\} \\
A_6(G) = \emptyset & A_6^{-1}(G) = \{1,2,4,5\}
\end{array}
$$

We thus compute the following maximal edges.

$$\boxed{(1,3)(1,6)}$$

Step 2 can be computed with a constant number of communication rounds.

Step 3

In this step we first compute for each $v \in V$ the subsets $W(v)$ formed by vertices w such that (v, w) is a maximal edge. From the maximal edges obtained in Step 2 we get

$$W(1) = \{3, 6\} \quad W(2) = \emptyset$$
$$W(3) = \emptyset \qquad W(4) = \emptyset$$
$$W(5) = \emptyset \qquad W(6) = \emptyset$$

Construct now the subgraph H of G_R induced by the vertices that are simultaneously descendants of v and ancestrals of any $w \in W(v)$. This can be done by determining the transitive closure [9] G_R and through the intersection of the vertices that leave v with those that enter each of $w \in W(v)$. The paths $v - w$ in G_R taken from a certain vertex v are exactly the source-sink paths in H. These paths can be obtained through the parallel unrestricted depth search algorithm (see Section 4).

For our example the subgraph H is shown in Fig. 4.

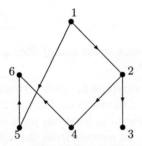

Fig. 4. Graph H

$$(1, 2)(1, 5)(2, 3)(2, 4)(4, 6)(5, 6)$$

Let us now perform an unrestricted search on the digraph H.

The unrestrited search in this acyclic graph can be done in $O(\log p)$ communication rounds.

Observe that digraph H is acyclic. Using the parallel unrestricted search algorithm of [2] we determine the maximal paths of H. For our example the maximal paths in H are:

$$C_1 = \{1, 2, 3\}$$
$$C_2 = \{1, 2, 4, 6\}$$
$$C_3 = \{1, 5, 6\}$$

The source-sink paths of H obtained in the previous step form maximal cliques of G.

6 Correctness

Lemma 1. *Let $G = (V, E)$ be a weakly connected digraph. Let $v \in V$ and a kl-path in G such that $k = v$ and $l = suc[v]$. Then the kl-path is: (i) a simple path; (ii) a cicle; or (iii) a path connected to a cycle.*

Proof: [3]

Lemma 2. *Let $G = (V, E)$ be an acyclic weakly connected digraph. Then each kl-path is a simple maximal path.*

Proof: [3]

Theorem 2. *Let $G = (V, E)$ be an acyclic weakly connected digraph. All maximal paths of G, starting at the root vertex r, will be found by the algorithm at least once.*

Proof: [3]

Theorem 3. *Let $G = (V, E)$ be an acyclic weakly connected digraph. None of the maximal paths of G, starting at the root vertex r, will be found by the algorithm more than once.*

Proof: [3]

Lemma 3. *Graph G_R obtained at the end of Step 1 of the parallel algorithm is the transitive reduction of G.*

Proof: At each iteration we compute for each vertex v the set L_v formed by the union of the adjacency lists of the vertices that arrive at v. The set L_v contains the vertices for which there exists at least one path of length $\geq 2^i$ to v. At each iteration i, for each vertex v, we remove, from the adjacency lists of the vertices that arrive at v, those vertives that belong to L_v. After $\lceil \log n \rceil$ iterations, all the vertices that are ancestral of v belong to L_v. Thus the adjacency lists will contain only vertices v_j for which there does not exist any path joining u_j to v different of (u_j, v).

Theorem 4. *The parallel algorithm computes all the maximal cliques correctly.*

Proof: By the previous lemma, the algorithm computes the transitive reduction correctly. By using the unrestricted search of an acyclic digraph, we determine all the maximal paths in the digraph H. The correctness result follows from the fact that the S_1−orientation G is locally transitive.

Acknowledgments

The first author is supported by PRONEX-SAI. The second author is supported by FAPESP (Fundação de Amparo à Pesquisa do Estado de São Paulo) Proc. No. 98/06138-2, CNPq Proc. No. 52.3778/96-1 and 46.1230/00-3, and CNPq/NSF Collaborative Research Program Proc. No. 68.0037/99-3. The third author is partially supported by the Conselho Nacional de Desenvolvimento Científico e Tecnológico, CNPq, and Fundação de Amparo à Pesquisa do Estado do Rio de Janeiro, FAPERJ, Brazil.

References

1. A. Bouchet.: Reducing Prime Graphs and Recognizing Circle Graphs. Combinatorica **7** (1987) 243–254
2. E.N. Cáceres.: Parallel Algorithms for Graph Problems. (In Portuguese.) PhD thesis, COPPE - UFRJ, Rio de Janeiro - RJ, (1992)
3. E. N. Cáceres and J. L. Szwarcfiter.: In preparation.
4. E. Cáceres, F. Dehne, A. Ferreira, P. Flocchini, I. Rieping, A. Roncato, N. Santoro, and S.W. Song.: Efficient Parallel Graph Algorithms For Coarse Grained Multicomputers and BSP. In: Proceedings ICALP '97 - 24th International Colloquium on Automata, Languages, and Programming. P. Degano, R. Gorrieri, A. Marchetti-Spaccamela (editors). Lecture Notes in Computer Science **1256** (1997) 390–400
5. W. H. Cunningham.: Decomposition of Directed Graphs. SIAM J. Alg. and Disc. Methods **3** (1982) 214–228
6. W. H. Cunningham and J. Edmonds.: A Combinatorial Decomposition Theory. Canad. J. Math. **32** (1980) 734–765
7. F. Dehne (Ed.), Coarse grained parallel algorithms. In: Special Issue of Algorithmica **24** (1999) 173–426
8. C. P. Gabor, W.L. Hsu, and K. J. Supowit.: Recognizing Circle Graphs in Polynomial Time. J. of Assoc. Comput. Mach. **36** (1989) 435–474
9. R. M. Karp and V. Ramachandran.: Parallel Algorithms for Shared-Memory Machines. In: J. van Leeuwen (ed.): Handbook of Theoretical Computer Science Vol. A. Chapter 17, The MIT Press/Elsevier (1990) 869–941
10. W. Naji.: Graphes des Cordes, Caractérisation et Reconnaissance. Disc. Math. **54** (1985) 329–337
11. J. L. Szwarcfiter and M. Barroso.: Enumerating the Maximal Cliques of Circle Graph. In: F.R.K. Chung, R.L. Graham, and D.F. Hsu (eds.): Graph Theory, Combinatorics, Algorithms and Applications, SIAM Publications (1991) 511–517
12. L. G. Valiant.: The Complexity of Enumeration and Reliability Problems. SIAM J. Comp. **8** (1979) 410–421
13. L. G. Valiant.: A Bridging Model for Parallel Computation. Communication of the ACM **33** (1990) 103–111

Parallel Models and Job Characterization for System Scheduling*

X. Deng[1], H. Ip[1], K. Law[1], J. Li[2], W. Zheng[3], and S. Zhu[1]

[1] Department of Computer Science, City University of Hong Kong
Tat Chee Avenue, Kowloon, Hong Kong, P.R. China
{csdeng, cship, cskckl, cszhusf}@cityu.edu.hk
[2] Department of Mathematics, Yunnan University
Kunming 650091, P.R. China
jianping@public.km.yn.cn
[3] Department of Computer Science, Tsinghua University
Beijing 100084, P.R. China
zwm-dcs@tsinghua.edu.cn

Abstract

In this work, we study job characterization in multi-programmed multiprocessor system by taking into consideration the parallel job models. We first introduce an example to illustrate the issues involved. Then we focus on two popular system scheduling policies: round-robin for single processor systems, and equi-partition for multiprocessor systems. We analytically study effect of job parallelization on the overall performance of the system, and also present simulation results. Through these studies, we discuss scheduling policies of parallel jobs in relation to the ratio of number of jobs and the number of processors in the multiprocessor system, and propose new paradigm for parallel algorithm designs.

1 Introduction

In contrary to single processor computers, the multiprocessor system environment creates many complications in design and analysis of architecture and algorithms. Different types of parallel architecture resulted in architecture-dependent algorithm designs in the early stages of parallel computing (and still do.) Later, several parallel models (e.g., [13, 18, 2]) emerged that proposed architecture independent parallel computing, and hence made it possible to have a general purpose parallel computer environment. In a general purpose parallel computer system, users submit jobs and the system schedule their execution to make full use of the system's computing power. Even though users may have a good estimation of the processing time of their submitted jobs, it is often not be taken into consideration by the system scheduler for

---

*Research is partially supported by a CERG grant (CityU 1074/00E) of Hong Kong RGC and a research grant of City U of Hong Kong (7001215).

various reasons: e.g., user programs may contain bugs that defy users' estimation; users may lie to gain advantage according to scheduling policies; and the resulted scheduler may cause extra overhead if all job information is taken into consideration. In fact, recent studies on parallel computer systems have favored simple policies (e.g., equi-partition and dynamic equi-partition [15, 19, 9, 10, 6, 8]) that originated from the Round-Robin policy in single processor systems. For such policies, no specific job information is required. We should consider such schedule policies in our discussion. In particular, we view a sequential algorithm for a problem as generating a single process, and parallel algorithm for a problem as generating k processes (k is determined by the particular algorithm and no more than the number of processors). Each process is assigned with an equal processing power when severed by the system as in the Round-Robin policy.

Despite the fact that it may be difficult to collect, or utilize all job information in a general purpose parallel system, there is still the possibility that some simple characterization of parallel jobs can be exploited to improve system performance [14, 3, 1, 11, 12]. To one extreme, some studies considered utilizing all job information in parallelization of jobs [13, 16, 17]. To understand job characterization in multiprocessor system, one must look into the type of jobs are presented to such systems, i.e., to study parallel algorithms developed to attack problems that demand high performance computing. While design of parallel algorithms is often dependent on the parallel models that they are going to be applied, the measurement of speedup is common to most of the models. The theoretical model of PRAMs, for example, call a parallel algorithm to be optimal if the product of the execution time T_p and the number of processors P is of the same magnitude as the best sequential time T_1: $P \times T_p = O(T_1)$ (we should call the ratio $\frac{P \times T_p}{T_1}$ the inefficiency.) In more recent models advocating architecture independent parallel algorithm designs by reducing communication cost ([18, 2, 5, 4]), extra works are often done to distribute data to be processed by individual processes (each executed in one single processor), with interleaved communication phases to coordinate and synchronize the processes.

While system scheduling calls for efficient utilization of computing resources, parallelism achieves speedup at the cost of introducing inefficiency (see, e.g., [7]). A comprehensive study of parallel system scheduling therefore should take into the consideration of the trade-off between the benefit of speedup and the cost of inefficiency when introducing parallelism for task processing. We focus on this factor of inefficiency in our study of scheduling policies. In particular, we are interested in decisions to execute a task with a parallel algorithm or a sequential algorithm in a general purpose parallel computer system environment, with n jobs and P processors. Very often, for the same parallel algorithm (may work for different number of processors), the inefficiency resulted from communication delays would be the same, no matter what is the number of processes the task is divided into. For example, the inefficiency that is resulted from communication delays (for the parallel system communication), the number of communication rounds (as advocated in architecture independent parallel algorithm designs), and the extra work necessary for efficient synchronization of separated computing processes are usually the same no matter how many processes are used, especially for new parallel algorithm design paradigms mentioned above. In design of algorithm for coarse grained multicomputers [4], for example, algorithms often require $O(\frac{T_1}{p})$ local computation time (the constant hidden in the big-O is independent of both T_1 and p). When the communication costs are sufficiently small (as a major requirement in design of coarse grained parallel algorithms), the inefficiency ratio is mainly determined by this constant in the local computation time, and thus independent of both T_1 and p.

We exploit the associated inefficiency for each job in the design of our scheduling policy. When inefficiency is bigger than one, it may not be helpful to execute jobs with parallel algorithms, especially when the number of jobs is relatively large. However, when the number of jobs is reduced to sufficiently small than the number of processors of the system, it may become necessary to execute these remaining jobs in parallel mode. However, we may have to start these jobs from the beginning since we may not know which are the jobs that we would like to execute and they may have been executed for some time in the system, under Round-Robin policy. This introduces a further inefficiency into the system performance. It is thus desirable to have jobs that could be execute in sequential mode first and then continue with parallel mode for the remaining part at any point in time. This poses a new challenge for parallel algorithm designs. We should not deal with the new challenge here but study the benefit of the existence of such algorithms.

Section 2 analyzes a special case of two jobs and three processors to illustrate the issues that might arise from our consideration. We consider the inefficiencies from generating two or three processes for each job and obtain analytic results. In Section 3, we study a general case of n jobs and p processors. We deal with the problem with a comparison of two policies: parallel processing and sequential processing. We aim at obtaining a policy that decides to introduce parallelism for job processing or to use sequential algorithms. In Section 4, we consider the more general case, where there are more jobs than threshold of introducing parallelism. In this situation, we consider the benefit of introducing parallelism, after many of the jobs finished execution, to the last few jobs remained uncompleted. Two models are considered: for one, only the remained portions are parallelized, and for another, jobs have to restart from the beginning in the parallel execution mode. The former is obviously more desirable but it would put harder requirement on parallel algorithm designs (and one that has not been noticed before to our best knowledge.) In Section 5, we conclude with discussion on the parallel algorithm design issues arisen from our study.

2 An Illustrative Example

In this section, we use our devised parallel algorithm to execute a task with two jobs $\mathcal{J}_1$, $\mathcal{J}_2$ on three identical parallel processors $\mathcal{M}_1$, $\mathcal{M}_2$, $\mathcal{M}_3$. Here, two jobs $\mathcal{J}_1$, $\mathcal{J}_2$ require respectively x_1, x_2 units of unknown processing time, but we shall know their exact values at the final execution.

Our devised parallel algorithm is as follows: For any job $\mathcal{J}_i$, we consider the two possibilities to process it, i.e., either $\mathcal{J}_i$ is parallelized in average into two small jobs requiring $\frac{\beta_2}{2} \cdot x_i$ units of processing time or it is serialized, then each part (including parallelized ones) will be parallelized to process one unit of processing time per each time on one of three processors. We may always assume $\beta_2 \leq 2$ below.

By using our devised algorithm, we get a first follwoing result of total completion time involving in either parallelizing one or two jobs in average into small parts or serializing two jobs, whose proof can be found in the full version.

Theorem 1 Suppose three identical parallel processors $\mathcal{M}_1$, $\mathcal{M}_2$, $\mathcal{M}_3$ and two jobs $\mathcal{J}_1$, $\mathcal{J}_2$ which require x_1, x_2 units of unknown processing time, respectively.

(i) If $1 \leq \beta_2 \leq \frac{6}{5}$, then we get the (determined) minimum total completion time by parallelizing each of two jobs $\mathcal{J}_1$, $\mathcal{J}_2$ in average into two small parts than any other ways;

(ii) If $\frac{6}{5} \leq \beta_2 \leq 2$ and it is assumed that two processing times x_1 and x_2 are both randomized variables with uniformly identical distribution, then the expected value of total completion time by parallelizing both jobs $\mathcal{J}_1$, $\mathcal{J}_2$ in average into two small parts is better than any other ways if $\frac{6}{5} < \beta_2 \leq \frac{18}{11}$ and the expected value of total completion time by only parallelizing one of two jobs in average into two small parts is better than any other ways if $\frac{18}{11} < \beta_2 \leq 2$.

In addition, we consider another possibility to process $\mathcal{J}_i$, i.e., $\mathcal{J}_i$ is parallelized in average into three small jobs requiring $\frac{\beta_3}{3} \cdot x_i$ units of processing time, where $\beta_3 \leq 3$. We can get following result whose proof can be found in the full version.

Theorem 2 Suppose three identical parallel processors $\mathcal{M}_1$, $\mathcal{M}_2$, $\mathcal{M}_3$ and two jobs $\mathcal{J}_1$, $\mathcal{J}_2$ which require at random x_1, x_2 units of processing time, respectively. The three variables Φ, Ψ and Υ are defined above. Then

(i) If $1 \leq \beta_2 \leq \frac{18}{13}$ and $\beta_3 \leq \frac{11}{10}\beta_2$, the minimum expected value of completion time of two jobs is obtained by parallelizing each of two jobs in average into three small parts with $\frac{\beta_3}{3}x_1$, $\frac{\beta_3}{3}x_2$ units of processing time, respectively;

(ii) If $1 \leq \beta_2 \leq \frac{18}{13}$ and $\frac{11}{10}\beta_2 \leq \beta_3 \leq 3$, the minimum expected value of completion time of two jobs is obtained by parallelizing each of two jobs in average into *two* small parts with $\frac{\beta_2}{2}x_1$, $\frac{\beta_2}{2}x_2$ units of processing time, respectively;

(iii) If $\frac{18}{13} \leq \beta_2 \leq 2$ and $\beta_3 \leq \frac{9}{20}\beta_2 + \frac{1}{10}$, the minimum expected value of completion time of two jobs is obtained by parallelizing each of two jobs in average into three small parts with $\frac{\beta_3}{3}x_1$, $\frac{\beta_3}{3}x_2$ units of processing time, respectively;

(vi) If $\frac{18}{13} \leq \beta_2 \leq 2$ and $\frac{9}{20}\beta_2 + \frac{1}{10} \leq \beta_3 \leq 3$, the minimum expected value of completion time of two jobs is obtained by parallelizing only one of two jobs in average into two small parts with $\frac{\beta_2}{2}x_1$ (or $\frac{\beta_2}{2}x_2$) units of processing time, respectively.

3 Decision to Process Job in Parallel or in Sequent

In this section, suppose that there are p identical parallel processors $\mathcal{M}_1$, $\mathcal{M}_2$, ..., $\mathcal{M}_p$ and n jobs $\mathcal{J}_1$, $\mathcal{J}_2$, ..., $\mathcal{J}_n$. We shall execute these n jobs on p identical parallel processors by following devised algorithm: For any job $\mathcal{J}_i$, we consider the two possibilities to process it, i.e., either $\mathcal{J}_i$ is parallelized in average into p small parts requiring $\frac{\alpha_i}{p} \cdot x_i$ units of processing time or it is serialized, then each part (including parallelized one) will be parallelized to process one unit of processing time per each time on one of p processors, where $\alpha_i \geq 1$ is called an expansion coefficient of job $\mathcal{J}_i$. If α_i is smaller, it is better to parallelize job $\mathcal{J}_i$ in average into p small parts requiring $\frac{\alpha_i}{p} \cdot x_i$ units of processing time, otherwise it is serialized. Roughly, if α_i satisfies $\alpha_i \leq \frac{p}{n}$ for each job $\mathcal{J}_i$, then it is better to parallelize each job in average into p small parts such that the completion time of these n jobs is better than that of serializing jobs (see Theorem 4); otherwise, we can parallelize partial jobs in average into p small parts such that the completion time of these n jobs is better than that of serializing jobs (see next section).

Firstly, if each of n jobs is serialized, we can get the total completion time of these n jobs to process these n jobs on p identical parallel processors as follows, whose proof can be found in the full version of this paper.

Theorem 3 Suppose that we process n jobs $\mathcal{J}_1$, $\mathcal{J}_2$, ..., $\mathcal{J}_n$ on p identical parallel processors $\mathcal{M}_1$, $\mathcal{M}_2$, ..., $\mathcal{M}_p$ by assigning an unit of processing time of each job to some processor per each time (i.e., each job is processed on at most one processor per each time), where n jobs require x_1, x_2, ..., x_n units of unknown processing times, respectively. Then

(i) If $n \leq p$, then the total completion time of these n jobs is $\sum_{i=1}^{n} x_i$;

(ii) If $p \leq n$, then the total completion time of these n jobs is $\sum_{i=1}^{n-p} \frac{2n-2i+1}{p} x_{j_i} + \sum_{i=n-p+1}^{n} x_{j_i}$, where $x_{j_1} \leq x_{j_2} \leq \cdots \leq x_{j_n}$. In particular, its total completion time is at most $\frac{n+p}{p} \sum_{i=1}^{n-p} x_{j_i} + \sum_{i=n-p+1}^{n} x_{j_i}$.

Secondly, if each of n jobs is parallelized into p small parts, we can get the total completion time of these n jobs as follows.

Theorem 4 Suppose that we parallelize each of n jobs $\mathcal{J}_1$, $\mathcal{J}_2$, ..., $\mathcal{J}_n$ in average into p small parts, each of which requires $\frac{\alpha_i}{p} \cdot x_i$ units of processing time, then assign each of which on p identical parallel processors $\mathcal{M}_1$, $\mathcal{M}_2$, ..., $\mathcal{M}_p$ by an unit per each time, where n jobs require x_1, x_2, ..., x_n units of unknown processing times, respectively. So the total completion time of these n jobs is as follows: $\frac{1}{p} \sum_{i=1}^{n} (2n - 2i + 1) \alpha_{j_i} x_{j_i} - \frac{n(n-1)}{2}$, where $\{j_1, j_2, \ldots, j_n\} = \{1, 2, \ldots, n\}$ satisfies $\alpha_{j_1} x_{j_1} \leq \alpha_{j_2} x_{j_2} \leq \cdots \leq \alpha_{j_n} x_{j_n}$. In particular, the total completion time of these n jobs is at most $\frac{n}{p} \sum_{i=1}^{n} \alpha_i x_i - \frac{n(n-1)}{2}$.

Proof. Since each job $\mathcal{J}_{j_i}$ is parallelized in average into p small jobs, each of which requires $\frac{\alpha_{j_i}}{p} \cdot x_{j_i}$ units of unknown processing time, we can suppose that the completion time of job $\mathcal{J}_{j_i}$ starting at the end of job $\mathcal{J}_{j_{i-1}}$ to the completion of job $\mathcal{J}_{j_i}$ is t_i for each $i \in \{1, 2, \ldots, n\}$ (here we put $x_{j_0} = 0$ and $\alpha_{j_0} = 0$ for convenience), then we get $t_1 = (\frac{\alpha_{j_1} x_{j_1}}{p} - 1)n + 1$ and $t_i = \frac{\alpha_{j_i} x_{j_i} - \alpha_{j_{i-1}} x_{j_{i-1}}}{p} \cdot (n - i + 1) + 1$ if $2 \leq i \leq n$. So the total completion time of n jobs $\mathcal{J}_1$, $\mathcal{J}_2$, ..., $\mathcal{J}_n$ is as follows

$$total = \sum_{j=1}^{n} \sum_{i=1}^{j} t_i = \frac{1}{p} \sum_{i=1}^{n} (2n - 2i + 1)\alpha_{j_i} x_{j_i} - \frac{n(n-1)}{2}.$$

Especially, by the fact $\frac{\alpha_{j_1} x_{j_1}}{p} \leq \frac{\alpha_{j_2} x_{j_2}}{p} \leq \cdots \leq \frac{\alpha_{j_n} x_{j_n}}{p}$, we easily get $\sum_{i=1}^{n} (2n - 2i + 1)\alpha_{j_i} x_{j_i} \leq \sum_{i=1}^{n} n\alpha_{j_i} x_{j_i} = n \sum_{i=1}^{n} \alpha_i x_i$. So we get $total \leq \frac{n}{p} \sum_{i=1}^{n} \alpha_i x_i - \frac{n(n-1)}{2}$, i.e., the total completion time is at most $\frac{n}{p} \sum_{i=1}^{n} \alpha_i x_i - \frac{n(n-1)}{2}$. □

For the case $\alpha_i = \alpha$ for each integer i, we get the following direct consequence as follows: the total completion time of these n jobs is $\frac{\alpha}{p} \sum_{i=1}^{n} (2n - 2i + 1) x_{j_i} - \frac{n(n-1)}{2}$, which is at most $\frac{n\alpha}{p} \sum_{i=1}^{n} x_i - \frac{n(n-1)}{2}$, where $x_{j_1} \leq x_{j_2} \leq \cdots \leq x_{j_n}$.

Here, we give some remarks: (i). For the case $n \leq p$ and $1 \leq \alpha \leq \frac{p}{n}$, we get $total < \sum_{i=1}^{n} x_i$, this means that it is better to parallelize each job in average into p small parts in this case such that the total completion times of these n jobs is smaller than that whose jobs are all serialized; (ii). In the preceding proof, we assume indeed that our devised parallel algorithm is processed on the order of $\mathcal{J}_{j_1}, \mathcal{J}_{j_2}, \ldots, \mathcal{J}_{j_n}$, but when the devised parallel algorithm will be processed on any order of these n jobs, then we can get the total completion time of these n jobs is at most $\frac{1}{p}\sum_{i=1}^{n}(2n-2i+1)\alpha_{j_i}x_{j_i}$, this means that the absolute deviation of our devised parallel algorithm is at most $\frac{n(n-1)}{2}$.

As an experimental result shown in Figure 1, when we assign n jobs $\mathcal{J}_1$, $\mathcal{J}_2$, ..., $\mathcal{J}_n$ to satisfy the uniform distributions, it is given out the ratio of $\sum_{i=1}^{n}x_i$ divided by $\frac{\alpha}{p}\sum_{i=1}^{n}(2n-2i+1)x_{j_i} - \frac{n(n-1)}{2}$ at the some different values of α and the number of jobs n. It is easy to see that the total completion time of parallelizing each of these n jobs in average into p small parts is less than that of serializing these n jobs if the ratio is greater than one. In addition, when we also assign n jobs $\mathcal{J}_1$, $\mathcal{J}_2$, ..., $\mathcal{J}_n$ to satisfy the negative exponent distribution or Poisson distribution respectively, we shall obtain the similar figure as Figure 1.

4   The Benefit of Parallelizing Partial Jobs

In this section, suppose that we have p identical parallel processors $\mathcal{M}_1$, $\mathcal{M}_2$, ..., $\mathcal{M}_p$ and n jobs $\mathcal{J}_1$, $\mathcal{J}_2$, ..., $\mathcal{J}_n$, where $n \le p$. Considering differently from parallelizing all jobs, we shall pay our attention to some possibilities of parallelizing partial jobs in average into p small parts. Roughly, by using the devised parallel algorithm in the proceeding section, we consider some possibilities of parallelizing partial jobs in average into p small parts in order to get the better total completion time than that of serializing these jobs. We consider two possibilities as follows: For some fixed integer s ($0 \le s \le n$), at the end of execution of the s shortest jobs, we have got the left $n - s$ jobs and we can parallelize each left job in average into p small parts. The first method is to parallelize each left job in average into p small parts, each final part of which requires $\frac{\alpha}{p} \cdot (x_i - x_s + 1)$ units of unknown processing time (here $x_s \ge 1$) , and the second one is to do so with each final part requiring $\frac{\alpha}{p} \cdot x_i$ units of processing time, where the parallelized job $\mathcal{J}_i$ requires x_i units of unknown processing time. We hope to obtain the better integer s such that the final total completion time of n jobs is minimum.

Theorem 5 Suppose that we have p identical parallel processors $\mathcal{M}_1$, $\mathcal{M}_2$, ..., $\mathcal{M}_p$ and n jobs $\mathcal{J}_1$, $\mathcal{J}_2$, ..., $\mathcal{J}_n$ which require respectively x_1, x_2, ..., x_n units of unknown processing time on some processors, where $n \le p$ and $x_1 \le x_2 \le \cdots \le x_n$. Let s be a fixed integer satisfying $0 \le s \le n$.

(i) At the end of s jobs $\mathcal{J}_1$, $\mathcal{J}_2$, ..., $\mathcal{J}_s$ executed and $x_s \ge 1$, suppose that each left job $\mathcal{J}_j$ requiring $x_j - (x_s - 1)$ units of unknown processing time is parallelized in average into p small jobs, each of which requires $\frac{\alpha}{p} \cdot (x_j - x_s + 1)$ units of processing time, then the total completion time of these n jobs $\mathcal{J}_1$, $\mathcal{J}_2$, ..., $\mathcal{J}_n$ is

$$\sum_{i=1}^{s} x_i + (n - s - \frac{\alpha(n-s)^2}{p})x_s + \frac{\alpha}{p}\sum_{j=s+1}^{n}(2n-2j+1)x_j + f(n,s),$$

where $f(n,s) = \frac{\alpha(n-s)^2}{p} - \frac{(n-s)(n-s-1)}{2}$;

(ii) At the end of s jobs $\mathcal{J}_1$, $\mathcal{J}_2$, ..., $\mathcal{J}_s$ executed (putting $x_0 = 0$ for convenience), suppose that each left job $\mathcal{J}_j$ requiring x_j units of unknown processing time is parallelized in average into p small jobs, each of which requires $\frac{\alpha}{p} \cdot x_j$ units of unknown processing time, i.e., we restart to process the $n - s$ jobs left by parallelizing each in average into p small parts, then the total completion time of n jobs $\mathcal{J}_1, \mathcal{J}_2, \ldots, \mathcal{J}_n$ is

$$\sum_{i=1}^{s} x_i + (n - s)x_s + \frac{\alpha}{p} \sum_{j=s+1}^{n} (2n - 2j + 1)x_j - \frac{(n - s)(n - s - 1)}{2}.$$

Proof. (i). We can get the total completion time of these n jobs in the two following stages:

Stage 1. At the end of the s executed jobs $\mathcal{J}_1, \mathcal{J}_2, \ldots, \mathcal{J}_s$, we get the total completion time of these s jobs as follows $total_1 = \sum_{i=1}^{s} x_i$.

Stage 2. At the end of the s executed jobs $\mathcal{J}_1, \mathcal{J}_2, \ldots, \mathcal{J}_s$ (here $x_s \geq 1$), we know that each left job $\mathcal{J}_j$ which requires $x_j - (x_s - 1)$ units of unknown processing time is parallelized in average into p small jobs, each of which requires $\frac{\alpha}{p} \cdot (x_j - x_s + 1)$ units of processing time. Put t_i to represent the completion time of job $\mathcal{J}_i$ starting at the end of job $\mathcal{J}_{i-1}$ to the completion of job $\mathcal{J}_i$ for each $i \in \{s + 1, s + 2, \ldots, n\}$ (note $t_s = x_s$ for convenience), by using similar arguments in the proof of preceding theorem, we easily get $t_{s+1} = (\frac{\alpha(x_{s+1} - x_s + 1)}{p} - 1)(n - s) + 1$ and $t_j = \frac{\alpha(x_j - x_{j-1})}{p} \cdot p(n - j + 1) + 1$ if $s + 2 \leq i \leq n$. Then we get the total completion time of $\mathcal{J}_{s+1}, \mathcal{J}_{s+2}, \ldots, \mathcal{J}_n$ as follows $total_2 = \sum_{i=s+1}^{n} \sum_{j=s}^{i} t_j$.

So the total completion time of n jobs $\mathcal{J}_1, \mathcal{J}_2, \ldots, \mathcal{J}_n$ is as follows

$$
\begin{aligned}
total &= \sum_{i=1}^{s} x_i + \sum_{i=s+1}^{n} \sum_{j=s}^{i} t_j = \sum_{i=1}^{s} x_i + (n - s)t_s + \sum_{j=s+1}^{n} (n - j + 1)t_j \\
&= \sum_{i=1}^{s} x_i + (n - s - \frac{\alpha(n - s)^2}{p})x_s + \frac{\alpha}{p} \sum_{i=s+1}^{n} (2n - 2i + 1)x_i \\
&\quad + \frac{\alpha(n - s)^2}{p} - \frac{(n - s)(n - s - 1)}{2}
\end{aligned}
$$

(ii). With the similar arguments in (i), we only consider the following stage: at the end of s executed jobs $\mathcal{J}_1, \mathcal{J}_2, \ldots, \mathcal{J}_s$, each left job $\mathcal{J}_j$ which also requires x_j units of processing time is parallelized in average into p small jobs and restart to be processed, each of which requires $\frac{\alpha}{p} \cdot x_j$ units of processing time. Put t_i to represent the completion time of job $\mathcal{J}_i$ starting at the end of job $\mathcal{J}_{i-1}$ to the completion of job $\mathcal{J}_i$ for each $i \in \{s + 1, s + 2, \ldots, n\}$ (note $t_s = x_s$ for convenience), we get $t_{s+1} = (\frac{\alpha x_{s+1}}{p} - 1)(n - s) + 1$ and $t_j = \frac{\alpha(x_j - x_{j-1})}{p} \cdot (n - j + 1) + 1$ if $s + 2 \leq i \leq n$. Then we get the total completion time of $\mathcal{J}_{s+1}, \mathcal{J}_{s+2}, \ldots, \mathcal{J}_n$ as follows $total_2 = \sum_{i=s+1}^{n} \sum_{j=s}^{i} t_j$.

So the total completion time of n jobs $\mathcal{J}_1, \mathcal{J}_2, \ldots, \mathcal{J}_n$ is as follows

$$total = \sum_{i=1}^{s} x_i + (n - s)x_s + \frac{\alpha}{p} \sum_{i=s+1}^{n} (2n - 2i + 1)x_i - \frac{(n - s)(n - s - 1)}{2}$$

$\square$

We give some remark: In the proof of Theorem 5, we assume indeed that our devised parallel algorithm is processed on the order of $\mathcal{J}_1, \mathcal{J}_2, \ldots, \mathcal{J}_n$, but when the devised parallel algorithm will be processed on any order of these n jobs, then we can get the total completion time of these n jobs is at most $\sum_{i=1}^{s} x_i + (n - s - \frac{\alpha(n-s)^2}{p})x_s + \frac{\alpha}{p}\sum_{j=s+1}^{n}(2n - 2j + 1)x_j$ in the first method (or $\sum_{i=1}^{s} x_i + (n - s)x_s + \frac{\alpha}{p}\sum_{j=s+1}^{n}(2n - 2j + 1)x_j$ in the second method, respectively), this means that the absolute deviation of our devised parallel algorithm is at most the absolute value of $\frac{\alpha(n-s)^2}{p} - \frac{(n-s)(n-s-1)}{2}$ in the first method (or $\frac{(n-s)(n-s-1)}{2}$ in the second method, respectively).

In the Figures 2 and 3, we give out some experimental results of the ratios of $\sum_{i=1}^{n} x_i$ divided by the total completion time of the two models stated in Theorem 5 when we assign n jobs $\mathcal{J}_1, \mathcal{J}_2, \ldots, \mathcal{J}_n$ to satisfy the uniform distributions. It is shown that the first model that the remained portions are parallelized is better than the second model that restarts from the beginning in the parallel execution mode. On the other hand, it would be better not to parallelize these n jobs if the value of α is increased. In addition, when we also assign n jobs $\mathcal{J}_1, \mathcal{J}_2, \ldots, \mathcal{J}_n$ to satisfy the negative exponent distribution or Poisson distribution respectively, we shall obtain the similar figures as Figures 2 and 3.

5 Remarks and Conclusion

In this work, we study the tradeoff of parallelism and efficiency in terms of minimizing the mean completion time of jobs for jobs with known inefficiency but unknown execution time. Our characterization of parallel jobs are based observation on parallel algorithms designs. Our results show that best performance would be achieved if jobs can be executed in single process mode first and then continue in parallel mode without creating much extra inefficiency. This may lead to interesting questions and new models for design of parallel algorithms.

We are particularly interesting to find parallel algorithms that allow for sequential execution with time $T_1 + o(T_1)$ or parallel execution with time $T_p + o(T_p)$ for problems with best known sequential algorithm of time T_1 and best known parallel algorithm of time T_p with p processors. Most interesting solutions would have sequential time τ_1 for duration in sequential mode and parallel time τ_p for duration in parallel mode such that the remaining portion of the sequential execution mode, $T_1 - \tau_1$ is as efficiently parallelized as the original problem. That is, $\frac{p\tau_p}{T_1 - \tau_1}$ is roughly the same as $\frac{pT_p}{T_1}$.

References

[1] T. Brecht, and K. Guha, *Using Parallel Program Characteristics in Dynamic Processor Allocation Policies*, to appear in Performance Evaluation.

[2] D. Culler, R. Karp, D. Patterson, A. Sahay, K. Schauser, E. Santos, R. Subramonian, and T. von Eicken, *LogP: Towards a Realistic Model of Parallel Computation*, Proceedings of Fourth ACM SIGPLAN Symposium on Principles and Practice of Parallel Programming, San Diego, May, 1993, pp. 1–12.

[3] S. Chiang, R. Mansharamani, and M. Vernon, *Use of Application Characteristics and Limited Preemption for Run-to-Completion Parallel Processor Scheduling Policies*, Proceedings of the 1994 ACM Sigmetrics Conference on Measurement and Modeling of Computer Systems, 1994, pp. 33–44.

[4] F. DEHNE, X. DENG, P. DYMOND, A. FABRI, AND A. KHOKHAR, *A Randomized Parallel 3D Convex Hull Algorithm for Coarse Grained Multicomputers*, Theory of Computing Systems, Vol. 30, 1997, pp. 547-558, a special issue for selected papers presented at the 7th ACM Symposium on Parallel Algorithms and Architectures, 1995, Santa Barbara.

[5] F. DEHNE, A. FABRI, AND A. RAU-CHAPLIN, *Scalable parallel computational geometry for coarse grained multicomputers*, International Journal on Computational Geometry, Vol. 6, No. 3, 1996, pp. 379 - 400.

[6] XIAOTIE DENG, NIAN GU, TIM BRECHT, KAICHENG LU, *Preemptive Scheduling of Parallel Jobs on Multiprocessors*, SIAM J. Comput. 30, (2000), pp.145-160.

[7] D. EAGER, J. ZAHORJAN, AND E. LAZOWSKA, *Speedup Versus Efficiency in Parallel Systems*, IEEE Trans. on Computers, Vol. 38., No. 3, (1989), pp. 408–423.

[8] JEFF EDMONDS, DONALD D. CHINN, TIM BRECHT, X. DENG, *Non-clairvoyant Multiprocessor Scheduling of Jobs with Changing Execution Characteristics* , Proceedings of the 22nd ACM Symposium on Theory of Computing, May, 1997, pp. 120–129.

[9] R. MANSHARAMANI AND M. VERNON, *Qualitative Behavior of the EQS Parallel Processor Allocation Policy*, Technical Report CS TR 1192, Computer Sciences Department, University of Wisconsin, Madison, Madison, WI, November, 1993.

[10] R. MOTWANI, S. PHILLIPS, AND E. TORNG *Non-Clairvoyant Scheduling*, Proceedings of the 4th Annual ACM-SIAM Symposium on Discrete Algorithms, Austin, Texas, January, 1993, pp. 422–431.

[11] T. NGUYEN AND R. VASWANI AND J. ZAHORJAN, *Using Runtime Measured Workload Characteristics in Parallel Processor Scheduling*, Proceedings of the IPPS'96 Workshop on Job Scheduling Strategies for Parallel Processor Scheduling, Honolulu, HI, 1996, pp. 93–104.

[12] E. PARSONS, AND K. SEVCIK, *Multiprocessor Scheduling for High-Variability Service Time Distributions*, Job Scheduling Strategies for Parallel Processing, Lecture Notes in Computer Science, 949: edited by G. Feitelson and L. Rudolph, Springer-Verlag, 1995, pp. 127–145.

[13] C. PAPADIMITRIOU, AND M. YANNAKAKIS, *Towards an Architecture-Independent Analysis of Parallel Algorithms*, Proceedings of the 20th ACM Symposium on Theory of Computing, 1988, pp. 510–513.

[14] K. SEVCIK, *Characterizations of Parallelism in Applications and their use in Scheduling*, Proceedings of the 1989 ACM Sigmetrics Conference on Measurement and Modeling of Computer Systems, May, 1989, pp. 171–180.

[15] A. TUCKER AND A. GUPTA, *Process Control and Scheduling Issues for Multiprogrammed Shared-Memory Multiprocessors*, Proceedings of the Twelfth ACM Symposium on Operating Systems Principles, 1989, pp. 159–166.

[16] J. TUREK, W. LUDWIG, J. L. WOLF, L. FLEISCHER, P. TIWARI, J. GLASGOW, U. SCHWIEGELSHOHN, P. YU, *Scheduling Parallelizable Tasks to Minimize Average Response Time*, Proceedings of the 6th Annual Symposium on Parallel Algorithms and Architectures, June, 1994, pp. 200–209.

[17] J. TUREK, U. SCHWIEGELSHOHN, J. WOLF, P. YU, *Scheduling Parallel Tasks to Minimize Average Response Time*, Proceedings of the 5th SIAM Symposium on Discrete Algorithms, 1994, pp. 112–121.

[18] L. VALIANT, *A Bridging Model for Parallel Computation*, CACM , Vol. 33, No. 8, (1990), pp. 103–111.

[19] J. ZAHORJAN AND C. MCCANN, *Processor Scheduling in Shared Memory Multiprocessors*, Proceedings of the 1990 ACM Sigmetrics Conference on Measurement and Modeling of Computer Systems, Boulder, CO, May, 1990, pp. 214–225.

Optimization

Heuristic Solutions for the Multiple-Choice Multi-dimension Knapsack Problem

Md Mostofa Akbar[1, 2], Eric G. Manning[3], Gholamali C. Shoja[2], Shahadat Khan[4]

[2] Department of CS, PANDA Lab, UVic, Victoria, BC, Canada
{mostofa,gshoja}@csc.uvic.ca

[3] Department of CS and ECE,PANDA Lab, UVic, Victoria, BC, Canada
Eric.Manning@engr.UVic.ca

[4] Eyeball.com, Suite 409-100 Park Royal, West Vancouver, B.C. Canada
shahadat@eyeball.com

Abstract.

The Multiple-Choice Multi-Dimension Knapsack Problem (MMKP) is a variant of the 0-1 Knapsack Problem, an NP-Hard problem. Hence algorithms for finding the exact solution of MMKP are not suitable for application in real time decision-making applications, like quality adaptation and admission control of an interactive multimedia system. This paper presents two new heuristic algorithms, M-HEU and I-HEU for solving MMKP. Experimental results suggest that M-HEU finds 96% optimal solutions on average with much reduced computational complexity and performs favorably relative to other heuristic algorithms for MMKP. The scalability property of I-HEU makes this heuristic a strong candidate for use in real time applications.

1 Introduction

The classical 0-1 Knapsack Problem (KP) is to pick up items for a knapsack for maximum total value, so that the total resource required does not exceed the resource constraint R of the knapsack. 0-1 classical KP and its variants are used in many resource management applications such as cargo loading, industrial production, menu planning and resource allocation in multimedia servers.

Let there be n *items* with *values* $v_1, v_2, \ldots, v_n$ and the corresponding resources required to pick the items are $r_1, r_2, \ldots, r_n$ respectively. The items can represent *services* and their associated values can be *utility* or *revenue* earned from that service. In mathematical notation, the 0-1 knapsack problem is to find $V=$ maximize $\sum_{i=1}^{n} x_i v_i$, where $\sum_{i=1}^{n} x_i r_i \leq R$

and $x_i \in \{0,1\}$.

The MMKP is a variant of the KP. Let there be n *groups of items*. Group i has l_i items. Each item of the group has a particular value and it requires m resources. The objective of the MMKP is to pick exactly one item from each group for

---

[1] Supported by grants from Canadian Commonwealth Scholarship Program.

maximum total value of the collected items, subject to m resource constraints of the knapsack. In mathematical notation, let v_{ij} be the value of the j th item of the i th group, $\vec{r}_{ij} = (r_{ij1}, r_{ij2}, \cdots, r_{ijm})$ be the required resource vector for the j th item of the i th group and $\vec{R} = (R_1, R_2, \cdots, R_m)$, be the resource bound of the knapsack. Then the problem is to find $V = $ maximize $\sum\limits_{i=1}^{n}\sum\limits_{j=1}^{l_i} x_{ij} v_{ij}$, so that, $\sum\limits_{i=1}^{n}\sum\limits_{j=1}^{l_i} x_{ij} r_{ijk} \leq R_k$, $k = 1$, $2, \ldots, m$ and $\sum\limits_{j=1}^{l_i} x_{ij} = 1$, $x_{ij} \in \{0,1\}$, the picking variables. Here V is called the *value of the solution.*

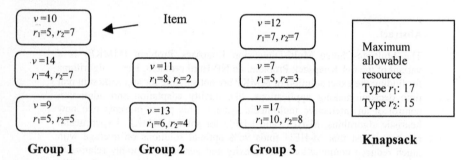

Fig. 1. Knapsack Problem.

Fig. 1 illustrates the MMKP. We have to pick exactly one item from each group. Each item has two resources, r_1 and r_2. The objective of picking items is to achieve maximum total value of the picked items subject to the resource constraint of the knapsack, that $\sum (r_1$ of picked items$) \leq 17$ and $\sum (r_2$ of picked items$) \leq 15$.

Many practical problems in resource management can be mapped to the MMKP. The Utility Model for adaptive multimedia systems proposed in [4, 5] is actually an MMKP. Users submitting their requests to a multimedia system can be represented by the groups of items. Each level of *QoS* (*Quality of Service*) of a user's requested session is equivalent to an item; each session is equivalent to a group of items. The values of the items are equivalent to offered prices for the session. The multimedia server is equivalent to the knapsack with limited resources, e.g. CPU cycles, I/O bandwidth and memory. The proposed exact solutions in [1, 2] are so computationally expensive that they are not feasible to apply in real time applications such as multimedia systems. Hence heuristic or approximate algorithms for solving the MMKP have an important role in solving real time problems.

In this paper, we present two heuristic algorithms for solving the MMKP, which are suitable for application in real time multimedia problems. Some related work on KP and MMKP will be described in section 2. In section 3 the heuristic algorithms M-HEU and I-HEU will be presented with complexity analysis. Some experimental results showing comparisons with other methods of solving the MMKP will be given in section 4. Section 5 concludes the paper, mentioning some of the possible applications of this algorithm in real time applications.

2 Related Work

There are different algorithms for solving variants of knapsack problems [8]. Actually MMKP is the combination of MDKP and MCKP. The Multi-Dimension Knapsack Problem (MDKP) is one kind of KP where the resources are multidimensional, i.e. there are multiple resource constraints for the knapsack. The Multiple Choice Knapsack Problem (MCKP) is another KP where the picking criterion for items is restricted. In this variant of KP there are one or more groups of items. Exactly one item will be picked from each group.

There are two methods of finding solutions for an MMKP: one is a method for finding exact solutions and the other is heuristic. Finding exact solutions is NP hard. Using the branch and bound with linear programming (BBLP) technique, Kolesar, Shih, Nauss and Khan presented exact algorithms for 0-1 KP, MDKP, MCKP and MMKP respectively [2, 4, 5, 10, 12]. Although the use of linear programming to determine the feasibility of picking any item of any group reduces the time requirement in the average case, it is not feasible to apply in all practical cases.

A greedy approach has been proposed [5, 8, 13] to find near optimal solutions of knapsack problems. For a 0-1 KP as described in section 1, items are picked from the top of a list sorted in descending order on v_i/r_i [8]. To apply the greedy method to the MDKP Toyoda proposed a new measurement called *aggregate resource consumption* [13]. Khan [4] has applied the concept of aggregate resource consumption to pick a new item in a group to solve the MMKP. His heuristic HEU selects the lowest-valued items by utility or revenue of each group as initial solution. It then upgrades the solution by choosing a new item from a group which has the highest positive Δa_{ij}, the change in aggregate consumed resource (the item which gives the best revenue with the least aggregate resource). If no such item is found then an item with the highest $(\Delta v_{ij})/(\Delta a_{ij})$ (maximum value gain per unit aggregate resource expended) is chosen. Here,

$$\Delta a_{ij} = \sum_k \left(r_{i\rho[i]k} - r_{ijk} \right) \times C_k \Big/ \left| \vec{C} \right| , \text{ increase in aggregate consumed resource.} \quad (1)$$

r_{ijk}=amount of the kth resource consumption of the jth item of the ith group.
$\rho[i]$= index of selected item from the ith group and C_k= amount of the kth resource consumption.
$\Delta v_{ij} = v_{i\rho[i]} - v_{ij}$, gain in total value.
This heuristic for MMKP provides solutions with utility on average equal to 94% of the optimum, with a worst case time complexity of $O\left(mn^2(l-1)^2\right)$. Here, n = number of groups, l =number of items in each group (assumed constant for convenience of analysis) and m =resource dimension.

Magazine and Oguz [7] proposed another heuristic based on Lagrange Multipliers to solve the MDKP. Moser's [9] heuristic algorithm also uses the concept of graceful degradation from the most valuable items based on Lagrange Multipliers to solve the MMKP. This algorithm is also suitable for real time application of MMKP problems and it performs better than HEU in terms of optimality when both methods reach a solution [4]. It has been observed from experiments that Moser's method does not always give a solution when there is a solution obtainable by selecting the least valuable items from each group. HEU does not have that difficulty because it starts from the least valuable items. But, HEU will fail to find a solution when the bottommost elements do not give feasible solution, while some items with higher values but with less resource consumption do.

3 Proposed Heuristic Algorithm for MMKP

3.1 Modified HEU (M-HEU)

We have to sort the items of each group in non-decreasing order according to the value associated with each item. Hence, we can say that in each group the bottom items are *lower-valued items* than the top ones. The items at the top can be defined as *higher-valued items* than those in the bottom. If a particular pick of items (one from each group) does not satisfy the resource constraint, we define the solution *infeasible*. A *feasible solution* is a solution that satisfies the resource constraints. For any resource k, C_k/R_k can be defined as infeasibility factor f_k. The kth resource is feasible if the infeasibility factor $f_k \leq 1$, otherwise it is infeasible.

We find that HEU requires additional steps to find a feasible solution if the lowest quality items from each group represent an infeasible solution. HEU finds a solution by only *upgrading* the selected items of each group. There might be some higher-valued items in the MMKP, which make the solution infeasible, but if some of the groups are downgraded we can get a feasible solution. This method of upgrading followed by downgrading may increase the total value of the solution.

Steps in the Algorithm
Step 1: Finding a feasible solution
Step 1.1: Select the lowest-valued items from each group.
Step 1.2: If the resource constraints are satisfied *then* notify "This MMKP has a solution". So go to step 2, *else* the solution is infeasible.
Step 1.3: Find out the resource k_m, which has the highest infeasibility factor. This is the *most infeasible resource*. Select a high-valued item from any group, which decreases f_{k_m}, does not increase the infeasiblity factor, of the other resources, does not make any feasible resource requirement into an infeasible one, and gives the highest Δa_{ij}, which has been defined by (1).
Step 1.4: If an item is found in *step 1.3 then* go to *step 1.2 else* notify "No Solution Found".
Step 2: Upgrading the selected feasible solution
/*This step is identical to the iteration performed in HEU*/
Step 2.1: Find a higher valued item from a group than the selected item of that group subject to the resource constraint which has the highest positive Δa_{ij}. If no such item is found then an item with the highest $(\Delta v_{ij})/(\Delta a_{ij})$ is chosen.
Step 2.2: If no such item is found in *step 2.1 then* go to *step 3 else* look for another items in *step 2.1*.
Step 3: Upgrading using one upgrade followed by at least one downgrade

Step 3.1: If there are higher-valued items than the selected item in any group *then* find such a higher-valued item (whatever is the resource constraint) which has the highest value of $\left(\Delta v_{ij}\right)/\left(\Delta a' r_{ij}\right)$. Here,

$$\Delta a'_{ij} = \sum_k \frac{r_{i\rho[i]k} - r_{ijk}}{R_k - C_k} = \text{ the ratio of increased resource requirement to} \qquad (2)$$

available resource, where, R_k= kth resource constraint.

Step 3.2: Find a lower-valued item than the selected item of the groups such that the downgrade still gives higher total value than the total value achieved in Step 2 and has the highest value of $\left(\Delta a''_{ij}\right)/\Delta v_{ij}$

$$\Delta a''_{ij} = \sum_k \frac{r_{i\rho[i]k} - r_{ijk}}{C_k - R_k} = \text{ the ratio of decreased resource requirement to} \qquad (3)$$

overconsumed resource.

Step 3.3: If an item is found in *step 3.2 then*

> If the item satisfies the resource constraint *then* look for a better solution in *step 2 else* go to *step 3.2* for another downgrade.

Else

> Revive the solution we found at the end of *step 2* and terminate.

Endif

3.2 Incremental Solution of the MMKP (I-HEU)

If the number of groups in the MMKP is very large then it is not efficient to run M-HEU once per minute, as a real time system, for example a multimedia system with 10000 sessions might well require. An *incremental* solution is a necessity to achieve better computation speed. By changing the step of finding a feasible solution (Step 1) we can use M-HEU to solve the MMKP *incrementally*, starting from an already solved MMKP with a smaller number of groups. The changed sub-steps in *step 1* are as follows.

Step 1.1: Select the lowest-valued item (according to value) from each new group.

Step 1.3: This step is similar to the *step 1.3* of M-HEU except that here we have to find any item instead of *"Higher-valued Item"*.

I-HEU does the feasibility test in step 1 with the lowest-valued items. I-HEU will require pretty much the same time if it finds the near optimal solution after doing a lot of upgrading and downgrading in the older groups as well as new groups. It will give the best performance when the solution is determined by upgrading or downgrading only the new groups of items. Typically, it is unlikely that it will require abrupt changes in the already solved groups while running step 2 and 3 of I-HEU. Thus we can expect some time advantage with this incremental method over M-HEU.

3.3 Analysis of the Algorithm

Non Regenerative Property: The three steps of M-HEU never regenerate a solution that has been found previously. The straightforward reasons for this convergence property are as follows:

- *Step 1* never makes any feasible resource requirement infeasible or infeasible resource requirement more infeasible.
- *Step 2* always upgrades the solution vector with increased total value.
- *Step 3* upgrades one item followed by downgrading one or more items for an increase in total value, thereby excluding the possibility of regenerating a previously determined solution or of infinite looping.

Complexity of the Algorithm: The computational complexity of step 1 will be the worst when there is no feasible solution or there is a feasible solution located at the highest valued item of each group and at each iteration only one item of one group moves one level up. Initially the selected items of each group is the lowest-valued item of each group. So the total upgrade in group i is $(l_i - 1)$. The total number of

upgrades in step 1 is $\sum_{i=1}^{n}(l-1)$. For convenience of analysis we assume that all groups

contain the same number of items, i.e., $l_i{=}l$. So step 1.3 and step 1.4 will be executed $(nl - n)$ times each.

Total floating point operations in step 1 is $\sum_{j=0}^{(l-1)n-1}\{(nl - n - j)\times(6m+1)+2m\}$

$$\text{Complexity in step 1, } T_1 = 3n^2(l-1)^2 m + \frac{n^2(l-1)^2}{2} + 5n(l-1)m + \frac{n(l-1)}{2} \qquad (4)$$

Step 2 requires the highest computation when there is a feasible solution located at the bottom most item of each group (initial solution), step 2.1 upgrades one level of one group in every execution and upgrading continues until it reaches the highest valued item of each group. So the analysis of step 2 is like step 1.

$$\text{Complexity in step 2, } T_2 = 3n^2(l-1)^2 m + 2n^2(l-1)^2 + 3n(l-1)m + 2n(l-1) \qquad (5)$$

We present an upper bound for the computational complexity of step 3. The situation will be the worst when an upgrade is done by step 3.1 and it is followed by all possible down grades by step 3.2 (called from step 3.3). Then it jumps to step 2 to do all possible $(n-1)l$ upgrades before going to step 3 again.
Upper bounds of the computational complexity by one upgrade in step 3.1 and all downgrades in step 3.2 are $n(l-1)\times(4m+1)$ and $n(l-1)\times\{(4m+5)\times(l-1)n+2m\}$ respectively.

$$\text{Upper bound in step 3, } T_3 = n^3(l-1)^3(7m+7)+n^2(l-1)^2(9m+2) \qquad (6)$$

Steps 1 and 2 share the job of upgrading. So the combined worst case complexity will be expressed by (5) i.e., $O(mn^2(l-1)^2)$. *Step 3* is executed when *step 2* fails to upgrade the solution. The available resource is very small in this step compared to *step 2*. We expect that this step require less iteration than the previous steps and we can improve the solution with less effort. This is analogous to the hill climbing approach in a plateau, for classical AI problems [14]. Please refer to [16] for a detailed analysis.

4 Experimental Results

4.1 Test Pattern Generation

The knapsack problem has been initialized by the following pseudo random numbers: r_{ijk} = kth weight of jth item of ith group = random(R_c). Value per unit resource p_k = random (P_c). Value for each item $v_{ij} = \sum_k r_{ijk} p_k + \text{random}(V_c)$. Here, R_c, P_c and V_c are the upper bound of any kind of resource requirement, unit price of any resource and the extra value of an item after its consumed resource price. The value of each item is not directly proportional to the resource consumption. The function random(i) gives an integer from 0 to (i-1), following the uniform distribution.

The total constraint for kth resource type $R_k = R_c \times n \times 0.5$, where n = number of groups. If we want to generate a problem for which the solution is known, the following reinitializations have to be done.

ρ_i = Selected item of ith group = random(C_i), C_i= number of items in ith group.

$R_k = \sum_i r_{i\rho_i k}$, i.e., exactly equal to the sum of the resources of the selected items.

The values associated with the selected items are $v_{i\rho_i} = \sum_k r_{i\rho_i k} \times p_k + V_c$.

This initialization ensures maximum value per unit resource for the selected items. Furthermore the total resource constraint is exactly the same as the aggregate of selected item resources. Hence there is no chance of selecting other items for maximization of total value.

4.2 Test Results

We implemented BBLP, Moser's heuristic, M-HEU and I-HEU for solving MMKP, using the C programming language. We tested the programs in a Pentium dual processor 200 MHZ PC with 32 MB RAM running Linux OS (RedHat Package 5.0).

Table 1 shows a comparison among BBLP, Moser's heuristic and our M-HEU. Here, 10 sets of data have been generated randomly for each of the parameters n, l and m. We used the constants R_c=10, P_c=10 and V_c=20 for generation of test cases. Data is not initialized for a predefined maximum total value for the selected items. The column BBLP gives the average value earned from BBLP. The columns *Moser* and *M-HEU* give the average standardized value earned in those heuristics with respect to exact solution, where solutions were found. The column $\bar{b}, \bar{m}, \bar{h}$ shows the number of data sets where BBLP, Moser and M-HEU *fail* to find the solution. We find that Moser's algorithm cannot always find a feasible solution when there is a solution found by M-HEU. In Table 1 row 1, 2, 3 and 7 shows that Moser's method failed to find a solution when the algorithms could. ∂_m and ∂_h are the standard deviation of standardized total value achieved in the 10 sets of data, given to indicate stability. The main observation from this table is the time requirement of the heuristic solutions compared with BBLP. This time requirement of BBLP increases dramatically with the size of the MMKP, because of the exponential computational complexity of BBLP.

It is impractical to test the performance of BBLP for larger MMKP. In order to determine the percentage of optimality (standardized value earned with respect to BBLP) achieved by the heuristics we used the technique described in the last subsection "Test Pattern Generation". Now if we look at table 1 and 2 we find that our proposed M-HEU always performs better than Moser's heuristic in finding feasible solutions, and in achieving optimality. From table 2 it is also observed that M-HEU performs better in terms of time requirement for larger problem sets. We can also conclude that the stability of the solution performance is almost the same in both the cases.

Table 3 shows the comparative performances of I-HEU and M-HEU for different *batch* sizes. Each batch contains a particular number of groups. For each set of parameters the program starts with no groups and continues until the number of groups reaches 100, after the arrival of several batches. The result from I-HEU is used to solve the MMKP for the next batch. We run M-HEU and I-HEU separately and the average time requirements per batch are shown in the table. The columns headed by $\bar{h}_m$ and $\bar{h}_i$ show the number of cases where M-HEU and I-HEU could not find feasible solutions respectively. The column I-HEU/M-HEU gives the ratio of total values achieved by two heuristics. We find that the performances of I-HEU and M-HEU are almost the same in achieving optimal solutions. This means the incremental approach in computation does not degrade the solution quality and we get better computational speed as observed from the test data. The main reason for the difference in solution quality is different starting points. M-HEU starts from scratch whereas I-HEU starts from an almost-done solution. That is why we are not getting the same result although we are doing the same thing in *step 2* and *step 3* of the algorithms.

Table 1. Comparison among BBLP, Moser's Heuristic and HEU

row no.	n	l	m	BBLP	Moser	M-HEU	t_{BBLP} (ms)	t_{Moser} (ms)	t_{M-HEU} (ms)	$\bar{b}, \bar{m}, \bar{h}$	∂_m	∂_h
1	5	5	5	621.60	0.94	0.97	49	0.57	0.53	0 1 0	0.039	0.032
2	7	5	5	946.10	0.93	0.95	400	1.09	1.06	0 1 0	0.010	0.021
3	7	7	5	964.20	0.93	0.96	1161	2.02	1.88	0 1 0	0.031	0.021
4	9	5	5	1163.40	0.95	0.97	1328	1.49	1.70	0 0 0	0.013	0.016
5	9	7	5	1110.20	0.92	0.96	8298	3.47	2.89	0 0 0	0.029	0.016
6	9	9	5	1135.00	0.93	0.95	13884	5.52	4.19	0 0 0	0.011	0.025
7	11	5	5	1341.10	0.93	0.96	4013	2.37	2.49	0 2 0	0.023	0.014
8	11	7	5	1431.20	0.94	0.96	15436	4.89	4.03	0 0 0	0.013	0.017
9	11	9	5	1394.10	0.93	0.96	38309	8.46	6.37	0 0 0	0.018	0.021
10	13	5	5	1648.50	0.94	0.96	12619	3.32	3.62	0 0 0	0.016	0.011
11	13	7	5	1545.30	0.94	0.95	51466	7.04	5.77	0 0 0	0.013	0.016
12	13	9	5	1387.40	0.93	0.95	55429	10.46	8.02	0 0 0	0.018	0.020
13	15	5	5	1803.10	0.94	0.97	39225	3.99	4.47	0 0 0	0.021	0.016
14	15	7	5	2007.30	0.94	0.95	84015	8.42	7.59	0 0 0	0.007	0.015
15	15	9	5	1766.60	0.93	0.95	139150	14.08	11.21	0 0 0	0.020	0.012
16	17	5	5	1995.40	0.93	0.97	49175	5.27	5.79	0 0 0	0.009	0.010
17	17	7	5	2148.90	0.93	0.95	107793	10.89	9.60	0 0 0	0.018	0.013
18	17	9	5	1811.00	0.93	0.96	199418	17.69	14.46	0 0 0	0.012	0.012
19	19	5	5	2218.50	0.94	0.96	75893	6.41	7.22	0 0 0	0.016	0.008
20	19	7	5	2104.40	0.93	0.96	124633	12.89	11.63	0 0 0	0.009	0.006
21	19	9	5	1880.60	0.92	0.96	270580	21.33	19.15	0 0 0	0.025	0.010
22	21	5	5	2578.50	0.95	0.96	64656	7.47	8.19	0 0 0	0.011	0.019

row no.	n	l	m	BBLP	Moser	M-HEU	t_{BBLP} (ms)	t_{Moser} (ms)	$t_{M\text{-}HEU}$ (ms)	$\bar{b},\bar{m},\bar{h}$	∂_m	∂_h
23	21	7	5	2174.60	0.93	0.96	114337	16.28	15.39	0 0 0	0.015	0.016
24	21	9	5	2791.50	0.94	0.95	463615	27.67	22.33	0 0 0	0.010	0.009

Table 2. Comparison between Moser's Heuristic and M-HEU

n	l	m	$\bar{m}$	$\bar{h}$	Max value	Moser	M-HEU	t_{Moser} (ms)	t_{HEU} (ms)	∂_m	∂_h
100	5	5	0	0	11574.9	0.94	0.96	137.4	134.6	0.0165	0.0077
100	15	5	0	0	9925.1	0.92	0.95	1410.4	806.0	0.0211	0.0133
100	25	5	0	0	10773.5	0.93	0.96	4002.5	1798.2	0.0121	0.0112
100	5	15	2	0	34942.6	0.94	0.95	263.2	248.4	0.0118	0.0095
100	15	15	0	0	34673.2	0.93	0.95	2859.2	1686.2	0.0078	0.0092
100	25	15	0	0	34533.5	0.93	0.95	8705.6	3488.7	0.0099	0.0093
100	5	25	9	0	59607.2	0.94	0.95	325.1	331.0	-	0.0035
100	15	25	1	0	59467.8	0.92	0.93	5026.2	2688.3	0.0101	0.0082
100	25	25	0	0	57338.3	0.92	0.94	14922.8	5715.8	0.0132	0.0083
200	5	5	0	0	22587.8	0.94	0.96	539.2	527.4	0.0104	0.0103
200	15	5	0	0	22360.8	0.94	0.96	5625.3	3404.3	0.0122	0.0095
200	25	5	0	0	22979.8	0.93	0.96	17643.3	8794.0	0.0172	0.0128
200	5	15	1	0	74289.9	0.94	0.95	1123.8	1192.4	0.0058	0.0091
200	15	15	0	0	71805.7	0.94	0.95	13901.0	7858.2	0.0080	0.0075
200	25	15	0	0	70007.3	0.94	0.95	37617.8	16023.4	0.0109	0.0081
200	5	25	6	0	113000.5	0.93	0.95	1747.3	1793.8	0.0069	0.0068
200	15	25	0	0	122391.5	0.93	0.94	22427.8	11907.6	0.0043	0.0075
200	25	25	0	0	116578.0	0.93	0.95	60908.8	23828.6	0.0067	0.0087
300	5	5	0	0	31790.7	0.94	0.95	1198.9	1162.5	0.0171	0.0103
300	15	5	0	0	31687.1	0.93	0.95	14641.0	7853.2	0.0221	0.0104
300	25	5	0	0	31908.5	0.93	0.96	43271.1	16561.2	0.0213	0.0091
300	5	15	0	0	102715.7	0.94	0.95	2981.1	2926.5	0.0048	0.0078
300	15	15	0	0	98948.5	0.93	0.95	31634.6	17635.6	0.0065	0.0081
300	25	15	0	0	97911.6	0.93	0.94	86546.2	36148.7	0.0075	0.0082
300	5	25	0	0	177995.2	0.94	0.96	5183.2	4475.2	0.0078	0.0054
300	15	25	0	0	167864.3	0.93	0.95	51186.4	26481.2	0.0089	0.0069
300	25	25	0	0	169079.8	0.93	0.95	132720.0	53770.6	0.0065	0.0072

Table 3. Comparison of I- HEU and M-HEU

Batch Size	l	m	I-HEU/ M-HEU	$t_{M\text{-}HEU}$	$t_{I\text{-}HEU}$	$\bar{h}_m$	$\bar{h}_i$
1	5	5	1.001	44.2	2.4	1	2
1	10	5	0.998	118.1	7.4	0	0
1	15	5	0.999	319.2	7.7	0	0
1	5	10	1.001	90.7	5.1	2	2
1	10	10	1.003	216.5	13.3	0	0
1	15	10	1.004	436.6	19.1	0	0
1	5	15	1.002	81.3	7.5	2	2
1	10	15	1.002	271.2	21.4	1	1
1	15	15	1.000	688.3	27.4	2	2
6	5	5	0.995	48.9	4.2	0	0
6	10	5	1.000	130.3	7.2	0	0
6	15	5	0.999	222.0	11.5	0	0
6	5	10	0.995	74.4	8.1	0	0
6	10	10	0.998	199.8	13.2	0	0
6	15	10	1.000	300.5	22.4	0	0
6	5	15	0.991	84.1	16.9	0	0
6	10	15	0.995	301.3	17.4	0	1
6	15	15	1.001	500.5	30.7	0	0
11	5	5	0.997	61.3	6.5	0	0
11	10	5	1.001	154.8	15.4	0	0
11	15	5	1.003	248.4	36.4	0	0
11	5	10	1.010	88.5	17.8	0	0
11	10	10	0.997	272.5	34.8	0	0
11	15	10	0.988	507.4	51.9	0	0
11	5	15	0.992	143.6	18.7	0	0
11	10	15	1.003	364.4	36.6	0	0
11	15	15	1.003	594.6	71.7	0	0

5 Concluding Remarks

The new heuristics M-HEU and I-HEU perform better than other algorithms considered here. M-HEU is applicable to real time applications like admission control and QoS adaptation in multimedia systems. We include a dummy QoS level with null resource requirement and zero revenue, which is therefore lower-valued than all other

QoS levels. Now selection of that null QoS level by M-HEU or I-HEU indicates rejection, comprising effect an admission control for the underlying system. On the other hand, the dummy QoS level is always feasible because it does not take any resource. So, there will be no problem regarding infeasible solution in this practical problem. Due to the quadratic complexity we can not claim too much about M-HEU's scalability property. However, I-HEU appears to be very effective indeed, as it offers almost the same result with much less time requirement. It therefore could be used with improved performance in a multimedia server system with thousands of admitted sessions. The QoS manager can execute I-HEU once per minute as some sessions are dropped and some new ones seek admission. M-HEU could be applied occasionally, e.g., once per hour for further improvement of the solution.

Only the worst case analysis of M-HEU and I-HEU has been presented here. The average case analysis of the algorithms is a good research topic for future work. The question of distributed algorithms for solving the MMKP is also an interesting unsolved problem. Parallel and distributed versions of MMKP with better computational complexity can improve the scalability and fault tolerance of adaptive multimedia.

References

1. R. Armstrong, D. Kung, P. Sinha and A. Zoltners. A Computational Study of Multiple Choice Knapsack Algorithm. *ACM Transaction on Mathematical Software*, 9:184-198 (1983).
2. P. Koleser, A Branch and Bound Algorithm for Knapsack Problem. *Management Science*, 13:723-735 (1967).
3. K. Dudziniski and W. Walukiewicz, A Fast Algorithm for the Linear Multiple Choice Knapsack Problem. *Operation Research Letters*, 3:205-209 (1984).
4. S. Khan. *Quality Adaptation in a Multi-Session Adaptive Multimedia System: Model and Architecture*. PhD thesis, Department of Electrical and Computer Engineering, University of Victoria (1998).
5. S. Khan., K. F. Li and E.G. Manning. The Utility Model for Adaptive Multimedia System. In *International Workshop on Multimedia Modeling*, pages 111-126 (1997).
6. M. Magazine, G. Nemhauser and L. Trotter. When the Greedy Solution Solves a Class of Knapsack Problem. *Operations Research*, 23:207-217 (1975)
7. M. Magazine and O. Oguz. A Heuristic Algorithm for Multidimensional Zero-One Knapsack Problem. *European Journal of Operational Research*, 16(3):319-326 (1984).
8. S. Martello and P. Toth. Algorithms for Knapsack Problems. *Annals of Discrete Mathematics*, 31:70-79 (1987).
9. M. Moser, D. P. Jokanovic and N. Shiratori. An Algorithm for the Multidimensional Multiple-Choice Knapsack Problem. *IEICE Transactions on Fundamentals of Electronics*, 80(3):582-589 (1997).
10. R. Nauss. The 0-1 Knapsack Problem with Multiple Choice Constraints. *European Journal of Operation Research*, 2:125-131(1978).
11. W.H. Press, S.A. Teukolsky, W. T. Vetterling and B.P. Flannery. *Numerical Recipes in C: The Art of Scientific Computing*. Cambridge University Press, Cambridge, UK, second edition (1992).
12. W. Shih. A branch and Bound Method for Multiconstraint Knapsack Problem. *Journal of the Operational Research Society*, 30:369-378 (1979).
13. Y. Toyoda. A Simplified Algorithm for Obtaining Approximate Solution to Zero-one Programming Problems. *Management Science*, 21:1417-1427 (1975)
14. G. F. Luger and W. A. Stubblefield. Artificial Intelligence, *Structures and Strategies for Complex Problem Solving*, Second edition, The Benjamin/Cummings Publishing Company, Inc., 1993.
15. R. K. Watson. *Applying the Utility Model to IP Networks: Optimal Admission & Upgrade of Service Level Agreements*. MASc Thesis, Dept of ECE, University of Victoria, 2001, to appear.
16. M. Akbar, E.G. Manning, G. C. Shoja, S. Khan, "*Heuristics for Solving the Multiple-Choice Multi-Dimension Knapsack Problem*", Technical Report DCS-265-1R, Dept. of CS, UVic, March 2001.

Tuned Annealing for Optimization

Mir M. Atiqullah[1] and S. S. Rao[2]

[1] Aerospace and Mechanical Engineering Department
Saint Louis University
St. Louis, MO 63103
[2] Department of Mechanical Engineering
University of Miami
Coral Gables, FL 33146

Abstract. The utility and capability of simulated annealing algorithm for general-purpose engineering optimization is well established since introduced by Kirkpatrick et. al[1]. Numerous augmentations are proposed to make the algorithm effective in solving specific problems or classes of problems. Some proposed modifications were intended to enhance the performance of the algorithm in certain situations. Some specific research has been devoted to augment the convergence and related behavior of annealing algorithms by modifying its parameters, otherwise known as cooling schedule. Here we introduce an approach to tune the simulated annealing algorithm by combining algorithmic and parametric augmentations. Such tuned algorithm harnesses the benefits inherent in both types of augmentations resulting in a robust optimizer. The concept of 'reheat' in SA, is also used as another tune up strategy for the annealing algorithm. The beneficial effects of 'reheat' for escaping local optima are demonstrated by the solution of a multimodal optimization problem. Specific augmentations include handling of constraints, fast recovery from infeasible design space, immunization against premature convergence, and a simple but effective cooling schedule. Several representative optimization problems are solved to demonstrate effectiveness of tuning annealing algorithms.

Keywords

Simulated annealing, design optimization, constrained optimization, tuned annealing, cooling schedule.

1 Introduction

In pursuit of high quality solutions, design optimization algorithms not only must find near optimum end result but also demonstrate efficiency in terms of computation. As a stochastic algorithm, the simulated annealing is well known for its capability to find the globally optimal solution. Guided probabilistic moves are key to finding global optimum by simulated annealing while overcoming local optima in the design space. The basic features of the annealing algorithm can be highlighted by the following pseudocode:

Program annealing
Set initial probability of accepting randomly generated worse design
Randomly generate a new solution by probabilistic move from current solution
If better than before accept it as next solution
If worse, accept it only with a certain probability and decrease it for next iteration..
Repeat the process by generating a new solution, until no improvement possible.
End program.

When used with limited probabilistic control, the objective function is improved over many such steps. But often its iterative and slow convergence can be prohibitive for problems with many design variables and lengthy function evaluations.

Many global optimization methods reported in the lierature[2,3] are stochastic in nature and converge asymptotically to the global optimum. When the modality of the design domain is not overly complex, these methods proved efficient. Many practical design optimization problems involve a large number of design parameters with objective functions characterized by many local minima. While the algorithm draws parallel to the physical heat treatment annealing of metals, the acceptance probability used for accepting occasional worse design draws its parallel to the temperature. The algorithm and logic used to progressively decrease this temperature is widely known as cooling schedule. Because of its discrete nature, the annealing algorithm can overcome non-smoothness and discontinuities in the design space. The convergence and global characteristics of simulated annealing (SA) are strongly affected by the structure of annealing and parameters of the cooling schedule. The analyses of several innovative cooling schedules are discussed and a parametric schedule is proposed, which is adaptive to problem at hand and the nature of the design domain and follows a 'gaussian' type decrement. Two combinatorial test problems are used as a platform for comparing effectiveness of various cooling schedules with that of the tuned algorithm.

No design should be considered complete without performing some form of optimization of the initial design. Traditional design approach used localized or specific performance based design improvement without considering globality of the process. This is mainly because of the inability of the traditional gradient based optimizers to find the global optimum.

The goal of this paper is to demonstrate the simplicity and the utility of the tuned annealing algorithm. Section two gives the algorithmic steps and the implementation details of tuned annealing algorithm. Numerical examples include a welded beam design, a 32 variable part stamping problem, a 200 variable part stamping problem, and a 25-bar space truss optimization, providing an spectrum of highly nonlinear and multimodal design problems.

2 Tune ups #1 and #2: Improve Feasibility and Handle Constraints

Simulated annealing responds to changes only in the objective function. Since almost all engineering design problems are highly constrained, SA would not be suitable for solving such problems. To take advantage of the global solution capability of SA and to alleviate this difficulty, it is customary to incorporate the constraint functions $g(x)$ to the design objective using a penalty function $P(x)$ such that:

$$P(\bar{x}) = F(\bar{x}) + r_k \sum_{j=1}^{m} G_j\left(g_j(\bar{x})\right) \tag{1}$$

where, $G_j = \max\left[0, g_j(\bar{x})\right]$ for inequality constraints and $G_j = abs\left[g_j(\bar{x})\right]$ for equality constraints. The factor r_k determines the relative degree of penalty. Two inherent problems plague the penalty function approach that usually slows the SA down and often leads the solution towards sub optimal solution. First, design problems with a large number of constraints with a large difference in the numerical values of the constraint functions pose special difficulty for the SA algorithm. Second, highly constrained problems may even pose difficulty in finding a feasible solution in the first place.

Ideally all constraints should be treated with equity and any design solution violating any constraint should be considered in comparable terms. To guide the annealing process to handle infeasible design space a constraint navigational strategy has been used successfully[2]. This approach takes the constraints into account explicitly, alleviates such scaling problems, and enhances the convergence through fast achievement of feasibility when the starting solution is infeasible (and random). Each iteration of such tuned SA algorithm can be described by the following pseudo-code:

> *Randomly perturb the variables: obtain a new design.*
> *Evaluate the objective function and the constraints.*
> *IF no constraint is violated proceed with regular annealing.*
> *ELSEIF the number of constraints violated increases, probabilistically accept the design and*
> > *go to the next iteration.*
> *ELSEIF the number of constraints violated decreases, unconditionally accept the design and*
> > *go to the next iteration.*
> *ELSEIF the number of constraints violated remains the same*
> *AND the amount of violation increases, probabilistically accept the design*
> *ELSE*
> > *accept the design.*
> *ENDIF*
> *Proceed to the next iteration.*

This tune up helps get rid of penalty function as well as scaling of constraints as often necessary in untuned algorithm.

3 Tune Up #3: Reheat

It is one of the most ambiguous parameters in annealing to determine with some certainty when to stop the random search process and delare the solution as optimal. While various cooling schedules are proposed[3,4] using statistical methods to sense closeness to optimality, numerical implementation of those are virtually impractical due to the requirement of prohibitively long Markov chains i.e. sequence of random probing of the design space. As a result algorithms have to be terminated after finite annealing process with uncertainty in the optimality of results. The situation is further complicated when real valued variables make up the design space and certain discretization is used to simulate the real space. For convergence purposes, the randomly selected steps in the random directions must be gradually decremented in magnitude. As the SA algorithm approaches the end of a Markov chain (long sequence of random steps), the steps become too small to continue searching far from the current solution, which practically halts the SA. As a compromise between excessively long annealing process and uncertain optimal result, a reheat strategy[2] has been used. Basically multiple SA algorithm runs are executed in a sequence such that the one SA algorithm picks up a solution as its starting point, which is left off by the previous one. It is essentially a re-annealing strategy with a stopping criteria built into it. The pseudocode for the reheat is given below:

> *Program Reanneal*
> *Preset annealing and reannealing parameters*
> *Anneal for a preset number of iterations. Results in set A*
> *Reset temperature to original (or determine for the neighborhood of the current solution A)*
> *Reanneal for the preset number of iterations. Reusults in set B.*
> *If B is better than A, replace A by B. Continue to Reanneal.*
> *If B is close or worse than A, terminate reannealing. B is the current solution.*
> *End Reanneal.*

Usually the only problems that may improve successively with each reanneal are those with multimodal objectives functions. The added benefit is that by setting shorter markov chains in the annealing and using multiple reannealing, optimal results may be found without lengthy annealing run, even for the unimodal functions with just single optimal solution.

4 Tune Up #4: Simple Cooling Schedule

An annealing algorithm can be made quite robust by selecting and implementing a proper cooling schedule irrespective of the algorithmic modifications discussed earlier. Much research has been devoted to this aspect. A cooling schedule is defined by a set of parameters governing the finite time behavior of the SA algorithm. The aim in the selection of the parameters is to closely follow the asymptotic behavior of the homogeneous algorithm using inhomogeneous implementation in finite time. The cooling schedules which use information from the objective evaluation and use it to adapt their annealing strategy, adjusts themselves for the design space at hand. For virtually all engineering design problems with multimodal objectives and large number of constraints and design variables, the capability to find globally optimal solution is extremely valuable. Several well-known schedules are discussed followed by the introduction of a simple schedule. Our schedule draws power from the adaptive probing of the design space. Moreover the decrement of the *temperature*, which is the control parameter of the cooling schedule, is designed to follow certain heuristics as well as characteristics of random search. While much simpler, our schedule performed better than or as well as several published schedules tested. The structure of any cooling schedule can be described by the following three parameters:

Initial temperature: The value of the initial temperature does not affect the adaptive characteristics of the cooling schedule but is critical for enabling the SA algorithm to overcome local optima. Initially, a quasi-equilibrium state can be assumed by choosing the initial value of the temperature such that virtually all transitions have guaranteed probability of acceptance. Many of the so-called adaptive schedules[4,5,6,7] draw their success from probing the neighborhood of the current design and determining the initial temperature. Starting with a too high temperature will unnecessarily prolong the already long SA process. As the algorithm progresses, the temperature must approach a value of zero, so that no worse solutions will be accepted. Then the algorithm virtually will not achieve any more significant improvement in the objective function. This tapering convergence is again much linked with any decreasing step length, specially if real valued design space is handled by discrete stepping of the SA.

Length of Markov chain: The number of transitions attempted at any specific temperature is called the length L_k of the Markov process at the kth step of temperature decrement. For finite time implementation, the chain length is governed by the notion of *closeness* of the current probability distribution a_{L_k,t_k} at temperature t_k to the stationary distribution q_{t_k}. The adaptive schedules have taken different approaches with different assumptions and preset conditions to determine when such *closeness* is achieved[8].

Decrement rule for temperature: The way the temperature is decremented is directly related to the notion of quasi-equilibrium of the probability distribution of configurations. It is intuitively obvious that a large decrement in the temperature t_k, at the k-th Markov chain, will necessitate a large number of transitions L_{k+1} at the (k+1)-th Markov chain before a quasi-equilibrium state is restored. Most adaptive cooling schedules follow the strategy of small decrements in temperature t_k to avoid long Markov chains L_k, albeit at the cost of increased number of temperature steps.

4.1 An Adaptive Schedule

The cooling schedules can be divided into two broad groups, static and adaptive. The schedules, which follow a predetermined course of decrement, are termed static while those using some statistical analysis of visited cost/design objective to control their decrement are known as dynamic. The static cooling schedules generally disregard the dynamic behavior of the problem at hand, often place too many restrictions on the annealing process. On the other hand, dynamic schedules, being computationally intensive, increase the computational effort many folds. To combine the beneficial characteristics of both classes, a simple hybrid schedule is proposed with two control parameters. The initial temperature t_0 should be high enough so that all configurations are equally admissible for acceptance at the initial temperature. In our approach, we include all moves for such estimations. In earlier works, the cost decreasing moves were not considered in the computation of the initial acceptance probability χ_0.

Thus, the augmented initial acceptance ratio is defined as

$$\chi_0 = \frac{\text{no. of accepted moves}}{\text{no. of proposed moves}} \approx 1.00 \approx \exp\left\{-\frac{(\overline{\Delta C} + 3\sigma_{\Delta C})}{t_0}\right\} \tag{2}$$

which leads to the new rule for the computation of the initial temperature:

$$t_0 = \frac{\left(\Delta C + 3\sigma_{\Delta C}\right)}{\ln\left(1/\chi_0\right)} \tag{3}$$

Experiments with several design problems, with arbitrary initial starting configurations (designs), suggest that the value of t_0 is increased by 50% or more when Eqn. (3) is used compared when standard deviation is not used.

The time to arrive at a quasi-equilibrium is related to the size of neighborhood ($\Re$). We propose by saying that the chain can be terminated if either the number of acceptances or rejections reach a certain number $\Lambda|\Re|$, where Λ is a multiplication factor. That is,

$$L_k = \begin{cases} m1 + \Lambda|\Re|; \{m2 = \Lambda|\Re|, m1 < m2\} \\ m2 + \Lambda|\Re|; \{m1 = \Lambda|\Re|, m2 < m1\} \end{cases} \tag{4}$$

where m1 and m2 are the cost decreasing and cost increasing moves experienced by the algorithm.

The progression of the cooling process, and simultaneously the annealing, can be divided into three segments, viz., global optimum locating (jumping over mountains), descending the mountain (and jumping over smaller hills), and local (mostly greedy) search. A cooling strategy should reflect these segments in the correct sequence.

At the onset, any rapid decrement of the temperature should be discouraged to avoid any 'quenching' effect. In the third (last) stage of temperature decrement, the algorithm is rarely expected to jump over hills and is assumed to be already in the region of the global solution. During this stage (last third of the Markov chain), the temperature value and decrement rates should be maintained at lower values to result in a flat convergence pattern. During the middle part of annealing, the algorithm should perform most of the necessary decrements in temperature. While annealing, the value of the cost function is assumed to follow a Gaussian pattern especially at higher temperatures. The above three tier cooling can be incorporated into one Gaussian-like temperature decrement over the entire annealing process. The following formula is used to compute the annealing temperature t_k during a Markov chain k:

$$t_k = t_0 \cdot a^{-\left[\frac{k}{f \cdot k \max}\right]^b} \tag{5}$$

where a and f are the control parameters and $kmax$ is the maximum number of chains to be executed. The exponent b can be computed once a and f are selected and the final temperature t_f (some small number) is set. At the final temperature decrement step (last Markov Chain), $t_k = t_f$ and $k = k\max$. Equations (5.59) and (5.55) yield,

$$t_f = t_0 \cdot a^{-\left(\frac{1}{f}\right)^b} \tag{6}$$

An interesting feature of the decrement rule, Eqn. (5), is that it can be tailored to go through any given temperature t_k $(t_o > t_k > t_f)$ during a given Markov chain k. The difficulty lies in the selection of the values for the parameters a and f. When the algorithm is in the kth

Markov chain and the parameter f equals $(k/kmax)$ the corresponding temperature is given by $t_K \frac{t_0}{\alpha}$. This indicates that the temperature attained in the kth chain is equal to the $\left(\frac{1}{a}\right)$th fraction of the initial temperature t_0. For a typical schedule, the temperature will be reduced to half of the initial temperature at about one-third the maximum number of allowed Markov chains, i.e., $a = 2$ and $f = 1/3$. This will allow 2/3 of the time to be devoted to finding the optimum in the current region.

Using a predetermined small number for the final temperature with a parametric form of the decrement rule, an upper limit is chosen for the number of iteration. As such, the algorithm is terminated if any of the following criteria are met in the order listed below:

(i) The final cost value in five consecutive Markov chains did not improve.

(ii) Similar to above. Five consecutive Markov chains did not improve the average cost $\overline{C}$ beyond a specified small fraction ε i.e.,

$$\frac{\overline{C}_{k-1} - \overline{C}_k}{\overline{C}_{k-1}} < \varepsilon \tag{7}$$

Here, the algorithm is assumed to have arrived at/very close to the optimum or have converged and, hence, it is terminated. The value of ε is set from past experience based on the cost values, scale factors, the accuracy desired and the computational effort involved.

(iii) The algorithm did not terminate in $kmax$ Markov chains. At this point, the temperature reaches a value of t_f.

If proper stopping criteria are used, it is unusual to have the algorithm stopped by this method, indicating insufficient annealing for the given problem situation.

5 Numerical Examples

The effects of the tune-ups are demonstrated by the following examples.

5.1 Welded Beam Design

Welded joints can save time and money, if designed properly. In this example, the total cost of a welded beam is minimized subject to various constraints. There are four dimensions that must be determined such that the cost is minimum while satisfying the constraints, shown in Figure 1. This nonlinear optimization problem was used by many researchers and the formulation is adopted from Reklaitis[9]. The optimization problem is:

Find the 4 dimensions $\{W,t,l,T\}$

that minimize the cost $F = 1.1047t^2 l + 0.04811 WT(14.0 + l)$

subject to $\tau_{weld} \leq \tau_d$, $\sigma_{weld} \leq \sigma_d$, $t \leq T$, $P_c \geq F$,

$t \geq 0.125$, and $\delta_{tip} \leq 0.25$,

where, τ_d = allowable shear stress in the weld, σ_d = allowable normal stress in the beam, P_c = buckling load, δ_{tip} = end deflection, and F = 6000 lb.

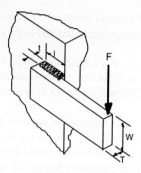

Fig. 1 : Welded beam design parameters and loading.

The derivations of the stress and deflection equations are found in [9]. The values of τ_d , σ_d and L are assumed to be 13600 psi, 30000 psi and 14 inches respectively. The values of G and E of the material are chosen as 12E6 and 30E6 psi respectively. The last physical constraint requires that all the dimensions be non-negative, i.e. $t, l, W, T \geq 0$. This problem is solved using the regular as well as tuned SA algorithm implemented on a Sun SparcStation. The results are shown in Table 1 along with those reported by others using

Table 1: Results of Welded Beam Optimization.

Variable	Regul. SA	Tuned-SA	[10]	[11] Soln.1	[11] Soln.2
t	.1525	0.2471	0.2536	0.2918	0.2489
l	11.64	6.1451	7.141	5.2141	6.173
W	8.5576	8.2721	7.1044	7.8446	8.1789
T	0.2489	0.2495	.2536	0.2918	.2533
Constraints violated	none	none	3	1	none
Objective Value	2.9265	2.4148	2.3398	2.606	2.4331

10,11 See the reference section for the sources.

different solution methods. It can be seen that the solution by the Tuned SA is superior than the one found by the regular SA algorithm. Moreover, the present solution compared favorably with others [9] in the table except those reported in [10, 11] which violated 3 and 1 constraints respectively. Considering that the SA is inherently a discrete method, the present results demonstrate the utility of the tuned algorithm as a robust optimizer.

5.2 Two-Variable Multimodal Function Optimization

The minimization of a two dimensional function that has fifteen stationary points is considered. This function, a form of the well-known *Camelback* function, is given by Equation 8.

$$f(x, y) = 4x^2 - 2.1x^4 + \frac{x^6}{3} + xy + 4y^4 - 4y^2 \tag{8}$$

There are 15 stationary points of this function with global optimums at (0.0898, -0.7127) and (-0.0898, 0.7127), and f = -1.0316. The starting point was x= -2, y= -1, f=5.7333 (same for implementation with and without the reheat) and the final solutions. The 'reheat tuned algorithm found one of the two global minima whereas the simple algorithm got trapped near one of the local minima on its search path. Both algorithms ran for a total of 500 iterations. The simple SA found the solution x*= -1.73, y*= -0.18, f*=2.29. While the 'reheat' tuned SA found the solution x*= -0.087, y*=0.72, f*=-1.03. The 'reheat' strategy was initiated at the midpoint (at 251st iteration) where the tuned algorithm temporarily accepted worse designs and got out of the local minimum finding better solution. In computationally intensive applications such as structural design optimization, the *reheat* strategy may prove to be an acceptable compromise between a local solution requiring limited computation and the global solution that might require a prohibitively large amount of computation.

5.3 Optimal Layout of 16 Circular Parts

Flat, thin components are cut out or stamped for consumer or industrial products from sheet metal, fabric, plastic, or other sheet stock. In most cases, like automobile body components, various size parts are cut out from rolls or plates. This example deals with the relative positioning of the parts to be cutout such that they are packed as close as possible and waste is minimized. Thus the objective is minimizing the rectangular area from which the parts are cut. The parts are assumed to be circular with constraints related to inclusion and intersection/overlap detection, as shown in Figure 2. The overlap constraints state that the center distance between any pair of circular parts must be greater than or equal to the sum of the radii of the respective parts. For a pair of circular parts 1 and 2, overlap is given by: $\left(\sqrt{(x2-x1)^2 + (y2-y1)^2} - (r1+r2) \right)$. Position and radii are given by (xi,yi) and ri respectively. As long as this is positive, there is no overlap. A total of $n(n+1)/2$ overlap constraints must be checked for satisfaction. Furthermore, another set of constraints specifies that the parts should be packed as close as possible to the axes but should not overlap. A set of 16 circles was randomly placed between the positions $0 \leq x_i, y_i \leq 25$ in., initially with 13 overlaps and an objective function value of 661.03 in^2. The optimization problem is solved using the tuned SA algorithm. In the final position of the parts generated by the SA, there were no overlap constraint violations and the pack was within 0.01 units from the axes. The objective was reduced to 584.46 in^2.

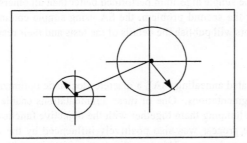

Fig. 2. Overlap calculation between 2 circles.

5.4 Optimal Layout of 100 Circular Parts

The circular part compaction is highly nonlinear and multimodal. Problems with large number of parts is becoming popular as test problems [12]. This example deals with 100 circular parts. It has more than 12 times as many design variables, the constraints increased from 136 to 5050, thus increasing the computational complexity of the problem. The numerical results are summarized in Table 2.

Table 2 : Initial and final data for the 100 part layout problem.

Items	Initial Design	Final Design
Constraints violated	44	1
Amount of violation	15899.75	.002199
Objective value	273148.9	73898.87
% waste	82.98%	37.1%

In this example, the design objective improved quickly during the early iterations but gradually tapered off in the latter iterations. This is an indication that the algorithm approached the vicinity of the optimum quickly but its characteristic slow convergence takes a relatively many more iterations before the convergence criteria are satisfied. The constraint violation of .002199 in the final design indicates existence of active constraint.

5.5 Optimal Design of a Space Truss

This example involves a 3-D space truss with 25 members grouped into 8 sets of sizes. Details of the formulation are available in many literature [7]. Dhingra and Bennage [13] used it to demonstrate an implementation of Tabu Search method. The objective is to minimize weight subject to buckling constraints under two loading conditions. Each set of truss members could assume any of the 30 allowable sizes between 0.1 to 2.6 in^2, thus allowing a total of 30^8 possible configurations. Loads varied from $-10,000$ lbs to $10,000$ lbs. The weight was minimized from 969.01 lbs to 481.33 lbs using the tuned SA algorithm which was slightly better than that reported in [13]. The simple cooling schedule as described earlier, was compared with several prominent schedules published in the literature through solution of a 8-circle part placement problem (16 variable, each may assume any of 45 values) and this 25-bar space truss optimization problem (8 variable, 30 allowable sizes) . In the first problem, the simple algorithm performed better than all others in terms of optimum found. In solving the second problem, the SA using simple cooling schedule came out second. The authors will publish the details of the tests and their results soon.

6 Conclusion

The simulated annealing (SA), a discrete stochastic optimization method is tuned up with several augmentations. One of these augmentations enabled handling of design constraints without lumping them together with the objective function and/or scaling. The convergence of the process was also positively influenced by the approach because of superior infeasibility reduction. This is particularly critical in problems where no solution exists that could be used as a starting point, as required by many algorithms. Furthermore, a simple cooling algorithm, which is both simpler and equally effective as other complex schedules, was implemented. The design of a welded beam and part layout optimization problems are solved using the tuned up SA, which yielded encouraging results. The 100 part layout optimization problem is taken as a representative, computationally large, optimization problem. This work showed that implementing several simple algorithmic developments could add robustness to the annealing algorithm for optimization.

References

1. Kirkpatrick, S., Gelatt Jr., C.D. and Vecchi, M.P., "Optimization by Simulated Annealing", *Science*, Vol. 220, pp. 671-680,1983.

2. Atiqullah, Mir, S. S. Rao, "Simulated Annealing and Parallel Processing: An Implementation for Constrained Global Design Optimization." *Engineering Optimization*, vol. 32, pp. 659-685, 2000.

3. Hajek, B., "Cooling Schedules for Optimal Annealing", Mathematics of Operations Research, Vol. .13, N0.4, pp. 563-571, 1988.

4. Aarts, E.H.L. and van Laarhoven, P.J.M., "Statistical Cooling : A general Approach to Combinatorial Optimization Problems," *Philips Journal of Research*, Vol. 40, pp. 193-226, 1985.

5. Aarts, E.H.L. and van Laarhoven, P.J.M.,"A New Polynomial Time Cooling Schedule," Proc. *IEEE International Conf. on Computer Aided Design*, Santa Clara, pp 206-228, 1985.

6. Huang, M.D., Romeo, F, and Sangiovanni-Vincentelli, A.L, "An Efficient General Cooling Schedule for Simulated Annealing," in *Proceedings of IEEE International Conference on Computer-Aided Design*, pp. 381-384, Santa Clara, November 1986.

7. Atiqullah, Mir M., "Global design optimization using Stochastic methods and Parallel processing," *Ph.D. Dissertation*, School of mechanical Engineering, Purdue University, West Lafayette, 1995.

8. Romeo, F. and Sangiovanni-Vincentelli, A.L., "Probabilistic Hill Climbing Algorithms : Properties and Applications," *Proceedings of the Chapel Hill Conference and VL SI*, pp. 393-417, May 1985.

9. Reklaitis, G.V., Ravindran, A., Ragsdell, K.M., " Engineering Optimization: Methods and Application," John Wiley and Sons, New York, 1983.

10. Ragsdell, K.M. and Phillips, D.T.; "Optimal Design of a Class of Welded Structures using Geometric Programming", ASME Journal of Engineering for Industry, Vol. 98, No 3, pp. 1021-1025, 1976.

11. Deb Kalyanmoy; "Optimal Design of a Class of Welded Structures via Genetic Algorithms", 31st AIAA/ASME/ASCE/AHS/ASC Structures, Structural Dynamics and Materials Conference. Long Beach, CA. April 2-4, 1990.

12. S.S. Rao, and E.L. Mulkay, "Engineering Design Optimization Using Interior-Point Algorithms," AIAA Journal, Vol. 38, No. 11, November 2000.

13. W.A. Bennage, and A.K. Dhingra, "Optimization of Truss Topology Using Tabu Search," International Journal for Numerical Methods in Engineering, Vol. 38, pp. 4035-4052, 1995.

A Hybrid Global Optimization Algorithm Involving Simplex and Inductive Search

Chetan Offord[1] and Željko Bajzer[2]

[1] Biomathematics Resource, Mayo Clinic
200 1st St. SW, Rochester, MN 55905, USA
offord.chetan@mayo.edu
[2] Biomathematics Resource and Department of Biochemistry & Molecular Biology,
Mayo Clinic, Mayo Medical and Mayo Graduate School
200 1st St. SW, Rochester, MN 55905, USA
bajzer@mayo.edu

Abstract. We combine the recently proposed inductive search and the Nelder-Mead simplex method to obtain a hybrid global optimizer, which is not based on random searches. The global search is performed by line minimizations (Brent's method) and plane minimizations (simplex method), while the local multidimensional search employs the standard and a modified simplex method. Results for the test bed of the Second International Competition of Evolutionary Optimization and for another larger test bed show remarkable success. The algorithm was also efficient in minimizing functions related to energy of protein folding.

1 Introduction

Nelder-Mead simplex method [1] for minimization has been used in various fields and especially for optimization in chemistry [2], [3]. It is known as a robust method which does not use derivatives, and has the advantage of easy implementation [4]. It is also known that the simplex search works best for low-dimensional problems and that the position and size of the starting simplex are important for the success of the search. Within the context of global optimization, the simplex method was successfully used as a component of hybrid algorithms that involve adaptive random search [5] or genetic algorithms [6].

Here we propose a hybrid algorithm (SIH) for global minimization based on a combination of inductive search [7] and multiple simplex searches. Two general characteristics of the proposed algorithm are different from standardly accepted paradigms. First, we do not use random numbers in our search as is done in many current global optimizers. This makes our algorithm immune from variations generated by random numbers. Second, besides using line minimizations (frequently employed in classical minimization methods) we also use multiple plane minimizations, i.e. we try to find the minimum in pairs of variables while the rest are kept constant. This feature makes it possible to capture some interdependencies among variables as they approach their values at the minimum, and yet it is less expensive in terms of function calls than an extensive n-dimensional search.

2 The Algorithm

The search domain for a given objective function $f(\mathbf{x})$ is defined by a n-dimensional box $\{\mathbf{x}|\mathbf{x} = (x_1, \ldots, x_n) \in \mathbf{R}^n,\ x_i \in [b_i^l, b_i^u],\ i = 1, \ldots, n\}$, where b_i^l and b_i^u are lower and upper limits for the variable x_i respectively.

Phase 1. Inductive search follows Bilchev and Parmee [7]. This search requires that the objective function $f(\mathbf{x})$ is also an explicit function of space dimension n. Thus, it is assumed that the following sequence of functions can be defined:

$$\phi(1, x_1),\ \phi(2, x_1, x_2),\ \phi(3, x_1, x_2, x_3),\ \ldots \phi(n, x_1, \ldots, x_n) \equiv f(x_1, \ldots, x_n)$$
$$\equiv f(\mathbf{x}) \qquad (1)$$

For many functions usually used in testing global minimization algorithms, this is naturally true, as they are defined for any dimension n, i.e. they are also explicit functions of n. However, in many applications this is not true; therefore we define the sequence of functions as follows:

$$\phi(i, x_1, \ldots, x_i) = f(x_1, \ldots, x_i, x_{i+1}^s, x_{i+2}^s, \ldots, x_n^s), \quad i = 1, \ldots, n \qquad (2)$$

where $(x_1^s, \ldots, x_n^s)$ is the point obtained by an initial n-dimensional search with the simplex algorithm based on the implementation in [4]. The initial simplex for that search is defined by points $\mathbf{p}_j = \mathbf{c} + 0.5[\mathbf{e}_{j-1}(\mathbf{d} \cdot \mathbf{e}_{j-1}) - \mathbf{d}/(n+1)]$, $j = 1 \ldots, n+1$, where $\mathbf{e}_0 = \mathbf{0}$, and $\mathbf{e}_i, i = 1, \ldots, n$ are unit vectors of $\mathbf{R}^n$, $\mathbf{d} = (b_1^u - b_1^l, \ldots, b_n^u - b_n^l)$ determines the dimensions of the n-dimensional box, and $\mathbf{c} = \mathbf{d}/2$ is its central point. Points $\mathbf{p}_j$ define a simplex with its center of gravity in the central point of the search domain. The induction algorithm now goes as follows:

1. The minimum of $\phi(1, x_1)$ at $x_1 = \xi_1$ is found by a line minimization. We used Brent's method as implemented in [4]; see details below.
2. The minimum of $\phi(2, \xi_1, x_2)$ at $x_2 = \xi_2$ is found by a line minimization.
3. The current "minimum" of $\phi(2, x_1, x_2)$ at (ξ_1, ξ_2) is now improved by a two-dimensional simplex minimization (algorithm ESIMP2 described below) to obtain a better or equal value at point (ξ_1^1, ξ_2^1).
4. The minimum of $\phi(3, \xi_1^1, \xi_2^1, x_3)$ at $x_3 = \xi_3$ is found by line minimization.
5. The current best point of $\phi(3, x_1, x_2, x_3)$ at $(\xi_1^1, \xi_2^1, \xi_3)$ is then improved by a three-dimensional simplex minimization (algorithm SIMP(i, r), $i = 3$, described below) to obtain a better or equal value at point $(\xi_1^2, \xi_2^2, \xi_3^2)$.
6. The process described in steps 4 and 5 is now repeated for functions (2), $i = 4, \ldots, n$, i.e. each time a line minimization based on the previous best point is performed and subsequently improved by an i-dimensional minimization (algorithm SIMP(i, r)). At the end, for $i = n$, the full function $f(x_1, \ldots, x_n)$ is minimized by a n-dimensional simplex algorithm with the best point $(\zeta_1, \ldots, \zeta_n)$.

The line minimization by Brent's method initially requires a bracketing triplet [4]. For a given variable x_i we choose the triplet (b_i^l, η_i, b_i^u), where for the function $g(x)$ to be minimized $g(\eta_i)$ is smaller than both $g(b_i^l)$ and $g(b_i^u)$. The point η_i is determined by evaluating the function at n_1 equidistant points within the interval $[b_i^l, b_i^u]$. If by chance $\eta_i = b_i^l$ or $\eta_i = b_i^u$, then the line search is omitted and this point is considered the result of the line minimization. With some experimentation we found that optimal n_1 is around 200 and chose $n_1 = 176$.

The simplex minimizations use the stopping criterion as in [4] with the tolerance $\epsilon = 10^{-7}$. In a simplex search, if a coordinate x_i of a vertex moves outside $[b_i^l, b_i^u]$, it is returned back within that interval to the closest border by a small distance defined as $\delta_i = 0.0001 d_i |\sin(2.2n_2)|$. Here n_2 is the total number of previous boundary crossings. This is designed to keep the point close to the border, but in various positions, to prevent degeneracy of the simplex.

Phase 2. While Phase 1 can yield the global minimum, sometimes the search must be extended further. We accomplish this by combining plane and n-dimensional simplex minimizations, while always utilizing the position of the current best minimum.

Plane minimizations. These are performed in sequence for each possible pair of variables; the remaining variables are resting on the values of the previous best point (initially the result from Phase 1). The sequence of pairs of variables is given by (x_j, x_{j+i}), $i = 1, \ldots, n-1$, $j = 1, \ldots, n-i$, and the procedure for extensive plane minimizations (algorithm ESIMP2) is described below. When the minimization for a given pair of variables results in a significant relative decrease of the function value ($> \sqrt{\epsilon}$), the plane minimizations are repeated for all pairs previously achieving a significant decrease. The first pair to achieve a significant decrease after Phase 1 is simply remembered and the minimization is not repeated. The described repetition procedure is also employed after the plane minimization for the last pair of variables in the sequence. We consider this last effort by repetition as a natural endpoint of SIH.

Multidimensional minimizations. When the plane minimizations for Int($n/2$) consecutive pairs (not counting repeated plane minimizations) are completed, we then interrupt the sequence of plane minimizations and perform two n-dimensional simplex minimizations to take into account all variables simultaneously. Both minimizations are initialized as SIMP(n, r) with the current best point and with $r = r_m = 0.7 + (m_1 - 1)0.01$, where m counts the number of n-dimensional minimizations and $m_1 = m \bmod 30$ (to avoid $r_m = 1$, which would introduce initial simplex points on the border of the search domain). The two n-dimensional minimizations differ in the basic simplex algorithm. The first is designed to avoid a rapid contraction towards the best point, so that the space is searched more exhaustively. This is achieved by contracting only the worst point towards the best point when in the standard simplex method all points are contracted towards the best point. The second n-dimensional minimization employs the standard method [4].

Sub-algorithm ESIMP2 . The search in a plane consists of five minimizations. The first four simplex minimizations are intended to search the rectangular domain of the plane extensively and independently of the previous best minimum. They are started with the initial triangles as depicted in Fig. 1. The fifth two-dimensional search is initialized by the best point found before the four plane minimizations, and the two points which yielded the lowest function values in those four minimizations. In the fifth plane minimization we use the non-standard simplex method with the slow contraction to the best point in order to find a better minimum in the neighborhood of the current one.

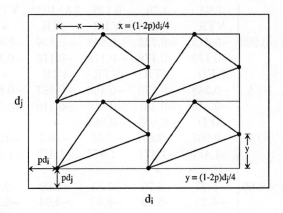

Fig. 1. Initial simplexes (triangles) for plane minimizations. In the ESIMP2 sub-algorithm we used $p = 0.12$

Sub-algorithm SIMP(i, r). Here we use the implementation of simplex method by Press et al., [4]. It is used in Phase 1 for i-dimensional minimizations, $i = 3, \ldots, n$ and in Phase 2 for $i = n$. The algorithm starts with the following initial simplex: $\mathbf{s}_j = (s_1^j, \ldots, s_n^j)$, $s_i^j = \zeta_i + r\sigma(b_i^u + b_i^l - 2\zeta_i) \max(b_i^u - \zeta_i, \zeta_i - b_i^l)$, $j = 1, \ldots, n + 1$, where $\sigma(x) = 1$ for $x > 0$ and $\sigma(x) = -1$ for $x \leq 0$. The point $(\zeta_1, \ldots, \zeta_n)$ is the current best minimum. For Phase 1, $r = 0.7$ and for Phase 2 as defined above.

3 Testing and Comparisons

To compare the efficiency of SIH with other algorithms, we found most suitable the published results of the Second International Competition on Evolutionary Optimization [8], [6]. In Table 1 we compare the results of SIH on the test bed, consisting of seven unconstrained minimizations with the results obtained by Differential Evolution (DE) [8], and the Simplex Genetic Hybrid (SGH) algorithm [6]. These appear to be the only results of the competition which were published.

Table 1. Comparisons for minimization of test functions: Generalized Rosenbrock (RO, $M = 5000$), Odd Square (OS, $M = 5000$), Modified Langerman (ML, $M = 3750$), Shekel's Foxholes (SF, $M = 16000$), Epistatic Michalewicz (EM, $M = 1250$, and Storn's Chebyshev Polynomials (CP, for $n = 9$, $M = 1500$ and for $n = 17$, $M = 10000$). $\phi(1, x)$ is not defined for RO function; therefore we used sequence (2). The functions and the corresponding search domains were obtained from K. Price; see also [8].

$F(\mathbf{x})$	Alg.	n	VTR	M	$2M$	$4M$	$8M$	$16M$	BVAT
RO	DE	10	10^{-6}	30.54	4.844	0.059	VTR	VTR	0
	SGH			7.93	3.75	0.122	$3.8 \cdot 10^{-4}$	VTR	$1.8 \cdot 10^{-9}$
	SIH			VTR	VTR	VTR	VTR	•	$8.5 \cdot 10^{-20}$
OS	DE	10	-0.999	-0.415	-0.735	-0.843	-0.859	-0.870	-0.873
	SGH			-0.173	-0.178	-0.178	-0.178	-0.178	-0.635
	SIH			NFD	VTR	VTR	VTR	•	-1.143
ML	DE	10	N/A	-0.348	-0.176	-0.460	-0.687	-0.834	-0.965
	SGH			-0.315	-0.487	-0.510	-0.510	-0.510	-0.655
	SIH			NFD	-0.965	-0.965	•		-0.965
SF	DE	10	N/A	-0.967	-2.67	-7.50	-10.1	-10.2	-10.208
	SGH			-1.477	-1.477	-1.477	-1.477	-1.477	-1.706
	SIH			-10.208	•				-10.208
EM	DE	10	-9.66	-5.77	-6.79	-7.81	-8.90	-9.44	-9.660
	SGH			-6.29	-6.29	-8.62	-8.94	-8.94	-9.556
	SIH			NFD	NFD	NFD	-8.34	VTR	-9.660
CP	DE	9	10^{-6}	$5.88 \cdot 10^4$	$4.12 \cdot 10^3$	46.4	0.0223	VTR	0
	SGH			9924	4442	3089	2018	835	11.85
	SIH			NFD	NFD	NFD	VTR	VTR	0
CP	DE	17	10^{-6}	$4.9 \cdot 10^6$	$3.5 \cdot 10^4$	11.8	$3.9 \cdot 10^{-5}$	VTR	0
	SGH			$6.4 \cdot 10^6$	$1.4 \cdot 10^6$	$3.9 \cdot 10^5$	$2.3 \cdot 10^5$	$1.2 \cdot 10^5$	$3.0 \cdot 10^4$
	SIH			NFD	NFD	NFD	2445	573	573

Corresponding to the difficulty in the minimization of a given function, each minimization is characterized by a different number M of function evaluations for the first check point. Subsequent check points are given as multiples of M. For each check point the corresponding value of the function is displayed. Value to reach (VTR) is a value presumed to lie within the basin of attraction which contains the global minimum, and if it is reached, the search is considered successful. The values obtained at different check points for DE and SGH correspond to the average of best values reached in 20 independent searches, differed by realization of random numbers. The best value attained BVAT is the lowest result achieved in any of the 20 searches at the last check point ($16M$). In the case of our non-random algorithm, we present the result of a single search in each column and the value (BVAT) as either the value obtained when SIH completed the search (all plane minimizations in Phase 2 performed, signified by •) or when the last check point had been reached, whichever happened first. In some cases the first few check points were reached before the function attained full dimensionality in the Phase 1; this is signified by NFD. The results of Table 1 show that our algorithm performed better than the DE algorithm in all cases but the last one, and much better than the SGH algorithm. We note that K. Price has communicated to us that the current version of DE is more efficient than the version from Table 1.

Table 2. Comparison of achievement of various algorithms after 20000 function calls for functions described in [9]: Odd Square (OS' [F29]), Ackley (AC [F20]), Shekel's Foxholes (SF [F14]), Griewank (GR [F16]), Katsuuras (KA [F18]), Rastrigin (RA [F19]), Rosenbrock (RO' [F2]), Epistatic Michalewicz (EM [F28]), Neumaier no. 3 (N3 [F5]), Hyper-Ellipsoid (HE [F4]), Storn's Chebyshev Polynomials (CP [F30]), Neumaier no. 2 (N2 [F26]), Modified Langerman (ML' [F27]), Shekel-7 (S7 [F7]), Colville (CO [F25]).

$F(\mathbf{x})$	n	SIH	ASA	CS	DE	GE3	GEIII	VFSR	RAND
OS'	10	-2.65	-0.015	-0.038	-0.137	-0.407	-0.772	-0.027	-0.089
AC	30	0.205^*	2.77	6.47	6.46	2.09	1.40	2.78	5.88
OS'	5	-1.81	-0.409	-0.500	-0.847	-0.727	-0.898	-0.127	-0.577
SF	5	-10.4	-2.67	-3.47	-3.18	-10.4	-2.67	-3.29	-2.14
GR	10	0	0.176	0.066	1.39	0.178	0.636	0.067	20.6
KA	10	1.00	10.5	68.5	417	1.00	316	9.70	199
RA	10	0	2090	$4 \cdot 10^5$	10^5	803	2490	1910	$7 \cdot 10^5$
RO'	10	$2 \cdot 10^{-21}$	2.97	$4 \cdot 10^{-10}$	429	8.18	8.85	0.147	5140
EM	5	-4.69	-3.72	-2.79	-4.12	-4.64	-3.82	-3.84	-3.24
SF	10	-10.2	-1.47	-10.2	-0.859	-1.42	-1.41	-1.47	-0.337
EM	10	-9.66	-7.66	-7.38	-4.11	-7.55	-6.65	-9.03	-5.29
N3	25	-2900^*	$2 \cdot 10^4$	-2900	$2 \cdot 10^5$	$3 \cdot 10^4$	$2 \cdot 10^3$	$3 \cdot 10^4$	$9 \cdot 10^5$
N3	30	-4930^*	10^5	-4930	$2 \cdot 10^6$	$4 \cdot 10^4$	$2 \cdot 10^4$	10^5	$2 \cdot 10^6$
HE	30	$4 \cdot 10^{-12*}$	14.7	10^{-21}	193	9.71	5.02	15.0	1060
N3	15	-665	1280	-665	$2 \cdot 10^4$	240	-178	520	$5 \cdot 10^4$
N3	20	-1520^*	9600	-1520	$3 \cdot 10^5$	-609	892	2490	$2 \cdot 10^5$
CP'	9	0	3400	3.94	$5 \cdot 10^5$	8660	722	2220	$7 \cdot 10^4$
N2	4	$4 \cdot 10^{-21}$	0.247	0.217	0.227	0.267	0.239	0.238	0.282
ML'	5	-0.940	-0.965	-0.482	-0.965	-0.501	-0.513	-0.908	-0.510
S7	4	-10.4	-2.75	-10.4	-10.4	-5.08	-10.4	-5.12	-2.83
RO'	5	$7 \cdot 10^{-20}$	0.515	$9 \cdot 10^{-11}$	$8 \cdot 10^{-7}$	2.80	3.59	0.338	10.6
N3	10	-210	-205	-210	2140	-39.1	-50.7	-208	2890
ML'	10	-0.806	-0.274	-0.003	-0.040	-0.513	-0.011	-0.513	-0.005
CO	4	10^{-20}	0.121	$2 \cdot 10^{-11}$	$4 \cdot 10^{-7}$	0.385	3.29	0.198	25.0

Further testing is based on a very detailed and thorough work of Janka [9] in which he compared the efficiency of eight global minimization algorithms, four additional variants, and a simple random search, on the test bed of thirty functions F1-F30, some in multiple dimensions. The algorithms we included here are: Adaptive Simulated Annealing (ASA) [10], [11], Very Fast Simulated Re-annealing (VSFR) [12], [11], clustering with single linkage (CS) [13], [14], Differential Evolution (DE) [8], and Genocop algorithms (GE3, GEIII) [15], [16]. Janka also considered algorithms SIGMA, PGAPack, which however were outperformed by the above mentioned ones. We chose to compare Janka's best variants of ASA, CS and VSFR with SIH. It is noteworthy that the actual implementation of the algorithms Janka used for his study, were based on the software available via the Internet; he did not consult the authors for possible optimal settings [9].

In Table 2 we compare the results of Janka for problems which he classified as the most difficult minimizations [9] with the results of SIH. Some of the problems are the same as in Table 1 and they are designated with the same abbreviation. Some of the problems are essentially the same as in Table 1, except for small differences in the definition of the function and/or search domain; these are denoted with a prime. The rounded values obtained after 20000 function calls are shown. Those underlined represent the function values obtained within the basin of attraction containing the global minimum. With an asterisk, we denoted the situation when the corresponding function is defined in such a way that sequence (1) can be used, but we had to use sequence (2) because Phase 1 of our algorithm had not reached the full dimensionality before 20000 function calls. This happened for problems with $n \geq 20$. Table 2 clearly reveals that our algorithm significantly outperformed all others. The global optimum was found in all but three cases. In two of those problems (AC, ML' for $n = 10$), our values were the lowest anyway and only in one case (ML' for $n = 5$) our value was the second lowest.

Our final comparison is based on a very recent work by Mongeau et al. [17], who compared the efficiency of public domain software for global minimization, i.e. Adaptive Simulated Annealing (ASA), clustering algorithm GLOBAL, genetic algorithm GAS, multilevel random search GOT, integral global optimization INTGLOB, and interactive system for universal functional optimization (UFO). We compare the results of the two highest dimensional functions: protein folding energy minimizations in 15 (pf5) and 18 dimensions (pf6). Table 3 indicates that SIH is comparable to UFO (variant with the best results) and outperformed all others.

Table 3. Comparisons for minimization of protein folding functions pf5 and pf6 as defined in [17]. Displayed is the number of function evaluations required to attain the minima -9.10 for pf5 and -12.71 for pf6. Algorithms which did not reach the minima in 17000 function evaluations for pf5 and 25000 for pf6, are marked by N.R.

$F(\mathbf{x})$	n	SIH	UFO	INTGLOB	ASA	GLOBAL	GOT	GAS
pf5	15	1336	1300	12200	N.R.	13700	N.R.	N.R.
pf6	18	6256	7700	N.R.	N.R.	22300	N.R.	N.R.

4 Concluding Remarks

The developed algorithm takes advantage of the well-known and liked Nelder-Mead simplex minimization and of inductive search, the novel idea of searching from the simple (e.g., minimizing one dimensional correlate of the function) to the complex (e.g., minimizing full n-dimensional function). Inductive search requires that the object function can be considered not only as a function of n variables, but also as a function of n. We have proposed a simple method of how in the context of minimization this can be achieved, when the objective function does not have such a property.

Another feature of inductive search is the combination of global searches in one dimension and subsequent local multidimensional searches. We carried this idea in Phase 2 of our algorithm by employing global 2-dimensional searches and local n-dimensional searches. Phase 2 proved being necessary in some difficult cases.

While the current tests have shown that our algorithm is quite efficient, obviously more extensive testing, especially in scientific applications, should be performed. For difficult problems we anticipate that Phase 2 should be repeated several times, each time with different starting simplexes.

Acknowledgements. We thank Dr. K. Price for providing us with the test bed of the Second International Competition of Evolutionary Optimization and for helpful suggestions. We are also grateful to Mr. E. Janka for making available his M.Sc. thesis and for additional explanations. This work is supported by Mayo Foundation and in part by GM34847.

References

1. Nelder, J.A., Mead, R.: A Simplex Method for Function Minimization. Comput. J. **7** (1965) 308–313
2. Jurs, P.C.: Computer Software Applications in Chemistry. 2nd edn. John Wiley, New York (1996)
3. Walters, F.H., Parker, L.R.Jr., Morgan, S.L., Deming, S.N.: Sequential Simplex Optimization. CRC Press, Boca Raton (1991)
4. Press, W.H., Teukolsky, S.A., Vetterling, W.T., Flannery, B.P.: Numerical Recipes. 2nd. edn. Cambridge University Press, New York (1992)
5. Huzak, M., Bajzer, Ž.: A New Algorithm for Global Minimization Based on the Combination of a Adaptive Random Search and Simplex Algorithm of Nelder and Mead. Croat. Chem. Acta **69** (1996) 775–791
6. Yen, J., Lee, B.: A Simplex Genetic Algorithm Hybrid. In: Proceedings of 1997 IEEE International Conference on Evolutionary Computation. IEEE Inc., Piscataway, NJ (1997) 175–180
7. Bilchev, G., Parmee, I.: Inductive Search. In: Proceedings of 1996 IEEE International Conference on Evolutionary Computation. IEEE Inc., Piscataway, NJ (1996) 832–836

8. Price, K.V.: Differential Evolution vs. the Functions of the 2nd ICEO. In: Proceedings of 1997 IEEE International Conference on Evolutionary Computation. IEEE Inc., Piscataway, NJ (1997) 153–157

9. Janka, E.: Vergleich Stochastischer Verfahren zur Globalen Optimierung. M.Sc. thesis, Vienna (1999); http://www.solon.mat.univie.ac.at/~vpk/; janka@utanet.at

10. Ingber, L.: Simulated Annealing: Practice Versus Theory. Math. Comput. Model. **18** (1993) 29–57

11. Ingber, L.: http://www.ingber.com

12. Ingber, L., Rosen, B.: Genetic Algorithms and Very Fast Simulated Re-annealing: A Comparison. Math. Comput. Model. **16** (1992) 87-100

13. Boender, C, Romeijn, H.: Stochastic Methods. In: Horst, R., Pardalos, P. (eds): Handbook of Global Optimization. Kluwer, Dordrecht (1995) 829–869

14. Csendes, T.: http://www.inf.u-szeged.hu/~csendes/

15. Michalewicz, Z.: Genetic Algorithms + Data Structures = Evolution Programs. 3rd ed. Springer, Berlin (1966)

16. Michalewicz, Z.: http://www.coe.uncc.edu/~zbyszek/

17. Mongeau, M., Karsenty, H., Rouzé, V., Huriart-Urruty, J.-B.: Comparison of Public-domain Software for Black Box Global Optimization. Optim. Meth. and Software **13** (2000) 203–226

Applying Evolutionary Algorithms to Combinatorial Optimization Problems

Enrique Alba Torres[1] and Sami Khuri[2]

[1] Universidad de Málaga, Complejo Tecnológico,
Campus de Teatinos, 29071 Málaga, Spain.
eat@lcc.uma.es
WWW home page: http://polaris.lcc.uma.es/~eat/
[2] Department of Mathematics & Computer Science, San José State University,
One Washington Square, San José, CA 95192-0103, U.S.A.
khuri@cs.sjsu.edu
WWW home page: http://www.mathcs.sjsu.edu/faculty/khuri

Abstract. The paper describes the comparison of three evolutionary algorithms for solving combinatorial optimization problems. In particular, a generational, a steady-state and a cellular genetic algorithm were applied to the maximum cut problem, the error correcting code design problem, and the minimum tardy task problem. The results obtained in this work are better than the ones previously reported in the literature in all cases except for one problem instance. The high quality results were achieved although no problem-specific changes of the evolutionary algorithms were made other than in the fitness function. The constraints for the minimum tardy task problem were taken into account by incorporating a graded penalty term into the fitness function. The generational and steady-state algorithms yielded very good results although they sampled only a tiny fraction of the search space.

1 Introduction

In many areas, such as graph theory, scheduling and coding theory, there are several problems for which computationally tractable solutions have not been found or have shown to be non-existent [12]. The polynomial time algorithms take a large amount of time to be of practical use. In the past few years, several researchers used algorithms based on the model of organic evolution as an attempt to solve hard optimization and adaptation problems. Due to their representation scheme for search points, Genetic Algorithms (GA) are the most promising and easily applicable representatives of evolutionary algorithms for the problems discussed in this work.

The goal of this work is two-fold. First, a performance comparison of three evolutionary algorithms for solving combinatorial optimization problems is made. These algorithms are the generational genetic algorithm (genGA), the steady-state genetic algorithm (ssGA) [13], and the cellular genetic algorithm (cGA) [8]. Second, the paper reports the improvement achieved on already known results for similar problem instances. We compare the results of our experiments to those of [2] and [6].

The outline of the paper is as follows: Section 2 presents a short overview of the basic working principles of genetic algorithms. Section 3 presents the maximum cut problem, the error correcting code design problem, and the minimum tardy task problem. The problem's encoding, the fitness function, and other specific particularities of the problem are explained. The experimental results for each problem instance are described in Section 4. The paper summarizes our findings in Section 5.

2 The Evolutionary Algorithms

Our genGA, like most GA described in the literature, is generational. At each generation, the new population consists entirely of offspring formed by parents in the previous generation (although some of these offspring may be identical to their parents). In steady-state selection [13], only a few individuals are replaced in each generation. With ssGA, the least fit individual is replaced by the offspring resulting from crossover and mutation of the fittest individuals. The cGA implemented in this work is an extension of [10]. Its population is structured in a toroidal 2D grid and the neighborhood defined on it always contains 5 strings: the one under consideration and its north, east, west, and south neighboring strings. The grid is a 7×7 square. Fitness proportional selection is used in the neighborhood along with the one–point crossover operator. The latter yields only one child: the one having the larger portion of the best parent. The reader is referred to [1] for more details on panmictic and structured genetic algorithms.

As expected, significant portions of the search space of some of the problem instances we tackle are infeasible regions. Rather than ignoring the infeasible regions, and concentrating only on feasible ones, we do allow infeasibly bred strings to join the population, but for a certain price. A penalty term incorporated in the fitness function is activated, thus reducing the infeasible string's strength relative to the other strings in the population. We would like to point out that the infeasible string's lifespan is quite short. It participates in the search, but is in general left out by the selection process for the succeeding generation.

3 Combinatorial Optimization Problems

In this paper, we apply three evolutionary algorithms to instances of different NP-complete combinatorial optimization problems. These are the maximum cut problem the error correcting code design problem, and the minimum tardy task problem. These problems represent a broad spectrum of the challenging intractable problems in the areas of graph theory [9], coding theory [7], and scheduling [4]. All three problems were chosen because of their practical use and the existence of some preliminary work in applying genetic algorithms to solve them [2], [3], and [6].

The experiments for graph and scheduling problems are performed with different instances. The first problem instance is of moderate size, but nevertheless, is a challenging exercise for any heuristic. While the typical problem size for the first instance is about twenty, the subsequent problem instances comprise of populations with strings of length one hundred and two hundred, respectively. In the absence of test problems of significantly large sizes, we proceed by introducing scalable test problems that can be scaled up to any desired large size, and more importantly, the optimal solution can be computed. This allows us to compare our results to the optimum solution, as well as to the existing best solution (using genetic algorithms). As for the error correcting code design problem, we confine the study to a single complex problem instance.

3.1 The Maximum Cut Problem

The maximum cut problem consists in partitioning the set of vertices of a weighted graph into two disjoint subsets such that the sum of the weights of the edges with one endpoint in each subset is maximized. Thus, if $G = (V, E)$ denotes a weighted graph where V is the set of nodes and E the set of edges, then the maximum cut problem consists in partitioning V into two disjoint sets V_0 and V_1 such that the sum of the weights of the edges from E that have one endpoint in V_0 and the other in V_1, is maximized. This problem is NP-complete since the satisfiability problem can be polynomially transformed into it [5].

We use a binary string $(x_1, x_2, \ldots, x_n)$ of length n where each digit corresponds to a vertex. Each string encodes a partition of the vertices. If a digit is 1 then the corresponding vertex is in set V_1, if it is 0 then the corresponding vertex is in set V_0. Each string in $\{0,1\}^n$ represents a partition of the vertices. The function to be maximized is:

$$f(x) = \sum_{i=1}^{n-1} \sum_{j=i+1}^{n} w_{ij} \cdot [x_i(1 - x_j) + x_j(1 - x_i)]. \tag{1}$$

Note that w_{ij} contributes to the sum only if nodes i and j are in different partitions.

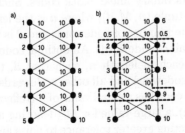

Fig. 1. Example of a maximum cut for the graph structure proposed for generating test examples. The problem size is $n = 10$, the maximum cut value is $f^* = 87$.

In this work, we consider the randomly generated sparse graph "cut20-0.1" and the randomly generated dense graph "cut20-0.9" found in [6]. In order to obtain larger problem instances, we make use of the scalable weighted graph with $n = 10$ nodes shown in Figure 1a. The cut-set that yields the optimal solution can be computed from the construction. The dotted line partition of Figure 1b is represented by the bit string 0101001010 (or its complement) with objective function value $f^* = 87$ and yields the optimum cut-set.

This graph can be scaled up, for any even value of n, to form arbitrarily large graphs with the same structure and an even number of nodes. The construction of a graph with n nodes consists in adding vertex pairs at the bottom of the graph and connecting them vertically by one edge of weight 1 per vertex and diagonally by one edge of weight 10 per vertex. According to this construction, the optimal partition is easily described by the concatenation of a copy of the $n/4$-fold repetition of the bit pattern 01, followed by a 0, then another copy of the $n/4$-fold repetition of the bit pattern 01, and finally a 0. Alternatively, one could take the complement of the described string. The string has objective function value $f^* = 21 + 11 \cdot (n - 4)$ for $n \geq 4$. One might be tempted to believe that such regularity in the formulation of the problem instance might favor the workings of genetic algorithms. In order to defuse any doubts, we introduce a preprocessing step which consists in randomly renaming the vertices of the problem instance. As a consequence, consecutive bit positions no longer correspond to vertices that are close to each other within the graph itself.

For the experiments reported here, a graph of size $n = 100$, "cut100", is used.

3.2 The Error Correcting Code Design Problem

The error correcting code design problem (ECC) consists of assigning codewords to an alphabet that minimizes the length of transmitted messages and that provides maximal correction of single uncorrelated bit errors, when the messages are transmitted over noisy channels. Note that the two restrictions are conflicting in nature. On one hand, we would like to assign codewords that are as short as possible, and on the other hand, good error correction is achieved by adding redundant bits so as to maximize the Hamming distance between every pair of codewords.

This study considers binary linear block codes. Such codes can be formally represented by a three-tuple (n, M, d), where n is the length (number of bits) of each codeword, M is the number of codewords and d is the minimum Hamming distance between any pair of codewords. An optimal code consists in constructing M binary codewords, each of length n, such that d, the minimum Hamming distance between each codeword and all other codewords, is maximized. In other words, a good (n, M, d) code has a small value of n (reflecting smaller redundancy and faster transmission), a large value for M (denoting a larger vocabulary) and a large value for d (reflecting greater tolerance to noise and error). As n increases, the search space of possible codes grows exponentially.

Linear block codes can either be polynomially generated, such as the Bose, Chaudhuri, and Hocquenghem (BCH) codes [7], or non-polynomially generated, by using some heuristic. Genetic algorithms can be used to design such codes [3]. Other researchers have used hybrids (e.g., simulated annealing and genetic algorithms) and parallel algorithms to achieve good codes [2].

In this study, we consider a problem instance that was tackled by [2], where $n = 12$ and $M = 24$, and use their fitness function with all three evolutionary algorithms. However, we do not parallelize our genetic algorithms, as is the case in their work. The function to be minimized is:

$$f(C) = \frac{1}{\sum_{i=1}^{M} \sum_{j=1; i \neq j}^{M} \frac{1}{d_{ij}^2}} \tag{2}$$

where d_{ij} represents the Hamming distance between codewords i and j in the code C (of n codewords, each of length M).

Note that for a code where $n = 12$ and $M = 24$, the search space is of size $\binom{4096}{24}$, which is approximately 10^{87}. It can be shown that the optimum solution for $n = 12$ and $M = 24$ has a fitness value of 0.0674. The optimum solution is illustrated in [2].

3.3 The Minimum Tardy Task Problem

The minimum tardy task problem is a task-scheduling problem. It is NP-complete since the partitioning problem can be polynomially transformed into it [5]. The following is a formal definition of the minimum tardy task problem [12]:

Problem instance:

Tasks:	1	2	...	n	,	$i > 0$
Lengths:	l_1	l_2	...	l_n	,	$l_i > 0$
Deadlines:	d_1	d_2	...	d_n	,	$d_i > 0$
Weights:	w_1	w_2	...	w_n	,	$w_i > 0$

Feasible solution: A one-to-one scheduling function g defined on $S \subseteq T$, $g : S \longrightarrow Z^+ \cup \{0\}$ that satisfies the following conditions for all $i, j \in S$:

1. If $g(i) < g(j)$ then $g(i) + l_i \leq g(j)$ which insures that a task is not scheduled before the completion of an earlier scheduled one.

2. $g(i) + l_i \leq d_i$ which ensures that a task is completed within its deadline.

Objective function: The tardy task weight $W = \sum_{i \in T-S} w_i$, which is the sum of the weights of unscheduled tasks.

Optimal solution: The schedule S with the minimum tardy task weight W.

A subset S of T is feasible if and only if the tasks in S can be scheduled in increasing order by deadline without violating any deadline [12]. If the tasks are not in that order, one needs to perform a polynomially executable preprocessing step in which the tasks are ordered in increasing order of deadlines, and renamed such that $d_1 \leq d_2 \leq \cdots \leq d_n$.

A schedule S can be represented by a vector $x = (x_1, x_2, \ldots, x_n)$ where $x_i \in \{0, 1\}$. The presence of task i in S means that $x_i = 1$, while its absence is represented by a value of zero in the ith component of x. We use the fitness function described in [6] which allows infeasible strings and uses a graded penalty term.

For our experiments, we use three problem instances: "mttp20" (of size 20), "mttp100" (of size 100) and "mttp200" (of size 200). The first problem instance can be found in [6]. The second and third problem instances were generated by using a scalable problem instance introduced in [6].

4 Experimental Runs

We performed a total of 100 experimental runs for each of the problem instances. Whenever no parameter setting is stated explicitly, all experiments reported here are performed with a standard genetic algorithm parameter setting: Population size $\mu = 50$, one-point crossover, crossover rate $p_c = 0.6$, bit-flip mutation, mutation rate $p_m = 1/n$ (where n is the string length), and proportional selection. All three algorithms were run on a uniprocessor machine. These were the settings used with the same problem instances reported in [2] and [6].

What follows is the convention used to present the results of the experimental runs. The first column for each evolutionary algorithm gives the best fitness value encountered during the 100 runs. The second column for each evolutionary algorithm records the number of times each one of these values is attained during the 100 runs. The values given in the first row of the table are the average number of iterations it took to obtain the maximum value. The first value recorded under $f(x)$ is the globally optimal solution. For example, Table 1 reports that genGA obtained the global optimal (whose value is 10.11981) ninety two times out of the 100 runs. The table also indicates that the optimum value was obtained after 2782.1 iterations when averaged over the 100 runs.

4.1 Results For The Maximum Cut Problem

Table 1. Overall best results of all experimental runs performed for "cut20-01".

ssGA		genGA		cGA	
avg = 626.4		avg = 2782.1		avg = 7499	
$f(x)$	N	$f(x)$	N	$f(x)$	N
10.11981	79	10.11981	92	10.11981	16
9.76	21	10.05	1	10.0	24
		10.0	1	9.765	53
		9.89	1	9.76	7
		9.76	5		

We notice that genGA performs better than ssGA for the sparse graph (see Table 1), while ssGA gives better results for the dense graph (see Table 2). Due to the very small population size ($n = 50$), in which neighborhoods cannot develop properly, the cGA did not produce results as good as the two other algorithms. In other words, we believe that with such a small population size, cGA is still mainly in the exploration stage rather than the exploitation stage. As for "cut100", all three algorithms were unable to find the global optimum. When compared to [6], our ssGA and genGA performed better for the sparse and dense graphs. As for "cut100", our algorithms were not able to improve on the results of [6].

Overall, these are good results especially when we realize that the evolutionary algorithms explore only about 1% of the search space.

Table 2. Overall best results of all experimental runs performed for "cut20-0.9".

ssGA		genGA		cGA	
avg = 2007.1		avg = 4798		avg = 7274	
$f(x)$	N	$f(x)$	N	$f(x)$	N
56.74007	72	56.74007	50	56.74007	2
56.04	10	56.73	19	56.5	12
55.84	16	56.12	12	55.5	59
55.75	2	56.04	9	54.5	24
		55	10	53.5	3

Table 3. Overall best results of all experimental runs performed for "cut100".

ssGA		genGA		cGA	
$f(x)$	N	$f(x)$	N	$f(x)$	N
1077	0	1077	0	1077	0
1055	9	1055	0	1055	0
1033	19	1033	8	1033	4
1011	36	1011	9	1011	11
989	22	989	14	989	8
967	7	967	9	967	4
≤ 945	7	≤ 945	60	≤ 945	73

4.2 Results For The ECC Problem

For the ECC problem, ssGA outperformed both genGA and cGA. For this problem instance, cGA produced results comparable to those of ssGA. But as can be seen from the average values in Table 4, ssGA is substantially faster than cGA. Our algorithms performed better than the one reported in [2]. We believe that our algorithms outperformed theirs mainly because we used symmetric encoding, where once a string is processed, we assume that its complement too has been taken care of, thus producing substantial time savings.

4.3 Results For The Minimum Tardy Task Problem

We notice that while ssGA outperforms the two other heuristics for the 20-task problem instance, genGA gives much better results for the 100-task and 200-task problems. For "mttp20", the local optimum of 46 differs from the global one by a Hamming distance of three. Compared to the results of [6] for "mttp20", ssGA performs much better, genGA is comparable, while cGA's performance is worse.

With "mttp100", the global optimum (200) is obtained with the unique string composed of 20 concatenations of the string $b=11001$. The second best solution of 243, is obtained by the strings that have 11000 as prefix (with tasks three, four and five contributing a total of 60 units towards the fitness value). This prefix is followed by the substring 11101 (contributing 3 units towards the fitness value) and 18 copies of $b=11001$ (each contributing 10 units towards the fitness value). Since there are 19 ways of placing 11101 among the 18 substrings 11001, there are 19 strings of quality 243 ($60 + 3 + (18 \times 10)$). A second accumulation of results is observed for the local optimum of 329, which is obtained by the schedule represented by the string with prefix: 001001110111101. The string is then completed by concatenating 17 copies of $b=11001$. This string too is unique. Compared to the results reported in [6] for "mttp100", both ssGA and genGA significantly outperform them. Once more, cGA lags behind.

For "mttp200", genGA is a clear winner among the three evolutionary algorithms. This problem instance was not attempted by [6].

Table 4. Overall best results of all experimental runs performed for the ECC problem instance.

ssGA		genGA		cGA	
avg = 7808		avg = 35204		avg = 30367	
$f(x)$	N	$f(x)$	N	$f(x)$	N
0.067	40	0.067	22	0.067	37
0.066	0	0.066	11	0.066	1
0.065	17	0.065	18	0.065	21
0.064	25	0.064	33	0.064	27
0.063	13	0.063	16	0.063	13
0.062	5	0.062	0	0.062	1

Table 5. Overall best results of all experimental runs performed for "mttp20".

ssGA		genGA		cGA	
avg = 871.4		avg = 2174.7		avg = 7064.2	
$f(x)$	N	$f(x)$	N	$f(x)$	N
41	86	41	73	41	23
46	10	46	11	46	7
51	4	49	8	49	9
		51	3	51	9
		56	1	53	6
		57	1	54	1
		61	1	56	12
		65	2	≥ 57	33

Table 6. Overall best results of all experimental runs performed for "mttp100".

ssGA		genGA		cGA	
avg = 43442		avg = 45426		avg = 15390	
$f(x)$	N	$f(x)$	N	$f(x)$	N
200	78	200	98	200	18
243	4	243	2	243	18
326	1			276	2
329	17			293	6
				316	1
				326	1
				329	37
				379	9
				≥ 429	8

Table 7. Overall best results of all experimental runs performed for "mttp200".

ssGA		genGA		cGA	
avg = 288261.7		avg = 83812.2		avg = 282507.3	
$f(x)$	N	$f(x)$	N	$f(x)$	N
400	18	400	82	400	6
443	8	443	9	443	7
476	2	476	2	493	1
493	1	493	2	529	34
516	1	496	1	543	1
529	42	529	3	579	10
579	3	629	1	602	1
602	1			629	8
665	23			665	17
715	1			≥ 679	15

5 Conclusion

This work explored the applications of three evolutionary algorithms to combinatorial optimization problems. The algorithms were the generational, the steady-state and the cellular genetic algorithm. The primary reason behind embarking on the comparison testing reported in this work was to see if it is possible to predict the kind of problems to which a certain evolutionary algorithm is or is not well suited. Two algorithms, ssGA and genGA, performed very well with the maximum cut problem, the error correcting code design problem and the tardy task scheduling problem. The third algorithm, namely cGA, usually lagged behind the other two. For all problem instances except one, genGA and ssGA outperformed previously reported results.

Overall, our findings confirm the strong potential of evolutionary algorithms to yield a globally optimal solution with high probability in reasonable time, even in case of hard multimodal optimization tasks when a number of independent runs is performed.

We subscribe to the belief that one should move away from the reliance on individual problems only in comparing the performance of evolutionary algorithms [11]. We believe researchers should instead create test problem generators in which random problems with certain characteristics can be generated automatically and methodically. Example characteristics include multimodality, epistasis, the degree of deception, and problem size. With this alternative method, it is often easier to draw general conclusions about the behavior of an evolutionary algorithm since problems are randomly created within a certain class. Consequently, the strengths and weaknesses of the algorithms can be tied to specific problem characteristics.

It is our belief that further investigation into these evolutionary algorithms will demonstrate their applicability to a wider range of NP-complete problems.

References

1. E. Alba and J. M. Troya. A survey of parallel distributed genetic algorithms *Complexity* pages 31–52, vol. 4, number 4, 1999.
2. H. Chen, N. S. Flann, and D. W. Watson. Parallel genetic simulated annealing: a massively parallel SIMD algorithm. *IEEE transactions on parallel and distributed systems*, pages 805–811, vol. 9, number 2, February 1998.
3. K. Dontas and K. De Jong. Discovery of maximal distance codes using genetic algorithms. *Proceedings of the Tools for Artificial Intelligence Conference*, pages 805–811, Reston, VA, 1990.
4. P. Brucker, Scheduling Algorithms, Springer-Verlag, 2nd edition, 1998.
5. R. M. Karp. Reducibility among combinatorial problems. In R. E. Miller and J. W. Thatcher, editors, *Complexity of Computer Computation*, pages 85–103. Plenum, New York, 1972.
6. S. Khuri, T. Bäck, and J. Heitkötter. An evolutionary approach to combinatorial optimization problems. *Proceedings of the 22nd Annual ACM Computer Science Conference*, pages 66–73, ACM Press, NY, 1994.
7. S. Lin and D. J. Costello, Jr. Error Control Coding: Fundamentals and Applications', Prentice Hall, 1989.
8. B. Manderick and P. Spiessens. Fine-grained parallel genetic algorithms, *Proceedings of the 3rd ICGA*, pages 428–433, Morgan Kaufmann, 1989.
9. C. H. Papadimitriou, Computational Complexity, Addison Wesley, 1994.
10. J. Sarma and K. De Jong. An analysis of the effect of the neighborhood size and shape on local selection algorithms. *Lecture Notes in Computer Science*, vol. 1141, pages 236–244, Springer-Verlag, Heidelberg, 1996.
11. W. Spears. Workshop on test problems generators. *Proceedings of the International Conference on Genetic Algorithms*, Michigan, July 1997.
12. D. R. Stinson. *An Introduction to the Design and Analysis of Algorithms*. The Charles Babbage Research Center, Winnipeg, Manitoba, Canada, 2nd edition, 1987.
13. G. Syswerda, A Study of Reproduction in Generational and Steady-State Genetic Algorithms. *Proceedings of FOGA*, pages 94–101, Morgan Kaufmann, 1991.

Program and Visualization

Session chair:

Brian J. d'Auriol (University of Texas at El Paso, USA)

Exploratory Study of Scientific Visualization Techniques for Program Visualization

Brian J. d'Auriol, Claudia V. Casas, Pramod Kumar Chikkappaiah,
L. Susan Draper, Ammar J. Esper, Jorge López, Rajesh Molakaseema,
Seetharami R. Seelam, René Saenz, Qian Wen, Zhengjing Yang

Department of Computer Science, The University of Texas at El Paso, El Paso, TX
79968, USA

Abstract. This paper presents a unique point-of-view for program visualization, namely, the use of scientific visualization techniques for program visualization. This paper is exploratory in nature. Its primary contribution is to re-examine program visualization from a scientific visualization point-of-view. This paper reveals that specific visualization techniques such as animation, isolines, program slicing, dimensional reduction, glyphs and color maps may be considered for program visualization. In addition, some features of AVS/Express that may be used for program visualization are discussed. Lastly, comments regarding emotional color spaces are made.

1 Introduction

Program Visualization has been defined as the use of various graphical techniques to enhance the human understanding of computer programs [1]. In order to achieve a good visualization of a program, there are three basic visualization stages that need to be taken into account: extraction or data collection, abstraction or analysis, and presentation or display of the result of the analysis [2]. Often, the goals of program visualization reflect the needs of program understanding, design, debugging, maintaining, testing, or code re-use.

Similar to the field of program visualization is scientific visualization. Scientific visualization provides for visual illustration of data and related properties that are often obtained from scientific (or other) disciplines. Scientific visualization helps users to gain a better understanding of complex results. In the past, scientific visualization enjoys more mature techniques, procedures and tools than commonly available in program visualization. Two other differences between the fields include the purpose for the visualizations and dimensional or spatial structuring of the data to be visualized. Whereas scientific visualization is often used to understand the nature of a phenomena, system or application area, program visualization often seeks to provide clarification of a process or procedure so that modifications to the process can be made to generate a 'better' process. In typical scientific visualizations, the data exist in some identifiable structure or coordinate space, e.g., temperature data in a volume. With respect to programs, there is often a lack of any identifiable structure, that is, the nature of the data to be visualized is not static so much as it is transformable. Program visualization is often about understanding how the data associated with a program is transformed during the execution of the program.

This paper is exploratory in nature. The focus is on program visualization from a scientific visualization context. Its primary contribution is to re-examine program visualization from a different point-of-view; to explore how structure may be added to data associated with a program; and to explore how a well known scientific visualization tool, AVS/Express, might be used in program visualization.

This paper is structured as follows. Section 2 reviews program visualization elements while related work in program visualization is described in Section 3. Section 4 presents a proposed approach to accomplish program visualization from a scientific visualization point-of-view. Section 5 presents a summary of AVS/Express. Section 6 presents various techniques, both scientific and program, that are exploratory in nature but considered potentially useful for program visualizations. An exploratory application example is presented in Section 7. Conclusions are given in Section 8.

2   Program and AVS/Express Visualization Elements

In order to visualize a program, characteristics of the program need to be identified. Characteristics can be categorized as either static or dynamic. Static data that can be visualized in a program involve abstractions such as the following: data structures and attributes, procedures, code metrics and types of operations. Execution statistics are part of the dynamic visualization aspects of a program. Execution statistics involve issues such as: the amount of computations performed, amount of storage space consumed, program control flow and data flow. Debuggers, tracers and other profiling tools are examples of dynamic visualization enabling tools. Animation is frequently used in program visualizations since it is often of interest to study a program's behavior during its execution.

Some AVS/Express features that are under consideration for use in program visualization are now briefly described [3, 4]; later sections may provide additional detail. In general, unstructured point data can be represented by coordinate values, multivariate data can be represented by glyphs and line data can be represented by points that describe lines. Colored streamlines with animated markers indicate flow direction. Plotting of surface grids allow for meshes. Arbitrary cut-away 'cuts' provides for a sectioned model. Line graphing facilities are available. Isosurfaces, 3D surfaces with coloring, velocity vectors and contour plots can be used. Discrete data can be interpolated between the points so as to provide more continuous color visualization of the data. Additional techniques include exterior and edge finding; contours, isolines, isosurfaces and isovolumes; slices and cross-sections; colors; 3D scatter data and solid contours in 3D; and city scapes, ribbon plots, and surface plots.

The primary data structure in AVS/Express are fields. Fields combine a structure with data where a structure may be one of: uniform mesh, rectilinear mesh, structured mesh, or unstructured mesh. However, it is not apparent how field structure information can be meaningfully associated with program related data. This subject is addressed in subsequent sections.

3 Related Work

Program visualization is important in the area of parallel and distributed computing. Here, visualization assists in the understanding of such aspects as concurrency, scaling, communication, synchronization and shared variables, see for example [2]. Program visualization is also incorporated into software fault detection [5]. When a failure occurs in a software testing process, it is necessary for fault detection to inspect a large number of software processes from various viewpoints. Program visualization is also used in education, see for example [6].

BALSA is an algorithm animation system that was first developed for educational purposes at Brown University [7]. Although BALSA offers several features that gives it a great value in education, its use is limited by the necessity that BALSA uses a code annotation method. In such a method, the programmer indicates 'interesting events' as annotations in the program code in order to mark areas or information that will be used in the visualization. This requires the visualizer to know the program code in order to produce a good visualization of it. The benefit of such a system is to a third party.

A good introduction to the area and much additional related work can be found in [8].

4 A Program Visualization Approach

Given the three stage visualization process of data collection, abstraction and presentation, the proposed approach is: a) identifying relevant program-related information, b) abstracting that information into a form suitable for visualization and c) incorporating specific visualization components for the presentation. Unique in this work is the adoption of scientific visualization techniques and framework for the third stage, namely, the use of a scientific visualization tool for the presentation. In this work, AVS/Express is used in this context.

However, since the data suitable for a scientific visualization has structure, specifically, the data exists in a coordinate space, a requirement is imposed on either the first or second stages to incorporate such a coordinate structure on the program-related information. For example, assume a linear scale and each program statement mapped to equidistant points on the line, then the information 'program statement' has been given a (simple) structure. The abstraction stage could augment the information in this example, by say, adding color to distinguish between several types of statements. Such an example (see Section 7), though simple, is suitable for presentation by AVS/Express.

Two key questions are apparent: 'In which stage ought the structure be incorporated?' and 'What would constitute a useful structure?' The answers to these questions lie outside the focus of this paper, and indeed, are motivations for work-in-progress. None-the-less, preliminary comments can be made regarding the first question; furthermore, this paper is devoted to explorations that address these questions.

Two possibilities exist for the addition of the structure to the data: firstly, in the first stage, the information can be extracted according to a pre-defined structure format; or secondly, in the second stage, a suitable structure can be added to the data. Two examples follow. For the first case, let a 2-D table be identified as relevant, then a coordinate (x, y) can identify an element in the table. For the second case, a control flow graph can be identified based on a standard program analysis where each node in the graph can be identified by an (x, y) coordinate.

Once a coordinate framework has been established, one may imagine the visualization process whereby a) a user creates geometric objects, and b) data is mapped to some geometric features or attributes of these objects.

5 AVS/Express

AVS/Express is a visualization tool developed by Advanced Visual Systems for visualizing scientific data. It uses object-oriented technology, provides for component software development, allows end-user programming and is a visual development tool. The software provides for visualization of complex data and allows applications to incorporate interactive visualization and graphics objects [3]. AVS/Express features a Data Viewer interface, that is, a "point-and-click" interface [4]. This is used for viewing text, graphs, images, geometries and volumes or manipulating display elements such as cameras and light sources. Users have control over the rendering of and interaction with objects in the Data Viewer.

The three programming interfaces in AVS/Express are the Network Editor, the V Command Processor and the Application Programming Interface. As the Network Editor is the principal interface, it is detailed further below.

The Network Editor [3] is a visual programming environment. In the Network Editor, rectangular boxes represent each object of an application. Each object has predefined input and output ports that provide interconnections between objects. Different port attributes, e.g. datatypes, are represented by different colors on the sides of each box. Macros are easily created and stored in libraries for re-use. Connections are made, broken, or re-arranged by "mouse drag and drop" operations. The Network Editor allows the user to modify applications 'on-the-fly', with no need to stop, recompile and start again. In the Network Editor objects display their hierarchy and relations in a flow graph-type manner where users may connect objects visually and graphically.

AVS/Express tool kits provide predefined application program components that in addition, enable users to create their own components or incorporate other developed components. Some of the available tool kits include: User Interface Kit, Data Visualization Kit, Image Processing Kit, Graphics Display Kit, Database Kit, AVS5 Compatibility Kit and Annotation and Graphing Kit [3]. Brief descriptions of some of these kits are given below [3, 9]. The User Interface Kit provides the ability to build platform independent or customizable graphical user interfaces. The Image Processing Kit is an extensive library with over 40 image processing functions for analyzing and manipulating images. The Database kit allows for multidimensional visualization applications linked to commercial relational database management systems.

6   Visualization Techniques

6.1   Animation

Animation refers to a process of dynamically generating a series of frames of a scene in which each frame is an alteration of the previous frame [10]. Animation is often used in two different ways in science and visualization: firstly, it is to visualize data to assist in understanding some phenomenon, and secondly, is used to convey information for teaching, recording, etc. In both contexts, animation can be useful for programmers. For example, it can be used to directly support algorithm or data structures teaching. In addition, it can be also used in the analysis of programs for debugging or algorithm enhancement purposes.

Algorithm animation is the process of abstracting the data, operations, and semantics of computer programs, and then creating animated graphical views of those abstractions. A number of program visualization systems such as XTANGO and POLKA [8] incorporated animation. A system called Eliot is an interactive animation environment. Eliot provides a library of visual data types which are ordinary data types with a set of pre-defined visualizations. The user can select one visualization for each data object to be animated. Eliot constructs an animation where the objects as well as their operations are animated based on these selections. Eliot can be used in algorithm design, visual debugging and learning programming [11]. See [12, 13] for further examples. In addition, there are a several web-based algorithm animation systems, see for example Animator [14].

Animation in AVS/Express [15, 16] includes the following four methods. The Loop module which outputs a sequence of integers or floats to drive 'down stream' modules. The loop controls allow the user to set starting value, ending value and other necessary control information. Animations with read_Field can be accomplished by using the temporal data option. The Key frame Animator interacts with modules mainly through the GUI interface. The particle advector animates massless particles along a vector field.

6.2   Isolines

An isoline is a curve that has the property that every point on it has the same value of the *isoline function*. Isolines are a well known visualization component in scientific visualization. Isolines can be used to assist in the visualization of control flow, static data structure, control flow, etc. For example, an isoline function defined as the (byte) size of every data structures would provide visual information about memory usage (the assumption is made in this and subsequent isoline examples that size exists in a pre-defined coordinate space). In visualizing data flow, isolines could also be used to indicate the data elements that have the same value during the execution of the program; the changes to isolines could provide information regarding how the data values change; the pattern of such a change could indicate program behaviors. Another example of incorporating isolines in a program visualization is the nesting level of program statements; such would indicate areas of nested code.

6.3 Program Slicing

Program Slicing has been used to analyze, debug, and understand computer programs [17]. Program slicing is a "decomposition technique that extracts from program statements relevant to a particular computation" [17]. A typical computer program consists of several smaller sub-programs. Program slicing is aimed at identifying these sub-programs and analyzing dependencies between them. There is much research reported in program slicing, see for example [17]. Program visualization may benefit from the ideas and methods in slicing by visualizing slices of a program then combining these visualizations to produce a whole program visualization. The first two stages of program visualization include data collection and data preprocessing where statements are analyzed to extract data for visualization. Interestingly, part of this work is completed during program slicing, hence, overhead due to program slicing may be reduced by the necessary requirements of the proposed program visualization method.

6.4 Dimension Reduction

For higher dimensional data (i.e., the coordinate space is greater than three dimensions), it is often necessary to reduce the 'higher' dimensions to three or two for visual rendering purposes. Reduction of dimension can be done by two methods: focusing and linking [18]. Focusing involves selecting subsets of the data; reduction of dimension by typically a projection or in some cases, by some general manipulation of the layout of information on the screen. Whereas focusing conveys only partial information about the data (i.e., the subset(s)), linking provides for contextual referencing of the data with respect to other data subsets. Essentially, this is accomplished by sequencing several visualizations over time or by showing the visualizations in parallel simultaneously. Specific techniques include dimensional stacking and hierarchical axis.

6.5 Glyphs in AVS/Express as a Means of Representing Higher Dimensional Discrete Data

A glyph is a single graphical object (e.g., sphere, diamond, square) that represents a multivariate data object/value at a single location in space. Multiple data attributes may be mapped to appearance attributes of the glyph (e.g. size, color, texture). A glyph can be used to represent a higher dimensional data set. See [19] for details regarding glyph representation.

In AVS/Express, the module GLYPH located in the main library, mappers section, can be used to represent higher dimensional data. It has three input ports one of which is called the in_field port that takes any mesh plus node data, another is the in_glyph port which takes as input an unstructured mesh that forms the geometric object used as a glyph. The parameters of the glyph module provide for size, color orientation of the glyph. AVS/Express is capable of representing data sets of multi-discrete data of up to 10 dimensions (i.e, position=3; color=3; shape/size=1, and orientation=3 — refer to [19]).

Using glyphs in AVS/Express may be accomplished as follows. The input field in_field takes as input one or many field type data structures that are output from other modules. Specific AVS/Express information include: a) Glyph Component: will determine the scale/size of the glyph; b) Map Component: selection of the data component from the input field(s) to color the glyph; and c) Mode: determination of how the module will portray the data values. By choosing a vector, a vector type object indicating the orientation of the glyph may be obtained.

6.6 Color Maps

Color sequencing in maps is important for proper understanding of the data attributes. Pseudosequencing is a technique of representing continuously varying map values using a sequence of colors. For example, geographers use a well-defined color sequence to display height above sea level: lowlands are green (which illustrates images of vegetation); browns represent mountains, etc. Such a sequence follows a (logical) perceptual sequence often made understandable with training.

In pseudosequencing a map, there are generally two important issues. The first is perceiving the shape of features. The second is the classification of values on the basis of color. If the color sequencing is poor, the understanding of the map tends to be more difficult. One important factor is that colors used in maps follow a perceptual (or human understanding) sequence. Several alternatives include the luminance sequence for representing large numbers of details; and a chromatic sequence. when representing little detail. In general, a perceptual ordered sequence will result from a series of colors that monotonically increases or decreases with respect to one or more of the color opponent channels. Also, in general, the larger the area that is color coded, the more easily colors can be distinguished. When large areas of color coding are used with map regions, the colors should be of low saturation and differ only slightly. (See [19])

6.7 Emotional Space Color Order

The feelings that humans experience such as happiness, sadness, loving, etc. are emotional behaviors. Emotional behaviors are powerful, in part, because of its impact. An interesting point to note is that color evokes emotion [20]. When humans talk about color for example, color becomes a descriptor of the physical world. Individuals have different interpretation of the different colors, which means that color is also a visual response to physical data; an emotional response to expectation and violation of appearance. It is of interest to consider that a color-based visualization may be used to provide emotional stimuli regarding user-feelings about the program (as for example, the common situation whereby a programmer firstly encounters a poorly written code fragment and has a negative emotional response).

The emotions that are experienced can be modeled as emotional space. One example of this is the relationship between color and sound: the voice of a child may be connected to the color yellow [21]. There are two important issues related to this idea of emotion in color communication. Firstly, it is difficult to reproduce a color message in different media, i.e., to select an alternative stimuli requires some 'emotion metric'. Secondly, composing a presentation of color is difficult. The message creators sometimes generate the unintended emotional message for lacking of guidance in selecting colors. To translate the relationship into the desired emotional response, a message creator needs to manipulate the stimuli properly.

According to Mendlers theory of emotion [20], the emotion system or the human feeling of emotion has an internal (human) sense-of-order or expectation. Emotion can be 'increased' or 'decreased' in the processing of the information conveyed by the colors. This information needs to have color relationships that fits some sense-of-order. To define an sense-of-order is insufficient to induce an emotional response. Additionally, resolution is required. Resolution is the effect of perceived color with respect to the sense-of-order.

For useful application to program visualization as presented in this paper, it is of interest to note if there are models that represent human sense-of-order interaction. Several tests have been done [20] regarding the questions like: do color interactions exist? What is the space definition of the color? What needs to be determined along a color dimension? How to measure a violation of the color order? Some reported results indicate that a color change in each dimension can be used as a frame work to set up the color expectation. When an expectation differs from the perceptions, an emotion may be generated.

7 Example Application

Figure 1 presents an example visualization of static, program-related data (this visualization was not done in AVS/Express). The horizontal axis refers to the time of the program's execution; the vertical axis, to the amount of resources consumed; and the depth axis, to the nesting of statements. Each statement is depicted by a rectangle placed on the horizontal-depth plane (the width of the rectangle is constant since no data has been mapped to this parameter) The colors are mapped as follows: gold – input statement, yellow – output statement, green – computational statement and light-blue – control statement (if gray-shaded, the classification is ordered by increasing grayness in the order of output, input, control and computation statements). The visualization shown indicates a typical program with the following characteristics: the program starts with an input statement followed by three computational statements at a nesting level of one. The next statement is a control statement at nesting level of two. Two subsequent computational statements execute in a nesting level of three, these statements consume the most resources out of any other statement. the program completes by an output statement, input statement, computational statement and lastly, an output statement.

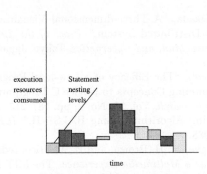

Fig. 1. A program visualization example.

This example illustrates several of the aspects described in this paper. First, program statements have been structured in a 1-D space (the horizontal axis). Second, two additional dimensions represent statement properties. Third, a colorization based on a statement category is used (however, it should be pointed out that the selected color map used in this example is *ad hoc* in nature).

8 Conclusion

This paper presents a unique point-of-view for program visualization, namely, the use of scientific visualization techniques for program visualization. This study confirms that the key aspect in integrating scientific visualization techniques in program visualization is the incorporation of structure, in particular, coordinate based data, into program-related data. This paper explores basic and fundamental issues and techniques from both scientific and program visualization in the pursuit of integrating the techniques. This paper reveals a number of interesting aspects that may be fruitful to consider in the future. Specific visualization techniques are presented and explored for utility in program visualization. This paper also describes certain features of AVS/Express that may be used for program visualization. Lastly, some additional comments regarding emotional color spaces are made. Future work to continue the investigation into scientific and program visualizations is indicated.

References

1. B. A. Price, R. M. Baecker, and I. S. Small, "A Principled Taxonomy of Software Visualization," *Journal of Visual Languages and Computing*, Vol. 4, pp. 211–266, 1993.

2. E. Kraemer and J. Stasko, "Issues in Visualization for the Comprehension of Parallel Programs," *Proc. of the 3rd Workshop on Program Comprehension*, Washington, D.C., Nov. 1994, pp. 116–125, Nov. 1994.

3. Advanced Visual Systems Inc., *Using AVS/Express*, r. 4.0 ed., 1998. Part No. 320-0321-05 Rev. A.

4. Advanced Visual Systems Inc., *Visualization Techniques*, r. 4.0 ed., 1998. Part No. 320-0324-05 Rev. A.

5. H. Amari and M. Okada, "A Three-dimensional Visualization Tool for Software Fault Analysis of a Distributed System," *Proc. of the 1999 IEEE International conference on Systems, Man, and Cybernetics*, Tokyo, Japan, Oct., 12-15, pp. 194–9, Oct. 1999. Vol. 4.

6. P. Smith and G. Webb, "The Efficacy of a Low-level Program Visualization Tool for Teaching Programming Concepts to Novice C Programmers," *Journal of Educational Computing Research*, Vol. 22, No. 2, pp. 187–215, 2000.

7. M. Brown, "Exploring Algorithms Using BALSA-II," *IEEE Computer*, Vol. 21, No. 5, pp. 14–36, 1988.

8. J. Stasko, J. Domingue, M. H. Brown, and B. A. Price, (eds.), *Software Visualization, Programming as a Multimedia Experience*. The MIT Press, 1998.

9. Advanced Visual Systems Inc., *AVS/Express Toolkits*, r. 4.0 ed., 1998. Part No. 320-0323-05 Rev. A.

10. N. Magnenat-Thalmann and D. Thalmann, *Computer Animation Theory and Practice*. Springer Verlag, 2 ed., 1991. ISBN: 038770051X.

11. S. Lahtinen, E. Sutinen, and J. Tarhio, "Automated Animation of Algorithms with Eliot," *Journal of Visual Languages & Computing*, Vol. 9, pp. 337–49, June 1998.

12. C. D. R. Baecker and A. Marcus, "Software Visualization for Debugging," *CACM*, Vol. 40, pp. 44–54, April 1997.

13. O. Astrachan and S. Rodger, "Animation, Visualization, and Interaction in CS1 assignments," *ACM. Sigcse Bulletin*, March 1998, Vol. 30, pp. 317–21, March 1998.

14. http://www.cs.hope.edu/~alganim/animator/Animator.html.

15. http://www.osc.edu/~kenf/Visualization/XPsciviz/sv-schedule.html.

16. http://www.ncsc.org/training/materials/express_class/XPsciviz/XPia/xpia11-frame.html.

17. D. W. Binkley and K. B. Gallagher, "Program Slicing," *Advances in Computers*, Vol. 43, 1996.

18. R. Wegenkittl, H. Loffelmann, and E. Groller, "Visualizing the Behaviour of Higher Dimensional Dynamical Systems," *Proc. of the 8th IEEE Visualization '97 Conference*, Oct, 19-24, 1997, pp. 119–125, Oct 1997.

19. C. Ware, *Information Visualization Perception for Design*. Morgan Kaufmann Publishers, 2000.

20. B. Burling and W. R. Bender, "Violating Expectations of Color Order," *Proc. SPIE Vol. 2657, Human Vision and Electronic Imaging,*, April 1996, B. E. Rogowitz and J. P. Allebach, (eds.), pp. 63–71, April 1996.

21. T. Miyasato, "Generation of Emotion Spaces Based on the Synesthesia phenomenon," *Proc. of the IEEE Third Workshop Multimedia Signal Processing*, 1999, pp. 463–7, 1999. Cat. No. 99TH8451.

Immersive Visualization Using AVS/Express

Ian Curington
Advanced Visual Systems Ltd.
Hanworth Lane, Chertsey, Surrey, KT16 9JX UK
ianc@avs.com

Abstract. To achieve benefit and value from high performance computing (HPC), visualization is employed for a direct understanding of computational results. The more closely coupled the visualization system can be to the HPC data source, the better the chance of fostering discovery and insight. A software architecture is described, with discussion of the dynamic object manager allowing scalable high-performance applications and computational steering. High performance parallel graphics methods are described, including interactive HPC data exploration in immersive environments.

1. Introduction

Visualization systems that aim to address complex data visualization problems, large data sources, while at the same time providing interactive immersive presentation, must be scalable, highly configurable and provide mechanisms for efficient data access and high performance display. The AVS/Express visualization software system addresses these requirements by using a runtime dynamic object manager. Methods of managing large data access, computational steering, parallel multi-pipe rendering and immersive menu and interactive device control systems are implemented using the object manager framework.

2. High Performance Framework

The AVS/Express visualization system is based on a framework for rapid prototyping and deployment of complex applications. Although visualization is the primary content of the software object library, it has been used in a wide variety of other areas.

2.1 Scalable Light-Weight Object Manager

At the heart of the framework is the object manager (OM). This performs the roles of scheduling, object state control, object notification and event management. An object is an abstract class, as small as a single byte or string and as large as a complete application container. As most applications have a large number of objects, the OM is highly tuned for efficient management. A current application for the European Space Agency (ESA) has as many as 750,000 visible objects for example.

Object relationships are hierarchical, both in class inheritance structure and in instance application design structure. Objects are moveable between processes and hosts. Applications often use dynamic object control so that only those objects used by the current application state are instanced.

The OM and visualization module implementations allow for full 64-bit address space support, including the layers that pass data between processes. This allows the system to scale to very large data problems.

2.2 Dynamic Data File Cache Management

A set of "file objects" are included that help to manage data file access during visualization. These dynamically map parts of files to memory arrays. This way references to the arrays by the rest of the system trigger dynamic file access, allowing files larger than virtual memory to be processed. This system is similar to an application specific cache management layer. Data sizes of 200 Gbytes have been processed using the file object system.

3. Computational Steering

As much of the data used in high-performance visualization systems originates from simulation, there is a natural interest in coupled simulation-visualization systems.

Using a modular visualization system, this can be achieved in a simple and easy way. By making minor changes to MPI parallel simulation programs, they can be steering enabled, whereby the visualization system can monitor simulation progress and provide interactive user control over simulation parameters [6]. New insights may quickly be gained by such continual monitoring and guiding the progress of computational simulations that were only previously analyzed in their final state.

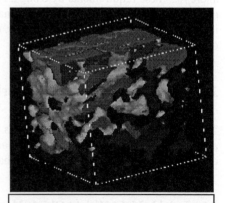

Figure 1 Result of a steered simulation to determine the spinodal point of an immiscible fluid.

Another system for computational steering, ViSIT [4] is a set of utility functions intended to support online-visualization and application steering. The current system has C and FORTRAN bindings for the application end (or 'simulation') and supports AVS on the visualization side. The idea is that the simulation is 'more important' than the visualization. Therefore the simulation triggers all actions. The simulation actively establishes a connection and sends or receives data.

Given this structure, a ViSIT-enabled simulation can establish a connection to a visualization/steering application, transfer 1,2,3 or 4D-data to and from that application, then shut the connection down again. On the AVS visualization side, the interactive user can receive and send data from and to the simulation while allowing data that arrives from the simulation to trigger visualization modules.

4. Parallel Rendering

In AVS/Express, the very flexible graphics architecture is exploited to minimize memory requirements and data structure conversions during rendering for large data problems. A thin layer OpenGL renderer provides a graphics display path, without dependence on scene tree, dynamic scene creation, or any requirement to hold the scene content in virtual memory. For small data sets and high speed interaction, stored scene tree mechanisms are used, but not required.

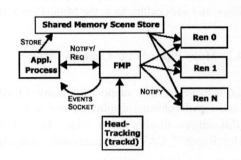

Figure 2 Frame Manager Process implements both pipeline and multi-pipe rendering parallelism.

Data source methods, and visualization methods are able to register user-defined draw methods providing a "chunking" mechanism to build up the scene using a multi-pass or data segmentation model. Visualization procedures are only executed as needed during the incremental updates, and do not need access to the entire model. In addition, this architecture allows procedural rather than stored data structure sources. By coupling this system to file access APIs, very large data sources may be managed in an efficient manner.

4.1 Multi-Pipe SDK

The SGI Multi-Pipe SDK is designed as a new problem-solving API [1]. The design is to allow OpenGL applications to easily migrate to multi-pipe computing and graphics environments. The system supports full portability, so the same application can run on a single display low-end desktop system, through to an Onyx2 with multiple Rendering Engines. Unlike Inventor or Performer, the Multi-Pipe SDK does not impose a scene-graph structure, so dynamic update of scene contents, as often happens in visualization applications, is handled in an efficient way. The system provides a transparent, call-back driven programming interface, and takes care of inter-process communication, parallel execution, and display configuration issues. The system is used to implement full parallel rendering within AVS/Express Multi-Pipe Edition [2].

The AVS/Express system is a full-featured visualization system and application development environment. In addition to over 850 visualization operators and interfaces, it has a generalized viewer and interaction system for the display of visualization data. The viewer is split into device (graphics API) independent and dependent layers. The mapping between these layers uses a "virtual renderer" dynamic binding structure. In this way, multiple renderers can be supported in the same runtime environment. The AVS/Express Multi-Pipe Edition extends the virtual renderer base class, and adds callbacks to the Multi-Pipe SDK subsystem.

4.2 Multi-Pipe / Multi-Channel Display

To achieve flexible support for many virtual display environments, full multi-pipe and/or multi-channel display control is mapped onto the parallel rendering process. This allows display on a wide class of VR systems, such as Immersadesk™, HoloBench™, CAVE™ or panoramic multi-channel displays.

4.3 Pipeline Parallelism

To implement full parallel rendering, an intermediate Frame Management Process (FMP) was introduced to separate the handling of the rendering from the application process [5]. When a frame is ready (fully tokenized) the application notifies the FMP which in turn notifies the MPU processes and is, itself, blocked. The application continues with its tasks and produces the next frame. Note that the application process is eventually blocked, so that there is no build-up of frames should the production of frames be faster than the consumption rate of the renderers. Meanwhile all rendering processes run in parallel for simultaneous update of all display channels.

4.4 VR Menu System

A menu system for virtual reality and immersive stereo applications is included that allows for rapid prototyping, configuration changes, and 3D-layout behavior management. The VRMenu user-interface toolkit exploits 3D geometric graphics systems and allows interactive immersive data exploration without leaving the environment to modify control parameters.

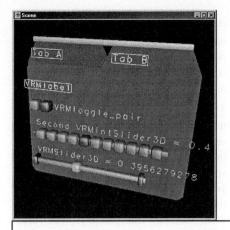

Figure 3 3D Menu manipulation tools – tab panels, buttons, choice lists, sliders and region crop controls.

A primary goal of interactive data visualization in an immersive environment is to get inside the data space for exploration and discovery. Ideally, no other distractions should be introduced, allowing full attention to be directed to the detail in the visualization. The VRMenu system is used to replace the 2D GUI with a 3D geometric equivalent, so that key parameters in the visualization can be controlled within the immersive environment without interruption.

4.5 Trackers / Controllers

An important part of the immersive experience is tracking the 3D viewer position, correcting the perspective stereo views for optimal effect. The head-tracking process (using *Trackd* software from VRCO Inc.) passes viewer location information to the FMP [7]. For head-tracking support use of the FMP means that the camera can now be decoupled from the application, the FMP piloting the center of the view with camera offsets without interruption of the application. If a new frame is ready then the FMP will immediately use it, otherwise it will re-render the previous frame according to the new camera position. Figure 2 shows that the head-tracking data feeds directly into the FMP, which will trigger the renderers to redraw with the updated camera, without interrupting the main AVS/Express Multi-Pipe Edition application process.

5. Visualization Methods

A recently developed method of using a moving texture alpha-mask to represent scientific data is used for the purpose of visualizing continuous fluid dynamics fields [3]. The method combines stream tubes and particle animation into one hybrid technique, and employs texture surface display to represent time, flow velocity and 3D

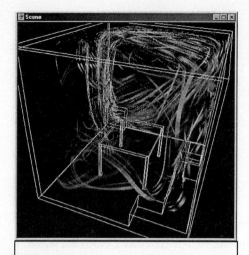

Figure 4 Convection Heat Flow in Room using Texture Wave Ribbons

flow structure in one view. The technique exploits texture support in high-performance graphics systems and achieves high graphics efficiency during animation.

This approach combines the particle tracing technique with the geometric stream ribbon and tube technique including the advantages of both. The spatial information is shown using geometry, while velocity variations over time are shown using an animated texture with variable transparency. The color used in the texture map is a direct replacement to scalar color assignment, yielding high quality color contours and avoids artifacts introduced by RGB interpolation during graphics display

One approach is to apply an offset to the *u-v* texture coordinates, creating an animation to make the texture crawl along the stream path. Here however, the u-v texture coordinates remain static, so the color and flow variables remain referenced to each other for analytical flow visualization, while the content of the texture alpha-mask is dynamically changed to produce an animation. In this way a continuous periodic function can be applied, rather than a direct positional shift. The center of highest opacity in the alpha-mask is used to create a time-wave pulse function. As phase is adjusted, the positions of the pulse function peaks move along the time path.

6. Conclusion

The AVS/Express visualization system has been described, highlighting the characteristics that address large data visualization. Through efficient data access and high performance parallel display, the underlying framework scales to handle large problems. Application architectures have been discussed for directly coupled high-performance computing applications with visualization for monitoring and steering of those applications through visual interfaces. Finally, the use of immersive visualization techniques has been shown to add value in the exploration of large data problems.

7. References

1. Bouchaud, P. "Writing Multipipe Applications with the MPU", SGI EMEA Developer Program document, 1998.
   http://www.sgi.com/software/multipipe/sdk/

2. Curington, I. "Multi-Pipe Rendering Framework for Visualization Applications" Proceedings of SIGRAD '99, Stockholm December 1999

3. Curington, I. "Animated Texture Alpha-Masks for Flow Visualization" Proceedings of IEEE IV2000, Information Visualization conference, London July 2000

4. W. Frings, T. Eickermann, "visit - VISualization Interface Toolkit", Zentralinstitut fuer Angewandte Mathematik, Forschungszentrum Juelich GmbH http://www.kfa-juelich.de

5. Lever, Leaver, Curington, Perrin, Dodd, John, Hewitt, "Design Issues in the AVS/Express Multi-Pipe Edition. IEEE Computer Society Technical Committee on Computer Graphics, Salt Lake City, October 2000.

6. Martin, Love "A Simple Steering Interface for MPI Codes" *(to appear)*

7. VRCO Inc., Trackd tracker/controller device drivers http://www.vrco.com

VisBench: A Framework for Remote Data Visualization and Analysis

Randy W. Heiland, M. Pauline Baker, and Danesh K. Tafti

NCSA, University of Illinois at Urbana-Champaign,
Urbana, Illinois
{heiland, baker, dtafti}@ncsa.uiuc.edu

Abstract. Computational researchers typically work by accessing a compute resource miles from their desk. Similarly, their simulation output or other data is stored in remote terabyte data servers. VisBench is a component-based system for visualizing and analyzing this remote data. A time-varying CFD simulation of heat exchange over a louvered fin provides sample data to demonstrate a workbench-oriented version of VisBench. An analysis technique (POD) for spatiotemporal data is described and applied to the CFD data.

1 Introduction

Computational and experimental scientists routinely access remote resources necessary for their research. While it was once customary for a researcher to download remote data to a local workstation in order to visualize and analyze it, this scenario is usually impractical today. Data files have become too large and too numerous from a typical simulation run at a high-performance computing center. This has led to data being stored in terabyte servers located at the HPC sites.

Experimental scientists face the same dilemma of burgeoning remote data. For example, the Sloan Digital Sky Survey has already captured nearly two years worth of data and the Large Hadron Collider at CERN and National Ignition Facility at LLNL will each produce vast amounts of data when they become fully operational.

The goal of the NSF PACI program is to build the Grid[1] – a distributed, metacomputing environment connected via high-speed networks, along with the necessary software to make it usable by researchers. As part of this effort, we present a software framework that is being used to remotely visualize and analyze data stored at NCSA.

Some goals of our project, called *VisBench*, include:

- minimize data movement,
- take advantage of (remote) HPC resources for visualization and analysis,
- provide application-specific *workbench* interfaces and offer a choice of clients, ranging from lightweight (run anywhere) to more specialized,
- be prudent of the cost of necessary software.

The notion of an application-specific workbench means offering high-level functionality that is pertinent to an application domain. We will demonstrate this with an analysis technique that VisBench offers through its CFD interface. The last goal stems from the fact that, as leading-edge site of the Alliance[1], NCSA will make software recommendations to our partners.

We now present an overview of VisBench, followed by an example of it being used to visualize and analyze remote CFD data.

2 VisBench: A Component-Based Workbench

VisBench adopts the software component model whereby application components are connected in a distributed object fashion. This model allows us to "plug-in" various software components – taking advantage of their individual strengths. A component may be anything from a small piece of code that performs one specific function to an entire software package that provides extensive functionality.

We have initially taken a coarse-grained approach; our framework consists of just a few, large components. In this paper we will present just three: a general-purpose visualization component, a general-purpose data analysis component, and a graphical user interface component. For visualization, we use an open source software package called the Visualization Toolkit (VTK)[2]. VTK is a relative newcomer in the visualization community and is being rapidly adopted by many groups. For data analysis, we use MATLAB, a commercial software package that has been in existence for over twenty years. For the user interface, we have written a graphical user interface (GUI) client using Java Swing.

In order to connect these different components together and have them easily interoperate, one needs *middleware*. We have initially chosen CORBA as the middleware for VisBench. Figure 1 shows a schematic of the VisBench framework discussed in this paper.

2.1 Visualization

We selected VTK as our core visualization component for a number of reasons. As a modern object-oriented software package (written in C++), it offers over 600 classes that perform numerous 2-D and 3-D visualization algorithms, rendering, and interaction techniques. The quality of example renderings, especially for scientific datasets, and the ability to interact in 3-D were important factors. So too was its ability to stream very large datasets through various filters within a visualization pipeline. Its open source model allows for valuable contributions from users. For example, a group at LANL has provided extensions for parallel filters using MPI. The Stanford Graphics group will be offering a parallel rendering version of VTK, and the Vis&VE group at NCSA has provided functionality for easily using VTK in the CAVE – a future VisBench client. The level of activity and excitement within the VTK community is very encouraging.

---

[1] National Computational Science Alliance, one of two partnerships within PACI

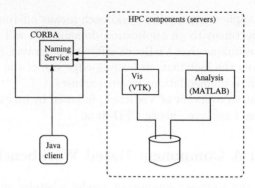

Fig. 1. VisBench design and some existing components

One feature lacking in the VTK distribution is a graphical user interface. There is, however, an option to automatically wrap each class in the Python or Tcl scripting language. Our visualization component (server) consists of an embedded Python interpreter that has access to all the VTK classes, providing a mechanism for dynamically constructing a visualization pipeline. A VisBench user does not need to know the Python language though. The Java GUI (client) transparently converts all actions into appropriate Python-VTK commands. This has the added benefit of generating a script, associated with an interactive session, which can then be run in batch mode if necessary.

2.2 Data Analysis

The MATLAB software package is familiar to nearly every scientist and engineer who performs numerical calculations. It has earned wide respect and is trusted for its numerical accuracy. Using its Application Program Interface, we have written a core analysis component for VisBench. As with the visualization component, the intent is to provide high level point-and-click workbench-related functionality. Of course, if users are capable of writing their own MATLAB scripts, these too can be entered via VisBench and executed on the remote analysis server.

Besides providing extensive core mathematical functionality, MATLAB has another distinct advantage over similar systems – there are numerous, freely available scripts that cover nearly every application domain. It is straightforward to incorporate such scripts into VisBench.

2.3 User Interface

In order to make the remote VisBench engines accessible to users working on a wide variety of platforms, our primary VisBench client is written in Java. The client makes extensive use of the Java Swing package for the graphical user interface. A sample VisBench session would proceed as follows:

- from the client, connect to a running server (a factory server that forks off a separate server for each user),
- using a Reader widget, read a (remote) data file; associated metadata will be displayed in the widget,
- create filters for the data – for example, slicing planes, isosurfaces, streamlines, vector glyphs, etc.,
- geometry associated with the filters is rendered on the back-end VTK server and the resulting image compressed and sent back to the client where it gets displayed,
- interactively view the results and make necessary changes.

An example of the CFD workbench interface is shown in Figure 2. This Java client is a very lightweight client. One needs only a Java Virtual Machine in order to run it. (The Java 2 platform is required since we use the Swing package and the CORBA bindings).

Workbench interfaces are being designed as a collaborative effort between application scientists and computer scientists. We envision these workbenches evolving into problem-solving environments.

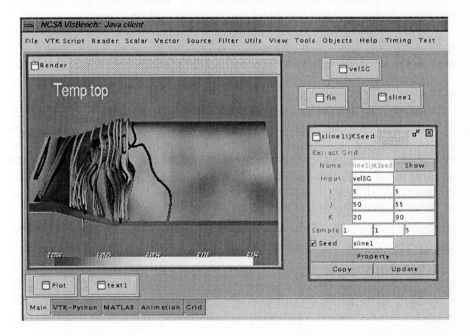

Fig. 2. An example of the CFD Java client

2.4 Middleware

Middleware is the software that allows distributed objects to communicate and exchange data with each other. There are three mainstream middleware solutions: COM, CORBA, and JavaBeans. COM (Component Object Model) is from Microsoft and is intended for Windows platforms and applications. CORBA (Common Object Request Broker Architecture) is a vendor-independent specification defined by a not-for-profit consortium, the Object Management Group. JavaBeans, from Sun, is for Java platforms and applications.

We chose CORBA as the middleware for VisBench for two primary reasons. It provides language interoperability – something quite valuable in scientific computing where there is a mix of Fortran, C/C++, and, with growing frequency, Java. It makes the most sense based on our current high-end hardware architectures and operating systems – primarily IRIX and Linux. We selected the ACE ORB (TAO), an open source university research project, as our CORBA implementation. An Object Request Broker (ORB) provides a mechanism for transparently communicating requests from a client to a *servant object*. Within the scope of VisBench that is being presented here, there are only two servant objects – the VTK and MATLAB components.

The interfaces to CORBA objects are specified using the Interface Definition Language (IDL). The IDL is independent of any programming language and allows for applications in different languages to interoperate. The IDL for a particular object contains the interfaces (methods) for its operations.

CORBA provides the specifications for a number of different *services*. The most commonly used is the Naming Service. The Naming Service provides a mechanism for mapping object names to object references. A servant object "advertises" itself via the Naming Service. A client then connects to the Naming Service, obtains object references, and invokes operations on those objects.

VisBench has less communication overhead than other HPC distributed component models. In the Common Component Architecture[3], for example, the goal is to construct a distributed computational pipeline requiring the transmission of large amounts of data. The current VisBench model assumes that the simulation (or experimental) data will be directly accessible from the machine hosting the VisBench server and that a relatively small amount of visualization, analysis, or user data will be exchanged between client and server.

3   POD Analysis

When trying to understand spatiotemporal data, visualization – especially an animation, is an extremely useful tool. However, sometimes one needs more quantitative information – for example, when trying to compare results from a parameter study.

We present a fairly complex analysis technique that is sometimes used in CFD research – particularly in the study of turbulence. The technique has its roots in statistics and is based on second-order statistical properties that result in a set of optimal eigenfunctions. In certain application domains, the algorithm is known as the Karhunen-Loève decomposition. Lumley[4] pioneered the idea that it be used to analyze data from turbulent flow simulations and suggested that it could provide an unbiased identification of coherent spatial structures. Within this context, the algorithm is known as the *proper orthogonal decomposition* (POD).

We use the POD as an example of providing high-level analysis functionality in the VisBench environment. Because of its computational demands, the POD is most often used for analyzing only 1-D or 2-D time-varying data. Since VisBench provides an analysis component running on a remote HPC resource, we will demonstrate the POD on time-varying 3-D data. An outline of the algorithm is presented.

Assume we are given a set of numeric vectors (real or complex):

$$\{\mathbf{X}_i\}_{i=1}^M$$

where $\mathbf{X} = [x_1, x_2, \ldots, x_N]$.

The mean is computed as:

$$\overline{\mathbf{X}} = \frac{1}{M} \sum_{i=1}^M \mathbf{X}_i$$

Hence, if we have a time series of spatial vectors, the mean will be the time average.

We shall operate on a set of caricature vectors with zero mean:

$$\hat{\mathbf{X}}_i = \mathbf{X}_i - \overline{\mathbf{X}}, \qquad i = 1, \ldots, M$$

Using the method of snapshots[5], we construct an approximation to a statistical covariance matrix:

$$\mathbf{C}_{ij} = \langle \hat{\mathbf{X}}_i, \hat{\mathbf{X}}_j \rangle, \qquad i, j = 1, \ldots, M$$

where $\langle \cdot, \cdot \rangle$ denotes the usual Euclidean inner product.

We then decompose this $M \times M$ symmetric matrix, computing its (non-negative) eigenvalues, λ_i, and its eigenvectors, ϕ_i, $i = 1, \ldots, M$, which form a complete orthogonal set.

The orthogonal eigenfunctions of the data are defined as:

$$\Psi^{[k]} = \sum_{i=1}^M \phi_i^{[k]} \hat{\mathbf{X}}_i, \qquad k = 1, \ldots, M$$

where $\phi_i^{[k]}$ is the i-th component of the k-th eigenvector. It is these eigenfunctions which Lumley refers to as *coherent structures* within turbulent flow data.

The *energy* of the data is defined as the sum of the eigenvalues of the covariance matrix:

$$E = \sum_{i=1}^M \lambda_i$$

Taking the ratio of an eigenvalue (associated with an eigenfunction) to the total energy, we calculate an energy percentage for each eigenfunction.

$$E_k = \frac{\lambda_k}{E}$$

Sorting the eigenvalues (and associated eigenvectors) from largest to smallest, we can order the eigenfunctions from most to least energetic. In a data mining context, we refer to the plot of descending eigenfunction energy percentages as a *quality of discovery* plot. Ideally, one wants just a few eigenfunctions that cumulatively contain most of the energy of the dataset. This would be an indication that these eigenfunctions, or coherent structures, are a good characterization of the overall data. (In a dynamical systems context, this would constitute a relatively low-dimensional phase space of the given system).

Furthermore, since the eigenfunctions span the space of the given data, it is possible to reconstruct an approximation to any given vector as:

$$\mathbf{X} \approx \overline{\mathbf{X}} + \sum_{i=1}^{K} a_i \Psi^{[i]}$$

taking the first $K(K < M)$ most energetic eigenfunctions, where the coefficients are computed by projecting the caricature vectors onto the eigenfunctions:

$$a_i = \left(\frac{\hat{\mathbf{X}} \cdot \Psi^{[i]}}{\Psi^{[i]} \cdot \Psi^{[i]}} \right)$$

The POD has been used to analyze near wall turbulence and other PDE simulations[6][7], as well as experimental data.

4   Example: Turbulent flow over fins

A HPC simulation of turbulent flow and heat transfer over a multilouvered fin[8] provides a source of example data for demonstrating VisBench. Figure 3a shows the multilouvered fin model. The flow is from left-to-right, i.e., hitting the leading (upper) edge of a fin and traveling down to the trailing edge. For the simulation, the computational domain is comprised of one-half of one full fin (due to symmetry) and the region around it, as shown in Figure 3b. It is periodic back-to-front (leading-edge boundary to trailing-edge boundary) and bottom-to-top. There is a wall at the flat base where much of the heat is retained.

The computational domain consists of a topologically regular grid which follows the geometry of the fin – having higher resolution in the area of interest (at the twisted junction). The grid size is $98 \times 98 \times 96$ for this particular model.

The simulation writes files of computed values at regular time intervals. From the VisBench client, a user selects some subset (M) of these remote files, representing a particular regime of the simulation. The user then selects a particular scalar or vector field within each file and, after connecting to an analysis server,

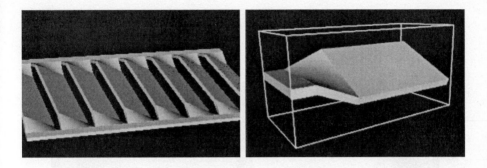

Fig. 3. a) The multilouvered fin and b) the computational domain

initiates the POD analysis. We have currently analyzed the temperature, pressure, streamwise vorticity, and velocity fields.

Figure 4 shows the results of applying the POD to the temperature field (for $M \approx 200$). The mean is shown at the top followed by the first three most energetic eigenfunctions. The first eigenfunction contained 32% of the energy, the second, 12%, and the third, 8%.

The first eigenfunction reveals "hot-spots" along the flat base (top and bottom), a circular region near the upper part of the top, and two streaks along the length of the bottom of the fin. The second eigenfunction visually appears to be orthogonal to the first (which it is, mathematically) – the base of the fin is now "cool" and there is a hot-spot near the upper region of the twisted junction (both top and bottom). Although we show only the surface of the fin (color-mapped with temperature), it should be noted that the analysis was indeed performed over the entire 3-D domain and has been visualized using various filters.

By encapsulating the POD algorithm in the analysis component, interactive viewing from the visualization component, and relevant graphical controls from a client, we have illustrated how VisBench can serve as an application workbench.

5 Summary

We have presented a framework for performing visualization and analysis of remote data. The VisBench model consists of a variety of servers running on remote HPC machines with access to data archived on those machines. An application-dependent (workbench) client with a graphical user interface can simultaneously communicate with different servers via CORBA. We have shown a Java client used for visualizing and analyzing CFD data. The client is lightweight in that it requires only the Java 2 runtime environment. The amount of data exchanged between client and visualization server is minimal – GUI-generated command strings to the server and compressed (JPEG) images to the client.

An earlier version of VisBench was demonstrated at Supercomputing '99 – a Java client on the show floor in Portland visualizing CFD data at NCSA. In addition, VisBench and Condor (www.cs.wisc.edu/condor) teamed together to

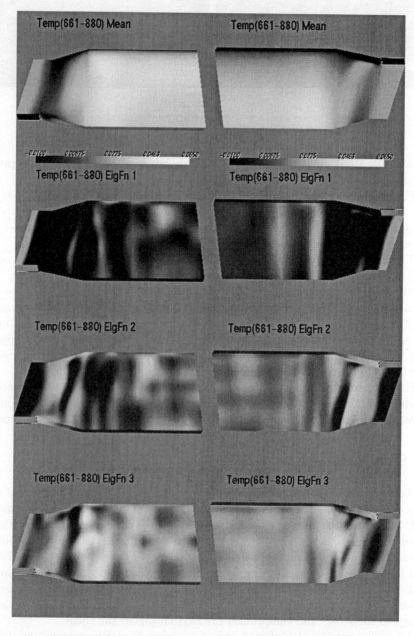

Fig. 4. POD results of temperature on fin (top and bottom)

demonstrate frame-based parallel rendering using the Grid. At Supercomputing '00 in Dallas, we demonstrated both the Java client (for a Chemical Engineering workbench) and a geometry client that received geometry from a VisBench server (at NCSA) and displayed it locally on a tiled wall.

We have not yet collected performance data for the different scenarios that VisBench offers. The visualization server that typically runs at NCSA (a multiprocessor SGI Onyx) takes advantage of hardware rendering and has shown a 50x speedup over software rendering. This is highly desirable for an image-based client, such as the Java client presented here or a client in a web browser. To offer an idea of performance for the Java client, the rendering in Figure 2 would update at about 2 frames/second at the Supercomputing conferences. For a geometry client, the rendering speed will be determined by the client machine's hardware. Obviously, VisBench will always be network limited. We have not yet incorporated parallel VTK filters into the visualization server, nor are we using parallel rendering. These two areas will receive more attention as we transition to Linux clusters. The analysis component is still considered proof of concept and the POD implementation, in particular, needs to be optimized.

Much of our effort has gone toward providing utilities that the application scientists have requested – remote file selection, animation, camera path editing, and providing VRML files on the server machine (which doubles as a web server).

We acknowledge the help of many members, past and present, of the NCSA Vis&VE group, as well as members of the Condor team, the Alliance Chemical Engineering team, and the Extreme! Computing team at Indiana University. For more information on VisBench, we refer readers to the web page at visbench.ncsa.uiuc.edu.

References

1. Foster, I., Kesselman, C. (eds.): The Grid: Blueprint for a New Computing Infrastructure. Morgan-Kaufmann, San Francisco (1998)
2. Schroeder, W., Martin, K., Lorensen, W.: The Visualization Toolkit: An Object-Oriented Approach to 3D Graphics, 2nd ed.. Prentice-Hall, Old Tappan, N.J. (1998)
3. Armstrong, R., Gannon, D., Geist, A., Keahey, K., Kohn, S., McInnes, L., Parker, S., Smolinski, B.: Toward a Common Component Architecture for High-Performance Scientific Computing. Proceedings of the Eighth IEEE International Symposium on High Performance Distributed Computing (1999) 115-124
4. Lumley, J.L.: The structure of inhomogeneous turbulent flow. In: Yaglom, A.M., Tatarski, V.I. (eds.): Atmospheric Turbulence and Radio Wave Propagation 25 (1993) 539-575
5. Sirovich, L.: Turbulence and the dynamics of coherent structures: Part I-III. Quarterly of Applied Mathematics, XLV(3) (1987) 561-590
6. Aubry, N., Holmes, P., Lumley, J.L., Stone, E.: The dynamics of coherent structures in the wall region of a turbulent boundary layer. J. Fluid Mech. 192 (1988) 115-173
7. Berkooz, G., Holmes, P., Lumley, J.L.: The proper orthogonal decomposition in the analysis of turbulent flows. Annual Review of Fluid Mechanics 25 (1993) 539-575
8. Tafti, D.K., Zhang, X., Huang, W., Wang, G.: Large-Eddy Simulations of Flow and Heat Transfer in Complex Three-Dimensional Multilouvered Fins. Proceedings of FEDSM2000: 2000 ASME Fluids Engineering Division Summer Meeting (2000) 1-18

The Problem of Time Scales in Computer Visualization

Mark Burgin[1], Damon Liu[2], and Walter Karplus[2]

[1] Department of Mathematics, University of California, Los Angeles
Los Angeles, CA 90095
mburgin@math.ucla.edu

[2] Computer Science Department, University of California, Los Angeles
Los Angeles, CA 90095
{damon, karplus}@cs.ucla.edu

Abstract. A new approach to temporal aspects of visualization is proposed. Empirical and theoretical studies of visualization processes are presented in a more general context of human-computer interaction (HCI). It makes possible to develop a new model of HCI as a base for visualization. This model implements a methodology of autonomous intelligent agents. Temporal aspects of the model are rooted in the system theory of time. This theoretical background provides for obtaining properties of time coordination. In particular, it is proved that some theoretically advanced models for computation and visualization cannot be realized by physical devices. Applications of the model, its implications for visualization and problems of realization are considered. The new approach does not refute or eliminate previous research in this area, but puts it in a more general and adequate context, systematizing it and providing efficient facilities for the development of computer visualization.

Keywords. Visualization, interaction time, system time, intelligent agent, virtual reality.

1   Introduction

Utilization of computers is based on three main components: computers as the technical component, users as human component, and human-computer interaction (HCI). Visualization is an important part of this interaction. At the very beginning, the main emphasis was on computers. Engineers and computer scientists tried to make computers more powerful, more efficient, and more flexible. At the same time, teaching computer skills and later computer science were organized.

The problem of visualization in HCI was understood much later [18]. In spite of this, research in visualization as a tool for HCI has been spectacularly successful, and has fundamentally changed computing. The main emphasis has been traditionally made on the means and forms of representation. Virtually all software written today employs visualization interface toolkits and interface builders.

An aspect of visualization that has usually been downplayed until now involves *time*. However, time dependencies and temporal constraints are an important aspect of action, and failure to meet them leads to an important class of human errors; many of the errors associated with safety critical systems have a significant temporal component. For example, as is stressed in [9], the goal of physics-based graphics modeling tools is a compromise between getting visual realism and reasonable computational times. Another example of a situation where time is a critical parameter is given in [8]. The scenario concerns an air traffic controller whose task is to schedule the arrival of aircraft at an airport, while maintaining an adequate separation distance between them. Thus, in many cases time plays a crucial role, and cannot be relegated to a secondary consideration or to be treated implicitly. For example, in most problem-solving systems, time is hidden in a search process, while in formal grammars time is overshadowed by the sequences of transformations. Besides, time is important for business systems. People expect their computers to interact with them quickly; delays chase user away.

Various authors considered the problem of time in HCI from different points of view. However, in each case only a partial perspective has been considered: functioning as interface software, behavior of a user, or perception of a user. Thus, in the method of the state transition diagrams [11], interaction is represented as a net of tasks where time is not given explicitly. Grammatical approach [12] also gives some temporal relations without including time as a separate parameter. In [12], the main concern is user behavior during interactions with computer systems.

Our aim is to elaborate a complete model for computer visualization with an explicit structure of time dependencies and relations. Here we consider a single level interaction model. For example, we take into account temporal aspects of computational processes on the level of data but disregard operations with bits of information as well as time in physical processes in computer chips. Multilevel models are more complicated and will be considered in the future.

The conventional approach assumes that time is a one-dimensional quantity, made up of points, where each point is associated with a value [12]. The points are ordered by their values. The common concepts of later and earlier correspond to larger and smaller values of time, respectively. This view is compatible with the traditional psychological, thermodynamic, and cosmic views of time [13].

Another approach is based on temporal intervals rather than the standard view of mapping time to points on the real number line [1]. Time intervals are primary objects of observation and objects of learning by children.

Our approach to time is essentially different being based on the system theory of time [6]. The main distinctions are: multiplicity of time, non-linearity in a general case and multiplicity of formal representations.

Multiplicity of time means that each system has its own time scale. Only interaction between systems makes explicit various relations between times in each of the systems in question. In addition to this, processes going on in the same system may also develop in different time scales.

Non-linearity of time means that the order between time elements (points or intervals) is not necessarily linear. For example, let us consider a cartoon which relates a story that goes on in 1900 and covers a one year period of time. Then time in this cartoon will be cyclic because each display of this cartoon repeats all events in the same sequence. Another example is given by a complex computer game, like Civilization, which is played on two computers that have no connections. In this case, we cannot say that one event in the first game happened earlier than another event in the second game. Thus, there is no linear time scale for both games.

Multiplicity of formal representations means that different formal structures may be corresponded to time. For example, time is usually represented by a real line (as in physics or biology), system of intervals on the real line [1], system relations (as in temporal and dynamic logics), and natural numbers (as in theory of algorithms and theoretical models of computers: Turing machines, Petri nets, RAM, inductive Turing machines [7], etc.). To explicate time in visualization to a full extent, we need a model for human-computer interaction which represents the time issues in visualization.

2 The M-Model of Human-Computer Interaction

HCI is a process of information transmission and reception, which are called *communication.* Consequently, HCI is a kind of communication. Usually two types of communication are considered: *technical communication,* which is the base of information technology, and *social communication* [2]. HCI, as the foundation on which society is structured, is both technical and social, including both human beings and technical systems. Consequently, it demands a specific approach.

The communication process as a total event has been a subject of many studies. Many models or structural descriptions of a communication event have been suggested to aid the understanding of the general organization of the process. Models provide clues that permit predictions of behavior and thus stimulate further research.

There are *static* and *dynamic* models. Static models represent the system in which the communication goes on. Dynamic models feature its functioning.

The simplest static model consists of three elements: *sender, receiver,* and *channel* (Figure 1a). The simplest dynamic model also consists of three elements: *sender, receiver,* and *message* (Figure 1b).

We use the more detailed model from [5], which incorporates both static and dynamic features of the process. It is presented in Figure 2. Connection here consists of three components: the *communication media (channel), message,* and *communication means.* Communication means include language for coding and decoding the message, transmission and reception systems, etc.

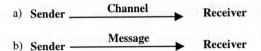

Figure 1: The simplest models of a communication event.
a) Static model; b) Dynamic model.

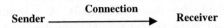

Figure 2: Synthetic model of communication.

All structures (1a, 1b, 2) are special cases of a more general structure that is called a *named set* or a *fundamental triad* [5]. Besides, these structures are models of a single communication act. They are "elementary particles" of a communication process when the sender and the receiver exchange their roles. A complete model of communication as a whole process is complex and is built from such elementary models. Thus, the cyclic representation of communication is given in Figure 3.

Figure 3: The cyclic representation of communication process.

At the beginning, HCI was correlated with a general schema of communication. However, the spectacular development of computer interface changed this situation, and HCI became a mediated communication. Actually, in many cases communication in society is mediated by technical devices such as telephones, facsimile machines, satellites, computers, and so on. But these technical devices are only examples of advanced communication channels, while the human-computer communication possesses a mediator of a different type. We call this mediator the *communication* or *interaction* space. Thus, we come up with the following schema (cf. Figure 4) which represents the *mediator model*, or simply, *M-model* of HCI.

The communication space is not simply a channel because both participants of the communication, the user and the computer, are represented in the corresponding communication space by special objects, and these objects interact with each other. We call these objects *communication representatives* of the user and the computer, respectively. One real object or person (a computer, program or user) may have several communication representatives (e.g., avatars of the user). Communication functions are delegated to these representatives. Real parties of the interaction process (users and computer programs) control these representatives, but their behaviors have to be described separately. An efficient tool for such description is given by multi-party grammars which portray, in a formal way, actions of several parties [12].

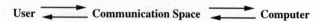

User ⟷ Communication Space ⟷ Computer

Figure 4: The relationship among user, computer, and communication space.

For the majority of software, communication space is the screen of the monitor of a computer. On the screen, as a communication space, a user is represented by a cursor and/or special signs (such as >) denoting commands together with the fields for these commands. A computer (a program) is represented by different icons and/or special signs denoting computer messages to the user together with the fields for these messages. However, current software does not capture the needs and preferences of the end users.

Only in some cases, the communication space is a virtual reality. In the future, VR will be the main type of communication space. The way to achieve high quality HCI is based upon personalization of the user representatives [16] and intellectualization of the computer representatives [19] in the communication space.

3 Multiagents to Enhance Computer Visualization

Visualization, according to [10], can go on different levels. The lowest one is *geometric modeling*, when only geometric forms are reproduced. Then goes the *physics-based modeling*, when the physical rules governing interaction of the visualized systems are taken into account and incorporated in visualization programs. The highest level is the *behavioral modeling.*

Agents may be used on each level but they become essentially efficient and even necessary only on the third level: behavioral modeling. This modeling always includes animation and also has three levels. On the first, *predetermined level* rules visualization and animation of a system **R** but ignores the changes in the environment of this system, in particular, the interaction of **R** with other systems. The next, *stimulus-reaction level* is based on a direct reaction of the visualized system to the current state of its environment. The highest, *cognitive level* presupposes learning of the visualization program. Means for such learning have been developed in artificial intelligence. The interaction of computer graphics and artificial intelligence is now on the merge of explosive growth as a new breed of highly autonomous, quasi-intelligent graphic characters begins to populate the domains of animation production, game development, and multimedia content creation, as well as distributed multi-user virtual worlds, e-commerce, and other Web-enabled activities [10].

In stratified visualization, image is stratified, i.e., separated into several strata, allowing to compute data for visualization of these strata by different devices (processors). The two natural ways of image stratification are localization and multilayer imaging. *Localization* means that the whole picture is divided into several parts and visualization information for each part is processed on a separate device. Another kind of localization is the systemic localization when a separate system, for example, a character in animation, is visualized autonomously. *Multilayer approach* presupposes that we have a dynamic picture that is changing in a different mode for some of its components. For example, in animation the whole picture is treated as composed of some active moving beings (people, animals, etc.) and things (cars, planes, etc.), and the background that changes much slowly. This makes it possible to animate moving things and beings separately from the background. Thus, two strata are utilized. Often, disjoint moving things and different beings are also animated separately, giving birth to many other strata.

Logical means for interpretation and investigation of stratified visualization are provided by logical theories of possible worlds [14], which were formalized by logical varieties [4]. Under certain conditions, this can considerably enhance the efficiency of computer visualization.

Structural means for realization of stratified visualization are provided by autonomous intelligent agents and multiagents. Here multiagent means a multiagent system [15] that acts as a single agent. In nature, a hive of bees or a colony of ants are examples of multiagents. Usually agents are considered as semi-autonomous computer programs that intelligently assist a user with computer utilization. They act on behalf of a user to carry out tasks.

Levels of behavioral modeling correspond to the types of software agents that are used for such modeling. According to [20], we have the following types of autonomous agents: *reflexive, model based, goal based,* and *utility based* agents. This linear typology is developed into a three dimensional classification with the structural, functional, and operational dimensions.

Along the structural dimension, we have three grades: 1) agents only with a stimulus-reaction system (SRS); 2) agents with SRS and a world model (WM); 3) agents with SRS, WM, and self-model (SM).

Along the functional dimension, we have the following grades: 1) action; 2) perception and reaction; 3) perception, reaction, and reasoning; 4) perception, reaction, reasoning, and mission orientation; 5) perception, reaction, reasoning, mission orientation, and evaluation; 6) perception, reasoning, mission orientation, evaluation, and self-transformation.

Operational dimension is determined by algorithmic classes that realize agents. For example, finite automata agents (level 1) are at the beginning of this dimension while Turing machine or RAM agents (level 2) are much higher. Super-recursive agents (level 3) are even higher [7].

In this three dimensional space, reflexive agents have the coordinates (1, 2, 1), model based agents have the coordinates (2, 2, 1) or (2, 3, 2), goal based agents have the coordinates (3, 4, 2) or (3, 4, 3), and utility based agents have the coordinates (3, 5, 2) or (3, 4, 3).

In the context of the M-model of HCI, we have agents of three types: 1) agents for a user (users) or U-agents; 2) agents for the software system or S-agents; 3) agents for the interacting media or M-agents.

As an example, let us consider visualization in a computer game, in which a player competes with different characters (dragons, tigers, bandits, etc.) who try to prevent the player achieving the goal of the game. As a rule, the user is represented on the screen by some character. According to the multiagent technique, this character is visualized by a U-agent. Characters who are competing with the user are visualized by a S-agent or S-multiagent. The background of this competition is realized by a M-agent.

Cognitive technology for agents design [10] makes it possible to create an intelligent virtual reality in the game. The same is true and even more urgent for training programs based on simulation of different characters. As examples, we can take military or police training programs which, to be realistic and thus efficient, have to employ technique of intelligent agents and cognitive visualization that can reflect adequately temporal aspects.

4 Dynamics and Time in Computer Visualization

According to the system theory of time [6], each system induces three categories of time: 1) *Internal* or *system* time; 2) *External* or *environmental* time; 3) *Intermediary* or *connective* time.

Let **R** be some system.

Definition 1. *Internal time of* **R** *is the time in which the whole system* **R** *or some of its subsystem functions.*

Definition 2. *External time for* **R** *is an internal time of its environment as a system.*

Definition 3. *Connective time for* **R** *is the time in which the processes of interaction of* **R** *with its environment goes on.*

Remark 1. According to these definitions, internal time of any system **P** that exists outside **R** is an external time for **R**. It implies that there are, as a rule, many different external times for a system as there are many systems that interact with **R**.

The existence of a variety of times is compatible with a logical approach to system representation that is based on logical varieties instead of calculi [4]. If time is included in the description, the variety of representation consists of dynamic logics (implicit representation of time) or temporal logics (explicit representation of time).

With respect to the general time classification, which is given above, we have three distinct times related to visualization:

1. System time of a user.
2. System time of a computer or of a separate computer program.
3. System time of the interaction space, which is the intermediary time for both the user and computer.

As a result, the interaction space is used for synchronization of the system times of a user and a computer. These times are essentially different. The system time of a user is a psychological time of a human being [6]. It is determined by the processes in the psyche. As interaction with a computer demands some intelligence, conscious processes play a more important role in the user's internal time than unconscious processes. If we compare conscious processes in the brain with information processing of a computer, we see that system time of the computer, which is defined by this processing, is much faster than the psychological time of a human being. That is the reason why computers perform calculations better than most people.

It should be recognized that any computer has not a single time but many, because different programs are functioning mostly in their own time scale. This is especially evident when a computer is utilized in the time-sharing mode.

Each computer has one *gauge time,* which is used for the synchronization of all computations and consequently, of all other internal times in this computer. This is necessary for the organization of digital computations, where time is discrete. It is assumed that analog computations go in continuous time [21], but actually their time is only fuzzy continuous [3].

The gauge time of a computer is the time of its clock, signals of which gauge time in all subsystems and processes in the computer. This is similar to the situation in society: each person has his or her own internal or inner time, but all synchronize their activities by physical time which is measured by clocks and calendars. The interaction space provides a similar possibility. Namely, it is useful to have a model time in the interaction space. The aim is to synchronize all processes in the interaction space. Now the time of computer clocks is used as the model time of the interaction space. However, in the future, a variety of virtual reality implementations will bring up with many different kinds of time. Even now we have some differences. While the physical time is linear (at least, it is assumed so), the time of computer clocks is cyclic. This distinction caused the famous Y2K problem, which troubled so many people at the end of 20th century. The cycle of time in computers is equal to a hundred years. Computer games also have a cyclic time but with much shorter cycles.

An important problem arises from existence of distinct time scales in different systems. This is the problem of time coordination. For example, computer time is much faster than human time. Consequently, if a computer displays on the screen the results of arithmetic calculations as they were produced, the user would not be able to discern these results because human visual perception is much slower than modern computers.

For a system $\mathbf{R}$ to interact properly with another system $\mathbf{Q}$, it is necessary to have a time coordination function $t_{c_{\mathbf{RQ}}}$. This function maps homomorphically time $T_{\mathbf{Q}}$ in $\mathbf{Q}$ into time $T_{\mathbf{R}}$ in $\mathbf{R}$. Here, homomorphically means that $t_{c_{\mathbf{RQ}}}$ preserves a chosen structure in $\mathbf{Q}$. For example, our conventional time has the structure of an order set [13], i.e., if we have two moments of time, then one of them is less (earlier) than the other. Homomorphism means that the relation to be earlier, or equivalently, the dual relation to be later, is preserved. Dynamic properties of systems and their investigation involve topological properties of time. With respect to time coordination functions, it implies continuity, or more generally, fuzzy continuity of such functions. Properties of fuzzy continuous functions are studied in [3]. These properties are important for time coordination.

Here we consider one example that has important practical implications. Many theories of super-recursive algorithms are based on infinite computations [7]. Practical realization of such computations, which took place at the finite interval of the user time, might extend enormously power of computers [22]. However, their realization is a very complicated problem which depends on existence of the time coordination functions of definite type. In particular, we have the following results.

Let T_U and T_C be inner times of user and computer, correspondingly. We suppose that both times are linear, and are subclasses of the real line.

Theorem 1. *If there are only fuzzy continuous time coordination functions $t_{c_{CU}} : T_U \to T_C$, then it is impossible to realize infinite computations.*

A proof of this theorem is based on the following property of fuzzy continuous functions, which is proved in [3].

Theorem 2. *A real function is fuzzy continuous if and only if it is bounded.*

Note that realization of super-recursive algorithms essentially extends the capacity of intelligent agents for visualization.

5 Conclusion

A model for the dynamic representation of human-computer interface developed in this paper is used for explication of the temporal aspects of computer visualization. The results obtained show that there are several time scales and consequently, multiple times in virtual reality. This important feature has to be taken into account in visualization design as well as in software design.

These peculiarities of the interaction space and its time characteristics are utilized in the Virtual Aneurysm system, developed at UCLA [17].

At the same time, the problem of different time scales is urgent not only for visualization but also for HCI design in general. Development of HCI tools and research in this area made clear that this is a critical factor for improving quality and usability of HCI [12]. Consequently, the dynamic model of multi-system interaction proposed in this paper is also aimed at further development of HCI.

References

1. J.F. Allen, "Towards a general theory of action and time," *Artificial Intelligence*, vol. 23, pp. 123–154, 1984.
2. K. Brooks (Ed.), *The Communicative Arts and Sciences of Speech*, Merrill Books, Columbus, Ohio, 1967.
3. M. Burgin, "Neoclassical analysis: fuzzy continuity and convergence," *Fuzzy Sets and Systems*, vol. 70, no. 2, pp. 291–299, 1995.

4. M. Burgin, "Logical tools for inconsistent knowledge systems," *Information: Theory and Applications,* vol. 3, no. 10, pp. 13–19, 1995.
5. M. Burgin, "Fundamental structures of knowledge and information," *Academy for Information Sciences,* Kiev, 1997. (in Russian)
6. M. Burgin, "Time as a factor of science development," *Science and Science of Science,* vol. 5, no. 1, pp. 45–59, 1997.
7. M. Burgin, "Super-recursive algorithms as a tool for high performance computing," *Proc. of the High Performance Computing Symposium,* pp. 224–228, San Diego, California, 1999.
8. B. Fields, P. Wright, and M. Harrison, "Time, tasks and errors," *SIGCHI Bulletin,* vol. 28, no. 2, pp. 53–56, Apr. 1996.
9. N. Foster and D. Metaxas, "Modeling water for computer animation," *ACM Communications,* vol. 43, no. 7, pp. 60–67, July 2000.
10. J. Funge, "Cognitive modeling for games and animation," *ACM Communications,* vol. 43, no. 7, pp. 40–48, July 2000.
11. M. Green, "The University of Alberta user interface management system," *Computer Graphics,* vol. 19, pp. 205–213, 1985.
12. H.R. Hartson and P.D. Gary, "Temporal aspects of tasks in the user action notation," *Human-Computer Interaction,* vol. 7, pp. 1–45, 1992.
13. S.W. Hawking, *A Brief History of Time,* Toronto: Bantam, 1988.
14. J. Hintikka, *Knowledge and Belief,* Cornell University Press, Ithaca, 1962.
15. M.N. Hunks and L.M. Stephens, "Multiagent systems and societies of agents," in *Multiagent Systems: A Modern Approach to Distributed Artificial Intelligence,* G. Weiss ed., Cambridge/London, The MIT Press, 1999.
16. J. Kramer, S. Noronha, and J. Vergo, "A user-centered design approach to personalization," *ACM Communications,* vol. 43, no. 8, pp. 45–48, Aug. 2000.
17. D. Liu, M. Burgin, and W. Karplus, "Computer support system for aneurysm treatment," *Proc. of the 13th IEEE Symposium on Computer-Based Medical Systems,* Houston, Texas, pp. 13–18, June 2000.
18. B.A. Meyers, "A brief history of human computer interaction technology," *ACM interactions,* vol. 5, no. 2, pp. 44–54, Mar. 1998.
19. M. Minsky, "Commonsense-based interfaces," *ACM Communications,* vol. 43, no. 8, pp. 67–73, Aug. 2000.
20. S.J. Russell and P. Norvig, *Artificial Intelligence: A Modern Approach,* Prentice Hall, New Jersey, 1995.
21. H.T. Siegelmann, *Neural Networks and Analogy Computation,* Birkhauser, Boston, 1999.
22. I. Stewart, "The dynamics of impossible devices," *Nonlinear Science Today,* vol. 1, no. 4, pp. 8-9, 1991.

Making Movies: Watching Software Evolve through Visualisation

James Westland Cain[1] and Rachel Jane McCrindle[2]

[1]Quantel Limited, Turnpike Road, Newbury, Berkshire, RG14 2NE, UK
james.cain@quantel.com
[2]Applied Software Engineering Research Group, Department of Computer Science, The
University of Reading, Whiteknights, PO Box 225, Reading, Berkshire, RG6 6AY, UK
r.j.mccrindle@reading.ac.uk

Abstract. This paper introduces an innovative visualisation technique for
exposing the software defects that develop as a software project evolves. The
application of this technique to a large-scale industrial software project is
described together with the way in which the technique was modified to enable
integration with the software configuration management process. The paper
shows how a number of forces acting on the project can be equated to changes
in the visualisations and how this can be used as a measure of the quality of the
software.

Keywords. Software Visualisation, Software Configuration Management, Large
Scale Software Engineering, Software Evolution.

1 Introduction

Even the simplest of designs in evolving software systems can over many years lead
to unnecessarily complex and inflexible implementations [11], which in turn can lead
to vast amounts of effort expenditure during their enhancement and maintenance.
Large-scale commercial developments rarely have simple design issues to solve and
resultantly are particularly prone to maintenance problems. Such problems are further
exacerbated by a decaying knowledge of the code as the software ages and original
programmers move onto new assignments [1], and by the huge variance in
effectiveness and quality of output of replacement programmers joining the team [12].
These factors make the maintenance and in particular the comprehension of a system
difficult and the associated impact analysis of a proposed change a time-consuming
and error prone process.

Maintenance and evolution of large-scale software systems can be aided through
good initial design [14, 15, 16, 17] coupled with definition of [18, 21] and compliance
to [4, 6, 19] programming heuristics. Software visualisation is another technique that
has been shown to have an important role to play in program comprehension and
which can be readily applied to legacy systems as well as newly developed code [3].
In order to expose the causes of software degradation and thereby assist in the
comprehension of large-scale software throughout its evolution we have developed
the Visual Class Multiple Lens (VCML) technique and toolset. VCML uses a
combination of spatial and temporal visualisations to expose and track the increasing

complexity, brittleness and ultimate degradation of a system over time [5]. This paper gives a brief overview of the key concepts of the VCML approach and discusses how we have extended the work by combining it with a configuration management tool in order that the evolution of an entire project can be visualised. The resultant software evolution 'movies' enable trends and deficiencies within a system to be rapidly identified and readily tracked. Examples of movies produced from the visualisation of a large-scale commercial system are also given and explained.

2 Visual Class Multiple Lens (VCML) Technique

As projects grow the maintenance overhead starts to cause inertia. It becomes harder for the developers to add new features without breaking the existing source code. More time has to be devoted to defect tracking and bug fixing and regression testing becomes paramount. It therefore becomes especially important to understand and maintain as far as is practicable the structure of the code. VCML uses a combination of visualisation techniques with a variety of lenses to show the structure, quality and compliance to good design principles and programming heuristics of the code.

2.1 Visualisations

Spatial visualisations [1, 2, 4, 10, 20] represent graphically the structure of source code whilst temporal visualisations can be used to visualise software release histories [8]. As both spatial and temporal visualisations have proved their worth to the program comprehension process in isolation, it is reasonable to infer that further advances could be made if both approaches were used in combination. That is, there is a need for a tool to visualise source code changes over time in order to check that the design intent is still realised by the source code. Gall et al. [8] for example, have visualised software release histories through the use of colour and the third dimension. They chose to represent the system at module level rather than the source code level and at this level of granularity found that the system could scale well using time as the z-axis. Koike et al. [10] also use the Z-axis in their visualisation to represent time. However they chose to represent a system at source level and found that this strategy led to high complexity for even fairly small amounts of source code and that the resultant visualisations did not scale well. Ball et al. [1] also represent changes at source code level over time with their visualisations, and by using colour to infer activity areas, so that reengineering hot spots become apparent. Our approach takes this a stage further by combining heuristic codification into spatial and temporal visualisation of software source code through the use of the VCML technique.

2.2 Lenses

A number of *lenses* have been defined that change the visual representation of a defined characteristic of a class to be proportional to its importance in the system. The importance can be based upon any metric such as the number of *references* to the class or the number of other classes that the class itself *uses*. This allows a sense of scale, in terms of importance or weighting, to be superimposed upon the graphical representation of the classes produced by the reverse engineering tools. [3]. This technique allows class size to be determined through metrics gathered from reverse engineering source code. The lenses presented in this paper are drawn using a radial view with information culling in order to project metrics into the visualisation.

Within our current visualisations we use *size* to represent the number of references to a class and *colour* to represent the number of classes that each class uses. The lenses are orthogonal in nature and may therefore be used in combination as a Visual Class Multiple Lens (VCML) view to expose how the different underlying metrics interact and affect compile time coupling [5]. For example, in order to assess some of the finer details of the code an overall visualisation may be refined such that the low *uses* count (blue) classes can be filtered out as shown by viewing Figure 2 (lens applied) rather than Figure 1.

2.3 A Visualisation of Piecemeal Growth

Version control systems such as MKS Source Integrity, Rational Clear Case and Microsoft Source Safe are used in industry for team development of software. They all have command line interfaces that allow for automation of source code check-out. Batch files can be made that check out images of software as it exists at a specified date and time. Therefore, using such systems, it is possible to check out source code trees at time lapsed intervals and to evaluate how the source code has changed. In our approach we are combining this output with the lenses in order to assess the evolving characteristics of a software system. Currently, the VCML are based on exhibiting characteristics of the code based on:

Increased size: the size in a VCML view indicates the number of *references* to the class throughout the source code. As the project develops the system often becomes highly coupled to a number of key classes. Thus if there is a need to change a large class there is a significant overhead in terms of recompilation and regression testing. The rest of the project is tightly coupled to the class definition. As time elapses in a project there is a tendency for classes to grow in size. The more large classes there are, the more likely a programmer will have to make a change that affects large parts of the source code. At the very least recompilation will be required and at a higher cost regression testing for all affected modules may be required. Even more costly however, is a failure to realise that regression testing is required on such a scale.

Red shift: the colour in a VCML view indicates the number of classes that the class *uses*. As the project develops class implementations get fleshed out and resultantly classes tend to get less cohesive and generally causing a 'red shift' in the VCML view.

3 Concatenating VCML Views into movies

Previously, VCML views have only ever been presented in sequences of three or four images [5] and with a time span of some six months between each view. Whilst this gives a useful indication of how a software system has evolved over time the level of granularity is insufficient to enable an analysis of software evolution to be correlated with events occurring within the project team during development.

For example, factors such as staffing, skill level of developers and project management strategies have been shown to have a huge impact on software evolution and its resultant quality [7, 12]. By taking for example, two-weekly snap shots of the software source code and making a VCML view for each, it is possible to generate 'movies' that illustrate the gradual change in software attributes over time. Through this approach it is also possible to identify increases in the rate of change of attributes in the VCML views and to attempt to correlate these with changes in the dynamics of the project team.

3.1 Layout Algorithm
Repeatability [9] is key to being able to make movies out of the individual images. If the layout algorithm is not deterministic and repeatable then the classes will be drawn in different locations for each source code snap shot, and the movies would show the classes 'jumping about' between each frame thereby preventing comparisons from being made and trends from being identified. To solve this VCML views use a radial tree view so that classes are generally always drawn in the same position.

VCML views also use a non Euclidean geometry, somewhat similar to the Focus + Context approach [13], in that the size of each element in the tree varies according to the interest that the viewer places on the element. The difference between VCML views and Focus + Context techniques is that in VCML views an attribute of each element is determining its own importance whilst Focus + Context techniques demand that the user determines where to investigate further in order to obtain more detail. VCML views thus allow the user to discover which elements are important without demanding viewer interaction.

3.2 Generating The Pictures In Theory
In order to generate the movie pictures, it was postulated that image collection could be automated through the aforementioned use of scripting (batch file) techniques. Command line driven source code check-out facilities from a configuration management database enables complete image trees to be checked-out, with each image tree containing the code in its entirety at a particular date. Therefore by using simple script files it should be possible to check out images of the code at regular (e.g. monthly or bi-weekly) intervals across the entire duration of the project, compile each check-out and save the browser database generated by the compiler. Once this operation has taken place the intermediate files can be deleted to recover disk space.

This approach would result in a directory of browser database files spanning the history of the project at regular intervals each of which can then be individually rendered by the Visual Class tool into a set of VCML views. Thus there is a fairly automated process that can generate a stack of images representing the evolution of the software, which can then be concatenated together to make a small animation or movie.

3.3 Generating The Pictures In Practice
In practice however, the above method had several problems which needed to be overcome. For example, one of the source code history files was found to have been corrupted about one year prior to archiving for the purposes of this research. This meant that about six months of source code images would not compile. Actually, the

corruption was relatively easy to fix by hand but as it has to be done for each image, was very labour intensive.

A second and more serious problem was that compilers, system header files and operating systems themselves evolve over time and resultantly most of the image trees would not link. This meant that it was not possible to build an executable from the source code. This did not matter for the purposes of the research, as only a compile is required to generate the information necessary to create a browser database. However, it did mean that the script files had to invoke the utility to generate the browser databases separately from the invocation of the compiler.

Finally Visual Class does not inherently have the ability to save bitmaps, even though this can easily be added at a future date. In the meantime however we are using screen dumps as an alternative approach. Another problem was that Visual Class tends to leak memory on repeated loading of browser databases; so each script invocation only generated four bitmaps, which although not prohibitive is inconvenient and time consuming (this too will be fixed in future versions of the tool).

4 The Movies

Swift is a large-scale commercial piece of software in daily use and which is continually evolving to meet the demands of its Industry. It has well over 100,000 lines of original in-house code as well as incorporating third party COTS (Components-off-the-Shelf) such as MFC (Microsoft Foundation Classes). Development of Swift began in the Spring of 1997 utilising a single developer and growing to eight fulltime personnel by the Summer of 1999. Resultantly, as well as the added complexity introduced through the required updates, the code became a mix of the coding styles and ideologies of the individual programmers.

Figure 1 shows twelve VCML views of the Swift project at two-month intervals. Each diagram has been generated by taking a snapshot of the source code, compiling it to generate a browser database, and loading the database in Visual Class.

Immediately it can be seen that the internal complexity of the Swift project is increasing over time. The first frame show 549 classes, the last shows 947. That is, the number of classes has almost doubled, and more importantly the proportion of large classes has grown even more rapidly. This can be seen as a visualisation of the compile time coupling within Swift.

Figure 2 also shows twelve VCML views of the Swift project at two-month intervals. It is different from Figure 1 because the lens threshold has been set to twenty in each image. This means that any class that uses less than twenty other classes in its implementation are not drawn. The first image has 6 classes, of which only one is sizable. Two years later, the final image has eleven classes, most of which are large. These views also show that there is a 'red shift' over time. This tends to indicate that the major classes in the project are becoming less cohesive as the system evolves.

4.1 Colour change analysis

From the outset CWnd is a prominent class in Figure 2's images. Over the two years it gradually grows in size but it does not change colour. This is because CWnd is at the heart of the Microsoft Foundation Classes, a COTS component that was used in the implementation of Swift. As the MFC version was not changed during the project development then the implementation of CWnd did not change. Thus the colour of CWnd is constant over time.

Although it is initially hardly visible, CMainFrame is present on every image in Figure 2. It only however becomes prominent towards the end of the first year's development. Over the duration of the project it has changed colour from yellow to red. This class is derived from an MFC class, but implemented by the Swift team. It is central to the application's implementation, as it represents the main window of the application, handling most of the menu invocations. Therefore it is both highly dependent on the MFC classes and acts as a conduit for message passing around the application. The change from yellow to red indicates that the developers are adding increasing functionality to CMainFrame, gradually turning it into a 'god' class [21].

4.2 Size change analysis

CWnd grows steadily in each image in Figure 2. This makes sense, as there were increasingly more windows and dialogs implemented in Swift as time went by, and each window and dialog implementation would inherit directly or indirectly from CWnd. As the MFC implementation was never changed this is not a problem because CWnd was never the cause of a recompilation.

CMainFrame grows markedly from the beginning to the end of the visualisation. It is being used by an increasing number of classes. This indicates that as time progresses an increasing amount of code will need to be recompiled if it ever changes, and as it grows the compile time coupling with the rest of Swift is increasing.

4.3 Size Increase and Red Shift

CMainFrame is both growing in size and moving across the colour spectrum towards red. The combination of these attributes means that CMainFrame is likely to change frequently and each time it does it demands the recompilation of more and more dependent classes.

Given this issue also applies to a host of other classes such as CBlock, CTransmit, CBound, CSwit32Doc and CSwift32App, one can see that a good proportion of the development work was being focused on these key classes. Each time they change most of the application needs recompilation. Towards the end of the two-year period these classes were being changed daily. Each time this occurred the developer would have to set the compiler running for about an hour.

Worse still, each time they were checked into the revision control system, all the other developers would have to recompile their code bases. Each change could cost the team as a whole six hours compilation time. The effect of this was that people were reluctant to check in their changes. They would keep code in private branches for many days or weeks. When they came to checking in their changes the head revision of the code base was often quite different to the version that they had checked out, so large merges of code were common.

Merging source code files is an arduous and labour intensive process, where conflicting code changes have to be resolved by comparing three files (the old checked out version, the modified local version and the head revision). Most commercial software configuration management tools supply tools to aid with these types of merging. Even so the merging process can easily reintroduce bugs that have already been fixed. This happens because the programmer may choose an old piece of code over a head revision to enable their code changes to still make sense and compile.

5 Program comprehension

Movies can show a great amount of detail regarding system behaviour over time, for example, it is possible to equate the speed up in growth of the in big red classes to the numbers of developers joining the project and their experience. The Swift project team changed markedly over the two-year time frame. The first six moths of the project had only one developer, the designer of the product. The second six months saw the addition of another senior engineer. During the third six-month period two junior developers were recruited. During the final six months the team had six developers, with one of the senior developers becoming the project manager and the original designer joining a different department (though he was available for consultation).

During the last six months many bugs that had been fixed seemed to reappear in new versions (patches were being issued to customers approximately every three weeks) and productivity seemed to slow even though there were more developers working on the system.

The last three images in Figure 1 show a marked increase in the rate of growth of the classes and the red shift. This shows that the novice programmers were placing more and more code in 'god' like classes. This breaks one of the heuristics advocated by [21] primarily because programmers new to C++ tend to write in a 'C' like manner that does not scale well. This then leads to large recompile times and the software becomes brittle and prone to bugs reoccurring.

Pressure on the project team to deliver software releases every three weeks meant that 'fire-fighting' fixes were common. In the short term these worked and got the product 'out of the door', but in the long term one can see from the VCML views during the last six months, that software quality was being eroded by these practices.

6 Conclusion

The VCML view can show how some classes can dominate a project when they both use many classes in their implementations and have many references to them in the rest of the source code. The concatenation of the views into movies allows the visualisation of the effects of code changes over time. The novel technique of making VCML movies illuminates the forces that cause the need for refactoring as projects grow. The visualisation presented here shows how the code from a commercial project became more brittle as junior programmers took over the software maintenance. The views show how problems of program comprehension can cause very hard project management problems.

VCML views made into movies allow for the visualisation of the project evolution on a macro scale. They show that software quality can be visualised to allow for analysis of hotspots, areas of code that are impeding software development and those which need to be refactored.

We are currently working on visualising another large industrial project and the effects on the technique before and after major code refactoring takes place. We are also working on correlating the VCML movies with 'bug' reports from the two products.

Acknowledgements

We would like to acknowledge the financial and project support provided by Quantel Limited and Softel Limited, without which this work would not have been possible.

References

[1] Ball T., Eick S., *Software Visualisation in the Large*, IEEE Computer, April 1996.
[2] Burkwald S., Eick S., Rivard K., Pryce J., *Visualising Year 2000 Program Changes*, 6th IWPC June 1998.
[3] Cain J., McCrindle R., *Program Visualisation using C++ Lenses*, Proceedings 7th International Workshop on Program Comprehension, 20-26, Pennsylvania, May 1999.
[4] Cain J., McCrindle R., *Visual Class Tool for Assessing and Improving C++ Product Quality*, 3rd Annual IASTED International Conference Software Engineering and Applications, Arizona, October 1999.
[5] Cain J., McCrindle R., *The Spatial and Temporal Visualisation of Large Software Systems*, submitted for peer review, January 2001.
[6] Ciupke O., *Automatic Detection of Design Problems in Object-Oriented Reengineering*, Technology of Object-Oriented Languages and Systems, August 1999.
[7] Coplien, J., A Generative Development-Process Pattern Language in Coplien J., & Schmidt D., (eds.), *Patterns Languages of Program Design*, Addison Wesley, 1995.
[8] Gall H., Jazayeri M., Riva C., *Visualizing Software Release Histories: The Use of Colour and The Third Dimension*, IEEE Int. Conf. on Software Maintenance, September 1999.
[9] Herman, I., Melançon, G., Marshall, M. S., *Graph Visualization and Navigation in Information Visualization: A Survey*, 24 IEEE Transactions on Visualization and Computer Graphics, vol. 6, no. 1, January-March 2000.
[10] Koike H., Chu H., *How Does 3-D Visualisation Work in Software Engineering? Empirical Study of a 3-D Version/Module Visualisation System*, The 20th International Conference on Software Engineering, 19 - 25 April 1998.
[11] Fowler, M., *Refactoring: Improving the Design of Existing Code,* Addison Wesley, 1999.
[12] Landauer, T.K., The Trouble with Computers: Usefulness, Usability, & Productivity, London: MIT Press, 1995
[13] Lamping, J., Rao, R., Pirolli P., *A Focus+Context Technique Based on Hyperbolic Geometry for Visualizing Large Hierarchies*, ACM Computer-Human Interaction, 1995
[14] Martin, R., C., *The Open-Closed Principle, Engineering Notebook*, January 1996.
[15] Martin, R., C., *The Liskov Substitution Principle, Engineering Notebook*, March 1996.
[16] Martin, R., C., *The Dependency Inversion Principle, Engineering Notebook*, May 1996.
[17] Martin, R., C., *The Interface Segregation Principle, Engineering Notebook*, August 1996.
[18] Meyers, S., *Effective C++,* 2nd Edition, Addison Wesley, 1997
[19] Meyers, S., Klaus, M., *A First Look at C++ Program Analysers, Dr. Dobbs Journal*, Feb. 1997.
[20] Rao, R., Card, S., Exploring *Large Tables with the Table Lens*, ACM Computer Human Interaction, 1995.
[21] Riel, A. J., *Object-Oriented Design Heuristics*, Addison Wesley, 1996.

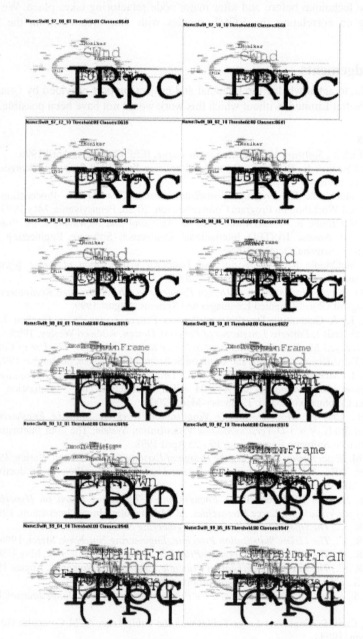

Figure 1. Twelve images with a threshold of zero, at approximately two-month intervals
(animated at http://www.reading.ac.uk/~sssmccri)

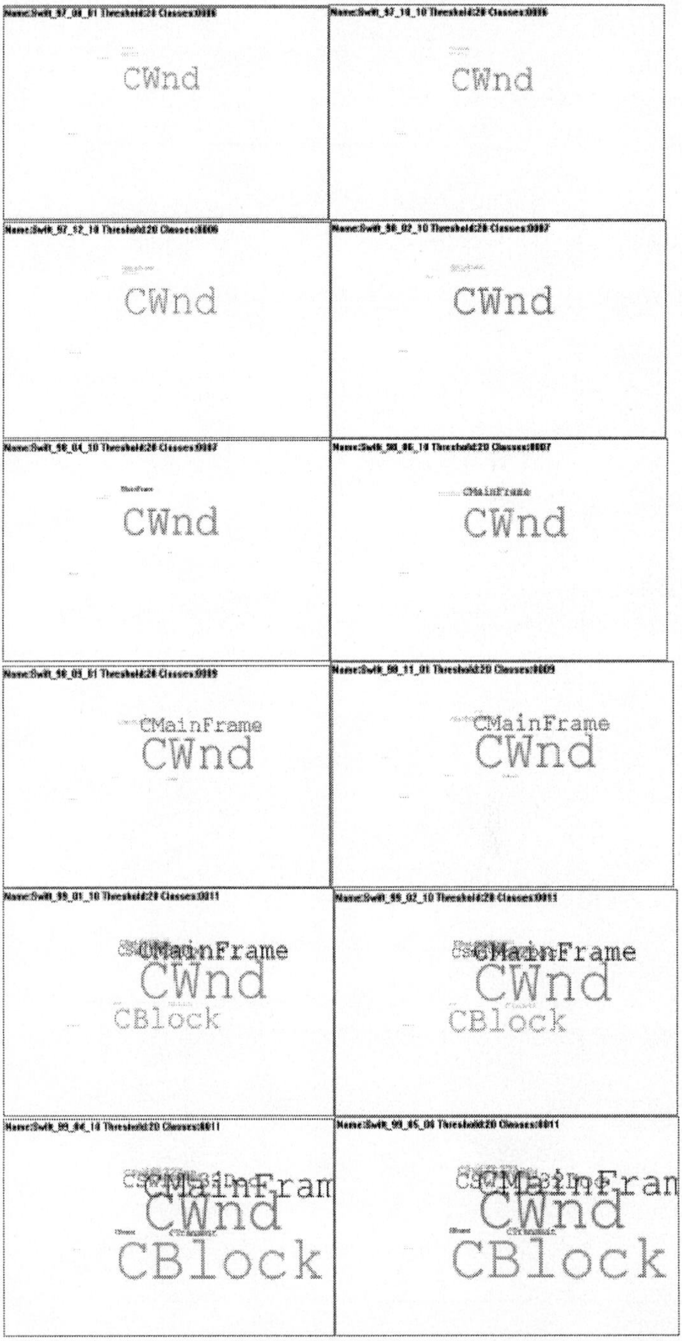

Figure 2. Twelve images with a threshold of twenty, at approximately two-month intervals (animated at http://www.reading.ac.uk/~sssmccri)

Tools and Environments for Parallel and Distributed Programming

Session chairs:

Jens Volkert (Johannes Kepler Universitat Linz, Austria)
Dieter Kranzlmueller (Johannes Kepler Universitat Linz, Austria)

Performance Optimization for Large Scale Computing: The Scalable VAMPIR Approach

Holger Brunst[1], Manuela Winkler[1], Wolfgang E. Nagel[1], and
Hans-Christian Hoppe[2]

[1] Center for High Performance Computing
Dresden University of Technology
D-01062 Dresden, Germany

[2] PALLAS GmbH
Hermühlheimer Str. 10
D-50321 Brühl, Germany

{brunst,winkler,nagel}@zhr.tu-dresden.de,
hch@pallas.com

Abstract. Performance optimization remains one of the key issues in parallel computing. Many parallel applications do not benefit from the continually increasing peak performance of todays massively parallel computers, mainly because they have not been designed to operate efficiently on the 1000s of processors of todays top of the range systems. Conventional performance analysis is typically restricted to accumulated data on such large systems, severely limiting its use when dealing with real-world performance bottlenecks. Event based performance analysis can give the detailed insight required, but has to deal with extreme amounts of data, severely limiting its scalability. In this paper, we present an approach for scalable event-driven performance analysis that combines proven tool technology with novel concepts for hierarchical data layout and visualization. This evolutionary approach is being validated by implementing extensions to the performance analysis tool Vampir.
Keywords: performance visualization, application tuning, massively parallel programming, scalability, message passing, multi-threading.

1 Introduction

Todays microprocessor technology provides powerful basic components normally targeted at the workstation market. Theoretically, this single processor technology enables an enormous peak performance when joined to a multiprocessor system consisting of 5000 or more processors. The multiplied peak performance gives the impression to most people that performance – and moreover performance optimization – is no longer an important issue. This assumption contradicts reality [12, 20] when dealing with parallel applications. Only a few highly specialized and optimized scientific programs scale well on parallel machines that provide a couple of thousands processors. In order to widen the range of applications that benefit from such powerful computational resources, performance analysis and optimization is an essential. The situation is such that performance analysis and optimization for large scale computing itself poses crucial difficulties. In the following we will present an analysis/optimization approach that combines existing tool technology with new hierarchical concepts. Starting with a brief summary of evolving computer architectures and related work, we will present the concepts and a prototyped tool extension of our scalable performance analysis approach.

2 Evolving Architectures

In the near future, parallel computers with a shared memory interface and up to 32 or more CPU nodes (SMP Systems) are expected to become the standard system for scientific numerical simulation as well as for industrial environments. In contrast to the homogeneous MPP systems that were more or less exclusively designed and manufactured for the scientific community, SMP systems mostly offer an excellent price/performance ratio as they are typically manufactured from standard components off the shelf (COTS). The comparably low price also makes SMP systems affordable for commercial applications and thus widens the community of parallel application developers needing tool support.

In cases where outstanding computational performance is needed, SMP systems can be coupled to clusters interconnected by a dedicated high performance network. Projects like ASCI [1, 3, 4] funded by the Department of Energy or the LINUX based Beowulf [2, 19] show that the coupling of relatively small SMP systems to a huge parallel computer consisting of approximately 10000 CPUs is feasible with current hardware technology. On the other hand the system and application software running on this type of machine still presents a lot of unanswered questions. It is a known fact that the different communication layers in such a hybrid system make it highly complicated to develop parallel applications performing well. Within this scope, *scalability* is one of the key issues.

Dealing with scalability problems of parallel applications in most cases requires performance analysis and optimization technology that is capable of giving detailed insight into an application's runtime behavior. Accumulative trace data analysis cannot fulfill this requirement as it typically does not explain the cause of performance bottlenecks in detail. We experienced that program event tracing is required for the location and solution of the majority of program scalability problems. Being aware of this implies that scalability also needs to be introduced into event based performance analysis and optimization tools as tracing the behavior of 1000-10000 processing entities generates an enormous amount of performance data. Today, the optimization tools are quite limited with respect to the number of processing entities and the amount of trace data that can be handled efficiently. The next section will give a brief summary on the current state of the art of performance tool development and its limits in order to provide a better understanding of our activities related to large scale performance analysis.

3 Current Analysis / Optimization Tools

Large machines with more than 1000 processing entities produce an enormous amount of trace data during a tracing session. Finding performance bottlenecks and their origin requires appropriate tools that can handle these GBytes of information efficiently. This implies data reduction, selection, archiving, and visualization on a large scale. Most of todays tools [6, 22] cannot yet fulfill this task. We will now classify todays tools regarding their task, capabilities, and drawbacks.

3.1 Tools for Accumulated Data Analysis

This type of tool lists measures - typically in a long table - like summarized function dwell, communication rates, performance registers, etc. Representatives of this category are *prof, gprof, iostat, vmstat* known from the UNIX world. This type of data presentation is disadvantageous when dealing with large, complex, parallel applications. Nevertheless, it can help to identify a general performance problem. Once detected, more complex tools are needed to find out what's going on 'inside' the application.

Based on the same information acquisition methods as the text based tools, tools like *Xprofiler* [23] from IBM, *APPRENTICE* [8] from Cray Research or *Speedshop* [21] from SGI have standard spreadsheet constructs to graphically visualize accumulated information. Although this allows for a quicker overview of the total application's behavior, the identification of a problem's cause remains unresolved in most cases.

3.2 Tools for Event Trace Analysis

The tools that can directly handle program traces are typically capable of showing accumulated performance measures in all sorts of diagrams for arbitrary time intervals of a program run. This is already a major difference to the above mentioned tools which typically only show a fixed time interval which in most cases is the overall program duration. In addition to this, they offer a detailed insight into a program's runtime behavior by means of so-called timeline diagrams. These diagrams show the parallel program's states for an arbitrary period of time. Performance problems that are related to imbalanced code or bad synchronization can easily be detected with the latter as they cause irregular patterns. When it comes to performance bottlenecks caused by bad cache access or bad usage of the floating point units, the timelines are less useful. Similar performance monitors can be found in almost every CPU now, and would be very helpful when combined with these CPU timeline diagrams. Representatives of this tool category are *Jumpshot* [13], *Vampir* [17], and *Xpvm* [24].

4 New Scalability Issues

In the scope of ASCI, detailed program analysis over a long period of time (not just a few seconds as is possible today) has been identified as an important requirement. Independently of the method of data acquisition, this implies an enormous amount of data to be generated, archived, edited and analyzed. The data extent is expected to be more than one TByte for a moderate size (1000 CPUs, 1 hour runtime) program run, where 0.5 MByte/s trace data per CPU is a rule of thumb observed in our daily work if pretty much every function call is traced. Obviously, this amount of data is far too large to be handled by an analysis sub-system significantly smaller than the parallel master platform. Dynamic instrumentation as possible with DYNINST [7], which was developed in the scope of the Paradyn [16] project, connected to a visualization tool would allow to deselect uninteresting code parts during runtime and thereby reduce the trace data to a moderate size of 10-100 GByte/h.

The hierarchical design of large scale parallel systems will have an impact on the way this data is stored. Nobody would really expect a single trace file as the outcome of a one hour tracing session on 1000 CPUs. Distributed files each representing a single CPU or a cluster of CPUs are likely to be used as they allow for best distributed I/O performance. Efficient access to these files forms the basis of our work and will be discussed in the following.

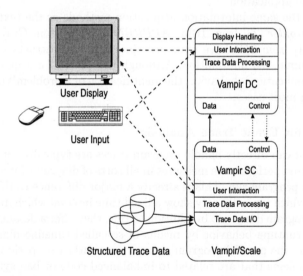

Fig. 1. A Scalable Optimization Tool Layout

4.1 Distributed Trace Data Preparation

The scalability requirements that arouse from the enormous amount of trace data to be archived, synchronized, edited, visualized and analyzed require considerable effort and time when incorporated into a fully new tool. As for smaller systems there are already a couple of good tools available on the market; we recommend a parallel data layer extension that acts in between the trace data base and the analysis tool. This data layer is responsible for all time consuming operations like for example:

- post-mortem synchronization of asynchronous trace data
- creation of statistics
- filtering of certain events, messages, CPUs, clusters
- summarizing the event history for long time periods

Outsourcing these activities from the visualization tool (which is typically a sequential program) allows for fast trace data access with limited changes to the user interface of the visualization tool. Our major goal was to keep the existing interface constant and to add scalability functionality in a natural way. Figure 1 illustrates the structure of this approach for the performance analysis tool Vampir.

The trace data can be distributed among several files, each one storing a 'frame' of the execution data as pioneered by jumpshot [13]. Frames can correspond to a single CPU, a cluster of CPUs, a part of the execution or a particular level of detail in the hierarchy (see 4.2). The frames belonging to a single execution are tied together by means of an index file. In addition to the frame references, the index file contains statistics for each frame and a 'performance thumbnail' which summarizes the performance data over time.

The analysis tool reads the index file first, and displays the statistics and the rough performance thumbnail for each frame. The user can then choose a frame to look at in more detail, and the frame file will be automatically located, loaded and displayed. Navigation across frame boundaries will be transparent.

4.2 Hierarchical Visualization

When it comes to visualization of trace data, dealing with 1000-10000 CPUs poses additional challenges. The limited space and resolution of a computer display allows the visualization of at most 200-300 processing entities at any one time. Acting beyond this limit seems to be useless especially when scrolling facilities are involved, as users become confused by too much data. Therefore, we propose a hierarchical approach where, based on the distributed system's architecture, the user can navigate through the trace data on different levels of abstraction. This works for both event trace oriented displays (timelines) and statistic displays for accumulated data. The hierarchy in figure 2a allows the visualization of at least 10000 processing entities where a maximum of 200 independent objects need ever be drawn simultaneously on the same display. This prediction is based on a hypothetical cluster consisting of 64 SMP nodes with 200 processes each (and no threads). As our model provides 3 layers which can each hold at maximum 200 objects, almost any cluster configuration can be mapped in an appropriate way.

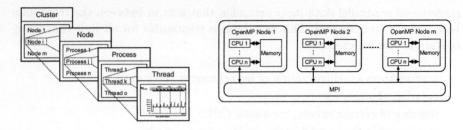

(a) Access Hierarchy   (b) SMP Cluster with MPI/OpenMP

Fig. 2. Hierarchical View on Event Trace Data

4.3 Performance Monitors

Current processor architectures usually offer performance monitor functionality, mostly in the form of special registers that contain various performance metrics like the number of floating-point operations, cash misses etc. The use of these registers is limited because there is no relation to the program structure: an application programmer typically does not know which parts of the application actually cause bad cache behavior. To help here, the use of profiling techniques is necessary: cyclic sampling and association of the sampled values to code locations (à la prof) or combination of sampling with subroutine entry/exit events (à la gprof) will provide the insight into which parts of a program need further optimization.

For the optimization of large scale applications, performance monitors gain additional significance. As stated above, dealing with large amounts of performance data requires multiple abstraction layers. While the lower layers are expected to provide direct access to the event data by means of standard timeline and statistic views, the upper layers must provide aggregated event information. The aggregation needs to be done in a way that provides clues to performance bottlenecks caused in lower layers. Measures like cache performance, floating point performance, communication volume, etc. turned out to have good summarizing qualities with respect to activities on lower layers. We suggest introducing a chart display with n graphs representing n ($n < 64$) nodes as the entry point to a scalable performance optimization approach.

4.4 Hybrid Programming Paradigms: MPI + OpenMP

Parallel computers are typically programmed by using a message passing programming paradigm (MPI [5, 15] PVM [10], etc.) or a shared memory programming paradigm (OpenMP[9, 14, 18] Multi Threading, etc.). Large machines are now often realized as clusters of SMPs. The hybrid nature of these systems is leading to applications using both paradigms at the same time to gain best performance. Figure 2b illustrates how MPI and OpenMP would cooperate on a SMP cluster consisting of m nodes with n CPUs each.

So far, most tools support either message passing or shared memory programming. For the hybrid programming paradigm, a tool supporting both models equally well is highly desirable. The combination of well established existing tools for both realms by means of a common interface can save development effort and spare the users the inconvenience of learning a completely new tool. In this spirit, Vampir and the GuideView tool by KAI [11] will be combined to support analysis of hybrid MPI/OpenMP applications. First results will be illustrated in the final paper.

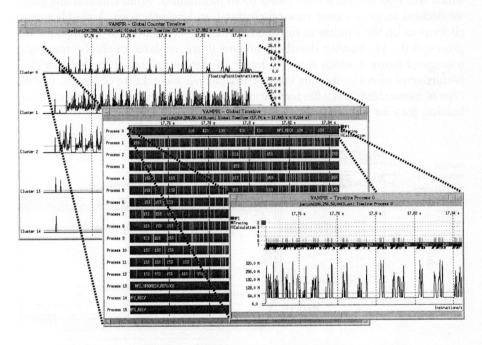

Fig. 3. Hierarchical View on 256 CPUs

5 Vampir - Scalability in Practice

The previous section introduced new ideas for performance analysis tools targeted towards the optimization of next generation applications. We already depend on such tools as part of our daily work, which is why we have put quite some effort into developing a prototype which implements many of the ideas mentioned above. The prototype is based on Vampir 2.5 [17], which is a commercial tool for performance analysis and visualization accepted in the field.

Based on real applications that were tuned at our center, we will now present one possible realization of a scalable performance analysis tool.

5.1 Navigation on GBytes of Trace Data

In section 4.2 we introduced the idea of a hierarchical view on event trace data. The reason for this was the large amount of event trace data generated by a large scale application running on hundreds of processing nodes over a longer period of time. Figure 3 illustrates the impact of this approach on trace data navigation for a test case generated on 256 (16 × 16) processors. The leftmost window depicts MFLOPS rates on the cluster level[1] for a time period of 0.1 s which was selected via a time based zoom mechanism. From this starting point we decided to get a closer view on cluster 1 and its processes. A simple mouse click opens up the window in the middle which depicts the state[2] changes for the processes 0 - 15. Further details to process 0 are available in the bottom right window of figure 3 which shows a detailed function call stack combined with a performance monitor showing the instructions per second rate of process 0. This type of hierarchical trace file navigation permits an intuitive access to trace files holding data for hundreds of clustered CPUs.

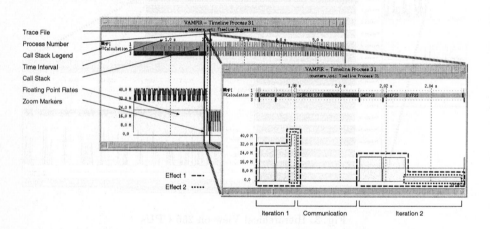

Fig. 4. Performance Monitor combined with Call Stack

5.2 Finding Hot Spots by Means of Performance Monitors

Performance monitors can be of much use when dealing with performance bottlenecks of unknown origin. The following example was taken from a performance

---

[1] Cluster 3 - 12 were filtered to adapt to the small figure size in the paper

[2] Depending on the type of instrumentation, a state is a certain function or code block

optimization session, which we recently carried out on one of our customer's programs. For some reason, his parallel program started off performing as expected but suffered a serious performance decrease in its MFLOPS rate after two seconds.

Figure 4 shows the program's call stack combined with the MFLOPS rates for a representative CPU over the time period of 6 seconds with a close-up of the time interval the program behavior changes. We see two similar program iterations separated by a communication step. The first one is twice as fast as the second one. We can also see that the amount of work carried out in both iterations is identical, as their integral surfaces are the same (effect 1). A third aspect can be found in the finalizing part (function DIFF2D) of each iteration. Obviously the major problem resides here, as the second iteration is almost 10 times slower than the first one (effect 2). We eventually discovered that the whole problem was caused by a simple buffer size limit inside the program which lead to repeated date re-fetching.

6   Conclusion

This paper has presented concepts for scalable event-based performance analysis on SMP clusters containing 1000s of processing entities. The key issues are the distributed storage and handling of event traces, the hierarchical analysis and visualization of the trace data and the use of performance monitor data to guide detailed event based analysis. The existing implementation of parts of these within the Vampir framework has been discussed. Further extensions to Vampir that are currently being worked on will serve as a proof of concept, demonstrating the benefits of event based performance analysis to real-world users with large applications on the upcoming huge SMP systems.

References

[1] Accelerated Strategic Computing Initiative (ASCI).
   http://www.llnl.gov/asci.

[2] D. J. Becker, T. Sterling, D. Saverese, J. E. Dorband, U. A. Ranawak, and C. V. Packer. Beowulf: A Parallel Workstation for Scientific Computation. In *Proceedings, International Conference on Parallel Processing*, 1995.
   http://www.beofulf.org.

[3] Blue Mountain ASCI Machine.
   http://w10.lanl.gov/asci/bluemtn/bluemtn.html.

[4] Blue Pacific ASCI Machine.
   http://www.llnl.gov/asci/platforms/bluepac.

[5] S. Bova, C. Breshears, H. Gabb, R. Eigenmann, G. Gaertner, B. Kuhn, B. Magro, and S. Salvini. Parallel programming with message passing and directives. *SIAM News*, 11 1999.

[6] S. Browne, J. Dongarra, and K. London. Review of performance analysis tools for mpi parallel programs.
   http://www.cs.utk.edu/~browne/perftools-review.

[7] B. Buck and J. K. Hollingsworth. An API for Runtime Code Patching. Technical report, Computer Science Department, University of Maryland, College Park, MD 20742 USA, 1998.
http://www.cs.umd.edu/projects/dyninstAPI.

[8] Cray Research. *Introducing the MPP Apprentice Tool*, IN-2511 3.0 edition, 1997.

[9] D. Dent, G. Mozdzynski, D. Salmond, and B. Carruthers. Implementation and performance of OpenMP in ECWMF's IFS code. In *Proc. of the 5th SGI/CRAY MPP-Workshop*, Bologna, 1999.
http://www.cineca.it/mpp-workshop/abstract/bcarruthers.htm.

[10] A. Geist, A. Beguelin, J. Dongarra, W. Jiang, R. Manchek, and V. Sunderam. *PVM: Parallel Virtual Machine.* The MIT Press, 1994.
http://www.epm.ornl.gov/pvm.

[11] The GuideView performance analysis tool.
http://www.kai.com.

[12] F. Hoßfeld and W. E. Nagel. Per aspera ad astra: On the way to parallel processing. In H.-W. Meuer, editor, *Anwendungen, Architekturen, Trends, FOKUS Praxis Informationen und Kommunikation*, volume 13, pages 246–259, Munich, 1995. K.G. Saur.

[13] The Jumpshot performance analysis tool.
http://www-unix.mcs.anl.gov/mpi/mpich.

[14] Lund Institute of Technology. *Proceedings of EWOMP'99, 1st European Workshop on OpenMP*, 1999.

[15] Message Passing Interface Forum. *MPI-2: Extensions to the Message-Passing Interface*, August 1997.
http://www.mpi-forum.org/index.html.

[16] B. P. Miller, M. D. Callaghan, J. M. Cargille, J. K. Hollingsworth, R. B. Irvin, K. L. Karavanic, K. Kunchithapadam, and T. Newhall. The Paradyn Parallel Performance Measurement Tools. *IEEE Computer*, 28(11):37–46, November 1995.
http://www.cs.wisc.edu/~paradyn.

[17] W. E. Nagel, A. Arnold, M. Weber, H.-C. Hoppe, and K. Solchenbach. VAMPIR: Visualization and Analysis of MPI Resources. *Supercomputer 63*, XII(1):69–80, January 1996.
http://www.pallas.de/pages/vampir.htm.

[18] *Tutorial on OpenMP Parallel Programming*, 1998.
http://www.openmp.org.

[19] D. Ridge, D. Becker, P. Merkey, and T. Sterling. Beowulf: Harnessing the Power of Parallelism in a Pile-of-PCs. In *Proceedings, IEEE Aerospace*, 1997.
http://www.beofulf.org.

[20] L. Smarr. Special issue on computational infrastructure: Toward the 21st century. *Comm. ACM*, 40(11):28–94, 11 1997.

[21] The Speedshop performance analysis tool.
http://www.sgi.com.

[22] Pointers to tools, modules, APIs and documents related to parallel performance analysis.
http://www.fz-juelich.de/apart/wp3/modmain.html.

[23] The Xprofiler performance analysis tool.
http://www.ibm.com.

[24] The XPVM performance analysis tool.
http://www.netlib.org/utk/icl/xpvm/xpvm.html.

TRaDe: Data Race Detection for Java

Mark Christiaens[1] and Koen De Bosschere[1]

ELIS, Ghent University,
Sint-Pietersnieuwstraat 41, 9000 Gent, Belgium,
{mchristi,kdb}@elis.rug.ac.be

Abstract. This article presents the results of a novel approach to race
detection in multi-threaded Java programs. Using a topological approach
to reduce the set of objects needing to be observed for data races, we have
succeeded in reducing significantly the time needed to do data race detection. Currently, we perform data race detection faster than all known
competition. Furthermore, TRaDe can perform data race detection for
applications with a high, dynamically varying number of threads through
the use of a technique called 'accordion clocks'.

1 Introduction

The technique of multi-threaded design has become a permanent part of almost
every serious programmer's toolbox. When applying a multi-threaded design to
a programming task, a task is split into several threads. Each of these subtasks
can be executed on a separate processor, increasing parallelism and speed. A
group of threads can be dedicated to respond to external events, so a very fast
response time can be obtained using this design.

Together with the advantages of multi-threaded design, a new type of bug
has appeared which is very hard to handle: a data race. As each thread performs
its task in the program, it modifies data structures along the way. It is up to
the programmer to ensure that the operations of threads happen in an orderly
fashion. If the threads do not behave properly, but modify a data structure in
a random order, a data race occurs. An example of a data race can be seen in
Figure 1.

Data Races are very hard to locate. They are non-deterministic, since their
occurrence is dependent on the execution speed of the threads involved. On
top of that, data races are non-local. Two pieces of code whose only relation
is that they both utilize the same data structures, independently can function
perfectly but when executed simultaneously, they will wreak havoc. Finding or
even suspecting a data race can take days of debugging time.

2 State of the Art in Data Race Detection

In general, finding data races is at least an NP-hard problem [9], so usually,
techniques are proposed to detect data races in one particular execution and

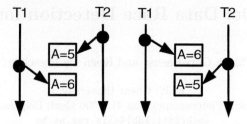

Fig. 1. On the left we see thread 2 accessing a common object, A, and writing the value 5. This is followed by thread 1, writing the value 6 to A. The result of this operation is that A contains the value 6. On the right, thread 1 for some reason executes faster which results in the same events happening but in reverse order. This is possible since there is no synchronisation between thread 1 and 2. A now contains the value 5.

not in all possible executions of a program. Several systems for detecting data races for general programs have been devised [10, 11]. These techniques can find some types of data races (their definitions of data race differ) but have the disadvantage of being very time consuming. A time overhead of a factor of 30 is not uncommon. This time overhead is caused by the fact that *every* read and write operation performed by the threads is observed.

Recently, modern programming languages with a more object oriented philosophy have seen the light. One of these is the Java programming language designed by Sun [8, 13] at the beginning of the 1990s. Java was designed with multi-threading in mind. Several language constructs are dedicated to the manipulation of threads. Since it is so easy to make a multi-threaded program in Java, the data race problem manifests itself even more frequently than in the usual C-programs.

It is therefore no coincidence that for this language, already two tools exists that detect data races in particular executions of Java programs: AssureJ and JProbe [6, 7]. Here too, the problem of overhead pops up. A time overhead of at least a factor 10 is common. This is smaller than for general programs because each thread in Java has a local stack. The values on this stack can only be manipulated by the thread owning the stack. Therefore all read and write operations performed on this stack need not be observed to find data races.

3 Topological Race Detection

Java has a specific property that can be exploited to make race detection even more efficient. The only way an object can be manipulated is through a reference to that object. Pointers do not exist so the problem that through a pointer every location in memory can be modified, does not exist.

It is therefore possible to determine exactly which objects can be reached by a thread at a certain point in an execution and which cannot. Objects that are reachable by only one thread cannot be involved in a data race since no other thread can simultaneously be modifying that object.

In Java we have exploited this fact as follows. Initially, at an object's construction through invoking one of the bytecodes `new`, `newarray`, `anewarray` and `multianewarray`, the only reference to it is stored on the stack of the thread creating the object. No other threads can access this reference, therefore at this point this object cannot be involved in a data race.

From that point forward, this reference can be manipulated by the thread, maybe even be stored in an object or a class. We have devised a system, called TRaDe (Topological RAce Detection), to detect dynamically when an object becomes reachable by more than one thread i.e. the object becomes global. It constantly observes all manipulations of references which alter the interconnection graph of the objects used in a program (hence the name "topological" of the method).

Using this classification mechanism, we can detect data races more efficiently. Only for objects that are reachable by more than one thread, is full data race detection necessary. This means that we save execution time *and* memory by not analyzing the read and write operation to these objects and by not maintaining data structures for race detection. During the last year we have implemented our ideas by modifying the interpreter (and shutting down the JIT compiler) contained in the source of the JDK1.2.1 from Sun so that it detects data races. To measure the performance of this implementation, we collected a set of applications consisting of a number of large programs to be used as a benchmark (see Table 1).

Table 1. Description of the Java benchmarks used

Name	Description
SwingSet	A highly multi-threaded demo of the Swing widget set
Jess	A clone of the expert system shell CLIPS
Resin	Web server entirely written in Java
Colt	Open source libraries for high performance scientific and technical computing
Raja	Ray tracer

In Table 2, we have measured the execution time and memory consumption of the benchmarks while doing race detection with TRaDe, AssureJ and JProbe. For comparison, we have added the execution time when executing with a modern JIT compiler, the Hotspot 1.0.1 build f and with the interpreter version on which TRaDe itself is based.

The benchmarks were performed on a Sun Ultra 5 workstation with 512 MB of memory and a 333 MHz UltraSPARC IIi with a 16 KB L1 cache and a 2 MB L2 cache. JProbe could not complete all benchmarks. In this case, the benchmarks were run on a larger system to get an indication of JProbe's performance. The larger system was a Sun Enterprise 450 with 2048 MB of memory, 4X400 Mhz UltraSPARC IIi processors with 16 KB L1 direct mapped cache and 4 MB L2 direct mapped cache. These measurements are indicated with a †.

As can be seen in Table 2, using TRaDe, we can outperform all known competition. TRaDe is approximately a factor 1.6 faster than AssureJ, the fastest competition. AssureJ is the closest competitor in terms of speed and usually beats us in terms of memory consumption. We must note however that we have found AssureJ not to detect all data races. When a thread is being destroyed, AssureJ apparently prematurely removes all information relating to this thread although future activities of other threads can still cause a data race with activities of this thread.

Table 2. Execution time and memory consumption of the Java benchmarks. At the bottom, averages and normalized averages (with TRaDe=1) are given for all systems except JProbe who couldn't perform all benchmarks.

	TRaDe		AssureJ		JProbe		HotSpot		Interpreter	
	s	MB	s	MB	s	MB	s	MB	s	MB
SwingSet	98.3	126.6	160.6	73.3	>1200†	>650 †	20	41.8	15	29.8
Jess	370.3	12.1	610	17.5	>3600†	>650†	22	19.8	76	8
Resin	56.5	27.6	68	27.3	193	226.17	11.8	27.9	10	13.7
Colt	132.5	25	187.8	21.2	471.6	71	27.8	23.6	40.4	13.5
Raja	204.8	19.5	372	17.8	1945†	1037 †	14.6	24.5	42.2	11.3
average	172.48	42.16	279.68	31.42			19.24	27.51	36.72	15.24
normalized	1	1	1.62	0.75			0.11	0.65	0.21	0.36

4 Accordion Clocks

A second problem we have tackled in TRaDe is the problem of the dynamic creation of threads. Many applications continuously create new threads: a browser updating a picture in a web page, an editor checking for modified files on disk, etc. To explain why this is a problem, we have to delve a little deeper in how data race are detected.

In parallel shared-memory programs there are two types of races: data races and synchronisation races. An example of a data race was already given in Figure 1.

In Figure 2, we see the other type of race, a *synchronisation race*. A synchronisation race occurs when two threads perform a synchronisation operation (like obtaining a mutex, a P operation on a semaphore, ...) on the same synchronisation object in a non-determined order.

The set of operations that are performed by a thread between two consecutive synchronisation operations is called a segment (e.g. S_1, S_2, S_3, S_4, ... in Figure 2).

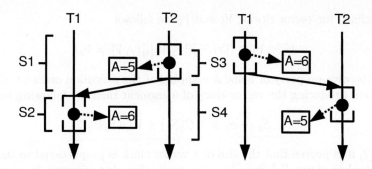

Fig. 2. The threads enter a mutex (indicated by the opening and closing brackets) before modifying the common variable. The final result of the execution is determined by which thread enters the mutex first. The order in which the threads access the mutex is indicated by the solid arrows.

Synchronisation primitives induce a partial order on the segments. For example, in Figure 2, thread T_2 is able to enter the mutex first. Thread T_1 must wait until T_2 leaves the mutex. Therefore, segment S_3 is ordered before S_2. We write $S_3 \overset{syn}{\to} S_2$.

In addition, segments are partially ordered simply by the fact that they are sequentially executed by their thread. Segment S_1 for example is executed before S_2 by thread T_1. We write $S_1 \overset{seq}{\to} S_2$.

Using these two partial orders, we can define the "execution order" on two segments S_x and S_y as the transitive closure of the union of the two previous relations:

$$S_x \to S_y \equiv S_x \overset{syn}{\to} S_y \vee S_x \overset{seq}{\to} S_y \vee \exists S_z.(S_x \to S_z \wedge S_z \to S_y) \tag{1}$$

Two segments are said to be "parallel" if there is no execution order between them i.e.

$$S_x \parallel S_y \equiv \neg((S_x \to S_y) \vee (S_y \to S_x)) \tag{2}$$

A data race occurs when operations from two parallel segments access a common variable and at least one operation is a write operation. If we define the set of locations written to resp. read from by a segment S as $W(S)$ and $R(S)$ then a data race occurs between two segments, S_x and S_y, when the following two conditions hold:

$$S_x \parallel S_y$$
$$(R(S_x) \cup W(S_x)) \cap W(S_y) \neq \emptyset \vee (R(S_y) \cup W(S_y)) \cap W(S_x) \neq \emptyset \tag{3}$$

Checking in practice whether or not two segments are ordered can be done using a construct called a "vector clock" [12], [4],[5]. A vector clock is essentially just an array of integers (one for every thread in the program) that is updated each time a synchronisation operation occurs. A comparison function,

$<$, is defined for vector clocks, V_1 and V_2, as follows

$$V_1 < V_2 \equiv ((\forall i . V_1[i] \leq V_2[i]) \wedge V_1 \neq V_2) \tag{4}$$

It has the nice property of being a mapping of the execution order i.e. if VC is a function producing the vector clock of a segment then the following holds:

$$S_x \rightarrow S_y \equiv VC(S_x) < VC(S_y) \tag{5}$$

In [3] it is proven that the size of a vector clock is proportional to the maximum number of parallel threads in the multi-threaded program. In general, we can't just reduce the size of the vector clock when a thread dies.

Why this is the case can be seen in Figure 3. We see a simple FTP-server which creates a new slave thread for each file request. Depending on the size of the file to transfer, the bandwidth of the network connection to the client and many other factors, this slave thread can take a variable amount of time to finish. When the slave thread has finished transferring the file, it is destroyed.

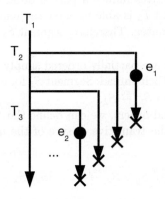

Fig. 3. An FTP-server spawning new threads for every file request.

Clearly, over a long period of time, only a small fraction of the slave threads will run concurrently. Still, the size of the vector clock will be equal to the total number of threads executed by the FTP-server. Indeed, if we would try to remove the information about the execution of thread T_2 when the thread has finished, we would run the risk of no longer detecting all data races.

Take for example event e_1 on thread T_2. Even when thread T_2 has finished execution, this event can still cause a race with another event e_2 on thread T_3. Indeed, if T_2 had run a little slower, e_2 would have been executed before e_1, which clearly constitutes a data race. The cause for this data race is that T_2 simply dies when it has finished its task without synchronizing with any other thread. Therefore, the risk for data races with event e_1 will continue to exist indefinitely and we can therefore never remove T_2's index from the vector clock.

We have devised a new approach called 'accordion clocks' which allows us to surmount the problem of the ever growing size of vector clocks. The idea behind accordion clocks is simple. Suppose a thread has finished its execution. In general we cannot yet remove its index from the vector clocks since we might still need it to verify if another thread accesses the data structures touched by the dead thread in a not properly synchronized manner. If however we detect that a thread has ended *and* there are no data structures left which were touched by this thread then we can safely remove the index from the vector clock since no races are possible anymore.

Accordion clocks do just that. Initially, at program startup, all accordion clocks contain one index since only one thread is active. As threads are created during program execution the accordion clocks are expanded transparently to include new indices for the new threads. If a thread dies, this is noted and periodically all objects on the Java heap are checked to see whether there are any left which were once accessed by this deceased thread. If there are none left, the index corresponding to the deceased thread is removed transparently from all the accordion clocks in the system. The time overhead of this check is comparable to the time overhead of a full garbage collection cycle.

We have implemented accordion clocks in TRaDe and tested them using the benchmarks described in Table 4. The results can be seen in Figure 4. The total memory used for storing all data structures to do data race detection is shown as it evolves in time. We have divided the memory consumption in two part: the top part indicates the memory used for accordion clocks and vector clocks. The bottom part is used for other per object data structures to perform data race detection. On the left, we see the execution of the benchmarks without using accordion clocks. On the right accordion clocks are used.

It is clear that after a short period of time the memory used for storing accordion clocks becomes overwhelming. This prohibits searching data races in long running applications. When using accordion clocks, the memory consumed for storing clocks levels off after a while. For some applications, like Forte and Java2D, memory consumption seems to level off to a constant cost. For Swing, accordion clocks are clearly beneficial but some work will have to be done to stop the other data structures expanding.

5   Conclusions

In this article we have shown that, using a topological approach, it is possible to more efficiently detect data races in Java. We have implemented this technique in a Java virtual machine and have found it superior in execution speed to competing systems. Furthermore, we have found that a large part of the memory consumption while doing data race detection is due to storing vector clocks. We have solved the problem of the increasing size of vector clocks by introducing accordion clocks. Accordion clocks can grow and shrink as threads are created and destroyed while maintaining correct data race detection.

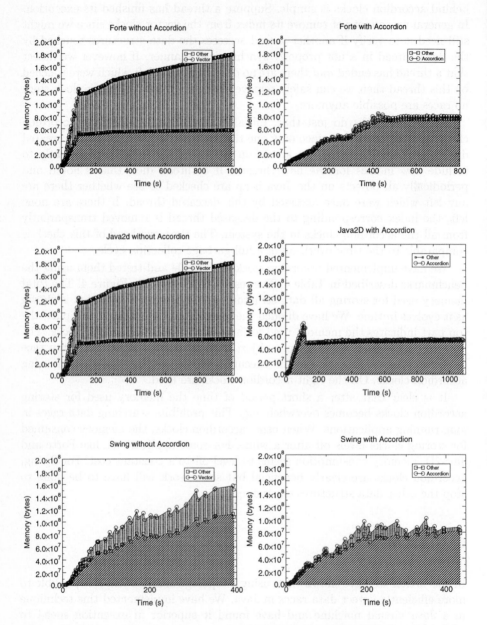

Fig. 4. Evolution of memory consumption during an execution of TRaDe without (on the left) accordion clocks and with (on the right) accordion clocks.

Name	Description
SwingSet Demo	A highly multi-threaded demo, included with the JDK 1.2, of the Swing widget set. It demonstrates buttons, sliders, text areas, ... Every tab in the demo was clicked twice during our tests. Immediately thereafter it was shut down.
Java2D Demo	A highly multi-threaded demo, included with the JDK 1.2. It demonstrates the features of the Java 2D drawing libraries and is highly multi threaded. Every tab in the demo was clicked 5 times. Immediately thereafter it was shut down.
Forte for Java	A full-blown Java IDE with object browsers, visual construction of GUI code, ... To exercise the benchmark, we created a small dialog while doing race detection.

Table 3.

6 Future Work

We are currently exploring whether it is possible to do data race detection by instrumenting the bytecode of class files. Recently, a lot of research has been devoted to escape analysis [2, 1]. This is a technique allowing to detect *at compile time* that an object will be local to a thread. We believe that this technique could be used to reduce the number of memory accesses that need to be observed thereby speeding up data race detection even further.

Acknowledgments

Mark Christiaens is supported by the GOA project 12050895. Our thanks go to Michiel Ronsse, with whom we had many stimulating discussions.

References

1. J. Aldrich, C. Chambers, E. G. Sirer, and S. Eggers. Static analyses for eliminating unnecessary synchronization from Java programs. In *Static Analysis Symposium 99*, September 1999.
2. Jeff Bogda and Urs Hölzle. Removing unnecessary synchronisation in Java . Technical report, Department of Computer Science, University of California, Santa Barbara, CA 93106, april 1999.
3. Bernadette Charron-Bost. Concerning the size of logical clocks in distributed systems. *Information Processing Letters*, 1(39):11–16, July 1991.
4. C.J. Fidge. Partial orders for parallel debugging. In *Proceedings of the ACM SIGPLAN and SIGOPS Workshop on Parallel and distributed debugging*, pages 183–194, May 1988.
5. Dieter Haban and Wolfgang Weigel. Global events and global breakpoints in distributed systems. In *21st Annual Hawaii International Conference on System Sciences*, volume II, pages 166–175. IEEE Computer Society, January 1988.

6. KL Group, 260 King Street East, Toronto, Ontario, Canada. *Getting Started with JProbe.*

7. Kuck & Associates, Inc., 1906 Fox Drive, Champaign, IL 61820-7345, USA. *Assure User's Manual,* 2.0 edition, march 1999.

8. Tim Lindholm and Frank Yellin. *The Java Virtual Machine Specification.* The Java Series. Addison-Wesley, 1997.

9. R. H. B. Netzer. *Race Condition Detection for Debugging Shared-Memory Parallel Programs.* PhD thesis, University of Wisconsin-Madison, 1991.

10. Michiel Ronsse and Koen De Bosschere. JiTI: Tracing Memory References for Data Race Detection. In E. D'Hollander, F.J. Joubert, and U. Trottenberg, editors, *Proceedings of ParCo97: Parallel Computing: Fundamentals, Applications and New Directions,* volume 12 of *Advances in Parallel Computing,* pages 327–334, Bonn, February 1998. North Holland.

11. Stefan Savage, Michael Burrows, Greg Nelson, Patrick Sobalvarro, and Thomas Anderson. Eraser: A dynamic data race detector for multi-threaded programs. In *Operating Systems Review,* volume 31, pages 27–37. ACM, October 1997.

12. Reinhard Schwarz and Friedman Mattern. Detecting causal relationships in distributed computations: in search of the holy grail. *Distributed Computing,* 7(3):149–174, 1994.

13. Bill Venners. *Inside the Java Virtual Machine.* McGraw-Hill, New York, New York, USA, second edition, 1999.

Automation of Data Traffic Control on DSM Architectures

Michael Frumkin, Haoqiang Jin, Jerry Yan[*]

Numerical Aerospace Simulation Systems Division
NASA Ames Research Center

Abstract. The distributed shared memory (DSM) architecture simplifies development of parallel programs by relieving a user from the tedious task of distributing data across processors. Furthermore, it allows incremental parallelization using, for example, OpenMP or Java threads. While it is easy to demonstrate good performance on a few processors, achieving good scalability still requires a good understanding of data flow in the application. In this paper we discuss ADAPT, an Automatic Data Alignment and Placement Tool, that detects data congestions in FORTRAN array oriented codes and suggests code transformations to resolve them. We then show how ADAPT suggested transformations, including data blocking, data placement, data transposition and page size control improve performance of the NAS Parallel Benchmarks.

1 Significance of Data Traffic Control

Few programming constructs allow explicit specification of data location in computer memory[1]. That leaves data placement decisions to the compiler and the operating system. The time for accessing data varies across different computer architectures, making code tuning machine dependent. Some memory problems such as cache trashing and false sharing are not difficult to identify and overcome with array dimension padding and variable privatization. Other data traffic problems such as data locality, excessive TLB misses and thread interference are difficult to diagnose and cure. Even the best implementations of Computational Fluid Dynamics (CFD) codes can achieve only about 20% of machine peak performance[14], spending 80% of the time waiting for data.

Two main factors contribute to the low efficiency of these codes. First, the distribution of Floating Point Instructions (FPI) does not present an optimal mix to keep all the functional units busy. There is no obvious cure for this. Second, many operations are stalled waiting for data, see Example 1 in Section 2. Solving this problem is the subject of the paper.

Approach to data traffic optimization. The first step in addressing this challenge is to identify code constructs causing data congestion, and the type of

---

[*] M/S T27A-2, NASA Ames Research Center, Moffett Field, CA 94035-1000; e-mail: {frumkin,hjin,yan}@nas.nasa.gov

[1] By explicit constructs, we mean statements such as the "register" qualifier in C or data distribution directives in High Performance Fortran.

congestion. We use four congestion metrics in this paper: Primary Data Cache (PDC) misses, Secondary Cache (SC) misses, Translation Lookaside Buffer (TLB) misses, and Cache Invalidations (CI). The second step involves choosing a proper remedy to resolve congestions and applying it to the code. This remedy can be either code transformation, or changing program environment such as page size or page placement/migration mechanisms.

Methods for controlling data traffic. These problems have been studied elsewhere [4, 10, 11] and techniques have been developed to improve data traffic. These techniques include data grouping for cache optimization, optimization of the page size for reduction of TLB misses, placement of pages to resolve contention in memory channel and data transposition for reduction of cache invalidations and TLB misses. Even with these techniques in hand, it is not easy for the user to identify the appropriate transformations to be applied to various part of the code. For example, it has been known that computations of the rhsz and zsolve operators in BT and SP of the NAS Parallel Benchmarks (OpenMP version) exhibit performance which is 3-4 times worse than of corresponding operators in the x-direction [7]. Neither, the reason for this, nor a mechanism for fixing the problem has been known in spite of NPB being de facto standards for reporting improvements in compilers and tools [12].

Many code transformations to improve data traffic are already implemented in compilers. These include loop interchange, loop fusion, software pipelining, prefetching, and modification of variable declarations (such as padding and privatization). Nevertheless, compilers cannot perform expensive analysis involving full interprocedural data dependency analysis. Many types of analysis are not possible at compile time at all.

Another approach to identifying data traffic problems relies on hardware counters that gather event statistics during program execution. Tools for collecting and analyzing the events, such as perfex and Performeter, have been built on top of these counters [6, 13]. These tools allow a user to instrument code and identify code constructs with performance anomalies. However, these tools are diagnostic in nature and leave analysis of the problems and identification of the remedies to the user.

In this paper we present ADAPT, a tool that combines both static code analysis and run-time testing. ADAPT analyzes data-to-data and data-to-computations affinity in the source code to report potential data traffic problems. The tool also inserts diagnostic statements in the code for those expressions that influence data traffic but cannot be evaluated at compile time. These statements are evaluated at run time and issue warnings about poorly performing code constructs. We demonstrate ADAPT's ability to identify problematic data traffic constructs and to suggest cures for these problems on three simulated CFD applications BT, SP and LU of the NAS Parallel Benchmarks (NPB). ADAPT was able to resolve data traffic problems with the rhsz and zsolve operators and helped to improve performance of the codes by 27% on average.

2 Automated Detection of Data Traffic Problems

To be able to control data traffic in an application, the user needs information on data movement across the memory hierarchy. While details of such movement can be complicated and machine dependent, they can be formulated in terms of cache parameters, array offsets, and data access strides, and can be characterized by simple metrics such as cache misses and invalidations. In a few cases, such as accessing shared data, data traffic depends on the cache coherency protocol and is sensitive to variations in execution order across different threads.

A tool that detects and corrects data traffic congestions greatly reduces the burden on a user to tune his code manually. Such a tool can reduce data traffic by increasing data reuse, avoid memory contention by optimizing initial data placement, and reduce the number of TLB misses and thread interference by optimizing page size. A typical scenario in which such a tool is applied is shown in the following example.

Example 1. The first two loop nests in `lhsz` of SP from NPB (serial version) are shown in Figure 1, left pane. The attempt to optimize the first nested loop actually slows down the computations since it increases the number of accessed memory pages and poorly utilizes the primary cache, see `lhsz` curve in Figure 2. Merging the first and the second nested loops and recalculating the expressions, see right pane in Figure 1, decreases the number of pages accessed, improves cache utilization, and reduces the total execution time, in spite of increasing the total number of FPI, see `lhsz_t` curve in Figure 2. We have implemented such a tool by adding features to ADAPT (Automatic Data Alignment and Placement Tool) [2]. Originally, ADAPT was designed for the automatic insertion of HPF directives into FORTRAN code[2]. The tool is able to identify data-to-data and to data-to-computations affinity and to express such affinities through HPF ALIGN and DISTRIBUTE directives. ADAPT's ability to extract these affinities is the key to enabling it with automatic data traffic diagnostic capabilities.

Data-to-data affinity. Two data items have an affinity if both are required by the same instruction (directly or indirectly) during program execution. For a pair of arrays used in the same loop nest statement, the affinity relation is a correspondence between array elements referenced by the same loop index. Grouping of affine data items and packing the groups into a continuous stream improves program performance by hiding the memory latency. In general, affinity is a many-to-many relation leaving many degrees of freedom to group affine data items. In [3], it is shown that the efficiency of grouping affine array elements depends on the geometry of the array interference lattice, defined as a set of solutions to the Cache Miss Equation [4].

The affinity relation can be deduced for each array pair in each nest statement. A control dependence results in affinity relations between the arrays involved in the control statement and all arrays in the basic blocks immediately dominated by the statement. The most common case observed in our CFD appli-

---

[2] ADAPT is built on top of CAPTools [9]. It uses a CAPTools generated data base and CAPTools code analysis and utilities.

lhsz

```
do j=1,ny
  do i=1,nx
    do k=1,nz
      cv(k)=ws(i,j,k)
      rhon(k)=SFunction(rho(i,j,k))
    end do
    do k=2,nz-1
      lhs(i,j,k,1)=0.0d0
      lhs(i,j,k,2)=-dttz2*cv(k-1)
                  -dttz1*rhon(k-1)
      lhs(i,j,k,3)= 1.0d0
                  +c2dttz1*rhon(k)
      lhs(i,j,k,4)= dttz2*cv(k+1)
                  -dttz1*rhon(k+1)
      lhs(i,j,k,5)=0.0d0
    end do
  end do
end do
```

lhsz_t

```
   do k=2,nz-1
do j=1,ny
  do i=1,nx
    lhs(i,j,k,1)=0.0d0
    lhs(i,j,k,2)=-dttz2*ws(i,j,k-1)
      -dttz1*SFunction(rho(i,j,k-1))
    lhs(i,j,k,3)=1.0d0
      +c2dttz1*SFunction(rho(i,j,k))
    lhs(i,j,k,4)=dttz2*ws(i,j,k+1)
      -dttz1*SFunction(rho(i,j,k+1))
    lhs(i,j,k,5)=0.0d0
  end do
end do
  end do
```

Fig. 1. *Data Traffic Optimization.* The original code (left), taken from `lhsz.f` of NPB2.3-serial, saves floating point instructions used in `SFunction`. However, such loop ordering creates a large number of TLB and PDC misses. By rearranging computations (right pane) these problems are resolved, improving execution time in spite of increase in the number of FPI. The profiles of both codes are shown in Figure 2.

cations is one-to-many affinity relations between arrays resulting from difference operators on structured discretization grids. These relations can be described by a stencil (i.e. by a set of vectors with constant elements) and we call them *stencil relations*. In order to deduce the affinity for arrays used in different statements of the same nest, ADAPT uses the chain rule, see [2]. The union of affinity relations over all directed paths leading from an array u to an array q forms the nest affinity relation between q and u. The relation lists all elements of u used for computation of each element of q and is a one-to-many mapping.

Data-to-computations affinity. We represent a program by a bipartite graph called the program affinity graph. Let C be the set of program statements, and let D be the program data, i.e. the set of (virtual) memory locations referenced in the program. We say a memory location d has an affinity with a statement c if the datum at address d is either an operand or a result of c. The program affinity graph has C and D as the vertices of the parts with appropriately directed arcs connecting statements with data affine to them. Many program properties can be expressed in terms of the affinity graph. For example, a statement $c2$ depends on a statement $c1$ if there is a directed path from $c1$ to $c2$. Otherwise, $c1$ and $c2$ are independent and can be executed in any order.

The analysis of the affinity graph can be simplified by indexing the statements inside the nests and the memory locations used by the arrays. In this

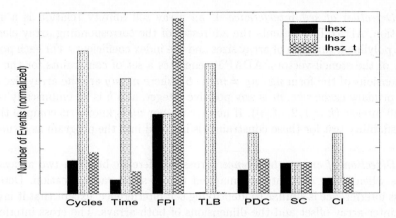

Fig. 2. The effect of TLB (Table Lookaside Buffer) and PDC (Primary Data Cache) usage optimization on the performance of `lhsz` nest. The performance of a similar operator in x-direction (`lhsx`) is given as a reference. The horizontal axis shows types of events measured with the use of hardware counters. The vertical axis shows a normalized number of measured events. As above, FPI stands for Floating Point Instructions, SC for Secondary Cache misses, and CI for secondary Cache Invalidations.

case, the arcs connecting data and statements can be expressed as $(I; \mathtt{idx}(I))$ where I is a loop nest index tuple, and $\mathtt{idx}(I)$ is a memory address of an array element referenced in iteration I. In our application domain (CFD applications on structured grids), the index function is usually a linear function of I with symbolic coefficients known at compile time. There are only a few nests in our applications where this is not a case. These include the core of the FFT algorithm, where $\mathtt{idx}(i, j, k) = i + 2 \cdot j \cdot k$; nests interpolating data between distinct grids, where $\mathtt{idx}$ function is read from a file; nests accessing specially enumerated grid points where the $\mathtt{idx}$ function is stored in a precomputed array. The tool identifies nests with nonlinear access functions, without any further analysis.

Some properties of the data traffic can be deduced using only symbolic analysis of the $\mathtt{idx}$ function (see the thread noninterference condition below). Others require knowledge of the actual numerical values of the coefficients. If the properties of the data traffic can be expressed by a symbolic expression but cannot be verified without knowing the numerical values of the coefficients, ADAPT inserts the expression in the code and the user receives a warning at run time. We call this a *run time test*.

Checking cache unfriendly access patterns. In general, cache friendly computations involve good temporal and spatial locality [5] and cannot be expressed in simple closed-form terms [4]. However, some necessary conditions for cache friendly computations can be formulated and checked. The first condition is simple: array elements accessed in the iterations of the innermost loop cover contiguous memory locations. Otherwise, nonunit stride in memory access would cause underutilization of data loaded into the cache.

Detection of self interference. If an array self affinity relation is a stencil relation, ADAPT represents the addresses of the corresponding array elements as a polylinear function of array sizes and the index coefficients. For each possible pair of the stencil vectors, ADAPT generates a set of constraints for the array dimensions of the form $nx \cdot ny \neq m/h \cdot S$, where nx, ny are the array sizes, S is the primary cache size, m is any positive integer, and h is an empirically chosen small integer ($h = 1, 2$, cf.[3]). If neither nx nor ny is known at compile time, a satisfiability test for these constraints is inserted into the program as a run time test.

Detection of cross interference. Cross interference between two arrays happens when affine elements are mapped to the same cache location. Detecting cross interference is similar to identifying self interference, except that it involves the inter-array offset and the dimensions of both arrays. The cross interference constraints are represented by a set of polylinear inequalities: $\mathtt{idx}_a(i + e_1, j + e_2, k + e_3) + \mathit{off}_{ab} - \mathtt{idx}_b(i + d_1, j + d_2, k + d_3) \neq m/h \cdot S$, where e_i, d_i, $i = 1, 2, 3$ are components of the stencil vectors. Evaluation of these inequalities requires knowledge of off_{ab} and can be done, for example, if both arrays are in the same common block or are segments of the same (bigger) array.

Detection of high TLB misses. TLB misses (as in Example 1) usually result from large memory strides due to iterations of an inner loop of a nest. For each array in the nest our TLB miss test checks a sufficient conditions for TLB misses in the inner loop: the number of iterations of the innermost nest exceeds TLB_SIZE, and the distance between the first and last address accessed in the inner loop exceeds PAGE_SIZE*TLB_SIZE. If both conditions are met, ADAPT issues a warning about high TLB misses in the inner loop of the nest. If both conditions can not be proved to be false, the tool inserts a run time test.

Checking thread noninterference. This condition can be formulated as absence of overlap of the address spaces accessed by different threads. If the noninterference condition is satisfied, then the memory accessed by a thread can be placed in the memory of the processor running the thread, improving data locality. This condition is checked only for "read/write" arrays since thread interference would cause cache lines invalidations. In the case of "read" arrays, this condition is not checked since read arrays are copied into secondary cache and threads do not affect each other, provided that initial placement of the array has been done correctly.

Interference of threads depends on the data sharing protocol implemented in the DSM computer. For example, if data coherence is supported at the level of secondary cache lines, as it is on the Origin 2000, then read/write interference can happen if two threads are accessing two different words on the same cache line. For a software performance tool, it is possible to be aware of the data sharing protocol and adjust CI estimates accordingly. We have implemented a more general approach: for each nest and each thread, ADAPT evaluates the interference as a ratio of the number of memory locations adjacent to the memory locations accessed by other threads to the total number of memory locations accessed by the thread. For example, in Figure 1 the *interference ratio* for lhs is

$P/(nx \cdot ny)$ in `lhsz` and $P/(nx \cdot ny \cdot nz)$ in `lhsz_t`, which correlates well with the CI numbers in Figure 2. This interference indicator is similar to the surface-to-volume ratio used to estimate cache utilization [3], and to the communications-to-computations ratio used to characterize MPI programs.

Detection of the data sources and the initial data placement. Page placement on a DSM computer is commonly controlled by a simple policy such as "First Touch" or "Round Robin". More sophisticated page placements can be implemented with special tools, such as `dplace`, see [13]. An inappropriate initial data placement – for example, concentrating data on a single processor – can cause memory contention during execution and may hamper application scalability. Therefore, we enabled ADAPT with a capability to detect data initialization constructs in the code and to insert an equitable data placement directive before each construct.

3 Experiments

We conducted experiments on 16 processors of a 512 processor single image SGI Origin 2000 (MIPS R12000 CPU). We submitted jobs through the Portable Batch System (PBS), which dedicates requested resources to the job and minimizes interference with other jobs running on the machine. We used 16 processors, since this is the minimal number of processors where the slowdown due to memory traffic effects was well pronounced for the 64^3-point grids of NPB class A. The effects are similar up to 32 processors, after which the load imbalance becomes dominant.

The primary memory hierarchy on the Origin 2000 involves registers, primary data and instruction caches, secondary unified data and instruction cache, and the main memory. The access time to data located at different levels of memory is shown in Table 1, cf. [13]. The primary cache is directly mapped 2-way set

Table 1. Access Time (in machine cycles) to Data in Origin 2000 Memory.

Data Location	Access Time	Condition
registers	0	
L1 cache	2-3	L1 hit
L1 cache	8-10	L1 miss, L2 hit
L2 cache	75-250	L2 miss, TLB hit
L2 cache	~2000	L2 miss, TLB miss

associative, having 512 lines of 32 bytes each in each set. The secondary cache is shared by data and instructions and it is also directly mapped 2-way set associative containing 32K 128 byte lines. The main memory is split in modules of 768 MB per node (2 processors per node), totaling 196 GB of memory on a 512 processor machine. The TLB has 64 entries containing base addresses of 64 pairs of pages.

A cache coherency protocol guarantees that data accessed by different processors do not go stale. This protocol invalidates a line in secondary cache every time a processor requests exclusive ownership (usually for writing) of data mapped to the line. In this case, all copies cached in all other processors are invalidated, and each processor working with this line has to request a fresh copy of the line to resume computations. An implication of this protocol is that it will cause significant slowdown when two processors attempt to write data located within the same 128 byte segment of main memory.

As the test codes, we chose the OpenMP version of the PBN3.0-b2, a release of the NAS Parallel Benchmarks which includes optimized serial, OpenMP, HPF and Java versions. The suite is designed for demonstrating capabilities of compilers and tools applied to CFD codes [7].

We measured execution time at various levels of optimization. Since we used the -O3 flag in the compilation, special effort was required in many cases to prevent the compiler from undoing our optimizations. We specified compiler flags for suppressing prefetching: --LNO:prefetch=0 to obtain accurate numbers for the hardware counters, and flag OPTFLAGS=-OPT:reorg_common=OFF to enforce our own padding of arrays declared in the common blocks.

4 Experimental Results and Discussion

We applied the tool to the SP, BT and LU codes of the NPB [7] (optimized OpenMP version PBN-O). For each code ADAPT was able to generate the following data traffic optimization diagnostics.

- *Nests for initial data placements.* ADAPT detected all nests where data were initialized. In all cases, the arrays were initialized from array of smaller dimensions (or constants). The initial data placement was appropriately implemented in the original code and we essentially did not make any changes.
- *Nests with nonunit strides and advice on loop interchange.* Such nests were detected only in calculation of the so-called right hand side array by the subroutines rhs, exact_rhs, and erhs.
- *Nests with big strides and advise on loop interchange and data transpositions.* Nests with big strides were detected in the subroutines rhs, exact_rhs, erhs and in zsolve. Advice on changing *jik* into *kji* loops was issued for the first three subroutines. In zsolve (BT and SP only), a dependency in the *k* index prevented the parallelization of the *k* loop, and no loop interchange was carried out.
- *Nests with self or cross interference, and advice on padding.* No arrays with self interference were detected indicating that the existing paddings of the first and second dimensions were sufficient. The cross interference condition was presented in the form that array offsets cannot be equal to a multiple of the cache size plus a stencil vector offset.

Following ADAPT's advice, we implemented a number of changes in the original OpenMP code by hand. Almost all changes occurred in rhs, zsolve, buts, and blts. We classify these changes into 3 categories, as shown in Table

2: removing auxiliary arrays and applying nest fusion in rhs, loop interchange in rhs, removing auxiliary arrays in solvers: zsolve in BT and SP and buts, blts in LU. Both exact_rhs and erhs were outside the main iteration loop, so we did not make any changes to these subroutines. Incremental improvement in performance via data traffic optimization for each benchmark is tabulated in Table 2. The improvement was about 200% for rhs and 20% for zsolve. Average overall speedup was about 27% on 16 processors. Some optimizations that gave

Table 2. *Improving Benchmark Performance via Data Traffic Optimization.* Time (in seconds) was measured on 16 processors of a 400MHz Origin 2000 machine, SGI OpenMP compiler.

Appli-cation	Original Code	Data reuse and nest fuse	Loop interchange	Removing solver aux. arrays	Total speedup
BT.A	54.00	51.44	44.22	42.12	22%
SP.A	63.40	55.89	38.70	37.92	40%
LU.A	59.35	52.19	48.95	48.78	18%

performance improvements in simple test codes did not result in expected improvements in the benchmarks. Optimizing page size by providing each processor with one page from each array did not affect performance of the jobs running under PBS. It improved performance of jobs running outside PBS by 10%. We could not find a remedy for the thread interference warning (high CI number) in zsolve, because the nest has a dependency in the z-direction, preventing loop interchange.

5 Conclusions and Related Work

In this paper we presented ADAPT, a tool to identify data traffic problems in array-oriented Fortran codes. The tool advises the user on excessive cache and TLB misses, and possible thread interference, and suggests code transformations for improving data traffic. Some of the transformations are counterintuitive, since they reduce data traffic and overall computational time by increasing the number floating point instructions.

We demonstrated data traffic improvements using three simulated CFD applications BT, SP and LU from the NAS Parallel Benchmark suite. For some subroutines performing operations in the z-direction, code transformations improved performance by a factor of 3 and improved scalability of these subroutines. Overall data traffic optimizations improved the benchmark performance on 16 processors by 27% on average.

Research is being carried out in three main directions to enhance ADAPT's ability to control data traffic: reducing communications in MPI programs [9], optimizing data distributions in HPF programs [2], and improving spatial and temporal data locality for optimizing cache performance [4]. With the proliferation of the DSM architecture, data traffic control on DSM is becoming increasingly important. Some problems of data distributions and page migrations on DSM are the subject of recent papers [10, 11]. We have plans to implement some global data traffic control features in ADAPT and to test it on a wider class of CFD applications and on variety of different implementations of DSM. Finally, we plan to incorporate ADAPT with CAPO, a parallelization tool that automatically inserts OpenMP directives into FORTRAN codes [8].

Acknowledgements. This work was supported by the NASA High Performance Computing and Communications (HPCC) Program, RTOP #725-10-31. The authors appreciate Rob F. Van Der Wijngaart's effort in reviewing this paper.

References

1. D. Bailey, J. Barton, T. Lasinski, and H. Simon (Eds.). *The NAS Parallel Benchmarks*. NAS Technical Report RNR-91-002, www.nas.nasa.gov.
2. M. Frumkin, J. Yan. *Automatic Data Distribution for CFD Applications on Structured Grids*. The 3rd Annual HPF User Group Meeting, Redondo Beach, CA, August 1-2, 1999, 5 pp. Full version: NAS Technical report NAS-99-012, December 99. 27 pp.
3. M. Frumkin, R.F. Van der Wijngaart. *Efficient Cache Use for Stencil Operations on Structured Discretization Grids*. Submitted to JACM, see also NAS Technical Report NAS-00-15, www.nas.nasa.gov.
4. S. Gosh, M. Martonosi, S. Malik. *Cache Miss Equations: An Analytical Representation of Cache Misses*. ACM ICS 1997, pp. 317–324.
5. J.L. Hennessy, D.A. Patterson. *Computer Organization and Design*. Morgan Kaufmann Publishers, San Mateo, CA, 1994.
6. Innovative Computing Lab., UTK. *PAPI: Hardware Performance Counter Application Programming Interface*. http://icl.cs.utk.edu/papi.
7. H. Jin, M. Frumkin, J. Yan. *The OpenMP Implementation of NAS Parallel Benchmarks and Its Performance*, NAS Technical Report, NAS-99-011, 1999, 26 pp, www.nas.nasa.gov.
8. H. Jin, M. Frumkin, J. Yan. *Automatic Generation of OpenMP Directives and Its Application to Computational Fluid Dynamics Codes*, Springer LNCS, v. 1940, p. 440–456.
9. S.P. Johnson, C.S. Ierotheou, M. Cross. *Automatic Parallel Code Generation on Distributed Memory Systems*. Parallel Computing, V. 22 (1996), pp. 227-258.
10. D. S. Nikolopoulos, T. S. Papatheodorou, C. D. Polychronopoulos, J. Labarta, E. Ayguade. *Is Data Distribution Necessary in OpenMP?* Proceedings of Supercomputing 2000. Dallas, TX, Nov. 4-10, 2000. 14 pp.
11. D. S. Nikolopoulos, T. S. Papatheodorou, C. D. Polychronopoulos, J. Labarta, E. Ayguade. *Leveraging Transparent Data Distribution in OpenMP via User-Level Dynamic Page Migration*. Springer LNCS, v. 1940, p. 415–427.
12. *Proceedings of Supercomputing 2000*. Dallas, TX, Nov. 4-10, 2000.
13. SGI Inc. Technical Document. *Origin2000 and Onyx2 Performance Tuning and Optimization Guide*. http://techpubs.sgi.com.
14. J. Taft. *Performance of the Overflow-MLP CFD Code on the NASA Ames 512 CPU Origin System*. NAS Technical Report, NAS-00-05. www.nas.nasa.gov.

The Monitoring and Steering Environment

Christian Glasner, Roland Hügl, Bernhard Reitinger,
Dieter Kranzlmüller, and Jens Volkert

GUP Linz, Johannes Kepler University Linz,
Altenbergerstr. 69, A-4040 Linz, Austria/Europe,
kranzlmueller@gup.uni-linz.ac.at,
http://www.gup.uni-linz.ac.at/

Abstract. Monitoring and steering promises interactive computing and visualization for large-scale scientific simulations. This is achieved by retrieving data about a program's execution during runtime, analyzing these data with sophisticated graphical representations, and applying changes to parameters of the simulation in progress. Based on these requirements the MoSt environment offers some novel ideas: Firstly, instrumentation of the code is performed on-the-fly during execution without a priori preparation of the target application's code. Secondly, all functionality is provided in modules which communicate with each other via a specialized network protocol, allowing the user to adapt MoSt to specific situations. Thirdly, visual representations can be generated by arbitrary visualization packages, which integrate an interface to the monitoring environment. The latter may also apply Virtual Reality techniques in order to enhance the user's understanding. This paper offers an overview of MoSt, the strategy behind it, and its main modules.

1 Introduction

Computational science and engineering requires not only huge amounts of processing power from high-performance computing (HPC) architectures, but also dedicated concepts, methods, and solutions incorporated in sophisticated programming and program analysis tools. Such a class of tools are computational steering environments, which offer functionality for online modifications of executing scientific and engineering simulations. In principle, there are two questions being addressed by the tools in this area:

- How to analyze a program's state at some arbitrary point during execution?
- How to modify a program (or its parameters, respectively) during execution?

Corresponding to these two questions, two distinct activities can be defined: monitoring and steering. Monitoring identifies the process of observing a program's execution during runtime, which helps the user to learn about a program's behavior. Based on the observed data, steering defines the activities applied to modify a program's execution by manipulating key characteristics of its algorithms.

Several software tools based on these two activities have been developed [15]. Some well-known examples are the on-line monitoring and steering environment FALCON [5], the Computational Steering Environment CSE [6], the Visualization and Steering Environment VASE [1], the runtime interaction framework DAQV [8] and its successor DAQV-II [9], the visualization and steering library CUMULVS [7], and the scientific programming environment SCIRUN [16].

The work described in this paper is based on some ideas derived from analyzing the tools mentioned above. Especially three aspects have been investigated:

- Attaching the monitoring part to the target application during runtime without prior instrumentation
- Using Virtual Reality technology to improve the analysis tasks, especially the problem of finding a suitable starting point for in-depth investigations
- Re-using existing visualization packages for in-depth investigations by integrating them via dedicated interfaces and efficient networking protocols.

These aspects have been implemented in the Monitoring and Steering environment MOST. Originally intended only as a test-bed for the ideas described above, it has lately been extended to a complete environment, that it is available for usage in real-world computational science applications.

This paper is organized as follows: The next section briefly describes the main strategy behind MOST, as well as the main modules currently available. The instrumentation and monitoring module is described in Section 3. The observed data is visualized either in Virtual Reality (Section 4) or with conventional visualization tools (Section 5). Afterwards, the experiences with our MOST prototype are summarized together with an outlook on future ideas in this project.

2 Computational Steering Strategy

The strategy incorporated in the MOST environment can be described with a simple example. We assume, that an arbitrary scientific application is already executed on some kind of HPC system. A some time during its execution, the user decides to investigate the two questions given in Section 1, which initiates the following steps:

1. The interface of MOST is initialized on a graphical workstation.
2. With MOST, the user logs into the HPC system and attaches the monitoring part to the target application.
3. After attaching, the user is effectively in control over the target application and starts downloading program state data.
4. The observed data is presented to the user, which in turn tries to identify critical points for in-depth investigation.
5. At these critical points, in-depth analysis is carried out, revealing more and more information about the program's execution and increasing the user's knowledge.
6. Eventually, the user identifies some key parameters, and changes their values.

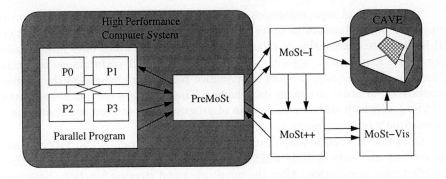

Fig. 1. Structure of the MoSt Environment

7. After the changes have been committed, MOST is detached from the target application and awaits future initiation.

This example execution of MOST demonstrates the different steps conducted during monitoring and steering. In practice, these steps are distributed among several independent modules, which communicate on a client-server basis via a dedicated networking protocol. The structure and interconnection of the MOST modules is shown in Figure 1. The following components can be identified:

- MOST++: user interface and control center
- PREMOST: instrumentation and monitoring module
- MOST-I: visualization of activity data in VR
- MOST-VIS: (conventional) data visualization

The main module of MOST is MOST++, which represents the control center and graphical user interface for all activities. All other modules are initiated via commands in MOST++, and status information as well as small amounts of analysis data (such as single scalar values) are displayed within it.

The PREMOST module performs the instrumentation and monitoring parts of MOST. Upon initiation of MOST++, it logs into the HPC system and retrieves information about the running processes. After selecting a set of processes, PREMOST attaches itself to the running code.

The next module initiated by MOST++ is MOST-I, which receives activity data about the target application. This data is visualized by MOST-I within a Virtual Reality (VR) environment (e.g. a CAVE). It is used to identify critical areas or hotspots in the program, which are qualified as starting points for in-depth investigation activities.

After critical areas have been detected with MOST-I, the user initiates the MOST-VIS module from the MOST++ interface. In contrast to MOST-I, which focuses on program activity data, MOST-VIS displays application dependant information, for example data structures processed by the scientific algorithms.

The graphical representations offered by MoST-Vis are used to improve program comprehension in order to understand an application's computation. It is therefore comparable to related work in this area, which offer functionality for model exploration, algorithm experimentation, and performance optimization [15]. Additionally, the user may induce parameter changes to the running application in order to change its behavior. These changes are conducted within MoST++, which allows modification of program data or even code patching of a program's original functions.

Of course, the actual modifications are again carried out by PreMoSt, which is actually the only module which needs to be executed on the same machine as the target application. After the monitoring and steering functions have been completed, PreMoSt can be completely removed from the HPC system, leaving the original application with only the applied changes.

The next sections offer some more details about the ideas included in MoST, with special focus on the main differences between MoST and other related monitoring and steering environments.

3   On-line Instrumentation with PreMoSt

One of the main goals of PreMoSt's design was to cope with the perturbation introduced when monitoring an arbitrary program. This is a general problem of on-line monitoring systems, because a program being observed is different from its counterpart without observation, and in some cases the observation target may even reveal completely different execution behavior [2]. Consequently, developers of monitoring tools must seek to generate as little monitor overhead as possible.

Most current approaches in this area apply dynamic monitoring [5], where the amount of overhead can be adjusted at runtime, such that less overhead is generated when less data is needed. Nevertheless, there is always a certain amount of overhead, no matter how few data is obtained. Consequently, the generated data has to be analyzed with care.

As an improvement, the approach incorporated in PreMoSt does not generate any monitor overhead if no analysis data are needed. This is especially useful for long-running scientific applications, where users are not constantly monitoring the program's execution. In concrete, PreMoSt is only attached to the target application if observation is actually requested.

The on-the-fly technique included in PreMoSt is based on the dynamic program instrumentation approach described in [11], which is available in the DyninstAPI [10]. Besides its main usage for performance tuning [13], DyninstAPI has also been applied to several other areas like debugging and program optimization.

DyninstAPI offers functionality to instrument an application that is already being executed on a computer system by inserting monitoring code on-the-fly. In addition, DyninstAPI can be used to establish connections between the program in memory and its source code by accessing the symbol table information stored in the program's object code.

Although DyninstAPI has proved very useful for our efforts, it still lacks some needed functionality. For example, access to raw memory locations are performed in DyninstAPI only via the symbol table information, which represents a certain restriction, especially for the example described in the next section. Therefore, some additional functionality had to be developed to fulfil the requirements of PREMOST.

4   "Feel the Program" with Most-I

After PREMOST has been attached to the program, a user needs to find out, what a target application is currently doing, or what parts of the program's data are actually being processed. The idea incorporated in MOST is to collect program activity data based on the number of accesses to a particular memory region. In concrete, the memory blocks containing the code- and data-segments of an arbitrary program are divided into areas with high activity and areas with low activity. Then, the normalized activity of an arbitrary area A consisting of n memory cells is computed as

$$activity(A) = \frac{\sum\limits_{i=1}^{n} \frac{\int\limits_{t_{min}}^{t_{max}} a_i(t)dt}{t_{max}-t_{min}}}{n}$$

In this formula, the activity $a(t)$ of a memory block at a certain time t is 0, if the memory block is not accessed, or 1, if it has been accessed recently. Furthermore, the activity over a given interval of time $[t_{min}, t_{max}]$ will yield higher values for memory location with more accesses.

Based on this formula, the activity of arbitrary areas or memory blocks, of distinct processes, or even a complete program can be determined. A problem is that each memory cell has to be obtained iteratively by the monitor in order to compute accurate activity values. However, as a first approximation memory blocks of coarser granularity can be selected for $a(t)$. Over time, the granularity of the cells can be refined for those blocks, that yield higher activities values in the first place. Please note, that the refinements are applied during the target program's execution, and the program's behavior is subject to change over time, Consequently, it is necessary to verify the focus of the refinement periodically. The hierarchical refinement process is described in more detail in [12].

The activity data obtained by PREMOST is then visualized with MOST-I. The main problem is to present the activity data of an executing program in a meaningful way to the user, so that the user understands what is happening in the program. Therefore, it was of major importance to investigate visualization techniques, which should guide the user through the activity data, where to guide means to support the user in discovering things, that were unknown before [14].

With the visualization, we want to improve the process of forming a mental image of the program's execution, which is utilized to aid the reasoning and understanding of program behavior [17].

This goal was addressed by Virtual Reality (VR) technology incorporated in MoSt-I, where the "I" stands for *Immersion*. Currently, the chosen environment is the CAVE (CAVE Automatic Virtual Environment), a multi-person, room-sized, high-resolution, 3D video and audio environment developed in 1992 at EVL (Electronic Visualization Laboratory, University of Illinois at Chicago). The task of MoSt-I is to generate the immersive model of the program's behavior in the CAVE. Then the user may interactively explore this virtual world and gain more and more knowledge about the program's characteristics.

The idea of using VR in MoSt-I is, that it allows to cause a "feeling" in the user about the program's behavior. By exploring the activity data presented in the CAVE, the user can intuitively investigate even huge amounts of data. This can mainly be attributed to the fact, that VR offers additional dimensions compared to traditional 2-dimensional displays. Some important characteristics in this context are: Firstly, the number of available dimensions is increased (e.g. 3D, immersion, sound). Secondly, important issues can be emphasized by combining attributes (e.g. color+sound). Thirdly, animation can be used to describe the variation of data over time.

A major difficulty for visualization is always to choose an appropriate visual representation. For MoSt-I several approaches have been investigated. The most suitable of these approaches seems to be the activity tunnel, which arranges the memory of parallel processes as stripes in a 3-dimensional tunnel [12]. Within this tunnel, the activity is visualized with attributes like color, shades, object size, and animation.

An example of the activity tunnel is presented in Figure 2. It shows the memory activity of 8 application processes, with different shades indicating different levels of activity. This picture has been taken within the CAVE environment and also contains the user who controls the application. In order to intensify the effect of VR onto the user, all interaction with the system is performed via voice input, which is shown in Figure 2 with the mounted headset.

In addition to the graphical representation, mappings to other sensory input are currently investigated. One example is sound, which was originally inspired by the following story about the Whirlwind computer (1950) [4]: *You even had audio output in the sense that you could hear the program because there was an audio amplifier on one of the bits of one of registers - so each program had a signature. You could hear the tempo of how your program was running. You could sense when it was running well, or when it was doing something surprising.*

Similar ideas are discussed in [3]. By integrating sound output in MoSt-I, we try to emphasize the activity of a program in the region of the tunnel, that is currently inspected by the user. Other ideas include the sonification of hardware performance counters, the program counter, and similar state data.

Fig. 2. Picture of user (with microphone) and the activity tunnel in the CAVE.

5 Data Visualization with MoSt-Vis

As described above, the visualization MoST-I is mainly used as a starting point for in-depth investigations. Therefore, after the user has decided where to perform program and data analysis, symbol table information is presented by MoST++. Based on this information, interesting data structures of the running program can be selected. The selected data is then extracted by PreMoST. Depending on the amount of data, the results of PreMoST are either printed in a window of MoST++, or forwarded to the visualization module MoST-Vis.

The graphical representations provided by MoST-Vis have to be chosen corresponding the the scientific applications data. Since most of these data is already visualized during post-mortem analysis, it seems beneficial to re-use existing visualization packages. This is achieved within MoST by developing corresponding input interfaces, such that PreMoST can forward its data to an arbitrary visualization tool. At present, input interfaces for IBM's Open Data Explorer (OpenDX) are available, while similar functions are currently being implemented for AVS/Express.

Some examples of data visualization are shown in Figure 3 to Figure 5. All screenshots have been produced with MoST while the target application was actually running. Figure 3 shows the scalar value of a chosen variable over time, with each curve corresponding to one process of the application. Figure 4 shows a two-dimensional heat-diagram of an array, that is currently processed by the target application. Figure 5 shows a 3-dimensional visualization of a crank shaft grid, that is currently used for computing the stress around this object.

The visualizations are constructed within MoST as follows: MoST++ reads the symbol table information and presents the set of available data structures

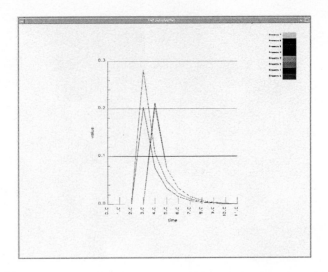

Fig. 3. Example visualization of scalar values on 8 processes over time

to the user. Upon choosing some of the data structures, PREMOST begins to download the corresponding data from the running program. In addition, the user selects a suitable graphical representation within MOST++ and specifies the mapping of the data structures onto the dimensions of the graphical output. These data and the mapping specification are forwarded to MOST-VIS which finally generates the output described above.

6 Conclusions

The MOST environment described in this paper offers some novel ideas for analyzing and controlling applications from computational science and engineering. Although originally intended only to verify some of our ideas, the first results with the tool where rather promising. This convinced us to develop the system into an environment, that can be offered interested scientists.

Most useful is the dynamic instrumentation module, which offers to inject commands on-the-fly. Therefore the application needs not be stopped and restarted, which allows to decrease the overall computation time, a factor that may be crucial for users of expensive HPC systems.

On contrary, the VR visualization of activity data may seem a little bit controversial in the first place. However, the problem of leading the user to an suitable starting place for in-depth investigations is very important, and there seems to be only limited support from related work in this area. Nevertheless, it is certainly necessary to assess the usefulness of the current approach for real-world applications, and to continue studying this important aspect.

Finally, the idea of re-using existing visualization tools by developing corresponding interfaces allowed to shift lots of concentration to the other aspects

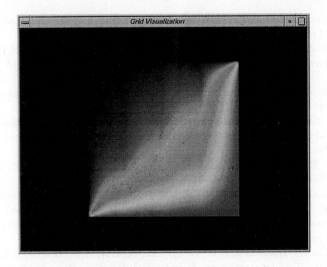

Fig. 4. Example heat diagram of 2-dimensional data array

Fig. 5. Example 3-dimensional data visualization of a crank shaft

of this project. The problem in this area is mainly related to the efficiency of the networking protocol and the underlying network infrastructure. Especially for tools with huge amounts of data, dedicated methods for filtering or preprocessing need to be developed.

References

1. Brunner, J.D., Jablonowski, D.J., Bliss, B., Haber, R.B., "VASE: The Visualization and Application Steering Environment", *Proc. Supercomputing '93*, Portland, OR, pp. 560-569 (November 1993).
2. Eisenhauer, G., Gu, W., Kraemer, E., Schwan, E., Stasko, J., "Online Displays of Parallel Programs: Problems and Solutions", *Proc. PDPTA '97*, Intl. Conference on

Parallel and Distributed Processing Techniques and Applications, Las Vegas, NV, pp.11-20 (July 1997).

3. Francioni, J.M., Jackson, J.A., "Breaking the Silence: Auralization of Parallel Program Behavior", *Journal of Parallel and Distributed Computing*, Vol. 18, pp. 181-194 (1993).

4. Frenkel, K.A., "An Interview with Fernando Jose Corbató", *Communications of the ACM*, Vol. 34, No. 9, pp. 83-90 (September 1991).

5. Gu, W., Eisenhauer, G., Schwan, K., Vetter, J., "Falcon: On-line Monitoring and Steering of Large-Scale Parallel Programs", Concurrency: Practice and Experience, Vol. 10, No. 9, pp. 699-736 (August 1998).

6. van Liere, R., Mulder, J.D., van Wijk, J.J., "Computational Steering", *Future Generation Computer Systems*, Vol. 12, No. 5, pp. 441-450 (April 1997).

7. Geist, G.A., Kohl, J.A., Papadopoulos, P.M., "CUMULVS: Providing Fault-Tolerance, Visualization, and Steering of Parallel Applications", *Intl. Journal of Supercomputer Applications and High Performance Computing*, Vol. 11, No. 3, pp. 224-236 (Fall 1997).

8. Harrop, Ch.W., Hackstadt, S.T., Cuny, J.E., Malony, A.D., Magde, L.S., "Supporting Runtime Tool Interaction for Parallel Simulations", *Proc. Supercomputing '98*, Orlando, FL (November 1998).

9. Hackstadt, S.T., Harrop, Ch.W., Malony, A.D., "A Framework for Interacting with Distributed Programs and Data", *Proc. HPDC-7, 7th IEEE Intl. Symposium on High Performance Distributed Computing*, Chicago, IL, pp. 206-214 (July 1998).

10. Hollingsworth, J.K., Buck, B., "DyninstAPI Programmer's Guide - Release 2.0", Computer Science Department, URL: http://www.cs.umd.edu/projects/dyninstAPI (2000).

11. Hollingsworth, J.K., Miller, B.P., Cargille, J., "Dynamic Program Instrumentation for Scalable Performance Tools", *Proc. 1994 Scalable High Performance Comp. Conf.*, Knoxville, TN (1994).

12. Kranzlmüller, D., Reitinger, B., Volkert, J., "Experiencing a Program's Execution in the CAVE", Proc. PDCS 2000, Conference on Parallel and Distributed Computing Systems, Las Vegas, NV, USA, pp. 259-265 (November 2000).

13. Miller, B.P., Callaghan, M.D., Cargille, J.M., Hollingsworth, J.K., Irvin, R.B., Karavanic, K.L., Kunchithapadam, K., Newhall, T., "The Paradyn Parallel Performance Measurement Tool", *IEEE Computer*, Vol. 28, No. 11, pp. 37-46 (November 1995).

14. Miller, B.P., "What to Draw? When to Draw? - An Essay on Parallel Program Visualization", *Journal of Parallel and Distributed Computing*, Vol. 18, No. 2, pp. 265-269 (June 1993).

15. Mulder, J.D., van Liere, R., van Wijk, J.J., "A Survey of Computational Steering Environments", *Future Generation Computer Systems*, Vol. 15, No. 1, pp. 119-129 (February 1999).

16. Parker, S.G., Johnson, C.R., "SCIRun: A Scientific Programming Environment for Computational Steering", *Proc. Supercomputing '95*, San Diego, CA, (December 1995).

17. Zhang, K., Ma, X., Hintz, T., "The Role of Graphics in Parallel Program Development", *Journal of Visual Languages and Computing*, Academic Press, Vol.10, No.3, pp. 215-243 (June 1999).

Token Finding Using Mobile Agents

Delbert Hart[1], Mihail E. Tudoreanu[2], and Eileen Kraemer[3]

[1] University of Alabama in Huntsville
Huntsville, AL 35899 USA
dhart@cs.uah.edu

[2] Washington University in St. Louis
St. Louis, MO, 63130 USA
renu@cs.wustl.edu

[3] University of Georgia Athens, GA, 30606 USA
eileen@cs.uga.edu

Abstract. One of the greatest challenges facing the software community today is the increasing complexity of software. Complexity limits understanding, making it difficult to evaluate the correctness, reliability, and performance of a system. Coupled with visualization, monitoring can provide users with insight into an application's behavior. Monitoring can also be used in conjunction with automated tools to adaptively tune performance. This paper presents a detailed look at how mobile agents that are embedded in a monitoring system can be used to find a token within a distributed system. Several strategies by which agents may accomplish this task are qualitatively compared. We then describe tests that were performed to evaluate trade-offs among the strategies considered, and discuss the results of those tests.

1 Introduction

Software systems are some of the most complex constructs created by man. This degree of complexity makes it difficult to evaluate their correctness and performance a priori. Monitoring provides a practical way of learning about distributed computations. To be effective, online monitoring must adapt to the changing needs of the user while minimizing the effect monitoring has on the application. In the case of distributed applications, lag, non-determinism and the lack of a global clock further complicate the monitoring task.

This paper explores the use of mobile agents to address the challenges of monitoring distributed computations. The use of mobile agents has a number of benefits: 1) Responsiveness - agents are able to react locally to conditions at the application processes. 2) Transience - the ability to deploy agents on demand helps to minimize the overall cost of monitoring. 3) Customization - agents may be encoded at run-time; and thus can make use of application-specific information, permitting efficient solutions. 4) Mobility - the ability to migrate between processes makes agents well suited to distributed applications in which properties are not necessarily bound to one process.

The idea of using mobile code for monitoring tasks has been proposed by many. Yet with the exception of agents within PathFinder [5], the only other general purpose monitoring system we are aware of that intends to support mobile code is the BRISK system[1], which has plans to employ a Scheme-based language for agents. Hence, actual experience with mobile code in general purpose monitoring systems is very limited. Other uses of agents for monitoring, such as in JAT[4] and WHERE[3], do not support the migration of agents between processes of the computation. Thus, these agents lack the ability react locally and to track non-local properties as easily.

General purpose mobile agent systems, such as AgentTcl[7], are not well-suited for the tasks of monitoring because they provide high level services such as name services, authentication, and/or network references that are not necessary for monitoring and incur extra perturbation. General purpose systems are typically designed to serve as stand-alone systems instead of being embedded within another application or library. For these reasons, we have designed and implemented a simple mobile code model, specialized for the purposes of monitoring and steering activities of distributed computations.

This model is used as the basis for implementing specific mobile agent modules that are part of a larger monitoring and steering framework, the PathFinder [5] exploratory visualization system. Within the wide range of environments and tasks in a monitoring system, agents may employ several strategies to perform the monitoring task. To illustrate the richness of agent solutions this paper presents an example monitoring task, finding a token in a distributed system. We provide a novel analysis of how mobile code can be used to address this problem. We then perform experiments designed to highlight the trade-offs between two specific strategies that represent the extremes of mobility versus non-mobility, providing insight as to when mobile agents are appropriate. Section 2 gives a brief description of the PathFinder exploratory visualization system and how mobile agents are supported. The token finding example and experiments are given in Sect. 3. The results of the experiments and discussion are provided in Sect. 4.

2 PATHFINDER

PathFinder manages the size and complexity of distributed systems by engaging the user as an active partner who guides both the collection of data from the application and the visual representation of the program's state and behavior. The key insight is that it is both unnecessary and inefficient to collect all possible data that an application can provide. Instead, one should collect only the data that supports the user's current interests. This selective monitoring, in conjunction with navigation tools to modify the viewer's perspective on the computation, provides a dynamic and interactive paradigm for monitoring distributed systems. The use of mobile code to realize these interactions is a natural extension of this exploratory visualization approach.

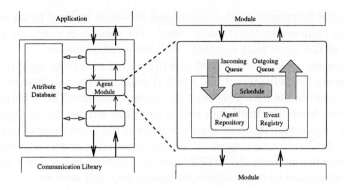

Fig. 1. An agent module can be one of many modules installed to provide monitoring and steering related services. The agent module provides the bridge between the application and the milieu.

2.1 Agent Model

The agent model has two kinds of entities: *agents* and *milieux* (agent servers). The distinguishing feature of this model is that it is specialized to be embedded within a monitoring library. This constrains the milieux to potentially sharing the thread of control with the application and utilizing an abstract asynchronous form of communication with other milieux. The former constraint arises from the languages being monitored. The latter is due to the fact that PathFinder is designed to be applicable to computations using any form of message passing. Thus, no assumptions about the interprocess communication is made other than its existence. Agents access information about the application via specialized agents called *avatar* agents and *event* agents, providing synchronous and asynchronous access, respectively. The milieu provides services that allow an agent to execute within it, to interact with other agents, and to move to other milieux. A milieu does not support any form of inter-milieux communication other than agent migration.

Each agent is an encoding of data and code. The code contained within an agent is stored as a set of handlers. These methods on the agent define how it reacts to events in the milieu. A distinguished handler of the agent is the *Arrival* handler which is automatically executed by the milieu when the agent arrives. An agent is considered to be active while one of its handlers is executing. If the agent migrates to another milieux all of its state information is included in the transfer. When an agent's handler completes execution the agent becomes dormant, waiting until one of its handlers is triggered.

The milieu (Fig. 1) interacts with the outside world via incoming and outgoing queues, which contain agents. When an agent arrives at a milieu, it is removed from the incoming queue and placed in the agent repository, where it remains while it is in the milieu. Internally, a milieu is driven by events. Events are signaled either by an agent or by the arrival of an agent via the incoming queue. Agents register to react to an event by indicating which of their handler(s) should be triggered. The milieu uses the event registry to schedule agent handlers in response to the signaled event. The handler is executed atomically, and receives the agent that signaled the event as an argument.

From within a milieu, agents are able to interact with other local (in the same milieu) agents and with the milieu itself. A milieu allows an agent to: 1) Access any other agent in the milieu through the agent repository. The fields or handlers of any agent, including itself, can be read from or written to. 2) Signal events. Further, an agent may insert or remove entries of the event registry pertaining to any agent. 3) Create a new agent. 4) Induce an agent to move to another milieu. For performance reasons the model does not address issues, such as security, that are orthogonal to using agents for monitoring and steering.

2.2 Implementation

One instantiation of the model uses Perl to construct the internals of the milieu (a PMilieu) and to serve as the base language for agents. Perl[8] is an object-oriented, imperative scripting language. The interface between the Perl portions of the PMilieu and the C coded portions consists of bindings of the incoming and outgoing queues, Perl modules that provide interfaces to resources present at the monitored process, and event agents that are generated by the agent module and inserted into the incoming queue of the milieu. SWIG[2] was used to create the extension modules that provide access to the resources of the process. Control is transferred to the PMilieu when an event occurs within the process. The application regains control after all of the agents have responded to the event, represented by an event agent.

Other implementations of agent modules for PathFinder support different styles of agents written in other languages. However, all of the agent modules are based on the same model, allowing reasoning about using mobile agents without going into specifics of the implementation and facilitating cooperation between agents in different modules.

3 Token Finding

3.1 Strategies

One use of mobile agents is to discover resources available in a wide-area network [6]. The resource discovery task is similar to finding a particular property in a distributed application. Throughout this section we will refer to the property as a token, but it could be any property over the local state of a process. A token enables the process holding it to perform some action(s), and can be used as a form of coordination among processes. Finding a token in a distributed system is a common task in monitoring. Data obtained by monitoring the process that holds the token can help to provide insight into the coordinated action. Mobile agents may be used to locate and continuously track a token as it moves from one process to another.

Each strategy begins with an initial process, called the coordinator, which initiates the token-finding process. If a process has an agent at it, then the process is said to be covered by the monitoring. Nodes are marked as clean when no token is present and no token is on the way to the process. We consider different strategies for token finding (Table 1).

wait The monitoring agent remains at a node until the token arrives there.

flood The coordinator sends an agent to one node. Upon receiving an agent, the node sends the token out on all of its links. This repeats until all nodes and links have been affected.

broadcast The coordinator sends a monitoring agent directly to all other nodes. **Broadcast** affects only one link per application node, in contrast to **flood** which affects all links.

search The monitoring agent moves from one node to another until the token is found. The movement may be directed or undirected. Only one copy of the agent exists.

entrapment Agents cooperate to partition the network, and then watch for the token at the boundary nodes, and slowly "tighten the net".

wavefront Similar to **entrapment**, the monitoring agent starts from one locations and sends out a wave of agents looking for the token.

time to live Similar to the **search** strategy, but the monitoring agent is assigned a maximum number of hops before it ends.

For these strategy the properties of growth, propagation, residue, and guarantees are examined. The properties differ in how the monitoring coverage changes over time. The *growth* property indicates how quickly the coverage can increase. The *residue* property indicates how long a node remains part of the coverage and the *propagation* property measures how far the coverage extends. Table 2 summarizes how the strategies in Table 1 differ.

Growth Each strategy begins the token finding process at some initial set of processes and then extends coverage until the token is found. The *growth* property describes how the coverage grows, how monitoring spreads from one process in the computation to another without intervention by the coordinator. In strategies such as **wait** and **broadcast**, processes do not interact with each other; hence, there is no growth. The **search** strategy also does not grow. Instead, it moves the monitoring agent from one process to another.

The **entrapment** strategy adds one process at a time to the coverage area, exhibiting *constant* growth. *Proportional* growth strategies scale their expansion based on the size of the computation or the local number of links at the process. The **flood** strategy activates monitoring at all neighbors of a process, exhibiting proportional growth.

A trade-off exists between the rate of growth of a strategy and the resource usage of the algorithm. A high growth rate will reduce the amount of lag between the request for finding the token and the time of its location. However, a computation may not be able to tolerate the resource demands of a high growth rate. Another consideration is that strategies with growth rates above a low constant, e.g. one, are difficult to halt before they have checked the whole network. Even if

Table 1. Example token finding strategies.

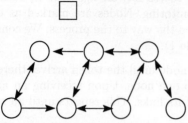

For simplicity, the examples will assume the network is fixed and FIFO. The circles represent nodes of the network and the arrows directional channels. The coordinator is shown as a square.

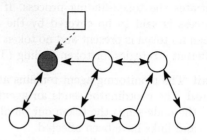

(wait) A simple strategy is to activate monitoring at a single node and wait for the token to arrive.

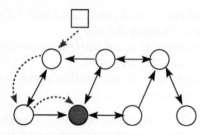

(search) Instead of waiting for the token to arrive the **search** strategy moves from node to node looking for the token. The search ends after a predetermined number of nodes have been visited.

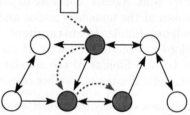

(entrapment) An entrapment strategy seeks to increasingly limit the token's possible locations until it is found. In this example the entrapment strategy has partitioned the network into two parts, watching the nodes in the middle to ensure that the token does not move into the left half while it is searching the right half.

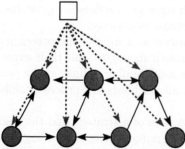

(broadcast) The coordinator directly activates monitoring in all of the nodes in the broadcast strategy. After a period of time the monitoring automatically deactivates.

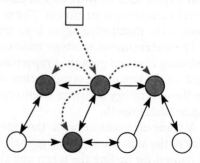

(flood) The flood strategy begins at one node and spreads on all links until the entire computation is looking for the token.

Table 2. Characteristics of different token finding strategies.

| | Growth | | | Residue | | Propagation | | Guaranteed |
	None/One	Constant	Proport.	expire	explicit	finite	infinite	to find
wait	x				x	n/a	n/a	no
search	x			n/a	n/a	x		no
entrapment		x			x		x	yes
broadcast	x			x		n/a	n/a	yes
flood			x		x		x	yes

the token is found almost immediately, there is no way to quickly communicate this information to halt the growth of coverage.

Residue Residue refers to how long artifacts from the strategy remain at a process. After initially determining that the token is not present at a process, most strategies will continue to watch the process for a time to ensure that the token does not arrive from one of the channels. Some strategies deactivate monitoring if it has been determined that the token can not reach the process undetected, such as in the case in which all channels arriving at the process have been flushed and the sources are being monitored.

Flushing lets a process know that a token is not in an incoming channel. Flushing depends upon the communication characteristics of the network. The **broadcast** strategy does an implicit flush of the channels. It knows that monitoring will be active at all processes by a certain time. Then it simply waits until all messages sent by that time would have been delivered. It assumes a maximum network delay exists, which is reasonable in many environments. Explicitly flushing a channel consists of sending a marker or message to indicate that the token is no longer in the channel. Once markers have been received on all channels, the process may safely deactivate its monitoring. This technique assumes that the channels are at least FIFO.

Explicit flushing has less lag, but a higher message overhead and typically involves a more complicated scheme to both clean-up and to keep message traffic at a reasonable level. Another trade-off to consider is between having long lasting residuals and the growth and propagation of the coverage. The slower growing and further propagating strategies often will have long lasting residuals. On the other hand a fast growing strategy may have a short residual (or none at all). One relevant factor in this decision is how much overhead the residual imposes on the process. That is, how expensive it is to watch the incoming messages and/or periodically check the computation's state for the token. Also of concern is how sensitive the computation is to message traffic.

Propagation & Guarantees Propagation is related to the rate of growth of a strategy and is related to how far the coverage extends from its point of origin. Some algorithms stop propagating after reaching some finite measure, such as a time limit or a hop-count. Other algorithms will continue until the token is found, no matter how long that takes. Performance or time bounds may exist on finding the token. For example, token finding may cease if the token is not found by the time the computation enters the next phase. The most common guarantee of a strategy is that if it is run to completion, then the token will be found. The **entrapment** and **flood** strategies guarantee that the token will be found, while strategies like **wait** and **search** do not make this guarantee. Other examples of guarantees made by a strategy might describe how resources will be utilized by it.

3.2 Train Simulation

To demonstrate the trade-offs between the different token finding strategies, a railroad simulation is used. The computation consists of a "smart" railroad that simulates algorithms that ensure the safe movement of trains. Components of the simulation are tracks, lights and switches. Each component represents a computing device capable of communicating with other components and making decisions that regulate the movement of trains. Each component is simulated by a separate process. Processes communicate only through message passing. A train moves from one component to the next via a message exchange between the two components. To allow user interaction with the simulation as it runs, each component performs a one second sleep after it processes incoming messages.

3.3 Experiments

The token in these experiments is the locomotive of the train. The experiments considered four different variables: agent strategy, distribution, topology, and number of components.

The test-bed consisted of a primary cluster of 22 Linux workstations and a remote cluster of 4 Linux workstations about 600 kilometers away. The machines were of various configurations and speeds. Two different configurations of the test-bed were used. The first configuration used all 22 machines in the primary cluster. The second configuration used 18 machines in the primary cluster and all 4 machines in the remote cluster. The coordinator for the search was on the remote cluster in the second configuration.

Two topologies of the railroad components were used. The first topology was of a general graph that was randomly generated. The graph was constructed to be connected. That is, any component is reachable from any component. Some track components were changed to switch components to create circuits and redundant paths that are typical of real world railroads. The second topology consisted of only straight track components. The tracks were configured so they formed a ring. The total number of components in the simulation was varied from 22 to 176.

Two different types of agents were used, based on the **broadcast** and **search** strategies. The residue in both cases was limited to the time required for the agent to check if the train was present at the component. Upon arrival at a component, an agent checks whether the locomotive is present and immediately dies or leaves that component depending on the strategy. When the locomotive was found the coordinator was signaled. For the **broadcast** strategy, the coordinating process sent the same agent to all railroad components. In the **search** strategy the coordinator sent an agent to a randomly chosen component. The search agent was provided with a list of all components and visited each one in order until the train was found. A timeout value was chosen after which point the train was considered to be not found.

4   Results and Discussion

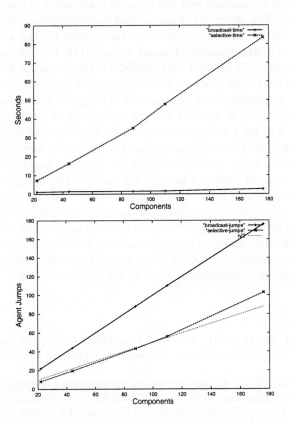

Fig. 2. (Top) The amount of time to find the token is plotted for various number of processes. The amount of time the **search** strategy took was linear in the number of components compared to the **broadcast** strategy which was relatively constant. (Bottom) For both strategies, the number of agent jumps is linear relative to the number of components. The **search** strategy uses approximately half the number of jumps of the **broadcast** strategy.

Approximately 150 tests were executed. No significant difference was found based on the topology or test-bed configuration. The one second delay in the application effectively hid any difference in the latency of agent transmittals between the two clusters. The train was not found by the agents during five tests – one a **broadcast** strategy and four **search** strategies. This happened because the train either was in transit when the agents were sent out or because the train moved to a component that had already been searched. If a longer residual had been provided for the **broadcast** strategy it would not have missed the token in that one instance.

The difference between the two strategies in terms of latency and number of agent transmittals is shown in Fig. 2. As expected, the latency for the **broadcast** strategy was close to constant and was linear in the number of components for the **search** strategy. The effect of the one second pause in the application processes can be easily noticed in the amplitude of the growth for the **search** strategy. Also, as expected, the number of agent transmittals for the **search** strategy was approximately half of what was needed for the **broadcast** strategy. Since the amount of overhead invoked by both types of agents was approximately the same, the perturbation of the **search** strategy is about half that of the **broadcast** strategy. The number of agent jumps by the **search** strategy was actually above half the number of components. This was due to the instances in which a token moved to a component that the agent had already checked.

Using different agent strategies allows for a simple way to customize what costs will be incurred during a monitoring session. The specific trade-off that we examined here was a fast, expensive, and bursty pattern of resource usage, versus a slower, cheaper, and more regular pattern. The choice that is appropriate will depend on the needs of both the user and the application. Utilizing mobile agents allows for this choice to be made easily at run-time. Further experiments are planned to quantify the cost of using agents in the monitoring activity and to further verify our qualitative analysis of the agent trade-offs.

References

[1] Bakic, A.M., Mutka, M.M., Rover, D.T.: BRISK: A Portable and Flexible Distributed Instrumentation System. Software: Practice and Experience. **12** (2000) 1353–1373

[2] Beazley, D.M.: SWIG: An easy to use tool for integrating scripting languages with C and C++. Proceedings of the 4th USENIX Tcl/Tk workshop. (1996) 129–139

[3] Brooks, C., Tierney, B., Johnston, W.: JAVA Agents for Distributed System Management. Technical report, Lawrence Berkeley National Laboratory. (1998)

[4] Frost, H.R.: Documentation for the Java(tm) Agent Template. Stanford University. (1996)

[5] Hart, D., Kraemer, E., Roman, G.C.: Consistency Considerations in the Interactive Steering of Computations. International Journal of Parallel and Distributed Systems and Networks. **3** (1999) 171–179

[6] Jun, K., Boloni, L., Palacz, K., Marinescu, D.C.: Agent-Based Resource Discovery. Proceedings of 9th Heterogeneous Computing Workshop. (2000)

[7] Rus, D., Gray, R., Kotz, D.: Transportable Information Agents. Readings in Agents. (1998) 283–291

[8] Schwartz, R., Wall, L.: Programming Perl. (1994)

Load Balancing for the Electronic Structure Program GREMLIN in a Very Heterogenous SSH-Connected WAN-Cluster of UNIX-Type Hosts

Siegfried Höfinger[1]

Dipartimento di Chimica "G. Ciamician",
Universita' degli Studi di Bologna,
Via F. Selmi 2,
I-40126, Bologna, Italy
sh@ciam.unibo.it
http://www.ciam.unibo.it

Abstract. Five far distant machines located at some French, Austrian and Italian research institutions are connected to a WAN-cluster via PVM 3.4.3. The secure shell protocol is used for connection and communication purposes between the different hosts. Operating-Systems, architectures and cpu-performances of all the 5 machines vary from LINUX-2.2.14/INTEL PPro-200MHz, over LINUX-2.2.13/INTEL PII-350MHz, OSF I V5.0/ALPHA EV6-500MHz, IRIX64 6.5/MIPS R10000-195MHz, up to IRIX64 6.5/MIPS R12000-300MHz. An initial benchmark run with the Hartree Fock program GREMLIN reveals a speed difference of roughly a factor 7x between the slowest and the fastest running machine. Taking into account these various speed data within a special dedicated load balancing tool in an initial execution stage of GREMLIN, may lead to a rather well balanced parallel performance and good scaling characteristics for this program if run in such a kind of heterogenous Wide Area Network cluster.

1 Introduction

Computer Science and Industry has made great progress in recent years and as a result of this, the average desktop personal computer as of today has become superior in many aspects to his supercomputer analogues. The other most rapid emerging field has been the internet and internet based technology, and therefore todays probably most potential computing resources might be lying in these huge number of ordinary internet computers, that are accessible in principal to everyone else on the net, but mainly remain idle and serve for minor computational tasks. Scientific research in many areas however suffers from limited access to computational resources and therefore great attention should be payed to development efforts especially focusing on parallel and distributed computing strategies and all the problems connected to them.

One such example for a really demanding scientific discipline is ab initio quantum chemistry, or electronic structure theory, which currently is about to enter the field of mainly application oriented sciences and bio-sciences as well, and thus experiences a never foreseen popularity, which all in all may be due to awarding the Nobel Price in Chemistry to J.A. Pople and W. Kohn in 1998.

In a previous article [1] we introduced one such quantum chemical program, which shall from hereafter be called **GREMLIN**, that solves the *time independent Schrödinger equation* [2] according to the *Hartree Fock Method* [3] [4]. One of the main features of this program had been the capability to execute the most expensive part in it in parallel mode on distributed cluster architectures as well as on shared memory multiprocessor machines [5]. In addition, what makes this application particularly attractive for a distributed computing solution, is its modest fraction in communication time, which on the other hand implies a principal possible extension to a *Wide Area Network* (**WAN**) cluster, where the individual "working" nodes are usually formed form a number of UNIX-type machines [^1] of usually hetereogenous architecture and the connection between them is simply realized from the ordinary low-bandwidth/high-latency internet.

Following previous results [1], a properly balanced distribution of the global computational work requires some basic interference with the theoretical concept of recursive *ERI* (Electron Repulsion Integrals) computation [6]. However, taking into account a system inherent, partial inseparability of the net amount of computational work, allows an estimation and decomposition into fairly equal sized fractions of node work, and from this adequate node specific pair lists may be built. The present article intends to describe, how one may extend this concept to an additional consideration of different node performance, since the previous study was based on multiprocessor machines made of equally fast performing CPUs.

1.1 Computational Challenge

Here we briefly want to recall, what makes ab-initio electronic structure calculation a real computational challenge. The main problem lies in the evaluation of ERIs, *the Electron Repulsion Integrals*, which are 6-dimensional, 4-center integrals over the basis functions φ.

$$ERI = \int_{r_1} \int_{r_2} \varphi_i(r_1)\varphi_j(r_1)\frac{1}{|r_2 - r_1|}\varphi_k(r_2)\varphi_l(r_2)dr_1 dr_2 \tag{1}$$

and the basis functions φ_i are expanded in a series over *Primitive Gaussians* χ_j

$$\varphi_i(r) = \sum_j d_{i,j}\,\chi_j(r) \ , \tag{2}$$

[^1]: although PVM 3.4.3 would support WIN32 like OSs as well

which typically are *Cartesian Gaussian Functions* located at some place (A_x, A_y, A_z) in space [2] [7] [8].

$$\chi_j(\boldsymbol{r}) = N_j(x - A_x)^l(y - A_y)^m(z - A_z)^n \, e^{-\alpha_j(\boldsymbol{r} - \boldsymbol{A})^2} \qquad (3)$$

Although somewhat reduced from numerical screening, the principal number of ERIs to be considered grows with the 4th power of the number of basis functions, which themselve is proportional to the number of atoms in the molecule. However, since the quality of the employed basis set must be kept high in order to enable quantitative reasoning, the according number of ERIs very soon exceeds conventional RAM and diskspace limits and thus becomes the only limiting factor at all. For example, a simple, small molecule like the amino acid alanine (13 atoms), that has been used as a test molecule throughout this present study, at a basis set description of aug-cc-pVDZ quality [9] [10] (213 basis functions of S, P and D type) leads to a theoretical number of approximately 260 x 10^6 ERIs, which requires about 2.1 GigaByte of either permanent or temporary memory and goes far beyond usual available computational resources.

Fortunately there is partially independence in the mathematical action of these many ERIs and one may solve the problem in a so called "Direct" way, which means, that a certain logical block of related ERIs is first calculated recursively [3], then the action of these block on all the corresponding Fock-matrix elements – from which there luckily are only a number of (*number of basis functions*)^2 – is considered, and then the procedure is repeated and a new block of ERIs overwrites the old one and thus only a small amount of working memory is permanently involved. Further complifying is the fact, that one has to respect a hierarchic structure in spawning the space to the final primitive cartesian gaussian functions χ_j, where, following the notation introduced in [1], a certain center $\boxed{\text{i}}$ refers to an according block of contracted shells → (j)...(k), from which each of them maps onto corresponding intervals of basis functions l...m and the later are expanded from primitive cartesian gaussian functions χ_j as seen from (2). Therefore, after defining a particular centre quartette $\boxed{\text{i1}}\,\boxed{\text{i2}}\,\boxed{\text{i3}}\,\boxed{\text{i4}}$, all the implicit dependencies down to the primitive cartesian gaussians χ_j must be regarded and as a consequence rather granular blocks of integrals must be solved all at once, which becomes the major problem when partitioning the global amount of integrals into equally sized portions.

1.2 Speed Weighted Load Balancing

Concerning parallelization, we follow a common downstream approach and define node specific pair lists, that assign a certain subgroup of centre quartettes

---

[2] An S-type basis function will consist of primitive gaussians with $l = m = n = 0$, a P-type however of primitives with $l + m + n = 1$, which may be solved at 3 different ways, either $l = 1$ and $m = n = 0$, or $m = 1$ and $l = n = 0$, or $n = 1$ and $l = m = 0$. D-type specification will likewise be $l + m + n = 2$ and similarly F-type $l + m + n = 3$.

[3] All complicated ERI-types ($l + m + n > 0$) may be deduced from the easier computed $(S_i, S_j | S_k, S_l)$ type.

to each of the individual nodes, which then shall work independently on their corresponding partial amount of global computational work. Ideally these pair lists are built in a way, such that each of the nodes needs the same time for executing its according fraction of net work. Suppose the total number of theoretical centre quartettes is represented from the area of a rectangle, like shown in Table 1, and one wants to distribute these many centre quartettes now onto a number of parallel executing nodes, then the simplest method would certainly be an arithmetic mean scheme (left picture in Table 1), where the theoretic number of centre quartettes is devided by the number of nodes and all of them get exactly this arithmetic mean fraction to work on. Due to the fact that several centres may now substantially differ in the number of deducable contracted shells $\rightarrow$ basis functions $\rightarrow$ primitive gaussians, this simple procedure has been shown to not be applicable for the case of distributing global computational work for recursive ERI calculation done in parallel [1]. In fact, instead, one had to take into account all these hierarchic dependencies down to the level of primitive gaussians χ_j in order to be able to estimate the real fraction one particular centre quartette actually had of the global amount of work measured in terms of theoretic establishable quartettes of primitive gaussians now. However, following this pathway led to considerable improvements in parallel performance and the according pair lists of center quartettes may then be symbolized like shown in the medium picture of Table 1. Note, that up to now, the indicated, individual 4 nodes are still considered to all operate at the same CPU speed and despite the fact, that the actual number of centre quartettes each node has to process has become apparently different now, the execution time for each of the nodes is now much more comparable – if not equal – to each other, which stands in great contrast to the arithmetic mean picture.

Going one step further and assuming different node performance next, would change the situation again. For example, let us hypothetically think of a parallel machine, where node II is twice as fast as node I, and node III and IV are running three times and four times as fast as I respectively. Then, we could equally well think of a parallel machine made up of 10 nodes of the speed of type I, divide the global amount of work (measured again at the innermost level of potential primitive gaussian quartettes) into 10 equal sized fractions, and let the fastest node (IV) work on 4 portions of that estimated unit metric, while node III and II get $\frac{3}{10}$ and $\frac{2}{10}$ of the global work and node I will just deal with the remaining $\frac{1}{10}$ of the global amount. The schematic representation of such a kind of partitioning is given in the right picture of Table 1. On the other hand, one could obtain a theoretical speed up factor of 2.5 ($= \frac{10}{4}$) for such a case [4], if at first instance communication time is said to be extremly small and bare serial execution intervals are neglected completely.

---

[4] compared to the situation where 4 equally fast performing CPUs operate on already load balanced pair lists

Table 1. Comparision between different partitioning schemes of the outermost loop over centre quartettes, represented from the partial areas of the 4 rectangles, that stand for node specific fractions of the global amount of theoretical combinations of centre quartettes. For the Speed Weighted Load Balancing, node II is assumed to be twice as fast as I, and nodes III and IV, are said to be three times and four times as fast as I.

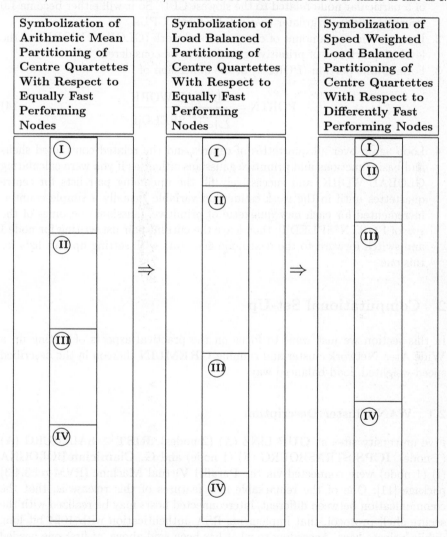

| Symbolization of Arithmetic Mean Partitioning of Centre Quartettes With Respect to Equally Fast Performing Nodes | Symbolization of Load Balanced Partitioning of Centre Quartettes With Respect to Equally Fast Performing Nodes | Symbolization of Speed Weighted Load Balanced Partitioning of Centre Quartettes With Respect to Differently Fast Performing Nodes |

1.3 Computational Implementation of Speed Weighted Load Balancing

As already explained within Sect. 1.2, we want to distribute the theoretical number of centre quartettes onto a number of nodes of different CPU performance. For this purpose we implement the following steps:

1. Determine the number of participating hosts (NMB) and their according relative speed factors (SPEED[I]). The speed factor is the relative performance of a particular node related to the slowest CPU. So it will either become 1.0 (weakest node), or greater than 1.0 for faster CPUs.
2. Estimate the net amount of computational work (GLOBAL WORK) at the level of quartettes of primitive gaussians to be considered.
3. Form a unit portion (PORTN) of the dimension of

$$\text{PORTN} = \frac{\text{GLOBAL WORK}}{\sum_{I=1}^{NMB} \text{SPEED[I]}} \tag{4}$$

4. Loop again over all quartettes of centres and the related contracted shells and basis functions and primitive gaussians either, as if you were calculating GLOBAL WORK, and successively fill the upcoming pair lists for centre quartettes until in the work estimation variable (usually a simple counter, incremented for each new quartette of primitive gaussians) becomes of the size of PORTN*SPEED[I]; then leave the current pair list writing for node I and switch forward to the next node and start with setting up pair lists for this one.

2 Computational Set-Up

In this section we just want to focus on the practical aspects of setting up a Wide Area Network cluster and running **GREMLIN** thereon in the described speed-weighted, load-balanced way.

2.1 WAN Cluster Description

Five university sites at **GUP** LINZ (A) (2 nodes), **RIST**[++] SALZBURG (A) (1 node), **ICPS** STRASBOURG (F) (1 node) and **G. Ciamician** BOLOGNA (I) (1 node) were connected via the **P**arallel **V**irtual **M**achine (PVM rel.3.4.3) package [11]. One of the remarkable nice features of this release is, that the communication between different, interconnected hosts may be realized with the secure shell protocol, that implements RSA authentication with 1024 bit long public/private keys. According to what has been said above, at first one needed to get an overview of the different node perfomance of all the individual hosts involved in the cluster. Therefore an initial benchmark run with the PVM version of **GREMLIN** (1 node at each location seperately) on a very small training

Table 2. Conditioning and description of the individual host performance in the WAN cluster. The data is due to a Hartree/Fock DSCF calculation on glycine/631g with the program **GREMLIN**. (Result: -271.1538 Hartree in 22 iterations)

Physical Location	Architecture/ Clock Speed/ RAM/2L-Cache	Operating System	real [s]	usr [s]	sys [s]	Rel. Speed
node I G.C. BOLOGNA Italy	INTEL Dual PPro 200 MHz 256 MB/512 KB	LINUX 2.2.14	2431	159	0	1.000
node II ICPS STRASBOURG France	MIPS R10000 200 MHz 20 GB/4 MB	IRIX 64 6.5	1186	9	2	1.934
node III GUP LINZ Austria	INTEL PII 350 MHz 128 MB/512 KB	LINUX 2.2.13	1167	60	1	2.054
node IV GUP LINZ Austria	MIPS R12000 300 MHz 20 GB/8 MB	IRIX 64 6.5	767	6	1	2.990
node V RIST SALZBURG Austria	ALPHA EV6 21264 500 MHz 512 MB/4 MB	OSF I V 5.0	341	6	1	6.823

system (glycine/631g, 10 centre, 55 basis functions) was performed, which led to the data shown in Table 2.

The timings were obtained with the simple UNIX-style **time a.out** command. According to the fact, that the PVM version of **GREMLIN** consists of a master-code and a node-code part, and since the node-code part got a different executable name, the mentioned **time**-command could easily distinguish between the parallel and the serial (diagonalization and pre-ERI work) fractions of the program execution. Thus to focus on the sections that really were running in parallel, one simply had to substract the usr+sys timings from the real one and could straightforwardly obtain the relative speed factors shown in Table 3. Note, that node I and III were lacking from special tuned **LAPACK** libraries, so their usr timings became significantly higher.

2.2 Estimation of Network Latency and Communication Time

To get a feeling for the time, that is lost through inter host communication — when nodes are receiving/sending data — we simply measured the bandwidth we got from the different host positions towards those node serving as the master machine in the WAN cluster later on (node III). For the real application of alanine/aug-cc-pVDZ (13 atoms, 213 basis functions) we had to expect a data transfer of the size of 1452 kB per iteration, which results in a net amount of 27.6

Table 3. Speed-Factor and Network-Latency table for the WAN cluster. Speed-Factors represent the relative performance of all the individual hosts in the WAN cluster with respect to the slowest performing CPU. Network bandwidth was obtained from measuring transfer rates between nodes and the future master-machine (node III).

Physical Location	Architecture/ Clock Speed/ RAM/2L-Cache	Operating System	Relative Speed Factor	Network Bandwidth [kB/s]	Exp.Total Comm. Time [s]
node I G.C. BOLOGNA Italy	INTEL Dual PPro 200 MHz 256 MB/512 KB	LINUX 2.2.14	1.000000	166	166
node II ICPS STRASBOURG France	MIPS R10000 200 MHz 20 GB/4 MB	IRIX 64 6.5	1.933617	608	45
node III GUP LINZ Austria	INTEL PII 350 MHz 128 MB/512 KB	LINUX 2.2.13	2.054250	—	—
node IV GUP LINZ Austria	MIPS R12000 300 MHz 20 GB/8 MB	IRIX 64 6.5	2.989474	918	30
node V RIST SALZBURG Austria	ALPHA EV6 21264 500 MHz 512 MB/4 MB	OSF I V 5.0	6.822822	592	47

MB for all the 19 iterations needed throughout the whole calculation. Network transfer rates and estimated total times spent on communication are also shown in Table 3.

3 Discussion

A final calculation of the above mentioned alanine/aug-cc-pVDZ (13 atoms, 213 basis functions) system on a successive increasing WAN cluster was performed and led to the execution timings and according Speed Up factors shown in Table 4. A similar, graphical representation of the Speed Up factors is shown in Fig. 1. Instead of strictly applying *Amdahl' s Law*, Speed Up $\leq \frac{1}{s+\frac{1-s}{N_{cpu}}}$, we tended to simply relate (real-usr) timings to each other, which was estimated to have almost no influence on relative values, and neglectable influence on absolute values.

Comparision of the final column of Table 3 to the 2nd column of Table 4 reveals a neglectable influence of communication time as well.

The 3rd column of Table 4 might be best suited to explain the actual heterogenity of the WAN cluster. In principle there should be one uniform amount of time spent on the diagonalization- and pre-ERI work, which basically is all what is reflected in the Usr Time. However, temporary network bottlenecks, OS-competion for CPU-time, temporary I/O management excess, CPU-time competition from interactive user operation — which all was allowed during program execution — led to that much more realistic, more variational picture.

The plot in Fig. 1 defines the number of machines in a cumulative way from left to the right on the abscissa, thus the always added new hosts are indicated at

Table 4. Execution timings and Speed Up factors for the DSCF Hartree Fock calculation of alanine/aug-cc-pVDZ with **GREMLIN** in a WAN cluster made of 1 to 5 nodes.

WAN Cluster Configuration	Real Time [s]	Usr Time [s]	Sys Time [s]	Theor. Speed Up $\sum$ SPEED[I]	Real Speed Up
master III nodes I	240 061	9 268	3	1.000	1.000
master III nodes I,II	90 280	9 261	8	2.934	2.847
master III nodes I,II,III	60 496	9 368	2	4.988	4.516
master III nodes I,II,III IV	45 014	9 923	3	7.977	6.577
master III nodes I,II,III IV,V	27 038	9 482	6	14.800	13.116

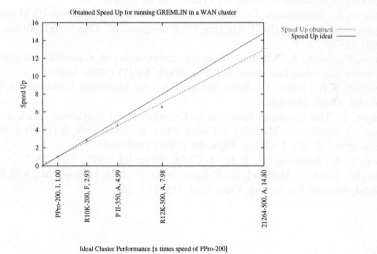

Fig. 1. Representation of the obtained and ideal Speed Up factors for the DSCF Hartree Fock calculation of alanine/aug-cc-pVDZ with **GREMLIN** in a ssh-connected WAN-cluster, made of up to 5 machines.

those final ideal speed level — relative to the slowest node — the cluster should ideally achieve at that very configuration.

3.1 Conclusion

Considering individual node performance in a heterogenous WAN cluster properly, may result in excellent parallel scalability for special dedicated applications, that are characterized from small communication time and large independent node intervals.

Acknowledgement

The author would like to thank Prof. Zinterhof from RIST[++] Salzburg, Prof. Volkert from GUP Linz and Dr. Romaric David from ICPS Strasbourg for providing access to their supercomputer facilities.

References

1. Höfinger, S., Steinhauser, O., Zinterhof, P.: Performance Analysis and Derived Parallelization Strategy for a SCF Program at the Hartree Fock Level. Lect. Nt. Comp. Sc. **1557** (1999) 163–172
2. Schrödinger, E.: Quantisierung als Eigenwertproblem. Ann. d. Phys. **79, 80, 81** (1926)
3. Hartree, D.R.: Proc. Camb. Phil. Soc., **24** (1928) 89
4. Fock, V.: Näherungsmethoden zur Lösung des Quantenmechanischen Mehrkörperproblems. Z. Phys. **61** (1930) 126 **62** (1930) 795
5. Höfinger, S., Steinhauser, O., Zinterhof, P.: Performance Analysis, PVM and MPI Implementation of a DSCF Hartree Fock Program. J. Comp. Inf. Techn. **8** (1) (2000) 19–30
6. Obara, S., Saika, A.: Efficient recursive computation of molecular integrals over Cartesian Gaussian functions. J. Chem. Phys. **84** (7) (1986) 3963–3974
7. Davidson, E.R., Feller, D.: Basis Set Selection for Molecular Calculations. Chem. Rev., **86** (1986) 681–696
8. Shavitt, I.: The Gaussian Function in Calculations of Statistical Mechanics and Quantum Mechanics. Methods in Comp. Phys. ac. New York, **2** (1963) 1–44
9. Dunning Jr., T. H.: J. Chem. Phys. **90** (1989) 1007–1023
10. Woon, D. E., Dunning Jr., T. H.: J. Chem. Phys. **98** (1993) 1358–1371
11. Geist, G., Kohl, J., Manchel, R., Papadopoulos, P.: New Features of PVM 3.4 and Beyond. Hermes Publishing, Paris Sept. (1995) 1–10

DeWiz - Modular Debugging for Supercomputers and Computational Grids

Dieter Kranzlmüller

GUP Linz, Johannes Kepler University Linz,
Altenbergerstr. 69, A-4040 Linz, Austria/Europe,
kranzlmueller@gup.uni-linz.ac.at,
http://www.gup.uni-linz.ac.at/

Abstract. Debugging is accepted as one of the difficult tasks of high performance software development, which can be attributed to the high complexity of parallel and distributed applications. Especially users of massively parallel supercomputers or distributed metacomputer systems experience huge obstacles, that are difficult if not impossible to overcome with existing error detection approaches. The prototype tool DeWiz presents an effort to improve this situation by applying the abstract event graph model as a representation of parallel program behavior. Besides its usability for different programming paradigms it permits analysis of data with various debugging activities like automatic error detection and sophisticated abstraction. In addition, DeWiz is implemented as a set of loosely connected modules, that can be assembled according to the user's needs and given priorities. Yet, it is not intended as a complete replacement but as a plug-in for well-established, existing tools, which may utilize it to increase their debugging functionality.

1 Introduction

The quest for ever increased computational performance seems to be a never-ending story mainly driven by so-called "grand challenge" problems of science and engineering, like simulations of complex systems such as weather and climate, fluid dynamics, and biological, chemical, and nuclear reactions. Since existing computing systems allow only insufficient calculations and restricted solutions in terms of processing speed, required memory size, and achieved numerical precision, new architectures and approaches are being developed to shift the performance barrier. The upper limit of this development is represented by supercomputer systems, which may be further coupled to metacomputers. Recent examples of realization projects are the US Accelerated Strategic Computing Initiative (ASCI) [6] and the multi-institutional Globus project [4]. While ASCI seeks to enable Teraflop computing systems far beyond the current level of provided performance, the Globus project tries to enable computational grids that provide pervasive, dependable, and consistent access to distributed high-performance computational resources.

Such systems achieve their level of performance due to their high degree of powerful parallel and distributed computing components. Yet, this complicates the software development task due to the required coordination of multiple, concurrently executing and communicating processes. As a consequence, big obstacles are experienced during all phases of the software lifecycle, which initiated many research efforts to improve the parallel programmers situation with manifold strategies and development tools.

One area of investigation is testing and debugging, which shares a great part in determining the reliability of the application and thus the quality of the software. The goal of debugging is to detect faulty behavior and incorrect results occurring during program execution, which is attempted by analyzing a program run and investigating process states and state changes. Obviously, the complexity of the program and the number of interesting state changes determines the amount of work needed to analyze the program's behavior. This means, that bigger systems are probably more difficult to debug than smaller systems. Yet, there are only a limited number of parallel debuggers, that are suitable for error detection of applications running on massively parallel and distributed systems. The biggest problem is that current debugging tools suffer from managing the amount of presented data, which stems from mainly two characteristics: Firstly, most parallel debugging tools are composed from combing several sequential tools and integrating them under a common user interface. These tools often lack support for detecting errors derived from parallelism. Secondly, many tools are based on textual representations, which may be inappropriate in many cases to display and manage the inherent complexity of parallel programs [13].

The work described in this paper differs from existing approaches due to the fact, that debugging activities are based on the event graph model instead of the underlying source code. It describes a parallel program's execution by occurring state changes and their interactions on concurrently executing processes, which allows to cope equally with programs based on message-passing and the shared memory paradigm. Furthermore, it can be applied for automatic error detection and to perform higher-level program abstraction.

These ideas are implemented in the tool prototype DeWiz, the Debugging Wizard. In contrast to other tools, DeWiz does not contain a graphical user interface for the analysis task, but instead offers its results to other existing debugging tools. By providing an adaptable input interface, traces from post-mortem debuggers and event-streams from on-line debuggers can be processed. Similarly, an output interface allows to use DeWiz as a plug-in for the user's preferred analysis tool. Another feature is its steering module, that offers a way for the user to describe the desired analysis more precisely. This covers the idea, that the user has knowledge about the program's expected behavior and may thus be able to identify different priorities for more or less critical errors and analysis tasks.

This paper is organized as follows. The next section discusses the target systems for our debugging model and the requirements imposed onto parallel debugging tools. Afterwards, the event graph model is introduced and some possibilities for analyzing a program's behavior are presented. This leads to the actual implementation of DeWiz in Section 4, which is described by its main features and its mode of operation.

2 Requirements to a Parallel Debugger

The main criterion for any tool developer is a definition of target systems, which in our case are high performance supercomputers. Since there exist different architectures and possibilities of programming them, a wide variety of strategies and tools have already been proposed. In the case parallel debuggers many promising approaches exist as academic or research prototypes, for example Mantis [11], P2D2 [5], PDBG [2], and PDT [1], or even as a commercial tool like Totalview [3]. A characteristic of these systems is, that every approach applies several instances of an existing sequential debugger in order to perform the debugging task on the participating processes. Although this may be useful in most cases, it introduces some obstacles especially on large scale computing systems like massively parallel machines or heterogenous clusters of supercomputers.

The problems experienced during debugging of supercomputer applications are mostly connected to the amount of data that has to be analyzed. Firstly, these programs tend to be long-lasting, from some hours to several days or even more. As a consequence, many state changes occur that have to be observed and processed by the debugging tool. In the worst case, debugging a supercomputer may require another supercomputer to perform the analysis task. Secondly, the execution time and the large number of participating processes leads to enormous interprocess relations, which cannot be comprehended by the user. Thirdly, a great amount of debugging activities has to be performed equally for different processes and repeated iterations of the target application.

Previous solutions were always based on down-scaling, which means that the program's size is reduced in terms of participating processes and executed numbers of iterations. While this may be successful in many cases, it also contains potential for critical errors, which may be experienced only in full-scale real-world applications. As a consequence, there may be some cases where the program has to be tested under regular conditions in its intended environment. In order to comply to this requirement, we have identified the following characteristics for a debugger of the above mentioned target systems:

- Heterogeneity: supporting different programming models and architectures.
- Usability: managing huge amounts of data and improving program understanding.
- Abstraction: reducing the amount of debugging data presented to the user.
- Automatization: performing repeated debugging activities without user interaction.

A strategy or tool supporting these characteristics may then be applicable to full-scale applications, and may allow to perform debugging activities impossible with existing solutions. In addition, it may also improve the error detection task on smaller-scale application sizes, especially if it can be combined with other tools in this area.

3   The Event Graph Model for Debugging

Our solution to fulfil the characteristics described above are based on the event graph model. An event graph is a directed graph G = (E,), where E is the non-empty set of vertices e of G, while is a relation connecting vertices, such that x y means that there is an edge from vertex x to vertex y in G. The vertices e are the events occurring during program execution, which change the state of the corresponding process [14]. The relation establishes Lamport's "happened before" ordering [10], which consists of the sequential order on one particular process and the order between disjunct processes whenever communication or synchronization takes place.

In principle, every change of process state is caused by an event, and there are huge numbers of events being generated during program execution. However, usually only a small subset of events is required during debugging, which allows to filter only interesting events for error detection and to reduce the number of state changes for investigation. The remaining events are then collected as the vertices of the event graph. One difficulty is to define, which events to collect and which state changes to ignore. In order to allow a large degree of flexibility, the events collected in the event graph are user-defined. For example, a user looking for communication errors may define point-to-point communication events as established by send and receive function calls to be the target of investigation.

During program analysis, it is not only important to know about the occurrence of an event, but also about its properties. These properties are called event attributes and represent everything that may be interesting for the investigator. Similar to the events, the set of attributes may be appointed by the user. For the above mentioned communication points, a user may identify the communication statement's parameters to be event attributes for the analysis. Another kind of attributes are source code pointers, which consist of filename and line number corresponding to the original function call or statement. These attributes are needed in order to establish a connection between the graph and the faulty source code.

With the event graph model defined as above it is possible to describe erroneous behavior. In principle, every error theory defines two groups of bugs in programs, failures and computational errors. While the former is clearly recognizable, e.g. through program break-downs or exceptions being taken, the latter always depends on the semantic contents of the results and requires a verification step. Thus, computational errors can only be detected by comparison of expected results with actually obtained results.

Integrating failure detection in the event graph model is relatively easy, since their occurrence usually leads to an end of the corresponding process. Thus, a failure is always the last event on a particular process, which is characterized by having only one approaching edge but no leaving edge. Therefore, a debugging tool can easily direct the programmer's attention to such places by analyzing the final event on each participating process.

On the other hand, computational errors may occur at both, edges and vertices of the event graph. Since the edges describe a set of state changes, and these

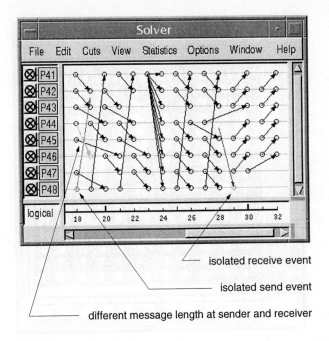

isolated receive event

isolated send event

different message length at sender and receiver

Fig. 1. Basic event graph analysis (e.g. errors in point-to-point communication)

state changes determine the results of the program, analyzing the edges may be required in order to detect incorrect behavior. Errors at vertices can be identified by describing the expected event attributes. If the event attributes obtained during execution differ from the expected data, incorrect or illegal operations have been detected. Please note, that incorrect operations may not necessarily result in errors, e.g. when the program is prepared to handle such unexpected events.

For instance, comparing expected and actual attributes of the communication events may expose isolated events or events with different message length. Isolated events are send events without corresponding receive events or vice versa. Events with different message length are revealed, if the size of the message data differs at sender and receiver. However, even if events with these characteristics are detected, they need not necessarily result in malign behavior. For example, isolated send events may have no effect, while isolated receive events may block the processes' execution forever.

An example of these basic debugging features is visible in Figure 1. It shows the execution of a program on 8 selected nodes from a possibly much larger execution. Some of the edges in the graph are highlighted to emphasize errors in the communication parameters of corresponding send and receive function calls.

Besides the basic analysis capabilities of checking event attributes, a more sophisticated way of analysis considers the shape of the graph itself. Often, a set of corresponding send and receive events resembles more or less complex

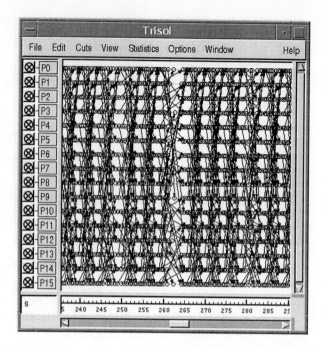

Fig. 2. Advanced event graph analysis (e.g. anomalous event graph pattern)

communication patterns, like broadcast, scatter, gather, and butterfly. Other possibilities are iterations in the observed algorithm, repeated function calls to selected communication statements, and grouping of processes (see [7] for an overview of some examples). These characteristics can be detected with simple pattern matching algorithms. As a result, this analysis allows to detect complete patterns, nearly complete patterns, and the absence of expected patterns in the event graph.

An example for communication pattern analysis is visible in Figure 2. This shows a finite element solver that was executed on 16 nodes. There have been 200 iterations to perform the computation, and only point-to-point communication events have been traced. In total, the trace contained around 20.000 events. Therefore, it is possible that the strange behavior during one of the operations could have failed to notice, especially if a smaller scale than seconds would have been selected. With pattern matching, this strange behavior would have been detected immediately. Please note, that this is a real example and we detected this error only by accident, before we developed this strategy.

The next step after detecting anomalous behavior is to direct the users attention to such places in the event graph. This is called automatic abstraction and means, that only a limited surrounding of the erroneous events is extracted. Therefore, instead of presenting all the data to the user, only critical sections of the event graph are displayed. A simplified operation of automatic abstraction is to evaluate the history of events that finally resulted in the detected bug. Therefore, it may only be necessary to display the communication partners of the corresponding process, and this only for a certain interval of time. As a consequence, in contrast to displaying huge numbers of events and several hundred processes, only a small subset is presented to the user, which still contains the interesting places for the debugging task.

4 DeWiz, the Debugging Wizard Prototype Tool

The event graph model as described above together with a preliminary set of error detection functions has been implemented in a tool prototype called DeWiz. Besides this functionality, several aspects have been applied during the design of DeWiz, which are as follows:

- Modularity: the debugging functionality must be adaptable to the users needs and the applications characteristics.
- Independence: the features of DeWiz must be applicable without a defined graphical user interface in mind, but instead as a plug-in to available debugging tools.
- Efficiency: due to the prospected amount of data, the tool must be implemented in order to facilitate all available performance by executing modules in parallel and applying parallelized algorithms during the analysis (e.g. pattern matching).

In addition, DeWiz contains a steering module, that allows to integrate the users knowledge into the debugging task. Within the proposed strategy it is able to allow the users to define:

- events interesting for monitoring
- expected behavior in terms of communication patterns
- priority between different kinds of errors

These aspects are available in a first tool prototype, whose operation during program analysis is displayed in Figure 3. The starting point is a target system, that is observed by an available monitoring tool. This monitoring tool is connected to DeWiz either on-line via event streams or post-mortem via tracefiles. In order to combine the tool with an available monitoring utility, the input interface has been adapted to process the given event data. This input interface forwards the data to the working modules of DeWiz, which perform the desired analysis.

At this point, the users knowledge interferes with the systems operation. A dedicated steering module allows to decide about the expected behavior and the priority of different errors. The user with the knowledge about the target system, called the debugging expert, enters this data via a configuration file. In the future, we will change this propriety form of steering with some kind of graphical user interface. The configuration given by the user determines the

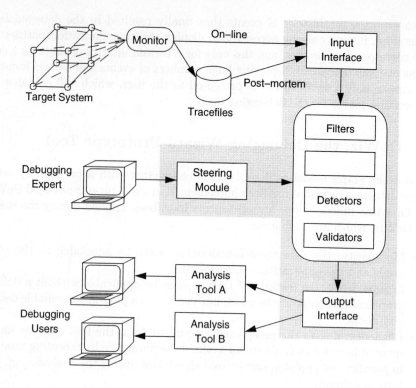

Fig. 3. Flowdiagram of DeWiz during opertation

arrangement of the working modules. Clearly, modules evaluating higher priority errors are executed earlier during the analysis task. In addition, the distribution of the modules to available processing elements can be defined in order to improve the analysis speed.

After the analysis has be completed, or critical errors have been detected, the results are forwarded to the output interface. Again, at this end of the tool, some customizations have to be carried out to provide the results of DeWiz to other existing analysis tools. Of course, since the results are delivered as event streams, any possible connection as well as concurrent connections to different tools are imaginable.

At present, the current prototype of DeWiz has been connected to the MAD environment [8], which has been developed at our department. Besides others, MAD contains an event monitoring utility that performs the program observation of MPI programs [11], and several analysis tools that visualize the evaluated analysis data. One of these analysis tools is ATEMPT, which visualizes a state-time diagram of the programs execution. The results of DeWiz are therefore mapped onto the display of ATEMPT, and allow to remove all analysis functionality from the graphical user interface. In addition, it is much easier to extend the modular approach of DeWiz, and as a side effect provide these extensions to other available debugging environments.

5 Conclusions and Future Work

The work described in this paper tries to provide a solution for the difficulties encountered during debugging of massively parallel supercomputers as well as distributed metacomputers. With the event graph model it is relatively easy to distinguish correct and incorrect behavior, to provide some means of abstraction and to perform tedious analysis tasks automatically.

The presented ideas have already been implemented in the DeWiz tool prototype, which has been successfully connected to an existing debugging environment. In addition, the modular composition of DeWiz allows to easily extend and improve its functionality, which is one of the current goals in this project. By providing more analysis features, the capabilities of DeWiz can be increased with benefits for the debugging user. In addition, we experienced that many analysis tasks are really time-consuming, especially with the large amounts of data to be processed. Thus, it seems necessary to further optimize and parallelize some of the existing modules in order to speed up the execution.

Another future goal in this project is the extension of the input and output interfaces. So far, we have only processed post-mortem tracefiles, and one of the next goals is to integrate an interface based on the OMIS monitoring standard [16]. This will allow us more flexibility in choosing the input connection and processing different event streams, and will probably deliver useful feedback for additional analysis modules.

Acknowledgements This work represents an extension of my PhD thesis [9] and covers aspects that have not been described before. Consequently, many of the ideas presented here have evolved from the many discussions with my PhD supervisor, Prof. Dr. Jens Volkert.

References

1. C. Clemencon, J. Fritscher, R. Rühl, "Visualization, Execution Control and Replay of Massively Parallel Programs within Annai's Debugging Tool", Proc. High Performance Computing Symposium, HPCS'95, Montreal, Canada, pp. 393-404 (July 1995).
2. J. C. Cunha, J. Loureno, J. Vieira, B. Mosco, and D. Pereira. , "A framework to support parallel and distributed debugging", Proc. of HPCN'98, High Performance Computing and Networking Europe, Amsterdam, Netherlands, (1998).
3. Etnus (Dolphin Interconnect Solutions Inc.): TotalView 4.1.0, Documentation: http://www.etnus.com/pub/totalview/tv4.1.0/totalview-4.1.0-doc-pdf.tar, Framingham, Massachusetts, USA, 2000.
4. I. Foster, C. Kesselman, "The Globus Project: A Status Report", Proc. IPPS/SPDP'98 Heterogenous Computing Workshop, pp. 4-18 (1998).

5. R. Hood, "The p2d2 Project: Building a Portable Distributed Debugger", Proc. SPDT'96, ACM SIGMETRICS Symposium on Parallel and Distributed Tools, Philadelphia, USA, pp. 127-136 (May 1996).
6. F. Hossfeld, "Teraflops Computing: A Challenge to Parallel Numerics?" Proc. 4th Intl. ACPC Conference, Springer, LNCS, Vol. 1557, Salzburg, Austria, pp. 1-12 (Feb. 1999).
7. D. Kranzlmüller, S. Grabner, J. Volkert, "Event Graph Visualization for Debugging Large Applications", Proc. SPDT'96, ACM SIGMETRICS Symposium on Parallel and Distributed Tools, Philadelphia, PA, pp. 108-117 (May 1996).
8. Kranzlmüller, D., Grabner, S., Volkert, J., *Debugging with the MAD Environment*, Parallel Computing, Vol. 23, No. 1–2, pp. 199–217 (Apr. 1997).
9. Kranzlmüller, D., *Event Graph Analysis for Debugging Massively Parallel Programs*, PhD Thesis, GUP Linz, Joh. Kepler University Linz, http://www.gup.uni-linz.ac.at/ dk/thesis, (Sept. 2000).
10. Lamport, L., *Time, Clocks, and the Ordering of Events in a Distributed System*, Communications of the ACM, pp. 558 - 565 (July 1978).
11. S.S. Lumetta, D.E. Culler, "The Mantis Parallel Debugger", Proc. of SPDT'96: SIGMETRICS Symposium on Parallel and Distributed Tools, Philadelphia, PA, pp. 118-126 (May 1996).
12. Message Passing Interface Forum, "MPI: A Message-Passing Interface Standard - Version 1.1", http://www.mcs.anl.gov/mpi/ (June 1995).
13. C.M. Pancake, "Visualization Techniques for Parallel Debugging and Performance-Tuning Tools", in: A.Y.Zomaya, "Parallel Computing: Paradigms and Applications", Intl. Thomson Computer Press, pp. 376-393 (1996).
14. M. van Rick, B. Tourancheau, "The Design of the General Parallel Monitoring System", Programming Environments for Parallel Computing, IFIP, North Holland, pp. 127-137 (1992).
15. M. Stitt, "Debugging: Creative Techniques and Tools for Software Repair", John Wiley & Sons, Inc., NY (1992).
16. R. Wismüller, "On-Line Monitoring Support in PVM and MPI", Proc. EuroPVM/MPI'98, LNCS, Springer, Vol. 1497, Liverpool, UK, pp. 312-319, (Sept 1998).

Fiddle: a Flexible Distributed Debugging Architecture

João Lourenço   José C. Cunha

{jml,jcc}@di.fct.unl.pt
Departamento de Informática
Faculdade de Ciências e Tecnologia
Universidade Nova de Lisboa
Portugal

Abstract. In the recent past, multiple techniques and tools have been proposed and contributed to improve the distributed debugging functionalities, in several distinct aspects, such as handling the non-determinism, allowing cyclic interactive debugging of parallel programs, and providing more user-friendly interfaces. However, most of these tools are tied to a specific programming language and provide rigid graphical user interfaces. So they cannot easily adapt to support distinct abstraction levels or user interfaces. They also don't provide adequate support for cooperation with other tools in a software engineering environment. In this paper we discuss several dimensions which may contribute to develop more flexible distributed debuggers. We describe Fiddle, a distributed debugging tool which aims at overcoming some of the above limitations.

1 Introduction

The debugging of parallel and distributed applications still largely relies upon ad-hoc techniques such as the ones based on *printf* statements. Such ad-hoc approaches have the main advantage of providing easy and flexible customization of the displayed information according to the user goals, but they are very limited concerning when, how and what information to display. This makes such approaches not acceptable in general.

Intensive research has been conducted in the field of distributed debugging in the recent past, with several techniques and tools being proposed [1–3], and a few debugging tools reaching a commercial status [28, 30]. Overall, such techniques and tools have contributed to improve the distributed debugging functionalities, in several distinct aspects, such as handling the non-determinism, allowing cyclic interactive debugging of parallel programs, and providing more user-friendly interfaces.

Most of the existing tools for the debugging of parallel and distributed applications are tied to a specific programming language and provide a rigid, albeit sometimes sophisticated, graphical user interface. In general, such tools cannot be easily adapted to support other abstraction levels for application development, or to allow access from other user interfaces. They also don't allow other software development tools having access to the distributed debugger interface, so that, for example, interesting forms of tool cooperation could be achieved in order to meet specific user requirements.

We claim there is a need to develop more flexible distributed debuggers, such that multiple abstraction levels, multiple distributed debugging methodologies, and multiple user and tool interfaces can be supported. In the rest of this paper we first discuss the main dimensions which should be addressed by a flexible distributed debugger. Then we introduce the Fiddle distributed debugger as an approach to meet those requirements, and briefly mention several types of tool interaction concerning the use of Fiddle in a parallel software engineering environment. Finally we discuss ongoing work and summarize the distinctive characteristics of this approach.

2 Towards Flexible Distributed Debugging

In order to better identify the main requirements for the design of a flexible distributed debugging tool, we consider the following main dimensions: i) Multiple abstraction levels; ii) Multiple programming methodologies; iii) Multiple user and tool debugging interfaces.

Multiple distributed debugging abstraction levels

The extra complexity of parallel and distributed applications has led to the development of higher-level programming languages. Some of them are textual languages, while others are based on graphical entities or more abstract concepts. Such higher-level models rely upon e.g., automatic translators to generate the low-level code which is executed by the underlying runtime software/hardware platform. These models have increased the gap to the target languages which are typically supported by a distributed debugger. The same happens concerning the abstractions of lower level communication libraries like PVM [5] or MPI [13], and operating system calls. In order to bridge such semantic gap, a debugger should be flexible and able to relate to the application level abstractions.

Many of the actual distributed debuggers support imperative languages, such as C and Fortran, and object-oriented languages, such as C++ or Java. They also allow to inspect and control the communication abstractions, e.g., message queues at lower-level layers [8]. But this is not enough. The distributed debugger should also be extensible so that it can easily adapt to new programming languages and models. The same reasoning applies to the requirement for supporting debugging at multiple abstraction levels, such as source code, visual (graphical) entities, or user defined program concepts.

Multiple distributed debugging methodologies

Among the diversity of parallel software development tools, one can find tools focusing on performance evaluating and tuning [4, 15, 18, 24, 25], and other focusing on program correction aspects [4, 14, 28, 30]. One can also find tools providing distinct operation modes: i) Tools which may operate on-line with a running distributed application in order to provide an interactive user interface [4, 11, 14, 30]; ii) Tools which can be used to enforce a specific behavior upon the distributed application by controlling its execution, e.g., as described in some kind of script file [22, 26, 27]; iii) Tools which may deal with the information collected by a tracing tool as described in a log file [4, 15, 16, 18, 25].

The on-line modes i) and ii) are mainly used for correctness debugging. The off-line or post-mortem mode iii) is mainly used for performance evaluation and/or visualization purposes, but is also applied to debugging based on a trace and replay approach [20]. The distributed debugger should be flexible in order to enable this diversity of operation modes, and their possible interactions.

The above mentioned distinct operation modes will be put into use by several alternative/complementary debugging methodologies. The basic support which is typically required of a distributed debugger is to allow interactive debugging of remote processes. This allows to act upon individual processes (or threads), but it doesn't help much concerning the handling of non-determinism. In order to support reproducible behavior, the distributed debugger should allow the implementation of a trace and replay approach for interactive cyclic debugging. Furthermore, the distributed debugger should provide support for a systematic state exploration of the distributed computation space, even if such a search is semi-automatic and user-driven [22].

All of the above aspects have great impact upon the design of a flexible distributed debugger architecture. On one hand, this concerns the basic built-in mechanisms which should be provided. On the other hand, it puts strong requirements upon the support for extensible services and for concurrent tool interaction. For example, a systematic state exploration will usually require the interfacing (and cooperation) between an interactive testing tool and the distributed debugger. In order to address the latter issue, we discuss the required support for multiple distributed debugging interfaces, whether they directly refer to human computer interfaces, or they are activated by other software tools.

Multiple user and tool interfaces

Existing distributed debuggers can be classified according to the degree of flexibility they provide regarding user and tool interfacing. Modern distributed debuggers such as TotalView provide sophisticated graphical user interfaces (GUI), and allow the user to observe and control a computation from a set of pre-defined views. This is an important approach, mainly if such graphical interfaces are supported at the required user abstraction levels and for the specific parallel and distributed programming models used for application development (e.g. PVM or MPI).

The main problem with the above approach is that typically the GUI is the only possible way to access the distributed debugger services. If there is no function-call based interface (API), which can be accessed from any separate software tool, then it is not possible to implement more advanced forms of tool interaction and cooperation.

The proposal for standard distributed debugging interfaces has been the focus of recent research [6]. In the meanwhile, several efforts have tried to provide flexible infrastructures for tool interfacing and integration. Most of these efforts are strongly related to the development of parallel software engineering environments [7, 10, 17]. A related effort, albeit with a focus on monitoring infrastructure, is the OMIS [23] proposal for a standard monitoring interface.

Beyond the provision of a well-defined API, a distributed debugger framework should support mechanisms for coordinating the interactions among multiple concurrent tools. This requirement has been recently recognized as of great importance to enable further development of dynamic environments based upon the cooperation among multiple concurrent tools which have on-line access to a distributed computation.

The architecture of a flexible distributed debugger

All of the above mentioned dimensions should have influence upon the design of the architecture of a distributed debugger:

- *Software architecture.* From this perspective distributed debuggers can be classified in two main types: monolithic and client-server. Several of the existing distributed debuggers and related software development tools belong to the first class. This has severe consequences upon their maintenance, and also upon their extensibility and adaptability. A client-server approach, as followed in [11, 14, 15], clearly separates the debugging user interface from the debugging engine. This allows changes to be confined to the relevant sub-part, which is the client if it deals with the user interface, or the server if it deals with the debugging functionalities. This eases the adaptability of the distributed debugger to new demands.
- *Heterogeneity.* Existing distributed debuggers provide heterogeneity support at distinct levels: hardware, operating system, programming language (e.g., imperative, object-oriented), and distributed programming paradigm (e.g., shared-memory, message passing). The first two are a common characteristic of the major debuggers. The imperative and object-oriented languages are also in general simultaneously supported by distributed debuggers [14, 15, 30]. The support for both distributed-memory and shared-memory models also requires a neutral distributed debugging architecture which can in general be achieved using the client-server approach.
- *User and tool interfacing.* In a monolithic approach, the interfacing defined (graphical or text based) is planned beforehand according to a specific set of debugging functionalities. So it has limited capabilities for extension and/or adaptation. On the other hand, a client-server approach allows fully customizable interfaces to be developed anew and more easily integrated into the distributed debugger architecture.

3 The Fiddle approach

Research following from our experiences in the participation of the EU Copernicus projects SEPP and HPCTI [10] has led to the development of Fiddle.

3.1 Fiddle Characteristics

Fiddle [21] is a distributed debugging engine, independent from the user interface, which has the following characteristics:

- *On-line interactive correctness debugging.* Fiddle can be interactively accessed by a human user through a textual or graphical user interface. Its debugging services can equally be accessed by any other software tool, acting as a Fiddle client;
- *Support of high-level user-definable abstractions.* Access to Fiddle is based on an API which is directly available to C and C++ programs. Access to Fiddle is also possible from visual parallel programming languages, allowing a parallel application to be specified in terms of a set of graphical entities (e.g., for application components and their interconnections). After code generation to a lower level language

such as C, it would not be desirable for the user to perform program debugging at such a low level. Instead, the user should be able to debug in terms of the same visual entities and high-level concepts which were used for application development. The Fiddle architecture enables such kind of adaptation, as it can incorporate a high-level debugging interface which understands the user-level abstractions, and is responsible for their mapping onto Fiddle basic debugging functions;

- *Many clients / many servers.* Multiple clients can connect simultaneously to Fiddle and have access to the same set of target processes. Fiddle uses smart daemons, spread over the multiple nodes where the target application is being executed, to do the heavy part of the debugging work, relying on a central daemon for synchronization and interconnection with the client tools. By distributing the workload by the various nodes, reaction time and global performance are not compromised. This software architecture also scales well as the number of nodes increases;

- *Full heterogeneity support.* Fiddle architecture is independent of the hardware and operating system platforms as it relies on node debuggers to perform the system dependent commands upon the target processes. Fiddle is also neutral regarding the distributed programming paradigm of the target application, i.e., shared-memory or distributed-memory, as the specific semantics of such models must be encapsulated as Fiddle services, which are not part of the Fiddle core architecture;

- *Tool synchronization.* By allowing multiple concurrent client tools to access a common set of target processes, Fiddle needs to provide some basic support for tool synchronization. This can be achieved in two ways: i) As all requests made by client tools are controlled by a central daemon, Fiddle is able to avoid certain interferences among these tools; ii) Tools can also be notified about events originated by other tools, thus allowing them to react and coordinate their actions (see Sec. 4). In general, however, the cooperation among multiple tools will require other forms of tool coordination, and no basic support is provided for such functionalities, as they are dependent on each application software development environment. However, some specific tool coordination services can be integrated into Fiddle so that, for example, multiple tools can get consistent views of a shared target application;

- *Easy integration in Parallel Software Engineering Environments.* The event-based synchronization mechanism provided by Fiddle can be used to support interaction and synchronization between Fiddle debugging client tools and other tools in the environment, e.g., on-line program visualization tools.

3.2 Fiddle Software Architecture

Fiddle is structured as a hierarchy of five functional layers. Each layer provides a set of debugging services which may be accessed through an interface library. Any layer may be used directly by a client tool which finds its set of services adequate for its needs. Layers are also used indirectly, as the hierarchical structure of Fiddle's software architecture implies that each layer $\mathcal{L}_i$ $_{(i>0)}$ is a direct client of layer $\mathcal{L}_{i-1}$ (see Fig. 1). In this figure, Layer 3_m has two clients (tools CT_2^{3m} and CT_1^{3m}), Layer 2_m has also two clients (tool CT_1^{2m} and Layer 3_m), and each of the remaining layers is shown with only one client, in the layer immediately above.

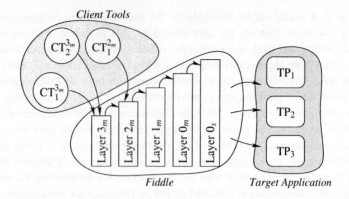

Fig. 1. Fiddle's Layered Architecture

A service requested by a Layer 3_m client tool will be processed and transferred to the successively underlying layers, until Layer 0_s is reached. At this point, the request is applied to the target process. The result of such service request is also successively processed and transferred to the upper layers until the client tool gets the reply.

There is a minimum set of functionalities common to all Fiddle layers, namely:

- *Inspect/control multi-threaded target processes.* Fiddle provides a set of debugging services to act upon threads within a multi-threaded process. These services will be active only if the node debugger being used on a specific target process is able to deal with multi-threaded processes;
- *Inspect/control multiple target processes simultaneously.* Any client tool may use Fiddle services to inspect and control multiple processes simultaneously. Except for Layer 0_s and Layer 0_m, the target process may also reside in remote nodes;
- *Support for client tool(s).* Some layers accept only one client tool while some others accept multiple client tools operating simultaneously over the same target application.

These common functionalities and the layer-specific ones described below are supported by the software architecture shown in Fig. 2.

Layer 0_s This layer implements a set of local debugging services. A client tool for this layer must be the only one acting upon the set of local single- or multi-threaded target processes.

The following components are known to this layer: i) *Target processes*: a subset of processes that compose the target application and are being executed in the local node; ii) *Node debuggers*: associated to each target process there is a node debugger which is responsible for acting upon it; iii) *Layer 0_s library*: provides single-threaded access to the debugging functionalities; iv) *Client tool*: this must be a single-threaded tool.

Layer 0_m This layer extends Layer 0_s to provide support for multi-threaded client tools. Client tools may use threads to control multiple target process and interact simultaneously with the user.

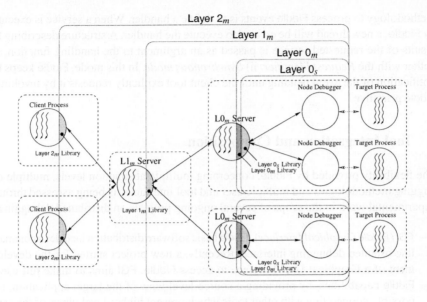

Fig. 2. The Fiddle software architecture

Layer 1ₘ This layer extends Layer 0ₘ to provide support for remote target processes. The target application can spread over multiple nodes and Fiddle can be used to debug local or remote target processes.

The components known to this layer are: 1. *Target processes*: running in the local or remote nodes; 2. *Node debuggers*: associated to each target process there is a node debugger; 3. *L0ₘ Server*: the daemon is a Layer 0ₘ client tool and acts as a gateway between the Layer 1ₘ client tool and the target application. 4. *Layer 1ₘ Library*: provides thread-safe access to local and remote debugging functionalities; 5. *Client tool*: Layer 1ₘ client tools may transparently access local and remote target processes, relying on *L0ₘ Server* daemons to act as tool representatives in each node.

Layer 2ₘ This layer extends Layer 1ₘ to provide support for multiple simultaneously client tools. These tools may concurrently request Fiddle's debugging services. The software architecture of Layer 2ₘ includes an instance of Layer 1ₘ, plus a daemon, the *L1ₘ Server*. This daemon is a Layer 1ₘ client tool, which multiplexes the service requests from the multiple client tools to be submitted to Layer 1ₘ. It also demultiplexes the corresponding replies back to the clients..

Layer 3ₘ This layer extends Layer 2ₘ to provide an event-based interaction between Fiddle and the client tools. In contrast to the lower layers, a thread that invokes a method in the *Layer 3ₘ Library* doesn't block waiting for its reply. Instead, it receives a *Request Identifier* which will be used later to process the reply. When the service is executed, its success status and data (the *reply*) is sent to the client as an event. These events may be processed by the client tool in two different ways: i) *Asynchronous mode*. The general methodology to process Fiddle events is to define a handler. When a service is executed by Fiddle, a new thread will be created to execute the handler. A structure describing the results of the requested service is passed as an argument to the handling function, together with the *Request Identifier*. ii) *Synchronous mode*. In this mode, Fiddle keeps the notification of the event pending until the client tool explicitly requests it by invoking a Fiddle primitive.

4 Tool Integration and Composition

The flexibility provided by Fiddle, concerning multiple abstraction levels, multiple debugging methodologies and multiple user and tool interfaces, is being assessed through experimental work in the scope of several ongoing projects on distributed debugging:

- *FGI (Fiddle Graphical Interface)*. Fiddle basic software distribution includes a command-line oriented debugging interface. Recently, a new project started for the development of a Graphical User Interface to access Fiddle. FGI aims to make full use of Fiddle capabilities, by providing a source-level view of the target application, and possibly cooperating with other tools which present higher-level views of the same application;
- *Deipa (Deterministic Execution and Interactive Program Analysis)*. The use of an interactive *testing tool* which partially automates the identification and localization of suspect program regions can improve the process of developing correct programs. Deipa is an interactive steering tool that uses Fiddle to drive the program execution path according to a specification produced by STEPS [19] (an interactive testing tool developed at TUG, Gdansk, Poland). Deipa cooperates with other Fiddle client tools, such as FGI, to allow localized program state inspection and control. With Deipa the user may direct the program execution to a specific testing point, and then another tool, such as FGI, can be used to inspect and fine-control the application processes. When satisfied, the user may reuse Deipa to proceed execution until the next testing point;
- *PADI (Parallel Debugger Interface)*. PADI [29] (developed at UFRGS, Rio Grande do Sul, Brasil) is a distributed debugging GUI, written in Java, which uses Fiddle as the distributed debugging engine, in order to implement debugging operations on groups or individual processes;
- *On-line Visualization and Debugging*. Research is under way to support the coordination of the on-line observation of a distributed application, as provided by a parallel program visualizer Pajé [15], and the debugging actions of Fiddle. Both tools were independently developed so they must now be adapted to achieve close cooperation. For example, whenever a process under debugging reaches a breakpoint and stops, then such a state transition must be accordingly updated by Pajé and reflected on its on-line visualization. On the other hand, if Pajé shows that a given process is blocked waiting for a message, we may be interested in selecting the process and having Fiddle automatically stopping the process, and selecting the source line containing the message reception code, and refreshing (or possibly opening) a source-level debugging GUI.

5 Conclusions and Future Work

The Fiddle distributed debugger is a result of our research on distributed debugging for the past 6 years [11,12]. It currently provides support for debugging multi-threaded/multi-process distributed applications. Its distinctive characteristics are the incremental design as a succession of layers, and its strong focus on the tool cooperation aspects. Fiddle current prototype fully implements Layer 0_s to Layer 2_m, while Layer 3_m is under development. It runs under Linux, uses GDB as the node debugger, and has a C/C++ API.

In order to evaluate and improve the support provided by Fiddle concerning flexible tool composition and integration in a Parallel Software Engineering Environment, there is experimental work under way. This is helping us to assess the design options, through the implementation of several case studies. Ongoing work also includes the development of a Java-based API.

As part of future work we plan to integrate Fiddle into the DAMS distributed monitoring architecture [9] as a *distributed debugging service*, in order to allow more generic and extensible tool cooperation and integration. We also plan to address large scale issues on distributed debugging for cluster computing environments.

Acknowledgments. The work reported in this paper was partially supported by the PRAXIS XXI Programme (SETNA Project), by the CITI (Centre for Informatics and Information Technology of FCT/UNL), and by the cooperation protocol ICCTI/French Embassy in Portugal.

References

1. *Proc. of the ACM Workshop on Parallel and Distributed Debugging*, volume 24 of *ACM SIGPLAN Notices*. ACM Press, January 1988.
2. *Proc. of the ACM Workshop on Parallel and Distributed Debugging*, volume 26 of *ACM SIGPLAN Notices*. ACM Press, 1991.
3. *Proc. of the ACM/ONR Workshop on Parallel and Distributed Debugging*, volume 28 of *ACM SIGPLAN Notices*. ACM Press, 1993.
4. Don Allen et al. The Prism programming environment. In *Proc. of Supercomputer Debugging Workshop*, pages 1–7, Albuquerque, New Mexico, November 1991.
5. A. Beguelin et al. A user's guide to PVM parallel virtual machine. Technical Report ORNL/TM-118266, Oak Ridge National Laboratory, Tennessee, 1991.
6. J. Brown, J. Francioni, and C. Pancake. White paper on formation of the high performance debugging forum. Available in "http://www.ptools.org/hpdf/meetings/mar97/-whitepaper.html", February 1997.
7. Christian Clémençon et al. Annai scalable run-time support for interactive debugging and performance analysis of large-scale parallel programs. In *Proc. of EuroPar'96*, volume 1123 of *LNCS*, pages 64–69. Springer, August 1996.
8. J. Cownie and W. Gropp. A standard interface for debugger access to message queue information in MPI. In *Proc. of the 6th EuroPVM/MPI*, volume 1697 of *LNCS*, pages 51–58. Springer, 1999.
9. J. C. Cunha and V. Duarte. Monitoring PVM programs using the DAMS approach. In *Proc. 5th Euro PVM/MPI*, volume 1497 of *LNCS*, pages 273–280, 1998.

10. J. C. Cunha, P. Kacsuk, and S. Winter, editors. *Parallel Program Development for Cluster Computing: Methodology, Tools and Integrated Environment.* Nova Science Publishers, 2000.

11. J. C. Cunha, J. Lourenço, and T. Antão. An experiment in tool integration: the DDBG parallel and distributed debugger. *Euromicro Journal of Systems Architecture*, 45(11):897–907, 1999. Elsevier Science Press.

12. J. C. Cunha, J. Lourenço, J. Vieira, B. Moscão, and D. Pereira. A framework to support parallel and distributed debugging. In *Proc. of HPCN'98*, volume 1401 of *LNCS*, pages 708–717. Springer, April 1998.

13. MPI Forum. *MPI-2: Extensions to the message-passing interface.* Univ. of Tennessee, 1997. http://www.mpi-forum.org/docs/mpi-20-html/mpi2-report.html.

14. R. Hood. The p2d2 project: Building a portable distributed debugger. In *Proc. of the 2nd Symposium on Parallel and Distributed Tools*, Philadelphia PA, USA, 1996. ACM.

15. J. C. Kergommeaux and B. O. Stein. Pajé: An extensible environment for visualizing multithreaded programs executions. In *Proc. Euro-Par 2000*, volume 1900 of *LNCS*, pages 133–140. Springer, 2000.

16. J. A. Kohl and G. A. Geist. The PVM 3.4 tracing facility and XPVM 1.1. In *Proc. of the 29th HICSS*, pages 290–299. IEEE Computer Society Press, 1996.

17. D. Kranzlmüller, Ch. Schaubschläger, and J. Volkert. A brief overview of the MAD debugging activities. In *Proc. of AADEBUG 2000*, Munich, Germany, August 2000.

18. D. Kranzlmuller, S. Grabner, and J. Volkert. Debugging massively parallel programs with ATTEMPT. In *High-Performance Computing and Networking (HPCN'96 Europe)*, volume 1067 of *LNCS*, pages 798–804. Springer, 1996.

19. H. Krawczyk and B. Wiszniewski. Interactive testing tool for parallel programs. In *Software Engineering for Parallel and Distributed Systems*, pages 98–109, London, UK, 1996. Chapman & Hal.

20. T. J. LeBlanc and J. M. Mellor-Crummey. Debugging Parallel Programs with Instant Replay. *IEEE Transactions on Cumputers*, C–36(4):471–482, April 1978.

21. J. Lourenço and J. C. Cunha. *Flexible Interface for Distributed Debugging (Library and Engine): Reference Manual (V 0.3.1).* Departamento de Informática da Universidade Nova de Lisboa, Portugal, December 2000.

22. J. Lourenço, J. C. Cunha, H. Krawczyk, P. Kuzora, M. Neyman, and B. Wiszniewsk. An integrated testing and debugging environment for parallel and distributed programs. In *Proc. of the 23rd EUROMICRO Conference*, pages 291–298, Budapeste, Hungary, September 1997. IEEE Computer Society Press.

23. T. Ludwing, R. Wismüller, V. Sunderam, and A. Bode. OMIS – On-line monitoring interface specification. Technical report, LRR-Technish Universiät München and MCS-Emory University, 1997.

24. B. P. Miller, J. K. Hollingsworth, and M. D. Callaghan. The Paradyn parallel performance measurement tools. *IEEE Computer*, 28(11):37–46, November 1995.

25. W. E. Nagel et al. VAMPIR: Visualization and analysis of MPI resources. *Supercomputer*, 12(1):69–80, January 1996.

26. M. Oberhuber. Managing nondeterminism in PVM programs. *LNCS*, 1156:347–350, 1996.

27. Michael Oberhuber. Elimination of nondetirminacy for testing and debugging parallel programs. In *Proc. of AADEBUG'95*, 1995.

28. Michael Oberhuber and Roland Wismüller. DETOP - an interactive debugger for PowerPC based multicomputers. pages 170–183. IOS Press, May 1995.

29. D. Stringhini, P. Navaux, and J. C. Kergommeaux. A selection mechanism to group processes in a parallel debugger. In *Proc. of PDPTA'2000*, Las Vegas, Nevada, USA, June 2000.

30. Dolphin ToolWorks. *TotalView.* Dolphin Interconnect Solutions, Inc., Framingham, Massachusetts, USA. http://www.etnus.com/Products/TotalView/.

Visualisation of Distributed Applications for Performance Debugging

F.-G. Ottogalli[1], C. Labbé[1], V. Olive[1], B. de Oliveira Stein[2],
J. Chassin de Kergommeaux[3], and J.-M. Vincent[3]

[1] France Télécom R&D - DTL/ASR, 38243 Meylan cedex, France
{francoisgael.ottogalli,cyril.labbe,vincent.olive}@rd.francetelecom.fr
[2] Universidade Federal de Santa Maria, Rio Grande do Sul, Brazil
benhur@inf.UFSM.br
[3] Laboratoire Informatique et Distribution, Montbonnot Saint Martin, France
{Jacques.Chassin-de-Kergommeaux,Jean-Marc.Vincent}@imag.fr

Abstract. This paper presents a method to perform visualisations of the behaviour of distributed applications, for performance analysis and debugging. This method is applied to a Java distributed application. Application level traces are recorded without any modification of the monitored applications nor of the JVMs. Trace recording includes records from the JVM, through the JVMPI, and records from the OS, through the data structure associated to each process. Recorded traces are visualised post mortem, using the interactive Pajé visualisation tool, which can be conveniently specialised to visualise the dynamic behaviour of distributed Java applications. Applying this method to the execution of a book server, we were able to observe a situation where both the computation or the communications could be at the origin of a lack of performances. The observation helped finding the origin of the problem coming in this case from the computation.

Keywords: performance analysis and debugging, distributed application, Java, JVMPI, meta-ORB.

1 Introduction

The aim of the work described in this article is to help programmers to analyse the executions of their distributed programs, for performance analysis and performance debugging. The approach described in the following includes two major phases: recording of execution traces of the applications and post mortem trace-based visualisations. The analysis of distributed applications is thus to be done by programmers, with the help of a visualisation tool displaying the execution behaviour of their applications.

In the presented example, a distributed application is executed by several Java[TM][1] Virtual Machines (JVM) cooperating through inter-objects method calls on a distributed infrastructure.

---

[1] Java and Java-based marks are trademarks or registered trademarks of Sun Microsystems, Inc. in the United States and other countries. The authors are independent of Sun Microsystems, Inc.

This work was carried out in the context of the Jonathan project [6] developed at France Télécom. Jonathan is a meta-ORB, that is a framework allowing the construction of object-oriented platforms such as ORBs (*Object Request Brokers*). Jonathan could be specialised to be CORBA or a Java RMI (*Remote Method Invocation*) style. In the following we will be concerned with the CORBA specialisation.

Performance traces are recorded at the applicative level of abstraction, by recording the method calls through the Java Virtual Machine Profiling Interface (JVMPI) [1, 14, 15]. Additional information needs to be recorded at the operating system level, in order to identify inter-JVMs communications. Execution traces are then passed to the interactive visualisation tool Pajé [4]. Pajé was used because it is an interactive tool – this characteristic being very helpful for performance debugging – which can be tailored conveniently to visualise the execution and communications between JVMs.

The existing Java monitoring and visualisation tools cannot be used conveniently for performance debugging of distributed Java applications. *hprof* [12] and *Optimizeit* [7] provide on-line cumulative information, without support for distributed applications (monitoring of communications). Although this information can be very useful to exhibit performance problems of sequential Java applications, it is of little use to help identifying the origin of performance problems in the distributed settings. The JaVis tool [10] can be used to trace the communications performed by Java Virtual Machines. However the recording is performed by modified JVMs while we decided to stick to standard JVMs.

The main outcome of this work is the ability to observe and visualise the dynamic behaviour of distributed Java applications **including communications**, without any modification of the application programs. Moreover it allows a temporal analysis of the hierarchy of methods invocations.

The organisation of this paper is the following. The recording of execution traces is described in Section 2. Section 3 describes the visualisation tool Pajé and its specialisation for the visualisation of the behaviour of distributed Java applications. Then comes the conclusion which also sketches future work.

2 Recording traces of the execution of Java applications on distributed platforms

Performance analysis is performed off-line, from execution traces which are ordered sets of events. An event is defined as a state change in a thread of a JVM. Two types of events will be considered: the infrastructure events, corresponding to state changes associated with the JVM machinery and to the standard JDK classes used, and the applicative events, associated with the execution of the methods of the application classes. In order to take into account the state changes of internal variables, it is possible to build specific recording classes, to be observed as well as applicative classes.

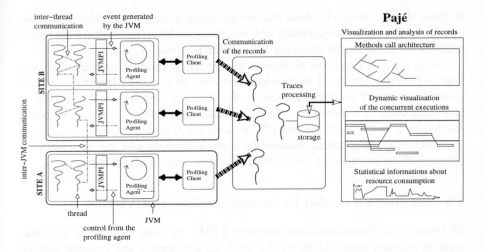

Fig. 1. *Global architecture for analysis and performance debugging.* Profiling agents are used to observe events and interact with the JVM. Profiling clients connect profiling agents to the applications dedicated to collect and process the data traces. Then, the data traces are visualised with Pajé.

To each recorded event a date, identification and location are assigned so that the execution path of the application as well as the communications can be reconstructed post-mortem by the analysis tools.

2.1 Application level recording traces (Figure 1)

Events are recorded using the JVM Profiling Interface (JVMPI) defined by SUN and implemented in the Linux_JDK_1.2.2_RC4. This functional interface allows event occurrences and control operations to be observed [14]. The JVMPI is accessed by implementing an observation agent [1, 15], loaded during the initialisation of the JVM. The following events are observed:

- loading or unloading of a class;
- beginning or termination of a method call;
- creation, destruction or transfer of an object in the execution stack;
- creation, suspension, destruction of a thread;
- use of monitors (waiting and acquisition);
- user defined events (program annotations to observe specific events).

Observing these events is used to reconstruct the execution from the JVM point of view. It is used to construct several representations of the execution. One, based on thread execution, displays the execution path of the methods (Figure 4). Another, not shown in this paper, represents the execution in terms of object and method execution.

Constructing an event trace requires to date, identify and localise each event. A unique identifier is associated to each loaded class, and each method defined in these classes. Similarly, Java threads are given a unique identifier. Since the events of a thread are sequentially ordered, it is possible to reconstruct the causal order of an execution by grouping the records by thread.

Using a JVMPI to observe events can potentially produce an important volume of recorded events. It is therefore necessary to provide **filtering** functions to limit the size of the recorded traces. For example (see Section4), filtering an execution trace by a mere exclusion of standard Java classes from the recording, divided by a factor of 32 the size of the execution trace.

Additional information, relating to communications and use of system resources will be recorded at the operating system level of abstraction.

2.2 Information needed for communications observations

To observe the communications between JVMs, we need to identify the links created between them. Calls to the methods performing the communications are recorded by the JVMPI. However, communication parameters, necessary to reconstruct the dynamic behaviour of the applications, are lost in our records.

Accessing these parameters through the JVMPI would not have been simple since calling parameters cannot be accessed without delving into the execution stack of the JVM; this would be the case for example with the communication parameters, when recording a communication event. The approach used instead to obtain the parameters is to observe the messages sent and received at **the operating system level of abstraction**. This choice was driven by two major reasons :

- to have a direct access to the parameters associated to the communications;
- to obtain information about operating system resources consumption [13].

We assume that "sockets" are used to establish connections between JVMs. The identification of the sockets used for inter-objects communications is performed at the operating system layer. In the case of the Linux operating system, the sockets used by a process or a thread are represented as typed "i-nodes". The data structures associated to these i-nodes can be used to identify a given link by the IP address of the remote host and the port number used.

Once all the data have been collected, as described in figure 1, they can be visualised with Pajé.

3 Visualisation using the Pajé generic tool

Pajé is an interactive visualisation tool for displaying the execution of parallel applications where a (potentially) large number of communicating threads of various life-times execute on each node of a distributed memory parallel system [2–4]. The main novelty of Pajé is an original combination of three of the most desirable properties of visualisation tools for parallel programs: extensibility, interactivity and scalability.

In contrast with passive visualisation tools [9] where parallel program entities – communications, changes in processor states, etc. – are displayed as soon as produced and cannot be interrogated, it is possible to inspect all the objects displayed in the current screen and to move back in time, displaying past objects again. Scalability is the ability to cope with a large number of visual objects such as threads. Extensibility is an important characteristic of visualisation tools to cope with the evolution of parallel programming interfaces and visualisation techniques. Extensibility gives the possibility to extend the environment with new functionalities: processing of new types of traces, adding new graphical displays, visualising new programming models, etc.

3.1 Adapting Pajé for visualising distributed Java executions

Extensibility is a key property of a visualisation tool. The tool has to cope with the evolutions of parallel programming models – since this domain is still evolving rapidly – and of the visualisation techniques. Several characteristics of Pajé were designed to provide a high degree of extensibility: modular architecture, flexibility of the visualisation modules and genericity. It is mainly the genericity of Pajé which made possible to specialise it for visualising distributed Java applications.

The Pajé visualisation tool can be specialised for a given programming model by inserting an instantiation program in front of a trace file. The visualisation to be constructed from the traces can be programmed by the user, provided that the types of the objects appearing in the visualisation are hierarchically related and that this hierarchy can be described as a tree (see Figure 2). This description is inserted in front of the trace file to be analysed and visualised.

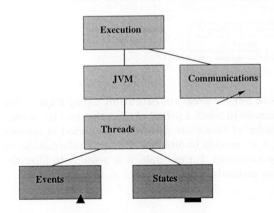

Fig. 2. *Example of type hierarchy for visualising Java distributed applications.* Intermediate nodes are containers while the leaves of the type hierarchy are the types of the elementary visual entities. Method calls are represented as state changes of the displayed threads.

3.2 Trace visualisation with Pajé

The Pajé command language, allowing users to discribe how traces should be visualised, is based on two generic types: containers and entities. While entities can be considered as elementary visual objects, containers are complex objects, each including a (potentially high) number of entities. An entity type is further specified by one of the following generic types: event, state, link or variable. Another generic type used to program Pajé is the value type which can describe the type of one or several fields of complex entity types. Java abstractions were mapped to the Pajé generic command language according to the type hierarchy of Figure 2.

The visualisation is similar to a space-time diagram, where the time varies along the x-axis while JVMs and threads are placed along the y-axis (see Figure 3). The execution of the Java methods are represented as embedded boxes in the frame of each of the threads. Inter-thread communications appear as arrows between the caller and the callee communication methods. It is also possible to visualise, on the same diagram, the evolution of several system variables and the use of monitors, in order to draw correlations between application programs and consumption of system resources (not shown in this paper).

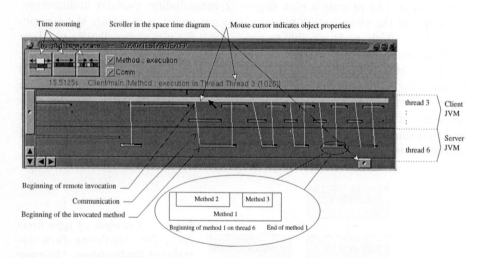

Fig. 3. *Visualisation of the Book Server program execution using Pajé.* The interactivity of Pajé allows programmers to select a period of execution and to "zoom" with respect to the time scale in order to have more details on the period of interest displayed on the screen. Similarly, it is possible to filter the displayed information, to restrict the amount of details to be visualised. For example, it is possible to filter the communications or the calls to some methods.

4   Observation of a book server execution

The main purpose of this observation is to highlight the ordering of events corresponding to the beginning and the end of methods execution as well as communications. This will help identifying performance bottlenecks in terms of processing and network resources consumption.

The target application, an electronic book server, is an important application of the "net economy". In addition it is representative of a whole set of on-demand data server applications in the domains of multimedia, video and music. Such applications can be characterised by massively concurrent accesses to the provided resources. In this example, the objective was to optimise the global performances of the application. One situation was identified where a bottleneck resulted from the excessive execution time of some methods and not from the communication delays. This drove us to conclude that optimisations should first concern these methods and their scheduling on the threads of the application.

4.1 Description of the experiment

The displayed trace represents the execution of two clients and of the "book" resource: these entities perform most of the computations and communications. Since our objective is to analyse and improve performances, these entities have to be analysed in priority.

The execution trace was realized during an experiment involving three different locations. The resources and the resource server are hosted by the first site while two clients are located on the two other sites and access the same resource simultaneously. The three computers are interconnected by a 10 MBit Ethernet network; they all run Linux - kernel 2.4.0 of Debian woody - and use the 1.2.2 RC4 JDK from SUN.

This execution environment does not have a hardware global clock. Therefore, a global clock had to be implemented by software [11, 8, 5].

4.2 Trace processing and interpretation

Trace files need to be processed to convert local dates into global dates using Maillet's algorithm [11]. Matching of Java and system threads is then performed, allowing system and application level traces to be merged.

Figure 4 exemplifies the functionalities provided by Pajé to visualise execution paths and communications. It is a detailed view showing the beginning of a communication.

Figure 5 displays a global view of the experiment described in Section 4.1. The execution of the book server, two clients as well as the communications between one of the clients and the server are visible. This representation makes it clear that the communication phase is very short with respect to the entire duration of the experiment.

Figure 6 is used to identify the origin and destination of communications (JVM, thread, method) as well as the interleaving of the computation and communication phases of threads.

Thus it is possible to observe a pattern (gray area in Figure 6) composed of four communications occurring five times during the visualised time interval.

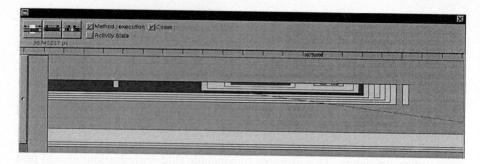

Fig. 4. *Visualisation of the execution path of a thread.* This visualisation represents the execution of two threads inside the same *JVM*. The embedding of clear and dark boxes represents the embedding of method calls. Dark boxes represent selected boxes. The line originating from the upper thread represents the beginning of a communication.

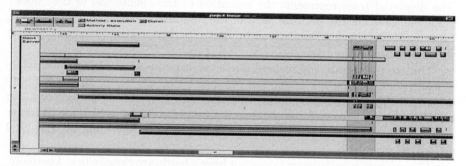

Fig. 5. *Visualisation of the execution of three JVMs.* This figure represents the execution of three *JVMs* on three different sites. An area, representing a period which includes several communications between two *JVMs*, was selected (in gray) to be magnified, using the zooming facilities of Pajé (Figure 6).

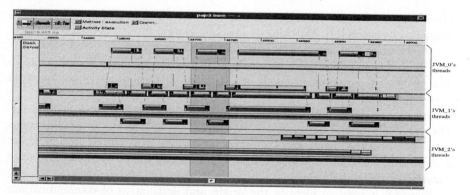

Fig. 6. *Observing a series of communication sequences between two JVMs.* *JVM* 0 represents the execution of the "book" resource while *JVM* 1 and 2 represent the executions of the first and second clients. Only communications between the "book" resource and the first client are displayed.

This pattern can be divided in two phases including each two communications: the first one from the client to the server and the second one from the server to the client. The execution time of this pattern is fairly constant, except for the forth occurrence. A trace analysis indicates that the whole execution is slowed down by the excessive duration of a thread (third from the top). The remaining of the sequence is indeed dependent on the sending of a message from this thread.

On the contrary, communication delays are low and constant with respect to computation times. In such a situation, improving the performances of this application is better done by optimising some specific methods that are identified to be time consuming than by trying to reduce the communication delays.

5 Conclusion

This paper presents a method to perform visualisations of the behaviour of distributed applications in the scope of performance analysis and performance debugging. Application level traces are recorded without any modification of the monitored applications nor of the JVMs. Trace recording includes recording of the method calls at the application level by the JVMPI as well recording communication information at the socket level of the operating system. Recorded traces are visualised post mortem, using the interactive Pajé visualisation tool, which can be conveniently specialised to visualise the dynamic behaviour of distributed Java applications. This method has been applied to a Java distributed application. It was thus possible to discriminate between two possible origins of a performance problem, ruling out the hypothesis of inefficient communications.

This work is still in progress and is amenable to several new developments, some of them concerning the tracing activity while the other extensions concern the visualisation tool.

First of all, the tracing overhead should be analysed in order to be able to assess the quality of the traced information and therefore of the visualisations of the executions of the traced programs. Further work will aim at evaluating the system resources required by on-line analysis of the traces: this analysis could help evaluating the system resources to be provided to analyse on-line distributed applications while keeping the analysis overhead low. Another perspective is the use of the observation and visualisation environment for system "black-box" analysis.

For the Pajé visualisation tool, the perspectives include several extensions in order to integrate the operating system resource consumption observations – such as the use of processors, memory or the bandwidth of the network – with the application visualisations and therefore help programmers to relate both. In addition, Pajé could be extended in order to take into account several problems related with the visualisation of embedded systems, such as the display of the synchronisation mechanisms.

References

1. Java virtual machine profiler interface (JVMPI). Technical report, Sun Miscrosystems, 1999.
2. J. Chassin de Kergommeaux and B. de Oliveira Stein. Pajé: an extensible and interactive and scalable environment for visualizing parallel executions. R.R. 3919, INRIA, april 2000. http://www.inria.fr/RRRT/publications-eng.html.
3. J. Chassin de Kergommeaux and B. de Oliveira Stein. Pajé: an extensible environment for visualizing multi-threaded programs executions. *Euro-Par 2000 Parallel Processing, LNCS* 1900, pages 133–140. Springer, 2000.
4. J. Chassin de Kergommeaux, B. de Oliveira Stein, and P. Bernard. Pajé, an interactive visualization tool for tuning multi-threaded parallel applications. *Parallel Computing*, 26(10):1253–1274, aug 2000.
5. J. Chassin de Kergommeaux, E. Maillet, and J.-M. Vincent. Monitoring parallel programs for performance tuning in distributed environments. *Parallel Program Development for Cluster Computing: Methodology, Tools and Integrated Environments*, chapter 6. Nova Science, 2001. To appear.
6. B. Dumant, F. Dang Trang, F. Horn, and J.-B. Stefani. Jonathan : an open distributed processing environment in java. In *Middleware'98: IFPI Int. Conf. on Distributed Systems Platforms and Open Distributed Processing*, Sept. 1998.
7. G. Freedman. Common java performance issues and solutions. Technical report, Intuitive System Inc., 1998.
8. Y. Haddad. *Performance dans les systèmes répartis : des outils pour les mesures.* PhD thesis, Université de Paris-Sud, centre d'Orsay, Septembre 1988.
9. M. T. Heath. Visualizing the performance of parallel programs. *IEEE Software*, 8(4):29–39, 1991.
10. I. Kazi and al. Javiz : A client/server java profiling tool. *IBM - Systems Journal*, 39(1):82, 2000.
11. E. Maillet and C. Tron. On efficiently implementing global time for performance evaluation on multiprocessor systems. *Journal of Parallel and Distributed Computing*, 28(1):84–93, 1995.
12. N. Meyers. A performance analysis tool. Tech. report, Sun Miscrosystems, 2000.
13. F.-G. Ottogalli and J.-M. Vincent. Mise en cohérence et analyse de traces logicielles. *Calculateurs Parallèles, Réseaux et Systèmes répartis*, 11(2), 1999.
14. Sun Microsystems. *Java Virtual Machine Profiler Interface (JVMPI)*, Feb 1999.
15. D. Viswanathan and S. Liang. Java virtual machine profiling interface. *IBM - Systems Journal*, 39(1):82, 2000.

Achieving *Performance Portability* with *SKaMPI* for High-Performance MPI Programs

Ralf Reussner and Gunnar Hunzelmann

Chair Computer Science for Engineering and Science
Universität Karlsruhe (T.H.)
Am Fasanengarten 5, D-76128 Karlsruhe, Germany.
{reussner | gunnar}@ira.uka.de

Abstract. Current development processes for parallel software often fail to deliver portable software. This is because these processes usually require a tedious tuning phase to deliver software of good performance. This tuning phase often is costly and results in machine specific tuned (i.e., less portable) software. Designing software for performance *and* portability in early stages of software design requires performance data for all targeted parallel hardware platforms. In this paper we present a publicly available database, which contains data necessary for software developers to design and implement portable and high performing MPI software.

1 Introduction

The cost of todays parallel software mostly exceeds the cost of sequential software of comparable size. This is mainly caused by parallel programming models, which are more complex than the imperative sequential programming model. Although the simplification of programming models for parallel and distributed systems is a promising research area, in the next years no change of the dominance of the currently message-passing or shared-memory imperative models for parallel and distributed systems is to be expected. One common problem of the mentioned programming models for parallel hardware is that parallelisation must be stated explicitly (despite a slightly better support for purely data-parallel programs through e.g., Fortran 90 or HPF [KLS+94]). (The need for manual parallelisation arises because compilers do not support parallelisation as good as vectorisation.)

Current software processes for the development of parallel software reflect this need of manual parallelisation in two phases [Fos94].

Design phase: Various ways of problem specific parallelisation are known, such as multi-grid methods, event-driven simulations, specialised numerical algorithms, etc. All these approaches are used during the design phase to create a design which either reflects the parallelism inherent in the problem, or applies parallel algorithms to speed up computations.

Tuning phase: Software development processes for parallel software contain a special tuning phase, which comes after the integration test of the software. In this tuning phase software is optimised for specific hardware platforms. That is, code is evaluated with measurements, and then several actions are applied: e.g., rearrangement of code, hiding communication latency, replacement of standardised communication operations by more specific ones, etc. These steps are often required to make the software as fast as necessary on the targeted platform. Unfortunately, portability to other platforms is usually lost.

Hence, machine specific performance considerations are not explicitly made during design stage, but only in the tuning phase. Some of the performance measurements done in the tuning phase are specific for the program's design, but other measurements concern the underlying MPI and hardware performance. In this paper we present a public database containing MPI performance data.

MPI as the most widely used standard for developing message passing software manifested *compiling portability*. This means, that the user is able to develop software which just has to be recompiled when changing the hardware platform, opposed to limited prortability when using vendor specific libraries. (Ommiting here all the remaining tiny peculiarities when trying to develop really portable software with the C programming language.)

Due to the above mentioned platform specific program tuning, this compiling portability usually does not help much when writing high performance software. What is really needed is *performance portability*. This term refers to the requirement, that portability for a parallel program does not only mean that it is compilable on serveral platforms, but also that it shows at least good performance on all these platforms without modifications. (A more quantitative notion of portability would take into account (a) the effort of making a program compilable, (b) the effort of showing good (or best) performance, and (c) the really achieved performance before and after modifications.)

In section 2 we present work related to MPI benchmarking. How performance data can be used during the design of parallel software is discussed in section 3. Section 4 contains a description of our performance database. An example is presented in section 5. Finally, section 6 concludes.

2   Related Work

Benchmarking in general and benchmarking communication operations on parallel platforms in particular require a certain caution to ensure the validity of benchmarked results for "normal application programs". Gropp et al. present guidelines for MPI benchmarking [GL99]. Hempel gives some additional advice and formulates as a general benchmarking principle, that benchmarks never should show better results when lowering the performance of the MPI implementation (the benchmarked entity in general) [Hem99]. Most interestingly, he describes a case, where making MPI point to point communication slower resulted in better results of the COMMS1 – COMMS3 suites of PARKBENCH [PAR94]. Currently no benchmark exactly fulfils all these requirements. The mechanisms applied by *SKaMPI* to tackle MPI benchmarking problems are described in [RSPM98,Reu99]. Some new algorithms to benchmark collective operations reliable are given in [dSK00]. The benchmarks which come probably closest to *SKaMPI*'s goals are the following two: A widely used MPI benchmark is the one shipped with the mpich[1] implementation of MPI; it measures nearly all MPI operations. Its primary goal is to validate mpich on the given machine; hence it is less flexible than *SKaMPI*, has less refined measurement mechanisms and is not designed for portability beyond mpich.

The b_{eff} benchmark of Rabenseifner measures network performance data from a perspective interesting for the application programmer and complements the measures included in *SKaMPI*. The results are publicly available on the web[2]. As intended for users of the Cray T3E and other machines installed at the HLRS in Stuttgart this database of the benchmark does not cover such a wide range of machines as *SKaMPI* does.

The low level part of the *PARKBENCH* benchmarks [PAR94] measure communication performance and have a managed result database[3] but do not give much information about the performance of individual MPI operations.

P. J. Mucci's[4] *mpbench* pursues similar goals as *SKaMPI* but it covers less functions and makes only rather rough measurements assuming a "quite dead" machine.

The *Pallas MPI Benchmark (PMB)*[5] is easy to use and has a simple well defined measurement procedure but has no graphical evaluation yet and only covers relatively few functions.

Many studies measure a few functions in more detail [GHH97,PFG97,RBB97,O.W96] but these codes are usually not publicly available, not user configurable, and are not designed for ease of use, portability, and robust measurements.

As one can imagine, the performance of a parallel computers (especially with a complex communication hardware) cannot be described by one number (even a single processor cannot be specified by one number). So many of the currently used benchmarks (e.g. [BBB+94,PAR94]) may be useful to rank hardware (like in the "top500" list[6]), but do not give much advice for program optimisation or even performance design.

3   Design of Parallel Software with MPI Performance Data Aware

When designing an MPI program for performance *and* portability the following questions arise:

Selection of point to point communication mode? MPI offers four modes for point to point communication: standard, buffered, ready, synchronous. Their performance depends on the implementation (esp. MPI_Send, where the communication protocol is intentionally unspecified) and the hardware support of these operations. (The ready mode is said to only be supported by the Intel Paragon... .) Additionally, MPI differentiates between blocking and non-blocking communication. Furthermore, there exist specialised operations like MPI_SendRecv for the common case of data exchange

---

[1] http://www.mcs.anl.gov/Projects/mpi/mpich/

[2] http://www.hlrs.de/mpi/b_eff

[3] http://netlib2.cs.utk.edu/performance/html/PDStop.html

[4] http://www.cs.utk.edu/~mucci/DOD/mpbench.ps

[5] http://www.pallas.de/pages/pmbd.htm

[6] http://www.top500.org

between two MPI processes. Also one is able to use wildcards like MPI_ANY_TAG or MPI_ANY_SOURCE which modify the behaviour of these operations. The MPI reference [GHLL+98] does a good job explaining the semantics of all these similar operations, but their performance depends highly on the MPI implementation and on hardware support.

Should compound collective operations be used? The MPI standard offers some compound collective operations (like MPI_Allreduce, MPI_Allgather, MPI_Reduce-scatter, MPI_Alltoall) which can be replaced by other, more primitive collective MPI operations (e.g., MPI_Bcast and MPI_Reduce). The compound collective operations are provided by the MPI standard, since it is possible to provide better algorithms for the compound operation than just putting some primitive collectives together. The question for the application software designer is: Is a compound operation worth the effort (e.g., designing particular data structures / partitionings) for using it?

Use of collective operations or hand made operations? Similar to compound collective operations made by more simple collective operations, also simple collective operations can be made solely out of point to point communicating operations. Again, the question arises, whether the MPI library provides a good implementation for the targeted platform. It might be worth the effort to reimplement collective operations for specific platforms with point to point communication in the application program.

Optimality of vendor provided MPI operations? The above question of the optimality of collective operations can be asked for all MPI operations provided by the vendor. Often a vendor provides some optimised MPI operations, while other operations perform suboptimal. Unfortunately these subsets of well-implemented MPI operations vary from vendor to vendor (i.e., from platform to platform).

Knowing the detailed performance of specific MPI operations helps to decide which MPI operation to choose when many similar MPI operations are possible. Of course, the best choice is the MPI operation performing well on all targeted platforms. If such a everywhere-nice operation does not exist, one can decide which operations on which platform must be replaced by hand-made code. Introducing static #ifdef PLAT-FORM_1 ... alternatives in the code simplifies the task to create portable software. Code selection during run-time is also feasible (by dynamically querying the platform), but introduces new costs during run-time.

To test the quality of provided MPI operations we reimplemented some operations with naive algorithms. Is the vendor provided implementation worse than these naive approaches, the application programmer can easily replace the vendor implementation. (Additionally, it says a lot about the quality of the vendor's implementation.)

How to use the database to answer the above question questions is shown in the next section, after a brief introduction of the database's terms.

4 The Public Result Database

The detailed design of the database is described in [Hun99]. Summarising the key concepts we present the following terms in a bottom-up order:

Single Measurement: A single measurement contains the MPI operation to perform. It has a right for its own, since it unifies calling different parameterised MPI operations in a unique way.

Measurement: A *measurement* is the accumulated value of repeated single measurement's results: Several single measurements are performed at the same argument (e.g., MPI_SendRecv at 1024 Kbytes). Their results are stored in an array. After reducing the influence of outliers by cutting (user defined) quartiles of that array, the average value is taken as the result of the measurement. The number of single measurements to perform before computing the result of the measurements depends on the standard error allowed for this measurement. Attributes of a measurement are: the single measurement to perform, the allowed standard error, the maximum and minimum number of repetitions.

Pattern: *Patterns* organise the way measurements are performed. For example collective operations must be measured completely different from point-to-point communicating MPI operations. *SKaMPI* currently contains four patterns: (a) for point-to-point measurements, (b) for measurements of collective operations, (c) for measuring communication structures arising in the master-worker scheme [KGGK94], and (d) for simple, i.e., one-sided operations like MPI_Commsplit. Besides grouping measurements the benefit of patterns lies in the comparability of measurements performed by the same pattern. Attributes of a pattern include the kinds and units of different arguments.

Suite of Measurements: A *suite of measurements* contains all measurements which measure the same MPI operation with the same pattern. (Note that the same MPI operation may be measured with two patterns, such as MPI_Send in the point-to-point pattern and the master worker pattern.) In a suite of measurements the measurements are varied over one parameter (such as the message length or the number of MPI processes). Informally spoken, the graph describing the performance of an MPI operation is given by the suite of measurements. The points of the graph represent the measurements. Important attributes of the suite are: the MPI operation, the pattern used, range of parameters, step-width between parameters (if fixed), scale of the axes (linear or logarithmic), use of automatic parameter refinement.

Run of Benchmark: A *run* is the top most relation, it includes several suites of measurements and their common data, such as: a description of the hardware (processing elements, network, connection topology), the date, the user running the benchmark, the operating system (and its version), and the settings of global switches of *SKaMPI*.

The structure of the relations used in the database are shown in figure 1. The *SKaMPI* result database has two web-user-interfaces . One is for downloading detailed reports of the various runs on all machines (http://liinwww.ira.uka.de/~skampi/cgi-bin/run_list.cgi.pl). The other interface (http://liinwww.ira.uka.de/~skampi/cgi-bin/frame_set.cgi.pl) is able to compare the operations between different machines according to the user's selection. In the following we describe this user-interface (see figure 2).

After loading the database's web site with a common netbrowser, querying is performed in three steps:

1. Choosing a run. Here you select one or more machine(s) you are interested in. Since on some machines several runs have been performed (e.g., with a different number

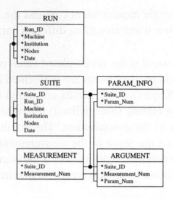

Fig. 1. Design of the *SKaMPI* result database

of MPI processes) you often can choose between several runs of one machine. This selection is performed in the upper left part of the user-interface (see figure 2).

2. Choosing the suites. After you selected some runs, the database is queried for the suites belonging to a run. The available runs are presented in a list for each selected run at the upper right part of the user-interface. There you now can select for each run the suites you are interested in (e.g., MPI_Reduce). Of course you may also select different suites on different machines (such as MPI_GatherSR in run A and MPI_Gather in run B [7].

3. After choosing the suites of interest the database is queried for all relevant measurements. The user-interface creates a single plot for all selected suites. The user is able to download an additional plot in encapsulated postscript (if selected in the previous step). There also exists the possibility to zoom into the plot.

5 Hamlet's Question: Using MPI_Gather or Not ?

As an example we discuss a special case of the question "Using collective operations or hand made operations?", as posed in section 3.[8] Here we look at the MPI_Gather operation, because one can replace it relatively simple by a hand-made substitute. The MPI_Gather operation is a collective operation which collects data from MPI processes at a designated root process. Two extremely naive implementations are provided by the *SKaMPI* benchmark. Both implementations can be programmed easily and fast by any application programmer.

The first implementation (MPI_GatherSR) simply uses point to point communication (implemented with MPI_Send - MPI_Recv) from all processes to the root process.

---

[7] An enumeration with detailed description of all operations measured by *SKaMPI* and all possible alternative naive implementations of MPI operations can be found in the *SKaMPI* user manual [Reu99]. This report is available online (http://liinwww.ira.uka.de/~reussner/ib-99-02.ps)

[8] and not by W. Shakespeare, as the heading might suggest.

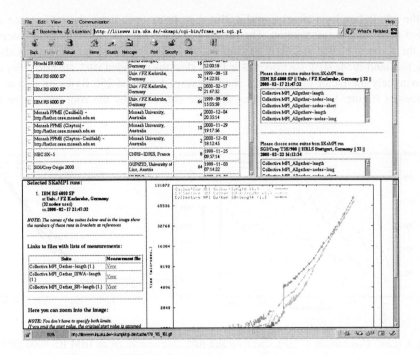

Fig. 2. Interactively querying the result database

Here the root process sequentially posts the receives. A receive is finished when the data arrived; each (but the first) receive has to wait until the previous receive finished (even if its data is ready to receive before).

The second implementation (`MPI_GatherISWA`) is only slightly more clever: the processes use the non-blocking `MPI_ISend` and the root process posts non-blocking receives (`MPI_IRecv`). This allows the root process to receive data from the other processes in the order the data arrives at the root process (assuming that the posting of the non-blocking receives is finished before any data arrived). The root process controls receiving with `MPI_Waitall`.

More sophisticated algorithms of gather operations usually perform much better. Their complexity depends on the topology of the underlying communication network; details can be found in e.g., [KGGK94,KdSF+00].

Consider you are developing an application for two of the most common parallel architectures: an IBM RS 6000 SP and a Cray T3E. Your implementation will make use of `MPI_Gather`. One of many questions is, whether `MPI_Gather` should be replaced by hand-made operations on all machines, or on one machine (which?), or never?

Looking at the *SKaMPI* result database for the Cray T3E shows, that the vendor provided `MPI_Gather` performs much better than the naive implementations. Concerning the results for an IBM RS 6000 SP one has to say that the IBM RS 6000 SP is a family of parallel computers rather than one single machine. Many different processors, connection networks and even topologies exist. Hence, generalising from these results to the whole family is clearly invalid. But this shows once more the importance detailed performance information. The results presented here are measured consistently on the Karlsruhe IBM RS 6000 SP from 1999 – 2000. [9] These measurements show that the hand-made MPI_GatherSR is faster than the provided MPI_Gather (by approximately the factor of 1.5) for a short message length of 256 Bytes on two to 32 processes (figure 3 (left)). Regarding the time consumed by the root process of the

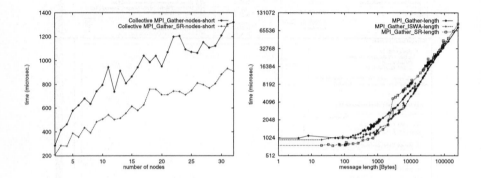

Fig. 3. (Left) Comparison of MPI_Gather and MPI_GatherSR on an IBM RS 6000 SP varied over the number of nodes at a fixed message length of 256 bytes. (Right) Comparison of MPI_Gather, MPI_GatherSR, and MPI_GatherISWA on an IBM RS 6000 SP with 32 processes varied over the message length

three implementations (MPI_Gather, MPI_GatherSR, MPI_GatherISWA) varied over the message length at the fixed number of 32 MPI processes (figure 3 (right)), we see that MPI_GatherSR is fastest if the message length is below 2048. For longer messages the used MPI_Send changes its internal protocol from a one-way direct-sending to a two-way request-sending protocol. For message lengths up to 16 KBytes the MPI_GatherISWA algorithm is fastest. For longer message lengths the difference between MPI_Gather and MPI_GatherISWA is not relevant; MPI_GatherSR is clearly the slowest. Till now, we looked at the time consumed by the root process. A detailed dicussion on timing collective operation lies beyond the scope of this paper (refer for example to [dSK00]). However, since MPI 1 collective operations are all blocking, we we might consider the non-blocking MPI_GatherISWA as useful, if our application allows the processes to perform computations without using (the still occupied) send buffer. The timing for all processes is shown in figure 4. From these results we can draw the conclusion, that we should provide a hand-made replacement for MPI_Gather for an IBM RS 6000 SP if we deal with short message lengths (below 10 KBytes). Whether we use the simple send-receive algorithm or the little more sophisticated non-blocking algorithm depends from the message length (smaller or greater 2 KBytes). If the processes can perform calculations without using the message buffers, we can hide communication latency by using the non-blocking MPI_GatherISWA. However, it is necessary to switch to the vendor provided MPI_Gather implementa-

---

[9] E.g., http://liinwww.ira.uka.de/~skampi/skampi/run2/l2h/

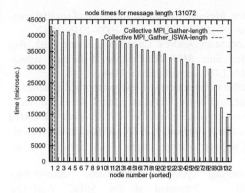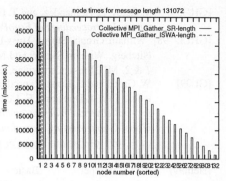

Fig. 4. Times consumed for each single MPI process for the different `MPI_Gather` implementations on an IBM RS 6000 SP.

tion on the Cray T3E (even when communication latency hiding would be possible, the vendor supplied `MPI_Gather` is to be preferred to the `MPI_GatherISWA`).

6 Conclusion

We presented a new process for the development of parallel software which moves the non-functional design considerations of performance and portability to the early stages of design and implementations. This lowers costs, even if only one platform is targeted. This process is supported by a publicly available MPI performance database. How performance data influences the design of MPI programs was discussed and an example for one particular case was presented.

To achieve an amount of standardised software in the parallel computing area more similar to that in the area of sequential programs, parallel software has to increase its portability. Considering this background, the creation of software development processes and tools supporting the engineering of more portable software is really crucial.

URL of the *SKaMPI* project
http://liinwww.ira.uka.de/~skampi

References

[BBB+94] D. Bailey, E. Barszcz, J. Barton, D. Browning, R. Carter, L. Dagum, R. Fatoohi, S. Fineberg, P. Frederickson, T. Lasinski, R. Schreiber, H. Simon, V. Venkatakrishnan, and S. Weeratunga. The NAS Parallel Benchmarks. Report RNR-94-007, Department of Mathematics and Computer Science, Emory University, March 1994.

[dSK00] B. R. de Supinski and N. T. Karonis. Accurately measuring mpi broadcasts in a computational grid. In *Proc. 8th IEEE Symp. on High Performance Distributed Computing (HPDC-8)*, Redondo Beach, CA, August 2000. IEEE.

[Fos94] Ian T. Foster. *Designung and Building Parallel Programs – Concepts and Tools for Parallel Software Engineering*. Addison Wesley, Reading, MA, 1994.

[GHH97]   V. Getov, E. Hernandez, and T. Hey. Message-passing performance of parallel computers. In *Proceedings of EuroPar '97 (LNCS 1300)*, pages 1009–1016, 1997.

[GHLL+98] William Gropp, Steven Huss-Lederman, Andrew Lumsdaine, Ewing Lusk, Bill Nitzberg, William Saphir, and Marc Snir. *MPI: The Complete Reference. Volume 1 & 2*. MIT Press, Cambridge, MA, USA, second edition, 1998.

[GL99]   W. Gropp and E. Lusk. Reproducible measurements of MPI performance characteristics. In J. J. Dongarra, E. Luque, and Tomas Margalef, editors, *Recent advances in parallel virtual machine and message passing interface: 6th European PVM/MPI Users' Group Meeting, Barcelona, Spain, September 26–29, 1999: proceedings*, volume 1697 of *Lecture Notes in Computer Science*, pages 11–18. Springer-Verlag, 1999.

[Hem99]   Rolf Hempel.   Basic   message   passing   benchmarks,   methodology and   pitfalls,   September   1999.   Presented   at   the   SPEC   Workshop (www.hlrs.de/mpi/b_eff/hempel_wuppertal.ppt).

[Hun99]   Gunnar Hunzelmann. Entwurf und Realisierung einer Datenbank zur Speicherung von Leistungsdaten paralleler Rechner. Studienarbeit, Department of Informatics, University of Karlsruhe, Am Fasanengarten 5, D-76128 Karlsruhe, Germany, October 1999.

[KdSF+00] N. T. Karonis, B. R. de Supinski, I. Foster, W. Gropp, E. Lusk, and J. Bresnahan. Exploiting hierarchy in parallel computer networks to optimize collective operation performance. In *Proceedings of the 14th International Conference on Parallel and Distributed Processing Symposium (IPDPS-00)*, pages 377–386, Los Alamitos, May 1–5 2000. IEEE.

[KGGK94]   Vipin Kumar, Ananth Grama, Anshul Gupta, and George Karypis. *Introduction to Parallel Computing: Design and Analysis of Algorithms*. Benjamin/Cummings, Redwood City, CA, 1994.

[KLS+94]   Charles H. Koelbel, David B. Loveman, Robert S. Schreiber, Guy L. Steele Jr., and Mary E. Zosel. *The High Performance Fortran handbook*. Scientific and engineering computation. MIT Press, Cambridge, MA, USA, January 1994.

[O.W96]   C. O.Wahl. Evaluierung von Implementationen des Message Passing Interface (MPI)-Standards auf heterogenen Workstation-clustern. Master's thesis, RWTH Aachen, Germany, 1996.

[PAR94]   PARKBENCH Committee, Roger Hockney, chair. Public international benchmarks for parallel computers. *Scientific Programming*, 3(2):iii–126, Summer 1994.

[PFG97]   J. Piernas, A. Flores, and J. M. Garcia. Analyzing the performance of MPI in a cluster of workstations based on Fast Ethernet. In *Recent advances in parallel virtual machine and message passing interface: Proceedings of the 4th European PVM/MPI Users' Group Meeting*, number 1332 in Lecture Notes in Computer Science, pages 17–24, 1997.

[RBB97]   M. Resch, H. Berger, and T. Bönisch. A comparison of MPI performance on different MPPs. In *Recent advances in parallel virtual machine and message passing interface: Proceedings of the 4th European PVM/MPI Users' Group Meeting*, number 1332 in Lecture Notes in Computer Science, pages 25–32, 1997.

[Reu99]   Ralf H. Reussner. SKaMPI: The Special Karlsruher MPI-Benchmark–User Manual. Technical Report 02/99, Department of Informatics, University of Karlsruhe, Am Fasanengarten 5, D-76128 Karlsruhe, Germany, 1999.

[RSPM98]   R. Reussner, P. Sanders, L. Prechelt, and M. Müller. SKaMPI: A detailed, accurate MPI benchmark. In V. Alexandrov and J. J. Dongarra, editors, *Recent advances in parallel virtual machine and message passing interface: 5th European PVM/MPI Users' Group Meeting, Liverpool, UK, September 7–9, 1998: proceedings*, volume 1497 of *Lecture Notes in Computer Science*, pages 52–59, 1998.

Cyclic Debugging Using Execution Replay

Michiel Ronsse, Mark Christiaens, and Koen De Bosschere

ELIS Department, Ghent University, Belgium
{ronsse,mchristi,kdb}@elis.rug.ac.be

Abstract. This paper presents a tool that enables programmers to use cyclic debugging techniques for debugging non-deterministic parallel programs. The solution consists of a combination of record/replay with automatic on-the-fly data race detection. This combination enables us to limit the record phase to the more efficient recording of the synchronization operations, and checking for data races during a replayed execution. As the record phase is highly efficient, there is no need to switch it off, hereby eliminating the possibility of Heisenbugs because tracing can be left on all the time.

1 Introduction

Although a number of advanced programming environments, formal methods and design methodologies for developing reliable software are emerging, one notices that the biggest part of the development time is spent while debugging and testing applications. Moreover, most programmers still stick to arcane debugging techniques such as adding print instructions or watchpoints or using breakpoints. Using this method, one tries to gather more and more detailed and specific information about the cause of the bug. One usually starts with a hypothesis about the bug that one wants to prove or deny.

Normally, a program is debugged using a program execution. Indeed, repeating the same program execution over and over will eventually reveal the cause of the error (cyclic debugging). Repeating a particular execution of a deterministic program (e.g. a sequential program) is not that difficult. As soon as one can reproduce the program input, the program execution is known (input and program code define the program execution completely). This turns out to be considerably more complicated for non-deterministic programs. The program execution of such a program can not be determined a-priori using the program code and the input only, as these programs make a number of non-deterministic choices during their execution, such as the order in which they enter critical sections, the use of signals, random generators, etc. All modern thread based applications are inherently non-deterministic because the relative execution speed of the different threads is not stipulated by the program code. Cyclic debugging can not be used as such for these non-deterministic programs as one cannot guarantee that the same execution will be observed during repeated executions. Moreover, the use of a debugger will have a negative impact on the non-deterministic nature of the program. As a debugger can manipulate the execution of the different threads of the application, it is possible that a significant discrepancy in execution speed arises, giving cause to appearing or disappearing. The existence of this kind of errors, combined with the primitive debugging tools used nowadays, makes debugging parallel programs a laborious task.

In this paper, we present our tool, RECPLAY, that deals with the non-determinism introduced by one cause of non-determinism that is specific for parallel programs: unsynchronized accesses to shared memory (the so-called *race conditions*[1]). RECPLAY uses a combination of techniques in order to allow the usage of standard debuggers for sequential programs for debugging parallel programs. RECPLAY is a so-called execution replay mechanism: information about a program execution can be traced (*record phase*) and this information is used to guide a faithful re-execution (*replay phase*). A faithful replay can only be guaranteed if and only if the log contains sufficient information about all non-deterministic choices that were made during the original execution (minimally the outcome of all the race conditions). This suffices to create an identical re-execution, the race conditions included. Unfortunately, this approach causes a huge overhead, severely slows down the execution, and produces huge trace files. An alternative approach we advocate in this paper is to record an execution as if it did not contain data races, and to check for the occurrence of data races during a replayed execution. As has been shown [CM91], replay will be guaranteed to be correct up to the race frontier, i.e., the point in the execution of each thread were a race event is about to take place.

2 The RECPLAY Method

As the overhead introduced by tracing all race conditions is far too high (it forces us to intercept all memory accesses), RECPLAY uses an approach based on the fact that there are two types of race conditions: synchronization races and data races. *Synchronization races* (introduced by synchronization operations) intentionally introduce non-determinism in a program execution to allow for competition between threads to enter a critical section, to lock a semaphore or to implement load balancing. *Data races* on the other hand are not intended by the programmer, and are most of the time the result of improper synchronization. By adding synchronization, data races can (and should) always be removed.

RECPLAY starts from the (erroneous) premise that a program (execution) does not contain data races. If one wants to debug such a program, it is sufficient to log the order of the synchronization operations, and to impose the same ordering during a replayed execution.[2] RECPLAY uses the ROLT method [LAV94], an ordering-based record/replay method, for logging the order of the synchronizations operations. ROLT logs, using Lamport clocks [Lam78], the partial order of synchronization operations. A timestamp is attached to each synchronization

---

[1] Technically, a race condition occurs whenever two threads access the same shared variable in an unsynchronized way, and at least one thread modifies the variable.

[2] Remember that RECPLAY only deals with non-determinism due to shared memory accesses, we suppose e.g. that input is refed during a replayed execution.

operation, taking the so-called clock condition into consideration: if operation a causally occurs before b in a given execution the timestamp $LC(a)$ of a should be smaller than the timestamp $LC(b)$ of b. Basically, ROLT logs information that can be used to recalculate, during replay, the timestamps that occurred during the recorded execution.

The ROLT method has the advantage that it produces small trace files and that it is less intrusive than other existing methods [Net93]. This is of paramount importance as an overhead that is too big will alter the execution, giving rise to Heisenbugs (bugs that disappear or alter their behavior when one attemps to isolate or probe it, [Gai86]. Moreover, the method allows for the use of a simple compression scheme [RLB95] which can further reduce the trace files. The information in the trace files is used during replay for attaching the Lamport timestamps to the synchronization operations. To get a faithful replay, it is sufficient to stall each synchronization operation until all synchronization operations with a smaller timestamp have been executed.

Of course, the premise that a program (execution) does not contain data races is not correct. Unfortunately, declaring a program free of data races is an unsolvable problem, at least for all but the simplest programs [LKN93]. Even testing one particular execution for data races is not easy: we have to detect whether the order in which two memory accesses occur during a particular execution is fixed by the program code or not. Unfortunately, this is only possible if the synchronization operations used reflect the synchronization order dictated by the program code. E.g. this is possible if the program only uses semaphores and the program contains no more than one $P()$ and one $V()$ operation for each semaphore. If this is not the case, it is impossible to decide whether the order observed was forced by the program code or not. However, for guaranteeing a correct replay, we do not need this information as we want to detect if *this* replayed execution contains a data race or not, as a data race would render the replay unreliable. And as *we are imposing a particular execution order* on the synchronization operations using the trace file, we know that the synchronization operations are forced in this order. However, this order is forced by RECPLAY, and not by the program itself.[3]

The online data race detection used by RECPLAY consists of three phases:

1. *collecting memory reference information* for each sequential block between two successive synchronization operations on the same thread (called segments). This yields two sets of memory references per segment: $S(i)$ are the locations that were written and $L(i)$ are the locations that were read in segment i. RECPLAY uses multilevel (see Figure 1) bitmaps for registering the memory accesses. Note that multiple accesses to the same variable in a segment will be counted as one, but this is no problem for detecting data races. The sets $L(i)$ and $S(i)$ are collected on a list.

---

[3] In fact, it is not necessary to re-execute the synchronization operations from the program, as RECPLAY forces an execution order (a total order) on the synchronization operations that is stricter than the one contained in the program (a partial order).

2. *detecting conflicting memory references* in concurrent segments. There will be a data race between segment i and segment j if either $(L(i) \cup S(i)) \cap S(j) \neq \emptyset$ or $(L(j) \cup S(j)) \cap S(i) \neq \emptyset$ is true. If the comparison indicates the existence of a data race, RECPLAY saves information about the data race (address and threads involved, and type of operations (load or store)). For each synchronization operation, RECPLAY will compare the bitmaps of the segment that just ended against the bitmaps of the parallel segments on the list. Moreover, RECPLAY will try to remove obsolete segments from the list. A segment becomes obsolete if it is no longer possible for future segments to be parallel with the given segment.

3. *identifying the conflicting memory accesses* given the traced information. This requires another replayed execution.

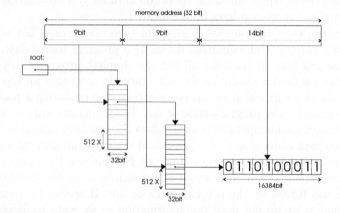

Fig. 1. RECPLAY uses a 3-level bitmap where each level is addressed using a different part of the address: the first two parts are used to address lists of pointers, while the last part of the address points to the actual bit. Such a bitmap favors programs with a substantial memory locality.

In our race detection tool, we use a classical logical vector clock [Mat89,Fid91] for detecting concurrent segments as segments x and y can be executed in *parallel* if and only if their vector clocks are not ordered (p_x is the thread on which segment x was executed):

$$x\|y \Leftrightarrow \begin{cases} (VC_x[p_x] \geq VC_y[p_x]) \text{ and } (VC_x[p_y] \leq VC_y[p_y]) \\ \text{or} \\ (VC_x[p_x] \leq VC_y[p_x]) \text{ and } (VC_x[p_y] \geq VC_y[p_y]) \end{cases}$$

This is possible thanks to the strong consistency property of vector clocks. For detecting and removing the obsolete segments, RECPLAY uses an even stronger clock: snooped matrix clocks [DBR97].

It is clear that data race detection is not a cheap operation. The fact that all memory accesses must be intercepted does indeed impose a huge overhead.

Fortunately, RECPLAY performs the data race detection during a replayed execution, making it impossible for the data race detector to alter the normal execution. Moreover, for each recorded execution, only one data race check is necessary. If no data races are found, it is possible to replay the execution without checking for data races. This will lead to a much faster re-execution that can be used for cyclic debugging.

3   Evaluation

program	normal runtime	record		replay		replay+detect	
		runtime	slow-down	runtime	slow-down	runtime	slow-down
cholesky	8.67	8.88	1.024	18.90	2.18	721.4	83.2
fft	8.76	8.83	1.008	9.61	1.10	72.8	8.3
LU	6.36	6.40	1.006	8.48	1.33	144.5	22.7
radix	6.03	6.20	1.028	13.37	2.22	182.8	30.3
ocean	4.96	5.06	1.020	11.75	2.37	107.7	21.7
raytrace	9.89	10.19	1.030	41.54	4.20	675.9	68.3
water-Nsq.	9.46	9.71	1.026	11.94	1.26	321.5	34.0
water-spat.	8.12	8.33	1.026	9.52	1.17	258.8	31.9

Table 1. Basic performance of RECPLAY (all times in seconds)

The RECPLAY system has been implemented for Sun multiprocessors running Solaris using the JiTI instrumentation tool we also developed [RDB00]. The implementation uses the dynamic linking and loading facilities present in all modern Unix operating system and instruments (for intercepting the memory accesses and the synchronization operations) on the fly: the running process is instrumented.

While developing RECPLAY, special attention was given to the probe effect during the record phase. Table 1 gives an idea of the overhead caused during record, replay, and race detection for programs from the SPLASH-2 benchmark suite [4]. The average overhead during the record phase is limited to 2.1% which is small enough to keep it switched on all the time. The average overhead for replay is 91% which can seem high, but is feasible during debugging. The automatic race detection is however very slow: it slows down the program execution about 40 times (the overhead is mainly caused by JiTI intercepting all memory accesses). Fortunately, it can run unsupervised, so it can run overnight and we have to run it only once for each execution.

The memory consumption is far more important during the data race detection. The usage of vector clocks for detecting the races is not new, but the

---

[4] All experiments were done on a machine with 4 processors and all benchmarks were run with 4 threads.

program	created	max. stored	compared
cholesky	13 983	1 915 (13.7%)	968 154
fft	181	37 (20.5%)	2 347
LU	1 285	42 (3.3%)	18 891
radix	303	36 (11.9%)	4 601
ocean	14 150	47 (0.3%)	272 037
raytrace	97 598	62 (0.1%)	337 743
water-Nsq.	637	48 (7.5%)	7 717
water-spat.	639	45 (7.0%)	7 962

Table 2. Number of segments created and compared during the execution, and the maximum number of segments on the list.

mechanism used for limiting the memory consumption is. The usage of multi-level bitmaps and the removal of obsolete segments (and their bitmaps) allows us to limit the memory consumption considerably. Table 2 shows the number of segments that was created during the execution, the maximum number on the list, and the number of parallel segments during a particular execution (this is equal to the number of segments compared). The average maximum number of segments on the list is 8.0%, which is a small number. Without removing obsolete segments, this number would of course be 100%. Figures 2 and 3 show the number of segments on the list and the total size of the bitmaps in function of the time (actually the number of synchronization operations executed so far) for two typical cases: lu and cholesky[5]. For lu, the number of segments is fairly constant, apart from the start and the end of the execution. The size of the bitmaps is however not that constant; this is caused by the locality of the memory accesses as can be seen in the third graph showing the number of bytes used by the bitmaps divided by the number of segments. The numbers for cholesky are not constant, but the correlation between the number of segments and the size of the bitmaps is much higher, apart from a number of peaks. The number of segments drops very quickly at some points, caused by *barrier* synchronization creating a large number of obsolete segments.

4 Related Work

In the past, other replay mechanisms have been proposed for shared memory computers. Instant Replay [LM87] is targeted at coarse grained operations and traces all these operations. It does not use any technique to reduce the size of the trace files nor to limit the perturbation introduced. It does not work for programs containing data races. A prototype implementation for the BBN Butterfly is described.

Netzer [Net93] introduced an optimization technique based on vector clocks. As the order of all memory accesses is traced, both synchronization and data

---

[5] These are not the runs used for Table 2

races will be replayed. It uses comparable techniques as ROLT to reduce the size of the trace files. However, no implementation was ever proposed (of course, the overhead would be huge as all memory accesses are traced, introducing Heisenbugs). We believe that it is far more interesting to detect data races than to record/replay them. Therefore, RECPLAY replays the synchronization operations only, while detecting the data races.

Race Frontier [CM91] describes a similar technique as the one proposed in this paper (replaying up to the first data race). Choi and Min prove that it is possible to replay up to the first data race, and they describe how one can replay up to the race frontier. A problem they do not solve is how to efficiently find the race frontier. RECPLAY effectively solves the problem of finding the race frontier, but goes beyond this. It also finds the data race event.

Most of the previous work, and also our RECPLAY tool, is based on Lamport's so-called *happens-before* relation. This relation is a partial order on all synchronization events in a particular parallel *execution*. If two threads access the same variable using operations that are not ordered by the happens-before relation and one of them modifies the variable, a data race occurs. Therefore, by checking the ordering of all events and monitoring all memory accesses data races can be detected for one *particular program execution*. Another approach is taken by a more recent race detector: Eraser [SBN+97]. It goes slightly beyond work based on the happens-before relation. Eraser checks that a *locking discipline* is used to access shared variables: for each variable it keeps a list of locks that were hold while accessing the variable. Each time a variable is accessed, the list attached to the variable is intersected with the list of locks currently held and the intersection is attached to the variable. If this list becomes empty, the locking discipline is violated, meaning that a data race occurred.

The most important problem with Eraser is however that its practical applicability is limited in that it can only process mutex synchronization operations and in that the tool fails when other synchronization primitives are build on top of these lock operations.

5 Conclusions

In this paper we have presented RECPLAY, a practical and effective tool for debugging parallel programs with classical debuggers. Therefore, we implemented a highly efficient two-level record/replay system that traces the synchronization operations, and uses this trace to replay the execution. During replay, a race detection algorithm is run to notify the programmer when a race occurs. After removing the data races, normal sequential debugging tools can be used on the parallel program using replayed executions.

References

[CM91]   Jong-Deok Choi and Sang Lyul Min. Race frontier: Reproducing data races in parallel-program debugging. In *Proc. of the Third ACM SIGPLAN Sym-*

posium on Principles & Practice of Parallel Programming, volume 26, pages 145–154, July 1991.

[DBR97] Koen De Bosschere and Michiel Ronsse. Clock snooping and its application in on-the-fly data race detection. In *Proceedings of the 1997 International Symposium on Parallel Algorithms and Networks (I-SPAN'97)*, pages 324–330, Taipei, December 1997. IEEE Computer Society.

[Fid91] C. J. Fidge. Logical time in distributed computing systems. In *IEEE Computer*, volume 24, pages 28–33. August 1991.

[Gai86] Jason Gait. A probe effect in concurrent programs. *Software - Practice and Experience*, 16(3):225–233, March 1986.

[Lam78] Leslie Lamport. Time, clocks, and the ordering of events in a distributed system. *Communications of the ACM*, 21(7):558–565, July 1978.

[LAV94] Luk J. Levrouw, Koenraad M. Audenaert, and Jan M. Van Campenhout. A new trace and replay system for shared memory programs based on Lamport Clocks. In *Proceedings of the Second Euromicro Workshop on Parallel and Distributed Processing*, pages 471–478. IEEE Computer Society Press, January 1994.

[LKN93] Hsueh-I Lu, Philip N. Klein, and Robert H. B. Netzer. Detecting race conditions in parallel programs that use one semaphore. Workshop on Algorithms and Data Structures (WADS), Montreal, August 1993.

[LM87] Thomas J. LeBlanc and John M. Mellor-Crummey. Debugging parallel programs with Instant Replay. *IEEE Transactions on Computers*, C-36(4):471–482, April 1987.

[Mat89] Friedemann Mattern. Virtual time and global states of distributed systems. In Cosnard, Quinton, Raynal, and Roberts, editors, *Proceedings of the Intl. Workshop on Parallel and Distributed Algorithms*, pages 215–226. Elsevier Science Publishers B.V., North-Holland, 1989.

[Net93] Robert H.B. Netzer. Optimal tracing and replay for debugging shared-memory parallel programs. In *Proceedings ACM/ONR Workshop on Parallel and Distributed Debugging*, pages 1–11, May 1993.

[RDB00] M. Ronsse and K. De Bosschere. Jiti: A robust just in time instrumentation technique. In *Proceedings of WBT-2000 (Workshop on Binary Translation)*, Philadelphia, 10 2000.

[RLB95] M. Ronsse, L. Levrouw, and K. Bastiaens. Efficient coding of execution-traces of parallel programs. In J. P. Veen, editor, *Proceedings of the ProRISC / IEEE Benelux Workshop on Circuits, Systems and Signal Processing*, pages 251–258. STW, Utrecht, March 1995.

[SBN+97] Stefan Savage, Michael Burrows, Greg Nelson, Patrick Sobalvarro, and Thomas Anderson. Eraser: A dynamic data race detector for multithreaded programs. *ACM Transactions on Computer Systems*, 15(4):391–411, November 1997.

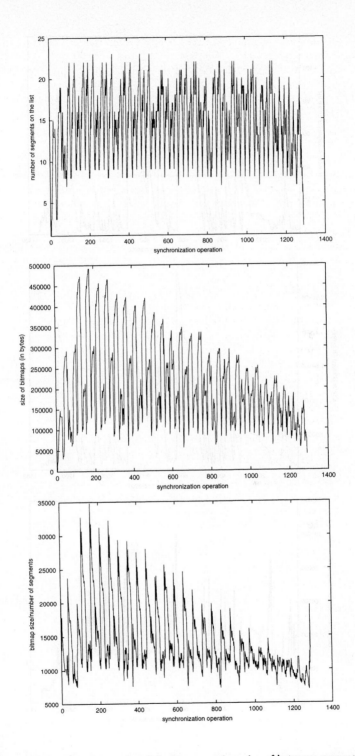

Fig. 2. Number of segments, size of the bitmaps and number of bytes per segment for lu.

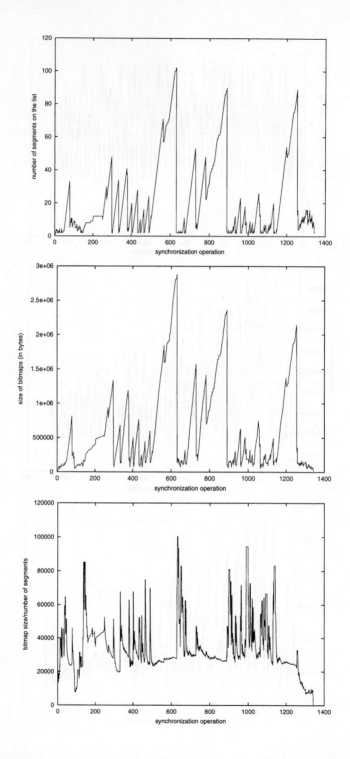

Fig. 3. Number of segments, size of the bitmaps and number of bytes per segment for cholesky.

Visualizing the Memory Access Behavior of Shared Memory Applications on NUMA Architectures

Jie Tao [*] , Wolfgang Karl, and Martin Schulz

LRR-TUM, Institut für Informatik,
Technische Universität München, 80290 München, Germany
E-mail: {tao,karlw,schulzm}@in.tum.de

Abstract. Data locality is one of the most important issues affecting the performance of shared memory applications on NUMA architectures. A possibility to improve data locality is the specification of a correct data layout within the source code. This kind of optimization, however, requires in depth knowledge about the run-time memory access behavior of programs. In order to acquire this knowledge without causing a probe overhead, as it would be caused by software instrumentation approaches, it is necessary to adopt a hardware performance monitor that can provide detailed information about memory transactions. As the monitored information is usually very low-level and not user-readable, a visualization tool is necessary as well. This paper presents such a visualization tool displaying the monitored data in a user understandable way thereby showing the memory access behavior of shared memory applications. In addition, it projects the physical addresses in the memory transactions back to the data structures within the source code. This increases a programmer's ability to effectively understand, develop, and optimize programs.

1 Introduction

Clusters with NUMA characteristics are becoming increasingly important and are regarded as adequate architectures for High Performance Computing. Such clusters based on PCs have been built for our SMiLE project (Shared Memory in a Lan–like Environment) [8]. In order to achieve the necessary NUMA global memory support, the Scalable Coherent Interface (SCI) [1, 5], an IEEE standardized System Area Network, is taken as the interconnection technology. With a latency of less than 2 μs and a bandwidth of more than 80 MB/s for process to process communication, SCI offers state-of-the-art performance for clusters. Shared memory programming is enabled on the SMiLE clusters using a hybrid DSM system, the SCI Virtual Memory (SCI-VM) [9, 14]. In this system, all communication is directly handled by the SCI hardware through appropriate remote memory mappings, while the complete memory management and the required global process abstraction across the individual operating system instances is handled by a software component. This creates a global virtual memory allowing a direct execution of pure shared memory applications on SCI-based clusters of PCs.

---

[*] Jie Tao is a staff member of Jilin University in China and is currently pursuing her Ph.D. at Technische Universität München in Germany.

Many shared memory applications, however, initially do not run efficiently on top of such clusters. The performance of NUMA systems depends on an efficient exploitation of memory access locality since remote memory accesses are still an order of magnitude slower than local memory accesses, despite the good latency offered by current interconnection technologies. Unfortunately, many shared memory applications without optimization exhibit a poor data locality when running on NUMA architectures and therefore do not achieve a good performance. It is necessary to develop performance tools which can be used to enable and exploit data locality. One example of such tools is a visualizer which allows the user to understand an application's memory access behavior, to detect communication bottlenecks, and to further optimize the application with the goal of better data locality.

Such tools, however, require detailed information about the low-level memory transactions in the running program. The only way to facilitate this without causing a high probe overhead is to perform the data acquisition using a hardware monitor capable of observing all memory transactions performed across the interconnection fabric. Such a monitoring device has been developed within the SMiLE project.

The hardware monitor traces memory references performed by the running program and gives the user a complete overview of the program's memory access behavior in the form of access histograms. Its physical implementation is based on SCI-based clusters; the general principles, however, could be applied to any NUMA architecture in the same way, therefore providing a general approach for the optimization of shared memory applications on top of such systems.

The information from the hardware monitor, however, is purely based on physical addresses, very detailed and low-level, and is therefore not directly suitable for the evaluation of applications. Additional tools are necessary in order to transform the user-unreadable monitored data in a more understandable and easy-to-use form. One example for such a tool, a visualizer, is presented in this paper. It offers both a global overview of the data transfer among processors and a description of given data structure or parts of arrays. This information can be used to easily identify access hot spots and to understand the complete memory access behavior of shared memory applications. Based on this information, the application can then be optimized using an application specific data layout potentially resulting in significant performance improvements.

The remainder of this paper is organized as follows. Section 2 introduces a tool environment for efficient shared memory programming. Section 3 describes the visualizer. In Section 4 a few related work are compared. The paper is then rounded up with a short summary and some future directions in Section 5.

2 Framework for On-line Monitoring

As many shared applications on NUMA systems do initially not achieve a good parallel performance primarily due to the poor data locality, a tool environment has been designed to enable extensive tuning. As shown in Figure 1, this environment includes three components: Data acquisition, an on-line monitoring interface called OMIS/OCM, and a set of tools.

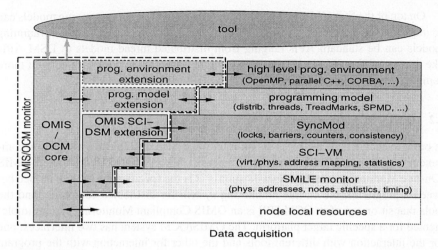

Fig. 1. Infrastructure of the on-line monitoring for efficient shared memory programming

2.1 Data acquisition

The data acquisition component is used to collect the data required by tools. It comprises a flexible multilayer infrastructure, including both system facilities and additional software packages, which prepares and delivers data to tools for performance analysis and program tuning. The system facilities are node local resources on the PCs and the hardware monitor. The node local resources include performance counters in the processors as well as the complete process and thread management services of the local operating systems. The hardware monitor [6, 7] provides the detailed information about the low-level transactions performed on the SCI network. It snoops a synchronous local bus on the network interface, over which all SCI packets from and to the nodes are transfered. The information carried by the packets is extracted and triggered by an event filter. Events of interest are counted. The information gathered by the hardware monitor is spilled to the corresponding buffer in the main memory via the PCI local bus.

On top of the hardware monitor, the SCI-VM handles further parts of the monitored data, e.g. the physical addresses, which are not suitable for a direct use of tools. As the current implementation of the hardware monitor uses physical addresses for identifying the access destination of a memory transaction event, the SCI-VM provides a virtual-to-physical address mapping list enabling tools to define events using physical addresses.

As the hardware monitor lacks the ability to monitor synchronization primitives, which generally have a critical performance impact on shared memory applications, the SyncMod module [15] is responsible for any synchronization within shared memory applications including locks and barriers. Information delivered by the SyncMod module ranges from simple lock and unlock counters to more sophisticated information like mean and peak lock times and peak barrier wait times. This information will allow the detection of bottlenecks in applications and forms the base for an efficient application optimization.

On top of the SCI-VM and the SyncMod, shared memory programming models can be implemented depending on the user's demands and preferences. Such programming models can be standard APIs ranging from distributed thread models to DSM APIs like the one of TreadMarks [2], but also form the basis for the implementation of more complex programming environments like CORBA [16] and OpenMP [12].

2.2 OMIS/OCM interface

In order to enable tools to access the information gathered by the data acquisition component, a system and tool independent framework, the OMIS/OCM, is adopted. OMIS (On-line Monitoring Interface Specification) [3] is the specification of an interface between a programmable on-line monitoring system for distributed computing and the tools that sit on top of it. OCM [19] is an OMIS Compliant Monitoring system implemented for a specific target platform. The OMIS/OCM system has two interfaces: one for the interaction with different tools and the other for interaction with the program and all run-time system layers. It comprises core of services that are fully independent of the underlying platform and the programming environment. It can be augmented by optional extensions to cover platform or programming environment specific services for a special tool. As we apply OMIS/OCM in our research work to implement a non-conflicting access from tools to the data acquisition component, the OMIS/OCM core has to be extended with various extensions allowing the definition and implementation of new services related to all of the components within the data acquisition multilayers. Using these services, information can be transfered from nodes, processes, and threads to the tools.

2.3 Tools

The information offered by the data acquisition component is still low-level and difficult to understand; tools are therefore required to manipulate the execution of programs and ease the analyzation and optimization of programs. Currently two tools are envisioned. A visualizer is intended to exhibit the monitored data in an easy-to-use way allowing programmers to understand the dynamic behavior of applications. As the SCI-VM offers all necessary capabilities to specify the data layout in the source code, programmers are enabled to optimize programs with a better data layout and further to improve data locality. Another tool, an adaptive run-time system, is intended to transparently modify the data layout at run-time. It detects communication bottlenecks and determines a proper location for inappropriately distributed data using the monitored information and also the information available from the SCI-VM, the programming models, and even the users. Incorrectly located data will be migrated during the execution of a program, resulting in a better data locality.

3 Visualizing the Run-time Memory Access Behavior

Data locality is the most important performance issue of shared memory programming on NUMA architectures and a visualization tool provides a direct way to understand the memory access behavior. Therefore, a first endeavor in the implementation of an appropriate tool environment focuses on the visualizer. Based on the monitored data, this toolset generates diagrams, tables, and other understandable representations to illustrate the low-level memory transactions. It aims at providing programmers with a valuable and easy-to-use tool for analyzing the run-time data layout of applications.

The visualizer is implemented using Java in combination with the Swing library to take full advantages of the advanced graphic capabilities of Swing minimizing the implementation efforts. It deals with the monitored data and generates diagrams, tables, curves, etc. The monitored data, however, is not directly applied to the visualizer since it is fine-grained and scattered. Some functions are provided to acquire corse-grained information from this original data. These functions are compiled to a shared library and connected to the visualizer using the Java Native Interface.

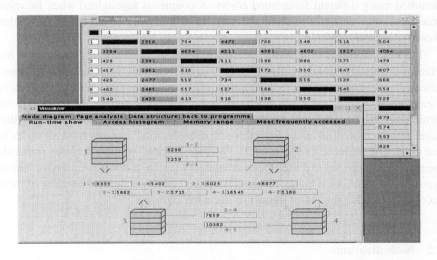

Fig. 2. The Run-time show window

The visualizer provides several different views of the acquired monitored data in order to show the various aspects of an application's memory access behavior. It shows the low-level data transfer among processors, the detailed access numbers to pages and memory regions, communication hot spots, the access behavior within a single page, and the projection back to the data structure within the source codes. Figure 2, 3, and 4 illustrate the common windows showing the monitored data acquired during the simulated execution of the RADIX program from the SPLASH2-Benchmark suite, with a working set size of 256KB running on a 4-nodes SCI-cluster. As the implementation of the hardware monitor is still on going, a simulation system [17] has been developed that simulates the execution of shared memory programs on hardware monitor equipped SCI clusters. The monitored data generated by the simulation system is exactly the one that will be provided by the future hardware monitor. We have applied the simulation system to the visualizer to enable first experiments.

3.1 Run-time show and detailed access numbers

The windows shown in Figure 2 and 3 provide some detailed information about the individual memory transactions. The "Run-time show" window, as illustrated in Figure 2, is used to reflect the actual transfer between nodes. Two different diagrams are available: a graph version for dealing with clusters with up to 4 PCs and a table version for larger configurations. The four components in the graph represent four memories located on the different nodes and the arrows between them represent the flow of data among nodes. Rectangles between components represent counters, which are used to store the number of data transfers between two nodes and are dynamically modified corresponding to the run-time data transfer. One counter is incremented every time a data transaction is performed between the two memories. For larger systems, a counter matrix is used to exhibit the node-to-node data transfer. Communication hot spots are identified using different foreground colors. A counter is highlighted when its value exceeds an user defined threshold and marked with another bright color if its value exceeds a second threshold.

The exact numbers of memory accesses to pages and regions are illustrated in the "Access histogram" table and the "Memory region" window. The "Access histogram" reflects the memory accesses to the whole global virtual memory, while the "Memory region" shows only a region of it. Each row of the "Access histogram" table displays the virtual page number , the node number, the access source, the number of accesses to that page, and the ratio of accesses from this source to the total accesses from all nodes to that page. The information shown in the "Memory region" window includes primarily the number of accesses to a user specified memory region on a node. Excessive remote accesses are highlighted in the "Access histogram" table and the exact sorting of these accesses is listed in the "Most frequently accessed" window.

3.2 Node diagrams

Further details about memory accesses to pages can be found in the "node diagram" window shown in Figure 3. This graph is used to exhibit the various aspects of memory transactions with page granularity. It uses colored columns to show the relative frequency of accesses performed by all nodes to a single page. One color stands for one node. Inappropriate data allocation can be therefore easily detected via the different height of the columns. Page 515 (virtual number), for example, is located on node 1 but accessed only by node 2. It is therefore incorrectly allocated and should be placed on node 2. This allocation can be specified by users within the source code using annotations provided by the SCI Virtual Memory. In order to support this work, a small subwindow is combined to the "Access histogram". The corresponding data structure of a selected page will be shown in this small upper window when a mouse button is pressed within the graphic area. The node diagrams therefore direct users to place data on the node that mostly requires it. Each diagram shows only a certain number of pages depending on the number of processors. A page list allows users to change the beginning of a diagram, enabling thereby an overall view of the data layout.

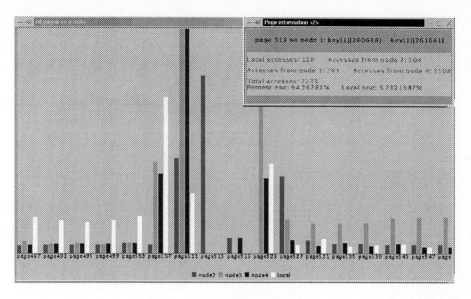

Fig. 3. An access diagram of node 1

3.3 Page analysis

The description above has shown that node diagrams provide information for a better allocation of individual pages which are frequently accessed by only one node. The allocation of pages accessed by more processors, however, is more difficult and requires a further analysis of the accesses within them. The "Section" diagram, "Read/write" curve, and "Phase" diagram, as shown together in Figure 4, provide the support for users to make a reasonable decision about this issue. The "Section" diagram exhibits the access behavior of different sections within a single page, the "Read/write" curve offers the comparison between the frequency of reads and writes to different sections, and the "Phase" diagram provides the per-phase statistics of memory accesses to a page. The section size can be specified by the user and phases are marked by global barriers within the applications. Corresponding data structures of selected sections can be displayed in the same way as done in the "Node digram".

3.4 Projecting back to the source code

The visualized memory access behavior with various granularities is not sufficient for the analyzation and optimization of data locality, as programmers do not use the virtual addresses explicitly in the source codes. Page numbers and memory addresses therefore have to be projected onto the corresponding data structures within the source codes of running applications. For this purpose, the visualizer offers a "Data structure" window to reflect this mapping. Instead of displaying only one page or section of interest as it is the case in the "Node diagram" and the "Page analysis" windows, this window shows all the shared variables occurring in a source code. In combination with the "Access histogram", the "Data structure" window provides users with a global overview

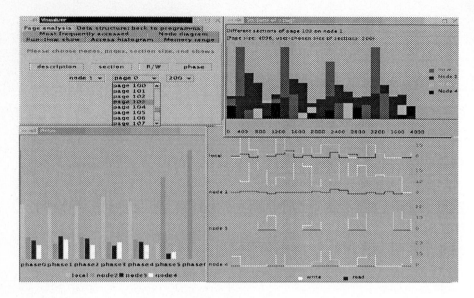

Fig. 4. The detailed access character of a page

of accesses to the complete working set of the application. The relationship between the virtual addresses and the variables, which forms the base of the projection, is currently accomplished using a set of macros available in the programming models. The SCI-VM delivers the information required for the transformation of physical to virtual addresses. As the programming models provide corresponding macros, this work will not cause much inconvenience for the users. On the other hand, many troubles and inaccuracies related to modifying compilers or extracting symbol information from binaries are avoided.

4 Related work

Few research projects have focused on providing hardware-supported monitoring for network traffic in clusters. The performance monitor Surfboard for the Princeton SHRIMP multicomputer [10] is one example. It is designed to measure all communications and respond to user commands. The monitored data is written to a DRAM as needed. Unlike our research work the SHRIMP project uses Surfboard primarily to measure the packet latency, packet size, and receive time. Another trace instrument [11] has been designed at Trinity College Dublin for SCI-based systems. Full traces are transfered into a relational database. Users are allowed to access this database only after the monitoring.This instrument is therefore intended for offline analysis, debugging and validation.

Visualization tools, however, have widely been investigated. VISTOP [18] is a state based visualizer for PVM applications. PROVE [4] is a performance visualization tool designed for an integrated parallel program development environment. Part of the work introduced in [13] is the visualization of the measured performance data obtained through the instrumentation of source code with calls to software probes. Unlike these performance visualization systems focusing on presenting process states and process communication, the visualization tool described in this paper is more oriented to visualizing the interconnect traffic. Based on a powerful performance monitor it provides programmers with a valuable support in optimizing the applications towards a better data locality at run-time.

5 Conclusion and Future Work

Data locality is one of the most important performance issues on NUMA architectures. It has been addressed intensively in recent years. Optimizing programs with the goal of a better data layout can significantly improve data locality. This requires the user, however, to understand a program's memory access behavior at run-time. This is a complex and difficult task since any communication on such architectures is performed implicitly and handled completely by the NUMA hardware. A scheme to ease this work is to monitor the interconnection traffic and then to visualize the memory access behavior, forming the base for a further locality optimization by programmers.

Such a hardware performance monitor and a visualizer have been developed for NUMA-characterized SCI-based clusters. This paper presents primarily the visualizer which shows the fine-grained monitored information about the memory accesses in a higher-level and more readable way. In addition, it maps the monitored data back to the source code enabling an exact analysis of a program's data layout. Using this visualized information, programmers are able to detect the communication bottlenecks and further optimize programs with a better data layout potentially resulting in significant performance improvements.

As some windows of the current visualizer are only considered for small clusters, the next step in this research work is to design more flexible data representations allowing a comprehensive show of monitored data even for large clusters. In addition, different kinds of applications will be analyzed and optimized using the visualized information. It can be expected that these applications will benefit from this optimization significantly.

References

1. IEEE Standard for the Scalable Coherent Interface(SCI). IEEE Std 1596-1992,1993, IEEE 345 East 47th Street, New York, NY 10017-2394, USA.
2. C. Amza, A. Cox, S. Dwarkadas, and etc. TreadMarks: Shared memory computing on networks of workstations. *IEEE Computer*, 29(2):18–28, Feb. 1995.
3. M. Bubak, W. Funika, R. Gembarowski, and R. Wismüller. OMIS-compliant monitoring system for MPI applications. In *Proc. 3rd International Conference on Parallel Processing and Applied Mathematics - PPAM'99*, pages 378–386, Kazimierz Dolny, Poland, September 1999.
4. P. Cacsuk. Performance visualization in the GRADE parallel programming environment. In *Proceedings of the 4th international conference on High Performance Computing in Asia-Pacific Region*, pages 446–450, Peking, China, 14-17, May 2000.

5. Hermann Hellwagner and Alexander Reinefeld, editors. *SCI: Scalable Coherent Interface: Architecture and Software for High-Performance Computer Clusters*, volume 1734 of Lecture Notes in Computer Science. Springer-Verlag, 1999.

6. R. Hockauf, W. Karl, M. Leberecht, M. Oberhuber, and M. Wagner. Exploiting spatial and temporal locality of accesses: A new hardware-based monitoring approach for DSM systems. In *Proceedings of Euro-Par'98 Parallel Processing / 4th International Euro-Par Conference Southampton*, volume 1470 of Lecture Notes in Computer Science, pages 206–215, UK, September 1998.

7. Wolfgang Karl, Markus Leberecht, and Michael Oberhuber. SCI monitoring hardware and software: supporting performance evaluation and debugging. In Hermann Hellwagner and Alexander Reinefeld, editors, *SCI: Scalable Coherent Interface: Architecture and Software for High-Performance Computer Clusters*, volume 1734 of Lecture Notes in Computer Science, chapter 24. Springer-Verlag, 1999.

8. Wolfgang Karl, Markus Leberecht, and Martin Schulz. Supporting shared memory and message passing on clusters of PCs with a SMiLE. In *Proceedings of the third International Workshop, CANPC'99*, volume 1602 of Lecture Notes in Computer Science, Orlando, Florida, USA(together with HPCA-5), January 1999. Springer Verlag, Heidelberg.

9. Wolfgang Karl and Martin Schulz. Hybrid-DSM: An efficient alternative to pure software DSM systems on NUMA architectures. In *Proceedings of the 2nd International Workshop on Software DSM (held together with ICS 2000)*, May 2000.

10. Scott C. Karlin, Douglas W. Clark, and Margaret Martonosi. SurfBoard–A hardware performance monitor for SHRIMP. Technical Report TR-596-99, Princeton University, March 1999.

11. Michael Manzke and Brian Coghlan. Non-intrusive deep tracing of SCI interconnect traffic. In *Conference Proceedings of SCI Europe'99*, pages 53–58, Toulouse, France, September 1999.

12. OpenMP Architecture Review Board. *OpenMP C and C++ Application, Program Interface*, Version 1.0, Document Number 004–2229–01 edition, October 1998. Available from http://www.openmp.org/.

13. Luiz De Rose and Mario Pantano. An approach to immersive performance visualization of parallel and wide-area distributed applications. In *Proceedings of the International Symosium on High Performance Distributed Computing*, Redondo Beach, CA, August 1999.

14. Martin Schulz. True shared memory programming on SCI-based clusters. In Hermann Hellwagner and Alexander Reinefeld, editors, *SCI: Scalable Coherent Interface: Architecture and Software for High-Performance Computer Clusters*, volume 1734 of Lecture Notes in Computer Science, chapter 17. Springer-Verlag, 1999.

15. Martin Schulz. Efficient coherency and synchronization management in SCI based DSM systems. In *Proceedings of the SCI-Europe, held in conjunction with Euro-Par 2000*, pages 31–36, Munich, Germany, August 2000.

16. J. Siegel. *CORBA - Fundamentals and Programming*. John Wiley & Sons, 1996.

17. Jie Tao, Wolfgang Karl, and Martin Schulz. Understanding the behavior of shared memory applications using the SMiLE monitoring framework. In *Proceedings of the SCI-Europe, held in conjunction with Euro-Par 2000*, pages 57–62, Munich, Germany, August 2000.

18. R. Wismüller. State based visualization of PVM applications. In *Parallel Virtual Machine –EuroPVM'96*, volume 1156 of Lecture Notes in Computer Science, pages 91–99, Munich, Germany, October 1996.

19. R. Wismüller. Interoperability support in the distributed monitoring system OCM. In *Proc. 3rd International Conference on Parallel Processing and Applied Mathematics - PPAM'99*, pages 77–91, Kazimierz Dolny, Poland, September 1999.

CUMULVS Viewers for the ImmersaDesk·

Torsten Wilde, James A. Kohl and Raymond E. Flanery, Jr.
Oak Ridge National Laboratory

Keywords: Scientific Visualization, CUMULVS,
ImmersaDesk, VTK, SGI Performer

Abstract. This paper will discuss the development of CUMULVS "viewers" for virtual reality visualization via ImmersaDesk/Cave systems. The CUMULVS (Collaborative, User Migration, User Library for Visualization and Steering) system, developed at Oak Ridge National Laboratory, is a base platform for interacting with high-performance parallel scientific simulation programs on-the-fly. It provides run-time visualization of data while they are being computed, as well as coordinated computational steering and application-directed checkpointing and fault recovery mechanisms. CUMULVS primarily consists of two distinct but cooperative libraries - an application library and a viewer library. The application library allows instrumentation of scientific simulations to describe distributed data fields, and the viewer library interacts with this application side to dynamically attach and then extract and assemble sequences of data snapshots for use in front-end visualization tools. A development strategy for integrating CUMULVS with the virtual reality visualization libraries and environments will be presented, including discussion of the various data transformations and the visualization pipeline necessary for converting raw CUMULVS data into fully rendered 3-dimensional graphical entities.

1. Introduction

Scientific simulation continues to proliferate as an alternative to expensive physical prototypes and laboratory experiments for research and development. Scientists often run software simulations on high-performance computers to obtain results in a fraction of the time, or at a higher resolution, versus traditional mainframe systems, PCs or workstations. Software experiments provide a cost-effective means for exploring a wide range of input datasets and physical parameter variations, but much infrastructure is required to enable the development of these parallel and distributed computer simulations. Teams of scientists often need to observe the ongoing progress of the simulation and share in the coordination of its control.

* Research supported by the Mathematics, Information and Computer Sciences Office, Office of Advanced Scientific Computing Research, Office of Science, U.S. Department of Energy, under contract No. DE-AC05-00OR22725 with UT-Battelle, LLC.

It is useful to visualize the ongoing simulation and interact with it at runtime, especially if the simulation runs for several days. For example, it could save time and money to discover that a simulation is heading in the wrong direction due to an incorrect parameter value, or because a given model does not behave as expected.

A proper visualization environment allows scientists to view and explore the important details of a simulation data set. For a 2-dimensional (2D) data set, a 2D visualization environment is sufficient. But for 3-dimensional (3D) problems, a 3D visualization environment is required to provide the best access to *all* information embedded in the data set. One of the most sophisticated ways to interact with a simulation in 3D is via a virtual reality environment. 3D vision is one of the core human senses, and scientists can use virtual reality to intuitively interact with a simulation using a fabricated "real-world" interface.

This paper describes work to build a 3D virtual reality viewer environment for high-performance applications that use CUMULVS as the interface to application data. The ImmersaDesk™ was chosen as the target display environment because it is less expensive and requires less space then a fully immersive CAVE™ environment. Yet the ImmersaDesk uses the same library interface as the CAVE™, so graphical views for an ImmersaDesk™ can also be displayed on a CAVE™ without modification. A variety of visualization software libraries are required to translate and convert raw data from CUMULVS into a form viewable on the ImmersaDesk™. Section 3 will describe these design choices, after a brief background section. Sections 4 and 5 present results and future work, respectively.

2. Background

2.1. CUMULVS

CUMULVS (Collaborative, User Migration, User Library for Visualization and Steering) [9,10] provides a fundamental platform for interacting with running simulation programs. With CUMULVS, a scientist can observe the internal state of a simulation while it is running via online visualization, and can "close the loop" and redirect the course of the simulation using computational steering. These interactions are realized using multiple independent front-end "viewer" programs that dynamically attach to, interact with and detach from a running simulation as needed. Each scientist controls his/her own viewer, and can examine the data field(s) of choice from any desired perspective and at any level of detail. A simulation program need not always be connected to a CUMULVS viewer; this proves especially useful for long-running applications that do not require constant monitoring. Similarly, viewer programs can disconnect and re-attach to any of several running simulation programs. To maintain the execution of long-running simulations on distributed computational resources or clusters, CUMULVS also includes an application-directed checkpointing facility and a run-time service for automatic heterogeneous fault recovery.

CUMULVS fundamentally consists of two distinct libraries that communicate with each other to pass information between application tasks and front-end viewers.

Together the two libraries manage all aspects of data movement, including the dynamic attachment and detachment of viewers while the simulation executes. The application or "user" library is invoked from the simulation program to handle the application side of the messaging protocols. A complementary "viewer" library supports the viewer programs, via high-level functions for requesting and receiving application data fields and handling steering parameter updates.

The only requirement for interacting with a simulation using CUMULVS is that the application must describe the nature of its data fields of interest, including their decomposition (if any) across simulation tasks executing in parallel. Using calls to the user library, applications define the name, data type, dimensionality / size, local storage allocation, and logical global decomposition structure of the data fields, so that CUMULVS can automatically extract data as requested by any attached front-end viewers. Given an additional periodic call to the stv_sendReadyData() service routine, CUMULVS can transparently provide external access to the changing state of a computation. This library routine processes any incoming viewer messages or requests, and collects and sends outgoing data frames to viewers.

When a CUMULVS viewer attaches to a running application, it does so by issuing a "data field request," that includes a set of desired data fields, a specific region of the computational domain to be collected, and the frequency with which data "frames" are to be sent back to the viewer. CUMULVS handles the details of collecting the data elements of the subregion, or "view region," for each data field. The view region boundaries are specified in global array coordinates, and a "cell size" is set for each axis of the data domain. The cell size determines the stride of elements to be collected for that axis, e.g. a cell size of 2 will obtain every other data element. This feature provides more efficient high-level overviews of larger regions by using only a sampling of the data points, while still allowing every data point to be collected in smaller regions where the details are desired.

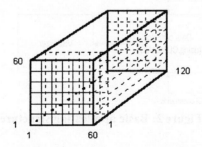

Figure 1: Data space of the viewer test application

2.2. Test Application

The simple test application for our viewer development calculates an expanding gas cloud. *Figure 1* shows the data space for this problem, which uses a rectangular mesh data topology. Potential visualization techniques for this kind of problem would include either data point visualization, contour rendering or volume rendering. For point visualization, each element of the data set is represented by a distinct geometrical object (such as a sphere). The properties of each object depend on the data values for each corresponding element, e.g. the color of each sphere can represent the temperature value at each point. This technique applies well to small data sets but requires significant computational power for larger data sets. Contour rendering visualizes the surface of an object, e.g. an isosurface can be drawn for a chosen contour value, identifying uniform values within the data set. Volume rendering additionally renders the interior of such contoured objects. Volume rendering is important for translucent objects, like water or gas clouds, where scattered light passing through and reflecting from the data can reveal various internal characteristics. Here, changing properties inside the data field need to be considered for proper rendering. The best visualization techniques for our test application would be either volume or contour rendering; we have chosen contour rendering for our experiments due to constraints in the available visualization libraries (Section 3.2).

3. CUMULVS Viewer Design for Immersive CAVE Environments

The goal of this work was to visualize data sets collected by CUMULVS in an immersive environment such as the ImmersaDesk or CAVE. CUMULVS already supports several graphical viewers, using AVS5 and Tcl/Tk, so a true 3D virtual reality viewer would be a useful addition. Because the ImmersaDesk used for this development was driven by an SGI Workstation, our viewer has been designed to run on the SGI Irix operating system.

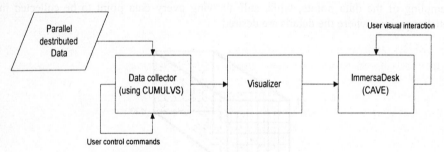

Figure 2: Basic application structure

3.1. Viewer Concept

Our basic viewer structure is shown in *Figure 2*. Depending on interaction from the user, the "Data Collector" requests the desired data set from the user application. This raw data is then converted into the data format as needed by the "Visualizer". In the Visualizer the data is converted into a useful visualization stream for the "ImmersaDesk" where the information is finally displayed.

To improve control over the viewer, the user can interact with the data collector, e.g. change the boundaries of the collected data fields, or change the frequency of data collection. This type of control will be referred to here as "User Control Commands." On the ImmersaDesk, the user can additionally interact directly with the virtual environment, e.g. zooming in/out of the data set and rotating the data set, independent of the ongoing data collection.

3.2. Viewer Pipeline

The purpose of the visualization pipeline is to create a scene, from the collected data sets, that can be properly interpreted by the ImmersaDesk (see *Figure 3*). The CAVE library [1], which is used to access the ImmersaDesk, recognizes OpenGL graphics primitives so the scene *could* be created in OpenGL. The problem with OpenGL, however, is that such programming is at the lowest abstraction level. The programmer has more flexibility in this format, but this approach adds significant time and complexity to the viewer development. For our development this is not cost effective, so we chose to utilize a higher-level toolkit with more built-in graphical objects to construct our viewer.

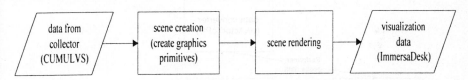

Figure 3: Basic visualization pipeline

With the increasing demand for data visualization, many so-called "visualization toolkits" have been developed, like VTK (Visualization Toolkit) from Kitware [3,5], WTK (World Tool Kit) from Sense8 [12], Open Inventor [13] and Java3D [14]. These toolkits help to create visualization pipelines by providing libraries for high-level graphics programming, or using direct visual programming. Two visualization toolkits that provide an applications programming interface (API), and are widely used, are VTK and WTK. VTK is freeware but doesn't support multi-process rendering to multiple channels, which is required for a CAVE environment. WTK is a commercial product that supports immersive environments in its latest version. Because our resulting CUMULVS viewer will be freely distributed, and could be used as a template for the creation of other viewers, our solution must avoid use of commercial products. Therefore VTK was chosen as our base visualization system.

However, without multi-process rendering to multiple channels, a 3D virtual reality (VR) scene can't be directly created for a CAVE environment using VTK. To solve this problem, SGI Performer [8] was integrated into our visualization pipeline. SGI Performer is a high-performance rendering package for SGI workstations with the capability for multi-process rendering to multiple channels. To translate VTK graphical objects into a form usable by SGI Performer, another tool called vtkActorToPF [4] is applied to convert a VTK Actor into an SGI Performer node. Using a special library called pfCAVE [2,6,7], SGI Performer can interface with the CAVE library, thereby completing our visualization pipeline!

The final structure of the Visualizer is shown in *Figure 4*. Raw data is converted into a VTK data structure (see section 3.3) in the "data converter." Next the visualization data is created, as a vtkActor, using VTK visualization techniques. The vtkActor data is converted into SGI Performer node data using vtkActorToPF. The view scene is created and rendered using SGI Performer and pfCAVE. Finally the virtual scene is displayed on the ImmersaDesk.

Because vtkActorToPF converts only a vtkActor into SGI Performer node data, we must use contour rendering as our visualization technique. VTK uses a different actor, "vtkVolume", to store volume-rendered information, and no appropriate translator from vtkVolume to SGI Performer currently exists. Therefore volume information cannot yet be transferred to SGI Performer in this way.

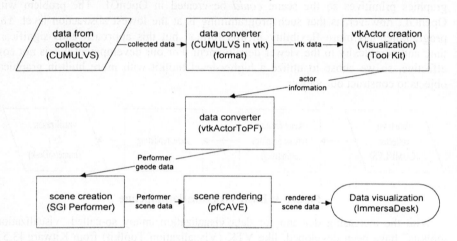

Figure 4: Final visualization pipeline

3.3. CUMULVS to VTK Data Conversion

Internally, VTK data types are derived from the organizing input data structure, and specify the relationship between cells and points [see also 11]. The first criterion is whether the structure is regular, i.e. has a single mathematical relationship between all composing points and cells, or irregular. Regular, or structured, data has a rectangular geometry and topology. The data in our test application are arranged on a rectangular lattice. The rows, columns and planes of the lattice are parallel to the global x-y-z coordinate system. The topology is built from voxels (cubes) that are arranged in a grid structure. Therefore our CUMULVS data set naturally maps to a structured point data set in VTK. If the data set topology and geometry of the source data set were unknown, it would be possible to use the unstructured grid data set as a general source object. But using this object requires more memory and computational resources. This is a motivation for writing individual viewers for different visualization problems, because it is very important in real time visualization to get the best possible performance. The structured point data set in VTK is represented using the data dimension, an origin point and the data spacing. The dimension of data is a 3-vector (n_x, n_y, n_z), specifying the number of points in the x, y and z directions. The origin point is the position in 3-dimensional space of the minimum x-y-z coordinate. Each voxel is identical in shape, with the spacing vector specifying the length in each of the x-y-z directions. VTK uses "Column Major" or FORTRAN storage order to store data elements. Because CUMULVS allows viewers to specify the target storage order of the data set to be collected, it is possible to correctly visualize data from both C and Fortran programs using our resulting viewer.

After determining the topology and size of the application data structure and creating a corresponding VTK equivalent, the data from the CUMULVS data collector can then be transferred into VTK. Due to the internal data organization in VTK, the CUMULVS data elements must be copied one-by-one into the VTK structured point data set. The data dimension and spacing must also be properly set to allow correct interpretation of the VTK data. The CUMULVS "cell size" regulates the number of collected data points along each axis. If we take a 1-dimensional data field with the boundaries $min_x = 1$ and $max_x = 10$, and a cell size $c_x = 1$, all 10 data points are collected. Converting this data into VTK, the dimension would be ($max_x - min_x + 1$) $= 10 = n_x$ and the spacing would be 1 because the number of data points is the same as the field size. If the CUMULVS cell size is changed to $c_x = 2$ then only half the data are collected, skipping every other element. In this case, VTK must have a dimension of $n_x = 5$ to account for the spacing of 2 (spacing equals the CUMULVS cell size). To solve this transformation problem in general, the VTK data dimension is calculated depending on the CUMULVS cell size for each data dimension, e.g. n_x / c_x. Because VTK requires an integer value as dimension the *ceiling* of this result is used. This same principle is applied along each axis for multi-dimensional spaces. Given this mapping the scalar data can be copied from CUMULVS directly into a vtkScalarArray, which is a VTK object for storing an array of scalars. After all the CUMULVS data is copied into it, the vtkScalarArray object is directly linked into the vtkStructuredPoints source object.

3.4. Final Viewer Structure

The final viewer structure is implemented in a single process as shown in *Figure 5*. The main event loop calls all handlers in a specific order. First the data set is collected, and then the visualization pipeline is executed. After that, any User Control Commands are handled and then the visualization handler is called again. This is done to improve the smoothness of the virtual reality interface.

4. Results

Figure 6 shows a sample view from the resulting viewer. To visualize our test application, nearly ½ million data elements are processed. The coloring of the sphere is done using a VTK "Elevation Color Mapper" that produces a color gradient based on the distance from the sphere origin. Otherwise the color of the entire sphere would be uniform. Unfortunately, the long visualization pipeline has some drawbacks in terms of the interactive response time when using this model. The time taken to handle the various visualization and rendering tasks creates delays in displaying view frames and reacting to user control requests. The most time consuming steps of the pipeline are copying the CUMULVS data into the VTK data structure and conversion into the proper formats for visualization. Because User Control Commands for data collection are separated from the visual interaction, multiple handler routines have to be called in the main viewer event loop. User interactivity was improved by adding a special User Control Command that stops the pipeline execution after the data set is collected and visualized. This makes it possible for the user to examine and manipulate the data set on the ImmersaDesk without interruption. A button press triggers collection of the next data set and restarts execution of the visualization pipeline. The paused mode can be repeatedly enabled using the User Control Command interface.

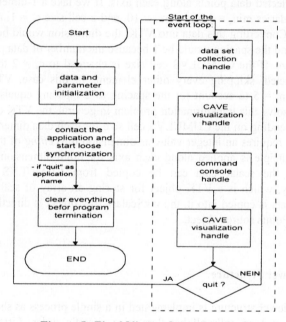

Figure 5: Final Viewer Structure

5. Summary/Future Work

Overall, this viewer is a good starting point for applying data collected by CUMULVS for use with custom 3D virtual reality viewers in CAVE-like environments. The essential portions of the visualization pipeline, including CUMULVS, VTK, vtkActorToPF and pfCAVE, are shown in implementation and behavior. The implemented visual interface is sufficient for most simple 3D visualizations and can be customized and adapted to a specific user application as needed. The use of VTK, vtkActorToPF, SGI Performer and pfCAVE as a visualization pipeline is a cost effective way to develop a 3D VR viewer using CUMULVS as a data collector.

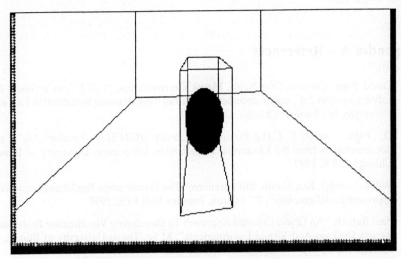

Figure 6: Screenshot of the CAVE simulator

There are several advantages to using VTK. It is "inexpensive" (VTK and vtkActorToPF are freeware, and pfCAVE comes with the CAVE library). It also enables fast prototyping, i.e. new viewers can be created by swapping out VTK objects in the visualization pipeline. On the other hand, the pipeline has a relatively long execution time – mostly due to the data conversions between different entities. Volume rendering is not currently possible because no translator from a vtkVolume to a SGI Performer scene node yet exists. Further, in order to solve specialized visualization problems, deep knowledge is required of all parts of the pipeline, including possible interactions. The documentation of the freeware is sometimes insufficient for new and inexperienced users, exacerbating this problem.

To illustrate the adaptability of our 3D CUMULVS viewer as a viewer template, a CUMULVS viewer for pure VTK was created at the end of this project. Using the VTK MyWindowRendererInteractor class provided by Mark Hoebeke, the visualization using pfCAVE was rewritten to support display using VTK only. The advantage of this viewer is that no expensive hardware is required, and because of the common use of VTK this viewer variation is a useful addition to the CUMULVS software package. The disadvantage of this viewer is that, because VTK is used at several places in the visualization pipeline, user visual interaction is currently only possible with data collection disabled.

Several improvements could be made to the viewer design in order to improve real time visualization and make the user interaction smoother. The structure could be converted from single process into multi-process, where the individual processes could execute independently. Distinct processes could be created for CUMULVS data collection, the steps of the visualization pipeline and the display on the ImmersaDesk. The problem here is that a communication interface for exchanging objects and/or data is required. Other improvements could be made to optimize the copying of the

CUMULVS data into VTK. Furthermore, the User Control Commands could be integrated into the visual user interface, eliminating one handler invocation in the viewer event loop.

Appendix A – References

[1] David Pape, Carolina Cruz-Neira, Marek Czernuszenko, "*CAVE User's Guide, CAVE Library version 2.6*", online documentation from the Electronic Visualization Laboratory, University of Illinois at Chicago, USA, 1997

[2] D. Pape, "*pfCAVE CAVE/Performer Library (CAVELib Version 2.6)*", online documentation from the Electronic Visualization Laboratory, University of Illinois at Chicago, USA, 1997

[3] Will Schroeder, Ken Martin, Bill Lorensen, "*The Visualization Toolkit an object-oriented approach to 3D graphics*", 2nd Edition, Prentice Hall PTR, 1998

[4] Paul Rajlich, "An Object Oriented Approach To Developing Visualization Tools Portable Across Desktop And Virtual Environments", M.Sc. Thesis University of Illinois, 1995 *vtkActorToPF*, http://hoback.ncsa.uiuc.edu/~prajlich/vtkActorToPF

[5] Kitware, *VTK*, http://www.kitware.com

[6] CAVERNUS user group, CAVE Research Network Users Society, http://www.ncsa.uiuc.edu/VR/cavernus

[7] Dave Pape, Online resort for CAVE programming, http://www.evlweb.eecs.uic.edu/pape/CAVE/prog

[8] Silicon Graphics Computer Systems, "*IRIS Performer™ Programmer's Guide*", document number: 007-1680-030

[9] G.A. Geist, J.A. Kohl, P.M. Papadopoulos, "*CUMULVS: Providing Fault-Tolerance, Visualization and Steering of Parallel Applications*", INTL Journal of High Performance Computing Applications, Volume II, Number 3, August 1997, pp. 224-236

[10] J.A. Kohl, P.M. Papadopoulos, "*CUMULVS user guide, computational steering and interactive visualization in distributed applications*", Oak Ridge National Laboratory, USA, Computer Science and Mathematics Division, TM-13299, 02/1999

[11] K.J. Weiler, "Topological Structures for Geometric Modeling", Ph.D. thesis, Rensselaer Polytechnic Institute, Troy, NY, May 1986

[12] Sense8, *World Tool Kit*, http://www.sense8.com/

[13] TGS Inc., *Open Inventor*, http://www.tgs.com/

[14] SUN Microsystems Inc., Java 3D™, http://java.sun.com/products/java-media/3D/

Simulation

N-Body Simulation on Hybrid Architectures

P.M.A. Sloot, P.F. Spinnato, and G.D. van Albada

Section Computational Science
Universiteit van Amsterdam
Kruislaan 403, 1098 SJ Amsterdam, The Netherlands
{sloot, piero, dick}@science.uva.nl
http://www.science.uva.nl/research/scs/index.html

Abstract. N-body codes are routinely used for simulation studies of physical systems, e.g. in the fields of computational astrophysics and molecular dynamics. Typically, they require only a moderate amount of run-time memory, but are very demanding in computational power. A detailed analysis of an N-body code performance, in terms of the relative weight of each task of the code, and how this weight is influenced by software or hardware optimisations, is essential in improving such codes. The approach of developing a dedicated device, GRAPE [9], able to provide a very high performance for the most expensive computational task of this code, has resulted in a dramatic performance leap. We explore on the performance of different versions of parallel N-body codes, where both software and hardware improvements are introduced. The use of GRAPE as a 'force computation accelerator' in a parallel computer architecture, can be seen as an example of a Hybrid Architecture, where Special Purpose Device boards help a general purpose (multi)computer to reach a very high performance[1].

1 Introduction

N-body codes are a widely used tool in various fields of computational science, from astrophysics to molecular dynamics. One important application is the simulation of the dynamics of astrophysical systems, such as globular clusters, and galactic clusters [10]. The core of an N-body code is the computation of the (gravitational) interactions between all pairs of particles that constitute the system. Many algorithms have been developed to compute (approximate) gravity interactions between a given particle i and the rest of the system. Our research is concerned with the simplest and most rigorous method [1], which computes the exact value of the gravity force that every other particle exerts on i. Unlike the well-known hierarchical methods [3,4], this method retains full accuracy, but it implies a computational load that grows as N^2, where N is the total number of particles. Consequently, the computational cost becomes excessive even for a few thousand particles, making parallelisation attractive [15,16].

---

[1] Parts of this work will be reported in the FGCS journal issue on HPCN 2000 [13].

The huge computational requirements of N-body codes also make the design and implementation of special hardware worthwhile. The goal of our research is the study of an emergent approach in this field: the use of Hybrid Computer Architectures. A hybrid architecture is a parallel general purpose computer, connected to a number of Special Purpose Devices (SPDs), that accelerate a given class of computations. An example of this model is presented in [11]. We have evaluated the performance of such a system: two GRAPE boards attached to our local cluster of a distributed multiprocessor system [2]. The GRAPE SPD [9] is specialised in the computation of the inverse square force law, governing both gravitational and electrostatic interactions:

$$\mathbf{F}_i = G \frac{m_j m_i}{|\mathbf{r}_j - \mathbf{r}_i|^3} (\mathbf{r}_j - \mathbf{r}_i) \tag{1}$$

(where m_i and m_j are star masses in the gravity force case, and charge values, in the Coulomb force case). The performance of a single GRAPE board can reach 30 GigaFlop/s. Though some fundamental differences, like electrostatic shielding, exist, this similarity in the force expression allows us in principle to use GRAPE for both classes of problems. Simulated gravothermal oscillations of globular clusters cores, and other remarkable results obtained by using GRAPE, are reported in [9].

Our research aims at understanding how hybrid architectures interact with a given application. For this purpose, we have used NBODY1 [1] as a reference code. It is a widely used code in the field of Computational Astrophysics. It includes all the relevant functionalities of a generic N-body code, without becoming overly complex. We have determined the scaling properties of various parallel versions of the code, with and without use of GRAPE boards. The data obtained are used for the realisation of a performance simulation model that will be applied to study a more general class of hybrid architectures and their interaction with various types of N-body codes [12].

2 Architecture description

The GRAPE-4 SPD is an extremely powerful tool for the computation of interactions that are a function of r^{-2}. Given a force law like (1), the main task of a GRAPE board is to evaluate the force that a given set of particles, the j-particles, exerts on the so called i-particles. This is done in a fully hardwired way, using an array of pipelines (up to 96 per board). Each pipeline performs, for each clock-cycle, the computation of the interaction between a pair of particles.

A GRAPE-4 system consisting of 36 boards was the first computer to reach the TeraFlop/s peak-speed [9]. GRAPE-4 is suitable for systems of up to $10^4 - 10^5$ particles, when running an N-body code whose computational complexity scales as N^2 (^{2}). More sophisticated algorithms, such as the Barnes and Hut tree-code,

---

^{2} Besides the $O(N^2)$ complexity due to force computation, another term due to the relaxation time of the system must be accounted for. This makes the total time complexity of a simulation run $\sim O(N^{8/3})$ for homogeneous systems.

reduce the computing cost to $O(N \cdot \log N)$, at the price of a decreased accuracy, and an increased code complexity [3, 16]. The latter codes change the work distribution between the GRAPE and the host, since many more computations not related to mere particle-particle force interactions must be done by the host. This can make the host become the system's bottleneck. This problem may be solved by using a high performance parallel machine as the host, leading to hybrid architectures.

We connected two GRAPE boards to two nodes of our local DAS (Distributed ASCI Supercomputer [2], unrelated to, and predating the American 'ASCI' machines) cluster. The DAS is a wide-area computer resource. It consists of four clusters in various locations across the Netherlands (one cluster is in Delft, one in Leiden, and two in Amsterdam). The entire system includes 200 computing nodes. A 6 Mbit/s ATM link connects remote clusters. The main technical characteristics of our DAS-GRAPE architecture are summarised in the table below:

local network	host	GRAPE	channel
Myrinet	PentiumPro 200 MHz	300 MFlop/s/pipe peak	PCI9080
150 MB/s peak-perf.	64 MB RAM	62, resp. 94 pipes per board	33 MHz clock
40 μs latency	2.5 GB disk	on-board memory for 44 000 j-particles	133 MB/s

3 Code description

We chose NBODY1 as the application code for our performance analysis work because it is a rather simple code, but includes all the main tasks which GRAPE has been designed to service. This allows us to evaluate the performance of our system. A number of modifications have been made to the code, in order to parallelise it, and to let it make full use of GRAPE's functionalities. We have built and tested the following versions of the code:

- BLK - a basic parallel version of NBODY1, enhanced by adding a block time-step scheme. This code does not make use of GRAPE.
- GRP - like BLK, but now using the GRAPE for the force calculations.

The codes were parallelised using MPI. They are described in more detail below. The basic program flow for an N-body code is given in fig. 1.

3.1 BLK: the BLOCK code

The original version of NBODY1 uses individual time-steps. Each particle is assigned a different time at which force will be computed. The time-step value Δt depends on the particle's dynamics [1]. Smaller Δt values are assigned to particles having faster dynamics (i.e. those particles which have large values in the higher order time derivatives of their acceleration). At each iteration, the code selects the particle that has the smallest $t + \Delta t$ value, and integrates only the orbit of that particle. This reduces the computational complexity, with respect to a code where a unique global time step is used. The individual time step approach reduces the overall time complexity to $O(N^{7/3})$, from the $O(N^{8/3})$ for global time step approach [7]. ([3]). An effect of individual times is that, for each particle, values stored in memory refer to a different moment in time, i.e. the moment of its last orbit integration. This means that an extrapolation of the other particles' positions to time t_i is needed, before the force on i is computed.

```
t = 0
while (t < t_end)
    find new i-particles
    t = t_i + Δt_i
    extrapolate particle positions
    compute forces
    integrate i-particle orbits
```

Fig. 1. Pseudocode sketching the basic NBODY1 tasks.

Since their introduction, N-body codes have evolved to newer versions, that include several refinements and improvements (*cf.* [15]). In the version of NBODY1 used in our study we implemented the so called *hierarchical block time step* scheme [8]. In this case, after computing the new Δt_i, the value actually used is the value of the largest power of 2 smaller than Δt_i. This allows more than one particle to have the same Δt, which makes it possible to have many i-particles per time step, instead of only one. Using this approach, force contributions on a (large) number of i-particles can be computed in parallel using the same extrapolated positions for the force-exerting particles, hereafter called j-particles. Moreover, when a GRAPE device is available, it is possible to make full use of the multiple pipelines provided by the hardware, since each pipeline can compute the force on a different particle concurrently.

parallelisation We let every PE have a local copy of all particle data (see fig. 2 for a pseudocode sketch). Each PE computes force contributions only from its own subset of j-particles, assigned to it during initialisation. A global reduction operation adds up partial forces, and distributes the result to all PEs. Then each PE integrates the orbits of all i-particles, and stores results in its own memory. To select the i-particles, each PE searches among only its j-particles, to determine a set of i-particle candidates. A global reduction operation is performed on the union of these sets in order to determine the real i-particles, i.e. those having the smallest time. The resulting set is scattered to all PEs for the force computation. Since every PE owns a local copy of all particle data, only a set of labels identifying the i-particles is scattered, reducing the communication time.

---

[3] These figures for the time complexity are valid for a uniformly distributed configuration. More realistic distributions show a more complicated dependence on N, although quantitatively only slightly different.

```
t = 0
while (t < t_end)
    find i-particle candidates among my j-particles
    global reduction to determine actual i-particles
    global communication to scatter i-particles
    t = tᵢ + Δtᵢ
    extrapolate particle positions
    compute partial forces from my j-particles
    global sum of partial force values
    integrate i-particle orbits
```

Fig. 2. Pseudocode sketching the tasks of a process of the parallel NBODY1 code.

3.2 GRP: the GRAPE code

The API for the GRAPE hardware consists of a number of function calls, the most relevant for performance analysis being those which involve communications of particles data to and from the GRAPE. These communication operations include: sending j-particle data to GRAPE, sending i-particle data to GRAPE, receiving results from GRAPE.

parallelisation The presence of the GRAPE boards introduces a certain degree of complexity in view of code parallelisation. The GRAPE-hosts obviously play a special role within the PE set. This asymmetry somehow breaks the SPMD paradigm that parallel MPI programs are expected to comply with. Besides the asymmetry in the code structure, also the data distribution among PEs is no longer symmetric. The force computation by exploiting GRAPE boards is done, similarly to the non-GRAPE code, by assigning an equal number of j-particles to each GRAPE, which will compute the partial force on the i-particle set, exerted by its own j-particles. After that, a global sum on the partial results, done by the parallel host machine will finally give the total force. The GRAPE does not automatically update the j-particles' values, when they change according to the system evolution. The GRAPE-host must take care of this task. Each GRAPE-host holds an 'image' of the j-particles set of the GRAPE board linked to it, in order to keep track of such update. Since all force computations and j-particles positions extrapolations are done on the GRAPE, the only relevant work to do in parallel by the PE set, is the search for i-particles candidates, which is accomplished exactly as in the code described in the previous sub-section, the position extrapolation of the $i$particles, and the orbit integration.

4 Results

Measurements for the evaluation of performance of the codes described in the previous section were carried out. They were intended to explore the scaling

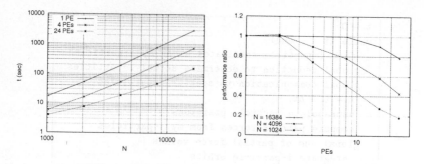

Fig. 3. *a*: Global timings for the parallel block time-step code. *b*: Parallel efficiency of the code.

behaviour of parallel N-body codes. Sample runs were made scaling both N, and the number of PEs n_{PE}; the former from 1024 to 16384, the latter from 1 to 24. NBODY1 does not need much run-time memory, just about 200 bytes per particle, but is heavily compute-bound [5]. Our timings were carried out in order to show the relative computational relevance of the various code tasks as a function of N and n_{PE}.

Our runs were started using a Plummer model distribution as initial condition (density of particles decreasing outward as a power of the distance from the cluster centre). The gravity force is modified by introducing a *softening parameter*, which is a constant term, having the dimension of a length, inserted in the denominator in eq. (1). It reduces the strength of the force in case of close encounters and thus prevents the formation of tightly-bound binaries. In this way very short time-steps and correspondingly long integration times are avoided. The use of a softening parameter is common practice in N-body codes. In our runs, this parameter was set equal to 0.004. As a reference, the mean inter-particle distance in the central core of the cluster, when $N = 16384$, is approximately equal to 0.037.

4.1   Block Time-step Code

The essential tasks of this version of the code are depicted in figure 1. As already sayed, the number of i-particles per iteration can be greater than one. This optimises the force computation procedure, also in view of the use of GRAPE, but, on the other hand, increases the communication traffic, since information about many more particles must be exchanged each time step. Fig. 3 shows total timings and performance of this code, performance being defined as:

$$P_n = \frac{t_1}{n_{PE} \cdot t_n},$$

with t_n the execution time when using n_{PE} PEs. Timings refer to 300 iterations of the code. The execution time grows as a function of N^2 because the number of

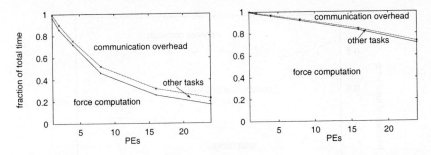

Fig. 4. Evolution of execution time shares for the BLK code. *a*: runs with 1024 particles; *b*: runs with 16384 particles.

i-particles, i.e. the number of force computations, grows approximately linearly with N ([4]). Since the computational cost for the force on each particle also grows linearly with N, the resulting total cost per time-step is $O(N^2)$. Fig. 3*b* shows a good performance gain for this code, affected anyway by a relevant communication overhead. This large overhead can be seen in Fig. 4, which also shows how the execution time shares evolve as a function of n_{PE}. These figures show that for the BLK code, almost all of the computational time is spent in the force computation task; the *j*-particle extrapolation, that takes roughly $25 \sim 30\%$ of the total time in the original code [13], is now reduced to a fraction of one percent.

4.2 GRAPE Code

The code-flow sketched in fig. 1 represents the actual working of GRP too. The only difference with BLK is now that forces are computed on the GRAPE, instead that on the host. We analysed the relative importance of the time spent in GRAPE computation, host computation, and mutual communication. For the parallel version, network communications overheads also have been analysed. The parallel code runs have been done by using only the DAS nodes connected to the GRAPE boards at our disposal, thus the maximum number of PEs in this case is two. We observed that the parallel performance of the GRP code is very poor. The large communication overhead that dominates the GRP code, as can be seen in fig. 5, can explain this. This figure shows, apart from the large communication overhead, that the time share spent in GRAPE computations (i.e. force computations) is quite low, resulting in a low efficiency of this code, in terms of GRAPE exploitation. One reason for that is of course the very high speed of the GRAPE. The GRAPE performs its task much faster than its host and the communication link between them. The figure clearly shows that for our

---

[4] In our runs we found that on average $2.5\% \sim 3\%$ of the particles are updated in every time-step for all values of N. This fraction may change as the system evolves.

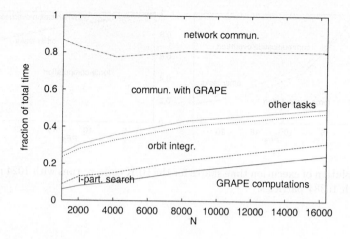

Fig. 5. Scaling of execution time shares for the parallel GRP code.

hardware configuration the capabilities of the GRAPE will only be fully utilised for problems of over 40 000 particles (for single GRAPEs) and approximately double than that for the parallel system. This number is, however, limited by the on-board memory for j-particles of GRAPE.

Measurements [14] show that most of the time spent in communication is due to software overhead in copy operations and format conversions; analogous measurements [6], performed on a faster host, showed a higher communication speed, linearly dependent on the host processor clock speed. Nevertheless, even though GRAPE boards are not exploited optimally, the execution times for the GRP code are much shorter than those for the BLK code. The heaviest run on two GRAPEs is about one order of magnitude faster than the analogous run of the BLK code on 24 PEs. A global comparison of the throughput of all codes studied in this work is given in the next subsection.

4.3 Code Comparison

In order to evaluate the relative performance of the two versions of the N-body code studied in this paper, runs were made, where a 8192 particle system, and a 32768 particle system were simulated for 7200 seconds. Initial conditions and values of numerical parameters were identical to the ones previously specified. The fastest hardware configuration was used in each case, i.e. 24 PEs for the BLK code runs, and 2 PEs (and 2 GRAPEs) for the GRP run. Figs. 6a,b show the evolution of the simulated time, as a function of the execution time. The performance of each code is measured as the time it takes before a simulation reaches a certain simulated time. The figures show that the GRP code outperforms the other code by a factor 8, for 8192 particles, and by a factor 20, for 32768 particles.

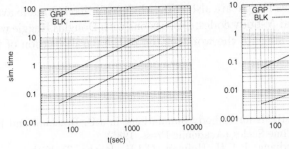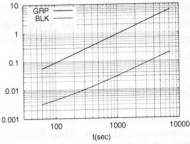

Fig. 6. Performance comparison for the two versions of the *N*-body code. *a*: runs with 8192 particles; *b*: runs with 32768 particles.

These figures clearly show the large performance gain obtained with GRAPE. Using only two PEs, an order of magnitude better performance was attained compared to the BLK code on 24 PEs. Due to the reduction in the time needed for the force calculation, the communication overhead for the GRP code accounts for approximately 50% of the total execution time. Hence an even larger relative gain may be expected for larger problems, as the relative weight of the communication overhead will become less. The difference in performance between the two cases shown in fig. 6 clearly illustrates this effect.

5 Discussion

The main conclusions from our work are, apart from the very good parallel performance of the BLK code, that the GRP code shows a dramatic performance gain, even at a low efficiency in terms of GRAPE boards utilisation. This low efficiency is mainly due to a very high communication overhead, even for the largest problem studied. This overhead can be strongly reduced with the use of a faster host, and by the development of an interface requiring fewer format conversions. The GRAPE hosts in the system that we studied have a 200 MHz clock speed. Nowadays standard clock speeds are 3 to 4 times faster; the use of a state-of-the-art processor would reduce the host and communication times significantly. An extremely powerful machine as GRAPE, in any case, can be exploited efficiently only when the problem size is large, thus attaining the highest SPD utilisation.

The measurements described in this paper have been used to validate and calibrate a performance simulation model for *N*-body codes on hybrid computers. The model will be used to study the effects of various software and hardware approaches to the *N*-body problem.

Acknowledgments Jun Makino has kindly made two GRAPE-4 boards available to us, without which this work would have been altogether impossible.

Sverre Aarseth and Douglas Heggie are also acknowledged for having made available to us the original serial N-body codes. Discussions with Douglas Heggie were of great value at an early stage of the work related to the parallelisation of the code.

References

1. S.J. Aarseth, Direct Methods for N-body Simulations, in J.U. Brackhill and B.I. Cohen, eds., Multiple Time Scales (Academic Press, 1985)
2. H.E. Bal, R.A.F. Bhoedjang, R.F.H. Hofman, C.J.H. Jacobs, T. Kielmann, J. Maassen, R. van Nieuwpoort, J. Romein, L. Renambot, T. Rühl, R. Veldema, K. Verstoep, A. Baggio, G. Ballintijn, I. Kuz, G. Pierre, M. van Steen, A.S. Tanenbaum, G. Doornbos, D. Germans, H. Spoelder, E.-J. Baerends, S. van Gisbergen, H. Afsarmanesh, G.D. van Albada, A.S. Belloum, D. Dubbeldam, Z.W. Hendrikse, L.O. Hertzberger, A.G. Hoekstra, K.A. Iskra, B.D. Kandhai, D.C. Koelma, F. van der Linden, B.J. Overeinder, P.M.A. Sloot, P.F. Spinnato, D.H.J. Epema, A. van Gemund, P.P. Jonker, A. Radulescu, C. van Reeuwijk, H.J. Sips, P.M. Knijnenburg, M. Lew, F. Sluiter, L. Wolters, H. Blom and A. van der Steen, The Distributed ASCI supercomputer project, Operating Systems Review 34, nr 4 (2000) 76–96.
3. J. Barnes and P. Hut., A Hierarchical $O(N \cdot \log N)$ Force-Calculation Algorithm, Nature 324 (1986) 446
4. H. Cheng, L. Greengard and V. Rokhlin, A Fast Adaptive Multipole Algorithm in Three Dimensions, J. Comput. Phys. 155 (1999) 468
5. P. Hut, The Role of Binaries in the Dynamical Evolution of Globular Clusters, in E.F. Milone and J.C. Mermilliod, eds, Proc. of Int. Symp. on the Origins, Evolution, and Destinies of Binary Stars in Clusters. ASP Conf. Series 90 (1996) 391
6. A. Kawai, T. Fukushige, M. Taiji, J. Makino and D. Sugimoto, The PCI Interface for GRAPE Systems: PCI-HIB, Publ. of Astron. Soc. of Japan 49 (1997) 607
7. J. Makino and P. Hut, Performance Analysis of Direct N-body Calculations. Astrophys. J. Suppl. 68 (1988) 833
8. J. Makino, A Modified Aarseth Code for GRAPE and Vector Processors, Publ. of Astron. Soc. of Japan 43 (1991) 859
9. J. Makino and M. Taiji, Scientific Simulations with Special-Purpose Computers, Wiley, 1998
10. G. Meylan and D. C. Heggie, Internal Dynamics of Globular Clusters, Astron. and Astrophys. Rev. 8 (1997) 1
11. P. Palazzari, L. Arcipiani, M. Celino, R. Guadagni, A. Marongiu, A. Mathis, P. Novelli and V. Rosato, Heterogeneity as Key Feature of High Performance Computing: the PQE1 Prototype, in Proc. of Heterogeneous Computing Workshop 2000, Mexico Cancun (May 2000). IEEE Computer Society Press (2000), http://www.computer.org/proceedings/hcw/0556/0556toc.htm
12. P.F. Spinnato, G.D. van Albada and P.M.A. Sloot, Performance Modelling of Hybrid Architectures, Computer Physcics Communications (2001), accepted.
13. P.F. Spinnato, G.D. van Albada and P.M.A. Sloot, Performance of N-body Codes on Hybrid Machines, Future Generation Computer Systems (2001), in press.
14. P.F. Spinnato, G.D. van Albada and P.M.A. Sloot, Performance Measurements of Parallel Hybrid Architectures. Technical Report, *in preparation*
15. R. Spurzem, Direct N-body Simulations, J. Comput. Appl. Math. 109 (1999) 407
16. M.S. Warren, and J.K. Salmon, A Portable Parallel Particle Program. Comp. Phys. Comm. 87 (1995) 266

Quantum Mechanical Simulation of Vibration-torsion-rotation Levels of Methanol

Yun-Bo Duan[1] and Anne B. McCoy[2]

[1]Institute for Computational Science and Engineering,
Ocean University of Qingdao, Shandong 266003, China

and

Department of Physics and Astronomy,
The University of British Columbia,
Vancouver, BC V6T 1Z1, Canada
yduan@physics.ubc.ca

[2]Department of Chemistry, The Ohio State University,
Columbus, OH 43210
mccoy@chemistry.ohio-state.edu

Abstract

Two kinds of vibration-torsion-rotation Hamiltonians, referred as a model Hamiltonian and quantum mechanical Hamiltonian, are constructed to investigate the vibration-torsion-rotational interaction in methanol. The model Hamiltonian is based on the formulation of reduction of Hamiltonian in which the CO-stretching mode ν_8, the large-amplitude torsion mode ν_{12} and the three degrees of freedom that correspond to the overall rotation of the molecule are considered simultaneously. This Hamiltonian is used to carry out an analysis of already published data for CH_3OH with $v_{co} \leq 1$ (CO-stretching vibrational quantum number), $v_t \leq 4$ (torsional quantum number), and $J \leq 5$. The relative locations of the CO-stretch vibrational ground state and the fundamental state are well reproduced for torsional states with $v_t \leq 4$ and $J \leq 5$. An effective potential energy surface that describes ν_8 and ν_{12} modes is obtained from this analysis. The quantum mechanical Hamiltonian is partitioned in the form $H_A + H_B + H_{int}$ by employing a body-fixed axis system and the Jacobi coordinates, where H_A and H_B are the rovibrational Hamiltonians of methyl group CH_3 and asymmetric rotor OH, and H_{int} represents their interactions. This Hamiltonian is used to carry out a pure quantum mechanical simulation of the CO-stretching-torsion-rotational states using the the potential function obtained from the model Hamiltonian analysis. The present analyses show that a variety of resonance interactions can affect states for energies larger than 1100 cm^{-1}.

1 Introduction

The interactions between large-amplitude internal motions, overall rotation and other vibrational motions in methanol that contains a three-fold internal rotational potential barrier lead to the complicated energy level patterns.[1,2] The investigation of vibration-torsion-rotation (VTR) effects in molecules with large-amplitude internal motions has assumed increasing prominence as modern experimental and computational technologies have progressed such that obtaining gas-phase spectra of a wide variety of internal-rotation-type molecules with very high resolution and thereby a full quantum mechanical description of the finer details associated with higher-order VTR effects through computer simulation are now possible. Indeed, these technologies have recently allowed great advances in understanding the observed spectra with high resolution in laboratory and interstellar space, such that many VTR effects and potential energy surfaces (PES) are now obtainable by inversion of experimental data or direct *ab initio* quantum chemical calculation. Paralleling these developments, many researchers have attempted to extend this success to investigation of Hamiltonian models and construction of more general potential energy surfaces, which are ultimately necessary to understand the properties of the molecular spectra, energy levels and transition intensities. Due to the extremely complicated energy level structure existing in the larger-amplitude internal-motion type molecules, for example methanol, the study of structure, dynamics and spectroscopy generally demands very accurate experimental and theoretical methods[3-10].

To date, most of theoretical treatments of this type molecule have been based on Hamiltonians that are obtained using the one-large-amplitude internal rotation model. These treatments are based on the results of applying low order perturbation theory, through contact transformation, to the Hamiltonian, written in terms of either the normal or local mode coordinates. As a result of this process, a model Hamiltonian is parameterized in terms of a set spectroscopic constants by fitting observed spectrum. Generally, the model Hamiltonians based on one-large-amplitude internal rotation model have met with a high degree of success, for both symmetric and asymmetric internal rotors, in determining the intramolecular potential energy hindering the internal rotation. The most advantage of these models is that they can be applied to highly excited states of larger polyatomic molecules. However, The models have their drawback. First, for a simple model of a rigid rotor with a rigid internal rotor, many effects observable in the spectra of the molecule depend on the interaction of both internal rotation and overall rotation with the other vibrations. These effects should be considered with analyses. Second, the perturbation theory is based on the usual normal mode coordinates and is assumed to provide an exact description of the energy level progressions only for non-degenerate case. All of these assumption are not entirely valid. Third, a correlation-free form of Hamiltonian obtained using contact transformation method is different for different molecules and even different isotopomers of the same molecule.[5] This indicates that the contact transformation procedure depends on the particular molecule under consideration. Finally, so many parameters in a practical application had to be used to achieve experimental precision that the physical picture of models used and the physical origin of the spectroscopic parameters fitted become ambiguous.[8-10] Evidently, the usual normal mode expansion of the Hamiltonian are no longer valid when the system of interest contains one or more large amplitude vibrational motions that are highly anharmonic. The usual perturbation expansions for the effects of anharmonicity and VTR coupling may not converge.

For methanol, in addition to the low-frequency, large amplitude torsion, eleven other vibrational degrees of freedom are needed to describe the dynamics of the molecule. Difficulties in locating the fundamentals arise from the overlap of bands due to the broad rotational envelopes, from complicated torsional structure, and from Coriolis couplings and resonances. Thus, how to provide a realistic model system in which interactions among rotation, torsion and other vibrational modes are clearly considered is an important challenge. As the fundamental frequency of the CO-stretch mode is the nearest one to that of the torsional mode in methanol, the understanding of interactions between the CO-stretching mode and the large amplitude hindered internal rotation is an intrinsic fundamental problem. In fact, the CO-stretch fundamental has been studied extensively because of interest in the rich optically pumped far infrared laser emission observed from methanol. The problem encountered now is the observed CO-stretch torsion-rotation energy structure is still not properly accounted for by existing models even though ν_8 is the second lowest frequency mode. In particular, the fact that the hindering barrier for torsional mode had to be increased by about 20 cm^{-1} relative to its value in the ground state indicates the presence of a significant torsion-CO-stretch interaction. Therefore, investigation of interaction between the CO-stretch fundamental and the large amplitude torsion is of considerable interest.

First, in the present work, we present a model Hamiltonian that accounts for the couplings among vibration, torsion and rotation for a molecule with an internal rotor. A five-dimensional VTR Hamiltonian for methanol is constructed based on the formulation of reduction of Hamiltonian in which the CO-stretching mode ν_8, the large-amplitude torsion mode ν_{12} and the three degrees of freedom that correspond to the overall rotation of the molecule are considered simultaneously. This Hamiltonian is used to carry out an analysis of already published data for CH$_3$OH with $v_{co} \leq 1$, $v_t \leq 4$, $J \leq 5$, and $|K| \leq 5$. The present analysis indicates that the height of the torsional barrier V_3 becomes 370 cm^{-1} by introducing torsion-CO-stretching-vibration interaction terms, this should be compared to the values for V_3 when only $v_{co} = 0$ or $v_{co} = 1$ are considered 373 cm^{-1} for ground vibrational state and 392 cm^{-1} for the CO-stretching fundamental state. Second, instead of taking the perturbation approach for investigating the structure, dynamics and spectroscopy of methanol, we focus on the variational approach that based on an internal coordinate Hamiltonian and underlying potential function. The variational approach is a perturbation-theory-free method[11] and it is possible to develop theoretical models applicable to all molecules in a certain class. This method has been extensively applied for weakly bond cluster molecules and chemical reaction dynamics. Employing a body-fixed axis system and the Jacobi coordinates, we derive a Hamiltonian for the vibrating-internally rotating-rotating CH_3OH molecule with a large amplitude internal motion-vibration-rotation interactions. The resulting Hamiltonian is used to carry out a pure quantum mechanical simulation for the VTR states of this kind of molecule.

2 Theory and Method

2.1 Perturbation Method - Model Hamiltonian

In the present model, we consider simultaneously the small amplitude CO-stretching mode and the torsion with the overall rotation in methanol. The five degree-of-freedom model arises by freezing the other vibrational modes except the two modes with lowest frequency. The distance between the centers of mass of methyl group CH_3 and asymmetric rotor OH, R, is used to describe the CO-stretching vibration. A frame-fixed axis system with its origin at the center of mass of CH_3OH is chosen in which the a axis is parallel to the axis of symmetry of the internal rotor, the b axis is parallel to the plane of symmetry, and the c axis is perpendicular to the plane of symmetry. The internal rotation coordinate γ is chosen as the torsional angle of the internal rotor with respect to the frame.

According to the perturbation theory, the reduced rovibrational Hamiltonian of methanol for a given total angular momentum J can be written as

$$H = H_V + H_T + H_R + H_{TR} + H_{VTR}, \qquad (2.1)$$

where H_V is the CO-stretching vibrational Hamiltonian, H_T is the torsional Hamiltonian, H_R is the overall rotation Hamiltonian, H_{TR} is the reduced torsion-rotational interaction Hamiltonian that is analogous to those described in Refs. 5 and 6 and is totally symmetric in the molecular symmetry (MS) group G_6 and invariant under time reversal, and H_{VTR} represents interactions among vibration, torsion, and rotation, which is obtained by multiplying each of the terms in H_R, H_T and H_{TR} and a power series in z and P_z.

The matrix elements of the complete CO-stretching-torsion-rotational Hamiltonian will be obtained by using the basis functions

$$|\Psi_{v_{co},v_t,\sigma,J,K}\rangle = |v_{co}\rangle|v_t,\sigma\rangle|JK\rangle, \qquad (2.2)$$

where the torsional symmetry label σ is equal to 0, +1, or -1, classifying the levels respectively as A, E_1, or E_2 symmetry species of MS G_6 group associated with the threefold nature of the hindering potential. The VTR energy levels are obtained by diagonalizing the matrix resulting from the Hamiltonian matrix.

2.2 Perturbation-free Method - Variational Calculation

In the present model, methanol is divided into two subsystems A and B, i.e., methyl group CH_3 and asymmetric diatomic OH. The five Jacobi vectors ($\mathbf{r}_{1A}, \mathbf{r}_{2A}$,

r_{3A}, r_B, R) specify the atomic positions in methanol, CH_3OH. To formulate our theory, three kinds of axis systems are introduced as follows: i) the space fixes (SF) axis system; ii) the body-fixed (BF) axis system (x, y, z) with its origin at the molecular center of mass; iii) the monomer fixed (MF) axis system (ξ, η, ζ) with its origin at the nuclear center of mass of the monomer. The ζ-axis is along with one of the instantaneous principal axes.

The coordinates used in the present work are defined as follows. In the BF coordinate frame, the z-axis is defined to lie along the Jacobi vector R that connects the centers of mass of the two fragments A and B, pointing from the monomer A to the monomer B. The symmetric plane, i,e., HCOH plane, of methanol coincides with the xz-plane with the x-axis points upward. This is the geometry with torsional angle setting to be zero. The R has the polar angles (α, β) with respect to the SF coordinate system. The MF frame of monomer A as follows, (i) the ζ-axis lies parallel to r_{3A}; (ii) r_{2A} lies in the $\xi\zeta$-plane. Thus, the six independent coordinates that describe the internal vibrations in CH_3 group are $(r_{1A}, r_{2A}, r_{3A}, \theta_1, \phi_1, \theta_2)$. The Euler angles $\Omega^{(\alpha)} = (\theta_\alpha, \phi_\alpha, \chi_\alpha)$ with $\alpha = A$ and B describe the orientation of the MF axis systems of the methyl group CH_3 and the OH group in the body fixed axes, where the dihedral angle $\phi = \phi_B - \phi_A$ describes the torsion of the fragments A and B about z-axis, χ_A describes rotation of CH_3 about its symmetric axis. The angle χ_B and the corresponding quantum number disappear for methanol due to a linear part, OH.

The full-dimensional rovibrational Hamiltonian of a molecule that can be divided into two monomers for a given total angular momentum J, in the BF axis system and the Jacobi coordinates above, can be written as

$$H = H_{rv}^{(A)} + H_{rv}^{(B)} + H_{int} \qquad (2.3)$$

where $H_{rv}^{(\alpha)}$, with $\alpha = A$ and B, represents the rovibrational kinetic operators of the CH_3 and OH, respectively. The $H_{rv}^{(A)}$ can be written as

$$H_{rv}^{(A)} = \sum_{i=1}^{2} \left[-\frac{\hbar^2}{2\mu_{iA}} \frac{\partial^2}{\partial r_{iA}^2} + \frac{L_{iA}^2}{2\mu_{iA} r_{iA}^2} \right] - \frac{\hbar^2}{2\mu_{3A}} \frac{\partial^2}{\partial r_{3A}^2} + \frac{(j_A - L_{12})^2}{2\mu_{3A} r_{3A}^2} \qquad (2.4)$$

where μ_{iA} is the reduced mass of two points of mass. The j_A is the total angular momentum operator of the CH_3. The $L_{12} = L_{1A} + L_{2A}$ is the total angular momentum operator of the hydrogen atoms $(H_1 H_2 H_3)$ in the CH_3. The rovibrational kinetic operator of OH, $H_{rv}^{(B)}$, is written as

$$H_{rv}^{(B)} = -\frac{\hbar^2}{2\mu_B} \frac{\partial^2}{\partial r_B^2} + \frac{j_B^2}{2\mu_B r_B^2}, \qquad (2.5)$$

where μ_B is the reduced mass of the diatomic OH and the j_B is the angular momentum operator of the OH.

The inter-monomer interaction Hamiltonian, H_{int}, is given by

$$H_{int} = -\frac{\hbar^2}{2\mu} \frac{\partial^2}{\partial R^2} + \frac{(J - j_{AB})^2}{2\mu R^2} + V, \qquad (2.6)$$

where μ represents the reduced mass of the two monomers in methanol and the $j_{AB} = j_A + j_B$ is the coupled internally rotational angular momentum. The potential V in Eq. (2. 6) is a function of twelve internally vibrational coordinates. The expression of the Hamiltonian operator shows that the Jacobi coordinates provide the exact kinetic operator and allow one to select a physical axis of quantization. Especially, the matrix elements of the Hamiltonian derived above can be conveniently evaluated in an adequate representation. It is worth to note that the MS group[11] that the molecule under consideration belongs to can be used to extremely simplify the evaluation of the matrix elements and help one classify the energy levels.

In the present work, we only discuss the case of the CO-stretch-torsion-rotation. That means all of the vibrational modes are frozen except the CO-stretching and torsional modes. Thus, the Hamiltonian given in Eq. (2. 3) becomes

$$
\begin{aligned}
H &= -\frac{\hbar^2}{2I_A}\frac{\partial^2}{\partial\phi_A^2} - \frac{\hbar^2}{2I_B}\frac{\partial^2}{\partial\phi_B^2} - \frac{\hbar^2}{2\mu}\frac{\partial^2}{\partial R^2} + \frac{(\mathbf{J}-\mathbf{j}_{AB})^2}{2\mu R^2} + V \\
&= -\frac{\hbar^2}{2\mu}\frac{\partial^2}{\partial R^2} + H_{ang}(R),
\end{aligned}
\tag{2.7}
$$

where I_A and I_B are the inertias of CH$_3$ and OH around the z axis, respectively.

To construct the matrix of the Hamiltonian in Eq. (2. 7), the total spectral representation is written as the direct product $\Re_{ang} \otimes \Re_{rad}$ with

$$
\Re_{rad} = \{|\psi_{v_{co}}\rangle, v_{co} = 1, N_R\},
\tag{2.8}
$$

where $\psi_{v_{co}}$ is the basis function in the radial degree of freedom of R. The angular basis $\Re_{ang}$ is taken as

$$
\Re_{ang} = \{|k_A, k_B\rangle\} \otimes \{|J, M, K\rangle\},
\tag{2.9}
$$

where the Wigner functions $\{|j, k, m\rangle\}$ are defined as

$$
\langle\phi, \theta, \chi|j, k, m\rangle = \sqrt{\frac{2j+1}{8\pi^2}} D_{mk}^{j*}(\phi, \theta, \chi)
\tag{2.10}
$$

with the Wigner rotation D-matrices defined by

$$
D_{mk}^{j}(\phi, \theta, \chi) = e^{-im\phi} d_{mk}^{j}(\cos\theta) e^{-ik\chi}.
\tag{2.11}
$$

By introducing $\gamma = \phi_B - \phi_A$ and $\Phi = \phi_B + \phi_A$, the Hamiltonian in Eq. (2. 7) and the corresponding basis vector become

$$
H = -\frac{\hbar^2}{2I_+}\left(\frac{\partial^2}{\partial\gamma^2} + \frac{\partial^2}{\partial\Phi^2}\right) - \frac{\hbar^2}{I_-}\frac{\partial^2}{\partial\gamma\partial\Phi} - \frac{\hbar^2}{2\mu}\frac{\partial^2}{\partial R^2} + \frac{(\mathbf{J}-\mathbf{j}_{AB})^2}{2\mu R^2} + V, \tag{2.12}
$$

and

$$
\langle\phi_A, \phi_B|k_A, k_B\rangle = \frac{1}{2\pi}e^{i(k_A\phi_A + k_B\phi_B)} = \langle\gamma, \Phi|m, K\rangle = \frac{1}{2\pi}e^{1/2i(K\Phi + m\gamma)}, \tag{2.13}
$$

where $I_{\pm} = I_A I_B/(I_A \pm I_B)$, $m = k_B - k_A$, and $K = k_A + k_B$. The matrix elements of the term, $-1/2\mu\partial^2/\partial R^2$, are of the analytical form in sinc basis function. In our calculation, the R dependence of the basis is given by a discrete variable representation (DVR).[12–14] Thus, the matrix representation of the the Hamiltonian is sparse. The energy levels are obtained by diagonalizing the Hamiltonian in Eq. (2. 12).

3   Results and Discussions

First, the model Hamiltonian given in Sec. 2. 1 is used to analyze all of published spectrum data listed in Ref. 2. The coordinate dependent terms are grouped to provide an effective potential,

$$V = \frac{1}{2}f_2 z^2 + \frac{V_3}{2}(1 - \cos 3\gamma) + \frac{V_6}{2}(1 - \cos 6\gamma)$$
$$+ z^2[^2V_3(1 - \cos 3\gamma) + {}^2V_6(1 - \cos 6\gamma)], \qquad (3.1)$$

which describes an essential part of the torsional problem if multiple CO-stretching states are to be treated simultaneously. The values of the parameters in Eq. (3. 1) are listed in Table 1. The data set that is used in the present fitting consists of 295 microwave (MW) and millimeter wave (MMW) transitions with $v_{co} = 0, v_t \leq 2$ and $J \leq 5$, 737 Fourier transform far-infrared (FIR) data involving energy levels with $v_{co} \leq 1, v_t \leq 4$ and $J \leq 5$ for all available K states of A and E torsional symmetries.[2] All of the MW and MMW lines have experimental uncertainties of 100 kHz and the experimental uncertainties of the Fourier transform FIR data are 0.0003 cm^{-1}= 9 MHz. To reflect the difference between the experimental uncertainties of the MW, MMW and FIR transitions, all of MW and MMW transitions were given a weight that is 100 times larger than that for the FIR data. Using 25 adjustable and 8 fixed parameters the fit converged with a root-mean-square (rms) deviation of 0.20 cm^{-1} for 1032 experimental data with $v_{co} \leq 1, v_t \leq 4, J \leq 5$, and $|K| \leq 5$. The 295 MW and MMW transitions were fit with a rms deviation of 2.1 MHz while 737 FIR data had a rms deviation of 0.238 cm^{-1}. The present results show that the relative location of the vibrational ground state and the CO stretch fundamental state can be well reproduced for torsional states with $v_t \leq 4$, $J \leq 5$, and $|K| \leq 5$ using the simple form of the effective potential in Eq. (3. 1). The first two columns of Table 1 provide the parameters and corresponding operators in the reduced torsion-rotational Hamiltonian for CH_3OH as well as the parameters and operators that were included in previous studies of this system. The values of the parameters that were obtained in the present fit are listed in the third column and their uncertainties (1σ) are given in parentheses. The fourth column provides the results of fit to the transitions with $v_{co}=1$ by Henningsen.[15] The fifth column provides the constants fitted by Duan et al,[9] where 470 MW and MMW transitions with $v_{co}=0$ were fit with a rms deviations of 0.35 MHz using 28 terms.

Table 1: Molecular constants used for representing the CO-stretch-torsion-rotation of CH_3OH

Operator	Constant[a]	Current fit $v_{co}=0, 1$	Fit[c] $v_{co}=1$	Fit[d] $v_{co}=0$
P_a^2	A	127520.580(4006)	127438.32	127536.083(6716)
P_b^2	B	24685.496(1149)	24479.01	24687.6247(336)
P_c^2	C	23764.049(1077)	23548.38	23762.4232(1040)
$\{P_a, P_b\}$	D_{ab}	-151.3(33)		-114.496(1590)
P_γ^2	$F(\text{cm}^{-1})$	27.737805(4606)	27.608226	27.6447930(8785)
	$\rho(\text{unitless})$	0.810211(7)		0.809883
$-P^4$	$\triangle_J$	0.0518(fixed)	0.049	0.05021(6)
$-P^2 P_a^2$	$\triangle_{JK}$	0.2793(947)	0.286	0.2793(982)
$-P_a^4$	$\triangle_K$	1.3053(1570)	1.14	1.6106(266)
$-2P^2(P_b^2 - P_c^2)$	δ_J	0.0018(fixed)		0.00169(7)
$P^2 P_\gamma^2$	G_V	-3.5046(1293)	-5.01	-3.5506(26)
$P^2(1 - \cos 3\gamma)$	F_V	-71.650(1219)	-196.2	-71.49904(3)
$P^2 P_a P_\gamma$	L_V	0.0942(fixed)	-0.068	0.0809(11)
$P_a^3 P_\gamma$	k_1	-3.116(150)	-1.44	-1.6339(1070)
$P_a^2 P_\gamma^2$	k_2	-63.833(1593)	-55.2	-66.445(1115)
$P_a P_\gamma^3$	k_3	-178.290(1519)	-161.1	-122.009(2275)
P_γ^4	k_4	-259.102(1302)	-256.3	-218.7(fixed)
$P_a^2(1 - \cos 3\gamma)$	k_5	343.33(1266)	410.9	325.93(1071)
$\{P_a, P_c\} \sin 3\gamma$	k_7'	-328.522(1250)		-274.52(1529)
$\{P_\gamma^2, P_b^2 - P_c^2\}$	c_1	1.6(fixed)	-1.2272	-1.6632(79)
$\{P_\gamma^2, \{P_a, P_b\}\}$	$\triangle_{ab}$	-2.891(1362)		-6.9643(6738)
$\{P_a, P_b\}(1 - \cos 3\gamma)$	d_{ab}	333.56(3341)		304.800(2591)
$P_a P_\gamma^5$	K_7	0.7658(889)		
$P_a^2 P_\gamma^4$	K_1	0.2327(fixed)		
$P_a^3 P_\gamma^3$	K_3	-0.2663(1703)		
$\{P_\gamma P_a, \{P_a^2, P_b^2 - P_c^2\}\}$	c_{10}	0.2183(fixed)		
$\{P_b^2 - P_c^2, P_a P_\gamma^3\}$	c_{12}	0.2109(fixed)		
$(P_b^2 - P_c^2)\{P_\gamma^2, 1 - \cos 3\gamma\}$	c_{19}	0.2419(fixed)		
$(1 - \cos 3\gamma)/2$	$V_3(\text{cm}^{-1})$	369.45509(8690)	392.35	373.37459(3352)
$(1 - \cos 6\gamma)/2$	$V_6(\text{cm}^{-1})$	6.3653(1379)	-0.52	-0.93196(4757)
z^2	$f_2(\text{SI})^b$	507.6181(864)		
$z^2(1 - \cos 3\gamma)$	$^2V_3(\text{SI})^b$	1.9856(371)		
$z^2(1 - \cos 6\gamma)$	$^2V_6(\text{SI})^b$	-3.7463(561)		
$z^2 P_\gamma^2$	$^2F \times 10^{-33}(\text{SI})^b$	5.381(25)		

[a] Parameters in MHz.
[b] Vibrational and related parameters in SI.
[c] Ref. 15.
[e] Ref. 9.

Using the fitted PES in Eq. (3. 1), we also calculated the CO-stretching-rotational energy levels based on the quantum mechanical Hamiltonian. Table 2 gives the comparison of the two kinds of calculated energy levels with experimental results for $J = 2$, where a total of 50 levels are arranged in order of frequency alone with the observed VTR energy levels. The table contains quantum numbers for the states, residuals (observed-fitted levels and observed-calculated levels), the symmetry (A or E) of the torsional sub-states, and the parity quantum number (+,-) that refer only to A torsional sub-states. The present results shown that the relative location of the vibrational ground state and the CO stretch fundamental state are well reproduced for torsional states with $v_t \leq 2$ using the quantum mechanical simulation. However, the rms of the deviation of the variational values with the experimental ones is about 2.2 cm^{-1} that is about 10 times larger than that in model Hamiltonian. The reason may result either from the approximated model or form the inaccurate potential

Table 2: Energy levels of CH_3OH in unit of cm^{-1a}

A/E	v_{co}	v_t	J	K	P	Exp.[b]	Fit[c]	E-F[d]	Cal.[e]	E-C[f]
						J=2				
A	0	0	2	0	+	4.84049	4.84047	.000	5.51544	-.675
E	0	0	2	-1		-.40543	-.40529	.000	-.00014	-.405
E	0	0	2	0		4.84073	4.84073	.000	5.51533	-.675
A	0	0	2	1	+	14.90430	14.90422	.000	15.45369	-.549
A	0	0	2	1	-	14.98778	14.98770	.000	15.47501	-.487
E	0	0	2	1		10.34657	10.34662	.000	10.46546	-.119
E	0	0	2	2		11.17828	11.17843	.000	10.64262	.536
E	0	0	2	-2		13.71898	13.71917	.000	13.46315	.256
A	0	0	2	2	-	31.04901	31.04893	.000	30.42347	.626
A	0	0	2	2	+	31.04905	31.04900	.000	30.44477	.604
E	0	1	2	1		198.29105	198.27993	.011	199.22743	-.936
A	0	1	2	2	+	212.41775	212.40593	.012	212.34853	.069
A	0	1	2	2	-	212.41775	212.44905	-.031	212.34853	.069
E	0	1	2	0		204.61844	204.60729	.011	205.43830	-.820
A	0	1	2	1	+	230.88932	230.87744	.012	230.53584	.353
A	0	1	2	1	-	230.89894	230.88712	.012	230.55709	.342
E	0	1	2	-2		249.12439	249.11268	.012	246.67994	2.444
E	0	1	2	2		273.64737	273.63810	.009	276.04858	-2.401
E	0	1	2	-1		283.01760	283.00952	.008	285.05823	-2.041
A	0	1	2	0	+	299.27974	299.27187	.008	300.65626	-1.377
A	0	2	2	0	+	358.04034	358.04295	-.003	359.30377	-1.263
E	0	2	2	-1		365.47979	365.48121	-.001	365.21149	.268
E	0	2	2	2		403.15497	403.15451	.000	399.82158	3.333
E	0	2	2	-2		443.95855	443.97235	-.014	449.40802	-5.449
E	0	2	2	1	+	479.37273	479.39099	-.018	483.96187	-4.589
A	0	2	2	1	-	479.39093	479.40920	-.018	483.98276	-4.592
E	0	2	2	0		506.01909	506.03283	-.014	508.81335	-2.794
E	0	2	2	1		551.00435	551.00722	-.003	551.13618	-.132
A	0	2	2	2	-	614.14008	614.13037	.010	610.81209	3.328
A	0	2	2	2	+	614.14008	614.13037	.010	610.81209	3.328
A	1	0	2	0	+	1038.68345	1038.77356	-.090	1040.03541	-1.352
E	1	0	2	-1		1033.91137	1033.51629	.395	1034.43474	-.523
E	1	0	2	0		1038.49648	1038.50936	-.013	1039.91482	-1.418
A	1	0	2	1	+	1048.75741	1048.64912	.108	1049.88171	-1.124
A	1	0	2	1	-	1048.84413	1048.73253	.112	1050.31464	-1.471
E	1	0	2	1		1043.83049	1043.94868	-.118	1044.98585	-1.155
E	1	0	2	2		1045.42780	1045.06618	.362	1045.41899	.009
E	1	0	2	-2		1047.77162	1047.53605	.236	1047.88661	-.115
A	1	0	2	2	+	1064.43044	1064.61015	-.180	1063.38838	1.042
A	1	0	2	2	-	1064.43044	1064.61015	-.180	1063.38838	1.042
E	1	1	2	1		1232.62536	1232.38543	.240	1234.32138	-1.696
A	1	1	2	2	-	1246.87386	1246.63343	.240	1247.47400	-.600
A	1	1	2	2	+	1246.87386	1246.63343	.240	1247.47400	-.600
E	1	1	2	0		1238.68106	1238.48414	.197	1240.41529	-1.734
A	1	1	2	1	+	1264.38316	1264.41266	-.029	1265.32102	-.938
A	1	1	2	1	-	1264.39137	1264.42377	-.032	1265.75318	-1.362
E	1	1	2	-2		1281.81559	1282.16204	-.346	1281.57984	.236
E	1	1	2	-1		*1305.26255	1313.74774	-8.485	1317.59291	-12.330
E	1	1	2	2			1305.75609		1309.68118	
A	1	1	2	0	+		1328.70054		1332.13530	

[a] Energy levels are arranged from lower to higher and are relative to the lowest one for A and E, respectively.
[b] Experimental results are taken from Ref. 2. The levels marked with '*' are not included in the fitted data set.
[c] The present fit results based on a model Hamiltonian.
[d] E-F = the experimental-the fitted values.
[e] The present calculated results based on the variational calculation.
[f] E-C = the experimental-the variational values.

function, which is being investigated.

In conclusion, we have used two kinds of the 5D Hamiltonian models, i.e., perturbation method and perturbation-free method, to investigate the CO-stretching-torsion-rotational states. The model Hamiltonian is used to analyze the observed data that involve the torsional and CO-stretching modes for CH_3OH to generate global values for the molecular parameters. The relative location of the vibrational ground state and the CO stretch fundamental state is well reproduced for torsional states with $v_t \leq 4$ and $J \leq 5$. We have proposed a Hamiltonian for calculating VTR energy levels of methanol through using the Jacobi vectors. The present results for the CO-stretching-torsion-rotation states shown that present formulation can be used to carry out a multi-dimensional quantum calculation for methanol or to fit the potential energy surface. On balance, the presented methods of analysis provide an important first step in an attempt to model various small amplitude vibrational interactions influenced by the internal rotation motion and to interpret observed spectrum data.

ACKNOWLEDGMENTS

The authors gratefully acknowledge to support from Office of Research at The Ohio State University and the Petroleum Research Fund administered by the American Chemical Society.

References

[1] C. C. Lin and J. D. Swalen, *Rev. Mod. Phys.* **31** (1959) 841-892.

[2] G. Moruzzi, B. P. Winnewisser, M. Winnewisser, I. Mukhopadhyay, and F. Strumia, *Microwave, Infrared and Laser Transitions of Methanol, CRC Press*, (1995).

[3] Y. B. Duan and K. Takagi, *Phys. Lett. A* **207** (1995) 203-208.

[4] Y. B. Duan, H. M. Zhang, and K. Takagi, *J. Chem. Phys.*, **104** (1996) 3914-3922.

[5] Y. B. Duan, L. Wang, I. Mukhopadhyay, and K. Takagi, *J. Chem. Phys.*, **110** (1999) 927-935.

[6] Y. B. Duan, L. Wang, and K. Takagi, *J. Mol. Spectrosc.* **193** (1999) 418-433.

[7] Y. B. Duan and A. B. McCoy, *J. Mol. Spectrosc.*, **199** (2000) 302-306.

[8] Y. B. Duan, L. Wang, X. T. Wu, I. Mukhopadhyay, and K. Takagi *J. Chem. Phys.*, **111** (1999) 2385-2391.

[9] Y. B. Duan, A. B. McCoy, L. Wang, and K. Takagi, *J. Chem. Phys.*, **112** (2000) 212-219.

[10] L. Wang, Y. B. Duan, I. Mukhopadhyay, D. S. Perry, and K. Takagi, *Chem. Phys.*, **263** (2001) 263-270.

[11] P. R. Bunker and P. Jensen, *Molecualr Symmetry and Spectroscopy*, NRC Research Press (1998), Ottawa.

[12] J. V. L. Lill, G. A. Parker, J. C. Light, *Chem. Phys. Lett.* **89** (1982) 483-489.

[13] J. C. Light and Z. Bacic, *J. Chem. Phys.* **87** (1987) 4008-4019.

[14] D. T. Colbert and W. H. Miller, *J. Chem. Phys.* **96** (1992) 1982-1992.

[15] J. O. Henningsen, *J. Mol. Spectrosc.* **85** (1981) 282-300.

Simulation-Visualization Complexes as Generic Exploration Environment

Elena V. Zudilova

Integration Technologies Department
Corning Scientific Center
4 Birzhevaya Linia, St. Petersburg 199034, RUSSIA
tel.: 7 (812) 3229525   fax: +7 (812) 3292061
http://www.corning.com
e-mail: ZudilovaEV@corning.com

Abstract. Simulation-visualization complexes combine tools for numerical simulation and data visual representation. They facilitate together the research process of investigating a phenomenon by decreasing necessary time and cost resources. The paper is devoted to the process of design and development of such kind of complexes. It introduces the approach that permits to combine simulation and visualization compounds together and represents at the same time a minimum of user discomfort related with the increasing functionality of a final complex. This goal can be achieved if the interface of the exploration complex will be designed and developed in accordance with such usability criterions as consistency, informative feedback and design simplicity.

1. Introduction

Today's existing simulation tools can not always satisfy researchers if, for instance, output data is too vast to be analyzed in numerical form. If it happens, the best approach for conducting effective analysis is to visualize the numerical data and then deal with the graphical interpretation of the obtained results.

The paper represents the approach of how to combine the features of simulation and visualization software in the framework of one generic exploration environment that can be obtained by the linkage of computational processes and the processes concerned with the graphical interpretation and interaction.

The introduction to numerical simulation and scientific visualization is provided in section 2 of the following paper. Section 3 is devoted to the integration of simulation and visualization compounds and shows what kind of feedback exists between them. Section 4 represents a view to the complex interface solution in generic exploration environment. Such interface usability criterions as feedback, consistency, visual and task simplicity are explained.

2. Introduction to Exploration Complexes

Supercomputer systems of different architectures have become very popular for solution of various mathematical and physical problems demanding large data volumes. The choice of this modern and powerful equipment is caused by the requirement to decrease the time interval necessary for simulation process and/or visualization of obtained results.

Generic simulation-visualization complexes are exploration complexes that contain both tools for numerical simulation, data visual representation and interaction capabilities that facilitate together the process of phenomenon investigating by shorten necessary time and cost resources.

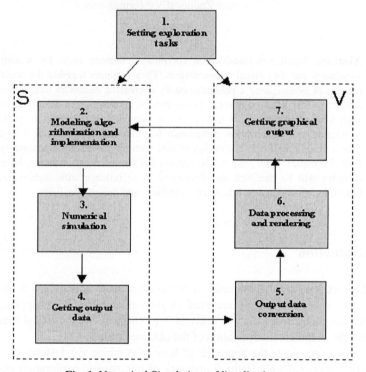

Fig. 1. Numerical Simulation ->Visualisation

Fig. 1 represents the main functional compounds of simulation-visualization process as a whole.

Three preparation stages usually precede numerical simulation: modeling, algorithmization and development of simulation software [11].

The stage of modeling is the most complicated as it covers the elaboration of mathematical models, development of numerical schemes for effective simulation, definition of the investigated zones and fields, selection of scales and criterions. [8]

Algorithmization stage [4] includes algorithm development, selection of hardware and software for further implementation of this algorithm, algorithm adaptation in accordance with selected hardware, if it is necessary. If it is a supercomputer of parallel architecture, then parallel algorithm should be developed.

As for the implementation stage, it means the cycle of scripting, testing and validation.

Supercomputer applications aimed to numerical simulation usually have command-line or zero-dimensional (0-D) user interface. Command line here is as usual a typical interface solution [7]. User may only vary simulation parameters, i.e. initial conditions, criterions, scale, etc. using a keyboard. Sometimes it is not effective and comfortable, especially when the computational processes are based on the complicated mathematical logic.

Only people participated in the development of this software or specially pre-trained persons can effectively interact with the computational processes and analyze output data.

If data generated by numerical simulation software is large and complicated, then the best way to analyze it is to present this data visually. The parameters of many physical and chemical processes are better observed in graphical form. Moreover, there are special information systems based on numerical simulation methods, where visualization can be considered as the only solution of representing output results, such as weather forecasting, medical diagnostics and different types of monitoring.

The problems concerned with data visual presentation are very urgent today because of the fast increasing of data volumes processed by different computer systems including supercomputers. The progress of visualization hardware and software is caused by continuous rising of requirements addressed to them. In the sphere of science and engineering these requirements include clear visual presentation and efficient real-time control of huge data volumes obtained through calculations based on complicated multidimensional models.

So today visualization software developers focus their efforts on the implementation of integrated interactive systems that consist of information management compounds, data structuring tools and visualization toolkits.

Fig. 2. Example of 3D visualization of electromagnetic flow

Output data volumes obtained as a result of numerical simulation are always represented in a special format. If this format is different from those that can be supported by visualization software, it is necessary to convert obtained data to one of

906

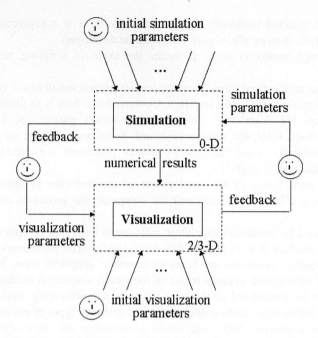

initial simulation
parameters

...

simulation
parameters

Simulation

feedback

0-D

numerical results

Visualization

feedback

visualization
parameters

2/3-D

...

initial visualization
parameters

Fig. 3. Traditional feedback scheme between simulation and visualization software

the available visualization formats. For this purpose a special data format converter should be included into simulation-visualization scheme.

There are many software applications to work with video, graphics and animation that can be used for the visualization of static and/or dynamic objects either existing or being simulated. In ideal case this software includes integrated systems of analyses, processing and data imaging based on the classical representation as graphs with the complicated data visualizations as 2D surfaces or 3D objects (Fig.2). With the help of the following systems users can obtain information from various databases, conduct mathematical and logical operations, edit or analyze them visually, and finally reproduce high-quality copies of images. The following functionality is available by so called 2D or 3D (2/3-D) graphical user interfaces [9].

Creating the Virtual Reality environment is the highest level of 3D visualization when the stereo virtual objects can be interacted during the presentation process. Virtual Reality technologies permit the real-time interaction with the models of 3D objects, including the effect of full presence in the real medium through the audio, video and even tactile components.

Virtual Reality is the modern concept that is widely used today not only in entertainment and education as it was several years ago but also in different science domains, such as biology, medicine diagnostics, molecular modeling, aerospace industry, electrostatics, etc.

3. Interaction between Simulation and Visualization Compounds

The main advantage of combination of simulation and visualization features in the framework of one exploration complex is the significant decrease of time resources necessary for the conduction of the experimental cycle. [3]

The experimental cycle can be shorten, first of all, by minimization of feedback processes, number of people involved in each process and their effort aimed to the maintenance of interaction–adaptation features. Fig. 3 and 4 illustrate how the situation changes when separate simulation and visualization software systems (Fig.3) are combined into one generic complex of the same purpose (Fig. 4).

Simulation-visualization complexes can be static or dynamic [1]. The static complexes deal with time independent data. Generated once, the following data does not change. Only visualization parameters can be varied for better observing the reproduced image.

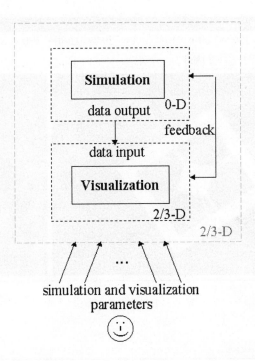

Fig. 4. Feedback scheme in generic simulation-visualization environment

As for dynamic complexes, the interaction between simulation and visualization compounds here is very complicated. Numerical data is generated by simulation software periodically and the visualization results are also updated with the same frequency. So the visualization reproduced currently is the graphical interpretation of numerical data generated at some moment. Visualization results are also can be modified by available visualization parameters.

Thus a common static complex can be interpreted as a special case of a dynamic complex. Ideal variant is when a dynamic exploration complex maintains all the features of static one.

IntelliSuite™ CAD for MEMS can be considered as an example of static simulation-visualization complex implemented as a standalone application. MEMS is an acronym of Micro Electro Mechanical Systems and is a microfabrication technology which exploits the existing microelectronics infrastructure to create complex machines with micron feature sizes. IntelliSuite is an integrated software complex which assists designers in optimizing MEMS devices by providing them access to manufacturing databases and by allowing them to model the entire device manufacturing sequence, to simulate behavior and to see obtained results visually without having to enter a manufacturing facility. [5]

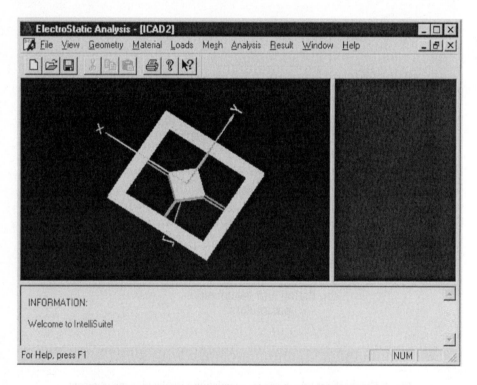

Fig. 5. IntelliSuite results: 3D visualization of an accelerometer

It provides the ability to simulate with high accuracy different classes of MEMS devices induced mechanically, electrostatically and electromagnetically and then to obtain the graphical presentation of the appearance of each simulated device. ACIS 3D Visualization Toolkit is used for the implementation for new visualization engine of IntelliSuite CAD for MEMS. Fig. 5 represents the prototype version of 3D interface of the Electrostatic Analysis Component of the following exploration complex.

There are also two different variants of implementation of generic exploration complexes. A simulation-visualization complex can be implemented as distributed software complex where numerical simulation is conducted on a separate computer system (or even systems) and visualization process is generated on a separate graphical server or workstation. There is another approach when it is implemented as standalone generic simulation-visualization software located on the same computer system possessing necessary computational and visualization resources. The solution about the implementation form is defined both by the complexity of the task to be simulated and the desired visualization features.

Sometimes UNIX workstation or even a simple PC is enough for obtaining sufficient results (simple CAD/CAM applications). But there are several situations when expected results can be provided only by means of distributed exploration complex, where, for instance, the numerical simulation is organized on supercomputer of Cray, HP Convex or Parsytec series and the visualization process is conducted on SGI Onix2 or Origin 2000 supercomputer (complicated run-time tasks). [2]

4. Complex Interface Solution

Integration of two interrelated software compounds of different purposes in a generic complex leads to necessity of design and development of a generic graphical interface that provides a user the ability to interact with each component and to influence on feedback between them. This interface permits to minimize the exploration cycle as a whole, as the features of each compound are now available to user via the same environment. And, moreover, the user interaction with the simulation compound becomes more friendly as 0-D interface of the simulation part is replaced by the graphical 2/3-D user interface of the entire generic simulation-visualization complex (see Fig. 3, 4).

User interface structures optimize access to application data and features. The human-computer interaction becomes highly efficient and productive by mapping the tasks to user goals.

The effective interface design starts from the structure of the product and integrates navigation, information design, visual design and technology. Complex interface solution for a generic simulation-visualization complex should be based on at least three main usability criterions: [6]

1. Consistency at all levels

If actions are executed in a specific way on one screen, users expect the action to be performed in the same way throughout the other screens. They expect certain features of the interface behavior in certain ways no matter with what compound of the exploration complex they work: with simulation or with visualization.

2. Informative feedback

Feedback closes the communication loop between the computer and the user, telling the user how their actions were processed and what the results of those actions are. In the absence of error messages, normal feedback lets the user know that the system is behaving in the expected manner.

3. Design simplicity:

Several types of simplicity contribute to a well-designed user interface:

- Visual simplicity is achieved by showing only the most important objects and controls.
- Verbal simplicity means usage of direct, active, positive language.
- Task simplicity is achieved when related tasks are grouped together, and only a few choices are offered at any one time.
- Conceptual simplicity is accomplished by using natural mappings and semantics, and by using progressive disclosure.

The main aim of the generic interface is to facilitate the work of users so that increasing functionality of simulation-visualization complex does not lead to rapid increasing of user discomfort while working with it.

5. Conclusion

The paper presents the approach of building the generic environment for numerical simulation and scientific visualization. Both tasks that are going to be linked are rather complicated. The paper provides a view of how it can be done with a minimum of user feeling that system complexity has been increased.

The main idea is to combine simulation and visualization features into the generic environment with a common graphical user interface through what a user will be able to manipulate simulation and visualization parameters wherever it is necessary with a minimal effort and in accordance with his expertise. This environment will combine both simulation, visualization and feedback capabilities between two these compounds that permits to deal with the results of numerical simulation in visualized form and to change the circumstances of conducting simulation processes.

Such usability criterions of interface design as: consistency, informative feedback and design simplicity permit to minimize demands on human memory. That will help to provide the success of the exploration complex among its further users. The ideal variant is even to provide the generic environment with adaptive user interface that permits a concrete user to have at the top level the most frequently used features [6].

References

1. Belleman R.G., Sloot P.M.A: The Design of dynamic Exploration Environments for Computational Steering Simulations, Proceedings of the 1st SGI Users' Conference, pp 57-72 (2000).
2. Bogdanov A.V., Gevorkyan A.S., Gorbachev Yu.E., Zudilova E.V., Mareev V.V., Stankova E.N.: Supercomputer Cluster of the Institute for High Performance Computing and Data Bases, Proceedings of CSIT Conference, Yerevan, Armenia, pp. 412-354 (1999).

3. Bogdanov A.V., Stankova E.N., Zudilova E.V. Visualization Environment for 3D Modeling of Numerical Simulation Results, Proceedings of the 1st SGI Users' Conference, pp. 487-494 (2000).

4. Bogdanov A.V., Zudilova E.V., Zatevakhin M.A., Shamonin D.P.: 3D Visualization of Convective Cloud Simulation. Proc. The 10th International Symposium on Flow Visualization, pp. 118-124 (2000).

5. He Y., Marchetti J., Maseeh F.: MEMS Computer-aided Design. Proc. of the 1997 European Design &Test Conference and Exhibition Microfabrication (1997).

6. Gavrilova T., Zudilova E., Voinov A. User Modeling Technology in Intelligent System Design and Interaction. Proc. of East-West International Conference on Human-Computer Interaction, EWHCI'95, Vol.2, pp. 115-128 (1995).

7. Gorbachev Y., Zudilova E: 3D-Visualization for Presenting Results of Numerical Simulation. Proceedings of HPCN Europe - 99, pp.1250-1253 (1999).

8. Stankova E., Zudilova E. Numerical Simulation by Means of Supercomputers. Proceedings of HPCN Europe - 98, pp. 901-903 (1998).

9. Upson C., Faulhaber J., Kamins D.: The Application visualization System: A Computational Environment for Scientific Visualization, IEE Computer Graphics and Applications. 9(4), pp..30-42 (1989).

10. Zudilova E.: Using Animation for Presenting the Results of Numerical Simulation. Proceedings of International Conference of Information Systems Analysis and Synthesis) / Conference of the Systemics, Cybernetics and Informatics, Orlando, Florida, USA, pp. 643-647 (1999).

11. Zudilova E., Shamonin D.: Creating DEMO Presentation on the base of visualization Model. Proceedings of HPCN Europe - 2000, pp.460-467 (2000).

Efficient Random Process Generation for Reliable Simulation of Complex Systems

Alexey S. Rodionov[1], Hyunseung Choo[2], Hee Yong Youn[2], Tai M. Chung[2], and Kiheon Park[2]

[1] Novosibirsk Institute of Computational Mathematics and Mathematical Geophysics
Siberian Branch of the Russian Academy of Sceiences, Novosibirsk, Russia
alrod@rav.sscc.ru
[2] School of Electrical and Computer Engineering
Sungkyunkwan University, Suwon, Korea
{choo, youn,tmchung,khpark}@ece.skku.ac.kr

Abstract. Generating pseudo random object is one of the key issues in computer simulation of complex systems. Most earlier systems employ independent and identically distributed random variables, while those of real processes often show nontrivial autocorrelation. In this paper a new approach is presented that generates random process for given marginal distribution and autocorrelation function. The proposed approach achieves significant speed-up and accuracy by using truncated distribution instead of order statistics. An experiment displays more than ten times speed-up for reasonable size system. Moreover, the proposed generator is simple and requires less memory.

1 Introduction

In developing a stochastic model to describe a real system, a model realistically replicating the actual operation should be chosen. Even though mathematical analysis is a useful tool for predicting the performance of a system wherever it is tractable, computer simulation is gaining more attention due to the relatively recent advent of fast and inexpensive computational power. It allows faithful reproduction of the behavior of real processes and comprehensive analysis based on the data collected.

For computer simulation of complex systems pseudo random objects of various nature need to be generated [1, 2]. In most general purpose discrete simulation independent and identically distributed random variables are usually used. However, the random numbers of real processes often show nontrivial autocorrelation. For example, observe that the two processes of a same exponential marginal distribution ($\lambda = 1$) in Figure 1 look quite different due to different autocorrelation functions. For an M/M/1 queue with the service rate of 1.5 and inter-arrival rate of 1, the average waiting time is 7.94. For the inter-arrival distribution of Figure 1(a) and (b), it is 12.14 and 7.62, respectively. The difference is significant based on the t-criteria with a significance level of 95%. This exemplifies the importance of simulation with dependent random numbers.

A number of researches related to this problem have been done, while they are for generating mostly correlated vectors and fields [3 − 6]. The method commonly employed for generating dependent random vectors [3, 7], however, is not suitable for long one-dimensional sequence due to the requirement on large size memory. As a result, other methods were developed to generate stable one-dimensional processes with some given properties and arbitrary length [8 − 10]. Based on the theoretical foundation on generating random processes with required properties [10, 12], we propose a new method for generating pseudo random processes with a given marginal distribution and autocorrelation function (ACF). Such processes often allow us to faithfully replicate the actual operation of the simulated system. For a 16-state randomized Markov chain about 14 times acceleration is achieved in comparison with the previous algorithm [12], while the storage requirement is much smaller. Moreover, the proposed generator is simple.

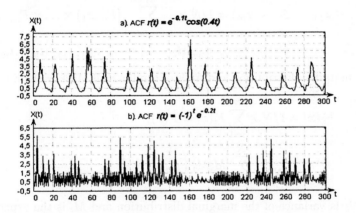

Fig. 1. Two different processes with the same marginal distribution.

2 Basic Concepts and Notation

Let $\xi_1, \xi_2, \ldots, \xi_N$ be a set of independent random variables with identical continuous distribution $F(x)$, and let $\xi_{(1)}, \xi_{(2)}, \ldots, \xi_{(N)}$ be the same set of random variables ranked in non-decreasing order. It is assumed that the ordering rule does not change by time as

$$P(\xi_{(i)}(t) < x) = P(\xi_{(i)} < x) = F_i(x).$$

We obviously have

$$\sum_{i=1}^{N} F_i(x) = N \times F(x).$$

Let Ψ be a homogeneous Markov chain independent of $\xi_i(t)$ with a set of states $i = 1, 2, \ldots, n$, a vector of initial probabilities, π_0, and two-stochastic matrix of transition probabilities:

$$(P(\Psi_{t+1} = j \mid \Psi_t = i)) = (p_{ij}) = P$$
$$\sum_{i=1}^{N} p_{ij} = 1; \quad \sum_{j=1}^{N} p_{ij} = 1; \quad i, j = 1, 2, \ldots, N. \tag{1}$$

Let the process

$$\xi_t = \xi_{\Psi_t}(t), t = 0, 1, 2, \ldots$$

be a randomized Markov chain (RMC).

If the initial probabilities of Markov chain are $\pi_0 = (1/N, \ldots, 1/N)$, then the finite-dimensional distribution of the process is:

$$P(\xi_{t_1} < x_1, \ldots, \xi_{t_m} < x_m) = 1/N \times \sum_{j_1, \ldots, j_m = 1}^{N} [F_{j_1}(x_1) \times \ldots \times F_{j_m}(x_m) \times$$
$$p_{j_1 j_2}(t_2 - t_1) \times \ldots \times p_{j_{m-1} j_m}(t_m - t_{m-1})] .$$

The marginal distribution and ACF of randomized Markov chain above are:

$$F_\xi(x) = 1/N \times \sum_{i=1}^{N} F_i(x) = F(x), \tag{2}$$

$$r'(\tau) = \frac{1}{N D_\xi} \sum_{i,j=1}^{N} M_i M_j \left(p'_{ij}(\tau) - \frac{1}{N} \right), \quad \tau = 1, 2, \ldots \tag{3}$$

where D_ξ is variance of the marginal distribution and M_i is the expectation of order statistics $\xi_{(i)}$ as:

$$M_i = \int_{-\infty}^{\infty} x \, dF_i(x).$$

To obtain $dF_i(x)$ of the expression above we use the well-known expression for probability elements of order statistics [11]:

$$dF_i(x) = \frac{\Gamma(n+1)}{\Gamma(i)\Gamma(N-i+1)} F^{i-1}(x) \left[1 - F(x)\right]^{N-i} f(x) \, dx$$
$$= \frac{N!}{(i-1)!(N-i)!} F^{i-1}(x) \left[1 - F(x)\right]^{N-i} f(x) \, dx \tag{4}$$
$$= (N-i+1) \binom{N}{i-1} F^{i-1}(x) \left[1 - F(x)\right]^{N-i} f(x) \, dx$$

Based on this we obtain an algorithm generating pseudo random process of a given distribution. It consists of two parts: obtaining a new state of the Markov chain and generating a random number according to the corresponding distribution.

The algorithm for simulating a Markov chain with a given matrix of transition probabilities is well-known. The following is the algorithm proposed for generating a Markov chain [9].

Algorithm G1.
Input: N is the number of states of the Markov chain; n is the required length of the process realized.
Output: an array $X(1 \times n)$ of a process realization. The algorithm consists of the following steps:

Step 1 (Preliminary). Put a sample of a non-correlated sequence with the given marginal distribution $F(x)$ into an urn with reset. The counter k is set to one. The initial state of Markov chain is gained uniformly from $1, 2, \ldots, N$.

Step 2. From an urn, ξ_i $(i = 1, \ldots, N)$ are extracted by a random fashion and then ranked in non-decreasing order. Let denote resulting variables as $\xi_{(i)}$ $(i = 1, \ldots, N)$.

Step 3. By the given two-stochastic matrix of transition probabilities the next state (say j) of the Markov chain is generated.

Step 4. ξ_j is copied to output as X_k, and k is increased by one.

Step 5. If $k > n$, the generation comes to an end. Otherwise, go to Step 2.

Since Algorithm G1 has several shortcomings [13], we propose an improved algorithm.

Algorithm G2.
Input and output are same as for Algorithm G1.

Step 1 (Preliminary). The initial state of Markov chain is gained uniformly from $1, 2, \ldots, N$. The counter k is set to one.

Step 2. By the given two-stochastic matrix of transition probabilities the next state (say j) of the Markov chain is generated.

Step 3. Generation of pseudo random number, ξ, by a truncated distribution $F_j(x)$ on j-th inter-fractile interval. ξ is copied to output as X_k, and k is increased by one.

Step 4. If $k > n$, the generation comes to an end. Otherwise, go to Step 2.

The cumulative distribution function $F_i(x)$ whose distribution is truncated on interval (x_{i-1}, x_i) is

$$
F_i(x) = \begin{cases} 0, & \text{for } x < x_{i-1} \\ \frac{F(x)-F(x_{i-1})}{F(x_i)-F(x_{i-1})}, & \text{for } x_{i-1} \le x < x_i \\ 1, & \text{for } x \ge x_i, \end{cases}
$$

If x_i is i-th fractile of the marginal distribution, we obtain a simpler expression

$$
F_i(x) = \begin{cases} 0, & \text{for } x < x_{i-1} \\ N(F(x) - \frac{i-1}{N}), & \text{for } x_{i-1} \le x < x_i \\ 1, & \text{for } x \ge x_i, \end{cases} \tag{5}
$$

The Expressions (2) and (3) are still true, but now M_i ought to be obtained using the distribution (5), so:

$$M_i = \int_{x_{i-1}}^{x_i} x \, dF_i(x) = x \, F_i(x) \Big|_{x_{i-1}}^{x_i} - \int_{x_{i-1}}^{x_i} F_i(x) \, dx = x_i - \int_{x_{i-1}}^{x_i} F_i(x) \, dx \qquad (6)$$

For obtaining the transition probability matrix we need to solve the system (3) of nonlinear equations of degree τ. The solutions were found only for some special kinds of two-stochastic matrix [10], whereas it is necessary to have a general solution for real problems that could be done only numerically. We next present how to determine transition probability matrix.

3 Determination of Transition Probability Matrix

In determining transition probability matrix, we need to minimize the difference between real ACF and ACF of simulated process by the elements of the two stochastic matrices. If we use the sum of squares of the differences as a distance, then we have to solve the following optimization task:

$$\Phi(\tau) = \sum_{\tau} (r(\tau) - r'(\tau))^2 \to min$$

$$r'(\tau) = \frac{1}{ND_\xi} \sum_{i,j=1}^{N} M_i M_j (P'_{ij}(\tau) - \tfrac{1}{N}) \qquad (7)$$

$$\sum_{i=1}^{N} P'_{ij} = 1; \quad \sum_{j=1}^{N} P'_{ij} = 1; \quad i,j = 1,2,\ldots,N \qquad (8)$$

$$\forall i,j \;\; 0 \le P'_{ij} \le 1,$$

where $P'_{ij}(\tau)$ is the element of the i-th row and j-th column of the derivable matrix in power τ. $r'(\tau)$ is the ACF of a randomized Markov chain corresponding to the matrix and given marginal distribution.

Molchan had stated [9] that the gradient optimization method is not suitable for solving the optimization problem of Expressions (7) and (8). Instead, he offered a heuristic method of stochastic optimization and a modification of it. Our experiments with various optimization methods revealed that the simple gradient method of "Quickest descent" with the use of penalty functions is the best, and thus it is adopted here.

First of all, let us free the task from the restrictions of two-stochasticity, Expression (8), by using the following substitutions:

$$P'_{iN} = 1 - \sum_{j=1}^{N-1} P'_{ij}, \quad P'_{Ni} = 1 - \sum_{j=1}^{N-1} P'_{ji}, \quad i = 1,\ldots,N-1, \qquad (9)$$

$$P'_{NN} = \sum_{i,j=1}^{N-1} P'_{ij} - N + 2. \qquad (10)$$

Then we can formally transform our optimization problem to a problem of unconstrained optimization by adding a penalty function:

$$Q(P'^{(k)}_{ij}, \rho^{(l)}_{ij}, \delta^{(m)}_{ij}) = \Phi(P'^{(k)}_{ij}) + \sum_{i,j=1}^{N-1} \left[\rho^{(l)}_{ij} U^{(k)}_{ij} (P'^{(k)}_{ij})^2 + \delta^{(l)}_{ij} V^{(k)}_{ij} (P'^{(k)}_{ij} - 1)^2 \right],$$

$$i, j = 1, \ldots, N-1,$$

where

$U^{(k)}_{ij} = \begin{cases} 1 & \text{if } P'_{ij} < 0, \\ 0 & \text{otherwise} \end{cases}$ (index operator),

$V^{(k)}_{ij} = \begin{cases} 1 & \text{if } P'_{ij} > 1, \\ 0 & \text{otherwise} \end{cases}$ (index operator),

$\rho^{(l)}_{ij} = 10^l$ (weight coefficients), where l is the amount of iterations from the moment of $P'^{(k)}_{ij} < 0$, $l \leq 100$;

$\delta^{(m)}_{ij} = 10^l$ (weight coefficients), where m is the amount of iterations from the moment of $P'^{(k)}_{ij} > 1$, $l \leq 100$.

Derivation dQ/dp_{kl} ($1 \leq k, l \leq N$) is defined by the following expressions (hereafter we will omit (k) and apostrophe at P for better readability of expressions):

$$\frac{dQ}{dp_{kl}} = \frac{d\Phi}{dp_{kl}} + 2 \sum_{i,j=1}^{N-1} \left[\rho^{(l)}_{ij} U_{ij} P_{ij} - \delta^{(l)}_{ij} V_{ij} (P_{ij} - 1) \right]$$

$$\frac{d\Phi}{dp_{kl}} = \frac{1}{ND_\xi} \sum_\tau \left[\frac{1}{ND_\xi} \sum_{i,j=1}^{N} M_i M_j \left(P_{ij}(\tau) - \frac{1}{N} \right) - r(\tau) \right] \sum_{i,j=1}^{N} M_i M_j \frac{dP_{ij}(\tau)}{dp_{kl}}$$

In [13], we show how the computation steps can be minimized. Experiments display very fast access to the minimum area in about 3 to 5 steps and abrupt deceleration of convergence. The deceleration rate greatly depends on the length of ACF; it is almost imperceptible for $n < 10$, while advancement to optimum slows down very much for $n > 30$. This might be due to the cancellations in the calculation of multiple sums and products of the numbers of very small magnitudes. The number of operations grows exponentially with τ. Use of special numerical methods and high-precision calculations could improve the situation. In case when high accuracy is needed, we propose the following optimization scheme.

Step 1. Input the initial two-stochastic matrix, say unitary matrix or matrix with equal elements.

Step 2. Find a matrix for ACF with a length restricted to a small number.

Step 3. Using the result of the previous step, find a matrix for ACF with the needed length.

Step 4. If the accuracy is not sufficient, use random optimization for more precise result.

For our experiments we choose one of the worst cases in the point of view of the difference from the examples given in [10]. Here the marginal distribution is exponential with $\lambda = 1$, $N = 10$, and ACF $r(\tau) = e^{-0.1\tau}\cos(0.4\tau)$,

The difference displayed by the algorithm above with 2500 random steps was about 10%. Higher accuracy can be achieved using the proposed optimization scheme. First, by sequential execution of the gradient optimization for an ACF length from 5 to 40 with step 5, we obtain a difference of 9.6% in not more than 10 steps for each length. Consequent execution of 500 steps of random search with one-dimensional optimization gives us the difference of 6.1%. The computation time was also significantly reduced as about 3 minutes on the 600MHz Pentium III PC. If the number of states is increased to 16, then a difference 2.4% is archived for the ACF, but the computation takes more than 20 minutes. The difference with gradient minimization is 14.6%. We next compare Algorithm G1 and G2 for generating random processes with exponential marginal distribution.

4 Case Studies

Algorithm G1 According to (4) the distribution density of ξ_i in our case is

$$f_i(x) = \lambda(N - i + 1)\binom{N}{i-1}\left(1 - e^{-\lambda x}\right)^{i-1}e^{-\lambda(N-i+1)x},$$

and, consequently, the expectation for ξ_i is

$$M_i = \int_0^\infty x f_i(x)\,dx = \int_0^\infty x\lambda(N-i+1)\binom{N}{i-1}\left(1 - e^{-\lambda x}\right)^{i-1}e^{-\lambda(N-i+1)x}\,dx =$$

$$\frac{1}{\lambda}(N - i + 1)\binom{N}{i-1}\int_0^\infty t\left(1 - e^{-t}\right)^{i-1}e^{-(N-i+1)t}\,dt =$$

$$\frac{1}{\lambda}(N - i + 1)\binom{N}{i-1}\sum_{j=0}^{i-1}\frac{(-1)^{i-j-1}}{N-j}\binom{i-1}{j}\int_0^\infty (N-j)te^{-(N-j)t}\,dt.$$

The last integral is expectation of the exponential distribution with parameter $N - j$. Hence the final result for M_i is:

$$M_i = \frac{N - i + 1}{\lambda}\binom{N}{i-1}\sum_{j=0}^{i-1}\frac{(-1)^{i-j-1}}{(N-j)^2}\binom{i-1}{j} = \frac{K_i}{\lambda},$$

where K_i is a coefficient that depends only on the values of i and N. N is decided as some not very large constant independent on the distribution parameter λ. When N is fixed, all K_i's can be easily calculated.

Algorithm G2 The cumulative distribution function $F_i(x)$ is determined by Expression (5) and expectations M_i by Expression (6). For the exponential marginal distribution considered, the i-th fractile of exponential distribution is equal to

$$x_i = -\frac{1}{\lambda}\ln\left(\frac{N - i}{N}\right), \quad 1 \le i \le N - 1, \tag{11}$$

where N is the number of intervals (for the uniformity of expressions, we also add the leftmost and rightmost values: $x_0 = 0$ and $x_N = \infty$). We obtain an expression for the cumulative distribution function of truncated exponential distribution on the interval between the fractiles from Expressions (5) and (11).

$$F_i(x) = N(1 - e^{-\lambda x}) - i + 1. \tag{12}$$

The corresponding expectations are

$$M_i = \frac{1}{\lambda}\left(1 + \ln \frac{N(N-i)^{N-i}}{(N-i+1)^{N-i+1}}\right) =$$

$$\frac{1}{\lambda}(1 + \ln N + (N - i)\ln(N - i) - (N - i + 1)\ln(N - i + 1)) = \frac{K_i^*}{\lambda},$$

$$i = 1, \ldots, N - 1$$

$$M_N = \frac{1}{\lambda}(1 + \ln N) = \frac{K_N^*}{\lambda}$$

As for Algorithm G1, the expectation M_i has inverse negative relationship with the distribution parameter λ. Note that calculating K_i^*'s in Algorithm G2 is much simpler than calculating K_i's in Algorithm G1, i.e., it does not require to calculate the sums and binomial coefficients.

For the generation process we need to produce the samples from the total population with exponential distribution for Algorithm G1, while from the truncated interval $[x_i - 1, x_i)$ for Algorithm G2. For detail, refer to [13]. As the Algorithm G2 does not need the sorting step, it is much faster than G1. For generating 100,000 numbers for a 16-state randomized Markov chain, 1.86 sec was taken with Algorithm G2. It was 25.04 sec with Algorithm G1, which is more than 13 times slower than the proposed approach. The accuracy of the proposed ACF approximation scheme is illustrated in Figure 2. Observe that the difference with the proposed algorithm is smaller than that of Algorithm G1. In Figure 3, we illustrate an example of ACF

$$r(\tau) = \begin{cases} 1 - 0.05\tau & \text{if } 0 < \tau \le 20, \\ 0 & \text{otherwise} \end{cases}$$

which shows the significance of proper choice of maximal ACF length, n. If we have $n = 20$, the ACF is well approximated up to $\tau = 20$ but the difference becomes large beyond that point. For $n = 25$, we achieve better result, and $n = 30$ allows a really good approximation. Yet on the interval $[1, \ldots, 20]$, all approximations are almost equally good. In this example, the number of states, N, is 16.

5 Conclusion

In this paper we have proposed an algorithm for generating randomized Markov chains with given marginal distribution and ACF. The proposed approach achieves significant speed-up and accuracy by using truncated distribution instead of order statistics. Randomized Markov chains can approximate different stationary

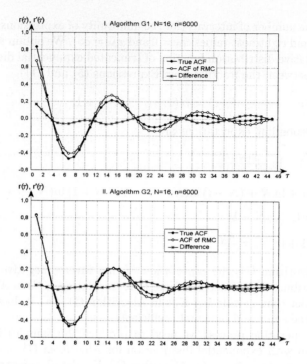

Fig. 2. Comparison of given and obtained ACFs.

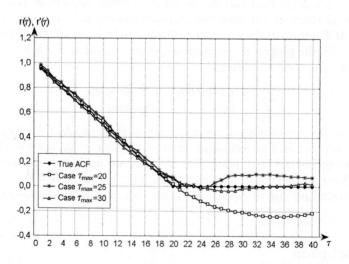

Fig. 3. Influence of an ACF length truncation on the approximation accuracy (based on algorithm G2).

random processes when only the marginal distribution and ACF are taken into account. The accuracy of an ACF approximation is good enough if the numbers of the states of the Markov chain is not large. Usually, not more than 20 states provide a sufficiently small difference between the approximated ACF and that of the randomized Markov chain. On the other hand, with the number of states more than 20, it is almost impossible to complete the optimization task for the transition probability matrix on modern PC in reasonable time. We have also presented an optimization approach using penalty functions. In the future researches we will try to improve the quality of optimization. A significant improvement will also be achieved by using parallel algorithms for matrix computation.

Acknowledgement This work was supported in part by Brain Korea 21 project and grant No. 2000-2-30300-004-3 from the Basic Research Program of Korea Science and Engineering Foundation.

References

1. Al-Shaer, E., Abdel-Wahab, H., Maly, K.: HiFi: A New Monitoring Architecture for Distributed Systems Management. The 19th IEEE Int'l Conf. on Distributed Computing Systems (1999) 171-178
2. Gribaudo, M., Sereno, M.: Simulation of Fluid Stochastic Petri Nets. Proceeding of the 8th MASCOTS (2000) 231-239
3. Dagpunar, J.: Principle of Random Variate Generation. Clarendon Press, Oxford (1988)
4. Blacknell, D.: New Method for the Simulation of Correlated $\mathcal{K}$-distributed Clutter. IEE Proc. Rad., Son. Nav., Vol. 141 (1994) 53-58
5. Ronning, G.: A Simple Scheme for Generating Multivariate Gamma Distributions with Non Negetive Covariance Matrix. Technometrics, Vol. 19 (1977) 179-183
6. Bustos, O.H., Flesia, A.G., Frery, A.C.: Simulation of Correlated Intensity SAR Images. Proc. XII Brazilian Symp. on Comp. Graph. and Image Proc. (1998) 10
7. Hammersley, J., Handscomb, D.: Monte-Carlo Methods. Methuen&Co, (1964)
8. Letch, K., Matzner, R.: On the Constraction of a Random Process with Given Power Spectrum and Probability Density function. Proceedings of the 1998 Midwest Symposium on Systems and Circuits, Vol. 41 (1999) 14-17
9. Molchan, S.I., Prelovskaja, A.A., Rodionov, A.S.: Program Complex for Generation of Random Processes with Given Properties. System modeling, Trans. of Computing Center of Siberian Branch of Academy of Sciences of USSR, Novosibirsk, ISSN 0134-630X, No. 13 (1998) 70-81
10. Molchan, S.I.: About One Approach to Generation of Random Processes with Given Properties. Simulation of Computing Systems and Processes, Perm's State Univercity, USSR (1986) 59-72
11. Wilks, S.: A Mathematical Statistics. Second edition, John Wiley&Sons (1962)
12. Feller, W.: An Introduction to Probability Theory and its Applications. John Wiley&Sons (1971)
13. Rodionov, A.S., Choo, H., Youn, H.Y., Chung, T.M., Park, K.: On Generating Random Variates for Given Autocorrelation Function. TR-2001-1, School of ECE, Sungkyunkwan University (2001)

Replicators & Complementarity: Solving the Simplest Complex System Without Simulation

Anil Menon

R & D Group
Cerebellum Software
600 Waterfront Dr.
Pittsburgh, PA 15222
anilm@acm.org

Abstract. Replicator systems are a class of first order, nonlinear differential equations, arising in an extraordinary variety of modeling situations. It is shown that finding the stationary points of replicator systems is equivalent to solving a nonlinear complementarity problem. One consequence is that it becomes possible to use replicator systems to solve very large instances of the NP-complete graph bisection problem. The methodological and philosophical import of their equivalence with complementarity problems (upto stationarity) is discussed.

1   Replicators

A *replicator* is a fundamental unit in evolutionary processes, representing a population *type*, and characterized by two attributes: $p_i(t)$, its proportion in the population at time t, and $f_i(t)$, its non-negative fitness at time t. A replicator's fitness is a measure of its significance to the future evolution of the population. The proportions of replicators in a population change as a result of their mutual interactions, and their relative fitnesses; the dynamics is described by a system of differential equations known as the *replicator equations*, given by:

Continuous replicator equations: For $i = 1 \ldots n$,

$$\frac{dp_i}{dt} = p_i(t)\ (f_i(\boldsymbol{p}) - \bar{f}(\boldsymbol{p})). \tag{1}$$

A *replicator system* consists of these, together with the relevant description of replicators, fitnesses, and proportion values. Examples of replicators include behavioral patterns in sociological modeling, primordial molecules in prebiotic models, species in a Lotka-Volterra system, strategies in certain N-person games, and reactants in autocatalytic reactions (see [8] for a comprehensive survey).

In fact, *any* first order differential equations model can be non-trivially related to a replicator model. To see this, consider the system,

$$\frac{dq_i}{dt} = g_i(q_1, \ldots, q_n, t) \quad i = 1, \ldots, n. \tag{2}$$

Define the variables, $z_i(t) = \exp(q_i(t))/\sum_{j=1}^{n} \exp(q_j(t))$. Differentiating the z_i w.r.t time, a replicator system is obtained:

$$\frac{dz_i}{dt} = z_i\left(\frac{dq_i}{dt} - \sum_{j=1}^{n} \frac{dq_j}{dt} z_j\right) \quad i = 1,\ldots,n. \tag{3}$$

(3) has the aspect that the "fitnesses" are the rates of change of the q_i variables. Assuming $\sum_{j=1}^{n} q_j(t) = $ constant, the system (3) equilibrates ($\frac{dz_i}{dt} = 0$ for all $i = 1,\ldots,n$) if and only if the system (2) does so.

Replicator systems are universal precisely because they seem to capture the essential relative growth dynamics of heterogeneous populations. In this sense, one could argue that replicator systems are the 'simplest complex systems'.

The outline of this paper is as follows: Section 2 demonstrates how heuristic arguments regarding graph bisection lead to a replicator system. Some of the problems associated with replicator optimization are discussed in Section 3. In Section 4, the connection between replicator systems and complementarity problems is introduced as a means to alleviate some of these problems. These ideas are applied to the graph bisection problem. Section 5 ends with a discussion.

2 Graph Bisection: Replicator Models

In the graph bisection problem, the nodes of a graph have to be assigned to either of two partitions (bins, colors, clusters), such that the total numbers of nodes assigned to each partition are the same, and the *cut-size* — the number of *cross edges* (edges whose end nodes are in different partitions) — is as small as possible. The graph bisection problem surfaces in a wide variety of contexts.

One of the earliest instances of the graph bisection problem occurs, appropriately enough, in biology. In the early 1900's, the zoologist H. V. Wilson observed a curious morphological phenomenon [15]. Wilson first noticed that sponges (the simplest metazoan organisms) could be disassociated into their constituent cells by straining them through a filter made of fine silk cloth. If the randomly mixed disassociated cells are left in the right physiological conditions, then after about a day, the cells form aggregates, and after $4-5$ days, reorganize themselves into a sponge, with histological organization characteristic of the original. There are many different cells types in a sponge: pinacocytes, archocytes, amebocytes, collencytes, collar cells etc. How all these different types (about 16 in all) reorganize themselves to cluster together in the right histological arrangement remains one of the key mysteries in morphogenesis [11, pp. 82-196].

One natural interpretation is that the histological organization of the sponge plays the role of a graph, with the sponge cells as nodes , and the adhesive and cohesive forces between cells the weighted edges of the graph. Minimizing the cut-size becomes minimizing the work of adhesion between different cell types. In fact, since the sponge has about 16 different types of cells, it can be seen as an instance of the graph partitioning problem (16 partitions)[1].

---

[1] It is tempting to solve the graph bisection problem by mimicking how the sponge solves its problem instance. Yet, two facts counsel against the impulse: first, cell

On account of the ubiquity of the graph bisection problem, there have been a great many proposals to solve it [5], as well pioneering attempts to solve it through replicator systems [14]. The development given below is believed to be original. It also has the advantage of revealing when the method is likely to fail.

The graph bisection problem is formally specified as follows: Let $G = (V, E)$ be an undirected graph, with node set V, edge set E, and no self loops. Let the total number of nodes be $n = |V|$. The nodes are assumed be numbered 1 through n. Each node is assigned a non-negative node-weight $c_i \in \mathbb{R}_+$, and similarly, each edge is assigned a non-negative edge-weight w_{ij} (if two edges are not connected, they are assigned a zero weight). For the weighted adjacency matrix $A = (w_{ij})_{n \times n}$, define the associated row-sum diagonal matrix $D = (d_{ij})_{n \times n}$ with the diagonal terms $d_{ii} = \sum_{j=1}^{n} w_{ij}$. The Laplacian matrix (Fiedler matrix, Kirchoff matrix, admittance matrix) of the graph is defined by $L = D - A$, is positive semidefinite. It can be shown that $x^t L x$ is the sum of the weights of those edges whose endpoints lie in different partitions (the so called *cut-size*). The graph bisection problem (GB) may then be represented by:

$$\underset{x \in \{0,1\}^n \ c^t x = \sum_{i=1}^{n} c_i/2}{\arg\min} \quad x^t L x. \tag{4}$$

The graph bisection problem is NP-complete and arises in a variety of contexts including load balancing, storage management, distributed directories, and VLSI design.

Interpret the graph as a network of individuals, where each edge in the graph represents a "friendship" relationship. The population is to be divided into two mutually exclusive groups A and B. For the ith individual, let x_i denote her *degree of confidence* in the proposition "I should be assigned to group A." It is assumed that $x_i \geq 0$, and that $\sum_{j=1}^{n} x_j > 0$ (i.e., at least one individual has a non-zero confidence in being assigned to A.) The ith individual is allowed to *continually* update her confidence degree based on its current value, and the values of those of her friends. What should the update rule look like? The following two guidelines are reasonable:

1. The *Inertia Heuristic*: The change in x_i should be proportional to its current value. After all, if individuals are quite confident about being assigned to a particular partition thenin the absence of any contrary information, their confidence can only increase with time. The Inertia Heuristic may be captured by the expression:

$$\frac{dx_i}{dt} \propto x_i. \tag{5}$$

2. The *Roman Rule Heuristic*[2]: If a majority of an individual's neighbors have very little confidence in the proposition that *they* should be assigned to the

---

sorting is *slow*. The time taken to re-organize is of the order of days. Second, simulations of theoretical models of cell sorting have proved to be somewhat disappointing, showing a tendency for the system to get stuck in local minima [11].

[2] In Rome behave as the Romans do. In this case, "In Rome, believe as the Romans do."

partition A (reflected by their low x_j values), then it is probably a good idea for an individual not to get over-confident. Yet, an individual may have many friends, and the opinions of some may be more reliable than those of others. Let w_{ij} indicate the *reliability* or *worth* of the opinion of jth friend of the ith individual. The Roman Rule is then captured by the expression,

$$\frac{dp_i}{dt} \propto \sum_{j=1}^{n} w_{ij} x_j. \tag{6}$$

The overall reliability is obtained by summing the individual reliabilities. Other aggregating procedures could be used, but summation has nice linear properties, behaves well under scaling, and is computationally inexpensive.

Combining these two heuristics the following update rule is obtained:

$$\frac{dx_i}{dt} = K x_i \left(\sum_{j=1}^{n} w_{ij} x_j\right) = x_i f_i(\boldsymbol{x}) \quad i = 1, \ldots, n. \tag{7}$$

In (7), K is a constant of proportionality. The above system could be simulated, starting with a random initialization of confidence degrees $(x_i(0))$, and stopping when no more changes in confidence degrees are seen. Then the nodes actually could be assigned to partitions based on the strength of their confidences. However, this leads to the problem of determining what constitutes a "strong" confidence, and what is a "weak" one. One resolution is to *normalize* (7), i.e., introduce the variables, $p_i = x_i / \sum_{j=1}^{n} x_j$. Transforming the x_i's to p_i's also takes care of scaling issues; if all the x_i's are multiplied a common factor, the p_i will be unaffected. Differentiating the variables p_i w.r.t time:

$$\frac{dp_i}{dt} = \frac{\left(\sum_{j=1}^{n} x_j\right)\frac{dx_i}{dt} - x_i \sum_{j=1}^{n} \frac{dx_j}{dt}}{\left(\sum_{j=1}^{n} x_j\right)^2}, \tag{8}$$

$$= p_i f_i(\boldsymbol{p}) - p_i \sum_{j=1}^{n} p_k f_k(\boldsymbol{p}), \tag{9}$$

$$= p_i (f_i(\boldsymbol{p}) - \bar{f}(\boldsymbol{p})). \tag{10}$$

The equations in (10) represents a replicator system, with "fitnesses" linearly proportional to replicator "proportions," (a Lotka-Volterra system). This system can be simulated, and after it reaches equilibrium, the stationary solution $\boldsymbol{p}^* = \{p_i^*\}$ is mapped to a binary feasible solution using the median assignment procedure[3]. Thus, heuristic arguments lead to a replicator system for graph bisection. Also, if the replicator system does not perform well for a given instance,

---

[3] This entails determining the median value p_m of the vector of proportions $\boldsymbol{p}^*$. If the ith component p_i^* is greater than the median it is assigned to A; if less, to B (ties are handled randomly). It can be shown that the median assignment rule is optimal with respect to several distance metrics [2].

the two heuristics may be culpable, and better heuristics (reflected in new fitness definitions) could be designed.

There are however, some chronic problems associated with techniques such as replicator optimization. A straightforward simulation of the above differential leads to a variety of problems. A brief discussion of these issues is given in the next section.

3 Replicator Optimization: Strategy and Problems

The use of replicators in discrete optimization consists of the following sequence of steps:

1. A discrete optimization problem is relaxed into a continuous optimization problem over the unit simplex.
2. Each relaxed variable is modeled as a replicator. Specifically, the ith discrete variable is conceptually replaced by a replicator; associated with the ith replicator are two values: its "proportion" p_i, representing the relaxed value of the ith variable, scaled to the real interval $(0, 1)$, and its non-negative real valued "fitness" f_i, the exact specification of which is problem dependent.
3. Starting from an initial set of values, the fitnesses and proportions of the replicators are changed as per the replicator equations. The procedure is carried out till some termination criteria are satisfied.
4. The convergent proportions are then mapped to suitable discrete values (this mapping is also dependent on the problem).

It can be shown that if the fitnesses of the replicators are gradients of the function to be maximized [8, pp. 242-243], then the replicator system acts as a gradient dynamical system in Shahshahani space (a Riemannian space named after the mathematiciam Shahshahani [12]). Accordingly, if these conditions are satisfied, replicator dynamics guarantee convergence to a local maximum of that function.

Unfortunately, the fact that replicator optimization is a gradient ascent method, implies that it also suffers from the same problems that have long plagued the application of gradient methods [1]. Amongst these are: (i) dependence of the convergence rates on the conditioning numbers of certain matrices associated with the computational surface, (ii) the average case convergence rates are typically closer to worst-case convergence rates, (iii) the fatal reliance on local information *alone*, and (iv) discrete gradient systems do not necessarily have the same convergence properties as their associated continuous gradient systems [7].

The definition of a gradient is metric dependent, and many of the above-mentioned problems can be alleviated by the use of a *variable* metric. Since replicators can be viewed as systems performing gradient ascent in Riemannian space, it is a *variable metric* gradient ascent method [8, pp. 242], and this may mitigate some of the above problems.

Another problem has to do with scalability. A replicator equation describes the time evolution of a *single* replicator. When mapped to problems such as graph bisection, this implies that the resulting system may consist of thousands of differential equations. Simulating such systems is very difficult, and the intricacies of simulation can become at least as complicated as the original problem itself. The next section gives a technique to solve for the equilibrium points, that does not depend on simulating the system of selection equations. It also becomes possible to sidestep the problems associated with gradient ascent procedures.

4   Replicators & Complementarity Problems

Let $D \subseteq \mathbb{R}^n$, and $g, h : \mathbb{R}^n \to \mathbb{R}^m$. The *complementarity problem*(CP) [9, pp. 23-24] is to,

$$\textbf{(CP)} \quad \text{Find a } \boldsymbol{x} \in D : g(\boldsymbol{x}) \geq \boldsymbol{0}, \ h(\boldsymbol{x}) \geq \boldsymbol{0}, \ (g(\boldsymbol{x}))^t h(\boldsymbol{x}) = 0, \quad (11)$$

$g(\boldsymbol{x}) \geq \boldsymbol{0}$ means that each component of $g(\boldsymbol{x})$ is greater than or equal to zero. A fascinating and thorough review of many instances of this problem may be found in [4]. Two special cases are to be distinguished:

Nonlinear Complementarity Problem:
  Here, $n = m$, $g(\boldsymbol{x}) = \boldsymbol{x}$, $h : \mathbb{R}^n \to \mathbb{R}^n$. Then,

$$\textbf{(NCP)} \quad \text{Find a } \boldsymbol{x} \in D : \boldsymbol{x} \geq \boldsymbol{0}, \ h(\boldsymbol{x}) \geq \boldsymbol{0}, \ \boldsymbol{x}^t h(\boldsymbol{x}) = 0. \quad (12)$$

Linear Complementarity Problem:
  Here, $n = m$, $g(\boldsymbol{x}) = \boldsymbol{x}$, and $h(\boldsymbol{x}) = M\boldsymbol{x} + \boldsymbol{r}$, where $M = (m_{ij})_{n \times n}$ and $\boldsymbol{r} \in \mathbb{R}^n$. Then,

$$\textbf{(LCP)} \quad \text{Find a } \boldsymbol{x} \in D : \boldsymbol{x} \geq \boldsymbol{0}, \ h(\boldsymbol{x}) = M\boldsymbol{x} + \boldsymbol{r} \geq \boldsymbol{0}, \ \boldsymbol{x}^t (M\boldsymbol{x} + \boldsymbol{r}) = 0.$$
$$(13)$$

The above LCP is denoted by $\text{LCP}(\boldsymbol{x}, M)$. A great variety of solution techniques exist for solving the LCP, ranging from those based on Lemke's classical complementarity pivot algorithm, through iterative methods such as successive overrelaxation (SOR), to those based on interior point methods such as Karmarkar's algorithm. A comprehensive discussion of the technology may be found in Cottle et. al. [3, pp. 383-506].

The relevance of nonlinear complementarity problems to this paper is seen by the following observation, a natural generalization of the observation of Y. Takeuchi and N. Adachi [13].

Observation A: Define $D = \mathbb{S}^n$, and consider the NCP problem of finding a $\boldsymbol{p} \in \mathbb{S}^n$ such that:

$$g(\boldsymbol{p}) = \boldsymbol{p} \geq \boldsymbol{0}, \quad h(\boldsymbol{p}) = [f_1(\boldsymbol{p}) - \bar{f}(\boldsymbol{p}), \ldots, f_n(\boldsymbol{p}) - \bar{f}(\boldsymbol{p})]^t \geq \boldsymbol{0},$$
$$\boldsymbol{p}^t h(\boldsymbol{p}) = 0. \quad (14)$$

Every solution $\boldsymbol{p} \in \mathbb{S}^n$ to the above problem is also an equilibrium point of the selection system $dp_i/dt = p_i[f_i(\boldsymbol{p}) - \bar{f}(\boldsymbol{p})]$.

Of course, not every equilibrium point of the selection equation is necessarily a solution to the corresponding NCP; because, an equilibrium point p^* for the selection equation has to satisfy $p^* \geq 0$, and $(p^*)^t(f_i(p^*) - \bar{f}(p)) = 0$, it need not satisfy $h(p) \geq 0$.

However, this is a minor technical problem. Notice that (14) implies that an equilibrium point p^* has to satisfy n equalities of the form,

$$p_i h_i(p) = p_i(f_i(p^*) - \bar{f}(p)) = 0. \tag{15}$$

Suppose there is an index i such that $h_i(p^*) < 0$. Consequently, $p_i^* = 0$. Now, in a replicator selection system, once a proportion goes to zero, it stays zero, and no longer takes part in the replicator dynamics. Since this is the case, the replicator fitness $f_i(p^*)$ can be defined to take any value, when $p_i^* = 0$. Whatever its value, the ith has no effect on the replicator dynamics. In particular, a new replicator system can be defined,

$$\frac{dp_i}{dt} = p_i(f_i'(p) - \bar{f}'(p)) \quad i = 1, \ldots, n, \tag{16}$$

$$f_i'(p) = \begin{cases} f_i(p) & \text{if } p_i > 0, \\ 0 & \text{if } p_i = 0. \end{cases} \tag{17}$$

The new replicator system has the same orbits and equilibrium points as the old one. It also satisfies the conditions that $h_i'(p) = f_i'(p) - \bar{f}'(p) \geq 0$. For this alternate system, its equilibrium points are also solutions to the corresponding NCP. The next two observations complete the connection between complementarity problems and replicator selection systems.

Observation B: When fitnesses are linearly dependent on replicator proportions, then the replicator system becomes the Lotka-Volterra equations.

To see this, suppose the fitnesses f_i were such that, $f_i(p) = \sum_{j=1}^{n} w_{ij}\, p_j$, so that the Lotka-Volterra equations is obtained, namely:

$$dp_i/dt = p_i \left(\sum_{j=1}^{n} w_{ij}\, p_j - \sum_{j,k} w_{jk} p_j\, p_k \right).$$

Using the Hofbauer transformation, that is, introducing non-negative real variables $z_1, \ldots, z_n$ such that, $p_i = z_i / \sum_{i=1}^{n} z_i$, the Lotka-Volterra system can be transformed into its *equivalent* classical form [8, pp. 134-135],

$$\frac{dz_i}{dt} = z_i\, [w_{in} + \sum_{j=1}^{n-1} w_{ij}\, z_i], \quad i = 1, \ldots, n - 1. \tag{18}$$

In vector form (18) becomes, $dz/dt = z^t\, y = z^t[c + A'z]$.

Observation C: Solving a linear complementarity problem is equivalent to finding the stationary points of an associated Lotka-Volterra system [13]. Precisely, every solution to the LCP$(\mathbf{c}, A')$, where: $\mathbf{c} = [w_{1,n}, \ldots, w_{n-1,n}]$ and $A' = [w_{i,j}]_{(n-1) \times (n-1)}$, is also a equilibrium point of the Lotka-Volterra system in (18).

These observations may be placed in a larger context. In a long series of papers, Grossberg, Cohen and others (see [6] for a review), studied a class of systems, the *competitive adaptation level systems*, characterized by the following dynamics,

$$\frac{dx_i}{dt} = a_i(\boldsymbol{x}) \left[b_i(x_i) - c(\boldsymbol{x})\right] \qquad i = 1, \ldots, n \tag{19}$$

where $\boldsymbol{x}^t = (x_1, x_2, \ldots, x_n)$ is a state vector describing the population of the n interacting individuals, and $a_i(\cdot), b_i(\cdot)$ and $c_i(\cdot)$ are all real valued functions. Under fairly mild conditions on the functions $a_i(\cdot), b_i(\cdot)$ and $c_i(\cdot)$, many stability results, both local and asymptotic, may be derived. An astonishing variety of neural networks and differential systems, amongst them Hebbian nets, the Boltzmann machine, and the Gilpin-Ayala, Eigen-Schuster and Lotka-Volterra systems are all special cases of (19).

The replicator system itself is a special case of (19). Clearly, the nonlinear complementarity problem is essentially identical to the problem of finding the stationary points of competitive adaptation level system. In particular, when replicator fitness are *linear* functions of replicator proportions, then finding the stationary points of the resulting replicator system (known as Lotka-Volterra systems) becomes identical to solving the linear complementarity problem. Thus, the methods used to solve nonlinear complementarity problems become applicable to finding the stationary points of replicator systems.

Earlier, heuristic arguments were used to show that to solve the graph bisection problem using replicator equations, the fitness functions of the replicators had to be linear functions of the replicator proportions. The above considerations lend support to these arguments. The fact that the graph bisection problem is a semidefinite integer programming problem implies that there exists an equivalent linear complementarity problem [3, pp. 4-5], and hence, a corresponding Lotka-Volterra system with fitnesses linearly proportional to replicator proportions.

Figure 1 (below) outlines the basic details of solving the graph bisection problems using complementarity. An experimental study of the algorithm and comparisons with spectral techniques can be found in [10].

5 Discussion

It is not atypical to find realistic graph bisection problems that have tens of thousands of nodes and edges; such graphs are found for example, in finite element models of complex 3D surfaces. A direct simulation of the corresponding replicator system would be rather challenging. Worse, discrete versions of the

---

GB_REP(L, b, x, *stepsize, maxgen, maxerr*)
{ Minimize $x^t L x$ such that $e^t x = b$, $x \in B^n$ }
{ where $b = n/2$, $L = (q_{ij})_{n \times n}$ }

1. Put $q^t = [0_n^t, -b]$, and $M = \begin{bmatrix} L & -e \\ e^t & 0 \end{bmatrix}$

2. Randomly initialize $z(0)$. Put $iter = 1$, $error = \infty$, $mincost = \infty$.

3. While (($error > maxerr$) and ($iter < maxiter$)) do
   (a) Compute $z(iter)$ from $z(iter - 1)$ using successive over-relation (SOR) on LCP(q, M).
   (b) Every *stepsize* generations do:
      i. Find $z_m = $ median $\{z_1, z_2, \ldots, z_n\}$
      ii. Assignment phase: For $i = 1, \ldots, n$ if $z_i > z_m$ put $x_i = 1$, else put $x_i = 0$
      iii. Compute $C(x) = x^t L x$
      iv. If $C(x) < mincost$, $mincost = C(x)$, $x^* = x$

4. Return x^*.

Fig. 1. Graph Partitioning : The Basic Procedure

replicator equations have stability issues not found in the continuous versions [7]. The formulation in terms of complementarity problems eliminates such concerns. Parallel implementations of the LCP are available, as well as a considerable body of experience in solving such large systems. Furthermore, the relationship between some of the most popular models in "soft computing" and some of the most classical of results in mathematical optimization is clearly seen. It is also possible that complementarity theory could benefit from some of the theoretical insights of replicator modeling.

The demonstrated connection with complementarity raises questions on the usefulness of replicator-based optimization. If finding the stationary points of a replicator system can be reduced to a complementarity problem, then why not just solve the bisection problem (or any other replicator problem) by framing it as a linear complementarity problem in the first place?

It should be kept in mind however, that languages are not just referential devices. While two languages may indeed refer to the same object, what it enables us to *do* with it, is to some extent, language dependent. Modeling a problem in terms of replicator theory leads to actions different from those actions induced by viewing the problem in terms of, say, nonlinear programming. Since simulation of the replicator systems is largely infeasible, the ends have to be achieved by analogy.

A subtler argument is that replicator theory provides a "pseudo-physics" for computational problems. Traditionally, physical, biological, and engineering problems have been attacked using linear and nonlinear programming, game theory, differential calculus and other tools. These tools are useful because their semantics are independent of the problem domain. So, for example, sponge reassambly may be modeled as a graph bisection problem, and spectral methods used to derive estimates on energy dissipation. The application of replicator-theoretic arguments to a computational problem, means that the problem gets endowed with a "pseudo-physics." This enables the engineer to reason in physical terms about abstract entities. The computational problem can be attacked at a conceptual level, rather than in purely algorithmic terms. The conceps of replicator theory suggest hypotheses and strategies, that more "applications-neutral" areas such as complementarity, may fail to inspire. Ultimately, the results of this paper are intended to "complement" the technology of replicator modeling, by providing effective means to achieve promised results.

References

1. H. Akaike. On a successive transformation of probability distribution and its application to the analysis of the optimum gradient method. *Annals of the Inst. of Statistical Math.*, X1:1–16, 1959.
2. T. F. Chan, P. Ciarlet Jr., and W. K. Szeto. On the optimality of the median cut spectral bisection graph partitioning method. *SIAM J. on Scientific Computing*, 18(3):943–948, 1997.
3. R. Cottle, J. S. Pang, and Stone R. E. *The Linear Complementarity Problem.* Academic Press, Boston, 1992.
4. M. C. Ferris and J. S. Pang. Engineering and economic applications of complementarity problems. *SIAM Review*, 39:669–713, 1997.
5. Per-Olof Fjällström. Algorithms for graph partitioning: A survey. *Linköping Electronic Articles In Computer and Information Science*, 3(10):1–37, 1998. http://www.ep.liu.se/ea/cis/1998/010/.
6. S. Grossberg. Nonlinear neural networks: principles, mechanisms and architectures. *Neural Networks*, 1:17–61, 1988.
7. J. Hofbauer. A Hopf bifurcation theorem for difference equations approximating a differential equation. *Monatshefte für Mathematik*, 98:99–113, 1984.
8. J. Hofbauer and K. Sigmund. *The Theory of Evolution and Dynamical Systems.* Cambridge University Press, Cambridge, 1988.
9. R. Horst and H. Tuy. *Global Optimization: Deterministic Approaches.* Springer-Verlag, Berlin, 1990.
10. A. Menon, K. Mehrotra, C. Mohan, and S. Ranka. Optimization using replicators. In *Proceedings of the Sixth International Conference on Genetic Algorithms*, pages 209–216, San Mateo, California, 1995. Morgan Kaufman.
11. G. D. Mostow. *Mathematical models of cell rearrangement.* Yale University Press, New Haven, 1975.
12. M. Shahshahani. A new mathematical framework for the study of linkage and selection. *Memoirs of the AMS*, 211, 1979.
13. Y. Takeuchi and N. Adachi. The existence of globally stable equilibria of ecosystems of the generalized Volterra type. *J. Math. Biology*, 10:401–415, 1980.
14. H.-M. Voigt, H. Mühlenbein, and H.-P. Schwefel. *Evolution and Optimization '89. Selected Papers on Evolution Theory, Combinatorial Optimization, and Related Topics.* Akademie-Verlag, Berlin, 1990.
15. H. V. Wilson. On some phenomena of coalescence and regeneration in sponges. *J. Expt. Zoology*, 5:245–258, 1907.

Soft Computing:
Systems and Applications

More Autonomous Hybrid Models in Bang^2

Roman Neruda*, Pavel Krušina, and Zuzana Petrová

Institute of Computer Science, Academy of Sciences of the Czech Republic,
P.O. Box 5, 18207 Prague, Czech Republic
roman@cs.cas.cz

Abstract. We describe a system which represents hybrid computational models as communities of cooperating autonomous software agents. It supports easy creation of combinations of modern artificial intelligence methods, namely neural networks, genetic algorithms and fuzzy logic controllers, and their distributed deployment over a cluster of workstations. The adaptive agents paradigm allows for semiautomated model generation, or even evolution of hybrid schemes.

1 Introduction

Hybrid models, including combinations of of artificial intelligence methods such as neural networks, genetic algorithms or fuzzy logic controllers, seems to be a promising and currently studied research area [2]. In our work [7] we have tested this approach in the previous implementation of our system with encouraging results on several benchmark test [7]. The models have included combinations as using genetic algorithm to set parameters of a perceptron network or fuzzy logic controller. Other example is setting learning parameters of back propagation (learning rate, decay) or genetic algorithm (crossover, mutation rate) by a fuzzy controller. Yet another example combination is using a fuzzifier or one back propagation step as a special kind of a genetic operator.

Recently we have turned our effort to more complex combinations, which have not been studied much yet, probably also because of the lack of a unified software platform that would allow for experiments with higher degree hybrid models. This is the motivation behind the design of the new version of our system called Bang^2 .

As before, the unified interface of the library of various AI computational components allows to switch easily e.g. between several learning methods, and to choose the best combination for application design. We have decided to allow components to run in distributed environment and thus to make use of parallel hardware architectures, typically a cluster of workstations. Second goal of Bang^2 design involves, beside creation of more complex models, also the semiautomated model generation and even the evolution of hybrid models.

For distributed and relatively complex system as Bang^2 it is favorable to make it very modular and to prefer the local decision making against global

* This work has been partially supported by GAASCR under grant no. B1030006.

intelligence. This lead us to the idea to take advantage of agent technology. Employing software agents simplifies the implementation of new AI components and even their dynamic changes. We also hope that some of the developed hybrid models will help the agents itself to become more adaptive and to behave more independently, which in turn should help the user of the system to build better models.

2 System architecture

Bang^2 consists of a population of agents living in the environment, which provides support for creation of agents, their communication, distribution of processes. Each agent provides and requires services (e.g. statistic agent provides statistic preprocessing of data and requires data to process). Agents communicate via special communication language encoded in XML. There are several special agents necessary for Bang^2 run (like the Yellow Pages agent who maintains information about all living agents and about services they provide). Most of the agents realize various computational methods ranging from simple statistics to advanced evolutionary algorithms.

Our notion of intelligent agent follows the excellent introductory work by Franklin [5]. Generally, an agent is an entity (a part of computer program with its own thread of execution in our case), which is autonomous, reacts to its environment (e.g. to user's commands or messages from other agents) in pursue of its own agenda. The agent can be adaptive, or intelligent in a sense that it is able to gather information it needs in some sophisticated way. Moreover, our agents are mobile and persistent. We do not consider other types or properties of agents that for example try to simulate human emotions, mood, etc.

Bang^2 environment is a living space for all the agents. It supplies resources and services the agents need and serves as a communication layer. One example of such an abstraction is a location transparency in communication between agents — the goal is to make the communication simple for the agent programmer and identical for local and remote case while still exploiting all the advantages of the local one. There should be no difference from the agent point of view between communication to local and remote agent. On the other hand, we want to provide an easy way how to select synchronous, asynchronous or deferred synchronous mode of operation for any single communication act. The communication should be efficient both for passing XML strings and binary data.

As the best abstraction from the agent programmer point of view we have chosen the CORBA-like model of object method invocation. This approach has several advantages in contrast to the most common model of message passing. Among them let us mention the fact that programmers are more familiar with concept of function calling then message sending and that the model of object method invocation simplifies the trivial but most common cases while keeping the way to the model of message passing open and easy.

We have three different ways of communication based on the way, how the possible answer is treated: synchronous, asynchronous and deferred synchronous (cf. 1). The synchronous way is basically a blocking call of the given agent returning its answer. In asynchronous mode, it is a non-blocking call discarding answer, while the deferred synchronous way is somewhere in between: it is a non-blocking call storing answer at a negotiated place.

Regarding the communication media, the standard way is to use the agent communication language, described in section 3. In order to achieve faster communication the agents can negotiate to use alternative binary interface (cf. 1) which does not employ the translation of binary data into XML textual representation and back.

Medium	XML strings	CData*	function parameters
Call	Sync	BinSync	UFastNX
Generality	High	Run-time	Hardwired
Speed	Normal	Fast	The fastest

Table 1. Communication functions properties: Sync is a blocking call of the given agent returning its answer, Async is non-blocking call discarding answer and Dsync is non-blocking call storing answer at negotiated place. BinSync and BinDsync are same as Sync and Dsync but the exchange binary data instead of XML strings. UFastNX is a common name for set of functions with number of different parameters of basic types usually used for proprietary interfaces.

From the programmer's point of view, an agent in Bang[2] is regular C++ classes derived from base class Agent which provide common services and connection to environment (Fig. 1). Agent behavior is mainly determined by its ProcessMsg function which serves as the main message handler. The ProcessMsg function parses the given message, runs user defined triggers via RunTriggers function and finally, if none is found, the DefaultBehavior function. The last mentioned function provides standard processing of common messages. Agent programmer can either override the ProcessMsg function on his own or (preferably) write trigger functions for messages he wants to process. Triggers are functions with specified XML tags and attributes. RunTriggers function calls matching trigger functions for a received XML message and fills up the variables corresponding to specified XML attributes with the values and composes the return statement from the triggers return values (see 3).

There are several helper classes and functions prepared for the programmers. Magic agent pointer, which is one of them, is an association of a regular pointer to Agent object with its name which is usable as a regular pointer to an agent class but has the advantage of being automaticly updated, when the targeted agent moves around.

The agent inner state is a general name for values of relevant member variables determining the mode of agent operation and its current knowledge. The control unit is its counterpart — program code manipulation with the inner state and performing agent behavior, it can be placed in all ProcessMsg functions or triggers.

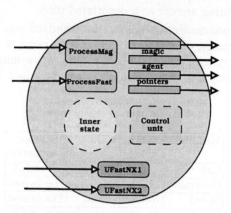

Fig. 1. The internal agent structure.

3 Communication language

Agents need a communication language for various negotiations and for data transfer between them. The language should be able to describe basic types of communication, such as requests, acceptance, denial, queries. Also, the language should be able to describe quite wide range of data formats, such as the arbitrary data set on one hand, or the inner state of a neural network on the other. The language should also be human readable, to some extent, although there might be other tools that can provide better means of communication with the user. Last but not least, we expect reliable protocol for message delivery (which is TCP/IP in our case).

Several existing agent communication languages for agents already try to solve these problems. ACL ([4]) and KQML ([3] — widely used, de facto standard) are lisp-based languages for definition of message headers. KIF (KQML group — [6]), ACL-Lisp (ACL group — [4]) are languages for data transfer. They both came out of predicate logic and both are lisp-based, enriched with keywords for predicates, cycles etc. XSIL [8] and PMML [1] are XML-based languages designed for transfer of complex data structures through the simple byte stream.

Messages in Bang[2] have adopted XML syntax. Headers are not necessary, because of the inner environment representation of messages — method invocation — the sender and receiver are known. The first XML tag defines the type of the message (similar to message types defined in an ACL header). Available message types are:

- *request* (used when an agent require another agent to do something),
- *inform* (information providing),
- *query* (information gathering),
- *ok* (reply, no error),
- *ugh* (reply, an error occurs).

The content of the message (everything between outermost tags) contains commands (type request), information provisions, etc. Some of them are understandable to all agents (ping, kill, move, ...), others are specific to one agent or a group of agents. Nevertheless, agent is always able to indicate whether he/she understands a particular message. For illustration of agent communication language messages see figure 2.

```
<broadcast><halt/></broadcast>
<inform>
<created myid="!000000000001"
  name="Lucy"
  type="Neural Net.MLP"/>
</inform>

<ok>Agent Lucy, id=!000000000001,
type=Neural Net.MLP created</ok>

<request><ping/></request>
```

Fig. 2. Example of Bang^2 language for agent negotiation.

There are two ways how to transfer data: as a XML string, or as a binary stream. The former is human readable, but may lack performance. This is not fatal in agents' negotiation stage (as above), but can represent a disadvantage during extensive data transfers. The latter way is much faster, but the receiver has to be able to decode it. Generally in Bang^2 , the XML way of data transfer is implicit and the binary way is possible after the agents make an agreement about format of transferred data. For illustration of agent data transfer language see figure 3.

4 Conclusions and future work

For now, the design and implementation of the environment is complete. The support agents, including the Yellow Pages and basic graphical user interface agent (written in Tcl/Tk) are ready. We have started to create a set of agents of different purpose and behavior to be able to start designing and experimenting with adding more sophisticated agent oriented features to the system. A general genetic algorithm agent and a RBF neural network agent has been developed and tested so far, more is coming soon.

```
<query><vector row="45"/></query>
<query><vector/></query>
<ok><data separator=",">
Here are binary data
</data></ok>
<query><bin><query>
<vector/>
</query></bin></query>
<ok session="5" funcnum="1"/>
```

Fig. 3. Example of Bang^2 language for data transfer.

For experimenting with the more sophisticated agent schemes, we will focus on mirroring agents, parallel execution, automatic scheme generating and evolving. Also the concept of an agent working as the other agent's brain by means of delegating the decisions seems to be promising. Another thing is the design of load balancing agent able to adapt to changing load of host computers and to changing communication/computing ratio. And finally we think about some form of inter Bang^2 -sites communication.

In the following we discuss some of these directions in more detail.

4.1 Task parallelization

There are two ways of parallelization: by adding an agent the ability to parallelize its work or by creating generic parallelization agent able to manage non-parallel agent schemes. Both have their pros and cons. The environment creates a truly virtual living space for agents, so the possibility for explicit inner parallel execution is there since the beginning. This approach will always be the most effective, but in general quite difficult to program.

On the other hand, the general parallelization agent can provide cheap parallelization in many cases. Consider an example of a genetic algorithm. It can explicitly parallelize by cloning fitness function agent and letting the population being fitness-ed simultaneously. Or on the other hand, the genetic algorithm can use only one fitness function agent, but be cloned together with it and share the best genoms with its siblings via a special purpose genetic operator. We can see this in figure 4, where agents of Camera and Canvas are used to automatize the sub scheme-cloning. Camera looks over the scheme we want to replicate and produces its description. Canvas receives such description and creates the scheme from new agents.

4.2 Agents scheme evolving

When thinking about implementing the task parallelization, we found it very useful to have a way of encoding scheme descriptions in a way which is understandable by regular agents. Namely, we think of some kind of XML description

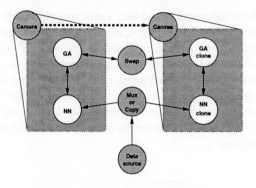

Fig. 4. Generic task parallelization.

of a scheme (which was in the previous version of the system represented in the custom Scheme Description Language). This lead to idea of agents not only creating and reading such a description, but also manipulating it.

Once we are able to represent a hybrid scheme, we can think of their automatic evolution by means of genetic algorithm. All we need is to create a suitable genetic operator package to plug into a generic GA agent. As a fitness, one can employ the part of generic task parallelization infrastructure (namely the Canvas, see fig. 5). For genetic evolving of schemes we use the Canvas for testing the newly modified schemes.

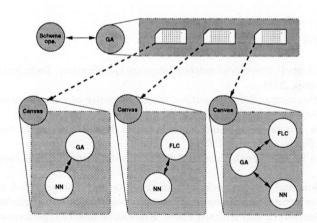

Fig. 5. Evolving a hybrid scheme by means of a genetic algorithm.

4.3 Agent as a brain of other agent

As it is now, the agent has some autonomous — or intelligent — behavior encoded in standard responses for certain situations and messages. A higher degree of intelligence can be achieved by adding some consciousness mechanisms into an agent. One can think of creating a planning agents, Brooks subsumption architecture agents, layered agents, or Franklin "conscious" agents.

Instead of hard-coding these mechanisms into an agent, we develop a universal mechanism via which a standard agent can delegate some or all of its control to a specialized agent that serves as its external brain. This brain will provide responses to standard situations, and at the same time it can additionally seek for supplementary information, create its own internal models, or adjust them to particular situations.

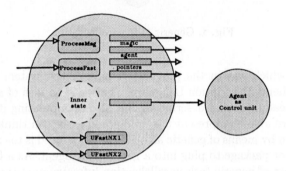

Fig. 6. Agent serving as an external brain of other agent.

References

1. PMML v1.1 predictive model markup language specification. Technical report, Data Mining Group, 2000.
2. Pietro P. Bonissone. Soft computing: the convergence of emerging reasoning technologies. *Soft Computing*, 1:6–18, 1997.
3. Tim Finnin, Yannis Labrou, and James Mayfield. KQML as an agent communication language. *Software Agents*, 1997.
4. Foundation for Intelligent Physical Agents. *Agent Communication Language*, October 1998.
5. Stan Franklin and Art Graesser. Is it an agent, or just a program?: A taxonomy for autonomous agents. In *Intelligent Agents III*, pages 21–35. Springer-Verlag, 1997.
6. Michael Genesereth and Richard Fikes. Knowledge interchange format, v3.0 reference manual. Technical report, Computer Science Department, Stanford University, March 1995.
7. Roman Neruda and Pavel Krušina. Creating hybrid AI models with Bang. *Signal Processing, Communications and Computer Science*, I:228–233, 2000.
8. Roy Williams. Java/XML for scientific data. Technical report, California Institute of Technology, 2000.

Model Generation of Neural Network Ensembles Using Two-Level Cross-Validation

S. Vasupongayya, R.S. Renner, and B.A. Juliano

Department of Computer Science
California State University, Chico
Chico, CA 95929-0410
{sang, renner, juliano}@ecst.csuchico.edu

Abstract. This research investigates cross-validation techniques for performing neural network ensemble generation and performance evaluation. The chosen framework is the Neural Network Ensemble Simulator (NNES). Ensembles of classifiers are generated using *level-one cross-validation*. Extensive modeling is performed and evaluated using *level-two cross-validation*. NNES 4.0 automatically generates unique data sets for each student and each ensemble within a model. The results of this study confirm that *level-one cross-validation* improves ensemble model generation. Results also demonstrate the value of *level-two cross-validation* as a mechanism for measuring the true performance of a given model.

1   Introduction

In a traditional neural network system, a model is represented by an individual network, one that has been trained on a single data set for a specific domain. Such a system can be replaced by an "ensemble" [3][5][6], a system model composed of multiple individual neural networks. In this study, the process of creating an ensemble consists of training each network in the ensemble individually using a unique training set, validating each network using a unique validation set, and combining all networks to form an ensemble using the *weighted contribution* combination method [10].

A neural network model, represented by either an individual neural network or an ensemble, is considered "good" if it is able to generalize over the entire domain and correctly predict or classify unseen data [14]. In order to generate good ensembles of unique neural networks, a sufficient amount of available data is needed for the training and validation processes. In reality, the available data are limited, so it is important to employ optimal usage of these data. Conventionally, researchers have worked around the limited data to achieve unique networks by using one or more of four methods: (1) changing the initial topology, (2) changing the tuning parameters, (3) using multiple learning algorithms, or (4) using different training data. According to Amari [1], with a fixed data set, the first three of these methods may lead to the *overfitting* or *overtraining* problem, because the training data are potentially biased.

Cross-validation is one of several techniques of using different training data to achieve unique networks [7][8]. *Cross-validation* rotates the training data and thus reduces the *bias* that leads to *overfitting*. There are two levels of *cross-validation* proposed in this study. The first one is *level-one cross-validation* (CV1), which potentially achieves unique networks by rotating the training and validation data. This research claims that CV1 will make *overfitting* to the entire training data set less likely.

Based on this research, CV1 as a method for *cross-validation* not only reduces the *overfitting* problem, it also eliminates the *bias* problem caused by the location of the test set in the sample data. Since it is commonly understood that a good neural network should be able to generalize over the entire domain and correctly classify unseen data, using one particular test set can lead to a misrepresentation in the performance measurement. The second level of *cross-validation* proposed in this study is called *level-two cross-validation* (CV2). CV2 eliminates the bias by grabbing a new test set each time a new ensemble is generated. The overall performance of the model is represented by the average performance over all ensembles in a given model.

2   Two-Level Cross-Validation

2.1.   Level-Two Cross-Validation

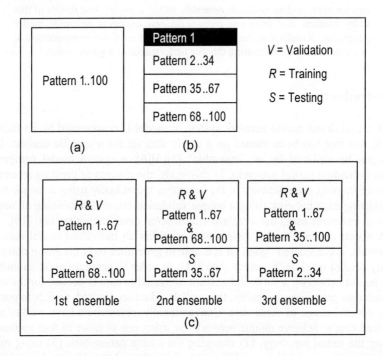

Fig. 1. An example of how the CV2 process rotates the test set for $P=100$ data patterns and $K_2=3$ ensembles to generate ensemble data sets: (a) An original example data file with 100 data patterns; (b) Data are divided into $K_2=3$ groups of equal size with 1 data pattern left; (c) The resulting ensemble data sets

By using the CV2 ensemble creation procedure, all the available P data patterns are divided into K_2 non-overlapping groups, where K_2 is the number of ensembles. Each ensemble gets a different test set consisting of P/K_2 data patterns and the rest of the available data are used as the training and validation set. If the division is uneven, the left over data patterns are always included in the training set for all ensembles.

Figure 1 illustrates the CV2 process for K_2=3 ensembles and P=100 data patterns. The test set for the first ensemble consists of 33 data patterns, numbered 68 to 100 in the original file. The remaining data, numbered 1 to 67 in the original file, are used as the training and validation set of the first ensemble. This way all data are used and each ensemble gets a test set of equal size.

2.2. Level-One Cross-Validation

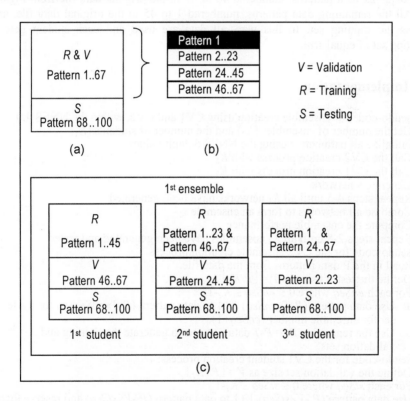

Fig. 2. An example of how the CV1 process rotates the validation set for P=67 data patterns and K_1=3 students to generate student data sets: (a) The data set for the first ensemble from Figure 1(c); (b) Training and validation data is divided into K_1=3 groups of equal size with 1 data pattern left; (c) The resulting student data sets

After the CV2 process creates each data set, the CV1 creation procedure is called for each K_2 ensemble data set. CV1 divides all the $P'=P-P/K_2$ training and validation data in a given data set into K_1 non-overlapping groups, where K_1 is the number of networks (students) in an ensemble. Each student is assigned one of the K_1 groups for its validation set. This group provides unique data, with a set size of P'/K_1, to validate each network. The remaining K_1-1 groups of the available data are used as the training set. If the division is uneven the extra patterns are assigned to the training set for all students. Each student gets a non-overlapping validation set, while the training set is overlapping.

Figure 2 illustrates the CV1 process for a training and validation set with $P'=P-P/K_2=100-33=67$ data patterns and $K_1=3$ students. The data set in Figure 2(a) is the data set for the first ensemble given in Figure 1(c). In this example, all $P'=67$ data patterns reserved for the training and validation set are divided into three groups, with one pattern remaining. The validation set for the first student consists of $P'/K_1=67/3=22$ data patterns, numbered 46 to 67 in the original data file from Figure 1(a). All the remaining data patterns, numbered 1 to 45 in the original data file, are used as the training set. In this way, all data are used and each student gets a validation set of equal size.

3   Implementation

The pseudo-code for ensemble creation using CV1 and CV2, within the NNES 4.0:
1. Get the number of ensembles (K_2) and the number of students (K_1)
2. Initialize all parameters using the NNES default values
3. Call the CV2 creation process with K_2
4. Call the CV1 creation process with K_1
5. Generate a network
6. Repeat steps 4-5 until all K_1 networks have been generated
7. Combine all networks to form an ensemble
8. Compute the ensemble performance
9. Repeat steps 2-8 until all K_2 ensembles have been generated and tested

The pseudo-code for the CV2 ensemble creation process:
1. Read in the P data patterns from the data file.
2. Define the test set size as $T = \lfloor P/K_2 \rfloor$.
3. For each $enSeq$ where $0 \leq enSeq \leq K_2$-1,
   a. Use data pattern $(P-T\times enSeq)-T+1$ to data pattern $(P-T\times enSeq)$ to form a test set for ensemble number $(enSeq+1)$.
   b. Use the remaining $P'=P-T$ data patterns to generate the training and validation sets.

The pseudo-code for the CV1 student creation process:
1. Define the validation set size as $V = \lfloor P'/K_1 \rfloor$.
2. For each $stSeq$ where $0 \leq stSeq \leq K_1$-1,
   a. Use data pattern $(P'-V\times stSeq)-V+1$ to data pattern $(P'-V\times stSeq)$ and reserve them for the validation set for student number $(stSeq+1)$.
   b. Use the remaining $P'-V$ data patterns in the training set.

4   The Model Considered

The Neural Network Ensemble Simulator (NNES) [10][13] is chosen as the development platform for CV1 and CV2. To support this research, NNES version 3.0 has been updated to version 4.0, which supports CV1 and CV2. Within the NNES 4.0

framework, the *cascade-correlation* learning algorithm is selected to create ensemble models [2]. The *weighted contribution* combination method is selected to combine networks to form an ensemble classifier [10]. The number (K_1) of networks desired for a given ensemble and the number of ensembles (K_2) for a given simulation model are entered by the user. Once the aforementioned parameters are entered, NNES default values are accepted for all remaining parameters.

Results presented in this research are generated from experimentation utilizing the *diabetes* data set from the PROBEN1 repository [9]. The PROBEN1 *diabetes* data set consists of three permutations of the same data set labeled *diabetes1*, *diabetes2* and *diabetes3*. For purposes of comparison to earlier experiments with NNES 3.0 [10][11][12], *diabetes1* is selected as the base data set. *Diabetes1* contains 576 data patterns for training and validation and an additional 192 for testing, for a total of 768 patterns.

Table 1. Details of each simulation in this research

Simulation Number	Number of Ensembles	Number of Students	Total Number of Networks
1	3	3	9
2	3	5	15
3	3	10	30
4	5	3	15
5	5	5	25
6	5	10	50
7	10	3	30
8	10	5	50
9	10	10	100
Total	54	54	324

5 Simulation Results

Nine simulations are run generating three sets of the 3-ensemble model, three sets of the 5-ensemble model, and three sets of the 10-ensemble model (see Table 1). Each student in all nine simulations is trained independently on *diabetes1* using CV1 and CV2. Performance evaluation focuses only on test data misclassification rates, and rate of disparity or improvement amongst models.

5.1. Simulation Results for the 3-Ensemble Model

According to the 3-ensemble model results presented in Table 2, while individual networks from Simulation 1, 2 and 3 misclassify the test data on average 31.57% of the time, the 3-ensemble offers an average misclassification rate of 27.72%. The 3-ensemble reflects an average misclassification reduction of 3.85, or

12.25% improvement from its average individual student. The ensemble set exhibiting the best performance in this model is the 3-ensemble of ten students, with an average misclassification rate of 26.56%. This rate reflects a 15.09% over the average of its individual networks.

Table 2. The 3-ensemble model result averages

Simulation #	Average misclassification rate of individual students	Misclassification rate of ensemble	Ensemble rate improvement compared to the average misclassification rate of its individual students	
			Improved	% Improved
1- 3ens/3std	31.36	27.08	4.28	13.65
2- 3ens/5std	32.08	29.52	2.57	8.00
3- 3ens/10std	31.27	26.56	4.71	15.09
Average	31.57	27.72	3.85	12.25

5.2. Simulation Results for the 5-Ensemble Model

According to the 5-ensemble model results presented in Table 3, while individual networks from Simulation 4, 5 and 6 misclassify the test data on average 30.63% of the time, the 5-ensemble offers an average misclassification rate of 26.38%. The 5-ensemble reflects an average misclassification reduction of 4.25, or 13.95% improvement from its average individual student. The ensemble set exhibiting the best performance in this model is the 5-ensemble of ten students, with an average misclassification rate of 26.09%. This rate reflects a 16.21% improvement over that of its average individual student. Table 3 depicts only the best, worst, and average ensembles for each model, where best and worst are defined by the model improvement factor.

Table 3. The 5-ensemble model results

Simulation#	Ens	Average misclass. rate of individual students	Misclass. rate of ensemble	Ensemble rate improvement compared to the average misclass. rate of individual students	
				Improved	%Improved
4- 5ens/3std	Best	26.38	20.87	5.51	20.89
	Worst	34.49	32.17	2.32	6.73
	Avg	30.26	26.26	4.00	13.53
5- 5ens/5std	Best	30.26	24.35	5.91	19.53
	Worst	31.13	30.43	.7	2.25
	Avg	3.47	26.78	3.69	12.10
6- 5ens/10std	Best	32.09	25.22	6.87	21.41
	Worst	31.39	28.70	2.69	8.57
	Avg	31.15	26.09	5.06	16.21
Average		30.63	26.38	4.25	13.95

5.3. Simulation Results for the 10-Ensemble Model

According to the 10-ensemble model results presented in Table 4, while individual networks from Simulation 7, 8 and 9 misclassify the test data on average 29.97% of the time, the 10-ensemble offers an average misclassification rate of 25.03%. The 10-ensemble reflects a misclassification reduction of 4.94, or 17.47% improvement from its average individual student. The ensemble set exhibiting the best performance in this model is the 10-ensemble of ten students, with an average misclassification rate of 24.38%. This rate reflects a 20.11% improvement over the average of its individual networks.

Table 4. The 10-ensemble model results

Simulation#	Ens	Average misclass. rate of individual students	Misclass. rate of ensemble	Ensemble rate improvement compared to the average misclass. rate of individual students	
				Improved	%Improved
7- 10ens/3std	Best	28.65	19.30	9.35	32.64
	Worst	26.90	28.07	-1.17	-4.35
	Avg	30.35	25.79	4.56	15.05
8- 10ens/5std	Best	22.46	12.28	10.18	45.33
	Worst	39.65	36.84	2.81	7.09
	Avg	29.72	24.91	4.81	17.24
9-10ens/10std	Best	22.63	12.28	10.35	45.74
	Worst	34.39	36.84	-2.45	-7.12
	Avg	29.84	24.38	5.46	20.11
Average		29.97	25.03	4.94	17.47

6 Discussion of Results

6.1. Results of CV1

From Tables 2,3, and 4, it can be concluded that all CV1 ensemble models have a significantly lower misclassification rate than the average of their independent counterparts. While individual networks misclassify the test data on average 30.72% of the time (calculation based on overall averages from tables 2,3,4), the CV1 models average 26.38%. This average represents a reduction of 4.35, or a 14.55% performance increase. The model exhibiting the best performance is the 10-ensemble model, with an average misclassification rate of 25.03%. This rate reflects a 17.47% improvement over the average of its individual networks.

These results confirms the proposition that CV1 may be used to generate ensembles that demonstrate a performance improvement over individual networks. Two ensembles in Simulation 7 and two ensembles in Simulation 9 show an individual student average which outperforms their respective ensembles by a small margin (see Table 4 for details). However, the average performance of the models in Simulation 7 and Simulation 9 are consistent with the CV1 proposition, reflecting a significant ensemble improvement of 15.05% and 20.11%, respectively. The most probable explanation for the performance degradation on these few ensembles can be attributed to a biased test set, based on its relative location. However, by rotating the test set for each ensemble of a given model CV2 reduces the impact of such bias.

6.2. Results of CV2

By convention, the unseen test data typically come from the bottom of the data set. Potential *bias* of a particular subset of the data, may inadvertently sabotage the testing process leading to inaccuracies in performance evaluation. A better measure of performance would test the model on as many different subsets of unseen data patterns as possible. Table 3 and Table 4 nicely illustrate the *bias* problem. In Table 4 Simulation 9, the ensemble performance varies by as much as 12.8 points, or 52.86%. The significance of these results provide support for the inclusion of cross-validation techniques in the model, for performance evaluation. The *bias* problem is clearly illustrated by the results presented for the 10-ensemble models with five or ten students, where the classification range for ensemble performance is as great as 24.56 percentage points. These two models have the best misclassification rate of 12.28% and the worst misclassification rate of 36.84%. Clearly, if the performance of a model were measured by a single test set it is likely not to be an accurate reflection of model performance [15][16]. These results and observations provide support for the significance of CV2 as both a valuable evaluation method and technique for experimentation efficiency in ensemble generation, testing, and analysis.

6.3. Secondary Findings

A secondary finding relates the misclassification rate of a given model to the number of ensembles in that model. Results show a steady decrease in misclassification rates when the number of ensembles in a given model is increased. This trend may be explained by the increase in available training data, based on the train-test split [4]. When the data set subdivisions increase in numbers they cause a decrease in the number of patterns per group. Further investigation is needed to provide conclusive evidence on the impact of the train-test split in CV1 and CV2 modeling. This investigation is left for future work.

7   Conclusions

The scope of this study represents only a small segment of the issues associated with ensemble model generation and evaluation. Limitations imposed on the experimental conditions provide avenues for future work. Deployment of future NNES versions will provide support for testing of multi-classification problems. Classification flexibility will encourage the continued evaluation of CV1 and CV2 within the NNES framework, as applied to other interesting domains. Another objective will seek expansion of the framework to include increased options for learning and combination methodologies. Future work will also explore the relationship between the number of ensembles in a given model and its relative performance.

In conclusion, although there is still much work to be done, significant progress has been made into the investigation of *two-level cross-validation* techniques for ensemble generation and evaluation. Ensembles generated using *level-one cross-validation* (CV1) are shown to provide a lower misclassification rate than their individual networks. Results support CV1 as a sound methodology for ensemble model generation. Likewise, simulation models using *level-two cross-validation* (CV2) provide a sound methodology for effectively evaluating the true performance of a model. Furthermore, ensemble investigations requiring large-scale experimentation have been simplified by this work and the deployment of CV1 and CV2 in NNES 4.0.

References

1. Amari, S., Murata, N., Muller, K.R., Finke, M., Yang, H.H.: "Asymptotic statistical theory of overtraining and cross-validation", *IEEE Transactions on Neural Networks*, Vol.8, No.15 (1997) 985-996

2. Fahlman, S.E., Lebiere, C.: "The cascade-correlation learning architecture," in *Advances in Neural Information Processing Systems 2*, D. S. Touretzky, Ed. Los Altos, CA: Morgan Kaufmann (1990) 524-532

3. Hansen, L.K., Salamon, P.: "Neural network ensembles," *IEEE Transactions on Pattern Analysis and Machine Intelligence*, vol. 12 (1990) 993-1001

4. Kearns, M.: "A bound on the error of cross validation using the approximation and estimation rates, with consequences for the training-test split," *Neural Computation*, Vol. 9 (1997) 1143-1162

5. Krogh, A., Sollich, P.: "Statistical mechanics of ensemble learning," *Physical Review*, vol. E 55 (1997) 811

6. Krogh, A., Vedelsby, J.: "Neural network ensembles, cross-validation, and active learning," in *Advances in Neural Information Processing System 7*, G. Tesauro, D. S. Touretzky, and T.K. Leen, Ed. Cambridge, MA: MIT Press (1995) 231-238

7. Leisch, F., Jain, L.C., Hornik, J.: "Cross-validation with active pattern selection for neural-network classifiers," *IEEE Transactions on Neural Networks*, Vol.9, No.1 (1998) 35-41

8. Nowla, S., Rivals, I., Personnaz, L.: "On cross validation for model selection," *Neural Computation*, Vol. 11 (1999) 863-871

9. Prechelt, L.: "PROBEN1--A set of neural network benchmark problems and benchmarking rules," Universitat Karlsruhe, Karlsruhe, Germany, Technical Report (1994) 21-94

10. Renner, R.S.: *Improving Generalization of Constructive Neural Networks Using Ensembles*, Ph.D. dissertation, The Florida State University (1999)

11. Renner, R.S.: "Systems of ensemble networks demonstrate superiority over individual cascor nets", in proceedings of the *International Conference on Artificial Intelligence* (2000) 367-373

12. Renner, R.S., Lacher, R.C.: "Combining Constructive Neural Networks for Ensemble Classification," in proceedings of the *Joint Conference on Intelligent Systems* (2000) 887-891

13. Renner, R.S., Lacher, R.C., Juliano, B.J.: "A Simulation Tool for Managing Intelligent Ensembles", in proceedings of the *International Conference on Artificial Intelligence* (1999) 578-584

14. Ripley, B.D.: *Pattern Recognition and Neural Networks*, Cambridge, MA: Cambridge University Press (1996)

15. Stone, M.: "Asymptotic for and against cross-validation," *Biometrika*, Vol. 64 (1977) 29-35

16. Stone, M.: "Cross-validation choice and assessment of statistical predictions," in *Journal of the Royal Statistical Society*, Vol. 36, No. 1 (1994) 111-147

A Comparison of Neural Networks and Classical Discriminant Analysis in Predicting Students' Mathematics Placement Examination Scores

Stephen J. Sheel[1], Deborah Vrooman[2], R.S. Renner[3], Shanda K. Dawsey[4]

[1]Department of Computer Science
Coastal Carolina University
Steves@coastal.edu

[2]Department of Mathematics
Coastal Carolina University
Vroomand@coastal.edu

[3]Department of Computer Science
California State University, Chico
Renner@ecst.csuchico.edu

[4]Coastal Carolina University
AVX Corporation
Dawsey@sccoast.net

Abstract. Implementing a technique that is efficient yet accurate for college student placement into the appropriate mathematics course is important. Coastal Carolina University currently groups students into entry-level mathematics courses based upon their scores on a mathematics placement examination given to incoming freshmen. This paper examines alternative placement strategies. Using multiple regression analysis, the accumulative high school grade point average, mathematics SAT, and the final grade in Algebra II were found to be the best predictors of success on a mathematics placement examination. Entry-level mathematics placement based on neural networks and discriminant analysis is contrasted with placement results from the mathematics placement test. It is found that a neural network can outperform classical discriminant analysis in correctly predicting the recommended mathematics placement. Consequently, a trained neural network can be an effective alternative to a written mathematics placement test.

1 Introduction

As technology advances, methodologies evolve to enhance our abilities to perform arduous tasks more expediently. Using modern computer technologies not only makes completing tasks more efficient, but also may achieve a higher degree of accuracy. For instance, classifying students into an appropriate entry-level course is often a time-consuming task. The traditional classification method is to administer a placement exam to each student for the purpose of measuring his or her ability in a particular subject and/or area. Testing students requires time and energy. The exam questions must be developed and analyzed, and a process for administrating and scoring the exams must be implemented. In administering the exam, there is the arrangement of a time and place for taking the exam, production of a physical copy of the exam, and the delegation of a proctor. Once the exams are given, the exams must be scored and students, based upon their scores, placed into the appropriate entry-level course. Similarly, a computer-based test requires design, programming, implementation, and scoring. Several tools can be used for predictive purposes in applied research. In studies where the criterion variable is nominal rather than continuous, neural networks and classical discriminant analysis can be used to predict group membership.

Currently at Coastal Carolina University, most incoming freshmen and transfer students take a mathematics placement test prior to course enrollment. Students are then placed into a mathematics course according to their performance. The entry-level mathematics courses are as follows: Math 130I -- Intensive College Algebra, Math 130 -- College Algebra, Math 131 -- Trigonometry and Analytic Geometry, and Math 160 -- Calculus I. The placement test created by the Department of Mathematics and Statistics is being utilized by all colleges at Coastal Carolina University, except the E. Craig Wall College of Business. Students entering the College of Business are assigned to a mathematics class according to their academic achievement in secondary education. Prior to summer 2000, the mathematics placement test was given in the form of a written instrument usually during a student orientation session in the summer before fall enrollment. In summer 2000, students applying to Coastal Carolina University had the option of taking a computer based mathematics placement test online over the Internet before arriving or taking it at the University.

Neural networks and discriminant analysis are two techniques that can be employed for membership classification. This study compares the effectiveness of placement based on neural networks and discriminant analysis with that of traditional testing. The results for both the neural network and discriminant analysis are compared to the results obtained by the mathematics placement test. For this study, results are contrasted for the mathematics placement test given before and after the advent of the computer based online testing. If a trained neural network or prediction equation based on discriminant analysis yields statistically similar results to the mathematics placement test, then one or the other could be used for entry-level placement in mathematics courses.

2 Related Applications in Education

Within the last decade, neural networks and discriminant analysis have been compared as predictors of success or failure in various disciplines. Gorr, Nagin, and Szczypula [5] conducted a comparative study of neural networks and statistical models for predicting college students' grade point averages. Discriminant analysis is included as one of the statistical models compared to a neural network. Their study concluded that the neural network serves as a better predictor than discriminant analysis. On the other hand, Wilson and Hardgrave [8] found that the discriminant analysis approach predicts graduate student success in a master's level business administration (MBA) program better than the neural network approach.

The use of neural networks as a predictor has increased over the past few years. Cripps [3] uses a neural network to predict grade point averages of Middle Tennessee State University students. According to Carbone and Piras [2], neural networks are instrumental in predicting high school dropouts. In Nelson and Henriksen's study [6], a neural network uses input from student responses on a mathematics placement examination given to incoming students at Ball State University and outputs the mathematics course in which each student should be placed. The implementation of neural networks as a prediction tool continues to increase.

3 Setting-up the Experiments

When assigning college students to an entry-level mathematics course on the basis of their high school performance, certain factors may contribute to student performance in mathematics. High school grade point average (GPA), class rank, Scholastic Aptitude Tests (SAT), grades earned in high school algebra I and II, geometry, and advanced mathematics courses are academic performance indicators that have the potential of affecting mathematics course placement in college. Multiple correlation coefficients adjusted to beta weights were used to analyze the variance of these predictors of scores on the mathematics placement test. The three factors chosen to have the most influence on mathematics placement test scores are high school GPA, SAT mathematics score, and final grade in high school algebra II.

The scale used for high school GPA and the final grade in high school algebra II is 0.0 to 4.0, with 0.0 representing the lowest possible score or an "F", 1.0 representing a "D", 2.0 representing a "C", 3.0 representing a "B", and 4.0 representing a perfect score or an "A". The SAT mathematics score serves as a measurement of a student's overall mathematical ability. The scale for the SAT mathematics score ranges from a low of 200 to a high of 800.

4 Neural Network Approach

An artificial neural network represents a connectionist structure whereby information is implicitly stored within the network connections and the solution is uncovered over a long series of computations. The neural network used in this study is a recurrent backpropagation network generated using a commercial simulation tool called *BrainMaker Professional* [1]. A three-layer feed-forward architecture was chosen: a three-unit input layer, ten-unit hidden layer, and a four-unit output layer. The first layer represents the inputs to the network (GPA, SAT, AlgII), and the output layer represents the classification (appropriate entry-level course). The hidden layers are purely computational units for the network.

Backpropagation is a *supervised learning* algorithm, which requires "training" the neural network. Training is performed using labeled data (data in which the target outputs are known). Values associated with network connections oscillate until the network stabilizes (reaches minimum error criteria), at which point the connection values (weights) are frozen (fixed). This fixed topology is then used to test the network against new unseen data representing student records that were not part of the training set.

In this experiment, the network is trained using data for entry-level college students grouped according to their performance on the mathematics placement test. The student's high school grade point average, SAT mathematics score, and high school algebra II score are used in the input layer. The output layer consists of four nodes representing courses in which the student may be placed into by the mathematics placement exam. A subset of the available data is held out for testing.

5   Classical Approach (Statistical Modeling)

Discriminant analysis is a statistical technique used to classify objects into distinct groups, based on a set of criteria or characteristics of the objects. Fisher's Linear Discriminant Analysis (FLDA) is a favorable classification rule that works well in situations where the groups to be discriminated are linearly separable. Objects are grouped according to their discriminant score, which is computed based on the object's observed values of the discriminating criteria or characteristics [4].

6   Data Collection

Academic information for students entering Coastal Carolina University from the fall semester of 1995 to the spring semester of 1997 was obtained through the Office of Institutional Advancement and used for the first data set. The second data set consists of students entering Coastal Carolina University in the fall of 2000 who took the computer based placement test. The data was divided into the following categories for each student: high school grade point average, high school rank, SAT verbal, SAT mathematics, algebra I, algebra II, geometry, advanced mathematics and mathematics placement examination scores. All student records with blank fields were deleted from the study. The first data set is composed of students who were correctly placed in an entry-level mathematics class using the traditional written instrument. Only records of students receiving a final grade of "C" or higher were selected. In the second data set, students with a final grade of "D" or higher were included in the analysis. The total number of records in the first data set is 458 and the second data set is 198.

7   Results and Conclusions

Using the first data representing the written placement test, the 458 student records are randomly ordered and assigned to two equal data partitions. The first partition is used to train the neural network and to create the discriminant analysis predictive equation. The second partition is used to test the trained neural network and the predictive discriminant analysis equation. Both methodologies use the overall high school GPA, the SAT mathematics score, and the final grade in Algebra II as input, and the mathematics placement result as the dependent variable.

The trained neural network correctly places 206 out of the 229 (90.0%) records in the testing data set when compared to the mathematics placement test. The predictive equations derived using discriminant analysis correctly places 155 out of 229 (67.7%) records in the testing data set. These experiments reflect a 22.2% positive difference in classification performance rate of the neural network and discriminant analysis. This difference represents an improvement of 68.9%, where improvement is measured by the reduction in classification error (e.g. (74-23)/74), see Table 1, row 1).

Using a chi-square contingency test, the null hypothesis of independence between the placement results and the methodology used is rejected. The placement results reflect a 0.01 level dependency on methodology. Hence, there is not a significant difference between the mathematics placement test and the trained neural network. This insignificance implicates the neural network as a valid replacement candidate for the mathematics placement test, at 90% accuracy.

Table 1. Classification results for comparison methods

	# Of Records	Nnet Misclass. Records	Nnet Class. Rate	D.A. Misclass. Records	D.A. Class. Rate	Improvement of nnet over D.A. (Misclassification Reduction)
Experiment 1	229	23	90.0%	74	67.7 %	68.9 %
Experiment 2	99	27	72.7 %	25	74.7 %	- 8.0 %
Average	—	50	81.4 %	99	71.2 %	49.5 %

In a similar manner, a second neural network was created to analyze the student placement data for the students taking the online computer based mathematics placement test. The trained neural network correctly places 72 of the 99 (72.7%) records in the testing data set. The predictive equations derived using discriminant analysis correctly places 74 of the 99 (74.7%) records. These experiments reflect a 2.0% negative difference in classification performance rate of the neural network and discriminant analysis, or an 8.0% increase in classification error (see Table 1, row 2). In this experiment, the results from neither method appear to be significant, and the dependency on the mathematics placement test is more significant.

The network generated for the first experiment outperformed the discriminant analysis technique by a significant amount. The network generated for the second experiment slightly under-performed the discriminant analysis technique, but not by a significant amount. When averaging the two experiments the results fall clearly in favor of the neural network, with an average misclassification rate of 81.4%, as compared to discriminant analysis with an average misclassification rate of 71.2%. The average improvement in terms of record classification error reduction is 49.5% (see Table 1, row 3). Due to the degree of difference between training data set sizes, one may speculate that the network in the first experiment was able to benefit from a larger training set size. The training size and the train-test split within this problem domain will be issues left for future research.

The results from these two experiments show that trained neural networks can be used with a minimal list of input parameters as a tool to effectively place students into entry-level mathematics courses. Such networks demonstrate an ability to improve upon discriminant analysis techniques, as shown here by as much as 68.9%. The neural network in the second data set performed as well as discriminant analysis, but was not statistically significant to the results of the actual mathematics placement test. When contrasted to the logistical difficulties of testing, scoring, and reporting the scores on a paper mathematics placement test, neural network methods also provide greater temporal efficiency and an opportunity for administrative cost reduction. Future work will explore the use of NNES ensemble networks [7], and alternative training methods to increase the performance of neural networks.

References

1. *BrainMaker Professional User's Guide and Reference Manual*, 4th Ed., California Scientific Software, Nevada City, CA (1993)

2. Carbone, V., Piras, G.: "Palomar project: Predicting school renouncing dropouts, using the artificial neural networks as a support for educational policy decisions", Substance *Use & Misuse*, Vol.33, No.3 (1998) 717-750

3. Cripps, A.: "Predicting grade point average using clustering and ANNs", *Proceedings of World Congress on Neural Networks* [CD-ROM], 486-490. Available: INSPEC Abstract Number C9807-7110-003 (1996)

4. Guertin, W.H., Bailey, J.P.: *Introduction to Modern FACTER ANALYSIS*, Ann Arbor, MI: Edward Brothers, Inc. (1970)

5. Gorr, W. L., Nagin, D., Szczypula, J.: "Comparative study of artificial neural network and statistical models for predicting student grade point averages" *International Journal of Forecasting*, Vol.10, No.2 (1994) 17-34

6. Nelson, C.V., Henriksen, L.W.: "Using neural net technology to enhance the efficiency of a computer adaptive testing application", *Mid-Western Educational Research Association Annual Meeting* [Microfiche], 1-40. Available: ERIC Number ED387122 (1994)

7. Renner, R.S., Lacher, R.C.: "Combining Constructive Neural Networks for Ensemble Classification," *Proceedings of the Fifth Joint Conference on Information Sciences* (2000) 887-891

8. Wilson, R.L., Hardgrave, B.C.: "Predicting graduate student success in an MBA program: Regression versus classification", *Educational and Psychological Measurement*, vol.55, No.2 (1995) 186-195

Neural Belief Propagation
Without Multiplication

Michael J. Barber

Institut für Theoretische Physik, Universität zu Köln, D-50937 Köln, Germany
mjb@thp.uni-koeln.de

Abstract. Neural belief networks (NBNs) are neural network models
derived from the hypothesis that populations of neurons perform statistical inference. Such networks can be generated from a broad class
of probabilistic models, but often function through the multiplication
of neural firing rates. By introducing additional assumptions about the
nature of the probabilistic models, we derive a class of neural networks
that function only through weighted sums of neural activities.

1 Introduction

It has been proposed [1] that cortical circuits perform statistical inference, encoding and processing information about analog variables in the form of probability density functions (PDFs). This hypothesis provides a theoretical framework
for understanding diverse results of neurobiological experiments, and a practical framework for the construction of neural network models that implement
information-processing functions. By organizing probabilistic models as Bayesian
belief networks [6, 7], this statistical framework leads to a class of neural networks called neural belief networks [2].

BBNs are directed acyclic graphs that represent probabilistic models (Fig. 1).
Each node represents a random variable, and the arcs signify the presence of
direct causal influences between the linked variables. The strengths of these influences are defined using conditional probabilities. The direction of a particular
link indicates the direction of causality (or, more simply, relevance): an arc points
from cause to effect.

NBNs can be used to represent diverse probabilistic models. These networks
update the PDFs describing a set of variables by pooling multiple sources of
evidence about the random variables in accord with the graph structure of the
BBN. There are two types of support that arise from the evidence: predictive
support, which propagates from cause to effect along the direction of the arc,
and retrospective support, which propagates from effect to cause, opposite to
the direction of the arc.

When there is more than one source of predictive support, their influences
are pooled multiplicatively [2]. This result is ubiquitous: except in the case of
tree-structured graphs, NBNs pool evidence through the multiplication of neural activation states. While there are some experimental [3] and theoretical [4]

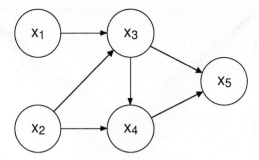

Fig. 1. A Bayesian belief network. Evidence about any of the random variables influences the likelihood of, or belief in, the remaining random variables. In a straightforward terminology, the node at the tail of an arrow is a parent of the child node at the head of the arrow, e.g. X_4 is a parent of X_5 and a child of both X_2 and X_3. From the structure of the graph, we can see the conditional independence relations in the probabilistic model. For example, X_5 is independent of X_1 and X_2 given X_3 and X_4.

indications of such multiplicative interactions in biological systems, their existence has not been conclusively established. Further, multiplication of neural activation states may be undesirable for practical implementation of such neural networks. Thus, we consider a variation of NBNs that pools the support from multiple sources without requiring the multiplication of neural activations.

2 Mean-Value Neural Belief Networks

We will attempt to find the set of marginal distributions $\{\rho(x_i; t)\}$ that best matches a desired probabilistic model $\rho(x_1, x_2, \ldots, x_D)$ over the set of random variables, which are organized as a BBN. One or more of the variables x_i must be specified as evidence in the BBN; to develop general update rules, we do not distinguish between evidence and non-evidence nodes in our notation.

The first assumption we make is that the populations of neurons only need to accurately encode the mean values of the random variables, rather than the complete PDFs. We take the firing rates of the neurons representing a given random variable X_i to be piecewise-linear functions of the mean value $\bar{x}_i(t)$ (Fig. 2)

$$a_j^i(t) = a_j^i(\bar{x}_i(t)) = [\alpha_j^i \bar{x}_i(t) + \beta_j^i]_+ , \tag{1}$$

where $[\,]_+$ denotes (half-wave) rectification and α_j^i and β_j^i are parameters describing the response properties of neuron j of the population representing random variable X_i. We can make use of (1) to directly encode mean values into neural activation states, providing a means to specify the value of the evidence nodes in the NBN.

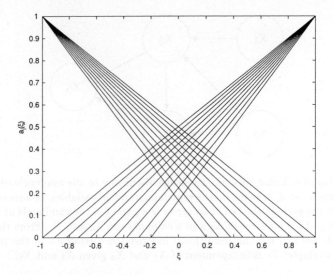

Fig. 2. The mean values of the random variables are encoded into the firing rates of populations of neurons. A population of twenty neurons with piecewise-linear responses is associated with each random variable. The neuronal responses a_j are fully determined by a single input ξ, which we interpret as the mean value of a PDF. The form of the neuronal transfer functions can be altered without affecting the general result presented in this work.

Using (1), we derive an update rule describing the neuronal dynamics, obtaining (to first order in τ)

$$a_j^i(t + \tau) = \left[\alpha_j^i\left(\bar{x}_i(t) + \tau\frac{d\bar{x}_i(t)}{dt}\right) + \beta_j^i\right]_+ . \qquad (2)$$

Thus, if we can determine how $\bar{x}_i$ changes with time, we can directly determine how the neural activation states change with time.

The mean value $\bar{x}_i(t)$ can be determined from the firing rates as the expectation value of the random variable X_i with respect to a PDF $\rho(x_i; t)$ represented in terms of some decoding functions $\{\phi_j^i(x_i)\}$. The PDF is recovered using the relation

$$\rho(x_i; t) = \sum_j a_j^i(t)\phi_j^i(x_i) . \qquad (3)$$

The decoding functions are selected so as to minimize the difference between the assumed and reconstructed PDFs [2]. In this case, we assume that the desired PDF has the form of a Dirac delta function $\delta(x - \bar{x})$, ensuring that the mean value is well represented.

With representations as given in (2) and (3), we have

$$\bar{x}_i(t) = \int x_i\rho(x_i; t)\,dx_i$$

$$= \sum_j a_j^i(t) \bar{x}_j^i \ , \tag{4}$$

where we have defined

$$\bar{x}_j^i = \int x_i \phi_j^i(x_i) \, dx_i \ . \tag{5}$$

We assume the PDFs $\rho(x_i; t)$ to be normally distributed $\rho(x_i; t) \equiv \rho(x_i; \bar{x}_i(t)) = N(x_i; \bar{x}_i(t), \sigma_{x_i}^2)$. Intuitively, we might expect that the variance $\sigma_{x_i}^2$ should be small so that the mean value is coded precisely, but we will see that the variances have no significance in the resulting neural networks.

Second, we make the key assumption that interactions between the nodes are linear

$$x_j = \sum_{x_i \in \mathrm{Pa}(x_j)} g_{ji} x_i \ . \tag{6}$$

To represent the linear interactions as a probabilistic model, we take the normal distributions $\rho(x_j \mid \mathrm{Pa}(x_j)) = N(x_j; \sum_{x_i \in \mathrm{Pa}(x_j)} g_{ji} x_i, \sigma_j^2)$ for the conditional probabilities.

We use the relative entropy [5] as a measure of the "distance" between the probabilistic model $\rho(x_1, x_2, \ldots, x_D)$ and the product of the marginals $\prod_i \rho(x_i; t)$. Thus, we minimize

$$E = \int \prod_i \rho(x_i; \bar{x}_i) \log \left(\frac{\prod_i \rho(x_i; \bar{x}_i)}{\rho(x_1, x_2, \ldots, x_D)} \right) dx_1 dx_2 \cdots dx_D \tag{7}$$

with respect to the mean values $\bar{x}_i$. By making use of the gradient descent prescription

$$\frac{d\bar{x}_k}{dt} = -\eta \frac{\partial E}{\partial \bar{x}_k} \tag{8}$$

and the fact that

$$\rho(x_1, x_2, \ldots, x_D) = \prod_i \rho(x_i \mid \mathrm{Pa}(x_i)) \tag{9}$$

for BBNs [6], we obtain the update rule for the mean values,

$$\frac{d\bar{x}_k}{dt} = -\eta \sum_{x_j \in \mathrm{Ch}(x_k)} \frac{g_{jk}}{\sigma_j^2} \left(\sum_{x_i \in \mathrm{Pa}(x_j)} g_{ji} \bar{x}_i - \bar{x}_j \right) \tag{10}$$

in the case of root nodes (nodes without parents), and

$$\frac{d\bar{x}_k}{dt} = \frac{\eta}{\sigma_k^2} \left(\sum_{x_j \in \mathrm{Pa}(x_k)} g_{kj} \bar{x}_j - \bar{x}_k \right)$$

$$- \eta \sum_{x_j \in \mathrm{Ch}(x_k)} \frac{g_{jk}}{\sigma_j^2} \left(\sum_{x_i \in \mathrm{Pa}(x_j)} g_{ji} \bar{x}_i - \bar{x}_j \right) \tag{11}$$

for all other nodes.

The update rule for the neural activities is obtained by combining (2), (4), and either (10) or (11) as appropriate. Clearly, the neural network operates by calculating weighted sums of neural activation states, and is not dependent upon multiplicative dynamics.

The foregoing provides an algorithm for generating and evaluating neural networks that process mean values of random variables. To summarize,

1. Establish independence relations between model variables. This may be accomplished by using a graph to organize the variables.
2. Specify the g_{ij} to quantify the relations between the variables.
3. Assign network inputs by encoding desired values into neural activities using (1).
4. Update other neural activities using (10) and (11).
5. Extract the expectation values of the variables from the neural activities using (4).

3 Applications

As a first example, we apply this strategy to the BBN shown in Fig. 1, with firing rate profiles as shown in Fig. 2. Specifying $x_1 = 1/2$ and $x_2 = -1/2$ as evidence, we find an excellent match between the mean values calculated by the neural network and the directly calculated values for the remaining nodes (Table 1).

We next focus on some simpler BBNs to highlight certain properties of the resulting neural networks (which will again utilize the firing rate profiles shown in Fig. 2). In Fig. 3, we present two BBNs that relate three random variables in different ways. The connection strengths are all taken to be unity in each graph, so that $g_{21} = g_{23} = g_{12} = g(13) = 1$.

With the connection strengths so chosen, the two BBNs have straightforward interpretations. For the graph shown in Fig. 3a, X_2 represents the sum of X_1 and X_3, while, for the graph shown in Fig. 3b, X_2 provides a value which is duplicated in X_1 and X_3. The different graph structures yield different neural networks; in particular, nodes X_1 and X_3 have direct connections for the neural network based on the graph in Fig. 3a, but no such direct weights exist in a

Table 1. The mean values decoded from the neural network closely match the values directly calculated from the linear relations. The coefficients for the linear combinations were randomly selected, with values $g_{31} = -0.2163$, $g_{32} = -0.8328$, $g_{42} = 0.0627$, $g_{43} = 0.1438$, $g_{53} = -0.5732$, and $g_{54} = 0.5955$.

Node	Direct Calculation	Neural Network
X_1	0.5000	0.5000
X_2	-0.5000	-0.5000
X_3	0.3083	0.3084
X_4	0.0130	0.0128
X_5	-0.1690	-0.1689

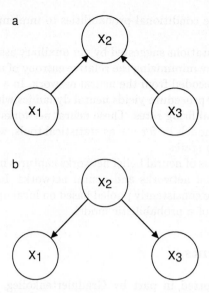

Fig. 3. Simpler BBNs. Although the underlying undirected graph structure is identical for these two networks, the direction of the causality relationships between the variables are reversed. The neural networks arising from the BBNs thus have different properties.

second network based on Fig. 3b. Thus, specifying $x_1 = -1/4$ and $x_2 = 1/4$ for the first network produces the expected result $\bar{x}_3 = -0.5000$, but specifying $x_2 = 1/4$ in the second network produces $\bar{x}_3 = 0.2500$ regardless of the value (if any) assigned to x_1.

To further illustrate the neural network properties, we use the graph shown in Fig. 3b to process inconsistent evidence. Nodes X_1 and X_3 should copy the value in node X_2, but we can specify any values we like as network inputs. When we assign $x_1 = -1/4$ and $x_3 = 1/2$, the neural network yields $\bar{x}_2 = 0.1250$ for the remaining value. This is a reasonable result, matching the least-squares solution to the inconsistent problem.

4 Conclusion

Neural belief networks are a general class of neural networks that can consistently mix multiple sources of evidence, but depend upon multiplicative neural dynamics that may be undesirable. To eliminate these possibly undesirable properties, we considered a more restricted class of related networks by making two auxiliary assumptions.

First, we assumed that only the mean values of the random variables need to be accurately represented, with higher order moments of the distribution being unimportant. We introduced neural representations of relevant probability density functions consistent with this assumption. Second, we assumed that the random variables of the probabilistic model are linearly related to one another, and chose appropriate conditional probabilities to implement these linear relationships.

Using the representations suggested by our auxiliary assumptions, we derived a set of update rules by minimizing the relative entropy of an assumed PDF with respect to the PDF decoded from the neural network. In a straightforward fashion, this optimization procedure yields neural dynamics which depend only on a weighted sum of neural firing rates. These neural networks can be implemented, for use as biological models or for use as statistical tools, without multiplication of neuronal activation states.

This restricted class of neural belief networks captures many of the properties of both Bayesian belief networks and neural networks. In particular, multiple sources of evidence are consistently pooled based on local update rules, providing a distributed version of a probabilistic model.

Acknowledgements

This work was supported in part by Graduiertenkolleg Azentrische Kristalle GK549 of the DFG. The author would like to thank J.W. Clark for valuable discussions.

References

[1] C.H. Anderson. Basic elements of biological computational systems. In J. Potvin, editor, *Proceedings of the 2nd IMACS Conference on Computational Physics*, pages 135–137, Singapore, 1994. World Scientific.

[2] M.J. Barber, J.W. Clark, and C.H. Anderson. Neural propagation of beliefs. Submitted to Neural Comput., 2001.

[3] F. Gabbiani, H.G. Krapp, and G. Laurent. Computation of object approach by a wide-field, motion-sensitive neuron. *J. Neurosci.*, 19(3):1122–41, February 1999.

[4] B. Mel. Information processing in dendritic trees. *Neural Comput.*, 6:1031–85, 1994.

[5] A. Papoulis. *Probability, Random Variables, and Stochastic Processes*. McGraw-Hill, Inc., New York, NY, third edition, 1991.

[6] Judea Pearl. *Probabilistic Reasoning in Intelligent Systems: Networks of Plausible Inference*. Morgan Kaufmann Publishers, Inc., San Mateo, CA, 1988.

[7] P. Smyth, D. Heckerman, and M.I. Jordan. Probabilistic independence networks for hidden Markov probability models. *Neural Comput.*, 9:227–69, February 1997.

Fuzzy Logic Basis
in High Performance Decision Support Systems

Bogdanov A., Degtyarev A., Nechaev Yu.

Institute for High Performance Computing and Databases
Fontanka 118, 198005 St.-Petersburg, Russia
bogdanov@hm.csa.ru, deg@fn.csa.ru, int@fn.csa.ru

Abstract. The problem of synthesis of the onboard integrated intellectual complexes (IC) of decision support acceptance in fuzzy environment is discussed. The approach allowing to formalize complex knowledge system within the framework of fuzzy logic basis is formulated. Interpretation of fuzzy models is carried out with use of high performance computing at processing information stream in problems of analysis and forecast of worst-case situations (disputable and extreme).

Introduction

The present research is connected with use of soft computing concept in onboard integrated real time IC functioning. Formulation of practical recommendations and decisions acceptance in complex dynamic object (DO) control in extreme conditions is carried out on the basis of measurement data, mathematical models and the structured knowledge base (KB). The fundamental basis of research is theoretical principles of formalization of knowledge used in the organization of designing and systems functioning having in view their discrepancy and uncertainty, and also their algorithmic processing in decisions acceptance or in choice of operations for purposeful actions support. Practical realization of this problem is connected with development of effective algorithms of the analysis of the information obtained from measuring system and from procedural components of KB. Aspects of fuzzy logic allow to formalize discrepancy and uncertainty in the description of information and procedural elements of soft computing processes. Artificial neural networks (ANN) and genetic algorithms (GA) allow to organize training during calculations performance and their high-performance realization.

1. Conceptual model and IC architecture

The fundamental idea of employing IC for solving complex problems of DO interacting with the environment is connected with application of the measuring means and methods of analyzing the original information and supercomputer technology.

The invariant kernel of the system includes the problem field, KB and DB of IC. For creation of this set conceptual model of IC is formed. The consideration of system in functional aspect allows to pick out the following components [3,5]:

$$<S_F, S_M, S_W> \tag{1}$$

where $S_F = \{S_{F1}, ..., S_{FN}\}$ is the set of functional subsystems; S_M is structural scheme of the system; S_W are conditions of formation of complete system;

$$S_F = <X, Y, A, P_F, T_F>, S_M = <E, C, \Phi, P_M, T_M>, S_W = <G, R, U_R, K_R, E_f>$$

where $X = X_j$ (j = 1,..., n) is the vector-set of input signals; $Y = Y_i$ (i = 1,..., m) is the vector-set of output signals; $A : \{X \rightarrow Y\}$ is the operator determining process of functioning of system S; $P_F = \{P_F^x, P_F^y, P_F^A\}$ is the full set of functional parameters; $P_F^x = \{\pi_F^x\}$; $P_F^y = \{\pi_F^y\}$; $P_F^A = \{\pi_F^A\}$ are the parameters for considered subsets; T_F is the set of the time moments, invariant to object of modeling; $E = E_v$ (v = 1,..., N_v) is the set of system components; $C = C_q$ (q = 1,..., Q_c) is the set determining relations between elements; $P_M = \{P_M^E, P_M^C, P_M^\Phi\}$ is the set of morphological parameters. $\Phi = \{\Phi_\lambda\}$ (λ=1,...,r) is the configuration set determining way of S_F formation. $P_M^E = \{\pi_M^E\}$; $P_M^C = \{\pi_M^C\}$; $P_M^\Phi = \{\pi_M^\Phi\}$ are the parameters for considered subsets; $T_M = \{t_k^M\}$ is the set of the time moments of dynamic structures; G are the purposes of functioning of object in realization of a task R; U_R are the principles and algorithms of management of object; K_R is an execution result quality; E_f is the efficiency determining at what price the purpose G is achieved.

Problem of selection of optimal requirements to quality operation index Q_l (l=1,...,L) and of characteristics of IC tools C_s (s=1,...,N) in conditions of incompleteness and indefiniteness of initial data is formulated on the basis on Zadeh' generalization principle [9]. In accordance with this principle possibilities of considered indexes realization are defined by the following:
for IC characteristics

$$\mu_S(C_S) = \max_{C_{S1},...,C_{Sh}:C_S=f_S(C_{S1}...C_{Sh})} \mu_{S1}(C_{S1}) \wedge ... \wedge \mu_{Sh}(C_{Sh}), C_S \in R^1 \tag{2}$$

and for quality indexes

$$\mu_{\Pi Q}(q_l) = \max_{C_{u1},...,C_{up}:q_l=Q_l(C_{u1}...C_{up})} \mu_{u1}(C_{u1}) \wedge ... \wedge \mu_{up}(C_{up}), q_l \in R^1 \tag{3}$$

with fuzzy sets

$$C_s = \{C_s, \mu_s(C_s)\}; \quad \Pi Q_l = \{q_l, \mu_{\Pi q_l}(q_l)\}.$$

It is possible to present optimal solution of problem for fuzzy goals Q and fuzzy possibilities PQ having in view conditions (2) and (3) as

$$\mu_{\Pi Ql}(q_l) \wedge \mu_{Ql}(q_l) \rightarrow \max \tag{4}$$

The realization of (1)–(4) is connected with the decision of the following functional tasks:

- diagnosing of the current condition of environment and dynamic object;
- recognition and classification of situation;
- the forecast of development of situation;
- search of control strategy;
- change of the current condition of dynamic object and forecast of its behavior under influence of the control decisions

Imitative modeling, procedure of the statistical and spectral analysis, filtration and forecast, identification and recognition of images are used as the mathematical tools for development of control algorithms.

The architecture of onboard IC is shown in the fig.1. The complex contains intelligence subsystems functioning on the basis of dynamic measurement data. Analysis and forecast of situations development and decisions acceptance are provided by means of the training system realizing adaptive component functions.

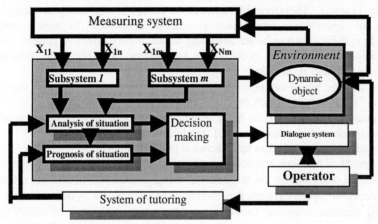

Fig. 1. Architecture of onboard intelligence complex

2. Fuzzy parallel systems

Development of IC software is based on the following artificial intellect paradigms. A basic element of information model of a considered subject domain is fuzzy production KB model. The conceptual model of IC functioning is formulated having in view requirements to information, computing and measuring technology. The formalized knowledge system is created by method of synthesis of control algorithm and decisions acceptance in conditions of uncertainty and incompleteness of the initial information.

Fuzzy logic inference is carried out with the help of modified modus ponens on the basis of the results of the current measurements [1], [8].

$$\text{«If } X, \text{ then } Y, \text{ else } Z\text{»} \ (X \rightarrow Y(Z)) \tag{5}$$

where X, Y, Z - are fuzzy sets defined on universal sets U, V, W. The sets X, Y, Z are interpreted as a fuzzy input and output of some system. The relationship between these sets is defined by fuzzy system model.

The result of given conditional fuzzy operator completing (5) is characterized by the expression

$$\Phi(X \to Y(Z)) = \{\mu_x/\Phi(Y), (1-\mu_x)/\Phi(Z)\} \tag{6}$$

where μ_x is truth degree of condition X.

Uniqueness of the defined group of operators is determined after calculating μ_x in accordance with given threshold of the truth degree $\gamma_0 \in [0,1]$:

$$\Phi(\bullet) = (\Phi(Y), \; if \; \mu_x \geq \gamma_0) \; or \; (\Phi(Z), \; if \; \mu_x < \gamma_0) \tag{7}$$

where $\Phi(\bullet)$ is a result of operators completing.

Specific feature of computing systems of new generation is orientation on knowledge processing and parallel functioning. Parallelism is inherent in knowledge processing systems. Therefore in perspective computer technologies design the problems of knowledge representation models adaptation to new computer architecture and hardware-software designing to suit highly effective parallel models of knowledge representation occur. Two approaches in fuzzy logic conclusion realization in onboard IC are used:

- Modus ponens based on minimax composition of entrance FS and the fuzzy relation "input-output" which describes a rule of decision acceptance;
- Situational inference based on fuzzy recognition of the situation by comparison of its description with the description of standard fuzzy situations.

Comparative analysis of these approaches is published in [12].

Development of methods and software of generation and alternatives choice on the basis of initial information on a situation being researched allows to rationally organize decision support system functioning. Analysis allows to determine set (A_0) of minimal dimension among known set of alternatives (A), each of which (a_n) is estimated by one or several generalized criteria of quality K_q. Then, for any alternative $a_i \in A$, the alternative $a_j \in A_0$, concerning which conditions

$$K_q(a_j \in A_0) > K_q(a_i \in A) \tag{8}$$

are true is put in conformity.

The considered class of systems is characterized by the fact that for situations set there is a certain number of allowable decisions. Then the set of situations is broken into classes in such manner that the unique decision with necessary practical recommendations is put in conformity for each of these classes. The mechanism providing functioning logic and decisions development in process of entry of information on object dynamics and on environment is based on decisions conclusion with the help of formal procedure:

$$F_i: S_k(t_i) \to U_j, \; k=1,\ldots,n; \; j=1,\ldots,J; \; i=1,\ldots,N, . \tag{9}$$

based on the initial data concerning control decisions $\{u_i\}$.

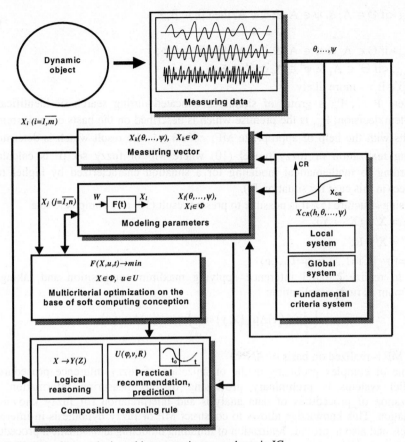

Fig. 2. Build-in data analysis and interpretation procedures in IC

In IC functioning in complex conditions inference rules usually contain the built-in procedures of analysis and interpretation of measurement data (fig.2). In this case situations when the current measurement data result in various interpretation may arise. Then expansion of analogical inference rule [4] on fuzzy systems may be used. The approach developed by authors allows to carry out a fuzzy inference in the following sequence:

- the built-in procedures are realized and their efficiency is checked by mathematical modeling methods ;
- according to modeling results membership functions of antecedent parts are constructed;
- inference is carried out on the basis of model (5)

As an example, let us consider fuzzy model of floating DO extreme situation identification by results of parameters θ, ψ, ζ measurement (where θ is rolling, ψ is pitching, ζ is heaving) in the case when antecedent part of rule is described by identical ratios

P_1: <if $\Theta \in A_1$ & $\psi \in A_2$ & $\zeta \in A_3$ then $\beta_1 \in B_1$>; $\qquad$ **(10)**

$\bullet \quad \bullet \quad \bullet$

P_n: <if $\Theta \in A_1$ & $\psi \in A_2$ & $\zeta \in A_3$ then $\beta_n \in B_n$>;

P_{n+1}: <if $\Theta' \in A_1$ & $\psi' \in A_2$ & $\zeta' \in A_3$ >

<$\beta' \in B_1$> - more likely.

Here $P_1, \ldots, P_n$ is group of situations allocated during search of identification problem decision; P_{n+1} is the premise which is described on the basis of measurement results with the help of appropriate MF; <$\beta' \in B_1$> is the result which is determined during realization of fuzzy model (10) where MF of fuzzy set β' is calculated according to mathematical modeling for a situation characterized by higher truth degree (in this case for situation P_1).

Using structure (5), it is possible to present result (10) as a composite rule

$Y' = X' \circ (X \to Y);$ $\qquad$ **(11)**

$Y' = X' \circ R;$

$R = [r_{ij}]$ ($i = 1, \ldots, m; j = 1, \ldots, n$)

and to realize parallel inference applying maximin composition and taking of minimum as fuzzy implication:

$$\mu_{Y'} = \bigcup_{x \in X} \mu_{X'}(x) \Lambda \mu_R(x, y) = \bigcup_{x \in X} \mu_{X' \cap X}(x) \Lambda \mu_Y(y) \qquad \textbf{(12)}$$

then MF is realized on basis of ANN.

One of complex problems in the organization of fuzzy inference procedure in parallel systems is preliminary processing of the measuring information and realization of procedures of data analysis and interpretation for fuzzy knowledge formation. This knowledge allows to construct MF of fuzzy hypothesis in inference models and also to provide realization of imitating modeling and statistical procedures of IC analysis in various operation conditions functioning especially in abnormal situations. Such calculations demand realization of high performance computing with the use of modern supercomputers [3]. Development of information processing algorithms in real time represents the important aspect at realization of IC designing computing technology requirements.

The paradigm of parallel calculations in realization of fuzzy logic inference assumes use of various models of approximate reasoning. As one of approaches it is possible to realize the algorithm offered in work [11]:

- *preprocessor processing*: construction of a matrix of parallelism and parallel sets Π_i ($i = 1, \ldots, n$) for all production models;
- *initialization*: input of initial condition S_0 and installation of current value $S = S_0$; input of target condition S_k; if $S = S_k$, then stop;
- *search of the decision (the basic production cycle)*: calculation of activation set for current condition S (performance of a phase of production cycle comparison); determination of parallel carried out productions subset from $\cap \Pi_i$, and obtaining new current condition S; if $S \neq S_k$, then stop, otherwise return to the beginning of decision search stage.

Interaction of rules in production system is fundamental concept in use of parallel processing. Parallel products activation is allowable at their mutual independence. Simultaneous performance of dependent rules may result in products interference $P=\{P_q\}$ (q=1,...,Q) on to independent subsets.

A fuzzy neural network is one of the effective models for design of real time systems KB. Such model was offered by authors [2] for realization of algorithm of control vector formation of ship movement in ship-born IC. In this paper fuzzy production model is used as knowledge representation model. The fuzzy logic inference by results of current measurements is carried out with the help of the modified modus ponens

$$\mu_{\sigma_j} = \max_c \min\left(\mu_{\sigma_j^c}, \min Poss_{c,i}\right) \tag{13}$$

where $Poss_{c,i}\left(Z_i^c, \acute{Z}_i\right) = \sup_z \min\left(\mu_{z_i^c}, \mu_{z_i'}\right)$ is a measure of an opportunity.

Transformation of the entrance fuzzy variables (result of procedure (13)) in to an accurate estimation u is realized with the use of defuzzification procedure. The organization of antecedent parts of each rule calculations is carried out in fuzzy multilayered ANN. Basic elements of a network are fuzzy «AND» and «OR» formal neurones. During preprocessing an input vector calculation of measures of an opportunity (13) is stipulated. Values of synaptic weights are formed during training according to algorithm of optimization problem decision offered in work [13]. Elements of control vector are calculated on the basis of production model. Calculation by modus ponens is carried out in blocks of "MAX" operation. Accurate values of vector elements are formed according to defuzzification method in blocks «df».

Neural network approach to the calculations organization according to intellectual algorithm of ship control allows to provide IC functioning in real time. Increase of intellectuality degree of onboard integrated IC is achieved due to use of soft computing concept. Principles of fuzzy logic allow to formalize inaccuracy and uncertainty in the description of information and procedural elements of soft computing processes. ANN allow to organize training during calculations and high performance realization. The practical application of the soft computing concept at design of onboard IC for underwater device is considered in [10]. The KB of IC is represented as fuzzy models. Tabulated values of FM are stored in a database. The fuzzy inference is carried out with the help of modus ponens and procedure of accurate interpretation (defuzzification). Complexity of the intellectual control law has resulted in necessity of use of the computing procedures focused on performance in multiprocessor computing environment on the base of neural network models. Realization of this technology has allowed to set off optimum structures from the suboptimum decisions found during GA work.

3. Decisions acceptance in fuzzy environment.

One of the most complex ways of on-board IC development is connected with development of tools ensuring functioning of integrated systems on the basis of a complex information flow. They include information from sensors of measuring system and results of mathematical modeling. The practical realization of the developed concept of analysis and forecast of dynamics of complex objects is carried out with using of supercomputer technologies ensuring the large resource of memory and speed. The functions of multiprocessor system are realized on the basis of the distributed architecture ensuring modularity of constructions and an opportunity of upgrading, reliability, distribution of loading and parallel processing of the information.

During IC functioning the control is carried out as a sequence of steps $u(\varphi,v,R) = u_i$ ($i=1,\ldots,m$). Local control goal is matched to each step.

As a result the dynamic object comes into certain condition $S(u) = S_j$ ($j=1,\ldots,n$) and has target signals $Y(U,R)=Y_k$ ($k=1,\ldots q$). Quality of IC work to obtain practical recommendations on r^{th} a step is

$$\Phi(r) = \Phi\left(S_k, S_f\right) \tag{14}$$

where $\Phi(r)$ is error function; S_k is technical condition of dynamic object (forecast); S_f is actual technical condition

The value of error function is defined by reliability of the algorithms incorporated in formalized knowledge system.

To increase of efficiency of IC functioning the methods of the theory of experiment design are used. The problem of experiment design in real-time systems is connected with solution of the following problems:

- construction of optimum structure of mathematical models for adaptive components realization;
- choice of optimum conditions of measurements ensuring a reliable estimation of dynamic object characteristics and parameters of environment;
- development of algorithms of measuring information analysis

Algorithm of decision-making process defines a sequence of the procedures connected to interpretation of situations, development and analysis of alternatives, formulation of criteria, risk calculation and estimation, choice of way out variant , realization and check of the decision.

Object structurization as decisions tree allows to present logic decision functions of recognition, of the analysis and forecasting of considered situations. Construction of decisions tree subseting A_k into M images assumes:

- variables·set $X \in X_k$ ($k=1,\ldots, N$) with value areas $D_1,\ldots, D_N$ and $D_1 \times \ldots \times D_N$
- variables set $x \in x_k(a)$ for object $a \in A_k$ ($k=1,\ldots,N$);
- separation of set D into pairs crossing subsets

$$E^1,\ldots,E^M, E^S = E_1^{S_1} \times \ldots \times E_k^{S_k} \; S_j \in \left\{1,\ldots,\ell_j\right\} S = 1,\ldots,M, j = 1,\ldots,k, k = 1,\ldots,N \tag{15}$$

In case of failure of initial information analysis with the help of traditional methods the decision is transferred to formation of the structure providing problem solution. The basis of such structure are the approximated methods and algorithms including approaches using nonconventional procedures (neural network logic basis, GA, etc.). In particular, algorithms of measuring system sensors parrying contain procedures of intelligence sensors creation basing on associative ANN and imitating modeling methods. Formation of the adaptive computing procedures which are taking into account volume and reliability of the initial data, makes one of the important trends in decision support systems design.

Process of decision support system synthesis is a sequence of hierarchical structure analysis operations. The elementary subtasks in this structure are analyzed preliminarily by method of hierarchies [7]. Realization of this method is connected to allocation of priorities (weights) of attributes with a view of choice of the best of them with the use of algebraic theory of matrixes and expert procedures:

$$W\pi = \lambda_{max}\pi; \quad \Pi = \begin{bmatrix} \pi_{11} & \cdots & \pi_{1m} \\ \vdots & \pi_{ij} & \vdots \\ \pi_{n1} & \cdots & \pi_{nm} \end{bmatrix} \times \begin{bmatrix} g_1 \\ \vdots \\ g_m \end{bmatrix}, \tag{16}$$

where W is the inverse symmetric matrix of values of paired comparisons of attributes concerning the given attribute; π is the normalized vector of attributes weights; λ_{max} is the greatest eigenvalue of matrix W; Π is the result of determination of global priorities of attributes $\Pi_1,\ldots, \Pi_N$; N is the number of attributes; π_{ij} (i=1,..., n, j=1,...,m) is the relative weight of i-th attribute on j-th attribute; g_j is the relative weight of j-th attribute.

During such analysis attributes, to which the minimal global priority values correspond are allocated. The obtained data allow to generate space of the alternatives determining DO condition and control decisions, providing task performance. The result is achieved by finding compromise between the allocated space of alternatives A_q^* and attributes dictionary volume $\{x_i\}$. Realization of developed information technology in the analysis of extreme situations is shown in the fig.3.

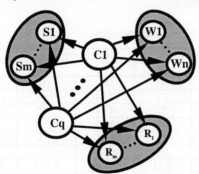

Fig.3 Fragment of a semantic network at the analysis of extreme situations

The semantic network presented here in which units frames are located, contains the following sets:

- S is the set of scripts containing the descriptions of dynamic stages and the appropriate updating in model parameters;
- W is the set of variants (DO conditions) described by information model;

- R is the set of conclusions describing results of calculations of the script on given variant;
- C is the set of links uniting in a semantic network triads < S, W, R > and containing references to these sets

4. Results of experiments.

Practical realization of the developed technology is carried out as the integrated onboard complex of a navigation safety.

The IC full-scale tests have been carried out aboard a tanker in the Baltic Sea, aboard a container ship in a voyage in the Mediterranean and in the Atlantic and on a small ship in the Black Sea. The most complete investigation of the IC operation was checked on the small ship, where the ship motions on waves were chosen according to specially developed program.

In the course of the test operation the dynamic characteristics measurements, this system adaptation and completion of the designing process, receiving new knowledge about the ship behavior on seas and its integration with the old one were carried out. The following factors were of special practical interest: testing the reliability of the applied methods for estimating and predicting the seas parameters, the transverse stability characteristics and the KB reaction to the standard and non-standard situations arising in the seaway. All this allowed us to correct the mathematical models of the interaction of the ship and environment, the mechanism of logical output and the procedure of decision making on the basis of fuzzy aims and limitations.

The test proved the possibility of practical evaluation and prediction of the dynamic characteristics, of the reliability of KB operation under various conditions of service. The selective data of wave prediction are shown in [6], seaworthiness experiments are shown in table 1

Table 1. Comparison of seaworthiness characteristics

Ship's type	Small ship		Container ship		Tanker	
Speed, kn	6	6	14.5	14.0	5	5
Course angle, deg	45	90	10	10	0	0
Draught, m	2.2	2.1	7.64	7.65	11.3	7.5
Wave length, m	20	26	43.4	67.5	22	23
Wave height, m	1.75	2.3	3.96	5.00	1.9	2.0
Rolling, deg	12	16	5.6	6.5	–	–
Pitching, deg	2.8	1.9	1.3	1.5	–	–
Actual GM, m	0.84	0.83	0.530	0.525	0.8	1.8
Critical GM, m	0.50	0.49	0.328	0.327	0.6	1.0
Heel on circ., deg	–	–	–	–	3.2	9.6
Speed lost, kn	3	1	2.3	2.8	–	–

Conclusion

The real-time program complex on the basis of multiprocessor technologies presented in the paper is intended for the safety control of ships and offshore structures.

Developed IC solves complex and difficult for formalization problems of analysis and forecast of seaworthiness characteristics in dynamic area on a basis continuously entering in "on-line" measuring information. Methods used in IC have statistical and experimental basis. They are easily realized in onboard supercomputer and allow the further perfection on the basis of full-scale tests.

The work is supported by grant RFBR 00-07-90227.

References

1. Averkin A.N. at all. Fuzzy sets in control and artificial intelligence models. M.: Nauka, 1986 (in Russian)
2. Nechaev Yu., Degtyarev A., Siek Yu., Decisions making in real-time intelligence systems using soft computing concept.//Artificial Intelligence N3, 2000, pp.525-533 (in Russian, English version now is available in http://www.maus-soft.com/iai)
3. Nechaev Yu., Gorbachev Yu. Realization of complex intelligence systems based on supercomputer technology. //Proc. of International conference "Intelligence Multiprocessor Systems", IMS'99, Taganrog, 1999, pp.78-85 (in Russian)
4. Polya G., Mathematics and plausible reasoning. Princeton University Press, 1954.
5. Nechaev Yu., Degtyarev A. Account of peculiarities of ship's non-linear dynamics in seaworthiness estimation in real-time intelligence systems.//Proc. of the Int. conf. STAB'2000, Launceston, Tasmania, Australia, Feb. 2000, vol.2, pp.688-701
6. Nechaev Yu. The problem of wave parameters identification in ship's intelligence system.//Proc. of 13th Int. conf. on hydrodynamics in ship design HYDRONAV'99, Sept. 1999, Gdansk, Poland, pp.201-208
7. Saaty T.L. A sealing method for priorities in hierarchical Structures //J.Math. Psychology, 1977. Vol.15. N3
8. Zadeh L.A. The concept of linguistic variable and its application to approximate reasoning. – N.Y.: Elsivieer P.C., 1973
9. Zadeh L Fuzzy logic, neural networks and soft computing//CACM,1994,v.37,N3,pp.77-84
10. Nechaev Yu.I., Siek Yu.L. Design of Ship-Board control system Based in the soft computing conception //Proc. of 11th Int. conf. on industrial and engineering applications of artificial intelligence and expert systems, IEA-98-AIE. Benicassim, Castellon, Spain, June,1998. Springer. Vol.11. pp.192--199.
11. Vagin V.N. Parallel inference in situation control systems //Proc. of the 1995 ISIC Workshop – 10th IEEE International Symposium on Intelligent Control. Aug.1995 Monterey (Cal.) USA, pp.109-116
12. Melikhov A.N., et al. Situation advised systems with fuzzy logic. M., Nauka, 1990 (in Russian)
13. Pedrycz W. Fuzzy neural networks with reference neurons as classifiers. // IEEE Trans. Neural Networks, vol. 3, 1992, pp. 770-775.

Scaling of Knowledge
in Random Conceptual Networks

Lora J. Durak, Alfred W. Hübler

Center for Complex Systems Research

Department of Physics, University of Illinois, Urbana, USA

February 8, 2001

Abstract

We use a weighted count of the number of nodes and relations in a conceptual network as a measure for knowledge. We study how a limited knowledge of the prerequisite concepts affects the knowledge of a discipline. We find that the practical knowledge and expert knowledge scale with the knowledge of prerequisite concepts, and increase hyper-exponentially with the knowledge of the discipline specific concepts. We investigate the maximum achievable level of abstraction as a function of the material covered in a text. We discuss possible applications for student assessment.

1 Introduction

Defining quantitative measures for knowledge and understanding has long been a challenging problem [1]-[4]. Assessment of understanding and knowledge occurs so frequently in every day life, and technology, and is of such considerable importance that a comprehensive mathematical theory of knowledge and understanding would seem to have been required long ago. Historically, there have been multiple attempts to define understanding and knowledge. One of the earlier written theoretical descriptions of knowledge can be found in Plato's dialogues. Much of Theaetetus is devoted to the rejection of Pythagora's view that all knowledge is perception[5]. Plato inherited from Socrates the view that there can be knowledge in the sense of objective universally valid knowledge, such as the properties of numbers and geometrical objects [6]. In the *Republic*, Plato introduces the "levels of knowledge" of the development of the human mind on it's way from ignorance to true knowledge[7].

Still true knowledge exists only if a foundation of axioms or a priori knowledge is assumed to be true. Defining a minimum set of a priori knowledge is still a subject of active research in metaphysics. Aristotle values the most abstract knowledge as wisdom[8], but he encourages to test knowledge with our senses[6]. St. Augustine emphasizes abstract knowledge too, in De Beata Vita [9] "only the wise man can be happy, and wisdom postulates knowledge of truth". For him 'a priori knowledge' has to be experiencable either by our own senses or others[9]. Leipnitz disagrees with the empiricists such as Locke that all our concepts are ultimately derived from experience [6]. Kant sides with Leipnitz and states: "That all of our knowledge begins with experience, there is no doubt... But though all our knowledge begins with experience, it does not follow that all arises out of experience"[10] and details in On the Form and Principles: "Since then, in metaphysics, we do not find empirical principles, but in the very nature of pure intellect, not as experiencable concepts, but as abstracted from the intrinsic laws of the mind (attending to its actions on the occasion of experience), and so as acquired" [6][11].

A recent and largely successful attempt of a quantitative theory of knowledge are formal languages in computer science, ranging from epsilon machines[12] and Bayesian networks to object oriented computer languages such as JAVA[13]. While expert systems have become a standard ingredient of manufacturing tools and consumer electronics, very little attempt has been made to assess and compare the amount of knowledge of these systems. The measure that comes closet to a measure of knowledge is Shannon's entropy[14], or average information content of a message. In information theory, the actual meaning of the message is unimportant. Instead the moment of surprise of the message, and the quality of transmission are measured.

Assessment of knowledge is a standard issue in education. National and International standardized test attempt to assess the knowledge of the students. Concept maps were developed by Ausubel [15], Novak and Gowin[16] to measure the change of the structure and organization of an individual's knowledge during the learning process (Anderson[17], Bruer [18], Stoddart [19]) in many areas [20] but little is known about the statistical and topological properties, in particular individual reproducibility and the predictive power of the test results. At this point, the philosophy of knowledge, the computer science approach to knowledge, and the assessment of knowledge in education appear to be quite disconnected.

In this paper, we try to bridge the gaps. We introduce several quantities to measure the amount of knowledge of an agent. One of these measures emphasis the value of the most abstract knowledge in line with Plato and St. Augustine, another measures emphasis the value of less abstract and a priori knowledge in line with Locke. From computer science, we adapt the definition of an agent, as an information processing system, including students, AI computer programs, and electronic courseware[21]. We study the scaling of these measures as a function of the prerequisite knowledge.

We classify knowledge in levels of abstraction, similar to Plato's classification, except that we use such levels only for classifying knowledge within a given discipline. A discipline is a given field of interest or a set of tasks with common properties, such as Algebra or Physics. We label knowledge as common knowledge if all of the agents know it. We don't enter into the discussion about a priori knowledge and derived common knowledge, therefore adapting Kant's view. However, we introduce a special name for knowledge that is required to describe the concepts of a given discipline, but is neither common knowledge, nor part of that discipline and call it prerequisite knowledge. We distinguish between common knowledge concepts and prerequisite concepts. Discipline specific concepts are for solving typical tasks within a discipline, whereas, prerequisite concepts solve other tasks, often more basic tasks. The collection of discipline specific concepts and relations, is our definition of knowledge. More specific, the collection of least abstract discipline specific knowledge we call know-how or practical knowledge[23], whereas the collection of most abstract discipline specific knowledge is labeled as wisdom or expert knowledge.

We assume that in each level of abstraction, knowledge is structured in terms of concepts, where each concept is referenced by an identifier, such as a name or a symbol and contains (i) an objective (ii) a definition complemented by (iii) a collection of "like-this"-examples [22] and "hands-on examples" [1], (iv) a collection of applications, and (v) a list of related concepts. We assume that each component, is given as one or several "trains of reasoning" [1]. A "train of reasoning" is a sequence of sentences in a spoken or a formal language that include cartoon-style illustrations or animations, which are comprehendible by the agent.

The objective specifies a task such as "this concept establishes a relation between force, mass and acceleration". This task is solved in the definition. The definition of an abstract concept, is typically a short sequence of sentences and illustrations that define a quantity, such as "density" or describe a relationship between concepts. For example, Newton's law describes a relation between mass, force, and acceleration.

The examples in a concept are problem-solution pairs, where the solution is a derived from the definition of the concept. The applications in a concept are problem-solution pairs which require repeated use of the concept or illustrate relations to other disciplines. Some of the applications are typically hands-on [1]. Applications relate the concept to common knowledge [24]. In contrast to the methods in the applications, the definition leaves the sequence of sentences and illustrations ambiguous wherever permissible and employs discipline specific concepts that are as abstract as possible. This briefness and high level of abstraction can make the definition appear ambiguous. We feel that it is important to employ the use of examples because they are less abstract and resolve some of that ambiguity.

An agent knows a concept, if it can tell the identifiers, reproduce the definition, give examples, applications, solve problems which are very similar to the given examples and applications, and can list related concepts. This implies that the agent knows a concept, if it knows all sub-concepts of the concept and can name, but does not necessarily know related concepts. Sub-concepts are those concepts, which are used in the definition, examples, and applications. In this paper, we study how the unknown prerequisite concepts limit knowledge. Knowing a concept is a prerequisite for understanding.[25] If the agent is able to give it's own interpretations and translations of the concept, the agent has an understanding of the concept.

If two concepts have a relation, i.e. similarities in their facts, methods, explanation, or typical application, they are called related. The concepts and their relations form a hierarchical conceptual network. If two concepts have a strong relation, this relation can be a concept by itself, an abstract concept[26].

2 A simple model of conceptual networks

We study conceptual networks for a given field of interest or discipline. A conceptual network is a cross-referenced library of concepts $C_i = 1 \ldots N[27]$. We extracted kinematics concepts from the highlighted equations in Serway and Faughn, College Physics[28]. Each concept C_i solves a typical task. The agent is described with a set of indicators, $c_i = 1, 2, 3, \ldots, N$, and $p_j, j = 1, 2, 3, \ldots, N_p$. c_i is set equal to 1 if the agent knows all the prerequisite concepts which are used in the description of the concept, otherwise it is set equal to zero. p_j is set equal to 1 if the agent knows the prerequisite concept P_j perfectly, otherwise it is set equal to zero $p_j = 0$, where $j = 1, 2, 3, \ldots$. $p_{i,D}$ is set equal to 1 if the agent knows the description of the concept, otherwise it is set equal to zero, $p_{i,D} = 0$. Then c_i, the knowledge of the agent, is defined as:

$$K = \sum_{i=1}^{N} w_i c_i, \tag{1}$$

where w_i is a measure for the importance or weight of each concept. If $< c_i >$ is the probability that an agent knows the concept C_i then the expectation value for the knowledge is:

$$< K >= \sum_{i=1}^{N} w_i < c_i > . \tag{2}$$

We assume that the agent knows common knowledge, whereas it knows prerequisite concepts $P_j, j = 1, 2, \ldots, N_p$ only with probability p.

A concept is called a base concept if it contains only prerequisite concepts and common knowledge. The number of base concepts is N_1. Abstract concepts of level $L = 2$ contain at least one base concept in the same discipline. If the base concept is replaced by it's own description in a level 2 concept, then this is called a substitution. For example, in a physics mechanics class, position $x(t)$ is a prerequisite concept, displacement $dx = x(t + dt) - x(t)$ is a base concept and $v(t) = dx/dt$ is a level 1 concept. If dx is replaced by $x(t + dt) - x(t)$

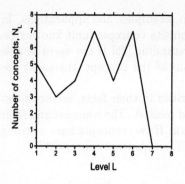 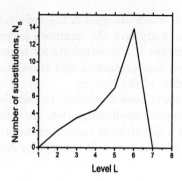

Figure 1: The number of concepts N_L as a function of the level of abstraction L in kinematics (left). The number of substitutions required N_S for a full substitution of a level L concept in kinematics.

in $v(t) = dx/dt = (x(t + dt) - x(t))/dt$, the displacement is substituted by its description of the velocity.

If a concept contains an abstract concept of level L_i and possibly some lower levels, it is called an abstract concept of level $L_i + 1$. The number of concepts of abstraction level L within a given discipline is called N_L. We assume the N_L is constant for $L \leq \hat{L}$ and $N_L = 0$ else . Fig. 1 shows the N_L versus the level L for kinematics[28]. The mean value is $N_L=5$ and $\hat{L} = 6$.

If the lower level concepts are recursively substituted by their descriptions, then the concept is called "fully-substituted". N_s is the average number of sub-concepts required to describe a level L concept in terms of prerequisite concepts and common knowledge concepts. We assume that N_S increases exponentially with L:

$$N_S = \alpha * \lambda^L. \tag{3}$$

This assumption is supported by the data shown in Fig. 1. For kinematics we find $\alpha = 1$ and $\lambda = 1.6$.

When a concept is fully substituted, it's description contains a certain number of prerequisite concepts N_P and a certain number of common knowledge concepts N_C. Fig. 2 suggests that N_P increases linearly for small L, $N_P = 0.5+2*L$ and is constant $N_P = 6.5$ for $L \leq 3$. In this paper, we assume that N_P is constant. The number of common knowledge concepts increases linearly with the level of abstraction:

$$N_C = N_{C,0} + n_C * L. \tag{4}$$

For kinematics we find $N_{C,0} = 1$, and $n_C = 0.6$, hence $N_C = 1 + 0.6 * L$.

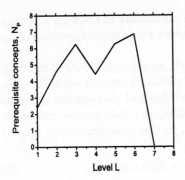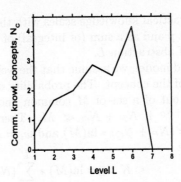

Figure 2: The number of prerequisite concepts in a fully-substituted concept N_P as a function of the level of abstraction L for kinematics (left). The number of common knowledge concepts in a fully-substituted concept N_C as a function of the level of abstraction L for kinematics.

The probability of knowing a concept with n sub-concepts is $p_{i_1}, p_{i_2}, p_{i_3}, \ldots, p_{i_n}$. For each agent the sequence of indicators $\{p_{i_1}, p_{i_2}, p_{i_3}, \ldots, p_{i_n}\}$ is a sequence of zeros and ones $\{0, 1, 1, 0, \ldots.0\}$ of length n. If we suppose that the indicators are statistically independent, then the probability that the agent knows a concept of level L is:

$$P_L = p^{N_P} * p_D^{N_s} = p^{N_P} * p_D^{\alpha * \lambda^L}, \tag{5}$$

where p is the probability of knowing a prerequisite concept, and p_D is the probability of knowing the description of a discipline-specific concept. In Eqn. 5, the first factor reflects the requirement that the agent knows all prerequisite concepts in the fully-substituted concept. The second term guarantees that the agent knows the descriptions of sub-concepts required for a full substitution. We define the maximum level of abstraction, which the agent can reach by the condition $N_L * P_L = 1$, where the agent has to know at least one level L concept. Solving this equation for L yields the maximum level of abstraction L_{max}.

$$L_{max} = \frac{\ln \frac{-\ln N_L * p^{N_P}}{a * \ln p_D}}{\ln \lambda}. \tag{6}$$

To determine the knowledge of an agent, one needs to model the importance of the concepts w_i. There are two simple models. In the first model we assume, that all concepts are equally important since each concept matches one task. Then $w_i = 1$, for all i. In this case, knowledge (practical knowledge) is :

$$< K >_P = \sum_{L=1}^{\hat{L}} N_L * P_L = p^{N_P} * \sum_{L=1}^{\hat{L}} p_D^{\alpha * \lambda^L}. \tag{7}$$

Hence, the practical knowledge scales with the probability of knowing a prerequisite concept p and as a sum (or integral) of hyper-exponential functions of p_D in the level of abstraction L.

In a second model, we assume that the weight w_i is proportional to information content of the concept. The probability of picking a concept with $N_P + N_C$ sub-concepts out of a set of M common knowledge and prerequisite concepts is $\rho = M^{N_P + N_C}$ if $N_P + N_C \ll M$. Therefore, the information content is $I = \ln(1/\rho) = (N_P + N_C) * \ln(M)$ and $w_i = I$. In this case, knowledge (expert knowledge) is:

$$< K >_E = \ln(M) * \sum_{L=1}^{\hat{L}} (N_P + N_C) * N_L * P_L \tag{8}$$

$$= \ln(M) * p^{N_P} \sum_{L=1}^{\hat{L}} (N_P + N_{C,0} + n_c * L) * N_L * p_D^{\alpha * \lambda^L}. \tag{9}$$

Hence, the expert knowledge scales with the probability of knowing a prerequisite concept p like the practical knowledge, but increases even faster than the practical knowledge as a function of p_D in the level of abstraction L.

3 Related Models

In the previous section, we assumed that the number of substitutions increases exponentially with the level of abstraction. In a linear graph, the number of substitions would increase linearly, such as $N_S = N_{S,0} * L$. In this case, the practical knowledge is:

$$< K >_P = \sum_{L=1}^{\hat{L}} N_L * P_L = p^{N_P} * \sum_{L=1}^{\hat{L}} p_D^{N_{S,0} * L} \approx p^{N_P} * p_D^{N_{S,0} * \hat{L} + 1}. \tag{10}$$

Consequently, the practical knowledge scales both with the prerequisite knowledge p and the knowledge of the field-specific concepts p_D.

In the previous section, we assumed that the number of prerequisite concepts N_P does not depend on the level of abstraction L. However, the experimental data indicate that at least for small L, N_P increases linearly in L, such as $N_P = N_{P,0} + n_P * L$. In this case, the practical knowledge increases rapidly as a function of p:

$$< K >_P = \sum_{L=1}^{\hat{L}} p^{N_{P,0} + n_P * L} * N_L * p_D^{\alpha * \lambda^L}. \tag{11}$$

Even if the experimental data for kinematics do not suggest this, one might assume that the number of abstract concepts N_L decreases with the level of abstraction L. We use $N_L = N_{L,0} * \delta^L$, an exponential model, where $0 < \lambda < 1$. In this case, the practical knowledge is:

$$< K >_P = \sum_{L=1}^{\hat{L}} N_{L,0} * \delta^L * p^{N_P} * p_D^{N_P}. \tag{12}$$

4 Discussion

If we assume the agent is a perfect learner, the probability of knowing a discipline-specific concept p_D is determined by the amount of material covered by a textbook or a course. The maximum level of abstraction has a logarithmic singularity at $p_D = 100\%$, see Eqn. 6. The means, that if the textbook or course is omitting just a very small fraction of the discipline concepts, this has a huge negative impact on the maximum level of abstration that the student or computer agent can achieve.

Knowledge scales only as a power-law of p, the probability of knowing knowing prerequisite concepts, see Eqn. 7 and Eqn. 8. This means that the knowledge does not decrease very rapidly, when p decreaes. Perhaps, this accommodates students who have learned a comparatively low level of prerequisite knowledge.

Currently, many printed textbooks are following a linear sequence of concepts through a conceptual network, where each new idea is deduced from the previous concept with very little reference to the branched topological structure of conceptual networks. In contrast, most electronic courseware makes the student aware of all the relations when a new concept is introduced; such as online C-computer langungage manuals, Windows NT Help, Yahoo Search Engine, Kaplan software and SAT English test preparation software. Courseware with intense cross-referencing exposes the student to concepts, which she or he may learn much later or never learn. In the past, there has been extensive discussion by Hamlyn[29] about the advantages and disadvantages of referring to unknown concepts for teaching. Cross-referencing is much easier in electronic courseware. In kinematics, 20 out of 25 concepts describe relations (80%). Considering that the largest part of knowledge is attributable to abstract knowledge, electronic courseware may help to improve learning environments since it supports relations.

The kinematics data suggest that substitutions required to substitute a concept of level L by prerequisite and common knowledge concepts increases exponential in L. If the number of subtitution would increase linearly, each concepts would be at a higher level of abstration. Representing concepts at a comparatively low level of abstraction, might be beneficial for some learners. In kinematics the higest level of abstration is $\hat{L} = 6$ (see Fig. 1). Humans can memorize 7 ± 2 chunks of information in short term memory[30] depending on their level of intelligence. Possibly, there is a connection between the level of abstraction in kinematics $\hat{L} = 6$ and human short term memory.

It is surprising, that both the number of concepts N_L at a given level of abstraction (see. Fig. 1) and the number of prerequisite concepts N_P required for a full-substitution of a concept (see Fig. 2) do not seem to depend on the level of abstraction L.

We give thanks to M. Osborne, E.A. Jackson, P. Melby, and D. Smyth for thoughtful discussion. This research was supported by the Office of Naval Research Grant No. N00014-96-1-0335.

References

[1] Hendley B.P. (ed.): Dewey, Russell, Whitehead, Philosophers as Educators, South. Illinois Univ. Press, Carbondale (1986) 16, 83.

[2] Peters, R.S. (ed.): The Philosophy of Education, Oxford Univ. Press, (1973).

[3] Grant, R.W., Tarcov N. (eds.): John Locke, Some Thoughts Concerning Education and Of the Conduct of the Understanding, Hackett Publishing Company, Inc., Indianapolis/Cambridge (1996).

[4] Osborne M.D.: Constructing Knowledge in the Elementary School Classroom: Teachers and Students, Falmer, New York (1999); Osborne M.D.: Teacher as Knower and Learner, Reflections on Situated Knowledge in Science Teaching, Journal of Research on Science Teaching **35/4** (1998) 427.

[5] Jowett B.: The Works of Plato (3rd edition), Oxford University Press, Oxford (1892) Theaetetus 152 c 5-7.

[6] Copleston F.: A History of Philosophy, **I** , Doubleday, NY (1993) 143, 200, 288.

[7] Jowett B.: The Works of Plato (3rd edition), Oxford University Press, Oxford (1892) Republic, 509 d 6 - 511 e 5.

[8] Ross Sir W.D.: Aristotle's Metaphysics, 2 vol, Oxford University Press, Oxford (1924) Metaphysics A, 980 a I.

[9] Quasten J. , Plumpe J.C. (eds.):Ancient Christian Writers: the Works of the Fathers in Translation, Westminster, Maryland (1946) De Beata Vita 2,10,and 14; 4; 27ff; De Trinitate, 15, 12, 21.

[10] Kant I.: Critique of Pure Reason, 1, translated by Meiklejohn J.M.D., 2.ed., Everyman's Library, London (1933).

[11] Kant I.:Gesammelte Schriften, Critical Edition, 22 vols., Prussian Academy of Sciences, Berlin (1902-42) On the Form and Principles 2, 8; W ., II, p. 395.

[12] Crutchfield J.P., Young, K.:Inferring statistical Complexity, Physical Review Letters **63** (1989) 105-108.

[13] Harold E.R.: JAVA Network Programming, O'Reilly, Cambridge (1997).

[14] Sloane N.J.A., Wyner A.D.(eds.): Claude Elwood Shannon: Collected Papers, IEEE Press, New York (1993).

[15] Ausubel D.: Ed. Psych.: A Cognitive View, Werbel and Pack, NY (1968).

[16] Novak J.D., Gowin D.B.: Learning how to Learn, Cambridge U. Press, (1984), Novak J.D.: Learning, Creating, and Using Knowledge, Lawrence Erlbaum Assoc., London (1998).

[17] Anderson, O.R.: Some Interrelationships between Constructivist Models of Learning and Current Neurobiological Theory, with Implications for Science Education, Journal of Reseach and Science Teaching **29/10** (1992) 1037-1058.

[18] Bruer, J. T.: Schools for Thought, MIT Press, Massachusetts (1993).

[19] Stoddart, T., Abrams R., Gasper, E., Canaday, D., Concept Maps as Assessment in Science Inquiry Learning - a Report of Methodology, International Journal of Science Education **22**, (2000) 1221-1246.

[20] Roth W.-M., Roychoudhury, A.: The Concept Map as a Tool for the Collaborative Construction of Knowledge: A Microanalysis of High School Physics Students, J Res. Sci. Teaching **30** (1993) 503-534; Stensvold, M.S., Wilson, J.T.: The Interaction of Verbal Ability with Concept Mapping in Learning from a Chemistry Laboratory Activity, Sci. Ed. **74** (1990) 473-480; Markow, P. G., Lonning, R.A.: Usefulness of Concept Maps in College Chemistry Laboratories, J. Res. Sci. Teaching **35** (1998) 1015-1029.

[21] Schwarz, E., Brusilovsky P., Weber G.: World-Wide Intelligent Textbooks in Ed. Telec. 1996, Patricia Carlson and Fillia Makedon eds., AACE, Charlottesville (1996) 302; Weber, G.: Individual Selection of Examples in an Intelligent Programming Environment. J Art. Int. Ed. **7(1)** (1996) 3-33; Weber, G.: Episodic Learner Modeling, Cog. Sci. **20**, (1996) 195-236.

[22] Wittgenstein, L: Philosophical Investigations, translated by Anscombe, G. E. M., Blackwell Publishers Ltd., Malden (1997) 490.

[23] Ryle, G.: The Concept of the Mind, Hutch. Univ. Lib., London (1949) 26-51.

[24] Whitehead, A. N.: An Introduction to Mathematics, Home University Library, London, (1911); rev. ed., Oxford University Press, New York (1958); Dewey, J.: The Relation of Theory to Practice in Education in The middle Works of John Dewey, Boydston, J. A. (ed.), Southern Illinois University Press, Carbondale (1977) **3** 249-272.

[25] Jackson E.A.: The Unbounded Vistas of Science: Evolutionary Limitations, Complexity **5** (2000) 35-44.

[26] Whitehead, A.N.: The Principles of Mathematics in Relation to Elementary Teaching, in Whitehead, A.N.: The Organization of Thought: Educational and Scientific, Williams and Norhgate, London (1917), reprint, Greenwood Press, Westport (1974), 101-2.

[27] Beland, A., Mislevy, R. J.: Probability-Based Inference in a Domain of Proportional Reasoning Tasks, Journal Educational Measurement **33** (1996) 3-27.

[28] Serway, R. A., Faughn, J. S.: College Physics, fourth ed., Saunders College Publishers, Fort Worth (1995).

[29] Hamlyn, D.Y.: Human Learning in The Philosophy of Education, by R.S. Peters ed. , Oxford University Press, Oxford (1973).

[30] Miller, G., 1956. The magic number seven, plus or minus two: some limits of our capacity for processing information. Psychological Review **63** (1956) 81-97.

Implementation of Kolmogorov Learning Algorithm for Feedforward Neural Networks

Roman Neruda, Arnošt Štědrý, and Jitka Drkošová*

Institute of Computer Science, Academy of Sciences of the Czech Republic,
P.O. Box 5, 18207 Prague, Czech Republic
roman@cs.cas.cz

Abstract. We present a learning algorithm for feedforward neural networks that is based on Kolmogorov theorem concerning composition of n-dimensional continuous function from one-dimensional continuous functions. A thorough analysis of the algorithm time complexity is presented together with serial and parallel implementation examples.

1 Introduction

In 1957 Kolmogorov [5] has proven a theorem stating that any continuous function of n variables can be exactly represented by superpositions and sums of continuous functions of only one variable. The first, who came with the idea to make use of this result in the neural networks area was Hecht-Nielsen [2]. Kůrková [3] has shown that it is possible to modify the original construction for the case of approximation of functions. Thus, one can use a perceptron network with two hidden layers containing a larger number of units with standard sigmoids to approximate any continuous function with arbitrary precision. In the meantime several stronger universal approximation results has appeared, such as [6] stating that perceptrons with one hidden layer and surprisingly general activation functions are universal approximators.

In the following we review the relevant results and show a learning algorithm based on Sprecher improved version of the proof of Kolmogorov's theorem. We focus on implementation details of the algorithm, in particular we proposed an optimal sequential version and studied its complexity. These results have also lead us to consider ways of suitable parallelization. So far, we have realized one parallel version of the algorithm running on a cluster of workstations.

2 Preliminaries

By $\mathcal{R}$ we denote the set of real numbers, $\mathcal{N}$ means the set of positive integers, $\mathcal{I} = [0, 1]$ and thus $\mathcal{I}^n$ is the n-dimensional unit cube. By $\mathcal{C}(\mathcal{I}^n)$ we mean a set of all continuous functions defined over $\mathcal{I}^n$.

* This research has been partially supported by GAASCR under grants B1030006 and A2030801, and by GACR under grant 201/99/0092.

Definition 1. *The sequence $\{\lambda_k\}$ is integrally independent if $\sum_p t_p \lambda_p \neq 0$ for any finite selection of integers t_p for which $\sum_p |t_p| \neq 0$*

Definition 2. *By* sigmoidal function *we mean any function $\sigma : \mathcal{R} \to \mathcal{I}$ with the following limits:* $\lim_{t \to -\infty} \sigma(t) = 0$ $\lim_{t \to \infty} \sigma(t) = 1$

Definition 3. *The set of* staircase-like functions *of a type σ is defined as: the set of all functions $f(x)$ of the form $f(x) = \sum_{i=1}^{k} a_i \sigma(b_i x + c_i)$, and denoted as $\mathcal{S}(\sigma)$.*

Definition 4. *A function $\omega_f : (0, \infty) \to \mathcal{R}$ is called a* modulus of continuity *of a function $f : \mathcal{I}^n \to \mathcal{R}$ if $\omega_f(\delta) = sup\{|f(x_1, \ldots, x_n) - f(y_1, \ldots, y_n)|; (x_1, \ldots, x_n), (y_1, \ldots, y_n) \in \mathcal{I}^n$ with $|x_p - y_p| < \delta$ for every $p = 1, \ldots, n\}$.*

In the following we always consider (neural) network to be a device computing certain function dependent on its parameters. Without the loss of generality we limit ourselves only to networks with n inputs with values taken from $\mathcal{I}$ and one real output. Thus we consider functions $f : \mathcal{I}^n \to \mathcal{R}$.

Moreover, we talk about two types of network architectures. One is the usual multilayer perceptron with two hidden layers. Perceptron units in each layer compute the usual affine combination of its inputs and weights (and bias) and then apply a sigmoidal activation function. An example can be the most common perceptron with logistic sigmoid. The (single) output unit computes just the linear combination.

The second type of network is a more general feedforward network where the units in different layers can compute different activation functions that can be more complicated than the logistic sigmoid. The description of a concrete form of such a network is subject to the section 4.

3 Kolmogorov theorem

The original Kolmogorov result shows that every continuous function defined on n-dimensional unit cube can be represented by superpositions and sums of one-dimensional continuous functions.

Theorem 1 (Kolmogorov). *For each integer $n \geq 2$ there are $n(2n + 1)$ continuous monotonically increasing functions ψ_{pq} with the following property: For every real-valued continuous function $f : \mathcal{I}^n \to \mathcal{R}$ there are continuous functions ϕ_q such that*

$$f(x_1, \ldots, x_n) = \sum_{q=0}^{2n} \phi_q \left[\sum_{p=1}^{n} \psi_{pq}(x_p) \right].$$ (1)

Further improvements by Sprecher provide a form that is more suitable for computational algorithm. Namely, the set of functions ψ_{pq} is replaced by shifts of a fixed function ψ which is moreover independent on a dimension. The overall quite complicated structure is further simplified by suitable parameterizations and making use of constants such as λ_p, β, etc.

Theorem 2 (Sprecher). *Let $\{\lambda_k\}$ be a sequence of positive integrally independent numbers. There exists a continuous monotonically increasing function $\psi : \mathcal{R}^+ \to \mathcal{R}^+$ having the following property: For every real-valued continuous function $f : \mathcal{I}^n \to \mathcal{R}$ with $n \geq 2$ there are continuous functions ϕ and a constant β such that:*

$$\xi(\mathbf{x}_q) = \sum_{p=1}^{n} \lambda_p \psi(x_p + q\beta) \tag{2}$$

$$f(\mathbf{x}) = \sum_{q=0}^{2n} \phi_q \circ \xi(\mathbf{x}_q) \tag{3}$$

Another result important for computational realization is due to Kůrková who has shown that both inner and outer functions ψ and ϕ can be approximated by staircase-like functions (cf. Definition 3) with arbitrary precision. Therefore, standard perceptron networks with sigmoidal activation functions can, in principle, be used in this approach. The second theorem of hers provides the estimate of units needed for approximation w.r.t. the given precision and the modulus of continuity of the approximated function.

Theorem 3 (Kůrková). *Let $n \in \mathcal{N}$ with $n \geq 2$, $\sigma : \mathcal{R} \to \mathcal{I}$ be a sigmoidal function, $f \in C(\mathcal{I}^n)$, and ε be a positive real number. Then there exists $k \in \mathcal{N}$ and functions ϕ_i, $\psi_{pi} \in \mathcal{S}(\sigma)$ such that:*

$$\left| f(x_1, \ldots, x_n) - \sum_{i=1}^{k} \phi_i \left(\sum_{p=1}^{n} \psi_{pi} x_p \right) \right| \leq \varepsilon \tag{4}$$

for every $(x_1, \ldots, x_n) \in \mathcal{I}^n$.

Theorem 4 (Kůrková). *Let $n \in \mathcal{N}$ with $n \geq 2$, $\sigma : \mathcal{R} \to \mathcal{I}$ be a sigmoidal function, $f \in C(\mathcal{I}^n)$, and ε be a positive real number. Then for every $m \in \mathcal{N}$ such that $m \geq 2n+1$ and $n/(m-n)+v < \varepsilon/\|f\|$ and $\omega_f(1/m) < v(m-n)/(2m-3n)$ for some positive real v, f can be approximated with an accuracy ε by a perceptron type network with two hidden layers, containing $nm(m + 1)$ units in the first hidden layer and $m^2(m+1)^n$ units in the second one, with an activation function σ.*

4 Algorithm proposal

Sprecher sketched an algorithm based on Theorem 2 that also takes into account Theorem 4 by Kůrková. Here we present our modified and improved version that addresses crucial computational issues.

The core of the algorithm consists of four steps:

For each iteration r:

1. Construct the mesh $\mathcal{Q}^n$ of rational points $\mathbf{d}_k$ dissecting the unit cube (cf. (5)).

2. Construct the functions ξ in the points $\mathbf{d}_k^q$ (see (6) and (11)).
3. Create sigmoidal steps θ described in (10).
4. Compile the outer functions ϕ_q^j (see (9)) based on θ and previous approximation error e_r.
5. Construct the r-th approximation f_r of original function f according to (14), and the r-th approximation error e_r according to (13).

4.1 The support set $\mathcal{Q}$

Take integers $m \geq 2n$ and $\gamma \geq m+2$ where n is the input dimension. Consider a set of rational numbers $\mathcal{Q} = \left\{ d_k = \sum_{s=1}^k i_s \gamma^{-s}; i_s \in \{0, 1, \ldots, \gamma - 1\}, k \in \mathcal{N} \right\}$. Elements of $\mathcal{Q}$ are used as coordinates of n-dimensional mesh

$$\mathcal{Q}^n = \{\mathbf{d}_k = (d_{k1}, \ldots, d_{kn}); d_{kj} \in \mathcal{Q}, j = 1, \ldots, n\}. \tag{5}$$

Note that the number k determines the precision of the dissection.

For $q = 0, 1, \ldots, m$ we construct numbers $\mathbf{d}_k^q \in \mathcal{Q}^n$ whose coordinates are determined by the expression

$$d_{kp}^q = d_{kp} + q \sum_{s=1}^k \gamma^{-s} \tag{6}$$

Obviously $d_{kp}^q \in \mathcal{Q}$ for $p = 1, 2, \ldots, n$. We will make use of $\mathbf{d}_k^q$ in the definition of functions ξ.

4.2 The inner function ψ

The function $\psi : \mathcal{Q} \to \mathcal{I}$ is then defined with the help of several additional definitions. For the convenience, we follow [11] in our notation.

$$\psi(d_k) = \sum_{s=1}^k \tilde{i}_s 2^{-m_s} \gamma^{-\rho(s-m_s)}, \tag{7}$$

$$\rho(z) = \frac{n^z - 1}{n - 1},$$

$$m_s = \langle i_s \rangle \left(1 + \sum_{l=1}^{s-1} [i_l] \cdot \ldots \cdot [i_{s-1}] \right), \tag{8}$$

$$\tilde{i}_s = i_s - (\gamma - 2)\langle i_s \rangle.$$

Let $[i_1] = \langle i_1 \rangle = 1$ and for $s \geq 2$ let $[i_s]$ and $\langle i_s \rangle$ be defined as:

$$[i_s] = \begin{cases} 0 \text{ for } i_s = 0, 1, \ldots, \gamma - 3 \\ 1 \text{ for } i_s = \gamma - 2, \gamma - 1 \end{cases}$$

$$\langle i_s \rangle = \begin{cases} 0 \text{ for } i_s = 0, 1, \ldots, \gamma - 2 \\ 1 \text{ for } i_s = \gamma - 1 \end{cases}$$

Figure 1 illustrates the values of ψ for $k = 1, 2, 3, 4; n = 2; \gamma = 6$.

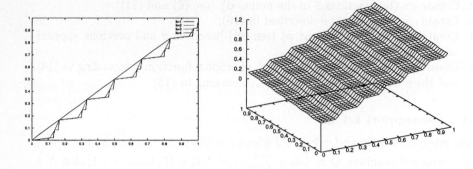

Fig. 1. a) Values of $\psi(d_k)$ for various k. b) Values of $\xi(\mathbf{d}_k)$ for $k = 2$.

4.3 The outer functions ϕ

The functions ϕ_q in equation (3) are constructed iteratively as functions $\phi_q(y_q) = \lim_{r \to \infty} \sum_{j=1}^{r} \phi_q^j(y_q)$. Each function $\phi_q^j(y_q)$ is determined by e_{j-1} at points from the set $\mathcal{Q}^n$. The construction is described in the following.

For $q = 0, 1, \ldots, m$ and $j = 1, \ldots, r$ we compute:

$$\phi_q^j \circ \xi(\mathbf{x}_q) = \frac{1}{m+1} \sum_{\mathbf{d}_k^q} e_{j-1}(\mathbf{d}_k) \, \theta(\mathbf{d}_k^q; \xi(\mathbf{x}_q)), \tag{9}$$

where $\mathbf{d}_k \in \mathcal{Q}$.

The real-valued function $\theta(\mathbf{d}_k; \xi(\mathbf{x}_q))$ defined for a fixed point $\mathbf{d}_k^q \in \mathcal{Q}^n$. The definition is based on a given sigmoidal function σ:

$$\theta(\mathbf{d}_k^q; y_q) = \sigma(\gamma^{\beta(k+1)}(y_q - \xi(\mathbf{d}_k^q)) + 1) \tag{10}$$
$$- \sigma(\gamma^{\beta(k+1)}(y_q - \xi(\mathbf{d}_k^q) - (\gamma - 2)b_k)$$

where $y_q \in \mathcal{R}$, and b_k is a real number defined as follows:

$$b_k = \sum_{s=k+1}^{\infty} \gamma^{-\rho(s)} \sum_{p=1}^{n} \lambda_p.$$

The functions $\xi(\mathbf{d}_k^q)$ are expressed by equation

$$\xi(\mathbf{d}_k^q) = \sum_{p=1}^{n} \lambda_p \psi(d_{kp}^q) \tag{11}$$

where $\psi(d_{kp}^q)$ are from (7), and coefficients λ_p are defined as follows.

Let $\lambda_1 = 1$ and for $p > 1$ let

$$\lambda_p = \sum_{s=1}^{\infty} \gamma^{-(p-1)\rho(s)} \tag{12}$$

Figure 1 shows values of $\xi(\mathbf{d}_k)$ for $k = 2$.

4.4 Iteration step

Let f is a known continuous function, $e_0 \equiv f$. The r-th approximation error function e_r to f is computed iteratively for $r = 1, 2, \ldots$

$$e_r(\mathbf{x}) = e_{r-1}(\mathbf{x}) - \sum_{q=0}^{m} \phi_q^r \circ \xi(\mathbf{x}_q), \tag{13}$$

where $\mathbf{x} \in \mathcal{I}^n$, $\mathbf{x}_q = (x_1 + q\beta, \ldots, x_n + q\beta)$, and $\beta = \gamma(\gamma - 1)^{-1}$.

The r-th approximation f_r to f is then given by:

$$f_r(\mathbf{x}) = \sum_{j=1}^{r} \sum_{q=0}^{m} \phi_q^j \circ \xi(\mathbf{x}_q). \tag{14}$$

It was shown in [11] that $f_r \to f$ for $r \to \infty$.

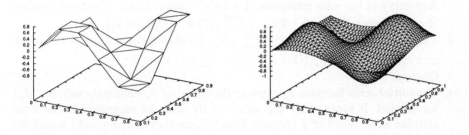

Fig. 2. Approximation of $\sin(6.28x) \cdot \sin(6.28y)$ for $k = 1$ and $k = 2$.

5 Time complexity

In the following time complexity analysis we consider our practical implementation of the above described algorithm, which introduces few further simplifications. First, the infinite sums are replaced by sums up to the sufficiently big constant K. We also keep the number $k(r)$ fixed to constant k during the iteration loop. Also note that all computations of f are performed on numbers taken from the dense support set Q.

The time complexity is derived with respect to the following basic operations: For time of the floating point multiplication and division the symbol t_m is used, while for the time of floating point addition and subtraction t_a is used.

In general, we first provide fine-grain analysis in the following lemmas, which is then summarized in the theorem 5 as an estimation of the total amount of computational time for one iteration step.

The following lemma summarizes the time needed for computing several terms that are independent of the iteration loop and can be performed in advance.

Lemma 1. *The times* $T_\lambda, T_b, T_\gamma, T_{e_0}$ *needed for computations of* $\lambda_p, b_k, \gamma^{\rho(k+1)}$, e_0 *are expressed by the following formulae:*

1. $T_\lambda = (2 + K)t_a + [p + 2 + 2K + (n^K - 1)/(n - 1)]t_m,$
2. $T_b = n \times T_\lambda + [2(K - k) + (n^K - n^k)/(n - 1)]t_m + (K - k)t_a,$
3. $T_\gamma = (k + 1 + n^k)t_m,$
4. $T_{e_0} = \gamma^{nk} \times (2kt_m + (k - 1)t_a).$

Proof. It is easy to derive these results by examining the corresponding definitions.

1. We compute $\lambda_p = \sum_{s=1}^\infty \gamma^{-(p-1)\rho(s)}$. The value γ^{1-p} is computed and saved. If we suppose that $\rho(s)$ is computed using $\rho(s - 1)$ then it costs $2t_m + 1t_a$. Any entry in the sum consumes $(1 + (\rho(s) - \rho(s - 1))t_m$. The total number of operations is $2t_a + (p + 2)t_m + \sum_{s=1}^K [(2 + n^{s-1})t_m + t_a].$
2. The term $b_k = \sum_{s=k+1}^\infty \gamma^{-\rho(s)} \sum_{p=1}^n \lambda_p$. To obtain the entries of the outer sum consumes $\sum_{s=k+1}^K [(2 + n^{s-1})t_m + t_a]$ and the time for λ_p.
3. Obvious.
4. The initial error function e_0 is set to the function f, while only values $f(\mathbf{d}_k)$ are needed. It means that the values of $\mathbf{d}_k$ must be expressed in the form suitable as an input of f (decadic form). Since they are originally stored via their ordinal values, we need $(2kt_m + (k - 1)t_a)$ computations for any $\mathbf{d}_k$.

In order to compute time T_S of one (serial) iteration of the algorithm, we need to express times for partial steps that are thoroughly described in section 4. The most inner terms in the computation are numbers $\mathbf{d}_k^q$ that are copmuted by means of their coordinates $d_{kp}^q = d_{kp} + q \sum_{s=1}^k \gamma^{-s}$.

Lemma 2. *Time* $T_{\mathbf{d}_k^q}$ *for computing* $\mathbf{d}_k^q$ *is* $T_{\mathbf{d}_k^q} = (k + n)t_a + (k + 1)t_m$.

Proof. To compute the sum costs $kt_a + kt_m$. Then, $t_m + nt_a$ operations are needed to complete the computation.

Quite complicated algorithm for computing Ψ is described in 4.2. Its complexity is estimated as follows.

Lemma 3. *For any coordinate* d_{kp} *the function* Ψ *requires* $T_\Psi = (1/6k^3 + 3k^2 + 10k + 1)t_a + [(n^k - 1)/(n - 1) + (k^2 + k)/2 + 1]t_m$.

Proof. The definition of m_s (cf. 8) plays the key role in computation of $\Psi(d_{kp}^q)$. The upper bound for the expression of this value is $1/2(s^2 + 3s + 4)t_a + st_a$. To simplify the estimation we consider the worst case $s - m_s = s$ as an input into $\rho(s - m_s)$. This computation then requires $2t_a + st_m$. To estimate the time consumed to express the function $\Psi(d_{kp})$ means to add the entries $1/2(s^2 + 5s + 14)t_a + (s + (n^s - 1)/(n - 1))t_m$ together.

The time T_ξ needed for function $\xi(d_k^q)$ is straightforwardly computed by means of the three above mentioned quantities.

Lemma 4. *Function* $\xi(d_k^q) = \sum_{p=1}^n \lambda_p \Psi(d_{kp}^q)$ *consumes for any* d_k^q *the amount* $T_\xi = n \times (T_\lambda + T_\Psi + T_{d_k^q})$

Next we analyze the time T_θ necessary to compute θ.

Lemma 5. *To evaluate the function* θ *costs* $T_\theta = [4K - k + 6 + n^k + (2n^K - n^k)/(n - 1) + n]t_m + [2K - k + 7]t_a + T_\sigma$.

Proof. $\theta(d_k^q, y_q) = \sigma(\gamma^{\rho(k+1)}(y_q - \xi(d_k^q)) + 1) - \sigma(\gamma^{\rho(k+1)}(y_q - \xi(d_k^q)) - (\gamma - 2)b_k))$ and according to our assumptions $y_q = \xi(d_k^q)$. The time T_θ is then the sum $T_\gamma + T_b + T_\sigma + 5t_a + 3t_m$. Using lemma 1 we directly obtain the result.

Lemma 6. *Time needed to compute the value of the outer function* Φ *for one particular* d_k^q *is the following.* $T_\Phi = [n^2 + n^k + 3n + 7 - k + (2n + 4)K + (n^K(n + 2) - n^k - n)/(n - 1)]t_m + [(2 + n)K + 2n - k + 7]t_a + T_\Psi + T_{d_k^q} + T_\sigma$

Proof. $T_\Phi = T_\xi + T_\theta$. Separate the operations that are computed just once and do not depend on the iteration loop and use lemmas 4 and 5, which gives the estimation.

Lemma 7. *The time for evaluation of the error function* e_r *in one particular* d_k *is* $T_e = (m + 1)t_a$.

Proof. If we assume that the values of $e_{r-1}(d_k)$ are saved and Φ_q^r have been already computed, then the computation of $e_r(d_k)$ costs $(m + 1)t_a$.

Lemma 8. *To generate the* $r - th$ *approximation to the function* f *in one particular value consumes* $T_f = mt_a$.

Proof. The values of f_r are recurrently computed by means previous iteration f_{r-1} and already computed Φ_q^r, thus only m additions is needed.

Our analysis of the serial case is summarized in the following theorem.

Theorem 5. *The computational requirements in one iteration of the sequential algorithm is estimated as follows.*

$$T_S = \mathcal{O}\left(m\, n\, \gamma^{nk}(n^k + k^3)\right). \tag{15}$$

Proof. Express T_S according to the algorithm description:

$$T_S = \gamma^{nk} \left[mT_\Phi + T_e + T_f \right].$$

The partial terms used have been specified by the preceeding lemmas. We do not take into account terms that can be pre-computed (cf. Lemma 1). Also, a particular sigmoidal function — step sigmoid — is considered, which can easily be computed by means of two comparisons. We have sacrificed the tightness of the bound to tractability of the result, and consider $t_a = t_m$ (cf. Discussion).

$$T_S \approx m \, n \, \gamma^{nk} \left[k^3 t_a + (n^k + k^2) t_m \right],$$

which proves (15).

6 Discussion

The overall time of the serial algorithm consists of initial computations of the constant terms which take $T_\lambda + T_b + T_\gamma + T_{e_0} + T_\lambda$ and $r \times T_S$. The assumption that the additive and multiplicative operations take approximately the same time is not a strong simplification if one considers current microporcessors such as Pentiums. Note also that the complexity analysis considers quite optimal sequential realization of the algorithm described in section 4. In particular, some formulae have been reformulated for speedup. Also, the mesh Q is represented in a way which allows efficient access, and we cache once computed values wherever possible.

From the point of view of parallel implementation, there is one straightforward approach employing $m + 1$ processors for computations of $\Phi_0^r, \ldots, \Phi_m^r$, which are mutually independent. This can reduce the T_S by a factor of m on one hand, but it requires additional communication to complete each iteration by exchanging the computed values. Although a finer analysis of this aproach has not been performed yet, our first experiments show that the communication requirements are tiny compared to the total computation time, especially for problems of higher dimension. Our parallel implementation has been done on a cluster of workstations in the PVM environment. Figure 3 shows a typical behaviour of the program running on six machines for two iterations of the algorithm.

In the future work we plan to focus on the exact analysis of the parallel implementation, and on employing other means of parallelization, such as clever partitioning of the input space.

References

1. F. Girosi and T. Poggio. Representation properties of networks: Kolmogorov's theorem is irrelevant. *Neural Computation*, 1:461–465, 1989.
2. Robert Hecht-Nielsen. Kolmogorov's mapping neural network existence theorem. In *Procceedings of the International Conference on Neural Networks*, pages 11–14, New York, 1987. IEEE Press.

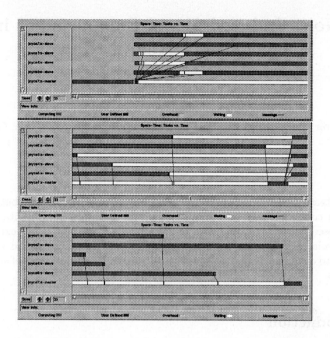

Fig. 3. Illustration of a typical parallel behaviour of the algorithm running on cluster of Pentium workstations under PVM.

3. Věra Kůrková. Kolmogorov's theorem is relevant. *Neural Computation*, 3, 1991.
4. Věra Kůrková. Kolmogorov's theorem and multilayer neural networks. *Neural Networks*, 5:501–506, 1992.
5. A. N. Kolmogorov. On the representation of continuous function of many variables by superpositions of continuous functions of one variable and addition. *Doklady Akademii Nauk USSR*, 114(5):953–956, 1957.
6. M. Leshno, V. Lin, A. Pinkus, and S. Shocken. Multilayer feedforward networks with a non-polynomial activation function can approximate any function. *Neural Networks*, (6):861–867, 1993.
7. G.G Lorentz. *Approximation of functions*. Halt, Reinhart and Winston, New York, 1966.
8. David A. Sprecher. On the structure of continuous functions of several variables. *Transactions of the American Mathematical Society*, 115:340–355, 1965.
9. David A. Sprecher. A universal mapping for Kolmogorov's superposition theorem. *Neural Networks*, 6:1089–1094, 1993.
10. David A. Sprecher. A numerical construction of a universal function for Kolmogorov's superpositions. *Neural Network World*, 7(4):711–718, 1996.
11. David A. Sprecher. A numerical implementation of Kolmogorov's superpositions II. *Neural Networks*, 10(3):447–457, 1997.

Noise-Induced Signal Enhancement in Heterogeneous Neural Networks

Michael J. Barber and Babette K. Dellen

Institut für Theoretische Physik, Universität zu Köln, D-50937 Köln, Germany
mjb@thp.uni-koeln.de   bd@thp.uni-koeln.de

Abstract. Neural networks can represent complex functions, but are often constructed of very simple units. We investigate the limitations imposed by such a simple unit, the McCulloch-Pitts neuron. We explore the role of stochastic resonance in units of finite precision and show how to construct neural networks that overcome the limitations of single units.

1 Introduction

Neural networks are often constructed of very simple model neurons. Typically, model neurons have a particular threshold or bias, and saturate to a fixed value for either strong or weak inputs, yielding a so-called sigmoidal activation or "squashing" function. While individual units are simple, it is well known that neural networks can represent complex functions. In this work, we will investigate the degree to which the simple dynamics of an individual unit limit the inputs that can be represented, and how these limitations can be overcome.

To do this, we consider the response of model neurons to a variety of signals. The model neurons have limited precision, which we implement by including noise in the systems. We consider noise that is intrinsic to the neurons, and thus has identical statistics for each of the units in the neural networks.

Noise is usually viewed as limiting the sensitivity of a system, but nonlinear systems can show an optimal response to weak or subthreshold signals when a non-zero level of noise is added to the system. This phenomenon is called stochastic resonance (SR) [2]. SR is seen in many systems, ranging from resonant cavities to neural networks to the onset of ice ages [5]. For example, a noise-free, subthreshold neuronal input can occasionally become suprathreshold when noise is added, allowing some character of the input signal to be detected.

Collins et al.[1] showed, in a summing network of identical Fitzhugh-Nagumo model neurons, that an emergent property of SR in multi-component systems is that the enhancement of the response becomes independent of the exact value of the noise variance. This allows networks of elements with finite precision to take advantage of SR for diverse inputs. To build upon the findings of Collins et al., we consider networks of simpler model neurons, but these are allowed to have different dynamics. In particular, we examine noisy McCulloch-Pitts (McP) neurons [3] with a distribution of thresholds. We construct heterogeneous networks that perform better than a homogeneous network with the same number of noisy McP neurons and similar network architecture.

2 Results

To investigate the effect of noise on signal transduction in networks of thresholding units, we constructed a network of noisy McCulloch-Pitts neurons. A McP neuron is a simple thresholding unit, which can be expressed mathematically as a step function. When the total input to a neuron (signal plus noise) exceeds its threshold, the neuron activates, firing an action potential or "spike." The presence of a non-zero level of noise can make the neuron more sensitive to weak signals; for example, a subthreshold signal can occasionally be driven above the neuronal threshold by the noise. The intrinsic neuronal noise is modeled in this work as independent, identically distributed Gaussian white noise that is added to the input signal. We use the standard deviation of the Gaussian as a measure of the noise strength.

The network architecture is very simple: an input layer of N noisy McP neurons is connected to a single linear output neuron. Each synaptic weight is of identical strength, so the output neuron calculates the mean value of the input neuron activation states. Each input unit is presented the same analog signal, but with a different realization of the intrinsic neuronal noise. The uncorrelated noises increase only by a factor of $\sqrt{N}$ on average across the population of neurons, while the coherent input signal is strengthened by a factor of N. This yields an increased signal-to-noise ratio and a more sensitive neural network.

We reconstruct the input signal by convolving the output neuron response with a linear filter [4]. The filter is generated to minimize a quadratic difference between the reconstructed signal and the input signal; the resulting filter for McP neurons is a low-pass filter. We quantitatively compare the reconstruction with the original signal, and make use of the standard correlation coefficient r^2 as a direct measure of the similarity between signal and reconstruction.

A single McP neuron shows a characteristic stochastic resonance profile in response to weak signals (generated here as a random walk with zero mean and a typical variance of one) with different noise intensities (Fig. 1a, green circles). As the noise increases, there is an initially enhanced network response, which then drops off as the noise further increases, overwhelming the signal. However, networks of McP neurons, all of which have identical thresholds, have a broad plateau of noise-enhanced responses (Fig. 1a, blue squares and red triangles), as well as an overall enhanced response to a signal. This is similar to the findings of Collins et al.[1].

The behavior seen in Fig. 1a appears to indicate that noise is a good thing to have in neurons, but it was calculated under the assumption of weak signals. To expand upon this assumption, we consider the ability of networks of McP neurons to reconstruct signals of different strength. In this instance, "strength" can be understood as the difference between the mean value of the input signal

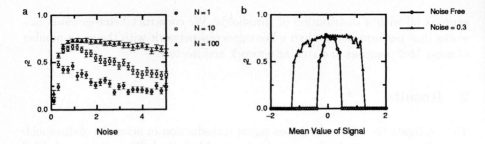

Fig. 1. Stochastic resonance is seen in networks of McCulloch-Pitts neurons. (a) With an increasing number N of neurons, the stochastic resonance peak broadens, losing the strong dependence on a particular noise level. The network responses shown are the average result for 50 zero-mean, random walk inputs; each signal is weak and normally subthreshold, with the neuronal thresholds equal to one. (b) The presence of noise widens the range over which signals are accurately detected and reconstructed. Here, we show a network of 100 neurons with thresholds of zero.

and the value of the neuronal thresholds, so we vary the signal mean while keeping the thresholds and the noise intensity fixed. In this manner, we see that the presence of noise broadens the range of signal strengths that the network can detect (Fig. 1b). This is the mirror image of the weak-signal stochastic resonance effect: strong input signals may be supra-threshold at all times, so that the McP neuron fires at each sampling time, but the presence of noise can drive the total signal below the threshold and improve the network sensitivity.

The network sensitivity increases as the number of input neurons increases (Fig. 2a). For large numbers of neurons, adding more units widens the range of signal detection, but does not significantly improve the peak reconstruction performance of the network.

The networks shown in Fig. 2a indicate that there is only a minimal widening of the range of signal detection for a large increase in the number of neurons. However, these networks were homogeneous, with each neuron having an identical threshold. A natural extension is to consider heterogeneous systems, where the thresholds are not uniform. We divide the neurons evenly into two subpopulations of different thresholds. The neurons in each subpopulation are identical, and the subpopulations have identical noise intensities, differing only in the thresholds. Again applying the signal reconstruction procedure, we see that the signal detection range can be more efficiently widened in this way than with a single homogeneous population of neurons (Fig. 2b). The use of multiple thresholds increases the range of signal detection without any significant loss in the quality of signal detection.

3 Conclusion

We have constructed a network of heterogeneous McP neurons that outperforms similar networks of homogeneous McP neurons. The network architectures are

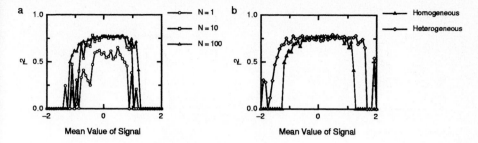

Fig. 2. Comparison of homogeneous and heterogeneous networks. (a) The range of effective signal detection and reconstruction increases gradually as the number of neurons increases. Every neuron in each of the networks shown here has an identical threshold, equal to zero. (b) Segregating the neurons into two groups with different thresholds leads to a greater range of effective signal detection and reconstruction than an increase in the number of neurons with uniform thresholds. The heterogeneous network has 10 neurons with a threshold of -0.65 and 10 neurons with a threshold of $+0.65$. The homogeneous network is identical to the 100 neuron network shown in (a).

identical, with the only difference being the distribution of neuronal thresholds. The heterogeneous system is sensitive to a wider range of signals than the homogeneous systems. Such networks of heterogeneous units are easily implemented, and could serve as simple models of many diverse natural and artificial systems.

A network of low-precision neurons (or other units) can take advantage of stochastic resonance to accurately encode and transmit an input. The collective properties of these systems can exceed the limitations of a single element. A careful consideration of properties of the elements, such as their thresholds, may yield even further improvements in the performance of the system.

Acknowledgements

We thank John Clark, Manfred Ristig, and Jürgen Hescheler for valuable discussion. This work was supported in part by Graduiertenkolleg Azentrische Kristalle GK549 of the DFG.

References

[1] J.J. Collins, C.C. Chow, and T.T. Imhoff. Stochastic resonance without tuning. *Nature*, 376:236–238, July 1995.
[2] L. Gammaitoni, P. Hänggi, P. Jung, and F. Marchesoni. Stochastic resonance. *Rev. Mod. Phys.*, 70(1):223–87, January 1998.
[3] J. Hertz, A. Krogh, and R.G. Palmer. *Introduction to the Theory of Neural Computation.* Addison-Wesley Publishing Company, Reading, MA, 1991.
[4] F. Rieke, D. Warland, R.R. de Ruyter van Steveninck, and W. Bialek. *Spikes: Exploring the Neural Code.* MIT Press, Cambridge, MA, 1997.
[5] K. Wiesenfeld and F. Moss. Stochastic resonance and the benefits of noise: From ice ages to crayfish and SQUIDs. *Nature*, 373:33–36, January 1995.

Phylogenetic Inference for Genome Rearrangement Data

Session chair:

Laura Salter (University of New Mexico, USA)

Evolutionary Puzzles:
An Introduction to Genome Rearrangement

Mathieu Blanchette

Department of Computer Science and Engineering
Box 352350
University of Washington
Seattle, WA 98195-2350 U.S.A.
206-543-5118
fax: 206-543-8331
blanchem@cs.washington.edu

Abstract. This paper is intended to serve as an introduction to genome rearrangement and its use for inferring phylogenetic trees. We begin with a brief description of the major players of the field (chromosomes, genes, etc.) and the types of mutations that can affect them, focussing obviously on genome rearrangement. This leads to a simple mathematical representation of the data (the order of the genes on the chromosomes), and the operations that modify it (inversions, transpositions, and translocations).

We then consider the problem of inferring phylogenetic (evolutionary) trees from genetic data. We briefly present the two major approaches to solve this problem. The first one, called distance matrix method, relies on the estimation of the evolutionary distance between each pair of species considered. In the context of gene order data, a useful measure of evolutionary distance is the minimum number of basic operations needed to transform the gene order of one species into that of another. This family of algorithmic problems has been extensively studied, and we review the major results in the field.

The second approach to inferring phylogenetic trees consists of finding a minimal Steiner tree in the space of the data considered, whose leaves are the species of interest. This approach leads to much harder algorithmic problems. The main results obtained here are based on a simple evolutionary metric, the number of breakpoints.

Throughout the paper, we report on various biological data analyses done using the different techniques discussed. We also point out some interesting open problems and current research directions.

1 Introduction

Understanding and classifying the incredible diversity of living beings has been the focus of much work, starting as early as in Ancient Greece. Since Darwin's thesis on evolution of species, we know that the diversity observed today is the result of a long process in which speciation (the event where two groups of organisms from one species slowly diverge until they form two different, though closely related, species) played a key role. The history of these speciation events can be represented by a phylogenetic tree, where each leaf is labeled with a contemporary species and where the internal nodes correspond to hypothetical speciation events. Figure 1 illustrates the phylogenetic tree relating a group of vertebrates.

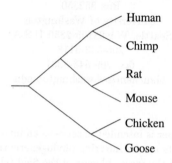

Fig. 1. The phylogenetic tree relating six species of vertebrates. Branch lengths are not to scale.

The problem of inferring the phylogenetic tree connecting a set of species is both of theoretical and practical importance. The problem has historically been addressed by considering morphologic, anatomic or developmental traits: species that have similar traits are likely to be closely related. However, for the last 30 years, most phylogenetic inference has been based on the DNA sequence of the species considered. The quantity of information contained in the genome of an organism is several orders of magnitude larger than that contained in its observable phenotype, and thus it potentially allows for a much more accurate inference. The DNA sequence of different organisms from different species differs because of mutations that occurred since their last common ancestor. It is by studying these mutations that one can hope to infer phylogenetic relationships.

This paper focuses on one specific type of mutations called genome rearrangement. In Section 2, we describe these mutations. In Section 3, we introduce the problem of inferring phylogenetic trees, and describe two classes of approaches that have been used.

2   Genome rearrangement

The genome of an organism consists of a long string of DNA, cut into a small number of segments called chromosomes. Genes are stretches of the DNA sequence that are responsible for encoding proteins. Each gene has an orientation, either forward or backward, depending in which direction it is supposed to be read. A chromosome can thus be abstracted as an ordered set of oriented genes. Most higher organisms' chromosomes are linear (their DNA sequence has a beginning and an end), but for lower organisms like bacteria, the chromosome is circular (their DNA sequence has no beginning or end).

The most common and most studied mutations operating on DNA sequences are local: they affect only a very small stretch on DNA sequence. These mutations include nucleotide substitutions (where one nucleotide is substituted for another), as well as nucleotide insertions and deletions. Most phylogenetic studies have been based on these types of mutations.

Genome rearrangement is a different class of mutation affecting very large stretches of DNA sequence. A genome rearrangement occurs when a chromosome breaks at two or more locations (called the breakpoints), and the pieces are reassembled, but in the "wrong" order. This results in a DNA sequence that has essentially the same features as the original sequence, except that the order of these features has been modified.

If the chromosome breaks occur in non-functional sequence, the rearrangement is unlikely to have any deleterious effects. On the other hand, a rearrangement whose breakpoints fall in functional sequence (e.g. genes) will almost certainly make the gene dysfunctional, rendering the offspring unlikely to survive. Consequently, almost all genome rearrangements that become fixed in future generations involve inter-genic breakpoints.

Figure 2 illustrates the three most common types of genome rearrangements. The first two, inversions and transpositions, affect only one chromosome at a time. The result of an inversion is to reverse the DNA segment between the two breakpoints. The order of the genes on this segment is reversed and their orientation is inverted. A transposition involves three breakpoints: the DNA segment between the two first breakpoints is spliced out and re-inserted somewhere else on the chromosome. A translocation involves two different chromosomes that exchange their ends.

The net effect of any genome rearrangement is to modify the order of features on a chromosome. The most important such features are genes, and most of the research done in this field has involved looking at genome rearrangement from the point of view of their effect on the gene order.

It is worth noticing that, compared to local mutations, genome rearrangements are extremely rare. However, over time, they accumulate and the order of the genes on each chromosome becomes more and more scrambled with respect to the original sequence. Thus, two closely related species will usually have similar gene orders (i.e. few genome rearrangements occurred during the evolution that separates them), whereas the gene orders of two more distant species will usually be less conserved.

3   Inferring phylogenetic trees

Inferring the phylogenetic (evolutionary) relationships among a set of species is a problem of both pure scientific and practical interest. This relationship is

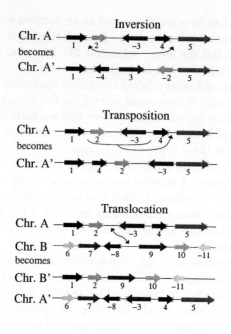

Fig. 2. Three most common types of genome rearrangements. A and B are chromosomes. Genes are numbered from 1 to 11.

usually depicted using a phylogenetic tree, whose leaves are labeled with the contemporary species under study, and whose internal structure indicates the order in which order these species diverged from each other. In some case, lengths will be associated with the branches of the tree. These lengths correspond to the amount of evolution that happened between the two endpoints of the branch.

By far the most common way to do phylogenetic inference is to study the evolution of DNA sequences from a local point of view. This approach has proven quite successful and has allowed us to infer the phylogenetic relationships among many species [14].

Gene arrangement can also be used to infer evolutionary relationships among species. Because genome rearrangements occur much more rarely than local mutations, it often allows to trace relationships between very distant species. Moreover, the fact that most genome rearrangements seem to have no effect at all on the fitness of the offspring makes our task easier. Consequently, one should see gene arrangement based studies as complimentary to sequence based studies.

3.1 Distance matrix methods

A first class of methods used to infer phylogenetic trees is based on the ability to estimate the amount of evolution that separates two species. Given species $S_1, S_2, ..., S_n$ with genomes $G_1, G_2, ..., G_n$, one can then compute the distance matrix D, with D_{ij} = evolutionary distance between G_i and G_j. Notice that D can be calculated using any type of data: morphologic similarity, DNA sequence, gene arrangement, etc. Once the matrix D is computed, one finds the phylogenetic tree T and the length of each of its branches such that the distance between S_i and S_j on the tree T is as close as possible to that specified by D, according to some criterion (for example, the sum of square of errors). This optimization problem is NP-hard, but very good heuristics exist for trees with a small number of species [14].

But one question remains: how to evaluate the amount of evolution between two species? In our case, this translates to "how many rearrangements have happened between the order of the genes in G_i and that in G_j?"

It is obviously impossible to answer this question precisely: nobody was there when these rearrangements occurred, and some early rearrangement may have been undone later during evolution. Much research has been devoted to estimating rearrangement distance. The most popular distance measure between two gene orders is the edit-distance, defined as the smallest number of rearrangements needed to transform one gene order into the other. This measure is bound to be an underestimation of the true number of events that occurred, but in many cases it appears to be very close to the actual distance. In particular, as long as the true distance between two gene orders is not too large compared to the number of genes involved, the edit-distance will be very close to the true distance.

The problem of computing the edit-distance between strings has challenged computer scientists for a long time. Most of the research done regarding edit-distance between genomes has been done under the rather strong assumption that there is a known one-to-one correspondence between the genes of the two genomes compared. That is, no gene is duplicated and both genomes contain exactly the same set of genes (this can be obtained by ignoring some genes, if needed). When the genome is made of only one chromosome, the two gene orders can thus be assimilated to mathematical permutations, and the problem becomes to transform one permutation into the other. If the orientation of the genes is known, a sign (+ or -) is associated with each element of the permutation. When the genomes considered are circular, the associated permutations are also circular (i.e. the last and first elements of the permutation are adjacent).

The problem of computing edit-distances between gene orders was first studied by Watterson [15] and by Sankoff [9], who considered edit-distance restricted to inversions only. The most important result on computing edit-distance between permutations is a polynomial time algorithm to find the inversion distance between two signed permutations, due to Hannenhalli and Pevzner [7]. The algorithm can actually be generalized to find the minimal sequence of inversions and translocations between genomes containing more than one chromosome. An exact description of that algorithm is beyond the scope of this paper. Given the two permutations, the algorithm builds an edge-colored graph called the breakpoint graph. The minimal inversion distance is then given by the sum of four terms, each measuring some property of the breakpoint graph (the simplest term is the number of connected components). The algorithm was later improved by [8] to run in time quadratic in the length of the sequences.

Notice that the knowledge of the orientation of each gene on the sequence is crucial to the time complexity of the algorithm. Indeed, the inversion distance problem for two unsigned permutations was shown to be NP-hard in [5]. No polynomial time algorithm for calculating the transposition distance is known, nor is any complexity result. A 3/2-approximation algorithm is described in [1].

Most algorithmic work has focussed on one type of rearrangement at a time, but nature doesn't have this restriction. In some groups of species, rearrangements occurring appear to be strongly biased towards one type of rearrangement, but most of the time, all three types of rearrangements can occur. Blanchette and Sankoff [4] have proposed a heuristic that computes the minimal-cost sequence of rearrangement between two gene orders, where each type of rearrangement is given a different cost. In nature, rearrangements involving short stretches of DNA seem to be more frequent than those involving longer segments. This could be taken into account when computing the minimal cost solution. Finally, it is believed that some inter-genic regions, called hot spots, are more prone to chromosome break than others. If this information was available, one could potentially use it to infer a more accurate evolutionary distance.

Edit-distance computation still offers many open and challenging problems. An important one is to generalize edit-distance computation to cases where genomes contain duplicated genes, and to the case where there is no clear one-to-one correspondence between the genes in the two genomes. The problem thus becomes an edit-distance problem on strings rather than on permutations.

Distance matrix methods have been used quite successfully to reconstruct phylogenetic trees. Many studies have used organelles (mitochondria and chloroplasts) genome to do their analysis. Organelles are small cellular structures that have their own genome, distinct from the nuclear genome. This genome is usually much smaller than the nuclear genome. For example, the mitochondrial genome of most metazoans have the same set of 37 genes. However, the order of these 37 genes varies a lot, and that makes them good candidates for phylogenetic inference (see, for example, [4] and [13]). Chloroplast genome have also been used for phylogenetic inference [6].

3.2 Reconstruction methods

The distance matrix based inference methods are attractive because all they require is the ability to compute pairwise distances. However, this approach also has its downsides. First, the tree inferred doesn't contain any information about what the ancestral genomes looked like. In fact, it is likely that there is no evolutionary scenario that can match the tree and the branch lengths inferred. Second, the fact the tree is inferred from a $n \times n$ real number matrix means that much of the data is left behind. One could potentially do a better job by doing the tree inference directly using the data, without going first through a distance matrix.

This is exactly what reconstruction methods do. The problem is now framed as a Steiner tree problem: find the evolutionary scenario that most economically explains the genomes observed. More precisely:

Given: a set of genomes $G = \{g_1, ..., g_n\}$, each located in some metric space $< S, d >$ (for example, the space of all possible gene orders under some edit-distance metric)

Find: the set of ancestral genomes $A = \{a_1, ..., a_{n-2}\}$, where $a_i \in S$ $\forall i$, and an unrooted tree $T = (G \cup A, E)$ with leaves G and internal nodes A, such that $\sum_{(v_i,v_j) \in E} d(v_i, v_j)$ is minimized.

This problem is NP-hard for most interesting metric spaces. The difficulty stems from two problems: i) there is a very large number of tree topologies, and ii) for each tree topology, there is a huge number of possible choices as to how locate the ancestral nodes. In fact, the very simplified Median problem, in which $n = 3$ and there is only one ancestral node, is NP-hard, even for signed inversions (for which the edit-distance computation is only of quadratic complexity). This leaves little hope to solve interesting problems with a larger number of species. However, good heuristics or approximation algorithms could be possible.

These rather depressing complexity results have motivated research towards finding new, simpler metric spaces in which the Steiner tree problem would be easier, while retaining as much biological relevance as possible. One such metric is the number of breakpoints between two permutations $A = a_1 a_2 ... a_n$ and $B = b_1 b_2 ... b_n$, which is defined as the smallest number of places where the chromosome A must be broken so that the pieces can be rearranged to form B. Notice that the pair of adjacent genes a_i, a_{i+1} in A needs to be broken only if a_i and a_{i+1} are not adjacent in B. When considering signed permutations, there is a breakpoint between a_i, a_{i+1} in A iff neither a_i follows a_{i+1} in B nor $-a_{i+1}$ follows $-a_i$ in B.

The number of breakpoints between two permutations can trivially be computed in time $O(n)$. This metric is not an edit-distance, but it has been shown to be closely related to the actual number of rearrangement events between two permutations [3]. Moreover, in many cases, different types of genome rearrangement occur with unknown probabilities, which makes the use of (weighted) edit-distance difficult. The breakpoint metric is not based on any specific type of rearrangement and thus can be applied without knowing anything about them.

Unfortunately, the Steiner tree problem under the number of breakpoint metric, and even the Median problem, is still NP-hard [11]. However, in this case, very good heuristics exist [12]. These heuristics rely on the ability to solve the Median problem by reducing it to a Traveling Saleman Problem (TSP), in which the number of cities is $2n$. TSP is itself an NP-hard problem, but it has been studied extensively, and very good heuristics have been developed [10]. This reduction allows one to solve quickly and optimally the Median problem for genomes containing less than a few hundred genes.

We then use our ability to solve the Median problem to iteratively assign near-optimal ancestral gene orders to the internal nodes of a fixed topology tree. Each tree topology is then evaluated in turn, and the topology requiring the smallest number of breakpoints is chosen. Efficient programs were developed using this heuristic (BPAnalysis [12], [3], GRAPPA [6]). These programs have been used successfully for several phylogenetic inference based on mitochondrial and chloroplast genomes, respectively.

Many interesting problems associated with reconstruction methods remain to be considered. A few of them are outlined here. First, no good heuristic is known for the Median problem for any of the important edit-distance metrics. There is interesting theoretical and practical work to be done in that direction.

When considering the breakpoint metric, many issues remain. One of the most important of them is how to solve the Median problem when some of the input genomes have some genes missing, in which case the TSP reduction breaks down. This is an extremely important case, because with many data sets, the gene content of each genome is quite variable. The current approach is to consider only genes that occur in each of the input genomes, but that obviously throws away a lot of information that may be valuable for inferring the correct phylogenetic tree.

The problem of generalizing algorithms from permutations (where there is a one-to-one correspondence between the genes of two genomes) to strings (where each gene can occur several times in one genome) is also of great interest in the context of breakpoint distances. In fact, even the problem of computing the number of breakpoints between two strings (defined as the smallest number of times you need to cut string A to be able to rearrange the pieces to form string B), hasn't been solved yet.

4 Conclusion

Genome rearrangement is a great source of interesting problems for computer scientists. Inferring phylogenetic trees based on genome rearrangement often translates into nice, clean algorithmic problems. Many of these problems remain open. But genome rearrangements are not just puzzles for computer scientists. Most algorithms developed in this field have been applied to real biological data and have given good insights about the evolution of the species considered. With the various sequencing projects in progress, new data sets will become available. For example, the genomes of more than 30 bacteria and archebacteria have now been completely sequenced. These genomes contain more than 1000 genes each, and will to take genome rearrangement studies to a whole new scale. The whole genome of various higher organisms (worm, fruit fly, human, mouse, etc.) is also completely sequenced or close to being completed. These genomes contain a few tens to thousands of genes, and promise new and interesting algorithmic problems.

5 Acknowledgements

I will like to sincerely thank Saurabh Sinha for his help with the preparation of this manuscript.

References

1. V. Bafna and P.A. Pevzner. Sorting by transpositions. *Proceedings of the 6th Annual ACM-SIAM Symposium on Discrete Algorithms (SODA 95)*, 614-623, 1995.
2. V. Bafna and P.A. Pevzner. Sorting by reversals: Genome rearrangements in plant organelles and evolutionary history of X chromosome. *Molecular Biology and Evolution*, 12: 239-246, 1995.
3. M. Blanchette, T. Kunisawa and D. Sankoff. Gene order breakpoint evidence in animal mitochondrial phylogeny. *Journal of Molecular Evolution*, 49, 193-203, 1998.
4. M. Blanchette, T. Kunisawa and D. Sankoff. Parametric genome rearrangement. *Gene*, 172, GC:11-17, 1996.
5. A. Caprara. Sorting by Reversals is Difficult. *Proceedings of the First Annual International Conference on Computational Molecular Biology (RECOMB 97)*, 75-83, 1997.
6. Cosner, M.E., Jansen, R.K., Moret, B.M.E., Raubeson, L.A., Wang, L.S., Warnow, T., and Wyman, S.. A new fast heuristic for computing the breakpoint phylogeny and a phylogenetic analysis of a group of highly rearranged chloroplast genomes.*Proc. 8th Int'l Conf. on Intelligent Systems for Molecular Biology ISMB-2000*, San Diego, 104-115, 2000.
7. S. Hannenhalli and P.A. Pevzner. Transforming men into mice (polynomial algorithm for genomic distance problem). *Proceedings of the IEEE 36th Annual Symposium on Foundations of Computer Science*, 581-592, 1995.
8. H. Kaplan, R. Shamir and R.E. Tarjan. Faster and Simpler Algorithm for Sorting Signed Permutations by Reversals. *Proceedings of the Eighth Annual ACM-SIAM Symposium on Discrete Algorithms (SODA 97)*, 1997.
9. J. Kececioglu and D. Sankoff. Exact and approximation algorithms for sorting by reversals, with application to genome rearrangement. *Algorithmica*, 13: 180-210, 1995.
10. E.L. Lawler, J.K. Lenstra, A.H.G. Rinnooy Kan, and D.B. Shmoys The Travelling Salesman Problem. John Wiley and Sons, 1985.
11. I. Pe'er and R. Shamir. The median problems for breakpoints are NP-complete. *Electronic Colloquium on Computational Complexity, Technical Report* 98-071, 1998.
12. Sankoff, D. and Blanchette, M. Multiple genome rearrangement and breakpoint phylogeny. *Journal of Computational Biology* 5, 555-570, 1998.
13. D. Sankoff, G. Leduc, N. Antoine, B. Paquin, B.F. Lang and R. Cedergren. Gene order comparisons for phylogenetic inference: Evolution of the mitochondrial genome. *Proceedings of the National Academy of Sciences USA*, 89: 6575-6579, 1992.
14. Swofford, D.L., G.J. Olsen, P.J. Waddell, and D.M. Hillis. *In Molecular Systematics* (2nd ed., D.M. Hillis, C. Moritz, and B.K. Mable, eds.). Sinauer Assoc. Sunderland, MA. Ch. 11 (pp. 407-514), 1996.
15. G. A. Watterson, W. J. Ewens, T. E. Hall et A. Morgan. The chromosome inversion problem. *Journal of Theoretical Biology*, 99: 1-7, 1982.

High-Performance Algorithm Engineering for Computational Phylogenetics

Bernard M.E. Moret[1], David A. Bader[2], and Tandy Warnow[3]

[1] Department of Computer Science, University of New Mexico,
Albuquerque, NM 87131, USA,
moret@cs.unm.edu, URL www.cs.unm.edu/~moret/
[2] Department of Electrical and Computer Engineering, University of New Mexico,
Albuquerque, NM 87131, USA,
dbader@eece.unm.edu, URL www.eece.unm.edu/~dbader/
[3] Department of Computer Sciences, University of Texas, Austin, TX 78712, USA,
tandy@cs.utexas.edu, URL www.cs.utexas.edu/users/tandy/

Abstract. Phylogeny reconstruction from molecular data poses complex optimization problems: almost all optimization models are NP-hard and thus computationally intractable. Yet approximations must be of very high quality in order to avoid outright biological nonsense. Thus many biologists have been willing to run farms of processors for many months in order to analyze just one dataset. High-performance algorithm engineering offers a battery of tools that can reduce, sometimes spectacularly, the running time of existing phylogenetic algorithms. We present an overview of algorithm engineering techniques, illustrating them with an application to the "breakpoint analysis" method of Sankoff *et al.*, which resulted in the GRAPPA software suite. GRAPPA demonstrated a million-fold speedup in running time (on a variety of real and simulated datasets) over the original implementation. We show how algorithmic engineering techniques are directly applicable to a large variety of challenging combinatorial problems in computational biology.

1 Background

Algorithm Engineering The term "algorithm engineering" was first used with specificity in 1997, with the organization of the first *Workshop on Algorithm Engineering (WAE 97)*. Since then, this workshop has taken place every summer in Europe and a parallel one started in the US in 1999, the *Workshop on Algorithm Engineering and Experiments (ALENEX99)*, which has taken place every winter, colocated with the *ACM/SIAM Symposium on Discrete Algorithms (SODA)*. Algorithm engineering refers to the process required to transform a pencil-and-paper algorithm into a robust, efficient, well tested, and easily usable implementation. Thus it encompasses a number of topics, from modelling cache behavior to the principles of good software engineering; its main focus, however, is experimentation. In that sense, it may be viewed as a recent outgrowth of *Experimental Algorithmics*, which is specifically devoted to the development of methods, tools, and practices for assessing and refining algorithms through experimentation. The online *ACM Journal of Experimental Algorithmics (JEA)*, at URL `www.jea.acm.org`, is devoted to this area and also publishes selected best papers from the WAE and ALENEX workshops. Notable efforts in algorithm engineering include the development of LEDA [19], attempts at characterizing the effects of caching on the behavior of implementations [1, 11, 16–18, 27, 29], ever more efficient implementation of network flow algorithms [7, 8, 13], and the characterization of the behavior of everyday algorithms and data structures such as priority queues [15, 32], shortest paths [6], minimum spanning trees [22], and sorting [21]. More references can be found in [20] as well as by going to the web site for the *ACM Journal of Experimental Algorithmics* at `www.jea.acm.org`.

High-Performance Algorithm Engineering

High-Performance Algorithm Engineering focuses on one of the many facets of algorithm engineering. The high-performance aspect does not immediately imply parallelism; in fact, in any highly parallel task, most of the impact of high-performance algorithm engineering tends to come from refining the serial part of the code. For instance, in the example we will use throughout this paper, the million-fold speed-up was achieved through a combination of a 512-fold speedup due to parallelism (one that will scale to any number of processors) and a 2,000-fold speedup in the serial execution of the code.

All of the tools and techniques developed over the last five years for algorithm engineering are applicable to high-performance algorithm engineering. However, many of these tools will need further refinement. For example, cache-aware programming is a key to performance (particularly with high-performance machines, which have deep memory hierarchies), yet it is not yet well understood, in part through lack of suitable tools (few processor chips have built-in hardware to gather statistics on the behavior of caching, while simulators leave much to be desired) and in part because of complex machine-dependent issues (recent efforts at cache-independent algorithm design [5, 12] may offer some new solutions). As another example, profiling a running program offers serious challenges in a serial environment (any profiling tool affects the behavior of what is being observed), but these challenges pale in comparison with those arising in a parallel or distributed environment (for instance, measuring communication bottlenecks may require hardware assistance from the network switches or at least reprogramming them, which is sure to affect their behavior).

Phylogenies

A phylogeny is a reconstruction of the evolutionary history of a collection of organisms; it usually takes the form of an evolutionary tree, in which modern organisms are placed at the leaves and ancestral organisms occupy internal nodes, with the edges of the tree denoting evolutionary relationships. Figure 1 shows two proposed phylogenies, one for several species of the *Campanulaceae* (bluebell flower) family and the other for Herpes viruses that are known to affect humans. Reconstructing phylogenies is a major component of modern research programs in many areas of biology and medicine (as well as linguistics). Scientists are of course interested in phylogenies for the usual reasons of scien-

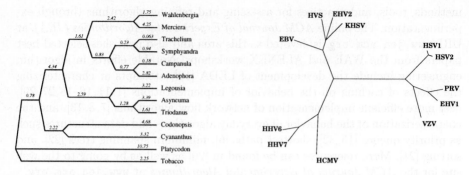

Fig. 1. Two phylogenies: some plants of the *Campanulaceae* family (left) and some Herpes viruses affecting humans (right)

tific curiosity. An understanding of evolutionary mechanisms and relationships is at the heart of modern pharmaceutical research for drug discovery, is helping researchers understand (and defend against) rapidly mutating viruses such as HIV, is the basis for the design of genetically enhanced organisms, etc. In developing such an understanding, the reconstruction of phylogenies is a crucial tool, as it allows one to test new models of evolution.

Computational Phylogenetics

Phylogenies have been reconstructed "by hand" for over a century by taxonomists, using morphological characters and basic principles of genetic inheritance. With the advent of molecular data, however, it has become necessary to develop algorithms to reconstruct phylogenies from the large amount of data made available through DNA sequencing, amino-acid and protein characterization, gene expression data, and whole-genome descriptions. Until recently, most of the research focused on the development of methods for phylogeny reconstruction from DNA sequences (which can be regarded as strings on a 4-character alphabet), using a model of evolution based mostly on nucleotide substitution. Because amino-acids, the building blocks of life, are coded by substrings of four nucleotides known as codons, the same methods were naturally extended to sequences of codons (which can be regarded as strings on an alphabet of 22 characters—in spite of the 64 possible codes, only 22 amino-acids are encoded, with many codes representing the same amino-acid). Proteins, which are built from amino-acids, are the natural next level, but are proving difficult to characterize in evolutionary terms. Recently, another type of data has been made available through the characterization of entire genomes: gene content and gene order data. For some organisms, such as human, mouse, fruit fly, and several plants and lower-order organisms, as well as for a large collection of organelles (mitochondria, the animal cells' "energy factories", and chloroplasts, the plant cells' "photosynthesis factories"), we have a fairly complete description of the entire genome, gene by gene. Because plausible mechanisms of evolution include gene rearrangement, duplication, and loss, and because evolution at this level (the "genome level") is much slower than evolution driven by mutations in the nucleotide base pairs (the "gene level") and so may enable us to recover deep evolutionary relationships, there has been considerable interest in the phylogeny community in the development of algorithms for reconstructing phylogenies based on gene-order or gene content data. Appropriate tools for analyzing such data may help resolve some difficult phylogenetic reconstruction problems; indeed, this new source of data has been embraced by many biologists in their phylogenetic work.[24, 25, 28] There is no doubt that, as our understanding of evolution improves, new types of data will be collected and well need to be analyzed in phylogeny reconstruction.

Optimization Criteria To date, almost every model of evolution proposed for modelling phylogenies gives rise to NP-hard optimization problems. Three main lines of work have evolved: more or less *ad hoc* heuristics (a natural consequence of the NP-hardness of the problems) that run quickly, but offer no quality guarantees and may not even have a well defined optimization criterion, such as the popular *neighbor-joining* heuristic [31]; optimization problems based on a *parsimony* criterion, which seeks the phylogeny with the least total amount of change needed to explain modern data (a modern version of Occam's razor); and optimization problems based on a *maximum likelihood* criterion, which seeks the phylogeny that is the most likely (under some suitable statistical model) to have given rise to the modern data. *Ad hoc* heuristics are fast and often rival the optimization methods in terms of accuracy; parsimony-based methods may take exponential time, but, at least for DNA data, can often be run to completion on datasets of moderate size; while methods based on maximum-likelihood are very slow (the point estimation problem alone appears intractable) and so restricted to very small instances, but appear capable of outperforming the others in terms of the quality of solutions. In the case of gene-order data, however, only parsimony criteria have been proposed so far: we do not yet have detailed enough models (or ways to estimate their parameters) for using a maximum-likelihood approach.

2   Our Running Example: Breakpoint Phylogeny

Some organisms have a single chromosome or contain single-chromosome organelles (mitochondria or chloroplasts), the evolution of which is mostly independent of the evolution of the nuclear genome. Given a particular strand from a single chromosome (whether linear or circular), we can infer the ordering of the genes along with the directionality of the genes, thus representing each chromosome by an ordering of oriented genes. The evolutionary process that operates on the chromosome may include inversions and transpositions, which change the order in which genes occur in the genome as well as their orientation. Other events, such as insertions, deletions, or duplications, change the number of times and the positions in which a gene occurs.

A natural optimization problem for phylogeny reconstruction from this type of data is to reconstruct the most parsimonious tree, the evolutionary tree with the minimum number of permitted evolutionary events. For any choice of permitted events, such a problem is computationally very intensive (known or conjectured to be NP-hard); worse, to date, no automated tools exist for solving such problems. Another approach is first to estimate leaf-to-leaf distances (based upon some metric) between all genomes, and then to use a standard distance-based heuristic such as *neighbor-joining* [31] to construct the tree. Such approaches are quite fast and may prove valuable in reconstructing the underlying tree, but cannot recover the ancestral gene orders.

Blanchette *et al.* [4] developed an approach, which they called *breakpoint phylogeny*, for the special case in which the genomes all have the same set of genes, and each gene appears once. This special case is of interest to biologists, who hypothesize that inversions (which can only affect gene order, but not gene content) are the main evolutionary mechanism for a range of genomes or chromosomes (chloroplast, mitochondria, human X chromosome, etc.) Simulation studies we conducted suggested that this approach works well for certain datasets (i.e., it obtains trees that are close to the model tree), but that the implementation developed by Sankoff and Blanchette, the BPAnalysis software [30], is too slow to be used on anything other than small datasets with a few genes [9, 10].

3 Breakpoint Analysis: Details

When each genome has the same set of genes and each gene appears exactly once, a genome can be described by an ordering (circular or linear) of these genes, each gene given with an orientation that is either positive (g_i) or negative $(-g_i)$. Given two genomes G and G' on the same set of genes, a *breakpoint* in G is defined as an ordered pair of genes, (g_i, g_j), such that g_i and g_j appear consecutively in that order in G, but neither (g_i, g_j) nor $(-g_j, -g_i)$ appears consecutively in that order in G'. The breakpoint distance between two genomes is the number of breakpoints between that pair of genomes. The breakpoint score of a tree in which each node is labelled by a signed ordering of genes is then the sum of the breakpoint distances along the edges of the tree.

Given three genomes, we define their *median* to be a fourth genome that minimizes the sum of the breakpoint distances between it and the other three. The *Median Problem for Breakpoints* (MPB) is to construct such a median and is NP-hard [26]. Sankoff and Blanchette developed a reduction from MPB to the Travelling Salesman Problem (TSP), perhaps the most studied of all optimization problems [14]. Their reduction produces an undirected instance of the TSP from the directed instance of MPB by the standard technique of representing each gene by a pair of cities connected by an edge that must be included in any solution.

BPAnalysis (see Figure 2) is the method developed by Blanchette and Sankoff

Initially label all internal nodes with gene orders
Repeat
    For each internal node v, with neighbors A, B, and C, do
        Solve the MPB on A, B, C to yield label m
        If relabelling v with m improves the score of T, then do it
   until no internal node can be relabelled

Fig. 2. BPAnalysis

to solve the breakpoint phylogeny. Within a framework that enumerates all trees, it uses an iterative heuristic to label the internal nodes with signed gene orders. This procedure is computationally very intensive. The outer loop enumerates all $(2n - 5)!!$ leaf-labelled trees on n leaves, an exponentially large value.[1] The inner loop runs an unknown number of iterations (until convergence), with each iteration solving an instance of the TSP (with a number of cities equal to twice the number of genes) at each internal node. The computational complexity of the entire algorithm is thus exponential in *each* of the number of genomes and the number of genes, with significant coefficients. The procedure nevertheless remains a heuristic: even though all trees are examined and each MPB problem solved exactly, the tree-labeling phase does not ensure optimality unless the tree has only three leaves.

4   Re-Engineering BPAnalysis for Speed

Profiling   Algorithmic engineering suggests a refinement cycle in which the behavior of the current implementation is studied in order to identify problem areas which can include excessive resource consumption or poor results. We used extensive profiling and testing throughout our development cycle, which allowed us to identify and eliminate a number of such problems. For instance, converting the MPB into a TSP instance dominates the running time whenever the TSP instances are not too hard to solve. Thus we lavished much attention on that routine, down to the level of hand-unrolling loops to avoid modulo computations and allowing reuse of intermediate expressions; we cut the running time of that routine down by a factor of at least six—and thereby nearly tripled the speed of the overall code. We lavished equal attention on distance computations and on the computation of the lower bound, with similar results. Constant profiling is the key to such an approach, because the identity of the principal "culprits" in time consumption changes after each improvement, so that attention must shift to different parts of the code during the process—including revisiting already improved code for further improvements. These steps provided a speed-up by one order of magnitude on the *Campanulaceae* dataset.

Cache Awareness   The original BPAnalysis is written in C++ and uses a space-intensive full distance matrix, as well as many other data structures. It has a significant memory footprint (over 60MB when running on the *Campanulaceae* dataset) and poor locality (a working set size of about 12MB). Our implementation has a tiny memory footprint (1.8MB on the *Campanulaceae* dataset) and good locality (all of our storage is in arrays preallocated in the main routine and retained and reused throughout the computation), which enables it to run almost completely in cache (the working set size is 600KB). Cache locality can be improved by returning to a FORTRAN-style of programming, in which storage is static, in which records (structures/classes) are avoided in favor of separate

---

[1] The double factorial is a factorial with a step of 2, so we have $(2n - 5)!! = (2n - 5) \cdot (2n - 7) \cdot \ldots \cdot 3$

arrays, in which simple iterative loops that traverse an array linearly are preferred over pointer dereferencing, in which code is replicated to process each array separately, etc. While we cannot measure exactly how much we gain from this approach, studies of cache-aware algorithms [1, 11, 16–18, 33] indicate that the gain is likely to be substantial—factors of anywhere from 2 to 40 have been reported. New memory hierarchies show differences in speed between cache and main memory that exceed two orders of magnitude.

Low-Level Algorithmic Changes Unless the original implementation is poor (which was not the case with BPAnalysis), profiling and cache-aware programming will rarely provide more than two orders of magnitude in speed-up. Further gains can often be obtained by low-level improvement in the algorithmic details. In our phylogenetic software, we made two such improvements. The basic algorithm scores every single tree, which is clearly very wasteful; we used a simple lower bound, computable in linear time, to enable us to eliminate a tree without scoring it. On the *Campanulaceae* dataset, this bounding eliminates over 95% of the trees without scoring them, resulting in a five-fold speed-up. The TSP solver we wrote is at heart the same basic include/exclude search as in BPAnalysis, but we took advantage of the nature of the instances created by the reduction to make the solver much more efficient, resulting in a speed-up by a factor of 5–10. These improvements all spring from a careful examination of exactly what information is readily available or easily computable at each stage and from a deliberate effort to make use of all such information.

5   A High-Performance Implementation

Our resulting implementation, *GRAPPA*,[2] incorporates all of the refinements mentioned above, plus others specifically made to enable the code to run efficiently in parallel (see [23] for details). Because the basic algorithm enumerates and independently scores every tree, it presents obvious parallelism: we can have each processor handle a subset of the trees. In order to do so efficiently, we need to impose a linear ordering on the set of all possible trees and devise a generator that can start at an arbitrary point along this ordering. Because the number of trees is so large, an arbitrary tree index would require unbounded-precision integers, considerably slowing down tree generation. Our solution was to design a tree generator that starts with tree index k and generates trees with indices $\{k + cn \mid n \in \mathcal{N}\}$, where k and c are regular integers, all without using unbounded-precision arithmetic. Such a generator allows us to sample tree space (a very useful feature in research) and, more importantly, allows us to use a cluster of c processors, where processor i, $0 \le i \le c - 1$, generates and scores trees with indices $\{i + cn \mid n \in \mathcal{N}\}$. We ran *GRAPPA* on the 512-processor Alliance cluster *Los Lobos* at the University of New Mexico and obtained a 512-fold speed-up. When combined with the 2000-fold speedup obtained through

---

[2] Genome Rearrangement Analysis through Parsimony and other Phylogenetic Algorithms

algorithm engineering, our run on the *Campanulaceae* dataset demonstrated a *million-fold* speed-up over the original implementation [2].

In addition, we made sure that gains held across a wide variety of platforms and compilers: we tested our code under Linux, FreeBSD, Solaris, and Windows, using compilers from GNU, the Portland group, Intel (beta release), Microsoft, and Sun, and running the resulting code on Pentium- and Sparc-based machines. While the gcc compiler produced marginally faster code than the others, the performance we measured was completely consistent from one platform to the other.

6   Impact in Computational Biology

Computational biology presents numerous complex optimization problems, such as multiple sequence alignment, phylogeny reconstruction, characterization of gene expression, structure prediction, etc. In addition, the very large databases used in computational biology give rise to serious algorithmic engineering problems when designing query algorithms on these databases. While several programs in use in the area (such as BLAST, see www.ncbi.nlm.nih.gov/BLAST/) have already been engineered for performance, most such efforts have been more or less *ad hoc*. The emergence of a discipline of algorithm engineering [20] is bringing us a collection of tools and practices that can be applied to almost any existing algorithm or software package to speed up its execution, often by very significant factors. While we illustrated the approach and its potential results with a specific program in phylogeny reconstruction based on gene order data, we are now in the process of applying the same to a collection of fundamental methods (such as branch-and-bound parsimony or maximum-likelihood estimation) as well as new algorithms.

Of course, even large speed-ups have only limited benefits in theoretical terms when applied to NP-hard optimization problems: even our million-fold speed-up with *GRAPPA* only enables us to move from about 10 taxa to about 13 taxa. Yet the very process of algorithm engineering often uncovers salient characteristics of the algorithm that were overlooked in a less careful analysis and may thus enable us to develop much better algorithms. In our case, while we were implementing the rather complex algorithm of Berman and Hannenhalli for computing the inversion distance between two signed permutations, an algorithm that had not been implemented before, we came to realize that the algorithm could be simplified as well as accelerated, deriving in the process the first true linear-time algorithm for computing these distances [3]. We would not have been tempted to implement this algorithm in the context of the original program, which was already much too slow when using the simpler breakpoint distance. Thus faster experimental tools, even when they prove incapable of scaling to "industrial-sized" problems, nevertheless provide crucial opportunities for exploring and understanding the problem and its solutions.

Thus we see two potential major impacts in computational biology. First, the much faster implementations, when mature enough, can alter the practice of research in biology and medicine. For instance pharmaceutical companies spend large budgets on computing equipment and research personnel to reconstruct phylogenies as a vital tool in drug discovery, yet may still have to wait a year or more for the results of certain computations; reducing the running time of such analyses from 2–3 years down to a day or less would make an enormous difference in the cost of drug discovery and development. Secondly, biologists in research laboratories around the world use software for data analysis, much of it rife with undocumented heuristics for speeding up the code at the expense of optimality, yet still slow for their purposes. Software that runs 3 to 4 orders of magnitude faster, even when it remains impractical for real-world problems, would nevertheless enable these researchers to test simpler scenarios, compare models, develop intuition on small instances, and perhaps even form serious conjectures about biological mechanisms.

Acknowledgments

This work was supported in part by NSF ITR 00-81404 (Moret and Bader), NSF 94-57800 (Warnow), and the David and Lucile Packard Foundation (Warnow).

References

1. Arge, L., Chase, J., Vitter, J.S., & Wickremesinghe, R., "Efficient sorting using registers and caches," *Proc. 4th Workshop Alg. Eng. WAE 2000*, to appear in LNCS series, Springer Verlag (2000).
2. Bader, D.A., & Moret, B.M.E., "GRAPPA runs in record time," *HPC Wire*, **9**, 47 (Nov. 23), 2000.
3. Bader, D.A., Moret, B.M.E., & Yan, M., "A fast linear-time algorithm for inversion distance with an experimental comparison," Dept. of Comput. Sci. TR 2000-42, U. of New Mexico.
4. Blanchette, M., Bourque, G., & Sankoff, D., "Breakpoint phylogenies," in *Genome Informatics 1997*, Miyano, S., & Takagi, T., eds., Univ. Academy Press, Tokyo, 25–34.
5. Bender, M.A., Demaine, E., and Farach-Colton, M., "Cache-oblivious search trees," *Proc. 41st Ann. IEEE Symp. Foundations Comput. Sci. FOCS-00*, IEEE Press (2000), 399–409.
6. Cherkassky, B.V., Goldberg, A.V., & Radzik, T., "Shortest paths algorithms: theory and experimental evaluation," *Math. Progr.* **73** (1996), 129–174.
7. Cherkassky, B.V., & Goldberg, A.V., "On implementing the push-relabel method for the maximum flow problem," *Algorithmica* **19** (1997), 390–410.
8. Cherkassky, B.V., Goldberg, A.V., Martin, P, Setubal, J.C., & Stolfi, J., "Augment or push: a computational study of bipartite matching and unit-capacity flow algorithms," *ACM J. Exp. Algorithmics* **3**, 8 (1998), www.jea.acm.org/1998/CherkasskyAugment/.
9. Cosner, M.E., Jansen, R.K., Moret, B.M.E., Raubeson, L.A., Wang, L.-S., Warnow, T., & Wyman, S., "A new fast heuristic for computing the breakpoint phylogeny and experimental phylogenetic analyses of real and synthetic data," *Proc. 8th Int'l Conf. Intelligent Systems Mol. Biol. ISMB-2000*, San Diego (2000).
10. Cosner, M.E., Jansen, R.K., Moret, B.M.E., Raubeson, L.A., Wang, L.S., Warnow, T., & Wyman, S., "An empirical comparison of phylogenetic methods on chloroplast gene order data in Campanulaceae," *Proc. Gene Order Dynamics, Comparative Maps, and Multigene Families DCAF-2000*, Montreal (2000).

11. Eiron, N., Rodeh, M., & Stewarts, I., "Matrix multiplication: a case study of enhanced data cache utilization," *ACM J. Exp. Algorithmics* **4**, 3 (1999), www.jea.acm.org/1999/EironMatrix/.

12. Frigo, M., Leiserson, C.E., Prokop, H., & Ramachandran, S., "Cache-oblivious algorithms," *Proc. 40th Ann. Symp. Foundations Comput. Sci. FOCS-99*, IEEE Press (1999), 285–297.

13. Goldberg, A.V., & Tsioutsiouliklis, K., "Cut tree algorthms: an experimental study," *J. Algs.* **38**, 1 (2001), 51–83.

14. Johnson, D.S., & McGeoch, L.A., "The traveling salesman problem: a case study," in *Local Search in Combinatorial Optimization*, E. Aarts & J.K. Lenstra, eds., John Wiley, New York (1997), 215–310.

15. Jones, D.W., "An empirical comparison of priority queues and event-set implementations," *Commun. ACM* **29** (1986), 300–311.

16. Ladner, R., Fix, J.D., & LaMarca, A., "The cache performance of traversals and random accesses," *Proc. 10th ACM/SIAM Symp. Discrete Algs. SODA99*, SIAM Press (1999), 613–622.

17. LaMarca, A., & Ladner, R., "The influence of caches on the performance of heaps," *ACM J. Exp. Algorithmics* **1**, 4 (1996), www.jea.acm.org/1996/LaMarcaInfluence.

18. LaMarca, A., & Ladner, R., "The influence of caches on the performance of sorting," *Proc. 8th ACM/SIAM Symp. Discrete Algs. SODA97*, SIAM Press (1997), 370–379.

19. Melhorn, K., & Näher, S. *The LEDA Platform of Combinatorial and Geometric Computing*. Cambridge U. Press, 1999.

20. Moret, B.M.E., "Towards a discipline of experimental algorithmics," *Proc. 5th DIMACS Challenge* (to appear), available at www.cs.unm.edu/~moret/dimacs.ps.

21. Moret, B.M.E., & Shapiro, H.D. *Algorithms from P to NP, Vol. I: Design and Efficiency*. Benjamin-Cummings, Menlo Park, CA, 1991.

22. Moret, B.M.E., & Shapiro, H.D., "An empirical assessment of algorithms for constructing a minimal spanning tree," in *Computational Support for Discrete Mathematics*, N. Dean & G. Shannon, eds., *DIMACS Monographs in Discrete Math. & Theor. Comput. Sci.* **15** (1994), 99–117.

23. Moret, B.M.E., Wyman, S., Bader, D.A., Warnow, T., & Yan, M., "A detailed study of breakpoint analysis," *Proc. 6th Pacific Symp. Biocomputing PSB 2001*, Hawaii, World Scientific Pub. (2001), 583–594.

24. Olmstead, R.G., & Palmer, J.D., "Chloroplast DNA systematics: a review of methods and data analysis," *Amer. J. Bot.* **81** (1994), 1205–1224.

25. Palmer, J.D., "Chloroplast and mitochondrial genome evolution in land plants," in *Cell Organelles*, Herrmann, R., ed., Springer Verlag (1992), 99–133.

26. Pe'er, I., & Shamir, R., "The median problems for breakpoints are NP-complete," *Elec. Colloq. Comput. Complexity*, Report 71, 1998.

27. Rahman, N., & Raman, R., "Analysing cache effects in distribution sorting," *Proc. 3rd Workshop Alg. Eng. WAE99*, in LNCS **1668**, Springer Verlag (1999), 183–197.

28. Raubeson, L.A., & Jansen, R.K., "Chloroplast DNA evidence on the ancient evolutionary split in vascular land plants," *Science* **255** (1992), 1697–1699.

29. Sanders, P., "Fast priority queues for cached memory," *ACM J. Exp. Algorithmics* **5**, 7 (2000), www.jea.acm.org/2000/SandersPriority/.

30. Sankoff, D., & Blanchette, M., "Multiple genome rearrangement and breakpoint phylogeny," *J. Comput. Biol.* **5** (1998), 555–570.

31. Saitou, N., & Nei, M., "The neighbor-joining method: A new method for reconstructing phylogenetic trees," *Mol. Biol. & Evol.* **4** (1987), 406–425.

32. Stasko, J.T., & Vitter, J.S., "Pairing heaps: experiments and analysis," *Commun. ACM* **30** (1987), 234–249.

33. Xiao, L., Zhang, X., & Kubricht, S.A., "Improving memory performance of sorting algorithms," *ACM J. Exp. Algorithmics* **5**, 3 (2000), www.jea.acm.org/2000/XiaoMemory/.

Phylogenetic Inference from Mitochondrial Genome Arrangement Data

Donald L. Simon and Bret Larget

Department of Mathematics/Computer Science
Duquesne University
Pittsburgh, PA 15282
{simon, larget}@mathcs.duq.edu
http://www.mathcs.duq.edu/profs/{simon.html, larget.html}

Abstract. A fundamental problem in evolutionary biology is determining evolutionary relationships among different taxa. Genome arrangement data is potentially more informative than DNA sequence data in cases where alignment of DNA sequences is highly uncertain. We describe a Bayesian framework for phylogenetic inference from mitochondrial genome arrangement data that uses Markov chain Monte Carlo (MCMC) as the computational engine for inference. Our approach is to model mitochondrial data as a circular signed permutation which is subject to reversals. We calculate the likelihood of one arrangement mutating into another along a single branch by counting the number of possible sequences of reversals which transform the first to the second. We calculate the likelihood of the entire tree by augmenting the state space with the arrangements at the branching points of the tree. We use MCMC to update both the tree and the arrangement data at the branching points.

1 Introduction

Determining the evolutionary history and relationships among a group of taxa is a fundamental problem in evolutionary biology. Phylogenies are branching tree diagrams that display these evolutionary relationships. Phylogenetic inference may involve estimation of the true evolutionary relationships among a set of taxa, estimation of times of speciation, estimation of ancestral data, and the assessment of uncertainty in these estimates.

Swofford et al.[15] provides an excellent overview of commonly used methods for phylogenetic analysis from aligned DNA sequence data. More recently, several authors have developed a Bayesian approach to phylogenetic inference from DNA sequence data (Rannala and Yang [10]; Yang and Rannala [16]; Mau, Newton, and Larget [8]; Larget and Simon [6]; Newton, Mau, and Larget [9]; Li, Pearl, and Doss [7]).

However, when the taxa of interest are quite distantly related, the difficulty in aligning sequences that have undergone substantial evolutionary change is great, and the sequences may no longer be phylogenetically informative. Several authors have argued that genome arrangement data is potentially more informative than DNA sequence data in comparing distantly related taxa, because large-scale genome rearrangements occur at much slower rates than nucleotide base substitution.

Mitochondrial genomes are circular and genes can appear on one of two strands. Nearly all animals have the same 37 mitochondrial genes. However, these genes are arranged differently in some species. These differences are a source of information to infer the evolutionary past.

Some recent papers which use genome arrangement data to make phylogenetic inferences include Smith *et al.*[11], Boore *et al.*[1], and Boore *et al.*[2]. The methods employed in these papers are not statistical.

Mitochondrial genome arrangement data is being collected at a rapidly increasing pace. In 1990, only about a dozen mitochondrial gene arrangements had been completely determined, but this number now exceeds one hundred (Boore [3]) and is expected to increase rapidly.

Despite this rapidly increasing amount of data, there is very little existing methodology for its use to reconstruct phylogenetic trees and to assess the uncertainty in the estimated evolutionary relationships. Most existing methodology is based on distance or parsimony methods. Sankoff *et al.*[12] is an early prominent example. In a recent article, Sankoff and Blanchette [13] describe a likelihood method for phylogenetic inference. This method, however, does not correspond to any particular mechanism for genome rearrangement.

This paper introduces a Bayesian framework for phylogenetic inference from genome arrangement data based on a likelihood model that assumes that reversals are the sole mechanism for genome rearrangement.

2 A model of mitochondrial genome rearrangement

A mitochondrial gene arrangement may be represented as a *signed circular permutation* where genes of the same sign are located on the same strand. There is a correspondence between signed circular permutations of size g and ordinary signed permutations of size $g - 1$ by choosing a single reference gene and a direction around the circle by which to list the remainder. There are several mechanisms through which mitochondrial genomes can rearrange. The 22 short (70–90 base pairs) genes code tRNAs. These short genes appear to rearrange by multiple mechanisms. The remaining 15 genes are longer (hundreds of base pairs). It is thought that inversions are the primary (and perhaps sole) mechanism by which large coding mitochondrial genes rearrange. For the remainder of this paper, we will ignore the 22 short tRNA genes and assume that reversals are the only mechanism of genome rearrangement. Given that an reversal occurs, our model assumes that all possible reversals are equally likely.

3   Representation and Modification of the Tree

We represent a phylogeny of s taxa with an unrooted tree topology τ and a $(2s-3)$-vector of branch lengths $\mathbf{b}_\tau$. The branch lengths represent the expected number of reversals. The observed data at each leaf is a signed circular permutation of g genes that we linearize and represent as a signed permutation of size $g-1$ by choosing a reference gene and direction around the circle to read the remaining genes. We represent the observed data at the leaf nodes with an array L of s signed permutations and augment our state space with an array I of $s-2$ signed permutations at the internal nodes.

We describe three algorithms for proposing new trees. The first method randomly selects one of the $s-2$ internal nodes and performs a random number of reversals on its permutation. The number of reversals is chosen randomly by a geometric distribution with parameter $p = 0.5$.

The second method is the LOCAL update method without the molecular clock used by BAMBE [6, 14], an MCMC program for the Bayesian analysis of aligned nucleotide sequence data. We randomly pick one of the $2s-3$ internal edges of the unrooted tree, designating its two nodes u and v. The other two neighbors of u are randomly labeled a and b and v's two other neighbors are randomly labeled c and d with equal probability. Set $m = \text{dist}(a, c)$. Our proposal changes m by multiplying edge lengths on the path from a to c by a random factor. We then detach either u or v with equal probability and reattach it along with its unchanged subtree to a point chosen uniformly at random on the path from a to c. Specifically, $m^* = m \times e^{\lambda_2(U_1 - 0.5)}$ where U_1 is a uniform$(0,1)$ random variable and λ_2 is a tuning parameter. Let $x = \text{dist}(a, u)$ and $y = \text{dist}(a, v)$ be distances in the current tree. If u is chosen to move, the proposal sets $x^* = U_2 \times m^*$ and $y^* = y \times m^*/m$. If v is chosen to move, $x^* = x \times m^*/m$ and $y^* = U_2 \times m^*$. In both cases U_2 is a uniform$(0,1)$ random variable. If $x^* < y^*$, the tree topology does not change while $\text{dist}(a, u^*) = x^*$, $\text{dist}(u^*, v^*) = y^* - x^*$, and $\text{dist}(v^*, c) = m^* - y^*$. If $x^* > y^*$, the tree topology does change as u^* becomes a neighbor of c and v^* becomes a neighbor of a while $\text{dist}(a, v^*) = y^*$, $\text{dist}(v^*, u^*) = x^* - y^*$, and $\text{dist}(u^*, c) = m^* - x^*$. The Hastings ratio in this case is $(m^*/m)^2$.

The third method both deforms the tree using the LOCAL update method above and changes the permutation at one of the internal nodes. Consider the case when during the LOCAL update method the node u is chosen to be moved. If the topology changes and the permutation at u is either the same or one reversal away from the permutation at its old neighbor a, then the permutation at u is copied from c, and a random reversal is applied with probability 0.5. If the topology does not change, or u is more than one reversal away from a, then a random reversal is applied to u's permutation. In the other case, when the node v is chosen to be moved, the same change is applied, except that the permutation may be copied from node a.

During one cycle the Markov chain procedure, each of the above three update methods is invoked in sequence. That is, the first method is used to change a node's permutation, then likelihood of the tree is calculated, and the new tree may be accept in place of the original tree. Then the LOCAL method without changing the permutation is invoked, the new tree is evaluated and possibly accepted. Finally, LOCAL with a permutation change is used, again possibly resulting in a tree being accepted.

4 Calculation of the likelihood

The likelihood of a tree with augmented internal data is the product of the likelihoods for each branch of the tree times the probability of a particular arrangement at an arbitrary root node.

Calculation of the likelihood along a single branch. Let $p(x, y; \beta)$ be the probability that gene arrangement x is converted to gene arrangement y after a Poisson(β) distributed random number of equally likely reversals. This probability is:

$$p(x, y; \beta) = \sum_{k=d(x,y)}^{\infty} \frac{e^{-\beta} \beta^k}{k!} \times \frac{\#_k(x, y)}{\binom{g}{2}^k} \tag{1}$$

where $d(x, y)$ is the length of a minimal sequence of reversals to convert x to y and $\#_k(x, y)$ is the number of different sequences of k reversals that transform x to y. Each term is the product of the Poisson probability of exactly k reversals and the probability that a sequence of k random reversals would transform arrangement x to y.

We calculate the first three terms of this sum exactly and use the approximation

$$\frac{\#_k(x, y)}{\binom{g}{2}^k} \approx \frac{1}{2^{g-1}(g-1)!} \tag{2}$$

for the remaining terms. This approximation is based on the approximately uniform distribution of gene arrangements after a large number of random reversals from any starting gene arrangement. The resulting approximation is

$$p(x, y; \beta) \approx \sum_{k=d(x,y)}^{d(x,y)+2} \left(\frac{e^{-\beta} \beta^k}{k!} \times \frac{\#_k(x, y)}{\binom{g}{2}^k} \right) + \left(1 - \sum_{k=0}^{d(x,y)+2} \frac{e^{-\beta} \beta^k}{k!} \right) \Big/ \left(2^{g-1}(g-1)! \right) \tag{3}$$

where x and y are the arrangements at the branch end-points, β is the branch length, $d(x, y)$ is the length of the smallest sequence of reversals from x to y $\#_k(x, y)$ is the number of sequences of k reversals that transform x to y, and there are g genes around the circle.

The likelihood of the data (L, I) for a given tree $(\tau, \mathbf{b}_\tau)$ is

$$p(L, I | \tau, \mathbf{b}_\tau) = \frac{1}{2^{g-1}(g-1)!} \times \prod_{i=1}^{2s-3} p(x_i, y_i; b_i) \tag{4}$$

where x_i and y_i are the gene arrangements from (L, I) that correspond to the end-points of branch i and b_i is the branch length of branch i.

Prior distributions. We assume a uniform prior distribution over the discrete set of unrooted tree topologies and a uniform distribution over the set of possible gene arrangements at an arbitrary root node. We assume the vector of branch lengths is uniformly distributed over the positive orthant $\mathbf{R}^{2s-3}$ subject to the constraint that the distance between any two leaves of the tree is less than a constant T. In the example below we use $T = 16$.

Calculation details. There is a rapid algorithm for calculating $d(x, y)$ (Hannenhalli and Pevzner [4], Kaplan, Shamir, and Tarjan [5]). The difficult part of the calculation is determining $\#_k(x, y)$. Each of these counts corresponds to an equivalent count between an arrangement and the identity arrangement. By brute force calculation, we have determined these counts for $k \leq d(x, y) + 2$ for $g \leq 10$. We take advantage of symmetries to reduce the necessary storage. The file that contains the three counts for all 4,674,977 possible arrangements up to symmetry for $g = 10$ requires nearly 250MB. To run our software on a data set with ten genes requires a computer with more than of 250MB of RAM to store and access the counts when needed. The time and space requirements to determine and store these counts increases exponentially. New approaches to quickly calculating or approximating these counts are necessary for the methods of this paper to be extended to gene arrangements with more genes.

5 Example Data

Due to the limitations of the current program, we tested the procedure on artificial data. We began with a known tree and a given signed permutation at the root. The permutation was of size 9, simulating a circular signed permutation of size 10. The permutations at the other nodes were then generated in a depth-first fashion by applying a number of random reversals to the parent of each node. The number of reversals was randomly chosen by a Poisson distribution with mean equal to the length of the branch to the parent. The data set given in Table 1 consists of the permutations at the leaves of the tree for one realization.

Table 1. Example Data Set

Taxon	Circular permutation
a	0,-5,-4,9,3,-8,2,7,-1,6
b	0,-5,-4,9,3,1,-7,-2,8,6
c	0,-7,-2,8,-3,-9,4,6,-1,5
d	0,-6,-2,-1,-8,-7,9,3,4,5
e	0,-5,-4,-3,-9,7,8,1,2,6
f	0,7,-8,-6,4,1,5,-3,-2,9

We completed four separate runs from randomly selected initial trees and obtained consistent results. Each run consisted of 10,000,000 cycles of which the first 100,000 were considered burn-in and discarded. Every tenth tree of the remaining data was sampled for inference. Each run took 90 minutes of CPU time on a 933 MHz Pentium III running Redhat Linux 7.0.

The combined samples yield a posterior probability of 0.480 for the best tree topology which agreed with the true tree topology. The estimated Monte Carlo standard error of this probability based on the four independent samples was 0.025.

Two other likely trees were generated by all four runs. The combined samples yield a posterior probability of 0.133 and 0.128, respectively, for the two trees, with an estimated Monte Carlo standard error of 0.01 for each.

6   Conclusions

We have demonstrated a Bayesian approach towards genome arrangement data may be successful. The usefulness of this method depends on the solution or an approximation of a solution to the problem of finding the number of paths between two permutations. A brute force method of pre-calculating the distance works on this small example, but clearly is not practical for full mitochondrial genomes.

A second limitation is that if there are a large of number of reversals between taxa the posterior distribution will be spread over a large number of phylogenies. However, this would seem to be a limitation of the data rather than the method. It is unlikely that any method of phylogenetic reconstruction would be successful with data that is not very informative. In this instance, it would be useful to combine the genome arrangement data with other data such as nucleotide sequence data. This is simple to do within the MCMC paradigm.

In comparison to Bayesian methods for phylogenetic reconstruction from nucleotide sequence data, our current approach to Bayesian analysis of genome arrangement data requires longer simulation runs to achieve the same Monte Carlo standard error. In BAMBE [6, 14], we were able to sum over all internal data whereas in this program such a summation would require enumerating all possible signed permutations which is far too large to be computed. Instead, we have data in the internal nodes which is updated at each cycle of the MCMC algorithm. This is not a limitation of this method, but one should be aware that longer runs are necessary.

Clearly this is a work in progress. We have only looked at small data sets in order to determine whether or not Bayesian methods can reconstruct phylogenies from genome arrangement data. We believe that with an appropriate approximation to the path counts this approach could be extended to the full mitochondrial genome and be computationally comparable to other methods which are not based on a likelihood model.

References

1. Boore, J. L., T. M. Collins, D. Stanton, L. L. Daehler, and W. M. Brown. Deducing arthropod phylogeny from mitochondrial DNA rearrangements. Nature **376** (1995) 163–165

2. Boore, J., D.V. Lavrov, and W.M. Brown. Gene translocation links insects and crustaceans. Nature **392** (1998) 667–678

3. Boore, J. Mitochondrial gene arrangement source guide, version 5.0. DOE Joint Genome Institute. http://www.jgi.doe.gov/Mitochondrial_Genomics.html (2000)

4. Hannenhalli, S. and Pevzner, P. Transforming Men into Mice (polynomial algorithm for genomic distance problem). 36th Annual IEEE Symposium on Foundations of Computer Science (1995) 581–592

5. Kaplan, H., Shamir, R. and Tarjan, R.E.. Faster and Simpler Algorithm for Sorting Signed Permutations by Reversals. SIAM Journal on Computing **29(3)** (1999) 880–892

6. Larget, B. and Simon, D.L. Markov chain Monte Carlo algorithms for the Bayesian analysis of phylogenetic trees. Mol. Biol. Evol. **16** (1999) 750–759

7. Li, S., Doss H., Pearl D. Phylogenetic tree construction using Markov chain Monte Carlo. J. Amer. Stat. Assoc. **95** (2000) 493–508

8. Mau, B., M.A. Newton, and B. Larget. Bayesian phylogenetic inference via Markov chain Monte Carlo methods. Biometrics. **55** (1999) 1–12

9. Newton, M., B. Mau, and B. Larget. Markov chain Monte Carlo for the Bayesian analysis of evolutionary trees from aligned molecular sequences. In F. Seillier-Moseiwitch (Ed.), Statistics in Molecular Biology and Genetics. IMS Lecture Notes-Monograph Series, **33** (1999) 143–162

10. Rannala, B., and Z. Yang. Probability distribution of molecular evolutionary trees: A new method of phylogenetic inference. J. Mol. Evol. **43** (1996) 304–311

11. Smith, A., and Roberts, G.. Bayesian computation via the Gibbs sampler and related Markov chain Monte Carlo methods. J. R. Statist. Soc. B. **55** (1993) 3–23

12. Sankoff, D., Leduc, G., Antoine, N., Paquin, B., Lang, B.F. and Cedergren, R. Gene order comparisons for phylogenetic inference: evolution of the mitochondrial genome. Proceedings of the National Academy of Sciences **89** (1992) 6575–6579

13. Sankoff, D. and Blanchette, M. Phylogenetic invariants for Genome Rearrangement. J. of Comp. Biol. **6** (1999) 431–445

14. Simon, D. and B. Larget. Bayesian analysis in molecular biology and evolution (BAMBE), version 2.02 beta. Department of Mathematics and Computer Science, Duquesne University (2000)

15. Swofford, D.L., Olsen, G.J., Waddell, P.J., and Hillis, D.M. Phylogenetic inference. In D. M. Hillis, C. Moritz, and B. K. Mable, editors, Molecular Systematics, 2nd Edition. Sinauer Associates, Sunderland, Massachusetts (1996)

16. Yang, Z. and Rannala, B. Bayesian Phylogenetic Inference Using DNA Sequences: A Markov Chain Monte Carlo Method. Mol. Biol. Evol. **14** (1997) 717–724

Late Submissions

Genetic Programming: A Review of Some Concerns

Maumita Bhattacharya and Baikunth Nath

School of Computing & Information Technology
Monash University (Gippsland Campus), Churchill 3842, Australia
{Email: Maumita,Baikunth.Nath@infotech.monash.edu.au}

Abstract: Genetic Programming (GP) is gradually being accepted as a promising variant of Genetic Algorithm (GA) that evolves dynamic hierarchical structures, often described as programs. In other words GP seemingly holds the key to attain the goal of "automated program generation". However one of the serious problems of GP lies in the "code growth" or "size problem" that occurs as the structures evolve, leading to excessive pressure on system resources and unsatisfying convergence. Several researchers have addressed the problem. However, absence of a general framework and physical constraints, viz, infinitely large resource requirements have made it difficult to find any generic explanation and hence solution to the problem. This paper surveys the major research works in this direction from a critical angle. Overview of a few other major GP concerns is covered in brief. We conclude with a general discussion on "code growth" and other critical aspects of GP techniques, while attempting to highlight on future research directions to tackle such problems.

1. Introduction

One of the major drawbacks of Genetic Programming (GP) coming in its way of becoming the dream "machine" for automated program generation is "Code Growth". If unchecked this results in consumption of excessive machine resources and slower convergence, thus making it practically unfit for complex problem domains needing large individuals as member programs in the search space. A good deal of research has gone into identifying causes and solutions to this problem [3,8,9,12,14,18]. However, as real life applications demand, absence of commonality in complexity of problem domain, operator structures, basic representations used in these works and the ultimate bounds on feasible population size, have failed to produce any clear-cut explanation of this problem. This paper analyses the "code growth" and a few other major GP concerns, research inputs in this direction and their shortcomings. An attempt has been made to highlight future research directions in this regard.

The discussion starts with a brief analysis of a few major genetic programming concerns relating to contribution and behavioral characteristics of important GP operators like crossover and mutation, and the issue of generality of genetic programming solutions (section 2 and 3). Detailed discussion on basic GP techniques and related operators can be found in [3].

The subsequent sections are devoted to a brief discussion on the "code growth" problem in general, while critically analyzing some of the research inputs in this direction. A brief summary of other GP constraints and probable future research directions are covered in section 5 and 6.

2. The Controversy Over GP Operators

The major issues related to the breeding operators are concerned with contributions and behavioral characteristics of these operators. The following discussion deals with a comparative analysis of the crossover and the mutation operator.

The Tussle Between Mutation and Crossover: In GP, since variable-length representation is involved, choosing a better operator means, picking the one, yielding the highest overall fitness before bloat stops all effective evolutions. While fitness and mean tree size are both highly domain-dependent factors, researches [17] have shown that a tenuous trend does exist [figure 1].

Inference 1: In general, population size more strongly determines whether crossover produces better fitness, whereas, number of generations more strongly determines whether it produces larger trees.

Inference 2: Crossover is apparently a more efficient approach for large populations with small run lengths. Mutation on the other hand, is more efficient for small population with long run lengths.

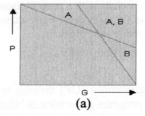
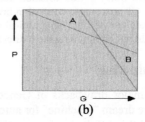

(a)   (b)

Figure 1. (a) Areas where Subtree Crossover has better fitness, or where Subtree Mutation has better tree size. Here, A: Area where Crossover has higher fitness and B: Area where Mutation has smaller tree size. (b) Areas where Subtree Crossover or Subtree Mutation is more efficient. Here, A: Area where Crossover is more efficient and B: Area where Subtree Mutation is more efficient. P and G denote the population size and the number of generations respectively.

Whither Crossover: Though an important breeding operator, crossover may not be the best choice in every situation. One of the "inadequacies" of crossover is realized in the form of loss of diversity in the population, and thereby forcing GP to rely on the ineffective mutation operator to create "fitter" solutions [2]. Let us consider GPs with function sets of function arity two. Let us consider the following:

d - the maximum tree depth; l - tree layer from root node to d ;

n_l - the number of nodes in layer $l = 2^l$, where $0 \leq l \leq d$;

N_l - the number of nodes in all layers from root to layer $l=$

$$\sum_{i=0}^{l} n_i = \sum_{i=0}^{l} 2^i = 2^{l+1} - 1;$$

N_d - the total number of nodes in a tree of depth $d = 2^{d+1} - 1;$

$P(any)_l$ - the likelihood of any node(s) in a tree, in layer l, experiencing some crossover activity, i.e. when crossover point occurs in any layer from 0 to $l =$

$$\frac{N_l}{N_d} = \frac{2^{l+1} - 1}{2^{d+1} - 1} \cong \frac{1}{2^{d-l}};$$

$P(layer)_l$ - the likelihood of crossover occurring within a layer l, i.e. when the two crossover points are in the same layer

$$= \frac{n_l^2}{N_d^2} = \frac{2^{2l}}{(2^{d+1} - 1)^2} \cong \frac{2^{2l}}{2^{2d+2}} = \frac{1}{2^{2(d-l+1)}}$$

$P(2\ legal\ offsprings)$ - the likelihood of crossover based on a random choice of crossover points, producing two legal offspring

$$= \frac{legal}{total} = \frac{\sum_{i=0}^{d} n_i^2}{N_d^2} \cong \frac{\sum_{i=0}^{d} 2^{2i}}{2^{2(d+1)}} \cong \frac{1}{4}, \text{ for large d;}$$

From the above calculations, analyzing $P(any)_l$, we find that the upper layers receive relatively little attention from the crossover operator, especially as the tree grows large. As subtree discovery and the spread of subtrees takes place at lower levels, chiefly involving the leaf nodes, immediately beneficial subtrees are quickly spread within the trees at the expense of other subtrees of less immediate benefit. This could lead to either faster convergence or loss of diversity in the population.

3. The Generality Issue

Generality of solutions can be treated as one of the major guiding principles for the search of representations using genetic programming. Here, generality can be expressed as how well a solution performs on unseen cases from the underlying distribution of examples. Generalization may be easier to achieve with small solutions, but it is hard to achieve, just by adding size component in the fitness function. Biasing search towards smaller programs, in the pursuit of finding compact solutions, also leads to biasing towards a model of lower dimensionality. A biased model has a fewer degrees of freedom and lower variance. We normally aim to achieve a low bias and low variance model, but in practical situations, there has to be a trade off between the two.

For standard GP, the structural complexity is simply the number of nodes in the tree representation. For GP, using subroutines, it is computed by adding up the structural complexities of all subroutines or ADFs. The average path length of an individual

i can be defined as $\overline{l(i)} = \dfrac{1}{n(i)} \sum\limits_{j=1}^{n(i)} dist(root(i), j)$,

Where, $n(i)$ is the number of leaves of the tree representing i and $dist(root(i), j)$ is the distance between two nodes. Researches [4] findings are as below:

• *The average path length of an offspring is crossover independent on an average, if only a single offspring is generated from each selected couple.*

• *Mutation, on the other hand is path length invariant on average over generations, only if newly generated subtrees have the same size as replaced subtrees.*

These size-operation links have enormous effect on the generality issue. Effective size offers an insight into the properties of the search space in a given problem domain. More importantly, this can be used for modularization purposes leading to a smaller complexity and increased generalization.

4. The "Code Growth" Problem

The problem of "code growth" arises, as the evolved structures appear to drift towards large and slow forms on an average. It is a serious problem as it causes enormous pressure on resources (in terms of storage and processing of extraneous codes) that is disproportional to the performance of the evolving programs. It even may reach a point where extraneous codes may invade the search space to such an extent that improving useful codes will become virtually impossible. This problem has been identified under various names, such as "bloat", "fluff" and "complexity drift" or increasing "structural complexity"[19]. The questions, which demand answers, are:

- *Cause of code growth - Is it a protective mechanism against destructive Crossover?*
- *Role played by "complexity drift" in selection & survival of structures.*
- *Role of other operators in producing code growth.*
- *Relationship between code growth and code shape.*
- *Relationship between code growth and code shape.*
- *Affectivity of parsimony pressure in controlling code growth.*

Most of the above mentioned concerns have been dealt with by Koza [3], Altenberg [6], Sean Luke[17], Blickle and Thiele [12], Nordin and Banzhaf [9], Nordin [11], McPhee and Miller [7], Soule et al. [14], Langdon et al. [18]. The common approach to address the problem has been either by enforcing a size or depth limit on programs or by an explicit size component in the GP fitness [3, 1, 5]. Alternate techniques have been proposed by Blickle [16], Nordin et al. [10]], Soule and Foster [14], Soule [15],

Angeline [8], Langdon [20]. However, as mentioned earlier, absence of a general framework makes it difficult to draw a generic conclusion about the "code growth" problem.

The Problem

Various research works have shown that program depth grows apparently linearly but rapidly with standard crossover. While being problem and operator dependent, on average program trees grow roughly by one level per generation, ultimately putting tremendous pressure on system resources [20]. It has been observed that if the program size exceeds some problem and fitness level dependent threshold limit, the distribution of their fitness value in the GP search space does not get affected by "complexity" or length. Hence the number of programs with a specific fitness is distributed like the total number of programs and most programs will have a depth close to $2\sqrt{\pi(\text{number of internal nodes})}$, ignoring terms $O(N^{1/4})$. Any stochastic search technique, inclusive of GP, finds it difficult to improve on the best trial solution and the subsequent runs are likely to produce either same or even worse trial solutions. The selection process ensures that only the better ones survive. Unless a bias is used, the longer programs with the current level of performance will have better chance to survive.

What Is Behind It

As has been mentioned earlier, several researchers have tried to find an answer to this question. Based on the chosen problem domain, types of operators various inferences have been drawn in this regard. While considering "code growth", its worthwhile to consider both the average program size and the size of the largest program. Both of these increase considerably, in a typical GP program [15, 20, 11]. In standard GP, only the crossover and the selection operators affect the population. A crossover operation affects only the individual programs within a population, but apparently does not have any direct effect on the average size of the population. On the other hand, the selection operator can easily change the average size of the population by preferentially selecting the larger or the smaller programs. Hence, the selection operator can only increase the average program size to the largest program size, and then by selecting several copies of the largest program. Thus, apart from other factors, the sustained "code growth" in GP can only possibly be explained by combined effect of these two operators or an interaction between the two [15]. Effect of inclusion of fitness over code growth has been discussed by Langdon and Poli [19]. According to them, random selection did not cause bloating. However, "code growth" does not directly contribute to fitness, as most of the excess codes are nothing but redundant data. Presently there are several distinct theories explaining the cause of "code growth", of which the major ones are [15, 10, 18]:

- *The Destructive Crossover Hypotheses*
- *The solution distribution theory*
- *Removal of bias theory*
- *The shape-size connection*

a) The Destructive Crossover Hypothesis

According to this hypothesis, on average, crossover has neutral or destructive effect on individuals as part of each parent gets discarded.

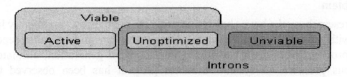

Figure 2. A Venn Diagram Depicting Various Types of Codes and Their Relations

However, programs having a large amount of redundant data, stands less chance of getting discarded, as crossover has less effect on unviable codes [figure 2] and hence ensures better chance of survival for the individual [15, 14,17]. The probability p_s, of an individual being selected and surviving crossover is:

$$p_s = \frac{f_i}{\overline{f}} (1 - p_c p_d)$$

Where, p_c and p_d are the crossover probability and the probability that crossover is destructive. Let f_i and $\overline{f}$ be the ith individual's fitness and the average fitness of the population respectively.

b) The solution distribution theory

This theory claims that "code growth" is partially caused by the distribution of semantically equivalent solutions in the phenotype or the solution space. In case of variable length representations, various semantically similar, but syntactically different individuals can represent a particular solution. A preferential choice towards larger solution will automatically lead to "code growth".

c) The Removal of bias theory

Subtree removal is the first step in both crossover and subtree mutation. This theory is based on the fact that changes in unviable code do not affect fitness and hence the source and size of a replacement branch during crossover, will not affect the fitness of the new program. Hence, there is no corresponding bias towards smaller (or larger) replacement branches. Fitness neutral operations will favor the removal of small branches and replacement with averaged sized branches creating a general increase in fitness neutral offspring. Besides, since crossover is either neutral or destructive in nature, these fitness-neutral offspring will be favored at the time of selection, thereby increasing the average program size, even in the absence of increase in fitness. Figure 3 depicts a generalized GP syntax tree [15]. The ancestors of any viable node are viable and if any node in a tree is viable, then the root node is definitely viable.

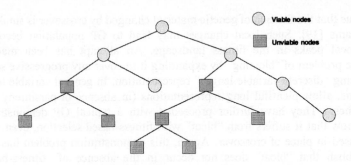

Figure 3. A Generalized GP Syntax Tree

d) The Shape-Size Connection

Apart from the above three causes of "code growth" a shape-size connection may also be a contributor to "complexity drift". Code growth leads towards a specific problem independent syntax tree shape. Syntax trees have been found to evolve into problem dependent shapes in absence of "bloating".

Average growth in program depth when using standard subtree crossover with standard random initial trees is near linear [20]. The rate is about 1 level per generation but varies between problems. When combined with the known distribution of number of programs, this yields a prediction of sub-quadratic, rising to quadratic, growth in program size. There is considerable variation between runs but on the average we observe sub-quadratic growth in size. This in a continuous domain problem, rises apparently towards a quadratic power law limit.

Discrete   mean length $\leq O$ (generations^2)

Continuous $\lim_{g \to \alpha}$ mean length$= O$ (generations^2)

However in large discrete programs, the whole population has the same fitness, even though its variety is 100%.

Most GP systems store each program separately and memory usage grows linearly with program size. i.e. in the range of O(generations$^{1.2}$) to O(generations^2). Run time is typically dominated by program execution time.

An Analysis of Some Major Research Inputs to "Code Growth" Problem

As has been mentioned earlier, the problem of "complexity drift" or "bloating" has been reported and analyzed by many researchers. The principal explanation for "bloat" has been the growth of "introns" or "redundancy". Bickle and Thiele[12, 16], Nordin et al.[9, 10], and McPhee et al.[7] have argued in favor of "introns", supporting its code protection property at times of "destructive" crossover. Major percentage of "code growth" seems to occur to protect existing, successful programs against the damaging effect of "destructive crossover"[15]. However, this code protection property does not guarantee selection of better training sets. Poli and

Langdon argue that the fraction of genetic material changed by crossover is smaller in longer programs [19]. Such local changes may lead to GP population becoming trapped at local peaks in the fitness landscape. An attempt has been made to generalize the problem of "bloating" by expanding it to cover any progressive search technique, using "discrete variable length" representation. In general variable length representations, allow plentiful long representations (in absence of parsimony bias) than short ones. They have further proceeded with a typical GP demonstration problem to show that it suffers from "bloat" with fitness-based selection, even when mutation is used in place of crossover. Again, this demonstration problem has been used to establish that "bloat" does not occur in the absence of "fitness-based" selection. However, generalization of this concept is perhaps questionable, as even fitness-free neutral selection might tend to favor longer programs and crossover operator may preferentially pick up smaller branches for replacement by averaged size branches. These two functions may have combined contribution to "code growth". Terence Soule [15] has suggested a size and shape relationship for "code growth". It is possible that sparser the shape of the syntax tree, higher the chances of upward drift in complexity. Whatever may be the cause of "code growth", it is obvious from the vast research inputs in this direction that standard GP technique does not essentially try to evolve "better" programs, rather the inclination being towards evolving programs those survive, regardless of their performances. Hence, having an explicit fitness function serves little purpose. Rather, the focus should be on basing reproduction on program behavior and assessing it dynamically. It is also important to explore the yet unknown forces that affect the complex, dynamic system of an evolving population.

Means to Control "Bloat"

Of the major mechanisms suggested for controlling "bloat"[19], the first and the most widely used one uses a universal upper bound either on tree depth [3] or program length. The second method exerts parsimony pressure (as discussed in Section 3) to control "code growth"[3, 1]. The third one involves manipulation of the genetic operations [19, 12, 16].

5. More Concerns for GP

The problem of "code growth" is not really any isolated problem for genetic programming. Effective identification of "redundancy", quantification of the destructive power of crossover operator, the "representation" problem, generalization of GP process to handle discrete, complex problem domains, affectivity of "building block" hypotheses and standard genetic operators are a few among a host of several other GP concerns. Besides, as a direct effect of measures taken to control "code growth", in the form of imposition of restriction on tree depth, may have adverse effect on GP performance. This could manifest itself by decreasing diversity in the population and thereby leading to a premature convergence to a sub-optimal solution. Diversity may tend to drop near the root node and thereby causing inability for GP to create "fitter" solution trees.

6. Conclusion

GP, as a stochastic process for automatic program generation has gained some early success. In many areas, including robot and satellite guidance, data mining, neural net design, circuit design, and economic strategy development, GP has produced satisfactory results. However, it has not yet scaled well to larger and more complex problems. "Code growth" has a major role to play in this. Our purpose has been to identify the typical characteristics of a few GP concerns and the "code growth" problem in particular, and explore future research direction to solve the problems.

It has been observed that due to enormous resource requirement, simpler problems needing small trees have been chosen to accommodate adequate number of nodes in search space. Hence, the problem remains un-addressed for complex, discrete domains requiring large candidate trees. Use of evolutionary techniques to decompose problems and evolve smaller trees to generate larger trees may be used. However, time and resource requirements will still pose a problem. Hybrid search techniques may be incorporated to hasten up search process. It has been further observed that use of different operators produces different "bloat" characteristics making it harder to identify general "bloating" trends. Absence of selection operator altogether, prevents bloating. Generation and use of new mutation and crossover operators or even altogether new operators may help reducing "bloating". Inter-generational processing on fluctuating numbers of individuals may be used to reduce redundancy. Representation is another critical issue in genetic programming. In case of "directed acyclic graph" representation, memory usage may be less, as in every generation individuals are deleted. Re-evaluation of unchanged codes may be avoided by caching relevant information using conventional storage structure and fast sear techniques. As has been mentioned earlier, unless checked "code growth" or "complexity drift" could even make it impossible to evolve useful codes. Hence, the problem needs understanding and control, for GP, which otherwise is a promising mechanism to solve a multitude of complex problems, to remain a viable technique.

References

[1] Byoung-Tak Zhang and Heinz Muhlenbein. *Balancing Accuracy and Parsimony in Genetic Programming*. Evolutionary Computation, 3(1): 17-38, 1995.

[2] Chris Gathercole and Peter Ross. *An Adverse Interaction Between Crossover and Restricted Tree Depth in Genetic Programming* - presented at GP 96

[3] John R.Koza. *Genetic Programming: On the Programming of Computers by Means of Natural Selection*.MIT Press, 1992.

[4] Justinian P. Rosca. *Generality Versus Size in Genetic Programming*. Proceedings of the Genetic Programming 1996 Conference (GP-96), The MIT Press.

[5] Justinian P. Rosca. *Analysis of Complexity Drift in Genetic Programming*. Proceedings of the Second Annual Conference on Genetic Programming, 1997.

[6]   Lee Altenberg. *Emergent Phenomena in Genetic Programming.* Evolutionary Programming – Proceedings of the Third Annual Conference, pp233-241. World Scientific Publishing, 1994.

[7]   Nicholas F. McPhee and J. D. Miller. *Accurate Replication in Genetic Programming.* Genetic Algorithms: Proceedings of the Sixth International Conference (ICGA95), pp303-309. Morgan Kaufmann.1995.

[8]   Peter J. Angeline. *Subtree Crossover Causes Bloat.* Genetic Programming 1998: Proceedings of the Third Annual Conference, pp745-752, Wisconsin. Morgan Kaufmann.1998.

[9]   Peter Nordin and Wolfgang Banzhaf. *Complexity Compression and Evolution.* ICGA95, pp310-317, Morgan Kaufmann. 1995.

[10]   Peter Nordin, Frank Francone, and Wolfgang Banzhaf. *Explicitly Defined Introns and Destructive Crossover in Genetic Programming.* Advances in Genetic Programming 2, pp111-134. MIT Press, 1996.

[11]   Peter Nordin. *Evolutionary Program Induction of Binary Machine Code and Its Applications.* PhD thesis, der Universitat Dortmund am Fachereich Informatik, 1997.

[12]   Tobias Blickle and L. Thiele. *Genetic Programming and Redundancy.* Proc. Genetic Algorithms within the Framework of Evolutionary Computation (Workshop at KI-94), Saarbrücken, Germany, 1994.

[13]   Terence Soule, J. A. Foster, and J. Dickinson. *Code Growth in Genetic Programming.* Genetic Programming 1996: Proceedings of the First Annual Conference, pp215-223, Stanford. MIT Press.1996.

[14]   Terence Soule and James A. Foster. *Code Size and Depth Flows in Genetic Programming.* Genetic Programming 1997, pp313-320. Morgan Kaufmann.1997.

[15]   Terence Soule. *Code Growth in Genetic Programming. PhD. Dissertation.* University of Idaho. May 15,1998.

[16]   Tobias Blickle. *Evolving Compact Solutions in Genetic Programming: A Case Study.* Parallel Problem Solving From Nature IV. LNCS 1141, pp564-573. Springer.1996.

[17]   Sean Luke. *Issues in Scaling Genetic Programming: Breeding Strategies, Tree Generation, and Code Growth.* PhD. Thesis, 2000.

[18]   William. B. Langdon, T. Soule, R. Poli, and J. A. Foster. *The Evolution of Size and Shape.* Advances in Genetic Programming 3, pp163-190. MIT Press.

[19]   William. B. Langdon and R. Poli. *Fitness Causes Bloat: Mutation.* EuroGP '98, Springer-Verlag.1998.

[20]   Willam.B. Langdon. *Quadratic Bloat in Genetic Programming.* Presented at GECCO'2000.

Numerical Simulation of Quantum Distributions: Instability and Quantum Chaos

G. Yu. Kryuchkyan, H. H. Adamyan, and S. B. Manvelyan
Institute for Physical Research, National Academy of Sciences,
Ashtarak-2, 378410, Armenia
Yerevan State University, Alex Manookyan 1, 375049,
Yerevan, Armenia

April 9, 2001

Abstract

Quantum state diffusion with moving basis approach is formulated for computation of Wigner functions of open quantum systems. The method is applied to two quantum nonlinear dissipative systems. Quantum dynamical manifestation of chaotic behavior, including emergence of chaos, as well as the formation of Hopf bifurcation in quantum regime are studied by numerical simulation of ensemble averaged Wigner functions.

1 Introduction and Models

All real experiments in quantum physics always deal with open systems, which are not even approximately isolated and significantly affected by the environment. Interaction with the environment leads to dissipation and decoherence, i.e. to irreversible loss of energy and coherence, and can monitor some of the system observables. Decoherence can destroy the quantum-mechanical interference or quantum superposition of states. All quantum systems suffer from decoherence, but it has very specific implications in area of quantum information and quantum computation.

The present paper is devoted to the problem of numerical simulation of dissipation and decoherence in open quantum system. We concentrated mainly on investigation of so-called quantum chaotic systems - quantum systems, which have chaotic dynamic in a classical limit. Their analysis requires the high-performance calculations as well as the development of new numerical algorithms.

One of the most practical tools for analyzing the time-evolution of an open quantum system is the master equation for the density matrix. However, the analytical studies in this area are very difficult and the solutions of master equations have been established only for relatively simple models. We remind the reader of a common approach to analysis of an open system. The starting point is the master equation in Lindblad form [1] for reduced density matrix $\rho(t) = Tr(\rho_{tot}(t))$ which is obtained from the total density matrix of the universe by tracing over the environment Hilbert space

$$\frac{\partial \rho}{\partial t} = -\frac{i}{\hbar} [H,\rho] + \sum_i \left(L_i \rho L_i^+ - \frac{1}{2} L_i^+ L_i \rho - \frac{1}{2} \rho L_i^+ L_i \right). \tag{1}$$

Here H is the Hamiltonian of the system and L_i are Lindblad operators which represent the effect of the environment on the system in a Markov approximation. Numerically there is often a large advantage in representing a system by quantum state rather than a density matrix. On this direction, there are many ways to "unraveling" the master equation into "trajectories" [2]. Our numerical analysis is based on the quantum state diffusion (QSD) approach which represents the density matrix into component stochastic pure states $|\Psi_\xi(t)\rangle$ describing the time-evolution along a quantum trajectory [3]. According to QSD the reduced density operator is calculated as the ensemble mean $\rho(t) = M(|\Psi_\xi(t)\rangle \langle \Psi_\xi(t)|)$, with M denoting the ensemble average. The corresponding equation of motion is

$$|d\Psi_\xi\rangle = -\frac{i}{\hbar} H |\Psi_\xi\rangle \, dt - \tag{2}$$
$$\frac{1}{2} \sum_i \left(L_i^+ L_i - 2 \langle L_i^+ \rangle L_i + \langle L_i \rangle \langle L_i^+ \rangle \right) |\Psi_\xi\rangle \, dt + \sum_i \left(L_i - \langle L_i \rangle \right) |\Psi_\xi\rangle \, d\xi_i.$$

Here ξ indicates the dependence on the stochastic process, the complex Wiener variables $d\xi_i$ satisfy the fundamental correlation properties:

$$M\left(d\xi_i \right) = 0, \; M\left(d\xi_i d\xi_j \right) = 0, \; M\left(d\xi_i d\xi_j^* \right) = \delta_{ij} dt, \tag{3}$$

and the expectation value equals $\langle L_i \rangle = \langle \Psi_\xi | L_i | \Psi_\xi \rangle$.

In what follows we apply this method for numerical calculations of quantum distributions averaged on trajectories. Among them the Wigner function play a central role because provides a wide amount of information about the system and also provides a pictorial view. Below we shall give the results of numerical calculations for two models of driven, dissipative nonlinear oscillators, which are relevant to some systems in quantum optics.

One of those is the model of nonlinear oscillator driven by two periodic forces, which described by the Hamiltonian

$$H = \hbar\omega_0 a^+ a + \hbar\chi(a^+ a)^2 + \hbar \left[(\Omega_1 \exp(-i\omega_1 t) + \Omega_2 \exp(-i\omega_2 t)) a^+ + h.c. \right], \tag{4}$$

where a, a^+ are boson annihilation and creation operators, ω_0 is an oscillatory frequency, ω_1 and ω_2 are the frequencies of driving fields, and χ is the strength of the anharmonicity. The couplings with two driving forces are given by Rabi frequencies Ω_1 and Ω_2.

The other is the model of two driven harmonic oscillators coupled by a nonlinear process, which is described by the following Hamiltonian

$$H = \hbar\omega_1 a_1^+ a_1 + \hbar\omega_2 a_2^+ a_2 + i\hbar\left(Ee^{-i\omega t}a_1^+ - E^*e^{-i\omega t}a_1\right) + i\hbar\frac{k}{2}\left(a_1^{+2}a_2 - a_1^2 a_2^+\right).$$
(5)

Here a_1, a_2 are the operators of the modes at the frequencies ω_1 and ω_2 respectively, k is the coupling coefficient between the modes, which in the case of an optical interaction can be proportional to the second-order nonlinear susceptibility $\chi^{(2)}$; E is a complex amplitude of the periodic force at the frequency ω.

2 MQSD Algorithm for Wigner Function

It seems naturally to use the Fock's state basis of two harmonic oscillators $\{|n\rangle_1 \otimes |m\rangle_2$; $n, m \in (0, 1, 2, ..., N)\}$ for numerical simulation. Unfortunately, in the most interesting cases, which are relevant to experiments, the effective number of Fock's states quickly becomes impractical. On this reason, it is very difficult to obtain a single quantum trajectory in such regimes, not to mention performing an ensemble averaging. However, it is possible to reduce significantly the number of need basis states by choosing an appropriate basis. It was demonstrated in Ref.[4] considering quantum state diffusion with moving basis (MQSD). Its advantages for computation was also shown in [4]. In this paper we develope the MQSD method for calculation of distribution functions including Wigner functions.

At first, we shortly describe the application of MQSD method for the numerical simulation of Wigner function using the standard definition based on the density matrix [2]. We have for each of the modes of the harmonic oscillators

$$W_i(\alpha) = \frac{1}{\pi^2} \int d^2\gamma Tr(\rho_i D(\gamma)) \exp(\gamma^*\alpha - \gamma\alpha^*),$$
(6)

where the reduced density operators for each of the modes are constructed by tracing over the other mode

$$\rho_{1(2)} = Tr_{2(1)}(\rho), \quad \rho = M(|\Psi_\xi\rangle\langle\Psi_\xi|).$$
(7)

We use the expansion of the state vector $|\Psi_\xi(t)\rangle$ in the basis of excited coherent states of two modes as

$$|\Psi_\xi(t)\rangle = \sum a_{nm}^\xi(\alpha_\xi, \beta_\xi)|\alpha_\xi, n\rangle_1 |\beta_\xi, m\rangle_2,$$
(8)

where

$$|\alpha, n\rangle_1 = D_1(\alpha)|n\rangle_1, \quad |\beta, m\rangle_2 = D_2(\beta)|m\rangle_2$$
(9)

are the excited coherent states centered on the complex amplitude $\alpha = \langle a_1 \rangle$, $\beta = \langle a_2 \rangle$. Here $|n\rangle_1$ and $|m\rangle_2$ are Fock's number states of the fundamental and second-harmonic modes, and D_1 and D_2 are the coherent states displacement operators

$$D_i(\alpha) = \exp(\alpha a_i^+ + \alpha^* a_i). \tag{10}$$

As it is known, the MQSD method is very effective for numerical simulation of quantum trajectories. However, the problem is that in this approach the two mode state vector $|\Psi_\xi(t)\rangle$ is expressed in the individual basis depending on the realization. It creates the additional definite difficulties for calculation of the density matrix at each time of interest on the formula (7), which contains the averaging on the ensemble of all possible states. In practical point of view, it is useful to operate with state vector $|\Psi_\xi(t)\rangle$ reduced to a basis which is the same for all realizations of the stochastic process $\xi(t)$. To avoid this we express the density operators as

$$\rho_i(t) = \sum_{nm} \rho_{nm}^{(i)}(t) \, |\sigma, n\rangle \, \langle \sigma, m| \tag{11}$$

in the basis of excited Fock states with an arbitrary coherent amplitude σ. It gives for the Wigner function (6)

$$W_i(\alpha + \sigma) = \sum_{nm} \rho_{nm}^{(i)} W_{nm}(r, \theta), \tag{12}$$

where (r, θ) are polar coordinates in the complex phase-space $X = \operatorname{Re}\alpha = r\cos\theta$, $Y = \operatorname{Im}\alpha = r\sin\theta$ and the coefficients W_{nm} are

$$W_{mn}(r,\theta) = \left\{ \begin{array}{l} \frac{2}{\pi}(-1)^n \sqrt{\frac{n!}{m!}} e^{i(m-n)\theta}(2r)^{m-n} e^{-2r^2} L_n^{m-n}(4r^2), \ m \geq n \\ \frac{2}{\pi}(-1)^m \sqrt{\frac{m!}{n!}} e^{i(m-n)\theta}(2r)^{n-m} e^{-2r^2} L_m^{n-m}(4r^2), \ n \geq m \end{array} \right\}, \tag{13}$$

where L_m^{n-m} are Laguerre polynomials.

As to the density matrix elements, they equal to

$$\rho_{nm}^{(1)}(\sigma_1) = M\left\{ \sum \langle n| D_1(\alpha_\xi - \sigma_1) |q\rangle \langle k| D_1^+(\alpha_\xi - \sigma_1) |m\rangle \, a_{qp}^{(\xi)}(\alpha_\xi, \beta_\xi) a_{kp}^{(\xi)*}(\alpha_\xi, \beta_\xi) \right\}, \tag{14}$$

$$\rho_{nm}^{(2)}(\sigma_2) = M\left\{ \sum \langle n| D_2(\beta_\xi - \sigma_2) |q\rangle \langle k| D_2^+(\beta_\xi - \sigma_2) |m\rangle \, a_{pq}^{(\xi)}(\alpha_\xi, \beta_\xi) a_{pk}^{(\xi)*}(\alpha_\xi, \beta_\xi) \right\}. \tag{15}$$

In short, the MQSD algorithm for numerical simulation is the following. In the initial basis centered at $\alpha = \beta = 0$ each set of stochastic $d\xi_i$ determines a quantum trajectory through Eqs.(2),(3). Then the state vectors (8) are calculated using a moving basis.

3 Pictorial View of Hopf Bifurcation

Now we apply the above numerical algorithm to calculate the ensemble-averaged Wigner function for the system of nonlinearly coupled, driven oscillators in contact with its environment. This model is accessible for experiments and is realized at least in the second-harmonic generation (SHG) in an optical cavity [5]. Intracavity SHG consists in transformation, via a $\chi^{(2)}$ nonlinear crystal, of an externally driven fundamental mode with the frequency ω_1 into the second-harmonic mode with the frequency $\omega_2 = 2\omega_1$ ($\omega_1 + \omega_2 \to \omega_2$). The system is described by both the Hamiltonians (5), and the Eqs.(1), (2). The Lindblad operators are $L_i = \sqrt{\gamma_i} a_i$, $(i = 1, 2)$, and γ_1, γ_2 are the cavity damping rates.

In the classical limit, this system is characterized by Hopf bifurcation which connect a steady-state regime to a temporal periodic regime (so-called self-pulsing instability [6]). In quantum treatment all quantum-mechanical ensemble-averaged quantities reach stable constant values at an arbitrary time exceeding the transient time. Therefore, the question has been posed what is the quantum-mechanical counerpart of a classical instability. The Wigner functions for the fundamental and second-harmonic modes have been analyzed in the vicinity of the Hopf bifurcation in Ref. [7], where it was shown that they describe "quantum instability". In this paper we expand our results [7] on the above Hopf-bifurcation range using MQSD algorithm. Below Hopf bifurcation the Wigner functions of both modes are Gaussian in the phase-space plane centered at $x = y = 0$. With increasing E we enter into critical transition domain, in the vicinity of Hopf bifurcation, where a spontaneous breaking of the phase symmetry occurs. The Wigner functions above bifurcation point at $E/E_{cr} = 2$ averaged over 1000 quantum trajectories are shown in Fig.1.

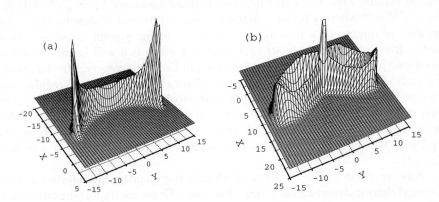

Fig.1.The Wigner functions of the second-harmonic mode (a) and the fundamental mode (b) beyond the Hopf bifurcation for the parameters: $\gamma_1 = \gamma_2 = \gamma$, $k/\gamma = 0.3$, $E_{cr}/\gamma = 20$.

It is obvious, that the appearance of two side humps of Wigner functions correspond to two most probable values of phases. Generally speaking, the contour pots in (x, y) plane of the Wigner functions correspond to phase trajectories of the system in the classical limit. We show these results on Fig.2.

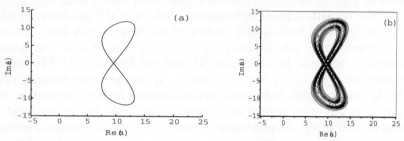

Fig.2. Phase-space trajectory (a) and contour plot of Wigner function (b) for the fundamental mode.

4 Quantum Chaos on Wigner Functions

Our second example is the model of dissipative nonlinear oscillator driven by two periodic forces of different frequencies and interacting with heat bath. This model presents an example of open time-dependent quantum system showing dissipative chaos in classical limit. The model was proposed to study quantum stochastic resonance in the authors' previous paper [8], where it is shown, in particular, that the model is available for experiments. It is described by Hamiltonian (4) and Eqs. (1), (2), with the Lindblad operators $L_1 = \sqrt{(N+1)\gamma}a$, $L_2 = \sqrt{N\gamma}a^+$, where γ is the spontaneous decay rate and N denotes the mean number of quanta of a heat bath. For $\Omega_2 = 0$ this system is reduced to a single driven dissipative anharmonic oscillator which is a well known model in nonlinear physics (see Ref. [9] and [8] for full list of references). In the classical limit the considered double driven oscillator exhibits regions of regular and chaotic motion. Indeed, our numerical analysis of the classical equations of motion in the phase-space shows that classical dynamics of the system is regular in domains $\delta \ll \gamma$ and $\delta \gg \gamma$, where $\delta = \omega_2 - \omega_1$, and also when $\Omega_1 \ll \Omega_2$ or $\Omega_2 \ll \Omega_1$. The dynamics is chaotic in the range of parameters $\delta \sim \gamma$ and $\Omega_1 \simeq \Omega_2$.

Now we come to the question of what is the quantum manifestation of a classical chaotic dynamics on Wigner function? These are important but rather difficult question of relevant to many problems of fundamental interest [10]. Our numerical analysis will show that the quantum dynamical manifestation of chaotic behavior does not appear on ensemble averaged oscillatory excitation numbers, but is clearly seen on the probability distributions. In Figs.3 we demonstrate moving our system from regular to chaotic dynamics by plotting the Wigner function of the system's quantum states for three values of Ω_2 : $\Omega_2/\gamma = 1$ (a), $\Omega_2/\gamma = \Omega_1/\gamma = 10.2$ (b), $\Omega_2/\gamma = 20$ (c), in a fixed moment of

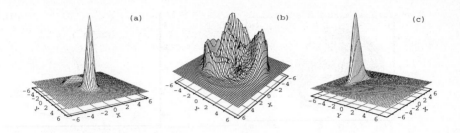

Fig.3. Transition from regular to chaotic dynamics on the Wigner func-
tions.The parameters are: $\chi/\gamma = 0.7$, $\Delta/\gamma = -15$, $\Omega_1/\gamma = \Omega_2/\gamma = 10.2$,
$\delta/\gamma = 5$. The averaging is over 2000 trajectories.

time. The values of Δ/γ ($\Delta = \omega_0 - \omega_1$), χ/γ, and Ω_1/γ are chosen to lead to
bistability in the model of single driven oscillator ($\Omega_2 = 0$).

We can see that for the case of a weak second force Fig.3(a) the Wigner
function has two humps, corresponding to the lower and upper level of excitation of anharmonic oscillator in the bistability region. The hump centered
close to $x = y = 0$ describes the approximately coherent lower state, while
the other hump describes the excited state. The graphs in Fig.3 are given at
an arbitrary time, exceeding the damping time. As calculations show, for the
next time intervals during the period of modulation $t = 2\pi/\delta$, the hump corresponding to the upper level rotates around the central peak. When we increase
the strength of the second force, the classical system reaches to a chaotic dynamics. The Wigner function for the chaotic dynamics is depicted in Fig.3(b).
Further increasing Ω_2, the system returns to the regular dynamics. The corresponding Wigner function at an arbitrary time exceeding the transient time is
presented in Fig.3(c). It contains only one hump, rotating around the centre of
the phase-space within the period. As we see, the Wigner function reflecting
chaotic dynamics (Fig.3(b)), has a complicated structure. Nevertheless, it is
easy to observe that its contour plots in (x, y) plane are generally similar to
the corresponding classical Poincaré section. Now we will consider this problem
in more detail, comparing the results for contour plots with classical strange
attractors on the classical maps, for the same sets of parameters. The results
are shown in Fig.4.

It can be seen in Fig.4.(a) that for the deep quantum regime ($\chi/\gamma = 0.7$,
$\Delta/\gamma = -15$, $\delta/\gamma = 5$), the contour plot of the Wigner function is smooth and
concentrated approximately around the attractor (Fig.4(b)). Nevertheless, the
different branches of the attractor are hardly resolved in Fig.4.(a). It can also be
seen, that in this deep quantum regime, an enlargement of the Wigner function
occurs in contrast to the Poincaré section.

Taking a smaller χ/γ, the contour plot of the Wigner function approaches
the classical Poincaré section. This can be seen in Figs.4.(d),(c). For the last
case the correspondence is maximal, and some details of the attractor (Fig.4(d))
are resolved much better in Fig.4(c).

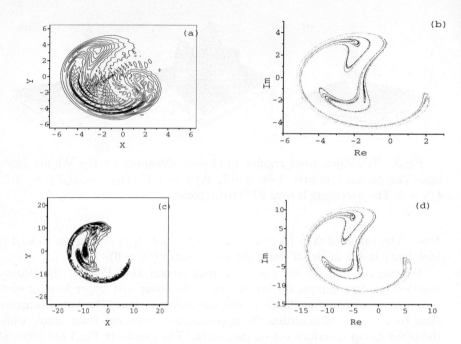

Fig.4. Contour plot of Wigner function (a), (c) corresponding to quantum chaos and Poincaré sections (b), (d) (approximately 20000 points) for the classical complex amplitude of double driven anharmonic oscillator, plotted at times of constant phase $\delta t = 2\pi n$, $(n = 1, 2, 3, ...)$. The parameters for the cases of (a), (c) are the same as in (b), (d) respectively.

5   Conclusion

We have formulated the effective method for computation of quantum distributions based on MQSD approach to open quantum systems. Using this approach we have numerically investigated the Wigner functions for two dissipative quantum systems showing both instability and chaos in a classical limit. We have shown how quantum instability and quantum chaos are displayed on the Wigner functions.

References

[1] C. W. Gardiner, Handbook of Stochastic Methods, (Springer, Berlin, 1986); D.F. Walls and G. J. Milburn Quantum Optics (Springer, Berlin, 1996).

[2] H. Y. Carmichael, An Open System Approach to Quantum Optics, Lecture Notes in Physics (Springer-Verlag, Berlin, 1993); M. B. Plenio and P. L. Knight, Rev. Mod. Phys. **70,** 101 (1998).

[3] N. Gisin and I.C. Percival, J. Phys. A **25**, 5677 (1992); A **26**, 2233 (1993); A **26**, 2245 (1993).

[4] R. Schack, T. A. Brun, and I. C. Percival, J. Phys. A **28**, 5401 (1995).

[5] P. D. Drummond, K. J. McNeil, and D. F. Walls, Opt. Acta **27**, 321 (1980); **28**, 211 (1981).

[6] S. Schiller and R. Byer, J. Opt. Soc. Am., B **10**, 1696 (1993).

[7] S. T. Gevorkyan, G. Yu. Kryuchkyan, and N. T. Muradyan, Phys. Rev. A **61**, 043805 (2000).

[8] H.H. Adamyan, S.B. Manvelyan, and G.Yu. Kryuchkyan, Phys.Rev, A **63**, 022102 (2001).

[9] P.D. Drummond and D. Walls, J. Phys. A **13** 725 (1980).

[10] *Quantum Chaos*, Eds. G. Casati and B. Chirikov (Cambridge University Press, 1995); F. Haake, "Quantum Signatures of chaos", (Springer-Verlag, Berlin, 2000).

Identification of MIMO Systems by Input-Output Takagi-Sugeno Fuzzy Models

Nirmal Singh[1], Renu Vig[2] and J. K. Sharma[1]

[1]Dept. of Mech. & Prod. Eng, BCET, Gurdaspur-143521, India
[2]Dept. of Computer Science, TTTI, Sector -26, Chandigarh-160026, India
[1]Beant College of Engineering & Technology, Gurdaspur-143521, India

Abstract: The extraction of fuzzy information out of raw data is very important and contains saving potential for science and engineering applications. The fuzzy models from this information can be used for different purposes in various applications because of their transparency. A number of techniques have been introduced to construct fuzzy models from measured data. Relatively little attention has been devoted to the identification of Multi Input Multi Output (MIMO) fuzzy models from input-output data. This paper concentrates on the fuzzy modeling and identification of MIMO systems by TS (Takagi-Sugeno) fuzzy models. In this paper product-space fuzzy clustering with adaptive distance measure named as Gustafson-Kessel (GK) algorithm is used. With this approach, the knowledge of the physical structure can be easily incorporated in the structure of the model. Implementation in the MATALAB is briefly described with simulation examples.

1. Introduction

At the computational level, fuzzy models can be regarded as flexible mathematical structures that can approximate a large class of complex non-linear systems to a desired degree of accuracy. Significant non-linearity and time delays exhibited by the automotive systems are common problems. Recently, a great deal of research activity has focused on the development of methods to build or update fuzzy models. Conventional system theory relies on crisp mathematical models of systems, such as algebraic and differential or difference equations. For some simple systems, mathematical models can be obtained. This is because the physical laws governing the systems are well understood. For a complex, dynamic and non-linear system, however, the gathering of an acceptable degree of knowledge needed for physical modeling is a difficult, time-consuming and expensive or even impossible task. [1].

Model based predictive control from input output data using TS fuzzy model is discussed in [3]. The modeling of a gas turbine aero engine based on multiobjective genetic algorithm is described in [3]. A progressive fuzzy modeling for control of an automotive engine air inlet is proposed in [4]. Feed forward scheme of a dynamic model with look up table of air flow rate mapped against engine crankshaft speed and throttle plate angle is used to predict the air flow into the engine with fuzzy c-mean clustering [5]. The c-mean clustering for fuzzy identification based engine fault detection and isolation is also discussed in [9]. So far most attention has been devoted to Single Input Single Output (SISO) or Multi Input Single Output (MISO) systems.

Relatively little attention has been devoted to the identification of MIMO fuzzy models from input-output data. The problem of developing a general methodology for intelligent system design has always been demanding. Fuzzy modeling is a framework in which different modeling and identification methods are combined, providing, on the one hand, a transparent interface with the designer or the operator and, on the other hand, a flexible tool for non-linear system modeling and control, comparable with other non-linear techniques. The rule-based character of fuzzy models allows for a model interpretation in a way that is similar to the one humans use. Such a compound system requires the identification of the local models from input-output data.

This work addresses a method for the fuzzy modeling and identification, which identifies the fuzzy rules describing the system behaviour from input-output data using TS models. In this paper product-space fuzzy clustering with adaptive distance measure i.e., GK algorithm is used. It gives a classical fuzzy model with local linear models in the consequent of each rule. The contribution of this paper is three-fold: First, it is shown that the fuzzy modeling & identification technique based on product-space fuzzy clustering can be extended to uncertain non-linear dynamical systems in a straight forwarded way. Second, a MATLAB implementation is presented which implements the presented techniques. Third, illustrative simulation examples of non-linear dynamic systems are given to prove that the approach is very simple and intuitively appealing.

2. TS Fuzzy Model Structure

TS fuzzy models are suitable to model a large class of non-linear systems [1]. Consider a MIMO system with n_i inputs: $u \in U \subset R^{n_i}$, and n_o outputs: $y \in Y \subset R^{n_o}$. This system will be approximated by a collection of coupled MISO discrete time fuzzy models. Denote q^{-1} the backward shift operator: $q^{-1}y(k) \overset{def}{=} y(k-1)$ where y is a signal sampled in discrete time instants k. Denote by the capital letters A and B polynomials in q, e.g., $A = a_0 + a_1 q + a_2 q^2 + \Lambda$ .. Given two integers, $m \le n$, define an ordered sequence of delayed samples of the signal y as: $\{ y(k)\}_m^n \overset{def}{=} [y(k-m), y(k-m-1), \Lambda , y(k-n)]$. The MISO models are of the input-output NARX (Non linear Auto Regressive with Exogenous input) type:

$$y_l(k+1) = F_l(x_l(k)), \quad l = 1, 2, \Lambda , n_o, \tag{1}$$

where the regression vector $x_l(k)$ is given by:

$$x_l(k) = [\{ y_1(k)\}_0^{n_{yl1}}, \{ y_2(k)\}_0^{n_{yl2}}, \Lambda, \{ y_{n_o}(k)\}_0^{n_{yln_o}},$$

$$\{ u_1(k)\}_{n_{dl1}}^{n_{ul1}}, \{ u_2(k)\}_{n_{dl2}}^{n_{ul2}}, \Lambda, \{ u_{n_i}(k)\}_{n_{dln_i}}^{n_{uln_i}}].$$

Here n_y and n_u are the number of delayed outputs and inputs, respectively, and n_d is the number of pure transport delays from the input to the output. n_y is a $n_o \times n_o$ matrix, and n_u, n_d are $n_o \times n_u$ matrices. F_l are rule based fuzzy models of the TS type:

$$R_{li} = \text{If } x_l(k) \text{ is } S_{li} \quad \text{then } y_{li}(k+1) = A_{li}y(k) + B_{li}u(k) + C_{li}, \quad i = 1, 2, \Lambda, K_l \quad (2)$$

Here S_{li} is the antecedent fuzzy set of the ith rule, and $A_{li} = [A_{li1}, \Lambda, A_{lin_o}]$, $B_{li} = [B_{li1}, \Lambda, B_{lin_i}]$, are vectors of polynomials. K_l is the number of rules in the lth model. The fuzzy sets S can be defined by multivariate membership functions $w(x(k)): R^{p_l} \rightarrow [0,1]$, where $p_l = \sum_{j=1}^{n_o} n_{ylj} + \sum_{j=1}^{n_i} n_{ulj} + 1$ is the dimension of the antecedent space. Alternatively, the antecedent of (2) can be given in the conjunctive form:

$$R_{li} = \text{If } x_{l1}(k) \text{ is } S_{li1} \text{ and } \Lambda \text{ and } x_{lp}(k) \text{ is } S_{lip}$$
$$\text{then } y_{li}(k+1) = A_{li}y(k) + B_{li}u(k) + C_{li}, i = 1, 2, \Lambda, K_l. \quad (3)$$

Both these forms can be constructed from data, as shown in next section. The output of the TS model is computed by the weighted mean:

$$y_l(k+1) = \frac{\sum_{i=1}^{K_l} w_{li}(x_l(k))y_{li}(k+1)}{\sum_{i=1}^{K_l} w_{li}(x_l(k))}. \quad (4)$$

3. Identification Method

The structure of the model, i.e., the matrices n_y, n_u and n_d are determined by the user on the basis of prior knowledge and/or by comparing several candidate structures in terms of the prediction error or other suitable criteria. Once the structure of the model is fixed, the parameters of the SISO, MISO and SIMO models (2) are estimated. These parameters are the antecedent membership functions and the consequent polynomials. An additional structural parameter is the number of rules, which can be specified or sought by the user. The identification GK algorithm proceeds in four steps as under, the theoretical details can be seen in [1-2]:

- Form the input-output sequences, $\{(u(k),y(k))\}_{k=1}^{N_d}$, form the non-linear regression problem of (1), using the user specified structural parameters n_y, n_u and n_d.
- Partition the data into the set of local linear sub models by using fuzzy clustering in the Cartesian product space $X \times Y$.
- Compute the antecedent membership function from the cluster parameters.
- Given the antecedent membership functions, estimate the consequent parameters by the least squares method.

4. MATLAB Implementation

The identification algorithm presented in the previous section was implemented in MATLAB. The core of the identification routine is the GK clustering algorithm. The function in the algorithm to construct a SISO, MISO, SIMO and MIMO fuzzy model from data is named *fmclust* and it has the following synopsis:

$$FM = fmclust(U,Y,c,FMtype,m,tol,seed,ny,nu,nd)$$

The data sequences are given in columns of the matrices U and Y, respectively. The number of required clusters, c is a scalar for MISO systems and a vector for MIMO systems (each MISO model may have a different number of clusters. The remaining parameters are optional. *FMtype* specifies whether the antecedent membership functions of the resulting model are computed analytically in the antecedent product space or are derived by projection. Product-space membership functions give faster but often less accurate models. In the algorithm, the parameter m is the fuzziness exponent. When $m \phi 1$, with the default value $m = 2$. Larger values imply fuzzier (more overlapping) clusters. Currently, there is no theoretical basis for an optimal choice for the value of m. However, it is established that the algorithm converges for any $m \in (1,\infty)$. The termination tolerance for the clustering algorithm can be given in *tol* (default $tol = 0.01$). The fuzzy partition matrix is initialised at random. In order to obtain reproducible results, the random generators may be seeded by supplying the seed parameter. Default value is sum $100 * clock$. The ny, nu and nd parameters are the delay matrices defined in section (2).

The output of *fmclust* is the FM model matrix, which contains the parameters of the obtained fuzzy model (under MATLAB , FM is defined as a structure). The obtained fuzzy model can be validated by using function *fmsim* with the following synopsis:

$$[Y_m,VAF] = fmsim(U,Y,FM)$$

This function simulates the fuzzy model FM from the input data U and plots the simulated output Y_m along with the true output Y. The first u_y values of Y are used to initialize Y_m. The output argument percentile Variance Accounted For *(VAF)* between two signals is the performance index of the model is given by:

$$VAF = 100\% \left[1 - \frac{var(Y - Y_m)}{var(Y)} \right] \tag{5}$$

The *VAF* of two equal signals is 100%. If the signals differ, *VAF* is lower. The *VAF* index is used to assess the quality of a model, by comparing the true output with the output of the fuzzy model. As additional output arguments, the degrees of fulfilment and the local outputs of the individual rules can be obtained.

The output obtained on the screen for approximation of $y = sin(x)$ is shown in Figures 1–2 with VAF and three rules are as:

If x is S_1 then y = 4.3137x + 0.1329,
If x is S_2 then y = −5.0666x + 2.2756
If x is S_3 then y = 5.4982x − 4.9211

5   Simulation Examples

5.1   Identification without data clustering

In the this example, fuzzy modelling & identification of a simple non-linear dynamic system without data clustering is described by a first-order difference equation:

$$y(k+1) = y(k) + u(k)e^{-3[y(k)]} \tag{6}$$

We use a stepwise input signal to generate with this equation a set of 300 input–output data pairs. The complexity of the system behaviour is typically not uniform. Some operating regions can be well approximated by a model, while other regions require fine partitioning. Comparatively fuzzy data clustering can capture the non-uniform behaviour of the system to obtain an efficient representation of identification, which follows in the next section. The identified model is shown in Fig 3.

5.2   Identification of internal combustion engine models

In the this example, fuzzy modelling & identification of the simple SISO, MISO and SIMO models are implemented with the above said algorithm by the TS fuzzy model with three rules. This example presents a model of a four-cylinder spark ignition automotive IC engine air inlet system, which demonstrates identification capabilities to model an IC engine. In the throttle body, the control input is the angle of the throttle plate [11]. The rate at which the model introduces air into the intake manifold can be expressed as the product of two functions: one, an empirical function of the throttle plate angle only; and the other, a function of the atmospheric and manifold pressures. In cases of lower manifold pressure (greater vacuum), the flow rate through the throttle body is sonic and is only a function of the throttle angle.

$$\dot{m}_{at} = f(\alpha)g(p_m)$$ (7)

$$f(\alpha) = 0.5642 - 0.05231\alpha + 0.10299\alpha^2 - 0.00063\alpha^3$$

where $\dot{m}_{at} = $ Mass flow rate into manifold (g/s), $\alpha = $ Throttle angle (deg), $p_m = $ Manifold pressure (bar)

$$g(p_m) = \begin{cases} 1, & p_m \le \dfrac{P_{amb}}{2} \\ \dfrac{2}{P_{amb}}\sqrt{p_m P_{amb} - p_m^2}, & \dfrac{P_{amb}}{2} \le p_m \le P_{amb} \\ -\dfrac{2}{p_m}\sqrt{p_m P_{amb} - P_{amb}^2}, & P_{amb} \le p_m \le 2P_{amb} \\ -1, & p_m \ge 2P_{amb} \end{cases}$$

$P_{amb} = $ Ambient atmospheric pressure (bar). The difference in the incoming and outgoing mass flow rates represents the net rate of change of air mass with respect to time. This quantity, according to the ideal gas law, is proportional to the time derivative of the manifold pressure.

$$\dot{p}_m = \frac{RT}{V_m}(\dot{m}_{at} - \dot{m}_{ao})$$ (8)

where $R = $ Specific gas constant, $T = $ Temperature, $V_m = $ Manifold volume m^3 $m_{ao} = $ Mass flow rate of air out of the manifold (g/s), $\dot{p}_m = $ Rate of change of manifold pressure (bar/s). The mass flow rate of air that the model pumps into the cylinders from the manifold is described in (9) by an empirically derived equation. This mass rate is a function of the manifold pressure and the engine speed [12].

$$\dot{m}_{ao} = -0.366 + 0.08979np_m - 0.0337np_m^2 + 0.0001n^2 p_m$$ (9)

where $n = $ engine speed (rad/s). In this example, we develop simple dynamic models for the relationship between the throttle angle, speed of an engine and manifold pressure for automotive IC engine air inlet. We will obtain a fuzzy model of the following structure: $y(k+1) = f(y(k), u(k))$, where $y(k)$ and $u(k)$ are the state and the input at time k, respectively, and f is a static function, called the state-transition function.

For SISO model the input is the throttle plate angle $u = [\alpha]^T$ and the output is the engine crankshaft speed $y = [n]^T$. For MISO model the inputs are $u = [\alpha, n]^T$ and the output is manifold pressure $y = [\, p_m \,]^T$. For SIMO model input is $u = [\, \alpha \,]^T$ and outputs are $y = [\, n, p_m \,]^T$. Now we load the data set, split it in halves and use the first half for identification and the second half for validation. The objective is to control the air inlet subsystem of an engine by implementing the above exposed algorithm. In the this automotive application, fuzzy modelling & identification of the throttle angle and the speed of an engine functions is implemented with the above said algorithm by the TS fuzzy model with three rules . The output obtained on the screen is shown in Figs 4 –5. The output obtained for MISO and SIMO model are shown in Figs 6 –7.

5.3 Identification of Water Tank Application

A tank with a pipe flowing in and a pipe flowing out is shown in Fig 8. The water that flows in, the outflow rate depends on the diameter of the outflow pipe (which is constant) and the pressure in the tank (which varies with the water level). The system has some very non-linear characteristics. A controller for the water level in the tank needs to know the current water level and it needs to be able to set the valve. The input is the water level and its output will be the rate at which the valve is opening or closing. The system is represented by the equation:

$$\overset{\&}{h} = 1/S\,(\,F_{in} - \beta\sqrt{h}\,)$$
(10)

where S is the horizontal cross sectional area, h is the level of the liquid in the tank, β is a flow coefficient , and F_{in} is the inlet liquid flow rate. The identification is shown in Fig.9-10 and the rules of identified fuzzy model are as:

R_1 : If F_{in} is S_{11} and h is S_{12} then $\overset{\&}{h} = 0.8912 F_{in} + 0.0949h - 0.0186$

R_2 : If F_{in} is S_{21} and h is S_{22} then $\overset{\&}{h} = 0.9285 F_{in} + 0.0958h - 0.0310$

R_3 : If F_{in} is S_{31} and h is S_{32} then $\overset{\&}{h} = 0.9451 F_{in} + 0.0945h - 0.0403$

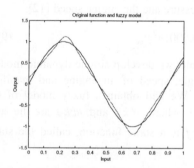

Fig.1. Static function and its approx. by a TS fuzzy model (VAF=99.67%)

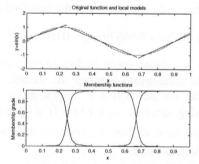

Fig. 2. Obtained local models (top), membership functions (bottom).

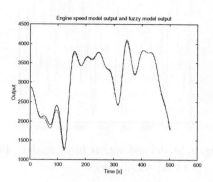

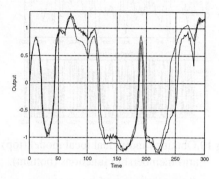

Fig 3. System / model (solid/dashed line)

Fig.4. SISO model and its approx.
by fuzzy model (VAF=99.67%)

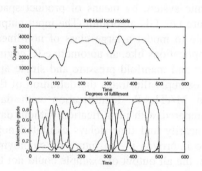

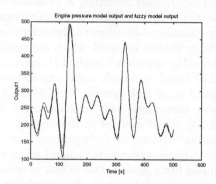

Fig. 5. Individual local models (top) and
membership functions (bottom)

Fig.6. MISO model and its approx. by a
fuzzy model (VAF=98.51%)

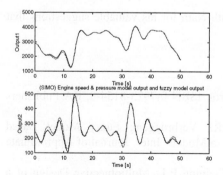

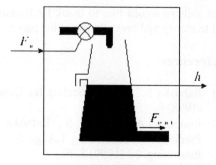

Fig.7. SIMO model and its approx. by
fuzzy model (VAF=99.67% and 57%)

Fig 8. Water tank

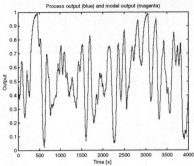
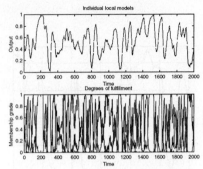

Fig. 9. Model and approx fuzzy model (VAF=99.58%) **Fig 10.** Obtained individual local model (top) and membership functions (bottom).

6 Conclusion

In this paper, fuzzy modelling & identification is presented for approximation of non-linear SISO, MISO, SIMO & MIMO dynamic system by means of product space fuzzy clustering with adaptive distance measure i.e., GK algorithm. The input-output TS fuzzy models are used, which are suitable to model a large class of non-linear systems. The approach is applied to simple applications like, an automotive IC engine air inlet based on throttle angle, engine speed and manifold pressure and single and cascaded water tank systems. The Identified example illustrates the definition of the various parameters with implementation in MATLAB. Identification with data clustering shows superior performance as compared with identification without data clustering. Because of the significant non-linearity and time delays exhibited non-linear dynamic systems, the control might be improved in the future by employing identified fuzzy models. Due to lack of space, the all output of example could not be presented here, however the simulation examples demonstrate the superior performance.

Acknowledgements

The authors would like to thank Dr Baikunth Nath for his valuable suggestions that led to clarity and presentation of the paper.

References

[1] Babuska R.: Fuzzy Modeling for Control, ISBN 0-7923-8154-8, KA Publishers (1998)

[2] Roubus J.A., Mollov S., Babuska R., Verbruggen.: Fuzzy Model Based Predictive Control Using Takagi- Sugeno Models. Inter. Journal of Approximate Reasoning, 22 (1999) 3-30.

[3] Bica B., Akat G., Chipperfield A.J., Fleming P.J.: Multiobjective Design of a Fuzzy Controller for a Gas Turbine Aero Engine. UKACC , Conf. on Control, 1(4) (1998) 901-906

[4] Bortolet P. Boverie S.: Fuzzy Modeling and Control of an Automotive Engine Air Inlet. SAE 980797, Int. Congress and Exposition, Detroit, Michigan, USA (1998)

[5] Copp D.G., Burnham K.J., Lockett F.P.: Model Comparison For Feed forward Air/Fuel Ratio Control. UKACC Int. Conference on Control,1(4) (1998) 670-675

[6] Costa Branco P.J., Dente J.A.: The Application of Fuzzy Logic in Automatic Modeling of Elctromechanical Systems. Fuzzy Sets & Systems , 95 (1998) 273-293

[7] Hendricks E., Chevalier A., Jensen M., Sorenson S. C., Trumpy D., Asik J.: Modeling of the intake manifold filling dynamics. SAE 960037, Int. Congress and Exposition, Detroit, Michigan, USA (1996)

[8] Jones J.D., Hua Y.: A Fuzzy Knowledge Base to Support Routine Engineering Design. Fuzzy Sets & Systems , 98 (1998) 267-278

[9] Laukonen E.G., Passino K.M., Krishnaswami V., Luh G.C., Rizzoni.: Fault Detection and Isolation for an Experiment Internal Combustion Engine via Fuzzy Identification. IEEE Trans. On Control Systems Technology, 3(3) (1995) 347-355

[10] Li X., Yurkovich S.: IC Engine Air/Fuel Ratio Prediction and Control Using Discrete Time Nonlinear Adaptive Techniques. American Control Conference, San Diego, CA (1999)

[11] Mathworks, Simulink, Stateflow.: Automotive Applications. The Mathworks (1998)

[12] Soliman A., Kim Y.Y., Rizzoni G.: A Fuzzy Decision-Making System for Automotive Diagnostics. SAE 980519, Int. Congress and Exposition, Detroit, Michigan, USA (1998)

[13] Sugeno M., Yashkawa T.: A Fuzzy Logic Based Approach to Qualitative Modeling. IEEE Trans. on Fuzzy Systems, 1(1), (1993) 7-31

[14] Takagi T., Sugeno M.: Fuzzy Identification of systems and its application to modeling and control. IEEE Trans. Systems, Man and Cybernetics, 15(1), (1985) 116-131

[15] Wang W., Nwagboso C., Zhou E., Lee C.: Intelligent Ignition Control Using Pressure Sensor with Fuzzy Logic in a Natural Gas Engine. IEE Int. Conference on Intelligent Vehicles (1998) 119-123

[16] Weeks R.W., Moskwa J.J.: Automotive Engine Modeling for Real Time Control Using MATLAB/SIMULINK. SAE 950417 (1995)

Control of Black Carbon, the Most Effective Means of Slowing Global Warming

Mark Z. Jacobson

Department of Civil and Environmental Engineering
Stanford University Terman Engineering Center, Room M-13
Stanford, CA 94305-4020 Phone: 650-723-6836 Fax: 650-725-9720
jacobson@ce.stanford.edu

Abstract

Under the Kyoto Protocol of 1997, no control of black carbon (BC) was considered. Recent studies, though, have suggested that BC and non-CO_2 greenhouse-gas emission controls might slow global warming. Yet, no study has compared the effects, over time, of theoretically reducing BC versus CO_2 or CH_4 emissions. In this study, a global model was used to compare the effects of such reductions. The model treated eight important feedbacks of aerosols to climate. Results suggest that the most efficient method of controlling global warming over the next 20-50 years, in terms of the magnitude and speed of a cooling, is control of fossil-fuel BC and associated organic matter. It is further shown that late-model diesel vehicles enhance global warming more than do equivalent gasoline vehicles, yet fuel-tax and carbon-tax laws favor diesel in many countries. It is concluded that control strategies for global warming should include control of BC along with control of CO_2 and other greenhouse gases.

Comparison of Two Schemes for the Redistribution of Moments for Modal Aerosol Model Applications

U. Shankar and A. L. Trayanov

MCNC-North Carolina Supercomputing Center,
P.O. Box 12889, Research Triangle Park, NC 27709-2889
uma@sol.ncsc.org

Abstract

The modal method for modeling atmospheric aerosol size-distributions takes advantage of their observed lognormal characteristics to formulate computationally efficient algorithms for many of the dynamical processes governing their time evolution (Whitby et al., 1991). Thus this approach has been particularly attractive in large-regional-to-global-scale model simulations that must be run over long periods, as in climate change studies. In previous applications of such models, two modes have been typically used to simulate the fine aerosol distribution, and one to represent the coarse particles, with the number of modes remaining fixed throughout the simulation. A drawback in some of these applications has been that the smaller of the fine modes, used to represent Aitken particles, tends to grow to sizes typical of the larger (accumulation) mode and thus cause mode overlap, unless the aerosol moments are redistributed between the modes. Two methods for moment redistribution are compared in the MCNC Multiscale Air Quality Simulation Platform (MAQSIP). In the first scheme the condensation growth rate of the moments is used to determine the initiation of moment transfer from the smaller to the larger of the fine modes to maintain two distinct modes. The second scheme uses a dynamic management of modes that takes account of all the dynamic processes that may lead to either two distinct fine modes, or a single mode, indistinguishable from the other. Size distribution parameters, as well as model performance parameters are compared between the two methods.

Whitby, E. R., McMurry, P. H., Shankar, U., and Binkowski, F. S. (1991). Modal Aerosol Dynamics Modeling, Rep. 600/3-91/020, Natl. Exposure Res. Lab., U. S. Environmental Protection Agency, Research Triangle Park, N. C. (Available as NTIS PB91-161729/AS from Natl. Tech. Inf. Serv., Springfield, VA.)

A Scale-Dependent Dynamic Model for Scalar Transport in the Atmospheric Boundary Layer

Fernando Port-Agel and Qiao Qin

Department of Civil Engineering and St Anthony Falls Laboratory,
University of Minnesota, Minneapolis, MN 55414
fporte@tc.umn.edu

Abstract

A key factor in large-eddy simulations (LES) of the atmospheric boundary layer is our limited ability to account for the physics that are not explicitly resolved in the simulations. These subgrid-scale (SGS) processes are particularly important in the near-ground region, where the characteristic eddy size is typically on the order of (or smaller than) the grid size. Numerical simulations and field experiments were performed to address open issues in SGS modeling. A new scale-dependent dynamic model was developed and used to optimize the value of the SGS model coefficient in the eddy-diffusion model at every position and time step based on the resolved field. Simulations with the scale-dependent dynamic model yield the expected trends of the coefficient as function of position, scale, and atmospheric stability. Furthermore, the new model gives improved predictions of mean profiles and turbulence spectra as compared with the traditional eddy-diffusion and dynamic models. Scale dependence of the model coefficient is further explored using high-resolution wind velocity and temperature measurements obtained in the surface layer using arrays of twelve three-dimensional sonic anemometers.

Advances in Molecular Algorithms

Session chairs:

Ben Leimkuhler (University of Leicester, UK)
Brian Laird (University of Kansas, USA)

MDT – The Molecular Dynamics Test Set

Eric Barth

Department of Mathematics, Kalamazoo College, Kalamazoo MI,49006, USA,
barth@kzoo.edu

Abstract

Over the past two decades, computational scientists have turned increasing attention to the field of molecular modeling. Advances have been made in the design of efficient algorithms for problems such as fast summation methods for computing non-bonded atomic interactions, long-time numerical integration of equations of motion, non-Newtonian dynamical formulations for simulation in a variety of statistical mechanical ensembles, and optimization algorithms. These projects have often involved mathematicians and/or computer scientists who, though adept at algorithm and software development, may possess limited or no physical or chemical knowledge.

The molecular dynamics test set is a collection of model problems for aiding numerical analysts, code developers and others in the design of computational methods for molecular dynamics (MD) simulation. Common types of calculations and desirable features of algorithms have been considered, and these were used to guide selection of representative models. By including essential features of certain classes of molecular systems,

but otherwise limiting the physical and quantitative details, it is hoped that the test set will help to facilitate cross-disciplinary algorithm and code development efforts.

Numerical Methods for the Approximation of Path Integrals Arising in Quantum Statistical Mechanics

Steve D. Bond

Departments of Chemistry and Mathematics, University of California - San Diego, La Jolla, CA 92093, USA

bond@ucsd.edu

Abstract

Discretizations of the Feynman-Kac path integral representation of the quantum mechanical density matrix are investigated. Each infinite-dimensional path integral is approximated by a Riemann integral over a finite-dimensional Sobolev space, by restricting the integration to a subspace of all admissible paths. Using this process, a wide class of methods can be derived, with each method corresponding to a different choice for the approximating subspace. The traditional "short-time" approximation and "Fourier discretization" can be recovered from this approach, using linear and spectral basis functions respectively. As an illustration, a new method is formulated using cubic elements and is shown to have improved convergence properties when applied to model problems. Experimental results will be discussed, including a double-well model and cluster simulations using the Blue Horizon.

The Multigrid N-Body Solver

David J. Hardy

Department of Computer Science, University of Illinois, Urbana, IL, 61801, USA
dhardy@uiuc.edu

Abstract

The multigrid N-body solver is a fast method for approximating the pairwise electrostatic forces due to a system of charged atoms. The electrostatic potential can be expressed as $U(r) = (1/2)q^T G(r)q$ where vector q contains the N charges and $N \times N$ matrix $G(r)$ has an element $C/\|r_j - r_i\|$ in its ijth position wherever there is an interaction between atoms i and j. The multigrid N-body solver uses a matrix splitting along with hierarchical interpolation to approximate the smooth dense $\tilde{G}(r)q$ product in $\mathcal{O}(N)$ time. The advantages of this algorithm are its simplicity, the fact that it maintains continuous forces, and that it enables cheaper multiple time stepping for molecular dynamics through a natural separation of spatial scales. The details of this method will be presented, and experimental results will be shown that compare this method with other popular fast electrostatics methods.

Do Your Hard-Spheres Have Tails? A Molecular Dynamics Integration Algorithm for Systems with Mixed Hard-Core/Continuous Potentials

Brian B. Laird

Department of Chemistry, University of Kansas, Lawrence KS, USA,
laird@ku.edu

Abstract

Integration algorithms for molecular-dynamics simulation generally fall into two mutually exclusive classes: those for purely collisional systems, such as hard spheres, and those for continuous potentials. However, there exist theoretically important model systems, such as the restricted primative model for electolytes, that have both collisional and continuous potential components and for which no satisfactory molecular dynamics algorithm has been developed. For this reason simulation studies of such systems have been limited to Monte Carlo studies, which give no information as to dynamical properties. We present a new molecular-dynamics algorithm for integrating the equations of motion for a system of particles interacting with mixed continuous/impulsive forces. This method, which we call Collision Verlet, is constructed using operator splitting techniques similar to those that have been used successfully to generate a variety molecular-dynamics integrators . In numerical experiments, the Collision Verlet method is shown to be superior to previous methods with respect to stability and energy conservation in long simulations.

An Improved Dynamical Formulation for Constant Temperature and Pressure Dynamics, with Application to Particle Fluid Models

Benedict J. Leimkuhler

Department of Mathematics and Computer Science, University of Leicester,
Leicester,LE1 7RH, UK
bll2@mcs.le.ac.uk

Abstract

A new fully dynamical scheme for constant temperature and pressure simulation of particle systems, based on a modification of the Nosé thermostat. A mechanical formulation for simultaneous control of temperature and pressure is also presented. This approach simplifies the construction of symplectic methods while providing a more intuitive perspective on the nature of controlled variable molecular dynamics. Moreover, the described method is Gallilean-invariant, hence angular momentum preserving, and the per-timestep simulation costs are similar to the per-timestep costs of microcanonical (N, V, E) simulation. The scheme is suited to molecular simulation and to large scale particle models of fluids such as smooth particle hydrodynamics and dissipative particle models.

Author Index

Lecture Notes in Computer Science

For information about Vols. 1–1976
please contact your bookseller or Springer-Verlag

UNIVERSITY OF LONDON
LIBRARY

Vol. 2020: D. Naccache (Ed.), Topics in Cryptology – CT-RSA 2001. Proceedings, 2001. XII, 473 pages. 2001

Vol. 2021: J. N. Oliveira, P. Zave (Eds.), FME 2001: Formal Methods for Increasing Software Productivity. Proceedings, 2001. XIII, 629 pages. 2001.

Vol. 2022: A. Romanovsky, C. Dony, J. Lindskov Knudsen, A. Tripathi (Eds.), Advances in Exception Handling Techniques. XII, 289 pages. 2001.

Vol. 2024: H. Kuchen, K. Ueda (Eds.), Functional and Logic Programming. Proceedings, 2001. X, 391 pages. 2001.

Vol. 2025: M. Kaufmann, D. Wagner (Eds.), Drawing Graphs. XIV, 312 pages. 2001.

Vol. 2026: F. Müller (Ed.), High-Level Parallel Programming Models and Supportive Environments. Proceedings, 2001. IX, 137 pages. 2001.

Vol. 2027: R. Wilhelm (Ed.), Compiler Construction. Proceedings, 2001. XI, 371 pages. 2001.

Vol. 2028: D. Sands (Ed.), Programming Languages and Systems. Proceedings, 2001. XIII, 433 pages. 2001.

Vol. 2029: H. Hussmann (Ed.), Fundamental Approaches to Software Engineering. Proceedings, 2001. XIII, 349 pages. 2001.

Vol. 2030: F. Honsell, M. Miculan (Eds.), Foundations of Software Science and Computation Structures. Proceedings, 2001. XII, 413 pages. 2001.

Vol. 2031: T. Margaria, W. Yi (Eds.), Tools and Algorithms for the Construction and Analysis of Systems. Proceedings, 2001. XIV, 588 pages. 2001.

Vol. 2032: R. Klette, T. Huang, G. Gimel'farb (Eds.), Multi-Image Analysis. Proceedings, 2000. VIII, 289 pages. 2001.

Vol. 2033: J. Liu, Y. Ye (Eds.), E-Commerce Agents. VI, 347 pages. 2001. (Subseries LNAI).

Vol. 2034: M.D. Di Benedetto, A. Sangiovanni-Vincentelli (Eds.), Hybrid Systems: Computation and Control. Proceedings, 2001. XIV, 516 pages. 2001.

Vol. 2035: D. Cheung, G.J. Williams, Q. Li (Eds.), Advances in Knowledge Discovery and Data Mining – PAKDD 2001. Proceedings, 2001. XVIII, 596 pages. 2001. (Subseries LNAI).

Vol. 2037: E.J.W. Boers et al. (Eds.), Applications of Evolutionary Computing. Proceedings, 2001. XIII, 516 pages. 2001.

Vol. 2038: J. Miller, M. Tomassini, P.L. Lanzi, C. Ryan, A.G.B. Tettamanzi, W.B. Langdon (Eds.), Genetic Programming. Proceedings, 2001. XI, 384 pages. 2001.

Vol. 2039: M. Schumacher, Objective Coordination in Multi-Agent System Engineering. XIV, 149 pages. 2001. (Subseries LNAI).

Vol. 2040: W. Kou, Y. Yesha, C.J. Tan (Eds.), Electronic Commerce Technologies. Proceedings, 2001. X, 187 pages. 2001.

Vol. 2041: I. Attali, T. Jensen (Eds.), Java on Smart Cards: Programming and Security. Proceedings, 2000. X, 163 pages. 2001.

Vol. 2042: K.-K. Lau (Ed.), Logic Based Program Synthesis and Transformation. Proceedings, 2000. VIII, 183 pages. 2001.

Vol. 2043: D. Craeynest, A. Strohmeier (Eds.), Reliable Software Technologies – Ada-Europe 2001. Proceedings, 2001. XV, 405 pages. 2001.

Vol. 2044: S. Abramsky (Ed.), Typed Lambda Calculi and Applications. Proceedings, 2001. XI, 431 pages. 2001.

Vol. 2045: B. Pfitzmann (Ed.), Advances in Cryptology – EUROCRYPT 2001. Proceedings, 2001. XII, 545 pages. 2001.

Vol. 2047: R. Dumke, C. Rautenstrauch, A. Schmietendorf, A. Scholz (Eds.), Performance Engineering. XIV, 349 pages. 2001.

Vol. 2048: J. Pauli, Learning Based Robot Vision. IX, 288 pages. 2001.

Vol. 2051: A. Middeldorp (Ed.), Rewriting Techniques and Applications. Proceedings, 2001. XII, 363 pages. 2001.

Vol. 2052: V.I. Gorodetski, V.A. Skormin, L.J. Popyack (Eds.), Information Assurance in Computer Networks. Proceedings, 2001. XIII, 313 pages. 2001.

Vol. 2053: O. Danvy, A. Filinski (Eds.), Programs as Data Objects. Proceedings, 2001. VIII, 279 pages. 2001.

Vol. 2054: A. Condon, G. Rozenberg (Eds.), DNA Computing. Proceedings, 2000. X, 271 pages. 2001.

Vol. 2055: M. Margenstern, Y. Rogozhin (Eds.), Machines, Computations, and Universality. Proceedings, 2001. VIII, 321 pages. 2001.

Vol. 2056: E. Stroulia, S. Matwin (Eds.), Advances in Artificial Intelligence. Proceedings, 2001. XII, 366 pages. 2001. (Subseries LNAI).

Vol. 2057: M. Dwyer (Ed.), Model Checking Software. Proceedings, 2001. X, 313 pages. 2001.

Vol. 2059: C. Arcelli, L.P. Cordella, G. Sanniti di Baja (Eds.), Visual Form 2001. Proceedings, 2001. XIV, 799 pages. 2001.

Vol. 2064: J. Blanck, V. Brattka, P. Hertling (Eds.), Computability and Complexity in Analysis. Proceedings, 2000. VIII, 395 pages. 2001.

Vol. 2068: K.R. Dittrich, A. Geppert, M.C. Norrie (Eds.), Advanced Information Systems Engineering. Proceedings, 2001. XII, 484 pages. 2001.

Vol. 2072: J. Lindskov Knudsen (Ed.), ECOOP 2001 – Object-Oriented Programming. Proceedings, 2001. XIII, 429 pages. 2001.

Vol. 2073: V.N. Alexandrov, J.J. Dongarra, B.A. Juliano, R.S. Renner, C.J.K. Tan (Eds.), Computational Science – ICCS 2001. Part I. Proceedings, 2001. XXVIII, 1306 pages. 2001.

Vol. 2074: V.N. Alexandrov, J.J. Dongarra, B.A. Juliano, R.S. Renner, C.J.K. Tan (Eds.), Computational Science – ICCS 2001. Part II. Proceedings, 2001. XXVIII, 1076 pages. 2001.

Vol. 2091: J. Bigun, F. Smeraldi (Eds.), Audio- and Video-Based Biometric Person Authentication. Proceedings, 2001. XIII, 374 pages. 2001.

Vol. 2092: L. Wolf, D. Hutchison, R. Steinmetz (Eds.), Quality of Service – IWQoS 2001. Proceedings, 2001. XII, 435 pages. 2001.

QUEEN MARY
UNIVERSITY OF LONDON
LIBRARY

WITHDRAWN
FROM STOCK
QMUL LIBRARY

Springer
Berlin
Heidelberg
New York
Barcelona
Hong Kong
London
Milan
Paris
Singapore
Tokyo

Lecture Notes in Computer Science 2074

Edited by G. Goos, J. Hartmanis and J. van Leeuwen

QMW Library

23 1192762 1

WITHDRAWN
FROM STOCK
QMUL LIBRARY

QA66.5 LEC/2074

DATE DUE FOR RETURN